FUNCTIONS

Linear Function: $f(x) = mx + b$

Two-point form: $y - y_1 = \left(\dfrac{y_2 - y_1}{x_2 - x_1}\right)(x - x_1)$

or $\begin{vmatrix} x & y & 1 \\ x_1 & y_1 & 1 \\ x_2 & y_2 & 1 \end{vmatrix} = 0$ (x_1, y_1) and (x_2, y_2) are any two points on the line.

Intercept form: $\dfrac{x}{a} + \dfrac{y}{b} = 1$ $(a, 0)$ is the x-intercept; $(0, b)$ is the y-intercept.

Quadratic Function: $f(x) = ax^2 + bx + c,\ a \neq 0$

Absolute Value Function: $f(x) = |x| = \begin{cases} x \text{ if } x \geq 0 \\ -x \text{ if } x < 0 \end{cases}$

Polynomial Function: $f(x) = a_n x^n + a_{n-1} x^{n-1} + a_{n-2} x^{n-2} + \cdots + a_2 x^2 + a_1 x + a_0,\ a_n \neq 0$

Rational Function: $f(x) = \dfrac{P(x)}{D(x)}$, where $P(x)$ and $D(x)$ are polynomial functions, $D(x) \neq 0$

Asymptotes:
Suppose $P(x)$ and $D(x)$ have no common factors, and let $P(x)$ have degree m with leading coefficient p and $D(x)$ have degree n with leading coefficient d. Then,

Vertical asymptote: $x = c$, where $D(c) = 0$

Horizontal asymptote: $y = 0$ if $m < n$; $y = \dfrac{p}{d}$ if $m = n$; none if $m > n$

Oblique asymptote: $y = mx + b$ if $m = n + 1$ and $f(x) = mx + b + \dfrac{r}{D(x)}$

Exponential Function: $f(x) = b^x,\ b > 0,\ b \neq 1$
Laws of exponents: Let a and b be nonzero real numbers and let m and n be any real numbers, except that the form 0^0 and division by zero are excluded.

Additive law: $b^m b^n = b^{m+n}$

Subtractive law: $\dfrac{b^m}{b^n} = b^{m-n}$

Multiplicative law: $(b^n)^m = b^{mn}$

Distributive laws: $(ab)^m = a^m b^m$ and $\left(\dfrac{a}{b}\right)^m = \dfrac{a^m}{b^m}$

Logarithmic Function: $f(x) = \log_b x,\ x > 0,\ b > 0,\ b \neq 1$
Laws of logarithms: Let $A > 0,\ B > 0$
Addition law: $\log_b AB = \log_b A + \log_b B$

Subtraction law: $\log_b \dfrac{A}{B} = \log_b A - \log_b B$

Multiplication law: $\log_b A^p = p \log_b A$
Log of both sides: $\log_b A = \log_b B$ is equivalent to $A = B$

Change of base: $\log_a x = \dfrac{\log_b x}{\log_b a}$

Precalculus

A Functional Approach to Graphing and Problem Solving

SIXTH EDITION

The Jones & Bartlett Learning Series in Mathematics

Geometry

Geometry with an Introduction to Cosmic Topology
Hitchman (978-0-7637-5457-0) © 2009

Euclidean and Transformational Geometry:
A Deductive Inquiry
Libeskind (978-0-7637-4366-6) © 2008

A Gateway to Modern Geometry: The Poincaré
Half-Plane, Second Edition
Stahl (978-0-7637-5381-8) © 2008

Understanding Modern Mathematics
Stahl (978-0-7637-3401-5) © 2007

Lebesgue Integration on Euclidean Space,
Revised Edition
Jones (978-0-7637-1708-7) © 2001

Precalculus

Precalculus: A Functional Approach to Graphing
and Problem Solving, Sixth Edition
Smith (978-1-4496-4916-6) © 2013

Precalculus with Calculus Previews, Fifth Edition
Zill/Dewar (978-1-4496-4912-8) © 2013

Essentials of Precalculus with Calculus Previews,
Fifth Edition
Zill/Dewar (978-1-4496-1497-3) © 2012

Algebra and Trigonometry, Third Edition
Zill/Dewar (978-0-7637-5461-7) © 2012

College Algebra, Third Edition
Zill/Dewar (978-1-4496-0602-2) © 2012

Trigonometry, Third Edition
Zill/Dewar (978-1-4496-0604-6) © 2012

Calculus

Multivariable Calculus
Damiano/Freije (978-0-7637-8247-4) © 2012

Single Variable Calculus: Early Transcendentals,
Fourth Edition
Zill/Wright (978-0-7637-4965-1) © 2011

Multivariable Calculus, Fourth Edition
Zill/Wright (978-0-7637-4966-8) © 2011

Calculus: Early Transcendentals, Fourth Edition
Zill/Wright (978-0-7637-5995-7) © 2011

Calculus: The Language of Change
Cohen/Henle (978-0-7637-2947-9) © 2005

Applied Calculus for Scientists and Engineers
Blume (978-0-7637-2877-9) © 2005

Calculus: Labs for Mathematica
O'Connor (978-0-7637-3425-1) © 2005

Calculus: Labs for MATLAB®
O'Connor (978-0-7637-3426-8) © 2005

Linear Algebra

Linear Algebra: Theory and Applications,
Second Edition
Cheney/Kincaid (978-1-4496-1352-5) © 2012

Linear Algebra with Applications, Seventh Edition
Williams (978-0-7637-8248-1) © 2011

Linear Algebra with Applications, Alternate
Seventh Edition
Williams (978-0-7637-8249-8) © 2011

Advanced Engineering Mathematics

A Journey into Partial Differential Equations
Bray (978-0-7637-7256-7) © 2012

Advanced Engineering Mathematics, Fourth Edition
Zill/Wright (978-0-7637-7966-5) © 2011

An Elementary Course in Partial Differential
Equations, Second Edition
Amaranath (978-0-7637-6244-5) © 2009

Complex Analysis

Complex Analysis for Mathematics and Engineering,
Sixth Edition
Mathews/Howell (978-1-4496-0445-5) © 2012

A First Course in Complex Analysis with Applications,
Second Edition
Zill/Shanahan (978-0-7637-5772-4) © 2009

Classical Complex Analysis
Hahn (978-0-8672-0494-0) © 1996

Real Analysis

Elements of Real Analysis
Denlinger (978-0-7637-7947-4) © 2011

An Introduction to Analysis, Second Edition
Bilodeau/Thie/Keough (978-0-7637-7492-9) © 2010

Basic Real Analysis
Howland (978-0-7637-7318-2) © 2010

Closer and Closer: Introducing Real Analysis
Schumacher (978-0-7637-3593-7) © 2008

The Way of Analysis, Revised Edition
Strichartz (978-0-7637-1497-0) © 2000

Topology

Foundations of Topology, Second Edition
Patty (978-0-7637-4234-8) © 2009

Discrete Mathematics and Logic

Essentials of Discrete Mathematics, Second Edition
Hunter (978-1-4496-0442-4) © 2012

Discrete Structures, Logic, and Computability,
Third Edition
Hein (978-0-7637-7206-2) © 2010

Logic, Sets, and Recursion, Second Edition
Causey (978-0-7637-3784-9) © 2006

Numerical Methods

Numerical Mathematics
Grasselli/Pelinovsky (978-0-7637-3767-2) © 2008

Exploring Numerical Methods: An Introduction to
Scientific Computing Using MATLAB®
Linz (978-0-7637-1499-4) © 2003

Advanced Mathematics

A Transition to Mathematics with Proofs
Cullinane (978-1-4496-2778-2) © 2013

Mathematical Modeling with Excel®
Albright (978-0-7637-6566-8) © 2010

Clinical Statistics: Introducing Clinical Trials,
Survival Analysis, and
Longitudinal Data Analysis
Korosteleva (978-0-7637-5850-9) © 2009

Harmonic Analysis: A Gentle Introduction
DeVito (978-0-7637-3893-8) © 2007

Beginning Number Theory, Second Edition
Robbins (978-0-7637-3768-9) © 2006

A Gateway to Higher Mathematics
Goodfriend (978-0-7637-2733-8) © 2006

For more information on this series and its titles,
please visit us online at http://www.jblearning.com.
Qualified instructors, contact your Publisher's
Representative at 1-800-832-0034 or info@jblearning
.com to request review copies for course consideration.

The Jones & Bartlett Learning International Series in Mathematics

Functions of Mathematics in the Liberal Arts
Johnson (978-0-7637-8116-3) © 2014

A Transition to Mathematics with Proofs
Cullinane (978-1-4496-2778-2) © 2013

Linear Algebra: Theory and Applications, Second Edition, International Version
Cheney/Kincaid (978-1-4496-2731-7) © 2012

Multivariable Calculus
Damiano/Freije (978-0-7637-8247-4) © 2012

Complex Analysis for Mathematics and Engineering, Sixth Edition, International Version
Mathews/Howell (978-1-4496-2870-3) © 2012

A Journey into Partial Differential Equations
Bray (978-0-7637-7256-7) © 2012

Association Schemes of Matrices
Wang/Hou/Ma (978-0-7637-8505-5) © 2011

Advanced Engineering Mathematics, Fourth Edition, International Version
Zill/Wright (978-0-7637-7994-8) © 2011

Calculus: Early Transcendentals, Fourth Edition, International Version
Zill/Wright (978-0-7637-8652-6) © 2011

Real Analysis
Denlinger (979-0-7637-7947-4) © 2011

Mathematical Modeling for the Scientific Method
Pravica/Spurr (978-0-7637-7946-7) © 2011

Mathematical Modeling with Excel®
Albright (978-0-7637-6566-8) © 2010

An Introduction to Analysis, Second Edition
Bilodeau/Thie/Keough (978-0-7637-7492-9) © 2010

Basic Real Analysis
Howland (978-0-7637-7318-2) © 2010

*For more information on this series and its titles, please visit us online at http://www.jblearning.com.
Qualified instructors, contact your Publisher's Representative at 1-800-832-0034 or info@jblearning.com
to request review copies for course consideration.*

Precalculus

A Functional Approach to Graphing and Problem Solving

SIXTH EDITION

Karl J. Smith, Ph.D.
Santa Rosa Junior College

Mathematics is a vast adventure in ideas; its history reflects some of the noblest thoughts of countless generations.

— Dirk J. Struik

JONES & BARTLETT
LEARNING

World Headquarters
Jones & Bartlett Learning
5 Wall Street
Burlington, MA 01803
978-443-5000
info@jblearning.com
www.jblearning.com

Jones & Bartlett Learning books and products are available through most bookstores and online booksellers. To contact Jones & Bartlett Learning directly, call 800-832-0034, fax 978-443-8000, or visit our website, www.jblearning.com.

Substantial discounts on bulk quantities of Jones & Bartlett Learning publications are available to corporations, professional associations, and other qualified organizations. For details and specific discount information, contact the special sales department at Jones & Bartlett Learning via the above contact information or send an email to specialsales@jblearning.com.

Production Credits

Chief Executive Officer: Ty Field
President: James Homer
SVP, Editor-in-Chief: Michael Johnson
SVP, Chief Technology Officer: Dean Fossella
SVP, Chief Marketing Officer: Alison M. Pendergast
Publisher: Cathleen Sether
Senior Acquisitions Editor: Timothy Anderson
Managing Editor: Amy Bloom
Senior Associate Editor: Megan R. Turner
Director of Production: Amy Rose
Senior Production Editor: Katherine Crighton
Production Editor: Tiffany Sliter

Production Assistant: Alyssa Lawrence
Senior Marketing Manager: Andrea DeFronzo
V.P., Manufacturing and Inventory Control: Therese Connell
Composition: Cenveo Publisher Services
Text Design: Anne Spencer
Cover and Title Page Design: Kristin E. Parker
Associate Photo Researcher: Lauren Miller
Cover and Title Page Image: © Nissan (2009). Nissan and the Nissan logo are registered trademarks of Nissan.
Printing and Binding: Courier Corporation
Cover Printing: Courier Corporation

Some images in this book feature models. These models do not necessarily endorse, represent, or participate in the activities represented in the images.

To order this product, use ISBN: 978-1-4496-4916-6

Library of Congress Cataloging-in-Publication Data
Smith, Karl J.
  Precalculus : a functional approach to graphing and problem solving/Karl Smith—6th ed.
    p. cm.
  Includes index.
  ISBN-13: 978-0-7637-5177-7 (hardcover)
  ISBN-10: 0-7637-5177-4 (ibid)
  1. Functions—Textbooks. I. Title.
  QA331.S6168 2010
  515—dc22
                              2009005347

6048
Printed in the United States of America
15 14 13 12 11   10 9 8 7 6 5 4 3 2 1

Dedication

This book is dedicated
to my daughter,
Melissa Ann Becker.

Missy graduated from UCLA
and is the principal at Meadow Elementary School in Petaluma, California.
She loves people, especially children,
and is now fulfilling her lifelong goal of working with children.
I am very proud of her, and dedicate this book
to her with love.

Table of Contents

A Note About Calculator Usage xiii

Preface xv

Problem Solving xix

Acknowledgments xxi

About the Author xxiii

Fundamental Concepts

Chapter 1 Algebraic and Geometrical Foundations 3

1.1 Real Numbers 4
1.2 Algebraic Expressions 12
1.3 Equations of Lines 20
1.4 Distance and Symmetry 29
1.5 Linear Inequalities and Coordinate Systems 36
1.6 Absolute Value Equations and Inequalities 42
*1.7 What Is Calculus? 50
1.8 Equations for Calculus 57
1.9 Inequalities for Calculus 65
   Chapter 1 Summary and Review 72

Functions and Graphing

Chapter 2 Functions with Problem Solving 77

2.1 Problem Solving 78
2.2 Introduction to Functions 92
2.3 Graph of a Function 103
2.4 Transformations of Functions 114
2.5 Piecewise Functions 123
2.6 Composition and Operations of Functions 129
2.7 Inverse Functions 136
2.8 Limits and Continuity 141
   Chapter 2 Summary and Review 152

Chapter 3 Polynomial Functions 159

3.1 Linear Functions 160
3.2 Quadratic Functions 169
3.3 Optimization Problems 177
3.4 Parametric Equations 187
3.5 Graphing Polynomial Functions 193
3.6 Polynomial Equations 203
   Chapter 3 Summary and Review 214

*optional section

Chapter 4 Additional Functions 219
   4.1 Rational Functions 220
   4.2 Radical Functions 230
   4.3 Real Roots of Rational and Radical Equations 236
   4.4 Exponential Functions 243
   4.5 Logarithmic Functions.. 254
   4.6 Logarithmic Equations.. 262
   4.7 Exponential Equations 268
       Chapter 4 Summary and Review 277

Trigonometry

Chapter 5 Trigonometric Functions 283
   5.1 Angles 284
   5.2 Fundamentals 295
   5.3 Trigonometric Functions of Any Angle 304
   5.4 Graphs of the Trigonometric Functions 312
   5.5 Inverse Trigonometric Functions 332
   5.6 Right Triangles.. 343
   5.7 Oblique Triangles... 355
       Chapter 5 Summary and Review........................ 369

Chapter 6 Trigonometric Equations and Identities 375
   6.1 Trigonometric Equations...................................... 376
   6.2 Proving Identities 385
   6.3 Addition Laws... 393
   6.4 Miscellaneous Identities 402
  *6.5 De Moivre's Theorem 413
       Chapter 6 Summary and Review 421

Advanced Algebra Topics and Analytic Geometry

Chapter 7 Analytic Geometry 429
   7.1 Parabolas 430
   7.2 Ellipses 439
   7.3 Hyperbolas 451
   7.4 Rotations 462
  *7.5 Polar Coordinates 468
   7.6 Graphing in Polar Coordinates 475
   7.7 Conic Sections in Polar Form 485
       Chapter 7 Summary and Review 492

*optional section

▨ Chapter 8   Sequences, Systems, and Matrices 499

   8.1   Sequences 500

   *8.2   Limit of a Sequence 509

   8.3   Series 518

   8.4   Systems of Equations 530

   8.5   Matrix Solution of a System of Equations 540

   8.6   Inverse Matrices 556

   8.7   Systems of Inequalities 570

       Chapter 8 Summary and Review 577

▨ Chapter 9   Vectors and Solid Analytic Geometry 583

   *9.1   Vectors in the Plane 584

   *9.2   Quadric Surfaces and Graphing in Three Dimensions 593

   *9.3   The Dot Product 603

   *9.4   The Cross Product 614

   *9.5   Lines and Planes in Space 624

       Chapter 9 Summary and Review 634

▨ Appendices 640

   Appendix A: Field Properties 640

   Appendix B: Complex Numbers 645

   Appendix C: Mathematical Induction 650

   Appendix D: Binomial Theorem 655

   Appendix E: Determinants and Cramer's Rule 661

   Appendix F: Library of Curves and Surfaces 670

   Appendix G: Answers 680

▨ Index 729

*optional section

A Note About Calculator Usage

One of my major criticisms about the calculator-based mathematics material I have seen published or presented at mathematics conferences is that the author of the material gets bogged down interpreting the precision of the calculator and the inputting of information into the calculator. Consequently, I lose track of the mathematics I'm trying to learn. Knowing about "pixel" size and how my calculator works does not enhance my *mathematical* understanding. Just as I am able to use my car to get to work without understanding the principles of the internal combustion engine, I believe we can use a calculator to advance our mathematical goals without understanding the details of "how" the calculations are accomplished. However, I do feel it is important to discuss how to interpret calculator output. For example, we need to recognize `2.409 -28` as a number in scientific notation, `2.409 × 10⁻²⁸`, which for most practical purposes is 0. Similarly, an output of `0.33333333` or `3.1415927` should be recognized as $\frac{1}{3}$ or π, respectively. Or, at a slightly higher level, if we are tracing a point on a curve and obtain successive coordinates `(11.982187, -.00239876)` and `(12.0498731, .0498710)`, we need to discuss the appropriate mathematics to be able to conclude that the *x*-intercept is probably $(12, 0)$. On the other hand, many real-world models do not have "nice" or rational intercepts and these problems are, from a practical standpoint, impossible to solve without the new technology. In such a case the `ZOOM` key can be used to approximate the intercept to any reasonable degree of accuracy. In other words, I believe the calculator to be an invaluable *tool* in understanding mathematics, but its use is not the ultimate goal of the material of this book.

You should notice that the notation I introduce in the book is consistent with that used by graphing calculators. I have attempted to make this book as calculator-independent as possible, and for the most part do not include little calculator keys telling you what keys to press, but instead expect you to know how to use the calculator you own.

I also have resisted the temptation to insert a cute little "calculator logo" on certain problems. Even though many problems unique to this book were designed with the calculator in mind, I believe the student should consider the calculator, and the graphing calculator, as useful *tools* for *any* problem.

Preface

Many colleges and universities offer a course called Precalculus, and there are many textbooks with that title, all of which tell the reader that the book is intended to prepare students for calculus. In order to understand why I wrote this book, and why I've added a subtitle—*A Functional Approach to Graphing and Problem Solving*—requires that we take a look at the reform issues that surround calculus as it is being taught today.

The major issue driving calculus reform today is the poor performance of students who are trying to master the concepts of calculus. Much of this failure can be attributed to the ways in which students are taught mathematics in high school. The high school mathematics textbooks that have been published over the last twenty years increasingly have focused on reducing mathematics into a series of small repeatable steps. Process is stressed over insight and understanding. In these same books we often find problems that are matched to worked-out sample problems that encourage students to memorize a series of problem-solving algorithms. Unfortunately, this reduces mathematics to taxonomy, and students show up for calculus generally willing to work hard, but for many, this translates into working hard at rote memorization! The task of getting students to think conceptually falls to the teachers of calculus and precalculus.

This book was written to provide sound development, stimulating problems, and a well-developed pedagogy. A common complaint in calculus is a lack of preparation in algebra and trigonometry. It is not uncommon for students who have had success in algebra, trigonometry, and analytic geometry, to find calculus too difficult. Part of the reason for these difficulties is that precalculus topics are divided into small compartments that students can master one at a time, but when confronted with a mixture of those same problems (which is what occurs in calculus), they become confused and unable to succeed. Students seem to have trouble seeing the forest because they are focused on the trees. Understanding the review material called *Practice for Calculus—Supplementary Problems: Cumulative Review* is necessary for success in calculus. These problems are presented as a true mixture of problems taken from calculus.

This book was written with two major themes in mind: (1) problem solving and (2) functions and graphs. In addition to developing some prerequisite skills in algebra, geometry, and problem solving, a precalculus text today should require students to develop verbal skills in a mathematical setting. This is not just because real mathematics wields its words precisely and compactly—verbalization should help students think conceptually. Included in this text are many opportunities for writing exercises, including individual and group projects, and WHAT IS WRONG? and IN YOUR OWN WORDS type problems. Classroom experience using these problems has been positive. At first, some students resist writing sentences, paragraphs, or papers in a mathematics classroom, but by the end, they have a better understanding of the subject.

Part of the process of *thinking conceptually* involves using a variety of learning models that include **numerical**, **algebraic**, and **geometric** solutions for a given problem. Whenever possible, more than one of these models have been presented. An important part of conceptualization is to use the available technology. Thus you will find *Computational Windows* throughout the book. However, it is imperative to know that the technology is used to help understand the mathematics, not the mathematics to better understand the technology. Students might want to read *A Note About Calculator Usage* immediately following the Table of Contents. It should be noticed that the technology in this book is platform neutral because specific calculators and computer programs frequently change. It is NOT necessary for the student (or the instructor) to use a graphing calculator to use this book, but it is important to recognize that technology must play a role in the learning process. Nevertheless, the focus of this book remains problem solving and functions.

To the Instructor

A great deal of flexibility is possible in the instructor's selection of topics presented in this book. More material than can be used in a single semester or quarter has been provided so that instructors can select material appropriate for their school or class.

Mathematics is loved by many, disliked by a few, admired and respected by all. Because of their immense power and reliability, mathematical methods inspire confidence in persons who comprehend them and awe in those who do not.

—Hollis R. Cooley, *Mathematics*, New York: New York University Press, 1963, p. 6

Some of the material may not be considered typical. Even though I had taught precalculus for over 20 years, had previously published a precalculus book, and spent the past several years writing a calculus textbook, I saw the inadequacy of the usual precalculus content. There are tremendous gaps in a typical student's preparation for calculus. Sometimes it is just terminology, such as the use of words like "if and only if," or "contrapositive"; sometimes it is a topic long forgotten, such as similar triangles (see p. 304); and sometimes it is a missing topic, such as solving equations by equating the coefficients (see p. 64).

Most often the problem is simply a lack of depth of development of the algebraic and trigonometric ideas as usually presented in precalculus. My calculus students easily mastered the new calculus ideas but had trouble with the algebraic manipulation. The complicated nature of the precalculus material in calculus was a stumbling block. There must be a closer tie between the precalculus preparation and the exact form of the problems that students will encounter in calculus. To bind the material of this book to calculus, a quick overview of calculus has been presented in Section 1.7. This allows the use of some of the terminology that students will encounter, such as limit, derivative, and integral.

The instructor will also notice that an intuitive introduction to limits and continuity has been included in Chapter 2. The intuitive introduction to these ideas in precalculus will help to ensure the student's success in calculus, since the limit concept is so essential to understanding in calculus. Parametric equations are presented early, not only because it is an important topic in calculus, but also because it is an important tool in using a graphing calculator.

To reinforce the precalculus practice for the problems that students will encounter in calculus, PROBLEMS FROM CALCULUS have been included. This designation has been used for problems that have been taken from calculus textbooks, and the problems have been rephrased to focus on precalculus content. For example, you will find factoring problems which I have found in calculus as derivatives of products and quotients and powers. Of course, students are not asked to find the derivative, but the form is given that they will find after determining the derivative (see p. 52). In Section 2.1, related rate problems from calculus are given (with the calculus part removed).

Finally, included at the end of each chapter (after Chapter 1), is a cumulative review of those problems that calculus students frequently encounter—those that require precalculus skills. It may be noticed that the problems are presented in a somewhat random order, both in topic and in difficulty; this is typical of the way calculus students will need to remember precalculus topics. In the final review at the end of Chapter 9 (see pp. 636 and 638), two problem sets are given that, if correctly worked by a student, will ensure success in calculus.

Instructor Resources

- A *Complete Solutions Manual* is available electronically for qualified instructors of this textbook.

- *PowerPoint Presentations* feature the key labeled figures as they appear in the text. This useful tool allows instructors to easily display and discuss figures and problems found within the text.

- A *Test Bank* is available through the Publisher as .rtf files for upload to instructors' Learning Management System.

- *WebAssign*™ has been developed by instructors for instructors. It is a premier independent online teaching and learning environment that has guided several million students through their academic careers since 1997. With WebAssign instructors can create and distribute algorithmic assignments using questions specific to this textbook. Instructors can also grade, record, and analyze student responses and performance instantly; offer more practice exercises, quizzes and homework; and upload additional resources to share and communicate with their students.

(continued)

- *eBook* When this textbook's course area has been registered at WebAssign, instructors and students have access to this textbook in electronic format for the life of the sixth edition.

Please contact your Jones & Bartlett Learning Account Specialist for information on, access to, and online demonstrations of the instructor resources and services described above at www.jblearning.com, or 1-800-832-0034.

To the Student

Of course, the selection of topics in this book is designed to prepare you for calculus. Calculus is a difficult subject, and there is no magic key to success—it will require hard work. This book should make your calculus journey easier, because it reviews the concepts and skills you will need for calculus, builds your problem-solving skills, and helps you form good study habits. You will need to read the book, work the examples in the book using your own pencil and paper, and make a commitment to do your mathematics homework on a daily basis.

☠ Read that last sentence once again. It's the best hint you will see about building success in mathematics. ☠

I have written this book so that it will be easy for you to know what is important.

- Important terms are presented in **boldface type**.
- Important ideas, definitions, and procedures are enclosed in boxes .
- *Common pitfalls, helpful hints, and explanations are shown using this font.*
- ☠ WARNINGS are given to call your attention to common mistakes.
- Color is used in a functional way to help you "see" what to do next.

Success in this course is a joint effort by the student, the instructor, and the author. The student must be willing to attend class and devote time to the course *on a daily basis*. There is no substitute for working problems in mathematics. If you can successfully complete this book and work the problems in the final cumulative review, you will be successful when you enroll in calculus.

- *The Student Survival Guide and Solutions Manual* was developed by the author as a study guide that offers students math study hints and a complete solution for every odd-numbered exercise in the textbook. This title is available for purchase in print and online format at www.jblearning.com

- *The Student Companion Website* features a complete online student tutorial that follows the order of the textbook. This text-specific content was created by the author and is available at go.jblearning.com/SmithPrecalculus

- *A WebAssign™ online homework and assessment student access card* is available bundled with this text or it can be purchased separately for students enrolled in a precalculus course requiring this sixth edition and access to WebAssign. Once registered through WebAssign, students will also have access to this sixth edition of the textbook in eBook format.

(continued)

Student Resources *(continued)*

- *CourseSmart:* This precalculus text is available for online subscription at CourseSmart .com. The eBook has many features designed to make studying more efficient, such as highlighting, online search, note taking, and print capabilities. For purchase information please visit www.jblearning.com/elearning/econtent/coursesmart

- *Exploring Mathematics: Solving Problems with the TI-84 Plus Graphing Calculator* by Jeffrey Gervasi of Porterville College is a useful manual that provides students with instruction on how to use a graphing calculator to solve precalculus problems. This manual can be purchased online at www.jblearning.com or in print format.

Problem Solving

Problems: Students *learn* mathematics by *doing* mathematics. Therefore, the problems and applications are perhaps the most important feature of any mathematics book. The problems in this book extend from routine practice to challenging. The problem sets are divided into Level 1 Problems (routine), Level 2 Problems (requiring independent thought), and Level 3 Problems (challenging). Application problems and *Modeling Applications* are integrated into almost every section of the book, and they are the cornerstone for mastering the prerequisite material for calculus.

Mathematical Modeling: Many books pay lip service to mathematical modeling, and some use it simply to mean "word" or "story" problems. Even though word problems are important in this book, it is not what is meant by mathematical modeling. It is not until Chapter 2 that this idea is formally introduced, but once introduced, it is a major thread of subsequent development. Simply stated, *mathematical modeling* is a *process* of creating a mathematical representation of a real-world phenomenon in order to gain better understanding of that phenomenon. It is an attempt to match observation in and about this world with symbolic mathematical statements. In the process, the student is taught to examine critically, evaluate, and modify. As an example, you might wish to look forward to Example 7, Section 4.7. In this problem, we seek to predict when the world's population will reach 7 billion. We make some assumptions and predict a date of February 27, 2010. Was this prediction correct? Other assumptions in the problem lead to other dates. What will be the ultimate test of our modeling ability for this problem? Find out when that population was actually reached and refine the model to match that date. Throughout the book you will find problems designated as MODELING APPLICATIONS.

Functions: A quick look at the Table of Contents will show that the central topic of this course is the notion of functions. After some algebraic and geometrical foundations, functions are defined in Chapter 2, and in that chapter some of the general properties of functions are discussed. Next, polynomial functions, rational functions, radical functions, exponential functions, logarithmic functions, and trigonometric functions are defined and discussed. For each different type of function, its properties and graph are considered and then investigated solving equations involving that type of function.

Algebra: It has been said that calculus is about 90% algebra and 10% calculus. There *are many* problems in this book that are designed to give the student the good old-fashioned algebraic practice that is needed (without relying on technology). The problems in this book provide a bridge between the (relatively) easy problems in algebra (in factoring, for example) and the (relatively) difficult problems in calculus.

Acknowledgments

I would like to thank the publishing staff at Jones & Bartlett Learning: Cathleen Sether, Publisher; Tim Anderson, Senior Acquisitions Editor; Amy Rose, Director of Production; Kat Crighton, Senior Production Editor; Tiffany Sliter, Production Editor; Megan Turner, Senior Associate Editor; Melissa Elmore, Digital Products Manager; and Andrea DeFronzo, Senior Marketing Manager.

The reviewers for this and all previous editions greatly helped to shape this precalculus textbook. Special thanks are given for the detailed reviews of this sixth edition that were submitted by Patricia Jones of Methodist College, Beth Long of Pellissippi State Technical Community College, Dr. Joanne Peeples of El Paso Community College, and Charles Ashbacher of Mount Mercy College.

The following math educators graciously participated in our precalculus workshop during the 2010 AMATYC conference in Boston, Massachusetts. Their advice influenced this new edition:

Prof. Susan McLoughlin, Union County College
Don Davis, Lakeland Community College
Sandy Spears, Jefferson Community and Technical College
Jack Keating, Massasoit Community College
Jane Marie Wright, Suffolk Community College
Sofya Antonova, Collin College
Jerry Chen, Suffolk Community College
Heather C. Knuth, Mass Bay Community College

Finally, I would like to thank the accuracy checkers for this edition: Melanie Fulton, John Fulton, Susan Reiland, Jenny Bagdigian, and Christopher Underwood.

Karl J. Smith

About the Author

Karl Smith has gained a reputation as a master teacher and guest lecturer. He has been a department chairperson, and is past president of the American Mathematical Association of Two-Year Colleges. He has served on numerous boards, including the Conference Board of Mathematical Sciences (which serves as a liaison between the National Academy of Sciences, the federal government, and the mathematical community in the United States), the Council of Scientific Society Presidents, the Board of the California Mathematics Council for Community Colleges, and was invited to participate in the National Summit on Mathematics Education. He received his B.A. and M.A. from the University of California at Los Angeles, and his Ph.D. from Southeastern University. In his spare time he enjoys running and swimming.

One of the most important goals of this book is to make mathematics real and alive for students. For that reason, personal notes of many mathematicians are included in the book.

Other Titles by Karl J. Smith

Calculus (with Monty Strauss and Jerry Bradley), 4th Edition
Biocalculus (with Wayne Getz and Sebastian Schreibner)
The Nature of Mathematics, 12th Edition
Mathematics: Its Power and Utility, 10th Edition
Trigonometry for College Students, 7th Edition
Essentials of Trigonometry, 4th Edition

Chapter Objectives

The material in this chapter is reviewed in the following list of objectives. After completing this chapter, review this list again, and then complete the self-test.

1.1
1.1 Be familiar with the counting numbers, whole numbers, integers, rational numbers, and real numbers.

1.2 State and use the trichotomy property.

1.3 Know the definition of absolute value; simplify absolute value expressions.

1.4 Find distance in one dimension.

1.5 Graph numbers on a number line.

1.6 Be able to convert from linear graphs, interval notation, and inequality notation.

1.2
1.7 Be familiar with the terminology of polynomials, including the definition and laws of exponents.

1.8 Simplify polynomials.

1.9 Factor polynomials.

1.3
1.10 Know what it means to graph an equation and to write the equation of a graph. Be able to graph by plotting points.

1.11 Find the slope of a line, given two points.

1.12 Graph lines using the slope and y-intercept.

1.13 Given information about a line, write the equation.

1.14 Graph lines given data points and find the equation to model an applied problem.

1.15 Determine whether given lines or line segments are parallel or perpendicular.

1.4
1.16 Know the Pythagorean theorem and the distance formula. Find the distance between points in two dimensions.

1.17 Know the equation for a circle; graph circles, and find the equation of a circle given the center and the radius.

1.18 Given information about a circle, write the equation.

1.19 Find the midpoint of a given segment. Find the perpendicular bisector of a line segment.

1.20 Describe symmetry about a line; draw curves showing symmetry with respect to a line; the x-axis; y-axis; and the origin. Test equations for symmetry.

1.5
1.21 Graph inequalities with one variable in both one and two dimensions.

1.22 Find values for which $y = 0$, $y > 0$, and $y < 0$ by looking at a graph.

1.23 Solve linear inequalities.

1.24 Formulate a mathematical problem numerically, algebraically, and geometrically.

1.6
1.25 Define absolute value.

1.26 Solve absolute value equations.

1.27 Solve absolute value inequalities.

1.28 Translate statements using absolute value.

1.7
1.29 Intuitively, what is a limit?

1.30 Intuitively, what is a derivative?

1.31 Intuitively, what is an integral?

1.32 Intuitively, what is calculus?

1.8
1.33 Solve quadratic equations by factoring.

1.34 Solve quadratic equations by completing the square.

1.35 Know the quadratic formula. Solve quadratic equations over the set of real numbers.

1.36 Solve equations by equating numerical coefficients.

1.9
1.37 Solve quadratic inequalities.

1.38 Solve inequalities in factored form.

Algebraic and Geometrical Foundations

1

Mathematics is the gate and key of the sciences . . . Neglect of mathematics works injury to all knowledge, since whoever is ignorant of it cannot know the other sciences or the things of this world.

—Roger Bacon (1894)

Chapter Sections

1.1 Real Numbers
One-Dimensional Coordinate System
Properties of Inequality
Absolute Value and Distance on a Number Line

1.2 Algebraic Expressions
Terminology
Operations with Polynomials
Factoring Polynomials

1.3 Equations of Lines
Two-Dimensional Coordinate System
Slope
Intercepts
Parallel and Perpendicular Lines

1.4 Distance and Symmetry
Distance and Midpoint Formulas
Symmetry

1.5 Linear Inequalities and Coordinate Systems
Formulating a Mathematical Problem
Relationship between One- and Two-Dimensional Coordinate Systems
Linear Inequalities in One Variable

1.6 Absolute Value Equations and Inequalities
Absolute Value Equations
Absolute Value Inequalities

1.7 What Is Calculus?
The Limit: Zeno's Paradox
The Derivative: The Tangent Problem
The Integral: The Area Problem

1.8 Equations for Calculus
Quadratic Equations
Completing the Square
Quadratic Formula
Equations in Quadratic Form
Equal Coefficients

1.9 Inequalities for Calculus
Quadratic Inequalities
Factored Inequalities

Chapter 1 Summary and Review
Self Test
Practice for Calculus
Chapter 1 Group Research Projects

▶ CALCULUS PERSPECTIVE

It may have been your experience that most mathematics courses you have taken start at about the same place, namely, a discussion of numbers and properties of numbers, and you have probably found that most of the material you need in a particular mathematics course is contained in that first chapter. The primary notable exception to those experiences will occur when you enroll in your first calculus course. The purpose of this book is to give you that preparation.

The development of calculus in the 17th century by Newton and Leibniz was the result of their attempt to answer some fundamental questions about the world and the way things work. These investigations led to two fundamental concepts of calculus—namely, the idea of a *derivative* and that of an *integral*. To understand these ideas, you must first have a firm foundation of elementary mathematics, including algebra, functions, graphing, and trigonometry. This course is designed to give you that necessary background, and as we travel through *this* course, we will constantly have our eye on our final goal: success in calculus. For example, the definitions for both the derivative and the integral depend on a notion defined in calculus called the *limit*. The formal definition of a limit involves both an absolute value and an inequality,

$$0 < |x - a| < \delta$$

In this first chapter, we introduce inequalities such as this as well as review the fundamental concepts necessary for the study of calculus.

1.1 Real Numbers

You should be familiar with various sets of numbers, as shown in Table 1.1.

TABLE 1.1	Sets of Numbers			
Name	**Symbol**	**Set**	**Examples**	**Comments**
Counting numbers or natural numbers	$\mathbb{N}$	$\{1, 2, 4, 5,...\}$	$72;\ 2{,}345;\ 950;\ \sqrt{25};\ \sqrt{1}$	$\sqrt{a}$ is the nonnegative real number b so that $b^2 = a$. It is called the **principal square root** of a.
Whole numbers	$\mathbb{W}$	$\{0, 1, 2, 3,...\}$	$0;\ 72;\ \dfrac{10}{2};\ \dfrac{16{,}425}{25}$	$\dfrac{a}{b}$ means $a \div b$. It is called the **quotient** of a and b. Some quotients are whole numbers.
Integers	$\mathbb{Z}$	$\{...,-2, -1, 0, 1, 2, 3,...\}$	$-5;\ 0;\ \dfrac{-10}{2};\ 956;\ -82;\ -\sqrt{25}$	
Rational numbers	$\mathbb{Q}$	$\dfrac{p}{q},\ p$ and q are integers, $q \neq 0$	$\dfrac{1}{3};\ \dfrac{4}{17};\ 0.863214;\ 0.8666...\ ;\ 8.\overline{6}$ $\dfrac{-11}{3};\ -15;\ \sqrt{\dfrac{1}{4}};\ 3.1416$	An overbar indicates repeating decimals.
Irrational numbers	$\mathbb{Q}'$	Numbers whose decimal representation does not terminate or repeat.	$5.12345678910...\ ;\ \sqrt{2};\ \sqrt{3};$ $4.313313331...\ ;\ \pi;\ \dfrac{\pi}{2}$	π is the ratio of the circumference of a circle to its diameter. Write $\pi \approx 3.1416$ to indicate that the numbers are **approximately equal**.
Real numbers	$\mathbb{R}$	All rational and irrational numbers.	All above examples are real numbers.	There is another set of numbers called the **complex numbers** (see Appendix B).

When adding real numbers, the result is called the **sum** and the numbers added are called **terms**. When we are multiplying real numbers, the result is called the **product** and the numbers multiplied are called the **factors**. The result from subtraction is called the **difference**; the result from division is the **quotient**.

Although it is assumed you are familiar with the basic operations of real numbers, it is worth noting some basic definitions:

Subtraction is defined as the opposite operation of addition:

$$a - b = a + (-b) \quad \text{for real numbers } a \text{ and } b$$

Multiplication is defined as repeated addition:

$$ma = \underbrace{a + a + \cdots + a}_{m \text{ terms}} \quad \text{for counting number } m$$

Division is defined as the opposite operation of multiplication:

$$a \div b = \frac{a}{b} = a \cdot \frac{1}{b} \quad \text{where } \frac{1}{b} \text{ is the **reciprocal**}$$

of a nonzero number b and a is any real number.

If an expression has more than one operation, they must be carried out according to an **order of operations agreement**. First, carry out all operations enclosed in parentheses. Next, perform multiplications and divisions as they occur by working from left to right, and finally, perform additions and subtractions as they occur by working from left to right. It is assumed that you are familiar with this fundamental agreement from beginning algebra.

One-Dimensional Coordinate System

The real numbers, $\mathbb{R}$, can most easily be visualized by using a **one-dimensional coordinate system** called a **real number line** (Figure 1.1).

A **one-to-one correspondence** is established between all real numbers and all points on a real number line:

- Every point on the line corresponds to precisely one real number; and
- Every real number corresponds to precisely one point.

A point associated with a particular number is called the **graph** of that number. Numbers associated with points to the right of the starting position, called the **origin**, are called **positive real numbers**, and those to the left are called **negative real numbers**. Numbers are called **opposites** if they are plotted an equal distance from the origin. The **opposite of a real number** a is denoted by $-a$. Notice that if a is positive, then $-a$ is negative; if a is negative, then $-a$ is positive. ☠ It is a common error to think of $-a$ as a negative number; it might be negative, but it might also be positive. If, say, $a = -3$, then $-a = -(-3) = 3$, which is positive. ☠

EXAMPLE 1   Plotting real numbers on a number line

Graph the following numbers on a real number line: $-2, 2.5, 5, -\sqrt{2}, \pi, \frac{2}{3}, 1.31331133311\ldots$.

Solution   When graphing, the exact positions of the points are usually approximated.

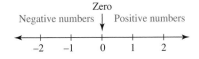

Zero

Negative numbers | Positive numbers

Figure 1.1 A real number line

Historical Note

The numeration system we use evolved over a long period of time. It is often called the *Hindu-Arabic* system because its origins can be traced back to the Hindus in Bactria (now Afghanistan). Later, in A.D. 700, India was invaded by the Arabs who used and modified the Hindu numeration system, and, in turn, introduced it to Western civilization. The Hindu Brahmagupta stated the rules for operations with positive and negative numbers in the seventh century A.D. There are some indications that the Chinese had some knowledge of negative numbers as early as 200 B.C. On the other hand, the Western mathematician Girolamo Cardan (1501–1576) was calling numbers such as (–1) absurd as late as 1545.

Properties of Inequality

There are certain relationships between real numbers with which you should be familiar.

LESS THAN

$a < b$ is read "a is less than b" and means $b - a$ is positive. On a number line, the graph of a is to the left of the graph of b.

GREATER THAN

$a > b$ is read "a is greater than b" and means $b < a$ or $a - b$ is positive. On a number line, the graph of a is to the right of the graph of b.

EQUAL TO

$a = b$ is read "a is equal to b" and means that a and b represent the same real number. On a number line, the graphs of both a and b are the same point. An **equation** is two expressions connected by an equal sign.*

If a and b represent any real numbers, then they are related by a property called **property of comparison**, or **trichotomy**.

TRICHOTOMY

Given any two real numbers a and b, exactly one of the following holds:

$$a = b \quad a > b \quad a < b$$

* We discuss equations in Section 1.8.

The operative word here is *exactly* one holds; this means that one and only one can be true at one time. Trichotomy establishes the order on a real number line. If you let $b = 0$, for example, then $a = 0$, $a < 0$, or $a > 0$. Using the definition of less than and greater than, it follows that

$a > 0$ if and only if a is positive;

$a < 0$ if and only if a is negative.

There are three other relationships that we need to know.

> **LESS THAN OR EQUAL TO**
>
> $a \leq b$ is read "a is less than or equal to b" and means that either $a < b$ or $a = b$ (but not both).
>
> **GREATER THAN OR EQUAL TO**
>
> $a \geq b$ is read "a is greater than or equal to b" and means that either $a > b$ or $a = b$ (but not both).
>
> **BETWEEN**
>
> $a < b < c$ is read "b is between a and c" and means both $a < b$ and $b < c$.
> Additional "between" relationships are also used:
>
> $a \leq b \leq c$ means $a \leq b$ and $b \leq c$
> $a \leq b < c$ means $a \leq b$ and $b < c$
> $a < b \leq c$ means $a < b$ and $b \leq c$
>
> In this book, we use the word *between* to mean strictly between, namely, $a < b < c$.

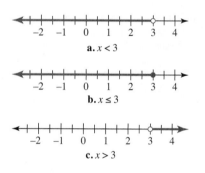

a. $x < 3$

b. $x \leq 3$

c. $x > 3$

d. $x \geq 3$

Figure 1.2 Inequality notation on $\mathbb{R}$

Graphs of linear inequality statements with a single variable are drawn on a one-dimensional coordinate system ($\mathbb{R}$). For example, $x < 3$ denotes the interval shown in Figure 1.2a. Notice that $x \neq 3$, and this fact is shown by an open circle at the endpoint of the ray. Compare this with $x \leq 3$ (Figure 1.2b) in which the endpoint $x = 3$ is included. Notice that to sketch the graph you darken (or color) the appropriate portion. Graphs of $x > 3$ and $x \geq 3$ are also shown.

There is a very useful notation to describe the graphs and intervals shown in Figure 1.2. This notation, called **interval notation**, uses brackets and parentheses to denote intervals algebraically. We will find this quite useful in calculus. The infinity symbol, ∞, is used to denote a ray, or $\mathbb{R}$ (the entire real number line). This notation is summarized in Table 1.2.

TABLE 1.2   Graph, Inequality, and Interval Notation

One-Dimensional Graph	Inequality Notation	Interval Notation
	$a < x < b$	(a, b)
	$a \leq x < b$	$[a, b)$
	$a < x \leq b$	$(a, b]$
	$a \leq x \leq b$	$[a, b]$
	$x > a$	(a, ∞)
	$x \geq a$	$[a, \infty)$
	$x < b$	$(-\infty, b)$
	$x \leq b$	$(-\infty, b)$
	$\mathbb{R}$ All real x values	$(-\infty, -\infty)$

EXAMPLE 2   Graphing an inequality and writing interval notation

Write each inequality in interval notation, and graph the inequality.

  a. $-6 \leq x < 5$ **b.** $7 < x < 8$ **c.** $-9 \leq x \leq -2$ **d.** $-5 < x \leq 0$ **e.** $x > 2$

Solution

  a. $[-6, 5)$

  b. $(7, 8)$

  c. $[-9, -2]$
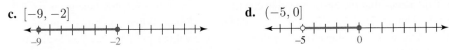

  d. $(-5, 0]$
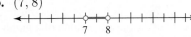

  e. $(2, \infty)$

■

EXAMPLE 3   Graphing an inequality and writing inequality notation

Write each interval using inequality notation, and graph the inequality.

  a. $[1, 3]$ **b.** $(2, 5]$ **c.** $[0, 4)$ **d.** $(-1, 4)$ **e.** $(2, 5] \cup [8, 10]$

Solution

We choose to let x be the variable for these solutions.

  a. Both endpoints are included: $1 \leq x \leq 3$.

  b. The left endpoint is excluded, and the right endpoint is included: $2 < x \leq 5$.

  c. The left endpoint is included, and the right endpoint is excluded: $0 \leq x < 4$.

  d. Both endpoints are excluded: $-1 < x < 4$.

  e. The symbol $\cup$ means *union* and is used for disjoint segments; note that the points 5, 8, and 10 are included and 2 is excluded: $2 < x \leq 5$ or $8 \leq x \leq 10$.

■

Absolute Value and Distance on a Number Line

An important idea in mathematics involves the notion of **absolute value**. If a is a real number, then its graph on a number line is some point, call it A. The distance between A and the origin is the geometric interpretation of the *absolute value of a*.

The **absolute value** of a number a is denoted by $|a|$ and is defined by

$$|a| = a \quad \text{if } a \geq 0$$
$$|a| = -a \quad \text{if } a < 0$$

>> IN OTHER WORDS If a number x is positive, then the absolute value of x is x itself. If x is negative, then the absolute value of a number x is the opposite of x. The absolute value of 0 is 0. This means that the absolute value can never be a negative number (see property 1, Table 1.3). That is $|5| = 5$, $|-5| = 5$, $|0| = 0$, and $|-\pi| = \pi$. When you find the numerical value of an absolute value (written without absolute value signs), we say you **evaluate** the absolute value.

STOP There are five important processes in algebra. *Evaluate* is the first of those five.

EXAMPLE 4   Evaluating an absolute value expression

Evaluate: **a.** $|\pi - 3|$ **b.** $|\pi - 4|$ **c.** $|\pi - \pi|$ **d.** $|x^2 + 8|$ **e.** $|-5 - x^2|$

Solution

  a. We note that $\pi - 3 > 0$, so $|\pi - 3| = \pi - 3$.
  b. Note that $\pi - 4 < 0$, so $|\pi - 4| = -(\pi - 4) = 4 - \pi$.
  c. Since $\pi - \pi = 0$, $|\pi - \pi| = 0$. *Note:* Absolute value is not always positive.
  d. $x^2 + 8 > 0$, so we find $|x^2 + 8| = x^2 + 8$.
  e. Since $-5 - x^2$ is negative for all real values of x,

$$|-5 - x^2| = -(-5 - x^2) = x^2 + 5$$

■

Since $|a|$ can be interpreted as the distance between the point A whose coordinate is a and the origin, it is a straightforward derivation to show that the distance between any two points on a number line can be expressed as an absolute value.

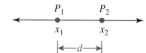

Let x_1 and x_2 be the coordinates of two points P_1 and P_2, respectively, on a number line. The **distance** d between P_1 and P_2 is

$$d = |x_2 - x_1|$$

Notice that if a is the coordinate of P_1 and 0 is the coordinate of P_2, then $d = |0 - a| = |-a|$; and if 0 is the coordinate of P_1 and a is the coordinate of P_2, then $d = |a - 0| = |a|$. Since d is the same distance for both of these calculations,

$$|-a| = |a| \quad \text{This is property 2, Table 1.3.}$$

It also follows that $d = |x_2 - x_1| = |x_1 - x_2|$.

Absolute value is related to square roots. In Table 1.1, we defined $\sqrt{a}$ as being that number so that

$$\sqrt{a}\sqrt{a} = a$$

In $\mathbb{R}$ we require that $a \geq 0$, but if $a < 0$, we might have $\sqrt{a^2}$ since $a^2 > 0$. However, in this case, if we write $\sqrt{a^2} = a$, we have a problem since a positive number $(\sqrt{a^2})$ is equal to a negative number a. To handle this, we note that

$$\text{if } a \geq 0, \sqrt{a^2} = a \quad \text{and if } a < 0, \sqrt{a^2} = -a$$

We summarize this with the following property.

SQUARE ROOT OF A PERFECT SQUARE

For all real numbers x,

$$\sqrt{x^2} = |x|$$

» IN OTHER WORDS This means that if $x \geq 0$, then $\sqrt{x^2} = x$, and if $x < 0$, then $\sqrt{x^2} = -x$. For example, $\sqrt{2^2} = |2| = 2$, and $\sqrt{(-2)^2} = |-2| = 2$.

EXAMPLE 5   Distance between points on $\mathbb{R}$, a number line

Find the distance between the points whose coordinates are given:

a. -10 and 5 **b.** 305 and -85 **c.** π and $\sqrt{5}$ **d.** π and $\sqrt{3}$

Solution

a. $d = |5 - (-10)| = |15| = 15$

b. $d = |-85 - (305)| = |-390| = 390$

c. $d = \left|\sqrt{5} - \pi\right| = \sqrt{5} - \pi$, since $\sqrt{5} - \pi > 0$

d. $d = \left|\sqrt{3} - \pi\right|$

$\quad = -\left(\sqrt{3} - \pi\right)$, since $\sqrt{3} - \pi < 0$

$\quad = \pi - \sqrt{3}$

■

Several properties of absolute value that you will need for a calculus course are summarized in Table 1.3.

TABLE 1.3	Properties of Absolute Value*	
Property	**Comment**
1. $|a| \geq 0$	Absolute value is nonnegative.
2. $|-a| = |a|$	The absolute value of a number and the absolute value of its opposite are equal.
3. $|a|^2 = a^2$	If an absolute value is squared, the absolute value can be dropped.
4. $|ab| = |a||b|$	The absolute value of a product is the product of the absolute values.
5. $\left|\dfrac{a}{b}\right| = \dfrac{|a|}{|b|}, b \neq 0$	The absolute value of a quotient is the quotient of the absolute values.
6. $-|a| \leq a \leq |a|$	Any number is between the absolute value of that number and its opposite, inclusive.
7. Let $b \geq 0$; $|a| = b$ if and only if $a = \pm b$.	This property is useful in solving absolute value equations; it is proved in Section 1.6.
8. Let $b \geq 0$; $|a| < b$ if and only if $-a < a < b$	This is used in solving absolute value inequalities.
9. Let $b \geq 0$; $|a| > b$ if and only if $a > b$ or $a < -b$	This is used in solving absolute value inequalities.
10. $|a + b| \leq |a| + |b|$	This is called the **triangle inequality**.

* Let a and b be any real numbers.

☠ "p if and only if q" is used to mean that both a statement and its converse are true; that is: If p, then q, *and* if q, then p. For example, property 8 has two parts: (1) If $|a| < b$, then $-b < a < b$. (2) If $-b < a < b$, then $|a| < b$. ☠

Property 7 is sometimes stated as $|a| = |b|$ if and only if $a = \pm b$. Since $|b| = \pm b$ (for any real number b), it follows that this property is equivalent to property 7. Also, properties 8 and 9 are true for $\leq$ and $\geq$ inequalities. Specifically, if $b > 0$, then

$$|a| \leq b \quad \text{if and only if} \quad -b \leq a \leq b$$

and

$$|a| \geq b \quad \text{if and only if} \quad a \geq b \text{ or } a \leq -b$$

PROBLEM SET 1.1

LEVEL 1

In Problems 1–6, classify each number as

$\mathbb{N} = \{natural\ numbers\}$

$\mathbb{W} = \{whole\ numbers\}$,

$\mathbb{Z} = \{integers\}$,

$\mathbb{Q} = \{rational\ numbers\}$,

$\mathbb{Q}' = \{irrational\ numbers\}$,

$\mathbb{R} = \{real\ numbers\}$.

Note from Table 1.1 that each number listed may be in none of these sets or perhaps more than one of them.

1. a. -4 **b.** $\dfrac{11}{2}$ **c.** $\sqrt{49}$ **d.** $\sqrt{50}$

2. a. 2π **b.** $\sqrt{\dfrac{1}{8}}$ **c.** $0.\overline{5}$ **d.** 0.5656

3. a. $\dfrac{\pi}{3}$ **b.** $\sqrt{2}$ **c.** $-\sqrt{2}$ **d.** $\sqrt{\dfrac{1}{100}}$

4. a. 9 **b.** $\sqrt{16}$ **c.** $\dfrac{0}{4}$ **d.** $\dfrac{4}{0}$

5. a. $0.\overline{6}$ **b.** $\sqrt{\dfrac{1}{9}}$ **c.** $\sqrt{1,000}$ **d.** $\dfrac{\pi}{6}$

6. a. 0.82333333 **b.** $0.8233333\ldots$
  c. $0.82333333^{1/2}$ **d.** $0.8232332333233\ldots$

Classify each number in Problems 7–12 as rational or irrational.

7. a. π **b.** $\dfrac{22}{7}$

  c. 3.141592654 **d.** $\dfrac{1}{17}$

8. a. 0.05882535294 **b.** $0.12112111211112\cdots$
  c. $\sqrt{2.7777\cdots}$ **d.** $\sqrt[5]{97.65625}$

9. a. $\sqrt{1+\sqrt{2}}$ **b.** $\dfrac{\sqrt{2}}{3}$

  c. 0.4714045208 **d.** $0.12345678910111213\cdots$

10. a. $0.33333333\cdots$ **b.** 0.333333333

  c. $\sqrt{\dfrac{2+\sqrt{5}}{2}}$ **d.** $\sqrt[3]{571.787}$

11. a. 1.414213562^2 **b.** $\left(\sqrt[5]{176+80\sqrt{5}}-1\right)^2$
12. a. 1.732050808^2
  b. $\sqrt{169+15\sqrt{77}}-\sqrt{\left(5\sqrt{11}-13\sqrt{2}\right)\left(3\sqrt{7}-13\sqrt{12}\right)}$

Illustrate the trichotomy property in Problems 13–20 by replacing the question mark by $=$, $>$, or $<$.

13. a. $-10\ ?\ -3$ **b.** $6-9\ ?\ 9-6$

14. a. $-\dfrac{8}{3}\ ?\ -2.66$ **b.** $\dfrac{3}{4}+\dfrac{4}{5}\ ?\ \dfrac{31}{20}$

15. a. $-4\ ?\ -8$ **b.** $\pi\ ?\ \dfrac{22}{7}$

16. a. $\sqrt{2}\ ?\ 1.41$ **b.** $\pi\ ?\ 3.1416$

17. a. $\sqrt{2.1}\ ?\ \sqrt{2}$ **b.** $-\sqrt{3}\ ?\ -\sqrt{3.1}$

18. a. $\dfrac{\pi}{2}\ ?\ 1.57$ **b.** $\dfrac{1}{3}\ ?\ 0.333333333$

19. a. $\sqrt{5}\ ?\ 2.236067978$ **b.** $\sqrt{1,000}\ ?\ 31.6227766$

20. a. $\dfrac{355}{113}\ ?\ \pi$ **b.** $\dfrac{355}{113}\left(1-\dfrac{0.003}{3,533}\right)\ ?\ \pi$

WHAT IS WRONG, *if anything, with each statement in Problems 21–26? Explain your reasoning.*

21. $\sqrt{2} = 1.414213562$
22. $\pi = 3.141592654$
23. $-n$ is a negative number.
24. Absolute value is always positive.
25. If we say x is between 5 and 6, then in this book we mean $5 \leq x \leq 6$.
26. If x is a real number, then x is either positive or negative.

Write each expression in Problems 27–32 without using the symbol for absolute value.

27. a. $|\pi-2|$ **b.** $|\pi-5|$

  c. $|x^2|$ **d.** $|-3-x^2|$

28. a. $\left|\sqrt{2}-1\right|$ **b.** $\left|\sqrt{2}-2\right|$

  c. $|-1-x^2|$ **d.** $|x^4+1|$

29. a. $|2\pi-6|$ **b.** $|2\pi-7|$

  c. $|x^4+x^2|$ **d.** $|-4-x^6|$

30. a. $\left|\frac{\pi}{6}-1\right|$ **b.** $\left|\frac{2\pi}{3}-1\right|$

  c. $-|x^2|$ **d.** $-|1+x^2|$

31. a. $\dfrac{x+y+|x-y|}{2}$, where $x = y$

  b. $\dfrac{x+y+|x-y|}{2}$, where $x > y$

  c. $\dfrac{x+y+|x-y|}{2}$, where $x < y$

32. a. $\dfrac{2x+y-|x-y|}{2}$, where $x = y$

  b. $\dfrac{2x+y-|x-y|}{2}$, where $x > y$

  c. $\dfrac{2x+y-|x-y|}{2}$, where $x < y$

Find the distance between the points whose coordinates are given in Problems 33–38.

33. a. (8) and (15) **b.** (-103) and (6)

34. a. $(-\pi)$ and (-3) **b.** (-143) and (-120)

35. a. (-23) and (-96) **b.** $\left(-\sqrt{5}\right)$ and (-3)

36. a. $(-\pi)$ and $\left(-\sqrt{2}\right)$ **b.** $\left(\sqrt{3}\right)$ and $(-\pi)$

37. a. $\left(\dfrac{4}{15}\right)$ and $\left(\dfrac{8}{10}\right)$ **b.** $\left(\sqrt{8}\right)$ and $\left(\sqrt{2}\right)$

38. $\left(-2\sqrt{3}-\sqrt{5}\right)$ and $\left(\sqrt{5}+\sqrt{3}\right)$

Graph each number given in Problems 39–42 on a real number line.

39. $-3; -\sqrt{3}; \dfrac{5}{4}; 2; 0.123456789101112\cdots$

40. $-4; -\dfrac{-5}{3}; \sqrt{2}; -1.343443444\cdots$

41. $0; \dfrac{\pi}{6}; \dfrac{\pi}{-3}; \dfrac{\pi}{2}; \pi$

42. $-\pi; -\dfrac{\pi}{2}; \dfrac{\pi}{3}; 0; 0.5$

For each inequality in Problems 43–46, graph and write in interval notation.

43. a. $3 < x < 7$ **b.** $-4 < x < -1$
   c. $-2 \le x \le 6$ **d.** $-3 < x \le 0$

44. a. $-6 \le x \le 0$ **b.** $-3 < x \le 3$
   c. $-\dfrac{\pi}{2} < x \le \dfrac{\pi}{2}$ **d.** $-\pi < x < \pi$

45. a. $x \le -3$ **b.** $x \ge -\pi$
   c. $x < 0$ **d.** $2 < x$

46. a. $3 < x \le 6$ **b.** $-\pi < x$
   c. $10 > x$ **d.** $-0.01 \le x \le 0.01$

For each interval in Problems 47–50, graph and write in inequality notation (use x as the variable).

47. a. $[-3, 2]$ **b.** $(-2, 2)$
   c. $(-\infty, 3]$ **d.** $(-\infty, 6) \cup (6, \infty)$

48. a. $(-3, 4]$ **b.** $[-3, 4)$
   c. $(-\infty, 1) \cup (1, \infty)$ **d.** $(2, \infty)$

49. a. $(-2, 0] \cup (3, 5)$ **b.** $(-\infty, 2) \cup (2, \infty)$
   c. $(-\infty, -3) \cup (0, 3]$ **d.** $[-5, -1) \cup (0, 5]$

50. a. $(-\infty, 0) \cup (0, \infty)$ **b.** $[-8, 3) \cup [5, \infty)$
   c. $[-4, 3)$ **d.** $[-2, \infty)$

LEVEL 2

Write each statement in Problems 51–54 symbolically, and then tell whether the statement is true or false. If it is false, give a counterexample.

51. a. The absolute value of a sum is the sum of the absolute values.
   b. The absolute value of a product is the product of the absolute values.

52. a. The absolute value of a difference is the difference of the absolute values.
   b. The absolute value of a quotient is the quotient of the absolute values.

53. a. The difference of squares is the square of the difference.
   b. The sum of squares is the square of the sum.

54. a. The square root of a sum of squares is the sum of the numbers.
   b. The square root of a difference of squares is the difference of the numbers.

LEVEL 3

55. IN YOUR OWN WORDS Give an example for each of the following:
   a. The sum of two rational numbers which is rational.
   b. The sum of two irrational numbers which is irrational.
   c. Is it possible for the sum of two irrational numbers to be rational?

56. IN YOUR OWN WORDS Give an example for each of the following:
   a. The product of two rational numbers which is rational.
   b. The product of two irrational numbers which is irrational.
   c. Is it possible for the product of two irrational numbers to be rational?

57. The *additive inverse of a number a* is a number $-a$ so that

$$a + (-a) = (-a) + a = 0$$

Find the additive inverse of each of the given numbers.

   a. -8 **b.** $\dfrac{\pi}{2}$

   c. $\sqrt{3}+5$ **d.** $\dfrac{1}{\sqrt{5}-2}$

58. The *multiplicative inverse of a number a $(a \ne 0)$* is a number $1/a$ so that

$$a\left(\dfrac{1}{a}\right) = \left(\dfrac{1}{a}\right)a = 1$$

Find the multiplicative inverse of each of the given numbers.

   a. -8 **b.** $\dfrac{\pi}{2}$

   c. $\sqrt{3}+5$ **d.** $\dfrac{1}{\sqrt{5}-2}$

59. Journal Problem (*Function*, Vol. 5, No. 3, June 1981, Problem 5.3.3 by Colin A. Wratten.) Prove

$$\sqrt{2+\sqrt{2}} + \sqrt{2-\sqrt{2}} < 2\sqrt{2}$$

60. Let $a = \sqrt{5+\sqrt{21}} + \sqrt{8+\sqrt{55}}$ and $b = \sqrt{7+\sqrt{33}} + \sqrt{6+\sqrt{35}}$.
   a. Use your calculator to show that $a \approx b$.
   b. Trichotomy says exactly *one* of the following is true: $a = b$, $a > b$, $a < b$. Without using technology, show which it is for this a and b.

1.2 Algebraic Expressions

Terminology

A **numerical expression** is a number or a grouping of numbers by valid mathematical operations (such as addition, subtraction, multiplication, nonzero division, extraction of roots, or raising to integral or fractional powers). If the numerical expression also includes a variable, it is called an **algebraic expression**. In your previous courses, you have used the laws of exponents extensively, and they must also be fully understood for the study of calculus.

RATIONAL EXPONENTS

If b is any real number and n is any natural number, then

$$b^n = \underbrace{b \cdot b \cdot b \cdots b}_{n \text{ factors}}$$

If $b \neq 0$, then $b^0 = 1$ and $b^{-n} = \dfrac{1}{b^n}$.

If $b > 0$, then $b^{1/n} = \sqrt[n]{b}$.

If m is also a natural number where m/n is reduced, then

$$b^{m/n} = (b^{1/n})^m = \sqrt[n]{b^m} = \left(\sqrt[n]{b}\right)^m$$

The number b is called the **base**, n is called the **exponent**, and b^n is called the **nth power** of b.

This definition sets the stage for much of our work in algebra. It is used in conjunction with five laws of exponents. You may recall from your previous work, to multiply expressions with the same base, add the exponents; to divide, subtract the exponents; to raise a power to a power, multiply exponents; to raise a product or quotient to a power, raise each factor to that power. These laws are summarized in the following box.

LAWS OF EXPONENTS

Let a and b be real numbers, and let m and n be integers. Then the five rules listed below govern the use of exponents except that the form 0^0 and division by zero are excluded.

Additive law of exponents:	$b^m \cdot b^n = b^{m+n}$
Subtractive law of exponents:	$\dfrac{b^m}{b^n} = b^{m-n}$
Multiplicative law of exponents:	$(b^n)^m = b^{mn}$
Distributive laws of exponents:	$(ab)^m = a^m b^m$ and $\left(\dfrac{a}{b}\right)^m = \dfrac{a^m}{b^m}$

We have called numbers being added *terms*, but we now need a slightly more general definition. A **term** is a constant, a variable, or a product of constants and variables. Thus, $10x$ is a term, but $10 + x$ is not (because the terms 10 and x are connected by addition and not multiplication). A finite sum of terms with *whole number exponents* on the variables is called a *polynomial*. For example,

POLYNOMIALS: $\qquad 6x, \quad 2x^2y + z^4, \quad \dfrac{1}{2}x + 4, \quad 0, \quad 12\dfrac{1}{3}, \quad \dfrac{x+3}{5}$

NOT POLYNOMIALS: $\quad \dfrac{1}{x}, \quad 2 + \sqrt{x}, \quad \dfrac{\sqrt[5]{x+1}}{5}, \quad x^{2/3}$

The general form of a polynomial with a single variable can be defined.

POLYNOMIAL

For any counting number n, an nth-degree **polynomial** is an expression which can be written in the form

$$a_n x^n + a_{n-1} x^{n-1} + \cdots + a_2 x^2 + a_1 x + a_0, \qquad a_n \neq 0$$

In an expression of the form $a_n x^n$, a_n is called the **coefficient** of x^n. If a_n is the coefficient of the greatest power of x, it is called the **leading coefficient** of the polynomial. If x is nonzero, the polynomial is said to have **degree** n. Notice that if $n = 2$ (degree 2), then the polynomial is called **quadratic**; if $n = 1$ (degree 1), then the polynomial is said to be **linear**; and if $n = 0$ (degree 0), then it is called a **constant**.

Each $a_k x^k$ of a polynomial is a term. Polynomials are frequently classified according to the number of terms. A **monomial** is a polynomial with one term; a **binomial** has two terms; and a **trinomial** has three terms. It is customary to arrange the terms of a polynomial in order of decreasing powers of the variable. **Similar terms** are terms with the same variable and the same degree. Two polynomials are **equal** if and only if coefficients of similar terms are the same. If some of the numerical coefficients are negative, they are usually written as subtractions of terms. Thus,

$$6 + (-3)x + x^3 + (-2)x^2$$

would customarily be written as

$$x^3 - 2x^2 - 3x + 6$$

To **evaluate** a polynomial means to replace the variable(s) with given numerical values and write the resulting numerical expression as a single number.

EXAMPLE 1   Evaluate a polynomial

Evaluate the given polynomial for $a = 2$ and $b = -3$.

a. $a^2 + b^2$ **b.** $(a + b)^2$ **c.** $-b^2$ **d.** $(-a)^2$

Solution

a. $\begin{aligned} a^2 + b^2 &= (2)^2 + (-3)^2 \\ &= 4 + 9 \\ &= 13 \end{aligned}$ **b.** $\begin{aligned} (a + b)^2 &= [2 + (-3)]^2 \\ &= (-1)^2 \\ &= 1 \end{aligned}$

c. $\begin{aligned} -b^2 &= -(-3)^2 \\ &= -9 \end{aligned}$ **d.** $\begin{aligned} (-a)^2 &= (-2)^2 \\ &= 4 \end{aligned}$ ∎

Operations with Polynomials

A numerical expression is said to be **simplified** if it is written as a single number after carrying out all indicated operations according to the order of operations agreement. A polynomial is said to be **simplified** if it is written with all similar terms combined and is expressed in the form of a polynomial, as given in the previous box. The distributive property shows you how to add polynomials by adding similar terms.

DISTRIBUTIVE PROPERTY

For real numbers a, b, and c,

$$a(b + c) = ab + ac$$

Since similar terms contain not only the same variables but also the same exponent (or exponents) on the variables, the distributive property can be applied. For example,

$$\underset{\downarrow}{5}\,\underset{\downarrow}{x} + \underset{\downarrow}{3}\,x = (5+3)x = 8x$$

If you also freely use the commutative and associative properties, you can simplify more complicated expressions using the idea of similar terms:

$$(5x^2 + 2x + 1) + (3x^3 - 4x^2 + 3x - 2) = \underbrace{3x^3}_{\text{3rd-deg. terms}} + \overbrace{\underbrace{5x^2 + (-4)x^2}_{\text{2nd-deg. terms}}}^{\text{Similar terms}} + \overbrace{\underbrace{2x + 3x}_{\text{1st-deg. terms}}}^{\text{Similar terms}} + \overbrace{\underbrace{1 + (-2)}_{\text{0-deg. terms}}}^{\text{Similar terms}}$$

$$= 3x^3 + x^2 + 5x - 1$$

There are five important processes in algebra. *Simplify* is the second of these. (Remember, the first was *evaluate*.)

EXAMPLE 2   Adding and subtracting polynomials

Simplify:

   a. $(4x - 5) + (5x^2 + 2x + 1)$ **b.** $(4x - 5) - (5x^2 + 2x + 1)$

   c. $(5x^2 + 2x + 1) - (3x^3 - 4x^2 + 3x - 2)$

Solution

   a. $(4x - 5) + (5x^2 + 2x + 1) = 4x - 5 + 5x^2 + 2x + 1$

$$= 5x^2 + (4x + 2x) + (-5 + 1)$$

$$= 5x^2 + 6x - 4$$

You will usually do the steps for a simplification like this mentally and will generally go directly from the problem to its answer.

   b. When adding polynomials, you simply add similar terms. However, when you subtract polynomials, you first subtract *each term* of the polynomial being subtracted and then combine similar terms.

$$\underset{\underset{\text{Subtracting polynomials}}{\uparrow}}{(4x - 5) - (5x^2 + 2x + 1)} = (4x - 5) \underset{\underset{\text{Subtract \textbf{each} term of the second polynomial}}{\uparrow\qquad\uparrow\qquad\uparrow}}{- (5x^2 - 2x - 1)}$$

$$= -5x^2 + 2x - 6 \quad \text{Combine similar terms}$$

   c. We complete this subtraction by using the definition (that is, by adding the opposite).

$$(5x^2 + 2x + 1) - (3x^3 - 4x^2 + 3x - 2) = (5x^2 + 2x + 1) + (-1)(3x^3 - 4x^2 + 3x - 2)$$

$$= 5x^2 + 2x + 1 - 3x^3 + 4x^2 - 3x + 2$$

$$= -3x^3 + 9x^2 - x + 3$$

Once again, we have shown more steps than you need to show in your work. You can probably do these steps in your head.   ■

EXAMPLE 3   Multiplying polynomials

Simplify:

   a. $5(2x^2 - 4x + 3)$ **b.** $(x - 5)(x + 3)$ **c.** $(4x - 5)(3x^3 - 4x^2 + 3x - 2)$

Solution

   a. $5(2x^2 - 4x + 3) = 10x^2 - 20x + 15$ Distributive property (distribute 5)

b. $(x-5)(x+3) = (x-5)x + (x-5)3$ Distributive property [distribute $(x-5)$]

$\qquad\qquad\quad = x^2 - 5x + 3x - 15$ Distributive property (again)

$\qquad\qquad\quad = x^2 - 2x - 15$ Combine similar terms (simplify).

Since multiplication of binomials is so common, we generally carry out the steps of this example by using the acronym FOIL, where F stands for the product of the *first* terms, O + I for the sum of the products of the *outer* and *inner* terms, and finally, the product of the *last* terms. If we use FOIL, this example is done in one step.

$$(x-5)(x+3) = x^2 - 2x - 15$$
$$\uparrow \qquad \uparrow \qquad \uparrow$$
$$\text{F} \quad \text{O + I} \quad \text{L}$$
$$x(x)3x + (-5x)(-5)3$$

c. Use the distributive property (twice):

$(4x-5)(3x^3 - 4x^2 + 3x - 2) = (4x-5)(3x^3) + (4x-5)(-4x^2) + (4x-5)(3x) + (4x-5)(-2)$

$\qquad = 4x(3x^3) + (-5)(3x^3) + 4x(-4x^2) + (-5)(-4x^2) + 4x(3x) + (-5)(3x) + 4x(-2) + (-5)(-2)$

$\qquad = 12x^4 - 15x^3 - 16x^3 + 20x^2 + 12x^2 - 15x - 8x + 10$

$\qquad = 12x^4 - 31x^3 + 32x^2 - 23x + 10$

Factoring Polynomials

A **factor** of a given algebraic expression (perhaps of some specified type, such as a polynomial) is another algebraic expression that divides evenly (that is, without a remainder) into the given expression. The process of **factoring** involves resolving the given expression into factors. The procedure we will use is to carry out a series of tests for different types of factors, as summarized in Table 1.4.

TABLE 1.4	Factoring Types	
Type	**Form**	**Comments**
1. Common factor	$ax + ay + az = a(x + y + z)$	This simply uses the distributive property. It can be applied with any number of terms.
2. Difference of squares	$x^2 - y^2 = (x - y)(x + y)$	Note that the sum of two squares cannot be factored in the set of real numbers.
3. Difference of cubes	$x^3 - y^3 = (x - y)(x^2 + xy + y^2)$	This can be proved by multiplication.
4. Sum of cubes	$x^3 + y^3 = (x + y)(x^2 - xy + y^2)$	Unlike the sum of squares, the sum of cubes can be factored.
5. Perfect squares	$x^2 + 2xy + y^2 = (x + y)^2$ $x^2 - 2xy + y^2 = (x - y)^2$	The middle term is twice the product of the first and last terms.
6. Perfect cubes	$x^3 + 3x^2y + 3xy^2 + y^3 = (x + y)^3$ $x^3 - 3x^2y + 3xy^2 - y^3 = (x - y)^3$	The numerical coefficients of the terms are 1 3 3 1 or 1 −3 3 −1
7. Trinomials	$acx^2 + (ad + bc)xy + bdy^2$ $\quad = (ax + by)(cx + dy)$	This procedure is used to factor a trinomial into binomial factors.
8. Grouping	See Example 6c.	Sometimes grouping works.
9. Irreducible	Cannot be factored over the set of integers.	

It is assumed that you are familiar with factoring the basic forms from your earlier algebra courses, although we will review all types of factoring with the examples of this section.

There are five important processes in algebra. *Factoring* is the third of these. (Remember, the first two are *simplify* and *evaluate*.)

In this book, we will factor *over the set of integers*, which means that all the numerical coefficients should be integers. If the original polynomial or expression has fractional coefficients, we will use common factoring first to factor out the fractional part.

Factoring over the set of integers also rules out factoring $x^2 - 3$ as $\left(x - \sqrt{3}\right)\left(x + \sqrt{3}\right)$. In this book, a polynomial is said to be *completely factored* if all fractional coefficients are eliminated by common factoring (as shown by Example 3) and if no further factoring is possible over the set of integers.

We now review the types of factoring you have done in your previous algebra courses.

EXAMPLE 4   Elementary factoring types

a. Common monomial factors   Factor: $a^2b + 5a^3b^2 + 7a^2b^3$
b. Common polynomial factors   Factor: $5x(3a - 5b) + 9y(3a - 5b)$
c. Difference of squares   Factor: $3x^2 - 75$
d. Difference of cubes   Factor: $27x^3 - 1$
e. Sum of cubes   Factor: $(x + 3y)^3 + 8$
f. Providing integral coefficients   Factor: $\frac{1}{36}x^2 - 9y^4$

Solution

a. The common factor is a^2b:

$$a^2b + 5a^3b^2 + 7a^2b^3 = a^2b(1 + 5ab + 7b^2)$$

b. The common factor is $(3a - 5b)$:

$$5x(3a - 5b) + 9y(3a - 5b) = (5x + 9y)(3a - 5b)$$

c. First common factor, then factor as a difference of squares:

$$3x^2 - 75 = 3(x^2 - 25)$$
$$= 3(x - 5)(x + 5)$$

d. $27x^3 - 1 = (3x - 1)(9x^2 + 3x + 1)$

e. $(x + 3y)^3 + 8 = [(x + 3y) + 2][(x + 3y)^2 - (x + 3y)(2) + (2)^2]$
$$= (x + 3y + 2)(x^2 + 6xy + 9y^2 - 2x - 6y + 4)$$

f. First common factor, then factor the difference of squares:

$$\frac{1}{36}x^2 - 9y^4 = \frac{1}{36}(x^2 - 324y^4)$$
$$= \frac{1}{36}[x^2 - (18y^2)^2]$$
$$= \frac{1}{36}(x - 18y^2)(x + 18y^2)$$

∎

☐ COMPUTATIONAL WINDOW   ⬓◻✕

Algebraic software contains factoring commands. The answers that you will obtain for the problems in this section will generally be the same as shown in the examples of this section, except for the order of the factors. For example, instead of $(5x + 9y)(3a - 5b)$ the software might show $(3a - 5b)(5x + 9y)$. Sometimes the software will show an unusual form. For Example 4e, we obtained the following form:

$$(x + 3y)^3 + 8 \ \boxed{\text{FACTOR}} \ (x + 3y + 2)(x^2 + x(6y - 2) + 9y^2 - 6y + 4)$$

In factoring trinomials, we use the process of FOIL, as illustrated in the following example.

EXAMPLE 5   Factoring trinomials

Factor:

  a. $x^2 - 8x + 15$ **b.** $6x^2 + x - 12$ **c.** $6w^2 - 9w - 15$ **d.** $6(x+y)^2 - 9(x+y) - 15$

Solution

                    $x \cdot x$ is F

                $\downarrow$ $\downarrow$

  a. $x^2 - 8x + 15 = (x - 5)(x - 3)$ *Select the last terms so that $O + I = -8$.*

                   $\uparrow$ $\uparrow$

                $(-5)(-3)$ is 15

  b. $6x^2 + x - 12 = (2x + 3)(3x - 4)$

  c. $6x^2 - 9w - 15 = 3(2w^2 - 3w - 5)$
                     $= 3(2w - 5)(w + 1)$

  d. This is the same as part **c** where $w = x + y$. Thus,
$$6(x+y)^2 - 9(x+y) - 15 = 3[2(x+y) - 5][(x+y) + 1]$$
$$= 3(2x + 2y - 5)(x + y + 1)$$

Factoring is a skill that must be nurtured and practiced over long periods of time in a variety of contexts. Even though you may have mastered the basic techniques of factoring, you will continue to see those techniques used in more elaborate extended applications throughout this and subsequent courses.

EXAMPLE 6   Advanced factoring

Factor:

  a. $4x^4 - 13x^2y^2 + 9y^4$ **b.** $\dfrac{9a^2}{b^2} - (a + 3b)^2$ **c.** $9x^3 + 18x^2 - x - 2$ **d.** $x^6 - 1$

Solution

  a. $4x^4 - 13^2y^2 + 9y^4 = (x^2 - y^2)(4x^2 - 9y^2)$
                         $= (x - y)(x + y)(2x - 3y)(2x + 3y)$ *Difference of squares*

  b. $\dfrac{9a^2}{b^2} - (a + 3b)^2 = \dfrac{1}{b^2}[9a^2 - b^2(a + 3b)^2]$ *Common factor to eliminate fractions*

                    $= \dfrac{1}{b^2}[3a - b(a + 3b)][3a + b(a + 3b)]$ *Difference of squares*

                    $= \dfrac{1}{b^2}(3a - ab - 3b^2)(3a + ab + 3b^2)$

  c. $9x^3 + 18x^2 - x - 2 = (9x^3 + 18x^2) - (x + 2)$ *Grouping*
                       $= 9x^2(x + 2) - (x + 2)$ *Common factoring*
                       $= (9x^2 - 1)(x + 2)$ *Common factoring*
                       $= (3x - 1)(3x + 1)(x + 2)$ *Difference of squares*

  d. You can consider this as a difference of squares or a difference of cubes. The difference of cubes does not lead to a completely factored form.

$$x^6 - 1 = (x^3)^2 - (1)^2$$
$$= (x^3 - 1)(x^3 + 1)$$
$$= (x - 1)(x^2 + x + 1)(x + 1)(x^2 - x + 1)$$

In calculus, one of the most commonly used types of factoring is common factoring. The following simplification uses common factoring and is required in calculus.

EXAMPLE 7   Advanced common factoring

Factor:

a. $\dfrac{A^5B^4 - A^4B^5}{A^{10}}$ **b.** $\dfrac{A^{2/3}B - A^{-1/3}B^2}{A^{4/3}}$

Solution

a. $\dfrac{A^5B^4 - A^4B^5}{A^{10}} = \dfrac{A^4B^4(A-B)}{A^{10}}$ *The common factor is the base with the smaller exponent.*

$\phantom{\dfrac{A^5B^4 - A^4B^5}{A^{10}}} = \dfrac{B^4(A-B)}{A^6}$ *Leave rational expressions in factored form.*

b. $\dfrac{A^{2/3}B - A^{-1/3}B^2}{A^{4/3}} = \dfrac{A^{-1/3}B(A-B)}{A^{4/3}}$ *Note the smaller exponent on the common factor A is −1/3 and the smaller exponent on the common factor B is 1. Also note $A^{2/3} = A^{-1/3}A$.*

$\phantom{\dfrac{A^{2/3}B - A^{-1/3}B^2}{A^{4/3}}} = A^{-5/3}B(A-B)$ *Note that $A^{-1/3}/A^{4/3} = A^{-5/3}$. For division, subtract exponents.*

EXAMPLE 8   Common factoring on a calculus problem

Simplify: $\dfrac{x^{2/3} \cdot 2(x^2-9) \cdot 2x - \frac{2}{3}x^{-1/3}(x^2-9)^2}{(x^{2/3})^2}$

Solution

You need to recognize the numerator as two terms separated by a minus sign:

$$x^{2/3} \cdot 2(x^2-9) \cdot 2x - \frac{2}{3}x^{-1/3}(x^2-9)^2$$

Looking at these terms (see highlighted regions) we see that the factors 2, x, and $(x^2 - 9)$ are in both regions. We choose as the common factor each of these base numbers with an exponent equal to the smaller of the exponents of these factors that appear highlighted—namely, $2, x^{-1/3}$, and $(x^2 - 9)^1$. In addition, since there is a fraction in the second region, we also factor out $\frac{1}{3}$:

$$\frac{x^{2/3} \cdot 2(x^2-9) \cdot 2x - \frac{2}{3}x^{-1/3}(x^2-9)^2}{(x^{2/3})^2} = \frac{\frac{2}{3}x^{-1/3}(x^2-9)[3x \cdot 2x - (x^2-9)]}{(x^{2/3})^2}$$

$$= \frac{\frac{2}{3}x^{-1/3}(x^2-9)(5x^2+9)}{x^{4/3}}$$

$$= \frac{2}{3}x^{-5/3}(x-3)(x+3)(5x^2+9)$$

PROBLEM SET 1.2

LEVEL 1

Evaluate the given polynomial for the given values in Problems 1–4.

1. Given $a = 3$, $b = -2$
   a. $(a+b)^2$ **b.** $a^2 + b^2$
   c. $(a-b)^3$ **d.** $a^3 - b^3$

2. Given $x = -1$, $y = -4$
   a. x^2y^2 **b.** $x^2 - y^2$
   c. $x(x-y)$ **d.** $y(x+2y)$

3. Given $x^2 - 2xy + y^2$
   a. $x = 1$, $y = 2$ **b.** $x = -2$, $y = 2$
   c. $x = -3$, $y = -1$ **d.** $x = 4$, $y = -5$

4. Given $(a+b)^3$
   a. $a = 1$, $b = 1$ **b.** $a = 10$, $b = -5$
   c. $a = -2$, $b = 4$ **d.** $a = 3$, $b = -2$

Simplify the expressions in Problems 5–8.

5. a. $(8x-4) + (3x^2 - 5x + 5)$
   b. $(2x^2 + 3x + 1) + (x^2 - 4x - 8)$
   c. $x + (x+3) + 2(x-5)$
   d. $(x^2 - 5) + 4(x-1)$

6. a. $(9x^2 - 3x + 2) + (16 - 5x^2)$
   b. $(8x^2 - 2x + 1) + (4x^2 + 3x - 5)$
   c. $2(x - 3) + x(x - 5)$
   d. $4(x^2 + x - 1) + 2(x - 4)$

7. a. $(8x - 4) - (3x^2 - 5x + 5)$
   b. $(2x^2 + 3x + 1) - (x^2 - 4x - 8)$
   c. $x + (x + 3) - 2(x - 5)$
   d. $(x^2 - 5) - 4(x - 1)$

8. a. $(9x^2 - 3x + 2) - (16 - 5x^2)$
   b. $(8x^2 - 2x + 1) - (4x^2 + 3x - 5)$
   c. $2(x - 3) - x(x - 5)$
   d. $4(x^2 + x - 1) - 2(x - 4)$

Simplify the expressions in Problems 9–16.

9. a. $(x + 2)(x + 1)$ **b.** $(y - 2)(y + 3)$
   c. $(x + 1)(x - 2)$ **d.** $(y - 3)(y + 2)$
10. a. $(2x + 1)(x - 1)$ **b.** $(2x - 3)(x - 1)$
   c. $(x + 1)(3x + 1)$ **d.** $(x + 1)(3x + 2)$
11. a. $(5x - 4)(5x + 4)$ **b.** $(3y - 2)(3y + 2)$
   c. $(a + 2)^2$ **d.** $(b - 2)^2$
12. a. $(u + v)^3$ **b.** $(s - 2t)^3$
   c. $(1 - 3n)^3$ **d.** $(2x - y)^3$
13. a. $2(x^2 - 3x + 4)$ **b.** $6(x - 4) + 2(x + 1)$
   c. $(3x - 1)(x^2 + 3x - 2)$ **d.** $(2x + 1)(x^2 + 2x - 5)$
14. a. $3(x^2 - 4x + 1)$
   b. $2(x^2 - 1) - 3(x + 1)$
   c. $(5x + 1)(x^3 - 2x^2 + 3x - 5)$
   d. $(4x - 1)(x^3 + 3x^2 - 2x - 4)$
15. a. $(x + 1)(x - 3)(2x + 1)$ **b.** $(2x - 1)(x + 3)(3x + 1)$
   c. $(x - 2)(x + 3)(x - 4)$ **d.** $(x + 1)(2x - 3)(2x + 3)$
16. a. $(x - 2)^2(x + 1)$ **b.** $(x - 2)(x + 1)^2$
   c. $(x - 5)^2(x + 2)$ **d.** $(x - 3)^2(3x + 2)$

Factor completely, if possible, the expressions in Problems 17–28.

17. a. $me + mi + my$ **b.** $a^2 - b^2$
   c. $a^2 + b^2$ **d.** $a^3 - b^3$
18. a. $a^3 + b^3$ **b.** $x^2 + 2xy + y^2$
   c. $x^2 - 2xy + y^2$ **d.** $u^2 + 2uv + v^2$
19. a. $a^3 + 3a^2b + 3ab^2 + b^3$
   b. $p^3 - 3p^2q + 3pq^2 - q^3$
   c. $-c^3 + 3c^2d - 3cd^2 + d^3$
   d. $x^2y + xy^2$
20. a. $(a + b)x + (a + b)y$ **b.** $(4x - 1)x + (4x - 1)3$
   c. $x^2 - 2x - 35$ **d.** $2x^2 + 7x - 15$
21. a. $3x^2 - 5x - 2$ **b.** $6y^2 - 7y + 2$
   c. $8a^2b + 10ab - 3b$ **d.** $2s^2 - 10s - 48$
22. a. $4y^3 + y^2 - 21y$ **b.** $12m^2 - 7m - 12$
   c. $12p^4 + 11p^3 - 15p^2$ **d.** $9x^2y + 15xy - 14y$
23. a. $(x - y)^2 - 1$ **b.** $(2x + 3)^2 - 1$
   c. $(5a - 2)^2 - 9$ **d.** $(3p - 2)^2 - 16$
24. a. $\dfrac{4}{25}x^2 - (x + 2)^2$ **b.** $\dfrac{4x^2}{9} - (x + y)^2$
   c. $\dfrac{x^6}{y^8} - 169$ **d.** $\dfrac{9x^{10}}{4} - 144$
25. a. $(a + b)^2 - (x + y)^2$ **b.** $(m - 2)^2 - (m + 1)^2$
   c. $2x^2 + x - 6$ **d.** $3x^2 - 11x - 4$

26. a. $6x^2 + 47x - 8$ **b.** $6x^2 - 47x - 8$
   c. $6x^2 + 49x + 8$ **d.** $6x^2 - 49x + 8$
27. a. $4x^2 + 13x - 12$ **b.** $9x^2 - 43x - 10$
   c. $9x^2 - 56x + 12$ **d.** $12x^2 + 12x - 25$
28. a. $x^6 + 9x^3 + 8$ **b.** $x^6 - 6x^3 - 16$
   c. $\left(x^2 - \dfrac{1}{4}\right)\left(x^2 - \dfrac{1}{9}\right)$ **d.** $\left(x^2 - \dfrac{1}{9}\right)\left(x^2 - \dfrac{1}{16}\right)$

LEVEL 2

WHAT IS WRONG, *if anything, with each statement in Problems 29–36? Explain your reasoning.*

29. $(5x)^3 = 5x^3$ **30.** $(2x^2y^3)^2 = 4x^4y^5$
31. $(x + y)^3 = x^3 + y^3$ **32.** $(x^2 + y^2) = (x + y)^2$
33. x^3 means $x + x + x$
34. $b^n = \underbrace{b \cdot b \cdot b \cdot \cdots \cdot b}_{n \text{ factors}}$ for all n

35. $(x^{-1} + y^{-1})^{-1} = x + y$ **36.** $\sqrt{x^2 + y^2} = x + y$

PROBLEMS FROM CALCULUS *Many processes in calculus require considerable factoring ability. Factor Problems 37–46, all of which were taken from a calculus textbook.*

37. $-2(2x^2 - 5x)^{-3}(4x - 5)$
38. $-4(x^4 + 3x^3)^{-2}(4x^3 + 9x^2)$
39. $\dfrac{(x^2 + 3)(3)(7x + 11)^2(7) - (7x + 11)^3(2x)}{(x^2 + 3)^2}$
40. $(2x - 5)^{-1}(4)(x + 5)^3 - (x + 5)^4(4)(2x - 5)^{-2}$
41. $3(x + 1)^2(x - 2)^4 + 4(x + 1)^3(x - 2)^3$
42. $4(x - 5)^3(x + 3)^2 + 2(x - 5)^4(x + 3)$
43. $3(2x - 1)^2(2)(3x + 2)^3 + (2x - 1)^3(3)(3x + 2)^2(3)$
44. $(2x - 3)^3(3)(1 - x^2)^2(-1) + 3(2x - 3)^2(1 - x^2)^3(2)$
45. $4(x + 5)^3(x^2 - 2)^3 + (x + 5)^4(3)(x^2 - 2)^2(2x)$
46. $5(x - 2)^4(x^2 + 1)^3 + (x - 2)^5(3)(x^2 + 1)^2(2x)$

LEVEL 3

47. PROBLEM FROM CALCULUS The expression $\dfrac{x^2 - 9}{x - 3}$ is not defined when $x = 3$. In this problem, we investigate the behavior of this expression for values of x close to 3. Suppose we choose x so that

$$|x - 3| < 0.1$$

(but so that $x \neq 3$).
   a. Find some x_0 satisfying this inequality.
   b. Calculate the value of the expression for x_0: $\dfrac{x_0^2 - 9}{x - 3}$
   c. Repeat parts **a** and **b** for two other values. Can you form a conclusion about the value of the given expression?
   d. How could factoring have been used to simplify your work in part **b**?

48. PROBLEM FROM CALCULUS The expression $\dfrac{x^3 - 27}{x - 3}$ is not defined when $x = 3$. In this problem, we investigate the behavior of this expression for values of x close to 3. Suppose we choose x so that

$$|x - 3| < 0.01$$

(but so that $x \neq 3$).

a. Find some x_0 satisfying this inequality.

b. Calculate the value of the expression for x_0: $\dfrac{x_0^3 - 27}{x_0 - 3}$

c. Repeat parts **a** and **b** for two other values. Can you form a conclusion about the value of the given expression?

d. How could factoring have been used to simplify your work in part **b**?

49. Journal Problem *(Mathematics and Computer Education, Fall 1984; Problem 208 by Charles W. Trigg.)*
Factor (over the reals):

$$6x^5 + 15x^4 + 20x^3 + 15x^2 + 6x + 1$$

50. Factor $(x+3)^2 - (x+3) - 6$

Simplify the expressions in Problems 51–56 in simplified polynomial form.

51. $(2x - 1)^2(3x^4 - 2x^3 + 3x^2 - 5x + 12)$
52. $(x - 3)^3(2x^3 - 5x^2 + 4x - 7)$
53. $(2x + 1)^2(5x^4 - 6x^3 - 3x^2 + 4x - 5)$
54. $(3x^2 + 4x - 3)(2x^2 - 3x + 4)$
55. $(2x^3 + 3x^2 - 2x + 4)^3$
56. $(x^3 - 2x^2 + x - 5)^3$

Factor completely, if possible, the expressions in Problems 57–60.

57. a. $x^{2n} - y^{2n}$ **b.** $x^{3n} - y^{3n}$
58. a. $x^{3n} + y^{3n}$ **b.** $x^{2n} - 2x^n y^n + y^{2n}$
59. $z^5 - 8z^2 - 4z^3 + 32$ **60.** $x^5 + 8x^2 - x^3 - 8$

1.3 Equations of Lines

Two-Dimensional Coordinate System

There is a one-to-one correspondence between $\mathbb{R}$ (the real numbers) and points on a number line, called a one-dimensional coordinate system. A **two-dimensional coordinate system** can be introduced by considering two perpendicular coordinate lines in a plane. Usually one of the coordinate lines is horizontal with the positive direction to the right (called the **x-axis**); the other is vertical with the positive direction upward (called the **y-axis**). These coordinate lines are called **coordinate axes**, and the point of intersection is called the **origin**. Notice from Figure 1.3 that the axes divide the plane into four parts called the **first, second, third**, and **fourth quadrants**. To **plot a point** (a, b) means to locate the point with coordinates (a, b) in the plane and to represent it by a (somewhat oversized) dot.

René Descartes (1596–1650)

$\maltese$ **Historical Note**

The Cartesian coordinate system is named after the philosopher René Descartes (1596–1650). Legend tells us that Descartes thought of his coordinate system while he was lying in bed watching a fly crawl around on the ceiling of his bedroom, and he wished to describe the path of the fly. The idea of a coordinate system ties together the two great branches of mathematics, algebra and geometry, making it truly one of the most revolutionary and profound ideas in history. The historian David Burton says, "Descartes's achievements in mathematics, philosophy, optics, meteorology, and science leave no doubt that he was the most dominant thinker of the 1600s."

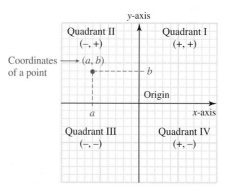

Figure 1.3 Cartesian coordinate system

Points of a plane are denoted by ordered pairs. The term **ordered pair** refers to two real numbers represented by (a, b), where a is called the **first component**, or **abscissa**, and b is the **second component**, or **ordinate**. The order in which the components are listed is important, since $(a, b) \neq (b, a)$, if $a \neq b$.

The plane determined by the x- and y-axes is called a **coordinate plane**, **Cartesian plane**, or **xy-plane**, and we denote this plane by $\mathbb{R}^2$.

The idea of graphing, or plotting, a point—first introduced in Section 1.1—is now extended to graphing points specified by ordered pairs. One of the most common methods for specifying a graph is to give an equation involving two variables (usually x and y). The procedure for graphing an unfamiliar relationship is to compute a sufficient number of ordered pairs **satisfying** the equation, plot the points represented by the ordered pairs, and then sketch the line or curve through the points. The **domain** in this two-dimensional setting is the set of all meaningful replacements for the first component (usually x). Any set of ordered pairs is called a **relation**.

$\maltese$ The domain is NOT x, but rather a set from which x is chosen. $\maltese$

EXAMPLE 1   **Graphing by plotting points**

Graph: $y = \left|\frac{1}{2}x - 2\right|$

Solution   Complete the graph by plotting points.

x	y		x	y
-3	$3.5 \leftarrow$ Let $x = -3$, then			
-2	3	$y = \left|\frac{1}{2}(-3) - 2\right| = 3.5$	5	0.5
-1	2.5		6	1
0	2		7	1.5
1	1.5		8	2
2	1		9	2.5
3	0.5		10	3
4	0		11	3.5

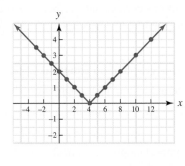

Other ordered pairs are shown on the graph. We say this is the graph of the relation defined by the equation $y = \left|\frac{1}{2}x - 2\right|$. ∎

EXAMPLE 2   **Graphing by plotting points**

Graph: $y = x^3 - 3x^2 - 5x - 6$

Solution   Complete the graph by plotting points.

x	y	x	y
-2	-16	2	-20
-1	-5	3	-21
0	-6	4	-10
1	-13	5	19

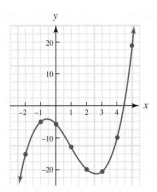

∎

COMPUTATIONAL WINDOW   ⬛▭⊡✕

You might have access to a graphing calculator. Such calculators are very easy to use. For Example 2, enter the function $\boxed{Y=}$ X^3 − 3X^2 − 5X − 6 then press the $\boxed{\text{GRAPH}}$ key, as shown on the left.

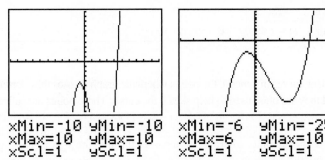

```
xMin=-10 yMin=-10    xMin=-6  yMin=-25
xMax=10  yMax=10     xMax=6   yMax=10
xScl=1   yScl=1      xScl=1   yScl=1
```

You can change the scale (as shown on the right) by using the $\boxed{\text{WINDOW}}$ key.

In later sections of this chapter, we will concentrate on the graphs of particular types of equations and find much more efficient methods for graphing than plotting points.

However, before beginning the development of these procedures, it is important that you clearly understand the relationship between an equation and its graph. This relationship is very explicit: There is a one-to-one correspondence between ordered pairs satisfying an equation and coordinates of points on the graph of the equation.

> **GRAPH**
>
> If we refer to the **graph of an equation**, or the **equation of a graph**, we mean that
> - Every point on the graph has coordinates that satisfy the equation.
> - Every ordered pair satisfying the equation has coordinates that lie on the graph.

 There are five important processes in algebra. *Graphing* is the fourth of these. (Remember, the first three are *evaluate*, *simplify*, and *factor*.)

Slope

In algebra, you studied several forms of the equation of a line. We begin with the concept of the steepness of a line. Consider the line shown in Figure 1.4.

The difference $x_2 - x_1$ is called the **horizontal change**, or Δx, and $|\Delta x|$ is the length of the horizontal side of the triangle. Similarly, the difference in the vertical coordinates, $y_2 - y_1$, is the **vertical change**, or Δy, and $|\Delta y|$ is the length of the vertical side of the triangle. The slope is a measure of the steepness of the line.

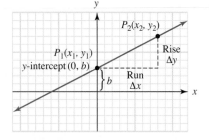

Figure 1.4 Line passing through $P_1(x_1, y_1)$ **and** $P_2(x_2, y_2)$

☠ Δx is one symbol and is pronounced "delta ex." ☠

> **SLOPE**
>
> The **slope** of a line passing through distinct points $P_1(x_1, y_1)$ and $P_2(x_2, y_2)$ such that $x_1 \neq x_2$ is
>
> $$\textbf{Slope} = \frac{\Delta y}{\Delta x} = \frac{y_2 - y_1}{x_2 - x_1}$$

> **» IN OTHER WORDS** Another way of remembering the slope of a line is to pick two points on the line (any two points will work) and find the RISE (up is positive) and the RUN (right is positive), and calculate the slope using RISE/RUN.

If you are moving from left to right on a graph, a positive slope is uphill and a negative slope is downhill. If $x_1 = x_2$, then $\Delta x = 0$ and the line is a **vertical line**, and we say the slope does not exist (or the slope is undefined). If $m = 0$, then $\Delta y = 0$, and the line is a **horizontal line**.

EXAMPLE 3   Finding the slope of a line given two points

Find the slope of the line passing through the points $(-2, 3)$ and $(4, -6)$.

Solution $m = \dfrac{\Delta y}{\Delta x} = \dfrac{-6 - 3}{4 - (-2)} = \dfrac{-9}{6} = -\dfrac{3}{2}$ ∎

Intercepts

When graphing lines, or curves in general, the points where the graph crosses the coordinate axes are usually easy to find, and they are often used to help sketch the curve. These points are called the **intercepts**.

> **INTERCEPTS**
>
> The **x-intercepts** are those points where a curve passes through the *x*-axis. They are found by setting $y = 0$ and solving the resulting equation for *x*. The **y-intercepts** are those points where a curve passes through the *y*-axis. They are found by setting $x = 0$ and solving for *y*.

When graphing lines, the *y*-intercept is used more often than the *x*-intercept. Every nonvertical line has exactly one *y*-intercept, so we speak about *the y*-intercept of a line. This point is almost universally denoted by $(0, b)$. Since the *y*-intercept always has a first component of 0, this notation is often shortened by simply saying *the y-intercept is b*. Remember, this means the line crosses the *y*-axis at the point $(0, b)$.

EXAMPLE 4   **Drawing lines with given intercept and given slope**

Draw the lines with intercept $(0, 3)$ and slope:

a. $m = \frac{1}{2}$ **b.** $m = 6$ **c.** $m = -1$

Solution

a. Plot $(0, 3)$; $m = \frac{1}{2}$, count out RISE $= 1$ and RUN $= 2$.

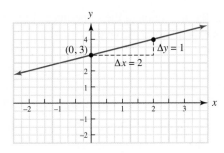

b. Plot $(0, 3)$; $m = 6$, which is $m = \frac{6}{1}$; count out RISE $= 6$ and RUN $= 1$.

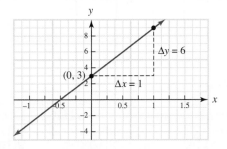

c. Plot $(0, 3)$; $m = -1$, which is $m = \frac{-1}{1}$.

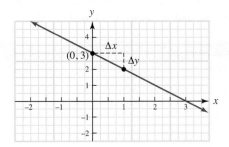

It is assumed that you know how to graph lines, but we review the procedure. If the line is not vertical, begin by solving the equation for the variable *y*. The resulting equation is known as the **slope-intercept form** of the equation of a line.

SLOPE-INTERCEPT FORM

The graph of the equation $y = mx + b$ is a line having slope *m* and *y*-intercept *b*.

EXAMPLE 5   Sketching lines using the slope and the intercept

Graph: **a.** $y = -\frac{2}{3}x + 75$ **b.** $5x - 4y = 12$

Solution

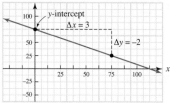

a. By inspection, the y-intercept is $(0, 75)$, and the slope is $-\frac{2}{3}$. Choose the scale to fit the y-intercept and write $m = \frac{-2}{3}$.

Plot $(0, 75)$, count out RISE $= -2$ (down 2) and RUN $= 3$ (over 3). The graph is shown.

This is an example of *negative slope* (downhill). Note the slope and the scale:

$$-\frac{2}{3} = \frac{-2}{3} \cdot \frac{25}{25}$$

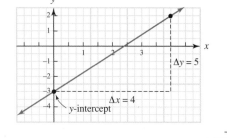

b. Solve* for y: $y = \frac{5}{4}x - 3$. By inspection of the slope-intercept form, $b = -3$ and $m = \frac{5}{4}$. The graph is shown. This is an example of *positive slope* (uphill).

COMPUTATIONAL WINDOW   _ □ X

You can easily graph the lines in Example 5. Press Y = to enter the equations, and then press RANGE or WINDOW to enter the same scaling numbers that you would if not using a graphing calculator (see Example 5a). These graphs are shown:

a.

```
\y1B(-2/3)x+75
xMin=-50  yMin=-25
xMax=150  yMax=150
xScl=25   yScl=25
```

b.

```
\y1B(5/4)x-3
xMin=-2   yMin=-6
xMax=6    yMax=4
xScl=1    yScl=1
```

In algebra you also studied several forms of the equation of a line. The derivation of some of these is reviewed in the problems, and the forms are stated in Table 1.5 for review.

TABLE 1.5	Forms of the Equation of a Line		
Form	**Equation**	**Variables defined**	**When used**
Standard form	$Ax + By + C = 0$	A, B, and C are any integers; A and B are not both zero; and $A > 0$; (x, y) is any point on the line.	A common form in which equations are often represented. It is the final form you will be asked to use for most problems.
Slope-intercept form	$y = mx + b$	m is the slope; b is the y-intercept; so $(0, b)$ is on the line.	Used when graphing a line and when finding the equation of a line when given the slope and intercept.
Point-slope form	$y - k = m(x - h)$	(h, k) is a given point on line.	Used when finding the equation of a line when given the slope and a point, or two points.
Horizontal line	$y = k$		Lines with slope 0.
Vertical line	$x = h$		Line with no slope (that is, slopes that are undefined).

*In this book, we assume you know how to solve linear equations. We discuss solving equations in Section 1.8.

EXAMPLE 6   Finding the equation of a line

Find the standard-form equation of the line using the given information.

a. y-intercept 8, slope 5   **b.** Slope -3, passing through $(2, -4)$

c. Passing through $(2, 3)$ and $(5, 7)$ **d.** No slope; passing through $(7, -3)$

Solution

a. Since you are given the slope and the y-intercept, use the slope-intercept form: $y = mx + b$, where $b = 8$ and $m = 5$: $y = 5x + 8$. In standard form, subtract y from both sides to obtain

$$5x - y + 8 = 0$$

b. Use the point-slope form, where $h = 2$, $k = -4$, and $m = -3$:

$$y - k = m(x - h) \qquad \text{Point-slope form}$$
$$y - (-4) = -3(x - 2) \qquad \text{Substitute.}$$
$$y + 4 = -3x + 6 \qquad \text{Simplify.}$$
$$3x + y - 2 = 0 \qquad \text{Add 3x and (-6) to both sides.}$$

c. Use the point-slope form, after first finding m: $\dfrac{7 - 3}{5 - 2} = \dfrac{4}{3}$.

$$y - k = m(x - h) \qquad \text{Point-slope form}$$
$$y - 3 = \frac{4}{3}(x - 2) \qquad \text{Use either point; we choose (2, 3).}$$
$$3y - 9 = 4x - 8 \qquad \text{Multiply both sides by 3 and distribute 4.}$$
$$4x - 3y + 1 = 0 \qquad \text{Subtract 3y and add 9 to both sides.}$$

d. This is a vertical line, so the equation has the form $x = h$ when it passes through (h, k). Thus, $x = 7$ is the equation. In standard form,

$$x - 7 = 0$$

☠ Do not confuse no slope (vertical line) with zero slope (horizontal line). ☠

Line segments are perhaps more common in applications than lines. The procedure is the same as that for graphing lines, with additional concern over the domain of the variable, as illustrated by the following example.

EXAMPLE 7   Demand application

Suppose that demand for a portable CD-ROM is linear, and the sales, y, are related to the price, x. Market research shows that at a price of $75, the sales are 4.5 million, but at a price of $100, the sales drop to 2 million units. Draw the demand line, and write an equation representing this information, if the price cannot drop below $50, nor can it be greater than $110.

Solution   It is not necessary to write the equation to draw the graph. Simply plot the points $(75, 4.5)$, $(100, 2)$, where the variable x is measured in dollars and the variable y is measured in millions of units. The graph is shown in Figure 1.5.

To find the equation, we first find the slope,

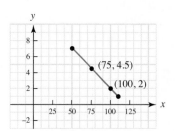

x is between 50 and 110.

Figure 1.5 Demand curve

$$m = \frac{y_2 - y_1}{x_2 - x_1} \qquad \text{Slope formula}$$
$$= \frac{2 - 4.5}{100 - 75} \qquad \text{The points are (75, 4.5) and (100, 2).}$$
$$= \frac{-2.5}{25}$$
$$= -\frac{1}{10}$$

The equation is found using the point-slope form:

$$y - k = m(x - h) \qquad \text{Point-slope form}$$

$$y - 2 = -\frac{1}{10}(x - 100) \quad \text{Use either point; we use (100, 2).}$$

$$10y - 20 = -x + 100 \qquad \text{Multiply both sides by 10 and distribute } -1.$$

$$x + 10y - 120 = 0 \qquad \text{Standard form; add } x \text{ and } -100.$$

The domain is limited so that the price x is between 50 and 110.

$$x + 10y - 120 = 0 \qquad 50 \le x \le 110$$

Parallel and Perpendicular Lines

Slope is essentially a rate, and it is important in the study of rates in calculus. One theorem involving slope states a condition for parallel and perpendicular lines.

Let L_1 and L_2 be two nonvertical lines with slopes m_1 and m_2, respectively. Then,

PARALLEL LINES

$$L_1 \parallel L_2 \quad \text{if and only if} \quad m_1 = m_2$$

PERPENDICULAR LINES

$$L_1 \perp L_2 \quad \text{if and only if} \quad m_1 m_2 = -1$$

>> IN OTHER WORDS Lines are parallel if and only if their slopes are equal, or if both lines are vertical. Lines are perpendicular if and only if their slopes are negative reciprocals or if one line is vertical and the other is horizontal.

EXAMPLE 8   Showing that two lines are perpendicular

Show that the lines $2x + y + 4 = 0$ and $x - 2y - 2 = 0$ are perpendicular.

Solution
Let m_1 and m_2 be the slopes of the two lines:

m_1: $2x + y + 4 = 0$ $\qquad\qquad\qquad\qquad$ m_2: $x - 2y - 2 = 0$

$\qquad\qquad y = -2x - 4$ $\qquad\qquad\qquad\qquad\qquad\quad 2y = x - 2$

We see $m_1 = -2$. $\qquad\qquad\qquad\qquad\qquad\qquad y = \frac{1}{2}x - 1 \quad$ We see $m_2 = \frac{1}{2}$.

$$m_1 m_2 = -2\left(\frac{1}{2}\right) = -1$$

Since the product of the slopes is -1, we know the lines are perpendicular.

EXAMPLE 9   Using slope to show that a triangle is a right triangle

Show that $\triangle ABC$ is a right triangle if $A(-2, -3)$, $B(-5, 1)$, and $C(6, 3)$ are the vertices. This triangle is shown in Figure 1.6.

Solution   We find the slopes of the sides:

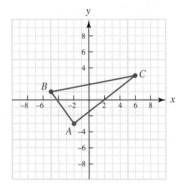

Figure 1.6 Graph of $\triangle ABC$

$$m_{AB} = \frac{\Delta y}{\Delta x} \qquad\qquad m_{BC} = \frac{\Delta y}{\Delta x} \qquad\qquad m_{AC} = \frac{\Delta y}{\Delta x}$$

$$= \frac{1-(-3)}{-5-(-2)} \qquad\qquad = \frac{3-1}{6-(-5)} \qquad\qquad = \frac{3-(-3)}{6-(-2)}$$

$$= -\frac{4}{3} \qquad\qquad\qquad = \frac{2}{11} \qquad\qquad\qquad = \frac{3}{4}$$

Since $m_{AB}m_{AC} = -\frac{4}{3}\left(\frac{3}{4}\right) = -1$, we see that two sides are perpendicular and $\triangle ABC$ is a right triangle. You could also use the distance formula to find the lengths of the sides and show that these lengths satisfy the Pythagorean theorem, which is stated at the beginning of the following section. ∎

PROBLEM SET 1.3

LEVEL 1

In Problems 1–8, find points that satisfy the given equation, plot the points, and sketch the graph.

1. $y = |3x - 4|$
2. $y = |2x + 5|$
3. $x + y = 3$
4. $x + 2y = 4$
5. $y = x^2 + 3$
6. $x = y^2 - 4$
7. $y = x^3 - x^2 + x$
8. $y = x^3 - 2x^2 - x + 2$

For the points given in Problems 9–16:

a. *Graph the line passing through the points;*
b. *Find the slope of the line.*

9. $(-5, 7), (-2, 3)$
10. $(4, -6), (1, -2)$
11. $(2, 1), (7, 13)$
12. $(1, 3), (16, 11)$
13. $(a, b), (b, a)$
14. $(a, b), (a+b, b-a)$
15. $(h, k), (h+a, k+a)$
16. $(x, y), (x+\Delta x, y+\Delta y)$

Find the slope of the line passing through the points given in Problems 17–20. Assume $h \neq 0$.

17. $(x, x^2), (x+h, (x+h)^2)$
18. $(x, x^3), (x+h, (x+h)^3)$
19. $\left(x, \sqrt{x}\right), \left(x+h, \sqrt{x+h}\right)$
20. $\left(x, \frac{1}{x}\right), \left(x+h, \frac{1}{x+h}\right)$
21. IN YOUR OWN WORDS Describe a procedure for graphing lines. Include the possibility of both horizontal and vertical lines.
22. IN YOUR OWN WORDS We defined slope as $m = \text{RISE/RUN}$, but many applications give slope as a percent. For example, a stepped stairway should have a slope of not less than 31.25% and not greater than 112.5%. Discuss slope written as a percent, and then convert these stairway specifications into an inequality using simplified slopes.

WHAT IS WRONG, *if anything, with each statement in Problems 23–28? Explain your reasoning.*

23. The slope of a horizontal line does not exist.
24. The line with equation $y = \frac{5}{3}x + 1$ passes through the point $(5, 3)$.
25. A line with slope 5, another with slope $\frac{1}{5}$, are perpendicular.
26. To graph a line with equation $y = 2x + 3$, begin by plotting the y-intercept 3, and then counting out the slope $m = 2/1$.
27. The slope $m = -\frac{2}{3}$ would be counted out on a graph by going down 2 and back 3.

28. Suppose a line has y-intercept of $(0, 9)$. The slope up 7 and back 9 defines the same line as down 7 and over 9.

Graph the lines whose equations are given in Problems 29–34 by finding the slope and the y-intercept.

29. **a.** $y = 3x + 5$ **b.** $y = -4x + 3$
30. **a.** $y = \frac{3}{4}x - 2$ **b.** $y = \frac{1}{5}x + 1$
31. **a.** $x - 4 = 0$ **b.** $y - 2 = 0$
32. **a.** $y = -\frac{1}{2}x + 125$ **b.** $y = \frac{3}{5}x - \frac{4}{7}$
33. **a.** $5x - 4y - 8 = 0$ **b.** $x - 3y + 2 = 0$
34. **a.** $10x - 25y + 50 = 0$ **b.** $2x - y - 1,200 = 0$

Find the equation of the line satisfying the given conditions in Problems 35–44. Give your answer in standard form.

35. **a.** y-intercept, 6; slope 5
    b. y-intercept, 0; slope 0
36. **a.** y-intercept, −3; slope −2
    b. y-intercept, 5; slope 0
37. **a.** slope, 3; passing through $(2, 3)$
    b. slope, −1; passing through $(-4, 5)$
38. **a.** slope, $\frac{1}{2}$; passing through $(3, 2)$
    b. slope, $\frac{2}{5}$; passing through $(5, -2)$
39. **a.** passing through $(4, -2)$ and $(4, 5)$
    b. passing through $(-4, -1)$ and $(4, 3)$
40. **a.** passing through $(5, 6)$ and $(1, -2)$
    b. passing through $(5, 6)$ and $(7, 6)$
41. passing through $(2, 4)$ parallel to $2x + 3y - 6 = 0$
42. passing through $(-1, -2)$ parallel to $x - 2y + 4 = 0$
43. passing through $(-1, -2)$ perpendicular to $x - 2y + 4 = 0$
44. passing through $(2, 4)$ perpendicular to $2x + 3y - 6 = 0$

LEVEL 2

Problems 45–51 provide some real-world examples of line graphs. Use the given information to write a standard-form equation of the line described by the problem.

45. The demand for a certain product is related to the price of the item. Suppose a new line of stationery is tested at two stores. It is found that 25 boxes are sold within a month if they are priced at $10, and 15 boxes priced at $20 are sold in the same time. Let x be the price and y be the number of boxes sold.

46. An important factor that is related to the demand for a product is the supply. The amount of the stationery in Problem 45 that can be supplied is also related to the price. At $10 each, 10 boxes can be supplied; at $20 each, 20 boxes can be supplied. Let x be the price and y be the number of boxes sold.

47. It costs $90 to rent a car if you drive 100 miles and $140 if you drive 200 miles. Let x be the number of miles driven and y the total cost of the rental. Use this equation to find how much it would cost if you drove 394 miles.

48. Suppose it costs $100 for maintenance and repairs to drive a three-year-old car 1,000 miles and $650 for maintenance and repairs to drive it 6,500 miles. Let x be the number of miles and y be the cost for repairs and maintenance. How much would it cost to drive 7,450 miles?

49. Charles's law for gasses states that if the pressure remains constant, then

$$V = V_0\left(1 + \frac{T}{273}\right)$$

where V is the volume (in.³), V_0 is the initial volume (in.³), and T is the temperature (in degrees Celsius).
  a. Sketch the equation for $V_0 = 100$ and $T \geq 273$.
  b. What is the temperature needed for the volume to double?

50. A particle starts at $P(0, 0)$ and its coordinates change every second by increments $\Delta x = 3$, $\Delta y = 5$. Find its new position after three seconds. Write the standard-form equation of the line described by this problem.

51. A particle starts at $P(-3, 5)$ and its coordinates change every second by increments $\Delta x = 5$, $\Delta y = -2$. Find its new position after two seconds. Write the equation of the line described by this problem.

52. PROBLEM FROM CALCULUS*

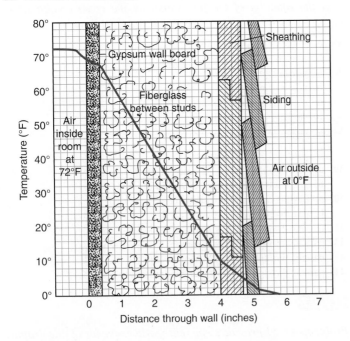

Find the rate of change of temperature in degrees/in. for
  a. gypsum wall board
  b. fiberglass insulation

---
*From *Calculus and Analytic Geometry*, 7th Edition, Thomas/Finney; Reading, MA, Addison Wesley, 1988, p. 10.

  c. wood sheathing
  d. Which of the materials is the best insulator? Explain.

LEVEL 3

53. Consider the line passing through the point $(2, 1)$ forming a triangle in the first quadrant. What is the slope m of this line in terms of b, where $(0, b)$ is the y-intercept?

54. Show that $A(2, 0)$, $B(4, 4)$, $C(0, 6)$, and $D(-2, 2)$ are the vertices of a square, if possible.

55. Show that $R(-2, -2)$, $B(3, -1)$, $U(4, 4)$, and $S(-1, 3)$ are the vertices of a rhombus, if possible. (A rhombus is an equilateral parallelogram.)

56. Show that $T(0, 0)$, $R\left(\sqrt{3}, 1\right)$, and $I\left(\sqrt{3}, -1\right)$ are the vertices of an equilateral triangle, if possible.

57. Show that $A(6, 0)$, $B(4, 4)$, $C(-6, -1)$, and $D(-2, -4)$ are the vertices of a trapezoid with two right angles, if possible. (A trapezoid is a quadrilateral with two sides parallel.)

58. Begin with the slope-intercept form and derive the point-slope form of the equation of a line.

59. Begin with the point-slope form and derive the following two-point form of the equation of a line.

$$y - y_1 = \left(\frac{y_2 - y_1}{x_2 - x_1}\right)(x - x_1)$$

where $P_1(x_1, y_1)$ and $P_2(x_2, y_2)$ are the given points.

60. The equation of a line with intercepts $(a, 0)$ and $(0, b)$ is

$$\frac{x}{a} + \frac{y}{b} = 1$$

Begin with the two-point form and derive this equation.

1.4 Distance and Symmetry

Distance and Midpoint Formulas

The *distance* formula for points in the coordinate plane can be derived by using the **Pythagorean theorem** from geometry.

PYTHAGOREAN THEOREM

A triangle with sides a and b and hypotenuse c (the side opposite the right angle) is a right triangle if and only if

$$a^2 + b^2 = c^2$$

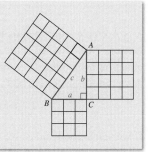

Let $P_1(x_1, y_1)$ and $P_2(x_2, y_2)$ be any two distinct points in a plane. If $x_1 = x_2$, then $\overline{P_1P_2}$ is a *vertical line segment*, as shown in Figure 1.7**a**. If $y_1 = y_2$, then $\overline{P_1P_2}$ is a *horizontal line segment*, as shown in Figure 1.7**b**.

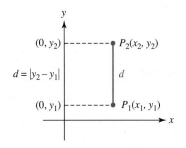

a. Vertical segment, $x_1 = x_2$

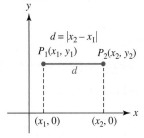

b. Horizontal segment, $y_1 = y_2$

Figure 1.7 Distance formula

With these special cases, it is easy to find the distance d between P_1 and P_2 because these distances correspond directly to distances on a one-dimensional coordinate system, as discussed in Section 1.1. Study Figure 1.7 and see Problems 57 and 58 for details.

We now focus on the general case in which P_1 and P_2 do not lie on the same horizontal or vertical line. Draw a line through P_1 parallel to the x-axis, and another through P_2 parallel to the y-axis. These lines intersect at a point Q with coordinates (x_2, y_1), as shown in Figure 1.8.

Thus,

$$d^2 = (\Delta x)^2 + (\Delta y)^2 \qquad \text{Pythagorean theorem}$$

$$d = \sqrt{(\Delta x)^2 + (\Delta y)^2} \qquad \text{Square root property (d nonnegative)}$$

$$= \sqrt{(x_2 - x_1)^2 + (y_2 - y_1)^2} \qquad \text{Substitution}$$

We summarize this result in the following box.

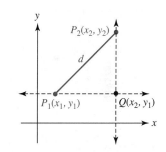

Figure 1.8 Distance between points in $\mathbb{R}$

DISTANCE FORMULA

The **distance** between any distinct points (x_1, y_1) and (x_2, y_2) is given by

$$d = \sqrt{(\Delta x)^2 + (\Delta y)^2}$$

where $\Delta x = x_2 - x_1$ and $\Delta y = y_2 - y_1$.

EXAMPLE 1   Distance between points in the plane

Find the distance between the points $(-5, 1)$ and $(4, -2)$.

Solution $\Delta x = x_2 - x_1 = 4 - (-5) = 9; \Delta y = y_2 - y_1 = -2 - 1 = -3$

$$d = \sqrt{9^2 + (-3)^2} = \sqrt{90} = 3\sqrt{10}$$

A **circle** is defined as the set of all points equidistant from a fixed point in a plane, called its **center**. That distance is the **radius** of the circle, and the distance formula can be used to develop an equation for a circle. Let the center be the point with coordinates (h, k), let the radius be r, and let any point on the circle be (x, y). Then from the distance formula,

$$\sqrt{(x - h)^2 + (y - k)^2} = r$$

If we square both sides, we obtain the following formula.

> **CIRCLE FORMULA**
>
> The **standard-form equation** of the circle with center (h, k) and radius r is
>
> $$(x - h)^2 + (y - k)^2 = r^2$$
>
> The **unit circle** is the circle with center $(0, 0)$ and radius 1:
>
> $$x^2 + y^2 = 1$$

EXAMPLE 2   Sketch the graph of a circle given an equation

Find the center and radius of the circle represented by $(x - 2)^2 + (y + 5)^2 = 16$, and sketch.

Solution   By inspection, $h = 2$, $k = -5$, and $r = \sqrt{16} = 4$. This is a circle with center at $(2, -5)$ and radius 4, as shown in Figure 1.9.

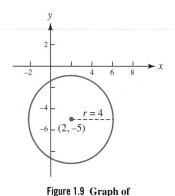

Figure 1.9 Graph of
$(x - 2)^2 + (y + 5)^2 = 16$

Another useful concept involving the coordinates of points is the midpoint of a segment determined by two points. The midpoint M of a segment determined by $P_1(x_1, y_1)$ and $P_2(x_2, y_2)$ can be found by considering the changes, Δx and Δy.

In Figure 1.10, if M is half the distance from P_1 to P_2, then the changes in x and y should be just half as great as Δx and Δy. If these changes are added to the coordinates of the initial point P_1, the midpoint is

$$M = \left(x_1 + \frac{\Delta x}{2}, y_1 + \frac{\Delta y}{2} \right)$$

$$= \left(x_1 + \frac{x_2 - x_1}{2}, y_1 + \frac{y_2 - y_1}{2} \right)$$

$$= \left(\frac{2x_1 + x_2 - x_1}{2}, \frac{2y_1 + y_2 - y_1}{2} \right)$$

$$= \left(\frac{x_1 + x_2}{2}, \frac{y_1 + y_2}{2} \right)$$

> **MIDPOINT FORMULA**
>
> The **midpoint** M between points (x_1, y_1) and (x_2, y_2) is given by
>
> $$M = \left(\frac{x_1 + x_2}{2}, \frac{y_1 + y_2}{2} \right)$$

> **» IN OTHER WORDS** You simply average the coordinates of the two endpoints: Add corresponding coordinates, and divide by 2 to find the coordinates of the midpoint.

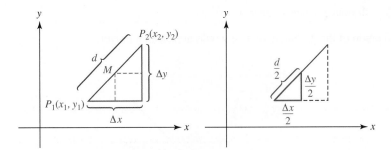

Figure 1.10 Midpoint formula

EXAMPLE 3   Finding the midpoint of a given segment

Find the midpoint between $(-5, 4)$ and $(1, 3)$.

Solution

$$M = \left(\frac{x_1 + x_2}{2}, \frac{y_1 + y_2}{2} \right) = \left(\frac{-5+1}{2}, \frac{4+3}{2} \right) = \left(-2, \tfrac{7}{2} \right)$$

EXAMPLE 4   Find the equation of a circle given its graph

Find the equation of the circle with a diameter having endpoints $(8, 4)$ and $(-2, 6)$, as shown in Figure 1.11.

Solution

The center of the circle is the point at the center of the diameter—that is, the midpoint:

$$\text{Center: } \left(\frac{8 + (-2)}{2}, \frac{4+6}{2} \right) = (3, \ 5)$$

The radius is the distance from the center to a point on the circle:

$$r = \sqrt{(8-3)^2 + (4-5)^2} = \sqrt{25+1} = \sqrt{26}$$

The desired equation is

$$(x-3)^2 + (y-5)^2 = 26$$

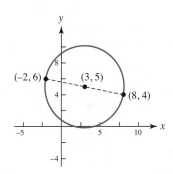

Figure 1.11 Circle with a given diameter

Symmetry

The *idea* of symmetry is the *idea* of mirror images. A graph or curve is **symmetric with respect to a line**, for example, if the graph is the same on both sides of that line, as shown in Figure 1.12.

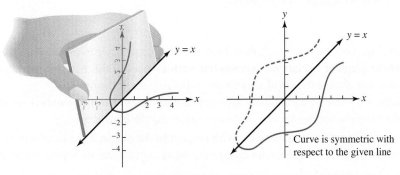

Figure 1.12 Symmetry with respect to a given line

The concept of symmetry is very useful in some physical applications—for example, Newton's law of universal gravitation or Coulomb's law uses symmetry in their formulation.

"Circuits and Symmetry" by Gray Haardeng-Pedersen, *Quantum*, July/August, 1995, p. 28.

EXAMPLE 5   Drawing a symmetric curve

Draw the reflection of the given curve as it would appear in a mirror.

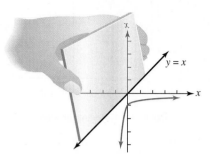

Solution   The answer is shown as a dashed curve. Your paper would look like the graph shown at the right.

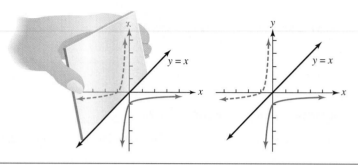

You will, of course, not use a mirror to test symmetry, but it is a worthwhile mental model when looking for curves that are symmetric to a line. Symmetry with respect to a line plays a role in analyzing inverse functions, which we will study in the next chapter. We begin with a definition about symmetric points.

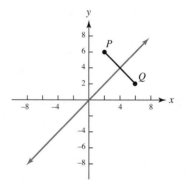

Figure 1.13 Points *P* and *Q* are symmetric about $y = x$

> **LINE SYMMETRY**
>
> Two points P and Q are said to be **symmetric about the line L** if the line segment $\overline{PQ}$ is perpendicular to L and the points P and Q are equidistant from L.

> **» IN OTHER WORDS** Points P and Q are symmetric about the line $y = x$ if $y = x$ is the perpendicular bisector of line segment $\overline{PQ}$. In Figure 1.13, we say that P and Q are **reflections** of each other about the line $y = x$. The line $y = x$ is called the **axis of symmetry**.

When graphing curves, we will be interested in symmetry with respect to one (or both) of the coordinate axes or with respect to the origin.

> **SYMMETRY**
>
> If the point P has coordinates (x, y), then the perpendicular distance to the x-axis is $|y|$. This means that the graph of a relation is **symmetric with respect to the x-axis** if substitution of $-y$ for y does not change the set of coordinates satisfying the equation.
>
> Similarly, the graph of a relation is **symmetric with respect to the y-axis** if substitution of $-x$ for x does not change the set of coordinates satisfying the equation.
>
> The graph of a relation is **symmetric with respect to the origin** if the simultaneous substitution of $-x$ for x and $-y$ for y does not change the set of coordinates satisfying the relation.

These concepts are illustrated in Figure 1.14.

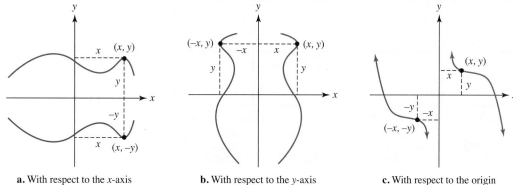

a. With respect to the *x*-axis   **b.** With respect to the *y*-axis   **c.** With respect to the origin

Figure 1.14 Symmetry with respect to the coordinate axes

EXAMPLE 6   Symmetry with respect to each of the coordinate axes

Draw the given curve so that it is:

a. Symmetric with respect to the *x*-axis

b. Symmetric with respect to the *y*-axis

c. Symmetric with respect to the origin

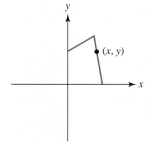

Solution

a. With respect to the *x*-axis   **b.** With respect to the *y*-axis   **c.** With respect to the origin

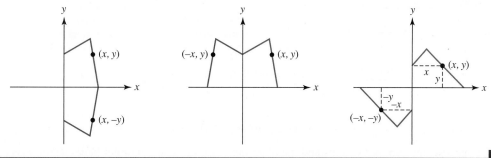

If a curve has any two of the three types of symmetry, it will also have the third. (You are asked to show this in Problem 59.) If you find that a curve has one type of symmetry, but not a second, then it cannot have the third type either.

EXAMPLE 7   Testing symmetry

Test for symmetry with respect to the *x*-axis, *y*-axis, and origin. The calculator graph for each of these curves is shown. Notice the input calculator function has a form that is solved for the variable *y*. We will discuss solving such equations in Section 1.8.

a. $y = \dfrac{3x^2 + 1}{x^4}$

b. $xy = 2$

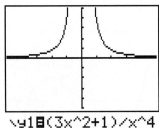

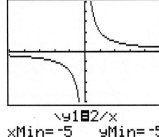

```
\y1B(3x^2+1)/x^4
xMin=-5     yMin=-5
xMax=5      yMax=5
xScl=1      yScl=1
```

```
\y1B2/x
xMin=-5     yMin=-5
xMax=5      yMax=5
xScl=1      yScl=1
```

c. $2x - y^2 - 2 = 0$

d. $9x^2 - 16y^2 = 144$

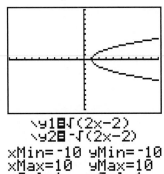

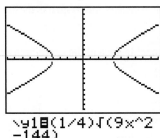

```
\y1B√(2x-2)
\y2B -√(2x-2)
xMin=-10  yMin=-10
xMax=10   yMax=10
xScl=1    yScl=1
```

```
\y1B(1/4)√(9x^2
-144)
\y2B(-1/4)√(9x^2
-144)
xMin=-10  yMin=-10
xMax=10   yMax=10
xScl=1    yScl=1
```

Solution

a. x-axis: $-y = \dfrac{3x^2 + 1}{x^4}$ Substitute −y for y. This is different from the original equation, so it is not symmetric with respect to the x-axis.

y-axis: $y = \dfrac{3(-x)^2 + 1}{(-x)^4}$ Substitute −x for x.

$= \dfrac{3x^2 + 1}{x^4}$ This is the same as the original equation, so it is symmetric with respect to the y-axis.

Because one symmetry holds and the other does not, the graph will not have the third type of symmetry. Therefore, the curve (as verified by looking at the calculator graph) is symmetric with respect to the y-axis.

b. x-axis: $x(-y) = 2$ This is a different equation.

y-axis: $(-x)y = 2$ This is a different equation.

origin: $(-x)(-y) = 2$ Substitute −x for x and −y for y.

$xy = 2$ This is the same equation.

This curve is symmetric with respect to the origin, which can be seen by looking at the calculator graph.

c. x-axis: $2x - (-y)^2 - 2 = 0$ Substitute −y for y.

$2x - y^2 - 2 = 0$ This is the same equation.

y-axis: $2(-x) - y^2 - 2 = 0$ This is a different equation.

Because one symmetry holds and the other does not, the graph will not have the third type of symmetry. Therefore, the curve (as verified by looking at the calculator graph) is symmetric with respect to the x-axis.

d. x-axis: $\quad 9x^2 - 16(y)^2 = 144$ $\quad$ This is the same equation.

$\quad y$-axis: $\quad 9(-x)^2 - 16y^2 = 144$ $\quad$ This is the same equation.

This curve is symmetric with respect to the x-axis, the y-axis, and the origin. ∎

PROBLEM SET 1.4

LEVEL 1

1. IN YOUR OWN WORDS Describe the concept of symmetry.
2. IN YOUR OWN WORDS Compare and contrast the formulas for the distance between two points and the midpoint of the segment connecting those points.

Find the distance between the points whose coordinates are shown in Problems 3–6. Also find the midpoint of each segment connecting the points whose coordinates are given.

3. **a.** $(5, 1)$ and $(8, 5)$
   b. $(1, 4)$ and $(13, 9)$
   c. $(7x, 5x)$ and $(3x, 2x)$, $x < 0$
4. **a.** $(-2, 4)$ and $(0, 0)$
   b. $(3, 6)$ and $(-2, 5)$
   c. $(x, 5x)$ and $(-3x, 2x)$, $x < 0$
5. **a.** $(4, 5)$ and $(3, -1)$
   b. $(0, 0)$ and $(5, -2)$
   c. $(7x, 5x)$ and $(3x, 2x)$, $x > 0$
6. **a.** $(-2, 1)$ and $(-1, -5)$
   b. $(-3, -1)$ and $(-6, -3)$
   c. $(x, 5x)$ and $(-3x, 2x)$, $x > 0$

Draw a reflection of each curve given in Problems 7–10. To do this, draw coordinate axes on your paper. Next, draw the line $y = x$ and the curve as shown. Finally, imagine a mirror on your paper, and draw the curve on your paper as it would look in this mirror.

7.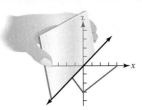

8.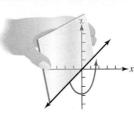

9.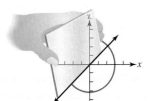

10.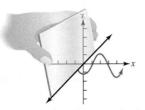

In Problems 11–14, draw a curve so that it is symmetric to the given curve with respect to

a. The x-axis; $\quad$ **b.** The y-axis; and $\quad$ **c.** The origin.

11.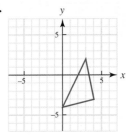

12.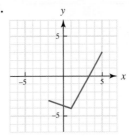

13.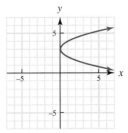

14.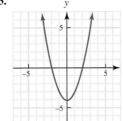

Graph each relation in Problems 15–26. Also, test for symmetry with respect to both axes and the origin.

15. $x + y + 3 = 0$ $\qquad$ 16. $3x - y + 1 = 0$
17. $2x + 3y + 6 = 0$ $\qquad$ 18. $4x - 5y + 2 = 0$
19. $y = -x^2$ $\qquad$ 20. $2x^2 + y = 0$
21. $x^2 + y^2 = 1$ $\qquad$ 22. $(x-1)^2 + (y+2)^2 = 9$
23. $y = -|x|$ $\qquad$ 24. $y = -|x+2|$
25. $y = 2|x|$ $\qquad$ 26. $y = |x| + 2$

In Problems 27–32, test for symmetry with respect to both axes and the origin. Check your answers by looking at the calculator graphs for each curve. If you will be using a graphing calculator, take note of the input values.

27. $y = x^2$ $\qquad\qquad$ **28.** $y = x^3$

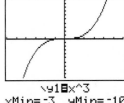

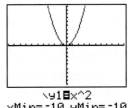

29. $y = -\dfrac{\sqrt{x+3}}{x-1}$

30. $xy + 6 = 0$

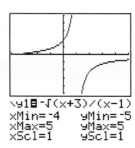

```
\y1B -√(x+3)/(x-1)
xMin=-4      yMin=-5
xMax=5       yMax=5
xScl=1       yScl=1
```

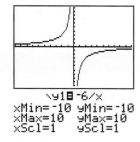

```
\y1B -6/x
xMin=-10  yMin=-10
xMax=10   yMax=10
xScl=1    yScl=1
```

31. $x^2 + y^2 = 9$

32. $x^2 - y^2 = 16$

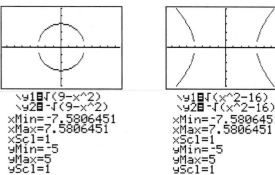

```
\y1B√(9-x^2)
\y2B -√(9-x^2)
xMin=-7.5806451
xMax=7.5806451
xScl=1
yMin=-5
yMax=5
yScl=1
```

```
\y1B√(x^2-16)
\y2B -√(x^2-16)
xMin=-7.5806451
xMax=7.5806451
xScl=1
yMin=-5
yMax=5
yScl=1
```

In Problems 33–40, test for symmetry with respect to both axes and the origin.

33. $y = \dfrac{x^3 - 1}{x - 1}$

34. $y = -\dfrac{\sqrt{9 - x^2}}{x + 3}$

35. $2\,|\,x\,| - |\,y\,| = 5$

36. $|\,y\,| = 5 - 3\,|\,x\,|$

37. $x^2 + 2xy + y^2 = 4$

38. $x^{1/2} + y^{1/2} = 4$

39. $y = x^4 - 2x^2 + 3$

40. $y = (x - 3)^2(x - 4)^2$

LEVEL 2

Find an equation of the circle that satisfies the conditions given in Problems 41–46; sketch each circle.

41. Center $(5, -1)$, radius 4

42. Center $(-1, 3)$, radius π

43. Endpoints of a diameter $(1, -4)$ and $(3, 6)$

44. Endpoints of a diameter $\left(\sqrt{2}, \pi\right)$ and $\left(3\sqrt{2}, 5\pi\right)$

45. Center (a, b), tangent to the x-axis

46. Center (c, d), tangent to the y-axis

47. Graph $y = x - 2$ and $y = |\,x - 2\,|$. Explain how the graph of $y = |\,x - 2\,|$ can be obtained from the graph of $y = x - 2$ by means of reflection.

48. Show that $P(4, 5)$ and $Q(5, 4)$ are symmetric with respect to the line $y = x$ by using the definition of symmetry.

Find a formula for the set of points (x, y) satisfying the conditions specified in Problems 49–50.

49. The distance from (x, y) to $(2, 3)$ is 7.

50. The distance from (x, y) to $(-3, -5)$ is 5.

51. Find the points on the x-axis that are 8 units from the point $(2, 4)$.

52. Find the points on the y-axis that are 8 units from the point $(2, 4)$.

LEVEL 3

Find a formula for the set of points (x, y) satisfying the conditions specified in Problems 53–56.

53. The distance from (x, y) to $(4, 0)$ plus the distance from (x, y) to $(-4, 0)$ equals 10.

54. The distance from (x, y) to $(3, 0)$ plus the distance from (x, y) to $(-3, 0)$ equals 10.

55. The distance from (x, y) to $(5, 0)$ minus the distance from (x, y) to $(-5, 0)$, in absolute value, is 6.

56. The distance from (x, y) to $(0, 5)$ minus the distance from (x, y) to $(0, -5)$, in absolute value, is 8.

57. Let $P_1(x_1, y_1)$ and $P_2(x_2, y_2)$ be two points such that $y_1 = y_2$. Show that the distance from P_1 to P_2 is $|\,x_2 - x_1\,|$.

58. Let $P_1(x_1, y_1)$ and $P_2(x_2, y_2)$ be two points such that $x_1 = x_2$. Show that the distance from P_1 to P_2 is $|\,y_2 - y_1\,|$.

59. Show that if a curve has any two of the three types of symmetry discussed in this section, then it must also have the third type of symmetry.

60. Show that if a curve has one type of symmetry discussed in this section but not the second, then it cannot have the third type either.

1.5 Linear Inequalities and Coordinate Systems

Formulating a Mathematical Problem

To understand difficult concepts in mathematics, we will find it worthwhile to consider, if possible, a numerical representation, an algebraic representation, and a geometrical representation. Let us illustrate what we mean with an easy familiar example. When cannons were introduced in the 13th century, their primary use was to demoralize the enemy. Cannons existed nearly three centuries before enough was known about the behavior of projectiles to use them with any accuracy. Consider a scale drawing (graph) of the path of a cannonball fired in a particular way, as shown in Figure 1.15.

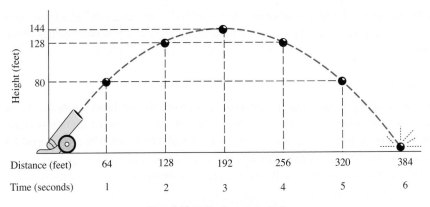

Figure 1.15 **Path of a cannonball**

We can describe the path (1) numerically, (2) algebraically, and (3) geometrically. A numerical analysis could list the time, horizontal distance, and height as shown in Table 1.6.

TABLE 1.6	Data for Cannonball's Path	
Time (in sec)	**Horizontal Distance (ft)**	**Height Distance (ft)**
1	64	80
2	128	128
3	192	144
4	256	128
5	320	80
6	384	0

EXAMPLE 1 Cannonball example of formulating a mathematics problem

For what horizontal distance(s) is the height of the cannonball described in Figure 1.15 greater than 128 ft? Give numerical, geometric, and algebraic solutions.

Solution   (1) *Numerical solution*

From the table, we see that it looks like the height is above 128 ft whenever t is between 2 and 4 (that is, $2 < t < 4$), or when x is between 128 and 256 ($128 < x < 256$). We can illustrate this relationship in $\mathbb{R}$:

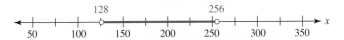

(2) *Geometric solution*

We can plot points from Table 1.6, and then we graph the line $y = 128$, as shown in Figure 1.16.

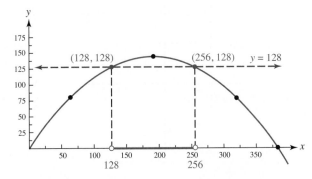

Figure 1.16 **Geometric solution to cannonball example**

We look for values on the real number line that correspond to values for which the height of the cannonball is greater than 128. We see that the solution $128 < x < 256$ corresponds to the numerical solution.

(3) *Algebraic solution*

⚠ As we progress through this course, we will often compare and contrast numerical, geometric, and algebraic methods.

We need to develop algebraic methods that allow us to solve this problem algebraically, so the best that we can do right now is to outline procedures that will be developed later in this chapter. From Table 1.6, we need to find the equation of the path. An algebraic analysis (developed later in the text) might describe the path of the cannon using an equation, such as $y = -\frac{1}{256}x^2 + \frac{3}{2}x$, where x represents the horizontal distance (in ft) and y represents the height (in ft). We must find the values of x that cause y to be greater than 128.

$$y > 128$$

$$-\frac{1}{256}x^2 + \frac{3}{2}x > 128$$

We will discuss problems such as this in Section 1.9. The solution (see Problem 58, Section 1.9) agrees with the numerical and geometric solutions.

Relationship between One- and Two-Dimensional Coordinate Systems

It is worthwhile to consider the relationship between one- and two-dimensional coordinate systems. In **one dimension**, denoted by $\mathbb{R}$, we have one variable, say x, and thus some relation involving the x-variable is given. In **two dimensions**, denoted by $\mathbb{R}^2$, we have two variables, usually x and y. Set y equal to that expression and solve (graph) in $\mathbb{R}^2$. We then look for the **x-intercept(s)** (that is, the places where the curve crosses the x-axis), and the intersection points are the solution for x in $\mathbb{R}$. We demonstrate this idea in the following examples.

EXAMPLE 2   Graphing an inequality in $\mathbb{R}$ and $\mathbb{R}^2$

Graph $-6 \leq x < 5$ in one and in two dimensions.

Solution   *In one dimension* the solution is shown in Figure 1.17.

Figure 1.17 Graph of $-6 \leq x < 5$

Note that the endpoint $x = -6$ is a (solid) point and the endpoint $x = 5$ is an excluded point.

In *two dimensions*, we see that $x = -6$ is a (solid) line and the boundary $x = 5$ is dashed to show that it is not included. The solution is the part of the number line that is the intersection of the two-dimensional regions $x \geq -6$ and $x < 5$, as shown in Figure 1.18.

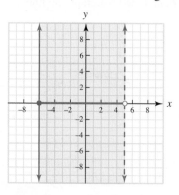

Figure 1.18 Graph of $-6 \leq x < 5$

Why would we want to solve an inequality such as shown in Example 2 in two dimensions? How can a two-dimensional solution be easier to consider when the one-dimensional solution is so easy? The answer is tied to the power of graphing calculators. We begin with an example a bit more difficult than Example 2.

■ COMPUTATIONAL WINDOW ▬▢▨

EXAMPLE 3

Use a graphing calculator to find the values of x that make each statement true:

a. $x^2 - 4 = 0$ **b.** $x^2 - 4 < 0$ **c.** $x^2 - 4 > 0$

Solution

Let $y = x^2 = 4$. On a graphing calculator, this is entered as $Y1 = X^2 - 4$. The graph is shown in Figure 1.19.

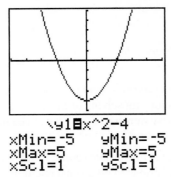

```
\y1∎x^2-4
xMin=-5     yMin=-5
xMax=5      yMax=5
xScl=1      yScl=1
```

Figure 1.19 Calculator graph

a. For $x^2 - 4 = 0$, we estimate the value of x for which $y = 0$. This is frequently called the **critical value** for y. We need to look at the scale for this graph, and we see that each tick mark on the x-axis represents one unit, so we determine that $y = 0$ when $x = -2$ or $x = 2$. These values are shown in Figure 1.20a.

b. For $x^2 - 4 < 0$, we look for values on the x-axis where the graph of $y = x^2 - 4$ is less than zero. We see this solution (Figure 1.20b) is $-2 < x < 2$.

c. For $x^2 = 4 > 0$, we look for values on the x-axis where the graph of $y = x^2 - 4$ is greater than zero. We see this solution (Figure 1.20c) is $x < -2$ or $x > 2$.

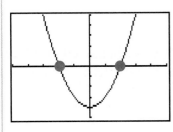 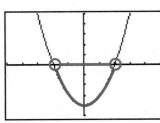 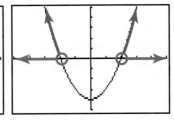

a. $y = 0$
$x = -2, x = 2$

b. $y < 0$
$-2 < x < 2$

c. $y > 0$
$x < -2$ or $x > 2$

Figure 1.20 Using a graphing calculator ■

Historical Note

© Karl Smith Library

Sophie Germain (1776–1831)

Sophie Germain is one of the first women recorded in the history of mathematics. She did original mathematical research in number theory. In her time, women were not admitted to first-rate universities and were not, for the most part, taken seriously, so she wrote at first under the pseudonym LeBlanc. The situation is not too different from that portrayed by Barbra Streisand in the movie *Yentl*. Even though Germain's most important research was in number theory, she was awarded the prize of the French Academy for a paper entitled "Memoir on the Vibrations of Elastic Plates." As we progress through this book we will profile many mathematicians in the history of mathematics, and you will notice that most of them are white males. Why? This issue is addressed in the group research projects at the end of this chapter.

Linear Inequalities in One Variable

A **linear inequality in one variable** is an inequality that can be written in one of the following forms:

$$ax + b < 0 \qquad ax + b \leq 0 \qquad ax + b > 0 \qquad ax + b \geq 0$$

where x is a variable, a is a nonzero real number, and b is a real number. To **solve a linear inequality** means to find all values that make the inequality true. Linear inequalities are solved by using the following principles stated for $<$, but they are also true for $\leq$, $>$, and $\geq$. These symbols are referred to as the **order** of the inequality.

INEQUALITY PROPERTIES

If a, b, and c are any real numbers, then

TRANSITIVITY

If $a < b$ and $b < c$, then $a < c$.

ADDITION

If $a < b$, then $a + c < b + c$.

POSITIVE MULTIPLICATION

If $a < b$ and $c > 0$, then $ac < bc$.

NEGATIVE MULTIPLICATION

If $a < b$ and $c < 0$, then $ac > bc$.

≫ IN OTHER WORDS The procedure and terminology for solving linear inequalities are identical to the procedure and terminology for solving linear equations, except for one fact: **If you multiply or divide by a negative number, the order of the inequality is reversed**.

Consider Example 4 in both one and two dimensions. If you have a graphing calculator, pay attention to how you can use two dimensions to give you a one-dimensional answer.

EXAMPLE 4   Solving a first-degree inequality

Solve $5x - 3 \geq 7$ graphically in   **a.** one dimension   **b.** two dimensions

Solution

a. In one dimension,

$$5x - 3 \geq 7 \qquad \text{Given}$$
$$5x - 3 + \mathbf{3} \geq 7 + \mathbf{3} \qquad \text{Add 3 to both sides.}$$
$$5x \geq 10 \qquad \text{Simplify.}$$
$$x \geq 2 \qquad \text{Divide both sides by 5.}$$

Answer: $[2, \infty]$

b. In two dimensions, let $y_1 = 5x - 3$ and $y_2 = 7$. The graph is shown in Figure 1.21. We are interested in finding the x-values for which $y_1 \geq y_2$.

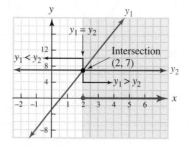

Figure 1.21   Graph of $5x - 3 \geq 7$ by graphing $y_1 = 5x - 3$ and $y_2 = 7$

The appropriate region is shown in color, and the desired solution for x on the x-axis is shown in color. The endpoint is the x-value for which y_1 and y_2 intersect.

EXAMPLE 5   Solving a *between* relationship

Solve $-5 \le 1 - 3x < 10$ in: **a.** one dimension   **b.** two dimensions

Solution

a. $-5 \le 1 - 3x < 10$ *Given*
   $-6 \le \quad -3x \quad < 9$ *Subtract 1 from all three parts.*
   $\quad 2 \ge \quad x \quad > -3$ *Divide both sides by −3 (reverse the order).*

The expression $2 \ge x > -3$ is not convenient for converting to interval notation, so it is rewritten as $-3 < x \le 2$. The solution is $(-3, 2]$. Graph:

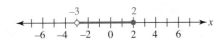

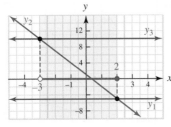

b. In two dimensions, we graph $y_1 = -5$, $y_2 = 1 - 3x$, and $y_3 = 10$. The graph is shown in Figure 1.22. The solution (in terms of x) is highlighted. We project this segment onto the x-axis to find the solution $(-3, 2]$.

Figure 1.22 **Two-dimensional solution** $(-3, 2]$

PROBLEM SET 1.5

LEVEL 1

1. IN YOUR OWN WORDS Describe a process for a one-dimensional solution to a linear inequality in one variable.
2. IN YOUR OWN WORDS Describe a process for a two-dimensional solution to a linear inequality in one variable.

In Problems 3–12, graph each inequality in

a. *one dimension; and* **b.** *two dimensions.*

Assume that the given intervals are for the variable x.

3. $x = 3$	**4.** $x = 0$
5. $x \le 3$	**6.** $x > -2$
7. $(-\infty, 4]$	**8.** $(-15, \infty)$
9. $[-10, 5] \cup (10, 20]$	**10.** $[-3, 1) \cup [2, 5]$
11. $-25 < x < 20$	**12.** $-3 \le x < 4$

WHAT IS WRONG, *if anything, with each statement in Problems 13–20? Explain your reasoning.*

13. If $5 > x > 1$, then $5 > x$ and $x > 1$, so $x > 5$.
14. If $1 \le x \le 5$, then $-1 \le -x \le -5$.
15. If $2x > -20$, then $x < -10$ because -20 is a negative number.

16. If $\dfrac{x}{y} \le 2$, then $x \le 2y$.

17. $x^2 > x$
   $\quad x > 1$ *Divide both sides by x.*
   Solution: $(1, \infty)$.
18. $2x = 3x$
   $\quad 2 = 3$ *Divide both sides by x.*
   No solution.
19. $x^2 + 2x = 3x$
   $\quad x + 2 = 3$ *Divide both sides by x.*
   $\quad\quad x = 1$ *Subtract 2 from both sides.*
20. $x - x^2 > -x$
   $\quad -1 + x < 1$ *Divide both sides by −x.*
   $\quad\quad x < 2$ **Add 1 to both sides.**

For each of the graphs and equations given in Problems 21–34, find the values of x for which

   a. $y = 0$ **b.** $y > 0$ **c.** $y < 0$

21.
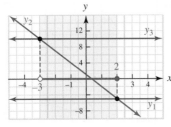
$\searrow y1 = x + 2$
xMin=-10   yMin=-10
xMax=10   yMax=10
xScl=1   yScl=1

22.
$\searrow y1 = 3x - 12$
xMin=-5   yMin=-20
xMax=10   yMax=10
xScl=1   yScl=1

23.
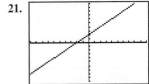
$\searrow y1 = 5x - 100$
xMin=-20   yMin=-150
xMax=50   yMax=100
xScl=10   yScl=10

24.
$\searrow y1 = -8x + 96$
xMin=-5   yMin=-100
xMax=15   yMax=120
xScl=4   yScl=20

25.
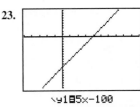
$\searrow y1 = x^2 - 9$
xMin=-10   yMin=-10
xMax=10   yMax=10
xScl=1   yScl=1

26.
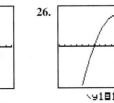
$\searrow y1 = 16 - x^2$
xMin=-10   yMin=-20
xMax=10   yMax=20
xScl=1   yScl=5

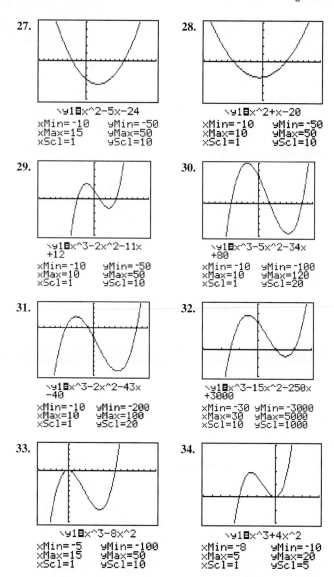

27.

$\y1\blacksquare x^2-5x-24$
xMin=-10   yMin=-50
xMax=15   yMax=50
xScl=1   yScl=10

28.

$\y1\blacksquare x^2+x-20$
xMin=-10   yMin=-50
xMax=10   yMax=50
xScl=1   yScl=10

29.

$\y1\blacksquare x^3-2x^2-11x$
+12
xMin=-10   yMin=-50
xMax=10   yMax=50
xScl=1   yScl=10

30.

$\y1\blacksquare x^3-5x^2-34x$
+80
xMin=-10   yMin=-100
xMax=10   yMax=120
xScl=1   yScl=20

31.

$\y1\blacksquare x^3-2x^2-43x$
-40
xMin=-10   yMin=-200
xMax=10   yMax=100
xScl=1   yScl=20

32.

$\y1\blacksquare x^3-15x^2-250x$
+3000
xMin=-30   yMin=-3000
xMax=30   yMax=5000
xScl=10   yScl=1000

33.

$\y1\blacksquare x^3-8x^2$
xMin=-5   yMin=-100
xMax=15   yMax=50
xScl=1   yScl=10

34.

$\y1\blacksquare x^3+4x^2$
xMin=-8   yMin=-10
xMax=5   yMax=20
xScl=1   yScl=5

LEVEL 2

Solve the inequalities in Problems 35–54. Write your answers using interval notation.

35. $3x - 9 \geq 12$
36. $-x > -36$
37. $-3x \geq 123$
38. $3(2 - 4x) \leq 0$
39. $5(3 - x) > 3x - 1$
40. $9x + 7 \geq 5x - 9$
41. $5(x - 1) + 3(2 - 4x) > 8$
42. $4(1 - x) - 7(2x - 5) < 3$
43. $2[3 - 3(x - 5)] < 1 - x$
44. $2(2x - 9) \geq 3[(2 - x) - 1]$
45. $-3 \leq -x < -2$
46. $-5 \leq 5x \leq 25$

47. $-8 < 5x < 0$
48. $-5 < 3x + 2 \leq 5$
49. $-7 \leq 2x + 1 < 5$
50. $-5 \leq 2x + 1 < 5$
51. $9 < 1 - 2x < 15$
52. $-4 < 1 - 5x \leq 11$
53. $-5 \leq 3 - 2x < 18$
54. $-3 \leq 1 - 2x \leq 7$

55. Boyle's law for a certain gas states that $PV = 400$, where P is the pressure (lb/in.²) and V is the volume (in.³). If $10 \leq V \leq 50$, what are the corresponding values of P?

56. Hooke's law for a certain spring states that $F = 5.5x$, where F is the force (lb) required to stretch this spring x inches beyond its natural length. If $5 \leq F \leq 20$, what are the corresponding values of x?

LEVEL 3

For Problems 57–60, show three methods of solution:

   a. *numerically* **b.** *algebraically* **c.** *geometrically*

57. In 2005 Stephanie Smith, known as the human cannonball, was fired out of a cannon in Melbourne. If the human cannonball follows the path described in Figure 1.15, for what horizontal distances is the height greater than 80 ft? (For the algebraic solution, show set-up only.)

© Dean Curtis, *Springfield News Leader*/AP Photos

58. If Amex Automobile Rental charges $35 per day and 45¢ per mile, how many miles can you drive and keep the cost under $125 per day?

59. A saleswoman is paid a salary of $300 plus a 40% commission on sales. How much does she need to sell in order to have an income of at least $2,000?

60. A certain experiment requires that the temperature be between 20° and 30°C. If Fahrenheit and Celsius degrees are related by the formula $C = \frac{5}{9}(F - 32)$, what are the permissible temperatures in Fahrenheit?

1.6 Absolute Value Equations and Inequalities

To model certain situations, we must be able to solve absolute value equations and inequalities. Consider, for example, the following problems:

- Suppose you are dealing with a substance that is unstable at or near room temperature. When storing this substance, you would want to avoid a temperature of or near 70°F. A safety factor of 10°F is necessary, so you want to avoid temperatures between 60° and 80°F. Find a mathematical model for this situation.

- One of the fundamental concepts in calculus is that of a limit, and to consider the definition of limit in calculus, you must have a mathematical model to describe the situation when "a number *x is close to* a number *c*." What do we mean by two points being "close to one another"?

Recall from Section 1.1 that we expressed absolute value algebraically as $|x| = x$ if $x \geq 0$ and $|x| = x$ if $x < 0$. We now consider absolute value geometrically, as a distance.

ABSOLUTE VALUE

Absolute value can symbolically be represented geometrically:

$$|x - a| = b \qquad x \text{ is } b \text{ units from } a$$

$$|x - a| > b \qquad x \text{ is more than } b \text{ units from } a$$

$$|x - a| < b \qquad x \text{ is less than } b \text{ units from } a$$

Absolute Value Equations

Since $|x - a|$ can be interpreted as the distance between *x* and *a* on the number line, an equation of the form

$$|x - a| = b$$

has two values of *x* that are a given distance from *a* when represented on a number line. For example, $|x - 5| = 3$ states that *x* is 3 units from 5 on a number line. Thus, *x* is either 2 or 8.

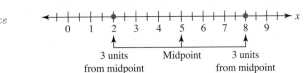

You can verify this conclusion by proving the following property (property 7, Table 1.3).

EQUATION PROPERTY

The **absolute value equation property** translates an absolute value equation into two equivalent equations.

$$|a| = b \text{ where } b \geq 0 \qquad \text{if and only if} \qquad a = b \text{ or } a = -b$$

There are two parts to a proof using the words "if and only if."

> **1.** If $|a| = b$ then $a = b$ or $a = -b$.
>
> **2.** If $a = b$ or $a = -b$, then $|a| = b$.

We will prove part 1 here and leave the proof of part 2 as a problem (see Problem 56). That is, suppose $|a| = b$; we now wish to show that $a = b$ or $a = -b$. Begin with the trichotomy law to compare the real number *a* with the real number 0:

$$a > 0 \qquad \text{or} \qquad a = 0 \qquad \text{or} \qquad a < 0 \qquad \text{Only one can be true.}$$

If $a > 0$ or if $a = 0$, then $|a| = a$ and

$$|a| = b \qquad \text{Given}$$
$$a = b \qquad \text{Substitute } a \text{ for } |a|.$$

If $a < 0$, then $|a| = -a$ and

$$|a| = b \qquad \text{Given}$$
$$-a = b \qquad \text{Substitute } -a \text{ for } |a|.$$
$$a = -b \qquad \text{Multiply both sides by } -1.$$

Thus if $|a| = b$ $(b \geq 0)$, then $a = b$ or $a = -b$.

EXAMPLE 1   Solving an absolute value equation

Solve: $|x| = 2$

Solution $x = 2$ or $x = -2$ ∎

EXAMPLE 2   Solving an absolute value equation with simplification

Solve: $|2x + 1| = 7$

Solution   Solve two separate equations:

$$2x + 1 = 7 \qquad\qquad 2x + 1 = -7$$
$$2x = 6 \qquad\qquad 2x = -8$$
$$x = 3 \qquad\qquad x = -4$$

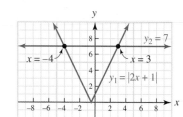

There are two values. You can check this solution graphically as shown in the margin. We have graphed $y_1 = |2x + 1|$ and $y_2 = 7$ and have noted that the points of intersection are $(3, 7)$ and $(-4, 7)$, so we see that $x = 3$ or $x = -4$ is the solution.   ∎

EXAMPLE 3   Absolute value equation with an empty solution set

Solve: $|5 - 2x| = -3$

Solution   There is no solution since the absolute value is always nonnegative. The graphs of $y_1 = |5 - 2x|$ and $y_2 = -3$ are shown. Notice that there is no intersection.

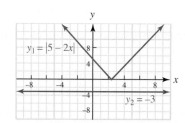

∎

If you are solving an absolute value equation with absolute values on both sides of the equation, the following property proves useful.

DOUBLE ABSOLUTE VALUE

The **double absolute value property** of equations removes absolute value signs:

$$|a| = |b| \qquad \text{if and only if} \qquad a = b \text{ or } a = -b$$

You are asked to prove this in Problem 60.

EXAMPLE 4   Solving a double absolute value equation

Solve: $|x - 1| = |2x - 5|$

Solution   Use the double absolute value property to write

$$x - 1 = 2x - 5 \qquad\qquad x - 1 = -(2x - 5)$$
$$-x = -4 \qquad\qquad\qquad 3x = 6$$
$$x = 4 \qquad\qquad\qquad\quad x = 2$$

■

Absolute Value Inequalities

Since the equation $|x - 5| = 3$ states that x is 3 units from 5, the inequality $|x - 5| < 3$ states that x is any number less than 3 units from 5.

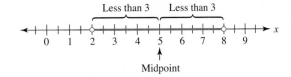

On the other hand, $|x - 5| > 3$ states that x is any number greater than 3 units from 5.

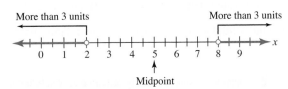

These concepts are summarized by four absolute value properties in which a and b are real numbers and $b > 0$ (properties 8 and 9, Table 1.3, page 9).

INEQUALITY PROPERTY

The **absolute value inequality property** translates an absolute value inequality into two equivalent inequalities.

$\lvert a \rvert < b$	if and only if	$-b < a < b$
$\lvert a \rvert \le b$	if and only if	$-b \le a \le b$
$\lvert a \rvert > b$	if and only if	$a > b$ or $a < -b$
$\lvert a \rvert \ge b$	if and only if	$a \ge b$ or $a \le -b$

There are four separate statements, all with similar proofs. We will prove the first one, namely, $|a| < b$ if and only if $-b < a < b$. Since this is an "if and only if" statement, the proof requires two parts:

Part 1:   If $|a| < b$, then $-b < a < b$,

Part 2:   If $-b < a < b$, then $|a| < b$.

To prove these parts, we will make use of the definition of absolute value:

$$|a| = a \quad \text{if } a \ge 0$$
$$|a| = -a \quad \text{if } a < 0$$

Part 1: Assume $|a| < b$, Prove $-b < a < b$.

If $a \ge 0, |a| = a$ and $|a| < b$, then by substitution, $a < b$.

If $a < 0, |a| = -a$ and $|a| < b$, then by substitution, $-a < b$.

Multiply both sides by -1 to obtain $-b < a$. Thus, $-b < a < b$.

Part 2:   Assume $-b < a < b$. Prove $|a| < b$.

If $a \geq 0$, and $a < b$ (given), then $|a| < b$ since $|a| = a$ when $a \geq 0$.

If $a < 0$, and $-b < a$ (given), then $b > -a$ (multiply both sides by -1).

Then, by substitution, $b > |a|$ because $|a| = -a$ when $a < 0$.

Thus, $|a| < b$.

EXAMPLE 5   Solving an absolute value inequality that is *less than*

Solve: $|2x - 3| \leq 4$

Solution   Write the given inequality without absolute value signs:

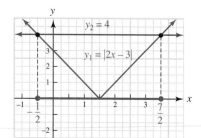

Let $y_1 = |2x - 3|$ and $y_2 = 4$; show solution on the x-axis where $y_1 \leq y_2$.

$$-4 \leq 2x - 3 \leq 4 \qquad \text{Absolute value inequality property}$$

$$-1 \leq \quad 2x \quad \leq 7 \qquad \text{Add 3 to all parts.}$$

$$-\frac{1}{2} \leq \quad x \quad \leq \frac{7}{2} \qquad \text{Divide by 2.}$$

This is the set of values between $-\frac{1}{2}$ and $\frac{7}{2}$ (inclusive), namely, $\left[-\frac{1}{2}, \frac{7}{2}\right]$, as shown in the margin.

Notice the solid dots imply brackets in the notation for the solution.   ∎

EXAMPLE 6   Absolute value inequality with an empty solution set

Solve: $|4 - x| < -3$

Solution   Absolute values must be nonnegative, so the solution set is empty.   ∎

EXAMPLE 7   Absolute value inequality with an order reversal

Solve: $|3 - 2x| < 7$

Solution   Write the inequality without absolute value signs:

$$-7 < 3 - 2x < 7 \qquad \text{Absolute value inequality property}$$

$$-10 < \quad -2x \quad < 4 \qquad \text{Subtract 3 from all parts.}$$

$$5 > \quad x \quad > -2 \qquad \text{Divide by } -2 \text{ (reverse order).}$$

$$-2 < \quad x \quad < 5 \qquad \text{Return to proper form.}$$

The solution set is the interval $(-2, 5)$.   ∎

For absolute value with *greater than* $(>, \geq)$ inequalities, the procedure is different. If $|x| < k$, then $|x| \not\geq k$. That is, the solution of $|x| \geq k$ is all the points that do *not* satisfy $|x| < k$. We call $|x| < k$ the **complement** of $|x| \geq k$. This provides an approach to solving absolute value inequalities that involve *greater than* relationships.

EXAMPLE 8   Solving an absolute value inequality that is *greater than*

Solve: $|2x - 5| > 3$

Solution   Find the solution of the complement, $|2x - 5| \leq 3$.

$$-3 \leq 2x - 5 \leq 3 \qquad \text{Absolute value inequality property}$$

$$2 \leq \quad 2x \quad \leq 8 \qquad \text{Add 5 to all three parts.}$$

$$1 \leq \quad x \quad \leq 4 \qquad \text{Divide all three parts by 2.}$$

This has the solution [1, 4]. Thus, the solution to the original inequality is the complement of this interval [1, 4] (i.e., every point *not* in this interval), namely,

$$(-\infty, 1) \cup (4, \infty)$$

∎

EXAMPLE 9   Greater than absolute value inequality with an order reversal

Solve: $|1 - 3x| \geq 2$

Solution   Solve the complementary problem: $|1 - 3x| < 2$.

$-2 < 1 - 3x < 2$	*Absolute value inequality property*
$-3 < \;-3x\; < 1$	*Subtract 1.*
$1 > \quad x \quad > -\dfrac{1}{3}$	*Divide by –3 (reverse order).*
$-\dfrac{1}{3} < \quad x \quad < 1$	*Rewrite in proper form.*

The solution is $\left(-\frac{1}{3}, 1\right)$. Thus, the solution to the original inequality is the complement, namely, $\left(-\infty, -\frac{1}{3}\right] \cup [1, \infty)$.

∎

EXAMPLE 10   Writing an inequality as an absolute value

Suppose by law a 90-pound bag of cement must weigh between 88 and 92 pounds. Express this relationship as an absolute value.

Solution   Let w be the actual weight of the bag of cement. Also recall that "between" implies strictly between:

$$88 < \quad w \quad < 92$$

$$88 - 90 < \; w - 90 \; < 92 - 90$$

$$-2 < w - 90 < 2$$

Thus, $|w - 90| < 2$.

∎

In Example 10, where did 90 come from? It is the **average value** of the endpoints. Note that the average of 88 and 92 is

$$\frac{88 + 92}{2} = 90$$

EXAMPLE 11   Writing an inequality as an absolute value

Rewrite $-37 < x < 73$ as an absolute value.

Solution   Find the average value of the endpoints (even though they are not included in the solution set of this example):

$$\frac{-37 + 73}{2} = 18$$

Now, subtract this average from each member:

$$-37 - 18 < x - 18 < 73 - 18$$

$$-55 < x - 18 < 55$$

Thus, $|x - 18| < 55$.

∎

A second application of absolute value inequalities concerns a mathematical statement of the situation when "a number x *is close to* a number c." What do we mean by two points being "close to one another"? We say that x is close to the number c if $|x - c| < \delta$ for some very small number δ (δ is the lowercase Greek letter delta).

EXAMPLE 12   Translating a "close to" statement

Translate the following statement into mathematical symbols: If x is close to 3 (within $\frac{1}{10}$ unit), then $6x + 1$ is close to 19 (within $\frac{3}{5}$ unit).

Solution   "If x is close to 3 (within $\frac{1}{10}$ unit)" is translated: $|x - 3| < \frac{1}{10}$. "Then $6x + 1$ is close to 19 (within $\frac{3}{5}$ unit)" is translated:

$$|(6x + 1) - 19| < \frac{3}{5}$$

To see this is true, note that

$$|(6x + 1) - 19| = |6x - 18|$$

$$= 6|x - 3|$$

$$< 6\left(\frac{1}{10}\right) \qquad \text{Since } |x - 3| < \tfrac{1}{10}$$

$$= \frac{3}{5}$$

In calculus, it is common to use absolute value equations and inequalities to specify distances. In particular, the set of all points x within a specified distance of δ units from a real number c is summarized in Table 1.7.

TABLE 1.7	Absolute Value as a Distance		
Symbol	**Description**	**Interval**	**Graph**
$|x - c| < \delta$	All points within δ units of c	$(c - \delta, c + \delta)$	
$|x - c| = \delta$	All points exactly δ units from c	Two points (not an interval) $x = c + \delta, x = c - \delta$	
$|x - c| > \delta$	All points more than δ units from c	$(-\infty, c + \delta) \cup (c + \delta, \infty)$	

PROBLEM SET 1.6

LEVEL 1

WHAT IS WRONG, *if anything, with each statement in Problems 1–6? Explain your reasoning.*

1. If $5 > x > 1$, then $5 > x$ and $x > 1$, so $x > 5$.
2. If $1 \le x \le 5$, then $-1 \le -x \le -5$.
3. If $\dfrac{x}{y^2} \le 2$, then $x \le 2y^2$.
4. If $|x - 5| \le 2$ then x is more than 2 units from 5.
5. If $|x - 3| > 4$ then x is within 3 units of 4.
6. If $|x - 2| = 3$, then x is 3 units from 2.

Use the definition of $|x|$ in Problems 7–12 to rewrite each equation as two linear equations; then solve.

7. **a.** $|x| = 5$ **b.** $|x| = -10$
8. **a.** $|x - 3| = 6$ **b.** $|x - 1| = 3$

9. **a.** $|x - 9| = 15$ **b.** $|2x + 4| = -8$
10. **a.** $|2x + 4| = -12$ **b.** $|3 - 2x| = 7$
11. **a.** $|5x + 4| = 6$ **b.** $|5 - 3x| = 14$
12. **a.** $|x - 3| = |x + 2|$ **b.** $|2x + 6| = |x - 3|$

In Problems 13–20, describe what values of x will ensure that:

a. $y_1 = y_2$ **b.** $y_1 < y_2$ **c.** $y_1 > y_2$

13.   14.

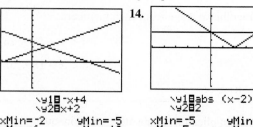

15.

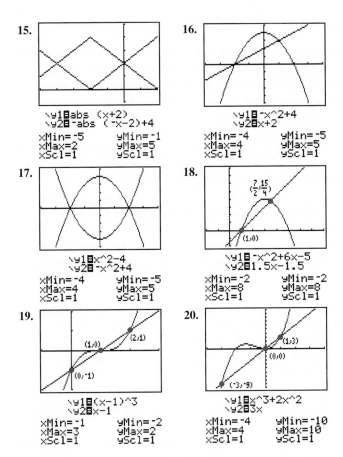

16.

```
\y1☐abs (x+2)
\y2☐ -abs (-x-2)+4
xMin=-5      yMin=-1
xMax=2       yMax=5
xScl=1       yScl=1
```

```
\y1☐ -x^2+4
\y2☐x+2
xMin=-4      yMin=-5
xMax=4       yMax=5
xScl=1       yScl=1
```

17.

18.

$\left(\frac{7}{2}, \frac{15}{4}\right)$

$(1,0)$

```
\y1☐x^2-4
\y2☐ -x^2+4
xMin=-4      yMin=-5
xMax=4       yMax=5
xScl=1       yScl=1
```

```
\y1☐ -x^2+6x-5
\y2☐1.5x-1.5
xMin=-2      yMin=-2
xMax=8       yMax=8
xScl=1       yScl=1
```

19.

$(1,0)$ $(2,1)$

$(0,-1)$

20.

$(1,3)$

$(0,0)$

$(-3,-9)$

```
\y1☐(x-1)^3
\y2☐x-1
xMin=-1      yMin=-2
xMax=3       yMax=2
xScl=1       yScl=1
```

```
\y1☐x^3+2x^2
\y2☐3x
xMin=-4      yMin=-10
xMax=4       yMax=10
xScl=1       yScl=1
```

LEVEL 2

Solve the inequalities in Problems 21–34.

21. a. $|x| < 2$ **b.** $|x| \le -2$
22. a. $|x| \le 8$ **b.** $|x| > -8$
23. $|x-3| > 7$ **24.** $|x-5| < 12$
25. $|x+3| \ge 4$ **26.** $|2x-5| \le 11$
27. $|2x-7| \le 13$ **28.** $|2x-11| < 9$
29. $|11-2x| < 5$ **30.** $|5-2x| > 15$
31. $|2-5x| > 1$ **32.** $|5-4x| \ge 9$
33. $|2x+31| \ge 19$ **34.** $|2x+19| \ge 31$

Express Problems 35–40 as absolute value statements.

35. The number x is three units from ten.
36. The number y is seven units from twenty.
37. The number five is d units from forty.
38. Eighteen is within b units of twenty-seven.
39. The number k is at least three units from negative four.
40. The number y is at least s units from position r.

PROBLEMS FROM CALCULUS *Problems 41–46 are taken from a calculus book. Express each as an absolute value statement.*

41. If x is within 1 unit of 5, then $y = 4x$ is within 4 units of 20.
42. If x is within 0.005 unit of 5, then $y = 4x$ is within 0.02 unit of 20.
43. If x is within δ units of 5, then $y = 4x$ is within ε units of 20.

44. If x is within 1 unit of 2, then

$$y = \frac{x^2 - 2x + 2}{x - 4}$$

is within 0.25 unit of -1.

45. If x is within 0.0006 unit of 2, then

$$y = \frac{x^2 - 2x + 2}{x - 4}$$

is within 0.0001 unit of -1.

46. If x is within δ units of 2, then

$$y = \frac{x^2 - 2x + 2}{x - 4}$$

is within ε units of -1.

In Problems 47–55, choose and define a variable and express the information in an absolute value statement.

47. The length of a machined part is to be 3.7 cm, with a tolerance of 0.04 cm.

© Bruce Amos/ShutterStock, Inc.

48. A survey predicts a market of 5.2 million, with an allowable error of 0.3 million.
49. A manufactured item must measure between 9 in. and 9.3 in.
50. A certain product must weigh between 32 lb and 36 lb.
51. A salesperson travels no more than 150 mi per day and sometimes as little as 90 mi.
52. A library has had a circulation of as many as 10,000 volumes and as few as 6,400 volumes.
53. An economist estimates the price index will grow by 11%, give or take 1%.
54. A mechanic estimates that repairs will run about $250, $30 more or less.
55. An air traffic controller requires you to stay in an air corridor between 5,000 ft and 8,000 ft.

LEVEL 3

56. Prove that if b is a nonnegative real number, and if $a = b$ or $a = -b$, then $|a| = b$.
57. IN YOUR OWN WORDS If $|a| \le b$ show that $b \le a \le b$.
58. IN YOUR OWN WORDS If $|a| > b$ show that either $a > b$ or $a < -b$.
59. IN YOUR OWN WORDS If $|a| \ge b$ show that either $a \ge b$ or $a \le -b$.
60. Prove that $|a| = |b|$ if and only if $a = b$ or $a = -b$.

ELEMENTARY MATHEMATICS

1. Slope of a line

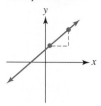

2. Tangent line to a circle

3. Area of a region bounded by line segments

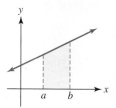

4. Average position and velocity

5. Average of a finite collection of numbers

Figure 1.23 Topics from elementary mathematics

1.7 What Is Calculus?

The purpose of this book is to prepare you for calculus, and in order to relate some of the material of this course to a course in calculus, we need to give you a quick overview of calculus, so that when we use words such as *derivative, differential calculus, integral, integral calculus,* or *limit,* you will have some concept of what we are talking about. In this section, we informally introduce you to this terminology.

If there is an event that marked the coming of age of mathematics in Western culture, it must surely be the essentially simultaneous development of calculus by Newton and Leibniz in the 17th century. Before this remarkable synthesis, mathematics had often been viewed as merely a strange but harmless pursuit, indulged in by those with an excess of leisure time. After the calculus, mathematics became virtually the only acceptable language for describing the physical universe. This view of mathematics and its association with the scientific method has come to dominate the Western view of how the world ought to be explained. This domination is so complete that it is virtually impossible for us to understand how earlier cultures explained what happened around them.

What distinguishes calculus from precalculus courses (algebra, geometry, and trigonometry) is the transition from static or discrete applications (see Figure 1.23) to those that are dynamic or continuous (see Figure 1.24). For example, in precalculus mathematics we consider the slope of a line, but in calculus we define the (nonconstant) slope of a nonlinear curve. In precalculus mathematics, we find average changes in quantities such as the position and velocity of a moving object, but in calculus, we can find instantaneous changes in the same quantities. In precalculus mathematics, we find the average of a finite collection of numbers, but in calculus, we can find the average value of infinitely many values over an interval.

You might think of calculus as the culmination of all of your mathematical studies. To a certain extent that is true, but it is also the beginning of your study of mathematics as it applies to the real world around us. Calculus is a three-semester or four-quarter course that *begins* your college work in mathematics. All your prior work in mathematics is considered elementary mathematics, with calculus the dividing line between elementary mathematics and mathematics as it is used in a variety of theoretical and applied topics. It is the mathematics of motion and change. The development of calculus in the 17th century by Newton and Leibniz was the result of their attempt to answer some fundamental questions about the world and the way things work. These investigations led to two fundamental concepts of calculus—namely, the idea of a *derivative* and that of an *integral.* The breakthrough in the development of these concepts was the formulation of a mathematical tool called a *limit.*

1. **Limit:** The limit is a mathematical tool for studying the *tendency* of a function as its variable *approaches* some value. Calculus is based on the concept of limit. In this book we shall introduce the limit of a function informally.

2. **Derivative:** The derivative is defined as a certain type of limit, and it is used initially to compute rates of change and slopes of tangent lines to curves. The study of derivatives is called *differential calculus.* Derivatives can be used in sketching graphs and in finding the extreme (largest and smallest) values of functions.

3. **Integral:** The integral is found by taking a special limit of a sum of terms, and the study of this process is called *integral calculus.* Area, volume, arc length, work, and hydrostatic force are a few of the many quantities that can be expressed as integrals.

First, let us return to a description about the foundation of calculus:

One might naturally suppose that an event so momentous must involve ideas so profound that average mortals can hardly hope to comprehend them. In fact, nothing could be further from the truth. The essential ideas of calculus—the derivative and the integral—are quite straightforward and had been known prior to either Newton or Leibniz. The contribution of Newton and Leibniz was to recognize that the idea of finding tangents (the derivative) and the idea of finding areas (the integral) are related and that this relation can be used to give a simple and unified description of both processes.

Let us begin by taking an intuitive look at each of these three essential ideas of calculus.

The Limit: Zeno's Paradox

Zeno (ca. 500 B.C.) was a Greek philosopher who is known primarily for his famous paradoxes. One of those concerns a race between Achilles, a legendary Greek hero, and a tortoise. When the race begins, the (slower) tortoise is given a head start, as shown in Figure 1.25.

Is it possible for Achilles to overtake the tortoise? Zeno pointed out that by the time Achilles reaches the tortoise's starting point, $a_1 = t_0$, the tortoise will have moved ahead to a new point t_1. When Achilles gets to this next point, a_2, the tortoise will be at a new point, t_2. The tortoise, even though much slower than Achilles, keeps moving forward. Although the distance between Achilles and the tortoise is getting smaller and smaller, the tortoise will apparently always be ahead.

Of course, common sense tells us that Achilles will overtake the slow tortoise, but where is the error in reasoning in the previous paragraph that always gives the tortoise the lead? The error is in the assumption that an infinite amount of time is required to cover a distance divided into an infinite number of segments. This discussion is getting at an essential idea in calculus—namely, the notion of a limit.

Consider the successive positions for both Achilles and the tortoise:

↓ Starting position

Achilles: $a_0, a_1, a_2, a_3, \ldots$

Tortoise: $t_0, t_1, t_2, t_3, \ldots$

After the start, the positions for Achilles, as well as those for the tortoise, form sets of positions that are ordered with positive integers. Such ordered listings are called *sequences* (see Section 8.1).

For Achilles and the tortoise we have two sequences, $\{a_1, a_2, a_3, \ldots, a_n, \ldots\}$ and $\{t_1, t_2, t_3, \ldots, t_n, \ldots\}$, where $a_n < t_n$ for all values of n. The idea of the limit of a sequence is essential to a calculus course, and we discuss this idea in Section 8.1. In Section 2.8, we introduce the idea of the limit of a *function*, a concept also used in calculus to define the other two basic concepts of calculus: the derivative and the integral. Even if the solution to Zeno's paradox using limits seems unnatural at first, do not be discouraged. It took over 2,000 years to refine the ideas of Zeno and provide conclusive answers to those questions about limits. The following example will provide an intuitive preview of a limit.

CALCULUS

1. Slope of a curve

2. Tangent line to a general curve

3. Area of a region bounded by curves

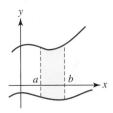

4. Instantaneous changes in position and velocity

5. Average of an infinite collection of numbers

Figure 1.24 Topics from calculus

t_2 t_1 Tortoise, t_0 Achilles, a_0

a_3 a_2 a_1

Figure 1.25 Achilles and the tortoise

EXAMPLE 1   An intuitive preview of a limit

The sequence $\dfrac{1}{2}, \dfrac{2}{3}, \dfrac{3}{4}, \dfrac{4}{5}, \ldots$ can be described by writing a *general term*: $\dfrac{n}{n+1}$, where $n = 1, 2, 3, 4, \ldots$* Can you guess the limit, L, of this sequence? We will say that L is the number that the sequence

---

*You might wish to review sequences and general terms in Section 8.1.

Check **www.mathnature.com** for some links to interactive sites illustrating the concept that a sequence of secant lines approaches a tangent line.

with general term $\frac{n}{n+1}$ tends toward as becomes large without bound. We will define a notation to summarize this idea:

$$L = \lim_{n \to \infty} \frac{n}{n+1}$$

Solution   As you consider larger and larger values for *n*, you find a sequence of fractions:

$$\frac{1}{2}, \frac{2}{3}, \frac{3}{4}, \ldots, \frac{1,000}{1,001}, \frac{1,001}{1,002}, \ldots, \frac{9,999,999}{10,000,000}, \ldots$$

It is reasonable to guess that the sequence of fractions is approaching the number 1. This number is called the **limit**.   ∎

The Derivative: The Tangent Problem

A *tangent line* (or, if the context is clear, simply say, "tangent") to a circle at a given point *P* is a line that intersects the circle at *P* and only at *P* (see Figure 1.26a). This characterization does not apply for curves in general, as you can see by looking at Figure 1.26b. If we wish to have one tangent line, which line shown should be called the tangent to the curve?

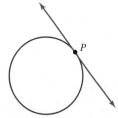 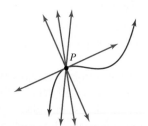

a. At each point *P* on a circle, there is one line that intersects the circle exactly once.

b. At a point *P* on a curve, there may be several lines that intersect that curve only once.

Figure 1.26 Tangent line

At each point *P* on a curve, there may be several lines that intersect the curve only once. To find a tangent line, begin by considering a line that passes through two points *P* and *Q* on the curve, as shown in Figure 1.27a. This line is called a **secant line**.

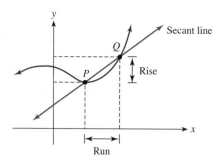 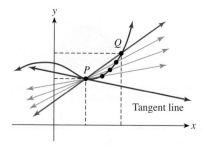

Figure 1.27 Secant line

Why would the slope of a tangent line be important? That is what we investigate in the first half of a calculus course, and it forms the definition of derivative, which is the foundation for what is called **differential calculus**.

EXAMPLE 2   Finding a tangent line

Let $y = \frac{1}{2}x^2 + 1$. Find the tangent line at $x = 2$.

Historical Note

Solution Sketch the curve $y = \frac{1}{2}x^2 + 1$ as shown in Figure 1.28.

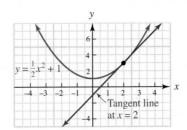

Figure 1.28 Tangent line for a given curve

We note that when $x = 2$, $y = \frac{1}{2}(2)^2 + 1 = 3$; plot the point $P(2, 3)$. Let Q_1 be the point $(8, 33)$. The line passing through PQ_1 is called a secant line.

$$m = \frac{33 - 3}{8 - 2} = \frac{30}{6} = 5$$

Now, let the point Q_1 slide along the curve toward the point P. We calculate the successive slopes of secant lines at selected points:

$$Q_2(4, 9) : m = \frac{9 - 3}{4 - 2} = \frac{6}{2} = 3$$

$$Q_3(3, 5.5) : m = \frac{5.5 - 3}{3 - 2} = 2.5$$

$$Q_4(2.5, 4.125) : m = \frac{4.125 - 3}{2.5 - 2} = \frac{1.125}{0.5} = 2.25$$

It appears that the slopes of the secant lines are getting closer to a line with slope 2. We verify this with some additional calculations.

$$Q_5(2.1, 3.205) : m = \frac{3.205 - 3}{2.1 - 2} = 2.05$$

$$Q_6(2.01, 3.02005) : m = \frac{3.02005 - 3}{2.01 - 2} = 2.005$$

$$Q_7(2.001, 3.0020005) : m = \frac{3.0020005 - 3}{2.001 - 2} = 2.0005$$

☠ Notice that at $x = 2$, the slope is not defined because of division by zero. ☠

In calculus, it will be shown that the tangent line at $P(2, 3)$ is the line passing through P with slope 2. The equation of this line is

$$y - k = m(x - h) \qquad \text{Point-slope form}$$

$$y - 3 = 2(x - 2) \qquad \text{Substitute } m = 2, (h, k) = (2, 3).$$

$$y - 3 = 2x - 4$$

$$y = 2x - 1 \qquad \text{Use this form to graph the line.}$$

$$2x - y - 1 = 0 \qquad \text{Standard form equation}$$

The tangent line is shown in Figure 1.28. ∎

Sir Isaac Newton (1643–1727)

At the time of their respective break-through publications using calculus (around 1685), there was a bitter controversy throughout Europe about whose work had been done first, along with accusations that each stole the idea from the other. Part of the explanation for this controversy is that each had done his actual work earlier, and another part can be attributed to the rivalry between mathematicians in England who championed Newton and those in Europe who supported Leibniz. The fact is, however, that the intellectual climate for the invention of calculus was ripe and inevitable.

The Integral: The Area Problem

You probably know the formula for the area of a circle with radius r:

$$A = \pi r^2$$

The Egyptians were the first to use this formula over 5,000 years ago, but the Greek Archimedes (ca. 300 B.C.) showed how to derive the formula for the area of a circle by using a limiting process. Consider the areas of inscribed polygons, as shown in Figure 1.29.

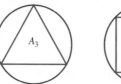   

Figure 1.29 Approximating the area of a circle

Since we know the areas of these polygons, we can use them to estimate the area of a circle. Even though Archimedes did not use the following notation, here is the essence of what he did, using a method called "exhaustion":

Let A_3 be the area of the inscribed equilateral triangle;

A_4 be the area of the inscribed square;

A_5 be the area of the inscribed regular pentagon.

How can we find the area of this circle? As you can see from Figure 1.29, if we consider the area of A_3, then A_4, then A_5, ..., we should have a sequence of areas such that each successive area more closely approximates that of the circle. We write this idea as a limit statement:

$$A = \lim_{n \to \infty} A_n$$

In calculus, limits are used in yet a different way to find the area of regions enclosed by curves. For example, consider the area shown in color in Figure 1.30.

We can approximate the area by using rectangles. If A_n is the area of the nth rectangle, then the total area can be approximated by finding the sum

$$A_1 + A_2 + A_3 + \cdots + A_{n-1} + A_n$$

This process is shown in Figure 1.31.

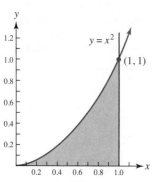

Figure 1.30 Area under a curve

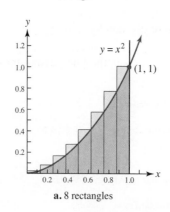 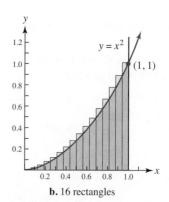

a. 8 rectangles   **b.** 16 rectangles

Figure 1.31 Approximating the area using rectangles

Notice that in the first figure the area of the rectangles approximates the area under the curve, but there is some error. In the second figure, there is a little less error than in the first approximation. As more rectangles are included, the error decreases.

EXAMPLE 3 Using rectangles to approximate the area under a curve

Consider the graph of $y = x^2$ and the region bounded by this curve, the x-axis, and the line $x = 1$, as shown in Figure 1.30. Approximate this area by using

a. Two rectangles   **b.** Four rectangles

Solution

a. For two rectangles, the base of each rectangle is 0.5 (total length 1 divided into two parts).

> If $x = 0.5$, then the height of the rectangle is $y = (0.5)^2 = 0.25$.
> Thus, $A_1 = \ell w = (0.25)(0.5) = 0.125$.

> If $x = 1$, then the height of the rectangle is $y = 1^2 = 1$.
> Thus, $A_2 = (1)(0.5) = 0.5$.

The area under the curve is approximated by

$$A_1 + A_2 = 0.625$$

b. For four rectangles (see Figure 1.32), the base of each rectangle is 0.25 (total length 1 divided into four parts).

> If $x = 0.25$, then the height of the rectangle is $y = (0.25)^2 = 0.0625$.
> Thus, $A_1 = \ell w = (0.0625)(0.25) = 0.015625$.

> If $x = 0.5$, then the height of the rectangle is $y = (0.5)^2 = 0.25$.
> Thus, $A_2 = (0.25)(0.25) = 0.0625$.

> If $x = 0.75$, then the height of the rectangle is $y = (0.75)^2 = 0.5625$.
> Thus, $A_3 = (0.5625)(0.25) = 0.140625$.

> If $x = 1$, then the height of the rectangle is $y = 1^2 = 1$.
> Thus, $A = (1)(0.25) = 0.25$.

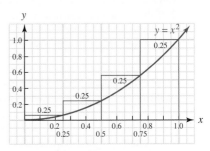

Figure 1.32 Estimating the area using four rectangles

The area under the curve is approximated by

$$A_1 + A_2 + A_3 + A_4 = 0.015625 + 0.0625 + .0140625 - 0.25$$
$$= 0.46875$$

If you have access to a calculator or a computer, you can verify the entries in Table 1.8.

Table 1.8 shows the result of approximating the area under the curve $y = x^2$ and bounded by the x-axis and the line $x = 1$. The first line (2 subintervals) shows the answer for Example 3a, and the second line (4 subintervals) shows the answer for Example 3b. Table 1.8 says if we repeated Example 3 for 1,024 rectangles we would obtain the sum of the area of the rectangles to be approximately equal to 0.3338218. In calculus, you will solve the area problem to find that the actual area is $\frac{1}{3}$.

The area problem leads to a process called *integration*, and the study of integration forms what is called **integral calculus**. Similar reasoning allows us to calculate such things as volumes, the length of a curve, the average value, or amount of work required for a particular task.

TABLE 1.8	Estimate for Area Under $y = x^2$
Number of terms	**Estimate over [0, 1]**
2	0.625
4	0.46875
8	0.3984375
16	0.3652344
32	0.3491211
64	0.3411865
128	0.3372498
256	0.3358900
512	0.3343105
1,024	0.3338218

PROBLEM SET 1.7

1. **IN YOUR OWN WORDS** An analogy to Zeno's tortoise paradox can be made as follows.

> A woman standing in a room cannot walk to a wall. To do so, she would first have to go half the distance, then half the remaining distance, and then again half of what still remains. This process can always be continued and can never be ended.

Draw an appropriate figure for this problem and then present an argument using sequences to show that the woman will, indeed, reach the wall.

2. **IN YOUR OWN WORDS** Zeno's paradoxes remind us of an argument that might lead to an absurd conclusion:

> Suppose I am playing baseball and decide to steal second base. To run from first to second base, I must first go half the distance, then half the remaining distance, and then again half of what remains. This process is continued so that I never reach second base. Therefore, it is pointless to steal base.

Draw an appropriate figure for this problem and then present a mathematical argument using sequences to show that the conclusion is absurd.

3. Consider the sequence 0.3, 0.33, 0.333, 0.3333, ... What do you think is the appropriate limit of this sequence?
4. Consider the sequence 0.9, 0.99, 0.999, 0.9999, ... What do you think is the appropriate limit of this sequence?
5. Consider the sequence 0.2, 0.27, 0.272, 0.2727, ... What do you think is the appropriate limit of this sequence?
6. Consider the sequence 3, 3.1, 3.14, 3.141, 3.1415, 3.14159, 3.141592, ... What do you think is the appropriate limit of this sequence?

Copy the figures in Problems 7–12 on your paper. Draw what you think is an appropriate tangent line for each curve at the point P by using the secant method.

7.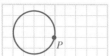
8.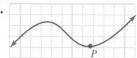

9.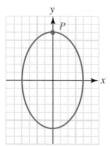
10.

11.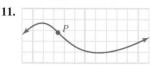
12.

*In Problems 13–18, **guess** the requested limits.*

13. $\lim\limits_{n \to \infty} \dfrac{2n}{n+4}$
14. $\lim\limits_{n \to \infty} \dfrac{2n}{3n+1}$

15. $\lim\limits_{n \to \infty} \dfrac{3n}{n^2+2}$
16. $\lim\limits_{n \to \infty} \dfrac{3n^2+1}{2n^2-1}$

17. $\lim\limits_{n \to \infty} \dfrac{3n^2}{n^2+2}$
18. $\lim\limits_{n \to \infty} \dfrac{2n^3-1}{n^3+1}$

Estimate the area in each figure shown in Problems 19–24.

19.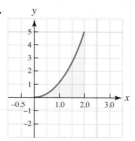
20.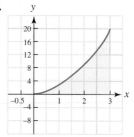

21.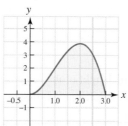
22.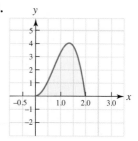

23.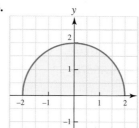
24.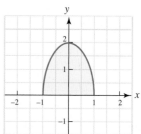

PROBLEMS FROM CALCULUS *In Problems 25–48, you are given a point P on the curve. Calculate the slope of a secant line, which passes through the point P and a point Q_1. Repeat for points Q_2, Q_3, and Q_4, each point on the curve closer to the given point P. Notice that the slope values seem to be getting close to a single number. In calculus, this limiting value is called the slope of the curve at the point P. State the slope of the tangent line and then draw the tangent line to the given curve at the point P.*

In Problems 25–30, sketch the curve $y = x^2$. Plot $P(2, 2^2)$ on this curve. Calculate the slope of the lines passing through the given points.

25. $Q_1(3, 3^2)$
26. $Q_2(2.5, 2.5^2)$
27. $Q_3(2.1, 2.1^2)$
28. $Q_4(2.01, 2.01^2)$

29. What is the slope of $y = x^2$ at P?

30. Find the equation of the tangent to $y = x^2$ at $(2, 4)$.

In Problems 31–36, sketch the curve $y = x^3$. Plot $P(2, 2^3)$ on this curve. Calculate the slope of the lines passing through the given points.

31. $Q_1(3, 3^3)$

32. $Q_2(2.5, 2.5^3)$

33. $Q_3(2.1, 2.1^3)$

34. $Q_4(2.01, 2.01^3)$

35. What is the slope of $y = x^3$ at P?

36. Find the equation of the tangent to $y = x^3$ at $(2, 8)$.

In Problems 37–42, sketch the curve $y = \sqrt{x}$. Plot $P\left(4, \sqrt{4}\right)$ on this curve. Calculate the slope of the lines passing through the given points.

37. $Q_1\left(9, \sqrt{9}\right)$

38. $Q_2\left(7, \sqrt{7}\right)$

39. $Q_3\left(5, \sqrt{5}\right)$

40. $Q_4\left(4.1, \sqrt{4.1}\right)$

41. What is the slope of $y = \sqrt{x}$ at P?

42. Find the equation of the tangent to $y = \sqrt{x}$ at $\left(4, \sqrt{4}\right)$.

In Problems 43–48, sketch the curve $y = 1/x$. Plot $P(2, 1/2)$ on this curve. Calculate the slope of the lines passing through the given points.

43. $Q_1(4, 1/4)$

44. $Q_2(2.5, 1/2.5)$

45. $Q_3(2.1, 1/2.1)$

46. $Q_4(2.01, 1/2.01)$

47. What is the slope of $y = 1/x$ at P?

48. Find the equation of the tangent to $y = \frac{1}{x}$ at $\left(2, \frac{1}{2}\right)$.

49. Suppose the circle in Figure 1.29 has radius 1. Table 1.9 shows the approximate areas of inscribed polygons.

TABLE 1.9	Area of Inscribed Polygons, $r = 1$
Polygon	**Approximate Area**
Triangle	1.299
Quadrilateral	2.000
Pentagon	2.378
Dodecagon	3.000
100 sides	3.1395

What is the limit of the approximate area? Reconcile this with the area found by using the formula for the area of a circle.

50. Suppose the circle in Figure 1.29 has radius 2. Table 1.10 shows the approximate areas of inscribed polygons.

TABLE 1.10	Area of Inscribed Polygons, $r = 2$
Polygon	**Approximate Area**
Triangle	5.196
Quadrilateral	8.000
Pentagon	9.511
Dodecagon	12.000
100 sides	12.558

What is the limit of the approximate area? Reconcile this with the area found by using the formula for the area of a circle.

PROBLEMS FROM CALCULUS *Consider the graph of $y = x^2$ bounded by the x-axis and the line $x = 2$. Approximate the area under the curve by using rectangles as described in Problems 51–56.*

51. Use two rectangles of width 1 unit each.

52. Use four rectangles of width 0.5 unit each.

53. Use eight rectangles of width 0.25 unit each.

54. Use the results of Problems 51–53 to see if you can make a guess about the area under the curve.

55. Calculate the sum of the areas of the rectangles shown in Figure 1.31a.

56. Calculate the sum of the areas of the rectangles shown in Figure 1.31b.

LEVEL 3

PROBLEMS FROM CALCULUS *First sketch the region under the graph on the interval $[a, b]$ in Problems 57–60. Then approximate the area of each region for the given value of n. Use the right endpoint of each rectangle to determine the height of that particular rectangle.*

57. $y = 4x + 1$ on $[0, 1]$ for
    a. $n = 4$ **b.** $n = 8$

58. $y = 4x^2 + 1$ on $[0, 1]$ for $n = 4$

59. $y = \dfrac{2}{x}$ on $[1, 2]$ for $n = 4$

60. $y = \dfrac{1}{x^2}$ on $[1, 2]$ for $n = 4$

1.8 Equations for Calculus

We have assumed that you have solved certain types of equations in your previous mathematics courses. An **equation** is a statement of equality between two mathematical expressions. Equations might be *true* (as with $4 + 1 = 5$), *false* (as with $4 + 1 = 11$), or *open* (as with $4 + x = 5$). That is, an open equation may be true or false depending on the replacement for the variable or variables. To **solve** an equation means to find all values of the variable (or variables) that make the equation true. A value that makes the equation true is called a **root** of the equation.

There are five important processes in algebra. *Solve* is the last of those five. (Remember, the first four are *evaluate, simplify, factor,* and *graph.*)

You will need to solve a variety of different types of equations in calculus, including the following:

- Linear equations (assumed for this course)
- Quadratic equations (reviewed in this section)
- Equations in quadratic form (reviewed in this section)
- Polynomial equations (factorable forms, and by equating coefficients reviewed in this section others considered in Section 3.6)
- Rational equations (Section 4.3)
- Radical equations (Section 4.3)
- Logarithmic equations (Section 4.6)
- Exponential equations (Section 4.7)
- Trigonometric equations (Section 6.1)

As you can see from this extensive list, the ability to solve equations is essential for the study of calculus.

Quadratic Equations

A **quadratic equation in one variable** is an equation that can be written in the form

$$ax^2 + bx + c = 0 \quad (a \neq 0)$$

where x is a variable and a, b, and c are real numbers. Quadratic equations can be solved by several methods. The simplest method, factoring, can be used if the quadratic expression $ax^2 + bx + c$ is factorable over the integers. This method depends on the following property of zero.

ZERO FACTOR THEOREM

$AB = 0$ if and only if $A = 0$ or $B = 0$ (or both).

» IN OTHER WORDS After getting a zero on one side, if the nonzero side can be factored, then the equation can be solved by setting *each* factor (separately) equal to zero to solve these simpler equations.

Thus, if the product of two factors is zero, then at least one of the factors is zero. If a quadratic expression is factorable, this property provides a method of solution.

EXAMPLE 1   Solving a quadratic equation by factoring

Solve: $x^2 = 2x + 15$

Solution

$$x^2 = 2x + 15 \qquad \text{Given}$$
$$x^2 - 2x - 15 = 0 \qquad \text{Obtain a 0 on one side.}$$
$$(x + 3)(x - 5) = 0 \qquad \text{Factor.}$$
$$x + 3 = 0 \quad \text{or} \quad x - 5 = 0 \qquad \text{Zero factor theorem}$$
$$x = -3 \qquad x = 5 \qquad \text{Solve each linear equation.}$$

The solution is $-3, 5$.

You can also use a calculator to graph the equation and find an approximate solution. The steps, of course, depend on the brand and model of calculator (see the front matter of this book). With a graphing calculator, you will input the equation and look at the graph (as shown in Figure 1.33). Using the ROOT feature, we find the roots (or approximate values, depending on the accuracy of your calculator). In order to find all the roots by calculator, you need to know how many roots to expect. Look at the places where the graph crosses the x-axis.

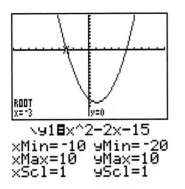 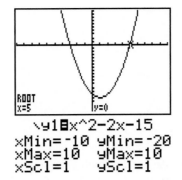

Figure 1.33 Roots using a calculator

We now summarize with the following procedure for solving factorable equations.

SOLVING BY FACTORING

To solve an equation that can be expressed as a product of factors:

Step 1 Rewrite all nonzero terms on one side of the equation.

Step 2 Factor the expression.

Step 3 Set each of the factors equal to zero.

Step 4 Solve each of the factor equations.

Step 5 The solution is the union of the solution sets of the factor equations.

EXAMPLE 2   Solving an equation given in factored form

Solve: $(3x - 2)(x + 1)^2(5x + 1) = 0$

Solution Set each factor equal to zero (you can do this by inspection) to find the solution:

$$x = \frac{2}{3}, -1, -\frac{1}{5}$$

Note that $x = -1$ is the solution to two of the factored equations, but we do not list the solution $x = -1$ twice. Sometimes this situation is described by saying that the root $x = -1$ has *multiplicity* of 2 because it appears as a factor twice.

Completing the Square

When the quadratic equation is not factorable, other methods must be employed. One such method depends on the square root property.

SQUARE ROOT PROPERTY

If P is an expression containing a variable and if $P^2 = Q$, then

$$P = \pm\sqrt{Q}$$

The equation $x^2 = 4$ could be rewritten as $x^2 - 4 = 0$, factored, and solved. However, the square root property can be used, as illustrated below.

Using the square root property: Using factoring:

$$x^2 = 4 \qquad \text{Given}$$

$$x^2 = 4$$

$$x = \pm\sqrt{4}$$

$$x^2 - 4 = 0$$

$$x = \pm 2$$

$$(x - 2)(x + 2) = 0$$

$$x = 2, -2$$

The square root property can be derived by finding the square root of a perfect square. Every quadratic may be expressed in the form $P^2 = Q$ by isolating the variable terms and **completing the square**, as illustrated in the following example.

EXAMPLE 3   Completing the square

Solve $4x^2 = 4x + 7$ by completing the square.

Solution

$$4x^2 = 4x + 7 \qquad \text{Given}$$

$$4x^2 - 4x = 7 \qquad \text{Rewrite to obtain variable terms on one side of the equation.}$$

$$x^2 - x = \frac{7}{4} \qquad \text{Divide by 4 to obtain a leading coefficient of 1.}$$

$$x^2 - x + \left(-\frac{1}{2}\right)^2 = \frac{7}{4} + \left(-\frac{1}{2}\right)^2 \qquad \text{Add the square of half the coefficient of the } x\text{-term to both sides.}$$

$$\left(x - \frac{1}{2}\right)^2 = \frac{7+1}{4} = 2 \qquad \text{Factor left side (it is a perfect square); simplify right side.}$$

$$x - \frac{1}{2} = \pm\sqrt{2} \qquad \text{Square root property}$$

$$x = \frac{1}{2} \pm \sqrt{2} \qquad \text{Solve for } x \text{ and simplify.}$$

$$= \frac{1 \pm 2\sqrt{2}}{2}$$

■

We can give a geometrical representation using areas to show why this process is called *completing the square*. Suppose we wish to complete the square for the algebraic expression $x^2 + 6x$. We begin by representing this as the area of a rectangle with sides x and $x + 6$:

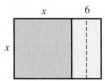

Area is $x\,(x + 6) = x^2 + 6x$

Cutting a piece and moving it does not change the area.

↑ Cut this piece in half and move the cut-off piece to the bottom.

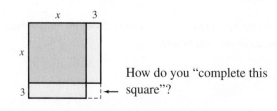

How do you "complete this square"?

Area is $x\,(x + 6) = x^2 + 6x$

Answer: Take one-half the coefficient of the first-degree term and square it.

Adding a 3-by-3 square completes the square figure, just as adding a term of $3^2 = 9$ completes the square trinomial.

Quadratic Formula

The process of completing the square is often cumbersome. However, if *any* quadratic equation $ax^2 + bx + c = 0$, $a \neq 0$, is considered, completing the square can be used to derive a formula for *x* in terms of the coefficients *a*, *b*, and *c*. The formula can then be used to solve all quadratic equations, even those that are nonfactorable.

QUADRATIC FORMULA

If $ax^2 + bx + c = 0$, $a \neq 0$, then

$$x = \frac{-b \pm \sqrt{b^2 - 4ac}}{2a}$$

Proof:

$ax^2 + bx + c = 0$	Given
$ax^2 + bx = -c$	Variables on one side
$x^2 + \left(\dfrac{b}{a}\right)x = -\dfrac{c}{a}$	Divide by a $(a \neq 0)$.
$x^2 + \left(\dfrac{b}{a}\right)x + \left(\dfrac{b}{2a}\right)^2 = -\dfrac{c}{a} + \left(\dfrac{b}{2a}\right)^2$	Complete the square.
$\left(x + \dfrac{b}{2a}\right)^2 = \dfrac{-4ac + b^2}{4a^2}$	Factor and simplify.
$x + \dfrac{b}{2a} = \pm\sqrt{\dfrac{b^2 - 4ac}{4a^2}}$	Square root property.
$x = -\dfrac{b}{2a} \pm \dfrac{\sqrt{b^2 - 4ac}}{2a}$	
$\quad = \dfrac{-b \pm \sqrt{b^2 - 4ac}}{2a}$	

EXAMPLE 4   Solving a quadratic equation by formula

Solve: $5x^2 + 2x - 2 = 0$

Solution   From the quadratic formula, where $a = 5$, $b = 2$, and $c = -2$:

$5x^2 + 2x - 2 = 0$	Given; $a = 5, b = 2$, and $c = -2$.
$x = \dfrac{-2 \pm \sqrt{2^2 - 4(5)(-2)}}{2(5)}$	Use $x = \dfrac{-b \pm \sqrt{b^2 - 4ac}}{2a}$
$= \dfrac{-2 \pm \sqrt{44}}{2(5)}$	
$= \dfrac{-2 \pm 2\sqrt{11}}{2(5)}$	
$= \dfrac{-1 \pm \sqrt{11}}{5}$	

EXAMPLE 5   Solving a quadratic equation that does not have real roots

Solve: $5x^2 + 2x + 2 = 0$

Solution

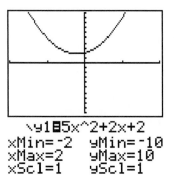

$$x = \frac{-2 \pm \sqrt{4 - 4(5)(2)}}{2(5)} = \frac{-2 \pm \sqrt{-36}}{10}$$

Since the square root of a negative number is not a real number, the solution set is *empty over the reals*. Notice that the graph shown in the margin does not intersect the real number line $y = 0$ (the *x*-axis), so the solution set is empty over the set of real numbers.

  If you are assuming the domain of complex numbers (see Appendix B), then you can complete the solution *over the complex numbers*, obtaining only nonreal roots.

$$x = \frac{-2 \pm \sqrt{-36}}{10} = \frac{-2 \pm 6i}{10} = \frac{-1 \pm 3i}{5} = -\frac{1}{5} \pm \frac{3}{5}i$$

◼

`\y1⊟5x^2+2x+2`
`xMin=-2   yMin=-10`
`xMax=2    yMax=10`
`xScl=1    yScl=1`

  Since the quadratic formula contains a radical, the sign of the radicand (the number under the square root sign) will determine whether the roots will be real or nonreal. If the radicand is negative, there will be no real solutions, but if it is positive, there will be two real solutions. When will there be exactly one solution? Only when the number under the radical is zero. This radicand is called the *discriminant* of the quadratic, and its properties are summarized in the box.

DISCRIMINANT

If $ax^2 + bx + c = 0$, $a \neq 0$, then $b^2 - 4ac$ is called the **discriminant**.

If $b^2 - 4ac < 0$, there are *no real* solutions.

If $b^2 - 4ac = 0$, there is *one real* solution.

If $b^2 - 4ac > 0$, there are *two real* solutions.

We now summarize a procedure for solving quadratic equations.

SOLVE A QUADRATIC

To solve a quadratic equation:

Step 1 Write the equation in the form

$$ax^2 + bx + c = 0$$

Step 2 Factor the quadratic expression, if possible. If it factors, then use the zero factor theorem.

Step 3 If $b = 0$, then consider using the square root property.

Step 4 If the quadratic is not yet factored, use the quadratic formula. The discriminant will tell you the nature of the roots.

Equations in Quadratic Form

Many equations are not quadratic but can be solved as if they were quadratic. We illustrate with two examples.

EXAMPLE 6   Solving an equation in quadratic form by factoring

Solve: $4x^4 + 3 = 13x^2$

Solution

$$4x^4 + 3 = 13x^2$$

$$4x^4 - 13x^2 + 3 = 0$$

$$(4x^2 - 1)(x^2 - 3) = 0$$

$$4x^2 - 1 = 0 \quad \text{or} \quad x^2 - 3 = 0$$

$$x^2 = \frac{1}{4} \qquad\qquad x^2 = 3$$

$$x = \pm\frac{1}{2} \qquad\qquad x = \pm\sqrt{3}$$

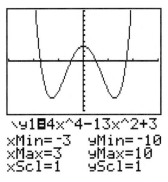

\y1■4x^4-13x^2+3
xMin=-3   yMin=-10
xMax=3   yMax=10
xScl=1   yScl=1

The solution is: $\frac{1}{2}, -\frac{1}{2}, \sqrt{3}, -\sqrt{3}$.

The graph reinforces the answer by noting where the curve crosses the x-axis, namely, at $x = -\frac{1}{2}$ and $x = \frac{1}{2}$; it also appears to cross near 1.75 and -1.75.

EXAMPLE 7   Solving an equation in quadratic form by quadratic formula

Solve: $x^4 + 6x^2 - 3 = 0$ over $\mathbb{R}$.

Solution

$$x^4 + 6x^2 - 3 = 0$$

$$(x^2)^2 + 6(x^2) - 3 = 0$$

$$x^2 = \frac{-6 \pm \sqrt{36 - 4(1)(-3)}}{2}$$

$$x^2 = \frac{-6 \pm 4\sqrt{3}}{2}$$

$$x^2 = -3 \pm 2\sqrt{3}$$

$$= \pm\sqrt{-3 + 2\sqrt{3}}$$

$$= \pm\sqrt{-3 + 2\sqrt{3}} \quad \text{Since } x = \pm\sqrt{-3 - 2\sqrt{3}} \text{ is not a real root.}$$

$$= \pm 0.6812500386$$

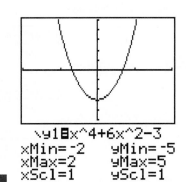

The graph in the margin supports these two real roots.

If you are including answers throughout the set of complex numbers, then the four roots are

$$\pm\sqrt{-3 + 2\sqrt{3}}, \pm\sqrt{3 + 2\sqrt{3}}i.$$

\y1■x^4+6x^2-3
xMin=-2   yMin=-5
xMax=2   yMax=5
xScl=1   yScl=1

Equal Coefficients

A rarely taught method for solving higher-order equations, which is used in calculus, is the method of equating coefficients. We state the result for a fourth-degree equation, but it is true for any degree.

> **EQUATING COEFFICIENTS**
>
> If $Ax^4 + Bx^3 + Cx^2 + Dx + E = ax^4 + bx^3 + cx^2 + dx + e$,
> then
> $$A = a,\ B = b,\ C = c,\ D = d,\ \text{and}\ E = e$$

» IN OTHER WORDS If you find two higher-order equations that are identical except for the numerical coefficients, then the coefficients of similar terms are also equal to each other.

EXAMPLE 8   Method of equal coefficients

Solve: $10x^4 + 5x^3 - 2x^2 + x = (t^2 + 1)x^4 + (u^2 - 3u + 1)x^3 + vx^2 - ux$

Solution   Equate the coefficients:

x^4 terms	x^3 terms	x^2 terms	x terms
$t^2 + 1 = 10$	$u^2 - 3u + 1 = 5$	$-2 = v$	$1 = -w$
$t^2 = 9$	$u^2 - 3u - 4 = 0$		$w = -1$
$t = \pm 3$	$(u - 4)(u + 4) = 0$		
	$u = 4, -1$		

The solution for the variables is $t = \pm 3$; $u = 4, -1$; $v = -2$; $w = -1$. ∎

PROBLEM SET 1.8

LEVEL 1

WHAT IS WRONG, *if anything, with each statement in Problems 1–8? Explain your reasoning.*

1. If $5x^2 = 125x$, then $x = 25$ (divide both sides by $5x$).
2. If $x^2 = -1$, then the solution set is empty in the set of real numbers.
3. If $(x + 3)(x - 2) = 4$, then $(x + 3) = 4$ and $(x - 2) = 4$.
4. If $(2x + 1)(x - 5) = 2$, then $x = -\frac{1}{2}, 5$.
5. The quadratic formula says
$$x = \frac{-b \pm \sqrt{b^2 - 4ac}}{2a}$$
6. If $ax^2 + bx + c = 0$, then
$$x = \frac{-b \pm \sqrt{b^2 - 4ac}}{2a}$$
7. If $bx^2 + ax + c = 0$, $b \neq 0$, then
$$x = \frac{-b \pm \sqrt{b^2 - 4bc}}{2a}$$
8. If $bx^2 + ax + c = 0$, $b \neq 0$, then
$$x = \frac{-a \pm \sqrt{a^2 - 4bc}}{2b}$$
9. IN YOUR OWN WORDS Describe a process for solving a linear equation.
10. IN YOUR OWN WORDS Describe a process for solving a quadratic equation.

Solve Problems 11–14 for x by factoring. If a root has multiplicity greater than 1, so state.

11. **a.** $(x - 2)(x + 1) = 0$ **b.** $(x + 3)(x + 4) = 0$
12. **a.** $(x - 2)(x - 3) = 0$ **b.** $(x + 1)(x - 3) = 0$
13. **a.** $(x + 3)(2x - 1)(3x + 1) = 0$
    b. $(x - 2)(3x + 1)(2x + 3) = 0$
14. **a.** $(2x - 1)(3x - 2)(x + 4)^2 = 0$
    b. $(3x + 2)(5x - 3)(x - 2)^3 = 0$

Solve Problems 15–20 for x by using the square root property. This may require that you complete the square.

15. **a.** $(x + a)^2 = 16$ **b.** $(x + a)^2 = b\ (b > 0)$
16. **a.** $(x + a)^2 = -16$ **b.** $(x + a)^2 = -b\ (b < 0)$
17. $x^2 + 2x - 15 = 0$ 18. $x^2 - 8x + 12 = 0$
19. $x^2 + 7x - 18 = 0$ 20. $2x^2 = 12 - 5x$

Solve Problems 21–30 for x by any appropriate method. If a root has multiplicity greater than 1, so state.

21. $6x^2 = 5x$ 22. $6x^2 = 12x$
23. $10x^2 - 3x - 4 = 0$ 24. $6x^2 + 7x - 10 = 0$
25. $16x^2 + 24x + 9 = 0$ 26. $9x^2 = 2 - 7x$
27. $3x(4 + 3x) = 13$ 28. $13 = x(4 - x)$
29. $x^2 + 2(3x + 5) = 0$ 30. $2x^2 + 1 = 2x$

LEVEL 2

Solve Problems 31–40 for x in terms of the other variable.

31. $2x^2 + x - w = 0$ **32.** $2x^2 + ux + 5 = 0$

33. $3x^2 + 2x - (y + 2) = 0$ **34.** $3x^2 + 5x + (4 - y) = 0$

35. $4x^2 - 4x + (1 - t^2) = 0$ **36.** $2x^2 + 3x + 4 - y = 0$

37. $y = 2x^2 + x + 6$

38. $4x^2 - (3t + 10)x + (6t + 4) = 0, t > 2$

39. $(x - 3)^2 + (y - 2)^2 = 4$ **40.** $(x + 2)^2 - (y + 1)^2 = 9$

Solve Problems 41–46 and state each solution set with roots correct to the nearest hundredth.

41. $3x = 75(x - 2)(x + 2)$

42. $(3x + 2)(2x + 1) = 2(2x - 1)(9x + 8)$

43. $2(3 - 4x)(2x + 1) = (6x - 1)(2x - 1)$

44. $2(3 - 4x)(2x + 1) = (6x - 1)(2x - 11)$

45. $x^3 - (x - 2)^3 = 92$

46. $(x - 1)(2x - 1)(3x - 1) = 6x^3 - 82$

PROBLEMS FROM CALCULUS *Solve Problems 47–52 using the method of equal coefficients.*

47. $Ax^5 + Bx^4 + Cx^3 + Dx^2 + Ex + F$
$= 5x^5 + 3x^3 - 3x + 8$

48. $Ax^7 + Bx^6 + Cx^5 + Dx^4 + Ex^3 + Fx^2 + Gx + H$
$= 8x^7 - 4x^3 + 2x + 9$

49. $sx^4 - (t + 4)x^3 - 7ux^2$
$= 16x^4 + 12x^3 + 5x^2$

50. $2x^2 + 5x = sx^3 + t^2x^2 - 2ux - vx^3 - x^2$

51. $8x^3 - 5x^2 + 2px - 8$
$= m^2x^3 - x^3 + nx^2 + 5x^2 - 8$

52. $px^4 + (q^2 - 4q + 1)x^3 + r^2x^2 + sx$
$= 5x^4 + 6x^3 + 9x^2 + 9x$

53. In Section 1.5, we described the path of a cannonball by $x^2 + 256y - 384x = 0$, where x is the horizontal distance (in ft) and y is the height (in ft). Find the horizontal distance when the height is 144 ft.

54. In Section 1.5, we described the path of a cannonball by $x^2 + 256y - 384x = 0$, where x is the horizontal distance (in ft) and y is the height (in ft). Find the horizontal distance at which the cannonball comes to rest.

55. In calculus, it is shown that if an object is thrown upward from the top of a 192-ft building with an initial velocity of 16 ft/s, then the height h above the ground t seconds later is

$$h = 192 + 16t - 16t^2$$

During what time interval will the ball be in the air?

56. In calculus, it is shown that if an object is thrown upward from the top of a 64-ft building with an initial velocity of 16 ft/s, then the height h above the ground t seconds later is

$$h = 64 + 16t - 16t^2$$

During what time interval will the ball be at least 32 ft above the ground?

LEVEL 3

57. Show that if r_1 and r_2 are roots of the quadratic equation $ax^2 + bx + c = 0$, $a \neq 0$, then $r_1 + r_2 = -b/a$ and $r_1 r_2 = c/a$.

58. Show that if r_1 and r_2 are roots of the quadratic equation $ax^2 + bx + c = 0$, $a \neq 0$, then

$$r_1^2 + r_2^2 = \frac{b^2 - 2ac}{a^2}$$

59. Show that if r_1 and r_2 are roots of the quadratic equation $ax^2 + bx + c = 0$, $a \neq 0$, then

$$\frac{1}{r_1^2} + \frac{1}{r_2^2} = \frac{b^2 - 2ac}{c^2}$$

60. Reconsider the equation $ax^2 + bx + c = 0$, $a \neq 0$. Isolate the variable terms, *multiply both sides by* $4a$, complete the square on the resulting terms, and solve for x. Use this procedure to prove the *quadratic formula.*

1.9 Inequalities for Calculus

One of the central themes of differential calculus is the ability to graph a variety of different types of relations. In calculus, we will define that portion of a graph that is cupped upward as *concave up*, and a portion that is cupped downward as *concave down*. To find those regions of differing concavity, you will need to know how to solve inequalities. In addition to linear inequalities (Section 1.5), the inequalities you will need for calculus include quadratic inequalities and inequalities in factored form.

Quadratic Inequalities

A restatement of the rules of signs for multiplication proves very useful in developing a method for solving polynomial inequalities that can be factored.

PROPERTY OF PRODUCTS

For expressions P and Q,

     $P \cdot Q = 0$, if and only if either $P = 0$ or $Q = 0$.
     $P \cdot Q < 0$, if and only if P and Q have opposite signs.
     $P \cdot Q > 0$, if and only if P and Q have the same sign.*

*The words "*p* if and only if *q*" mean "if *p*, then *q*," and also "if *q*, then *p*."

To apply the rules of signs to the solution of inequalities, look first at the associated equality and decide for what values of the variable the expression is zero. These values are called the **critical values** for the inequality. Then, by the trichotomy law (Section 1.1), PQ must be greater than or less than zero everywhere else. The signs of the factors can then be examined on the intervals between critical values. Finally, the interval or intervals that satisfy the inequality can be selected. A **quadratic inequality** is a region of the plane whose boundary is the graph of a quadratic equation.

EXAMPLE 1   Solving a quadratic inequality (one-interval solution)

Solve: $x^2 - x < 12$

Solution   Find the critical values.

$$x^2 - x - 12 = 0$$
$$(x + 3)(x - 4) = 0$$
$$x = -3, 4$$

These critical values divide the number line into three intervals.

One dimension   In one dimension ($\mathbb{R}$), we take one factor at a time and look at its critical value. That factor is positive or negative in *each* of the intervals (it cannot be 0 because then there would be a critical value at that point).

Step 1   Consider the factor $x + 3$, with critical value: $x = -3$ (Solve: $x + 3 = 0$). Label the parts of $\mathbb{R}$ that are positive and that are negative (see Figure 1.34a).

Step 2   Consider the next factor $x - 4$, with critical value: $x = 4$ (Solve: $x - 4 = 0$). Label the parts of $\mathbb{R}$ that are positive and that are negative (see Figure 1.34a).

Step 3   Since the factors are to be multiplied, use the rules of multiplication for each region on $\mathbb{R}$. The first region of $\mathbb{R}$. ($x < -3$): note Neg · Neg = Pos; the second region ($-3 < x < 4$): note Neg · Pos = Neg; and the third region ($x > 4$): Pos · Pos = Pos.

Step 4   Shade in the appropriate portion of the number line indicated by the original inequality: For this example, we see "< 0," which means that we are looking for the portion on the number line labeled "Neg." Highlight this portion of $\mathbb{R}$ as shown in Figure 1.34a.

Step 5   Finally, because the endpoints are excluded, make each of the endpoints an open dot.

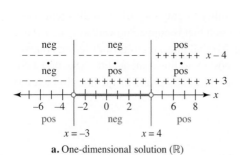

a. One-dimensional solution ($\mathbb{R}$)

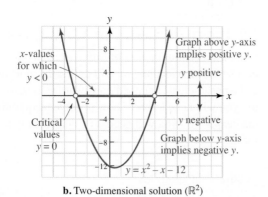

b. Two-dimensional solution ($\mathbb{R}^2$)

Figure 1.34 Solution for $x^2 - x < 12$

Two dimensions   In two dimensions, graph $y = x^2 - x - 12$. (You can do this by hand or using a graphing calculator.) Note from Figure 1.34b that the portion where the graph is below the *x*-axis is the portion for which the inequality is negative ($y < 0$), and the portion for which the graph is above the *x*-axis is the portion for which the inequality is positive ($y > 0$). We see that the graph is negative on the interval $(-3, 4)$. ∎

EXAMPLE 2   Solving a quadratic inequality (two-interval solution)

Solve: $2x^2 \geq 5 - 9x$

Solution

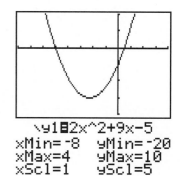

$$2x^2 \geq 5 - 9x \qquad \text{Given}$$
$$2x^2 + 9x - 5 \geq 0 \qquad \text{Write with a zero on one side}$$
$$(2x - 1)(x + 5) \geq 0 \qquad \text{Factor, if possible.}$$

\y1◻2x^2+9x-5
xMin=-8   yMin=-20
xMax=4   yMax=10
xScl=1   yScl=5

The graphical (two-dimensional) solution is shown in the margin. You can use the root feature to find the critical values, and then note that you want the portion of the x-axis for which the curve is above the y-axis.

If you do not have a graphing calculator, you can use a number line to solve the inequality for the critical values $x = \frac{1}{2}$ and $x = -5$, which come from $(2x - 1)(x + 5) = 0$; the steps are shown in Figure 1.35.

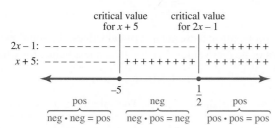

Figure 1.35 Steps in solving $2x^2 \geq 5 - 9x$

We are looking for the positive intervals. Since the solution is separated into two intervals, we use a union symbol (∪) to specify the answer in interval notation. The solution is $(-\infty, -5] \cup [\frac{1}{2}, \infty)$. ∎

EXAMPLE 3   Solving a quadratic inequality that does not factor

Solve: $x^2 + 2x - 4 > 0$

Solution   The polynomial on the left cannot be factored, so we consider $x^2 + 2x - 4$ as a single factor. To find the critical values, we find the values for which the factor $x^2 + 2x - 4$ is zero (by the quadratic formula):

$$x^2 + 2x - 4 = 0$$
$$x = \frac{-2 \pm \sqrt{4 - 4(1)(-4)}}{2}$$
$$= \frac{-2 \pm 2\sqrt{5}}{2}$$
$$= -1 \pm \sqrt{5}$$

Plot the critical values and check the sign of the factor in each interval. We are looking for values that are positive (i.e., ">0"). For example, if $x = -10$ (which is on the interval $\left(-\infty, -1 - \sqrt{5}\right)$), we see that $x^2 + 2x - 4 = (-10)^2 + 2(-10) - 4 > 0$. Similarly, if $x = 0$, $x^2 + 2x - 4 = 0^2 + 2(0) - 4 < 0$, and if $x = 10$, $x^2 + 2x - 4 > 0$. These steps are shown in Figure 1.36.

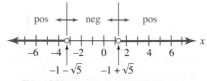

Figure 1.36 Graph of the solution of
$$x^2 + 2x - 4 > 0$$

The solution is $\left(-\infty, -1 - \sqrt{5}\right) \cup \left(-1 + \sqrt{5}, \infty\right)$.

EXAMPLE 4   Solving a quadratic inequality with solution $\mathbb{R}$

Solve: $x^2 + 5x + 12 > 0$

Solution   The quadratic does not factor, so we look for a critical value by solving $x^2 + 5x + 12 = 0$ by the quadratic formula.

$$x^2 + 5x + 12 = 0$$

$$x = \frac{-5 \pm \sqrt{25 - 4(1)(12)}}{2}$$

$$= \frac{-5 \pm \sqrt{-23}}{2}$$

This equation has no real roots. Since there are no critical values, the solution of the original inequality is either all real values or no real values. Let $x = 0$ be a test value, for which we see that

$$0^2 + 5(0) + 12 > 0$$

is true. Thus, all real values satisfy the inequality. The solution is $(-\infty, \infty)$.

Factored Inequalities

Higher-order inequalities, sometimes called **polynomial inequalities**, can be solved using the same techniques, provided the inequalities can be written in factored form.

EXAMPLE 5   Solving a polynomial inequality

Solve: $(x - 1)^2(x + 1)(3 - x) > 0$

Solution   The critical values are $x = 1$, $x = -1$, and $x = 3$. Since the product is greater than zero, we are looking for those regions on the number line for which the product of the factors is positive (see Figure 1.37a). We see that the solution is $(-1, 1) \cup (1, 3)$.

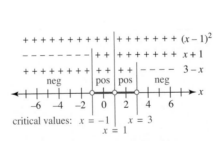

a. One-dimensional solution ($\mathbb{R}$)

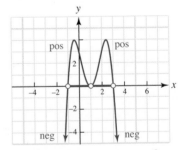

b. Two-dimensional solution ($\mathbb{R}^2$)

Figure 1.37 Solution for $(x - 1)^2(x + 1)(3 - x) > 1$

For a graphical solution, graph $y = (x - 1)^2(x + 1)(3 - x)$ and look for the portions above the x-axis ($y > 0$), as shown in Figure 1.37b. The graph is above the x-axis (y is positive, or $y > 0$) on the interval

$$(-1, 1) \cup (1, 3)$$

Compare this solution to the solution for $(x - 1)^2(x + 1)(3 - x) \geq 0$. Trace through the steps shown here to see the solution to this inequality is simply $[-1, 3]$.

EXAMPLE 6   Solving a rational inequality in factored form

Solve: $\dfrac{1}{2x + 6} \leq \dfrac{3}{2 - 2x}$

Solution   To solve an inequality with degree greater than 1, we wish to obtain a zero on one side. ☠ Do not multiply both sides by expressions that involve variables (because you do not know whether the expression is positive or negative). ☠

$$\frac{1}{2x + 6} \leq \frac{3}{2 - 2x} \qquad \textit{Given}$$

$$\frac{1}{2(x + 3)} - \frac{3}{2(1 - x)} \leq 0 \qquad \textit{Subtract } \frac{3}{2 - 2x} \textit{ from both sides to obtain a zero on the right side.}$$

$$\frac{(1 - x) - 3(x + 3)}{2(x + 3)(1 - x)} \leq 0 \qquad \textit{Obtain a common denominator and simplify.}$$

$$\frac{-4x - 8}{2(x + 3)(1 - x)} \leq 0$$

$$\frac{-4(x + 2)}{-2(x + 3)(x - 1)} \leq 0$$

$$\frac{2(x + 2)}{(x + 3)(x - 1)} \leq 0$$

We begin by using technology to graph

$$y = \frac{2(x + 2)}{(x + 3)(x - 1)}$$

as shown in Figure 1.38a. The **critical values for a rational inequality** are values for which either the numerator or denominator is zero. For this example, we see the critical values are $x = -2, -3, 1$. The inclusive inequality includes the critical values, but that is overridden by any values that cause division by zero. The geometric solution seems to be $(-\infty, -3) \cup [-2, 1)$.

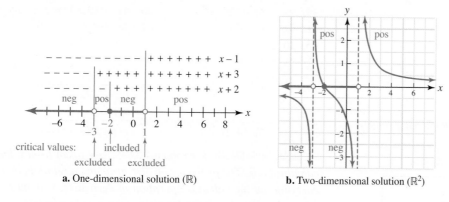

a. One-dimensional solution ($\mathbb{R}$)   **b.** Two-dimensional solution ($\mathbb{R}^2$)

Figure 1.38 Solution for $\dfrac{2}{2x + 6} \leq \dfrac{3}{2 - 2x}$

If you do have a graphing calculator, plot the critical values (values for which any of the factors is equal to zero) of -2, -3, and 1. These are included (because it is $\leq$), but *values that cause division by zero* ($x = -3$ and $x = 1$) need to be *excluded*. The details are shown in Figure 1.38b. The answer is $(-\infty, -3) \cup [-2, 1)$.

PROBLEM SET 1.9

LEVEL 1

WHAT IS WRONG, *if anything, with each statement in Problems 1–4? Explain your reasoning.*

1. If $x(x+8) < 0$, then $x < 0$ and $x + 8 < 0$.

2. If $x(x+2) \geq 0$, then $x + 2 \geq 0$ (divide both sides by x).

3. If $ax^2 + bx + c < 0$, then the critical values are

$$x = \frac{-b \pm \sqrt{b^2 - 4ac}}{2a}$$

4. Solve $3x^2 < 8 - 10x$

$$3x^2 < 8 - 10x$$
$$3x^2 + 10x - 8 < 0$$
$$(3x - 2)(x + 4) < 0$$

$$\begin{array}{ccc} 3x - 2 < 0 & \text{or} & x + 4 < 0 \\ 3x < 2 & & x < -4 \\ x < \dfrac{2}{3} & & \end{array}$$

Solve each inequality in Problems 5–24. State the solution using interval notation. You can solve these problems algebraically or geometrically. If you use a calculator for a graphical solution, show a sketch of the graph on your paper.

5. $x(x+1) < 0$
6. $x(x-1) \geq 0$
7. $(x-5)(x-2) \geq 0$
8. $(x+2)(x-3) \leq 0$
9. $(x-3)(x+7) < 0$
10. $(x+1)(x+5) > 0$
11. $(x+2)(3-x) \leq 0$
12. $(2-x)(x+3) > 0$
13. $(1-3x)(x-4) < 0$
14. $(2x+1)(3-x) > 0$
15. $x(x-3)(x+4) \leq 0$
16. $x(x+3)(x-4) \geq 0$
17. $\dfrac{x+2}{x} < 0$
18. $\dfrac{x}{x+8} > 0$
19. $\dfrac{x-3}{x+4} \leq 0$
20. $\dfrac{x+3}{6-x} \geq 0$
21. $(x-2)(x+3)(x-4) \geq 0$
22. $(x+1)(x-2)(x+3) \leq 0$
23. $(x+1)(2x+5)(7-3x) > 0$
24. $(x-2)(3x+2)(3-2x) < 0$

LEVEL 2

Solve each inequality in Problems 25–30. You should use geometrical methods.

25. $(x-5)(x+3)(x-1) \geq 0$

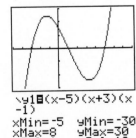

26. $x(x-2)(x+3) \leq 0$

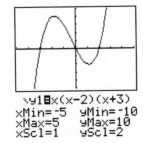

27. $x^3 + 2x^2 - 5x - 6 < 0$

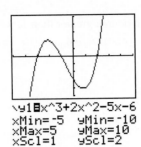

28. $x^3 - 3x^2 - 10x + 24 > 0$

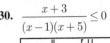

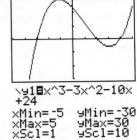

29. $\dfrac{10x}{(x-3)^2} \geq 0$

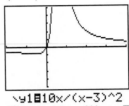

30. $\dfrac{x+3}{(x-1)(x+5)} \leq 0$

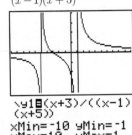

Solve each inequality in Problems 31–50. State the solution using interval notation. You can solve these problems algebraically or geometrically. If you use a calculator, show a sketch of your graph.

31. $x^2 - 3x + 2 > 0$
32. $x^2 + 3x + 2 < 0$
33. $x^2 - x - 2 \leq 0$
34. $x^2 - x - 2 \geq 0$
35. $5x^2 - 4x - 1 < 0$
36. $2x^2 - 3x - 35 > 0$
37. $x^2 + 5 \geq 6x$
38. $2x^2 + x \leq 6$
39. $2x^2 + 5x < 12$
40. $4x^2 + 10 > 13x$
41. $6x^2 + 6 \leq 13x$
42. $4x^2 + 23x + 15 > 0$
43. $10(x^2 + 1) \geq 29x$
44. $25x < 12(1 + x^2)$
45. $x(x-3) > 340$
46. $x(x+5) < 300$
47. $\dfrac{x(2x-1)}{5-x} > 0$
48. $\dfrac{x}{(2x+3)(x-2)} < 0$
49. $\dfrac{1}{x(x-3)(x+2)} \leq 0$
50. $\dfrac{1}{x(x+1)(5-x)} \geq 0$

PROBLEMS FROM CALCULUS *Problems 51–52 are taken from a leading calculus book. The problems involve graphing a relation, and after carrying out the calculus operation of derivative, you must determine when the given expression is positive and when it is negative.*

51. $y = 12x^3 - 12x^2 - 24x$
$\quad = 12x(x-2)(x+1)$
   a. When is $y > 0$?   **b.** When is $y < 0$?

52. $y = 3x^2(x-2)^2 + x^3 \cdot 2(x-2)$
   a. When is $y > 0$?   **b.** When is $y < 0$?

LEVEL 3

Solve each inequality in Problems 53–57.

53. $\dfrac{2}{x-2} \le \dfrac{3}{x+3}$

54. $\dfrac{1}{x+1} \ge \dfrac{2}{x-1}$

55. $\dfrac{x-4}{x^2-5x+6} \le 0$

56. $\dfrac{x-7}{x^2-4x-21} < 0$

57. $\dfrac{3-x}{1-3x} \ge \dfrac{x+3}{2x+1}$

58. In Example 1, Section 1.5, we solved

$$-\frac{1}{256}x^2 + \frac{3}{2}x > 128$$

numerically and graphically. Solve this inequality algebraically.

59. The property of trichotomy tells us that exactly one of the following is true:

(1) $\sqrt{ab} > \dfrac{a+b}{\sqrt{2}}$

(2) $\sqrt{ab} < \dfrac{a+b}{\sqrt{2}}$

(3) $\sqrt{ab} = \dfrac{a+b}{\sqrt{2}}$

Which one of these statements is true?

60. The property of trichotomy tells us that exactly one of the following is true:

(1) $\sqrt{\dfrac{a^2+b^2}{2}} > \dfrac{a+b}{2}$

(2) $\sqrt{\dfrac{a^2+b^2}{2}} < \dfrac{a+b}{2}$

(3) $\sqrt{\dfrac{a^2+b^2}{2}} = \dfrac{a+b}{2}$

Which one of these statements is true?

CHAPTER 1 SUMMARY AND REVIEW

T*he calculus is the greatest aid we have to the appreciation of physical truth.*

W. F. Osgood (1907)

 Take some time getting ready to work the review problems in this section. First, look back at the definition and property boxes. You will maximize your understanding of this chapter by working the problems in this section only after you have studied the material.

SELF TEST   *All of the answers for this self test are given in the back of the book.*

1. **Simplify**
   a. IN YOUR OWN WORDS Describe the meaning of this word.
   b. $(4x-1)(2x+3)$ c. $(x-2)^3$
   d. $(5x^3 - 2x^2 + 5) - (1 - 4x^2)$

2. **Factor**
   a. IN YOUR OWN WORDS Describe the meaning of this word.
   b. $6x^2 + 5x - 6$ c. $x^4 - 10x^2 + 9$
   d. $\left(x^3 - \frac{1}{8}\right)(8x^3 + 8)$

3. **Evaluate**
   a. IN YOUR OWN WORDS Describe the meaning of this word.
   b. $2x^2 + 3x - 4$ for $x = -3$
   c. $\left| 6 - \sqrt{50} \right|$ d. $|x^2 + 2\pi|$

4. **Graph**
   a. IN YOUR OWN WORDS Describe the meaning of this word.
   b. $[2, 4) \cup [7, \infty)$ c. $2x - 3y + 6 = 0$
   d. $y = x^2 + 2x - 3$ (use technology or plot points)

5. **Solve**
   a. IN YOUR OWN WORDS Describe the meaning of this word.
   b. $2x^2 - 4x - 30 = 0$ c. $x(14 - x)(x + 30) < 0$
   d. $\dfrac{(x-5)(x+2)}{x-1} \geq 0$

6. Write the standard-form equation of the lines with the given information.
   a. Passing through $(1, 3)$ and $(-5, -9)$.
   b. A horizontal line passing through $(45, 16)$.

7. Solve $x^2 - 5x + 1 = 0$ by completing the square.

Consider the graph of $y = |x^3 - 2x|$ shown in Figure 1.39 for Problems 8–10.

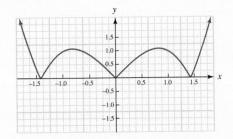

Figure 1.39 Graph of $y = |x^3 - 2x|$

8. Solve by geometrical methods: $|x^3 - 2x| \leq 0$.
9. Tell whether it is symmetric with respect to the *x*-axis, *y*-axis, or the origin.
10. a. Find the point on the curve where $x = 1$, and then find the slope of the line through this point and the origin.
    b. Repeat for $x = \sqrt{2}$.

STOP STUDY HINTS *Compare your solutions and answers to the self test.*

Practice for Calculus—*Supplementary Problems*

Evaluate the expressions in Problems 1–6.

1. $|x^2 + 2|$ 2. $|\pi - 7|$
3. $-a^2$ for $a = -3$ 4. $2x^2 - x + 1$ for $x = 2$
5. $(a - b)(a^2 + ab + b^2)$ for $a = 3, b = 2$
6. $-c^2 - d^2$ for $c = 2, d = 3$

Simplify the expressions in Problems 7–12.

7. $3x(1 - x^2) + x(2x^2 - 3x + 1)$
8. $3x(1 - x^2) - x(2x^2 - 3x + 1)$
9. $(2x + 1)(5x^3 + 4x^2 - 35x - 15)$
10. $(2x + 1) + (5x^3 + 4x^2 - 35x - 15)$
11. $(2x + 1) - (5x^3 + 4x^2 - 35x - 15)$
12. $(2x + 1)^3$

Factor the expressions in Problems 13–18.

13. $10x^2 - 5x - 50$ 14. $\dfrac{4x^2}{y^2} - (2x + y)^2$
15. $x^4 - 26x^2 + 25$ 16. $\left(x^3 - \frac{1}{64}\right)(64x^3 + 64)$
17. $6x^3 + 3x^2 - 6x - 3$ 18. $3x^{-4}y^2 - 4x^{-5}y^3$

Graph the lines in Problems 19–24 by using the slope and the y-intercepts.

19. $y = \frac{2}{3}x - 4$ 20. $y = -\frac{3}{5}x$
21. $2x + 5y = 15$ 22. $3x - 4y = -5$
23. $x = y$ 24. $2x = 3y$

Solve the inequalities in Problems 25–30 and leave your answers in notation.

25. $5x + 12 \leq 7$

26. $-10 \leq -x$

27. $5 \leq 2x + 3 \leq 23$

28. $3 \leq 1 - 2x \leq 11$

29. $-0.0001 \leq x + 3 \leq 0.0001$

30. $-0.001 < 2x - 5 < 0.001$

By plotting points, graph the relations defined by the equations in Problems 31–34.

31. $3x + y - 5 = 0$

32. $y = -\frac{4}{5}x$

33. $y = -\frac{4}{5}x^2$

34. $y = -\frac{4}{5}|x|$

Write the standard-form equation of each line in Problems 35–38.

35. Slope is -5 and passes through $(4, -3)$.

36. Passing through $(-2, 1)$ and $(5, -6)$.

37. Passing through $(4, 5)$ and perpendicular to the line $2x - 3y + 8 = 0$.

38. A vertical line through $(-50, -250)$.

Solve each equation or inequality in Problems 39–55 in $\mathbb{R}$.

39. $(5 - x)(7 - 3x) = 0$

40. $2x^2 - 17x + 30 = 0$

41. $x(2 - x)(x + 1) > 0$

42. $x(2 - x)(x + 1)^2 > 0$

43. $3x^2 + 2x + 1 = 0$

44. $x^2 = 144$

45. $|x| = -5$

46. $|2 - 5x| = 17$

47. $|2x - 5| < 8$

48. $|4 - x| \geq 2$

49. $3 + 5x \geq 2x^2$

50. $\dfrac{x + 4}{x - 5} \leq 0$

51. Solve $x^2 - 3x + 1 = 0$ by completing the square.

52. What is the slope of a line perpendicular to $6x - 4y + 3 = 0$?

53. Is the triangle with vertices $A(2, -7), B(4, 3)$, and $C(-3, -2)$ a right triangle?

54. Solve using geometrical methods: $x^3 + 5x^2 - 250x + 1,000 \geq 0$

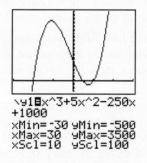

```
\y1=x^3+5x^2-250x
+1000
xMin=-30 yMin=-500
xMax=30  yMax=3500
xScl=10  yScl=100
```

55. Solve using geometrical methods: $\dfrac{x + 1}{x^2 - 4} \leq 0$

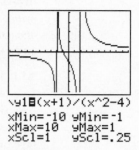

```
\y1=(x+1)/(x^2-4)
xMin=-10 yMin=-1
xMax=10  yMax=1
xScl=1   yScl=.25
```

For each curve in Problems 56–59:

a. *Tell whether it is symmetric with respect to the x-axis, the y-axis, or the origin.*

b. *Determine where $y = 0$.*

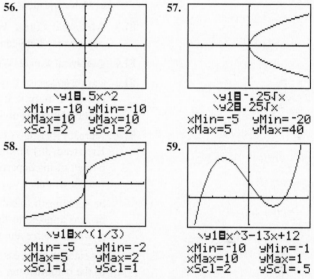

56.
```
\y1=.5x^2
xMin=-10 yMin=-10
xMax=10  yMax=10
xScl=2   yScl=2
```

57.
```
\y1=-.25√x
\y2=.25√x
xMin=-5  yMin=-20
xMax=5   yMax=40
```

58.
```
\y1=x^(1/3)
xMin=-5  yMin=-2
xMax=5   yMax=2
xScl=1   yScl=1
```

59.
```
\y1=x^3-13x+12
xMin=-10 yMin=-1
xMax=10  yMax=1
xScl=2   yScl=.5
```

60. PROBLEM FROM CALCULUS Sketch the curve $y = (x + 2)^2$. Plot $P(-2, 0)$ on this curve. Calculate the slope of the line passing through the given points.

a. $Q_1(2, 16)$

b. $Q_2(0, 4)$

c. $Q_3(-1, 1)$

d. $Q_4(-1.5, 0.25)$

e. What is the slope of the tangent line?

f. Write the equation of the tangent line, and sketch the tangent line at P.

CHAPTER 1 Group Research Projects

Working in small groups is typical of most work environments, and this book seeks to develop skills with group activities. At the end of each chapter, we present a list of suggested projects, and even though they could be done as individual projects, we suggest that these projects be done in groups of three or four students.

G1.1 Look at your college's catalog. Count the number of mathematics courses offered.
   a. How many are college level?
   b. How many are precalculus level?
   c. How many are calculus?
   d. How many are postcalculus?
   e. Are these results what you expected? Comment.

G1.2 Write out formulas for solving:
   a. Linear equations
   b. Quadratic equations (the answer for this is discussed in this chapter).
   c. Cubic equations

G1.3 **Historical Quest** Write a brief history of calculus. Include some documentation on the controversy between Newton and Leibniz in the 17th century.

G1.4 **Historical Quest** Write a paper on the history of the quadratic formula.

G1.5 RULES OF DIVISIBILITY It is well known that a number is divisible by 2 if the last digit of the number is 0, 2, 4, 6, or 8; it is also well known that a number is divisible by 3 if the sum of the digits of the number is divisible by 3. Find rules of divisibility for the numbers 1 to 12, inclusive.

G1.6 Journal Problem (*The Journal of Recreational Mathematics*, Problem 1036 by Friend H. Kierstead, Jr.) Professor Dimwit of the University of Lower Slobovia squandered the annual budget of his department to purchase a handheld calculator that had, in addition to the usual four functions, a square root key and a memory. To test his prowess with the electronic marvel, he set himself to calculate the successive powers of $3 + \sqrt{8}$. After the first few calculations, he was amazed to find that the calculator gave him nothing but integers for the succeeding powers, when even a fool knows that all powers of such a number should be irrational. So he repeated the process with $4 + \sqrt{15}$, with no better luck. At this point, he returned the calculator to the manufacturer with an angry letter claiming that the machine was defective. You are the manager of the manufacturer's Repair Department. How do you convince the good professor that his calculator is working correctly?

G1.7 **Historical Quest** The history of mathematics has many more white males recorded than it does females or minorities. Write a paper discussing some of the reasons this seems to be the case. You might check *Outstanding Women in Mathematics and Science*, 1991 National Women's History Project, 7738 Bell Road, Windsor, CA 95492, or *Black Mathematicians and Their Works*, edited by Newell et al. Ardmore, PA, Dorrance & Company, 1980.

G1.8 Journal Problem (*The Mathematical Gazette*, December 1982, Problem 66G by C. C. Goldsmith.) Note that $x^2 + 13x - 30$ and $x^2 - 13x + 30$ both factor over the integers. Prove that if $x^2 + px - q$ and $x^2 - px + q$ both factor over the integers, then the integers p and q are the hypotenuse and area of a right triangle with sides of integer length.

G1.9 Journal Problem (*Quantum,* January/February, 1996, p. 35; problem by Andrey Yegorov).

Prove

$$a^2 + b^2 + c^2 \geq ab + bc + ca$$

where *a*, *b*, and *c* are arbitrary real numbers.

G1.10 Journal Problem (*Parabola*, Vol. 1, No. 1, 1982, Problem H82.3, p. 13.)

Prove that when *x*, *y*, and *z* are real and not all equal,

$$x^2 + y^2 + z^2 > yz + zx + xy$$

and deduce that when also $x + y + z = 1$ then

$$yz + zy + xy \leq \frac{1}{3}$$

Chapter Objectives

The material in this chapter is previewed in the following list of objectives. After completing this chapter, review this list again, and complete the self-test.

2.1
- 2.1 In your own words, explain the process of problem solving, and describe mathematical modeling.
- 2.2 State the substitution property.
- 2.3 Define similar triangles.
- 2.4 State the proportional property of similar triangles.
- 2.5 Solve applied problems.

2.2
- 2.6 Define a function. Describe a function by rule, table, equation, graph, mapping, or set of ordered pairs. Use the vertical line test.
- 2.7 Be able to classify examples as one-to-one. Use the horizontal line test.
- 2.8 Distinguish between f and $f(x)$ notation. Use functional notation.
- 2.9 Find the difference quotient for a given function f.
- 2.10 Classify functions.
- 2.11 Use functions to model problems.

2.3
- 2.12 Know the graphs of the standard functions in Table 2.1, page 104.
- 2.13 Find the domain and range for a curve whose equation is given.
- 2.14 Determine when functions are equal.
- 2.15 Be able to find the x- and y-intercepts for a curve whose equation is given.
- 2.16 Find points satisfying specified conditions.
- 2.17 Determine where a function is increasing (graph rising), where it is decreasing (graph falling), where it is constant (graph horizontal), and locate the turning points.
- 2.18 Classify functions as even, odd, or neither.

2.4
- 2.19 Find the shift (h, k) when given an equation $y - k = f(x - h)$.
- 2.20 Given $y = f(x)$ and (h, k), write $y - k = f(x - h)$.

- 2.21 Given a function defined by $y = f(x)$, draw the graph of $y - k = f(x - h)$.
- 2.22 Given a function defined by $y = f(x)$, draw the graph's reflections, compressions, and dilations.

2.5
- 2.23 Define a piecewise function.
- 2.24 Graph a piecewise function.
- 2.25 Graph an absolute value function; translate an absolute value function.
- 2.26 Graph a greatest integer function; translate a greatest integer function. Graph a rounding up function.
- 2.27 Model problems using piecewise functions.

2.6
- 2.28 Find the composition of functions.
- 2.29 Express a given function as the composite of two functions using an inner and an outer function.
- 2.30 Find the sum, difference, product, and quotient functions.
- 2.31 Use composition to write functional iteration.

2.7
- 2.32 Given two functions, decide whether they are inverses.
- 2.33 Given a one-to-one function, find its inverse.
- 2.34 Graph a function, its inverse, and the line $y = x$ on the same coordinate axes.
- 2.35 Evaluate a function and its inverse by looking at a graph.

2.8
- 2.36 State the informal definition of the limit of a function.
- 2.37 Estimate limits by graphing (geometrical method).
- 2.38 Estimate limits by table (numerical method).
- 2.39 Evaluate limits of polynomials (analytic method).
- 2.40 Define continuity, and describe the concept in your own words.
- 2.41 Find suspicious points.
- 2.42 State and use the root location property.

Functions with Problem Solving

2

*S*olving problems is a practical art, like swimming, or skiing, or playing the piano; you can learn it only by imitation and practice . . . if you wish to learn swimming you have to go into the water, and if you wish to become a problem solver you have to solve problems.

—George Pólya
Mathematical Discovery, Vol. 1
New York: John Wiley and Sons, 1962, p. v

Chapter Sections

2.1 Problem Solving
Pólya's Problem-Solving Procedure
Inductive Reasoning
Deductive Reasoning
Mathematical Modeling
Word Problems
Rate Problems
Similar Triangle

2.2 Introduction to Functions
Definitions
Horizontal and Vertical Line Tests
Functional Notation
Using Functional Notation in Problem Solving
Classification of Functions

2.3 Graph of a Function
Domain and Range
Intercepts
Properties of Functions

2.4 Transformations of Functions
Reflections
Dilations and Compressions
Dilations and Compressions in the y-direction
Dilations and Compressions in the x-direction

2.5 Piecewise Functions
Definitions
Absolute Value Functions
Greatest Integer Function

2.6 Composition and Operations of Functions
Operations with Functions
Functional Iteration

2.7 Inverse Functions
The Idea of Inverse
Inverse Functions
Graph of f^{-1}

2.8 Limits and Continuity
Intuitive Notion of a Limit
Limits by Graphing or by Table
Intuitive Notion of Continuity
Definition of Continuity
Continuity on an Interval

Chapter 2 Summary and Review
Self Test
Practice for Calculus
Chapter 2 Group Research Projects

▶ CALCULUS PERSPECTIVE

In this chapter, we introduce a concept that is not only basic to this course but also basic to a major portion of a calculus course, namely, the concept of a function. In the calculus perspective from Chapter 1, we mentioned that the concept of a limit is central to the study of calculus. In this chapter, we present an intuitive introduction to limits in a particularly easy-to-understand setting. In this chapter, we introduce the notions of problem solving, functions, properties of functions, as well as limits and continuity.

2.1 Problem Solving

Pólya's Problem-Solving Procedure

We begin this study of problem solving by looking at the *process* of problem solving. As a mathematics teacher, I often hear the comment, "I can do mathematics, but I can't solve word problems." There *is* a great fear and avoidance of "real-life" problems because they do not fit into the same mold as the "examples in the book." It is easier for instructors to teach "word problems" when they fit into some mold, but by definition, that does not constitute genuine problem solving. Few practical problems from everyday life come in the same form as those you study in school. All the meaningful problems in calculus involve problem-solving skills. To compound the difficulty, learning to solve problems takes time. All too often, the mathematics curriculum is so packed with content that the real process of problem solving is slighted and because of time limitations becomes an exercise in mimicking the instructor's steps, instead of developing an approach that can be used long after the final examination is over.

The model for problem solving that we will use was first published in 1945 by the great, charismatic mathematician George Pólya. His book *How to Solve It* (Princeton University Press, 1971) has become a classic and has been reprinted several times since 1945.

In Pólya's book, you will find this problem-solving model as well as a treasure trove of strategy, know-how rules of thumb, good advice, anecdotes, history, and problems at all levels of mathematics. His problem-solving model is as follows:

PROBLEM-SOLVING GUIDE

Pólya's problem-solving guideline for problem solving:

Step 1   You have to *understand the problem.*

Step 2   *Devise a plan.* Find the connection between the data and the unknown. Look for patterns, relate to a previously solved problem or a known formula, or simplify the given information to obtain an easier problem.

Step 3   *Carry out the plan.*

Step 4   *Look back.* Examine the solution obtained.

Pólya's original statement of this procedure is reprinted on the inside front cover of this book.

Problem solving is a difficult task to master, and you are not expected to master it after one section of this book (or even after one mathematics course). Don't think you can avoid problem solving by skipping this section. Problem solving is one of the major threads with which this course is woven. You will be challenged to solve problems in nearly every section of this book.

One of the most important aspects of problem solving is to relate new problems to old ones. The problem-solving techniques outlined here should be applied when you are faced with a new problem. When you are faced with a question similar to one you have already worked, you can apply previously developed techniques.

We begin our journey toward problem solving by briefly introducing the two principal types of reasoning used in mathematics.

Inductive Reasoning

The type of reasoning—first observing patterns and then predicting answers for more complicated problems—is called **inductive reasoning**. It is a very important method of reasoning in problem solving and in using Pólya's method. With inductive reasoning, the results are not certain, only probable. These predictions can be checked or otherwise verified. One method of *proving* such conjectures is called **mathematical induction**, which we will discuss in Appendix C.

EXAMPLE 1   **Using patterns**

What is the sum of the first 100 positive consecutive odd numbers?

Solution

Step 1:   *Understand the problem.* Do you know what the words mean?

Odd numbers are 1, 3, 5, . . . , and *sum* means to add:

$$1 + 3 + 5 + \cdots + ?$$

The first thing you need to understand is what the last term will be, so you will know when you have reached 100 consecutive odd numbers.

$1 + 3$ *is two terms.*

$1 + 3 + 5$ *is three terms.*

$1 + 3 + 5 + 7$ *is four terms.*

It seems as if the last term is always one less than twice the number of terms. Thus, the sum of the first 100 consecutive terms is

$$1 + 3 + 5 + \cdots + 195 + 197 + 199$$

This is one less than 2(100).

Step 2:   *Devise a plan.* The plan we will use is to look for a pattern:

$1 = 1$ One term

$1 + 3 = 4$ Sum of two terms

$1 + 3 + 5 = 9$ Sum of three terms

Do you see a pattern yet?   If not, continue:

$1 + 3 + 5 + 7 = 16$

$1 + 3 + 5 + 7 + 9 = 25$

Step 3:   *Carry out the plan.* It looks like the sum of two terms is 2^2; of three terms, 3^2; of four terms, 4^2; and so on. The sum of the first 100 consecutive odd numbers therefore seems to be 100^2.

Step 4:   *Look back.* Does $100^2 = 10{,}000$ seem correct?   ∎

Deductive Reasoning

Another method of reasoning used in mathematics is called **deductive reasoning**. This method of reasoning produces results that are *certain* within the logical system being developed. That is, deductive reasoning involves reaching a conclusion by using a formal structure based on a set of **undefined terms** and a set of accepted, unproved **axioms** or **premises**. The conclusions are said to be proved and are called **theorems**.

The most useful axiom in problem solving is the principle of substitution.

SUBSTITUTION PROPERTY

If $a = b$, then a may be substituted for b in any mathematical statement without affecting the truth or falsity of the given mathematical statement.

≫ IN OTHER WORDS If two quantities are equal, you can remove one of the quantities and replace it with the other.

The simplest way to illustrate the substitution property is to use it in evaluating a formula.

EXAMPLE 2   Evaluating a formula

A water tank is in the shape of an inverted circular cone resting on top of a cylinder. The water is stored in the cylindrical part with base radius 2 m and height 4 m. How much water is in the tank when the water is 3 m deep?

Solution   Sometimes some outside information from a previous course is required when solving problems. For this example, you need the formula for the volume of a cone:

$$V = \frac{1}{3}\pi r^2 h$$

Substitute the known values into the formula:

$$V = \frac{1}{3}\pi \underset{\underset{h=3}{\uparrow}}{\left(\overset{\overset{r=2}{\downarrow}}{2^2}\right)}(3) \qquad \text{In this book, we will use arrows to show substitution.}$$

$$= 4\pi$$

The volume is $4\pi \, \text{m}^3$. Notice that we state our answer in sentence form (with units). Also notice that we do not round our answer unless a rounded answer is somehow indicated in the statement of the problem, or our only alternative is to work with a calculator or rounded results. ∎

The formal study of deductive reasoning is beyond the scope of this course. However, there are certain principles and terminology associated with proof and deductive reasoning that you will encounter in calculus. For example, in Chapter 1 we discussed the meaning of the words *if and only if*, and how a mathematical statement

$$p \text{ if and only if } q$$

requires two proofs:

$$(1) \text{ if } p, \text{ then } q \quad (2) \text{ if } q, \text{ then } p$$

There is some associated terminology. If p and q are any propositions, then the statement "if p, then q," written $p \to q$, is called a **conditional** and is translated several (equivalent) ways:

Conditional Translation	Example
If p, then q.	If you are 18, then you can vote.
q, if p.	You can vote, if you are 18.
p, only if q.	You are 18 only if you can vote.
All p are q.	All 18-year-olds can vote.

In addition to these translations for the conditional, there are related statements.

LOGICAL STATEMENT

Given the conditional $p \to q$, we define
the **converse** is: $q \to p$;
the **inverse** is: not $p \to$ not q;
the **contrapositive** is: not $q \to$ not p.

EXAMPLE 3   Logical statements

Let p be the proposition: "It is a 300ZX" and q be the proposition: "It is a car." The given statement, symbolized by $p \rightarrow q$, is "If it is a 300ZX, then it is a car." State the converse, inverse, and contrapositive.

Solution   Converse: $q \rightarrow p$ If it is a car, then it is a 300ZX.

Inverse: not $p \rightarrow$ not q If it is not a 300ZX, then it is not a car.

Contrapositive: not $q \rightarrow$ not p If it is not a car, then it is not a 300ZX. ∎

As you can see from Example 3, not all these statements are equivalent in meaning. The contrapositive and the original statement always have the same truth values, as do the converse and the inverse. We accept the following law, which is frequently used not only in calculus but in all of mathematics.

> **LAW OF CONTRAPOSITION**
>
> A conditional may always be replaced by its contrapositive without having its truth value affected.

Mathematical Modeling

Problem solving depends not only on the substitution property but also on translating statements from English to mathematical symbols. On a much more advanced level, this process is called **mathematical modeling** and involves matching mathematical skills with knowledge about the real world. A characteristic of mathematical modeling is that it is *iterative* (requires assumptions, feedback, and revision). It often involves solving problems with either too little information (some further research is needed) or too much information (you have information that is not needed to solve the problem).[*]

Mathematical models are formed, modified by experimentation and the accumulation of data, and are then used to predict some future occurrence in the real world. Such mathematical models are continually being revised and modified as additional relevant information becomes known.

Some mathematical models are quite accurate—particularly those used in the physical sciences. For example, one of the first models we consider in calculus is the path of a projectile. Other models predict the time of sunrise and sunset or the distance that an object falls in a vacuum. Other modeling, however, is less accurate—in particular when examples from the life sciences and social sciences are chosen. Only recently has modeling in these disciplines become precise enough to be included in a mathematics course.

What, precisely, is a mathematical model? At the low end of the spectrum, mathematical modeling can mean nothing more than real-life word problems. At the high end of the spectrum, mathematical modeling can mean choosing appropriate mathematics to solve a problem that has previously been unsolved.

The first step of what we call mathematical modeling involves *abstraction.*

With the method of abstraction, certain assumptions about the real world are made, variables are defined, and appropriate mathematics is developed. The next step is to simplify the mathematics or derive related mathematical facts from the mathematical model.

Historical Note

Mathematical modeling plays an important role in sporting events. For example, in 1992, preparing for the America's Cup competition included the following mathematical models:
- Race Modeling Programs
- Computational Fluid
- Computer-Aided Design
- Experimental Design
- Dynamic Modeling Program

Scientific American, May 1992

---

[*]In a classroom setting, it is not practical to give too little information in a problem (unless it is designated as a research problem). However, in the real world, it is common to need additional information to find a solution. Since problem solving involves a long learning process, we will only occasionally give you "extra" information. Our goal in this book is to build your problem-solving skills to bring you to a level of competency that will allow you to know when you have too much or too little information.

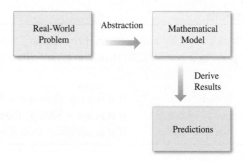

The results derived from the mathematical model should lead us to some predictions about the real world. The next step is to gather data from the situation being modeled and then to compare those data with the predictions. If the two do not agree, then the gathered data are used to modify the assumptions used in the model.

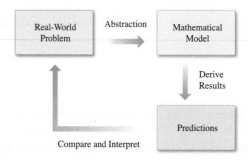

Mathematical modeling is an ongoing process. As long as the predictions match the real world, the assumptions made about the real world are regarded as correct, as are the defined variables. On the other hand, as discrepancies are noticed, it is necessary to construct a closer and a more dependable mathematical model.

Throughout this book, we will designate certain problems MODELING APPLICATION to designate problem-solving or mathematical modeling examples. You may have noticed Example 1 of this section was so designated. Mathematical modeling necessarily involves abstraction from the real world, deriving results, making predictions, and then comparing and interpreting in the real world. As an example, consider a problem we have adapted from calculus.

EXAMPLE 4   A volume problem from calculus   **MODELING APPLICATION**

A container is to be constructed from an 11 in. by 16 in. sheet of cardboard. Squares will be cut from the corners of the sheet and discarded as waste. The domain for the variable must be restricted so that the area of the base of the container exceeds (or is equal to) the area of wasted cardboard. We will revisit this problem in Chapter 3 when we consider optimization problems to ask for the maximum volume.

Solution

Step 1:   *Understand the problem.* Let s be the length of the side of a square that is cut from the cardboard.

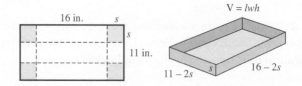

The domain requires that all sides be nonnegative. That is,

$$s \geq 0 \qquad 16 - 2s \geq 0 \qquad 11 - 2s \geq 0$$
$$s \leq 8 \qquad\qquad s \leq 5.5$$

Step 2: *Devise a plan.* We will do a numerical, algebraic, and if we have a graphing calculator, geometrical analysis.

Step 3: *Carry out the plan.*

Numerical (tabular) approach: To get some idea about the problem, a numerical approach is often helpful. For this approach, let us assume that the dimensions must be integers. With this modeling assumption, there are few values for s, and we can easily calculate the domain conditions (as well as the volume) of the six possible boxes. This is particularly easy to do if you use a calculator or computer.

Value of s	Length $16 - 2s$	Width $11 - 2s$	Height s	Area of Base (LENGTH)(WIDTH)	Area of Waste $4s^2$	Conditions Satisfied	Volume
0	16	11	0	176	0	Yes	0
1	14	9	1	126	4	Yes	126
2	12	7	2	84	16	Yes	168
3	10	5	3	50	36	Yes	150
4	8	3	4	24	64	No	96
5	6	1	5	6	100	No	30

It looks like the domain for integer values s is $0 \leq s \leq 3$. We also note that for integer values the maximum volume is 168 in.^{3}, which occurs for a box with dimensions 12 in. $\times$ 7 in. $\times$ 2 in.

Algebraic approach: We note from the question that the area of the base of the container should exceed (or be equal to) the area of the wasted cardboard. Thus:

$$\text{AREA OF BASE} \geq \text{AREA OF WASTE}$$
$$(\text{LENGTH OF BASE})(\text{WIDTH OF BASE}) \geq 4(\text{AREA OF CORNER})$$

$$(16 - 2s)(11 - 2s) \geq 4s^2$$
$$176 - 54s + 4s^2 \geq 4s^2$$
$$-54s \geq -176$$
$$s \leq \frac{176}{54} = \frac{88}{27}$$

This means the domain for this problem is $0 \leq s \leq 88/27$. By the way, the volume of the box at the endpoint $s = 0$ is 0, and at the endpoint $s = 88/27$ it is:

$$\text{VOLUME} = (\text{WIDTH})(\text{LENGTH})(\text{HEIGHT})$$
$$= (11 - 2s)(16 - 2s)(s)$$
$$\approx 138.4894579$$

We will see in calculus that the maximum value of *this* box occurs at a value where the derivative is zero or at one of the endpoints of the domain. This means the maximum value could possibly be between $s = 0$ and $s = 88/27$. We will need to wait until we will consider optimization problems in Chapter 3 to know for sure if this is the case.

Geometrical approach: You can use a graphing calculator for the inequality:

$$\text{AREA OF BASE} \geq \text{AREA OF WASTE}$$

$$(16 - 2s)(11 - 2s) \geq 4s^2$$

Since calculators use the variables x and y, you will input $\boxed{\text{Y1}} = (16 - 2\text{X})(11 - 2\text{X})$ and $\boxed{\text{Y2}} = 4\text{X}^{\wedge}2$. Then, look for the intersection of these curves, which is found using the intersection capabilities of the calculator. These steps can be illustrated:

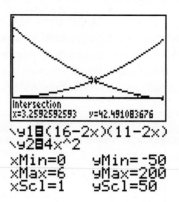

We see that the intersection is at $s \approx 3.2592592593$.

Step 4: *Look back.* The algebraic solution yields an exact answer ($88/27$), and the graphical method an approximate answer (3.2592592593). These values are both approximately equal to 3.26, so we say they agree. This seems to agree with the intuitively appealing numerical solution for integral values. ∎

Word Problems

Not all word problems are modeling applications, and not all modeling applications are word problems. To do problem solving outside of a classroom setting, you need to have a great deal of practice, sometimes with rather mundane or even trivial questions. The reason for this procedure is to allow you to gain confidence and to build your problem-solving skills as you progress through the course. You might say, "I want to learn how to become a problem solver, and textbook problems are not what I have in mind; I want to do *real* problem solving." But to become a problem solver, you must first learn the basics, and word problems are part of a textbook for good reason. We start with these problems *to build a problem-solving **procedure** that can be expanded to apply to problem solving in general.* As we do this, however, we will firmly focus on our goal: to build the skills you need to succeed in calculus.

We will now rephrase Pólya's problem-solving guidelines in a setting that is appropriate to solving word problems. This procedure is summarized in the following box.

PROBLEM SOLVING WITH WORD PROBLEMS

Pólya's problem-solving guideline for problem solving can be amplified to solve word problems.

Step 1   *Understand the problem.* This means read the problem. Note what it is all about. Focus on processes rather than numbers. You cannot work a problem you do not understand. A sketch may help in understanding the problem.

Step 2   *Devise a plan.* Write down a verbal description of the problem using operation signs and an equal or inequality sign. Note the following common translations.

Symbol	Verbal Description
=	is equal to; equals; are equal to; is the same as; is; was, becomes; will be; results in
+	plus; the sum of; added to; more than; greater than; increased by
−	minus; the difference of; the difference between; is subtracted from; less than; smaller than; decreased by; is diminished by
×	times; product; is multiplied by; twice (2×); triple (3×)
÷	divided by; quotient of

Step 3 *Carry out the plan.* In the context of word problems, we need to proceed deductively by carrying out the following steps.

Choose a variable. If there is a single unknown, choose a variable. If there are several unknowns, you can use the substitution property to reduce the number of unknowns to a single variable. Later, we will consider word problems with more than one variable.

Substitute. Replace the verbal phrase for the unknown with the variable.

Solve the equation. This is generally the easiest step. Translate the symbolic statement (such as $x = 3$) into a verbal statement. Probably no variables were given as part of the word problem, so $x = 3$ is not an answer. Generally, word problems require an answer stated in words. Pay attention to units of measure and other details of the problem.

Step 4 *Look back.* Be sure your answer makes sense by checking it with the original question in the problem. *Remember to answer the question that was asked.*

EXAMPLE 5   Area problem

Two rectangles have the same width, but one is 40 ft² larger in area. The larger rectangle is 6 ft longer than it is wide. The other is only 1 ft longer than its width. What are the dimensions of the larger rectangle?

Solution

Numerical analysis: If one rectangle is 40 ft² larger in area than the other, then the difference of their areas must be 40. Also, the larger rectangle is 6 ft longer than it is wide. For example, if we start with a width (which must be the same for both rectangles) of 1, then 2, then 3, . . . , we find

First (Larger) Rectangle			Second (Smaller) Rectangle			
Length	Width	Area	Length	Width	Area	Difference of Areas
7	1	7	2	1	2	5
8	2	16	3	2	6	10
9	3	27	4	3	12	15
10	4	40	5	4	20	20
12	6	72	7	6	42	30
13	7	91	8	7	56	35
14	8	112	9	8	72	40

We have apparently found the solution numerically, but it was a great deal of work, and we would have had more difficulty with this method if the answer had not been integral. The larger rectangle is 8 ft wide by 14 ft long.

Algebraic analysis: The areas differ by 40, and each area is the product of length and width. We also note that the widths of both rectangles are the same.

$$\text{AREA OF LARGER} - \text{AREA OF SMALLER} = 40$$

$$(\text{LENGTH OF LARGER})(\text{WIDTH}) - (\text{LENGTH OF SMALLER})(\text{WIDTH}) = 40$$

$$(\text{WIDTH} + 6)(\text{WIDTH}) - (\text{WIDTH} + 1)(\text{WIDTH}) = 40$$

Let W be the width of each rectangle. Then,

$$(W + 6)(W) - (W + 1)(W) = 40$$

$$W^2 + 6W - W^2 - W = 40$$

$$5W = 40$$

$$W = 8, \ W + 6 = 14$$

The larger rectangle is 8 ft wide by 14 ft long.

Geometrical analysis: A sketch will frequently simplify a problem. In this case, since the widths are the same, the difference in area can be seen in the larger rectangle.

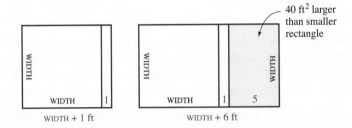

The difference is shown as the shaded region. The area of this region is stated in the problem:

$$\text{DIFFERENCE IN AREAS} = 40$$

$$(5)(\text{WIDTH}) = 40$$

Let W be the width of each rectangle. Then

$$5W = 40$$

$$W = 8 \text{ and } W + 6 = 14$$

The larger rectangle is 8 ft wide by 14 ft long. ∎

Rate Problems

There are two applications of word problems that are particularly important in calculus. One type of application involves rates, and the other type of application involves similar triangles. A dictionary definition of *rate* is "the quantity of a thing in relation to the units of something else." This definition is quite general; yet, rate is too often limited to rate–time–distance relationships. A typing rate is the number of words typed per unit of time, most prices are based on cost per unit, and interest earned is a part of the principal invested. In calculus, you will find rates to be a frequent application. In everyday usage, we might speak of miles per hour, or mph, but in scientific and mathematical applications, we write this as "mi/h." If it is included as part of the problem calculation, it is written as a fraction

$$\frac{\text{miles}}{\text{hour}}$$

interpreted as a division of miles by hours. Similarly, feet per second in everyday usage is abbreviated as fps; in mathematics it is abbreviated as ft/s and if used as part of the problem calculation is thought of as the fraction feet divided by seconds. However, most elementary applied problems do not require that we include the units of measurement as part of the calculations, but rather are used only in the statement of the solution to the problem. This usage is illustrated in the following example.

EXAMPLE 6   Rate–time–distance application   `MODELING APPLICATION`

An office worker takes 55 minutes to return from the job each day. This person rides a bus that averages 30 mi/hr and walks the rest of the way at 4 mi/h. If the total distance is 21 miles from office to home, what is the distance walked home each day?

Solution

Step 1:   *Understand the problem.* The total trip is composed of two distinct distances. These distances are the product of rate and time. This is the familiar formula $d = rt$.

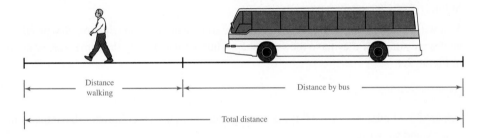

Step 2:   *Devise a plan.* Don't be too eager to choose a variable. Sometimes forcing an incorrect or poorly defined variable at the beginning of a problem often results in disaster. We will use

BUS DISTANCE + WALK DISTANCE = TOTAL DISTANCE

We know that the total distance is 21 mi, but we do not know the other two distances, so we use the principle of substitution for

BUS DISTANCE = (BUS RATE)(BUS TIME)

WALK DISTANCE = (WALK RATE)(WALK TIME)

Step 3:   *Carry out the plan.* By substitution, we have:

$$(\text{BUS RATE})(\text{BUS TIME}) + (\text{WALK RATE})(\text{WALK TIME}) = 21$$

$$\uparrow \qquad \qquad \qquad \qquad \uparrow$$
$$30 \qquad \qquad \qquad \qquad 4$$

We know that BUS TIME + WALK TIME $= \dfrac{55}{60}$ hr

We write 55 min as 55/60 hr. This means we can substitute BUS TIME $= \dfrac{55}{60} - $ WALK TIME

$$30\left(\frac{55}{60} - \text{WALK TIME}\right) + (4)(\text{WALK TIME}) = 21$$

Let W be the time walked, in hours. Thus,

$$30\left(\frac{55}{60} - W\right) + 4W = 21$$

$$\frac{55}{2} - 30W + 4W = 21 \qquad \text{Distributive property}$$

$$27.5 - 26W = 21 \qquad \text{Similar terms}$$

$$-26W = -6.5 \qquad \text{Subtract 27.5 from both sides.}$$

$$W = 0.25 \qquad \text{Divide both sides by } -26.$$

The worker walks 0.25 hr, but the question asks for distance, so

$$\text{WALK DISTANCE} = 4(0.25) = 1$$

The worker walks 1 mile daily.

Step 4:   *Look back.* If the walker walks 1 mi, then the bus ride is 20 mi and lasts (55 min − 15 min) = 40 min or 2/3 hr. (2/3)(30) = 20 mi, so the solution checks.   ∎

Similar Triangles

You will need the idea of *similar triangles* for some calculus and trigonometry problems. Two triangles are similar if they have the same shape (but not necessarily the same size, or area), as shown in Figure 2.1.

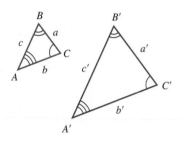

Figure 2.1 Similar triangles

SIMILAR TRIANGLES

$\triangle ABC$ is similar to $\triangle A'B'C'$ if the corresponding angles are congruent (have the same measure), and we write

$$\triangle ABC \sim \triangle A'B'C'$$

» IN OTHER WORDS Since the sum of the measures of any triangle is 180°, it follows that two triangles are similar if two angles of one are equal to two angles of the other.

   You will also need to remember the *proportional property of similar triangles*; namely, if two triangles are similar, then the corresponding sides are proportional.

SIMILAR TRIANGLE PROPERTY

If $\triangle ABC$ is similar to $\triangle A'B'C'$ with sides of lengths as shown in Figure 2.1, then

$$\frac{a}{c} = \frac{a'}{c'} \qquad \frac{a}{b} = \frac{a'}{b'} \qquad \frac{b}{c} = \frac{b'}{c'} \qquad \frac{c}{a} = \frac{c'}{a'} \qquad \frac{b}{a} = \frac{b'}{a'} \qquad \frac{c}{b} = \frac{c'}{b'}$$

The similar triangle property has a wide variety of applications, some of which are provided in the problem set. We conclude this section with two applications adapted from calculus problems.

EXAMPLE 7   Streetlight problem from calculus

A person 6 ft tall is standing 7 ft from the base of a streetlight. If the light is 20 ft above ground, how long is the person's shadow due to the streetlight?

Solution   Let x denote the length (in feet) of the person's shadow, as shown in Figure 2.2. (*Note:* $x > 0$.)

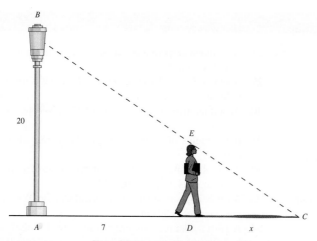

Figure 2.2 Streetlight calculus problem

Note that $\angle A$ and $\angle D$ are right angles, and $\angle C$ is common to both $\triangle ABC$ and $\triangle DEC$, so we have two angles of one triangle congruent to two angles of the other triangle, thus $\triangle ABC \sim \triangle DEC$. Since the triangles are similar, we have proportional sides, namely:

$$\frac{6}{x} = \frac{20}{7+x}$$
$$6(7+x) = 20x$$
$$42 + 6x = 20x$$
$$42 = 14x$$
$$3 = x$$

The shadow's length is 3 ft.   ■

The last example of this section, which is the famous inverted cone problem from calculus, uses similar triangles.

EXAMPLE 8   Inverted cone problem from calculus

A water tank is in the shape of an inverted cone 20 ft high with a circular base whose radius is 5 ft. How much water is in the tank when the water is 8 ft deep?

Solution   The volume of a cone is $V = \frac{1}{3}\pi r^2 h$. We are given $h = 8$ and need to find r. Once again, we use similar triangles as shown in Figure 2.3.

$\triangle ABC \sim \triangle DEC$ (can you see why?); thus,

$$\frac{5}{20} = \frac{r}{8}$$
$$40 = 20r$$
$$2 = r$$

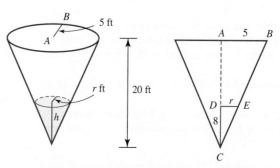

Figure 2.3 Inverted cone problem from calculus

The desired volume is $V = \frac{1}{3}\pi(2)^2(8) = \frac{32}{3}\pi$. That is, the desired volume is about 33.5 ft³.   ■

PROBLEM SET 2.1

1. IN YOUR OWN WORDS Describe Pólya's problem-solving procedure.
2. IN YOUR OWN WORDS Compare and contrast inductive and deductive reasoning.
3. IN YOUR OWN WORDS State the substitution property.
4. IN YOUR OWN WORDS What is meant by mathematical modeling?
5. IN YOUR OWN WORDS Describe a process for solving word problems.
6. IN YOUR OWN WORDS What is the proportion property of similar triangles?

Write the converse, inverse, and contrapositive of the statements in Problems 7–12.

7. If you break the law, then you will go to jail.
8. I will go Saturday, if I get paid.
9. If a polygon has three sides, then it is a triangle.
10. If $a^2 + b^2 = c^2$, then the triangle with sides a, b, and c is a right triangle.
11. If $5 + 10 = 15$, then $15 - 10 = 2$.
12. If $8x = 16$, then $x = 2$.

Problems 13–20 give a verbal description of some of the formulas you will need to use in a calculus course. Write each formula in symbolic form.

13. The area A of a parallelogram is the product of the base b and the height h.
14. The area A of a triangle is one-half the product of the base b and the height h.
15. The area A of a rhombus is one-half the product of the diagonals p and q.
16. The area A of a trapezoid is the product of one-half the height h and the sum of the bases a and b.
17. The volume V of a cube is the cube of the length s of an edge.
18. The volume V of a rectangular solid is the product of the length l, the width w, and the height h.
19. The volume V of a cone is one-third the product of pi, the square of the radius r, and the height h.
20. The volume V of a sphere is the product of four-thirds pi and the cube of the radius r.

Solve Problems 21–26. Because you are practicing a procedure, you must show all of your work.

21. What is the sum of the first 20 positive consecutive odd numbers?
22. What is the sum of the first 1,000 positive consecutive odd numbers?
23. The sum of two consecutive odd integers is 48. What is the smaller integer?
24. The sum of two consecutive even integers is 30. What is the larger number?
25. The sum of four consecutive integers is 74. What are the integers?
26. What is the sum of the first 1,000 positive consecutive even numbers?

PROBLEMS FROM CALCULUS *Problems 27–36 use formulas you will need for calculus. State the formula you need, and then use that formula to answer the question.*

27. A rectangular field is 100 ft long and 75 ft wide. What length of fencing is necessary to enclose its perimeter?

28. How much carpeting is necessary to cover an area that is 6 yd wide and 12 yd long?
29. A square field is 540 ft on each side. What is the distance around the field, and what is its area?
30. An automotive tire has a radius of 15 in. What is the circumference of the tire?
31. If the radius of a circular region is 15 in., what is the area of the region?
32. An airplane maintains a constant speed of 670 mi/h for a 3-hr flight. How far does the plane travel?
33. Air is being pumped into a spherical balloon. What is its volume when the diameter is 50 cm?
34. A 10-ft ladder rests against a vertical wall. If the bottom of the ladder is 3 ft from the wall, how high up the building (to the nearest foot) does the ladder reach?

35.

Sand is being dumped from a conveyor belt so that it forms a pile in the shape of a cone whose base diameter and height are always equal. What is the volume of the pile when it is 10 ft high?

36. A boat is pulled into a dock by a rope attached to the bow of the boat and passing through a pulley on the dock that is 1 m higher than the bow of the boat. How much rope is necessary to connect the pulley and the bow of the boat when the boat is 8 m from the dock?

Solve Problems 37–44. Because you are practicing a procedure, you must show all of your work.

37. Two rectangles have the same width, but one is 20 ft² larger in area. The larger rectangle is 6 ft longer than it is wide. The other is only 2 ft longer than its width. What are the dimensions of the larger rectangle?
38. Two triangles have the same height. The base of the larger one is 3 cm greater than its height. The base of the other is 1 cm greater than its height. If the areas differ by 3 cm², find the dimensions of the smaller figure.

39. A rectangle is 2 ft longer than it is wide. If you increase the length by a foot and reduce the width the same, the area is reduced by 3 ft². Find the width of the new figure.

40. The length of a rectangle is 3 m more than its width. The width is increased by 2 m, and the length is shortened by a meter. If the two figures have the same perimeter, what is it?

41. A businesswoman logs time in an airliner and a rental car to reach her destination. The total trip is 1,100 mi, the plane averaging 600 mi/h and the car 50 mi/h. How long is spent in the automobile if the trip took a total of 5 hr 30 min?

42. Barry hitchhikes back to campus from home, which is 82 mi away. He makes 4 mi/h walking, until he gets a ride. In the car, he makes 48 mi/h. If the trip took 4 hours, how far did Barry walk?

43. Two joggers set out at the same time from their homes 21 mi apart. They agree to meet at a point in between in an hour and a half. If the rate of one is 2 mi/h faster than the other, find the rate of each.

44. Two joggers set out at the same time but in opposite directions. If they were to maintain their normal rates for 4 hr, they could be 68 mi apart. If the rate of one is 1.5 mi/h faster than the other, find the rate of each.

PROBLEMS FROM CALCULUS *Problems 45–55 are modeled from problems taken from a variety of calculus textbooks.*

45. A rectangular area is to be fenced. If the space is twice as long as it is wide, for what dimensions is the area numerically greater than the perimeter?

46. A rectangular area three times as long as it is wide is to be fenced. For what dimensions is the perimeter numerically greater than the area?

47. A small manufacturer of high-gaming computers determines that the price of each item is related to the number of items, x, produced per day. The manufacturer knows that (1) the maximum number that can be produced is 10 items; (2) the price should be $400 - 25x$ dollars; (3) the overhead (the cost of producing x items) is $5x^2 + 40x + 600$ dollars; and (4) the daily profit is then found by subtracting the overhead from the revenue:

$$\text{PROFIT} = \text{REVENUE} - \text{COST}$$
$$= (\text{NUMBER OF ITEMS})(\text{PRICE/ITEM}) - \text{COST}$$
$$= x(400 - 25x) - (5x^2 + 40x + 600)$$
$$= 400x - 25x^2 - 5x^2 - 40x - 600$$
$$= -30x^2 + 360x - 600$$

What is the number of items produced if the profit is zero?

48. Current postal regulations do not permit a package to be mailed if the combined length, width, and height exceeds 72 in. What are the dimensions of the largest permissible package with length twice the length of its square end?

49. Suppose you throw a rock at 48 ft/s from the top of the Sears Tower in Chicago and the height in feet, h, from the ground after t sec is given by

$$h = -16t^2 + 48t + 1,454$$

a. What is the height of the Sears Tower?

b. How long will it take (to the nearest tenth of a second) for the rock to hit the ground?

50. If an object is shot up from the ground with an initial velocity of 256 ft/s, its distance in feet above the ground at the end of t sec is given by $d = 256t - 16t^2$ (neglecting air resistance). Find the length of time for which $d \geq 240$.

51. Find the length of time the projectile described in Problem 50 will be in the air.

52. Many materials, such as brick, steel, aluminum, and concrete, expand because of increases in temperature. This is why fillers are placed between the cement slabs in sidewalks. Suppose you have a 100-ft roof truss securely fastened at both ends, and assume that the buckle is linear. (It is not, but this assumption will serve as a worthwhile approximation.) Let the height of the buckle be x ft. If the percentage of swelling is y, then, for each half of the truss,

$$\text{NEW LENGTH} = \text{OLD LENGTH} + \text{CHANGE IN LENGTH}$$
$$= 50 + (\text{PERCENTAGE})(\text{LENGTH})$$
$$= 50 + \left(\frac{y}{100}\right)(50)$$
$$= 50 + \frac{y}{2}$$

These relationships are shown in Figure 2.4.

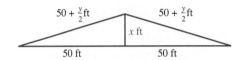

Figure 2.4 Buckling and expansion

By the Pythagorean theorem,

$$x^2 + 50^2 = \left(50 + \frac{y}{2}\right)^2$$
$$x^2 + 50^2 = \frac{(100 + y)^2}{4}$$
$$4x^2 + 4 \cdot 50^2 = 100^2 + 200y + y^2$$
$$4x^2 - y^2 - 200y = 0$$

Solve this equation for x and then calculate the amount of buckling (to the nearest half-inch) for the following materials:

a. brick; $y = 0.03$

b. steel; $y = 0.06$

c. aluminum; $y = 0.12$

d. concrete; $y = 0.05$

53. A 1-mi length of pipeline connects two pumping stations.

© Alaska Stock LLC/Alamy Images

Special joints must be used along the line to provide for expansion and contraction due to changes in temperature. However, if the pipeline were actually one continuous length of pipe fixed at each end by the stations, then expansion would cause the pipe to bow. Approximately how high would the middle of the pipe rise if the expansion was just 1 in. over the mile? (You may make the same assumption as we did in Problem 52, namely, that the buckle is linear.)

54. Consider the following pattern:

$$9 \times 1 - 1 = 8$$
$$9 \times 21 - 1 = 188$$
$$9 \times 321 - 1 = 2,888$$
$$9 \times 4,321 - 1 = 38,888$$

a. Use this pattern and inductive reasoning to specify the next equation in the sequence.

b. Predict the answer to

$$9 \times 987,654,321 - 1$$

c. Predict the answer to

$$9 \times 10,987,654,321 - 1$$

55. Consider the following pattern:

$$123,456,789 \times 9 = 1,111,111,101$$
$$123,456,789 \times 18 = 2,222,222,202$$
$$123,456,789 \times 27 = 3,333,333,303$$

a. Use this pattern and inductive reasoning to specify the next equation in the sequence.

b. Predict the answer to

$$123,456,789 \times 63$$

c. Predict the answer to

$$123,456,789 \times 81$$

LEVEL 3

Problems 56–60 were in the May 1989 issue of The Mathematics Teacher.

56. **Historical Quest** (From Bhaskara, ca. A.D. 1120.) "In a lake the bud of a water lily was observed, one cubit above the water, and when moved by the gentle breeze, it sunk in the water at two cubits' distance." Find the depth of the water.

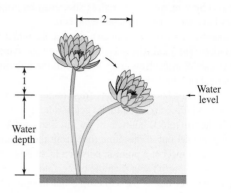

57. **Historical Quest** (From Bhaskara, ca. A.D. 1120.) "One third of a collection of beautiful water lilies is offered to Mahadev, one fifth to Huri, one sixth to the Sun, one fourth to Devi, and the six which remain are presented to the spiritual teacher." Find the total number of lilies.

58. **Historical Quest** (From Bhaskara, ca. A.D. 1120.) "One fifth of a hive of bees flew to the Kadamba flower; one third flew to the Silandhara; three times the difference of these two numbers flew to an arbor, and one bee continued flying about, attracted on each side by the fragrant Keteki and the Malati." Find the number of bees.

59. **Historical Quest** (From Brahmagupta, ca. A.D. 630.) "A tree one hundred cubits high is distant from a well two hundred cubits; from this tree one monkey climbs down the tree and goes to the well, but the other leaps in the air and descends by the hypotenuse from the high point of the leap, and both pass over an equal space." Find the height of the leap.

60. **Historical Quest** "Ten times the square root of a flock of geese, seeing the clouds collect, flew to the Manus lake; one eighth of the whole flew from the edge of the water amongst a multitude of water lilies; and three couples were observed playing in the water." Find the number of geese.

Tower of Pisa

2.2 Introduction to Functions

Definitions

In the previous chapter, we considered a Cartesian coordinate system to easily see the relationship between two variables. We now introduce an algebraic characterization for certain relationships.

Suppose we drop an object from a tall structure (such as the Leaning Tower of Pisa). The distance the object falls is dependent (among other things) on the length of time it falls. If we let the variable d be the *distance* the object has fallen (in feet) and t the *time* it has fallen (in seconds), and if we disregard air resistance, the formula (from physics and calculus) is

$$d = 16t^2$$

where 16 is a constant determined by the force of gravity acting on the object. Using the formula, we can calculate the height of the tower by timing the number of seconds it takes for the object to hit the ground. If it takes 3 seconds for the object to hit the ground, then the height of the tower (in feet) is

$$d = 16t^2$$
$$= 16(3)^2$$
$$= 144$$

The formula $d = 16t^2$ gives rise to a set of data for $0 \leq t \leq 15$:

Time (in sec)	0	1	2	3	4	. . .	14	15
Distance (in ft)	0	16	64	144	256	. . .	3,136	3,600

For every nonnegative value of t, there is a corresponding value for d. We can represent the data in the table as a set of ordered pairs in which the first component represents a value for t and the second component represents a corresponding value for d:

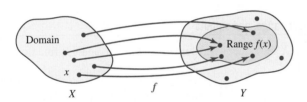

First component (values for t)
$$(x, \ y)$$
Second component (values for d)

The determined value (distance in this example) is called the **dependent variable**, and the specified variable is the **independent variable**. For this example (see preceding table), we have a set of ordered pairs: $(0, 0)$, $(1, 16)$, $(2, 64)$, . . . , $(14, 3136)$, $(15, 3600)$. The **domain** of a function is the set of values of the independent variable for which it is defined. For this example, the domain is defined as $0 \leq t \leq 15$, which means that t is any value between 0 and 15 (including the endpoints). Thus, other ordered pairs (not shown in the table) are $(0.5, 4)$, $(12.75, 2601)$, ... The set of all corresponding values of the dependent variable (second components) is called the **range**. A visual representation of a function is shown in Figure 2.5.

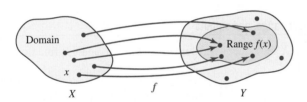

Figure 2.5 A function as a mapping

Historical Note

Gottfried Leibniz (1646–1716)

We first met Leibniz in Chapter 1 as one of the inventors of calculus. However, as you might guess, he is one of the giants in the history of mathematics. For example, the word *function* was used as early as 1694 by Leibniz to denote any quantity connected with a curve. Leibniz was one of the most universal geniuses of all time, and as a teenager, he came up with many of the great ideas in mathematics. However, his ideas were not fully accepted at the time because teenagers did not command much attention in intellectual circles. He was refused a doctorate at the University of Leipzig because of his youth, even though he had completed all the requirements. Among other things, Leibniz invented the calculus, exhibited an early calculating machine that he invented, and distinguished himself in law, philosophy, and linguistics. His ideas on functions were generalized by other mathematicians, including P. G. Lejeune-Dirichlet (1805–1859).

FUNCTION

A **function** is a rule that assigns to each element of the domain a single element in its range.

» IN OTHER WORDS To each x in the domain, there corresponds exactly one y in the range.

The pairs of corresponding values assigned in the definition of a function may be viewed as ordered pairs (x, y). This allows a rewording of the definition: *A function is a set of ordered pairs (x, y) in which no two different ordered pairs have the same first element x.*

It is customary to give functions letter names, such as f, g, f_1, or F. If y is the value of the function f corresponding to x, it is written $y = f(x)$ and is read "y is equal to the value of the function f at x," or, more briefly, "y equals f at x," or "y equals f of x."

Let $D = \{1, 2, 3, 4\}$ be the domain of a function called f. Think of the function f as a machine (*a function machine* as shown in Figure 2.6) that accepts an input x from D and produces an output $f(x)$, pronounced "f of x."

Input value x

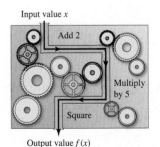

Output value $f(x)$

Figure 2.6 Function machine

☠ The symbol $f(x)$ does NOT mean multiplication; it is a single symbol representing the second component of the ordered pair (x, y). ☠

The function machine description has the advantage of being easy to understand, but it is awkward to use. We might also describe this function in a variety of ways, as shown in Example 1.

EXAMPLE 1   Alternate descriptions for a function

Describe the function machine in Figure 2.6 for the domain $D = \{1, 2, 3, 4\}$ using a rule, stating an equation, showing a mapping, using a set of ordered pairs, writing a table, and drawing a graph.

Solution

RULE   *For each input value, add two, then multiply by five, and finally square the result to find the output.* The function in this example would probably not be defined by such a verbal rule; nevertheless, a verbal rule is often the best way we have to describe a certain function.

GRAPH

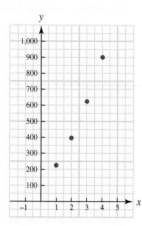

EQUATION $f(x) = [5(x + 2)]^2$ MAPPING

TABLE

x	$f(x)$
1	225
2	400
3	625
4	900

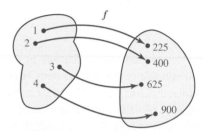

SET OF ORDERED PAIRS $\{(1, 225), (2, 400), (3, 625), (4, 900)\}$

EXAMPLE 2   Domain, range, and outputs

Given a mapping from X to Y, name the domain and range, and use functional notation to name the outputs for each input.

a.

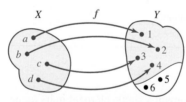

$X = \{a, b, c, d\}; Y = \{1, 2, 3, 4, 5, 6\}$

b.

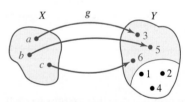

$X = \{a, b, c\}; Y = \{1, 2, 3, 4, 5, 6\}$

c.

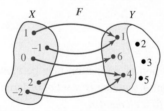

$X = \{1, -1, 0, 2, -2\}; Y = \{1, 2, 3, 4, 5, 6\}$

Solution

a. Domain $= \{a, b, c, d\}$; Range $= \{1, 2, 3, 4\}$;

The range consists only of those elements of Y that are actually used as outputs. However, the domain and X must be the same.

$f(a) = 1, f(b) = 2, f(c) = 3, f(d) = 4$

b. Domain $= \{a, b, c\}$; Range $= \{3, 5, 6\}$;

$g(a) = 3, g(b) = 5, g(c) = 6$

c. Domain $= \{1, -1, 0, 2, -2\}$; Range $= \{1, 4, 6\}$;

$F(0) = 6, F(1) = 1, F(-1) = 1, F(2) = 4, F(-2) = 4$

☠ Different elements in the domain may have the same value in the range. ☠ ■

Notice that Example 2c showed some repeated outputs. The ultimate example of repeated outputs is a function defined by $f(x) = c$ for *all* values of x. Such a function is called a **constant function**. If the outputs are always different (that is, if there are no repeated outputs), then the function is called *one-to-one*.

ONE-TO-ONE FUNCTIONS

If a function f maps X into Y so that for any distinct (different) elements x_1 and x_2 in X, $f(x_1) \neq f(x_2)$ then f is said to be a **one-to-one** function of X into Y.

EXAMPLE 3   A mapping that is not a function

Draw a picture of a mapping that is not one-to-one.

Solution

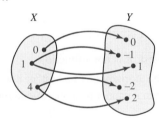

☠ Do not use $f(x)$ notation unless f is a function. ☠ This is not a function because 1 and 4 are associated with more than one image; because it is not a function, it follows that the mapping is not one-to-one. ■

Horizontal and Vertical Line Tests

There are two tests that involve sweeping a line across a graph. The first tells us whether a graph represents a function, and the second tells us whether a graph represents a one-to-one function.

VERTICAL LINE TEST

Every vertical line passes through the graph of a function in at most one point. This means if you sweep a vertical line across a graph and it simultaneously intersects the curve at more than one point, then the curve is not the graph of a function.

HORIZONTAL LINE TEST

Every horizontal line passes through the graph of a one-to-one function in at most one point. This means that if you sweep a horizontal line across the graph of a function and it simultaneously intersects the curve at more than one point, then the curve is not the graph of a one-to-one function.

EXAMPLE 4   Horizontal and vertical line tests

Use the vertical line test to determine whether the given curve is the graph of a function, and if it is the graph of a function, use the horizontal line test to determine whether it is one-to-one. Name the probable domain and range by looking at the graphs.

a. 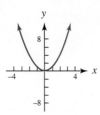   **b.** 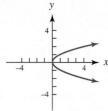   **c.**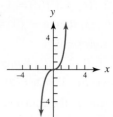

d. 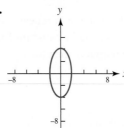   **e.**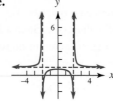

Solution

a.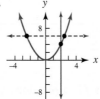

Passes the vertical line test; it is a function.

Does not pass the horizontal line test; it is not one-to-one.

Domain: $\mathbb{R}$

Range: $y \geq 0$

b.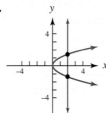

Does not pass the vertical line test; it is not a function.

If it is not a function, then the horizontal line test is not needed because if it is not a function, then it cannot be a one-to-one function.

Domain: $x \geq 0$

Range: $\mathbb{R}$

c.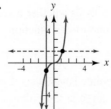

Passes the vertical line test; it is a function.

Passes the horizontal line test; it is one-to-one.

Domain: $\mathbb{R}$

Range: $\mathbb{R}$

d.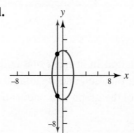

Does not pass the vertical line test; it is not a function.

If it is not a function, then the horizontal line test is not needed because if it is not a function, then it cannot be a one-to-one function.

Domain: $-1 \leq x \leq 2$

Range: $-4 \leq y \leq 4$

e.

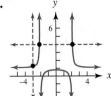

Passes the vertical line test; it is a function.

Does not pass the horizontal line test; it is not one-to-one.

Domain: $\mathbb{R}, x \neq \pm 2$

Range: $y \leq \frac{3}{4}$ or $y > 1$

Functional Notation

One of the most useful inventions in all the history of mathematics is the notation $f(x)$, called **functional notation**. Remember

x is a member of the domain.
$$\downarrow$$
$$\underline{f(x)}$$
$$\uparrow$$
f(x) is a number.

☠ A *function* is denoted by f; $f(x)$ is a *number* associated with x. ☠

Sometimes functions are defined by expressions such as

$$f(x) = x^2 + 1 \quad \text{or} \quad g(x) = (x+1)^2$$

To emphasize the difference between f and $f(x)$, some books use $f : x \to x^2 + 1$. In this book, however, we write "$f(x) = x^2 + 1$" to mean, "let f be the function defined by $f(x) = x^2 + 1$; this denotes the set of all ordered pairs (x, y) so that $y = x^2 + 1$."

EXAMPLE 5   Evaluating functions

Let $f(x) = x^2 + 1$ and $g(x) = (x+1)^2$.
Find **a.** $f(1)$ **b.** $g(1)$ **c.** $f(-3)$ **d.** $g(-3)$ **e.** $f(w)$ **f.** $g(w)$
   g. $f(w+h)$ **h.** $g(w+h)$

☠ Note that $f \neq g$ since $x^2 + 1 \neq (x+1)^2$. ☠

Solution

a. $f(1) = 1^2 + 1 = 2$

b. $g(1) = (1+1)^2 = 2^2 = 4$

c. $f(-3) = (-3)^2 + 1 = 9 + 1 = 10$

d. $g(-3) = (-3+1)^2 = (-2)^2 = 4$

e. $f(w) = w^2 + 1$

f. $g(w) = (w+1)^2 = w^2 + 2w + 1$

g. $f(w+h) = (w+h)^2 + 1$
$$= w^2 + 2wh + h^2 + 1$$

h. $g(w+h) = (w+h+1)^2$
$$= w^2 + wh + w + wh + h^2 + h + w + h + 1$$
$$= w^2 + 2wh + h^2 + 2w + 2h + 1$$

In calculus, functional notation is used in the definition of derivative. The value

$$\frac{f(x+h) - f(x)}{h}$$

is called a **difference quotient**.

EXAMPLE 6   Difference quotient

Let $f(x) = x^3$. Find the difference quotient $\dfrac{f(x+h) - f(x)}{h}$.

Solution

$$\frac{f(x+h)-f(x)}{h} = \frac{(x+h)^3 - x^3}{h}$$

$$= \frac{x^3 + 3x^2h + 3xh^2 + h^3 - x^3}{h}$$

$$= \frac{3x^2h + 3xh^2 + h^3}{h}$$

$$= 3x^2 + 3xh + h^2$$

☠ When finding a difference quotient, do not start with $f(x)$. Find $f(x+h)$ first, then subtract $f(x)$ and simplify; finally, divide by h. ☠

■

EXAMPLE 7   Difference quotient of a polynomial function

Let $f(x) = x^2 + 2x + 3$. Find $\dfrac{f(x+h)-f(x)}{h}$.

Solution

$$\frac{f(x+h)-f(x)}{h} = \frac{[(x+h)^2 + 2(x+h) + 3] - [x^2 + 2x + 3]}{h}$$

$$= \frac{x^2 + 2xh + h^2 + 2x + 2h + 3 - x^2 - 2x - 3}{h}$$

$$= \frac{2xh + h^2 + 2h}{h}$$

$$= 2x + h + 2$$

■

Using Functional Notation in Problem Solving

Functional notation can be used to work a wide variety of applied problems.

EXAMPLE 8   Falling object problem revisited

At the beginning of this section, we used the formula $d = 16t^2$ to represent the distance (in ft) that an object falls (neglecting air resistance) after t seconds. This relationship can be represented by $f(t) = 16t^2$. Use functional notation to represent each of the given ideas.

  a. The distance the object will fall in one second.
  b. The distance the object will fall in the next two seconds.
  c. The distance the object will fall during the h seconds following the second second.
  d. The average distance the object will fall in the first 5 seconds.
  e. The average distance the object will fall in the next 5 seconds (after the first 5 seconds).
  f. The average distance the object will fall in h seconds after the first x seconds.

Solution

  a. $f(1) = 16(1)^2 = 16$; the object will fall 16 ft.
  b. $f(3) = 16(3)^2 = 144$ ft is the distance the object will fall in the first 3 seconds; $f(3) - f(1)$ is the distance the object will fall in the next 2 seconds (after the first second). That is,

$$f(3) - f(1) = 144 - 16 = 128$$

In the next two seconds, the object falls 128 ft.

c. $f(2) = 16(2)^2 = 64$ ft is the distance the object will fall in the first two seconds. $f(2 + h)$ is the distance the object will fall in the next h seconds, so the distance in the h seconds following the second second is

$$f(2 + h) - f(2) = 16(2 + h)^2 - 64$$
$$= 16(4 + 4h + h^2) - 64$$
$$= 64 + 64h + 16h^2 - 64$$
$$= 16h^2 + 64h$$

Thus, the object will fall $(16h^2 + 64h)$ ft.

d. $f(5) = 16(5)^2 = 400$ ft is the distance the object will fall in the first 5 seconds. Thus, the average distance for the first 5 seconds is

$$\frac{f(5) \text{ ft}}{5 \text{ sec}} = \frac{400 \text{ ft}}{5 \text{ sec}} = 80 \text{ ft/s}$$

e. The distance the object travels in the *next* 5 seconds is

$$f(5 + 5) - f(5) = 16(10)^2 - 16(5)^2$$
$$= 1{,}600 - 400$$
$$= 1{,}200$$

The average distance is

$$\frac{f(10) - f(5)}{10 - 5} = \frac{1{,}200}{5} = 240 \text{ ft/s}$$

f. In the first x seconds the object travels $f(x)$ ft; in the *next* h seconds, the object travels $f(x + h) - f(x)$. The average distance for the h seconds is

$$\frac{f(x + h) - f(x)}{(x + h) - x} = \frac{16(x + h)^2 - 16x^2}{h} \qquad \text{\textit{Recognize the difference quotient.}}$$
$$= \frac{16(x^2 + 2xh + h^2) - 16x^2}{h}$$
$$= \frac{16x^2 + 32xh + 16h^2 - 16x^2}{h}$$
$$= \frac{32xh + 16h^2}{h}$$
$$= 32x + 16h$$

■

EXAMPLE 9   Calculus example, writing a function

Suppose you need to fence a rectangular play zone for children, to fit into a right-triangular plot with sides measuring 4 m and 12 m, as shown in Figure 2.7.

Write the area of the play zone as a function of the length of the play zone. We will continue with this example in Section 3.3.

Solution   Let x and y denote the length and width of the inscribed rectangle. The appropriate formula for the area is $A = lw = xy$. We wish to find a formula for this area. To write this as a single variable, x in this example, we note that $\triangle ABC \sim \triangle ADF$, which means that corresponding sides of these triangles are proportional; therefore,

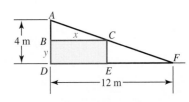

Figure 2.7 Play Zone

$$\frac{4-y}{4} = \frac{x}{12}$$

$$4 - y = \frac{1}{3}x$$

$$y = 4 - \frac{1}{3}x$$

We now write A as a function of x alone:

$$A(x) = xy = x\left(4 - \frac{1}{3}x\right) = 4x - \frac{1}{3}x^2$$

∎

Classification of Functions

If you have looked at the table of contents for this book, you will see that one of the unifying concepts of this book is that of a function. As a preview of what is to follow, we will define many of the functions you will encounter in this course and in your calculus course.

POLYNOMIAL FUNCTION

A **polynomial function** is a function of the form

$$f(x) = a_n x^n + a_{n-1} x^{n-1} + \cdots + a_2 x^2 + a_1 x + a_0$$

where n is a nonnegative integer and $a_n, \ldots, a_2, a_1, a_0$ are constants. If $a_n \neq 0$, the integer n is called the **degree** of the polynomial. The constant a_n is called the **leading coefficient** and the constant a_0 is called the **constant term** of the polynomial function. In particular,

A **constant function** is zero degree: $f(x) = a$

A **linear function** is first degree: $f(x) = ax + b$

A **quadratic function** is second degree: $f(x) = ax^2 + bx + c$

A **cubic function** is third degree: $f(x) = ax^3 + bx^2 + cx + d$

A **quartic function** is fourth degree: $f(x) = ax^4 + bx^3 + cx^2 + dx + e$

We will consider polynomial functions in Chapter 3.

A second important algebraic function is a *rational function*.

RATIONAL FUNCTION

A **rational function** is the quotient of two polynomial functions, $p(x)$ and $d(x)$:

$$f(x) = \frac{p(x)}{d(x)}, \qquad d(x) \neq 0$$

When we write $d(x) \neq 0$ we mean that all values c for which $d(c) = 0$ are excluded from the domain of d. Here are some examples of rational functions, written in different ways.

$$f(x) = x^{-1} \qquad f(x) = \frac{x-5}{x^2 + 2x - 3} \qquad f(x) = x^{-3} + \sqrt{2}x$$

We will consider rational functions in Chapter 4.

If r is any nonzero real number, the function $f(x) = x^r$ is called a **power function** with exponent r. You should be familiar with the following cases:

Integral powers ($r = n$, a positive integer): $f(x) = x^n = \underbrace{x \cdot x \cdots x}_{n \text{ factors}}$

Reciprocal powers (r is a negative integer): $f(x) = x^{-n} = \dfrac{1}{x^n}$ for $x \neq 0$

Roots ($r = \frac{m}{n}$ is a positive rational number): $f(x) = x^{m/n} = \sqrt[n]{x^m} = \left(\sqrt[n]{x}\right)^m$ for $x \geq 0$ if n even, $n \neq 0$ ($\frac{m}{n}$ is reduced)

Power functions can also have irrational exponents (such as $\sqrt{2}$ or π), but such functions must be defined in a special way and are not introduced until Chapter 4.

A function is called **algebraic** if it can be constructed using algebraic operations (such as adding, subtracting, multiplying, dividing, or taking roots) starting with polynomials. Any rational function is an algebraic function.

Functions that are not algebraic are called **transcendental**. The following functions are transcendental functions:

Exponential functions are functions of the form $f(x) = b^x$, where b is a positive constant. We will study these functions in Chapter 4.

Logarithmic functions are functions of the form $f(x) = \log_b x$, where b is a positive constant. We will also study these functions in Chapter 4.

Trigonometric functions are the functions sine, cosine, tangent, secant, cosecant, and cotangent. We will define these functions in Chapter 5.

PROBLEM SET 2.2

LEVEL 1

Determine whether the sets given in Problems 1–4 are functions. If it is a function, state its domain.

1. a. $\{(6, 3), (9, 4), (7, -1), (5, 4)\}$
   b. $\{6, 9, 7, 5\}$
   c. $y = 5x + 2$
   d. $y = -1$ if x is a rational number, and $y = 1$ if x is an irrational number.
2. a. $\{(3, 6), (4, 9), (-1, 7), (4, 5)\}$
   b. $\{10, 20, 30, 40\}$
   c. $y \leq 5x + 2$
   d. $y = -1$ if x is a positive integer, and $y = 1$ if x is a negative integer.
3. a. $\{(x, y) \mid y = \text{closing price of Xerox stock on July 1 of year } x\}$
   b. $\{(x, y) \mid x = \text{closing price of Xerox stock on July 1 of year } y\}$
4. a. (x, y) is a point on a circle with center $(2, 3)$ and radius 4.
   b. (x, y) is a point on a line passing through $(2, 3)$ and $(4, 5)$.

For each verbal description in Problems 5–8, write a rule in the form of an equation and then state the domain.

5. For each number x in the domain, the corresponding range value y is found by multiplying by three and then subtracting five.
6. For each number x in the domain, the corresponding range value y is found by squaring and then subtracting five times the domain value.
7. For each number x in the domain, the corresponding range value y is found by taking the square root of the difference of the domain value subtracted from five.
8. For each number x in the domain, the corresponding range value y is found by adding one to the domain value and then dividing that result into five added to five times the domain value.
9. Let $P(x)$ be the number of prime numbers less than x. Find
   a. $P(10)$ **b.** $P(-10)$ **c.** $P(100)$

10. Let $S(x)$ be the exponent on a base 2 that gives the result x. Find
   a. $S(32)$ **b.** $S\left(\frac{1}{8}\right)$
   c. $S\left(\sqrt{2}\right)$

In Problems 11–18, let $f(x) = 5x - 1$ and $g(x) = 3x^2 + 1$. Find the requested values.

11. a. $f(0)$ **b.** $f(2)$
   c. $f(-3)$ **d.** $f\left(\sqrt{5}\right)$
12. a. $f(w)$ **b.** $g(w)$
   c. $g(t)$ **d.** $g(v)$
13. a. $f(t)$ **b.** $f(p)$
   c. $f(t + 1)$ **d.** $g(t + 1)$
14. a. $f(x + 2)$ **b.** $g(x + 2)$
   c. $f(t + h)$ **d.** $g(t + h)$
15. a. $f(t^2 - 3t) - g(t + 2)$
   b. $f(t^2 + 2t + 1) - g(t + 3)$
16. a. $\dfrac{f(t + 3) - f(t)}{3}$ **b.** $\dfrac{g(t + 2) - g(t)}{2}$
17. a. $\dfrac{f(t + h) - f(t)}{h}$ **b.** $\dfrac{g(t + h) - g(t)}{h}$
18. a. $\dfrac{f(x + h) - f(x)}{h}$ **b.** $\dfrac{g(x + h) - g(x)}{h}$

In Problems 19–24, use the vertical line test to determine whether the curve is a function and if it is the graph of a function, use the horizontal line test to determine whether it is one-to-one. Also state the probable domain and range.

19.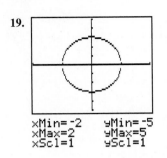

xMin=-2   yMin=-5
xMax=2   yMax=5
xScl=1   yScl=1

20.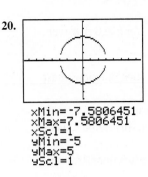

xMin=-7.5806451
xMax=7.5806451
xScl=1
yMin=-5
yMax=5
yScl=1

21.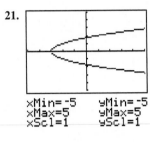

xMin=-5   yMin=-5
xMax=5   yMax=5
xScl=1   yScl=1

22.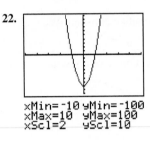

xMin=-10 yMin=-100
xMax=10   yMax=100
xScl=2   yScl=10

23.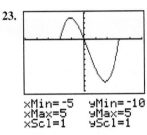

xMin=-5   yMin=-10
xMax=5   yMax=5
xScl=1   yScl=1

24.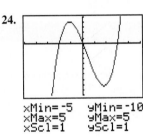

xMin=-5   yMin=-10
xMax=5   yMax=5
xScl=1   yScl=1

WHAT IS WRONG, *if anything, with each statement in Problems 25–34? Explain your reasoning.*

25. $f(x+2) = f(x) + f(2)$ **26.** $f(2x) = 2f(x)$

27. If $f(x) = 3x^2 + 5$, then $f(2)$ is a function.

28. If $f(x) = 3x^2 + 5$, then $f(x)$ is a function.

29. If $f(x) = 3x^2 + 5$, then f is a function.

30. The horizontal line test is used to determine whether a graph represents a function.

31. The horizontal line test is used to determine whether a graph represents a one-to-one function.

32. The vertical line test is used to determine whether a graph represents a function.

33. The vertical line test is used to determine whether a graph represents a one-to-one function.

34. If $f(x) = 3x^2$, then $\dfrac{f(x+h) - f(x)}{h} = \dfrac{3x^2 + h - 3x^2}{h}$

LEVEL 2

In Problems 35–40, find the difference quotient for the given function f.

35. $f(x) = 4x^2$ **36.** $f(x) = 6x^2$

37. $f(x) = x^2 + 3$ **38.** $f(x) = 6x^2 + 2x$

39. $f(x) = x^2 - x + 3$ **40.** $f(x) = x^2 + 2x - 3$

41. If $S(x) = \dfrac{3x+2}{4x-1}$, find $S\left(\dfrac{1}{x}\right)$.

42. Let $R(x) = 3x^2 + 3x^{-2} - x - x^{-1}$. Show $R\left(\dfrac{1}{x}\right) = R(x)$.

43. Let $f(x) = ax + b, a \neq 0$. Find a and b so that
$$f(x + y) = f(x) + f(y)$$

44. Let $f(x) = ax + bx + c, a \neq 0$. Find a, b, and c so that
$$f(x + y) = f(x) + f(y)$$

45. If $f(x) = ax + b$, $a \neq 0$, evaluate $f\left(-\dfrac{b}{a}\right)$.

46. Let $g(x) = ax^2 + bx + c$, $a \neq 0$. Find
$$g\left(\frac{-b + \sqrt{b^2 - 4ac}}{2a}\right)$$

47. a. Let $Q(x) = \dfrac{x+a}{x-a}$. Does
$$Q(2a + 3a) = Q(2a) + Q(3a)?$$

b. Give an example of a function E for which
$$E(2a + 3a) = E(2a) + E(3a)$$

48. a. Let $T(x) = 2^x$. Does
$$T(a + b) = T(a) \cdot T(b)?$$

b. Give an example of a function D for which $D(a + b) \neq D(a) \cdot D(b)$.

49. Let d be a function that represents the distance an object falls (neglecting air resistance) from rest in the first t seconds. Find the distance the object falls for the given intervals of time if $d(t) = 16t^2$.

a. From $t = 2$ to $t = 6$. *Hint*: This is $d(6) - d(2)$.

b. From $t = 2$ to $t = 4$.

c. From $t = 2$ to $t = 3$.

d. From $t = 2$ to $t = 2 + h$.

e. From $t = x$ to $t = x + h$.

f. Give a physical interpretation for
$$\frac{d(t + h) - d(t)}{h}$$

50. Suppose the total cost (in dollars) of manufacturing q units of a certain item is given by
$$C(q) = q^3 - 30q + 400q + 500$$
on $[0, 30]$.

a. What is the cost of manufacturing 20 units?

b. Compute the cost of manufacturing the 21st unit.

51. An efficiency study of the morning shift at a certain factory indicates that an average worker who arrives on the job at 8:00 A.M. will have assembled
$$f(x) = -x^3 + 6x + 15x^2$$
units x hours later $(0 \leq x \leq 8)$.

a. How many units will such a worker have assembled by 10:00 A.M.?

b. How many units will such a worker assemble between 9:00 A.M. and 10:00 A.M.?

52. It is estimated that t years from now the population of a certain suburban community will be
$$P(t) = 20 - \frac{6}{t+1}$$
thousand people.

a. What will the population of the community be nine years from now?

b. By how much will the population increase during the ninth year?

c. What will happen to the size of the population in the "long run"?

53. Find the area of a square as a function of its perimeter.

54. Find the area of a circle as a function of its circumference.

PROBLEMS FROM CALCULUS *Functions are, of course, central to the study of calculus. Problems 55–60 are adapted from a leading calculus textbook.*[*]

55. A manufacturer wants to design an open box having a square base (length x) and a surface area of 108 square inches. Write the volume as a function of the length of a side of the base.

56. Write the distance between a point (x, y) on the graph of $y = 4 - x^2$ and the point $(0, 2)$ as a function of x.

57. A rectangular page is to contain 24 square inches of print where x is the height of the printed portion. The margins at the top and bottom of the page are each $1\frac{1}{2}$ inches. The margins on each side are 1 inch. Write the area of the paper as a function of x.

58. Two posts, one 12 feet high and the other 28 feet high, stand 30 feet apart. The top of each post is fastened by wire to a single stake, running from ground level to the top of each post. Write the length of the wire as a function of the distance, x, the stake is located from the 12-ft post.

59. Four feet of wire is to be used to form a square and a circle. Write the total area (sum of the area of the square and the area of the circle) as a function of the length, x, of the side of the square.

60. A hospital patient receives an intravenous glucose solution from a cylindrical bottle of radius 8 cm with height 20 cm. Suppose the fluid level drops 0.25 cm/min. (*Note*: The volume of a right circular cylinder of radius r and height h is $\pi r^2 h$.)

a. Write a formula for the amount S of solution in cubic centimeters (cm³) that has entered the patient's vein when the height of the removed fluid is h cm.

b. Write a formula for the height of the fluid (in cm) t minutes after the full bottle is hooked up to a patient.

c. Write a formula for S as a function of t.

d. How long does it take for all the fluid to enter the patient's vein?

2.3 Graph of a Function

Graphs have visual impact. They also reveal information that may not be evident from verbal or algebraic descriptions. To represent a function $y = f(x)$ geometrically as a graph, it is traditional to use a Cartesian coordinate system on which units for the independent variable x are marked on the horizontal axis and units for the dependent variable y are marked on the vertical axis.

> **GRAPH OF A FUNCTION**
>
> The **graph** of a function f consists of all points whose coordinates (x, y) satisfy $y = f(x)$, for all x in the domain of f.

One of the principal tasks of this book is to discuss efficient techniques involving calculus that you can use to draw accurate graphs of functions. In beginning algebra, you began sketching lines by plotting points, but you quickly found out that this is not a very efficient way to draw more complicated graphs, especially without the aid of a graphing calculator or computer. Table 2.1 includes a few common graphs you have probably encountered in previous courses. We will assume that you are familiar with their general shape and know how to sketch each of them.

We will use the functions in Table 2.1 as a basis for discussion in this chapter as we look at *properties* of functions, and then in subsequent chapters of the book, we will use the properties of this chapter to help us graph functions in general.

Even if you do not now have access to a graphing calculator or computer software that graphs, you will no doubt be using this technology in the future. Many have a misconception that if they only had this technology they would not need to study graphing in a mathematics course. Quite the contrary is true. Even the best software will often come up with a blank screen when an equation or a curve is input. Graphing calculators require input in the form [Y=], which means that you are expected to input equations that are functions.

Domain and Range

In this book, unless otherwise specified, the domain of a function is the set of real numbers for which the function is defined. We call this the **domain convention.**

If a function f is **undefined** at x, it means that x is not in the domain of f. The most frequent exclusions from the domain are those values that cause division by 0 and negative values under a

☠ Notice this agreement about the domain of a function that will be used throughout the book. ☠

*From Section 3.7, pp 213–216, of *CALCULUS*, Fifth Edition, by Larson, Hostetler, and Edwards.

TABLE 2.1	Directory of Curves	
Identity Function $y = x$	**Standard Quadratic Function** $y = x^2$	**Standard Cubic Function** $y = x^3$
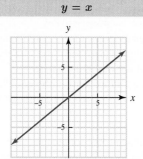	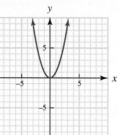	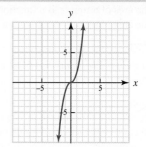
Absolute Value Function $y = |x| = \sqrt{x^2}$	**Square Root Function** $y = \sqrt{x}$	**Cube Root Function** $y = \sqrt[3]{x}$
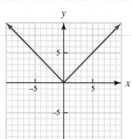	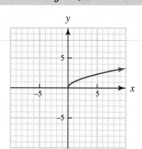	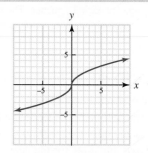
Standard Reciprocal $y = \dfrac{1}{x}$	**Standard Reciprocal Squared** $y = \dfrac{1}{x^2}$	**Standard Square Root Reciprocal** $y = \dfrac{1}{\sqrt{x}}$
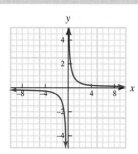	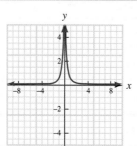	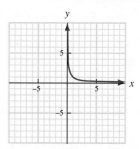

square root. In applications, the domain is often specified by the context. For example, if x is the number of people on an elevator, the context requires that negative numbers and nonintegers be excluded from the domain; therefore, x must be an integer such that $0 \le x \le c$, where c is the maximum capacity of the elevator.

EXAMPLE 1   Domain of a function

Find the domain for the given functions.

   a. $f(x) = 2x - 1$

   b. $g(x) = 2x - 1,\ x \ne -3$

   c. $h(x) = \dfrac{(2x - 1)(x + 3)}{x + 3}$

   d. $F(x) = \sqrt{12 + x - x^2}$

Solution

a. All real numbers; $D = (-\infty, \infty)$.

b. All real numbers except -3.

c. Because the expression is meaningful for all $x \neq -3$, the domain is all real numbers except -3.

d. F is defined whenever $12 + x - x^2$ is nonnegative:

$$12 + x - x^2 \geq 0$$
$$(4 - x)(3 + x) \geq 0$$

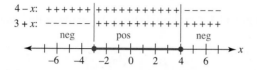

We see that x is nonnegative when $-3 \leq x \leq 4$, so $D = [-3, 4]$. ■

EQUAL FUNCTIONS

Two functions and f and g are **equal** if and only if

1. f and g have the same domain.
2. $f(x) = g(x)$ for all x in the domain.

In Example 1, the functions g and h are equal, but the functions f and h are not. A common mistake is to "reduce" the function h to the function f:

☠ WRONG: $h(x) = \dfrac{(2x - 1)(x + 3)}{x + 3} = 2x - 1 = f(x)$ ☠

RIGHT: $h(x) = \dfrac{(2x - 1)(x + 3)}{x + 3} = 2x - 1,\ x \neq -3;$ therefore, $h(x) = g(x)$.

Even though the usual graphing procedure is to find the domain, draw the graph, and then use the graph to determine the range, it is sometimes necessary to find both the domain and the range. We summarize these procedures:

FINDING THE DOMAIN

To find the domain, **solve for y** and look for exclusions for x.

FINDING THE RANGE

To find the range, **solve for x** and look for exclusions for y.

With this procedure, it is not necessary that the given relation be a function.

EXAMPLE 2   Finding the domain and range of a relation

Find the domain and range for:

a. $x^2 + y^2 = 3$

b. $y = \sqrt{3 - x^2}$

c. $xy^2 - y^2 - 1 = 0$

Solution

a. *Domain:* Solve for y and look for exclusions for x.

$$x^2 + y^2 = 3$$
$$y^2 = 3 - x^2$$
$$y = \pm\sqrt{3 - x^2}$$

Solve the inequality $3 - x^2 \geq 0$ to find critical values $\pm\sqrt{3}$. From a number line, we find

$$D = \left(-\sqrt{3}, \sqrt{3}\right)$$

Range: Solve for x and look for exclusions for y. The range is the same as the domain (the steps are identical), namely,

$$D = \left(-\sqrt{3}, \sqrt{3}\right)$$

b. *Domain:* Solve for y and look for exclusions for x. The radicand must be nonnegative, so

$$3 - x^2 \geq 0$$

Critical values are $x = \pm\sqrt{3}$, and from a number line we find $D = \left[-\sqrt{3}, \sqrt{3}\right]$.
Range: Solve for x and look for exclusions for y.

$$y = \sqrt{3 - x^2}$$
$$y^2 = 3 - x^2$$
$$x^2 = 3 - y^2$$
$$x = \sqrt{3 - y^2} \qquad \text{Positive only because } y \geq 0.$$

☠ Compare this step in part **b** with the similar step in part **a**. These steps are often confused. Do you see why the "$\pm$" is needed in part **a** but is not needed here in part **b**? ☠

Critical values are $y = \pm\sqrt{3}$, and from a number line, we find $R = \left[0, \sqrt{3}\right]$.

c. *Domain:* Solve for y and look for exclusions for x.

$$xy^2 - y^2 - 1 = 0$$
$$y^2(x - 1) = 1$$
$$y^2 = \frac{1}{x - 1}$$
$$y = \frac{\pm 1}{\sqrt{x - 1}}$$

By inspection, $D = (1, \infty)$.

Range: Solve for x and look for exclusions for y.

$$xy^2 - y^2 - 1 = 0$$
$$xy^2 = y^2 + 1$$
$$x = \frac{y^2 + 1}{y^2}$$

For this equation, we see that x is real for all y except $y = 0$. Thus, $R = (-\infty, 0) \cup (0, \infty)$. ■

Intercepts

As we discussed in Section 1.3, the points where a graph intersects the coordinate axes are called *intercepts*. We restate the definition here in functional notation.

> **INTERCEPTS**
>
> If the number zero is in the domain of f and $f(0) = b$, then the point $(0, b)$ is called the
> **y-intercept** of the graph of f. If a is a real number in the domain of f such that $f(a) = 0$, then
> $(a, 0)$ is an **x-intercept** of f. Any number x such that $f(x) = 0$ is called a **zero of the func-
> tion**.

Functions can have several x-intercepts (or no x-intercepts) but can have at most one y-intercept.
(Do you see why?)

EXAMPLE 3   Finding the intercepts

Find the intercepts and determine whether each is a function.

a. $yx^2 + y - 1 = 0$ $\qquad\qquad$ **b.** $x^2 - xy^2 + 4y^2 - 1 = 0$

c. $|x| + |y| = 4$ $\qquad\qquad\qquad$ **d.** $x^{2/3} + y^{2/3} = 16$

e. $f(x) = -x^2 + x + 2$

Solution

a. If we solve for y, we see that this is a function because for each x there is exactly one y-value.

y-intercept: $\qquad$ Let $x = 0$: $\qquad yx^2 + y - 1 = 0$
$$0 + y - 1 = 0$$
$$y = 1$$

The y-intercept is $(0, 1)$. **STOP** *Functions can have at most one y-intercept.*

x-intercept(s): $\qquad$ Let $y = 0$: $\qquad yx^2 + y - 1 = 0$
$$0 + 0 - 1 = 0 \qquad \text{False}$$

A false equation means that there is no point; in this example, there are no x-intercepts.

b. If we solve for y, we see that for each x there are two possible values of y. Thus, this is not a
function.

y-intercept(s): $\qquad$ Let $x = 0$: $\qquad x^2 - xy^2 + 4y^2 - 1 = 0$
$$0^2 - 0 + 4y^2 - 1 = 0$$
$$y^2 = \frac{1}{4}$$
$$y = \pm\frac{1}{2}$$

The y-intercepts are $\left(0, \dfrac{1}{2}\right), \left(0, -\dfrac{1}{2}\right)$.

x-intercept(s): $\qquad$ Let $y = 0$: $\qquad x^2 - xy^2 + 4y^2 - 1 = 0$
$$x^2 - 0 + 0 - 1 = 0$$
$$x = \pm 1$$

The x-intercepts are $(1, 0), (-1, 0)$.

c. This is not a function.

y-intercept(s): $\qquad$ Let $x = 0$: $\qquad |x| + |y| = 4$
$$|0| + |y| = 4$$
$$y = \pm 4$$

The y-intercepts are $(0, 4), (0, -4)$.

x-intercept(s): $\qquad$ Let $y = 0$; the x-intercepts are found similarly to be $(4, 0), (-4, 0)$.

d. This is not a function.

y-intercept(s):   Let $x = 0$:

$$x^{2/3} + y^{2/3} = 16$$
$$0^{2/3} + y^{2/3} = 16$$
$$(y^{2/3})^{3/2} = (4^2)^{3/2}$$
$$|y| = 4^3$$
$$y = \pm 64$$

The *y*-intercepts are $(0, 64)$, $(0, -64)$.

x-intercept(s):   Let $y = 0$; by symmetry, the *x*-intercepts are $(64, 0)$, $(-64, 0)$.

e. This is given in function notation, so it is obviously a function.

y-intercept: Let $x = 0$:

$$f(x) = -x^2 + x + 2$$
$$f(0) = -0^2 + 0 + 2$$
$$= 2$$

The *y*-intercept is $(0, f(0)) = (0, 2)$.

x-intercept(s):   Let $y = f(x) = 0$; factoring, we find that

$$-x^2 + x + 2 = 0$$
$$x^2 - x - 2 = 0$$
$$(x + 1)(x - 2) = 0$$
$$x = -1 \text{ or } x = 2$$

The *x*-intercepts are $(-1, 0)$ and $(2, 0)$.   ∎

Sometimes when you are graphing a curve, you want to find a point other than an intercept so that the point is in a certain region or with certain properties. For example, if you want to know one point on the line defined by the equation $2x + 3y - 4 = 0$, where $x > 5$, then you can choose any *x*-value satisfying $x > 5$ and find the corresponding *y*-value. Consider the following example.

EXAMPLE 4   Finding points satisfying specified conditions

Find a point on the curve defined by the equation $y = \dfrac{2x^2 - 3x + 5}{x - 3}$ that also satisfies the specified conditions, if it exists.

a. $x > 5$

b. passes through the line $y = -4$

c. passes through the line $y = 2x + 1$

Solution

a. Choose any value of $x > 5$, say, $x = 10$:

$$y = \frac{2(10)^2 - 3(10) + 5}{10 - 3} = \frac{175}{7} = 25$$

One possible point is $(10, 25)$.

b. Solve

$$-4 = \frac{2x^2 - 3x + 5}{x - 3}$$
$$-4(x - 3) = 2x^2 - 3x + 5 \qquad x \neq 3$$
$$2x^2 + x - 7 = 0$$
$$x = \frac{-1 \pm \sqrt{1 - 4(2)(-7)}}{2(2)}$$
$$= \frac{-1 \pm \sqrt{57}}{4}$$
$$\approx 1.6, -2.1$$

The given curve passes through the line $y = -4$ at approximately the points $(1.6, -4)$ and $(-2.1, -4)$.

c. Solve

$$2x + 1 = \frac{2x^2 - 3x + 5}{x - 3}$$
$$(2x + 1)(x - 3) = 2x^2 - 3x + 5 \qquad x \neq 3$$
$$2x^2 - 5x - 3 = 2x^2 - 3x + 5$$
$$-2x = 8$$
$$x = -4$$

The curve intersects the line $y = 2x + 1$ when $x = -4$ and when $y = 2(-4) + 1 = -7$. That is at the point $(-4, -7)$. ∎

Properties of Functions

Several different properties of functions are useful in a variety of ways.

> **PROPERTIES OF FUNCTIONS**
>
> Let S be a subset of the domain of a function f. Then:
>
> f is **increasing** on S if $f(x_1) < f(x_2)$ whenever $x_1 < x_2$ in S;
> f is **decreasing** on S if $f(x_1) > f(x_2)$ whenever $x_1 < x_2$ in S;
> f is **constant** on S if $f(x_1) = f(x_2)$ for every x_1 and x_2 in S.
>
> If the value a separates an interval over which f is increasing from an interval over which f is decreasing, then $(a, f(a))$ is a **turning point**.

☠ Note the terminology; we say that the function is increasing and the graph is rising. We say that the function is decreasing and the graph is falling. ☠

These properties are illustrated with the following example.

EXAMPLE 5   Properties of a function

Let $y = (x - 5)^2 - 4 = x^2 - 10x + 21$ with the graph as shown in Figure 2.8.

a. Where are the intercepts?

b. Where is the turning point?

c. Where is the function increasing?

d. Where is the function decreasing?

Solution   *Note*: We state the intervals over which f is increasing, decreasing, or constant in terms of x, that is, the S that is a subset of the domain of f.

a. y-intercepts: If $x = 0$, then $y = 0^2 - 10(0) + 21 = 21$
    x-intercepts: If $y = 0$, then

$$x^2 - 10x + 21 = 0$$
$$(x - 7)(x - 3) = 0$$
$$x = 3, 7$$

The intercepts are $(0, 21)$, $(3, 0)$, and $(7, 0)$.

b. By inspection from the equation $y + 4 = (x - 5)^2$, we note the turning point is the point $(5, -4)$.

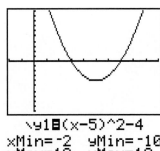

\y1目(x-5)^2-4
xMin=-2   yMin=-10
xMax=10   yMax=10
xScl=1   yScl=1

Figure 2.8 Graph of $y = (x - 5)^2 - 4$

c. We see the graph is rising to the right of the turning point, so we say the function is increasing on $(5, \infty)$.

d. We see the graph is falling to the left of the turning point, so we say the function is decreasing on $(-\infty, 5)$. ∎

The information in Example 5 is frequently used in calculus, so we have superimposed the correct terminology on the graph in Figure 2.9.

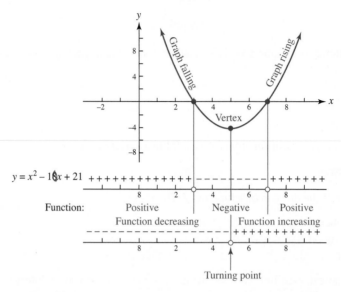

Figure 2.9 **Terminology associated with the properties of functions**

In the last section, we defined a *difference quotient*. There are several applications of difference quotient, and you might recall that this was the average distance we found in Example 8, Section 2.2. Another application involves the notion of an *average rate of change*.

AVERAGE RATE OF CHANGE

Let f be a function defined on some interval $[a, b]$. Then the **average rate** of change from $(a, f(a))$ to $(b, f(b))$ is the difference quotient

$$\frac{\Delta y}{\Delta x} = \frac{f(b) - f(a)}{b - a}$$

You might note that the average rate of change of a function f between two points is the slope of the line (called the *secant line*) connecting these points. That is, in calculus this is often stated in terms of a starting point x, and an incremental change h so that $a = x$ and $b = x + h$, so that

$$\frac{\Delta y}{\Delta x} = \frac{f(x + h) - f(x)}{h}$$

EXAMPLE 6   **Average rate of change of a function**

Consider the function (from Example 5) defined by $f(x) = (x - 5)^2 - 4$

a. Find the average rate of change over the interval $[2, 6]$.

b. What is the equation of the secant line passing through $(2, f(2))$ and $(6, f(6))$?

Solution

a. $f(2) = (2-5)^2 - 4 = 5$ The average rate of change is: $\dfrac{\Delta y}{\Delta x} = \dfrac{f(6) - f(2)}{6 - 2}$

$f(6) = (6-5)^2 - 4 = -3$

$$= \dfrac{-3 - 5}{4}$$

$$= -2$$

b. The slope of the secant line is $m = -2$ (part **a**). We can use either one of the given points. We choose $(2, 5)$:

$$y - k = m(x - h) \quad \text{Point-slope form}$$
$$y - 5 = -2(x - 2) \quad \text{Substitute known values.}$$
$$y - 5 = -2x + 4 \quad \text{Simplify.}$$
$$2x + y - 9 = 0 \quad \text{Standard form}$$

■

Another classification of functions is related to the symmetry of its graph. A function whose graph is symmetric with respect to the y-axis is called **even**. A function whose graph is symmetric with respect to the origin is called **odd**. If the function is found to be even or odd, then the symmetry of its graph helps in the graphing of the function. This concept can be used to reduce (by half) the amount of work necessary on many problems.

EVEN AND ODD FUNCTIONS

A function f is called

   even if $f(-x) = f(x)$ and

   odd if $f(-x) = -f(x)$

for all x in the domain of f.

Just as not every real number is even or odd (2 is even, 3 is odd, but 2.5 is neither), not every function is even or odd.

EXAMPLE 7   Even and odd functions

Classify the given functions as even, odd, or neither.

   a. $f(x) = x^2$
   b. $g(x) = x^3$
   c. $h(x) = x^2 + 5x$

Solution

a. $f(x) = x^2$ is *even* because

$f(-x) = (-x)^2 = x^2 = f(x)$

The graph at the right shows that the graph of the even function $f(x) = x^2$ is symmetric with respect to the y-axis.

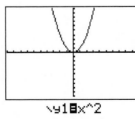

\y1☐x^2

xMin=-10 yMin=-10
xMax=10 yMax=10
xScl=1 yScl=1

b. $g(x) = x^3$ is *odd* because

$g(-x) = (-x)^3 = -x^3 = -g(x)$

The graph at the right shows that the graph of the odd function $g(x) = x^3$ is symmetric with respect to the origin.

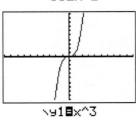

\y1☐x^3

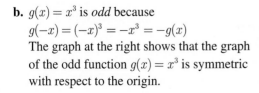

xMin=-10 yMin=-10
xMax=10 yMax=10
xScl=1 yScl=1

c. $h(x) = x^2 + 5x$ is *neither* because

$h(-x) = (-x)^2 + 5(-x) = x^2 - 5x$

Note that $h(-x) \neq h(x)$ and $h(-x) \neq -h(x)$.

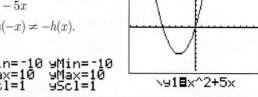

```
xMin=-10  yMin=-10
xMax=10   yMax=10
xScl=1    yScl=1
```

$\backslash y1\blacksquare x^2+5x$

PROBLEM SET 2.3

LEVEL 1

Sketch each of the functions in Problems 1–2, and classify each as odd, even, or neither.

1. **a.** identity function
   b. absolute value function
   c. standard reciprocal function
   d. standard square root reciprocal function
2. **a.** standard quadratic function
   b. square root function
   c. standard reciprocal squared function
   d. standard cubic function
3. IN YOUR OWN WORDS What is the graph of a function?
4. IN YOUR OWN WORDS How do you find the domain and range of a function?
5. IN YOUR OWN WORDS Distinguish between an x-intercept and a zero of a function.
6. IN YOUR OWN WORDS In the book we state that "functions can have several x-intercepts (or no x-intercepts) but can have at most one y-intercept. Explain why this is true.

State whether the functions f and g defined in Problems 7–10 are equal.

7. **a.** $f(x) = \dfrac{3x^2 + x}{x}$;

   $g(x) = 3x + 1$

   b. $f(x) = \dfrac{2x^2 - 7x - 4}{x - 4}$;

   $g(x) = 2x + 1,\ x \neq 4$

8. **a.** $f(x) = \dfrac{(3x + 1)(x - 4)}{x - 4}$;

   $g(x) = 3x + 1$

   b. $f(x) = \dfrac{3x^2 - 5x - 2}{3x + 1}$;

   $g(x) = x - 2$

9. **a.** $f(x) = \dfrac{2x^2 - x - 6}{x - 2}$;

   $g(x) = 2x + 3,\ x \neq 2$

   b. $f(x) = \dfrac{3x^2 - 5x - 2}{x - 2},\ x \neq 2$;

   $g(x) = 3x + 1$

10. **a.** $f(x) = \dfrac{(3x + 1)(x - 2)}{x - 2},\ x \neq 6$

    $g(x) = \dfrac{(3x + 1)(x - 6)}{x - 6},\ x \neq 2$

    b. $f(x) = \dfrac{(5x - 1)(x + 4)}{x + 4},\ x \neq -4$

    $g(x) = \dfrac{(5x - 1)(x - 2)}{x - 2},\ x \neq 2$

Find the domain for the functions defined by the equations in Problems 11–16.

11. **a.** $f(x) = 2x - 3$
    b. $g(x) = 2x - 3,\ x \neq 1$
12. **a.** $f(x) = \dfrac{(2x + 1)(x - 1)}{x - 1}$
    b. $g(x) = 2x + 1$
13. **a.** $f(x) = \dfrac{(3x + 1)(x - 3)}{x^2 + 2}$
    b. $g(x) = x^2 + 3x - 5$
14. **a.** $f(x) = \sqrt{3x + 9}$
    b. $g(x) = \sqrt{x^2 - 4}$
15. $f(x) = \sqrt{2 - x - x^2}$
16. $g(x) = \sqrt{2 + x - x^2}$

PROBLEMS FROM CALCULUS *Graphs similar to those shown in Problems 17 and 18 are common in calculus. Specify the coordinates of the requested points by looking at the given graphs.*

17. **a.** Point R
    b. Point S

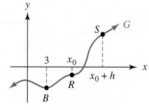

Graph of G

18. a. Point T
b. Point U

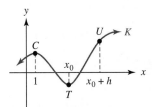

Graph of K

LEVEL 2

Find the domain, intercepts, and turning points of the functions defined by the graphs indicated in Problems 19–30. Also give the intervals for which the function is constant, where it is increasing, and where it is decreasing.

19.

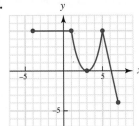

20.

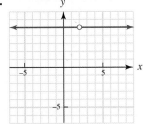

21.

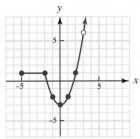

22.

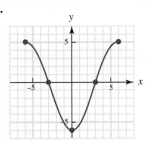

23.

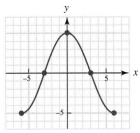

24.

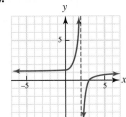

25.

26.

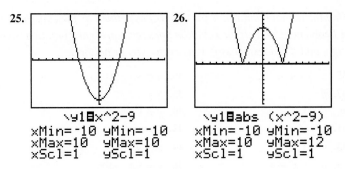

```
\y1■x^2-9
xMin=-10  yMin=-10
xMax=10   yMax=10
xScl=1    yScl=1
```

```
\y1■abs (x^2-9)
xMin=-10  yMin=-10
xMax=10   yMax=12
xScl=1    yScl=1
```

27.

28.

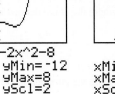

```
\y1■x^4-2x^2-8
xMin=-4   yMin=-12
xMax=4    yMax=8
xScl=1    yScl=2
```

```
\y1■x^3+x+2
xMin=-3   yMin=-10
xMax=3    yMax=10
xScl=1    yScl=1
```

29.

30.

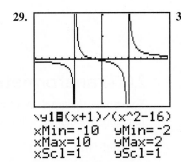

```
\y1■(x+1)/(x^2-16)
xMin=-10  yMin=-2
xMax=10   yMax=2
xScl=1    yScl=1
```

```
\y1■√(9-x^2)
xMin=-7.5 yMin=-5
xMax=7.5  yMax=5
xScl=1    yScl=1
```

PROBLEMS FROM CALCULUS *In calculus, the average rate of change of a function f between x and $x + h$ is defined to be the quantity*

$$\frac{f(x+h) - f(x)}{h}$$

Find the average rate of change from 2 to 2 + h for the functions in Problems 31–34.

31. Identity function, $f(x) = x$.
32. Standard quadratic function, $f(x) = x^2$.
33. Standard reciprocal function, $f(x) = 1/x$.
34. Consider the average rate of change for the standard quadratic function, $f(x) = x^2$.
   a. Which is larger, the average rate of change from 2 to 3 or from 10 to 11?
   b. What is the average rate of change from 2 to 2.1?
   c. What is the average rate of change from 2 to 2.01?
   d. What is the average rate of change from 2 to 2.001?
   e. What value does the sequence of calculation seem to be approaching?

In calculus, this value is called the *instantaneous rate of change.*

Find the domain and range for the graphs defined by the equations in Problems 35–52. If the equation does not represent a function, so state, and if it does, classify it as even, odd, or neither.

35. $y = x + 4$

36. $y = \sqrt{x - 4}$

37. $y = x^3$

38. $y = \sqrt{x}$

39. $y = 8x^3$

40. $y = \sqrt[3]{x}$

41. $xy = 1$

42. $y = x^2 + 4$

43. $y = x^2 - 8$

44. $y = \sqrt{x^2 - 9}$

45. $y = \sqrt{x^2 - 4}$

46. $y = \sqrt{x^3 - 9x}$

47. $y = \sqrt{x^2 + x - 12}$

48. $y = \dfrac{3}{x^2 + 1}$

49. $xy + 6 = 0$

50. $y = \dfrac{x^2 - 4}{x + 1}$

51. $2|x| - |y| = 5$

52. $|y| + 3|x| = 5$

53. Find the points (if any) where the curve defined by the equation

$$y = \frac{5x^2 - 8x}{2x + 1}$$

crosses the horizontal line $y = 3$.

54. Find the points (if any) where the curve defined by the equation

$$y = \frac{5x^2 - 8x}{2x + 1}$$

crosses the vertical line $x = -4$.

55. Find the points (if any) where the curve defined by the equation

$$y = \frac{2x^3 + 2x}{x^2 - 2}$$

crosses the vertical line $x = -1$.

56. Find the points (if any) where the curve defined by the equation

$$y = \frac{5x^2 - 8x}{2x - 1}$$

crosses the horizontal line $y = -4$.

57. Find the points (if any) where the curve defined by the equation

$$y = \frac{x^3 + 2x^2 - 2x}{x^2 - 2}$$

crosses the line $y = x + 1$.

58. Find the points (if any) where the curve defined by the equation

$$y = \frac{3x^3 + 4x^2 + 3}{3x^2 + 1}$$

crosses the horizontal line $y = x + 2$.

LEVEL 3

59. If f is increasing throughout its domain, prove that f is one-to-one.

60. If a function f is decreasing throughout its domain, prove that f is one-to-one.

2.4 Transformations of Functions

Sometimes the graph of a function can be sketched by translating or reflecting the graph of a related function. We call these translations and reflections *transformations* of a function.

Translations

We begin by considering two examples. Graph the functions $y = x^2$ and $y - 2 = (x - 6)^2$ by plotting points, as shown in Figure 2.10.

x	$y = x^2$
0	0
1	1
-1	1
2	4
-2	4
3	9
-3	9

a. Graph of $y = x^2$

x	$y = (x - 6)^2 + 2$
0	38
1	27
2	10
3	11
4	6
5	3
6	2
7	3
8	6
9	11

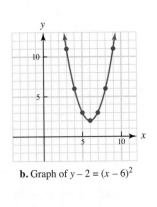

b. Graph of $y - 2 = (x - 6)^2$

Figure 2.10 Graphing by plotting points

Notice that the graphs in Figure 2.10 are identical, except they are in different locations. You also should have noticed (if you did the arithmetic) that the first table of values was much easier to calculate than the second. When two curves are congruent (have the same size and shape) and have the same orientation, we say that one can be found from the other by a **shift** or **translation**.

TRANSLATION

The graph defined by the equation

$$y - k = f(x - h)$$

is said to be a **translation** of the graph defined by $y = f(x)$. The translation (shift, as shown in Figure 2.10) is

to the right if $h > 0$

to the left if $h < 0$

up if $k > 0$

down if $k < 0$

» IN OTHER WORDS The procedure for graphing a translated graph is a two-step process:

(1) Plot (h, k). The numbers h and k are directed distances.

Horizontal translation $|h|$ units; to the right if h is positive and to the left if h is negative.

Similarly, the vertical translation is $|k|$ units; up if k is positive and down if k is negative.

(2) Graph the curve $y = f(x)$ using (h, k) as the starting point.

EXAMPLE 1　Translations of a standard curve

Given the standard quadratic function $y = x^2$ (see Table 2.1). Graph the given curves by translation.

a. $y - 4 = x^2$　**b.** $y + 6 = x^2$　**c.** $y = (x - 5)^2$　**d.** $y = (x + 5)^2$

Solution　Begin with the graph of $y = x^2$, as shown in Figure 2.11. The vertex of this standard quadratic function is $(0, 0)$. We call this the starting point.

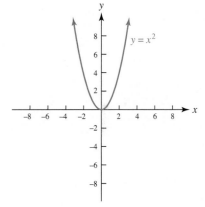

Figure 2.11 Graph of $y = x^2$

a. Write $y - 4 = x^2$ as $y - k = (x - h)^2$, and compare it to $y = x^2$ to see that $(h, k) = (0, 4)$. Draw the curve shown in Figure 2.11 with the starting point shifted up 4 units.

b. Write $y + 6 = x^2$ as $y - k = (x - h)^2$, and compare it to $y = x^2$ to see that $(h, k) = (0, -6)$ Draw the curve shown in Figure 2.11 with the starting point shifted down 6 units.

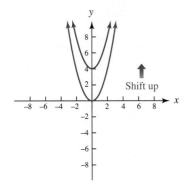

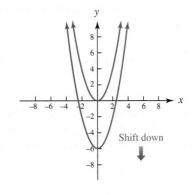

Compare the equations in parts **a** and **b** with the shift up and shift down directions of the graph.

c. Write $y = (x - 5)^2$ as $y - k = (x - h)^2$, and compare it to $y = x^2$ to see that $(h,\ k) = (5,\ 0)$. Draw the curve shown in Figure 2.11 with the starting point shifted to the right 5 units.

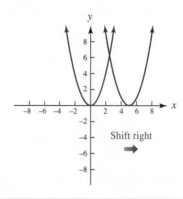

Shift right ➡️

d. Write $y = (x + 5)^2$ as $y - k = (x - h)^2$, and compare it to $y = x^2$ to see that $(h,\ k) = (-5,\ 0)$. Draw the curve shown in Figure 2.11 with the starting point shifted to the left 5 units.

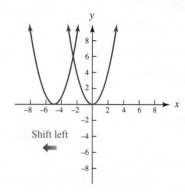

Shift left ⬅️

Compare the equations in parts **c** and **d** with the shift right and shift left directions of the graph. ∎

EXAMPLE 2   Translations of different standard curves

Graph: **a.** $y - \frac{1}{2} = \left(x - \frac{3}{2}\right)^2$ **b.** $f(x) = |x - 3| + 2$ **c.** $y = \sqrt{x - 2} - 3$

Solution   We begin by looking at Table 2.1.

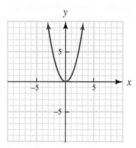

a. $y = x^2$

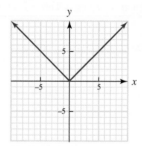

b. $y = |x|$

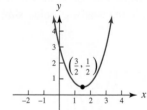

c. $y = \sqrt{x}$

a. First, plot $\left(\dfrac{3}{2}, \dfrac{1}{2}\right)$. Translate the standard quadratic function to this point, as shown at the right.

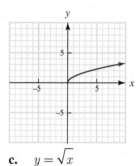

⚠️ Translate the graph, do NOT calculate values.

b. This equation can be written as $y - 2 = |x - 3|$, which is identical to the graph of the function $y = |x|$ from Table 2.1. Identify and plot the point $(h,\ k) = (3,\ 2)$. The graph is shown at the right.

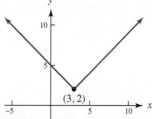

c. Rewrite as $y + 3 = \sqrt{x - 2}$. This graph is the same as the square root function $y = \sqrt{x}$ translated to the point $(h, k) = (2, -3)$. The graph is shown at the right.

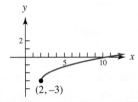

Reflections

In Chapter 1, we introduced the notion of symmetry. In terms of functions and functional notation, we restate those notions of symmetry with respect to the coordinate axes as **reflections.**

REFLECTION

A **reflection in the x-axis** of the graph of $y = f(x)$ is the graph of

$$y = -f(x)$$

A **reflection in the y-axis** of the graph of $y = f(x)$ is the graph of

$$y = f(-x)$$

» IN OTHER WORDS The graph is reflected in the x-axis if we replace y by $-y$ in its equation; it is reflected in the y-axis if we replace x by $-x$.

This procedure is illustrated in Figure 2.12, in which we have sketched the graph of $y = x^2$ and then reflected that graph.

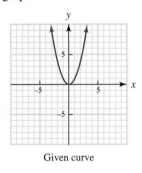

Given curve

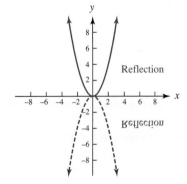

Figure 2.12 Reflection of $y = x^2$

EXAMPLE 3   Graphing with a reflection

Graph $y = -\sqrt{x}$.

Solution   The graph of this function is a reflection in the x-axis of the graph of $y = \sqrt{x}$, as shown in Figure 2.13.

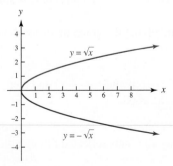

Figure 2.13 Reflection in the x-axis of $y = \sqrt{x}$

EXAMPLE 4   Reflections of the absolute value function

Graph **a.** $y = -|x|$ and **b.** $y = |-x|$.

Solution

a. Recall the graph of $y = |x|$ from Table 2.1. Replacing y by $-y$ reflects the graph of $y = |x|$ in the x-axis, as shown at the right.

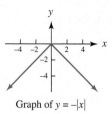

Graph of $y = -|x|$

b. Because x has been replaced by $-x$, the graph of $y = |-x|$ will be a reflection of $y = |x|$ in the y-axis. Note, too, that the graphs of these functions are identical. This is because (see Table 1.2, property 2) $|-a| = |a|$ for any real number a.

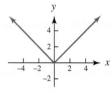

Graph of $y = |-x|$

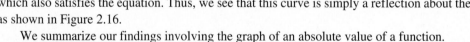

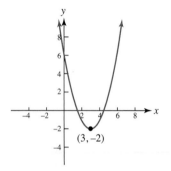

Figure 2.14 Graph of $y = x^2$ translated to $(3, -2)$

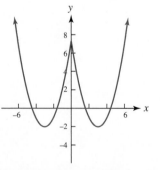

Figure 2.16 Graph of $y = (|x| - 3)^2 - 2$

An application of this reflection property helps us to graph curves such as

$$y = |(x - 3)^2 - 2|$$

If $y = f(x)$, where

$$f(x) = (x - 3)^2 - 2$$

the graph we seek is of the form $y = |f(x)|$. We begin by graphing

$$f(x) = (x - 3)^2 - 2$$

which is the standard quadratic function shown in Table 2.1 translated to the point $(3, -2)$ as shown in Figure 2.14.

Notice that part of the graph is above the x-axis and part is below the x-axis. Since the absolute value of a positive value leaves the positive value unchanged, the absolute value on $y = |f(x)|$ will leave the graph unchanged above the x-axis. However, the portion below the x-axis will be reflected above the x-axis because of the definition of absolute value. The desired graph is shown in Figure 2.15.

Contrast the graph shown in Figure 2.15 with the graph of the function

$$y = (|x| - 3)^2 - 2$$

For this curve, we also begin with the standard quadratic function shown in Table 2.1 and translate to the point $(3, -2)$. In this case, we note that for each value of x there is a corresponding value $-x$, which also satisfies the equation. Thus, we see that this curve is simply a reflection about the y-axis as shown in Figure 2.16.

We summarize our findings involving the graph of an absolute value of a function.

ABSOLUTE VALUE GRAPHS

The graph of $y = |f(x)|$ is found by graphing $y = f(x)$ and then reflecting all points of the graph that are below the x-axis through the x-axis.

The graph of $y = |f(x)|$ is an even function, so the graph is found by graphing $y = f(x)$ and then *in addition to those points* drawing all points reflected through the y-axis.

Dilations and Compressions

You may be familiar with the dilation of an eye or a compression fracture in your spine. To understand these concepts, we begin by observing that a curve can be compressed or dilated in either the x-direction, the y-direction, or both, as shown in Figure 2.17.

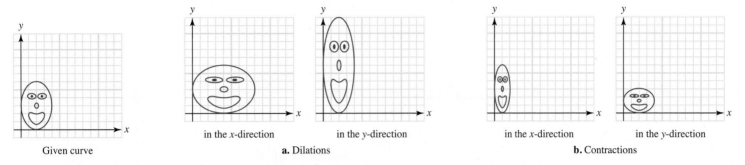

Given curve   in the x-direction   in the y-direction   in the x-direction   in the y-direction

a. Dilations   **b.** Contractions

Figure 2.17 Dilations and contractions of a given curve

DILATIONS AND CONTRACTIONS

To sketch the graph of $y = af(x)$, replace each point (x, y) with (x, ay)

If $a > 1$, then we call the transformation a **y-dilation.**

If $0 < a < 1$, then we call the transformation a **y-compression.**

To sketch the graph of $y = f(bx)$, replace each point (x, y) with $\left(\frac{1}{b}x, y\right)$

If $0 < b < 1$, then we call the transformation an **x-dilation.**

If $b > 1$, then we call the transformation an **x-compression.**

Dilations and Compressions in the y-direction

We are interested in modifications of a known function, which we call f. Consider the graph of $y = f(x)$ as shown in Figure 2.18**a**. The graph of $y = 2f(x)$ has the same shape except that each y-value is double the corresponding y-value of f. On the other hand, the graph of $y = \frac{1}{2}f(x)$ has the same shape except that each y-value is one-half the corresponding y-value of f. These graphs are shown in Figures 2.18**b** and **c**.

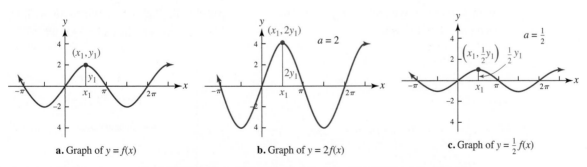

a. Graph of $y = f(x)$ **b.** Graph of $y = 2f(x)$ **c.** Graph of $y = \frac{1}{2}f(x)$

Figure 2.18 Dilations and compressions in the y-direction for a given function

Dilations and Compressions in the *x*-direction

To describe a dilation and compression in the x-direction, we will consider the function $y = f(x)$ and examine the effect of $y = f(bx)$. If we take a particular value of y, then it follows that the value bx is plotted in the same y-value as the x-value in the original curve. This means that to graph $y = f(bx)$, replace each point (x, y) on the graph of $y = f(x)$ with the point $\left(\frac{1}{b}x, y\right)$ See Figure 2.19.

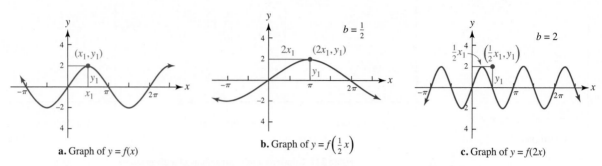

a. Graph of $y = f(x)$ **b.** Graph of $y = f\left(\frac{1}{2}x\right)$ **c.** Graph of $y = f(2x)$

Figure 2.19 Dilations and compressions in the *x*-direction for a given function

EXAMPLE 5   Compressions and dilations

Graph each function and describe each as a dilation or compression.

a. $y = 5\sqrt{x}$ **b.** $y = \sqrt{5x}$ **c.** $y = \left|\frac{1}{5}x\right|$ **d.** $y = \frac{1}{10}x^2$

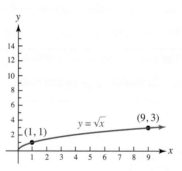

Solution   Compare parts **a** and **b** with the standard square root curve from Table 2.1, which we repeat here in the margin. Note that we have plotted two points in particular, namely, $(1, 1)$ and $(9, 3)$.

a. $y = 5\sqrt{x}$ is a y-dilation when compared with $f(x) = \sqrt{x}$. We see $a = 5 > 1$; (x, y) is replaced by $(x, 5y)$.

b. $y = \sqrt{5x}$ is an x-compression when compared with $f(x) = \sqrt{x}$. We see $b = 5$, so $\frac{1}{b} = \frac{1}{5}$; (x, y) is replaced by $\left(\frac{1}{5}x, y\right)$.

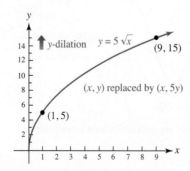

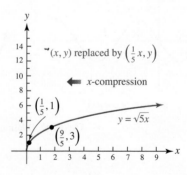

c. $y = \left|\frac{1}{5}x\right|$ is an x-dilation when compared with $f(x) = |x|$. We see $b = \frac{1}{5}$, so $0 < b < 1$; (x, y) is replaced by $(5x, y)$ because $\frac{1}{b} = 5$.

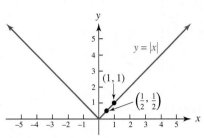

Graph of $y = |x|$ from Table 2.1

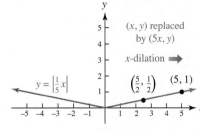

Graph of $y = \left|\frac{1}{5}x\right|$ as a dilation

d. $y = \frac{1}{10}x^2$ is a y-compression when compared with $f(x) = x^2$. We see $a = \frac{1}{10}$, so $0 < a < 1$; (x, y) is replaced by $\left(x, \frac{1}{10}y\right)$.

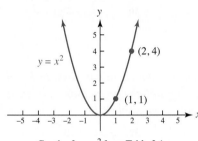

Graph of $y = x^2$ from Table 2.1

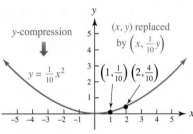

Graph of $y = \frac{1}{10}x^2$ as a compression

PROBLEM SET 2.4

LEVEL 1

Each of the graphs in Problems 1–10 is a translation of one of the curves from Table 2.1. Write the equation of the curves illustrated in Problems 1–10.

1.

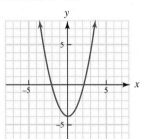

2.

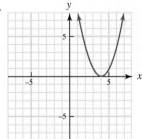

5.

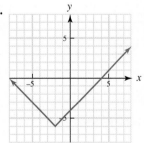

6.

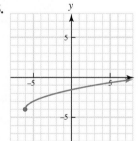

3.

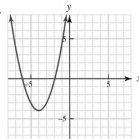

4.

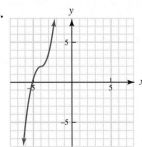

7.

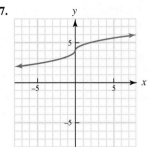

8.

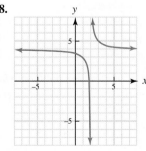

9. 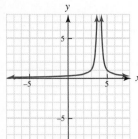   **10.**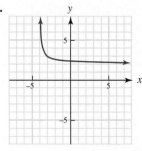

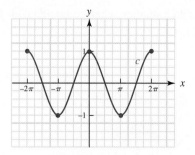

Figure 2.21 Graph of c

Let f, g, and s be the functions whose graphs are shown in Figure 2.20. Graph the functions indicated by the equations in Problems 11–20.

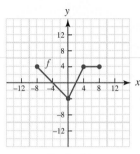

Graph of f

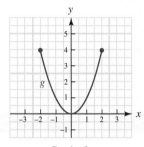

Graph of g

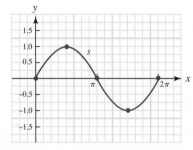

Graph of s

Figure 2.20 Graphs of curves for Problems 11–20

11. $y + 4 = f(x)$ **12.** $y - 4 = g(x)$
13. $y + \frac{1}{2} = s(x)$ **14.** $y - \pi = f(x)$
15. $y = g(x - 3)$ **16.** $y = s\left(x + \frac{\pi}{2}\right)$
17. $y = f(x + 4)$ **18.** $y - 4 = g(x - 3)$
19. $y - 1 = s\left(x + \frac{\pi}{2}\right)$ **20.** $y - \pi = f(x + 4)$

Let $y = c(x)$ be the function whose graph is given in Figure 2.21. Graph the curves indicated by the equations in Problems 21–30.

21. $y = 3c(x)$ **22.** $y = \frac{1}{2}c(x)$
23. $y = -c(x)$ **24.** $y = -2c(x)$
25. $y = c(2x)$ **26.** $y = c\left(\frac{1}{2}x\right)$
27. $y = c\left(x - \frac{\pi}{2}\right)$ **28.** $y + 1 = c\left(x - \frac{\pi}{2}\right)$
29. $y = 2c(x + 1)$ **30.** $y = \frac{3}{2}c(x - 2) + \sqrt{2}$

LEVEL 2

Graph the curves defined by the equations given in Problems 31–48. Do not plot points to graph these equations, but treat them as transformations of graphs from Table 2.1.

31. $y - 3 = (x - 2)^2$ **32.** $y = (x + 3)^2$
33. $y = x^2 - 1$ **34.** $y - 1 = |x - 7|$
35. $y = \sqrt{x} + 4$ **36.** $y = \sqrt{x + 4}$
37. $y = |x + \pi|$ **38.** $y = |x| - 6$
39. $y - \frac{\pi}{4} = (x + \pi)^2$ **40.** $y + 2 = \left(x + \sqrt{3}\right)^2$
41. $y + \pi = \sqrt{x - 3}$ **42.** $y = -3|x + 5|$
43. $y = 4\sqrt{x - 2}$ **44.** $y = -2(x + 4)^2$
45. $y + \sqrt{3} = \left| x - \sqrt{2} \right|$ **46.** $y - \sqrt{2} = \left(x + \sqrt{5}\right)^2$
47. $y - \sqrt{2} = \sqrt{x + 5}$ **48.** $y - 3 = -\frac{1}{2}(x + 2)^2$

LEVEL 3

Graph the equations in Problems 49–60. Do not graph these by plotting points.

49. $y = |(x - 4)^2 - 9|$ **50.** $y = |(x - 2)^2 - 4|$
51. $y = (|x| - 2)^2 - 4$ **52.** $y = (|x| - 4)^2 - 9$
53. $y + 9 = (x + 8)^2$, such that $-14 \leq x < -8$
54. $y - 2 = (x + 3)^2$, such that $-7 < x \leq -2$
55. $y + 12 = \left(x + \frac{25}{2}\right)^2$, such that $y > -10$
56. $y + 3 = (x + 3)^2$, such that $y < 6$
57. $y + 12 = (x - 8)^2$, such that $y < 4$
58. $y - 5 = 2|x - 1|$, such that $-4 \leq x \leq 2$
59. $y - 2 = \dfrac{1}{(x - 3)^2}$, such that $y < 5$
60. $y - 4 = \frac{1}{10}x^2$, such that $-2 \leq x \leq 2$

2.5 Piecewise Functions

Definition

There are many everyday examples of functions that cannot be defined in terms of a single equation. For example, suppose electricity is \$0.325/kwh for the first 1,000 kwh and then drops to \$0.088/kwh for usage between 1,000 and 3,000 kwh (including 3,000 kwh). A graph of this function is shown in Figure 2.22.

We see that the graph passes the vertical line test, so it is a function. To write an equation for the electric charges, C, we need to break up the domain into "pieces." If x is the kilowatt hours (kwh) and y is the price charged for the electric usage, then we find the domain: We see that $x \geq 0$. Two equations must be written, one for $0 \leq x \leq 1,000$ and another for $1,000 < x \leq 3,000$:

$$y = 0.325x \qquad \text{for } 0 \leq x \leq 1,000$$

$$y = 0.088x + 325 \qquad \text{for } 1,000 < x \leq 3,000$$

Such a function is called a *piecewise function* and is usually defined using a brace:

$$C(x) = \begin{cases} 0.325x & \text{if} \quad 0 \leq x \leq 1,000 \\ 0.088x + 325 & \text{if} \quad 1,000 < x \leq 3,000 \end{cases}$$

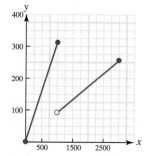

Figure 2.22 Electric charges function

> **PIECEWISE FUNCTION**
>
> A function whose domain D can be separated into a finite number of pieces such that the function has a different definition for each piece of the domain is called a **piecewise function.**

EXAMPLE 1   Graphing a piecewise function

Graph: $f(x) = \begin{cases} 3 - x & \text{if } -5 \leq x < 2 \\ x - 2 & \text{if } 2 \leq x \leq 5 \end{cases}$

Solution   The domain is $[-5, 5]$, and we graph the line segments for each of the separate parts:

Graph $y = 3 - x$ on $[-5, 2]$.
Graph $y = x - 2$ on $[2, 5]$.

The graph is shown in Figure 2.23. ∎

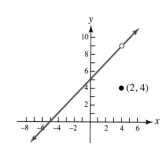

Figure 2.23 Graph of f

EXAMPLE 2   Example from calculus

Graph: $g(x) = \begin{cases} \dfrac{x^2 + 3x - 10}{x - 2} & \text{if } x \neq 2 \\ 4 & \text{if } x = 2 \end{cases}$

Solution   The first piece is a rational function with a deleted point. We see (if $x \neq 2$)

$$g(x) = \frac{x^2 + 3x - 10}{x - 2} = \frac{(x - 2)(x + 5)}{x - 2} = x + 5$$

The second piece is a single point, namely, $(2, 4)$. The graph is shown in Figure 2.24. ∎

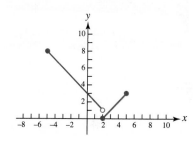

Figure 2.24 Graph of g

Consider an example very much like Example 2.

$$G(x) = \begin{cases} \dfrac{x^2 + 3x - 10}{x - 2} & \text{if } x \neq 2 \\ 7 & \text{if } x = 2 \end{cases}$$

By looking at Figure 2.24, we see that the point $(2, 7)$ for G "plugs the hole" in the deleted point from the first piece of the curve. This concept will be used in calculus when you study the topic of *continuity*.

Absolute Value Function

In Section 1.1, we defined absolute value as

$$|x| = \begin{cases} x & \text{if } x \geq 0 \\ -x & \text{if } x < 0 \end{cases}$$

The **absolute value function** is defined by $f(x) = |x|$. In Section 1.3, Example 1, we graphed $y = \left|\frac{1}{2}x - 2\right|$ by comparing it with a standard absolute value function. We now can treat the absolute value function as a piecewise function. That is, for $y = \left|\frac{1}{2}x - 2\right|$, we note that if $\frac{1}{2}x - 2 \geq 0$, then $x \geq 4$ graph

$$f(x) \begin{cases} \frac{1}{2}x - 2 & \text{if } x \geq 4 \\ -\left(\frac{1}{2}x - 2\right) & \text{if } x < 4 \end{cases}$$

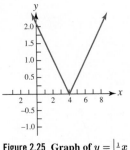

Figure 2.25 Graph of $y = \left|\frac{1}{2}x - 2\right|$

The graph is shown in Figure 2.25.

You might also notice that if we write

$$= \left|\frac{1}{2}x - 2\right| = \left|\frac{1}{2}(x - 4)\right| = \frac{1}{2}|x - 4|$$

we can consider the graph of y as a translation of $f(x) = \frac{1}{2}|x|$ where $(h, k) = (4, 0)$.

Greatest Integer Function

© CBS/Landov

The Price Is Right (1972–present) is one of several games dubbed *pricing games* where the contestants guess the price of an item, and the contestant coming the *closest without exceeding the true value* wins the item. This, basically, is the idea of another function, the *greatest integer function, or step function.* This rule provides a means by which to assign an integral value to the function.

Consider the following:

$$G(x) = \begin{cases} 0 & \text{if } 0 \leq x < 1 \\ 1 & \text{if } 1 \leq x < 2 \\ 2 & \text{if } 2 \leq x < 3 \\ 3 & \text{if } 3 \leq x < 4 \\ 4 & \text{if } 4 \leq x < 5 \end{cases}$$

Note that the domain for this function is $[0, 5)$ and the range is $\{0, 1, 2, 3, 4\}$. Here is the evaluation of some specific values:

$$G(3) = 3, \quad G\left(\frac{10}{3}\right) = 3, \quad G\left(\sqrt{10}\right) = 3, \quad G(\pi) = 3, \quad G(3,9999) = 3, \quad G(4) = 4$$

The graph is shown in Figure 2.26.

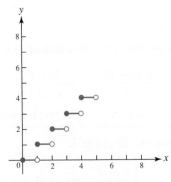

Figure 2.26 Graph of G

Notice that each part of the graph is simply a constant function and that the result looks somewhat like the steps in a stairway. As can be seen, the function is easy enough to understand, but if the domain were very large, it would be cumbersome to write, so the following notation is used.

GREATEST INTEGER FUNCTION

The **greatest integer function**, denoted by $f(x) = [\![x]\!]$, is defined by

$$[\![x]\!] = n \quad \text{if } n \le x < n+1$$

where n is an integer.

EXAMPLE 3 Evaluating a greatest integer function

Let $f(x) = [\![x]\!]$, $F(x) = [\![2x]\!]$, $g(x) = 2[\![x]\!]$, $G(x) = [\![x]\!]^2$, and $h(x) = [\![x^2]\!]$.
Evaluate these functions for $x = 5, 5.6, -5.6, \pi$, and $\frac{3}{4}$.

Solution   We arrange the answer in tabular form; make sure you see how to find each value listed.

x	$[\![x]\!]$	$[\![2x]\!]$	$2[\![x]\!]$	$[\![x]\!]^2$	$[\![x^2]\!]$
5	5	10	10	25	25
5.6	5	11	10	25	31
-5.6	-6	-12	-12	36	31
π	3	6	6	9	9
$\frac{3}{4}$	0	1	0	0	0

EXAMPLE 4 Graphing a greatest integer function

Graph: $f(x) = x + [\![x]\!]$ on $[-3, 3)$

Solution   Use the definition of the greatest integer function:

$$f(x) = \begin{cases} x - 3 & \text{if } -3 \le x < -2; \text{ that is } [\![x]\!] = -3 \text{ on } [-3, 2) \\ x - 2 & \text{if } -2 \le x < -1; \text{ that is } [\![x]\!] = -2 \text{ on } [-2, -1) \\ x - 1 & \text{if } -1 \le x < -0; \text{ that is } [\![x]\!] = -1 \text{ on } [-1, 0) \\ x & \text{if } 0 \le x < 1; \text{ that is } [\![x]\!] = 0 \text{ on } [0, 1) \\ x + 1 & \text{if } 1 \le x < 2; \text{ that is } [\![x]\!] = 1 \text{ on } [1, 2) \\ x + 2 & \text{if } 2 \le x < 3; \text{ that is } [\![x]\!] = 2 \text{ on } [2, 3) \end{cases}$$

Each of these linear functions is graphed for its respective domain, as shown by the solid line segments in Figure 2.27.

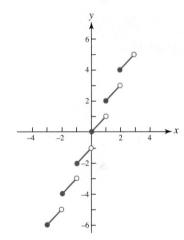

Figure 2.27 Graph of f

EXAMPLE 5 Rounding up instead of rounding down

The greatest integer function $y = [\![x]\!]$ "rounds down" because

$$y = 0 \quad \text{on } [0, 1)$$
$$y = 1 \quad \text{on } [1, 2)$$
$$y = 2 \quad \text{on } [2, 3)$$
$$\vdots$$

Write an expression using the greatest integer function that "rounds up":

$$y = 1 \quad \text{on } (0, 1]$$
$$y = 2 \quad \text{on } (1, 2]$$
$$y = 3 \quad \text{on } (2, 3]$$
$$\vdots$$

Solution   We begin by comparing the graphs of the "rounding down—greatest integer function"— and the "rounding up" function as shown in Figure 2.28.

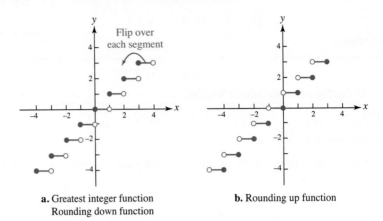

a. Greatest integer function
Rounding down function

b. Rounding up function

Figure 2.28 Comparison of rounding down and rounding up functions

We find that if we replace y by $-y$ and x by $-x$ in the equation $y = [\![x]\!]$, we obtain

$$-y = [\![-x]\!]$$

$$y = -[\![-x]\!]$$

In graphing this function, we find that it matches the graph shown in Figure 2.28b. ∎

The "rounding down" function is called the greatest integer function, and we give the name **ceiling function** to the "rounding up" function. Why would we want to consider a "rounding up" function? Most real-life situations require rounding up. Also, most calculations in business today are done by using computer programs, and the means of translating rounding problems for computer use is to use "rounding up" or "rounding down" functions.

EXAMPLE 6   Rental charges   MODELING APPLICATION

The Rental Store charge for a special drill is $10 for 4 hours or less usage. Additional charges are $3.00 for each additional hour or fraction thereof. Write a function that expresses the rental charge, and graph the function for a person who rents the drill for 10 hours or less.

Solution

Step 1: *Understand the problem.* "Let's see; if I rent the drill for 2 hours, my charge is $10. If I rent it for 5 hours, my charge is $10 + $3 = $13; if I rent the drill for 10 hours my charge is $10 + $3(6) = $28 because I have used 6 hours additional to my flat fee for 4 hours."

Step 2: *Devise a plan.* We will write a function to represent the charges, and we expect this function to be a step function. Then we will complete the problem by graphing the function.

Step 3: *Carry out the plan.* Let x be the number of hours the drill is rented, and let C be a function representing the cost (in dollars).

  On [0, 4], $C(x) = 10$

  On (4, ∞), the cost is dependent on the time.

   If x is an integer greater than 4, then the cost for these hours is $3 times $(x - 4)$. The reason we subtract 4 is that the first 4 hours are covered in the flat charge.

We consider

$$C(x) = \text{FLAT FEE} + \text{HOURLY FEE (in excess of 4 hours)}$$
$$= 10 + 3(x - 4)$$

For example, for $x = 5$, the charge is

$$C(5) = 10 + 3(5 - 4) = 13$$

If x is not an integer, then we must use a greatest integer function, and this is a rounding up function, so the desired function is

$$C(x) = 10 - 3[\![-(x - 4)]\!] \quad \text{From Example 5}$$
$$= 10 - 3[\![4 - x]\!]$$

For example, if $x = 4.5$ (4 hours 30 minutes) we have

$$C(4.5) = 10 - 3[\![4 - 4.5]\!]$$
$$= 10 - 3[\![-0.5]\!]$$
$$= 10 - 3(-1)$$
$$= 13$$

The desired function is a piecewise function

$$C(x) = \begin{cases} 10 & \text{if } 0 < x \le 4 \\ 10 - 3[\![4 - x]\!] & \text{if } 4 < x \le 10 \end{cases}$$

The graph of $y = C(x)$ is shown in Figure 2.29.

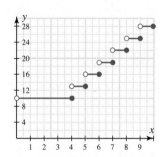

Figure 2.29 Rental charges

Step 4: *Look back.* We can try several examples to see whether the charges seem to be correct.

If I use the drill 4 hours or less, the cost is $10.

If I use the drill 8 hours 15 minutes, the cost is

$$C(8.25) = 10 - 3[\![4 - 8.25]\!]$$
$$= 10 - 3[\![-4.25]\!]$$
$$= 10 - 3(-5)$$
$$= 25$$

These amounts seem correct, and the function makes sense. ∎

PROBLEM SET 2.5

LEVEL 1

In Problems 1–16, find:

  a. $f(1)$ **b.** $f(5.3)$ **c.** $f(\pi)$
  d. $f\left(-\frac{1}{2}\right)$ **e.** $f(-5.3)$

  1. $f(x) = |x|$ **2.** $f(x) = -|x|$
  3. $f(x) = |x + 2|$ **4.** $f(x) = |x| - 2$
  5. $f(x) = |x| + x$ **6.** $f(x) = |x| - x$

  7. $f(x) = \begin{cases} -1 & \text{if } x < 0 \\ 0 & \text{if } x = 0 \\ 1 & \text{if } x > 0 \end{cases}$

  8. $f(x) = \begin{cases} 1 & \text{if } x < \pi \\ 2 & \text{if } x = \pi x > 0 \\ 3 & \text{if } x > \pi \end{cases}$

  9. $f(x) = \begin{cases} -x & \text{if } x < -3 \\ 2x & \text{if } -3 \le x \le 3 \\ x^2 & \text{if } x \ge 3 \end{cases}$

  10. $f(x) = \begin{cases} 10x & \text{if } x \le 0 \\ x^3 & \text{if } 0 < x < 3 \\ 0 & \text{if } x \ge 3 \end{cases}$

11. $f(x) = \begin{cases} x - 1 & \text{if } x \geq 2 \\ -x + 3 & \text{if } x < 2 \end{cases}$

12. $f(x) = \begin{cases} x - 1 & \text{if } x \geq -1 \\ -x - 3 & \text{if } x < -1 \end{cases}$

13. $f(x) = [\![x]\!]$ 14. $f(x) = [\![x]\!] + 1$
15. $f(x) = [\![x]\!] + x$ 16. $f(x) = 2[\![x]\!]$

WHAT IS WRONG, *if anything, with each statement in Problems 17–24?*
Explain your reasoning.

17. If $f(x) = |x|$, then f is a positive function.
18. If $f(x) = [\![x]\!]$, then f is a positive function.
19. If $f(x) = [\![x]\!]$, then x is an integer.
20. $[\![8]\!] = 8$ and $[\![8.1]\!] = 8$
21. $[\![-8]\!] = -8$ and $[\![-8.1]\!] = -8$
22. $|[\![x]\!]| = [\![|x|]\!]$
23. $|x|^2 = |x^2|$
24. $[\![x]\!]^2 = [\![x^2]\!]$

LEVEL 2

Graph the functions given in Problems 25–40.

25. $f(x) = \begin{cases} x - 2 & \text{if } x \geq 2 \\ 2 - x & \text{if } x < 2 \end{cases}$

26. $f(x) = \begin{cases} x + 1 & \text{if } x \geq -3 \\ -x - 3 & \text{if } x < -3 \end{cases}$

27. $g(x) = \begin{cases} -2 & \text{if } x \leq -2 \\ 0 & \text{if } -2 < x < 2 \\ 2 & \text{if } x \geq 2 \end{cases}$

28. $r(x) = \begin{cases} -4 & \text{if } x < 1 \\ 3 & \text{if } x = 1 \\ -2 & \text{if } x > 1 \end{cases}$

29. $f(x) = \begin{cases} x - 1 & \text{if } x \geq 2 \\ -x + 3 & \text{if } x < 2 \end{cases}$

30. $g(x) = \begin{cases} x - 1 & \text{if } x \geq -1 \\ -x - 3 & \text{if } x < -1 \end{cases}$

31. $f(x) = |x| + 2$
32. $g(x) = 3 - |x|$
33. $h(x) = |x + 3|$
34. $k(x) = |x - 2|$
35. $f(x) = |2x| - 3$
36. $G(x) = |3x| + 2$
37. $m(x) = |x|^2$

38. $n(x) = |x^2|$
39. $f(x) = |x - 2| + 1$
40. $g(x) = |x + 1| - 2$

PROBLEMS FROM CALCULUS *Problems 41–50 are found in calculus.*

41. If $f(x) = |x|$, find $\dfrac{f(x+h) - f(x)}{h}$, where $x \geq 0$, $h \geq 0$.

42. If $f(x) = |x|$, find $\dfrac{f(x+h) - f(x)}{h}$, where $x < 0$, $h < 0$.

43. If $f(x) = |x + 2|$, find $\dfrac{f(x+h) - f(x)}{h}$, where $x < -2$, $h < 0$.

44. If $f(x) = |x + 2|$, find $\dfrac{f(x+h) - f(x)}{h}$, where $x \geq -2$, $h \geq 0$.

45. A national fraternity allows one delegate for each 500 state members (or fraction thereof) of the fraternity. Suppose a state has n members. How many state representatives are allowed?

46. A salesperson receives a $500 bonus for each $10,000 worth of sales over an established base. What is the bonus for d dollars in sales over the base?

47. If the charges for a taxi are $2.50 plus 25¢ per each $\frac{1}{2}$ mile or fraction thereof, write a function that gives the cost of a taxi ride of x miles.

48. The charge for a certain telephone call is 75¢ for the first 3 minutes and 25¢ for each additional minute or fraction thereof. Write a function that gives the cost of a call lasting x minutes.

49. The telephone company charges $1.00 for the first 3 minutes for a certain call and 35¢ for each additional minute or fraction thereof. Write a function that gives the cost of a call lasting x minutes.

50. A measure of the disorder or randomness in a physical system is called *entropy* and is measured in calories per Kelvin. (Note: To convert from degree Celsius to Kelvin add 273.15.) An approximate model for the entropy of one mole of water under one atmosphere pressure is given by

$$S = \begin{cases} 10 + 0.04T & \text{if } T \leq 0 \\ 10 + 0.05T & \text{if } 0 < T < 100 \\ 12 + 0.03T & \text{if } T \geq 100 \end{cases}$$

where T is the temperature in degrees Celsius and S is in calories per Kelvin. Graph S.

LEVEL 3

Graph the functions given in Problems 51–60.

51. $b(x) = [\![x]\!] + 2$ 52. $d(x) = [\![x + 2]\!]$
53. $f(x) = 2[\![x]\!] + 1$ 54. $g(x) = [\![x]\!] - 2$
55. $h(x) = [\![x]\!] - x$ 56. $j(x) = [\![x]\!] + |x|$
57. $r(x) = |x - 2| + |x|$ 58. $s(x) = |x + 1| - |x|$
59. $F(x) = |x| - |x - 1|$ 60. $G(x) = |x| - |x + 2|$

2.6 Composition and Operations of Functions

There are many situations in which a quantity is given as a function of one variable that, in turn, can be written as a function of a second variable. This is known as *functional composition.*

Composition

Suppose, for example, that your job is to ship x packages of a product via Federal Express to a variety of addresses. Let x be the number of packages to ship, let f be the weight of the x objects, and let g be the cost of shipping. Then

The weight is a function of the number of objects: $f(x)$.

The cost is a function of the weight: $g[f(x)]$.

This process of evaluating a function of a function illustrates the idea of *composition of functions.*

> **COMPOSITION**
>
> The **composite function** $f \circ g$ is defined by
>
> $$(f \circ g)(x) = f[g(x)]$$
>
> for each x in the domain of g for which $g(x)$ is in the domain of f.

>> IN OTHER WORDS To visualize how functional composition works, think of $f \circ g$ in terms of an "assembly line" in which g and f are arranged in series, with output $g(x)$ becoming the input of f, as illustrated in Figure 2.30.

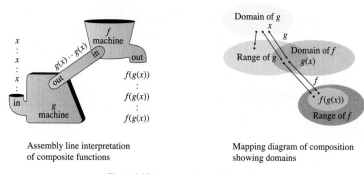

Assembly line interpretation of composite functions

Mapping diagram of composition showing domains

Figure 2.30 **Composition of functions**

EXAMPLE 1   Finding the composition of functions

If $f(x) = 3x + 5$ and $g(x) = \sqrt{x}$, find the composite functions $f \circ g$ and $g \circ f$.

Solution   The function $f \circ g$ is defined by $f[g(x)]$:

$$(f \circ g)(x) = f[g(x)] = f\left(\sqrt{x}\right) = 3\sqrt{x} + 5$$

The function $g \circ f$ is defined by $g[f(x)]$:

$$(g \circ f)(x) = g[f(x)] = g(3x + 5) = \sqrt{3x + 5}$$

∎

Example 1 illustrates that *functional composition is not commutative.* That is, $f \circ g$ is not, in general, the same as $g \circ f$.

Some attention must be paid to the domain of $g \circ f$. Let X be the domain of a function f, and let Y be the range of f. The situation can be viewed as shown in Figure 2.31.

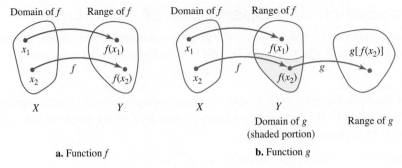

a. Function f **b.** Function g

Figure 2.31 **Two functions f and g viewed as mappings**

If part of f maps into the shaded portion of Y and part of f maps into the portion that is not shaded, as indicated in Figure 2.31**b**, then, as shown in Figure 2.32, the domain of $g \circ f$ is just the part of X that maps into the shaded portion of Y.

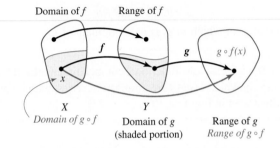

Figure 2.32 **Composition of two functions $g \circ f$. Notice that the domain of $g \circ f$ is the subset of X for which $g \circ f$ is defined (shaded portion of Y).**

EXAMPLE 2   Composition with focus on domains

Let $f = \{(0, 0), (-1, 1), (-2, 4), (-3, 9), (5, 25)\}$;
    $g = \{(0, -5), (-1, 0), (2, -3), (4, -2), (5, -1)\}$

a. What are the domains of f and g?

b. Find $g \circ f$ and state its domain.

c. Find $f \circ g$ and state its domain.

Solution

a. $D_f = \{0, -1, -2, -3, 5\}$; $D_g = \{0, -1, 2, 4, 5\}$

b. $\begin{matrix} f & & g & & g \circ f \end{matrix}$

$0 \to 0$ $\cdots$ $0 \to -5$ $0 \to -5$

$-1 \to 1$ $\cdots$ not defined, exclude -1 from the domain of $g \circ f$.

$-2 \to 4$ $\cdots$ $4 \to -2$ $-2 \to -2$

$-3 \to 9$ $\cdots$ not defined, exclude -3 from the domain of $g \circ f$.

$5 \to 25$ $\cdots$ not defined, exclude 5 from the domain of $g \circ f$.

Thus, $g \circ f = \{(0, -5), (-2, -2)\}$. The domain of the composite function $g \circ f$ is $\{0, -2\}$. Notice that -1, -3, and 5 are excluded from the domain of $g \circ f$ even though they are in the domain of f.

c.

g	f	$f \circ g$
$0 \to -5$ $\cdots$	not defined	
$-1 \to 0$ $\cdots$	$0 \to 0$	$-1 \to 0$
$2 \to -3$ $\cdots$	$-3 \to 9$	$2 \to 9$
$4 \to -2$ $\cdots$	$-2 \to 4$	$4 \to 4$
$5 \to -1$ $\cdots$	$-1 \to 1$	$5 \to 1$

Thus, $f \circ g = \{(-1, 0), (2, 9), (4, 4), (5, 1)\}$. The domain of the composite function $f \circ g$ is $\{-1, 2, 4, 5\}$.

EXAMPLE 3   An application of composite functions

Air pollution is a problem for many metropolitan areas.

Suppose that carbon monoxide is measured as a function of the number of people according to the information shown in Table 2.2.

TABLE 2.2 Carbon Monoxide Pollution	
Number of People	**Daily Level (ppm)**
100,000	1.41
200,000	1.83
300,000	2.43
400,000	3.05
500,000	3.72

Studies show that a refined formula for the average daily level of carbon monoxide in the air is

$$L(p) = 0.7\sqrt{p^2 + 3}$$

Further assume that the population of a given metropolitan area is growing according to the formula $p(t) = 1 + 0.02t^3$, where t is the time from now (in years) and p is the population (in hundred thousands). Based on these assumptions, what level of air pollution should be expected in four years?

Solution   The level of pollution is $L(p) = 0.7\sqrt{p^2 + 3}$, where $p(t) = 1 + 0.02t^3$. Thus, the pollution level at time t is given by the composite function

$$(L \circ p)(t) = L[p(t)] = L(1 + 0.02t^3) = 0.7\sqrt{(1 + 0.02t^3)^2 + 3}$$

In particular, when $t = 4$, we have

$$(L \circ p)(4) = 0.7\sqrt{[1 + 0.02(4)^3]^2 + 3} \approx 2.00 \text{ ppm}$$

In calculus, it is frequently necessary to express a function as the composite of two simpler functions.

EXAMPLE 4   Separating a function into two composite functions

Express each of the following functions as the composite of two functions u and g so that $f(x) = g[u(x)]$.

a. $f(x) = (x^2 + 2x)^2$

b. $f(x) = \sqrt{x^2 + 1}$

c. $f(x) = (x^2 + 5x + 1)^5$

d. $f(x) = \sqrt{5x^2 - x}$

☠ Consider Example 4 and the following paragraph carefully. ☠

Solution   We call f the *given function*, u the *inner function*, and g the *outer function*.

Given Function $f(x) = g[u(x)]$	Inner Function $u(x)$	Outer Function $g[u(x)]$
a. $f(x) = (x^2 + 2x)^2$	$u(x) = x^2 + 2x$	$g[u(x)] = [u(x)]^2$
b. $f(x) = \sqrt{x^2 + 1}$	$u(x) = x^2 + 1$	$g[u(x)] = \sqrt{u(x)}$
c. $f(x) = (x^2 + 5x + 1)^5$	$u(x) = x^2 + 5x + 1$	$g[u(x)] = [u(x)]^5$
d. $f(x) = \sqrt{5x^2 - x}$	$u(x) = 5x^2 - x$	$g[u(x)] = \sqrt{u(x)}$

∎

There are often other ways to express a composite function, but the most common procedure is to choose the function u to be the "inside" portion of the given function f. Notice that in parts **a** and **c** the "inside" portion is the portion inside the parentheses.

Operations with Functions

In algebra, you spend a great deal of time learning the algebra of real numbers. Operations with functions follows similar straightforward definitions.

FUNCTIONAL OPERATIONS

Let f and g be functions with domains D_f and D_g, respectively. Then, $f + g$, $f - g$, fg, and f/g are defined for the domain $D_f \cap D_g$:

$$(f + g)(x) = f(x) + g(x)$$

$$(f - g)(x) = f(x) - g(x)$$

$$(fg)(x) = f(x)g(x)$$

$$(f/g)(x) = \frac{f(x)}{g(x)} \qquad \text{provided } g(x) \neq 0$$

» IN OTHER WORDS Remember the operation symbols on the left of the equations, namely, $f + g$, $f - g$, fg, and f/g are operations on *functions* that are being defined; the operation on the right is defined for operations on *numbers*.

EXAMPLE 5   Operations with functions

If f and g are defined by $f(x) = x^2$ and $g(x) = x + 3$, find $f + g$, $f - g$, fg, and f/g. Evaluate these functions for $x = -1$ and $x = 5$, and state the domain for each.

Solution   The domain of both f and g is the set of real numbers, so the domains of $f + g$, $f - g$, and fg are also the set of real numbers. The domain of f/g is the intersection of the domains of f and g for which values causing $g(x) = 0$ are excluded. These are the usual domains for functions we consider in precalculus and calculus.

$$(f + g)(x) = f(x) + g(x) = x^2 + (x + 3) = x^2 + x + 3 \qquad D: (-\infty, \infty)$$
$$(f + g)(-1) = (-1)^2 + (-1) + 3 = 3$$
$$(f + g)(5) = 5^2 + 5 + 3 = 33$$

$$(f-g)(x) = f(x) - g(x) = x^2 - (x+3) = x^2 - x - 3 \qquad D:(-\infty, \infty)$$
$$(f-g)(-1) = (-1)^2 - (-1) - 3 = -1$$
$$(f-g)(5) = 5^2 - 5 - 3 = 17$$

$$(fg)(x) = f(x) \cdot g(x) = x^2(x+3) = x^3 + 3x^2 \qquad D:(-\infty, \infty)$$
$$(fg)(-1) = (-1)^3 + 3(-1)^2 = 2$$
$$(fg)(5) = 5^3 + 3(5)^2 = 200$$

$$(f/g)(x) = \frac{f(x)}{g(x)} = \frac{x^3}{x+3} \qquad D:(-\infty, -3) \cup (-3, \infty)$$
$$(f/g)(-1) = \frac{(-1)^2}{(-1)+3} = \frac{1}{2}$$
$$(f/g)(5) = \frac{5^2}{5+3} = 3.125$$

∎

Functional Iteration

In calculus, you will use a process in which the result of one step is used in the following step. This process, called **iteration**, is easy to describe in terms of composition. For example, if $f(x) = \sqrt{x}$, then $f(4) = \sqrt{4} = 2$. If we now take *this* answer and again evaluate f to find $f(2) = \sqrt{2}$, we have carried out an *iterative step*, which we could describe as $f(f(x))$ or as $f \circ f$.

Suppose we consider a problem from *The American Mathematical Monthly* (Vol. 92, Jan. 1985, pp. 3–23).* Define a function f, with domain the positive integers as follows:

$$f(x) = \begin{cases} 3x+1 & \text{if } x \text{ is odd} \\ x/2 & \text{if } x \text{ is even} \end{cases}$$

Let x_0 be some number in the domain of a function f. The **iterates** of x_0 are the numbers $f(x_0)$, $f(f(x_0)), f(f(f(x_0))), \ldots$. We can write this using composition: $f, f \circ f, f \circ f \circ f, \ldots$. The article asserts that for every positive integer, the iterates eventually return to 1.

EXAMPLE 6 Iteration conjecture   **MODELING APPLICATION**

Show that the iterates return to 1 for the case $x = 3$.

Solution

Step 1: *Understand the problem.* Let's see if we can understand what this problem is about. We calculate the first few values of f: (Remember the domain is the set of positive integers.)

$$f(1) = 3(1) + 1 = 4 \qquad f(2) = \frac{2}{2} = 1$$

$$f(3) = 3(3) + 1 = 10 \qquad f(4) = \frac{4}{2} = 2$$

$$\vdots \qquad\qquad\qquad \vdots$$

$$f(99) = 3(99) + 1 = 298 \qquad f(100) = \frac{100}{2} = 50$$

---

*"The $3x + 1$ Problem and Its Generalizations," by Jeffrey C. Lagarias.

Step 2: *Devise a plan.* Calculate the iterates for the smallest member of the domain.

The first iterate of $x_0 = 1$: $f(1) = 3(1) + 1 = 4$

the second iterate of $x_0 = 1$: $f(f(1)) = f(4) = \frac{4}{2} = 2$

the third iterate of $x_0 = 1$: $f(f(f(1))) = f(f(4)) = f(2) = 1$

We see for the first member, the iterates return to 1. The second number, $x = 2$, iterates to 1 in the first step (since 2 is even).

Step 3: *Carry out the plan.* Calculate the iterates for the number $x = 3$.

$f(3) = 10$; $f(10) = \frac{10}{2} = 5$; $f(5) = 3(5) + 1 = 16$; $f(16) = \frac{16}{2} = 8$;
$f(8) = \frac{8}{2} = 4$; $f(4) = \frac{4}{2} = 2$; $f(2) = \frac{2}{2} = 1$. Ah ha, we ended back at 1! Thus,
$f(f(f(f(f(f(f(3))))))) = 1$.

Step 4: *Look back.* The hypothesis that all values return to 1 is hardly proved. We showed it true for 1, 2, 3; you are asked to show it true for 4 and 5 in the problem set (Problem 55) and then asked to present an argument that it is always true.

PROBLEM SET 2.6

LEVEL 1

In Problems 1–4 find the indicated values where $f(x) = 3x - 2$ and $g(x) = 2x^2 + 1$.

1. **a.** $(f + g)(4)$ **b.** $(fg)(2)$
2. **a.** $(f - g)(3)$ **b.** $(f/g)(1)$
3. **a.** $(f \circ g)(2)$ **b.** $(g \circ f)(2)$
4. What is the domain of $(f + g)$, $(f - g)$, (fg), and (f/g)?

In Problems 5–8, find the indicated values, where $f(x) = \dfrac{x-2}{x+1}$ and $g(x) = x^2 - x - 2$.

5. **a.** $(f + g)(2)$ **b.** $(fg)(102)$
6. **a.** $(f - g)(5)$ **b.** $(f/g)(99)$
7. **a.** $(f \circ g)(1)$ **b.** $(g \circ f)(1)$
8. What is the domain of $(f + g)$, $(f - g)$, (fg), and (f/g)?

In Problems 9–12, find the indicated values, where $f(x) = \dfrac{2x^2 - x - 3}{x - 2}$ and $g(x) = x^2 - x - 2$.

9. **a.** $(f + g)(-2)$ **b.** $(f/g)(2)$
10. **a.** $(f - g)(2)$ **b.** $(f/g)(102)$
11. **a.** $(f \circ g)(0)$ **b.** $(g \circ f)(0)$
12. What is the domain of $(f + g)$, $(f - g)$, (fg), and (f/g)?

In Problems 13–16, find the indicated values, where

$$f = \{(0, 1), (1, 4), (2, 7), (3, 10)\}$$

and

$$g = \{(0, 3), (1, -1), (2, 1)(3, 3)\}$$

13. **a.** $(f + g)(1)$ **b.** $(fg)(2)$
14. **a.** $(f - g)(3)$ **b.** $(f/g)(0)$
15. **a.** $(f \circ g)(2)$ **b.** $(g \circ f)(2)$
16. What is the domain of $(f + g)$, $(f - g)$, (fg), and (f/g)?

LEVEL 2

WHAT IS WRONG, *if anything, with each statement in Problems 17–20? Explain your reasoning.*

17. If $f(x) = \sqrt[3]{x}$ and $u(x) = 2x^2 + 1$, then $f[u(x)] = 2\sqrt[3]{x} + 1$.
18. If $f(x) = \dfrac{1}{2x^2 + 1}$ and $u(x) = \sqrt{x - 1}$, then $f[u(x)] = \dfrac{1}{2x - 1}$.
19. If $f(x) = 3x^2 + 5x + 1$ and $u(x) = x^3$, then $f[u(x)] = (3x^2 + 5x + 1)^3$.
20. If $f(x) = \dfrac{x-1}{x+2}$ and $g(x) = \dfrac{2x+1}{x-3}$, then $(f \circ g)$ is the function

defined by $\left(\dfrac{x-1}{x+2} \right)\left(\dfrac{2x+1}{x-3} \right)$.

21. Given $f(x) = x^2 - 1$ and $(f - g)(x) = 2x + 1$. Find $g(x)$.
22. Given $f(x) = x^3 + 2$ and $(f + g)(x) = 4x - 6$. Find $g(x)$.
23. Given $f(x) = x^{-1}$ and $(fg)(x) = x$ $(x \neq 0)$. Find $g(x)$.
24. Given $f(x) = \dfrac{x-2}{x+3}$ and $(f/g)(x) = \dfrac{x+1}{x+2}$. Find $g(x)$.
25. Given $f(x) = \sqrt{x}$ and $(f \circ g)(x) = \sqrt{x^3 + 1}$. Find $g(x)$.
26. Given $f(x) = x^4$ and $(f \circ g)(x) = (2x + \sqrt{5})^4$. Find $g(x)$.

PROBLEMS FROM CALCULUS *In Problems 27–38, express f as a composition of two functions u and g so that $f(x) = g[u(x)]$.*

27. $f(x) = (x^2 + 1)^2$ 28. $f(x) = (x^2 - 1)^3$
29. $f(x) = (2x^2 - 1)^4$ 30. $f(x) = (x^2 + 4)^{3/2}$
31. $f(x) = (3x^2 + 4x - 5)^3$ 32. $f(x) = (2x^2 - x + 1)^2$
33. $f(x) = \sqrt{5x - 1}$ 34. $f(x) = \sqrt{x^2 - 1}$
35. $f(x) = \sqrt[3]{x^2 - 4}$ 36. $f(x) = \sqrt[4]{x^3 - x + 1}$
37. $f(x) = (x^2 - 1)^3 + \sqrt{x^2 - 1} + 5$
38. $f(x) = |x + 1|^2 + 6$

In Problems 39–42, find the sum, difference, product, and quotient of the given functions. Also state the domain for each.

39. $f(x) = 2x - 3$ and $g(x) = x^2 + 1$

40. $f(x) = \dfrac{x-2}{x+1}$ and $g(x) = x^2 - x - 2$

41. $f(x) = \dfrac{2x^2 - x - 3}{x-2}$ and $g(x) = x^2 - x - 2$

42. $f(x) = 4x + 2$ and $g(x) = x^3 + 3$

In Problems 43–46, find $f \circ g$ and $g \circ f$ for the given functions.

43. $f(x) = 2x - 3$ and $g(x) = x^2 + 1$

44. $f(x) = \dfrac{x-2}{x+1}$ and $g(x) = x^2 - x - 2$

45. $f(x) = \dfrac{2x^2 - x - 3}{x+1}$ and $g(x) = x^2 - x - 2$

46. $f(x) = 4x + 2$ and $g(x) = x^3 + 3$

47. If $f(x) = x^2$, $g(x) = 2x - 1$, and $h(x) = 3x + 2$, find:
   a. $(f \circ g) \circ h$ **b.** $f \circ (g \circ h)$

48. If $f(x) = x^2$, $g(x) = 3x - 2$, and $h(x) = x^2 + 1$, find:
   a. $(f \circ g) \circ h$ **b.** $f \circ (g \circ h)$

49. If $f(x) = \sqrt{x}$, $g(x) = x^2$, and $h(x) = x + 2$ all within domain $(0, \infty)$, find
   a. $(f \circ g) \circ h$ **b.** $f \circ (g \circ h)$

50. If $f(x) = \sqrt{-x}$, $g(x) = x^2$, and $h(x) = x$ all with domain $(-\infty, 0)$, find:
   a. $(f \circ g) \circ h$ **b.** $f \circ (g \circ h)$

51. Consider the volume of a particular cone as a function of its height by the formula

$$V(h) = \frac{\pi h^3}{12}$$

Suppose the height is expressed as a function of time by letting $h(t) = 2t$.
   a. Find the volume for $t = 2$.
   b. Express the volume as a function of time by finding $V \circ h$.
   c. If the domain of V is $(0, 6]$, find the domain of h; that is, what are the permissible values for t?

52. The surface area of a spherical balloon is given by $S(r) = 4\pi r^2$. Suppose the radius is expressed as a function of time by $r(t) = 3t$.
   a. Find the surface area for $t = 2$.
   b. Express the surface area as a function of time by finding $S \circ r$.
   c. If the domain of S is $(0, 8)$, find the domain of r; that is, what are the permissible values for t?

53. If $f(x) = x^2$, then

$$f\left(\frac{1}{x}\right) = \left(\frac{1}{x}\right)^2 = \frac{1}{x^2} = \frac{1}{f(x)}$$

Give an example of a function for which

$$f\left(\frac{1}{x}\right) \neq \frac{1}{f(x)}$$

54. If $f(x) = x$, then $f(x^2) = [f(x)]^2$. Give an example of a function for which

$$f(x^2) \neq [f(x)]^2$$

55. In Example 6, an unproved conjecture was checked for $x = 1$, 2, and 3. Check this conjecture for $x = 4$ and $x = 5$. Make a statement about values for which x is even.

56. PROBLEM FROM CALCULUS* A buffalo herd at Yellowstone has population P, which is a function of the amount of grass cover available for the herd.

Grass yield is estimated in lb/ft² by weighing the harvest from a 20 ft² test plot. The number of grasshoppers affects this yield, Y. An approximation for Y is given by

$$g(N) = 1 - \frac{1}{20} N\left(\frac{1}{20} N - 1\right)$$

where N is the number of grasshoppers contained on the test plot on July 1. An approximation for the population of the buffalo herd on October 1 is

$$P = f(Y) = 1{,}000\sqrt{Y^2 + 3Y + 1}$$

Find a formula for P as a function of N.

LEVEL 3

57. Let $f(x) = 1 + \dfrac{1}{x}$, find:
   a. $(f \circ f)(x)$ **b.** $(f \circ f \circ f)(x)$
   c. $(f \circ f \circ f \circ f)(x)$
   d. Can you predict the output for further iterations of compositions?

58. Choose *any* positive x. Find a numerical value for $(f \circ f)(x)$, $(f \circ f \circ f)(x)$, and $(f \circ f \circ f \circ f)(x)$. If you continue this iterative procedure, predict the outcome for any x for each of the given functions.
   a. $f(x) = \sqrt{x}$ **b.** $f(x) = 2\sqrt{x}$
   c. $f(x) = 3\sqrt{x}$ **d.** $f(x) = k\sqrt{x}$

59. PROBLEMS FROM CALCULUS A process in calculus called the *Newton-Raphson method* uses an iterative process for approximating roots of a given equation. A leading calculus book shows that the solution to the equation $x^2 = 5$ (which we know has a solution $x = \sqrt{5}$) can be found iteratively by evaluating the function

$$f(x) = \frac{x^2 + 5}{2x}$$

Compute the iterates of this function for $x_0 = 2$ and compare with a calculator approximation of $\sqrt{5}$.

60. Journal Problem From *School Science and Mathematics*, Vol. 83, No. 1, Jan. 1983. Let $f(x) = \dfrac{(x^2 + 1)^2}{2x^2}$ and $g(t) = \sqrt{t} \pm \sqrt{t - 1}$, where t is a positive integer. Find $f[g(t)]$.

* From *Calculus*, James F. Hurley. Belmont, CA: Wadsworth, Inc., 1987, p. 61

2.7 Inverse Functions

The Idea of an Inverse

In mathematics, the ideas of "opposite operations" and "inverse properties" are very important. The basic notion of an opposite operation or an inverse property is to "undo" a previously performed operation. For example, pick a number, and call it x; then:

	x	*Think:* I pick 8.
Add 5:	$x + 5$	*Think:* Now, I have $8 + 5 = 13$.
	The next operation returns you to x:	*Think:* I want to find an operation to get back to my original number
Subtract 5:	$x + 5 - 5 = x$	*Think:* $13 - 5 = 8$, my original number.

We now want to apply this idea to functions. Pick a number in the domain of a function f; call this number a:

	a	*Think:* I'll pick 4 this time.

Now, evaluate f for the number you picked; suppose we let f be defined by $f(x) = 2x + 7$:

Evaluate f:	$f(a) = 2a + 7$	*Think:* $f(4) + 7 = 15$.
	The next operation returns us back to x:	*Think:* I want to find a function, called the *inverse function*, denoted by f^{-1} if it exists, so that $f^{-1}(a) = a$.
Let $f^{-1}(x) = \dfrac{1}{2}(x - 7)$		*Think:* Where did this come from? This is the topic of this section!

☠ The symbol f^{-1} means the inverse of f and does not mean $\frac{1}{f}$. ☠

Evaluate f^{-1}:	$f^{-1}(2a + 7) = a$	*Think:* $f^{-1}(2a + 7) = \dfrac{1}{2}(2a + 7 - 7)$
		$\qquad = \dfrac{1}{2}(2a)$
		$\qquad = a$

Of course, for f^{-1} to be an inverse function, it must "undo" the effect of f for *each and every member* of the domain. This may be impossible if f is a function such that two x values give the same y value. For example, if $g(x) = x^2$, then $g(2) = 4$ and $g(-2) = 4$, so we cannot find a function g^{-1} such that $g^{-1}(4)$ equals *both* 2 and -2 because that would violate the very definition of a function. We see it is necessary to limit the given function so that it is *one-to-one*. Recall the *horizontal line test* from Section 2.2 to determine whether a function is one-to-one.

Inverse Functions

For a given function f, we write $b = f(a)$ to indicate that f maps the number a in its domain into the corresponding number b in the range. If f has an inverse f^{-1}, it is the function that reverses the *inverse* of f and does not affect f in the sense that

$$f^{-1}(b) = a$$

This means that

$$(f^{-1} \circ f)(a) = a$$

Furthermore, for every b in the domain of f^{-1},

$$(f \circ f^{-1})(b) = b$$

Let f be a function with domain D and range R. Then the function f^{-1} with domain R and range D is the **inverse of f** if

$$f^{-1}[f(a)] = a \quad \text{for all } a \text{ in } D$$

$$f[f^{-1}(b)] = b \quad \text{for all } b \text{ in } R$$

» IN OTHER WORDS Start with some x value in the domain of a function f. You can think of this in terms of a function machine.

Input x – value
10,000

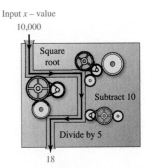

18

Output $f(x)$ from f machine is now input into another g machine

18

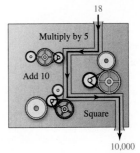

10,000
The output is x.

This machine is an inverse machine if it "undoes" the effect of the f machine; that is, it allows us to 'get back' to the answer x.

EXAMPLE 1   Showing that two given functions are inverses

Show that f and g defined by $f(x) = 5x + 4$ and $g(x) = \dfrac{x - 4}{5}$ are inverse functions.

Solution   We must show that f and g are inverse functions in two parts:

$$(g \circ f)(a) = g(5a + 4) \qquad \text{and} \qquad (f \circ g)(b) = f\left(\frac{b - 4}{5}\right)$$

$$= \frac{(5a + 4) - 4}{5} \qquad\qquad\qquad = 5\left(\frac{b - 4}{5}\right) + 4$$

$$= \frac{5a}{5} \qquad\qquad\qquad\qquad = (b - 4) + 4$$

$$= a \qquad\qquad\qquad\qquad\qquad = b$$

Thus, $(g \circ f)(a) = a$ and $(f \circ g)(b) = b$, so f and g are inverse functions. ∎

Once you are certain that a function g is the inverse of a function f, you can denote it by f^{-1}. This relationship is shown in Figure 2.33. Note that the range of f is the domain of f^{-1}.

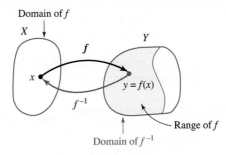

Figure 2.33 Inverse functions

Even though the definition of inverse functions shows us how to check to see whether two functions are inverses, it does not tell us how to *find* the inverse of a given function. To find the inverse, it is helpful to visualize a function as a set of ordered pairs. Suppose we pick a number, say 4, and evaluate a function f at 4 to find $f(4) = 15$. Then $(4, 15)$ is an element of f. Now the inverse function f^{-1} requires 15 to be changed back into 4; that is, $f^{-1}(15) = 4$ so that $(15, 4)$ is an element of f^{-1}. Thus, *if a function f has an element (a, b), then the inverse function f^{-1} must have element (b, a)*.

EXAMPLE 2   Inverse of a given function defined as a set of ordered pairs

Let $f = \{(0, 3), (1, 5), (3, 9), (5, 13)\}$; find f^{-1}, if it exists.

Solution   The inverse simply reverses the ordered pairs:

$$f^{-1} = \{(3, 0), (5, 1), (9, 3), (13, 5)\}$$

The inverse of a function may not exist. For example,

$$f = \{(0, 0), (1, 1), (-1, 1), (2, 4), (-2, 4)\}$$

and

$$g(x) = x^2$$

do not have inverses because if we attempt to find the inverses, we obtain relations that are not functions. In the first case, we find

Possible inverse of f: $\{(0, 0), (1, 1), (1, -1), (4, 2), (4, -2)\}$

This is not a function because not every member of the domain is associated with a single member in the range: $(1, 1)$ and $(1, -1)$, for example.

In the second case, if we interchange the x and y in the equation for the function g where $y = x^2$ and then solve for y, we find:

$$x = y^2 \quad \text{or} \quad y = \pm\sqrt{x} \text{ for } x \geq 0$$

But this is not a function of x, because for any positive value of x, there are two corresponding values of y, namely, $\sqrt{x}$ and $-\sqrt{x}$. These examples show why we impose the one-to-one condition.

Let us summarize the procedure for finding an inverse as illustrated in Example 3.

FINDING AN INVERSE

The procedure for finding an equation for a given inverse function, f.

Step 1   Let $y = f(x)$ be a given *one-to-one* function.

Step 2   Replace all x's and all y's (that is, interchange x and y) in the given equation.

Step 3   Solve for y. The resulting function defined by the equation $y = f^{-1}(x)$ is the inverse of f.

The domain of f and the range of f^{-1} must be equal as well as the domain of f^{-1} and the range of f. This property is evident in the following example (part **c**).

EXAMPLE 3   Finding the inverse using functional notation

a. If $s(x) = x^3 + 3$, find s^{-1}, if it exists.

b. If $u(x) = x^2$, find u^{-1}, if it exists.

c. If $t(x) = x^2$ on $(-\infty, 0]$ find t^{-1}, if it exists.

Solution

a. Note that s is a one-to-one function and thus will have an inverse.

$y = x^3 + 3$	Step 1: Given function.
$x = y^3 + 3$	Step 2: Interchange x and y.
$y^3 = x - 3$	Step 3: Solve for y.
$y = (x - 3)^{1/3}$	Thus, $s^{-1}(x) = (x - 3)^{1/3}$

b. Since u is not a one-to-one function, we say it has no inverse function.

c. With the given restriction on the domain, t is a one-to-one function, so the inverse exists.

$$y = x^2 \text{ for } -\infty < x \le 0 \qquad \text{Step 1: Given function}$$

$$x = y^2 \text{ for } -\infty < y \le 0 \qquad \text{Step 2: Interchange } x \text{ and } y.$$

$$y = -\sqrt{x} \qquad \text{Step 3: Solve for } y.$$

Thus, $t^{-1}(x) = -\sqrt{x}$

Note that y is negative, and $\sqrt{x}$ is positive, so the opposite is necessary. The implied domain here is $(0, \infty]$.

Graph of f^{-1}

The graphs of f and its inverse f^{-1} are closely related. In particular, if (a, b) is a point on the graph of f, then $b = f(a)$ and $a = f^{-1}(b)$, so (b, a) is on the graph of f^{-1}. It can be shown that (a, b) and (b, a) are reflections of one another in the line $y = x$. (See Figure 2.34.)

These observations yield the following procedure for sketching the graph of an inverse function.

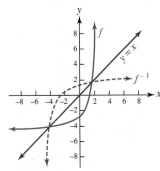

Figure 2.34 The graphs of f and f^{-1} are reflections in the line $y = x$

| INVERSE GRAPHS |

If f^{-1} exists, its graph may be obtained by reflecting the graph of f in the line $y = x$.

EXAMPLE 4   Graphing an inverse

Show that $t(x) = x^2$ on $(-\infty, 0]$ and $t^{-1}(x) = -\sqrt{x}$ (inverse functions from Example 3c) are symmetric with respect to the line $y = x$ by graphing each.

Solution   The graphs are shown in Figure 2.35.

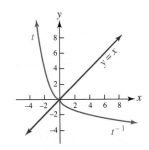

Figure 2.35 Graphs of t, t^{-1}, and $y = x$

EXAMPLE 5   Function and inverses from graph

Consider the function f defined by the graph in Figure 2.36. **a.** Find $f(5)$ **b.** Find $f^{-1}(6)$.

Solution   Use the graph as shown in Figure 2.36.

This is a member of the domain of f; locate this on the x-axis.
↓
a. $f(5) = 3$
↑
This is found by following the dots in Figure 2.36.

This is a member of the domain of f^{-1}; locate this on the y-axis.
↓
b. $f^{-1}(6) = 9$
↑
This is found by following the dashes in Figure 2.36.

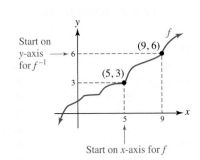

Figure 2.36 Graph of f

PROBLEM SET 2.7

LEVEL 1

WHAT IS WRONG, *if anything, with each statement in Problems 1–4?*
Explain your reasoning.

1. If $f(x) = x + 1$, then $f^{-1}(x) = \dfrac{1}{x+1}$.

2. **a.** If $f(3) = 5$, then $f^{-1}(5) = 3$.
   b. If $f(5) = 3$, then $f^{-1}(5) = \frac{1}{3}$.

3. Every function has an inverse because $f[f^{-1}(x)] = x$ for all x.

4. If $f^{-1}[f(a)] = a$, then f and f^{-1} are inverse functions.

Determine which pairs of functions defined by the equations in Problems 5–6 are inverses.

5. **a.** $f(x) = 3x$; $g(x) = \frac{1}{3}x$

   b. $f(x) = 5x + 3$; $g(x) = \dfrac{x-3}{5}$

   c. $f(x) = \dfrac{1}{x}$; $g(x) = \dfrac{1}{x}$

   d. $f(x) = x^2, x < 0$; $g(x) = \sqrt{x}, x > 0$

6. **a.** $f(x) = -5x$; $g(x) = \frac{1}{5}x$

   b. $f(x) = \frac{2}{3}x + 2$; $g(x) = \frac{3}{2}x + 3$

   c. $f(x) = \dfrac{1}{x+1}$; $g(x) = \dfrac{1}{x-1}$

   d. $f(x) = x^2, x \geq 0$; $g(x) = \sqrt{x}, x \geq 0$

Find the inverse function, if it exists, of each function given in Problems 7–12.

7. **a.** $f = \{(4,5),(6,3),(7,1),(2,4)\}$
   b. $f(x) = x + 3$
   c. $f(x) = 5x$
   d. $f(x) = 5$

8. **a.** $g = \{(1,4),(6,1),(4,5),(3,4)\}$
   b. $g(x) = 2x + 3$
   c. $g(x) = \frac{1}{5}x$ **d.** $g(x) = -3$

9. **a.** $f(x) = x^2 - 4$ **b.** $f(x) = \dfrac{1}{x-3}$

10. **a.** $g(x) = \sqrt{x} + 4$ **b.** $g(x) = \dfrac{2x+1}{x}$

11. $f(x) = \dfrac{2x-6}{3x+3}$ 12. $g(x) = \dfrac{3x+1}{2x-3}$

LEVEL 2

If s and c are defined by the graphs in Figure 2.37, find the values requested in Problems 13–16.

13. **a.** $s(2)$ **b.** $s(-2)$
    c. $s(0)$ **d.** $s(6)$
14. **a.** $s^{-1}(2)$ **b.** $s^{-1}(-2)$
    c. $s^{-1}(0)$ **d.** $s^{-1}(1)$
15. **a.** $c(0)$ **b.** $c^{-1}(1)$
    c. $c^{-1}(0)$ **d.** $c(8)$
16. **a.** $c^{-1}(-1)$ **b.** $c(4)$
    c. $c(12)$ **d.** $c^{-1}(-2)$

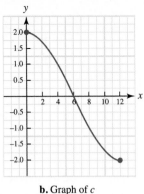

a. Graph of s **b.** Graph of c

Figure 2.37 Graphs for Problems 13–16

The functions s and c in Problems 17–24 are defined by the graphs in Figure 2.37.

17. What are the domain and range of the function s?
18. What are the domain and range of the function c?
19. Graph $y = s(x + 2)$.   20. Graph $y = c(x - 2)$.
21. Graph $y = s(x) + 2$.   22. Graph $y = c(x) - 2$.
23. Graph $y = s^{-1}(x)$.   24. Graph $y = c^{-1}(x)$.

In problems 25–26, use the function defined in Figure 2.38 to graph the requested functions.

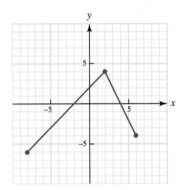

Figure 2.38 Graph of f

25. **a.** $y = f^{-1}(x)$ **b.** $y = f^{-1}(-x)$
26. **a.** $y = f^{-1}(x - 2)$ **b.** $y = f^{-1}(x) - 2$

For each of the Problems 27–38, graph the defined function and its inverse on the same coordinate axes, and use the graphs to decide whether the functions are inverses.

27. $f(x) = x + 2$; $f^{-1}(x) = x - 2$

28. $f(x) = 5x - 2$; $f^{-1}(x) = \dfrac{x+2}{5}$

29. $f(x) = x^2, x \geq 0$; $f^{-1}(x) = \sqrt{x}$
30. $f(x) = x^2, x \leq 0$; $f^{-1}(x) = -\sqrt{x}$
31. $f(x) = x^2 - 2, x \geq 0$; $f^{-1}(x) = \sqrt{x+2}$
32. $f(x) = |x-1|, x \geq 1$; $f^{-1}(x) = |x+1|, x \geq 0$
33. $f(x) = \frac{1}{4}x - 2$
34. $f(x) = \frac{1}{3}x + 1$

35. $f(x) = \frac{1}{3}x - \frac{5}{3}$

36. $f(x) = -\frac{1}{4}x + \frac{3}{4}$

37. $f(x) = 4x - 2$

38. $f(x) = 3x + 6$

39. If $f(x) = x^3 + 5x + 3$, find:

   a. $f[f^{-1}(5)]$ **b.** $f^{-1}[f(-2)]$

   c. $(f \circ f^{-1})(\pi)$ **d.** $(f^{-1} \circ f)\left(\sqrt{3}\right)$

40. If $f(x) = 5x + 1$

   a. Find $f^{-1}(x)$.   **b.** Find $f^{-1}(3)$.

   c. Find $\dfrac{1}{f(3)}$.   **d.** Does $f^{-1}(3) = \dfrac{1}{f(3)}$?

41. If $f(x) = x^4 - 3x^2 + 6$, find $f^{-1}(6)$.

42. If $F(x) = \dfrac{ax + b}{cx + d}$, find:

   a. $F^{-1}(x)$ **b.** $F[F^{-1}(x)]$

   c. $F[F(0)]$ **d.** $F^{-1}[F(0)]$

43. Show that $f(x) = \dfrac{5x + 3}{8x - 5}$ is its own inverse.

44. The function

$$C(F) = \frac{5}{9}(F - 32)$$

gives the temperature in Celsius when the Fahrenheit temperature (F) is known. Find the inverse function and give a verbal description.

45. The function

$$C(x) = 2.54x$$

gives the approximate number of centimeters when the length is x inches. Find the inverse function and give a verbal description.

46. Let $f(x) = \begin{cases} 2x - 3, x \le 1 \\ x - 2, x > 1 \end{cases}$

Sketch the graphs of f and f^{-1}.

47. Let $f(x) = \begin{cases} \frac{1}{2}x - 4, x \le 1 \\ x - 2, x > 1 \end{cases}$

Sketch the graphs of f and f^{-1}.

48. Let $f(x) = 10^x$.

   a. Graph f, f^{-1}, and the line $y = x$ on the same coordinate axes.

   b. Show that $f(x + y) = f(x)f(y)$.

49. Let $f(x) = 2^x$.

   a. Graph f, f^{-1}, and the line $y = x$ on the same coordinate axes.

   b. Show that $f(x - y) = \dfrac{f(x)}{f(y)}$.

50. Find the coordinates of A, B, C, and D in Figure 2.39.

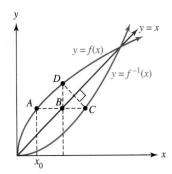

Figure 2.39 Graphs of f and f^{-1}

LEVEL 3

Determine which pairs of functions defined by the equations in Problems 51–56 are inverses.

51. $f(x) = 2x^2 + 1$, $x \ge 0$; $g(x) = \frac{1}{2}\sqrt{2x - 2}$, $x \ge 1$

52. $f(x) = 2x^2 + 1$, $x \le 0$; $g(x) = -\frac{1}{2}\sqrt{2x - 2}$, $x \ge 1$

53. $f(x) = (x + 1)^2$, $x \ge 1$; $g(x) = -1 - \sqrt{x}$, $x \ge 0$

54. $f(x) = (x + 1)^2$, $x \ge -1$; $g(x) = -1 + \sqrt{x}$, $x \ge 0$

55. $f(x) = |x + 1|$, $x \ge -1$; $g(x) = |x - 1|$, $x \ge 1$

56. $f(x) = |x + 1|$, $x \ge -1$; $g(x) = x - 1$, $x \ge 0$

Find the inverse for each function defined by the equations in Problems 57–59.

57. a. $f(x) = x^2$ on $[0, \infty)$ **b.** $f(x) = x^2$ on $(-\infty, 0]$

58. a. $f(x) = x^2 + 1$ on $(-\infty, 0]$ **b.** $f(x) = x^2 + 1$ on $[0, \infty)$

59. a. $f(x) = 2x^2$ on $[2, 10]$ **b.** $f(x) = 2x^2$ on $[-10, -1]$

60. If $f(x) = \dfrac{x + 1}{x - 3}$, find:

   a. $f^{-1}(x)$ **b.** $f(x^{-1})$ **c.** $[f(x)]^{-1}$

2.8 Limits and Continuity

Two essential ideas in understanding the nature of functions are limits and continuity. You will investigate these ideas at length in calculus, but because of their importance, we introduce them in this section as a foretaste of things to come.

Intuitive Notion of a Limit

The limit of a function f is a tool for investigating the behavior of $f(x)$ as x gets closer and closer to a particular number c. To visualize this concept, we return to the falling object example of Section 2.2.

EXAMPLE 1   Velocity as a limit   **MODELING APPLICATION**

A freely falling body experiencing no air resistance falls $s(t) = 16t^2$ feet in t seconds. Express the body's velocity at time $t = 2$ as a limit.

Solution

Step 1: *Understand the problem.* We know how to find the average velocity over a period of time, but here we need to define some sort of "mathematical speedometer" for measuring the *instantaneous velocity* of the body at time $t = 2$.

Step 2: *Devise a plan.* We first compute the *average velocity* $\overline{v}(t)$ of the body between time $t = 2$ and any other time t by the formula $\text{AVERAGE VELOCITY} = \dfrac{\text{DISTANCE TRAVELED}}{\text{ELAPSED TIME}}$.

Step 3: *Carry out the plan.*

$$\overline{v}(t) = \frac{\text{DISTANCE TRAVELED}}{\text{ELAPSED TIME}}$$

$$= \frac{s(t) - s(2)}{t - 2}$$

$$= \frac{16t^2 - 16(2)^2}{t - 2}$$

$$= \frac{16t^2 - 64}{t - 2}$$

As t gets closer and closer to 2, it is reasonable to expect the average velocity $\overline{v}(t)$ to approach the value of the required instantaneous velocity at time $t = 2$.

$$\lim_{t \to 2} \overline{v}(t) = \underbrace{\lim_{t \to 2} \frac{16t^2 - 64}{t - 2}}$$

This is the instantaneous velocity at $t = 2$.

Step 4: *Look back.* Notice that we cannot find the instantaneous velocity at time $t = 2$ by simply substituting $t = 2$ into the average velocity formula because this would yield the meaningless form $0/0$. ■

We now devote the remainder of this section to an intuitive introduction of how we can find the value of limits such as the one that appears in Example 1.

LIMIT OF A FUNCTION (informal definition)

The notation

$$\lim_{x \to c} f(x) = L$$

is read "the **limit** of $f(x)$ as x approaches c is L" and means that the functional value $f(x)$ can be made arbitrarily close to L by choosing x sufficiently close to c (but not equal to c).

> **» IN OTHER WORDS** If $f(x)$ becomes arbitrarily close to a single number L as x approaches c from either side, then we say that L is the limit of $f(x)$ as x approaches c.

Limits by Graphing or by Table

Figure 2.40 shows the graph of a function f and the number $c = 3$.

The arrowheads are used to illustrate possible sequences of numbers along the x-axis, approaching from both the left and the right. As x approaches $c = 3$, $f(x)$ gets closer and closer to 5. We write this as

$$\lim_{x \to 3} f(x) = 5$$

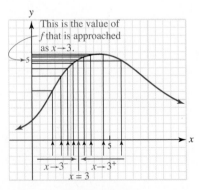

Figure 2.40 Limit as $x \to c$

As x approaches 3 from the left, we write $x \to 3^-$, and as x approaches 3 from the right, we write $x \to 3^+$. We say that the limit at $x = 3$ exists only if the value approached from the left is the same as the value approached from the right.

STOP *Pay attention to this notation.*

EXAMPLE 2   Estimating limits by graphing

Given the functions f, g, and h defined by the graphs in Figure 2.41, find the following limits by inspection, if they exist: **a.** $\lim\limits_{x \to 0} f(x)$ **b.** $\lim\limits_{x \to 1} g(x)$ **c.** $\lim\limits_{x \to 1} h(x)$

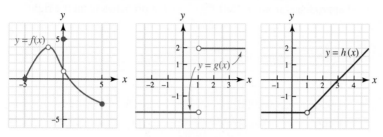

Figure 2.41 Limits from a graph

Solution

a. Take a good look at the given graph; notice the open circles on the graph at $x = 0$ and $x = -2$; also notice that $f(0) = 5$. To find $\lim\limits_{x \to 0} f(x)$, we need to look at both the left-hand and right-hand limits. Look at Figure 2.41 (left graph) to find

$$\lim_{x \to 0^-} f(x) = 1 \qquad \text{and} \qquad \lim_{x \to 0^+} f(x) = 1$$

so $\lim\limits_{x \to 0} f(x)$ exists and $\lim\limits_{x \to 0} f(x) = 1$. ☠ Notice here that the value of the limit as x approaches 0 is not the same as the value of the function at $x = 0$. ☠

b. Look at the center graph in Figure 2.41 to find

$$\lim_{x \to 1^-} g(x) = -2 \qquad \text{and} \qquad \lim_{x \to 1^+} g(x) = 2$$

so the limit as x approaches 1 does not exist.

c. Look at the graph at the right to find

$$\lim_{x \to 1^-} h(x) = -2 \qquad \text{and} \qquad \lim_{x \to 1^+} h(x) = -2$$

so $\lim\limits_{x \to 1} h(x) = -2$.

■

EXAMPLE 3   Velocity limit

In Example 1, we found the velocity of a falling object at time $t = 2$ as a limit:

$$\lim_{t \to 2} \frac{16t^2 - 64}{t - 2}$$

Find this limit using the graphical approach and the numerical approach (tabular).

Solution   *Graphical approach*

$$\overline{v}(t) = \frac{16t^2 - 64}{t - 2} = \frac{16(t^2 - 4)}{t - 2} = \frac{16(t - 2)(t + 2)}{t - 2} = 16(t + 2) \qquad t \neq 2$$

The graph of $\overline{v}(t)$ is a line with a deleted point, as shown in Figure 2.42.

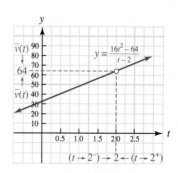

Figure 2.42 Graph of $\overline{v}(t)$

The limit can now be seen:

$$\lim_{t \to 2} \overline{v}(t) = 64$$

That is, the instantaneous velocity of the falling body at time $t = 2$ is 64 ft/s.

Numerical (tabular) approach

We begin by selecting sequences of numbers for $t \to 2^-$ and $t \to 2^+$:

t approaches from the left		t approaches from the right	
$t \to 2^-$		$t \to 2^+$	
t	$\overline{v}(t)$	t	$\overline{v}(t)$
1.950	63.200	2.100	65.600
1.995	63.920	2.015	64.240
1.999	63.984	2.001	64.016
2	undefined	2	undefined

That is, the pattern of numbers suggests

$$\lim_{t \to 2} \frac{16t^2 - 64}{t - 2} = 64$$

This tabular approach agrees with what we found with the graphical approach. ∎

Example 3 clearly shows that we cannot evaluate limits by substitution because if we attempt to substitute $t = 2$ into the velocity function $\dfrac{16t^2 - 64}{t - 2}$ we find that it does not exist (because we cannot divide by zero). However, the limit *does* exist and is 64. However, in calculus, you will prove that you may find the limit of a polynomial by direct substitution.

> **LIMIT OF A POLYNOMIAL**
>
> If P is a polynomial function, then
>
> $$\lim_{x \to c} P(x) = P(c)$$

EXAMPLE 4   Limit of a polynomial function

Evaluate   **a.** $\lim_{x \to 2}(2x^5 - 9x^3 + 3x^2 - 11)$ **b.** $\lim_{x \to -3} \dfrac{x^2 + x - 6}{x + 3}$

Solution

a. It would be difficult to graph this function, and a table of values would be lengthy. However, because this is a polynomial function, we can find the limit by direct substitution.

$$\lim_{x \to 2}(2x^5 - 9x^3 + 3x^2 - 11) = 2(2)^5 - 9(2)^3 + 3(2)^2 - 11 = -7$$

b. $\lim_{x \to -3} \dfrac{x^2 + x - 6}{x + 3} = \lim_{x \to -3} \dfrac{(x + 3)(x - 2)}{x + 3}$ Note that direct substitution gives an undefined expression (can't divide by 0).

$$= \lim_{x \to -3}(x - 2)$$ Because $x - 2$ is a polynomial, evaluate by substitution.

$$= 3 - 2$$

$$= 1$$

Notice from the preceding examples that when we write

$$\lim_{x \to c} f(x) = L$$

we do not require c itself to be in the domain of f, nor do we require $f(c)$, if it is defined, to be equal to the limit. Functions with the special property that $\lim_{x \to c} f(x) = f(c)$ are said to be *continuous* at $x = c$.

Intuitive Notion of Continuity

Continuity may be thought of informally as the quality of having parts that are in immediate connection with one another. This idea evolved from the vague or intuitive notion of a curve "without breaks or jumps" to a rigorous definition first given toward the end of the 19th century.

We begin with a discussion of *continuity at a point*. It may seem strange to talk about continuity *at a point*, but it should seem natural to talk about a curve being "discontinuous at a point," as illustrated by Figure 2.43.

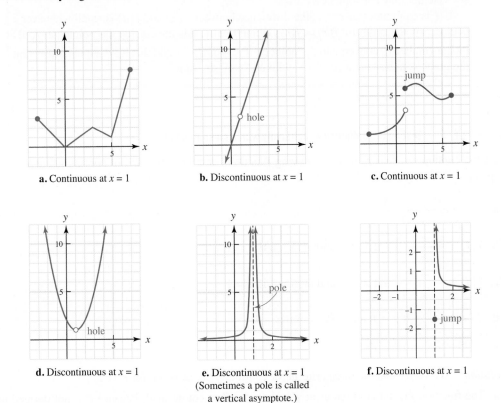

a. Continuous at $x = 1$ **b.** Discontinuous at $x = 1$ **c.** Continuous at $x = 1$

d. Discontinuous at $x = 1$ **e.** Discontinuous at $x = 1$ **f.** Discontinuous at $x = 1$
(Sometimes a pole is called
a vertical asymptote.)

Figure 2.43 Holes, poles, and jumps

Definition of Continuity

Let us consider the conditions that must be satisfied for a function f to be continuous at a point c. First, $f(c)$ must be defined. For example, the curves in Figures 2.43**b** and **e** are not continuous at $x = 1$ because they are not defined for $x = 1$. (An open dot indicates an excluded point.) A second condition for continuity at a point $x = c$ is that the function makes no jumps there. This means that if "x is close to c," then "$f(x)$ must be close to $f(c)$." This condition is satisfied if $\lim_{x \to c} f(x)$ exists. Looking at Figure 2.43, we see that the graphs in parts **c**, **d**, and **f** have a jump at the point $x = 1$. A third condition for continuity at point $x = c$ is that $\lim_{x \to c} f(x) = f(c)$. Note that in the curve in Figure 2.43**d**, $\lim_{x \to 1} f(x)$ exists but is not equal to $f(1)$. These considerations lead us to a formal definition of continuity of a function at a point.

Historical Note*

The idea of continuity evolved from the notion of a curve "without breaks or jumps" to a rigorous definition given by Karl Weierstraß. Our definition of continuity is a refinement of a definition first given by Bernhard Bolzano (1781–1848). Galileo and Leibniz had thought of continuity in terms of the density of points on a curve, but using today's standards, we would say they were in error because the rational numbers have this property, yet do not form a continuous curve. However, this was a difficult concept, which evolved over a period of time. Another mathematician, J. W. R. Dedekind (1831–1916), took an entirely different approach to conclude that continuity is due to the division of a segment into two parts by a point on the segment. As Dedekind wrote, "By this commonplace remark, the secret of continuity is to be revealed."

*From Carl Boyer, *A History of Mathematics* (New York, John Wiley & Sons, Inc., 1968), p. 607.

CONTINUITY AT A POINT

A function f is **continuous at a point** $x = c$ if

1. $f(c)$ is defined;

2. $\lim\limits_{x \to c} f(x)$ exists;

3. $\lim\limits_{x \to c} f(x) = f(c)$.

A function that is not continuous at c is said to have a **discontinuity** at that point.

>> IN OTHER WORDS Step 1 refers to the domain of the function and ignores what happens at points $x \neq c$, whereas step 2 refers to points close to c but ignores the point $x = c$. Continuity looks at the whole picture: at $x = c$ and at points close to $x = c$ and checks to see whether they are somehow "alike."

If f is continuous at $x = c$, the difference between $f(x)$ and $f(c)$ is small whenever x is close to c because $\lim\limits_{x \to c} f(x) = f(c)$ Geometrically, this means that the points $(x, f(x))$ on the graph of f converge to the point $(c, f(c))$ as $x \to c$ and this is what guarantees that the graph is unbroken at $(c, f(c))$ with no "gap" or "hole."

EXAMPLE 5   Testing the definition of continuity with a given function

Test the continuity of each of the following functions at $x = 1$. If it is not continuous at $x = 1$, explain.

a. $f(x) = \dfrac{x^2 + 2x - 3}{x - 1}$

b. $g(x) = \dfrac{x^2 + 2x - 3}{x - 1}$ if $x \neq 1$, and $g(x) = 6$ if $x = 1$

c. $h(x) = \dfrac{x^2 + 2x - 3}{x - 1}$ if $x \neq 1$, and $h(x) = 4$ if $x = 1$

d. $F(x) = \dfrac{x + 3}{x - 1}$ if $x \neq 1$, and $F(x) = 4$ if $x = 1$

e. $G(x) = 7x^3 + 3x^2 - 2$

Solution   We verify that the three criteria for continuity are satisfied for $c = 1$.

a. The function f is not continuous at $x = 1$ (hole; $f(c)$ not defined) because it is not defined at this point.

b. 1. $g(1)$ is defined; $g(1) = 6$.

   2. $\lim\limits_{x \to 1} g(x) = \lim\limits_{x \to 1} \dfrac{x^2 + 2x - 3}{x - 1}$

   $= \lim\limits_{x \to 1} \dfrac{(x - 1)(x + 3)}{x - 1}$

   $= \lim\limits_{x \to 1}(x + 3)$

   $= 4$

   3. $\lim\limits_{x \to 1} g(x) \neq g(1)$, so g is not continuous at $= 1$ (hole; $g(c)$ defined).

c. Compare h with g of part **b**. We see that all three conditions of continuity are satisfied, so h is continuous at $x = 1$.

d. 1. $F(1)$ is defined; $F(1) = 4$.

2. $\displaystyle\lim_{x\to 1} F(x) = \lim_{x\to 1}\frac{x+3}{x-1}$; the limit does not exist.

   The function F is not continuous at $x = 1$ (pole; $F(c)$ defined).

e. 1. $G(1)$ is defined; $G(1) = 8$.

2. $\displaystyle\lim_{x\to 1} G(x) = 7(1)^3 + 3(1)^2 - 2$ G is a polynomial function

   $\qquad\qquad = 8$

3. $\displaystyle\lim_{x\to 1} G(x) = G(1)$.

Because the three conditions of continuity are satisfied, G is continuous at $x = 1$. ∎

Continuity on an Interval

The function f is said to be **continuous on the open interval** (a, b) if it is continuous at each number in this interval. If f is also continuous from the right at a, we say it is **continuous on the half-open interval** $[a, b)$. Similarly, f is **continuous on the half-open interval** $(a, b]$ if it is continuous at each number between a and b and is continuous from the left at the endpoint b. Finally, f is **continuous on the closed interval** $[a, b]$ if it is continuous at each number between a and b and is both continuous from the right at a and continuous from the left at b.

EXAMPLE 6   Testing for continuity on an interval

Find the intervals on which each of the given functions is continuous.

a. $f_1(x) = \dfrac{x^2 - 1}{x^2 - 4}$

b. $f_2(x) = |x^2 - 4|$

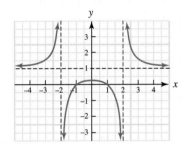
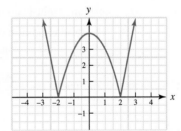

Solution

a. Function f_1 is not defined when $x^2 - 4 = 0$ or when $x = 2$ or $x = -2$. The curve is continuous on $(-\infty, -2) \cup (-2, 2) \cup (2, \infty)$.

b. Function f_2 is continuous on $(-\infty, \infty)$. ∎

Because we do not always have the graph of a function readily available, as we did in Example 6, and because the task of checking for continuity will focus on certain values, we consider a procedure involving identifying and then checking those values of concern. To help us describe the situation, we define a **suspicious point** as a point having an x value for which the definition of the function changes, or a value that causes division by zero for the given function.

For Example 6, the suspicious points can be listed:

a. $\dfrac{x^2 - 1}{x^2 - 4}$ has suspicious points for division by zero when and $x = 2$ and $x = -2$.

b. $|x^2 - 4| = x^2 - 4$ when $x^2 - 4 \geq 0$, and $|x^2 - 4| = 4 - x^2$ when $x^2 - 4 < 0$. This means the definition of the function changes when $x^2 - 4 = 0$, namely, $x = 2$ and $x = -2$ are suspicious points.

EXAMPLE 7 Checking continuity at suspicious points

Let $f(x) = \begin{cases} 3-x & \text{if } -5 \le x < 2 \\ x-2 & \text{if } 2 \le x < 5 \end{cases}$ and $g(x) = \begin{cases} 2-x & \text{if } -5 \le x < 2 \\ x-2 & \text{if } 2 \le x < 5 \end{cases}$

Find the intervals on which f and g are continuous.

Solution The domain for both functions is $[-5, 5)$; the functions are continuous everywhere on that interval except possibly at the suspicious points.

Examining f, we see $f(x) = 3 - x$ on $[-5, 2)$, which is a polynomial function and thus is continuous; $f(x) = x - 2$ on $[2, 5)$, which is a polynomial function and thus is continuous. The suspicious point on the real number line is the x value for which the definition of f changes—in this case, $x = 2$.

For g, we likewise see that the function is continuous except possibly at the suspicious point $x = 2$, the value where the definition of the function changes.

	Function f	Function g
Suspicious point(s): $x = 2$		$x = 2$
1.	$f(2) = 2 - 2 = 0$	$g(2) = 2 - 2 = 0$
	f is defined at $x = 2$.	g is defined at $x = 2$.
2.	$\lim\limits_{x \to 2^-} f(x) = \lim\limits_{x \to 2^-}(3 - x)$	$\lim\limits_{x \to 2^-} g(x) = \lim\limits_{x \to 2^-}(2 - x)$
	$= 1$	$= 0$
	$\lim\limits_{x \to 2^+} f(x) = \lim\limits_{x \to 2^+}(x - 2)$	$\lim\limits_{x \to 2^+} g(x) = \lim\limits_{x \to 2^+}(x - 2)$
	$= 0$	$= 0$
	Thus, $\lim\limits_{x \to 2^-} f(x)$ does not exist because the left- and right-hand limits are not equal.	Thus, $\lim\limits_{x \to 2} g(x) = 0$
3.	The third condition of continuity cannot hold at $x = 2$ because the limit does not exist.	$\lim\limits_{x \to 2^-} g(x) = g(2)$
Conclusion:	Continuous on $[-5, 2)$ and on $[2, 5)$ but not on $[-5, 5)$.	Continuous on $[-5, 5)$.

Root Location and Limits That Do Not Exist

In calculus, you will use continuity to prove an important result that we will need to approximate the roots of equations in algebra. This result asserts that if a function is continuous on a closed interval and has opposite signs someplace on that interval, then there is a root somewhere between its positive and negative values.

ROOT LOCATION THEOREM

If f is continuous on the closed interval $[a, b]$ and if $f(a)$ and $f(b)$ have opposite algebraic signs (one positive and the other negative), then $f(c) = 0$ for at least one number c on the open interval (a, b).

EXAMPLE 8 Using the root location theorem

Show that $x^4 = 5\sqrt{x} - 1 = 0$ for at least one number c on $[0, 2]$.

Solution Notice that the function $f(x) = x^4 - 5\sqrt{x} - 1$ is continuous on $[0, 2]$. Also notice that $f(0) = -1$ (it is negative) and $f(2) = 15 - 5\sqrt{2}$ (it is positive), so the conditions of the root location theorem apply.

Therefore, there is at least one number c on $(0, 2)$ for which $f(c) = 0$. This means that on $(0, 2)$, $c^4 - 5\sqrt{c} - 1 = 0$, as required. ∎

It may happen that a function f does not have a (finite) limit as $x \to c$. When $\lim_{x \to c} f(x)$ fails to exist, we say that $f(x)$ **diverges** as x approaches c. The following examples illustrate how divergence may occur.

EXAMPLE 9   A function that diverges

Evaluate $\lim_{x \to 0} \dfrac{1}{x^2}$.

Solution As $x \to 0$, the corresponding functional values of $f(x) = \frac{1}{x^2}$ grow arbitrarily large, as indicated in the table below:

x approaches from the left $x \to 0^-$		x approaches from the right $x \to 0^+$	
x	$f(x)$	x	$f(x)$
-0.1	$100 = 1 \times 10^2$	0.01	1×10^2
-0.05	$400 = 4 \times 10^2$	0.05	4×10^2
-0.001	1×10^6	0.001	1×10^6
0	undefined	0	undefined

The graph of f from Table 2.1 is shown in Figure 2.44.

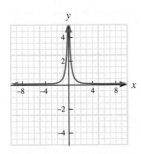

Figure 2.44 Graph of $y = \dfrac{1}{x^2}$

Geometrically, the graph of $y = f(x)$ rises without bound as $x \to 0$. Thus, $\lim_{x \to 0} \dfrac{1}{x^2}$ does not exist, so we say f diverges as $x \to 0$. ∎

PROBLEM SET 2.8

Describe each illustration in Problems 1–4 using a limit statement.

1.
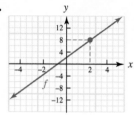

2.
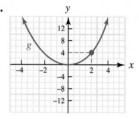

3.
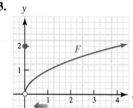

4.
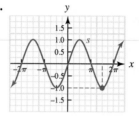

5. State the domain of each function, if possible, and determine whether it represents a continuous function.
   a. The temperature on a specific day at a given location considered as a function of time
   b. The humidity on a specific day at a given location considered as a function of time
   c. The selling price of AT&T stock on a specific day considered as a function of time
   d. The number of unemployed people in the United States during January 2007 considered as a function of time
   e. The charges for a taxi ride across town considered as a function of mileage
   f. The charges to mail a package as a function of its weight

6. Let $f(x) = \begin{cases} x^2 & \text{if } x > 2 \\ x+1 & \text{if } x \le 2 \end{cases}$
   Show that f is continuous from the left at 2 but not from the right.

Given the functions defined by the graphs in Problems 7–8, find the limits.

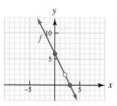

Graph of f

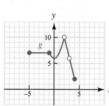

Graph of g

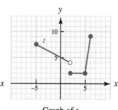

Graph of t

7. a. $\lim\limits_{x \to 3} f(x)$ b. $\lim\limits_{x \to 2} f(x)$
   c. $\lim\limits_{x \to -3} g(x)$ d. $\lim\limits_{x \to -1} g(x)$
   e. $\lim\limits_{x \to 2^-} t(x)$ f. $\lim\limits_{x \to 2^+} t(x)$

8. a. $\lim\limits_{x \to 0} f(x)$ b. $\lim\limits_{x \to -1} f(x)$
   c. $\lim\limits_{x \to 2} g(x)$ d. $\lim\limits_{x \to 3^+} g(x)$
   e. $\lim\limits_{x \to 4} t(x)$ f. $\lim\limits_{x \to -4} t(x)$

PROBLEMS FROM CALCULUS *Evaluate the limits in Problems 9–18. You may use the numerical, graphical, or algebraic method of solution. If the limit does not exist, explain why.*

9. $\lim\limits_{x \to 2^-}(x^2 - 4)$

10. $\lim\limits_{x \to 3^+}(x^2 - 4)$

11. $\lim\limits_{x \to -2}(x^2 + 3x - 7)$

12. $\lim\limits_{x \to 0}(x^3 - 5x^2 + 4)$

13. $\lim\limits_{x \to 3} \dfrac{x^2 + 3x - 10}{x - 2}$

14. $\lim\limits_{x \to 2} \dfrac{x^2 + 3x - 10}{x - 2}$

15. $\lim\limits_{x \to -3^+} \dfrac{1}{x - 3}$

16. $\lim\limits_{x \to -3^-} \dfrac{1}{x - 3}$

17. $\lim\limits_{x \to 0} \dfrac{\frac{1}{x+3} - \frac{1}{3}}{x}$

18. $\lim\limits_{x \to 0} \dfrac{\frac{1}{x} - \frac{1}{3}}{x - 3}$

Which of the functions defined in Problems 19–24 are continuous? If it is not continuous, explain why.

19.
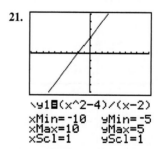

20.
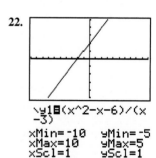

21.
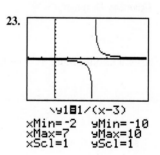

22.
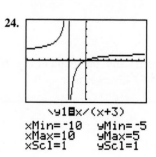

23.

24.

PROBLEMS FROM CALCULUS *In Problems 25–28, compute the one-sided limit.*

25. $\lim\limits_{x \to 2} (x^2 - 2x)$

26. $\lim\limits_{x \to 2^-} f(x)$, where $f(x) = \begin{cases} 3 - 2x & \text{if } x \le 2 \\ x^2 - 5 & \text{if } x > 2 \end{cases}$

27. $\lim\limits_{s \to 1^-} g(x)$, where $g(x) = \begin{cases} \dfrac{s^2 - s}{s - 1} & \text{if } s < 1 \\ \sqrt{1 - s} & \text{if } s \ge 1 \end{cases}$

28. $\lim\limits_{s \to 1^+} g(x)$, where $g(x) = \begin{cases} \dfrac{s^2 - s}{s - 1} & \text{if } s < 1 \\ \sqrt{1 - s} & \text{if } s \ge 1 \end{cases}$

PROBLEMS FROM CALCULUS *Identify all suspicious points and determine all points of discontinuity in Problems 29–38.*

29. $f(x) = x^3 - 3x + 5$

30. $f(x) = \dfrac{2x + 4}{x - 6}$

31. $f(x) = \dfrac{x}{x^2 - x}$

32. $f(x) = 3 - (5 + 2x)^3$

33. $f(x) = \sqrt{x} + \dfrac{5}{x}$

34. $f(x) = \sqrt[3]{x^2 - 1}$

35. $f(x) = \dfrac{1}{x} - \dfrac{3}{x + 1}$

36. $f(x) = \dfrac{x^2 - 1}{x^2 + x - 2}$

37. $f(x) = \begin{cases} x^2 - 2 & \text{if } x > 1 \\ 2x - 3 & \text{if } x \le 1 \end{cases}$

38. $f(x) = \begin{cases} 3x^2 - 2 & \text{if } x > 3 \\ 5 & \text{if } 1 < x \le 3 \\ 3x + 2 & \text{if } x \le 1 \end{cases}$

PROBLEMS FROM CALCULUS *In Problems 39–42, determine whether or not the given function is continuous on the prescribed interval.*

39. $f(x) = \dfrac{1}{x}$

    a. on $[1, 2]$ **b.** on $[0, 1]$ **c.** on $[-3, 0)$

40. $f(x) = \dfrac{1}{5 - x}$

    a. on $[0, 7)$ **b.** on $[0, 5]$ **c.** on $[-1, 1]$

41. $f(x) = \begin{cases} x^2 & \text{if } 0 \le x < 2 \\ 3x + 1 & \text{if } 2 \le x < 5 \end{cases}$

42. $g(t) = \begin{cases} 2t & \text{if } 0 < t \le 3 \\ 15 - t^2 & \text{if } -3 \le t \le 0 \end{cases}$

LEVEL 2

PROBLEMS FROM CALCULUS *In Problems 43–50, either evaluate the limit or explain why it does not exist.*

43. $\lim\limits_{x \to 1} \dfrac{1}{x - 1}$

44. $\lim\limits_{x \to 2} \dfrac{x^2 - 4}{x^2 - 4x + 4}$

45. $\lim\limits_{x \to 1} f(x)$, where $f(x) = \begin{cases} 2 & \text{if } x \ge 1 \\ -5 & \text{if } x < 1 \end{cases}$

46. $\lim\limits_{t \to -1} g(t)$, where $g(t) = \begin{cases} 2t + 1 & \text{if } t \ge -1 \\ 5t^2 & \text{if } t < -1 \end{cases}$

47. $\lim\limits_{t \to 5} f(t)$, where $f(t) = \begin{cases} t + 3 & \text{if } t \ne 5 \\ 4 & \text{if } t = 5 \end{cases}$

48. $\lim\limits_{t \to 2} g(t)$, where $g(t) = \begin{cases} t^2 & \text{if } t \ne 2 \\ 5 & \text{if } t = 2 \end{cases}$

49. $\lim\limits_{x \to 2} F(x)$, where $F(x) = \begin{cases} 2(x + 1) & \text{if } x < 2 \\ 4 & \text{if } x = 2 \\ x^2 - 1 & \text{if } x > 2 \end{cases}$

50. $\lim\limits_{x \to 3} G(x)$, where $G(x) = \begin{cases} 2(x + 1) & \text{if } x < 3 \\ 4 & \text{if } x = 3 \\ x^2 - 1 & \text{if } x > 3 \end{cases}$

In Problems 51–56, show that the given equation has at least one solution on the indicated interval.

51. $x^3 - 4x - 5 = 0$ on $[-5, 5]$
52. $x^5 - 4x + 6 = 0$ on $[-5, 5]$
53. $x^4 - 4x^2 = 0$ on $[-5, 5]$
54. $x^3 - 4x^2 = 0$ on $[-5, 5]$
55. $x^3 - 24x^2 + 188x - 465 = 0$ on $[-10, 10]$
56. $x^3 + 36x^2 + 424x + 1{,}657 = 0$ on $[-20, 20]$
57. A ball is thrown directly upward from the edge of a cliff and travels in such a way that t seconds later, its height above the ground at the base of the cliff (in feet) is

$$s(t) = -16t^2 + 40t + 24$$

    a. Compute the limit

$$v(t) = \lim_{x \to t} \frac{s(x) - s(t)}{x - t}$$

    to find the instantaneous velocity of the ball at time t.

    b. What is the ball's initial velocity?

    c. When does the ball hit the ground, and what is its impact velocity?

    d. When does the ball have velocity 0? What physical interpretation should be given to this time?

58. Tom and Sue are driving along a straight, level road in a car whose speedometer needle is broken but which has a trip odometer that can measure the distance traveled from an arbitrary starting point in tenths of a mile. At 2:50 P.M., Tom says he would like to know how fast they are traveling at 3:00 P.M., so Sue takes down the odometer readings listed in the table below, makes a few calculations, and announces the desired velocity. What is her result?

Time, t	2:50	2:55	2:59	
Odometer reading	33.9	38.2	41.5	

Time, t	3:00	3:01	3:03	3:06
Odometer reading	42.4	43.2	44.9	47.4

LEVEL 3

In Problems 59–60, find constants a *and* b *so that the given function will be continuous for all* x *throughout its domain.*

59. $f(x) = \begin{cases} ax + 12 & \text{if } x > 2 \\ 20 & \text{if } x = 2 \\ x^2 + bx + 5 & \text{if } x < 2 \end{cases}$ **60.** $g(x) = \begin{cases} ax + 3 & \text{if } x > 5 \\ 8 & \text{if } x = 5 \\ x^2 + bx + 1 & \text{if } x < 5 \end{cases}$

CHAPTER 2 SUMMARY AND REVIEW

*T*he *business of concrete mathematics is to discover the equations which express the mathematical laws of the phenomenon under consideration; and these equations are the starting-point of the calculus.*

Auguste Compte

Take some time getting ready to work the review problems in this section. First, look back at the definition and property boxes. You will maximize your understanding of this chapter by working the problems in this section only after you have studied the material.

STOP

SELF TEST   *All of the answers for this self test are given in the back of the book.*

1. If $f(x) = \dfrac{6x^2 + x - 2}{2x - 1}$, find:

   a. $f(2x)$ **b.** $2f(x)$

   c. $f\left(-\frac{2}{3}\right)$ **d.** $|f(3) - f(2)|$

   e. $f^{-1}(x)$

2. Let $f(x) = \sqrt{34 - 2x^2}$.

   a. What is the domain?   **b.** What is the range?

   c. If $f(x) = g[u(x)]$, find g and u.

   d. What is $f(-3)$?   **e.** Find $\lim\limits_{x \to -3} f(x)$.

3. If $16x^2 - x^2y^2 + y^2 - 1 = 0$

   a. What is the domain?   **b.** What is the range?

   c. What are the x-intercepts?   **d.** What are the y-intercepts?

   e. Does this equation represent a function? What or why not?

4. If $f(x) = \frac{1}{2}(x - 2)^2$

   a. Describe f by comparing it to $y = x^2$; classify it and then specify the shift, compression, or dilation, as appropriate.

   b. Graph f.   **c.** Find $\lim\limits_{x \to 0} f(x)$.

   d. Find $\dfrac{f(x+h) - f(x)}{h}$.   **e.** Find $\lim\limits_{h \to 0} \dfrac{f(x+h) - f(x)}{h}$.

5. If $f(x) = \begin{cases} x^2 + 1 & \text{if } x > 1 \\ |x + 1| & \text{if } x \le 1 \end{cases}$

   a. What is the domain of this function?

   b. What are the intercepts?   **c.** Graph f.

   d. Is f continuous at $x = 1$?

   e. If $y = f(x)$, graph $y - 1 = f(x + 2)$.

6. Suppose f is defined by the graph shown in Figure 2.45.

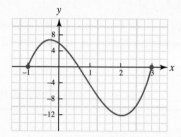

Figure 2.45 Graph of f

   a. Is f a function?

   b. If f one-to-one? Does it have an inverse?

   c. What are the domain and range of f?

   d. If $y = f(x)$, draw the graph of $y = \frac{1}{2}f(x)$.

   e. Graph $y + 2 = f(x - 1)$.

7. Consider the function defined by the graph in Figure 2.46.

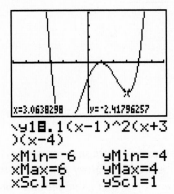

Figure 2.46 Function of f defined by a graph

a. Is f continuous on $[-6, 6]$?
b. Is f one-to-one?
c. What are the intercepts?
d. What are the approximate coordinates of the turning points for $x \geq 0$?
e. What is $\lim\limits_{x \to 0} f(x)$?

8. Let $f(x) = 5 - x^2$, and $g(x) = \dfrac{5 - x}{x + 2}$.

a. Find g^{-1}, if it exists.   **b.** Find $f \circ f$.
c. Find $g \circ f$.
d. Classify f and g as even, odd, or neither.
e. For which values is $f(x) \geq -20$?

9. a. An efficiency expert found that at a particular company that employs x workers $(x \geq 3)$, it takes

$$H = \frac{3x + 4}{2x - 5}$$

hours to complete a certain task. How many hours will the task take for 3 workers? How many hours will the task take for 10 workers? For 20 workers? If the task must be completed in 2 hours, how many workers are required? If each worker earns $25/hr, express the total labor cost, C, as a function of x.

b. If f and g are defined by

$$f(x) = \frac{6x^2 - x - 2}{3x - 2} \quad \text{and} \quad g(x) = 2x + 1$$

is it true that $f = g$? Why or why not?

10. An open box with a square base is to be built for $96. The sides of the box will cost $3/ft² and the base will cost $8/ft². Express the volume of the box as a function of the length of its base.

STOP STUDY HINTS *Compare your solutions and answers to the self test.*

Practice for Calculus—*Supplementary Problems: Cumulative Review Chapters 1–2*

Simplify each expression in Problems 1–4.

1. Find $|f(x) - L|$, where $f(x) = x^2 - 2$ and $L = 2$.
2. Find $|f(x) - L|$, where $f(x) = x^2 + 2$ and $L = 6$.
3. Find $|f(x) - L|$, where $f(x) = \dfrac{2x^2 - 3x - 2}{x - 2}$ and $L = 6$.
4. Find $|f(x) - L|$, where $f(x) = \dfrac{x^2 - 2x + 2}{x - 4}$ and $L = -1$.

Solve each inequality in Problems 5–12.

5. $-0.001 \leq x + 2 \leq 0.001$ **6.** $-0.001 \leq 2x + 5 \leq 0.001$
7. $|3x + 1| < 0.25$ **8.** $|x - 5| < 0.01$
9. $|5 - 3x| < 0.001$ **10.** $|1 - 8x| < 0.001$
11. $|2x - 1| < 0.0001$ **12.** $|5x + 3| < 0.0001$

Solve each equation for x in Problems 13–24.

13. $x^2 - 5x + 3 = 0$ **14.** $2x^2 - 5x - 3 = 0$
15. $x^2 + 9x + 20 = 0$ **16.** $x^2 - 3x + 1 = 0$
17. $|2x + 3| = 8$ **18.** $|5x + 1| = 10$
19. $|2 - 3x| = 11$ **20.** $|5 - 4x| = 20$
21. $x^2 = 4cy$ **22.** $x^2 + y^2 = 9$
23. $4x^2 + 3y^2 = 1$ **24.** $3x^2 - 4y^2 = 12$

Factor the expressions in Problems 25–32.

25. $1 - x^6$ **26.** $x^6 - 64$
27. $\left(x^2 - \frac{1}{9}\right)\left(x^2 - \frac{1}{25}\right)$ **28.** $x^4 - 41x^2 + 400$
29. $-4(3x^2 - 2x)^{-5}(6x - 4)$ **30.** $-2(2x^3 - 8x)^{-3}(6x - 8)$
31. $2(2x^2 + 3)(4x)(x^3 - 1)^3 + 3(2x^2 + 3)^2(x^3 - 1)^2(3x^2)$
32. $4(x^2 - 8)^3(2x)(5x^2 - 1)^3 + 3(x^2 - 8)^4(5x^2 - 1)(10x)$

In Problems 33–38, find $\dfrac{f(x + h) - f(x)}{h}$.

33. $f(x) = x + 5$ **34.** $f(x) = 6 - 3x$
35. $f(x) = 5x^2$ **36.** $f(x) = 6$
37. $f(x) = \frac{1}{x}$ **38.** $f(x) = 2x^2 + x$

*In the **graphs** in Problems 39–44, assume the dashed lines are parallel to the coordinate axes.*

39. What are the coordinates of P and Q?

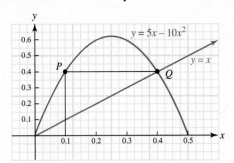

40. What are the coordinates of P, Q, and R?

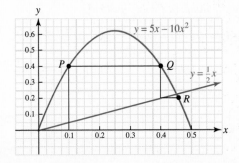

41. What are the coordinates of P and Q?

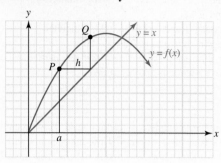

42. What are the coordinates of P and Q?

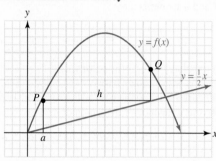

43. What are the coordinates of R?

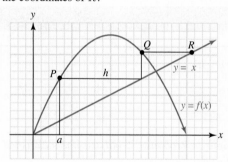

44. What is the second component of R?

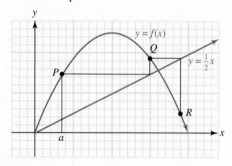

Evaluate the limits in Problems 45–52.

45. $\lim\limits_{x\to 3}(2x^2 - x - 10)$

46. $\lim\limits_{x\to 3}\dfrac{2x^2 - x - 10}{x + 2}$

47. $\lim\limits_{x\to 2}(2x^2 - x - 10)$

48. $\lim\limits_{x\to 2}\dfrac{2x^2 - x - 10}{x + 2}$

49. $\lim\limits_{h\to 0}\dfrac{[(x+h)+5] - (x+5)}{h}$

50. $\lim\limits_{h\to 0}\dfrac{[6 - 3(x+h)] - (6 - 3x)}{h}$

51. $\lim\limits_{h\to 0}\dfrac{(x+h)^2 - x^2}{h}$

52. $\lim\limits_{h\to 0^-}\dfrac{|2+h| - |2|}{h}$

In Problems 53–58, express each function as the composite of two functions. That is, express the given function f as $f(x) = g[u(x)]$, and then state the inner function u and the outer function g.

53. $f(x) = (3x^2 - 5x)^2$

54. $f(x) = \sqrt{x^2 + 9}$

55. $f(x) = (3x^4 - 1)^{3/2}$

56. $f(x) = |x^2 + 8|$

57. $f(x) = (x^2 + 3)^4 - (x^2 + 3)^{5/2}$

58. $f(x) = \sqrt[5]{x^2 + 3x + 1}$

59. To study the rate at which animals learn, a psychology student performed an experiment in which a rat was sent repeatedly through a laboratory maze. Suppose that the time (in minutes) required for the rat to traverse the maze on the nth trial was approximately

$$f(n) = 3 + \frac{12}{n}$$

a. What is the domain of the function f?

b. For what values of n does $f(n)$ have meaning in the context of the psychology experiment?

c. How long did it take the rat to traverse the maze on the third trial?

d. On which trial did the rat first traverse the maze in 4 minutes or less?

e. According to the function f, what will happen to the time required for the rat to traverse the maze as the number of trials increases?

60. Biologists have found that the speed of blood in an artery is a function of the distance of the blood from the artery's central axis. According to *Poiseuille's law*, the speed (cm/sec) of blood that is r cm from the central axis of an artery is given by the function

$$S(r) = C(R^2 - r^2)$$

where C is a constant and R is the radius of the artery.* Suppose that for a certain artery, $C = 1.76 \times 10^5$ cm/sec² and $R = 1.2 \times 10^{-2}$ cm.

a. Compute the speed of the blood at the central axis of this artery.

b. Compute the speed of the blood midway between the artery's wall and central axis.

*The law and the unit poise, a unit of viscosity, are both named for the French physician Jean Louis Poiseuille (1799–1869).

CHAPTER 2 Group Research Projects

Working in small groups is typical of most work environments, and this book seeks to develop skills with group activities. At the end of each chapter, we present a list of suggested projects, and even though they could be done as individual projects, we suggest that these projects be done in groups of three or four students.

G2.1 Let $S(x) = \dfrac{4^x - 4^{-x}}{2}$ and $C(x) = \dfrac{4^x + 4^{-x}}{2}$. Show that

$$[C(x)]^2 - [S(x)]^2 = 1$$

G2.2 Suppose $f(x) = \frac{1}{1-x}$. Define $f_1(x) = f \circ f$; $f_2(x) = f \circ f \circ f$; $f_3(x) = f \circ f \circ f \circ f$; .... Find $f_{100}(x)$.

G2.3 Use functional iteration to find $(f \circ f \circ f \circ f \dots)(x)$ when:

   a. $f(x) = \sqrt{1+x}$ **b.** $f(x) = \dfrac{1}{1+x}$ **c.** $f(x) = |x| - 1$

G2.4 𝔥istorical 𝔔uest Write a paper on George Pólya. Include, as part of this paper, a report on his book *How to Solve It* (Princeton University Press, 1971).

G2.5 Journal Problem (From Fourth U.S.A. Olympiad, May 6, 1975.) Prove that

$$[\![5x]\!] + [\![5y]\!] \geq [\![3x+y]\!] + [\![3y+x]\!]$$

G2.6 Journal Problem (From *The Mathematics Teacher*, December 1995, pp. 734–737, "Precalculus Explorations of Function Composition with a Graphing Calculator," by Lewis Lum.) This article describes a process called *carom paths*, as follows. Let

$$y_1 = f(x) = -2\sqrt{x-3}, \qquad y_2 = x, \qquad y_3 = g(x) = 3\sqrt{-2-x}$$

We describe the "carom path" as follows. Start at some value of x in the domain of $g \circ f$, say $x = 6$. Move vertically to the point on y_1 (point A), and bounce off horizontally to point B on y_2. Carom vertically to the point on y_3 (point C), and finally rebound horizontally to the point D on the graph of y_2. The graphs are shown in Figure 2.47.

We can calculate the coordinates of the indicated points.

   $A:\left(6, -2\sqrt{3}\right)$ since $f(6) = -2\sqrt{6-3} = -2\sqrt{3}$

   $B:\left(-2\sqrt{3}, -2\sqrt{3}\right)$ since $y = x$

   $C:\left(-2\sqrt{3}, 3\sqrt{2\sqrt{3}-2}\right)$ since $g\left(-2\sqrt{3}\right) = 3\sqrt{-2-\left(-2\sqrt{3}\right)}$

   $D:\left(3\sqrt{2\sqrt{3}-2}, 3\sqrt{2\sqrt{3}-2}\right)$ since $y = x$

   a. What are the points A, B, C, and D for the starting point $x = 12$?

   b. Attempt to plot the carom path of $x = 2$. Explain why the path cannot be completed. Find other points with the same problem as $x = 2$.

   c. Attempt to plot the carom path of $x = 3.5$. Explain why the path cannot be completed. Find other points with the same problem as $x = 3.5$.

   d. What is the domain for x that assures that the carom path exists?

G2.7 𝔥istorical 𝔔uest The notion of a function is not only central for this book but also fundamental to the study of calculus. In the article "The Mathematical Way of Thinking," Hermann Weyl begins by saying, "By the mathematical way of thinking I mean first that form of reasoning

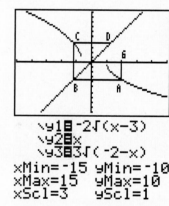

```
\y1🄱-2√(x-3)
\y2🄱x
\y3🄱3√(-2-x)
xMin=-15  yMin=-10
xMax=15   yMax=10
xScl=3    yScl=1
```

Figure 2.47 Carom path for f and g

through which mathematics penetrates into the sciences of the external world." He goes on to say that the average education of every person should include mathematics to teach everyone to think in terms of variables and functions. Do some research into the history of the idea of a function. Write a report of your research, and include a statement about whether you agree or disagree with Weyl's theses.

G2.8 "There are, in every culture, groups or individuals who think more about some ideas than do others. For other cultures, we know about the ideas of some professional groups or some ideas of the culture at large. We know little, however, about the mathematical thoughts of individuals in those cultures who are specially inclined toward mathematical ideas. In Western culture, on the other hand, we focus on, and record much about, those special individuals while including little about everyone else. Realization of this difference should make us particularly wary of any comparisons across cultures. Even more important, it should encourage finding out more about the ideas of mathematically oriented innovators in other cultures and, simultaneously, encourage expanding the scope of Western history to recognize and include mathematical ideas held by different groups within our culture or by our culture as a whole."*

Write a paper discussing this quotation.

G2.9 Journal Problem (From *The American Mathematical Monthly*, Vol. 92, January 1985, pp. 3–23.) Write a paper on the $3x + 1$ conjecture. See Example 6 on page 80 of "The $3x + 1$ Problem and Its Generalizations," by Jeffrey C. Lagarias.

G2.10 We introduced problem solving in this chapter. Here is an excerpt of a poem followed by three questions.

> *Yes, weekly from Southampton,*
> *Great steamers, white and gold,*
> *Go rolling down to Rio*
> *(Roll down—roll down to Rio!),*
> *And I'd like to roll to Rio*
> *Some day before I'm old!*
>
> Rudyard Kipling
> *Just So Stories*

a. How many steamers wending their way home will I see on this ocean?

b. On what days of the week will I see them?

c. How far from Southampton will I meet them?

Build a mathematical model to answer these questions. Not enough information is given, so here is a start. "Well, so weekly (say each Thursday) from Southampton great steamers go rolling down to Rio . . . It takes 14 days for a great white and gold steamer to cover the entire distance of 9,800 km (700 km per day) and arrive at Rio de Janeiro exactly at noon on Thursday. After a four-day stopover, the ship sets off on the return trip, and in a fortnight, at noon on Monday, it arrives at Southampton. Three days later—again on a Thursday—it leaves on its next voyage to Brazil . . . I wanted to roll to Rio, too, so I stepped onto an ocean liner at Southampton on Thursday and my voyage began."†

---

* From *Ethnomathematics* by Marcia Ascher (Pacific Grove: Brooks/Cole, 1991), pp. 188–189.
† From "Atlantic Crossings," by A. Rozental, *Quantum*, July/Aug 1993, pp. 46–47.

Chapter Objectives

The material in this chapter is previewed in the following list of objectives. After completing this chapter, review this list again, and then complete the self test.

3.1
- 3.1 Define a linear function.
- 3.2 Graph linear functions.
- 3.3 Describe the slope of the tangent line.
- 3.4 Find the equation of a tangent line.
- 3.5 Find the equation of a normal line.
- 3.6 Solve applied problems working with linear functions.

3.2
- 3.7 Define a quadratic function.
- 3.8 Graph quadratic functions of the form $y - k = a(x - h)^2$.
- 3.9 Graph a quadratic function of the form $y = ax^2 + bx + c$, $(a \neq 0)$ by finding the vertex; name the maximum or minimum value.
- 3.10 Model applied problems with a quadratic function.

3.3
- 3.11 Find the maximum or minimum value of a quadratic function.
- 3.12 Set up and solve an optimization problem.
- 3.13 Minimize a distance by looking at the square of the distance.
- 3.14 Find the maximum or minimum of a function numerically or geometrically.

3.4
- 3.15 Sketch a curve represented by parametric equations by plotting points.

- 3.16 Eliminate the parameter, if possible, to sketch a curve defined by parametric equations.

3.5
- 3.17 Use the division algorithm and carry out long division of polynomials.
- 3.18 Use synthetic division.
- 3.19 Use synthetic division and the remainder theorem to find points on the graph of a polynomial function.
- 3.20 Know the properties of polynomial graphs and use these properties to sketch the graphs of polynomial functions with degree greater than two.
- 3.21 Use inflection points to identify changes in concavity.

3.6
- 3.22 State the fundamental theorem of algebra and the number of roots theorem.
- 3.23 Find the zeros of polynomial functions by using the factor theorem, and state the multiplicity of each zero.
- 3.24 List the possible rational roots of a polynomial equation.
- 3.25 Use Descartes's rule of signs to determine the number of positive and negative real roots.
- 3.26 Solve polynomial equations.
- 3.27 Use the conjugate pair theorem to solve polynomial equations.

Polynomial Functions

3

*I*t is customary to ask of a piece of mathematics:

"Is it true?"... this is a mistaken question ... the appropriate questions are different ones:

- Is this piece of mathematics **correct**? That is, do the calculations follow the formal rules prescribed...?
- Is this piece of mathematics **illuminating**? That is, does it help understand what had gone before, either by further analysis or by abstraction or otherwise?
- Is this piece of mathematics **promising**? That is, though it is a novel departure from precedent or fashion, is there a reasonable chance that it will subsequently fit into the picture?
- Is this piece of mathematics **relevant**? That is, is it tied to something which is tied to human activities or to science?

—Saunders Mac Lane (1986)

Chapter Sections

3.1 Linear Functions
Lines and Linear Functions
Slope as a Rate
Slope of a Tangent

3.2 Quadratic Functions
Graphing Quadratic Functions
Modeling with Functions

3.3 Optimization Problems
Optimization Procedure
Optimization of Discrete Functions

3.4 Parametric Equations
Parameters
Lines in Parametric Form
Parametrization of a Curve

3.5 Graphing Polynomial Functions
Long Division of Polynomials
Synthetic Division
Graphing Polynomial Functions
Concavity

3.6 Polynomial Equations
Factoring Polynomials Using Synthetic Division
Fundamental Theorem of Algebra

Chapter 3 Summary and Review
Self Test
Practice for Calculus
Chapter 3 Group Research Projects

▶ CALCULUS PERSPECTIVE

This chapter, polynomial functions, is the foundation chapter for the study of different types of functions. Many of the problems you will encounter in mathematics will be modeled by a polynomial expression or a polynomial equation. Polynomials include lines, parabolas, cubics, as well as many higher-degree polynomials. In calculus, you will use something called a first derivative test, followed by (you guessed it) a second derivative test. In this book, we graph functions without calculus, but we will keep in mind that what we do in this book should build a foundation for the more powerful methods to follow later.

3.1 Linear Functions

Lines and Linear Functions

In Chapter 2, we considered the general definitions of functions as well as some of their general properties. In this chapter, we will narrow our focus to look at particular kinds of polynomial functions. You might want to review the classifications of functions in Section 2.3.

The first type of polynomial function we will consider is one that you have had some experiences within your previous courses and in Section 1.3. It is a *first-degree polynomial*.

> **LINEAR FUNCTION**
>
> A function f is a **linear function** if
> $$f(x) = mx + b$$
> where m and b are real numbers.

Notice that if $m = 0$, then $f(x) = b$, which is called a constant function. If the domain of a constant function is the set of real numbers, then the graph of $f(x) = b$ is a **horizontal line**, as shown in Figure 3.1**a**.

Let $P_1(x_1, y_1)$ and $P_2(x_2, y_2)$ be any points on a line and suppose that $x_1 = x_2$. Then the line is parallel to the y-axis and is called a **vertical line**, as shown in Figure 3.1**b**. Notice that vertical lines are not functions.

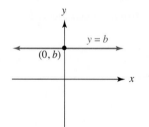
a. A horizontal line is a function.

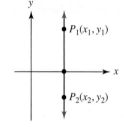
b. A vertical line is not a function.

Figure 3.1 Horizontal and vertical lines

EXAMPLE 1 Example from calculus: slope of a line

Let a function be defined by $y = f(x)$, and consider two points on the graph of f at $x_1 = a$ and $x_2 = a + h$. What is the slope of the line, called a *secant line*, which passes through these two points? The graph is shown in Figure 3.2.

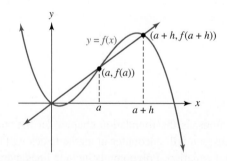

Figure 3.2 Slope of a line determined by two points on a curve

Solution The coordinates of the points are $(a, f(a))$ and $(a + h, f(a + h))$. We can now apply the definition of slope to find

$$\frac{f(a + h) - f(a)}{h}$$

Slope as a Rate

Consider the distance formula $d = rt$, where d is the distance traveled, r is the rate, and t is the time. This means, for example, that if a car is traveling at 55 mph, the distance traveled in x hours is

$$d(x) = 55x$$

The graph in Figure 3.3 shows this relationship.

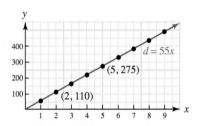

Figure 3.3 Graph of $d(x) = 55x$

It is easy to recognize this as a linear equation with y-intercept 0 and slope 55. To see why this *must* be the given 55 mph as a rate, we first note that in mathematics and physics, the rate of mph would usually be written as mi/h (miles per hour). Consider two points on the graph shown in Figure 3.3, say $(2, 110)$ and $(5, 275)$. We now find the slope, keeping track of the units:

$$m = \frac{275 \text{ mi} - 110 \text{ mi}}{5 \text{ hr} - 2 \text{ hr}} = \frac{165 \text{ mi}}{3 \text{ hr}} = 55 \frac{\text{mi}}{\text{hr}} \text{ or } 55 \text{ mi/h}$$

Thus, m is a rate of change, or *velocity*. Recall from the previous chapter how we found the average velocity. Notice that the calculation here is the same.

Another application of the slope as a rate is found in business and economics. The word *marginal* is used to mean *rate of change*. The following is an example.

EXAMPLE 2   **Example from calculus: linear equations in business**   `MODELING APPLICATION`

Suppose that it is known that the total cost (in dollars) for a product is given by the equation $C(x) = 18,500 + 8.45x$, where x is the number of items sold. Furthermore, suppose that the market demand is linear and that it has been found that at a selling price of \$15 a company will sell 7,500 items per month, but at a \$25 selling price, sales would drop to 7,000 items per month.

a. Find the marginal cost.

b. Graph the demand function.

c. Find the equation for the demand function.

d. Find the equation for the profit function.

Solution

`Step 1:` *Understand the problem.* The *marginal cost* is the additional cost to produce one more item. The demand is the number of items that could be sold at a given price. Since we are assuming that the demand is linear, we would expect that the higher price, the fewer items sold, and the lower the price, the more items sold. The profit is the revenue less the cost.

`Step 2:` *Devise a plan.* The first step is to decide on the variables. Apparently, the variables are the price, p, and the number of items sold per month, x. In mathematics, it is customary to put the independent variable (which in this problem is p) on the horizontal axis and the dependent variable (in this problem it is x) on the vertical axis. This is exactly the opposite of the way that economists do it. In 1890, a book by Alfred Marshall, *Principles of Economics*, the classic that is one of the foundations of modern price theory, made the break with mathematics and put the dependent variable, x, on the horizontal axis.[*]

---

[*] It is unfortunate that mathematicians and economists disagree about how to draw graphs like the one shown in Figure 3.3. Many mathematics professors will want to put p on the horizontal axis, but the simple fact is that Marshall's scheme is now used by everybody, although mathematicians must wonder about the odd ways of economists.

Also note that an economist would have called the number of items n instead of x, but by calling the number of items x instead of n, we are set to draw these graphs the same way that they are drawn in economics and, at the same time, make them more agreeable to mathematicians by using x for the horizontal axis.

Step 3:   *Carry out the plan.*

a. Since $C(x) = 18{,}500 + 8.45x$, we note that the p-intercept is $18{,}500$ and that $m = 8.45$; thus, the marginal cost is \$8.45. It is easy to see this if we find the slope for two points while keeping track of the units. For example,

If $x = 3{,}600$, then $C(3{,}600) = 18{,}500 + 8.45(3{,}600) = 48{,}920$.

If $x = 4{,}000$, then $C(4{,}000) = 18{,}500 + 8.45(4{,}000) = 52{,}300$.

Thus, the slope is

$$m = \frac{\$52{,}300 - \$48{,}920}{4{,}000 \text{ units} - 3{,}600 \text{ units}} = \frac{\$3{,}380}{400 \text{ units}} = \$8.45/\text{unit}$$

b. The demand is linear and passes through the points $(7500, 15)$ and $(7000, 25)$, where x is the number of units, and p is the price. We need to decide on a scale: For the x-axis, we choose one square for each 500 units. On the p-axis, we choose a scale with each square equal to 25 units. The graph is shown in Figure 3.4.

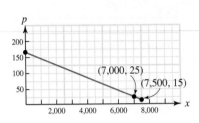

Figure 3.4 Demand graph

c. We use the slope-intercept form:

$$p - p_1 = m(x - x_1) \qquad \text{for ordered pairs } (x,\, p)$$

First, find m:

$$m = \frac{y_2 - y_1}{x_2 - x_1} \qquad \text{Slope formula}$$

$$= \frac{p_2 - p_1}{x_2 - x_1} \qquad \text{Slope formula for the variables in this example}$$

$$= \frac{25 - 15}{7{,}000 - 7{,}500}$$

$$= \frac{10}{-500}$$

$$= -0.02$$

Thus, the equation is

$$p - 15 = -0.02(x - 7{,}500) \qquad \text{Substitute } (7500, 15).$$
$$p - 15 = -0.02x + 150$$
$$p = -0.02x + 165$$

d. The profit function, P, is found by subtracting the cost from the revenue. We are given the cost function, but we need to find the revenue. The total revenue is xp:

$$\text{TOTAL REVENUE} = xp = x(-0.02x + 165) = -0.02x^2 + 165x$$

$$\text{TOTAL COST} = 18{,}500 + 8.45x$$

$$\text{TOTAL PROFIT} = \text{TOTAL REVENUE} - \text{TOTAL COST}$$

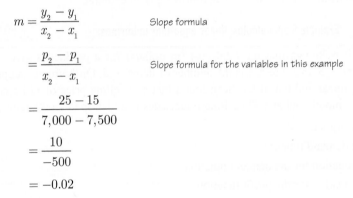

$$P(x) = (-0.02x^2 + 165x) - (18{,}500 + 8.45x)$$

$$= -0.02x^2 + 156.55x - 18{,}500$$

Step 4: *Look back.* The *marginal profit* is the rate of change of profit with respect to changes in the number of items. It will be shown in calculus that the marginal profit for this problem is

$$\lim_{h\to0}\frac{P(x+h)-P(x)}{h}=\lim_{h\to0}\frac{[(-0.02(x+h)^2+156.55(x+h)-18{,}500]-(-0.02x^2+156.55x-18{,}500)}{h}$$

$$=\lim_{h\to0}\frac{-0.02x^2-0.04xh-0.02h^2+156.55x+156.55h-18{,}500+0.02x^2-156.55x+18{,}500}{h}$$

$$=\lim_{h\to0}\frac{-0.04xh-0.02h^2+156.55h}{h}$$

$$=\lim_{h\to0}(-0.04x-0.02h+156.55)$$

$$=-0.04x+156.55$$

So we have found the marginal profit; what does this really mean, and how could it be used? Consider the numerical approach to answer this question. Look at Table 3.1.

TABLE 3.1	Comparison of Prices, Production Levels, Revenue, Cost, Profit, and Marginal Profit				
p Price	x Production	xp Total Revenue	$C(x)$ Total Cost	$P(x)$ Total Profit	$-0.04x+156.55$ Marginal Profit
100	3,250	$325,000	$45,962.50	$297,037.50	26.55
99	3,300	$326,700	$46,385.00	$280,315.00	24.55
98	3,350	$328,300	$46,807.50	$281,492.50	22.55
97	3,400	$329,800	$47,230.00	$282,570.00	20.55
96	3,450	$331,200	$47,652.50	$283,547.50	18.55
95	3,500	$332,500	$48,075.00	$284,425.00	16.55
94	3,550	$333,700	$48,497.50	$285,202.50	14.55
93	3,600	$334,800	$48,920.00	$285,880.00	12.55
92	3,650	$335,800	$49,342.50	$286,457.50	10.55
91	3,700	$336,700	$49,765.00	$286,935.00	8.55
90	3,750	$337,500	$50,187.50	$287,312.50	6.55
89	3,800	$338,200	$50,610.00	$287,590.00	4.55
88	3,850	$338,800	$51,032.50	$287,767.50	2.55
87	3,900	$339,300	$51,455.00	$287,845.00	0.55
86	3,950	$339,700	$51,877.50	$287,822.50	−1.45
85	4,000	$340,000	$52,300.00	$287,700.00	−3.45
84	4,050	$340,200	$52,722.50	$287,477.50	−5.45
83	4,100	$340,300	$53,145.00	$287,155.00	−7.45
82	4,150	$340,300	$53,567.50	$286,732.50	−9.45
81	4,200	$340,200	$53,990.00	$286,210.00	−11.45
80	4,250	$340,000	$54,412.50	$285,587.50	−13.45

Notice that as price decreases, the production increases. This makes sense because the lower the price, the greater the demand. Take a close look at the profit and marginal profit columns. The maximum profit occurs when the product is priced at $87. Also notice that as long as the marginal profit is positive the profit is increasing, and when the marginal profit is negative, the profit is declining. It looks like the "turning point" is when the marginal profit is 0.

Slope of a Tangent

We have just considered the slope as a rate. In calculus, the slope of a curve at a point is defined as the slope of the tangent to the curve at that point. But what do we mean by a tangent line? In Section 1.7, we discussed a *tangent to a circle* as well as the *slope of a secant line*. The slope of a tangent line can be thought of as the slope of a limiting line, as originally introduced in Section 1.7. We now return to this idea.

Consider the concept of a line *tangent* to a given curve (not necessarily a circle) at a given point. In general, it is not a simple matter to define a tangent to a curve at a point on the curve. We formulate a definition of such a line by specifying a property of this line. If the line has this property, it is called a **tangent line** to a curve at a point. That property is simply the *slope* of a tangent line at a given point $P_0(x_0, y_0)$. This is because the formula

$$\text{slope} = m = \frac{\Delta y}{\Delta x} = \frac{y_1 - y_0}{x_1 - x_0}$$

💀 Δx is a single symbol and does not mean delta times x. 💀

requires knowledge of not only the point of tangency (x_0, y_0) but of at least one other point (x_1, y_1) on the line as well.

The limit procedure for finding the slope of a tangent was originally developed by Pierre de Fermat and was later used by Isaac Newton. The brilliance of the Fermat-Newton approach was the use of the "dynamic" limit process to attack the "static" problem of finding slopes of tangents.

Suppose we wish to find the slope of the tangent to $y = f(x)$ at the point $P(x_0, f(x_0))$. The strategy is to approximate the tangent by other lines whose slopes can be computed directly. In particular, consider the line joining the given point P to the neighboring point Q on the graph of f as shown in Figure 3.5. This line is called a **secant** (a line that intersects, but is not tangent to, a curve). Compare the secants shown in Figure 3.5.

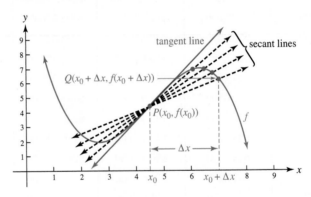

Figure 3.5 The secant $\overleftrightarrow{PQ}$

Notice that a secant is a good approximation to the tangent at point P as long as Q is close to P.

To compute the slope of a secant, first label the coordinates of the neighboring point Q, as indicated in Figure 3.5. In particular, let Δx denote the change in the x-coordinate between the given point $P(x_0, f(x_0))$ and the neighboring point $Q(x_0 + \Delta x, f(x_0 + \Delta x))$. The slope of this secant, m_{sec}, is easy to calculate:

$$m_{\text{sec}} = \frac{\Delta y}{\Delta x} = \frac{f(x_0 + \Delta x) - f(x_0)}{\Delta x}$$

To bring the secant closer to the tangent, let Q approach P *on the graph of f* by letting Δx approach 0. As this happens, the slope of the secant should approach the slope of the tangent at P. We denote the slope of the tangent by m_{tan} to distinguish it from the slope of a secant. These observations suggest the following definition.

> **SLOPE OF TANGENT LINE**
>
> At the point $P(x_0, f(x_0))$, the tangent line to the graph of f has **slope** given by the formula
>
> $$m_{\text{tan}} = \lim_{\Delta x \to 0} \frac{f(x_0 + \Delta x) - f(x_0)}{\Delta x}$$
>
> provided that this limit exists.

EXAMPLE 3   Slope of a tangent at a particular point

Find the slope of the tangent to the graph of $f(x) = x^2$ at the point $P(-1, 1)$.

Solution   Figure 3.6**a** shows the tangent to f at $x = -1$.

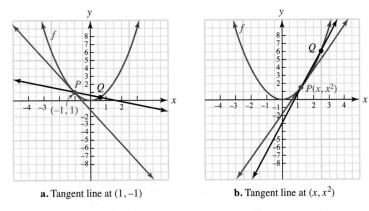

a. Tangent line at $(1, -1)$ **b.** Tangent line at (x, x^2)

Figure 3.6 Tangent lines to the graph of $y = x^2$

The slope of the tangent is given by

$$
\begin{aligned}
m_{\text{tan}} &= \lim_{\Delta x \to 0} \frac{f(-1 + \Delta x) - f(-1)}{\Delta x} \\
&= \lim_{\Delta x \to 0} \frac{(-1 + \Delta x)^2 - (-1)^2}{\Delta x} \qquad \text{Because } f(x) = x^2, f(-1 + \Delta x) = (-1 + \Delta x)^2. \\
&= \lim_{\Delta x \to 0} \frac{1 - 2\Delta x + (\Delta x)^2 - 1}{\Delta x} \\
&= \lim_{\Delta x \to 0} \frac{-2\Delta x + (\Delta x)^2}{\Delta x} \qquad \text{Factor out } \Delta x \text{ and reduce.} \\
&= \lim_{\Delta x \to 0} (-2 + \Delta x) \\
&= -2
\end{aligned}
$$

∎

In Example 3, we found the slope of the tangent $y = x^2$ at the point $(-1, 1)$. In Example 4, we perform the same calculation again, this time representing the given point algebraically as (x, x^2). This is the situation shown in Figure 3.6**b** for the slope of the tangent to $y = x^2$ at *any* point (x, x^2).

EXAMPLE 4   Slope of a tangent at an arbitrary point

Derive a formula for the slope of the tangent to the graph of $f(x) = x^2$, and then use the formula to compute the slope at $(4, 16)$.

Solution   Figure 3.6b shows a tangent at an arbitrary point $P(x, x^2)$ on the curve. From the definition of slope of the tangent,

$$
\begin{aligned}
m_{\text{tan}} &= \lim_{\Delta x \to 0} \frac{f(x + \Delta x) - f(x)}{\Delta x} \\
&= \lim_{\Delta x \to 0} \frac{(x + \Delta x)^2 - x^2}{\Delta x} \qquad \text{Because } f(x) = x^2, f(x + \Delta x) = (x + \Delta x)^2. \\
&= \lim_{\Delta x \to 0} \frac{x^2 + 2x\Delta x + (\Delta x)^2 - x^2}{\Delta x} \\
&= \lim_{\Delta x \to 0} \frac{2x\Delta x + (\Delta x)^2}{\Delta x} \qquad \text{Factor out } \Delta x \text{ and reduce.} \\
&= \lim_{\Delta x \to 0} (2x + \Delta x) \\
&= 2x
\end{aligned}
$$

At the point $(4, 16)$, $x = 4$, so $m_{\text{tan}} = 2(4) = 8$. ∎

The result of Example 4 gives a general formula for the slope of a line tangent to the graph of $f(x) = x^2$, namely, $m_{\text{tan}} = 2x$. The answer from Example 3 can now be verified using this formula; if $x = -1$, then $m_{\text{tan}} = 2(-1) = -2$.

EXAMPLE 5   Equation of a tangent

Find an equation of the tangent to the graph of $f(x) = \dfrac{1}{x}$ at the point where $x = 2$.

Solution   The graph of the function $y = \dfrac{1}{x}$, the point where $x = 2$, and the tangent at the point are shown in Figure 3.7.

First, find the slope of the tangent line:

$$
\begin{aligned}
m_{\text{tan}} &= \lim_{\Delta x \to 0} \frac{f(x + \Delta x) - f(x)}{\Delta x} \\
&= \lim_{\Delta x \to 0} \frac{\dfrac{1}{x + \Delta x} - \dfrac{1}{x}}{\Delta x} \qquad \text{Because } f(x) = \frac{1}{x}, f(x + \Delta x) = \frac{1}{x + \Delta x}. \\
&= \lim_{\Delta x \to 0} \frac{x - (x + \Delta x)}{x\Delta x(x + \Delta x)} \qquad \text{Simply the fraction.} \\
&= \lim_{\Delta x \to 0} \frac{-1}{x(x + \Delta x)} \\
&= \frac{-1}{x^2} \qquad \begin{array}{l}\text{Since the numerator is a constant, we look at the} \\ \text{denominator. The denominator is a polynomial; as} \\ \Delta x \to 0, x(x + \Delta x) \to x^2.\end{array}
\end{aligned}
$$

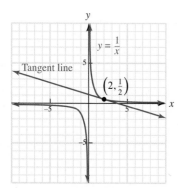

Figure 3.7 Tangent to $y = \dfrac{1}{x}$ **at** $\left(2, \dfrac{1}{2}\right)$

Next, find the slope of the tangent at $x = 2$: $m_{\tan} = -\dfrac{1}{4}$. Finally, find the point of tangency $(2, f(2)) = \left(2, \dfrac{1}{2}\right)$. The equation of the tangent can now be found by using the point-slope form:

$$y - \frac{1}{2} = -\frac{1}{4}(x - 2)$$

or in standard form, multiply both sides by 4 and simplify to obtain $x + 4y - 4 = 0$. ∎

EXAMPLE 6   A line that is perpendicular to a tangent

Find the equation of the line that is perpendicular to the tangent of $f(x) = \dfrac{1}{x}$ at $x = 2$ and intersects it at the point of tangency.

Solution   From Example 5, we found that the slope of the tangent is $-\dfrac{1}{4}$ and that the point of tangency is $\left(2, \dfrac{1}{2}\right)$. In Section 1.3, we saw that two lines are perpendicular if and only if their slopes are negative reciprocals of each other. Thus, the perpendicular line we seek has slope 4 (the negative reciprocal of $-\dfrac{1}{4}$), as shown in Figure 3.8.

The desired equation is

$$y - \frac{1}{2} = 4(x - 2) \quad \text{Point-slope formula}$$

∎

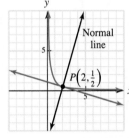

Figure 3.8 Graph of tangent and normal lines

The line we found in Example 6 has a name, which is defined in the following box.

> **NORMAL LINE**
>
> The **normal line** to the graph of f at the point P is the line that is perpendicular to the tangent to the graph at P.

PROBLEM SET 3.1

LEVEL 1

1. **IN YOUR OWN WORDS** What is a linear function?
2. **IN YOUR OWN WORDS** What is meant by a line tangent to a curve at a given point?

Find a linear function satisfying the conditions in Problems 3–6.

3. $f(2) = 5$ and $f(3) = 12$
4. $f(-2) = 6$ and $f(3) = 12$
5. $g(-1) = \dfrac{4}{5}$ and $g(3) = -\dfrac{2}{3}$
6. $g(1) = \dfrac{2}{3}$ and $g(3) = -\dfrac{4}{5}$

Find the velocity of the line passing through the points (t, m) given in Problems 7–14.

7. $(3,165), (4, 220)$
8. $(2, 136), (5, 340)$
9. $(3, 25), (8, 75)$
10. $(1, 8), (3, 24)$
11. $(1, 2), (6, 10)$
12. $(3, 6), (5, 6)$
13. $(0, 0), (5, 20)$
14. $(0, 0), (8, 320)$

PROBLEMS FROM CALCULUS *In Problems 15–18, use the graph to state the coordinates of A and B, and then find the slope of the line passing through A and B.*

15.

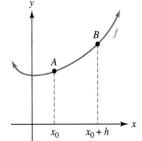

16.

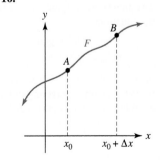

17.

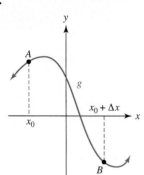

18.

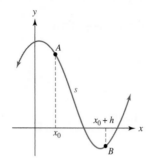

Find the slope of the lines tangent to the curves given in Problems 19–24 at the point c.

19. $f(x) = 2x$ at $c = 1$ **20.** $f(x) = -5x$ at $c = 8$

21. $f(x) = 2x^2$ at $c = 1$ **22.** $f(x) = 2 - x^2$ at $c = 0$

23. $f(x) = \dfrac{2}{x}$ at $c = 1$ **24.** $f(x) = -\dfrac{1}{x}$ at $c = \dfrac{1}{2}$

Find an equation for the tangent to the graph of the function at the specified point in Problems 25–30.

25. $f(x) = 3x - 7$ at $(3, 2)$ **26.** $g(x) = 5 - 2x$ at $(3, -1)$

27. $h(x) = 9 - x^2$ at $(0, 9)$ **28.** $F(x) = -x^2$ at $(1, -1)$

29. $f(s) = s^3$ at $s = 2$ **30.** $f(t) = \dfrac{1}{t+3}$ at $t = -1$

Find an equation of the normal line to the graph of the functions of Problems 25–30 at the specified point in Problems 31–36.

31. $f(x) = 3x - 7$ at $(3, 2)$ **32.** $g(x) = 5 - 2x$ at $(3, -1)$

33. $h(x) = 9 - x^2$ at $(0, 9)$ **34.** $F(x) = -x^2$ at $(1, -1)$

35. $f(s) = s^3$ at $s = 2$ **36.** $f(t) = \dfrac{1}{t+3}$ at $t = -1$

37. The distance, d, in feet, traveled by an object in t seconds is given by $d = 35t$.
   a. What is the velocity of the object?
   b. What is the distance covered in 10 sec?

38. A point moves along the x-axis and its x-component after t seconds is $x = 5t + 15$ cm.
   a. What is the velocity of the point?
   b. What is the distance covered in 1 minute?

39. Two points P and Q are moving (in ft/s) along the y-axis according to the equations:

$$P: y = 5t + 105$$
$$Q: y = 20t$$

   a. Which point has a "head start"?
   b. Which point is traveling faster?
   c. When does the point Q catch up to point P?

40. Two points P and Q are moving (in m/s) along the x-axis according to the equations:

$$P: x = 25t + 250$$
$$Q: x = 50t - 50$$

   a. Which point has a "head start"?
   b. Which point is traveling faster?
   c. When does the point Q catch up to point P?

41. Suppose you have a spring with a maximum stretch of 12 in. With no weight on the end, there is no stretch of the spring, but with a 10-lb weight the spring is stretched 6 in. Write a linear function expressing this relationship, and then use the function to find the weight if the spring is stretched 9.5 in. Also, draw the graph for a stretch between 0 and 12 in.

42. If a 250-kg weight stretches a certain spring 15 cm, write a linear function expressing this relationship, and then use the function to find the weight of an object that stretches the spring 8.4 cm. Assume that there is no stretch of the spring when there is no weight. Also draw the graph for a stretch between 0 and 21 cm.

43. The distance d, in miles, traveled by a car in t hours $(0 \le t \le 5)$ is given by the equation $d = 60t$.
   a. What is the car's velocity at time $t = 0$?
   b. What is the distance covered in 3 hours?
   c. What is the distance covered during the third hour?
   d. What is the car's velocity at time $t = 3$?

44. The distance d, in miles, traveled by a car in t hours $(0 \le t \le 5)$ is given by the equation $d = -2(t - 5)^2 + 50$.
   a. What is the car's velocity at time $t = 0$?
   b. What is the distance covered in 3 hours?
   c. What is the distance covered during the third hour?
   d. What is the car's velocity at time $t = 3$?

LEVEL 2

Use Table 3.1 to answer the questions in Problems 45–48. Assume the independent variable is the price of the item and the domain for price is [$80, $100].

45. If you lower the price,
   a. what happens to production?
   b. what happens to revenue?
   c. what happens to total cost?
   d. what happens to the total profit?

46. If you raise the price,
   a. what happens to production?
   b. what happens to revenue?
   c. what happens to total cost?
   d. what happens to the total profit?

47. a. What is the maximum total revenue? What price(s) provide the maximum revenue?
   b. What is the minimum total cost?
   c. Which provides more profit, setting a price that maximizes revenue or minimizes cost?

48. a. What is the minimum production?
   b. What is the maximum production?
   c. Explain why profit is what you need to maximize, and explain how you can use this table to maximize profit.

49. Suppose the cost, $C(x)$ in dollars, of producing x cribs is given by $C(x) = 200 + 150x$.
  a. Graph this cost function on $[0, 10]$.
  b. What is the marginal cost?
  c. Find $C(x+1) - C(x)$. Compare with your answer to part b.
50. Suppose the cost, $C(x)$ in dollars, of producing x barbecues is given by $C(x) = 50 + 0.15x^2$.
  a. Graph this cost function on $[0, 200]$.
  b. Find $C(200)$.
  c. What is the marginal cost?

Assume the depreciation for each item described in Problems 51–54 is linear. This is called straight-line depreciation.

51. A sand and gravel company purchases a piece of property for $215,000. It is estimated that the property will have a resale value of $35,000 after 10 years. What is the rate of depreciation?
52. An apartment building is purchased for $382,000 and has an effective life of 40 years. If the site is still worth $50,000 after that period, what is the rate of depreciation?
53. A $15,000 machine has an expected life of eight years and a scrap value of $1,000. If the machine is replaced after five years, what is its value at the time of replacement?
54. A sales representative depreciates her car over five years. If it is purchased for $48,400 and is presumed to have a scrap value of $5,950, what is its depreciated value after two years?
55. Find the coordinates of A, B, and C for the linear function $y = mx + b$ shown in Figure 3.9.

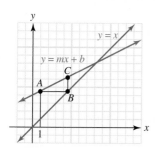

Figure 3.9 Graph of $y = mx + b$

56. Find the coordinates of A, B, and C for the linear function $y = mx + b$ shown in Figure 3.10.

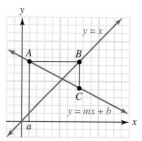

Figure 3.10 Graph of $y = mx + b$

57. Show that the linear function $f(x) = mx$ satisfies
  a. $f(x_1 + x_2) = f(x_1) + f(x_2)$
  b. $f(ax_0) = af(x_0)$
  c. Does the linear function $g(x) = mx + b$ satisfy either of the properties in parts a or b?
58. A function f is defined on the set of real numbers so that
$$f(x + y) = y(x) + f(y)$$
If $f(1) = 1$, find $f\left(\dfrac{1}{2}\right)$.
59. A function f is defined on the set of real numbers so that
$$f(x + y) = f(x) + f(y)$$
If $f(3) = 6$, find $f\left(\dfrac{3}{2}\right)$.
60. Consider the following functional property:
$$f\left(\frac{x + y}{2}\right) = \frac{f(x) + f(y)}{2}$$
  a. Does this property hold for the general linear function $f(x) = mx + b$?
  b. Does this property hold for the function $f(x) = x^2$?

3.2 Quadratic Functions

Graphing Quadratic Functions

The second type of polynomial function to be considered in this chapter is the quadratic function.

QUADRATIC FUNCTION

A function f is a **quadratic function** if
$$f(x) = ax^2 + bx + c$$
where a, b, and c are real numbers and $a \neq 0$

If $b = c = 0$, however, the quadratic function has the form $y = ax^2$ and has a graph called a **standard-position parabola**. We translated this function in Section 2.4. The graph of $y = x^2$, along with $y = ax^2$ (for various choices of a) is shown in Figure 3.11.

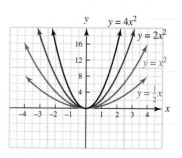

Figure 3.11 Graph of $y = ax^2$

Notice that for $f(x) = ax^2$,

$$f(-x) = a(-x)^2 = ax^2$$

so the graph is symmetric with respect to the y-axis. We also know that graphs of quadratic functions are continuous since they are polynomial functions. If $a > 0$, we say the parabola *opens upward*, and if $a < 0$, we say the parabola *opens downward*. The point $(0, 0)$ is the lowest point on the parabola if the parabola opens upward $(a > 0)$; $(0, 0)$ is the highest point if the parabola opens downward $(a < 0)$. This highest or lowest point is called the **vertex**. The parabola is symmetric with respect to the vertical line passing through the vertex.

> **MAXIMUM/MINIMUM**
>
> If $y = ax^2 + bx + c$, $a < 0$, then the **maximum** value of y is located at the vertex of the downward-opening parabola.
>
> Also, if $a > 0$, then the **minimum** value of y is located at the vertex of the upward-opening parabola.

Relative to a fixed scale, the magnitude of a determines the "wideness" of the parabola. Small values of $|a|$ yield "wide" parabolas while large values of $|a|$ yield "narrow" parabolas.

For graphs of parabolas of the form

$$y - k = a(x - h)^2$$

you simply translate the origin to the point (h, k) and then graph the parabola with the same shape, namely, $y = ax^2$, as shown in the following example. The point (h, k) is the vertex. Using functional notation, we can write

$$f(x) = a(x - h)^2 + k$$

EXAMPLE 1   Graphing a translated parabola

Graph $y + 5 = 3(x + 2)^2$ by:

a. plotting points   **b.** translation   **c.** using symmetry and the y-intercept

Solution

a. By plotting points:

x	y
0	7
1	22
-1	-2
-2	-5
-3	-2
-4	7

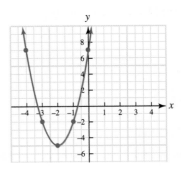

b. By inspection, $(h, k) = (-2, -5)$. Plot this point, and then plot the curve $y = 3x^2$ using the point $(-2, -5)$ as the starting point instead of the point $(0, 0)$.

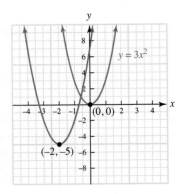

c. Note that $y + 5 = 3(x + 2)^2$ Vertex is $(-2, -5)$.

$$y = 3x^2 + 12x + 12 - 5$$

$$= 3x^2 + 12x + 7 \qquad \text{y-intercept is } (0, 7).$$

Plot the vertex, $(-2, -5)$, and then the y-intercept, $(0, 7)$. Using these two points, find a third point using symmetry, and then sketch the graph by drawing a smooth curve. The minimum value of y is -5.

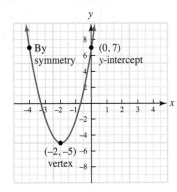

We are now ready to consider the general quadratic function

$$y = ax^2 + bx + c \qquad a \neq 0$$

Look at Example 1c and note that $y + 5 = 3(x + 2)^2$ was written as $y = 3x^2 + 12x + 7$. Our task now is to begin with this latter form and write it in factored form. This process is known as completing the square (review Section 1.8). We will show the general form as well as a particular example from Example 1c.

	General Form	Example
	$y = ax^2 + bx + c$	$y = 3x^2 + 12x + 7$

Step 1: Subtract c (the constant term) from both sides.

$$y - c = ax^2 + bx \qquad\qquad y - 7 = 3x^2 + 12x$$

Step 2: Factor a from the expression on the right (remember $a \neq 0$).

$$y - c = a\left(x^2 + \frac{b}{a}x\right) \qquad\qquad y - 7 = 3(x^2 + 4x)$$

Step 3: To complete the square on the number inside the parentheses, square one-half the coefficient of the x-term, and then add a times this number to both sides.

$$y - c + \frac{b^2}{4a} = a\left(x^2 + \frac{b}{a}x + \frac{b^2}{4a^2}\right) \qquad\qquad y - 7 + 12 = 3(x^2 + 4x + 4)$$

Step 4: Factor the right-hand side as a perfect square, and simplify the left-hand side:

$$y + \frac{-4ac}{4a} + \frac{b^2}{4a} = a\left(x + \frac{b}{2a}\right)^2 \qquad\qquad y + 5 = 3(x + 2)^2$$

$$y + \frac{b^2 - 4ac}{4a} = a\left(x + \frac{b}{2a}\right)^2$$

The vertex is:

$$\left(-\frac{b}{2a}, -\frac{b^2 - 4ac}{4a}\right) \qquad\qquad (-2, -5)$$

The form $y - k = a(x - h)^2$ is called the **general form of a parabola**. This equation is graphed by doing a translation as shown in the following example.

EXAMPLE 2   Graphing a parabola by completing the square

Sketch the graph of $y = 1 + 6x - 2x^2$.

Solution

$$y = 1 + 6x - 2x^2 \qquad \text{Given}$$

$$y - 1 = -2x^2 + 6x \qquad \text{Subtract 1 from both sides.}$$

$$y - 1 = -2(x^2 - 3x) \qquad \text{Factor the coefficient of the square term.}$$

$$y - 1 + \left[-2\left(-\frac{3}{2} \right)^2 \right] = -2\left[x^2 - 3x + \left(-\frac{3}{2} \right)^2 \right] \qquad \text{Complete the square.}$$

$$y - 1 - \frac{9}{2} = -2\left(x - \frac{3}{2} \right)^2 \qquad \text{Simplify.}$$

$$y - \frac{11}{2} = -2\left(x - \frac{3}{2} \right)^2$$

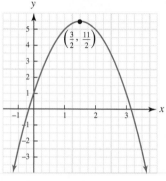

Graph by plotting the vertex at $\left(\dfrac{3}{2}, \dfrac{11}{2} \right)$ and then translating the graph of $y = -2x^2$ to this vertex location. Also note the parabola is narrow and opens down, and that the y-intercept is $(0, 1)$. The graph is shown in Figure 3.12. Notice that when plotting values like $\dfrac{3}{2}$ and $\dfrac{11}{2}$ it is convenient to choose two squares to represent 1 unit.

Figure 3.12 Graph of $y = 1 + 6x - 2x^2$

Modeling with Functions

One of the most useful techniques of problem solving is to define a function and then evolve the variables to obtain an equation involving only the defined variable. While many of the equations can be modeled by a linear or quadratic function, we will not limit ourselves to functions of only the first or second degree. A common problem in calculus is to maximize or minimize functions, a topic we will consider in the next section.

EXAMPLE 3   Product problem

The sum of two numbers is 100. Express the product of these numbers as a function of a single variable, and state the domain for the variable.

Solution   Let x and y be the numbers so that $x + y = 100$. Then the product, P, is $P = xy$. Write P as a function of a single variable by solving the first equation: $y = 100 - x$, so that

$$P = xy$$
$$= x(100 - x)$$

Therefore, we can write

$$P(x) = 100x - x^2$$

The domain of this function is $\mathbb{R}$; that is, $(-\infty, \infty)$.

Preview: In the next section, we will seek the maximum or minimum value of a function. For Example 3, we can find the maximum value numerically, geometrically, and algebraically.

Numerical Approach	Geometrical Approach	Algebraic Approach

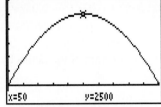

From the table, it looks like the maximum product is 2,500.

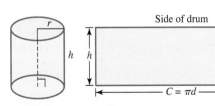

$\y1\triangleq100x-x^2$
xMin=0   yMin=-500
xMax=100 yMax=3000
xScl=10   yScl=500

From the graph, it looks like the maximum value is at the point $(50, 2{,}500)$.

$y = 100x - x^2$

$y = -(x^2 - 100x)$

$y - 50^2 = -\left[x^2 - 100x + \left(-\dfrac{100}{2}\right)^2\right]$

$y - 50^2 = -(x - 50)^2$

We see the vertex is $(50, 50^2)$, so the maximum value is $50^2 = 2{,}500$, which occurs at $x = 50$.

EXAMPLE 4   Geometric problem

Express the area as a function of x, the length of the base of the triangular region in the fourth quadrant formed by picking a point with coordinates (x, y) on the parabola $y = x^2 - 9$. State the domain for the area function.

Solution   Draw a picture, as shown in Figure 3.13.

To express the area of the triangular region, we use the formula $A = \dfrac{1}{2}bh$. The base of the triangle is x, and the height of the triangle is $-y$. (The height of the triangle is $|y| = -y$, since y is negative.) Also, since $P(x, y)$ is a point on the parabola $y = x^2 - 9$, we have the area as a function of x as

$$A(x) = \frac{1}{2}x[-(x^2 - 9)]$$

$$= -\frac{1}{2}x(x^2 - 9)$$

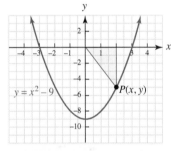

Figure 3.13 Area function

By looking at Figure 3.13, or by solving $-\dfrac{1}{2}x(x^2 - 9) \geq 0$, we see that the domain of this area function is $[0, 3]$.

EXAMPLE 5   Example from calculus: construction problem

Suppose we wish to design a cylindrical drum with radius r and height h. The top and bottom cost \$0.05/cm² and the curved sides cost \$0.10/cm². If the cost of the drum is $\$15\pi \approx \$47.12$, write:

a. the height as a function of the radius;

b. the volume as a function of the radius.

Also, state the domain for each function.

Solution   First, consider the drum in three parts (side, top, and bottom) to find the area:

Drum:	Sides (rolled out)	Top/bottom
Area:	$A = \pi dh = 2\pi rh$	$A = \pi r^2$
Cost:	$C = 0.10(2\pi rh)$	$C = 0.05(\pi r^2)$

Since the total cost is 15π, we have

$$0.10(2\pi rh) + 0.05\pi r^2 + 0.05\pi r^2 = 15\pi$$
$$0.2\pi rh + 0.1\pi r^2 = 15\pi$$

a. We find the height as a function of the radius by solving for h:

$$0.2\pi rh + 0.1\pi r^2 = 15\pi$$
$$0.2\pi rh = 15\pi - 0.1\pi r^2$$
$$h = \frac{150 - r^2}{2r}$$

Thus, $h(r) = \dfrac{150 - r^2}{2r}$; for the domain (i.e., the permissible values for r), we note that since

$h \geq 0$, $\dfrac{150 - r^2}{2r} \geq 0$. But since $2r > 0$, $150 - r^2 \geq 0$ or $r \leq \sqrt{150}$. Thus, the domain is

$\left(0, \sqrt{150}\right]$.

b. Since $V = Ah$, we have

$$V(r) = \pi r^2 h$$
$$= \pi r^2 \left[\frac{150 - r^2}{2r}\right]$$
$$= \frac{\pi r}{2}(150 - r^2)$$

The domain is $\left(0, \sqrt{150}\right]$. ∎

EXAMPLE 6　　**Example from calculus: sawmill problem***　　MODELING APPLICATION

Suppose that you need to cut a beam with maximal rectangular cross section from a circular log of radius 1 ft. (This is the geometric problem of finding the rectangle of greatest area that can be inscribed in a circle of radius 1.) Write the area of the beam as a function of x, and state the domain of this function.

Solution

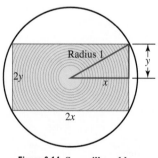

Figure 3.14 Sawmill problem

Step 1:　*Understand the problem.*　What do we mean by a "rectangular cross section"? We draw a figure to help us understand the problem. We also know wood logs are not circular, but we make that assumption for this problem.

Step 2:　*Devise a plan.*　This example illustrates a useful technique in choosing variables to avoid fractions. Instead of letting x and y denote the base and height, we find it convenient to let x and y denote half the base and half the height, respectively, of the beam, as shown in Figure 3.14. We will use the Pythagorean theorem to find the height, y.

Step 3:　*Carry out the plan.*

$$x^2 + y^2 = 1$$
$$y^2 = 1 - x^2$$
$$y = \pm\sqrt{1 - x^2} \quad \text{Square root property}$$

Since y is a length, select the positive value for y. Then,

$$A = (2x)(2y)$$
$$A(x) = 4x\sqrt{1 - x^2} \quad \text{Write the area as a function of } x.$$

---

*This example is from *Calculus, 5th Edition* by Edwards and Penney (Englewood Cliffs: Prentice Hall), p. 155.

Step 4: *Look back.* The domain of A is $[0, 1]$. We have now written the area as a function of x.

PROBLEM SET 3.2

LEVEL 1

Match each equation in Problems 1–4 to one of the given graphs. Compare with $y = x^2$.

1. a. $f(x) = 3x^2$

 b. $y = \dfrac{1}{3}x^2$

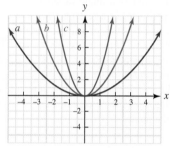

2. a. $f(x) = x^2 - 2$
 b. $y = (x - 2)^2$

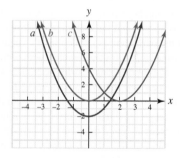

3. a. $y - 3 = (x + 4)^2$
 b. $y + 3 = (x - 4)^2$

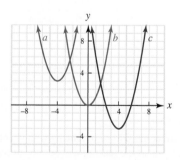

4. a. $y - 2 = \dfrac{1}{2}(x - 3)^2$
 b. $y - 2 = 2(x - 3)^2$

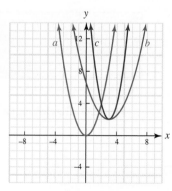

WHAT IS WRONG, if anything, with each statement in Problems 5–13? Explain your reasoning.

5. A function f defined by

$$f(x) = ax^2 + bx + c$$

is a quadratic function.

6. $f(x) = 5x^2 - 3x - 5x^2 + 2$ is a quadratic function.

7. The parabola $x - 5 = 3(y + 2)^2$ opens upward.

8. The vertex of the parabola

$$y - 3 = 2(x + 4)^2$$

is at the origin.

9. The graph of $y + 1 = \dfrac{1}{2}(x - 8)^2$ can be considered as a translation of the graph $y = x^2$.

10. To complete the square on

$$y + 25 = 2(x^2 + 20x)$$

add 10^2 to both sides.

11. To complete the square on

$$y = 3(x^2 + 8x)$$

you must add 16 to both sides.

12. The maximum value of

$$y - 3 = 4(x - 5)^2$$

is $y = 3$.

13. The minimum value of

$$y + 400 = -6(x + 4)^2$$

is $y = -400$.

In Problems 14–22, graph the quadratic function, and specify the vertex.

14. a. $y = -x^2$ **b.** $y = 2x^2$

15. a. $y = 5x^2$ **b.** $y = -\dfrac{1}{5}x^2$

16. a. $y = -2\pi x^2$ **b.** $y = -\dfrac{1}{10}x^2$

17. a. $y = (x - 3)^2$ **b.** $y = (x - 1)^2$

18. $f(x) = -2(x - 1)^2$ **19.** $f(x) = \dfrac{1}{4}(x - 1)^2$

20. $y - 2 = 3(x + 3)^2$ **21.** $y + 3 = \dfrac{2}{3}(x + 2)^2$

22. $y - 1 = \dfrac{1}{3}(x - 4)^2$

Adjust the scale on each of the graphs in Problems 23–28 so that they look like parabolas. Show the graph of each parabola.

23.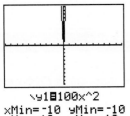

```
         \y1⊟100x^2
xMin=-10  yMin=-10
xMax=10   yMax=10
xScl=1    yScl=1
```

24.

```
         \y1⊟.01x^2
xMin=-10  yMin=-10
xMax=10   yMax=10
xScl=1    yScl=1
```

25.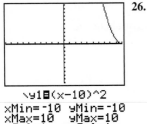

```
         \y1⊟(x-10)^2
xMin=-10  yMin=-10
xMax=10   yMax=10
xScl=1    yScl=1
```

26.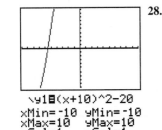

```
         \y1⊟x^2+10
xMin=-10  yMin=-10
xMax=10   yMax=10
xScl=1    yScl=1
```

27.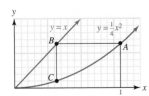

```
         \y1⊟(x+10)^2-20
xMin=-10  yMin=-10
xMax=10   yMax=10
xScl=1    yScl=1
```

28.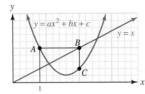

```
         \y1⊟-(x+10)^2+20
xMin=-10  yMin=-10
xMax=10   yMax=10
xScl=1    yScl=1
```

LEVEL 2

In Problems 29–30, find the coordinates of the points A, B, and C.

29. $f(x) = 0.25x^2$

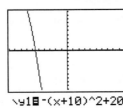

30. $f(x) = ax^2 + bx + c$

Sketch the graph of each equation given in Problems 31–44. State the maximum or minimum value of y.

31. $y = x^2 + 4x + 4$

32. $y = x^2 + 6x + 9$

33. $y = x^2 + 2x - 3$

34. $y = 2x^2 - 4x + 5$

35. $y = 2x^2 - 4x + 4$

36. $y = 2x^2 + 8x + 5$

37. $y = \frac{1}{2}x^2 + 2x - 1$

38. $y = \frac{1}{2}x^2 + 4x + 10$

39. $y = -3x^2 - 12x - 10$

40. $y = -3x^2 - 30x - 76$

41. $x^2 - 6x - 2y - 1 = 0$

42. $3x^2 - 12x - y + 10 = 0$

43. $x^2 + 2x + 2y - 3 = 0$

44. $2x^2 - 4x + 3y + 11 = 0$

45. What is the vertex of the parabola

$$y = ax^2 + bx + c, \quad a \neq 0$$

46. The sum of two numbers is 17. Express the product of these numbers as a function of a single variable.

47. The perimeter of a rectangle is 44 cm. Express the area of the rectangle as a function of a single variable.

48. Find the area of the triangular region in the third quadrant formed by picking a point with coordinates (x, y) on the parabola $y = x^2 - 100$ and the points $(0, 0)$, $(0, y)$.

49. Find the area of the triangular region in the first quadrant formed by picking a point with coordinates (x, y) on the parabola $y = 16 - x^2$ and the points $(0, 0)$, $(x, 0)$.

PROBLEMS FROM CALCULUS*

50. A rectangular box has a square base with edges at least 1 in. long. It has no top, and the total area of its five sides is 300 in.². Write the volume of the box as a function of a single variable.

51. A farmer has 600 m of fencing with which to enclose a rectangular pen adjacent to a long existing wall. She will use the wall for one side of the pen and the available fencing for the remaining three sides. Write the area of the pen as a function of a single variable.

52. A rectangle of fixed perimeter 36 is rotated about one of its sides, thus sweeping out a figure in the shape of a right circular cylinder, as shown in Figure 3.15. Write the volume of the cylinder as a function of a single variable.

Figure 3.15 Volume of a cylinder

53. A farmer has 600 yd of fencing with which to build a rectangular corral. Some of the fencing will be used to construct two internal divider fences, both parallel to the same two sides of the corral, as shown in Figure 3.16. Write the total area of the corral as a function of a single variable.

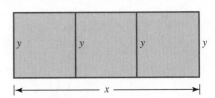

Figure 3.16 Cross fencing for a corral

---

* Problems 50–53 are adapted from Calculus, Fifth Edition *by Edwards and Penney.*

Let $P(x, y)$ be a point on the curve $y = \sqrt{x}$ (see Figure 3.17). Express the requested quantity in Problems 54–56 as a function of x.

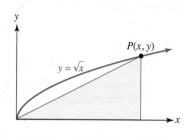

Figure 3.17 Graph of $y = \sqrt{x}$

54. Distance of $P(x, y)$ to $(0, 0)$
55. The area of the shaded triangle
56. The perimeter of the shaded triangle

LEVEL 3

57. A rectangle is inscribed in a circle of diameter 10 cm. Express the perimeter of the rectangle as a function of a single variable.
58. Express $\overline{|AB|}$ in Figure 3.18 as a function of x.

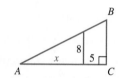

Figure 3.18 Similar triangles

59. Express the volume of a right circular cone with surface area 24π cm² as a function of its radius. A right circular cone along with the formulas for volume V and the lateral surface area S are given in Figure 3.19.

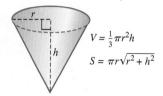

$$V = \frac{1}{3}\pi r^2 h$$
$$S = \pi r \sqrt{r^2 + h^2}$$

Figure 3.19 A right circular cone

60. Suppose the height and radius of a right circular cone (see Figure 3.19) are related by the equation $h = \sqrt{3}r$. Express the lateral surface area as a function of r.

3.3 Optimization Problems

It is common to ask for the best procedure, the greatest value, the least cost, or the shortest path. The process of developing something to the utmost extent is called **optimization**. Entire courses in mathematics are devoted to this topic, and in this section, we will lay the groundwork for a process that will continue when you take calculus.

Optimization Procedure

We will develop procedures to solve real-life problems that seek the maximum or minimum value. We will model our procedure after one attributed to George Pólya.

OPTIMIZATION PROCEDURE

Here are the steps for finding the maximum or minimum value.

Step 1 Understand the problem. Ask yourself if you can separate the given quantities and those you must find. What is unknown? Draw a picture to help you understand the problem.

Step 2 Choose the variables. Decide which quantity is to be optimized (that is, maximized or minimized) and call it Q (or some other suitable variable). Choose other variables for unknown quantities and label your diagram using these symbols.

Step 3 Express Q in terms of the variables defined in Step 2. Use the information given in the problem and the principle of substitution to rewrite Q in terms of a single variable, say, x.

Nothing takes place in the world whose meaning is not that of some maximum or minimum.

—Leonhard Euler

> **» IN OTHER WORDS** Q may begin as a formula involving several variables, but by using given information or known formulas, the goal is to write Q as a function of *one* variable so that $Q = f(x)$. This is the process we discussed in the last section.

Step 4   Find the domain $[a,\ b]$ for the function $Q = f(x)$.

Step 5   If the function is quadratic, the maximum and minimum values of f will occur at the vertex or endpoints $x = a$, $x = b$. If the function is not quadratic, you may be able to find the maximum or minimum, but in general, higher-order functions will require calculus to find the optimal value.

Step 6   Convert the result obtained in Step 5 back into the context of the original problem, making all appropriate interpretations. Be sure to answer the question asked.

One type of optimization problem involves making a decision based on a mathematical analysis of the problem. Typically, a product is being sold at some price, and a price increase is contemplated. It is known that an increase in price will increase the revenue if the number of items sold does not change; however, the price increase may decrease the number of items sold. The problem is to balance the increased revenue against the decrease in demand. For these problems, you will need the following formulas:

$$\text{PROFIT} = \text{REVENUE} - \text{COST}$$

$$\text{REVENUE} = (\text{NUMBER SOLD})(\text{PRICE PER ITEM})$$

$$\text{COST} = (\text{NUMBER SOLD})(\text{COST PER ITEM}) + \text{FIXED COSTS}$$

Sometimes the price per item is given in terms of what is known as a **demand function**, which is a given relationship between price and number of items sold. We illustrate these ideas with the following examples.

EXAMPLE 1   Determining the price increase to maximize profits

A manufacturer can produce a pair of earrings at a cost of \$3. The earrings have been selling for \$5 per pair, and at this price, consumers have been buying 4,000 per month. The manufacturer is planning to raise the price of the earrings and estimates that for each \$1 increase in the price, 400 fewer pairs of earrings will be sold each month. At what price should the manufacturer sell the earrings to maximize profit? Assume there are no fixed costs.

Solution

Step 1   If we sell 4,000 earrings at \$5/pair, then the revenue is \$5(4,000) = \$20,000. The cost of manufacturing these earrings is \$3(4,000) = \$12,000, so the profit is

$$\$20,000 - \$12,000 = \$8,000$$

If we raise the price by \$1, then we will sell 3,600 earrings at \$6/pair for revenue of \$21,600 and cost of \$10,800, for a profit of \$10,800. This is better than selling them at \$5/pair. The question is, how many price increases are necessary to maximize the profit?

Step 2   Let x denote the number of \$1 price increases, and let $P(x)$ represent the corresponding profit. (*Note*: This variable P is the quantity to be maximized; we called it Q in the optimization procedure.)

Step 3

$$\text{PROFIT} = \text{REVENUE} - \text{COST}$$

$$= (\text{NUMBER SOLD})(\text{PRICE PER PAIR}) - (\text{NUMBER SOLD})(\text{COST PER PAIR})$$

$$= (\text{NUMBER SOLD})(\text{PRICE PER PAIR} - \text{COST PER PAIR})$$

Recall that 4,000 pairs of earrings are sold each month when the price is $5 per pair, and 400 fewer pairs will be sold each month for each added dollar in the price. Thus,

NUMBER OF PAIRS SOLD $= 4,000 - 400($NUMBER OF $1 INCREASES$)$

$$= 4,000 - 400x$$

Knowing that the price per pair is $5 + x$, we can now write the profit as a function of x:

$$P(x) = (\text{NUMBER SOLD})(\text{PRICE PER PAIR} - \text{COST PER PAIR})$$
$$= (4,000 - 400x)[(5 + x) - 3]$$
$$= 400(10 - x)(2 + x)$$

Step 4 To find the domain, we note that $x \geq 0$. And $400(10 - x)$, the number of pairs sold, should be nonnegative, so $x \leq 10$. Thus, the domain is $[0, 10]$.

Step 5 Because this is a quadratic function, the optimal value is at the vertex. Algebraically, we complete the square:

$$y = 400(10 - x)(2 + x) \qquad \text{Given}$$
$$y = 400(20 + 8x - x^2)$$
$$y - 8,000 = -400(x^2 - 8x) \qquad \text{Subtract 8,000 from both sides.}$$
$$y - 8,000 - 6,400 = -400(x^2 - 8x + 4^2) \qquad \text{Complete the square.}$$
$$y - 14,400 = -400(x - 4)^2 \qquad \text{Factor.}$$

The graphical solution is shown in Figure 3.20, and because the parabola opens downward, the maximum value occurs at the vertex, $(4, 14,400)$.

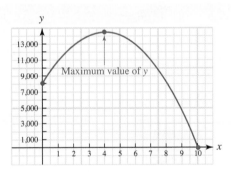

Figure 3.20 The profit function $P(x)$

Step 6 The maximum possible profit is $14,400, which will be generated if the earrings are sold for $9.00 per pair. ($5 plus four $1 price increases). ∎

EXAMPLE 2   Maximum profit

A manufacturer estimates that when x units of a particular commodity are produced each month, the total cost (in dollars) will be

$$C(x) = \frac{1}{8}x^2 + 4x + 200$$

and all units can be sold at a price of $p(x) = 49 - x$ dollars per unit. Determine the price that corresponds to the maximum profit.

Solution   The profit is found using the following formula:

$$\text{PROFIT} = \text{REVENUE} - \text{COST}$$

$$\text{PROFIT} = (\text{NUMBER OF ITEMS})(\text{PRICE PER ITEM}) - \text{COST}$$

$$\text{PROFIT} = x(49 - x) - \left(\frac{1}{8}x^2 + 4x + 200\right)$$

$$= 49x - x^2 - \frac{1}{8}x^2 - 4x - 200$$

$$= -\frac{9}{8}x^2 + 45x - 200$$

We recognize this profit function as a quadratic function whose graph is a parabola that opens downward. The maximum value occurs at the vertex. For this problem, we note that the domain for x is $[0, 49]$ because it does not make sense to produce a negative number of items, and it does not make sense if the price is negative. That is,

$$p(x) \geq 0$$

$$49 - x \geq 0$$

$$49 \geq x$$

We will solve this problem algebraically, numerically, and geometrically. In practice, it is not necessary that you use all three methods, but you may want to use a second method as a check on the first.

Numerical Approach	Geometrical Approach	Algebraic Approach

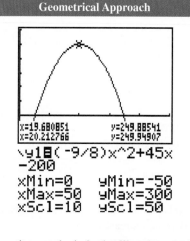

From the table, it looks like the maximum value is 250.

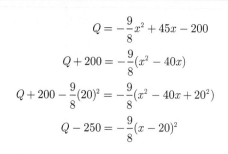

From the graph, it looks like the maximum value of 250 occurs when x is about 20.

$$Q = -\frac{9}{8}x^2 + 45x - 200$$

$$Q + 200 = -\frac{9}{8}(x^2 - 40x)$$

$$Q + 200 - \frac{9}{8}(20)^2 = -\frac{9}{8}(x^2 - 40x + 20^2)$$

$$Q - 250 = -\frac{9}{8}(x - 20)^2$$

We see the vertex is $(20, 250)$, so the maximum value is 250 and occurs when $x = 20$.

The price that corresponds to the maximum profit is

$$p(20) = 49 - 20 = 29 \text{ dollars}$$

∎

A common application is to minimize the distance between two points. Recall the distance formula; it involves a square root, and a square root function is not always the easiest to work with, so you will find the following result (which can be proved in calculus) helpful.

The distance d between two points $P(x_1,\ y_1)$ and $P_2(x_2,\ y_2)$, which is found by

$$d = \sqrt{(x_2 - x_1)^2 + (y_2 - y_1)^2}$$

is minimized when

$$d^2 = (x_2 - x_1)^2 + (y_2 - y_1)^2$$

is minimized.

EXAMPLE 3   Minimize a distance

What point on the curve $y = 2\sqrt{x}$ is closest to the point $(4,\ 0)$?

Solution   Draw a picture, as shown in Figure 3.21.

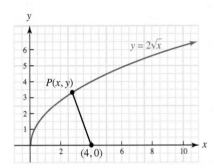

Figure 3.21 Graph of $y = 2\sqrt{x}$

We wish to minimize the distance d, from a point $P(x,\ y)$ on the graph to the point $(4,\ 0)$. We know (from the distance formula)

$$d = \sqrt{(x - 4)^2 + y^2}$$

The smallest value of d will occur when

$$d^2 = (x - 4)^2 + y^2$$

is minimized. Write this as a function $Q(x)$:

$$
\begin{aligned}
Q(x) &= (x - 4)^2 + \left(2\sqrt{x}\right)^2 && \text{Substitute } y = 2\sqrt{x}. \\
Q(x) &= x^2 - 8x + 16 + 4x && \text{Simplify.} \\
Q(x) - 16 &= x^2 - 4x && \text{Complete the square to find the vertex.} \\
Q(x) - 16 + 4 &= x^2 - 4x + 4 \\
Q(x) - 12 &= (x - 2)^2
\end{aligned}
$$

The vertex is $(2,\ 12)$, so the minimum value of Q is 12, which occurs when $x = 2$. If $x = 2$, then $y = 2\sqrt{2}$, so the minimum value of d^2 is 12, and the minimum value for d is $\sqrt{12} = 2\sqrt{3}$, which occurs for the point $\left(2,\ 2\sqrt{2}\right)$. The distance between $\left(2,\ 2\sqrt{2}\right)$ and $(4,\ 0)$ is

$$
\begin{aligned}
d &= \sqrt{(4 - 2)^2 + \left(0 - 2\sqrt{2}\right)^2} \\
&= \sqrt{4 + 4(2)} \\
&= 2\sqrt{3}
\end{aligned}
$$

EXAMPLE 4   Maximize a constrained area   `MODELING APPLICATION`

You need to fence in a rectangular play zone for children to fit into a right-triangular plot with sides measuring 4 m and 12 m. What is the maximum area for this play zone?

Solution

`Step 1:` *Understand the problem.* A diagram (as shown in Figure 3.22) will help us understand the problem. We modeled this as a function of x in Section 2.2.

`Step 2:` *Devise a plan.* We wish to maximize Q, where $Q = A(x) = 4x - \frac{1}{3}x^2$ (see page 99 for the formulation of this function). The domain for x is $0 \ge x \ge 12$. To maximize Q, we complete the square to find the vertex.

`Step 3:` *Carry out the plan.*

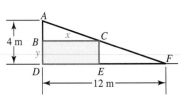

Figure 3.22 Children's play zone

$$Q = 4x - \frac{1}{3}x^2 \qquad \text{Given}$$

$$Q = -\frac{1}{3}(x^2 - 12x) \qquad \text{Factor } -\frac{1}{3}.$$

$$Q - 12 = -\frac{1}{3}(x^2 - 12x + 6^2) \qquad \text{Add } -\frac{1}{3}(6^2) \text{ to both sides.}$$

$$Q - 12 = -\frac{1}{3}(x - 6)^2 \qquad \text{Factor.}$$

The vertex of the parabola is $(6, 12)$, and because the parabola opens downward, we see the maximum is 12 when $x = 6$. This means that

$$y = 4 - \frac{1}{3}(6) = 2 \qquad \text{From Section 2.2, } y = 4 - \frac{1}{3}x.$$

The largest rectangular play zone that can be built in the triangular plot is a rectangle 6 m long and 2 m wide. Thus, the maximum area for this play zone is 12 m².

`Step 4:` *Look back.* We can check to see whether this result is reasonable by taking a geometric approach as shown in Figure 3.23.

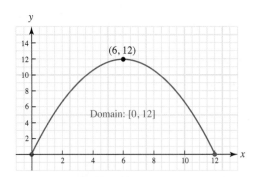

Figure 3.23 Maximum value of Q

The vertex is at $(6, 12)$, which corresponds to the algebraic solutions we found.

In calculus, you will prove that if a function is continuous on a closed interval $[a, b]$, there will be both a maximum and minimum value for the function on that interval. If the function is quadratic and the vertex is in the interval, then the maximum or minimum will occur at the vertex, and the other extreme value will occur at an endpoint. For continuous functions (not necessarily quadratic), you will develop, in calculus, a process for finding the extreme value within the interval by finding *turning points*. Recall, a *turning point* is a point on the graph that separates an interval over which a function *f* is increasing from an interval over which *f* is decreasing. A linear function has no turning points, and a quadratic function has one turning point. In calculus, it is shown that the graph of a polynomial function of degree *n* can have *at most* $n - 1$ turning points. However, if the function is increasing or decreasing throughout the entire interval, we can state the following important result.

OPTIMUM VALUES

If *f* is a continuous function that is increasing on $[a, b]$, then the minimum value of *f* is $f(a)$, and the maximum value is $f(b)$. If *f* is a continuous function that is decreasing on $[a, b]$, then the maximum value of *f* is $f(a)$, and the minimum value is $f(b)$.

Optimization of Discrete Functions

Sometimes the function to be optimized has practical meaning only when its independent variable is a positive integer. This can lead to certain problems because the theorems we have developed require continuous functions. If we model a function whose variable is defined as a positive integer, but as part of the modeling process we assume the function is defined for all real values (so that it is continuous), it may happen that the optimization procedure leads to a nonintegral or negative value of the independent variable, and additional analysis is needed to obtain a meaningful solution.

EXAMPLE 5   **Maximizing a discrete revenue function**   `MODELING APPLICATION`

A bus company will charter a bus that holds 50 people to groups of 35 or more. If a group contains exactly 35 people, each person pays \$60. In larger groups, everybody's fare is reduced by \$1 for each person in excess of 35. Determine the size of the group for which the bus company's revenue will be the greatest.

Solution

`Step 1:`   *Understand the problem.* The revenue is the amount of money collected:

REVENUE = (NUMBER OF PEOPLE IN THE GROUP)(FARE PER PERSON)

The question asks us to maximize this revenue.

`Step 2:`   *Devise a plan.* Let *x* be the number of people in excess of 35 who take the trip. Then

NUMBER OF PEOPLE IN THE GROUP = $35 + x$

FARE PER PERSON = $60 - x$

Let $R(x)$ be the revenue for the bus company:

$$R(x) = (35 + x)(60 - x) = 2{,}100 + 25x - x^2$$

Next, find the domain: We note that there must be at least 35 people ($x = 0$) and at most 50 people ($x = 15$); thus, $0 \le x \le 15$, *but because x represents the number of people, it must also be an integer.* We must now complete the square to find the maximum value.

`Step 3:`   *Carry out the plan.*

$$R(x) - 2{,}100 = -x^2 + 25x \qquad \text{\small Given}$$
$$R(x) - 2{,}100 - 12.5^2 = -(x^2 - 25x + 12.5^2) \qquad \text{\small Complete the square.}$$

The maximum occurs where $x = 12.5$. But *x* must be an integer, so $x = 12.5$ is not in the domain. To find the optimal *integer* solution, observe that *R* is increasing on $(0, 12.5)$ and decreasing on $(12.5, 15)$ as shown in Figure 3.24.

x	No.	Fare	Revenue
0	35	60	2,100
1	36	59	2,124
2	37	58	2,146
3	38	57	2,166
4	39	56	2,184
5	40	55	2,200
6	41	54	2,214
7	42	53	2,226
8	43	52	2,236
9	44	51	2,244
10	45	50	2,250
11	46	49	2,254
12	47	48	2,256
13	48	47	2,256
14	49	46	2,254
15	50	45	2,250

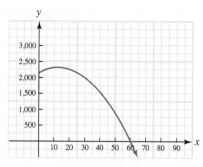

a. The continuous revenue function

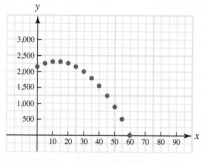

b. The discrete revenue function

Figure 3.24 Graphs of $R(x) = -x^2 + 25x + 2{,}100$

It follows that the optimal integer value of x is either $x = 12$ or $x = 13$. Because

$$R(12) = 2{,}256 \quad \text{and} \quad R(13) = 2{,}256$$

we conclude that the bus company's revenue will be greatest when the group contains either 12 or 13 people in excess of 35—that is, for groups of 47 or 48. In either case, the revenue will be $2,256.

Step 4: *Look back.* The graph of revenue as a function of x is a collection of discrete points corresponding to the integer values of x, as indicated in Figure 3.24b. To verify that this integer solution is correct, we can look at the table in the margin, which shows all possibilities from $x = 0$ (35 people) to $x = 15$ (50 people). For each number of travelers, the total revenue is calculated, and we see that the maximum value 2,256 is obtained when $x = 12$ and again when $x = 13$.

PROBLEM SET 3.3

LEVEL 1

1. IN YOUR OWN WORDS Describe an optimization procedure.
2. IN YOUR OWN WORDS Why does a tabular approach not always provide a maximum or minimum value?

Find the minimum value of y for the functions defined in Problems 3–8.

3. $y = x^2 - 10x + 27$
4. $y = x^2 + 8x + 4$
5. $y = x^2 - 2x + 2$
6. $y = 5x^2 - 20x + 6$
7. $x^2 + 70x - y + 1{,}265 = 0$
8. $4x^2 + 56x - y + 179 = 0$

Find the maximum value of y for the functions defined in Problems 9–14.

9. $y = -4x^2 - 8x - 1$
10. $y = -x^2 - 2x + 2$
11. $y = -3x^2 + 6x - 5$
12. $y = -5x^2 - 20x + 6$
13. $8x^2 - 16x + y + 655 = 0$
14. $6x^2 + 84x + y + 302 = 0$

For Problems 15–20, decide whether it makes better sense to find a maximum or a minimum value for y. Find this value, and state the value of x for which this optimal value of y is obtained.

15. $2x^2 + 2y + 16x + 8 = 0$
16. $3x^2 + 6x - 27y - 51 = 0$
17. $10x^2 + 100x + 20y + 90 = 0$
18. $10x^2 + 100x - 20y + 110 = 0$
19. $x^2 + 9y + 10x - 47 = 0$
20. $100x^2 - 120x - 25y + 41 = 0$
21. The sum of two numbers is 10. Find the largest possible product.

22. The difference of two numbers is 8. Find the smallest possible product.
23. The difference of two numbers is 6. Find the minimum value for the product of the numbers.
24. The sum of two numbers is 20. Find the smallest possible value for the sum of their squares.
25. Suppose the total cost of manufacturing x units of a certain commodity is

$$C(x) = 3x^2 + x + 48$$

dollars. Determine the minimum cost.
26. Suppose the total cost of producing x units is

$$C(x) = 5x^2 + 10x + 15{,}600$$

dollars. Determine the minimum cost.
27. A profit function, P, is

$$P(x) = 1{,}875 + 150x - 10x^2$$

Find the maximum profit.
28. A profit function, P, is

$$P(x) = 100x - 2x^2 - 600$$

Find the maximum profit.

LEVEL 2

Use numerical or graphical techniques to find the maximum and minimum values (to the nearest unit) on the given domains for the functions defined in Problems 29–34. Also state the x-value (correct to the nearest unit) that gives the requested optimal value.

29. Interval: $[-3, 2]$

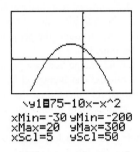

```
\y1075-10x-x^2
xMin=-30 yMin=-200
xMax=20  yMax=300
xScl=5   yScl=50
```

30. Interval: $[0, 30]$

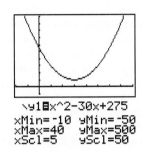

```
\y10x^2-30x+275
xMin=-10 yMin=-50
xMax=40  yMax=500
xScl=5   yScl=50
```

31. Interval: $[-6, 6]$

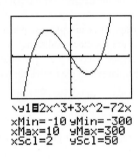

```
\y102x^3+3x^2-72x
xMin=-10 yMin=-300
xMax=10  yMax=300
xScl=2   yScl=50
```

32. Interval: $[-4, 8]$

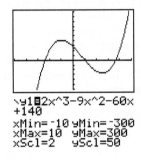

```
\y102x^3-9x^2-60x
+140
xMin=-10 yMin=-300
xMax=10  yMax=300
xScl=2   yScl=50
```

33. Interval: $[-6, 8]$

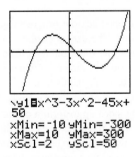

```
\y10x^3-3x^2-45x+
50
xMin=-10 yMin=-300
xMax=10  yMax=300
xScl=2   yScl=50
```

34. Interval: $[-4, 10]$

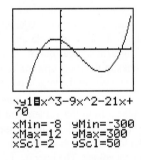

```
\y10x^3-9x^2-21x+
70
xMin=-8  yMin=-300
xMax=12  yMax=300
xScl=2   yScl=50
```

WHAT IS WRONG, *if anything, with each statement in Problems 35–37?*

35. If the function to be maximized or minimized is a second-degree function, then its graph is a parabola. If the parabola opens upward, then there will be a minimum value that must occur at the vertex. If the parabola opens downward, then there will be a maximum value that must occur at the vertex.

36. If the graph of a second-degree function to be maximized opens upward on some domain $[a, b]$, then the maximum value must occur at an endpoint.

37. If the graph of a second-degree function to be minimized opens downward on some domain $[a, b]$, then the minimum value must occur at an endpoint.

38. The highest bridge in the world is the bridge over the Royal Gorge of the Arkansas River in Colorado.

It is 1,053 ft above the water. If a rock is projected vertically upward from this bridge with an initial velocity of 64 ft/s, the height of the object h above the river at time t is described by the function

$$h(t) = -16t^2 + 64t + 1,053$$

What is the maximum height possible for a rock projected vertically upward from the bridge with an initial velocity of 64 ft/s? After how many seconds does it reach that height?

39. In 1974, Evel Knievel attempted a skycycle ride across the Snake River.

Suppose the path of the skycycle is given by the equation

$$d(x) = -0.0005x^2 + 2.39x + 600$$

where $d(x)$ is the height in feet above the canyon floor for a horizontal distance of x feet from the launching ramp, as shown in Figure 3.25. What was Knievel's maximum height?

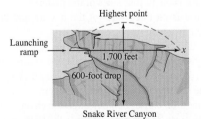

Figure 3.25 Evel Knievel's skycycle ride

40. The sum of the length and width of a rectangular area is 50 ft. Find the greatest area possible, and find the dimensions of the figure.

41. One side of a storage yard is against a building. The other three sides of the rectangular yard are to be fenced with 36 ft of fencing. How long should the sides be to produce the greatest area with the given length of fence? What is the area obtained?

42. Suppose the total cost of producing x units of a particular commodity is

$$C(x) = 2x + 9$$

and the selling price is

$$p(x) = \frac{1}{2}(74 - x)$$

Determine the level of production that maximizes the profit.

43. Suppose the total cost of producing x units of a particular commodity is

$$C(x) = \frac{2}{5}x^2 + 10x$$

and the selling price is

$$p(x) = \frac{1}{5}(100 - x)$$

Determine the level of production that maximizes the profit.

44. An arch (see Figure 3.26) has equation

$$y - 18 = -\frac{2}{81}x^2$$

where both x and y are measured in feet.

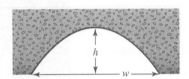

Figure 3.26 Stone archway

a. To relate the equation to the arch, you must superimpose a coordinate system over Figure 3.26. Where is the origin? How do x and y relate to h and w?

b. What is the maximum height, h?

c. What is the width, w, of the arch?

d. What is the height of the arch 9 ft from the center?

e. What is the height of the arch 18 ft from the center?

45. A farmer wishes to enclose a rectangular pasture with 360 ft of fence. Find the dimensions that give the maximum area in these situations:

a. The fence is on all four sides of the pasture.

b. The fence is on three sides of the pasture and the fourth side is bounded by a wall.

46. A truck is 250 mi due east of a sports car and is traveling west at a constant speed of 60 mi/h. Meanwhile, the sports car is going north at 80 mi/h. When will the truck and the car be closest to each other? What is the minimum distance between them?

47. A commuter train carries 600 passengers each day from a suburb to a city. It now costs $5 per person to ride the train. A study shows that 50 additional people will ride the train for each 25¢ reduction in fare. What fare should be charged to maximize the total revenue?

48. A store has been selling skateboards at the price of $40 per board, and at this price, skaters have been buying 45 boards a month. The owner of the store wishes to raise the price and estimates that for each $1 increase in price, 3 fewer boards will be sold each month. If each board costs the store $29, at which price should the store sell the boards to maximize the profit?

49. A tour agency is booking a tour and has 200 people signed up. The price of a ticket is $2,000 per person. The agency has booked a plane seating 250 people at a cost of $125,000. Additional costs to the agency are incidental fees of $500 per person. For each $10 that the price is lowered, a new person will sign up. How much should the per person price be lowered to maximize the profit to the tour agency?

50. A Florida citrus grower estimates that if 60 orange trees are planted, the average yield per tree will be 400 oranges. The average yield will decrease by 4 oranges for each additional tree planted on the same acreage. How many trees should the grower plant to maximize the total yield?

51. Farmers can get $2 per bushel for their potatoes on July 1, and after that the price drops 2¢ per bushel per day. On July 1, a farmer has 80 bu of potatoes in the field and estimates that the crop is increasing at the rate of 1 bu per day. When should the farmer harvest the potatoes to maximize revenue?

52. A viticulturist estimates that if 50 grapevines are planted per acre, each grapevine will produce 150 lb of grapes. Each additional grapevine planted per acre (up to 20) reduces the average yield per vine by 2 lb. How many grapevines should be planted to maximize the yield per acre?

LEVEL 3

53. IN YOUR OWN WORDS Show that of all rectangles with a given perimeter, the square has the largest area.

54. A woman plans to fence off a rectangular garden whose area is 64 ft². What should be the dimensions of the garden if she wants to minimize the amount of fencing used?

55. A carpenter wants to make an open-topped box out of a rectangular sheet of tin 24 in. wide and 45 in. long. The carpenter plans to cut congruent squares out of each corner of the sheet and then bend the edges of the sheet upward to form the sides of the box. What are the dimensions of the largest box that can be made in this fashion?

56. IN YOUR OWN WORDS Pull out a sheet of $8\frac{1}{2}$ in. by 11 in. binder paper. Cut squares from the corners and fold the sides up to form a container. Show that the maximum volume of such a container is about 1 liter.

57. A closed box with square base and vertical sides is to be built to house an ant colony. The bottom of the box and all four sides are to be made of material costing $1/ft², and the top is to be constructed of glass costing $5/ft². What are the dimensions of the box of greatest volume that can be constructed for $72?

58. According to postal regulations, the girth plus the length of a parcel sent by fourth-class mail may not exceed 108 in. What is the largest possible volume of a rectangular parcel with two square sides that can be sent?

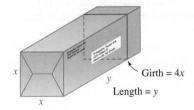

59. Example 4 in Section 2.1 considered the following question: A container is to be constructed from an 11 in. by 16 in. sheet of cardboard. Squares will be cut from the corners of the sheet and discarded as waste. The domain for the variable must be restricted so that the area of the base of the container exceeds the area of wasted cardboard. In Chapter 2, we found the domain to be $0 \geq s \geq 88/27$. Find the maximum volume.

60. A wire of length 1 yd is to be cut into two pieces, one of which will be bent to form a circle and the other, to form a square. How should the cut be made to maximize and also to minimize the sum of the areas enclosed by the two pieces?

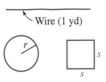

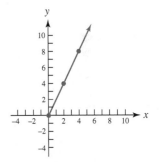

Let r be radius of circle and let s be length of side of square.

3.4 Parametric Equations

Until now, the curves we have discussed have been represented by a single equation or, in the case of piecewise-defined curves, a single equation for a single piece of the curve. However, many applications can best be described by specifying the x and y components of a graph in terms of another variable. For example, suppose we describe a particle moving in the plane so that its position is defined in terms of the length of time (in seconds) from some starting time, $t = 0$ (in more advanced work, this is often denoted by t_0). In particular, let its location be described by the functions

$$x = 2t \quad \text{and} \quad y = 4t$$

If $t = 0$, then $x = 0$ and $y = 0$, so we plot $(0, 0)$;

if $t = 1$, then $x = 2$ and $y = 4$, plot $(2, 4)$;

if $t = 2$, then $x = 4$ and $y = 8$, plot $(4, 8)$;

$$\vdots$$

The completed graph is shown in Figure 3.27.

Figure 3.27 Graph of $x = 2t$, $y = 4t$, $t \geq 0$

Parameters

Since this representation is not only convenient but also frequently used, we introduce some new terminology.

> **PARAMETER**
>
> Let t be a number in an interval I. Consider the curve defined by the set of ordered pairs (x, y), where
>
> $$x = f(t) \quad \text{and} \quad y = g(t)$$
>
> for functions f and g defined on I. Then the variable t is called a **parameter** and the equations $x = f(t)$ and $y = g(t)$ are called **parametric equations** for the curve defined by (x, y). The direction on the curve of increasing values of t is called the **orientation** of the curve.

EXAMPLE 1   Graphing parametric equations

Graph: $x = 1 + t$ and $y = 2t$ for $t \geq 0$.

Solution   These equations may be interpreted by saying that although the first component has a 1-unit "head start," the rate at which the second component is changing (with respect to t) is twice as fast as the first component. We can tabulate the values of x and y by choosing values for t ($t \geq 0$).

t	0	1	2	3	4	5
x	1	2	3	4	5	6
y	0	2	4	6	8	10

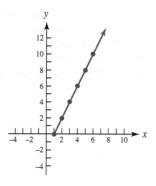

COMPUTATIONAL WINDOW

It is particularly easy to graph parametric equations using a calculator. In fact, we will see that the graphing calculator has added a new dimension to parametric equations and has provided motivation for an enhanced treatment of parametric equations. Most graphing calculators allow you to change from rectangular to parametric form by changing the MODE . Example 1 is shown.

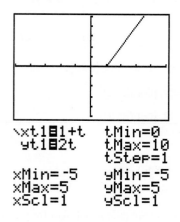

```
\xt1█1+t    tMin=0
 yt1█2t     tMax=10
            tStep=1

xMin=-5     yMin=-5
xMax=5      yMax=5
xScl=1      yScl=1
```

Lines in Parametric Form

Notice that the graph of Example 1 is a line. We can prove that the graph is a line by *eliminating the parameter*. We solve the first equation for t ($t = x - 1$) and substitute the result into the second equation:

$$y = 2t = 2(x - 1) = 2x - 2$$

which is a linear equation.

PARAMETRIC FORM OF A LINE

The graph of the parametric equations

$$x = x_1 + at \qquad y = y_1 + bt$$

is a line passing through (x_1, y_1) with slope $m = \dfrac{b}{a}$, $a \neq 0$.

It is easy to derive this result by solving one of the equations for t and substituting the result into the other equation:

$$x - x_1 = at \qquad\qquad y = y_1 + bt$$

$$\frac{x - x_1}{a} = t \qquad\qquad = y_1 + b\left(\frac{x - x_1}{a}\right)$$

$$y - y_1 = \frac{b}{a}(x - x_1)$$

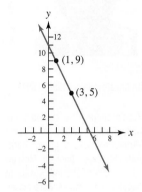

You recognize this as a line through (x_1, y_2) with slope $\dfrac{b}{a}$. In Section 1.3, we called a the run, which was denoted by Δx, and b the rise, denoted by Δy. Notice how easy this is to use with the following example.

EXAMPLE 2   Graphing a line given in parametric form

Graph $x = 3 - 2t$, $y = 5 + 4t$, for t any real number.

Solution   Recognize this as the parametric form of the equation of a line passing through $(3, 5)$ with $\Delta x = -2$ and $\Delta y = 4$. The slope is $\dfrac{b}{a} = \dfrac{4}{-2} = -2$. The graph is shown in Figure 3.28.

Figure 3.28 Graph of $x = 3 - 2t$, $y = 5 + 4t$

EXAMPLE 3   Eliminating the parameter

Graph $x = t + 2, y = t^2 + 2t - 1$ for t any real number by

  a. plotting points   **b.** eliminating the parameter

Solution

  a. By plotting points, we set up a table of values:

t	-4	-3	-2	-1	0	1	2	3
x	-2	-1	0	1	2	3	4	5
y	7	2	-1	-2	-1	2	7	14

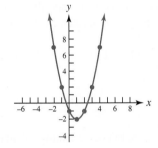

The points are plotted in Figure 3.29.

Figure 3.29 Graph of $x = t + 2$, $y = t^2 + 2t - 1$

  b. To eliminate the parameter, we can solve the first equation for t ($t = x - 2$) and substitute into the second equation:

$$y = t^2 + 2t - 1$$
$$= (x - 2)^2 + 2(x - 2) - 1$$
$$= x^2 - 4x + 4 + 2x - 4 - 1$$
$$= x^2 - 2x - 1$$

We recognize this as a parabola opening upward, which has y-intercept $(0, -1)$ and vertex at

$$x = -\frac{b}{2a} = -\frac{-2}{2} = 1.$$

It is not always easy to eliminate the parameter, as illustrated by the following example.

EXAMPLE 4   Eliminating the parameter with quadratic equations

Eliminate the parameter (t is any real number) for the parametric equations

$$x = t^2 - 3t + 1 \qquad y = -t^2 + 2t + 3$$

Solution   Instead of solving one equation first, which involves considerable (too much) algebra, we seek another method. We know that if we add equal values to equal values, the results are equal, so here we add one equation to the other:

$$+\begin{cases} x = t^2 - 3t + 1 \\ y = -t^2 + 2t + 3 \end{cases}$$

$$x + y = -t + 4$$

We can now solve this sum equation for the variable t:

$$t = 4 - x - y$$

This value for t can be substituted into either of the original parametric equations:

$$\begin{aligned} x &= t^2 - 3t + 1 \\ &= (4 - x - y)^2 - 3(4 - x - y) + 1 \\ &= 16 - 4x - 4y - 4x + x^2 + xy - 4y + xy + y^2 - 12 + 3x + 3y + 1 \\ &= x^2 + 2xy + y^2 - 5x - 5y + 5 \end{aligned}$$

∎

Parametrization of a Curve

Suppose we wish to graph the curve given in Example 4. Would you rather graph the curve given the rectangular coordinates $x = x^2 + 2xy + y^2 - 5x - 5y + 5$ (ouch!) or the parametric form

$$x = t^2 - 3t + 1 \qquad y = -t^2 + 2t + 3$$

⚠ Many curves are easier to graph in parametric form. In fact, this is one of the primary reasons for using parametric equations in the first place. You will see many examples as you work through the rest of this book.

If you were to use a graphing calculator or computer to help you sketch the curve given in Example 4, you could easily do so in parametric form (as shown in Figure 3.30).

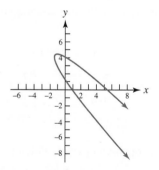

Figure 3.30 Graph of
$x = t^2 - 3t + 1, \ y = -t^2 + 2t + 3$

In rectangular form, you need to solve for y, and even then there would be two functions to graph (because it is quadratic in y). The process by which we change a rectangular-form equation into a parametric form is called **parametrization** of a curve. The easiest curve to parametrize is one that is a function.

EXAMPLE 5   Parametrization of a function

Find a parametrization of the curve C that is the portion of the parabola $y = x^2$ traversed from $(0, 0)$ to $(2, 4)$.

Solution   If $x = t$, then (by substitution) we have $y = t^2$. If we let $t = 0$, we have the point $(0, 0)$, and if we let $t = 2$, we find $(2, 4)$. So we see that this curve has the proper orientation if we let $0 \leq t \leq 2$. To see this in detail, we show a table of values:

Orientation: t increases from 0 to 2.

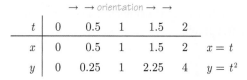

t	0	0.5	1	1.5	2	
x	0	0.5	1	1.5	2	$x = t$
y	0	0.25	1	2.25	4	$y = t^2$

The points are plotted in Figure 3.31.

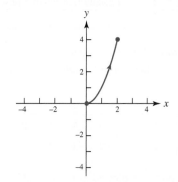

Figure 3.31 Graph of $y = x^2$ with orientation

EXAMPLE 6   Parametrization of an oriented line segment

Find a parametrization of the curve C that is the portion of the line $y = -2x + 3$ traversed from $(2, -1)$ to $(0, 3)$.

Solution   Graph the line segment as shown in Figure 3.32.

If $x = t$, then (by substitution), $y = -2t + 3$. If we let $t = 0$, we have the point $(0, 3)$, and if we let $t = 2$, we find $(2, -1)$. We see that this orientation is the wrong direction when we let $0 \leq t \leq 2$. We see this from a table of values:

Orientation: t increases from 0 to 2.

t	0	0.5	1	1.5	2	
x	0	0.5	1	1.5	2	$x = t$
y	3	2	1	0	−1	$y = -2t + 3$

→ From $(0, 3)$ to $(2, -1)$ → → Wrong direction

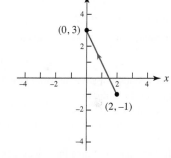

Figure 3.32 Graph of oriented line segment

To reverse the direction, use the parametric form of the equation of a line where the starting point is $(2, -1)$, and by looking at the graph (or the table), we see if $\Delta x = -2$, then $\Delta y = 4$. Thus, $x = 2 - 2t$, $y = -1 + 4t$, $0 \leq t \leq 1$. A table of values confirms that this is the correct orientation.

Orientation: t increases from 0 to 1.

t	0	0.25	0.5	0.75	1	
x	2	1.5	1	0.5	0	$x = 2 - 2t$
y	−1	0	1	2	3	$y = -1 + 4t$

→ From $(2, -1)$ to $(0, 3)$ → → Correct direction

A second curve that is easy to parametrize is one that is first degree in x and second degree in y; for example, $x = y^2 + 4y + 4$. This is not a function, but nevertheless, we can graph this curve. If we use a computer or a graphing calculator, we would need to solve for y to obtain $y = -2 \pm \sqrt{x}$. Since this is not a function, we could graph this by considering $y_1 = -2 + \sqrt{x}$ and $y_2 = -2 - \sqrt{x}$. However, it is much easier to parametrize this curve, as shown in the following example.

EXAMPLE 7   Parametrization of a first-degree equation in x

Graph $x = y^2 + y + 4$ by considering a parametrization.

Solution   Let $y = t$; then $x = t^2 + 4t + 4$, with no restrictions on the parameter. You can set up a table of values, or use a calculator to obtain the parametric graph.

t	-4	-3	-2	-1	0	1
x	4	1	0	1	4	9
y	-4	-3	-2	-1	0	1

The points are plotted in Figure 3.33.

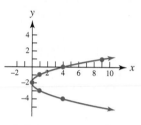

Figure 3.33 Graph of $x = y^2 + 4y + 4$

PROBLEM SET 3.4

LEVEL 1

WHAT IS WRONG, *if anything, with each statement in Problems 1–6? Explain your reasoning.*

1. A parametric equation is an equation with three variables. For example,
$$y = 3t + 4x + 5$$

2. The equation of a line in parametric form is $y = x_1 + mt$, where m is the slope of the line.

3. The equation of a line in parametric form is $x = a + (\Delta y)t$, $y = b + (\Delta x)t$, where (x, y) is a point on the line, Δy is the rise, and Δx is the run.

4. If you want to change the orientation of the line defined by the equations
$$x = 2 + t \quad \text{and} \quad y = 3t$$
oriented from $(2, 0)$ to $(7, 15)$ to the orientation $(7, 15)$ to $(2, 0)$, interchange the x and y values to obtain
$$y = 2 + t \quad \text{and} \quad x = 3t$$

5. To parametrize the curve $y = x^2$, let $t = x$ and substitute into the other equation: $y = t^2$. The orientation does not matter with this parabola because it is symmetric with respect to the y-axis.

6. To eliminate the parameter with the parametric equations
$$x = 2t^2 + 4t - 3, \quad y = -2t^2 + 5t + 3$$
solve the first equation for t and substitute into the second equation.

Write the parametric equations for the lines in Problems 7–14.

7. Slope $\dfrac{3}{2}$ passing through $(4, 5)$

8. Slope $\dfrac{1}{3}$ passing through $(-1, -2)$

9. Slope 3 passing through $(-5, 2)$
10. Slope -4, passing through $(6, -8)$
11. Slope $-\dfrac{1}{2}$ passing through the origin
12. Slope $-\dfrac{4}{5}$ passing through $(-1, -2)$
13. Horizontal line passing through $(-4, 5)$
14. Vertical line passing through $(-6, 8)$

Plot the curves in Problems 15–24 by plotting points, and then check your work by eliminating the parameter. Assume t is any real number.

15. $x = -1 + t$, $y = 3 - 2t$ 16. $x = -2 + t$, $y = 4 - 5t$

17. $x = t$, $y = 2 + \dfrac{2}{3}(t - 1)$ 18. $x = t$, $y = 3 - \dfrac{3}{5}(t + 2)$

19. $x = 2t$, $y = t^2 + t + 1$ 20. $x = 3t$, $y = t^2 - t + 6$
21. $x = t^2 - t + 6$, $y = 3t$ 22. $x = t^2 + t + 1$, $y = 2t$
23. $x = t^2 + 2t + 3$, $y = t^2 + t - 4$
24. $x = t^2 - 2t + 3$, $y = t^2 - t + 4$

Find an appropriate parametrization for the curves described in Problems 25–34. Be sure to consider the orientation, and give the domain for the parameter.

25. The line segment from $(0, 0)$ to $(4, 9)$
26. The line segment from $(2, 3)$ to $(-2, 5)$
27. The line segment from $(4, 9)$ to $(0, 0)$
28. The line segment from $(-2, 5)$ to $(2, 3)$
29. The parabola $y = x^2$ from $(3, 9)$ to $(0, 0)$
30. The parabola $y = x^2$ from $(0, 0)$ to $(3, 9)$
31. The graph of $y = \sqrt{x}$ from $(1, 1)$ to $(9, 3)$

32. The graph of $y = \sqrt{x}$ from $(9, 3)$ to $(1, 1)$

33. The graph of $y = \sqrt[3]{x}$ from $(27, 3)$ to $(0, 0)$

34. The graph of $y = \sqrt[3]{x}$ from $(0, 0)$ to $(27, 3)$

Graph the curves in Problems 35–46 by any convenient method. Assume t is any real number unless otherwise noted.

35. $x = -3 + 4t$, $y = -1 - 3t$ **36.** $x = -2 - 3t$, $y = -2 + 5t$

37. $x = 6 - 2t$, $y = -5 + 3t$ **38.** $x = 5 - t$, $y = -3 - 2t$

39. $x = 60t$, $y = 80t - 16t^2$ **40.** $x = 30t$, $y = 60t - 9t^2$

41. $x = t^2$, $y = t^3$ **42.** $x = t^3 + 1$, $y = t^2 - 1$

43. $x = t$, $y = t^3 + t^2$, $-1 \geq t \geq 1$

44. $x = t$, $y = t^3 - 2t^2 + t$, $0 \geq t \geq 1$

45. $x = t^3 + t^2$, $y = t$, $-1 \geq t \geq 1$

46. $x = t^3 - 2t^2 + t$, $y = t$, $0 \geq t \geq 1$

LEVEL 2

47. A particle begins at the point $(0, 0)$ at $t = 0$ on the graph of $y = 6x$ and moves in the direction of increasing x so that its x-coordinate changes 0.5 unit/sec. Find a set of parametric equations so that the coordinates are given at time t.

48. A particle begins at the point $(0, 1)$ at $t = 0$ on the graph of $y = 6x + 1$ and moves in the direction of increasing x so that its x-coordinate changes 2 units/sec. Find a set of parametric equations so that the coordinates are given at time t.

49. A particle begins at the point $(0, 1)$ at $t = 0$ on the graph of $y = 6x + 1$ and moves in the direction of decreasing x so that its x-coordinate changes 2 units/sec. Find a set of parametric equations so that the coordinates are given at time t.

50. A particle begins at the point $(0, 0)$ at $t = 0$ on the graph of $y = 6x$ and moves in the direction of decreasing x so that its x-coordinate changes 0.5 unit/sec. Find a set of parametric equations so that the coordinates are given at time t.

51. A particle begins at the point $(0, 0)$ at $t = 0$ on the graph of $y = x^2$ and moves in the direction of increasing x so that its x-coordinate changes 10 units/sec. Find a set of parametric equations so that the coordinates are given at time t.

52. A particle begins at the point $(0, 2)$ at $t = 0$ on the graph of $y = x^2 + 2$ and moves in the direction of decreasing x so that its x-coordinate changes 2 units/sec. Find a set of parametric equations so that the coordinates are given at time t.

53. One object follows path $f: x = 2t$, $y = 3t$, while another follows path $g: x = t$, $y = -\frac{1}{2}(x - 400) + 50$. Do these paths intersect? How about the objects?

54. One object follows path $F: x = t$, $y = 2t$, and another follows the path $G: x = t$, $y = -\frac{1}{2}(x - 100) + 25$. Do these paths intersect? How about the objects?

LEVEL 3

55. IN YOUR OWN WORDS Consider the parametric graphs $x = kt$, $y = t^2$ for $k = 1, 2, 3, \ldots$. What is the effect of the k-value?

56. IN YOUR OWN WORDS Consider the parametric graphs $x = t$, $y = (kt)^2$ for $k = 1, 2, 3, \ldots$. What is the effect of the k-value?

Sketch the curves in Problems 57–60 by introducing a parameter.

57. $x = y^2 - 4y + 1$

58. $y^2(x - 2) = 2$

59. $y^2(x + 4) = 2$

60. $xy^3 - y^2 + 2xy = 0$

3.5 Graphing Polynomial Functions

We have discussed graphing linear and quadratic functions, as well as a variety of other graphing techniques. For other curves, we have used a calculator. Sometimes, however, we need to locate exact coordinates for points on the graph (as opposed to the approximate coordinates we find with a calculator). To plot points on the graph of a polynomial function, and to find the roots of a polynomial equation (Section 3.6), we will need to know a process called *synthetic division*. Before we introduce this process, however, we will consider *long division* of polynomials.

Long Division of Polynomials

To check multiplication, we know we can divide. For example, if $3 \cdot 5 = 15$, then we know $15 \div 3 = 5$. The same is true for polynomials. For example,

$$(4x - 5)(5x^2 + 2x + 1) = 20x^3 - 17x^2 - 6x - 5$$

This product implies we can check by forming the quotient

$$\frac{20x^3 - 17x^2 - 6x - 5}{4x - 5}$$

If this result is $5x^2 + 2x + 1$, then the result checks.

We will illustrate a process called **long division**. The procedure calls for successive multiplications and subtractions, similar to the way we do long division for natural numbers.

$$
\begin{array}{r}
5x^2 \\
4x - 5 \overline{\smash{)}\ 20x^3 - 17x^2 - 6x - 5} \\
\underline{20x^3 - 25x^2}
\end{array}
$$

Divide the first terms and multiply to duplicate the highest-degree term.

$$
\begin{array}{r}
5x^2 \\
4x - 5 \overline{\smash{)}\ 20x^3 - 17x^2 - 6x - 5} \\
\underline{20x^3 - 25x^2} \\
8x^2 - 6x - 5
\end{array}
$$

Subtract; the first term will be eliminated.

$$
\begin{array}{r}
5x^2 + 2x \\
4x - 5 \overline{\smash{)}\ 20x^3 - 17x^2 - 6x - 5} \\
\underline{20x^3 - 25x^2} \\
8x^2 - 6x - 5 \\
8x^2 - 10x
\end{array}
$$

Repeat the process, making sure that the first terms are the same.

$$
\begin{array}{r}
5x^2 + 2x \\
4x - 5 \overline{\smash{)}\ 20x^3 - 17x^2 - 6x - 5} \\
\underline{20x^3 - 25x^2} \\
8x^2 - 6x - 5 \\
\underline{8x^2 - 10x} \\
4x - 5
\end{array}
$$

Subtract.

$$
\begin{array}{r}
5x^2 + 2x + 1 \\
4x - 5 \overline{\smash{)}\ 20x^3 - 17x^2 - 6x - 5} \\
\underline{20x^3 - 25x^2} \\
8x^2 - 6x - 5 \\
\underline{8x^2 - 10x} \\
4x - 5 \\
\underline{4x - 5} \\
0
\end{array}
$$

Repeat.

Subtract; remainder is 0.

Thus, we see that

$$
\frac{20x^3 - 17x^2 - 6x - 5}{4x - 5} = 5x^2 + 2x + 1
$$

Note that we can use multiplication to check division. For example, if $\frac{15}{3} = 5$, then $15 = 3 \cdot 5$. If there is a remainder from the division, we have

$$
\frac{17}{3} = 5 + \frac{2}{3}, \text{ then } 17 = 3 \cdot 5 + 2
$$

Consider positive integers P and D. If the result of P divided by D is an integer Q, then D is a *factor* of P. We summarize these observations:

If $\dfrac{P}{D} = Q$, then $P = QD$

If $\dfrac{P}{D} = Q + \dfrac{R}{D}$, then $P = QD + R$ $(R < D)$

Division of polynomials is similar and leads to a result called the **division algorithm**.

DIVISION ALGORITHM

If P and D are polynomials $[D(x) \neq 0]$, then there exist unique polynomials Q and R, such that

$$\frac{P(x)}{D(x)} = Q(x) + \frac{R(x)}{D(x)}$$

where Q is a unique polynomial and R is a polynomial such that the degree of R is less than the degree of D. The polynomial Q is the *quotient* and R is the *remainder*.

1. If $R(x) = 0$, then $D(x)$ is a factor of $P(x)$.
2. If the degree of D is greater than the degree of P, then $Q(x) = 0$.
3. If both sides are multiplied by $D(x)$, the division algorithm is then stated in *product form*:

$$P(x) = Q(x)D(x) + R(x)$$

EXAMPLE 1   Long division of polynomials

Divide: $\dfrac{x^4 - 5x^2 + 6}{x^2 - x - 2}$.

Solution   Set up the problem; note the use of 0 coefficients:

$$
\begin{array}{r}
x^2 + x - 2 \\
x^2 - x - 2 \overline{\smash{\big)}\, x^4 + 0x^3 - 5x^2 + 0x + 6} \\
\underline{x^4 - x^3 - 2x} \\
x^3 - 3x^2 + 0x + 6 \\
\underline{x^3 - x^2 - 2x} \\
-2x^2 + 2x + 6 \\
\underline{-2x^2 + 2x + 4} \\
2
\end{array}
$$

$$\frac{x^4 - 5x^2 + 6}{x^2 - x - 2} = x^2 + x - 2 + \frac{2}{x^2 - x - 2}$$ This is the form $\dfrac{P}{D} = Q + \dfrac{R}{D}$.

█ COMPUTATIONAL WINDOW   ▭▭✕

In many software programs, you can work problems like Example 1 by using an **EXPAND** command. After inputting the expression

$$(X^4 - 5X^2 + 6)/(X^2 - X - 2)$$

and executing the **EXPAND** command, you will obtain $-\dfrac{2}{3(x+1)} + \dfrac{2}{3(x-2)} + x^2 + x - 2.$

Notice that the order of the terms is not the same as we found in Example 1 and that the fractional part is written as a sum. If you simplify $-\dfrac{2}{3(x+1)} + \dfrac{2}{3(x-1)}$, you will obtain $\dfrac{2}{x^2 - x - 2}$.

Synthetic Division

If the divisor (the denominator of the fraction) is of the form $x - k$, the long division process can be shortened considerably. Since all variables can be aligned in columns of descending degree, the variables may be omitted and the coefficients written alone.

$$
\begin{array}{r}
x^2 + 2x + 1 \\
x - 2\overline{\smash{)}x^3 + 0x^2 - 3x - 2} \\
\underline{x^3 - 2x^2} \\
2x^2 - 3x - 2 \\
\underline{2x^2 - 4x} \\
x - 2 \\
\underline{x - 2} \\
0
\end{array}
\qquad
\begin{array}{r}
1 \quad\ 2 \quad\ 1 \\
1-2\overline{\smash{)}1 \quad\ 0 \ -3 \ -2} \\
\underline{1 \ -2} \\
2 \ -3 \ -2 \\
\underline{2 \ -4} \\
1 \ -2 \\
\underline{1 \ -2} \\
0
\end{array}
$$

Notice that when the division is written as at the right above, the *position* of the entries keeps track of the process. At the right, we highlight some entries in color. The entries we have highlighted are those that are repeated. Here is what it looks like without those entries:

$$
\begin{array}{r}
1 \qquad 2 \qquad 1 \\
1-2\overline{\smash{)}1 \qquad 0 \qquad -3 \qquad -2} \\
-2 \\
\rule{5cm}{0.4pt} \\
-4 \\
\rule{5cm}{0.4pt} \\
0
\end{array}
$$

You will notice that we have now highlighted still other terms in color. The first term (green 1) at the left is not necessary because it is *always* 1 (remember the divisor has the form $x - k$). It is not necessary to write the form on the right in such spread-out form, and we also move the terms shown in blue to the bottom. The division now has the following compact form:

$$
\begin{array}{r|rrrr}
-2 & 1 & 0 & -3 & -2 \\
 & & -2 & -4 & -2 \\
\hline
 & 1 & 2 & 1 & 0
\end{array}
$$

This compact form retains the essentials for the division. The top line of the synthetic division contains the constant k of the divisor $x - k$ (the -2 in this example) and the *coefficients* of the polynomial ($1, 0, -3, -2$ in this example). The bottom line contains the coefficients of the quotient and the remainder (the answer to the division). The process is usually simplified one more step, since the same result is obtained if we replace -2 by 2 and *add instead of subtract* at each step:

Find: $\dfrac{x^3 - 3x - 2}{x - 2}$.

The form is $x - k$; write k here.
↓

$$
\begin{array}{r|cccc}
2 & 1 & 0 & -3 & -2 \\
 & & 2^* & 4(2 \cdot 2) & 2(2 \cdot 1) \\
\hline
 & 1 & 2^* & 1\,(\text{Add.}) & 0\,(\text{Add.})
\end{array}
$$

← *These are the coefficients. Don't forget the 0 coefficient.*

This entry is found by multiplication: $2 \cdot 1 = 2$.

Add; $0 + 2 = 2$

↑
Bring down this entry.

To read the answer, look at the last row and put it back into polynomial form; the last digit (0 in this example) represents the remainder. The degree of the quotient (the answer) is 1 less than the degree of the polynomial you start with:

$$\frac{x^3 - 3x - 2}{x - 2} = x^2 + 2x + 1$$

Coefficients are the last row of the synthetic division.

EXAMPLE 2   Synthetic division

Divide: $\dfrac{x^4 + 6x^3 + 8x^2 - 6x - 9}{x + 3}$.

Solution

$$
\begin{array}{r|rrrrr}
-3 & 1 & 6 & 8 & -6 & -9 \\
   &   & -3 & -9 & 3 & 9 \\
\hline
   & 1 & 3 & -1 & -3 & 0
\end{array}
$$
We see the remainder is 0.

$$\frac{x^4 + 6x^3 + 8x^2 - 6x - 9}{x + 3} = x^3 + 3x^2 - x - 3$$

∎

EXAMPLE 3   Synthetic division with remainder

Divide: $\dfrac{6x^3 - 7x^2 - 10}{x - 2}$.

Solution

$$
\begin{array}{r|rrrr}
2 & 6 & -7 & 0 & -10 \\
  &   & 12 & 10 & 20 \\
\hline
  & 6 & 5 & 10 & 10
\end{array}
$$

$$\frac{6x^3 - 7x^2 - 10}{x - 2} = 6x^2 + 5x + 10 + \frac{10}{x - 2}$$

∎

Graphing Polynomial Functions

The following box summarizes some of the properties of polynomial functions that may help us in drawing their graphs.

> **PROPERTIES OF GRAPHS OF POLYNOMIAL FUNCTIONS**
>
> - The graph of a polynomial function is continuous.
> - The graph of a polynomial function is smooth. That is, there are no corners or abrupt changes.
> - The graph of an nth-degree polynomial function will have one y-intercept (where $x = 0$) and at most n x-intercepts.
> - The graph of an nth-degree polynomial function will have at most $n - 1$ turning points.

There is a close relationship between the graph of a function and the zeros of that function. We can use the graph to help us find the zeros, or we can use the zeros to help us draw a graph. We begin with a result called the **remainder theorem**.

> **REMAINDER THEOREM**
>
> When a polynomial defined by $f(x)$ is divided by $x - r$, the remainder is equal to $f(r)$.

>> **IN OTHER WORDS** We can divide $f(x)$ by $x - r$ and look at the remainder to find the functional value of f at $x = r$. To verify this theorem, recall the division algorithm:

$$\frac{P(x)}{D(x)} = Q(x) + \frac{R(x)}{D(x)} \qquad \text{or} \qquad P(x) = Q(x)D(x) + R(x)$$

In this context, $P(x) = f(x), D(x) = x - r$, and $R(x)$ is a constant since the degree of R must be less than the degree of D, which is 1. The division algorithm may be restated as

$$f(x) = Q(x)(x - r) + R$$

where R is the remainder. Now, by substitution,

$$f(r) = Q(r)(r - r) + R = R \qquad \text{Since } Q(r)(r - r) = Q(r) \cdot 0 = 0.$$

Because of the remainder theorem, we can use synthetic division to find points on the graph of a polynomial function, as illustrated in the following example.

EXAMPLE 4   Graphing a higher-degree polynomial function

Sketch the graph $f(x) = 3x^4 - 8x^3 - 30x^2 + 72x + 47$.

Solution   This graph has at most three turning points. We carry out synthetic division, but once again, we simplify the notation. Since the same polynomial is to be evaluated repeatedly, it is not necessary to recopy it each time. The work can be arranged as shown here:

Coefficients of polynomial:

x	3	-8	-30	72	47	Point:	
-4	3	-20	50	-128	559	$(-4, 559)$	
-3	3	-17	21	9	20	$(-3, 20)$	
-2	3	-14	-2	76	-105	$(-2, -105)$	
-1	3	-11	-19	91	-44	$(-1, -44)$	
0					47	$(0, 47)$	*y*-intercept
1	3	-5	-35	37	84	$(1, 84)$	
2	3	-2	-34	4	55	$(2, 55)$	
3	3	1	-27	-9	20	$(3, 20)$	
4	3	4	-14	16	111	$(4, 111)$	

You select the integers that are convenient. Plot the points and draw a smooth curve as shown in Figure 3.34.

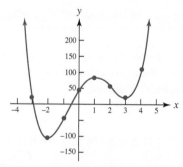

Figure 3.34 Graph of f with three turning points

■ COMPUTATIONAL WINDOW ▫ _▫◻︎☒

You might think that if you have a calculator you do not need to pay attention to these techniques, but if you input the function from Example 4 into a graphing calculator the graph might look like this:

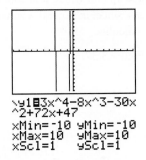

You can use the [TRACE] to obtain some idea about scale, or you can use the information from synthetic division to draw a proper graph:

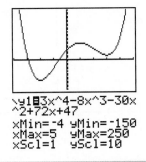

Concavity

In Section 2.3, we defined *increasing* and *decreasing functions*, which tell us where a graph is rising and where it is falling. But this gives only a partial picture of the graph. A portion of a graph that is cupped upward is called **concave up**, and a portion that is cupped downward is **concave down**. In Figure 3.35, the graph is concave up on the interval (A, B) and concave down on the interval (B, C).

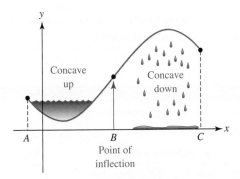

Figure 3.35 Concavity and point of inflection

Intuitively, a graph is concave up if it "holds water." A point where a graph changes concavity is called an **inflection point**. The study of concavity and the location of the inflection points is a topic you will consider in calculus, and one goal of this chapter is to focus on the terminology and properties of functions.

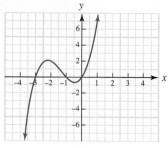

Figure 3.36 Graph of f

EXAMPLE 5   Maximum, minimum, inflection points

The graph of $f(x) = x^3 + 4x^2 + 3x$ is shown in Figure 3.36.

Label the following points or intervals: critical values; intervals where f is positive and where f is negative; turning points, and where the graph is rising and where the graph is falling; inflection points, and where the graph is concave up and where it is concave down. For this example, simply label the requested points and intervals, since, in general, you will not be able to find the exact coordinates of these points until you use calculus.

Solution   The answer is shown in Figure 3.37.

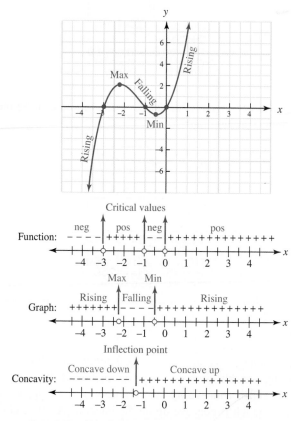

Figure 3.37 Important information about the graph of f

The information listed below is included to draw your attention to certain portions of Figure 3.37, but the figure completely answers the questions asked.

Property	Comment	Point or Interval
critical value(s)	Places where the graph crosses the x-axis.	$x = -3$, $x = -1$, $x = 0$
sign of function	positive (graph above x-axis)	interval: $(-3, -1) \cup (0, \infty)$
	negative (graph below x-axis)	interval: $(-\infty, -3) \cup (-1, 0)$
turning point(s)	Places where the graph changes from rising to falling.	$x \approx -2.2, -0.7$
rising		interval: $(-\infty, -2.2) \cup (-0.4, \infty)$
falling		interval: $(-2.2, -0.4)$
inflection point(s)	Place where concavity changes.	$x \approx -1.4$
concave down		interval: $(-\infty, -1.4)$
concave up		interval: $(-1.4, \infty)$

PROBLEM SET 3.5

Fill in the blanks for the synthetic divisions in Problems 1–4.

1. $\dfrac{3x^3 - 9x^2 + 11x - 10}{x - 2}$

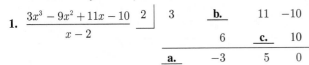

The quotient is **d.**

2. $\dfrac{4x^3 - 6x^2 - 8x - 5}{x - 3}$

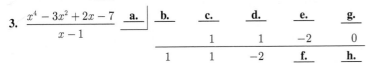

The quotient is **g.**

3. $\dfrac{x^4 - 3x^2 + 2x - 7}{x - 1}$

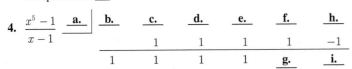

The quotient is **i.**

4. $\dfrac{x^5 - 1}{x - 1}$

The quotient is **j.**

Use synthetic division to find the quotient of the first polynomial divided by the second polynomial in Problems 5–8. Remember, if

$$\frac{P(x)}{D(x)} = Q(x) + \frac{R(x)}{D(x)}$$

then $Q(x)$ is called the quotient and $R(x)$ is called the remainder.

5. $x^3 + 2x^2 - x - 2; \; x - 1$
6. $x^3 - 4x^2 - 11x + 30; \; x - 2$
7. $x^3 - 8x^2 + x + 42; \; x + 2$
8. $x^3 - 8x^2 + x + 42; \; x - 7$

Use synthetic division to find the quotient and the remainder in Problems 9–12.

9. $\dfrac{2x^3 - 3x^2 + 4x - 10}{x - 2}$

10. $\dfrac{x^4 - 3x^3 - 4x^2 + 2x - 5}{x - 4}$

11. $\dfrac{5x^4 + 10x^3 - 20x^2 - 12x - 2}{x + 3}$

12. $\dfrac{x^5 - 3x^4 + 2x^2 - 5}{x + 2}$

In Problems 13–16, determine whether the given graph could be the graph of a polynomial, and if it cannot, tell why. If it is a polynomial, state the minimum degree. Also state the number of zeros.

13. **a.** 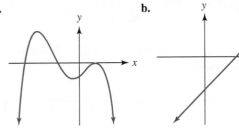 **b.**

14. **a.** 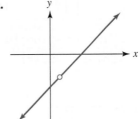 **b.**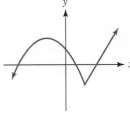

15. **a.** 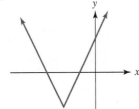 **b.**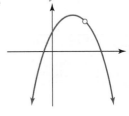

16. **a.** 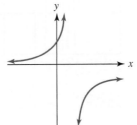 **b.**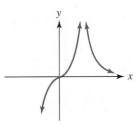

Use synthetic division to find the values specified for the functions in Problems 17–20.

17. $P(x) = x^4 - x^3 - 39x - 70$
  a. $P(0)$ **b.** $P(1)$
  c. $P(-1)$ **d.** $P(-5)$

18. $f(x) = 8x^4 - 6x^3 + 5x^2 + 4x - 3$
  a. $f(0)$ **b.** $f(1)$
  c. $f\left(-\dfrac{1}{2}\right)$ **d.** $f\left(\dfrac{1}{2}\right)$

19. $f(x) = 16x^4 + 64x^3 + 19x^2 - 81x + 18$
  a. $f(0)$ **b.** $f(-2)$
  c. $f(-3)$ **d.** $f\left(\dfrac{1}{4}\right)$

20. $f(x) = 4x^4 - 8x^3 - 43x^2 + 29x + 60$

  a. $f(-1)$ **b.** $f(1)$

  c. $f\left(-\dfrac{5}{2}\right)$ **d.** $f\left(\dfrac{3}{2}\right)$

WHAT IS WRONG, *if anything, with each graph in Problems 21–26? Explain your reasoning.*

21.

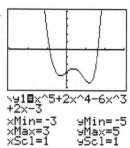

```
\y1█x^5+2x^4-6x^3
+2x-3
xMin=-3    yMin=-5
xMax=3     yMax=5
xScl=1     yScl=1
```

22.

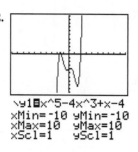

```
\y1█x^5-4x^3+x-4
xMin=-10   yMin=-10
xMax=10    yMax=10
xScl=1     yScl=1
```

23.

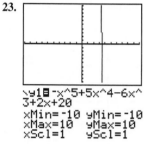

```
\y1█-x^5+5x^4-6x^
3+2x+20
xMin=-10   yMin=-10
xMax=10    yMax=10
xScl=1     yScl=1
```

24.

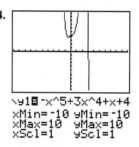

```
\y1█-x^5+3x^4+x+4
xMin=-10   yMin=-10
xMax=10    yMax=10
xScl=1     yScl=1
```

25.

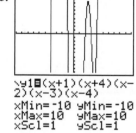

```
\y1█(x+1)(x+4)(x-
2)(x-3)(x-4)
xMin=-10   yMin=-10
xMax=10    yMax=10
xScl=1     yScl=1
```

26.

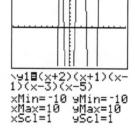

```
\y1█(x+2)(x+1)(x-
1)(x-3)(x-5)
xMin=-10   yMin=-10
xMax=10    yMax=10
xScl=1     yScl=1
```

LEVEL 2

PROBLEMS FROM CALCULUS *The graph of f is shown in Problems 27–32.*

a. Find the critical value(s) and state the intervals where *f* is positive and where it is negative.

b. Find the turning point(s) and state the intervals where the graph is rising and where the graph is falling.

c. Find the inflection point(s) and state the intervals where the graph is concave up and where it is concave down.

27. $f(x) = x^3 - 3x^2 - 18x + 40$

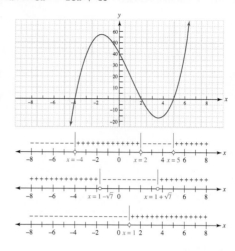

28. $f(x) = x^3 + 4x^2 - x - 4$

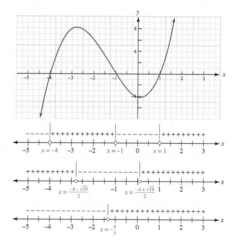

29. $f(x) = x^4 + 2x^3 - 41x^2 - 42x + 360$

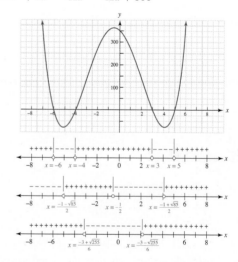

30. $f(x) = 3x^4 - 8x^3 - 48x^2$

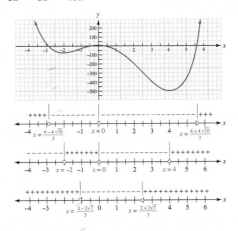

31. $f(x) = x^4 - 3x^3 - 31x^2 + 63x + 90$

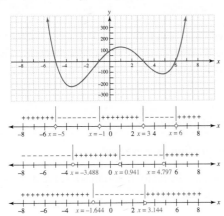

32. $f(x) = x^4 - 4x^3 - 20x^2 + 96$

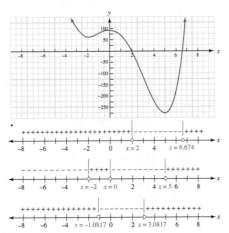

Sketch the graph of each polynomial function in Problems 33–58.

33. $f(x) = x^3 - 6x^2 + 9x - 9$
34. $f(x) = 2x^3 - 3x^2 - 36x + 78$
35. $f(x) = 4x^4 - 8x^3 - 43x^2 + 29x + 60$
36. $f(x) = 16x^4 + 64x^3 + 19x^2 - 81x + 18$
37. $f(x) = x^4 - 7x^2 - 2x + 2$
38. $f(x) = x^4 - 14x^3 + 58x^2 - 46x - 9$
39. $f(x) = x^6 - 4x^4 - 4x^2 + 4$
40. $f(x) = x^5 + 2x^4 - 5x^3 - 10x^2 + 4x + 8$
41. $f(x) = 3x^4 - x^3 - 14x^2 + 4x + 8$
42. $y = 5x^4 + 3x^3 - 22x^2 - 12x + 8$
43. $y = x^4 - x^3 - 3x^2 + 2x + 4$
44. $y = x^4 - 2x^2 - 4x + 3$
45. $y = (x - 1)(x + 1)(x + 3)$
46. $y = (x - 1)(x - 4)(x + 3)$
47. $y = (x + 1)(x + 3)(2x - 5)$
48. $y = (x - 1)(x - 4)(2x + 1)$
49. $y = (x + 1)(x + 2)(3x - 1)$
50. $y = x(x - 3)(x + 3)$
51. $y = 3x^2(x - 3)(x + 1)$
52. $y = 5x^2(x - 4)(x + 2)$
53. $y = x^2(x^2 - 1)$
54. $y = x^2(x^2 - 4)$
55. $f(x) = 3x^4 - x^3 + 5x^2 + x - 10$
56. $f(x) = 8x^4 + 12x^3 - 3x^2 + 4x + 20$
57. $f(x) = x^5 - 3x^4 + 2x^3 - 7x + 15$
58. $f(x) = x^5 - 5x^4 + 3x^3 + x^2$

LEVEL 3

Sketch the graph of each pair of graphs in Problems 59–60 on the same coordinate axes.

59. a. $y = x^3 + 3x^2 - x - 3$
　　b. $y = |x^3 + 3x^2 - x - 3|$
60. a. $y = -x^3 + 2x^2 + x - 2$
　　b. $y = -|x|^3 + 2|x|^2 + |x| - 2$

3.6 Polynomial Equations

Factoring Polynomials Using Synthetic Division

If $P(x) = a_n x^n + a_{n-1} x^{n-1} + \cdots + a_2 x^2 + a_1 x + a_0$, $a_n \neq 0$, then the *roots* or *solutions* of $P(x) = 0$ are values of x that satisfy this equation. Such an equation is called a **polynomial equation**. Recall from Section 2.3 that a is called a *zero* of a function P if $P(a) = 0$, and recall that we solved quadratic equations in Section 1.8.

If we wish to solve a polynomial equation, we begin by factoring the polynomial, if possible. From the division algorithm, we know that if $R = 0$ and

$$P(x) = Q(x)(x - r) + R$$

then $P(x) = Q(x)(x - r)$. This says that $(x - r)$ is a factor of $P(x)$. But since you are looking for values of x such that $P(x) = 0$,

$$0 = Q(x)(x - r)$$

Notice that this equation is satisfied by $x = r$ and leads to a result called the *factor theorem*. The polynomial equation $Q(x) = 0$ is called a **depressed equation** of the polynomial equation $P(x) = 0$.

FACTOR THEOREM

If P is defined by

$$P(x) = (x - r)Q(x) + R$$

and if $R = 0$, then $x - r$ is a factor of P.

» IN OTHER WORDS If P and D are polynomials (with $D(x) \neq 0$), then D is a factor of P if the remainder when dividing P by D is 0. Also recall the remainder theorem; when $P(x)$ is divided by $x - r$, then the remainder is equal to the value of P at $x = r$.

EXAMPLE 1   Determining a factor of a polynomial function

Factor $2x^3 + x^2 - 13x + 6$ by answering the question, is $\frac{1}{2}$ a zero?

Solution   If $\frac{1}{2}$ is a zero of the polynomial, then by the factor theorem we see that $x - \frac{1}{2}$ is a factor. We check for this factor by using synthetic division:

$$
\begin{array}{r|rrrr}
\frac{1}{2} & 2 & 1 & -13 & 6 \\
            &   & 1 & 1 & -6 \\
\hline
            & 2 & 2 & -12 & 0
\end{array}
$$

Since we have a zero remainder, we see that $x - \frac{1}{2}$ is a factor. We can use this result to complete the factoring:

$$2x^3 + x^2 - 13x + 6 = \left(x - \frac{1}{2}\right)(2x^2 + 2x - 12)$$

$$= \frac{1}{2}(2x - 1)(2)(x^2 + x - 6)$$

$$= (2x - 1)(x + 3)(x - 2)$$

After Example 1, you may be asking where the $1/2$ came from. If you have a graphing calculator, you might have found $1/2$ by estimating a zero of the polynomial. If you do not have a graphing calculator, or if the zeros are not easy to estimate, how could we have factored that polynomial? The answer to this question comes from a branch of algebra called *theory of equations*. It gives us a list of *possible* zeros for a polynomial and, consequently, a list of factors to *try* by using synthetic division.

RATIONAL ROOT THEOREM

If P is a function defined by

$$P(x) = a_n x^n + a_{n-1} x^{n-1} + \cdots + a_2 x^2 + a_1 x + a_0$$

with integer coefficients and a rational zero of the form $\dfrac{p}{q}$ (in lowest terms), then p is a factor of a_0 and q is a positive factor of a_n.

»» IN OTHER WORDS All rational zeros of the polynomial P are of the form p/q, where the constant term is divisible by p and the leading coefficient is divisible by q. As far as factoring is concerned, this means we should try $(x - r)$ as a factor where r is taken from the set of possible rational numbers p/q.

EXAMPLE 2   Factoring using the rational root theorem

Factor: $2x^4 - 5x^3 - 8x^2 + 25x - 10$

Solution   We note that p is a factor of the constant term: $\pm 1,\ \pm 2,\ \pm 5,\ \pm 10$; and q is a positive factor of the leading coefficient: 1, 2. We try factors of the form $(x - r)$, where r comes from a list of possible rational roots $\dfrac{p}{q}$: $\pm 1,\ \pm \dfrac{1}{2},\ \pm 2,\ \pm 5,\ \pm \dfrac{5}{2},\ \pm 10$. Check each of these using synthetic division.

	2	−5	−8	25	−10
1	2	−3	−11	14	4
−1	2	−7	−1	26	−36
2	2	−1	−10	5	0
−2	2	−5	0	5	
5	2	9	35	180	
−5	2	−11	45	−220	
$\frac{1}{2}$	2	0	−10	0	

$(x - 2)$ is a factor. → 2

$\left(x - \frac{1}{2}\right)$ is a factor. → $\frac{1}{2}$

If the remainder is not 0, then $x - r$ is not a factor. We will show the portion to show you that the value checked does not give us a factor. After a 0 remainder, use nonzero terms on next line.

Stop when the remaining polynomial is quadratic.

We can now factor the polynomial:

$$2x^4 - 5x^3 - 8x^2 + 25x - 10 = (x - 2)\left(x - \frac{1}{2}\right)(2x^2 - 10)$$

$$= (x - 2)\frac{1}{2}(2x - 1)2(x^2 - 5)$$

$$= (x - 2)(2x - 1)(x^2 - 5)$$

Look again at Example 2. After you obtain a zero remainder (as with the synthetic division by 2 in the example), you then use the coefficients on this line (and not the original coefficients) for subsequent division.

EXAMPLE 3   Repeated factors by using the rational root theorem

Factor: $2x^5 - 23x^3 + x^2 + 61x + 39$

Solution   All possible values of $\dfrac{p}{q}$: $\pm 1,\ \pm 3,\ \pm 13,\ \pm 39,\ \dfrac{1}{2},\ \pm \dfrac{3}{2},\ \pm \dfrac{13}{2},\ \pm \dfrac{39}{2}$.

We show the values from synthetic division, without showing those values we tried that did not work.

	2	0	−23	1	61	39	
−1	2	−2	−21	22	39	0	After a 0 remainder, use nonzero terms on the next line.
−1	2	−4	−17	39	0		
−3	2	−10	13	0			

Notice that if you try a value and it does not work, you do not need to try it again. However, if a value does give a 0 remainder once, there is no reason to expect that it cannot give a 0 remainder again.

$$2x^5 - 23x^3 + x^2 + 61x + 39 = (x+1)(x+1)(x+3)(2x^2 - 10x + 13)$$
$$= (x+1)^2(x+3)(2x^2 - 10x + 13)$$

Fundamental Theorem of Algebra*

In 1799, a 22-year-old graduate student named Karl Gauss proved in his doctoral thesis that every polynomial equation has at least one solution in the set of complex numbers. (Do not forget that the real numbers are a subset of the set of complex numbers.) This, of course, is an assumption that you have made throughout your study of algebra—from the time you first solved first-degree equations in beginning algebra. It is an idea so basic to algebra that it is called the **fundamental theorem of algebra**. To prove this theorem, it is necessary to allow the domain to be the set of complex numbers; when we speak of complex coefficients of a polynomial equation, we are including all those polynomial equations previously considered in this textbook.

> **FUNDAMENTAL THEOREM OF ALGEBRA**
>
> Every polynomial equation in a single variable with complex coefficients has at least one root.

If an equation has one solution, say, r, then $(x - r)$ is a factor, and the given polynomial is the product of $x - r$ and a polynomial of degree $n - 1$. Then that new equation, according to the fundamental theorem, has at least one solution. This root may now be used to obtain an equation of lower degree. The result of this process suggests the following theorem.

> **NUMBER OF ROOTS**
>
> Every polynomial equation of degree n in a single variable has n roots (where n is a counting number).

Of course, the roots need not be distinct or real. You need to pay attention to this last sentence; for example,

$$x^2 - 6x + 9 = 0$$

has two solutions (since it is second degree), but when we factor and solve, we find

$$(x - 3)(x - 3) = 0$$
$$x = 3$$

*Complex numbers (Appendix B) are required for portions of this section.

Notice that $x = 3$ comes from $x - 3 = 0$, but there are *two* identical factors, so there are *two* roots, namely, $x = 3$ (from the first factor) and $x = 3$ (from the second factor). When considering the theorems of this section, we will count *each* occurrence of the root. This means that for this example, there are two roots (which are not distinct). We say that $x = 3$ is a *root of multiplicity 2* (because it appears twice). If a factor $(x - r)$ occurs exactly k times in the factorization of $P(x)$, then r is called a **zero of multiplicity k.** If a factor $(x - r)$ occurs exactly k times in the factorization of $P(x)$ in the equation $P(x) = 0$, then r is called a **root of multiplicity k.** For example,

$$f(x) = (x - 1)(x - 3)(x - 4)(x - 1)(x - 3)(x - 1)$$

has *zeros* of 1 (multiplicity 3), 3 (multiplicity 2), and 4. Notice that this is a function, not an equation, so we use the word *zero*. On the other hand,

$$(x - 1)(x - 3)(x - 4)(x - 1)(x - 3)(x - 1) = 0$$

has *roots* of 1 (multiplicity 3), 3 (multiplicity 2), and 4.

We can use the number of roots theorem in conjunction with the rational root theorem along with another theorem called the **upper and lower bound theorem**, to know when we have found all of the roots of a given equation. If P is a polynomial function, then a real number a is an **upper bound** for the polynomial equation $P(x) = 0$ if there is no root, or solution, larger than a. A real number b is a **lower bound** if there is no solution less than b. Notice from Figure 3.38 that there may be several upper or lower bounds.

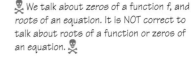 We talk about zeros of a function f, and roots of an equation. It is NOT correct to talk about roots of a function or zeros of an equation.

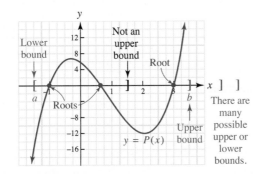

Figure 3.38 Upper and lower bounds for a polynomial function

UPPER AND LOWER BOUND THEOREM

Suppose P is a polynomial function with real coefficients and positive leading coefficient.

If $a > 0$ and in the synthetic division of $P(x)$ by $x - a$, all the numbers in the last row have either positive or zero coefficients, then a is an upper bound for the real roots of $P(x) = 0$.

If $b < 0$ and in the synthetic division of $P(x)$ by $x - b$, all the numbers in the last row alternate in sign (where a 0 is considered to be either positive or negative), then b is a lower bound for the real roots of $P(x) = 0$.

We use this upper and lower bound theorem along with the root location theorem (of Section 2.8) to locate the real roots. For example, if $x = 5$ is an upper bound for a polynomial equation, then we look for roots smaller than 5, and if $x = -1$ is a lower bound for that same polynomial equation, then we look for roots between -1 and 5.

Furthermore, if $f(3) = -2.34$ and $f(4) = 3.09$, then the root location theorem tells us that there must be a root between 3 and 4.

This theorem says that if the conditions of the theorem are satisfied, then the number is an upper (or lower) bound. It does not say that some number might exist that fails the test but is still an upper (or lower bound).

EXAMPLE 4   Using the upper and lower bound theorem

Solve: $4x^4 - 8x^3 - 43x^2 + 29x + 60 = 0$.

Solution   We begin with an approximate (graphical) solution, as shown in Figure 3.39.

It looks like there is a root near –1; other possibilities are between –2 and –3; between 1 and 2; and another root near 4. If we want to find the exact values of the roots, we find the possible rational roots.

$$p: \quad \pm 1, \pm 2, \pm 3, \pm 4, \ldots \qquad q: 1, 2, 4$$

$$\frac{p}{q}: \quad \pm 1, \pm \frac{1}{2}, \pm \frac{1}{4}, \pm 2, \ldots \quad \text{We will list the others as needed.}$$

We now turn to synthetic division, and we pick values to try first by looking at the graph (if we have a graphing calculator) and by the list of rational roots.

	4	−8	−43	29	60
−1	4	−12	−31	60	0
−3	4	−24	41	−63	
−2	4	−20	9	42	
$-\dfrac{5}{2}$	4	−22	24	0	

$x - 1$ is a factor.

Alternating signs with a negative value; lower bound.

Positive at $x = -2$ and negative at $x = -3$; thus, by the root location theorem, there is a root between −2 and −3. Since a root is between −2 and −3, select an appropriate value from the list of possible rational roots.

$x + \dfrac{5}{2}$ is a factor.

We can now complete the factorization:

$$4x^4 - 8x^3 - 43x^2 + 29x + 60 = 0 \quad \text{Given equation}$$

$$(x+1)\left(x+\frac{5}{2}\right)(4x^2 - 22x + 24) = 0 \quad \text{Factors from the synthetic division}$$

$$(x+1)\left(x+\frac{5}{2}\right)(2)(2x^2 - 11x + 12) = 0$$

$$(x+1)\left(x+\frac{5}{2}\right)(2)(2x - 3)(x - 4) = 0 \quad \text{Use factor theorem on this equation.}$$

The roots are $-1, -\dfrac{5}{2}, \dfrac{3}{2}$, and 4.   ■

EXAMPLE 5   Equation with roots of multiplicity

Solve: $8x^5 - 44x^4 + 86x^3 - 73x^2 + 28x - 4 = 0$.

Solution   You might use a calculator, if available, to help you make your selection of divisors in the synthetic division. We begin by listing the possible rational roots: $\pm 1, \pm \dfrac{1}{2}, \pm \dfrac{1}{4}, \pm \dfrac{1}{8}, \pm 2,$ ± 4. The number of roots theorem tells us to expect five solutions (but these roots are not necessarily real or distinct).

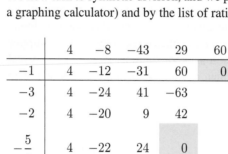

\y1■4x^4-8x^3-43x
^2+29x+60

xMin=-3 yMin=-150
xMax=5 yMax=100
xScl=1 yScl=50

Figure 3.39 Calculator graph

	8	−44	86	−73	28	−4	
−1	8	−52	138	−211	239	−243	−1 is a lower bound.
1	8	−36	50	−23	5	1	
2	8	−28	30	−13	2	0	$x - 2$ is a factor.
2	8	−12	6	−1	0		$x - 2$ is a factor again.
2	8	4	14	27			2 is an upper bound.
$\frac{1}{2}$	8	−8	2	0			$x - \frac{1}{2}$ is a factor.

The remaining polynomial in the quotient is quadratic, so we can finish by factoring or using the quadratic formula. We do this as part of the factoring process.

$$8x^5 - 44x^4 + 86x^3 - 73x^2 + 28x - 4 = 0 \quad \text{Given equation.}$$

$$(x-2)(x-2)\left(x - \frac{1}{2}\right)(8x^2 - 8x + 2) = 0 \quad \text{Factors from the synthetic division}$$

$$(x-2)^2 \frac{1}{2}(2x-1)(2)(4x^2 - 4x + 1) = 0$$

$$(x-2)^2(2x-1)(2x-1)(2x-1) = 0 \quad \text{Factor the quadratic.}$$

$$(x-2)^2(2x-1)^3 = 0 \quad \text{Use factor theorem on this equation.}$$

The roots are 2 and $\frac{1}{2}$. To reconcile with the number of roots theorem, we see that 2 has multiplicity two and $\frac{1}{2}$ has multiplicity three, for a total of five roots. ∎

In Example 5, the degree was 5 and we found 5 roots (taking into account the multiplicity of the roots). Suppose, however, you attempt to solve a polynomial equation that has fewer *real* roots than the degree. In such a situation, you will not know whether you cannot find the roots because, although they are real, they are *not rational* (that is, they are *x*-intercepts) or because they are simply not real numbers. This is summarized in Table 3.2.

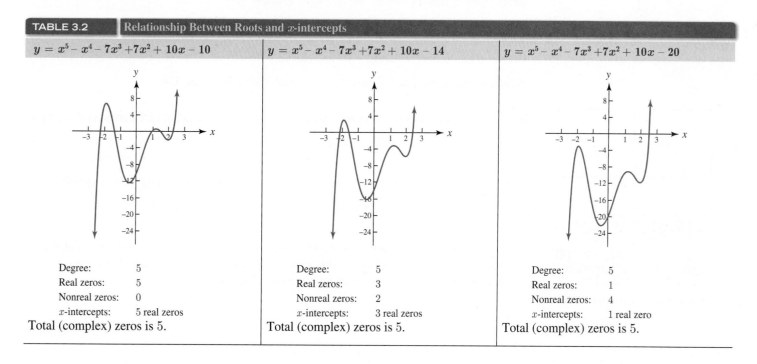

TABLE 3.2	Relationship Between Roots and x-intercepts	
$y = x^5 - x^4 - 7x^3 + 7x^2 + 10x - 10$	$y = x^5 - x^4 - 7x^3 + 7x^2 + 10x - 14$	$y = x^5 - x^4 - 7x^3 + 7x^2 + 10x - 20$
Degree: 5	Degree: 5	Degree: 5
Real zeros: 5	Real zeros: 3	Real zeros: 1
Nonreal zeros: 0	Nonreal zeros: 2	Nonreal zeros: 4
x-intercepts: 5 real zeros	x-intercepts: 3 real zeros	x-intercepts: 1 real zero
Total (complex) zeros is 5.	Total (complex) zeros is 5.	Total (complex) zeros is 5.

All the polynomials shown in Table 3.2 were chosen so that there are fewer rational roots than the degree of the polynomial. The unanswered question is, how many of the roots not found by the rational root theorem are real and how many are not real?

The following theorem will help you to answer this dilemma by telling you when you have found all the positive or negative real roots. When applying this theorem, remember that a root of multiplicity m is counted as m roots.

DESCARTES'S RULE OF SIGNS

Let P be a polynomial with real coefficients written in descending powers of x. Count the number of sign changes of the coefficients of $P(x)$ and of $P(-x)$.

Positive: The number of positive real zeros is equal to the number of sign changes or is equal to that number decreased by an even integer.

Negative: The number of negative real zeros is equal to the number of sign changes in $P(-x)$ or is equal to that number decreased by an even integer.

EXAMPLE 6   Using Descartes's rule of signs

Show that $x^9 - x^5 + x^4 + x^2 + 1 = 0$ has at least six nonreal complex solutions.

Solution $P(x) = x^9 - x^5 + x^4 + x^2 + 1$
$+ - + + +$ ← Two sign changes so that there are 2 or 0 positive real roots.

$P(-x) = -x^9 + x^5 + x^4 + x^2 + 1$
$- + + +$ ← One sign changes so that there is 1 negative real root.

The polynomial equation has one negative real root and at most two positive real roots. However, it has nine solutions since it is of degree 9; thus, there are at least six nonreal complex solutions. ∎

If you have a graphing calculator, you might be saying to yourself, "I have no need of Descartes's rule of signs because my calculator will show me the necessary graph." Indeed, the graphing calculator, along with more realistic problems, has *increased* the need for results such as Descartes's rule of signs. Consider the following example.

EXAMPLE 7   Graphing a function using Descartes's rule of signs

Graph: $f(x) = 0.01x^4 + 0.22x^3 + 0.1x^2 - 3.85x + 3$

Solution   Begin by noticing the calculator graph shown in Figure 3.40.

It is NOT correct. To draw the graph correctly, let us begin by checking Descartes's rule of signs:

$$f(x) = 0.01x^4 + 0.22x^3 + 0.1x^2 - 3.85x + 3$$
$+ + + - +$ ← Two sign changes

There are 2 or 0 positive roots (or x-intercepts); this is consistent with the calculator graph.

$$f(-x) = 0.01x^4 - 0.22x^3 + 0.1x^2 + 3.85x + 3$$
$+ - + + +$ ← Two sign changes

There are 2 or 0 negative roots (or x-intercepts); this is not consistent with the calculator graph. We need to look for a lower bound, which we can do with synthetic division. At the same time, the synthetic division gives us additional points to complete the graph.

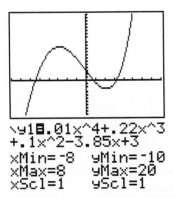

```
\y1◻.01x^4+.22x^3
+.1x^2-3.85x+3
xMin=-8    yMin=-10
xMax=8     yMax=20
xScl=1     yScl=1
```

Figure 3.40 This calculator graph is WRONG!

	0.01	0.22	0.1	-3.85	3
-10	0.01	0.12	-1.1	7.15	-68.5
-20	0.01	0.02	-0.3	2.15	-40
-30	0.01	-0.08	2.5	-78.85	2,368.5

The last row has alternating signs, so $x = -30$ is a lower bound. We use these points to help us complete the graph, as shown in Figure 3.41.

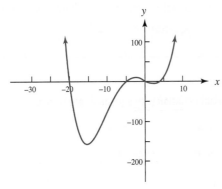

Figure 3.41 Graph using Descartes's rule of signs is CORRECT!

Many of the polynomial equations solved in this chapter have had rational solutions, but many others have irrational or nonreal solutions. If you know one such solution, then the next theorem tells us how to find another.[*]

> **CONJUGATE PAIR THEOREM**
>
> If $P(x) = 0$ is a polynomial equation with real coefficients, then when $a + bi$ is a root, $a - bi$ is also a root (a and b are real numbers).
>
> If $P(x) = 0$ is a polynomial equation with rational coefficients, then when $m + \sqrt{n}$ is a root, $m - \sqrt{n}$ is also a root (m and n are rational numbers and $\sqrt{n}$ is irrational).

EXAMPLE 8   Solving a polynomial equation with a known irrational root

Solve $x^4 - 4x - 1 = 4x^3$ given that $2 + \sqrt{5}$ is a root.

Solution   We use synthetic division (using the longer form because of the complexity of the numbers):

$$
\begin{array}{r|ccccc}
2+\sqrt{5} & 1 & -4 & 0 & -4 & -1 \\
 & & 2+\sqrt{5} & 1 & 2+\sqrt{5} & 1 \\
\hline
 & 1 & -2+\sqrt{5} & 1 & -2+\sqrt{5} & 0 \\
\end{array}
$$

---

[*]The rest of this section requires optional Appendix B (complex numbers).

By the conjugate pair theorem, $2 - \sqrt{5}$ must also be a root, so we begin where we left off from the first division:

$$
\begin{array}{r|ccccc}
2-\sqrt{5} & 1 & -2+\sqrt{5} & 1 & -2+\sqrt{5} \\
& & 2-\sqrt{5} & 0 & 2-\sqrt{5} \\
\hline
& 1 & 0 & 1 & \boxed{0}
\end{array}
$$

The remaining equation is $x^2 + 1 = 0$ with roots $x = \pm i$. Thus, the roots of the given polynomial equation are $2 \pm \sqrt{5}$, $\pm i$.

■

EXAMPLE 9   Solving a polynomial equation with a given nonreal root

Solve $x^4 + 2x^3 = 4x + 4$ given that $-1 - i$ is a root.

Solution   Rearrange terms to obtain $x^4 + 2x^3 - 4x - 4 = 0$.

$$
\begin{array}{r|ccccc}
-1-i & 1 & 2 & 0 & -4 & -4 \\
& & -1-i & -2 & 2+2i & 4 \\
\hline
& 1 & 1-i & -2 & -2+2i & \boxed{0}
\end{array}
$$

Since $-1 - i$ is a root, we know $-1 + i$ is also a root:

$$
\begin{array}{r|cccc}
-1+i & 1 & 1-i & -2 & -2+2i \\
& & -1+i & 0 & 2-2i \\
\hline
& 1 & 0 & -2 & \boxed{0}
\end{array}
$$

The roots are $-1 \pm i$, $\pm\sqrt{2}$.

■

PROBLEM SET 3.6

LEVEL 1

In Problems 1–6, find all the zeros of the polynomial and state the multiplicity of each zero.

1. a. $f(x) = (x-1)(x+4)^2$ **b.** $f(x) = (x+2)^2(x-3)^2$
2. a. $f(x) = x^3(2x-5)^2$ **b.** $f(x) = x^2(5x+1)^3$
3. a. $f(x) = (x^2-1)^2(x+6)$ **b.** $f(x) = (x-2)(x^2-9)^3$
4. a. $f(x) = (x^2+2x-15)^2$ **b.** $f(x) = (6x^2+7x-3)^2$
5. a. $f(x) = x^4 - 8x^3 + 16x^2$ **b.** $f(x) = x^4 + 6x^3 + 9x^2$
6. a. $f(x) = (x^3-9x)^2$ **b.** $f(x) = (x^3-25x)^2$

In Problems 7–12, use Descartes's rule of signs to state the number of possible positive and negative real roots.

7. a. $x^4 - 3x^3 + 7x^2 - 19x + 15 = 0$
  b. $3x^3 - 7x^2 + 5x + 7 = 0$
8. a. $2x^5 + 6x^4 - 3x + 12 = 0$
  b. $x^3 + 3x^2 - 4x - 12 = 0$
9. a. $x^3 + 2x^2 - 5x - 6 = 0$
  b. $2x^3 + x^2 - 13x + 6 = 0$

10. a. $2x^3 - 3x^2 - 32x - 15 = 0$
  b. $x^4 - 12x^3 + 54x^2 - 108x + 81 = 0$
11. a. $x^4 + 3x^3 - 20x^2 - 3x + 18 = 0$
  b. $x^4 - 13x^2 + 36 = 0$
12. a. $2x^2 + 6x - 3x^3 - 4 = 0$
  b. $5x^3 - 2x^4 + x^2 - 7 = 0$

Show that the equations in Problems 13–16 have no rational roots.

13. $x^3 - 2x^2 + 3x - 4 = 0$ **14.** $2x^3 + 5x^2 - 3x + 1 = 0$
15. $x^4 + 4x^3 - x^2 - 2x + 3 = 0$ **16.** $3x^4 - x^3 + 4x^2 + 2x - 2 = 0$

LEVEL 2

In Problems 17–22, decide whether the given value is a root of the given equation. If it is, then name another root.

17. $1+\sqrt{2}$; $x^3 - 2x^2 - x + 1 = 0$?
18. $2+\sqrt{5}$; $x^4 - 4x^3 - 5x^2 + 16x + 4 = 0$?

*19. $1+i;$ $x^3 - 4x^2 + 6x - 4 = 0?$

*20. $1-2i;$ $x^3 - x^2 + 3x + 5 = 0?$

*21. $1-2i;$ $x^4 - 2x^3 - 4x^2 + 2x - 5 = 0?$

*22. $1+2i;$ $x^4 - 7x^3 + 14x^2 + 2x - 20 = 0?$

Solve the polynomial equations in Problems 23–28 over the set of real numbers.

23. $x^4 - 4x^3 + 6x^2 - 4x + 1 = 0$

24. $x^4 + 2x^3 - 13x^2 - 14x + 24 = 0$

25. $x^4 - 20x^2 - 125 = 0$

26. $x^4 - 2x^3 - 13x^2 + 14x + 24 = 0$

27. $x^4 - 4x^3 - 2x^2 + 4x + 1 = 0$

28. $x^4 - 12x^3 - 13 = 6(3 - 2x - 5x^2)$

†*Solve the polynomial equations in Problems 29–34 over the set of complex numbers.*

29. $x^4 - 6x^2 - 27 = 0$ 30. $x^4 - 20x^2 - 125 = 0$

31. $x^5 - 4x^4 + 2x^3 + 2x^2 + x + 6 = 0$

32. $2x^4 - x^3 + 2x = 1$ 33. $2x^3 + 4x + 3 = 3x^2$

34. $x^4 - 2x^3 + 4x^2 + 2x - 5 = 0$

WHAT IS WRONG? *Use Descartes's rule of signs to determine what is wrong with each of the graphs shown in Problems 35–38. Use this information to assist in drawing a graph showing all of the zeros of the given function.*

35.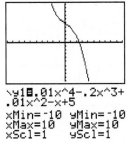

```
\y1⊟.01x^4-.2x^3+
.01x^2-x+5
xMin=-10  yMin=-10
xMax=10   yMax=10
xScl=1    yScl=1
```

36.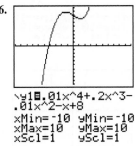

```
\y1⊟.01x^4+.2x^3-
.01x^2-x+8
xMin=-10  yMin=-10
xMax=10   yMax=10
xScl=1    yScl=1
```

37.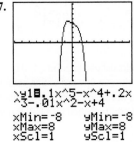

```
\y1⊟.1x^5-x^4+.2x
^3-.01x^2-x+4
xMin=-8   yMin=-8
xMax=8    yMax=8
xScl=1    yScl=1
```

38.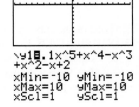

```
\y1⊟.1x^5+x^4-x^3
+x^2-x+2
xMin=-10  yMin=-10
xMax=10   yMax=10
xScl=1    yScl=1
```

‡*Solve the polynomial equations $P(x) = 0$ in Problems 39–51.*

39. $P(x) = x^3 - 8$ 40. $P(x) = x^3 - 64$

41. $P(x) = x^4 - 81$ 42. $P(x) = x^4 - 625$

43. $P(x) = x^4 - 25$ 44. $P(x) = x^4 - 64$

45. $P(x) = x^4 + 9x^2 + 20$ 46. $P(x) = x^4 + 10x^2 + 9$

47. $P(x) = x^4 + 13x^2 + 36$

48. $P(x) = (x^2 - 4x - 1)(x^2 - 6x + 7)$

49. $P(x) = (x^2 - 6x + 10)(x^2 - 8x + 7)$

50. $P(x) = (x^2 + 2x + 5)(x^2 - 3x + 5)$

51. $P(x) = (x^2 - 4x - 1)(x^2 - 3x + 5)$

52. IN YOUR OWN WORDS Draw the graphs of a fifth-degree equation that crosses the x-axis at the number of places indicated.

   a. two places

   b. one place

   c. four places

53. IN YOUR OWN WORDS Draw the graphs of a sixth-degree equation that crosses the x-axis at the number of places indicated.

   a. four places

   b. three places

54. Use Descartes's rule of signs and the root location theorem to show that the polynomial equation

$$x^3 + x^2 - 6x + 1 = 0$$

has exactly two positive real roots.

55. Use Descartes's rule of signs and the root location theorem to show that the polynomial equation

$$x^4 + x^3 - 8x^2 - 3x + 2 = 0$$

has exactly two negative real roots.

*56. Show that

$$x^5 - 10x + 2 = 0$$

has two positive real roots, one negative real root, and two roots that are not real.

LEVEL 3

57. Show that

$$x^6 + 2x - 1 = 0$$

has one positive real root, one negative real root, and four roots that are not real.

58. Solve $x^3 - 2x^2 + 4x - 8 = 0$.

59. Solve $x^4 - 7x^3 + 14x^2 + 2x = 0$ over the set of real numbers with your answer correct to the nearest hundredth.

60. Solve $x^4 - 4x^3 + 3x^2 + 8x - 10 = 0$ given that $2 + i$ is a root.

---

*Requires complex numbers (Appendix B).

†Requires complex numbers (Appendix B).

‡Requires complex numbers (Appendix B). However, if you have skipped complex numbers, you can work these problems by finding only the real roots.

CHAPTER 3 SUMMARY AND REVIEW

$\mathcal{S}$*cholarship lies in the direction of paying deference to the loyal continuous function rather than to the outlaws of mathematical society.*

E. D. Roe, Jr. (1910)

Take some time getting ready to work the review problems in this section. First, look back at the definition and property boxes. You will maximize your understanding of this chapter by working the problems in this section only after you have studied the material.

SELF TEST   *All of the answers for this self test are given in the back of the book.*

Graph the functions given in Problems 1–2.

1. a. $y - 1 = \frac{3}{5}(x + 2)^2$ **b.** $y = x^2 - 8x + 10$
   c. $x^2 + 4x - 2y - 2 = 0$ **d.** $x^2 - 8x - 3y + 3{,}616 = 0$

2. a. $f(x) = 3x^3 + 2x^2 - 12x - 8$
   b. $h(x) = x^4 - 14x^2 + x^3 - 14x$
   c. $x = 2 + 5t, \ y = -1 - 3t$
   d. $x = t^2 + 3t - 1, \ y = t^2 + 2t + 5$

3. a. Graph $f(x) = 2x^3 - 4x^2$.   **b.** What are the zeros of $f(x)$?
   c. State the number of roots theorem, and apply it to f.
   d. Reconcile your answers to parts **b** and **c**.

4. The function $f(x) = (x - 1)^2(x + 2)^3$ has zeros at $x = 1$ and $x = -2$. Thus, the two real roots for the polynomial equation, $f(x) = 0$, are $x = 1$ (root of multiplicity 2) and $x = -2$ (root of multiplicity 3). Draw the graph of a fifth-degree equation that crosses the x-axis in three places.

Find the zeros of the polynomial functions in Problems 5–6. State the multiplicity of each zero, show a list of possible rational roots, and use Descartes's rule of signs to determine the number of positive and negative real roots.

5. a. $3x^3 + 4x^2 - 35x - 12 = 0$
   b. $3x^3 + 2x^2 - 12x - 8 = 0$

6. a. $6x^4 - 13x^3 + 3x^2 + 9x - 5 = 0$
   b. $x^4 + 11x^3 - 25x^2 + 11x - 26 = 0$

7. Factor: $6x^4 - 61x^3 - 675x^2 + 6{,}837x - 12{,}155$
   (*Hint:* $12{,}155 = 5 \cdot 11 \cdot 13 \cdot 17$.)

8. WHAT IS WRONG, *if anything, with each of the graphs? Draw each graph correctly.*

a.
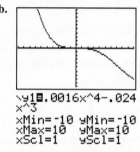

b.
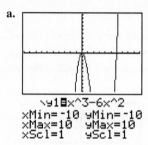

c.
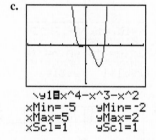

d.
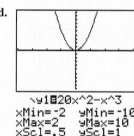

9. Which point on the curve $y^2 = 2x$ is closest to the point $(8, 0)$?

10. PROBLEM FROM CALCULUS If an object is thrown upward with a velocity of v_0 ft/s, its height h above the ground in t seconds is approximately $h = v_0 t - 16t^2$. What is the maximum height reached by a ball thrown upward with a velocity of 56 ft/s?

 STUDY HINTS *Compare your solutions and answers to the self test.*

Practice for Calculus—*Supplementary Problems: Cumulative Review Chapters 1–3*

Solve each equation for x in Problems 1–6.

1. $12x^2 + 8x - 15 = 0$ **2.** $6x^2 - 350x + 3{,}500 = 0$
3. $6x^2 - 100x + 250 = 0$ **4.** $x^2 + 50x - 10 = 0$
5. $y = 3x^2 + 2x + 1$
6. $y = ax^2 + bx + c$,   where $a \neq 0$

Solve each inequality in Problems 7–10.

7. $-0.00001 \leq 3x + 5 \leq 0.00001$
8. $-0.01 \leq 3 - 2x \leq 0.01$
9. $|2x + 3| \leq 0.1$
10. $|1 - 5x| \leq 0.001$

Factor each expression in Problems 11–16.

11. $3xy^2 - 5x^{-1}x^{-1}y^4 + x^{-1}y^{-2}$ **12.** $5(3x^3 - 4x)^5(9x^2 - 4)$

13. $-2(2x^2 - 3x)^{-3}(4x - 3)$ **14.** $-(3x^2 + 4x)^{-2}(6x + 4)$

15. $3(x^2 + 1)^2(2x)(3x + 2)^4 + 4(x^2 + 1)^3(3x + 2)^3(3)$

16. $4(x^2 + 3)^3(x + 4)^3(1) + 3(x^2 + 3)^4(x + 4)^2(1)$

Evaluate the limits in Problems 17–22.

17. $\lim_{x \to 0}(4x^3 + 5x^2 - 3x + 5)$ **18.** $\lim_{x \to 0}(2x^2 - 5x - 9)$

19. $\lim_{x \to 0}(5x^2 - 4x)$ **20.** $\lim_{x \to 0}(x^4 - 3x^3 + 5x + 8)$

21. $\lim_{h \to 0}\dfrac{(x + h)^3 - x^3}{h}$ **22.** $\lim_{h \to 0}\dfrac{2(x + h)^2 - 2x^2}{h}$

In Problems 23–28, find $\dfrac{f(x + h) - f(x)}{h}$ *and* **simplify**.

23. $f(x) = x^2 - 5x + 7$ **24.** $f(x) = 4x^2 + x - 9$

25. $f(x) = 4x^3 - 3x^2$ **26.** $f(x) = x^3 - 2x^2 + x + 1$

27. $f(x) = x^4$ **28.** $f(x) = 2x^4 - x$

Find an explicit relationship between x *and* y *in Problems 29–34 by eliminating the parameter. In each case,* **graph** *the path described by the parametric equations over the prescribed interval.*

29. $x = t + 1,\ y = t - 1;\ 0 \le t \le 2$

30. $x = -t,\ y = 3 - 2t;\ 0 \le t \le 1$

31. $x = 60t,\ y = 80t - 16t^2;\ 0 \le t \le 3$

32. $x = 30t,\ y = 60t - 9t^2;\ -1 \le t \le 2$

33. $x = t,\ y = 2 + \frac{2}{3}(t - 1);\ 2 \le t \le 5$

34. $x = t^2 + 1,\ y = t^2 - 1;\ 1 \le t \le \sqrt{2}$

Find

$$f(x_1)\Delta x + f(x_2)\Delta x + f(x_3)\Delta x + f(x_4)\Delta x$$

for $x_1 = 0.5$, $x_2 = 1$, $x_3 = 1.5$, $x_4 = 2$ *and for* $\Delta x = 0.5$. *The functions are provided in Problems 35–40.*

35. $f(x) = x^2$

36. $f(x) = 2x + 5$

37. $f(x) = 10 - 2x$

38. $f(x) = 3x^2 + x$

39. $f(x) = x^3$

40. $f(x) = 1 - x^3$

Find the slope of the line passing through the curve

$$y = x^3 - 5x^2 + x$$

and the points indicated in Problems 41–46.

41. $x = 2$ to $x = 5$

42. $x = 2$ to $x = 4$

43. $x = 2$ to $x = 3$

44. $x = 2$ to $x = 2.5$

45. $x = 2$ to $x = 2.1$

46. $x = 2$ to $x = 2 + h$

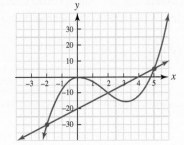

Find the slope of the line passing through the curve

$$y = x^5 - 4x^3 + x^2 - x$$

and the points indicated in Problems 47–52.

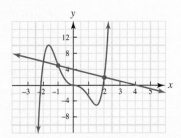

47. $x = -1$ to $x = 2$ **48.** $x = -1$ to $x = 1$

49. $x = -1$ to $x = 0$ **50.** $x = -1$ to $x = -0.1$

51. $x = -1$ to $x = -0.5$ **52.** $x = -1$ to $x = -1 + h$

Use a numerical or graphical approach to find the maximum and minimum values (correct to the nearest unit) of the function

$$f(x) = 12x^5 - 105x^4 + 100x^3 + 930x^2 - 1{,}800x$$

on the given interval in Problems 53–58.

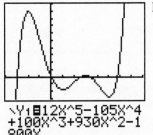   Detail: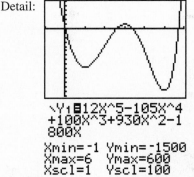

53. $[0, 4]$ **54.** $[0, 2]$

55. $[2, 5]$ **56.** $[-4, 6]$

57. $[-6, 6]$ **58.** $[-8, 8]$

59. A 10-ft pole is to be erected and held in the ground by four guy wires attached at the top. The guy wires are attached to the ground at a distance of 15 ft from the base of the pole. What is the exact length of one guy wire? How much wire should be purchased if it cannot be purchased in fractions of a foot?

60. A small manufacturer of software determines that the price of each program is related to the number of items produced. If x items are produced per day, and the maximum number that can be produced is 10 programs per day, then the price (in dollars) should be

$$p = 400 - 25x$$

It is also determined that the overhead (the cost, in dollars, of producing x items) is

$$C(x) = 5x^2 + 40x$$

Recall, that profit is the revenue less the cost, and a negative profit is called a loss.

a. If we assume that x is any real number, then for what values of x is the profit positive?

b. If we assume that x is any real number, then for what values of x does the manufacturer show a loss?

c. What is the domain for the profit function?

d. How many items should the manufacturer produce per day in order to maximize the profit, and what would be the expected daily profit?

CHAPTER 3   Group Research Projects

Working in small groups is typical of most work environments, and this book seeks to develop skills with group activities. At the end of each chapter, we present a list of suggested projects, and even though they could be done as individual projects, we suggest that these projects be done in groups of three or four students.

G3.1   𝔥𝔦𝔰𝔱𝔬𝔯𝔦𝔠𝔞𝔩 𝔔𝔲𝔢𝔰𝔱 Write a paper on the nature of a function and the history of the functional concept.

G3.2   Journal Problem (*Ontario Secondary School Mathematics Bulletin*, Vol. 17, No. 1, 1981.) Determine all polynomials

$$f(x) = ax^2 + bx + c$$

such that $f(a) = a, f(b) = b, f(c) = c$.

G3.3   For which real numbers c is there a line that intersects the curve

$$y = x^4 + 8x^3 + cx^2 + 10x + 5$$

in four distinct points?

G3.4   Consider the function defined by

$$f(x) = \frac{x^3 - 2.1x^2 + x - 2}{x^6 + 1}$$

    a. Find at least one zero (that is, a value x_0 so that $f(x_0) = 0$) in the interval $[-10, 10]$.

    b. Use technology to graph f to show at least one zero on this interval.

G3.5   SUM OF CUBES Write each of the numbers from 1 to 99 as a sum (or difference) of cubes. For example, $1 = 1^3, 2 = 1^3 + 1^3$, and $8 = 2^3$ are easy; however,

$$50 = 29^3 + 29^3 + 41^3 - 49^3$$

is not. It is not expected that you will find all of them, but the research project is to see how many of them you can find. Bonus points will be given for more original representations and negative points if you use only the sum of an appropriate number of 1^3 terms.

G3.6   Journal Problem (*Parabola*, Vol. 20, No. 1, 1984, Problem Q593.) Suppose your computer can produce the reciprocal of all available numbers and the sum of any two different ones. Which numbers can be produced if initially only the number 100 is in the computer, but new numbers may be stored in the computer's memory as they are calculated and become available.

G3.7   Journal Problem (*The Mathematics Student Journal.*[*]) Given that $f(11) = 11$ and

$$f(x + 3) = \frac{f(x) - 1}{f(x) + 1}$$

for all x, find $f(2012)$.

G3.8   Journal Problem (*Parabola*, Vol. 19, No. 1, 1983, p. 22.) Farmer Jones has to build a fence to enclose a 1,200-m² rectangular area $ABCD$. Fencing costs $3 per meter, but Farmer Smith has agreed to pay half the cost of fencing $\overline{CD}$, which borders the property. Given x is the length of side $\overline{BC}$, what is the minimum amount (to the nearest dollar) Jones has to pay?

---

[*]Volume 28, 1980, issue 3, p. 2; note the journal problem requests $f(1979)$, which, no doubt, was related to the publication date. We have taken the liberty of updating the requested value.

G3.9 Journal Problem (*The Mathematics Teacher*, March 1994, pp. 172–175, "What Manufacturers Say About a Max/Min Application," by Robert F. Cunningham, Trenton State College.) Use the fact that 12 fl oz is approximately 6.89π in.³ to find the dimensions of the 12-oz Coke® can that can be constructed using the least amount of metal. Compare these dimensions with a Coke from your refrigerator. What do you think accounts for the difference? The cited article discusses a similar question regarding tuna fish cans and the resulting responses.

G3.10 LETTER GRAPHING Consider graphing each letter of the alphabet for $0 \le x \le 4,\ 0 \le y \le 4$. For example, the graph of

$$(y - 2)(2x - y)(2x + y - 8) = 0$$

is shown in Figure 3.42. It looks like the letter "A."

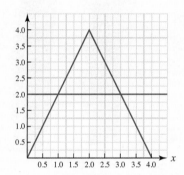

Figure 3.42 Graph of A

To see why these separate equations graph the letter A, use the zero product theorem to write three separate equations:

$$y - 2 = 0; \quad 2x - y = 0; \quad \text{and} \quad 2x + y - 8 = 0$$

Graph each of these lines within the domain and range limitations. Letters can be put together using translations. For example, the author's name, KARL, has the equations:

K: $\ x(x + y - 2)(2x - 3y + 6) = 0$

A: $\ (y - 2)(2x - y)(2x + y - 8) = 0$

R: $\ x(2y^2 + x - 12y + 16)(x + y - 2) = 0$

L: $\ xy = 0$

These letters are not translated. With translations and attention to domains and ranges, it is theoretically possible to write a single equation for the name KARL. Write an equation for each letter of the alphabet.

Chapter Objectives

The material in this chapter is previewed in the following list of objectives. After completing this chapter, review this list again, and then complete the self test.

4.1
4.1 Graph rational functions.
4.2 Find vertical, horizontal, and slant asymptotes.
4.3 Solve applied problems involving rational functions.

4.2
4.4 Graph radical functions.
4.5 Find difference quotients involving radical functions.
4.6 Solve applied problems involving radical functions.

4.3
4.7 Solve rational equations for real numbers.
4.8 Solve radical equations for real numbers.
4.9 Solve applied problems involving rational and radical equations.

4.4
4.10 Define an exponential function.
4.11 Sketch exponential functions.
4.12 Know (and use) the extended laws of exponents.
4.13 Solve applied problems using the compound interest formula, future value problems, present value problems, and effective annual yield.
4.14 Define the number e.
4.15 Solve applied problems using continuous compounding and inflation.

4.5
4.16 Define a logarithm; define a common logarithm; define a natural logarithm.
4.17 Define a logarithmic function.
4.18 Sketch logarithmic functions.
4.19 State and use the change of base theorem.
4.20 State the fundamental properties of logarithms. Show that the exponential and logarithmic functions are inverse functions.
4.21 Evaluate logarithms.
4.22 Solve applied problems involving logarithmic functions.

4.6
4.23 State the log of both sides theorem.
4.24 State the laws of logarithms; that is, the additive, subtractive, and multiplicative laws.
4.25 Solve logarithmic equations.
4.26 Solve applied problems involving logarithmic equations.

4.7
4.27 Solve exponential equations.
4.28 Solve applied problems involving exponential equations.

Additional Functions

4

*T*wo relatively simple problems—the determination of the diagonal of a square and that of the circumference of a circle—revealed the existence of new mathematical beings for which no place could be found within the rational domain.

—Tobias Dantzia (1991)

Chapter Sections

4.1 Rational Functions
Discontinuities of Rational Functions
Asymptotes
Graphs with Asymptotes

4.2 Radical Functions
Graph of Functions of the Form $y = x^n$
Graph of Functions of the Form $y = \sqrt{f(x)}$
Graphs of General Radical Functions

4.3 Real Roots of Rational and Radical Equations
Radical Equations
Higher-Order Equations

4.4 Exponential Functions
Irrational Exponents
Definition and Graphs of Exponential Functions
Real Exponents and Compound Interest
Continuous Compounding; The Number e

4.5 Logarithmic Functions
Solving for an Exponent; Definition of a Logarithm
Evaluating Logarithms
Logarithmic Functions
Exponential and Logarithmic Functions as Inverse Functions

4.6 Logarithmic Equations
Types of Logarithmic Equations
Laws of Logarithms
General Logarithmic Equations

4.7 Exponential Equations
Types of Exponential Equations
Growth and Decay
General Exponential Equations

Chapter 4 Summary and Review
Self Test
Practice for Calculus
Chapter 4 Group Research Projects

▶ CALCULUS PERSPECTIVE

Functions that are not continuous are common in calculus, and in this chapter we study two of this type, *rational* and *radical* functions. Rational functions can be used to model certain behaviors that increase without limit, and radical functions are sometimes used to model distances between points. In the latter part of this chapter, we introduce two very important functions, *exponential* and *logarithmic*. In calculus, exponential and logarithmic functions are important in solving equations called *differential equations*. In algebra, they are used in applications, including growth, decay, population studies, and earthquake intensity.

4.1 Rational Functions

An expression of the form $P(x)/D(x)$ is called a *rational function*.

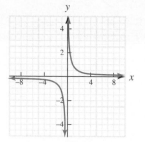

Figure 4.1 Graph of $y = \dfrac{1}{x}$;
Key points: $(1, 1)$, $(-1, -1)$

> **RATIONAL FUNCTION**
>
> A **rational function** f is the quotient of polynomial functions defined by $P(x)$ and $D(x)$ so that
>
> $$f(x) = \frac{P(x)}{D(x)} \quad \text{where} \quad D(x) \neq 0$$

You are familiar with the standard reciprocal function $(y = 1/x)$ from Table 2.1 (Figure 4.1), and we have discussed reflections, compressions, dilations, and shifting this function, as shown in Table 4.1.

TABLE 4.1	Variations of a Rational Function		
Equation:	**a.** $y = -\dfrac{1}{x}$	**b.** $y = \dfrac{2}{x}$	**c.** $y = \dfrac{1}{2x}$
Type:	Reflection	Dilation	Compression
Points:	$(1, -1), (-1, 1)$	$(2, 1), (-2, -1)$	$(\frac{1}{2}, 1), (-\frac{1}{2}, -1)$
Restriction:	$x \neq 0$	$x \neq 0$	$x \neq 0$
Equation:	**d.** $y = \dfrac{1}{x - 3}$	**e.** $y - 2 = \dfrac{1}{x}$	**f.** $y + 2 = \dfrac{1}{x - 3}$
Type:	Translation	Translation	Translation
Points:	$(4, 1), (2, -1)$	$(1, 3), (-1, 1)$	$(4, -1), (2, -3)$
Restriction:	$x \neq 3$	$x \neq 0$	$x \neq 3$

Graphs of rational functions differ from graphs of polynomials in two fundamental ways:

1. Rational functions are discontinuous at values for which the denominator is zero.
2. Rational functions approach lines, called *asymptotes*, which are lines with the property that the distance between the graph of the rational function and the line approaches zero as x becomes very large in absolute value.

Discontinuities of Rational Functions

We will consider each of these possibilities in this section. A rational function may be discontinuous at $x = c$ in two possible ways. The first is with a **deleted point**, that is, a value that causes division by zero. The other is a **vertical asymptote**, which is a line at $x = c$ with the property that the graph of the rational function either increases or decreases without limit at $x = c$.

EXAMPLE 1   Graph with a deleted point

Graph $y = \dfrac{x^3 + 4x^2 + 7x + 6}{x + 2}$.

Solution

Notice that there is a value of x that causes division by zero, namely, $x = -2$. This value is excluded from the domain and indicates the location of the deleted point.

$$y = \frac{x^3 + 4x^2 + 7x + 6}{x + 2} \qquad \text{Given}$$

$$= \frac{(x + 2)(x^2 + 2x + 3)}{x + 2} \qquad \text{Use synthetic division.}$$

$$= x^2 + 2x + 3, x \neq -2 \qquad \text{Simplify.}$$

$$y - 3 = x^2 + 2x \qquad \text{Complete the square; first subtract 3 from both sides}$$

$$y - 3 + 1 = x^2 + 2x + 1 \qquad \text{One-half of 2, squared and added to both sides}$$

$$y - 2 = (x + 1)^2$$

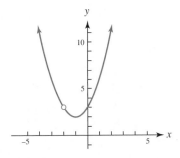

This parabola has vertex at $(-1, 2)$, opens upward, and is drawn with the deleted point, as shown in Figure 4.2.

Figure 4.2 Graph with deleted point

EXAMPLE 2   Graph with a vertical asymptote

Graph $y = \dfrac{x + 1}{x + 3}$.

Solution   First, find the x- and y-intercepts:

$$\text{If } x = 0, \quad \text{then} \quad y = \frac{0 + 1}{0 + 3} \qquad\qquad \text{If } y = 0, \quad \text{then} \quad 0 = \frac{x + 1}{x + 3}$$

$$= \frac{1}{3} \qquad\qquad\qquad\qquad 0 = x + 1$$

$$-1 = x$$

Plot the point: $\left(0, \dfrac{1}{3}\right)$. Plot the point: $(-1, 0)$.

Next, use long division to write:

$$y = \frac{x + 1}{x + 3}$$

$$y = 1 + \frac{-2}{x + 3}$$

$$y - 1 = \frac{-2}{x + 3}$$

Compare this with $y = \dfrac{1}{x}$. There are three modifications:

1. It is twice as far from the x-axis (numerator is 2).
2. It is reflected about the x-axis (negative).
3. It has been shifted to the left 3 units and up 1 unit.

The graph is shown in Figure 4.3.

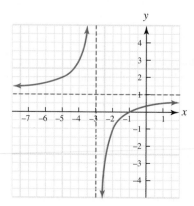

Figure 4.3 Graph of $y = \dfrac{x+1}{x+3}$

We will now consider the general class of rational functions of the form $y = 1/x^n$. We have seen (from Table 2.1) the standard reciprocal function $y = 1/x$, which is characteristic of n an odd number ($n = 1$), and the standard reciprocal squared function $y = 1/x^2$, which is characteristic of n an even number ($n = 2$). Some graphs and properties are shown in Table 4.2.

TABLE 4.2	Graphs of $y = \dfrac{1}{x^n}$	
Property	**n is a positive odd integer**	**n is a positive even integer**
Graphs:	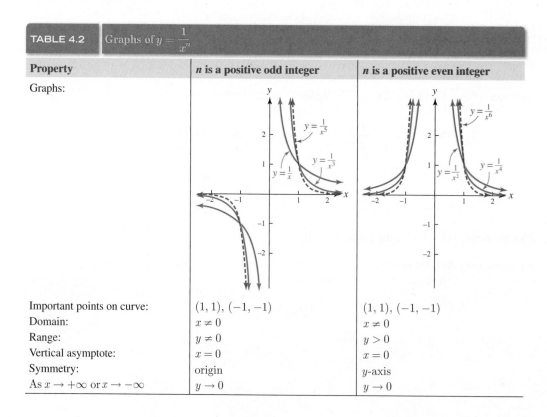	
Important points on curve:	$(1, 1)$, $(-1, -1)$	$(1, 1)$, $(-1, -1)$
Domain:	$x \neq 0$	$x \neq 0$
Range:	$y \neq 0$	$y > 0$
Vertical asymptote:	$x = 0$	$x = 0$
Symmetry:	origin	y-axis
As $x \to +\infty$ or $x \to -\infty$	$y \to 0$	$y \to 0$

EXAMPLE 3   Graphing a power of the reciprocal function

Graph $y = \dfrac{-3}{(x-2)^3}$.

Solution

We note that this function behaves like $y = 3/x^3$, moved two units to the right. We then reflect the graph of $y = 3/(x-2)^3$ about the x-axis to obtain the graph shown in Figure 4.4.

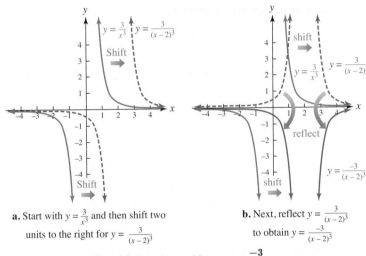

a. Start with $y = \frac{3}{x^3}$ and then shift two units to the right for $y = \frac{3}{(x-2)^3}$

b. Next, reflect $y = \frac{3}{(x-2)^3}$ to obtain $y = \frac{-3}{(x-2)^3}$

Figure 4.4 Steps in graphing $y = \dfrac{-3}{(x-2)^3}$

A rational function may have more than one vertical asymptote. Each value that causes division by zero is either a deleted point or locates a vertical asymptote.

EXAMPLE 4   Graphing a rational function with two vertical asymptotes

Graph $y = \dfrac{x-3}{x^3 - 6x^2 + 11x - 6}$.

Solution

The first step is to determine whether there are any common factors. Since the only factor in the numerator is $(x-3)$, we check the denominator using synthetic division:

$$
\begin{array}{r|rrrr}
3 & 1 & -6 & 11 & -6 \\
  &   & 3 & -9 & 6 \\
\hline
  & 1 & -3 & 2 & 0
\end{array}
$$

Thus, $x^3 - 6x^2 + 11x - 6 = (x-3)(x^2 - 3x + 2)$
$$= (x-3)(x-2)(x-1)$$

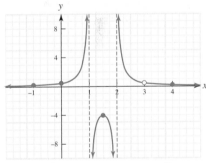

Figure 4.5 Graph of $y = \dfrac{x-3}{x^3 - 6x^2 + 11x - 6}$

From this we find

$$y = \frac{x-3}{x^3 - 6x^2 + 11x - 6} \qquad \text{Given.}$$

$$= \frac{x-3}{(x-3)(x-2)(x-1)} \qquad \text{Factor.}$$

$$= \frac{1}{(x-2)(x-1)} \qquad \text{Simplify; deleted point at } x = 3.$$

Since the denominator is 0 when $x = 2$ or $x = 1$, we see that there are two vertical asymptotes. Furthermore, since the denominator is quadratic, the graph should "behave" like $1/x^2$ for large values of $|x|$. We also need to plot a few values as shown in Figure 4.5. ■

Notice that the graph in Figure 4.5 has three separate branches. In general, the number of branches in the graph of a rational function is one more than the number of vertical asymptotes. We will now take a closer look at asymptotes, in general.

Asymptotes

As a preview of the ideas we plan to explore, let us examine the graph shown in Figure 4.6.

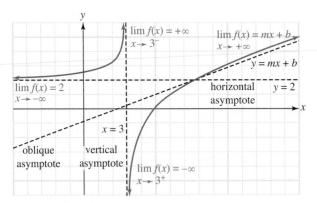

Figure 4.6 A typical graph with asymptotes

Suppose the graph shown in Figure 4.6 is the graph of a function we will call f. Notice that as x approaches 3 from either side, the corresponding functional values of f get large without bound (in absolute value) and the graph of f approaches the vertical line $x = 3$. This approach is through positive values ("up") as x approaches 3^- (from the left) and through negative values ("down") as x approaches 3^+ (from the right). We indicate the behavior of for x near 3 by writing*

$$\lim_{x \to 3^-} f(x) = +\infty \qquad \text{and} \qquad \lim_{x \to 3^+} f(x) = -\infty$$

and we describe the corresponding geometric behavior by saying that the line $x = 3$ is a *vertical asymptote* of the graph of f.

Note also that as x increases without bound (that is, as x moves toward the right on the x-axis), the graph of f follows the line $y = mx + b$. For this reason, the line $y = mx + b$ (where $m \neq 0$) is called an **oblique** (or **slant**) **asymptote**.

At the other end of the x-axis (as x decreases without bound), the graph approaches the line $y = 2$. We write $\lim_{x \to -\infty} f(x) = 2$ and say that the line $y = 2$ is a **horizontal asymptote** of the graph.

Graphs with Asymptotes

We shall now show how limits involving infinity can be used along with the curve-sketching techniques developed in previous sections to sketch graphs with asymptotes. These properties are proved in calculus.

---

*Many books simply write ∞ to mean $+\infty$. For now, we use $+\infty$ to remind you that $+\infty$ is not the same as $-\infty$.

Vertical Asymptote

Let $f(x) = \dfrac{P(x)}{D(x)}$ where P is a polynomial function with leading coefficient p and D is a polynomial function with leading coefficient d. Moreover, $P(x)$ and $D(x)$ have no common factors. The line $x = c$ is a *vertical asymptote* of the graph of f if either of the one-sided limits

$$\lim_{x \to +\infty} f(x) \qquad \text{or} \qquad \lim_{x \to -\infty} f(x)$$

is infinite. **The vertical asymptote is the line $x = c$ where $D(c) = 0$. That is, values that cause division by 0.**

Horizontal Asymptote

The line $y = L$ is a *horizontal asymptote* of the graph of f if

$$\lim_{x \to +\infty} f(x) = L \qquad \text{or} \qquad \lim_{x \to -\infty} f(x) = L$$

If $P(x)$ has degree m and $D(x)$ has degree n, then the **horizontal asymptote is $y = 0$ if $m < n$ and it is the horizontal line $y = p/d$ if $m = n$.**

Oblique Asymptote

The line $y = mx + b$ is an *oblique asymptote* of the graph of f if

$$f(x) = \frac{P(x)}{D(x)} = mx + b + \frac{r}{D(x)}$$

where $\displaystyle\lim_{x \to +\infty} \frac{r}{D(x)} = 0$ or where $\displaystyle\lim_{x \to -\infty} \frac{r}{D(x)} = 0$. **The oblique asymptote is $y = mx + b$.**

EXAMPLE 5   Graphing with vertical and horizontal asymptotes

Graph $y = \dfrac{x^2 + x - 2}{x^2 - x - 12}$.

Solution

Before beginning, take some time to think about the curve you are graphing. This curve is not symmetric with respect to the *x*- or *y*-axis or the origin. Look for common factors:

$$y = \frac{x^2 + x - 2}{x^2 - x - 12} \qquad \text{Given.}$$

$$= \frac{(x-1)(x+2)}{(x+3)(x-4)} \qquad \text{Factor; note there are no deleted points.}$$

Intercepts: *x*-intercepts $(y = 0)$:   *y*-intercepts $(x = 0)$:

$$0 = \frac{(x-1)(x+2)}{(x+3)(x-4)} \qquad y = \frac{(0-1)(0+2)}{(0+3)(0-4)}$$

$$0 = (x-1)(x+2) \qquad y = \frac{(-1)(2)}{(3)(-4)}$$

$$x = 1, -2 \qquad y = \frac{1}{6}$$

Plot the points $(1, 0)$, $(-2, 0)$, and $\left(0, \frac{1}{6}\right)$.

Vertical asymptotes: Values that cause division by 0; draw vertical asymptotes at $x = -3$ and $x = 4$, as shown in Figure 4.7**a**.

Horizontal asymptotes: Since the numerator and denominator have the same degree, and since $p = 1$, $q = 1$, we draw the horizontal asymptote $y = 1$, as shown in Figure 4.7**a**.

It looks like we need a few more points. First, we look for places where the curve crosses the asymptotes. A graph cannot cross a vertical asymptote (because of division by zero), but it can cross a horizontal or slant asymptote. In this case, we wish to know where the graph crosses the line $y = 1$:

$$y = \frac{x^2 + x - 2}{x^2 - x - 12}$$

$$1 = \frac{x^2 + x - 2}{x^2 - x - 12}$$

$$x^2 - x - 12 = x^2 + x - 2$$

$$-10 = 2x$$

$$-5 = x$$

Thus, the curve crosses the horizontal asymptote at $(-5, 1)$; plot this point. It looks like we need a couple more points: if $x = -4$, $y = \frac{5}{4}$, and if $x = 5$, $y = \frac{7}{2}$. After we have plotted these points, we have a grid, which is shown in Figure 4.7**b**.

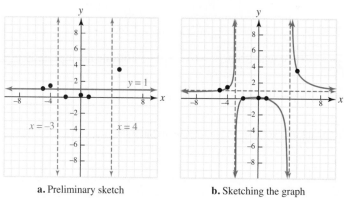

a. Preliminary sketch   **b.** Sketching the graph

Figure 4.7 Graph of $y = \dfrac{x^2 + x - 2}{x^2 - x - 12}$

EXAMPLE 6   Graphing with slant asymptote

Graph $f(x) = \dfrac{x^2}{x - 2}$.

Solution

Take a look at the function you are graphing before starting. The degree of the numerator is one more than the degree of the denominator, so we expect a slant asymptote. We also expect a vertical asymptote at $x = 2$. By synthetic division,

$$\begin{array}{r} 2\, \lfloor\ 1 \quad 0 \quad 0 \\ \ 2 \quad 4 \\ \hline 1 \quad 2 \quad 4 \end{array} \qquad \begin{aligned} y &= \frac{x^2}{x - 2} \\[4pt] &= x + 2 + \frac{4}{x - 2} \end{aligned}$$

This means for large $|x|$ the curve behaves like $y = x + 2$. The graph is shown in Figure 4.8.

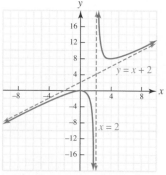

Figure 4.8 Graph of $y = \dfrac{x^2}{x - 2}$

EXAMPLE 7 **Inventory management** MODELING APPLICATION

A retailer buys 6,000 calculator batteries per year from a distributor and is trying to decide how often to order the batteries. The ordering fee is $20 per shipment; the storage cost is $0.96 per battery per year; and each battery costs the retailer $0.25. Suppose that the batteries are sold at a constant rate throughout the year and that each shipment arrives just as the preceding shipment has been used up. How many batteries should the retailer order each time to minimize the total cost?

Solution

Step 1: *Understand the problem.* We begin by writing the cost function:

TOTAL COST = STORAGE COST + ORDERING COST + COST OF BATTERIES

Step 2: *Devise a plan.* We need to find an expression for each of these unknowns. After writing this as a function of a single variable, we will seek to minimize this function.

Step 3: *Carry out the plan.* Assume that the same number of batteries must be ordered each time an order is placed; denote this number by x so that $C(x)$ is the corresponding total cost.

$$\text{STORAGE COST} = \begin{pmatrix} \text{AVERAGE NUMBER} \\ \text{IN STORAGE PER YR} \end{pmatrix} \begin{pmatrix} \text{COST OF STORING 1} \\ \text{BATTERY FOR 1 YR} \end{pmatrix}$$

$$= \left(\frac{x}{2}\right)(0.96)$$

$$= 0.48x$$

The average number of batteries in storage during the year is half of a given order, that is, $x/2$. Thus, the total yearly storage cost is the same as if the $x/2$ batteries were kept in storage for the entire year. This situation is illustrated in Figure 4.9. To find the total ordering cost, we can multiply the ordering cost per shipment by the number of shipments. We also note that because 6,000 batteries are ordered during the year, and because each shipment contains x batteries, the number of shipments is $6,000/x$.

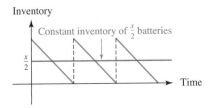

Inventory

Constant inventory of $\frac{x}{2}$ batteries

Time

Figure 4.9 Inventory graph

$$\text{ORDERING COST} = \begin{pmatrix} \text{ORDERING COST} \\ \text{PER SHIPMENT} \end{pmatrix} \begin{pmatrix} \text{NUMBER OF} \\ \text{SHIPMENTS} \end{pmatrix}$$

$$= (20)\left(\frac{6,000}{x}\right)$$

$$= \frac{120,000}{x}$$

The third component in finding the total cost is to formulate an expression for the cost of the batteries:

$$\text{COST OF BATTERIES} = \begin{pmatrix} \text{TOTAL NUMBER} \\ \text{OF BATTERIES} \end{pmatrix} \begin{pmatrix} \text{COST PER} \\ \text{BATTERY} \end{pmatrix}$$

$$= 6,000(0.25)$$

$$= 1,500$$

Thus, we can now formulate the total cost function:

$$\text{TOTAL COST} = 0.48x + \frac{120,000}{x} + 1,500$$

Let $C(x)$ represent the total cost. The goal is to minimize $C(x)$ on $(0, 6,000]$.

The analytic (algebraic) solution requires calculus, but we recognize the function C. It has a vertical asymptote at $x = 0$ and a slant asymptote $y = 0.48x + 1,500$. We graph the function as shown in Figure 4.10.

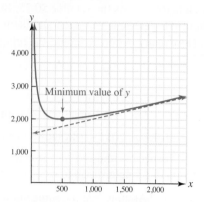

Figure 4.10 The total cost function

From the graph, it looks like the minimum value is at $x = 500$. We check some close-by values as well as the endpoints:

$$c(499) = 1,980.001 \qquad\qquad c(501) = 1,980.001$$

$$c(500) = 1,980.00 \qquad\qquad c(6,000) = 4,400.00$$

Step 4: *Look back.* It looks like the cost function has a minimum value of $1,980$ when $x = 500$. Thus, to minimize cost, the manufacturer should order the batteries in lots of 500.

PROBLEM SET 4.1

LEVEL 1

In Problems 1–14, graph the equations for each problem on the same coordinate axes, by comparing to one of the curves in Tables 4.1 and 4.2.

1. a. $y = \dfrac{1}{x}$ **b.** $y = \dfrac{3}{x}$

2. a. $y = \dfrac{1}{x}$ **b.** $y = -\dfrac{1}{x}$

3. a. $y = -\dfrac{1}{x}$ **b.** $y = -\dfrac{2}{x}$

4. a. $y = -\dfrac{1}{x}$ **b.** $y = -\dfrac{4}{x}$

5. a. $y = \dfrac{1}{x^2}$ **b.** $y = \dfrac{3}{x^2}$

6. a. $y = \dfrac{1}{x^2}$ **b.** $y = \dfrac{5}{x^2}$

7. a. $y = \dfrac{1}{x^2}$ **b.** $y = -\dfrac{1}{x^2}$

8. a. $y = -\dfrac{1}{x^2}$ **b.** $y = -\dfrac{2}{x^2}$

9. a. $y = -\dfrac{1}{x^2}$ **b.** $y = -\dfrac{4}{x^2}$

10. a. $y = \dfrac{1}{x}$ **b.** $y = \dfrac{1}{4x}$

11. a. $y = \dfrac{1}{x^2}$ **b.** $y = \dfrac{1}{(2x)^2}$

12. a. $y = \dfrac{1}{x-4}$ **b.** $y - 4 = \dfrac{1}{x}$

13. a. $y - 3 = \dfrac{1}{(x-2)^2}$ **b.** $y + 3 = \dfrac{1}{(x-2)^2}$

14. a. $y - 2 = \dfrac{1}{(x+4)^2}$ **b.** $y - 2 = \dfrac{1}{(x-4)^2}$

Graph the functions defined by each equation in Problems 15–24.

15. $y = \dfrac{(x+1)(x-2)(x+2)}{(x+1)(x-2)}$ **16.** $y = \dfrac{(x-3)(x+1)(x+5)}{(x+1)(x+5)}$

17. $y = \dfrac{(x+2)(x-1)(3x+2)}{x^2+x-2}$ **18.** $y = \dfrac{(x-3)(2x+3)(x+2)}{x^2-x-6}$

19. $y = \dfrac{x^2-x-12}{x+3}$ **20.** $y = \dfrac{x^2+x-2}{x-1}$

21. $y = \dfrac{x^2-x-6}{x+2}$ **22.** $y = \dfrac{2x^2-13x+15}{x-5}$

23. $y = \dfrac{(15x^2+13x-6)(x-1)}{3x^2-4x+1}$ **24.** $y = \dfrac{x^2+x-6}{x+3}$

Find the horizontal, vertical, and slant asymptotes, if any exist, for the functions given in Problems 25–32.

25. $y = \dfrac{1}{x}$ **26.** $y = \dfrac{1}{x+2}$

27. $y = \dfrac{2x^2+2}{x^2}$ **28.** $y = \dfrac{1}{x-4}$

29. $y = \dfrac{(x-1)(x+2)(x-3)(x+4)}{(x+2)(x-1)}$

30. $y = \dfrac{(x+3)(x-3)(x+1)(2x-3)}{(x-3)(2x-3)}$

31. $y = \dfrac{x^3-2x^2+x-2}{(x-2)(x^2+1)}$ **32.** $y = \dfrac{x^3-2x^2-x+2}{x^2-1}$

LEVEL 2

Graph each curve in Problems 33–42 by finding asymptotes.

33. $y = \dfrac{x+3}{x-2}$ **34.** $y = \dfrac{x-1}{x+1}$

35. $y = \dfrac{3x+5}{3x-2}$ **36.** $y = \dfrac{3x+5}{2x-3}$

37. $y = \dfrac{x^2+x-6}{x^2+2x-8}$ **38.** $y = \dfrac{x^2-x-2}{x^2-2x-3}$

39. $y = \dfrac{-x^2}{x-1}$ **40.** $y = \dfrac{x^2-x-12}{x-4}$

41. $y = \dfrac{2x^3-3x^2-32x-15}{x^2-2x-15}$ **42.** $y = \dfrac{(x-3)(x^2+1)}{x^3+3x^2+x+3}$

LEVEL 3

Graph each curve in Problems 43–56 by finding intercepts and asymptotes.

43. $y = \dfrac{x}{x^2+x-6}$ **44.** $y = \dfrac{-x}{x^2+x-6}$

45. $y = \dfrac{x^2}{x-4}$ **46.** $y = \dfrac{x^3}{(x-1)^2}$

47. $y = \dfrac{x^3}{x^2+4}$ **48.** $y = \dfrac{x^2}{x^2+1}$

49. $y = \dfrac{x^2+3x-2}{x^2+2x-8}$ **50.** $y = \dfrac{x^2}{20x-x^2-x^3}$

51. $y = \dfrac{6x^2-6x-12}{3x^2+4x+5}$ **52.** $y = \dfrac{4x^2+8x-12}{2x^2+3x+5}$

53. $y = \dfrac{x^2}{x^3-x^2-20x}$ **54.** $y = \dfrac{x^2}{20x-x^2-x^3}$

55. $y = \dfrac{x^3+x^2+2x+2}{x^2+1}$ **56.** $y = \dfrac{x^3+x^2+2x+2}{x^2+9}$

57. Consider a line with slope m passing through the parabola $y = x^2$ in the first quadrant and the point $(0, -9)$. Find the value of m so that m is a minimum.

58. Consider a line with slope m passing through the parabola $y = x^2$ in the first quadrant and the point $(0, -2)$. Find the value of m so that m is a minimum.

59. A store owner expects to sell 6,000 units of a certain commodity each year. The cost of each unit is \$10; it costs \$40 to order each new shipment of 500 units, and it costs \$2 to store each unit for a year. Assuming that the commodity is used at a constant rate throughout the year and that each shipment arrives just as the preceding shipment has been used up, find the number of units that should be ordered each time to minimize the cost.

60. A plastics firm has received an order from the city recreation department to manufacture 8,000 special Styrofoam kick boards for its summer swimming program. The firm owns 10 machines, each of which can produce 50 kickboards per hour. The cost of setting up the machines to produce the kickboards is \$800 per machine. Once the machines have been set up, the operation is fully automated and can be overseen by a single production supervisor earning \$35/hr.

   a. How many machines should be used to minimize the cost of production?

   HINT: Do not forget the answer should be a positive integer.

   b. How much will the supervisor earn during the production run if the optimal number of machines is used?

4.2 Radical Functions

We define a **radical function** as a function in which a variable is enclosed under a radical or, equivalently, a variable is raised to a fractional exponent. As before, we begin by looking at some common properties for graphs of radical functions and then modifications such as translations, reflections, dilations, and compressions. Finally, we graph some general radical functions using general graphical techniques.

Graph of Functions of the Form $y = x^n$

We begin with an example that examines the behavior of the function $y = x^n$, with special focus on the case where n is a rational number. Remember, if n is a rational number in reduced form p/q, then the expression x^n represents a radical expression. Furthermore, x^n is undefined when x is negative and q is even, so for convenience, we restrict x in our example to positive x-values.

EXAMPLE 1   Graph of $y = x^n$ for several choices of n

Sketch the graphs of $y = x^n$ for $x \geq 0$ and the given choice for n:

a. $n = 1$ **b.** $n = 2$ **c.** $n = 1/2$ **d.** $n = 1/10$ **e.** $n = 3/2$

Solution   The corresponding functions to graph are

a. $y = x$ **b.** $y = x^2$ **c.** $y = \sqrt{x}$ **d.** $y = x^{0.1}$ **e.** $y = x^{3/2}$

Parts **a** and **b** are included for reference since we are familiar with these graphs. All these curves pass through $(0, 0)$ and $(1, 1)$, and we note the behavior of these graphs is different for $0 < x < 1$ and for $x > 1$, as shown in Figure 4.11.

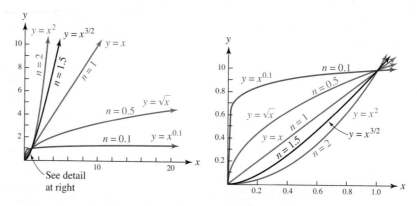

Figure 4.11 Graphs of functions of the form $y = x^n$

Notice that the concavity of the graphs for the form $y = x^n$ depend on n and are divided by the linear function $y = x$.

Function	Value of n	Concavity
$y = x^n$	$n > 1$	up
$y = x^n$	$0 < n < 1$	down

We are familiar (from Table 2.1) with the graphs of the square root and the cube root functions. We note that these functions are both concave down for $x > 0$, and that for $x < 0$ the square root function is concave down. The square root function (Figure 4.12**a**) does not exist for $x < 0$, but the cube root function (Figure 4.12**b**) is concave up for $x < 0$. These functions are copied from Table 2.1 for easy reference.

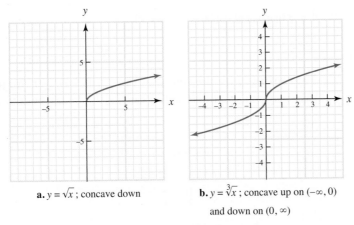

a. $y = \sqrt{x}$; concave down

b. $y = \sqrt[3]{x}$; concave up on $(-\infty, 0)$ and down on $(0, \infty)$

Figure 4.12 Square-root and cube-root functions

Some variations of the cubic function are considered in the following example.

EXAMPLE 2   Variations of the graph of $y = \sqrt[3]{x}$

Graph: **a.** $y = -\sqrt[3]{x}$ **b.** $y = \sqrt[3]{x} + 4$ **c.** $y = \sqrt[3]{x - 5}$ **d.** $y = -\sqrt[3]{x - 5} + 4$

Solution   We use the graph of the standard cube-root function (Figure 4.12b) as a reference.

a. $y = -\sqrt[3]{x}$ is a reflection of the standard cube-root function, as shown in Figure 4.13a.

b. $y = \sqrt[3]{x} + 4$ is a translation up 4 units, as shown in Figure 4.13b.

c. $y = \sqrt[3]{x - 5}$ is a translation 5 units to the right, as shown in Figure 4.13c.

d. $y = -\sqrt[3]{x - 5} + 4$ is a reflection and translation, as shown in Figure 4.13d.

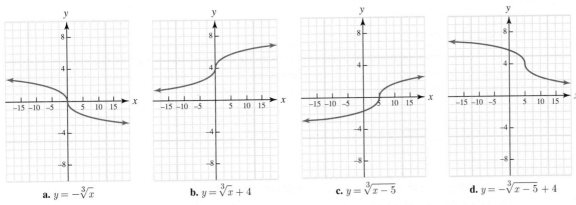

a. $y = -\sqrt[3]{x}$ **b.** $y = \sqrt[3]{x} + 4$ **c.** $y = \sqrt[3]{x - 5}$ **d.** $y = -\sqrt[3]{x - 5} + 4$

Figure 4.13 Variations of the graph of $y = \sqrt[3]{x}$

Graph of Functions of the Form $y = \sqrt{f(x)}$

When graphing a function of the form $y = \sqrt{f(x)}$ where f is a polynomial or a rational function, we pay particular attention to the domain, so that $f(x) \geq 0$.

EXAMPLE 3   Graph of a function of the form $y = \sqrt{f(x)}$

Graph $y = \sqrt{(x-2)(x-4)(x-6)}$.

Solution   We begin by looking at the domain: $(x-2)(x-4)(x-6) \geq 0$. We recognize $f(x) = (x-2)(x-4)(x-6)$ as a polynomial function, and we can find the domain by drawing a number line:

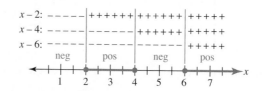

The domain is $[2, 4] \cup [6, \infty)$. The graph is shown in Figure 4.14. Notice that the graph of $y = \sqrt{f(x)}$ exists only where f is positive.

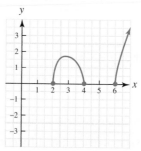

Figure 4.14 Graph of $y = \sqrt{f(x)}$

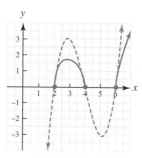

It would be incorrect to assume that the graph shown in Figure 4.14 is the same as the positive portions of the graph of the polynomial function. A comparison between the two is shown in the margin. Note the value for a particular value of x; the value of the square root function is the square root of the value of the polynomial function. ∎

Graphs of General Radical Functions

When considering the graph of a function containing a radical, the procedure requires more analysis. In addition to comparing it to known functions, you will need to pay attention to domain, range (if possible), the intercepts, and asymptotes.

EXAMPLE 4   Graphing a radical function

Graph $y = \frac{2}{3}\sqrt{x^2 - 9}$:

a. by comparing to $y = f(x)$ for f a quadratic function;
b. by finding the intercepts and asymptotes;
c. by calculator.

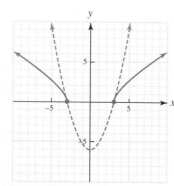

Solution

a. The function $f(x) = x^2 - 9$ is a parabola with vertex $(0, -9)$, so the graph of $y = \frac{2}{3}(x^2 - 9)$ has vertex at $(0, -6)$ since $\frac{2}{3}(-9) = -6$. The domain for the radical function is that part for which $y = \frac{2}{3}(x^2 - 9)$ is positive. Since y is equal to a square root, we see that the range includes all nonnegative values of y. The graph is shown in red where the graph of the related parabola $y = \frac{2}{3}(x^2 - 9)$ is shown as a dashed curve.

b. *x*-intercepts $(y = 0)$: *y*-intercepts $(x = 0)$:

$$0 = \tfrac{2}{3}\sqrt{x^2 - 9} \qquad y = \tfrac{2}{3}\sqrt{0^2 - 9}$$

$$9 = x^2 \qquad\qquad \text{No real numbers make this equation true.}$$

$$x = \pm 3$$

Asymptotes:

No vertical asymptotes since there are no values that cause division by zero.

As $x \to +\infty$ or as $x \to -\infty$, y increases without bound, so there are no horizontal asymptotes.

As $x \to +\infty$ or as $x \to -\infty$,

$$\sqrt{x^2 - 9} \approx \sqrt{x^2} = |x|$$

Thus, $y \to \tfrac{2}{3}|x|$ as $x \to +\infty$ or as $x \to -\infty$; the slant asymptote is $y = \tfrac{2}{3}|x|$.

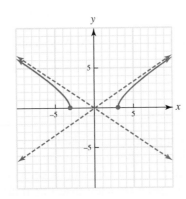

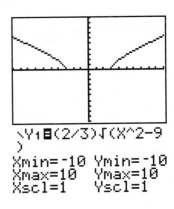

c. The calculator graph is shown in the margin.

EXAMPLE 5   Graphing a rational function that includes a radical

Graph $y = \dfrac{x}{\sqrt{x^2 - 4}}$.

Solution   We do not recognize this curve as a common type, so we look at the domain, intercepts, and asymptotes.

Domain: The domain is the set of all real numbers for which $x^2 - 4 > 0$. (Note that we exclude $x^2 - 4 = 0$ because of division by 0.) We solve this inequality to find the domain of $(-\infty, -2) \cup (2, \infty)$.

Intercepts: *y*-intercept $(x = 0)$: $y = \dfrac{0}{\sqrt{0^2 - 4}}$ does not exist in the set of real numbers, so there is no *y*-intercept.

x-intercept $(y = 0)$: $\dfrac{x}{\sqrt{x^2 - 4}} = 0$ when $x = 0$, but $x = 0$ is not in the domain, so there are no *x*-intercepts.

Asymptotes: For the vertical asymptotes, determine when the denominator can be zero:

$$x^2 - 4 = 0$$

$$(x - 2)(x + 2) = 0$$

$$x = \pm 2$$

☠ If the denominator had been $\sqrt{x^2 + 4}$, there would have been no vertical asymptotes because $x^2 + 4 \neq 0$ for all real values of x. ☠

Vertical asymptotes are $x = 2$, $x = -2$.

For the horizontal asymptotes, find $\displaystyle\lim_{x\to+\infty}\frac{x}{\sqrt{x^2-4}}$.

$$\frac{x}{\sqrt{x^2-4}}=\frac{x}{\sqrt{x^2\left(1-\dfrac{4}{x^2}\right)}}$$

$$=\frac{x}{|x|\sqrt{1-\dfrac{4}{x^2}}}$$

☠ If the denominator had been $\sqrt{4-x^2}$, there would have been no horizontal asymptotes because the domain is $(-2, 2)$. It would be impossible for $x\to+\infty$ and still remain in the domain $(-2, 2)$. ☠

Thus, $\displaystyle\lim_{x\to+\infty}\frac{x}{|x|\sqrt{1-\dfrac{4}{x^2}}}=1$ since x is positive and since $\dfrac{4}{x^2}\to0$ as $x\to+\infty$.

Also, $\displaystyle\lim_{x\to-\infty}\frac{x}{|x|\sqrt{1-\dfrac{4}{x^2}}}=-1$ since x is negative and since $\dfrac{4}{x^2}\to0$ as $x\to-\infty$.

Thus, $y=1$ and $y=-1$ are horizontal asymptotes. We can now sketch the function as shown in Figure 4.15.

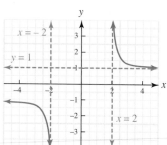

Figure 4.15 Graph of $y=\dfrac{x}{\sqrt{x^2-4}}$

EXAMPLE 6   Example from calculus: Minimize time of travel

A dune buggy is on the desert at a point A located 40 km from a point B, which lies on a long, straight road, from B through point D, as shown in Figure 4.16.

The driver can travel at 45 km/h on the desert and 75 km/h on the road. The driver will win a prize if she arrives at the finish line (point D) in less than 1 hr. If the distance from B to D is 28 km, is it possible for her to choose a route so that she can collect the prize?[*]

Solution   Suppose the driver heads for a point C, located x km down the road from B toward her destination, as shown in Figure 4.16. We want to minimize the time. We will need to remember the formula $d=rt$, or in terms of time, $t=\frac{d}{r}$.

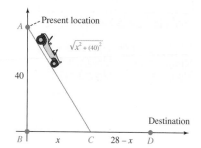

Figure 4.16 Path traveled by a dune buggy

$$\text{TIME}=\text{TIME FROM } A \text{ TO } C + \text{TIME FROM } C \text{ TO } D$$

$$=\frac{\text{DISTANCE FROM } A \text{ TO } C}{\text{RATE FROM } A \text{ TO } C}+\frac{\text{DISTANCE FROM } C \text{ TO } D}{\text{RATE FROM } C \text{ TO } D}$$

$$T(x)=\frac{\sqrt{x^2+1{,}600}}{45}+\frac{28-x}{75}$$

Let T be the time in the dune buggy as a function of x. The domain for x is $[0, 28]$. Calculus is required to find a minimum value analytic solution. However, we can find a minimum value graphically or numerically. The minimum seems to be close to $x\approx30$. This estimated value is not in the domain. However, in the domain $[0, 28]$, the function is continuous, so the minimum point will occur at one of the endpoints:

$$T(0)=\frac{\sqrt{0^2+1{,}600}}{45}+\frac{28-0}{75}\approx1.26 \qquad T(28)=\frac{\sqrt{28^2+1{,}600}}{45}+\frac{28-28}{75}\approx1.09$$

The driver can minimize the total driving time by heading directly across the desert from A to D. Even so, she takes more than an hour for the trip, so she does not win the prize.

---

[*]This problem is from *Calculus Fourth Edition* by Strauss, Bradley, and Smith (Englewood Cliffs, New Jersey: Prentice Hall, 2002, p. 241).

PROBLEM SET 4.2

LEVEL 1

1. Draw a curve that passes through $(0, 0)$ and is concave up on the interval $(0, +\infty)$.
2. Draw a curve that passes through $(0, 0)$ and is concave down on the interval $(0, +\infty)$.
3. Draw a curve that passes through $(0, 0)$ and is concave up on the interval $(-\infty, 0)$ and concave down on $(0, +\infty)$.
4. Draw a curve that passes through $(0, 0)$ and is concave down on the interval $(-\infty, 0)$ and concave up on $(0, +\infty)$.

WHAT IS WRONG, *if anything, with each statement in Problems 5–8? Explain your reasoning.*

5. The graph of $y = \sqrt{x^2 + 4}$ is the same as the graph of $y = x + 2$ except for their domains.
6. The graph of $y = x^2$ is concave up, so the graph of $y = \sqrt{x}$ is also concave up.
7. The graph of $y = x^3$ is concave up in the first quadrant, so the graph of $y = \sqrt[3]{x}$ is also concave up in the first quadrant.
8. The graph of $y = \sqrt{(x-1)(x-2)}$ is the same as the graph of the positive portions of the graph of $y = (x-1)(x-2)$.

In Problems 9–20, graph the equations for each problem on the same coordinate axes.

9. **a.** $y = \sqrt[4]{x}$ **b.** $y = -\sqrt[4]{x}$
10. **a.** $y = \sqrt[5]{x}$ **b.** $y = \sqrt[5]{x} - 3$
11. **a.** $y = \sqrt[3]{x}$ **b.** $y = \sqrt[3]{x-4} + 5$
12. **a.** $y = \sqrt[3]{x}$ **b.** $y = 10\sqrt[3]{x}$
13. **a.** $y = \dfrac{1}{\sqrt{x}}$ **b.** $y = 3 + \dfrac{1}{\sqrt{x}}$
14. **a.** $y = \sqrt{x}$ **b.** $y = \sqrt{x} + 6$
15. **a.** $y = x^{2/3}$ **b.** $y = x^{-2/3}$
16. **a.** $y = -x^{2/3}$ **b.** $y = -(x-2)^{2/3}$
17. **a.** $y = x^2 - 9$ **b.** $y = \sqrt{x^2 - 9}$
18. **a.** $y = x^2 + 9$ **b.** $y = \sqrt{x^2 + 9}$
19. **a.** $y = x^2 + x - 6$ **b.** $y = \sqrt{x^2 + x - 6}$
20. **a.** $y = x^2 - x - 6$ **b.** $y = \sqrt{x^2 - x - 6}$

Graph the function defined by each equation given in Problems 21–40.

21. $y = x^{3/5}$ on $[0, 32]$
22. $y = x^{2/3}$ on $[0, 1]$
23. $y = x^{1/4}$ on $[0, 25]$
24. $y = x^{1/4}$ on $[0, 1]$
25. $y = -2x^{1/4}$ on $[0, 16]$
26. $y = -2x^{1/4}$ on $[0, 1]$
27. $y = -\frac{1}{2}x^{3/5}$ on $[0, 32]$
28. $y = -\frac{1}{2}x^{3/5}$ on $[0, 1]$
29. $y = x^{3/5} - 2$
30. $y = (x-2)^{3/5}$
31. $y = x^{2/5} + 2$
32. $y = (x+2)^{2/5}$
33. $y = x^{1/5} - 1$
34. $y = (x-1)^{1/5}$
35. $y = \sqrt[4]{x-2} - 1$
36. $y = \sqrt[3]{x+3} - 1$

37. $y = \sqrt{x^2 - 1}$
38. $y = \sqrt{x^2 - 9}$
39. $y = \sqrt{x^2 + 1}$
40. $y = \sqrt{x^2 + 9}$

LEVEL 2

Graph the function defined by each equation given in Problems 41–56.

41. $y = \sqrt{(x-2)(x-3)}$
42. $y = \sqrt{x^2 - 6x + 8}$
43. $y = \sqrt{x^2 + 4}$
44. $y = \sqrt{4 - x^2}$
45. $y = \sqrt{10(x^2 - 1)}$
46. $y = \sqrt{0.1(x^2 - 9)}$
47. $y = \sqrt{x(x-2)(x-4)}$
48. $y = \sqrt{(x-3)(x-6)(x-9)}$
49. $y = \dfrac{x}{\sqrt{x^2 - 9}}$
50. $y = \dfrac{x}{\sqrt{x^2 + 9}}$
51. $y = \dfrac{x}{\sqrt{1 - x^2}}$
52. $y = \dfrac{x}{\sqrt{16 - x^2}}$
53. $y = \sqrt{\dfrac{3}{x^2 - 1}}$
54. $y = \dfrac{1}{\sqrt{1 - x^2}}$
55. $y = -\sqrt{\dfrac{x^2 + 1}{x}}$
56. $y = -\sqrt{\dfrac{x^2 + 4}{x}}$

57. Graph the relationship between the length and width of a rectangle, where the length is the dependent variable and the width is greater than or equal to 10 ft, but no larger than 100 ft. Give a graphical solution to the following possibilities:
    a. What is the width when the area is 500 ft²?
    b. What is the width when the area is $1,000\pi$ ft²?

LEVEL 3

58. PROBLEM FROM CALCULUS Missy Becker is at a point A on the north bank of a long, straight river 6 mi wide, as shown in Figure 4.17. Directly across from her on the south bank is a point B, and she wishes to reach a cabin C located 6 mi down the river from B. Given that Missy can row at 6 mi/h (including the effect of the current) and run at 10 mi/h, what is the minimum time (to the nearest minute) required for her to travel from A to C?

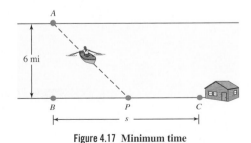

Figure 4.17 Minimum time

59. PROBLEM FROM CALCULUS Missy Becker is at a point A on the north bank of a long, straight river 6 mi wide, as shown in Figure 4.17. Directly across from her on the south bank is a point B, and she wishes to reach a cabin C located 4 mi down the river from B. Given that Missy can

row at 6 mi/h (including the effect of the current) and run at 10 mi/h, what is the minimum time (to the nearest minute) required for her to travel from A to C?

60. PROBLEM FROM CALCULUS Two towns A and B are 12.0 mi apart and are located 5.0 and 3.0 mi, respectively, from a long, straight highway, as shown in Figure 4.18.

A construction company has a contract to build a road from A to the highway and then to B. How long is the *shortest* (to the nearest tenth mile) road that meets these requirements?

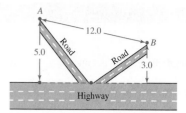

Figure 4.18 Shortest distance

4.3 Real Roots of Rational and Radical Equations

A **rational equation** is an equation with at least one variable in a denominator. The multiplication property for equality (from beginning algebra) specifies that both sides of an equation may be multiplied by any *nonzero* number without changing the solution of that equation.

SOLVING A RATIONAL EQUATION

To solve a rational equation:

Step 1 Exclude values that cause division by zero.

Step 2 Multiply both sides by the least common denominator and simplify.

Step 3 Solve the resulting equation.

Step 4 Check each solution to make sure it is not one of the excluded values; that is, division by zero is not permitted. Any value that is excluded in this manner is called an **extraneous root.**

» IN OTHER WORDS The procedure for solving rational equations is the same as that for solving other equations, except that values of the variables that cause division by zero are excluded from the solution set. It is *necessary* that you check for extraneous roots.

EXAMPLE 1   Solving a rational equation, root not excluded

Solve: $\dfrac{1}{x} - \dfrac{1}{4} = \dfrac{x-2}{4x}$.

Solution

Step 1: Set each denominator equal to 0 and solve (usually by inspection); any value that causes division by zero is *excluded*. For this example, $x \neq 0$.

Step 2: The least common denominator (LCD) is $4x$, so we multiply both sides by $4x$.

$$\frac{1}{x} - \frac{1}{4} = \frac{x-2}{4x} \qquad \text{Given.}$$

$$4x\left(\frac{1}{x} - \frac{1}{4}\right) = 4x\left(\frac{x-2}{4x}\right) \qquad \text{Multiply both sides by the LCD.}$$

Step 3: Solve the resulting equation.

$$4 - x = x - 2 \qquad \text{Simplify.}$$
$$6 = 2x$$
$$3 = x$$

Step 4: By noting the restriction, $x \neq 0$, $x = 3$ is verified as a solution.

EXAMPLE 2   Solving a rational equation, root excluded

Solve: $\dfrac{2x}{x-3} - 3 = \dfrac{2x-12}{3-x}$.

Solution

Step 1: Excluded value: $x \neq 3$.

Step 2:
$$\frac{2x}{x-3} - 3 = \frac{2x-12}{3-x}$$

$$(x-3)\left(\frac{2x}{x-3} - 3\right) = (x-3)\left(\frac{2x-12}{3-x}\right)$$

Step 3:
$$2x - 3(x-3) = (-1)(2x-12)$$
$$2x - 3x + 9 = -2x + 12$$
$$x = 3$$

Step 4: Checking the root $x = 3$ against the excluded value, we see that the solution is an extraneous solution. You might notice that if we attempted to check $x = 3$ in the original equation, we would find division by 0. Because there are no values that satisfy the given equation, we say the solution is *empty* and write this as $\varnothing$. ∎

EXAMPLE 3   One root included, one root excluded

Solve: $\dfrac{2x+1}{x+3} + \dfrac{3x-7}{2-x} = \dfrac{2x^2+10x-13}{6-x-x^2}$.

Solution

Step 1: Find (by inspection, if possible) the values that make any denominator equal to 0.

$$x+3 = 0 \qquad 2-x = 0 \qquad 6-x-x^2 = 0$$
$$x = -3 \qquad\quad 2 = x \qquad (3+x)(2-x) = 0$$
$$x = -3,\, 2$$

Thus, the excluded values are –3 and 2, so we write $x \neq -3,\, 2$.

Step 2:
$$\frac{2x+1}{x+3} + \frac{3x-7}{2-x} = \frac{2x^2+10x-13}{6-x-x^2}$$

$$(x+3)(2-x)\left(\frac{2x+1}{x+3} + \frac{3x-7}{2-x}\right) = (x+3)(2-x)\left(\frac{2x^2+10x-13}{(3+x)(2-x)}\right)$$

Step 3:
$$(2-x)(2x+1) + (x+3)(3x-7) = 2x^2 + 10x - 13$$
$$4x + 2 - 2x^2 - x + 3x^2 + 2x - 21 = 2x^2 + 10x - 13$$
$$x^2 + 5x - 19 = 2x^2 + 10x - 13$$
$$x^2 + 5x + 6 = 0$$
$$(x+2)(x+3) = 0$$
$$x = -2,\, -3$$

Step 4: Because $x = -3$ is extraneous, we see that the solution is -2. To distinguish the notation from solving the equation in step 3, we will show the solution to the given equation using set notation; namely, $\{2\}$. We use this notation to imply that we have checked the solution to step 3 by comparing it with the excluded values in step 1. ∎

The Greeks held that there was one most pleasing ratio of height to width in a rectangle. For a height h and a width w, the proportion the Greeks used is

$$\frac{h}{w} = \frac{w}{h+w}$$

This relationship is called the *divine proportion*. If $h = 1$ and this equation is solved for w (see Problem 53), the ratio h to w is a constant called the *golden section* or *golden ratio*. The Parthenon in Athens has been used as an example of a building with a height-to-width ratio that is almost equal to the golden ratio.

Parthenon

EXAMPLE 4   Golden rectangle application

MODELING APPLICATION

If the Parthenon is 101 feet wide, what is its height (to the nearest foot) if we assume the dimensions are in a golden ratio?

Solution

Step 1: *Understand the problem.* Since the Parthenon is built to satisfy the golden ratio, the height h and the width w satisfy the following proportion:

$$\frac{h}{w} = \frac{w}{h+w}$$

Step 2: *Devise a plan.* The width is 101 feet, so

$$\frac{h}{101} = \frac{101}{h+101}$$

We will solve this equation for h.

Step 3: *Carry out the plan.* There is only one unknown, which is written in variable form, so we now solve the equation for h:

$$h(h+101) = 101^2$$

$$h^2 + 101h - 101^2 = 0$$

$$h = \frac{-101 \pm \sqrt{101^2 - 4(1)(-101^2)}}{2(1)}$$

$$= \frac{-101 \pm 101\sqrt{1+4}}{2}$$

≈ 62.4 Disregard the negative solution, since distances are nonnegative.

Step 4: *Look back.* The Parthenon is about 62 feet high.

Radical Equations

We have encountered irrational numbers that involve radicals (square root, cube root, . . .), and now we consider expressions that involve variables as the radicand. Such expressions are called **radical expressions**. A **radical equation** is an equation that involves a radical expression. Solving equations requires the following property of powers.

PROPERTY OF POWERS

If P and Q are algebraic expressions in a variable x, and n is any positive integer, then the solution set of $P = Q$ is a subset of the solution set of $P^n = Q^n$. The equation $P^n = Q^n$ is called a **derived equation** of $P = Q$.

» IN OTHER WORDS Not all solutions of the derived equation may be solutions of the original equation. Whenever both sides of an equation are raised to a power, the solution *must be checked* in the original equation. Solutions that do not check are *extraneous solutions*.

To use this property of powers, a radical expression is first isolated on one side of the equation. This ensures that raising each side to a power will eliminate that radical. If there is more than one radical, the process is repeated. The following examples illustrate the procedure.

EXAMPLE 5   Radical equation with one radical

Solve: $\sqrt{3 - x} + 1 = x$.

Solution   First, isolate the radical, then square both sides:

$$\sqrt{3 - x} + 1 = x \qquad \text{Given.}$$

$$\sqrt{3 - x} = x - 1 \qquad \text{Isolate the radical.}$$

$$\left(\sqrt{3 - x}\right)^2 = (x - 1)^2 \qquad \text{Square both sides; this is the derived equation.}$$

$$3 - x = x^2 - 2x + 1$$

$$0 = x^2 - x - 2$$

$$0 = (x + 1)(x - 2)$$

$$x = -1, 2$$

Check   (This is required for radical equations as explained previously.)

$$\begin{array}{cc}
x = -1: & x = 2: \\
\sqrt{3 - (-1)} + 1 \stackrel{?}{=} -1 & \sqrt{3 - 2} + 1 \stackrel{?}{=} 2 \\
\sqrt{4} + 1 \stackrel{?}{=} -1 & \sqrt{1} + 1 \stackrel{?}{=} 2 \\
3 \neq -1 & 2 = 2 \\
x = -1 \text{ is extraneous.} & x = 2 \text{ is a root.}
\end{array}$$

Using the symbolism of writing the (checked) roots with set notation, we write the solution set to the original radical equation as $\{2\}$. ∎

EXAMPLE 6   Radical equation with two radicals

Solve: $\sqrt{2x + 5} + \sqrt{x + 2} = 1$.

Solution

$$\sqrt{2x + 5} + \sqrt{x + 2} = 1 \qquad \text{Given.}$$

$$\sqrt{2x + 5} = 1 - \sqrt{x + 2} \qquad \text{Isolate one of the radicals.}$$

$$\left(\sqrt{2x+5}\right)^2 = \left(1-\sqrt{x+2}\right)^2 \qquad \text{Square both sides}^*$$

$$2x+5 = 1-2\sqrt{x+2}+x+2 \qquad \text{Square and simplify.}$$

$$x+2 = -2\sqrt{x+2} \qquad \text{Isolate the remaining radical.}$$

$$(x+2)^2 = \left(-2\sqrt{x+2}\right)^2 \qquad \text{Square both sides again.}$$

$$x^2+4x+4 = 4x+8 \qquad \text{Square and simplify.}$$

$$x^2-4 = 0 \qquad \text{Now, solve this quadratic.}$$

$$(x-2)(x+2) = 0$$

$$x = 2, -2$$

Check: $x = 2$: $x = -2$:

$$\sqrt{2(2)+5}+\sqrt{2+2} \overset{?}{=} 1 \qquad \sqrt{2(-2)+5}+\sqrt{(-2)+2} \overset{?}{=} 1$$

$$\sqrt{9}+\sqrt{4} \overset{?}{=} 1 \qquad\qquad \sqrt{1}+\sqrt{0} \overset{?}{=} 1$$

$$3+2 \neq -1 \qquad\qquad\qquad 1 = 1$$

$$x = 2 \text{ is extraneous.} \qquad x = -2 \text{ is a root.}$$

The solution set is $\{-2\}$.

■

EXAMPLE 7   Radical equation with an index larger than 2

Solve: $\sqrt[3]{x^2+2} = 3$.

Solution

$$\sqrt[3]{x^2+2} = 3 \qquad \text{Given.}$$

$$\left(\sqrt[3]{x^2+2}\right)^3 = 3^3 \qquad \text{Cube both sides.}$$

$$x^2+2 = 27 \qquad \text{Simplify.}$$

$$x^2-25 = 0$$

$$(x+5)(x-5) = 0$$

$$x = -5, 5$$

Check: $x = -5$: $x = 5$:

$$\sqrt[3]{(-5)^2+2} \overset{?}{=} 3 \qquad \sqrt[3]{(5)^2+2} \overset{?}{=} 3$$

$$\sqrt[3]{27} \overset{?}{=} 3 \qquad\qquad \sqrt[3]{27} \overset{?}{=} 3$$

$$3 = 3 \qquad\qquad\qquad 3 = 3$$

$$x = -5 \text{ is a root.} \qquad x = 5 \text{ is a root.}$$

The solution set is $\{-5, 5\}$.

■

We return to Example 6 of Section 4.2.

---

*Remember, because of this step you *must* check the result.

EXAMPLE 8   Dune buggy problem revisited

A dune buggy is on the desert at a point A located 40 km from a point B, which lies on a long, straight road, as shown in Figure 4.16. The driver can travel at 45 km/h on the desert and 75 km/h on the desert road. If the distance from B to D is 28 km, at what point C would the driver meet the road so that the trip takes 1 hr 9 min?

VW Concept T: The dune buggy lives!

Solution   From Example 6 of Section 4.2, we find the domain $[0, \, 28]$ and the time function:

$$T(x) = \frac{\sqrt{x^2 + 1,600}}{45} + \frac{28 - x}{75} \qquad \text{Derived in Example 6 of Section 4.2}$$

$$\frac{69}{60} = \frac{\sqrt{x^2 + 1,600}}{45} + \frac{28 - x}{75} \qquad \text{Given} \qquad T(x) = 1 + \frac{9}{60} = \frac{69}{60}.$$

$$(900)\frac{69}{60} = (900)\left(\frac{\sqrt{x^2 + 1,600}}{45} + \frac{28 - x}{75}\right) \qquad \text{Multiply both sides by the LCD 900.}$$

$$1,035 = 20\sqrt{x^2 + 1,600} + 12(28 - x) \qquad \text{Simplify.}$$

$$1,035 = 20\sqrt{x^2 + 1,600} + 336 - 12x$$

$$12x + 699 = 20\sqrt{x^2 + 1,600} \qquad \text{Isolate the radical.}$$

$$144x^2 + 16,776x + 699^2 = 400(x^2 + 1,600) \qquad \text{Square both sides.}$$

$$0 = 256x^2 - 16,776x + 151,399 \qquad \text{Quadratic; obtain a 0 on one side.}$$

$$x = \frac{16,776 \pm \sqrt{(-16,776)^2 - 4(256)(151,399)}}{2(256)} \qquad \text{Quadratic formula}$$

$$x = \frac{2,097 \pm 5\sqrt{79,001}}{64}$$

$$x \approx 10.807, \, 54.724$$

We reject the value 54.724 because it is not in the domain. The dune buggy must head for a location about 10.8 km from B and 17.2 km from the point of destination, so the trip takes 1 hr 9 min. ∎

Higher-Order Equations

Sometimes we can use the techniques of this section to solve certain higher-degree equations, as illustrated with the following example.

EXAMPLE 9   Solving a sixth-degree equation

Solve: $x^6 + 7x^3 - 8 = 0$.

Solution

$$\text{Let } w = x^3.$$

$$x^6 + 7x^3 - 8 = 0$$

$$w^2 + 7w - 8 = 0$$

$$(w + 8)(w - 1) = 0$$

$$w = -8, 1$$

$$\text{If } w = -8, x^3 = -8$$

$$\sqrt[3]{x^3} = \sqrt[3]{-8}$$

$$x = -2$$

Check: $(-2)^6 + 7(-2) - 8 \overset{?}{=} 0$

$$64 - 14 - 8 \neq 0$$

$$\text{If } w = 1, x^3 = 1$$

$$\sqrt[3]{x^3} = \sqrt[3]{(1)^3}$$

$$x = 1$$

Check: $(1)^6 + 7(1) - 8 \overset{?}{=} 0$

$$1 + 7 - 8 = 0$$

Solution is $\{1\}$; -2 is extraneous.   ■

PROBLEM SET 4.3

LEVEL 1

WHAT IS WRONG, *if anything, with each statement in Problems 1–4? Explain your reasoning.*

1. $\sqrt{x^2} = x$

2. Solve $\sqrt{x^2} = x$.

$\sqrt{x^2} = x$ Given.

$x^2 = x^2$ Square both sides.

This equation is true for all real numbers.

3. Solve $\sqrt{x + 1} = x + 3$.

$\sqrt{x + 1} = x + 3$ Given.

$x + 1 = x^2 + 9$ Square both sides.

$x^2 - x + 8 = 0$ Get a 0 on one side.

There are no values satisfying this equation.

4. Solve $\dfrac{3x - 5}{5x - 5} + \dfrac{5x - 1}{7x - 7} = \dfrac{-9}{35 - 35x}$.

$$\frac{3x - 5}{5x - 5} + \frac{5x - 1}{7x - 7} = \frac{-9}{35 - 35x}$$

$$35(x - 1)\left[\frac{3x - 5}{5x - 5} + \frac{5x - 1}{7x - 7}\right] = 35(x - 1)\left[\frac{-9}{35 - 35x}\right]$$

$$7(3x - 5) + 5(5x - 1) = 9$$

$$21x - 35 + 25x - 5 = 0$$

$$49x - 40 = 9$$

$$49x = 49$$

$$x = 1$$

The solution is $x = 1$.

Solve the equations for real number solutions in Problems 5–30. State the solution set restrictions for rational equations and all extraneous roots found.

5. $\dfrac{x^2}{12} - \dfrac{x}{4} = \dfrac{1}{3}$

6. $\dfrac{x^2}{12} - \dfrac{x}{4} = \dfrac{5}{6}$

7. $\dfrac{3}{4x} + \dfrac{7}{16} = \dfrac{4}{3x}$

8. $\dfrac{7}{4x} + \dfrac{5}{18} = \dfrac{3}{36}$

9. $\dfrac{x + 3}{x - 2} = \dfrac{5}{x - 2}$

10. $\dfrac{x - 8}{x - 10} = \dfrac{2}{x - 10}$

11. $\dfrac{4}{y} - \dfrac{3(2 - y)}{y - 1} = 3$

12. $\dfrac{6}{y} - \dfrac{2(y - 5)}{y + 1} = 3$

13. $\dfrac{1}{x + 2} - \dfrac{1}{2 - x} = \dfrac{3x + 8}{x^2 - 4}$

14. $\dfrac{16}{x + 5} + \dfrac{4}{5 - x} = \dfrac{5 - 3x}{x^2 - 25}$

15. $\dfrac{x + 1}{x - 1} - \dfrac{x - 1}{x + 1} = \dfrac{5}{6}$

16. $\dfrac{5 - x}{x + 1} + \dfrac{x - 3}{x - 1} = \dfrac{8}{15}$

17. $\dfrac{x - 1}{x - 3} + \dfrac{x + 1}{x + 3} = \dfrac{1}{2}$

18. $\dfrac{x - 2}{x - 3} - \dfrac{x + 2}{x + 3} = \dfrac{5}{8}$

19. $2\sqrt{x} = x + 1$

20. $\sqrt{x} = x - 6$

21. $\sqrt{x + 2} = 3$

22. $\sqrt{2x + 3} = 3$

23. $\sqrt[3]{4x + 4} = 2$

24. $\sqrt[3]{3x + 1} = 1$

25. $x - \sqrt{x} - 2 = 0$

26. $x + \sqrt{x + 8} = -2$

27. $x = 6 - 3\sqrt{x - 2}$

28. $x - \sqrt{4x - 11} - 4 = 0$

29. $\sqrt{x - 3} = \sqrt{4x - 5}$

30. $\sqrt{x - 6} = \sqrt{3x + 2}$

LEVEL 2

Solve the equations for real number solutions in Problems 31–52.

31. $\dfrac{3}{2x + 1} + \dfrac{2x + 1}{1 - 2x} = 1 - \dfrac{8x^2}{4x^2 - 1}$

32. $\dfrac{x - 2}{x + 3} - \dfrac{1}{x - 2} = \dfrac{x - 4}{x^2 + x - 6}$

33. $\dfrac{x + 2}{3x - 1} - \dfrac{1}{x} = \dfrac{x + 1}{3x^2 - x}$

34. $\dfrac{x}{2x - 1} - \dfrac{1}{x} = \dfrac{x + 1}{2x^2 - x}$

35. $\dfrac{x - 2}{x + 2} + \dfrac{1}{x - 1} = \dfrac{4x - 1}{x^2 + x - 2}$

36. $\dfrac{x - 1}{x - 2} + \dfrac{x + 4}{2x + 1} = \dfrac{1}{2x^2 - 3x - 2}$

37. $\dfrac{x - 1}{2x - 1} + \dfrac{4 - x}{x + 1} = \dfrac{3x}{2x^2 + x - 1}$

38. $\dfrac{x - 3}{x - 2} + \dfrac{x - 1}{x} = \dfrac{22x - 110}{3x^2 - 15x}$

39. $\sqrt{x + 3} + \sqrt{x} = \sqrt{3}$

40. $1 + \sqrt{x + 2} = \sqrt{x}$

41. $\sqrt{4x + 1} - \sqrt{2x + 1} = 2$

42. $\sqrt{3x + 1} - \sqrt{2x - 2} = 2$

43. $\sqrt{x + 1} - \sqrt[4]{x + 1} = 0$

44. $\sqrt{x - 2} = \sqrt[4]{x^2 - 6x + 1}$

45. $\sqrt{x - 1} + 2 = 2\sqrt[4]{x - 1}$

46. $16x^4 - 17x^2 + 1 = 0$

47. $36x^4 + 4 = 25x^2$

48. $4x^{-4} - 35x^{-2} - 9 = 0$

49. $x^{-4} - 5x^{-2} + 6 = 0$

50. $(x^2 - 3x)^2 - 2(x^2 - 3x) - 8 = 0$

51. $(x^2 + 4x)^2 + 7(x^2 + 4x) + 12 = 0$

52. $x^{1/2} + 6x^{-1/2} + 5 = 0$

53. Solve $\dfrac{h}{w} = \dfrac{w}{h + w}$ for w if $h = 1$ and $w > 0$. This number is the *golden ratio.*

54. If a rectangle is to have sides with lengths in the golden ratio, what is the height if the length (the longer side) is one unit?

55. If a window is to be 5 ft wide, how high should it be, to the nearest tenth of a foot, to be a golden rectangle?

56. If a canvas for a painting is 18 in. wide, how high should it be, to the nearest inch, to be the divine proportion?

57. A photograph is to be printed on a rectangle in the divine proportion. If it is 9 cm high, how wide is it, to the nearest centimeter?

58. If the Parthenon is 60 ft high, what is its width, to the nearest foot, if we assume the building conforms to the golden ratio?

LEVEL 3

59. IN YOUR OWN WORDS In Example 4, we assumed the width of the Parthenon to be 101 ft and found the height to be 62.4 ft (assuming the golden ratio). If you worked Problem 58, you assumed the height to be 60 ft and found the width using the golden ratio to be 97 ft. Are the numbers from Example 4 and from Problem 58 consistent? Can you draw any conclusions? You might want to read "Misconceptions about the Golden Ratio" by George Markowsky in the January 1992 issue of *The College Mathematics Journal.*

60. Historical Quest This problem is from Brahmagupta, ca. 630. "Ten times the square root of a flock of geese, seeing the clouds collect, flew to the Manus lake; one-eighth of the whole flew from the edge of the water amongst a multitude of water lilies; and three couples were observed playing in the water." Find the number of geese in the original flock.

4.4 Exponential Functions

The linear, quadratic, polynomial, rational, and radical functions considered in the first part of this book are all called **algebraic functions**. An algebraic function is a function that can be expressed in terms of algebraic operations alone. If a function is not algebraic, it is called a **transcendental function**. In the remainder of this chapter, two examples of transcendental functions, *exponential* and *logarithmic* functions, are considered. In the next chapter, other types of transcendental functions are considered.

Irrational Exponents

To understand an exponential function, it is necessary to recall the definition of an exponent. In Chapter 1 (page 12), we defined an expression with a rational exponent. We need to define expressions such as $2^{\sqrt{3}}$. You might turn to a calculator to find $2^{\sqrt{3}} \approx 3.321997085$, but what does this mean? Why is your calculator programmed to provide *this* display? To define b^x for x an irrational number (or more generally, for x any real number), we require calculus, but the following theorem from calculus will help us to understand the evaluation of an expression with an irrational exponent.

Suppose b is a real number greater than 1. Then, for any real number x, there is a unique real number b^x. Moreover, if h and k are any two rational numbers such that $h < x < k$, then $b^h < b^x < b^k$

This squeeze theorem gives meaning to expressions such as $2^{\sqrt{3}}$. Consider the graph of the function $f(x) = 2^x$ by plotting points as shown in Figure 4.19**a**. The case where b is a real number $0 < b < 1$ is left for the problem set.

x	y
-3	0.125
-2	0.25
-1	0.5
0	1
1	2
2	4
3	8

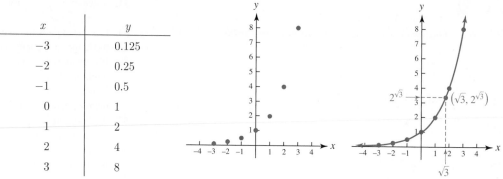

a. Plot selected points   **b.** Connect points with a smooth curve

Figure 4.19 Graph of $f(x) = 2^x$

If the points shown in Figure 4.19**a** are connected with a smooth curve (see Figure 4.19**b**), you can see that $2^{\sqrt{3}}$ is defined and is between 2^1 and 2^2. The number $2^{\sqrt{3}}$ can be approximated to any desired degree of accuracy:

Since $1 < \sqrt{3} < 1.8$ $2^1 < 2^{\sqrt{3}} < 2^2$

$1.7 < \sqrt{3} < 1.8$ $2^{1.7} < 2^{\sqrt{3}} < 2^{1.8}$

$1.73 < \sqrt{3} < 1.74$ $2^{1.73} < 2^{\sqrt{3}} < 2^{1.74}$

$\vdots$ $\vdots$

We can now assume that b^x is defined for all real numbers x because we have the squeeze theorem for irrational exponents and a direct definition for rational exponents. We will also accept that the usual laws of exponents hold for all real exponents.

Let a and b be nonnegative real numbers and let p and q be positive real numbers except that the form 0^0 and division by zero are excluded.

First law (Additive): $b^p \cdot b^q = b^{p+q}$

Second law (Subtractive): $\dfrac{b^p}{b^q} = b^{p-q}$

Third law (Multiplicative): $(b^q)^p = b^{pq}$

Fourth law (Distributive): $(ab)^p = a^p b^p$

Fifth law (Distributive): $\left(\dfrac{a}{b}\right)^p = \dfrac{a^p}{b^p}$

Definition and Graphs of Exponential Functions

We can now define an exponential function.

EXPONENTIAL FUNCTION

The function f is an **exponential function** if

$$f(x) = b^x$$

where b is a positive constant other than 1, and x is any real number. The number x is called the **exponent** and b is called the **base**.

The shape of a particular exponential graph depends on b, as illustrated by the following example.

EXAMPLE 1   Graphs of exponential functions

a. Graph $f(x) = 10^x$ $(b = 10;$ type $b > 1)$.   **b.** Graph $f(x) = 0.1^x$ $(b = 1/10;$ type $b < 1)$.

Solution

a. The function f is an increasing function, with a horizontal asymptote at $y = 0$. The graph, along with a table of values, is shown in Figure 4.20. Notice that it is often necessary to alter the scale for exponential functions.

x	y
-3	0.001
-2	0.01
-1	0.1
0	1
1	10
2	100
3	1,000

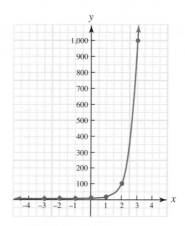

Figure 4.20 Graph of $f(x) = 10^x$

b. This function is a decreasing function with a horizontal asymptote at $y = 0$. The graph, along with a table of values, is shown in Figure 4.21.

x	y
-3	1,000
-2	100
-1	10
0	1
1	0.1
2	0.01
3	0.001

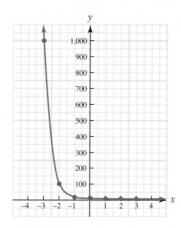

Figure 4.21 Graph of $f(x) = 0.1^x$

Example 1 is summarized in Table 4.3. We include these curves in our Library of Curves in Appendix F.

TABLE 4.3	Directory of Curves (Part II)
Exponential —Type I ($b > 1$) $y = b^x$	Exponential —Type II ($0 > b > 1$) $y = b^x$

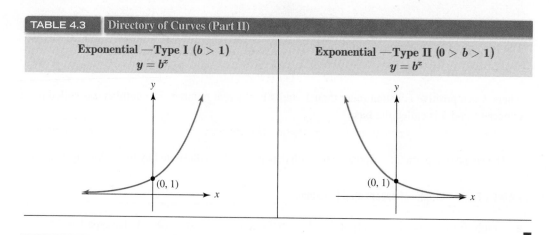

EXAMPLE 2   Variations of exponential graphs

Graph the curves defined by the given equations by comparing with the graph of $y = 2^x$.

a. $y = -2^x$ **b.** $y = 2^{-x}$ **c.** $y = (-2)^x$ **d.** $y = 2^x - 2$ **e.** $y = 2^{x-2}$

Solution   We begin each of these with the graph of $y = 2^x$.

a. Recall that -2^x means "opposite of 2 to the x power"; namely, $y = -(2^x)$. The graph of $y = -2^x$ is found by reflecting the graph of $y = 2^x$ in the x-axis.

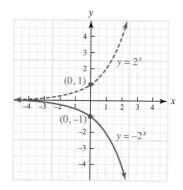

b. $y = 2^{-x}$ is obtained from the graph of $y = 2^x$ by reflecting in the y-axis. Also note that $y = 2^{-x}$ is the same as $y = \left(\frac{1}{2}\right)^x$, so this reflected graph has the shape of the exponential graph for $0 < b < 1$.

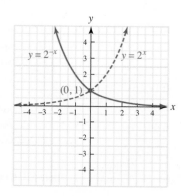

c. $y = (-2)^x$ is not defined since $b = -2$ is not a positive number.

d. $y = 2^x - 2$ can be written as $y + 2 = 2^x$; as compared with $y = 2^x$ is translated down 2 units.

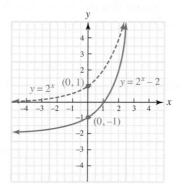

e. $y = 2^{x-2}$; compare with part **d**. This graph is translated to the right 2 units.

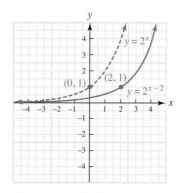

Real Exponents and Compound Interest

Compound interest provides an important application of an exponential function. If a sum of money, called the **principal** or **present value**, is denoted by P and invested at an annual interest rate of r, then the amount of money present in t years is called the **future value** and is denoted by A. It is found by

$$A = P + I$$

where I denotes the interest. **Interest** is an amount of money paid for the use of another's money. **Simple interest** is found according to the formula

$$I = Prt$$

For example, \$1,000 invested for 3 years at 8% simple interest would generate an interest of \$1,000(0.08)3 = \$240, so the amount present after 3 years is \$1,000 + \$240 = \$1,240.

Most applications pay interest on the interest, as well as on the principal, after a certain period of time. This type of interest is called **compound interest**. For example, \$1,000 invested at 8% annual interest compounded annually for 3 years can be found as follows:

First year $(t = 1)$:
$$A = P(1 + r)$$
$$= 1,000(1 + 0.08)$$
$$= 1,080$$
$$\downarrow$$

Second year $(t = 1)$:
$$A = P(1 + r) \qquad \text{One year's principal is previous year's total.}$$
$$= 1,080(1 + 0.08)$$
$$= 1,166.40$$
$$\downarrow$$

Third year $(t = 1)$
$$A = P(1 + r)$$
$$= 1,166.40(1 + 0.08)$$
$$= 1,259.71$$

Notice that the future value with simple interest is \$1,240.00, whereas with annual compounding the future value is \$1,259.71. Notice that the future value for each subsequent year is multiplied by $(1 + 0.08)$. This means that a formula for 3 years is (from the repeated calculations shown earlier)

$$A = \$1{,}000(1 + 0.08)^3$$

This formula assumes that the interest is compounded annually. Let n be the number of times the money is compounded each year. This means that

$$n = 1 \text{ if compounded annually}$$
$$n = 2 \text{ if compounded semiannually}$$
$$n = 4 \text{ if compounded quarterly}$$
$$n = 12 \text{ if compounded monthly}$$
$$n = 360 \text{ if compounded daily}$$

 In this book, we assume ordinary interest unless otherwise stated.

Note that when paying interest, banks use 360 for the number of days in a year; this is called **ordinary interest**. Unless instructed otherwise, use 360 for the number of days in a year. If 365 days are used, it is called **exact interest**.

COMPOUND INTEREST

$$A = P\left(1 + \frac{r}{n}\right)^{nt}$$

where A = future value; P = present value; r = annual interest rate; t = number of years; and n = number of times compounded per year. To simplify, we define two calculated values:

$$i = \text{rate per period} = \frac{r}{n} \qquad N = \text{number of periods} = nt$$

Thus, the compound interest formula is

$$A = P(1 + i)^N$$

EXAMPLE 3   Future value of an investment

If \$12,000 is invested for 5 years at 12% compounded quarterly, what is the future value?

Solution $P = \$12{,}000$, $r = 0.12$, $t = 5$, $n = 4$; calculate $i = \dfrac{r}{n} = \dfrac{0.12}{4} = 0.03$ and $N = nt = 4(5) = 20$. Then

$$A = P(1 + i)^N$$
$$= \$12{,}000(1 + 0.03)^{20}$$
$$\approx \$21{,}673.33 \qquad \text{Use a calculator, and round money problems to the nearest cent.}$$

Since the investment is growing (that is, the base of the exponent is positive) for $t > 0$, we can graph the future value of an investment. For example, the graph of the investment described in Example 3 is shown in Figure 4.22. This graph (or the graph of any exponential function) could not be drawn if it were not for the squeeze theorem for exponents.

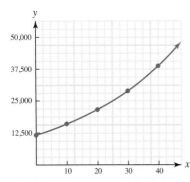

Figure 4.22 Growth of an investment

STOP In this book our convention is not to round until stating the final answer, and then round to the nearest cent.

EXAMPLE 4   Present value of an investment

You need to have \$50,000 in 8 years. If the rate of return is 9% and the investment is compounded monthly, how much must be deposited today so that you have the money at the required time?

Solution $A = \$50,000$, $r = 0.09$, $t = 8$, $n = 12$; calculate $i = \dfrac{r}{n} = \dfrac{0.09}{12}$ and $N = nt = 12(8) = 96$.
Then

$$A = P(1 + i)^N$$

$$50,000 = P\left(1 + \frac{0.09}{12}\right)^{96}$$

$$\frac{50,000}{\left(1 + \frac{0.09}{12}\right)^{96}} = P$$

Using a calculator, you can more easily calculate this by writing

$$P = 50,000\left(1 + \frac{0.09}{12}\right)^{-96} \approx 24,403.09$$

The present value needed is $\$24,403.09$.

∎

EXAMPLE 5 Effective annual yield formula

MODELING APPLICATION

Suppose your bank advertises an effective annual yield of 8.33%, but upon checking you find that your bank pays 8% interest compounded daily. What do they mean by **effective annual yield**, and how is it calculated?

Solution

> NEW, HIGHEST INTEREST RATE
> EVER ON INSURED SAVINGS
>
> **8.33%** annual yield on
> **8%** interest compounded daily
> Annual yield based on daily compounding when
> funds and interest remain on deposit a year. Note:
> Federal regulations require substantial interest
> penalty for early withdrawal of principal from
> Certificate Accounts.

Step 1: *Understand the problem.* We call a bank and are told that the effective annual yield is the simple interest rate that would yield the same future value at the end of 1 year. Daily compounding at 8% on a present value of P dollars means we have a future value of

$$A = P\left(1 + \frac{0.08}{360}\right)^{360}$$

On the other hand, compounded once, the future value at the effective rate is

$$A = P(1 + 0.0833)^1 \quad \text{Since simple interest for } t = 1$$

If we divide both sides of these equations by P, we find compounded daily:

$$\frac{A}{P} = \left(1 + \frac{0.08}{360}\right)^{360} \approx 1.08327744 \approx 1.0833$$

Effective rate:

$$\frac{A}{P} = \left(1 + \frac{0.08}{360}\right)^{360} = 1.0833$$

We can see these are the same, so that the bank advertisement is correct.

Step 2: *Devise a plan.* To find the effective annual yield formula, we consider the following. Since we do not know the present value ($P \neq 0$) but know that it is the same for both the compound interest calculation and the effective yield calculation, and since we want to find the same future value for both calculations, we see that we will work with the constant A/P. We also assume that the following values are known: $r = $ annual rate, $n = $ number of times compounded per year, and $t = 1$.

$$\underbrace{\frac{A}{P} \text{ for the compound calculation}}_{\left(1 + \frac{r}{n}\right)^n} = \underbrace{\frac{A}{P} \text{ for the effective yield calculation}}_{(1 + \text{EFFECTIVE RATE})}$$

Step 3: *Carry out the plan.* Let $Y = $ EFFECTIVE RATE:

$$\left(1 + \frac{r}{n}\right)^n = 1 + Y$$

$$Y = \left(1 + \frac{r}{n}\right)^n - 1$$

Step 4: *Look back.* We check the calculations for those given by the bank in the advertisement. An investment at 8% compounded daily $(n = 360)$ gives an effective rate of

$$Y = \left(1 + \frac{0.08}{360}\right)^{360} - 1 \approx 0.0832774398 \approx 8.33\%$$

EXAMPLE 6   Effective annual yield

What is the effective annual yield (correct to the nearest hundredth of a percent) for $4\frac{1}{2}\%$ compounded quarterly?

Solution   We see $r = 0.045$ and $n = 4$, and we know

$$Y = \left(1 + \frac{r}{n}\right)^n - 1$$

$$= \left(1 + \frac{0.045}{4}\right)^4 - 1$$

$$\approx 0.0457650863$$

The effective yield (or effective rate) is about 4.58%.

Continuous Compounding; The Number e

A reasonable extension of the current discussion is to ask the effect of more frequent compounding. To model this situation, consider the following contrived example. Suppose $1 is invested at 100% interest for 1 year compounded at different intervals. The compound interest formula for this example is

$$A = \left(1 + \frac{1}{n}\right)^n$$

where n is the number of times of compounding in 1 year. The calculations of this formula for different values of n are shown in the following table.

Number of Periods	Formula	Amount
Annual, $n = 1$	$\left(1 + \frac{1}{1}\right)^1$	$2.00
Semiannual, $n = 2$	$\left(1 + \frac{1}{2}\right)^2$	$2.25
Quarterly, $n = 4$	$\left(1 + \frac{1}{4}\right)^4$	$2.44
Monthly, $n = 12$	$\left(1 + \frac{1}{12}\right)^{12}$	$2.61
Daily, $n = 360$	$\left(1 + \frac{1}{360}\right)^{360}$	$2.71

Looking only at this table, you might (incorrectly) conclude that as the number of times the investment is compounded increases, the amount of the investment increases without bound. Let us continue these calculations for even larger n:

$n = 8,640$ (compounding every hour)   2.718124537

$n = 518,000$ (every minute)   2.718279142

$n = 1,000,000$ 2.718280469

$n = 10,000,000$ 2.71828169

$n = 100,000,000$ 2.718281815

The spreadsheet we are using for these calculations can no longer distinguish the values of $(1 + 1/n)^n$ for larger n. These values are approaching a particular number. This number, it turns out, is an irrational number, and it does not have a convenient decimal representation. (That is, its decimal representation does not terminate and does not repeat.) Mathematicians, therefore, have agreed to denote this number by using the symbol e. This number is called the **natural base** or **Euler's number**.

THE NUMBER e

As n increases without bound, the number e is the irrational number that is the limiting value of the formula

$$\left(1 + \frac{1}{n}\right)^n$$

> **» IN OTHER WORDS** Using limit notation, $\lim_{n \to +\infty} \left(1 + \frac{1}{n}\right)^n = e$. You must wait until you discuss the concept of limit in calculus for a formal definition of e, but the preceding discussion should be enough to convince you that
>
> $$e \approx 2.72$$
>
> Even though you will not find a bank that compounds interest every minute, you will find banks that use this limiting value to compound **continuously**.

CONTINUOUS COMPOUNDING

The future value, A, of an investment of P, **compounded continuously** at a rate of r for t years is found by using

$$A = Pe^{rt}$$

EXAMPLE 7   **Evaluating powers of e**

Approximate e, e^2, and e^{-3} using your calculator.

Solution   This example is a check to make sure you know how to use your calculator. Check the output of your calculator with those shown here. Look for the e^x key on your calculator. Some calculators will input the value of x first, whereas others input the exponent after using the e^x key.

$e \approx 2.718281828$ $e^2 \approx 7.389056099$ $e^{-3} \approx 0.0497870684$

EXAMPLE 8   **Inflation using continuous compounding**   MODELING APPLICATION

If your salary today is \$65,000 per year, what would you expect your salary to be in 20 years if you assume that inflation will continue at a constant rate of 8% over that time period?

Solution

Step 1: *Understand the problem.* Inflation is an example of continuous compounding. The problem with estimating inflation is that you must "guess" the value of future inflation, which in reality does not remain constant. However, if you look back 20 years and use an average inflation rate for the past 20 years—say, 8%—you may use this as a reasonable estimate for the next 20 years.

Step 2: *Devise a plan.* We will use the future value formula for compound interest.

Step 3: *Carry out the plan.* Let 65,000, $r = 0.08$, and $t = 20$ to find

$$A = Pe^{rt}$$
$$= 65,000e^{0.08(20)}$$
$$\approx 321,947.11$$

Step 4: *Look back.* The number (which we have rounded to the nearest cent) is $321,947.18. This seems very large, but it is based on our assumptions about inflation and our salary in 20 years. This assumes no salary increases other than those that result from inflation.

PROBLEM SET 4.4

LEVEL 1

Use a calculator to evaluate the expressions given in Problems 1–4.

1. a. e^3 **b.** e^{-2} **c.** $e^{0.15(5)}$

2. a. $e^{0.12}$ **b.** $e^{0.08}$ **c.** $856e^{0.05}$

3. a. $\left(1 + \dfrac{0.08}{12}\right)^{24}$ **b.** $\left(1 + \dfrac{0.05}{6}\right)^{72}$ **c.** $\left(1 + \dfrac{1}{1,000}\right)^{1,000}$

4. a. $\left(1 + \dfrac{0.12}{365}\right)^{365}$ **b.** $\left(1 + \dfrac{0.18}{360}\right)^{720}$ **c.** $\left(1 + \dfrac{1}{100,000}\right)^{100,000}$

If defined, simplify each of the expressions in Problems 5–12.

5. a. $9^{1/2}$ **b.** $-9^{1/2}$ **c.** $(-9)^{1/2}$

6. a. $64^{1/6}$ **b.** $(-64)^{1/6}$ **c.** $-64^{1/6}$

7. a. $(-32)^{1/5}$ **b.** $-32^{1/5}$ **c.** $32^{1/5}$

8. a. $-27^{2/3}$ **b.** $-8^{2/3}$ **c.** $27^{2/3}$

9. a. $[4^{1/2}]^2$ **b.** $[4^2]^{1/2}$
   c. $[(-4)^{1/2}]^2$ **d.** $[(-4)^2]^{1/2}$
   e. $[-4^{1/2}]^2$ **f.** $[-4^2]^{1/2}$

10. a. $[9^{1/2}]^2$ **b.** $[9^2]^{1/2}$
   c. $[(-9)^{1/2}]^2$ **d.** $[(-9)^2]^{1/2}$
   e. $[-9^{1/2}]^2$ **f.** $[-9^2]^{1/2}$

11. a. $[3^{1/2}]^2$ **b.** $[3^2]^{1/2}$
   c. $[(-3)^{1/2}]^2$ **d.** $[(-3)^2]^{1/2}$
   e. $[-3^{1/2}]^2$ **f.** $[-3^2]^{1/2}$

12. a. $[5^{1/2}]^2$ **b.** $[5^2]^{1/2}$
   c. $[(-5)^{1/2}]^2$ **d.** $[(-5)^2]^{1/2}$
   e. $[-5^{1/2}]^2$ **f.** $[-5^2]^{1/2}$

In Problems 13–22, sketch each pair of graphs on the same axes.

13. a. $y = 2^x$ **b.** $y = \left(\frac{1}{2}\right)^x$

14. a. $y = 3^x$ **b.** $y = 3^{-x}$

15. a. $y = 4^x$ **b.** $y = -4^x$

16. a. $y = 5^x$ **b.** $y = 5^x - 3$

17. a. $y = 2^x$ **b.** $y = 2^{x-3}$

18. a. $y = e^x$ **b.** $y = -e^x$

19. a. $y = e^x$ **b.** $y = e^x + 1$

20. a. $y = e^x$ **b.** $y = e^{x+1}$

21. a. $y = e^x$ **b.** $y = e^{x-3}$

22. a. $y = 2^x$ **b.** $y - 4 = 2^{x+3}$

LEVEL 2

Assume that $1,000 is invested for t years at r% interest compounded in the frequencies given in Problems 23–26. Give the future value.

23. $t = 25$, $r = 0.07$; annual compounding

24. $t = 10$, $r = 0.12$; semiannual compounding

25. $t = \frac{1}{3}$ (4 months), $r = 0.16$; monthly

26. $t = \frac{1}{2}$ (6 months), $r = 0.08$; monthly

27. What is the effective yield (rounded to the nearest hundredth of a percent) of 6% compounded quarterly?

28. What is the effective yield (rounded to the nearest hundredth of a percent) of 6% compounded monthly?

PROBLEM FROM CALCULUS *Problems 29–32 define three functions that are used in calculus.*

29. Graph $c(x) = \dfrac{e^x + e^{-x}}{2}$.

30. Graph $s(x) = \dfrac{e^x - e^{-x}}{2}$.

31. The functions c and s of Problems 29 and 30 are called the *hyperbolic cosine* and *hyperbolic sine* functions. These are defined by

$$\cosh x = \frac{e^x + e^{-x}}{2} \quad \text{and} \quad \sinh x = \frac{e^x - e^{-x}}{2}.$$

a. Show that $\cosh^2 x - \sinh^2 x = 1$.
b. Graph $y = \cosh x + \sinh x$.

32. Define the *hyperbolic tangent* function as

$$\tanh x = \frac{e^x - e^{-x}}{e^x + e^{-x}}$$

a. Show that $\tanh x = \dfrac{\sinh x}{\cosh x}$

  (see Problem 31).

b. Graph $\tanh x$.

33. Your first child has just been born. You want to give her a million dollars when she retires at age 65. If you invest your money at 12% compounded quarterly, how much do you need to invest today so your child will have a million dollars in 65 years?

34. An insurance agent wants to sell you a policy that will pay you $200,000 in 30 years. If you assume an average rate of inflation of 9% over the next 30 years, what is the value of that insurance payment in terms of today's dollars?

35. In 2007, the national debt passed $9 trillion. If the annual interest rate is 8%, what is the interest on the national debt *every second*? (Use a 365-day year.)

If P dollars are borrowed for N months at an annual rate of r, then the monthly payment, m, is found by the formula

$$m = \frac{Pi}{1 - (1+i)^{-N}}$$

where $i = r/n$. Use this information for Problems 36–39.

36. What is the monthly payment for a new car costing $34,560 with a down payment of $2,560? The car is financed for 4 years at 12%.

37. A purchase of $2,650 is financed at 21% for 2 years. What is the monthly payment?

38. A home loan is made for $215,000 at 8.4% for 30 years. What is the monthly payment?

39. A home is financed at $7\frac{1}{2}\%$ interest for 20 years. If the home price is $840,000 with 20% down, what is the monthly payment?

40. Radioactive argon-39 has a half-life of 4 min. This means that the time required for half the argon-39 to decompose is 4 min. If we start with 100 mg of argon-39, the amount A left after t minutes is given by

$$A = 100\left(\frac{1}{2}\right)^{t/4}$$

Graph this function for $t \geq 0$.

41. Carbon-14 used for archaeological dating has a half-life of 5,700 years. This means that the time required for half of the carbon-14 to decompose is 5,700 years. If we start with 100 mg of carbon-14, the amount A left after t years is given by

$$A = 100\left(\frac{1}{2}\right)^{t/5,700}$$

Graph this function for $t \geq 0$.

42. A graph of world population growth from 1 AD to 2000 is shown in Figure 4.23.

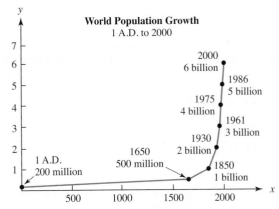

Figure 4.23 World population

In 1999, the world population was about 6 billion. If we assume a growth rate of 0.018%, the world population P (in billions) for t years after 2000 is given by the formula

$$P = 6e^{0.018t}$$

Graph this function for 2000–2010. Does your graph look like the one in Figure 4.23? Why or why not?

43. Compare the graphs for future value of simple and compound interest (annual compounding) for $1,000 invested at a 5% rate for x years where $0 \leq x \leq 20$.

44. Compare the graphs for future value of simple and compound interest (quarterly compounding) for $1,000 invested at an 8% rate for x years where $0 \leq x \leq 20$.

45. Use graphical methods to estimate $2^{\sqrt{2}}$.
46. Use graphical methods to estimate $3^{\sqrt{2}}$.

Sketch the curves defined by the equations in Problems 47–52 by plotting points.

47. $f(x) = 2^{x^2}$
48. $f(x) = 3^{-x^2}$
49. $f(x) = -4^{x^2}$
50. $f(x) = -5^{-x^2}$
51. $f(x) = \left(1 + \frac{1}{x}\right)^x$ for $x > 0$
52. $f(x) = \left(1 + \frac{1}{x}\right)^x$ for $x < 0$

LEVEL 3

WHAT IS WRONG, *if anything, with each of the proofs in Problems 53–54? Explain your reasoning.*

53. Find the error in the following "proof":

$2 = 2$
$\quad = 4^{1/2}$ *Definition of square root*
$\quad = [(-2)^2]^{1/2}$ *Substitute $(-2)^2 = 4$.*
$\quad = (-2)^{2(1/2)}$ *Theorem: $(b^q)^p = b^{pq}$.*
$\quad = (-2)^1$ *Since $2 \cdot \left(\frac{1}{2}\right) = 1$*
$\quad = -2$

Therefore, $2 = -2$.

54. The fourth law of exponents states $(ab)^p = a^p b^p$. For example,

$8 = 64^{1/2}$
$\quad = (4 \cdot 16)^{1/2}$
$\quad = 4^{1/2} \cdot 16^{1/2}$
$\quad = 2 \cdot 4$
$\quad = 8$

However, where is the error in the following steps, which show that 8 is equal to a number that is not defined?

$$8 = 64^{1/2}$$
$$= [(-4)(-16)]^{1/2}$$
$$= (-4)^{1/2}(-16)^{1/2}$$

Notice that $(-4)^{1/2}$ and $(-16)^{1/2}$ are not defined.

55. Find a counterexample to show that the squeeze theorem does not hold for $0 < b < 1$. State an alternate form of the squeeze theorem for this case.

56. Use graphical methods to determine which is larger:

a. $\left(\sqrt{3}\right)^{\pi}$ or $\pi^{\sqrt{3}}$

b. $\left(\sqrt{5}\right)^{\pi}$ or $\pi^{\sqrt{5}}$

c. $\left(\sqrt{6}\right)^{\pi}$ or $\pi^{\sqrt{6}}$

d. Consider $\left(\sqrt{x}\right)^{\pi} = \pi^{\sqrt{x}}$, where x is a positive real number. From parts **a–c**, notice that $\left(\sqrt{x}\right)^{\pi}$ is larger for some values of x, while $\pi^{\sqrt{x}}$ is larger for others. For $x = \pi^2$:

$$\left(\sqrt{x}\right)^{\pi} = \pi^{\sqrt{x}}$$

is obviously true (since $\pi^{\pi} = \pi^{\pi}$). Using a graphic method, find another value (approximately) for which the given statement is true.

57. Can an irrational number raised to an irrational power give an answer that is rational? *Hint*: Consider $\sqrt{2}^{\sqrt{2}}$. It is either rational or irrational.

PROBLEM FROM CALCULUS *In calculus, the number e is sometimes introduced using slopes. Problems 58–60 explore this idea.*

58. a. Draw the graph of $y = 2^x$ and plot the points $(0, 1)$ and $(2, 4)$, which are on the graph.

b. Consider the secant line passing through these points. Now, consider the slope of the secant line as the point $(2, 4)$ slides along the curve toward the point $(0, 1)$. Draw the line that you think will result when the point $(2, 4)$ reaches the point $(0, 1)$. This is the *tangent line* to the curve $y = 2^x$ at $(0, 1)$.

c. Using the tangent line, and the fact that the slope of a line is RISE/RUN, estimate (to the nearest tenth) the slope of the tangent line.

59. a. Draw the graph of $y = 3^x$ and plot the points $(0, 1)$ and $(2, 9)$, which are on the graph.

b. Consider the secant line passing through these points. Now, consider the slope of the secant line as the point $(2, 9)$ slides along the curve toward the point $(0, 1)$. Draw the line that you think will result when the point $(2, 9)$ reaches the point $(0, 1)$. This is the *tangent line* to the curve $y = 3^x$ at $(0, 1)$.

c. Using the tangent line, and the fact that the slope of a line is RISE/RUN, estimate (to the nearest tenth) the slope of the tangent line.

60. a. Draw a line passing through $(0, 1)$ with slope 1.

b. Compare the graphs of $y = 2^x$ (Problem 58) and $y = 3^x$ (Problem 59). Now, it seems reasonable that there exists a number between 2 and 3 with the property that the slope of the tangent through $(0, 1)$ is 1. Draw such a curve.

c. On the same coordinate axes, draw the graph of $y = e^x$. How does this curve compare with the curve you drew in part **b**? In calculus, the number between 2 and 3 with the property that the slope of the tangent through $(0, 1)$ is 1 is used as the number e.

4.5 Logarithmic Functions

Solving for an Exponent; Definition of Logarithm

In Chapter 1, we defined b^x for x rational, and in the previous section we used the squeeze theorem for exponents and discussed b^x for x irrational. This means we can consider b^x for all real numbers x and positive numbers b ($b \neq 1$). Suppose we wish to solve the equation $b = A$ for x.

Problem: **Solve $b^x = A$**   for x.

The algebraic techniques we have been using do not help us solve this equation. We begin by writing this exponential equation out in words, as if we were reading it out loud to someone else.

> **x is the exponent on the base b that yields A**

This can be rewritten

> **$x =$ exponent on base b to get A**

It appears that the equation is now solved for x, but this is simply a notational change. The expression "exponent on base b to get A" is called, for historical reasons, "the log of A to the base b." That is,

> **$x = \log A$ to the base b**

This phrase on the right is shortened to the notation

> **$x = \log_b A$**

The term **log** is an abbreviation for **logarithm**, but even the introduction of this notation is still a notational change only:

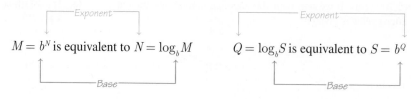

LOGARITHM

For positive b and A, $b \neq 1$

$$x = \log_b A \qquad \text{means} \qquad b^x = A$$

x is the called the **logarithm** and A is called the **argument**.

>> IN OTHER WORDS The statement $x = \log_b A$ should be read as "x is the log (exponent) on a base b that gives the value A."

 Do not forget that a logarithm is an exponent.

EXAMPLE 1   Changing from exponential form to logarithmic form

Write in logarithmic form: **a.** $5^2 = 25$ **b.** $\frac{1}{8} = 2^{-3}$ **c.** $\sqrt{64} = 8$

Solution

 a. In $5^2 = 25$, 5 is the base and 2 is the exponent, so we write

$$2 = \log_5 25$$

  Remember, the logarithmic expression "solves" for the exponent.

 b. In $\frac{1}{8} = 2^{-3}$, the base is 2 and the exponent is -3:

$$-3 = \log_2 \frac{1}{8}$$

 c. In $\sqrt{64} = 8$, the base is 64 and the exponent is $\frac{1}{2}$ (since $\sqrt{64} = 64^{1/2}$):

$$\frac{1}{2} = \log_{64} 8$$

EXAMPLE 2   Changing from logarithmic form to exponential form

Write each in exponential form: **a.** $\log_{10} 100 = 2$ **b.** $\log_{10} \frac{1}{1,000} = -3$ **c.** $\log_3 1 = 0$

Solution

 a. In $\log_{10} 100 = 2$, the base is 10 and the exponent is 2: $10^2 = 100$.

 b. In $\log_{10} \frac{1}{1,000} = -3$, the base is 10 and the exponent is -3: $10^{-3} = \frac{1}{1,000}$.

 c. In $\log_3 1 = 0$, the base is 3 and the exponent is 0: $3^0 = 1$.

Historical Note

In his history book, F. Cajori wrote, "The miraculous powers of modern calculation are due to three inventions: the Arabic notation, decimal fractions, and logarithms." Today, we would no doubt add a fourth invention to this list, namely, the handheld calculator. Nevertheless, the idea of a logarithm was a revolutionary idea that forever changed the face of mathematics. Simply stated, a **logarithm is an exponent**, and even though logarithms are no longer important as an aid in calculation, the logarithmic and exponential functions are extremely important in the study of advanced mathematics.

Evaluating Logarithms

To **evaluate** a logarithm means to find a numerical value for the given logarithm. The first ones we evaluate require that we know under what conditions $x = y$ when $b^x = b^y$. If $b = 1$, you cannot conclude that $x = y$ since $1^5 = 1^4$, but $5 \neq 4$. If $b \neq 1$, however, the statement can be proved true for all positive real numbers b and is called the **exponential property of equality**.

EXPONENTIAL PROPERTY OF EQUALITY

For positive real numbers b $(b \neq 1)$

$$b^x = b^y \text{ if and only if } x = y$$

Many logarithms can be evaluated by using the definition and the exponential property of equality.

EXAMPLE 3   Evaluating logarithms using the definition

Evaluate: **a.** $\log_2 64$ **b.** $\log_3 \frac{1}{9}$ **c.** $\log_9 27$ **d.** $\log_e 1$ **e.** $\log_{10} 0.1$

Solution

a. It is often necessary to supply a variable to convert from logarithmic to exponential form. We use N in this example.

Let $N = \log_2 64$ so that $2^N = 64$

$2^N = 2^6$

$N = 6$ *Exponential property of equality*

Thus, $\log_2 64 = 6$.

b. Let $N = \log_3 \frac{1}{9}$ so that $3^N = \dfrac{1}{9}$

$3^N = 3^{-2}$

$N = -2$ *Exponential property of equality*

Thus, $\log_3 \frac{1}{9} = -2$.

c. Let $N = \log_9 27$ so that $9^N = 27$

$(3^2)^N = 3^3$

$3^{2N} = 3^3$

$2N = 3$ *Exponential property of equality*

$N = \dfrac{3}{2}$

Thus, $\log_9 27 = \frac{3}{2}$.

d. Let $N = \log_e 1$ so that $e^N = 1$

$e^N = e^0$

$N = 0$ *Exponential property of equality*

Thus, $\log_e 1 = 0$.

e. Let $N = \log_{10} 0.1$ so that $10^N = 0.1$

$10^N = \dfrac{1}{10}$

$10^N = 10^{-1}$

$N = -1$ *Exponential property of equality*

Thus, $\log_{10} 0.1 = -1$. ■

Suppose we wish to find $\log_{10} 5.03$. If we let $N = \log_{10} 5.03$, then $10^N = 5.03$, and we are not closer to evaluating the logarithm because the exponential property of equality does not apply for this example. We can, however, estimate its value by noting

$$10^0 = 1$$
$$10^N = 5.03$$
$$10^1 = 10$$

The number N should be between 0 and 1 by the squeeze theorem for exponents. All is not lost because tables showing approximations for exponents such as this have been prepared. Base 10 is the most commonly used, so we call a logarithm to the base 10 a **common logarithm**, and we agree to write it without using a subscript 10; that is, log x is a *common logarithm*. A logarithm to the base e is called a **natural logarithm** and is denoted by ln x. The expression ln x is often pronounced "ell en x" or "lon x."

LOGARITHMIC NOTATIONS

Common logarithm: log x means $\log_{10} x$

Natural logarithm: ln x means $\log_e x$

Calculators have, to a large extent, eliminated the need for logarithm tables. You should find two logarithm keys on your calculator. One is labeled `LOG` for common logarithms, and the other is labeled `LN` for natural logarithms.

EXAMPLE 4   Evaluating logarithms using a calculator

Use a calculator to evaluate: **a.** $\log 5.03$ **b.** $\ln 3.49$ **c.** $\log 0.00728$

Solution   Be sure you use your own calculator to verify these answers.

   a. $\log 5.03 \approx 0.7015679851^{*}$
   b. $\ln 3.49 \approx 1.249901736$
   c. $\log 0.00728 \approx -2.137868621$

Example 4 shows fairly straightforward evaluations, since the problems involve common or natural logarithms and since your calculator has both `LOG` and `LN` keys. However, suppose we wish to evaluate a logarithm to some base *other than* base 10 or base *e*. The first method uses the definition of logarithm (as we did in Example 3), and the second method uses what is called the **change of base theorem**. Before we state this theorem, we consider its plausibility with the following example.

EXAMPLE 5   Looking for a pattern in evaluating logarithms

Evaluate the given expressions:

   a. $\log_2 8, \dfrac{\log 8}{\log 2}$, and $\dfrac{\ln 8}{\ln 2}$ **b.** $\log_3 9, \dfrac{\log 9}{\log 3}$, and $\dfrac{\ln 9}{\ln 3}$ **c.** $\log_5 625, \dfrac{\log 625}{\log 5}$, and $\dfrac{\ln 625}{\ln 5}$

Solution

   a. $\log_2 8 = x$ means (from the definition of logarithm)

$$2^x = 8$$
$$2^x = 2^3$$
$$x = 3$$

   Thus, $\log_2 8 = 3$. By calculator, $\dfrac{\log 8}{\log 2} \approx \dfrac{0.903089987}{0.3010299957} \approx 3$. Also, $\dfrac{\ln 8}{\ln 2} \approx \dfrac{2.079441542}{0.6931471806} \approx 3$.

   b. $\log_3 9 = x$ means $3^x = 3^2$ so that $x = \log_3 9 = 2$.

   By calculator, $\dfrac{\log 9}{\log 3} \approx \dfrac{0.9542425094}{0.4771212547} \approx 2$. Also, $\dfrac{\ln 9}{\ln 3} \approx \dfrac{2.197224577}{1.098612289} \approx 2$.

   c. $\log_5 625 =$ means $5^x = 5^4$ so that $\log_5 625 = 4$.

   By calculator, $\dfrac{\log 625}{\log 5} \approx \dfrac{2.795880017}{0.6989700043} \approx 4$ and $\dfrac{\ln 625}{\ln 5} \approx \dfrac{6.43775165}{1.609437912} \approx 4$.

You no doubt noticed that the answers to each part of Example 5 are the same. This result is summarized with Selfridge's theorem, which is proved on page 267 (in Section 4.6). We will refer to this theorem by its more common name, the **change of base theorem**.

CHANGE OF BASE THEOREM

$$\log_a x = \frac{\log_b x}{\log_b a}$$

---

*The number of digits shown may vary. Calculator answers are more accurate than were the old table answers, but it is important to realize that any answer (whether from a table or a calculator) is only as accurate as the input numbers. However, in this book, we will not be concerned with significant digits but instead will use all the accuracy our calculator gives us, rounding only once (if requested) at the end of the problem.

EXAMPLE 6   Evaluating logarithms with arbitrary bases

Evaluate (rounded to the nearest hundredth):   **a.** $\log_7 3$ **b.** $\log_3 3.84$

Solution

a. $\log_7 3 \quad \dfrac{\log 3}{\log 7} \approx \dfrac{0.4771212547}{0.84509804} \approx 0.5645750341 \approx 0.56$

This is all done by calculator and not on your paper.

b. $\log_3 3.84 = \dfrac{\log 3.84}{\log 3} \approx \dfrac{0.5843312244}{0.4771212547} \approx 1.224701726 \approx 1.22$

Calculator work

Logarithmic Functions

The notion of a logarithm can also be considered as a special type of function.

> **LOGARITHMIC FUNCTION**
>
> The function f defined on $(0,\infty)$ by
>
> $$f(x) = \log_b x$$
>
> where $b > 0$, $b \neq 1$, is called the **logarithmic function** with base b.

We begin our study of the logarithmic function, as we did the exponential function, by looking at some general properties of graphs of logarithmic functions.

EXAMPLE 7   Graph of a logarithmic function for $b > 1$

Sketch: $f(x) = \log_2 x$.

Solution   This function is an increasing function with a vertical asymptote $x = 0$. The equation $y = \log_2 x$ is equivalent (by definition) to the equation $x = 2^y$. However, since x is the independent variable, we construct a table of values using the change of base theorem. That is, $f(0.1) = \log_2 0.1 = \dfrac{\log 0.1}{\log 2} \approx -3.321928095$. A table of values and graph are shown in Figure 4.24. We show the values rounded to the nearest hundredth.

x	y
0.1	-3.32
0.5	-1
0.75	-0.42
1	0
2	1
3	1.58
4	2
8	3

Figure 4.24 Graph of $f(x) = \log_2 x$

EXAMPLE 8   Graph of a logarithmic function for $0 < b < 1$

Sketch: $f(x) = \log_{0.5} x$.

Solution   This function is a decreasing function with a vertical asymptote $x = 0$. The equation $y = \log_{0.5} x$ is equivalent to the equation $x = (0.5)^y$. A table of values follows; the values are, once again, rounded to the nearest hundredth. A graph is shown in Figure 4.25.

x	y
0.1	3.32
0.5	1
0.75	0.42
1	0
2	-1
3	-1.58
4	-2
8	-3

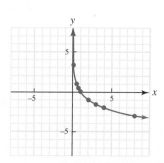

Figure 4.25 Graph of $f(x) = \log_{0.5} x$

Examples 7 and 8 are summarized in Table 4.4. We include these curves in our Library of Curves, Appendix F.

TABLE 4.4	Directory of Curves (Part III)

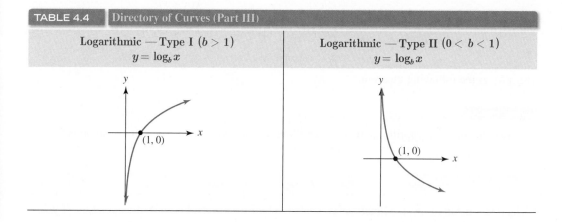

Logarithmic — Type I $(b > 1)$ $y = \log_b x$	Logarithmic — Type II $(0 < b < 1)$ $y = \log_b x$

Exponential and Logarithmic Functions as Inverse Functions

By looking at the graphs of this section, you may notice a relationship between the graphs of the exponential and logarithmic functions. For example, consider the graphs of $y = 2^x$, $y = \log_2 x$, along with the line $y = x$, as shown in Figure 4.26.

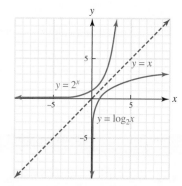

Figure 4.26 Graphs of $y = 2^x$, $y = \log_2 x$, and $y = x$

The graphs of $y = 2^x$ and $y = \log_2 x$ appear to be inverse functions. To prove that they are indeed inverses, we need to consider two properties of logarithms that follow from the definition of a logarithm.

PROPERTIES OF LOGS

1. $\log_b b^x = x$

2. $b^{\log_b x} = x$　　　$x > 0$

To justify property 1, remember the definition of logarithm:

$$b^M = N \text{ means } \log_b N = M$$

and let $M = x$ and $N = b^x$. Then since $b^x = b^x$, we have $\log_b b^x = x$.

To justify property 2, let $M = \log_b x$ and $N = x$ in the definition. Since $\log_b x = \log_b x$, then using the definition, we have $b^{\log_b x} = x$.

To show that $f(x) = \log_b$ and $g(x) = b^x$ are inverse functions, we need to show

$$(f \circ g)(x) = (g \circ f)(x) = x$$

for all x in the intersection of the domains of f and g:

$$
\begin{array}{ll}
(f \circ g)(x) = f[g(x)] & (g \circ f)(x) = g[f(x)] \\
\quad\quad = f(b^x) & \quad\quad = g(\log_b x) \\
\quad\quad = \log_b b^x & \quad\quad = b^{\log_b x} \\
\quad\quad = x \quad \text{by property 1} & \quad\quad = x \quad \text{by property 2}
\end{array}
$$

This proves the following theorem.

INVERSE THEOREM

The exponential and logarithmic functions with base b are inverse functions of one another.

COMPUTATIONAL WINDOW　　　_ □ ✕

The relationship of inverse functions is needed to find e on several brands of calculators. If your calculator has an e^x key, you do NOT need to read this. However, if your calculator has no key marked e^x but does have keys marked $\ln x$ and $\boxed{\text{INV}}$, how can you find e or e^x? Since $y = \ln x$ and $y = e^x$ are inverse functions, to find e (or e^1), press

This gives e^x.

The display is: 2.718281828. To find $e^{5.2}$ on such a calculator, press $\boxed{5.2}$ $\boxed{\text{INV}}$ $\boxed{\ln x}$ to obtain the display: 181.2722419.

PROBLEM SET 4.5

LEVEL 1

Write the equations in Problems 1–6 in logarithmic form.

1. a. $64 = 2^6$　　　　**b.** $100 = 10^2$

2. a. $1{,}000 = 10^3$　　**b.** $81 = 9^2$

3. a. $125 = 5^3$　　　　**b.** $a = b^c$

4. a. $m = n^p$　　　　　**b.** $\frac{1}{e} = e^{-1}$

5. a. $9 = \left(\frac{1}{3}\right)^{-2}$　　**b.** $8 = \left(\frac{1}{2}\right)^{-3}$

6. a. $\frac{1}{3} = 9^{-1/2}$　　　**b.** $\frac{1}{2} = 4^{-1/2}$

Write the equations in Problems 7–12 in exponential form.

7. a. $\log 10{,}000 = 4$　　**b.** $\log 0.01 = -2$

8. a. $\log 1 = 0$　　　　　**b.** $\log x = 2$

9. a. $\log_e e^2$　　　　　**b.** $\ln e^3 = 3$

10. a. $\ln x = 5$ **b.** $\ln x = 0.03$
11. a. $\log_2 \frac{1}{8} = -3$ **b.** $\log_2 32 = 5$
12. a. $\log_m n = p$ **b.** $\log_a b = c$

Evaluate the expressions in Problems 13–16 by using the definition of logarithm.

13. a. $\log_b b^2$ **b.** $\log_t t^3$
14. a. $\log_e e^4$ **b.** $\log_\pi \sqrt{\pi}$
15. a. $\log_\pi \frac{1}{\pi}$ **b.** $\log_2 8$
16. a. $\log_3 9$ **b.** $\log_{19} 1$

Evaluate the expressions in Problems 17–24 by using a calculator. Round your answers to four decimal places.

17. a. $\log 4.27$ **b.** $\log 1.08$
18. a. $\log 8.43$ **b.** $\log 9,760$
19. a. $\log 71,600$ **b.** $\log 0.042$
20. a. $\log 0.321$ **b.** $\log 0.0532$
21. a. $\ln 2.27$ **b.** $\ln 16.77$
22. a. $\ln 2$ **b.** $\ln 0.125$
23. a. $\ln 13$ **b.** $\ln 0.15$
24. a. $\ln 7.3$ **b.** $\ln 10.57$

In Problems 25–28, write each of the expressions in terms of common logarithms and then give a calculator approximation (four decimal places).

25. a. $\log_3 45$ **b.** $\log_5 91$
26. a. $\log_5 304$ **b.** $\log_2 1,513$
27. a. $\log_6 0.1$ **b.** $\log_4 3.05$
28. a. $\ln 10$ **b.** $\ln 1,000$

In Problems 29–32, write each of the expressions in terms of natural logarithms and then give a calculator approximation (four decimal places).

29. a. $\log_2 0.0056$ **b.** $\log_{8.3} 105$
30. a. $\log_\pi 100$ **b.** $\log_{\sqrt{2}} 8.5$
31. a. $\log_{1.08} 5,450$ **b.** $\log_{1.12} 12,450$
32. a. $\log e^2$ **b.** $\log e^8$

Sketch the graph of each equation given in Problems 33–44.

33. $y = \log x$ **34.** $y = \ln x$
35. $y - 3 = \ln x$ **36.** $y = \log x + 4$
37. $y = \log|x|$ **38.** $y = |\log x|$
39. $y = \log_3 x$ **40.** $y = \log_{1/3} x$
41. $y = \log_e x$ **42.** $y = \log_{\sqrt{2}} x$
43. $y = \log 10^x$ **44.** $y = \log_6 6^x$

LEVEL 2

WHAT IS WRONG, *if anything, with each statement in Problems 45–52? Explain your reasoning.*

45. $\log_b N$ is negative when N is negative.
46. $\log N$ is negative when $N > 1$.
47. In $\log_b N$, the exponent is N.
48. $\log 500$ is the exponent on 10 that gives 500.
49. A common logarithm is a logarithm in which the base is 2.
50. A natural logarithm is a logarithm in which the base is 10.

51. $\dfrac{\log A}{\log B} = \dfrac{\ln A}{\ln B}$
52. To evaluate $\log_5 N$, divide $\log 5$ by $\log N$.
53. An advertising agency conducted a survey and found that the number of units sold, N, is related to the amount a spent on advertising (in dollars) by the following formula:

$$N = 1,500 + 300 \ln a \qquad (a \geq 1)$$

  a. How many units are sold after spending $1,000?
  b. How many units are sold after spending $50,000?
54. The pH of a substance measures its acidity or alkalinity. It is found using the formula

$$\text{pH} = -\log[\text{H}^+]$$

where $[\text{H}^+]$ is the concentration (given in moles per liter) of hydrogen ions in an aqueous solution.
  a. What is the pH (to the nearest tenth) of a lemon for which $[\text{H}^+] = 2.86 \times 10^{-4}$?
  b. What is the pH (to the nearest tenth) of rainwater for which $[\text{H}^+] = 6.31 \times 10^{-7}$?
55. The Richter scale for measuring earthquakes was developed by Gutenberg and Richter. It relates the energy E (in ergs) to the magnitude of the earthquake M by the formula

$$M = \frac{\log E - 11.8}{1.5}$$

  a. A small earthquake is one that releases 15^{15} ergs of energy. What is the magnitude (to the nearest hundredth) of such an earthquake on the Richter scale?
  b. A large earthquake is one that releases 10^{25} ergs of energy. What is the magnitude (to the nearest hundredth) of such an earthquake on the Richter scale?
56. The intensity of sound is measured in decibels D and is given by

$$D = \log\left(\frac{I}{I_0}\right)^{10}$$

where I is the power of the sound in watts per cubic centimeter (W/cm^3) and $I_0 = 10^{-16}\ \text{W}/\text{cm}^3$ is the power of sound just below the threshold of hearing. Find the intensity (to the nearest decibel) of:
  a. A whisper, $10^{-13}\ \text{W}/\text{cm}^3$
  b. Normal conversation,

$$3.16 \times 10^{-10}\ \text{W}/\text{cm}^3$$

57. A learning curve describes the rate at which a person learns certain tasks. If a person sets a goal of typing N words per minute (wpm), the length of time t, in days, to achieve this goal is given by the formula

$$t = -62.5\ln\left(1 - \frac{N}{80}\right)$$

  a. How long would it take to learn to type 30 wpm?
  b. If we accept this formula, is it possible for anyone to learn to type 80 wpm?
58. In Problem 57, a formula for learning is given. Psychologists are also concerned with forgetting. In a certain experiment, students were asked to remember a set of nonsense syllables, such as "myt." The students then had to recall the syllables after t seconds. The equation that was found to model the forgetting experiment was

$$R = 80 - 27\ln t \qquad (t \geq 1)$$

where R is the percentage of students who remember the syllables after t seconds.

a. What percentage of the students remembered the syllables after 3 seconds?

b. What percentage of the students remembered the syllables after 15 seconds?

LEVEL 3

The time, t, in years, that it will take to pay off a loan of P dollars with monthly payments of m dollars at a rate r is given by the formula

$$t = \frac{\ln m - \ln(m - Pi)}{12\ln(1 + i)}$$

where $i = \frac{r}{12}$. *In particular, if we consider a typical home mortgage, say $150,000 at 8%, this formula becomes*

$$t \approx 12.5416\ \ln m - 12.5416\ \ln(m - 1{,}000)$$

Most home loans are for 30 years, but this may not be the best choice for the consumer. Use this model in Problems 59–60.

59. Draw a graph of the logarithmic equation for a typical home mortgage. Using the graph as a basis, comment on crossing different monthly payments and lengths for repayment of the loan.

60. What is the approximate length of a home mortgage of $150,000 at 8% with a monthly payment of $1,800?

4.6 Logarithmic Equations

As seen in the previous section, the most important use of logarithms today is modeling natural phenomena: Compound interest, growth, decay, learning, forgetting, Richter scale for earthquakes, pH (alkalinity and acidity), intensity of sound (decibels), and Newton's law of cooling can all be modeled using logarithms. Until now we have been limited to graphing exponential and logarithmic functions, along with some of the simplest evaluations. In this section, we shall discuss solving logarithmic equations, and in the next section, solving exponential equations.

Types of Logarithmic Equations

The key to solving logarithmic equations is the following theorem, which we will call the **log of both sides theorem**.

LOG OF BOTH SIDES THEOREM

If A, B, and b are positive real numbers with $b \neq 1$, then $\log_b A = \log_b B$ is equivalent to $A = B$.

The proof of this theorem is not difficult, and it depends on the two fundamental properties of logarithms given in the previous section.

Basically, all logarithmic equations in this book fall into one of four types:

Log Type I: The unknown is the logarithm; $\log_2 \sqrt{3} = x$.

Log Type II: The unknown is the base; $\log_x(5x + 6) = 2$.

Log Type III: The logarithm of an unknown is equal to a number; $\ln x = 5$.

Log Type IV: The logarithm of an unknown is equal to the logarithm of a number; $\log_5 x = \log_5 72$.

The following example illustrates the procedure for solving each of these types of logarithmic equations.

EXAMPLE 1   **Procedure for solving logarithmic equations**

Solve the following logarithmic equations:

a. $\log_2 \sqrt{3} = x$ **b.** $\log_x(5 + 6) = 2$ **c.** $\ln x = 5$ **d.** $\log_5 = \log_5 72$

Solution

a. Log Type I: $\log_2 \sqrt{3} = x$. If the logarithmic expression does not contain a variable, you can use your calculator to evaluate. Remember, this type was evaluated in Section 4.5. If it is a common logarithm (base 10), use the [LOG] key; if it is a natural logarithm (base e), use the [LN] key; if it has another base, use the change of base theorem:

$$\log_a N = \frac{\log N}{\log a} \qquad \text{or} \qquad \log_a N = \frac{\ln N}{\ln a}$$

For this example, we see

$$x = \log_2 \sqrt{3}$$

$$= \frac{\log \sqrt{3}}{\log 2}$$

$$\approx 0.7924812504$$

b. Log Type II: $\log_x(5x + 6) = 2$. If the unknown is the base, then use the definition of logarithm to write an equation that is not a logarithmic equation.

$$\log_x(5x + 6) = 2$$

$$x^2 = 5x + 6 \qquad \text{Definition of logarithm.}$$

$$x^2 - 5x - 6 = 0$$

$$(x - 6)(x + 1) = 0$$

$$x = 6, -1$$

When solving logarithmic equations, you must make sure your answers are in the domain of the variable. Remember that, by definition, the argument must be positive. For this example, $x = -1$ is not in the domain so the solution is $x = 6$.

c. Log Type III: The third and fourth types of logarithmic equations are the most common, and both involve the logarithm of an unknown quantity on one side of an equation. For the third type, use the definition of logarithm (and a calculator for an approximate solution, if necessary).

$$\ln x = 5$$

$$e^5 = x \qquad \text{Definition of logarithm}$$

This is the exact solution. An approximate solution is $x \approx 148.4131591$.

d. Log Type IV: When a log form occurs on both sides, use the log of both sides theorem.

$$\log_5 x = \log_5 72$$

$$x = 72$$

■

Historical Note

For years, the concept of the logarithm was used as a computational aid, but today it is more important in solving problems. It is an interesting historical note to observe that the first use of logarithms by the Babylonians was to solve problems and not to do calculations. The modern basis of logarithms was developed by Tycho Brahe (1546–1601) to disprove the Copernican theory of planetary motion. The name he used for the method was prostaphaeresis. In 1590, a storm brought together Brahe and John Craig, who, in turn, told Napier about Brahe's method. It was John Napier (1550–1617) who was the first to use the word *logarithm*. Napier was the Isaac Asimov of his day, having envisioned the tank, the machine gun, and the submarine. He also predicted that the end of the world would occur between 1688 and 1700. He is best known today as the inventor of logarithms, which, until the advent of the calculator, were used extensively with complicated calculations. Today, we use logarithms to solve both logarithmic and exponential equations.

☠ Make sure the log on both sides has the same base. ☠

Laws of Logarithms

Example 1 illustrates the procedures for solving logarithmic equations, but most logarithmic equations are not as easy as those illustrated in Example 1. You must usually do some algebraic simplification to put the problem into the form of one of the four types of logarithmic equations. You might also have realized that Log Type IV is a special case of Log Type III. For example, to solve

$$\log_5 x = \log_3 72$$

which looks like Example **1d**, we see that we cannot use the log of both sides theorem because the bases are not the same. We can, however, treat this as a Log Type III equation by using the definition of logarithm to write

$$x = 5^{\log_3 72}$$

This can be evaluated using a calculator. However, you might find it easier to visualize if we write

$$\log_5 x = \log_3 72 \approx 3.892789261$$

so that

$$x \approx 5^{3.892789261} \approx 525.9481435$$

To simplify logarithmic expressions, we remember that a logarithm is an exponent, so the laws of exponents lead us to corresponding laws of logarithms.

LAWS OF LOGARITHMS

Let A, B, and b be positive numbers and let p be any real number, $b \neq 1$.

First law (Additive):

$$\log_b(AB) = \log_b A + \log_b B$$

The log of a product is the sum of the logs of those numbers.

Second law (Subtractive):

$$\log_b\left(\frac{A}{B}\right) = \log_b A - \log_b B$$

The log of a quotient is the difference of the logs of those numbers.

Third law (Multiplicative):

$$\log_b(A)^p = p \log_b A$$

The log of a power is the product of the exponent times the log of the base.

The proofs of these laws of logarithms are easy. The first law of logarithms comes from the first law of exponents:

$$b^x b^y = b^{x+y}$$

Let $A = b^x$ and $B = b^y$, so that $AB = b^{x+y}$. Then from the definition of logarithm, these three equations are equivalent to

$$x = \log_b A, \qquad y = \log_b B, \qquad x + y = \log_b(AB)$$

Therefore, by putting these pieces together, we have

$$\log_b(AB) = x + y = \log_b A + \log_b B$$

Similarly, for the second law of logarithms,

$$\frac{A}{B} = \frac{b^x}{b^y}$$

$$= b^{x-y} \qquad \text{Second law of exponents}$$

Thus,

$$x - y = \log_b\left(\frac{A}{B}\right) \qquad \text{Definition of logarithm}$$

which means

$$\log_b\left(\frac{A}{B}\right) = \log_b A - \log_b B \qquad \text{Since } x = \log_b A \text{ and } y = \log_b B$$

The proof of the third law of logarithms follows from the third law of exponents, and you are asked to do this in the problem set. We can also prove this third law by using the second law of logarithms for p a positive integer:

$$\log_b A^p = \log_b \underbrace{A \cdot A \cdot A \cdots\cdots A}_{p \text{ factors}} \qquad \text{Definition of } A^p$$

$$= \underbrace{\log_b A + \log_b A + \log_b A + \cdots + \log_b A}_{p \text{ terms}} \qquad \text{Second law of logarithms}$$

$$= p \log_b A$$

When logarithms were used for calculations, the laws of logarithms were used to expand an expression such as $\log\left(\dfrac{6 \cdot 45.62^2}{84.2}\right)$. Calculators have made such problems obsolete. Today, logarithms are important in solving equations, and the procedure for solving logarithmic equations requires that we take an algebraic expression involving logarithms and write it as a single logarithm. We might call this *contracting* a logarithmic expression.

EXAMPLE 2   Using the laws of logarithms to contract

Write each of the given statements as a single logarithm.

   a. $\log x + 5 \log y - \log z$ **b.** $\log_2 3x - 2 \log_2 x + \log_2(x + 3)$

Be sure the bases are the same before using the laws of logarithms.

Solution

   a.
$$\log x + 5\log y - \log z = \log x + \log y^5 - \log z \quad\text{Third law}$$
$$= \log xy^5 - \log z \quad\text{First law}$$
$$= \log \frac{xy^5}{z} \quad\text{Second law}$$

   b.
$$\log_2 3x - 2\log_2 x + \log_2(x + 3) = \log_2 3x - \log_2 x^2 + \log_2(x + 3)$$
$$= \log_2 \frac{3x(x + 3)}{x^2}$$
$$= \log_2 \frac{3(x + 3)}{x}$$

■

EXAMPLE 3   Using the laws of logarithms to solve a logarithmic equation

Solve: $\log_8 3 + \frac{1}{2}\log_8 25 = \log_8 x$.

Solution

The goal here is to make this look like a Log Type IV logarithmic equation so that there is a single log expression on both sides.

$$\log_8 3 + \tfrac{1}{2}\log_8 25 = \log_8 x \quad\text{Given equation; note the bases are the same.}$$
$$\log_8 3 + \log_8 25^{1/2} = \log_8 x \quad\text{Third law of logarithms}$$
$$\log_8 3 + \log_8 5 = \log_8 x \quad 25^{1/2} = (5^2)^{1/2} = 5$$
$$\log_8(3 \cdot 5) = \log_8 x \quad\text{First law of logarithms}$$
$$15 = x \quad\text{Log of both sides theorem}$$

The solution is 15. (We check to be sure 15 is in the domain of the variable.)

■

When solving logarithmic equations, you must be mindful of extraneous solutions. The reason for this is that the logarithm function requires that the arguments be positive, but when solving an equation, we may not know the signs of the arguments. For example, if you solve an equation involving $\log x$ and $x = 3, -4$ is the solution, the value $x = -4$ must be extraneous because $\log(-4)$ is not defined.

General Logarithmic Equations

EXAMPLE 4   Logarithmic equations

Solve: **a.** $\log 15 + 2 = \log(x + 250)$ **b.** $\log 5 + \log(2x^2) = \log x + \log 15$

Solution

Use the laws of logarithms to combine the log statements. We have chosen two examples that are quite similar, but whose solutions require slightly different procedures. For part **a**, rewrite the given expression so that all parts involving logarithms are on one side. In part **b**, rewrite the given expression so that all logarithms involving the variable are on one side.

a. $\log 15 + 2 = \log(x + 250)$ *Given equation*

$2 = \log(x + 250) - \log 15$ *Subtract log 15 from both sides.*

$2 = \log \dfrac{x + 250}{15}$ *Second law of logarithms*

$10^2 = \dfrac{x + 250}{15}$ ☠ Log Type III: Use definition of logarithm. ☠

$1{,}500 = x + 250$ *Multiply both sides by 15.*

$1{,}250 = x$ *Subtract 250 from both sides.*

The solution is $x = 1{,}250$.

b. $\log 5 + \log(2x^2) = \log x + \log 15$ *Given equation.*

$\log(2x^2) - \log x = \log 15 - \log 5$ *Subtract log x and log 5 from both sides.*

$\log \dfrac{2x^2}{x} = \log \dfrac{15}{5}$ *Second law of logarithms*

$\log(2x) = \log 3$ *Simplify.*

$2x = 3$ ☠ Log Type IV: Use log of both sides theorem. ☠

$x = \dfrac{3}{2}$ *Divide both sides by 2.*

The solution is $x = \frac{3}{2}$.

EXAMPLE 5   Solving a logarithmic equation

Solve: $\ln x - \frac{1}{2}\ln 2 = \frac{1}{2}\ln(x + 4)$.

Solution

$\ln x - \frac{1}{2}\ln 2 = \frac{1}{2}\ln(x + 4)$ *Given equation*

$2\ln x - \ln 2 = \ln(x + 4)$ *Multiply both sides by 2.*

$2\ln x - \ln(x + 4) = \ln 2$ *Get the logarithms with variables on one side.*

$\ln x^2 - \ln(x + 4) = \ln 2$ *Third law of logarithms*

$\ln \dfrac{x^2}{x + 4} = \ln 2$ *Second law of logarithms*

$\dfrac{x^2}{x + 4} = 2$ *Log of both sides theorem*

$x^2 - 2x - 8 = 0$ *Multiply both sides by (x + 4) and get a 0 on one side.*

$(x - 4)(x + 2) = 0$ *Factor to solve the quadratic equation.*

$x = 4, -2$ *Zero factor theorem*

Since the domain requires that the argument is positive, we see that -2 is extraneous. The solution is $x = 4$.

EXAMPLE 6   Proof of the change of base theorem

Prove: $\log_a x = \dfrac{\log_b x}{\log_b a}$.

Proof:

Let $y = \log_a x$.

$$a^y = x \qquad \text{Definition of logarithm}$$

$$\log_b a^y = \log_b x \qquad \text{Log of both sides theorem}$$

$$y \log_b a = \log_b x \qquad \text{Third law of logarithms}$$

$$y = \frac{\log_b x}{\log_b a} \qquad \text{Divide both sides by } \log_b a \; (\log_b a \neq 0).$$

$$\log_a x = \frac{\log_b x}{\log_b a} \qquad \text{Substitution}$$

PROBLEM SET **4.6**

LEVEL 1

Solve the equations in Problems 1–6. Do not approximate these answers.

1. a. $\log_5 25 = x$ **b.** $\log_2 128 = x$
   c. $\log_3 81 = x$ **d.** $\log_4 64 = x$

2. a. $\log \frac{1}{10} = x$ **b.** $\log 10{,}000 = x$
   c. $\log 1{,}000 = x$ **d.** $\log \frac{1}{1{,}000} = x$

3. a. $\log_x 81 = 4$ **b.** $\log_x 84 = 2$
   c. $\log_x 28 = 2$ **d.** $\log_x 50 = 2$

4. a. $\ln x = 4$ **b.** $\ln x = \ln 14$
   c. $\ln 9.3 = \ln x$ **d.** $\ln 109 = \ln x$

5. a. $\log_3 x^2 = \log_3 125$ **b.** $\ln x^2 = \ln 12$
   c. $\log x^2 = \log 70$ **d.** $\log_2 8\sqrt{2} = x$

6. a. $\log_3 27\sqrt{3} = x$ **b.** $\log_x 1 = 0$
   c. $\log_x 10 = 0$ **d.** $\log_2 x = 5$

WHAT IS WRONG, *if anything, with each statement in Problems 7–16? Explain your reasoning.*

7. If $2 \log x = 8$, then $\log x - 2 = 4$.

8. $\ln \dfrac{x}{2} = \dfrac{\ln x}{2}$

9. $\log_b (A + B) = \log_b A + \log_b B$

10. $\log_b (AB) = (\log_b A)(\log_b B)$

11. $\dfrac{\log_b A}{\log_b B} = \log_b \dfrac{A}{B}$

12. $\dfrac{\log_b A}{\log_b B} = \log_b (A - B)$

13. $\dfrac{\log_b A}{\log_b B} = \log_b A - \log_b B$

14. If $2 \log_3 81 = 8$, then $\log_3 6{,}561 = 8$.

15. If $2 \log_3 81 = 8$, then $\log_3 81 = 4$.

16. If $\log_{1.5} 8 = x$, then $x^{1.5} = 8$.

In Problems 17–24, use the properties of logarithms to evaluate each of the expressions without using a calculator.

17. $\log 25 + \log 4$ **18.** $\log 60 - \log 6$

19. $\ln e + \ln e^2$ **20.** $5 \ln e$

21. $\log_6 72 - \log_6 12$ **22.** $\log_2 \sqrt[3]{32}$

23. $5^{\log_5 93}$ **24.** $\pi^{\log_\pi (1/2)}$

25. $\ln 2 + \frac{1}{2} \ln 9 = \ln x$ has solution $x = 6$.
   a. Check this solution by evaluating
$$\ln 2 + \tfrac{1}{2} \ln 9 \qquad \text{and} \qquad \ln 6$$
to show the left- and right-hand values are equal.
   b. Use the properties of logarithms to show $\ln 2 + \ln \sqrt{9}$ and $\ln 6$ have the same values.
   c. Multiply both sides by 2 to write
$$\ln 2 + \tfrac{1}{2} \ln 9 = \ln x$$
$$2 \ln 2 + \ln 9 = 2 \ln x$$
Show $\ln 4 + \ln 9$ and $\ln 6^2$ have the same values.

26. $\ln e = \ln \frac{\sqrt{2}}{x} - \ln e$ has solution $x = e^{-2} \sqrt{2}$.
   a. Check this solution by evaluating
$$\ln e \qquad \text{and} \qquad \ln \frac{\sqrt{2}}{e^{-2}\sqrt{2}} - \ln e$$
to show the left- and right-hand values are equal.
   b. Add $\ln e$ to both sides to obtain
$$2 \ln e = \ln \frac{\sqrt{2}}{x}$$
$$e^2 = \frac{\sqrt{2}}{x}$$
Show $x = e^{-2} \sqrt{2}$.

c. Note that $\ln e = 1$, so

$$1 = \ln\frac{\sqrt{2}}{x} - 1$$

$$2 = \ln\frac{\sqrt{2}}{x}$$

Use the definition of logarithm to show $x = e^{-2}\sqrt{2}$.

LEVEL 2

Contract the expressions in Problems 27–30. That is, use the properties of logarithms to write each expression as a single logarithm with a coefficient of 1.

27. a. $\log 2 + \log 3 + \log 4$ **b.** $\log 40 - \log 10 - \log 2$
   c. $2\ln x + 3\ln y - 4\ln z$ **d.** $5\ln x - 2\ln y + 3\ln z$
28. a. $3\ln 4 - 5\ln 2 + \ln 3$ **b.** $3\ln 4 - 5\ln(2+3)$
   c. $3\ln 4 - 5(\ln 2 + 3)$ **d.** $\ln 3 - 2\ln(4+8)$
29. a. $\log(x^2 - 9) - 2\log(x+3) + 3\log x$
   b. $\log(x^2 - 9) - 2[\log(x+3) + 3\log x]$
   c. $\ln(x-1) + \ln(x+2) - \ln(x-1)$
   d. $\frac{1}{3}[\ln(x-1) + \ln(x+2) - \ln(x-1)]$
30. a. $\frac{1}{2}\log(x-1) + \log(x+1) - \frac{3}{2}\log(x-1)$
   b. $\frac{1}{2}\log(3x+1) - [\log(x+1) - \frac{3}{2}\log(3x+1)]$
   c. $\ln(x^2 - 4) - \ln(x+2)$
   d. $\log(x^3 - 8) - \log(x^2 + 2x + 4)$

Solve the equations in Problems 31–55. Give exact answers only.

31. $\log 2 = \frac{1}{4}\log 16 - x$ **32.** $\frac{1}{2}\log x - \log 100 = 2$
33. $2\log x + 7 = 207$ **34.** $5\ln x - 6 = 104$
35. $\frac{1}{2}\log_b x = 3\log_b 5 - \log_b x$ **36.** $\log_8 5 + \frac{1}{2}\log_8 9 = \log_8 x$
37. $\ln x^2 = 1$ **38.** $2\ln x^2 = 4$
39. $\log x + \log(x-3) = 1$ **40.** $\log(x-21) - 2 = -\log x$
41. $\log 2 - \frac{1}{2}\log 9 + 3\log 3 = x$ **42.** $3\log 3 - \frac{1}{2}\log 3 = \log\sqrt{x}$
43. $\log 2 = \frac{1}{4}\log 16 - \log x$ **44.** $2\log x + 3\log 10 = 5$
45. $\log_b x - \frac{1}{2}\log_b 2 = \frac{1}{2}\log_b(3x-4)$
46. $\log_7 x - \frac{1}{2}\log_7 4 = \frac{1}{2}\log_7(2x-3)$
47. $\log 10 = \log\sqrt{1000} - \log x$

48. $\log 10^x - 2 = \log 100$ **49.** $\ln e^3 - \ln x = 1$
50. $\ln 1 + \ln e^x = 2$ **51.** $3\ln\frac{e}{\sqrt[3]{5}} = 3 - \ln x$
52. $5\ln\frac{e}{\sqrt[5]{2}} = 1 - \ln x$ **53.** $\log_x(x+6) = 2$
54. $\log(\log x) = 1$ **55.** $\ln[\log(\ln x)] = 0$
56. The "learning curve" describes the rate at which a person learns certain tasks. If a person sets a goal of typing N words per minute (wpm), the length of time t (in days) to achieve this goal is given by

$$t = -62.5\ln\left(1 - \frac{N}{80}\right)$$

Solve for N.
57. The "forgetting curve" for memorizing nonsense syllables is given by
$$R = 80 - 27\ln t \quad (t \geq 1)$$
where R is the percentage who remember the syllables after t sec. Solve for t.
58. The equation for the Richter scale relating energy E (in ergs) to the magnitude of the earthquake, M, is given by the formula

$$M = \frac{\log E - 11.8}{1.5}$$

Solve for E.

LEVEL 3

59. Prove the third law of logarithms using the third law of exponents. That is, prove $\log_b A^p = p\log_b A$.
60. Journal Problem *The College Mathematics Journal* (FFF #49; Vol. 23, 1992, p. 36). WHAT IS WRONG, if anything, with the solution to solving the given equation? Explain your reasoning.
$$2\log(x - 2) = \log(x - 3) + \log x$$

Solution:

$$\begin{aligned}
2\log(x-2) &= \log(x-3) + \log x \\
\log(2x-4) &= \log[x(x-3)] \\
\log(2x-4) &= \log(x^2 - 3x) \\
2x - 4 &= x^2 - 3x \\
x^2 - 5x + 4 &= 0 \\
(x-4)(x-1) &= 0 \\
x &= 4, 1
\end{aligned}$$

$x = 1$ is not a solution. Hence, $x = 4$.

4.7 Exponential Equations

In this section, we return to the question that prompted our discussion of logarithms in the first place: "How do you solve an exponential equation?" The most straightforward method for solving exponential equations applies the exponential property of equality stated in Section 4.5. Recall that if equal bases are raised to some power and the results are equal, then the exponents must also be equal. For example, solve $8 = 0.25^x$:

$$8 = 0.25^x \qquad \text{Given.}$$

$$2^3 = \left(\frac{1}{2^2}\right)^x \qquad \text{Write factors in exponent form.}$$

$$2^3 = (2^{-2})^x$$

$$2^3 = 2^{-2x}$$

$$3 = -2x \qquad \text{If the bases are the same, then the exponents are equal.}$$

$$-\frac{3}{2} = x$$

The unknown in this problem is t.

$$A = P(1+i)^N$$

$$2,000 = 1,250(1+0.04)^{2t}$$

$$1.6 = (1.04)^{2t}$$

$$2t = \log_{1.04} 1.6$$

$$t = \frac{1}{2}\log_{1.04} 1.6 \approx 5.991778214$$

We see that it is almost 6 years, but we need the answer to the nearest day. The time is 5 years + 0.991778214 year. Multiply 0.991778214 by 365 to find 361.9990482. This means that on the 361st day, we are still *a* bit short of $2,000, so the time necessary is 5 years 362 days.

◼ COMPUTATIONAL WINDOW ▭◻☒

The "work" in this example *seems* complicated, but it is not. If you have a calculator, you can do the entire thing in one step. Because there are many different brands of calculator, we do not usually show keystrokes, so we offer the following as one possible example:

| LOG | 1.6 | ÷ | LOG | 1.04 | ÷ | 2 | − | 5 | × | 365 | Display: 361.9990482

---

b. For continuous compounding, use the formula $A = Pe^{rt}$.

$$A = Pe^{rt}$$

$$2,000 = 1,250e^{0.08t}$$

$$1.6 = e^{0.08t}$$

$$0.08t = \ln 1.6$$

$$t = \frac{\ln 1.6}{0.08} \approx 5.87504566$$

To find the number of days, we once again subtract 5 and multiply by 365 to find that the time necessary is 5 years 320 days. (Note: 319.39 means that on the 319th day, you do not quite have the $2,000.) ◼

Growth and Decay

Example **3b** illustrates exponential growth. Growth and decay problems are common examples of exponential equations.

GROWTH/DECAY FORMULA

Exponential **growth** or **decay** can be described by the equation

$$A = A_0 e^{rt}$$

where r is the annual growth/decay rate, t is the time (in years), A_0 is the amount present initially (present value), and A is the future value. If r is positive, this formula models growth, and if r is negative, the formula models decay.

Note: You can use this formula as long as the units of time are the same. That is, if the time is measured in days, then the growth/decay rate is a daily growth/decay rate, as illustrated by the next example.

EXAMPLE 4   Radioactive decay

If 100.0 mg of neptunium-239 (^{239}Np) decays to 73.36 mg after 24 hours, find the value of r in the growth/decay formula for t expressed as days.

Solution
Since $A = 73.36$, $A_0 = 100.0$, and $t = 1$ (day), we have (from the growth/decay formula):

$$A = A_0 e^{rt}$$

$$73.36 = 100 e^{r(1)}$$

$$0.7336 = e^r$$

$$r = \ln 0.7336 \approx -0.309791358$$

EXAMPLE 5   Half-life of strontium-90

It has been determined that the half-life of strontium-90 is 28 years. How much (to the nearest gram) of a 50-gram sample will remain after **a.** 1 year? **b.** 10 years? **c.** 100 years?

Solution
We will use the growth/decay formula with $A/A_0 = 1/2$ for half-life when $t = 28$ to find the decay constant r:

$$\frac{1}{2} = e^{r(28)}$$

$$r = \frac{1}{28} \ln 0.5 \approx -0.0247552564$$

We use this approximation for r and the decay formula $A = 50 e^{-0.0247552564t}$ to answer the questions.

a. $t = 1$: $50 e^{-(0.0247552564)(1)} \approx 48.7774321$ *There are 49 grams remaining.*

b. $t = 10$: $50 e^{-(0.0247552564)(10)} \approx 39.03545913$ *There are 39 grams remaining.*

c. $t = 100$: $50 e^{-(0.0247552564)(100)} \approx 4.205938102$ *There are 4 grams remaining.*

EXAMPLE 6   Deaths from AIDS   `MODELING APPLICATION`

According to the AIDs Education Global Information System, in January 1992, the total number of AIDS-related U.S. deaths for all ages was 209,693. In January 2004 this number had risen to 525,060. Estimate the cumulative number of deaths from AIDS by January 2010.

Solution

Step 1: *Understand the problem.* We assume that the growth rate will remain constant over the years of our study, and we also assume that the growth takes place continuously. We wish to use the available information to predict the total impact of AIDS through 2009.

Step 2: *Devise a plan.* From January 1992 to January 2004 is 12 years, so $t = 12$. We will use the growth formula $A = A_0 e^{rt}$ to first find r for the known numbers, and then use this value of r to predict a future value for A when $t = 6$ years past 2004.

Step 3: *Carry out the plan.* First find r for $A_0 = 209,693$ for $t = 0$ and $A = 525,060$ for $t = 12$.

$$A = A_0 e^{rt}$$
$$525,060 = 209,693 e^{r(12)}$$

$$\frac{525,060}{209,693} = e^{12r}$$

$$12r = \ln\left(\frac{525,060}{209,693}\right)$$

$$r = \frac{1}{12}\ln\left(\frac{525,060}{209,693}\right) \approx 0.076488998787$$

Now use this value of r in the same formula where A is the cumulative number of U.S. AIDS deaths in January 2010, $A_0 = 525,060$ and $t = 6$:

$$A = A_0 e^{rt}$$
$$= 525,060 e^{6r}$$
$$= 830,847.72$$

Step 4: *Look back.* The expected number of cumulative U.S. deaths due to AIDS is about 830,848.

EXAMPLE 7 **Growth rate for world population** MODELING APPLICATION

On July 7, 1986, it was reported that the world population reached 5 billion. On October 12, 1999, the newspapers reported the world population reached 6 billion. If you assume the same growth rate as for the given period, when will the world population reach 7 billion?

Solution

Step 1: *Understand the problem.* We need to find a calendar date when the world population will reach 7 billion. We repeat Figure 4.23 to help us understand world population growth.

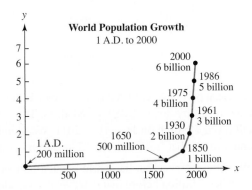

Step 2: *Devise a plan.* We will use the growth formula, $A = A_0 e^{rt}$. First, we will find the growth rate for the given dates and times. Then, we will use this value of r to begin with 6 billion in 1999 to find t when the population is 7 billion.

Step 3: *Carry out the plan.* We need to find t (in years) between July 7, 1986 and October 12, 1999. If we ignore leap years, it is 13 years from July 7, 1986 to July 7, 1999. From July 7 to October 12 is 97 days.

$$\frac{97}{365} \approx 0.265753424658 \qquad \text{so} \qquad t \approx 13.265753424658$$

If we use a calculator or a spreadsheet (that is, we do not ignore leap years), we find the actual number of days to be 4,845 (this includes leap-year days). This means

$$t = \frac{4,845}{365} \approx 13.2739726027$$

It is worthwhile to stop and comment on the nature of mathematical modeling. When modeling a real-world situation, there is often no one "correct" answer. Your answer is based on the assumptions you make. To answer the *mathematical* question, we needed to *make some assumptions* about what number to use. In this example, we need to *make some assumption* about the value of t. You will note that for this example we have calculated $t \approx 13.265753424658$ or $t \approx 13.2739726027$. For this problem we will use the value for t that represents the actual number of days. That is, we use $t = 4,845/365 \approx 13.2739726027$.

Use this value of t to find r where $A_0 = 5, A = 6$:

$$A = A_0 e^{rt}$$

$$6 = 5e^{rt}$$

$$\frac{6}{5} = e^{rt}$$

$$r = \frac{1}{t}\ln\frac{6}{5} \approx 0.01373526692$$

Now, if we assume a constant growth rate, we can predict populations past 6 billion. We want to find the time t, when $A_0 = 6$ and $A = 7$:

$$A = A_0 e^{rt}$$

$$7 = 6e^{rt}$$

$$\frac{7}{6} = e^{rt}$$

$$t = \frac{1}{r}\ln\frac{7}{6} \approx 11.22298392303$$

To find the fractional part of a year (in days), subtract 11 (i.e., the whole number part) and multiply the result by 365: $81.389 \approx 81$ days. This means that we predict the population will reach 7 billion on January 1, 2011. We found January 1 by looking at our desk calendar to see that from October 12 to December 31 is 80 days.

Step 4: *Look back.* We should expect that the world population reaches 7 billion on January 1, 2011. If it reaches 7 billion before that date, we can conclude that the rate is higher that we assumed, and if it reaches 7 billion after that date, the growth rate is lower than we assumed. ∎

General Exponential Equations

We complete our discussion with some additional examples of exponential equations.

EXAMPLE 8   Exponential with nonmatching bases

Solve: **a.** $5^{x+2} = 9^x$ **b.** $5^{x+2} = 8 \cdot 9^x$
Give both the exact answer and an approximate answer (to the nearest hundredth).

Solution

a. $5^{x+2} = 9^x$ *Given.*

   $x + 2 = \log_5 9^x$ *Definition of logarithm*

   $x + 2 = x \log_5 9$ *Third law of logarithms*

   $2 = x \log_5 9 - x$ *Terms involving x should be on one side.*

   $2 = x(\log_5 9 - 1)$ *Common factor*

   $x = \dfrac{2}{\log_5 9 - 1} \approx 5.48$ *Divide both sides by the coefficient of x to solve for x.*

b. This is similar to part **a**, except for the coefficient of 9^x.

   $5^{x+2} = 8 \cdot 9^x$ *Given.*

   $x + 2 = \log_5(8 \cdot 9^x)$ *Definition of logarithm*

   $x + 2 = \log_5 8 + \log_5 9^x$ *Second law of logarithms*

   $x + 2 = \log_5 8 + x \log_5 9$ *Third law of logarithms*

   $2 - \log_5 8 = x \log_5 9 - x$ *Terms involving x should be on one side.*

   $2 - \log_5 8 = x(\log_5 9 - 1)$ *Common factor*

   $x = \dfrac{2 - \log_5 8}{\log_5 9 - 1} \approx 1.94$ *Divide both sides by the coefficient of x to solve for x.*

∎

EXAMPLE 9   Exponential using the quadratic formula

Solve $\dfrac{10^x - 10^{-x}}{2} = 8$ correct to four decimal places.

Solution

$$\frac{10^x - 10^{-x}}{2} = 8$$ *Given.*

$$10^x - 10^{-x} = 16$$ *Multiply both sides by 2.*

$$10^x - \frac{1}{10^x} = 16$$ *Definition of exponent*

$$10^{2x} - 1 = 16 \cdot 10^x$$ *Multiply both sides by 10^x.*

$$10^{2x} - 16 \cdot 10^x - 1 = 0$$ *Treat this as a quadratic equation for which the variable is 10^x, $a = 1$, $b = -16$, and $c = -1$.*

$$10^x = \frac{16 \pm \sqrt{(-16)^2 - 4(1)(-1)}}{2(1)}$$ *Quadratic formula*

$$= \frac{16 \pm \sqrt{260}}{2}$$

$$= \frac{16 \pm 2\sqrt{65}}{2}$$

$$= 8 \pm \sqrt{65}$$

Thus, $10^x = 8 + \sqrt{65}$ or $10^x = 8 - \sqrt{65}$

$x = \log(8 + \sqrt{65}) \approx 1.2058$ $x = \log(8 - \sqrt{65})$

No values (argument must be positive)

PROBLEM SET 4.7

LEVEL 1

Solve the exponential equations in Problems 1–6. Give the exact answer.

1. a. $2^x = 128$ **b.** $3^x = 243$
2. a. $125^x = 25$ **b.** $216^x = 36$
3. a. $4^x = \frac{1}{16}$ **b.** $27^x = \frac{1}{81}$
4. a. $8^{3x} = 2$ **b.** $64^{2x} = 2$
5. a. $5^{2x+3} = 4$ **b.** $e^{2x+1} = 25$
6. a. $6^{5x-3} = 5$ **b.** $5^{3x-1} = 0.45$

Solve the exponential equations in Problems 7–12. Show the approximation you obtain with your calculator without rounding.

7. a. $10^x = 42$ **b.** $10^x = 0.0234$
8. a. $e^{1-2x} = 3$ **b.** $e^{1-5x} = 15$
9. a. $5^{-x} = 8$ **b.** $7^{-x} = 125$
10. a. $10^{5-3x} = 0.041$ **b.** $10^{2x-1} = 515$
11. a. $6 \cdot 8^x - 11 = 25$ **b.** $8\pi^x - 10 = 102$
12. a. $2 \cdot 3^x + 7 = 61$ **b.** $3 \cdot 5^x + 30 = 105$

LEVEL 2

Solve the exponential equations in Problems 13–40. State your answer rounded to four decimal places.

13. $5^x = 8^{2x}$ **14.** $10^x = 4^{2x}$
15. $3^x = 5^{x+3}$ **16.** $6^{x+2} = 10^x$
17. $\left(1 + \frac{0.08}{360}\right)^{360x} = 2$ **18.** $\left(1 + \frac{0.055}{12}\right)^{12x} = 2$
19. $100 = 64(10)^{0.005x^2}$ **20.** $850 = 55(10)^{0.08x^2}$
21. $100 = 64e^{0.005x^2}$ **22.** $850 = 55e^{0.08x^2}$
23. $100 = 64(4)^{0.005x^2}$ **24.** $850 = 55(4)^{0.08x^2}$
25. $10^{2x} - 18 \cdot 10^x + 80 = 0$ **26.** $10^{2x} - 10^x - 6 = 0$
27. $e^{2x} - e^x - 2 = 0$ **28.** $e^{2x} - 7e^x + 12 = 0$
29. $10^{2x} - 2 \cdot 10^x - 5 = 0$ **30.** $3e^{2x} + 2e^x - 6 = 0$
31. $e^x + e^{-x} = 10$ **32.** $e^x - e^{-x} = 100$
33. $\frac{10^x - 10^{-x}}{5} = 45$ **34.** $\frac{e^x - e^{-x}}{2} = 10$
35. $\frac{3^x - 3^{-x}}{4} = 8$ **36.** $\frac{4^x - 4^{-x}}{5} = 50$
37. $10^{5x+1} = e^{2-3x}$ **38.** $5^{2+x} = e^{3x+2}$
39. $x^2 5^x = 5^x$ **40.** $x^2 3^x = 9(3^x)$

Solve for the indicated variable in Problems 41–48.

41. $P = P_0 e^{rt}$ for t **42.** $I = I_0 e^{-rt}$ for r
43. $P = 14.7e^{-0.21a}$ for a **44.** $A = 250e^{-0.248t}$ for t
45. $A = A_0 \left(\frac{1}{2}\right)^{t/h}$ for t **46.** $A = A_0 e^{t/h}$ for h
47. $P_n = P_1 e^{0.11(1-n)}$ for n **48.** $N = 80(1 - e^{-0.16n})$ for n

49. The Arrhenius function is used to relate the viscosity (η) of a fluid (the fluid's internal friction, which is what makes it resist a tendency to flow) to its absolute temperature T:

$$\frac{1}{\eta} = Ae^{-E/RT}$$

where A is a constant specific to that fluid and R is the ideal gas constant. Solve this equation for T. The resulting formula is one you could use in conducting an experiment investigating the viscosity of different grades of motor oil at different temperatures.

50. The atmospheric pressure P in pounds per square inch (psi) is approximated by $P = 14.7e^{-0.21a}$, where a is the altitude above sea level in miles. If the atmospheric pressure of Denver is 11.9 psi, estimate Denver's altitude.

51. The half-life of ^{234}U, uranium-234, is 2.52×10^5 years. If 97.3% of the uranium in the original sample is present, what length of time (to the nearest thousand years) has elapsed?

52. The half-life of ^{22}Na, sodium-22, is 2.6 years. If 15.5 g of an original 100-g specimen remains, how long (to the nearest year) has elapsed?

53. If $850 is deposited into an account paying 8.5% interest compounded monthly, how long will it take (to the nearest month) to have $1,000?

54. If $850 is deposited into an account paying 8.5% interest compounded continuously (365-day year), how long will it take (to the nearest day) to have $1,000?

55. In Example 6, we used data from the AIDs Education Global Information System. The Centers for Disease Control and Prevention in Atlanta has slightly less recent data. They report that in January 2001, the cumulative number of U.S. deaths due to AIDS is 459,523. Use this number to estimate the cumulative number of deaths from AIDS by January 2010.

56. In Example 7, we calculated two values for the time between July 7, 1987, and October 12, 1999. We selected one of these values for the solution in the example. For this problem, select the other to estimate the date when the world population will reach 7 billion.

57. A healing law for skin wounds states that $A = A_0 e^{-0.1t}$, where A is the number of square centimeters of unhealed skin after t days when the original area of the wound was A_0. Draw a graph showing the healing of a 100-cm^2 wound. How many days does it take for half the wound to heal?

58. A satellite has an initial radioisotope power supply of 50 watts. The power output in watts is given by the equation $P = 50e^{-t/250}$, where t is the time in days. Draw a graph of the power output. If the satellite will operate if there is at least 10 watts of power, how long would we expect the satellite to operate?

LEVEL 3

59. PROBLEM FROM CALCULUS In calculus, it is shown that

$$e^x = 1 + x + \frac{x^2}{2} + \frac{x^3}{2 \cdot 3} + \frac{x^4}{2 \cdot 3 \cdot 4} + \cdots$$

a. What are the next two terms?

b. Approximate e using the first six terms.

c. Approximate $\sqrt{e}$ using the first four terms.

60. Journal Problem (*The College Mathematics Journal* FFF #49; Vol. 23 1992, p. 36). WHAT IS WRONG, if anything, with the solution to solving the given equation? Explain your reasoning.

$$e^{2x} - e^x - 3 = 0$$

Solution:

$$e^{2x} - e^x - 2 = 0$$
$$\log e^{2x} - \log e^x - \log 2 = 0$$
$$2x - x - \log 2 = 0$$
$$x = \log 2$$

CHAPTER 4 SUMMARY AND REVIEW

T he only way to learn mathematics is to do mathematics.

Paul Halmos (1967)

Take some time getting ready to work the review problems in this section. First, look back at the definition and property boxes. You will maximize your understanding of this chapter by working the problems in this section only after you have studied the material.

SELF TEST   *All of the answers for this self test are given in the back of the book.*

Graph each function in Problems 1–2.

1. a. $f(x) = \dfrac{1}{x-3} + 4$ **b.** $f(x) = \dfrac{x}{\sqrt{x^2 + 25}}$

  c. $y = 25e^{-x}$ **d.** $y = \log(3x - 1)$

2. a. $f(x) = \dfrac{3x^2 + 2x - 5}{x - 1}$ **b.** $f(x) = \dfrac{x}{\sqrt{x^2 - 25}}$

  c. $y = 25^{-1}10^x$ **d.** $y = \ln(x - 1)$

3. Without using a calculator or computer, simplify each expression:

  a. $\log 100 + \log \sqrt{10} + 10^{\log e}$

  b. $\log 4 + \log 10 + \log 25$

  c. $\ln e + \ln 1 + \ln e^{542} + \ln e^{\log 1{,}000}$

  d. $\log_8 4 + \log_8 16 + \log_8 8^{2.3}$

Solve for x each of the equations in Problems 4–6.

4. a. $\dfrac{5}{x} - \dfrac{x-5}{x-3} = \dfrac{x-6}{2x-12}$ **b.** $x - \sqrt{2x - 3} = 1$

  c. $10^x = 85$ **d.** $\log(x + 1) = 2 + \log(x - 1)$

5. a. $\dfrac{8}{x+6} - \dfrac{1}{x-4} = \dfrac{1}{3}$ **b.** $\sqrt{x+2} = \sqrt{x+3} - 1$

  c. $e^{3x+1} = 45$ **d.** $\log 2 + 2\log x = 5$

6. a. $435^x = 890$ (Round x to 4 places.)

  b. $\log_6 x = 4$

  c. $3 \ln \dfrac{e}{\sqrt[3]{5}} = 3 - \ln x$

  d. $A = P(1 + i)^x$

7. Write the equation $y = 14.8(2.5)^x$ as an equation of the form $y = ae^{bx}$.

8. Suppose that $1,000 is invested at 7% interest compounded monthly.

  a. How long before the value is $1,250?

  b. How long before the money doubles?

  c. What is the interest rate (compounded quarterly) if the money doubles in 5 years?

9. The half-life of arsenic-76 is 26.5 hr.

  a. Find the decay constant.

  b. If 85 mg of arsenic-76 are present at the start, how long would it take for 6 mg to be left?

10. In 2002, it was reported that the number of teenagers with AIDS doubles every 14 months. Find an equation to model the number of teenagers to be infected over the next 10 years.

STOP STUDY HINTS *Compare your solutions and answers to the self test.*

In Problems 1–6, find $\dfrac{f(x+h)-f(x)}{h}$ and **simplify**.

1. $f(x)=\dfrac{2x}{x-2}$ **2.** $f(x)=\dfrac{1}{x^2}$

3. $f(x)=e^x$ **4.** $f(x)=10^x$

5. $f(x)=\log x$ **6.** $f(x)=\ln x$

State the coordinates of P and Q, and then find the slope of the secant line passing through those points for the curves shown in Problems 7–10.

7.

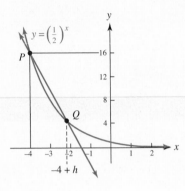

$y=\left(\tfrac{1}{2}\right)^x$

8.

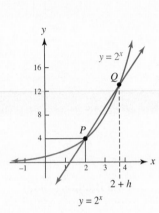

$y=2^x$

9.

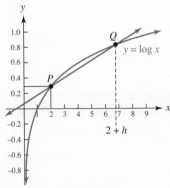

$y=\log x$

10.

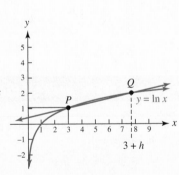

$y=\ln x$

Let $f(x)=\dfrac{e^x}{x-2}$ and $g(x)=\dfrac{x-2}{e^x}$ in Problems 11–14.

11. Evaluate $(f+g)(3)$.   **12.** Find $(f/g)(x)$.

13. Find $(f\circ g)(0)$.   **14.** Are f and g inverses?

Evaluate the functions given in Problems 15–20 at each of the endpoints of the intervals:

   a. $[1,2]$ **b.** $[9,10]$

Compute the average rate of change (rounded to the nearest hundredth).

15. $f(x)=x^2$ **16.** $f(x)=\dfrac{1}{x}$

17. $f(x)=e^{x+1}$ **18.** $f(x)=3^x$

19. $f(x)=\ln x$ **20.** $f(x)=\log x^2$

Factor each expression in Problems 21–24.

21. $8x^2-2x-15$

22. $6+3x-9x^2$

23. $5x^2y^3z+20x^{-2}yz^2+100xy^{-1}z^3$

24. $12xy^{-1}+3x^{-2}y+9xy^2$

Problems 25–26 show some graphs that you might see when looking at a graphing calculator. From what you see on these graphs, classify each as a polynomial, rational, logarithmic, or exponential function.

25. a.

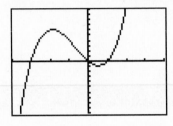

   b.

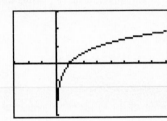

   c.

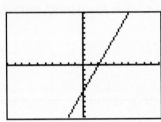

   d.

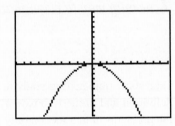

   e.

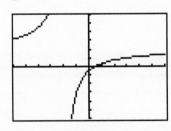

26. a.

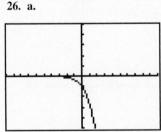

   b.

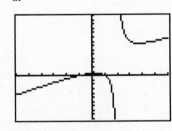

   c.

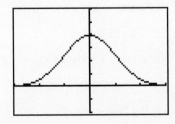

   d.

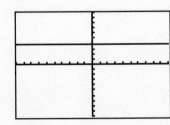

e.

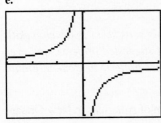

In Problems 27−32, express each function as the composite of two functions. That is, express the given function f as $f(x) = g[u(x)]$, and then state the inner function u and the outer function g.

27. $f(x) = e^{2x^2 + 1}$

28. $f(x) = \sqrt{3x^2 + 2x - 5}$

29. $f(x) = (x^2 + 3x)^2$

30. $f(x) = e^{3x^4}$

31. $f(x) = \log(3x^4 + 5x)$

32. $f(x) = \ln(x^3 - 3x^2)$

Graph the curves whose equations are given in Problems 33−48.

33. $y = \dfrac{2x + 1}{5 - x}$

34. $y = \dfrac{12}{x - 12}$

35. $y - 3 = |x + 2|$

36. $y = \dfrac{10}{x}$

37. $f(x) = -[\![2 - x]\!]$

38. $y = \sqrt{x(x^2 - 4)}$

39. $y = \dfrac{x^2}{x(x - 2)}$

40. $y = \dfrac{x^2 - 4x + 3}{x^2 - x - 6}$

41. $y = e^{2x}$

42. $y = 6^{-x^2}$

43. $y = \log(x - 1)$

44. $y = \sqrt{(x - 1)(x - 4)}$

45. $y = \sqrt{x^2 - 4x + 4}$

46. $y = \sqrt{4(x^2 + 1)}$

47. $y = \sqrt{0.5(x^2 + 9)}$

48. $y = x^3 - 3x^2 + 2$

Solve the equations in Problems 49−56.

49. $\dfrac{3}{x + 2} + \dfrac{x - 1}{x + 5} = \dfrac{5x + 20}{6x + 24}$

50. $\dfrac{x - 3}{x - 2} + \dfrac{x - 1}{x} = \dfrac{22x - 110}{3x^2 - 15x}$

51. $2 - \sqrt{3x + 1} = \sqrt{x - 1}$

52. $\sqrt{3x + 1} - \sqrt{2x - 1} = 1$

53. $\log 5 = \log x + \log(x + 4)$

54. $2 \ln \dfrac{e}{\sqrt{5}} = 2 - \ln x$

55. $75 = 60 + (100 - 60)10^{-0.05t}$

56. $675{,}000 = 550{,}000 e^{2x + 1}$

57. Solve $m = \dfrac{Pi}{1 - (1 + r)^{-n}}$ for n.

58. Solve $m = \dfrac{Pi}{1 - (1 + r)^{-n}}$ for r.

59. The half-life of arsenic-76 is 26.5 hours. If 55 mg of arsenic are present initially, how long before there are 30 mg?

60. If $2,500 is placed in a $2\frac{1}{2}$-year time certificate paying 9.5% compounded monthly, what is the future value of the account when the certificate matures in $2\frac{1}{2}$ years?

CHAPTER 4   Group Research Projects

Working in small groups is typical of most work environments, and this book seeks to develop skills with group activities. At the end of each chapter, we present a list of suggested projects, and even though they could be done as individual projects, we suggest that these projects be done in groups of three or four students.

G4.1 Before Hurricane Katrina in 2005, the entrance of the Aquarium of Americas in New Orleans had a gigantic building-size curve called a *logarithmic spiral*.

Courtesy of Audubon Aquarium of the Americas.

Find out how to construct a logarithmic spiral, and write a paper about what you learned. Why do you suppose it would appear on the front of an aquarium?

G4.2 If $x = \sqrt{7 + \sqrt{48}} + \sqrt{7 - \sqrt{48}}$, find the simplified value of x without using a calculator.

G4.3 Solve $\sqrt{x^2 + \sqrt{(x-1)^2}} = 1$.

G4.4 Discuss graphing using a logarithmic scale rather than a linear scale.

G4.5 Write an essay on earthquakes. In particular, discuss the Richter scale for measuring earthquakes. What is its relationship to logarithms?

G4.6 Write an essay on carbon-14 dating. What is its relationship to logarithms?

G4.7 From your local Chamber of Commerce, obtain the population figures for your city for the years 1980, 1990, and 2000. Find the rate of growth for each period. Forecast the population of your city for the year 2010. Include charts and graphs. List some factors, such as new zoning laws, that could change the growth rate of your city.

G4.8 If we assume that the world population grows exponentially, then it is also reasonable to assume that the use of some nonrenewable resources (such as petroleum) will also grow exponentially. In calculus, it is shown that for some constant k, under these assumptions, the formula for the amount of the resource, A, consumed from time $t = 0$ to $t = T$ is given by the formula

$$A = \frac{A_0}{k}\left(e^{yT} - 1\right)$$

where r is the relative growth rate of annual consumption.

a. Solve this equation for T to find a formula for life expectancy of a particular resource.

b. According to the Energy Information Administration, the annual world production (in billions of barrels per day) of petroleum is shown in the following table:

Year	1975	1980	1985	1990	1995	2000	2003
Quantity	52.42	62.39	52.97	60.90	61.85	66.03	67.00

Find an exponential equation for these data.

c. If in 1998, the world petroleum reserves were 2.8 trillion barrels, estimate the life expectancy for petroleum.

G4.9 Write an essay on John Napier. Include what he is famous for today, and what he considered to be his crowning achievement. Include a discussion of "Napier's bones."

G4.10 (*The Mathematics Teacher*, January 1996, p. 33.)* Neil Simon plans to open his newest play, *London Suite*, off-Broadway. His producer, Emanuel Azenberg, provided the financial comparison shown in Table 4.5.

TABLE 4.5 | Economics of a Broadway Play

Money Needed to Open			Weekly Budget		
Item	Broadway 1,000 seats $165/ticket	Off-Broadway 500 seats $120/ticket	Item	Broadway 1,000 seats $165/ticket	Off-Broadway 500 seats $120/ticket
Sets, costumes, lights	$1,071,000	$261,000	INCOME		
			Receipts	$750,000	$327,000
Loading	$525,000	$24,000	EXPENSES		
Rehearsal salaries	$306,000	$189,000	Rent/house crew	$135,000	$36,000
Director/ designer fees	$378,000	$183,000	Salaries	$162,000	$60,000
Advertising	$900,000	$363,000	Advertising	$90,000	$45,500
Administration	$705,000	$300,000	Lights/sound rental	$18,000	$10,500
TOTAL	$3,885,000	$1,320,000	Administration	$96,000	$33,000
			Royalties	$70,500	$42,000
			Extras	$48,000	$19,500
			TOTAL	$619,500	$246,500

a. Make a chart for half a year (26 weeks) showing the profit and loss for both a Broadway and off-Broadway production.

b. Find the break-even points for both productions. That is, how long does it take for the total profit to equal the initial expenditure?

c. Let P_1 be the profit or loss for the Broadway production X weeks after opening night expressed as a function (a percent) of the initial expenditures. Let P_2 be the function (a percent) for the off-Broadway production. Develop equations for P_1 and P_2 in terms of X. Use these equations to sketch the graphs of P_1 vs X and P_2 vs X.

d. Discuss what happens to the values of P_1 and P_2 as X is steadily increased from 0 to 52 (one year).

e. During the first year, is P_1 greater than P_2 at any time?

*Numbers have been updated.

© Randall Stevens/ShutterStock, Inc.

Chapter Objectives

The material in this chapter is reviewed in the following list of objectives. After completing this chapter, review this list again, and then complete the self test.

5.1

5.1 Know the definition and notation for angles, including positive and negative angles. Know the Greek letters. Know what it means for an angle to be in standard position.

5.2 Be familiar with the degree measure of an angle and be able to approximate the angle associated with a given degree measure without using any measuring devices.

5.3 Be familiar with the radian measure of an angle and be able to approximate the angle associated with a given radian measure without using any measuring devices.

5.4 Find the reference angle $\bar{\theta}$ for a given angle θ.

5.5 Change from radian measure to degree measure and from degree measure to radian measure; know the commonly used radian and degree measure equivalences.

5.6 Find angles coterminal with a given angle and less than one revolution.

5.7 Know the arc length formula and be able to apply it.

5.2 5.8 Know the unit circle definition of the trigonometric functions.

5.9 Evaluate trigonometric functions using the definition and using a calculator.

5.10 State and prove the eight fundamental identities.

5.11 Know the signs of the six trigonometric functions in each of the four quadrants. Name the function(s) that is (are) positive in the given quadrant.

5.12 Use the fundamental identities to find the other trigonometric functions if you are given the value of one function and want to find the others.

5.13 Graph with a unit circle parameter; parametrize part of a circle.

5.14 Use trigonometric functions to model projectile motion.

5.3 5.15 Know the ratio definition of the trigonometric functions.

5.16 Use the definition of the trigonometric functions to approximate their values for a given angle or for an angle passing through a given point.

5.17 Know and be able to derive the table of exact values.

5.18 Use the reduction principle, along with the table of exact values, to evaluate certain trigonometric functions.

5.4 5.19 Graph the trigonometric functions, or variations, by plotting points.

5.20 Graph one period of each of the standard trigonometric functions. Know the period and amplitude for each.

5.21 Graph the general trigonometric functions. Be able to state (h, k), a, b, and the period for each trigonometric function.

5.22 Graph a trigonometric curve from given data points.

5.5 5.23 Know the definition of the inverse trigonometric functions, especially the range of each.

5.24 Evaluate inverse functions using exact values and a calculator.

5.25 Graph the inverse cosine, sine, and tangent functions.

5.26 Evaluate functions of inverse trigonometric functions.

5.6 5.27 Know the right triangle definition of the cosine, sine, and tangent.

5.28 Solve right-triangle problems.

5.29 Find area of triangles, given two sides and an included angle.

5.30 Solve applied problems involving angle of elevation, angle of depression, and bearing.

5.7 5.31 State and prove the law of cosines.

5.32 State and prove the law of sines.

5.33 Know when to apply the laws of cosines and sines.

5.34 Solve oblique triangles.

5.35 Find the area of a triangle.

5.36 Solve applied problems involving oblique triangles.

Trigonometric Functions

5

T here is perhaps nothing which occupies, as it were, the middle position of mathematics, as trigonometry.

—J. F. Herbart (1890)

Chapter Sections

5.1 Angles
Measuring Angles
Relationship between Degree and Radian Measures
Reference Angles
Arc Length

5.2 Fundamentals
Definition of the Trigonometric Functions
Evaluation of Trigonometric Functions
Trigonometric Identities
Modeling with Trigonometric Functions

5.3 Trigonometric Functions of Any Angle
Ratio Definition of the Trigonometric Functions
Exact Values and the Reduction Principle
Angle of Inclination

5.4 Graphs of the Trigonometric Functions
Graphs of the Standard Trigonometric Functions
Graphs of the General Trigonometric Functions

5.5 Inverse Trigonometric Functions
Inverse Sine Function
Inverse Trigonometric Functions

Evaluating Inverses of the Reciprocal Functions
Functions of Inverse Functions

5.6 Right Triangles
Right Triangle Definition of the Trigonometric Functions
Trigonometric Substitutions
Solving Triangles
Areas

5.7 Oblique Triangles
SSS
SAS
SSA
ASA
AAA
Summary
Area of a Triangle

Chapter 5 Summary and Review
Self Test
Practice for Calculus
Chapter 5 Group Research Projects

▶ CALCULUS PERSPECTIVE

As you look at the structure of this book, you can see that it is organized around the various functions that are fundamental in your study of calculus. In this chapter, we will not only define and graph the trigonometric functions, but we will also consider limits of these functions.

The origins of trigonometry are obscure, but we know that it began before the development of calculus. Trigonometry traces its roots back to more than 2,000 years ago with the Mesopotamian, Babylonian, and Egyptian civilizations. The ancients used trigonometry in a practical way to measure triangles for surveying land. We consider this in the sections examining right and oblique triangles. However, today trigonometry is used more theoretically, not only in mathematics but also in electronics, engineering, and computer science. Trigonometry is the branch of mathematics that deals with the properties and applications of six functions: cosine, sine, tangent, secant, cosecant, and cotangent. These functions are defined in this chapter. As we will see, these are functions of angles or real numbers, so we begin our study in this chapter with angles.

5.1 Angles

In geometry, an angle is usually defined as the union of two rays with a common endpoint. In trigonometry and calculus, a more general definition is used.

> **ANGLE**
>
> An **angle** is formed by rotating a ray about its endpoint (called the **vertex**) from some initial position (called the **initial side**) to some terminal position (called the **terminal side**). The measure of an angle is the amount of rotation from the initial side to the terminal side.

If the rotation of the ray is in a counterclockwise direction, the measure of the angle is called **positive**, and if the rotation is in a clockwise direction, the measure is called **negative**. The notation $\angle ABC$ means the measure of an angle with vertex B and points A and C (different from B) on the sides; $\angle B$ denotes the measure of an angle with vertex at B, and a curved arrow is used to denote the direction and amount of rotation, as shown in Figure 5.1. If no arrow is shown, the measure of the angle is considered to be the smallest positive rotation.

TABLE 5.1	Commonly Used Greek Letters
Symbol	**Name**
α	alpha
β	beta
γ	gamma
δ	delta
θ	theta
μ	mu
ϕ or φ	phi
λ	lambda
ω	omega

Note: The lowercase Greek letter π always represents the irrational number approximately equal to 3.141592654.

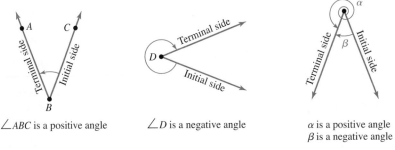

$\angle ABC$ is a positive angle $\angle D$ is a negative angle α is a positive angle
β is a negative angle

Figure 5.1 Examples of angles

Lowercase Greek letters are also used to denote the angles as well as the measure of angles. For example, θ may represent the angle or the measure of the angle called θ; you will know which is meant by the context in which it is used. Some commonly used Greek letters are shown in Table 5.1.

A Cartesian coordinate system may be superimposed on an angle so that the vertex is at the origin and the initial side is along the positive x-axis. In this case, we say the angle is in **standard position**. Angles in standard position having the same terminal side are **coterminal angles**. In Figure 5.2, β is coterminal with α. If the terminal side is on one of the coordinate axes, then θ is called a **quadrantal angle**. If θ is not a quadrantal angle, then it is said to be in a certain **quadrant** if its terminal side lies in that quadrant. In Figure 5.2, both α and β are called Quadrant II angles. We can state these facts as inequalities: $90° < \alpha < 180°$ and $-180° < \beta < -270°$.

Measuring Angles

Several units of measurement are used for measuring angles. Let α be an angle in standard position with a point P (not the vertex) on the terminal side. As this side is rotated through one revolution, the trace of the point P forms a circle. It is easy to understand measuring angles of one or more revolutions, but much of our work is with angles less than one revolution, so we will define measures that are smaller than one revolution. Historically, the most common scheme divides one revolution into 360 equal parts, with each part called a **degree**. Sometimes even finer divisions are necessary, so a degree is divided into 60 equal parts, each called a **minute** ($1° = 60'$), which is furthermore

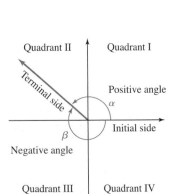

Figure 5.2 Standard-position angles α and β

divided into 60 equal parts, each called a **second** $(1' = 60'')$. For most applications, we write decimal parts of degrees instead of minutes and seconds. That is, $41.5°$ is preferred over $41° 30'$. If we write $\alpha = 41.5°$, we mean that the measure of the angle α is $41.5°$. A protractor for degree measure is shown in Figure 5.3**a**.

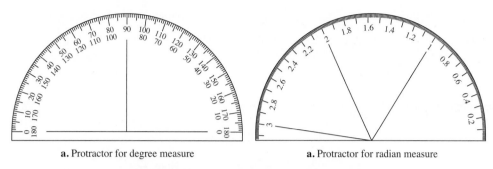

a. Protractor for degree measure   **a.** Protractor for radian measure

Figure 5.3 Protractors are devices for measuring angles

In calculus and scientific work, another measure for angles is defined; this method uses real numbers to measure angles. To understand this measure of angles, draw a circle with any nonzero radius r. Next, measure out an arc with length r. Figure 5.4**a** shows the case in which $r = 1$, and Figure 5.4**b** shows $r = 2$. Regardless of your choice for r, when the radius of the circle equals the length of the arc, the angle determined by this arc of length r is the same. (It is labeled θ in Figure 5.4.) This angle is used as a basic unit of measurement and is called a **radian**. A protractor for radian measure is shown in Figure 5.3**b**.

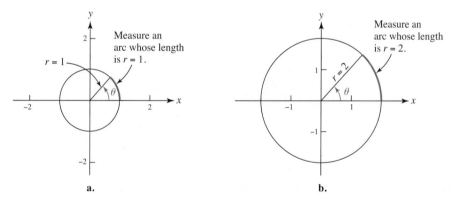

Figure 5.4 Radian measure; angles of measure 1

Since one revolution traces out the circumference of a complete circle, and we know that $C = 2\pi r$, we see that one revolution traces out exactly 2π radians. That is, the radian measure of one revolution is the ratio of the circumference $(2\pi r)$ divided by the radius of the circle (r), to give a measure of 2π. Thus, we say one revolution has a radian measure of 2π.

☠ When measuring angles in radians, we are using *real numbers*. Because radian measure is used so frequently, we agree that **radian measure is understood when no units of measure for an angle are indicated.** ☠

EXAMPLE 1 Drawing angles in radian measure

Sketch the angles:

a. $\frac{\pi}{2}$ **b.** $\frac{\pi}{3}$ **c.** $\frac{\pi}{6}$ **d.** $-\frac{5\pi}{6}$ **e.** $-\frac{3\pi}{2}$

Solution

a. Think of $\frac{\pi}{2}$ as $\frac{1}{2}(\pi)$.

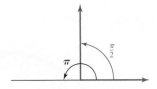

A right angle has a measure of $\frac{\pi}{2}$.

Note: $\frac{\pi}{2} + \frac{\pi}{2} = \pi$.

b. Think of $\frac{\pi}{3}$ as $\frac{1}{3}(\pi)$.

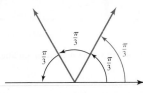

Note: $\frac{\pi}{3} + \frac{\pi}{3} + \frac{\pi}{3} = \pi$.

c. Think of $\frac{\pi}{6}$ as $\frac{1}{2}\left(\frac{\pi}{3}\right)$.

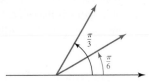

d. Think of $-\dfrac{5\pi}{6}$ as $5\left(\dfrac{\pi}{6}\right)$ in the negative direction.

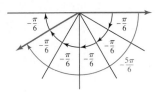

e. Think of $-\dfrac{3\pi}{2}$ as $3\left(\dfrac{\pi}{2}\right)$ in the negative direction.

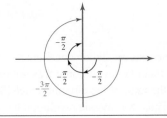

STOP Practice thinking in radian measure. You should memorize the approximate size of an angle of measure 1, in much the same way you have memorized the approximate size of an angle whose measure is 45°.

Relationship between Degree and Radian Measures

One revolution is measured by $360°$ or by 2π radians. Thus,

$$\text{Number of revolutions} = \frac{\text{Angle in degrees}}{360}$$

$$\text{Number of revolutions} = \frac{\text{Angle in radians}}{2\pi}$$

Therefore:

DEGREE/RADIAN FORMULA

$$\frac{\text{Angle in degrees}}{360} = \frac{\text{Angle in radians}}{2\pi}$$

>> IN OTHER WORDS To change from *degree measure to radian measure*, let θ be the angle measured in radians, and multiply both sides by 2π to obtain:

$$\theta = \frac{\pi}{180} \qquad (\text{DEGREE MEASURE OF ANGLE})$$

To change from *radian measure to degree measure*, let θ be the angle measured in degrees, and multiply both sides by 360 to obtain:

$$\theta = \frac{180}{\pi} \qquad (\text{RADIAN MEASURE OF ANGLE})$$

EXAMPLE 2   Change from degree measure to radian measure

Change to radians:

a. $45°$

b. $1°$

c. $123.45°$

Solution   In each case, let θ be the angle we seek, measured in radians.

a.
$$\frac{45}{360} = \frac{\theta}{2\pi}$$
$$\frac{45(2\pi)}{360} = \theta$$
$$\frac{\pi}{4} = \theta$$

An alternative method is to remember that π radians is $180°$. That is, because $45°$ is $\frac{1}{4}$ of $180°$, we know that the radian measure is $\frac{1}{4}$ of π or $\frac{\pi}{4}$.

The answer $\frac{\pi}{4}$ is called the *exact value* of the radian angle measure. If you use your calculator, θ is approximated as

$$\theta = \frac{\pi}{180}(45) \approx 0.78539816$$

b.
$$\frac{1}{360} = \frac{\theta}{2\pi} \qquad \textit{Substitute 1 for degree measure.}$$

$$\theta = \frac{\pi}{180} \qquad \textit{Exact answer}$$

The approximate value is $\frac{\pi}{180} \approx 0.0174532925$.

c. We find a decimal approximation correct to two decimal places.

$$\theta = \frac{\pi}{180}(123.45) \approx 2.15$$

Notice that this angle is in Quadrant II.

EXAMPLE 3   Change radian measure to degree measure

Change to degrees: **a.** $\dfrac{\pi}{9}$ **b.** 1   **c.** 4.30

Solution   In each case, let θ be the degree measure of the angle we seek.

a. $\theta = \dfrac{180}{\pi}\left(\dfrac{\pi}{9}\right)$ Substitute $\dfrac{\pi}{9}$ for radian measure.

$= 20$

The degree measure of the angle is $20°$.

b. $\theta = \dfrac{180}{\pi}(1)$ Substitute 1 for radian measure.

$= \dfrac{180}{\pi}$ This is the exact answer.

A calculator approximation is $\dfrac{180°}{\pi} \approx 57.30°$.

c. We find the decimal approximation correct to two decimal places.

$\theta = \dfrac{180}{\pi}(4.30)$ Substitute 4.30 for radian measure.

≈ 246.3718519 Calculator approximation

To two decimal places, the measure of the angle is $246.37°$. This angle is in Quadrant III.   ∎

For the more common measures of angles, it is a good idea to memorize the equivalent degree and radian measures shown in Table 5.2. These angles are shown in Figure 5.5.

TABLE 5.2	Relationship between Degree and Radian Measure of Angles
Degree	**Radians**
$0°$	0
$30°$	$\dfrac{\pi}{6} \approx 0.52$
$45°$	$\dfrac{\pi}{4} \approx 0.79$
$60°$	$\dfrac{\pi}{3} \approx 1.05$
$90°$	$\dfrac{\pi}{2} \approx 1.57$
$180°$	$\pi \approx 3.14$
$270°$	$\dfrac{3\pi}{2} \approx 4.71$
$360°$	$2\pi \approx 6.28$

← STOP

If you remember that

$$180° = \pi$$

you will easily be able to remember the other measures in this table.

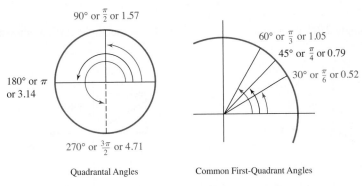

Figure 5.5 Graphs of commonly used angles in mathematics

Reference Angles

Table 5.2 lists angles with equal measure. That is, we say an angle with measure $180°$ is equal to an angle with measure π radians. Do not confuse equal angles with coterminal angles. Equal angles are always coterminal angles, but coterminal angles are not necessarily equal. For example, we say $\pi/2$ and $90°$ are equal angles, and these angles are coterminal. On the other hand, $\pi/2$ and $-3\pi/2$ are coterminal but are not equal.

We now introduce still another relationship between pairs of angles. If an angle θ is not a quadrantal angle, then we refer to its *reference angle*, which is denoted by $\overline{\theta}$ throughout this book.

> **REFERENCE ANGLE**
>
> If a standard-position angle θ is not a quadrantal angle, then the **reference angle** $\overline{\theta}$ is defined as the acute angle the terminal side of θ makes with the x-axis.

>> **IN OTHER WORDS** This means that $\overline{\theta}$ is always positive (regardless of θ) and is between $0°$ and $90°$ or between 0 and $\pi/2$. We can write:

In degree measure for θ: $0° < \overline{\theta} < 90°$ In radian measure for θ: $0 < \overline{\theta} < \dfrac{\pi}{2} \approx 1.57$

The procedure for finding the reference angles varies, depending on the quadrant of θ, as shown in Figure 5.6. Notice that the reference angle $\overline{\theta}$, shown in color, is always drawn to the x-axis and *never* to the y-axis. Also, remember that if θ is a quadrantal angle, there is no reference angle.

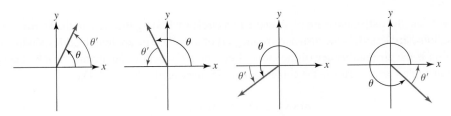

Figure 5.6 Reference angles

(STOP) Make sure you understand this idea.

EXAMPLE 4   Finding reference angles

Find the reference angle for each angle, and draw both the given angle and the reference angle.

a. $210°$ **b.** $150°$ **c.** $-\dfrac{5\pi}{3}$ **d.** 5 **e.** $812°$

Solution

a. Draw the angle $210°$, as shown:

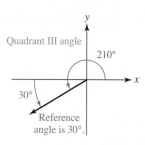

Reference angle is $210° - 180° = 30°$.

b. Draw the angle $150°$, as shown:

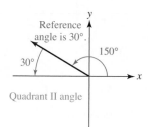

Reference angle is $180° - 150° = 30°$.

c. Draw the angle $-\dfrac{5\pi}{3}$, as shown:

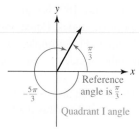

Reference angle is $2\pi - \dfrac{5\pi}{3} = \dfrac{\pi}{3}$.

d. Draw the angle 5 as shown:

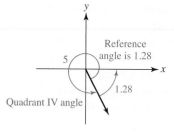

Reference angle is $2\pi - 5 \approx 1.28$.

e. Draw the angle $812°$. If the angle is more than one revolution, first find a nonnegative coterminal angle less than one revolution: $812°$ is coterminal with $92°$.

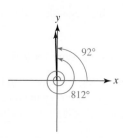

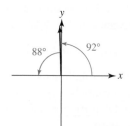

☠ It is NOT correct to say $812° = 92°$. It IS correct to say $812°$ and $92°$ are coterminal. ☠

Arc Length

We now relate the radian measure of an angle to a circle by finding the arc length. An **arc** is part of a circle; thus, **arc length** is the distance around part of a circle. The arc length corresponding to one revolution is the **circumference** of the circle. Let s be the length of an arc and let θ be the central angle (an angle whose vertex is the center of a circle) measured in radians. Then,

$$\text{ANGLE IN REVOLUTIONS} = \frac{s}{2\pi r}$$

since one revolution has an arc length (circumference of a circle) of $2\pi r$. Also,

$$\text{ANGLE IN RADIANS} = (\text{ANGLE IN REVOLUTIONS})\,(2\pi)$$

$$= \frac{s}{2\pi 4}(2\pi)$$

$$= \frac{s}{r}$$

This leads us to the **arc length formula**.

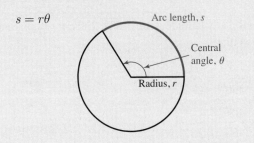

ARC LENGTH FORMULA

The **arc length** cut by a central angle θ (measured in radians) from a circle of radius r is denoted by s and is found by

$$s = r\theta$$

Arc length, s

Central angle, θ

Radius, r

EXAMPLE 5   Finding arc length

Find the length of an arc subtended (cut off) by a central angle of $36°$ in a circle with radius 5 in.

Solution   First, change the degree measure to radian measure:

$$\theta = \frac{\pi}{180}(36) = \frac{\pi}{5}$$

where θ is the radian measure of the angle. Thus,

$$s = r\theta \qquad \text{Arc length formula}$$

$$= 5\left(\frac{\pi}{5}\right) \qquad \text{Substitute known values.}$$

$$= \pi \qquad \text{This is the exact answer; the approximate answer is 3.14.}$$

The arc length is π inches. ∎

We conclude this section by considering an application. Locations on the earth are given by *latitude* and *longitude*. **Latitude** is measured north or south of the equator by angles between $0°$ and $90°$. **Longitude** is measured by angles between $0°$ and $180°$ either west or east of the Greenwich meridian, as shown in Figure 5.7. A **nautical mile** is defined to be s for a central angle of $1'$ on the earth.

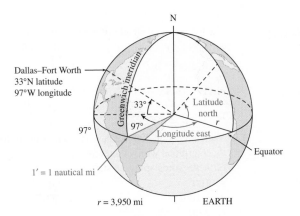

Assume that the radius of the earth is 3,950 mi or 6,370 km (kilometers).

Figure 5.7 Latitude and longitude on earth

EXAMPLE 6   Global distances

The Dallas–Fort Worth airport is located at approximately 33°N latitude and 97°W longitude. How far north (to the nearest hundred miles) is it from the equator?

Solution   The location is 33°N latitude, so $\theta = 33°$. Since $r = 3{,}950$ mi, it follows that $C = 2\pi r = 2\pi(3{,}950)$, so that

$$\frac{\theta}{360°} = \frac{s}{C} \qquad \text{Arc length formula}$$

$$\frac{33°}{360°} = \frac{s}{2\pi(3{,}950)} \qquad \text{Substitute known values.}$$

$$\left(\frac{33}{360}\right)2\pi(3{,}950) = s \qquad \text{Multiply both sides by } 2\pi(3{,}950).$$

$$s \approx 2{,}300 \qquad \text{Round to the nearest hundred; the calculator value is } 2275.03668.$$

The Dallas–Fort Worth airport is approximately 2,300 miles north of the equator. ■

PROBLEM SET 5.1

LEVEL 1

WHAT IS WRONG, *if anything, with each statement in Problems 1–6? Explain your reasoning.*

1. $30° = \dfrac{\pi}{3}$

2. $390° = 30°$

3. $-60° = 60°$

4. $-\dfrac{\pi}{4} = \dfrac{7\pi}{4}$

5. A circle with $r = 10$ has a circumference of 10π.

6. The arc length formula is $s = r\theta$, so the length of an arc of a circle of radius 10 with an angle of $10°$ is $s = (10)(10) = 100$.

7. From memory, give the radian measure for each of the angles whose degree measure is stated. Also sketch the angle.
   a. 30°   b. 90°
   c. 45°   d. 60°

8. From memory, give the radian measure for each of the angles whose degree measure is stated. Also sketch the angle.
   a. 120°   b. 660°
   c. −135°   d. −240°

9. From memory, give the degree measure for each of the angles whose radian measure is stated. Also sketch the angle.
   a. $\dfrac{\pi}{3}$ b. $\dfrac{\pi}{6}$
   c. $\dfrac{\pi}{2}$ d. $\dfrac{\pi}{4}$

10. From memory, give the degree measure for each of the angles whose radian measure is stated. Also sketch the angle.
   a. $\dfrac{5\pi}{6}$ b. $\dfrac{5\pi}{3}$
   c. $-\dfrac{\pi}{4}$ d. $-\dfrac{5\pi}{2}$

11. Pick the coin that most closely approximates the given dollar value. For example, $\pi/6$ is a decimal approximately equal to 0.52. This is most closely approximated by a half-dollar (Answer E).
   A. penny   B. nickel
   C. dime   D. quarter
   E. Kennedy half-dollar   F. Presidential dollar
   a. $\pi/3$ b. $\pi/12$
   c. $\pi/180$ d. $\dfrac{\pi}{4} - \dfrac{1}{4}$
   e. $\dfrac{1}{3}\left(\dfrac{\pi}{10}\right)$ f. $\dfrac{1}{50}(\pi)$

12. In degree measure, $50' + 40' = 90'$ or $1°30'$, a very small angle. In radian measure, $50 + 40 = 90$, a very large angle (over 14 revolutions). Here is another example in radian measure:

   $$\frac{\pi}{4} + \frac{\pi}{3} = \frac{3\pi}{12} + \frac{4\pi}{12} = \frac{7\pi}{12}$$

   Complete the given addition problems.
   a. $30' + 50'$ b. $30 + 50$
   c. $5°25' + 8°55'$ d. $\dfrac{\pi}{6} + \dfrac{\pi}{2}$
   e. $121°16'45'' + 16°55'50''$ f. $2\pi - \dfrac{\pi}{6}$
   g. $\pi + \dfrac{\pi}{3}$ h. $\dfrac{3\pi}{4} + 2\pi$

13. Sketch the angle 200°. Classify each of the following angles as equal to, coterminal with, or a reference angle of the sketched angle. If none of these words apply, so state.
   a. 160°   b. 20°
   c. −160°   d. 560°
   e. 200°

14. Sketch the angle $300°$. Classify each of the following angles as equal to, coterminal with, or a reference angle of the sketched angle. If none of these words apply, so state.
    a. $60°$ b. $-60°$
    c. $660°$ d. $120°$
    e. $-420°$

15. Sketch the angle $2\pi/3$. Classify each of the following angles as equal to, coterminal with, or a reference angle of the given angle. If none of these words apply, so state.
    a. $\dfrac{\pi}{3}$ b. $-\dfrac{\pi}{3}$
    c. $\dfrac{8\pi}{3}$ d. $-\dfrac{4\pi}{3}$
    e. $\dfrac{4\pi}{6}$

16. Sketch the angle $13\pi/6$. Classify each of the following angles as equal to, coterminal with, or a reference angle of the given angle. If none of these words apply, so state.
    a. $\dfrac{26\pi}{12}$ b. $\dfrac{25\pi}{6}$
    c. $-\dfrac{11\pi}{6}$ d. $\dfrac{\pi}{6}$
    e. $\dfrac{13\pi}{3}$

17. Sketch the angle $225°$. Classify each of the following angles as equal to, coterminal with, or a reference angle of the given angle. If none of these words apply, so state.
    a. $\dfrac{\pi}{4}$ b. $-\dfrac{\pi}{4}$
    c. $\dfrac{5\pi}{4}$ d. $\dfrac{3\pi}{4}$
    e. $\dfrac{11\pi}{4}$

18. Sketch the angle $5\pi/6$. Classify each of the following angles as equal to, coterminal with, or a reference angle of the given angle. If none of these words apply, so state.
    a. $30°$ b. $150°$
    c. $330°$ d. $60°$
    e. $-210°$

Sketch each angle in Problems 19–22 and change to radians using exact values.

19. a. $270°$ b. $480°$ c. $40°$
20. a. $150°$ b. $135°$ c. $20°$
21. a. $300°$ b. $-150°$ c. $85°$
22. a. $225°$ b. $-240°$ c. $250°$

Sketch each angle in Problems 23–26 and change to radians correct to the nearest hundredth.

23. a. $120°$ b. $-115°$ c. $100°$
24. a. $-60°$ b. $400°$ c. $23.7°$
25. a. $350°$ b. $525°$ c. $-45°$
26. a. $38.4°$ b. $-210°$ c. $-825°$

Sketch each angle in Problems 27–30 and change to degrees using exact values.

27. a. 5π b. $-\dfrac{5\pi}{2}$ c. $\dfrac{11\pi}{6}$

28. a. $-\dfrac{5\pi}{3}$ b. $\dfrac{5\pi}{3}$ c. -2π

29. a. $\dfrac{2\pi}{3}$ b. $\dfrac{4\pi}{3}$ c. $\dfrac{11\pi}{3}$
30. a. $-\dfrac{\pi}{4}$ b. $\dfrac{5\pi}{4}$ c. $-\dfrac{11\pi}{4}$

Sketch each angle in Problems 31–34 and change to decimal degrees correct to the nearest degree.

31. a. $\dfrac{2\pi}{9}$ b. $\dfrac{\pi}{2}$ c. $\dfrac{5\pi}{3}$
32. a. 2 b. -3 c. 0.5
33. a. -0.42 b. 0.4 c. 7
34. a. 12 b. -1.5 c. 4.712389

Find the exact value of the positive angle less than one revolution in Problems 35–36 that is coterminal with each of the given angles. Also, find the reference angle for each.

35. a. 3π b. $\dfrac{13\pi}{6}$
    c. $-\pi$ d. 7
    e. -7 f. $-\dfrac{5\pi}{4}$

36. a. $-\dfrac{\pi}{4}$ b. $\dfrac{17\pi}{4}$
    c. $\dfrac{11\pi}{3}$ d. -2
    e. 8 f. $-\dfrac{\pi}{6}$

Find the positive angle less than one revolution correct to four decimal places so that it is coterminal with each of the angles in Problems 37–40.

37. a. 9 b. -5 c. $\sqrt{50}$
38. a. -6 b. 6.2832 c. $3\sqrt{5}$
39. a. 30 b. -3.1416 c. $-\dfrac{5\pi}{3}$
40. a. -0.7854 b. 6.8068 c. $\dfrac{9\pi}{4}$

LEVEL 2

41. State inequalities (positive angles less than one revolution) that summarize each verbal statement. Use degree measure.
    a. α is in Quadrant I   b. β is in Quadrant II
    c. γ is in Quadrant III   d. φ is in Quadrant IV
42. State inequalities (less than one revolution) that summarize each verbal statement. Use radian measure.
    a. δ is a positive angle in Quadrant I or II (including $\dfrac{\pi}{2}$).
    b. ω is a positive angle in Quadrant II or III (including π).
    c. ϕ is in Quadrant IV or I (including 0).
    d. λ is a negative angle in Quadrant I or II (including $\dfrac{\pi}{2}$).

43. Make a statement about the angle θ.
    a. $180° < \theta < 360°$ b. $\dfrac{3\pi}{2} < \theta < 2\pi$
    c. $0° < 2\theta < 180°$ d. $0° < \dfrac{1}{2}\theta < 90°$
44. Make a statement about the angle θ.
    a. $-\dfrac{\pi}{2} < \theta < \dfrac{\pi}{2}$ b. $180° < \theta < 270°$
    c. $0 < 2\theta < -\pi$ d. $0° < \dfrac{1}{3}\theta < -90°$

45. If $0 \le \theta < \pi$, make a statement about
   a. 2θ **b.** 3θ
   c. $\dfrac{1}{2}\theta$ **d.** $\dfrac{1}{3}\theta$

46. If $-90° \le \theta < 90°$, make a statement about
   a. 2θ **b.** 3θ
   c. $\dfrac{1}{2}\theta$ **d.** $\dfrac{1}{3}\theta$

Find the length of the intercepted arc (to the nearest hundredth) if the central angle and radius are given in Problems 47 and 48.

47. **a.** $\theta = 1, r = 1$ m   **b.** $\theta = \dfrac{\pi}{3}, r = 4$ in.

48. **a.** $\theta = 2.34, r = 6$ cm   **b.** $\theta = \dfrac{2\pi}{3}, r = 15$ cm

49. How far (to the nearest hundredth cm) does the tip of an hour hand on a clock move in exactly 3 hours if the hour hand is 2.00 cm long?

50. A 50-cm pendulum on a clock swings through an angle of $100°$. How far (to the nearest cm) does the tip travel in one arc?

51. Through how many revolutions does a pulley with a 20.0-cm diameter turn when 1.00 m of cable is pulled through it without slippage? (See Figure 5.8.)

   HINT: 1 m = 100 cm; answer to the nearest hundredth revolution.

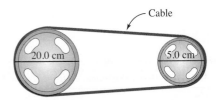

Figure 5.8 Pulley with two wheels

52. Answer the question in Problem 51 for a pulley with a 5.0-cm diameter.

53. If Columbia, South Carolina, is located at $34°$N latitude and $81°$W longitude, what is its approximate distance (to the nearest hundred miles) from Disneyland, which is located at $34°$N latitude and $118°$W longitude? (See Figure 5.9.) [Assume the radius of the earth is approximately 3,950 mi.]

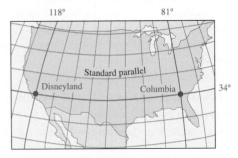

Figure 5.9 Heading to Disneyland

54. If New York City is at $40°$N latitude and $74°$W longitude, what is its approximate distance (to the nearest hundred miles) from the equator? [Assume the radius of the earth is 3,950 mi.]

55. A gas gauge (see Figure 5.10) is scaled from empty (E) to full (F) with an arc length of 5 cm with a radius of 2 cm. What is the angle (to the nearest degree) between the E and F readings?

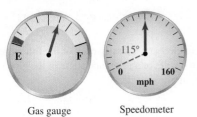

Gas gauge   Speedometer

Figure 5.10 Automobile gauges

56. A speedometer (see Figure 5.10) showing 80 mi/h forms an angle of 2 radians (approximately $115°$) with the mark for 0 mi/h. If the radius is 2 cm, find the length of the arc from 0 to 80.

57. How high will the weight in Figure 5.11 be lifted if the pulley of radius 10 in. is rotated through an angle of 4.5π? Express your answer to the nearest tenth of an inch.

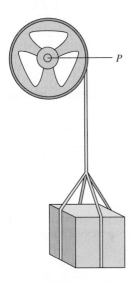

Figure 5.11 Pulley lift

LEVEL 3

58. A simple chain link from the drive sprocket (pedals) to the rear wheel of a bike is shown in Figure 5.12. If the sprocket has radius 4 cm and the drive sprocket has radius 10 cm, what is the number of rotations for the wheel for one rotation of the drive sprocket?

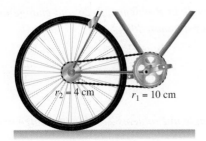

Figure 5.12 Drive sprocket

59. 𝕳𝖎𝖘𝖙𝖔𝖗𝖎𝖈𝖆𝖑 𝕼𝖚𝖊𝖘𝖙 The Hubble Space Telescope (HST) is a joint mission of the National Aeronautics and Space Administration and the European Space Agency (Figure 5.13).

Figure 5.13 Hubble Space Telescope

It was launched on April 25, 1990, and was upgraded on February 18, 1997. The reflecting telescope is deployed in low-earth orbit (600 km) with each orbit lasting about 95 minutes. The *linear velocity*, v, is calculated by the formula

$$v = \frac{r\theta}{t}$$

where r is the radius of the orbit and θ is the angle (in radians) per unit of time, t. If the radius of the earth is about 6,370 km, what is the linear velocity of the Hubble Space Telescope?

60. 𝕳𝖎𝖘𝖙𝖔𝖗𝖎𝖈𝖆𝖑 𝕼𝖚𝖊𝖘𝖙 In about 230 BC, a mathematician named Eratosthenes estimated the radius of the earth using the following information (see Figure 5.14). Syene and Alexandria in Egypt are on the same line of longitude. They are also 800 km apart. At noon on the longest day of the year, when the sun was directly overhead in Syene, Eratosthenes measured the sun to be 7.2° from the vertical in Alexandria. Because of the distance of the sun from the earth, he assumed that the sun's rays were parallel. Thus, he concluded that the central angle subtending rays from the center of the earth to Syene and Alexandria was also 7.2°. Using this information, find the approximate radius (to the nearest hundred km) of the earth.

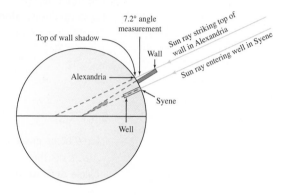

Figure 5.14 Earth's radius

5.2 Fundamentals

To introduce you to the trigonometric functions, we will consider a relationship between angles and circles. The **unit circle** is the circle centered at the origin with radius 1. The equation of the unit circle is

$$x^2 + y^2 = 1$$

Draw a unit circle with an angle θ in standard position, as shown in Figure 5.15.

Label the point $(1, 0)$ as B. A second point, A, is the intersection of the terminal side of the angle θ and the unit circle. If the coordinates of A are (a, b), we define two functions of θ:

$$c(\theta) = a \qquad \text{and} \qquad s(\theta) = b$$

The coordinates of point A, the point of intersection of the terminal side of θ and the unit circle, are found by evaluating c and s. For certain values, this evaluation is easy, as shown in the following example.

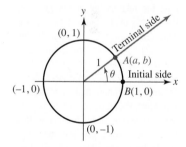

Figure 5.15 Unit circle with angle θ

EXAMPLE 1 Evaluating *c* and *s* using a unit circle

Find: a. $s(90°)$ **b.** $c\left(\frac{\pi}{2}\right)$ **c.** $c\left(\frac{3\pi}{2}\right)$ $s(270°)$ **e.** $c(-3.1416)$

Solution

a. $s(90°)$; this is the second component of the ordered pair (a, b) where $A(a, b)$ is the point of intersection of the terminal side of a 90° angle and a unit circle, as shown in Figure 5.16. By inspection, it is 1. Thus, $s(90°) = 1$.

b. $c\left(\frac{\pi}{2}\right) = 0$; for an angle of $\frac{\pi}{2}$ (a right angle), the first component of A is 0.

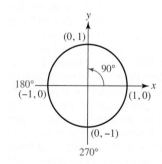

Figure 5.16 Angles with $\theta = 90°$ $\left(\text{or } \frac{\pi}{2}\right)$

c. $c\left(\dfrac{3\pi}{2}\right) = 0$; for an angle of $\dfrac{3\pi}{2}$, the first component of A is 0.

d. $s(270°) = -1$; for an angle of $270°$, the second component of A is -1.

e. $c(-3.1416) = -1$; for an angle of $-3.1416 \approx -\pi$, the first component of A is -1.   ∎

Since the title of this section is "Fundamentals," we might stop and ask, "Why is this worth knowing? When could I ever use this knowledge?" The answer is important: It sets the stage for the branch of mathematics called *trigonometry*. **Trigonometry** is the study of six functions and their relationships to one another, as well as the study of applications involving these functions.

Definition of the Trigonometric Functions

The function $c(\theta)$ is called the **cosine function**, and the function $s(\theta)$ is called the **sine function**. These functions, along with four others, make up the **trigonometric functions**.

 STOP This is an important definition. Spend some time learning this before going on. We list the cosine first because cosine always comes first (it is the **first** component); sine is second because it is the second component (also second alphabetically). Note that, with this definition, θ is no longer restricted as it was in Section 5.1.

> **UNIT CIRCLE DEFINITION**
> **TRIGONOMETRIC FUNCTIONS**
>
> Let θ be an angle with vertex at $(0, 0)$ drawn so that the initial side of θ is the positive x-axis, and $A(a, b)$ is the point of intersection of the terminal side of θ with the **unit circle**. Then the **unit circle definition** of the six trigonometric functions is defined as follows:
>
> **Cosine function**: $\cos \theta = a$ (read "cosine of θ is a")
>
> **Sine function**: $\sin \theta = b$ (read "sine of θ is b")
>
> **Tangent function**: $\tan \theta = \dfrac{b}{a}, a \neq 0$ (read "tangent of θ is b/a")
>
> **Secant function**: $\sec \theta = \dfrac{1}{a}, a \neq 0$ (read "secant of θ is $1/a$")
>
> **Cosecant function**: $\csc \theta = \dfrac{1}{b}, b \neq 0$ (read "cosecant of θ is $1/b$")
>
> **Cotangent function**: $\cot \theta = \dfrac{a}{b}, b \neq 0$ (read "cotangent of θ is a/b")

The angle θ is an important part of this definition. The words *cos*, *sin*, *tan*, and so on are meaningless without θ. You can speak about the cosine function, or you can speak about $\cos \theta$, but you cannot correctly write only *cos*. Also, take special note of the conditions on the tangent, secant, cosecant, and cotangent functions. These conditions exclude division by 0; for example, $a \neq 0$ means that $\sec \theta$ and $\tan \theta$ are not defined when θ is $90°$, $270°$, or any angle coterminal with either of these angles. Similarly, $b \neq 0$ means that $\csc \theta$ and $\cot \theta$ are not defined when θ is $0°$, $180°$, or any angle coterminal with these angles. Thus, *cosecant and cotangent are not defined for* $0°$ *or* $180°$. In this book, we assume all values that cause division by zero are excluded from the domain.

The angle θ in the unit circle definition is called the **argument** of the function. The argument, of course, does not need to be the same as the variable. For example, in $\cos(2\theta + 1)$, we say the *function* is $\cos(2\theta + 1)$, the *argument* is $2\theta + 1$, and the variable is θ.

Evaluation of Trigonometric Functions

In many applications, we will know the argument and will want to find one or more of its trigonometric functions. To help you see the relationship between the angle θ and the function in the definition, consider the following example.

EXAMPLE 2   Evaluating trigonometric functions using the definition

Use the definition to evaluate the following:

a. $\cos 110°$ **b.** $\sin 110°$ **c.** $\tan 110°$ **d.** $\sec 110°$ **e.** $\cos 200°$ **f.** $-\sin 50°$

g. $\sin(-50°)$ **h.** $\cos 4$

Solution   We draw a large circle with $r = 1$, as shown in Figure 5.17.

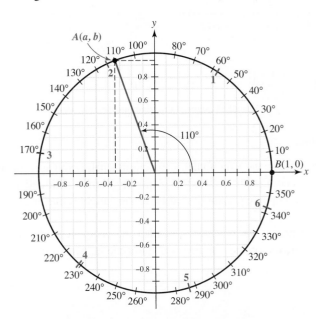

Figure 5.17 Approximate values for cosine and sine

Draw an angle where $\theta = 110°$, and estimate the coordinates of $A(a, b)$: $(a, b) \approx (-0.35, 0.95)$.

a. $\cos 110° \approx -0.35$ (cosine is the first component)

b. $\sin 110° \approx 0.95$ (sine is the second component)

c. $\tan 110° \approx \dfrac{0.95}{-0.35} \approx -2.7$

d. $\sec 110° \approx \dfrac{1}{-0.35} \approx -2.9$

e. Look at Figure 5.17, and *think* of an angle $\theta = 200°$. We are looking for the *first* component (remember *cosine is first, sine is second*); $\cos 200° \approx -0.9$.

f. By looking at Figure 5.17, we estimate the *second* component of the point determined by $\theta = 50°$; $\sin 50° \approx 0.75$. Thus, $-\sin 50°$ is the opposite of $\sin 50°$: $-\sin 50° \approx -0.75$.

g. From Figure 5.17, we estimate the *second* component of the point determined by $\theta = -50°$: $\sin(-50°) \approx -0.75$.

h. From Figure 5.17, we estimate the *first* component of the point determined by $\theta = 4$: $\cos 4 \approx -0.66$. ∎

Notice from Example 2 that some of the trigonometric functions for an angle, 110° for example, are positive and others are negative. In many applications, it is important to know whether a particular function is positive or negative for a particular angle. Consider the point (a, b) in the definition of the trigonometric functions. In Quadrant I, a and b are both positive, so all six trigonometric functions must be positive. In Quadrant II, a is negative and b is positive, so from the definition of trigonometric functions, all are negative except the sine and cosecant. In Quadrant III, a and b are both negative, so all the functions are negative except tangent and cotangent because those are ratios of two negatives. Finally, in Quadrant IV, a is positive and b is negative, so all are negative except cosine and secant. These results are summarized in Figure 5.18.

Easy-to-remember form:

$$A\ S_{\text{mart}}\ T_{\text{rig}}\ C_{\text{lass}}$$

🕱 Learn the signs of the trigonometric functions in each of the quadrants. 🕱

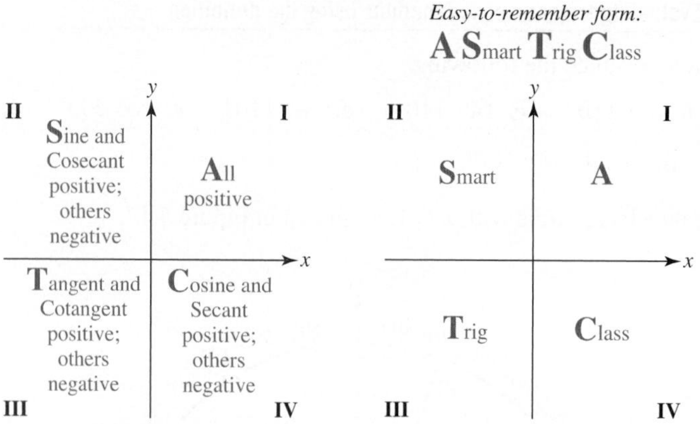

Figure 5.18 Signs of the trigonometric functions

Trigonometric Identities

Certain relationships among the trigonometric functions enable you to evaluate these functions. These relationships are commonly referred to as the **fundamental identities**. Recall from algebra that an **identity** is an open equation (has at least one variable) that is true for all values in the domain.

The first three identities we consider are called the **reciprocal identities**. Remember from arithmetic that 5 and $\frac{1}{5}$ are reciprocals, as are $\frac{2}{3}$ and $\frac{3}{2}$; in general, two numbers are reciprocals if their product is 1.

RECIPROCAL IDENTITIES

1. $\sec\theta = \dfrac{1}{\cos\theta}$ The cosine and secant are reciprocal functions.

2. $\csc\theta = \dfrac{1}{\sin\theta}$ The sine and cosecant are reciprocal functions.

3. $\cot\theta = \dfrac{1}{\tan\theta}$ The tangent and cotangent are reciprocal functions.

It is easy to prove that these are identities by looking at the definitions of secant, cosecant, and cotangent. Also, notice that each of these can be rewritten in other forms. For example, identity 1 can be written as:

$$\cos\theta\sec\theta = 1 \qquad \text{or} \qquad \cos\theta = \dfrac{1}{\sec\theta}$$

Since the method shown in Example 2 for evaluating the trigonometric functions is not practical, other procedures are often used. Today, we commonly use calculators.

▢ COMPUTATIONAL WINDOW   ▬▢✕

The procedure for evaluating the trigonometric functions is straightforward, but you should be aware of several details:

1. Note the unit of measure used: degree or radian. Calculators have a variety of ways of changing from radian to degree format, so consult your owner's manual to find out how your particular calculator does it. Most, however, simply have a switch (similar to an on/off switch) that sets the calculator in either degree or radian mode. *Remember:* If later in the course, when you are evaluating trigonometric functions, you suddenly start obtaining strange answers and have no idea what you are doing wrong, double-check to make sure your calculator is in the proper mode.

2. With most scientific calculators: `Input angle measure` `Input trig function` With most graphing calculators: `Input trig function` `Input angle measure`

3. You must remember which functions are reciprocals, since most calculators do not have $\sec\theta$, $\csc\theta$, and $\cot\theta$ keys. These calculators, however, do have a reciprocal key, which is labeled `1/x` or `x⁻¹`. For the secant, cosecant, and cotangent functions, evaluate the reciprocal function and then take the reciprocal of the result. *Note:* The key labeled `sin⁻¹` does NOT represent the reciprocal of sine. Also, `cos⁻¹` and `tan⁻¹` do NOT represent reciprocals.

EXAMPLE 3   Calculator evaluation of the trigonometric functions

Rework Example 2 using a calculator. Check your calculator answers with the answers shown here.

a. $\cos 110°$ **b.** $\sin 110°$ **c.** $\tan 110°$ **d.** $\sec 110°$ **e.** $\cos 200°$
f. $\sin 50°$ **g.** $\sin(-50°)$ **h.** $\cos 4$

Solution

   a. $\cos 110° \approx -0.3420201433$
   b. $\sin 110° \approx 0.9396926208$
   c. $\tan 110° \approx -2.747477419$
   d. $\sec 110° \approx -2.9238044$
   e. $\cos 200° \approx -0.9396926208$
   f. $-\sin 50° \approx -0.7660444431$
   g. $\sin(-50°) \approx -0.7660444431$ Do not confuse this with $-\sin 50°$; take the opposite of the angle, then evaluate the function.

   h. $\cos 4 \approx -0.6536436209$ If you obtained 0.9975640503, you forgot to change to radian mode. ∎

The next two identities we consider are called the **ratio identities**; they express the tangent and cotangent functions using sine and cosine.

RATIO IDENTITIES

4. $\tan\theta = \dfrac{\sin\theta}{\cos\theta}$

5. $\cot\theta = \dfrac{\cos\theta}{\sin\theta}$

STOP You will need to know these ratio identities to write the tangent and cotangent functions in terms of sines and cosines.

To prove these identities, we use the definition of the trigonometric functions:

$$\tan\theta = \frac{b}{a} \qquad \text{Definition of tangent}$$

$$= \frac{\sin\theta}{\cos\theta} \qquad \text{Substitution of values from the definition:}$$

$$\sin\theta = b \text{ and } \cos\theta = a.$$

The ratio identity of the cotangent is proved in exactly the same way (see Problem 38).

The last three identities are called the **Pythagorean identities** because they follow directly from the Pythagorean theorem. Since (a, b) is on a unit circle (see Figure 5.19), the Pythagorean theorem tells us

$$a^2 + b^2 = 1$$

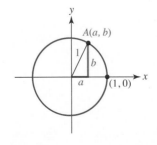

Figure 5.19 (a, b) **on a unit circle**

From the definition of sine and cosine (and by substitution), we obtain

$$(\cos\theta)^2 + (\sin\theta)^2 = 1$$

In trigonometry, we simplify the notation involving the square of a trigonometric function by writing the previous equation as follows:

PYTHAGOREAN IDENTITIES

6. $\cos^2\theta + \sin^2\theta = 1$
7. $1 + \tan^2\theta = \sec^2\theta$
8. $\cot^2\theta + 1 = \csc^2\theta$

 STOP   *Memorize the eight identities presented in this section. They are known as the **fundamental identities**, and a great deal of the work we do in trigonometry, and even in more advanced work, is dependent on them.*

To prove identities 7 and 8, we need only divide both sides of $a^2 + b^2 = 1$ by a^2 (for identity 7) or b^2 (for identity 8):

$$a^2 + b^2 = 1 \qquad \text{Pythagorean theorem}$$

$$\frac{a^2}{a^2} + \frac{b^2}{a^2} = \frac{1}{a^2} \qquad \text{Divide both sides by } a^2 (a \neq 0).$$

$$1 + \left(\frac{b}{a}\right)^2 = \left(\frac{1}{a}\right)^2 \qquad \text{Properties of exponents}$$

$$1 + (\tan\theta)^2 = (\sec\theta)^2 \qquad \text{Definition of } \tan\theta \text{ and } \sec\theta, \text{ and substitution}$$

$$1 + \tan^2\theta = \sec^2\theta \qquad \text{Notational change}$$

The proof of identity 8 is left as an exercise (see Problem 40).

These Pythagorean identities are frequently written in a different form. For example, we can solve identity 7 for $\tan\theta$:

$$1 + \tan^2\theta = \sec^2\theta \qquad \text{Identity 7}$$

$$\tan^2\theta = \sec^2\theta - 1 \qquad \text{Subtract 1 from both sides.}$$

$$\tan\theta = \pm\sqrt{\sec^2\theta - 1} \qquad \text{Do not forget the } \pm \text{ when you use the square root property from algebra.}$$

EXAMPLE 4   Using fundamental identities

Write all the trigonometric functions in terms of $\sin\theta$ by using the eight fundamental identities.

Solution

a. $\sin\theta = \sin\theta$ — Begin with the easiest function to find.

b. $\cos^2\theta + \sin^2\theta = 1$ — Use identity 6 to find $\cos\theta$.
$\cos^2\theta = 1 - \sin^2\theta$ — Subtract $\sin^2\theta$ from both sides.
$\cos\theta = \pm\sqrt{1 - \sin^2\theta}$ — Square root property

c. $\tan\theta = \dfrac{\sin\theta}{\cos\theta}$ — Identity 4

$= \dfrac{\sin\theta}{\pm\sqrt{1 - \sin^2\theta}}$ — Substitute from part **b***.

---

*In algebra, you needed to rationalize the denominator; however, since we work with many reciprocals in trigonometry, we generally relax the requirement that expressions with a radical in a denominator be rationalized.

d. $\cot\theta = \dfrac{1}{\tan\theta}$ Identity 3

$= \dfrac{1}{\dfrac{\sin\theta}{\pm\sqrt{1-\sin^2\theta}}}$ Substitute from part **c**.

$= \dfrac{\pm\sqrt{1-\sin^2\theta}}{\sin\theta}$ Invert and multiply.

e. $\csc\theta = \dfrac{1}{\sin\theta}$ Identity 2

f. $\sec\theta = \dfrac{1}{\cos\theta}$ Identity 1

$= \dfrac{1}{\pm\sqrt{1-\sin^2\theta}}$ Substitute from part **b**.

■

Modeling with Trigonometric Functions

In Section 3.4, we introduced the notion of a parameter, so we can graph a circle using θ as a parameter to define x and y.

EXAMPLE 5   Graphing with a unit circle parameter

Plot the curve represented by the parametric equations

$$x = \cos\theta, \qquad y = \sin\theta$$

Solution

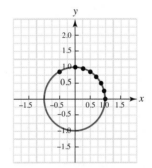

θ	0°	15°	30°	45°	60°	75°	90°	...	120°	...
x	1.00	0.97	0.87	0.71	0.50	0.26	0.00	c	−0.50	c
y	0.00	0.26	0.50	0.71	0.87	0.97	1.00	c	0.87	c

Figure 5.20 Unit circle

These points are plotted in Figure 5.20.

If the plotted points are connected, you can see that the curve is a circle with center at the origin and radius 1. We can also show this by eliminating the parameter:

$x^2 + y^2 = (\cos\theta)^2 + (\sin\theta)^2$ Substitute $x = \cos\theta$ and $y = \sin\theta$

$= \cos^2\theta + \sin^2\theta$ Notational change

$= 1$ Pythagorean identity

We recognize the equation

$$x^2 + y^2 = 1$$

as the equation of the unit circle, with center at $(0, 0)$ and radius 1. We can use these unit circle parameters for circles whose radii are not 1, say, r. Let $x = r\cos\theta$ and $y = r\sin\theta$, so that

$$x^2 + y^2 = (r\cos\theta)^2 + (r\sin\theta)^2 = r^2\cos^2\theta + r^2\sin^2\theta = r^2$$

which is a circle with center $(0, 0)$ and radius r.

■

EXAMPLE 6   Modeling projectile motion   `MODELING APPLICATION`

Graph the path of a projectile with initial velocity $v_0 = 50$ ft/s and angle $\theta = 45°$. Also, estimate how far downrange the projectile will hit, and the time until impact.

Solution

Step 1:   *Understand the problem.* We need some additional information to answer this question. Using Newton's laws of motion, as well as concepts from calculus, it can be shown that, if air resistance is neglected, the path of a projectile is given by the parametric equations

$$x = (v_0 \cos \theta)t, \qquad y = h_0 + (v_0 \sin \theta)t - 16t^2$$

where h_0 is the initial height off the ground, and v_0 is the initial velocity of the projectile in the direction of θ with the horizontal $(0° \le \theta \le 180°)$, as shown in Figure 5.21.

Step 2:   *Devise a plan.* The variables x and y represent the horizontal and vertical distances, respectively, measured in feet, and the parameter t represents the time in seconds after firing the projectile. We see $h_0 = 0$, $v_0 = 50$, and we use the parametric equations from Newton's laws.

Step 3:   *Carry out the plan.* We have

$$x = (50 \cos 45°)t, \qquad y = (50 \sin 45°)t - 16t^2$$

We can set up a table of values, or we can use a graphing calculator to show the values correct to the nearest tenth.

t	0	0.5	1	1.5	2	2.5
x	0.0	17.7	35.4	53.0	70.7	88.4
y	0.0	13.7	19.4	17.0	6.7	−11.6

Sketch the curve as shown in Figure 5.22.

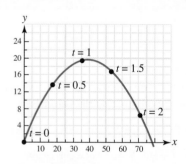

Figure 5.22 Path of a specific projectile

If we wish to estimate how far downrange the projectile will hit or when it will impact, we can estimate the point of intersection of the curve with the x-axis [or solve $y = 0$ for t, and find $x(t)$]. We obtain the following values:

t	2.1	2.2	2.3
x	74.2	77.8	81.3
y	3.7	0.34	−3.3

We see that the time until impact is between 2.2 and 2.3 seconds, and the point of impact is between 77.8 and 81.3 ft downrange.

Step 4:   *Look back.* We estimate that the impact will occur about 80 ft downrange after 2.2 sec.

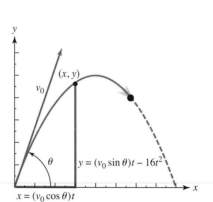

Figure 5.21 Path of a projectile

$y = (v_0 \sin \theta)t - 16t^2$

$x = (v_0 \cos \theta)t$

PROBLEM SET 5.2

Find the coordinates of the requested point in Problems 1–4. Assume that each circle is a unit circle.

1. Point P **2.** Point R

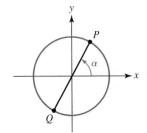 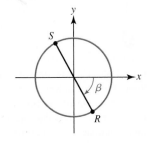

3. Point Q **4.** Point S

5. Name the reciprocal of each given trigonometric function.
  a. cosine   **b.** cotangent
  c. sine   **d.** secant
  e. cosecant   **f.** tangent

6. a. Solve $\cos^2\theta + \sin^2\theta = 1$ for $\cos\theta$.
  b. Solve $\cos^2\theta \sin^2\theta = 1$ for $\sin\theta$.
  c. Solve $\cot^2\theta + 1 = \csc^2\theta$ for $\cot\theta$.

WHAT IS WRONG, *if anything, with each statement in Problems 7–12? Explain your reasoning.*

7. $\sin 30° = -0.9880316241$
8. $\sin^2 + \cos^2 = 1$
9. Sine is positive in Quadrant IV.
10. Cosine is positive in Quadrant IV.
11. Both sine and cosine are negative in Quadrant II.
12. $\sin 2\theta = 2\sin\theta$

Use Figure 5.23 and the definitions of cosine and sine to evaluate the functions in Problems 13–16 to one decimal place. Check your answers using a calculator.

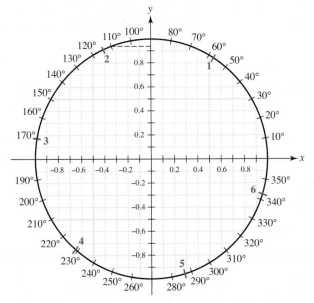

Figure 5.23 Using a unit circle definition to evaluate circle functions

13. a. $\cos 50°$ **b.** $\sin 70°$ **c.** $\sin(-150°)$
14. a. $\cos 200°$ **b.** $\cos 250°$ **c.** $\sin(-45°)$
15. a. $\cos 2$ **b.** $\sin 4$ **c.** $\tan(-6)$
16. a. $\sec 3$ **b.** $\cot 5$ **c.** $\csc(-3)$

Use a calculator to evaluate the functions given in Problems 17–22.

17. a. $\cos 50°$ **b.** $\cos 70°$
  c. $\sec 70°$ **d.** $\sin 20°$
  e. $\sin(-95°)$ **f.** $\cot 250°$

18. a. $\tan\dfrac{2\pi}{3}$ **b.** $\cot\dfrac{\pi}{3}$
  c. $\csc\dfrac{\pi}{6}$ **d.** $2\sin 15°$
  e. $\sin(2\cdot 15°)$ **f.** $5\sec 15°$

19. a. $\cos\left(\dfrac{103°}{2}\right)$ **b.** $\dfrac{\cos 103°}{2}$
  c. $\dfrac{\cot 103°}{2}$ **d.** $\cos\left(-\dfrac{5\pi}{4}\right)$
  e. $\sin\left(-\dfrac{5\pi}{4}\right)$ **f.** $-\cot\left(-\dfrac{5\pi}{4}\right)$

20. a. $\sin(-2)$ **b.** $-\sin 2$
  c. $\sec 2$ **d.** $\cos(-4)$
  e. $-\cos(-4)$ **f.** $-\csc(-4)$

21. a. $\dfrac{1}{\csc 2}$ **b.** $\dfrac{1}{\sec 3.5}$
  c. $\dfrac{1}{\cot 4.5}$ **d.** $\dfrac{1}{\sin 183°}$

22. a. $\dfrac{1}{\cos 215°}$ **b.** $\dfrac{1}{\tan 335°}$
  c. $\dfrac{\sin 50°}{\cos 50°}$ **d.** $\dfrac{\cos 5}{\sin 5}$

Tell whether each of the functions in Problems 23–28 is positive or negative. You should be able to do this without using a calculator.

23. a. sine, Quadrant I   **b.** cosine, Quadrant II
  c. tangent, Quadrant III   **d.** cotangent, Quadrant IV
24. a. secant, Quadrant I   **b.** cosecant, Quadrant II
  c. sine, Quadrant III   **d.** cosine, Quadrant IV
25. a. $\sin 1$ **b.** $\sin 2$
  c. $\sin 3$ **d.** $\sin 4$
26. a. $\sin 5$ **b.** $\sin 6$
  c. $\sin 7$ **d.** $\sin 8$
27. a. $\cos(-1)$ **b.** $\tan 2$
  c. $\cot 3$ **d.** $\sec 4$
28. a. $\csc 5$ **b.** $\cot 6$
  c. $\sin(-3)$ **d.** $\cos 10$

Specify in which quadrant(s) a standard-position angle θ could lie and satisfy the conditions stated in Problems 29–36.

29. a. $\cos\theta = 0.1234$ **b.** $\sin\theta > 0$
30. a. $\sin\theta = -0.85$ **b.** $\cos\theta < 0$
31. a. $\sec\theta = -1.45$ **b.** $\tan\theta < 0$
32. a. $\csc\theta = -1.35$ **b.** $\cot\theta > 0$
33. a. $\sin\theta < 0$ and $\tan\theta > 0$ **b.** $\sin\theta > 0$ and $\tan\theta < 0$
34. a. $\cos\theta > 0$ and $\tan\theta < 0$ **b.** $\cos\theta < 0$ and $\tan\theta > 0$

35. a. $\cos\theta = \sqrt{0.195}$ and $\tan\theta > 0$
 b. $\sin\theta = -\sqrt{1-\cos^2\theta}$ and $\tan\theta > 0$

36. a. $\csc\theta = \frac{3}{2}$ and $\tan\theta < 0$
 b. $\sec\theta = -\sqrt{1+\tan^2\theta}$ and $\sin\theta > 0$

LEVEL 2

Prove the fundamental identity in Problems 37–40 using the unit circle definition.

37. $\csc\theta = \dfrac{1}{\sin\theta}$

38. $\cot\theta = \dfrac{\cos\theta}{\sin\theta}$

39. $1 + \tan^2\theta = \sec^2\theta$
40. $\cot^2\theta + 1 = \csc^2\theta$

Write all the trigonometric functions in terms of the functions given in Problems 41–46. You do not need to rationalize denominators.

41. $\sin\theta$ **42.** $\cos\theta$
43. $\tan\theta$ **44.** $\cot\theta$
45. $\sec\theta$ **46.** $\csc\theta$

Use the fundamental identities to find the other trigonometric functions of θ using the information given in Problems 47–50. State the answers correct to the nearest hundredth.

47. a. $\cos\theta = \frac{5}{13}$; $\tan\theta > 0$ **b.** $\cos\theta = -\frac{5}{13}$; $\tan\theta < 0$

48. a. $\cos\theta = -\frac{5}{13}$; $\tan\theta > 0$ **b.** $\cos\theta = \frac{5}{13}$; $\tan\theta < 0$

49. a. $\tan\theta = \frac{5}{12}$; $\sin\theta > 0$ **b.** $\tan\theta = -\frac{5}{12}$; $\sin\theta > 0$

50. a. $\sin\theta = 0.65$; $\sec\theta > 0$ **b.** $\cot\theta = 1.25$; $\cos\theta < 0$

Sketch the curves represented by the parametric equations in Problems 51–56 by plotting points.

51. $x = 4\cos\theta,\ y = 3\sin\theta,\ 0° \le \theta \le 360°$
52. $x = 5\cos\theta,\ y = 2\sin\theta,\ 0° \le \theta \le 360°$
53. $x = \cos\theta,\ y = \sin\theta,\ 0° \le \theta \le 180°$
54. $x = 10\cos\theta,\ y = 10\sin\theta,\ 90° \le \theta \le 270°$
55. $x = 8\cos\theta,\ y = 8\sin\theta,\ 45° \le \theta \le 135°$
56. $x = 1 + \sin\theta,\ y = 1 - \cos\theta,\ 0° \le \theta \le 360°$

LEVEL 3

57. Graph the path of the projectile in Example 6 for an angle of 60° instead of 45°. How does increasing the size of this angle affect the point of impact as well as the time of flight?

58. Graph the path of the projectile in Example 6 for an angle of 30° instead of 45°. How does decreasing the size of this angle affect the point of impact as well as the time of flight?

59. A cannon is fired with an initial speed of 300 ft/s at an angle of 42° to the horizontal. Neglecting air resistance, sketch the path of the projectile. What is the time of impact (to the nearest tenth of a second)?

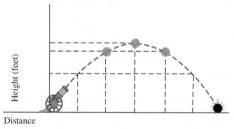

Distance

60. The shortest distance from home plate to an outfield fence at the old Fenway Park in Boston is 302 ft. If a batter hits a baseball 3 ft above the ground with an initial velocity of 100 ft/s (≈ 68 mi/h) and an angle of 45° with respect to the ground, will the baseball clear a 10-ft-high fence located 302 ft from home plate?

5.3 Trigonometric Functions of Any Angle

It is not always convenient to work with the unit circle definition of the trigonometric functions developed in the previous section. In this section, we use a result from geometry concerning similar triangles to generalize the unit circle definition. This ratio definition of the trigonometric functions will be derived in Example 1.

Ratio Definition of the Trigonometric Functions

Suppose you want to find the trigonometric functions of an angle whose terminal side passes through some known point, $(3, 4)$. To apply the unit circle definition, we need to find the point (a, b), as shown in Figure 5.24.

Let $(3, 4)$ be denoted by P and (a, b) by A. Let B be the point $(a, 0)$ and Q be the point $(3, 0)$; then $|\overline{OA}| = 1$, $|\overline{OP}| = \sqrt{3^2 + 4^2} = 5$. Now consider $\triangle AOB$ and $\triangle POQ$. Recall from geometry that two triangles are **similar** if two angles of one are congruent to two angles of the other. For these triangles, $\angle OBA$ is congruent to $\angle OQP$ because both are right angles, and $\angle O$ is congruent to $\angle O$ because equal angles are also congruent. Thus, these triangles are similar, which is denoted by

$$\triangle AOB \sim \triangle POQ$$

The important property of similar triangles is that corresponding parts of similar triangles are proportional. Thus,

$$a = \frac{a}{1} = \frac{3}{5} \quad \text{and} \quad b = \frac{b}{1} = \frac{4}{5}$$

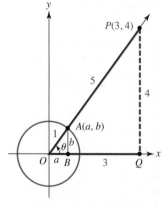

Figure 5.24 Finding (a, b) on a unit circle for the point $P(3, 4)$

Thus, $\cos\,\theta = \frac{3}{5}$, $\sin\,\theta = \frac{4}{5}$, and $\tan\,\theta = \frac{4/5}{3/5} = \frac{4}{3}$. The reciprocals are $\sec\,\theta = \frac{5}{3}$, $\csc\,\theta = \frac{5}{4}$, and $\cot\,\theta = \frac{3}{4}$.

EXAMPLE 1   Coordinates of $P(x, y)$ on the terminal side of angle

Let P be a point on the terminal side of a standard-position angle θ. Find the coordinates of P.

Solution   Consider the point P with coordinates (x, y), and let (a, b) be the point of intersection of the terminal side of θ and a unit circle, as shown in Figure 5.25.

Now we know (from the unit circle definition of the trigonometric functions) that $a = \cos\theta$ and $b = \sin\theta$. As before,

$$\Delta AOB \sim \Delta POQ$$

By the Pythagorean theorem, the hypotenuse is $r = \sqrt{x^2 + y^2}$, and since the triangles are similar, corresponding sides are proportional. In particular,

$$\frac{x}{r} = \frac{a}{1} \qquad \frac{y}{r} = \frac{b}{1}$$

Thus,

$$\frac{x}{r} = a \qquad \frac{y}{r} = b$$

$$\frac{x}{r} = \cos\theta \qquad \frac{y}{r} = \sin\theta$$

$$x = r\cos\theta \qquad y = r\sin\theta$$

In other words, the coordinates of P are $(r\cos\theta,\ r\sin\theta)$. ∎

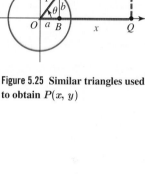

Figure 5.25 Similar triangles used to obtain $P(x, y)$

We find it worthwhile to continue the discussion started in Example 1. Since

$$\Delta AOB \sim \Delta POQ$$

we know there are six possible ratios of the three sides of these similar triangles:

$$a = \frac{a}{1} = \frac{x}{r} \qquad \frac{1}{a} = \frac{r}{x}$$

$$b = \frac{b}{1} = \frac{y}{r} \qquad \frac{1}{b} = \frac{r}{y}$$

$$\frac{b}{a} = \frac{y/r}{x/r} = \frac{y}{x} \qquad \frac{a}{b} = \frac{x}{y}$$

These ratios lead to an alternative definition of the trigonometric functions that allows you to choose *any* point (x, y). In practice, this definition of the trigonometric functions is more frequently used than the unit circle definition.

STOP   *Once again, we present a fundamental definition for the study of trigonometry. You should memorize this definition.*

RATIO DEFINITION TRIGONOMETRIC FUNCTIONS

Let θ be any angle in standard position, and let $P(x, y)$ be any point on the terminal side of the angle at a distance of r from the origin $(r \neq 0)$. Then, the **ratio definition of the trigonometric functions** is:

$$\cos\theta = \frac{x}{r} \qquad\qquad \sin\theta = \frac{y}{r} \qquad\qquad \tan\theta = \frac{y}{x} \quad (x \neq 0)$$

$$\sec\theta = \frac{r}{x} \quad (x \neq 0) \qquad \csc\theta = \frac{r}{y} \quad (y \neq 0) \qquad \cot\theta = \frac{x}{y} \quad (y \neq 0)$$

Note from the Pythagorean theorem that r is determined by x and y:

$$r^2 = x^2 + y^2 \qquad \text{Pythagorean theorem}$$

$$r = \pm\sqrt{x^2 + y^2} \qquad \text{Square root property}$$

$$r = \sqrt{x^2 + y^2} \qquad \text{Positive value since } r \text{ is a distance}$$

Note that since r is a distance, it is always positive; however, (x, y) is a point, so x and y could be positive, negative, or zero. This is why there are exclusions for certain parts of the definition.

EXAMPLE 2   Evaluating trigonometric functions given a point on the terminal side

Find the values of the six trigonometric functions for an angle θ in standard position with the terminal side passing through $(-5, -2)$.

Solution $x = -5, \quad y = -2, \quad$ and $\quad r = \sqrt{(-5)^2 + (-2)^2} = \sqrt{29}$. Thus,

$$\cos\theta = \frac{-5}{\sqrt{29}} \qquad\qquad \sec\theta = \frac{-\sqrt{29}}{5}$$

$$\sin\theta = \frac{-2}{\sqrt{29}} \qquad\qquad \csc\theta = \frac{-\sqrt{29}}{2}$$

$$\tan\theta = \frac{-2}{-5} = \frac{2}{5} \qquad\qquad \cot\theta = \frac{5}{2}$$

Exact Values and the Reduction Principle

There are certain angles for which we can find the **exact value** (as opposed to calculator approximations) of the trigonometric functions.

EXAMPLE 3   Exact values of trigonometric functions of $\frac{\pi}{4} = 45°$

Find the exact values of the trigonometric functions of $\frac{\pi}{4}$ (see Figure 5.26).

Solution

$$\cos\frac{\pi}{4} = \frac{x}{r} \qquad\qquad \text{From the ratio definition (this is true for any angle).}$$

$$= \frac{x}{\sqrt{x^2 + y^2}} \qquad r = \sqrt{x^2 + y^2} \text{ for any angle.}$$

$$= \frac{x}{\sqrt{x^2 + x^2}} \qquad \text{If } \theta = \frac{\pi}{4}, \text{ then } \theta \text{ bisects Quadrant I so that } x = y.$$

$$= \frac{x}{\sqrt{2x^2}} \qquad \text{Simplify.}$$

$$= \frac{x}{x\sqrt{2}} \qquad \sqrt{x^2} = |x| = x \text{ since } x \text{ is positive in Quadrant I.}$$

$$= \frac{1}{\sqrt{2}} \qquad \text{This can also be written as } \frac{1}{2}\sqrt{2}.$$

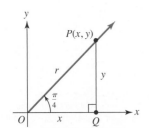

Figure 5.26 Right triangle with $\theta = \dfrac{\pi}{4} = 45°$

Similarly (remember that $y = x$):

$$\sin \frac{\pi}{4} = \frac{y}{r} = \frac{y}{\sqrt{x^2 + y^2}} = \frac{x}{\sqrt{x^2 + x^2}} = \frac{x}{\sqrt{2x^2}} = \frac{1}{\sqrt{2}}$$

This can, likewise, be written as $\frac{1}{2}\sqrt{2}$. By the ratio identity, we find

$$\tan \frac{\pi}{4} = \frac{\sin \dfrac{\pi}{4}}{\cos \dfrac{\pi}{4}} = \frac{\dfrac{1}{2}\sqrt{2}}{\dfrac{1}{2}\sqrt{2}} = 1$$

Finally, reciprocals give us the reciprocal functions:

$$\sec \frac{\pi}{4} = \sqrt{2}, \csc \frac{\pi}{4} = \sqrt{2}, \text{ and } \cot \frac{\pi}{4} = 1$$

EXAMPLE 4   Exact values of trigonometric functions of $\dfrac{\pi}{6} = 30°$

Evaluate the trigonometric functions of $30°$.

Solution   Consider not only the standard-position angle $30°$, but also the standard-position angle $-30°$. Choose $P_1(x, y)$ and $P_2(x, -y)$, respectively, on the terminal sides, as shown in Figure 5.27.

Since $\angle OQP_1$ and $\angle OQP_2$ are right angles, $\angle OP_1Q = 60°$ and $\angle OP_2Q = 60°$. Thus, $\triangle OP_1P_2$ is an equiangular triangle. (That is, all angles measure $60°$.) From geometry, we know that an equiangular triangle has sides the same length. Thus, $2y = r$. Notice the following relationship between x and y:

$$r^2 = x^2 + y^2 \qquad \textsf{Pythagorean theorem}$$

$$(2y)^2 = x^2 + y^2 \qquad \textsf{Since } 2y = r$$

$$3y^2 = x^2 \qquad \textsf{Simplify, and subtract } y^2 \textsf{ from both sides.}$$

$$\sqrt{3}\,|y| = |x| \qquad \textsf{Simplify; square root property}$$

$$x = \sqrt{3}y \qquad \textsf{Since } x > 0, y > 0 \textsf{ (first quadrant)}$$

$$\cos 30° = \frac{x}{r} = \frac{\sqrt{3}y}{2y} = \frac{\sqrt{3}}{2} \quad \sin 30° = \frac{y}{r} = \frac{y}{2y} = \frac{1}{2} \quad \tan 30° = \frac{y}{x} = \frac{y}{\sqrt{3}y} = \frac{1}{\sqrt{3}} \text{ or } \frac{1}{3}\sqrt{3}$$

$$\sec 30° = \frac{2}{\sqrt{3}} \text{ or } \frac{2}{3}\sqrt{3} \qquad \csc 30° = 2 \qquad \qquad \cot 30° = \sqrt{3}$$

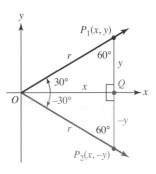

Figure 5.27 Right triangle with $\theta = 30°$, with reflection

---

EXAMPLE 5   Exact values of trigonometric functions of $\dfrac{\pi}{3} = 60°$

Find the exact values of the trigonometric functions of $\dfrac{\pi}{3}$ (see Figure 5.28).

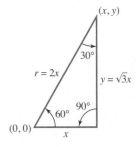

Figure 5.28 A right triangle with $\theta = \dfrac{\pi}{3}$

Solution   For an angle of $60°$ instead of $30°$, it can be shown that $r = 2x$, so that

$$r^2 = x^2 + y^2 \quad \text{Pythagorean theorem}$$

$$(2x)^2 = x^2 + y^2 \quad \text{Since } r = 2x$$

$$3x^2 = y^2 \quad \text{Simplify, and subtract } x^2 \text{ from both sides.}$$

$$\sqrt{3}\,|x| = |y| \quad \text{Simplify; square root property}$$

$$y = \sqrt{3}x \quad \text{Since } x > 0,\ y > 0 \text{ (first quadrant)}$$

$$\cos 60° = \frac{x}{r} = \frac{x}{2x} = \frac{1}{2} \qquad \sin 60° = \frac{y}{r} = \frac{\sqrt{3}x}{2x} = \frac{\sqrt{3}}{2} \qquad \tan 60° = \frac{y}{x} = \frac{\sqrt{3}x}{x} = \sqrt{3}$$

$$\sec 60° = 2 \qquad\qquad \csc 60° = \frac{2}{\sqrt{3}} = \frac{2}{3}\sqrt{3} \qquad\qquad 60° = \frac{1}{\sqrt{3}} = \frac{1}{3}\sqrt{3}$$

The results of Examples 3–5 are summarized in Table 5.3. You should now be able to verify all the entries in this table.

TABLE 5.3	**Table of Exact Values**						
Angle θ	0	$\dfrac{\pi}{6}$	$\dfrac{\pi}{4}$	$\dfrac{\pi}{3}$	$\dfrac{\pi}{2}$	π	$\dfrac{3\pi}{2}$
$\cos\theta$	1	$\dfrac{\sqrt{3}}{2}$	$\dfrac{\sqrt{2}}{2}$	$\dfrac{1}{2}$	0	-1	0
$\sin\theta$	0	$\dfrac{1}{2}$	$\dfrac{\sqrt{2}}{2}$	$\dfrac{\sqrt{3}}{2}$	1	0	-1
$\tan\theta$	0	$\dfrac{\sqrt{3}}{3}$	1	$\sqrt{3}$	undefined	0	undefined
$\sec\theta$	1	$\dfrac{2}{\sqrt{3}}$	$\dfrac{2}{\sqrt{2}}$	2	undefined	-1	undefined
$\csc\theta$	undefined	2	$\dfrac{2}{\sqrt{2}}$	$\dfrac{2}{\sqrt{3}}$	1	undefined	1
$\cot\theta$	undefined	$\sqrt{3}$	1	$\dfrac{1}{\sqrt{3}}$	0	undefined	0

 You should spend some time memorizing this table. These exact values are used throughout this book and will also be used in your future work in mathematics.

You can find exact values of the trigonometric functions that are multiples of those in Table 5.3 by using the idea of a reference angle and the **reduction principle**:

REDUCTION PRINCIPLE

If t represents any of the six trigonometric functions, then

$$t(\theta) = \pm t(\overline{\theta})$$

where $\overline{\theta}$ is the reference angle of θ and the plus or minus sign depends on the quadrant of the terminal side of the angle θ. This means to evaluate a trigonometric function of an angle $\overline{\theta}$ (that is not in the first quadrant):

(continues)

Go back and read this procedure box again, and ask questions if you do not understand what it is saying.

REDUCTION PRINCIPLE *(continued)*

Step 1 Find the reference angle for θ, namely, $\overline{\theta}$.

Step 2 Evaluate the trigonometric function for the reference angle, $\overline{\theta}$.

Step 3 Place a "+" in front of this evaluation if the function is positive in the quadrant of θ, and place a "−" in front of this evaluation if the function is negative in the quadrant of θ. Recall the easy-to-remember form for determining whether a trigonometric is positive or negative: "**A S**mart **T**rig **C**lass" (page 300).

EXAMPLE 6　Exact values by using the reduction principle

Evaluate:　**a.** $\cos 135°$　**b.** $\tan 210°$　**c.** $\sin\left(-\dfrac{7\pi}{6}\right)$　**d.** $\tan\dfrac{5\pi}{3}$　**e.** $\csc\left(-\dfrac{5\pi}{3}\right)$

Solution

a. $\cos 135° = \underset{\uparrow}{-\cos 45°} = -\dfrac{\sqrt{2}}{2}$

　　　　　　　reference angle

$\theta = 135°$, so θ is in Quadrant II; cosine is negative in Quadrant II.

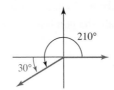

b. $\tan 210° = \underset{\uparrow}{+\tan 30°} = \dfrac{\sqrt{3}}{3}$

　　　　　　reference angle

$\theta = 210°$, so θ is in Quadrant III; tangent is positive in Quadrant III.

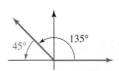

c. $\sin\left(-\dfrac{7\pi}{6}\right) = \underset{\uparrow}{+\sin\dfrac{\pi}{6}} = \dfrac{1}{2}$

　　　　　　　reference angle

$\theta = -\dfrac{7\pi}{6}$, so θ is in Quadrant II; sine is positive in Quadrant II.

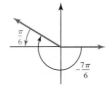

d. $\tan\dfrac{5\pi}{3} = \underset{\uparrow}{-\tan\dfrac{\pi}{3}} = -\dfrac{\sqrt{3}}{3}$

　　　　　　reference angle

$\theta = \dfrac{5\pi}{3}$, so θ is in Quadrant IV; tangent is negative in Quadrant IV.

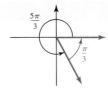

e. $\csc\left(-\dfrac{5\pi}{3}\right) = +\csc\dfrac{\pi}{3} = \dfrac{2}{\sqrt{3}}$ or $\dfrac{2}{3}\sqrt{3}$　*(Quadrant I)*

Angle of Inclination

There is a useful trigonometric formulation of slope. The *angle of inclination* of a line ℓ is defined to be the nonnegative angle ϕ $(0 \le \phi < \pi)$ formed between ℓ and the positively directed x-axis, as shown in Figure 5.29.

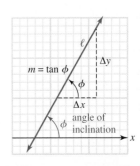

Figure 5.29 Angle of inclination

ANGLE OF INCLINATION

The **angle of inclination** of a line ℓ is the angle ϕ $(0 \le \phi < \pi)$ between line ℓ and the positive x-axis. Then the **slope** of line ℓ with inclination ϕ is

$$m = \tan \phi$$

We see that the line ℓ is *rising* if $0 < \phi < \frac{\pi}{2}$ and is *falling* if $\frac{\pi}{2} < \phi < \pi$. The line is *horizontal* if $\phi = 0$ and is vertical if $\phi = \frac{\pi}{2}$. Notice that if $\phi = \frac{\pi}{2}$, $\tan \phi$ is not defined; therefore, m is not defined for a vertical line. A **vertical line** is said to have **no slope**. ☠ Sometimes we say a vertical line has **infinite slope**. Pay special attention that if we say a line has no slope (vertical line) that is NOT the same as saying the line has 0 slope (horizontal line). ☠

To derive the trigonometric representation for slope, we need to find the slope of the line through $P(x_1, y_1)$ and $Q(x_2, y_2)$, where ϕ is the angle of inclination (see Figure 5.29.) From the definition of the tangent, we have

$$\tan \phi = \frac{y_2 - y_1}{x_2 - x_1} = \frac{\Delta y}{\Delta x} = m$$

EXAMPLE 7   Drawing a line with a given angle of inclination

Write the standard-form equation of a line passing through $(1, 2)$ with an angle of $60°$. Graph this line.

Solution   Note that $\tan 60° = \sqrt{3}$, so we draw the line as shown in Figure 5.30. Recall, the point-slope form of an equation is

$$y - y_1 = m(x - x_1)$$

$$y - 2 = \sqrt{3}(x - 1)$$

The standard form is $Ax + By + C = 0$, so we continue:

$$y - 2 = \sqrt{3}x - \sqrt{3}$$

$$\sqrt{3}x - y + 2 - \sqrt{3} = 0$$

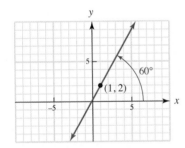

Figure 5.30 Graph of a line passing through $(1, 2)$ with a $60°$ angle of inclination

PROBLEM SET 5.3

LEVEL 1

Find the exact value in Problems 1–4.

1. a. $\tan \frac{\pi}{4}$ **b.** $\cos 0$

c. $\sin 60°$ **d.** $\cos 30°$

e. $\cos 270°$ **f.** $\tan \frac{\pi}{6}$

g. $\tan 180°$ **h.** $\sin 45°$

i. $\sin \pi$ **j.** $\sin \frac{\pi}{2}$

2. a. $\tan 0$ **b.** $\cos \frac{\pi}{4}$

c. $\cos 60°$ **d.** $\sin \frac{\pi}{6}$

e. $\sin 0°$ **f.** $\cos \pi$

g. $\tan \frac{\pi}{3}$ **h.** $\cos \frac{\pi}{2}$

i. $\sec \frac{\pi}{6}$ **j.** $\csc 0$

3. a. $\sec \frac{\pi}{4}$ **b.** $\sec 0°$

c. $\cot \frac{3\pi}{2}$ **d.** $\cot \pi$

e. $\sec \frac{\pi}{3}$ **f.** $\cot \frac{\pi}{6}$

g. $\cot 90°$ **h.** $\csc 60°$

i. $\cot \frac{\pi}{3}$ **j.** $\cot 0$

4. a. $\csc 30°$ **b.** $\csc \pi$

c. $\cot \frac{3\pi}{2}$ **d.** $\csc 90°$

e. $\sec \frac{3\pi}{2}$ **f.** $\sec 90°$

g. $\tan 90°$ **h.** $\sin \frac{3\pi}{2}$

i. $\tan \frac{3\pi}{2}$ **j.** $\cot \frac{\pi}{4}$

Use the reduction principle in Problems 5–9 to give the exact values in simplified form.

5. a. $\sin \frac{5\pi}{6}$ **b.** $\csc \left(-\frac{3\pi}{2}\right)$

c. $\cos(-300°)$ **d.** $\sin 390°$

6. a. $\sin \frac{17\pi}{4}$ **b.** $\cos(-6\pi)$

c. $\cos \frac{9\pi}{2}$ **d.** $\cos 495°$

7. a. $\sin(-765°)$ **b.** $\cos 300°$
  c. $\sin 120°$ **d.** $\tan 120°$
8. a. $\cot 240°$ **b.** $\sec 120°$
  c. $\tan 135°$ **d.** $\csc \frac{5\pi}{6}$
9. a. $\sin(-420°)$ **b.** $\csc(-6\pi)$
  c. $\sin(-390°)$ **d.** $\tan \frac{11\pi}{4}$

Find the values of the six trigonometric functions for angle θ in standard position whose terminal side passes through the points given in Problems 10–15. Draw a sketch showing θ and the reference angle.

10. $(3, 4)$ **11.** $(3, -4)$
12. $(-5, -12)$ **13.** $(-5, 12)$
14. $(-6, 1)$ **15.** $(-2, -3)$

WHAT IS WRONG, *if anything, with each statement? That is, decide whether each statement in Problems 16–27 is true or false by choosing various values for θ. If it is false, give a counterexample (that is, an example that shows that it is not true).*

16. $\sin\theta \csc\theta = 1$ **17.** $\cos\theta \csc\theta = 1$
18. $\sin^2\theta + \cos^2\theta = 1$ **19.** $(\sin\theta + \cos\theta)^2 = 1$

20. $\tan^2\theta - \sec^2\theta = 1$ **21.** $\cos\theta = \frac{\sin\theta}{\tan\theta}$

22. $\cos(-\theta) = \cos\theta$ **23.** $\sin(-\theta) = \sin\theta$

24. $\sin \frac{1}{2}\theta = \frac{1}{2}\sin\theta$ **25.** $\sin \frac{1}{2}\theta = \sqrt{\frac{1-\cos\theta}{2}}$

26. $\frac{\cos 2\theta}{2} = \cos\theta$ **27.** $\frac{\sin 5\theta}{5} = \sin\theta$

LEVEL 2

28. Find $\cos \frac{5\pi}{4}$ by using the procedure illustrated in Example 3.
29. Find $\cos 135°$ by choosing an arbitrary point (x, y) on the terminal side of $135°$ and applying the ratio definition of the trigonometric functions.
30. Find $\sin\left(-\frac{\pi}{4}\right)$ by choosing an arbitrary point (x, y) on the terminal side of $-\frac{\pi}{4}$ and applying the ratio definition of the trigonometric functions.
31. Find $\cos 210°$ by choosing an arbitrary point (x, y) on the terminal side of $210°$ and applying the ratio definition of the trigonometric functions.
32. Find $\sin 210°$ by choosing an arbitrary point (x, y) on the terminal side of $210°$ and applying the ratio definition of the trigonometric functions.

Substitute the exact values for the trigonometric functions in the given expressions in Problems 33–45 and simplify.

33. a. $2\cos \frac{\pi}{2}$ **b.** $\cos \frac{2\pi}{2}$
34. a. $\sin \frac{2\pi}{4}$ **b.** $2\sin \frac{\pi}{4}$
35. a. $\sin 30° + \cos 0°$ **b.** $\sin \frac{\pi}{2} + 3\cos \frac{\pi}{2}$
36. a. $\sin^2 60°$ **b.** $\cos^2 \frac{\pi}{4}$
37. a. $\sin^2 \frac{\pi}{6} + \cos^2 \frac{\pi}{6}$ **b.** $\sin^2 \frac{\pi}{2} + \cos^2 \frac{\pi}{2}$
38. a. $\sin^2 \frac{\pi}{3} + \cos^2 \frac{\pi}{3}$ **b.** $\sin^2 \frac{\pi}{6} + \cos^2 \frac{\pi}{3}$
39. a. $\sin \frac{\pi}{6}\csc \frac{\pi}{6}$ **b.** $\csc \frac{\pi}{2}\sin \frac{\pi}{2}$
40. a. $\cos\left(\frac{\pi}{4} - \frac{\pi}{2}\right)$ **b.** $\cos \frac{\pi}{4} - \cos \frac{\pi}{2}$
41. a. $\tan(2 \cdot 30°)$ **b.** $2\tan 30°$

42. a. $\csc\left(\frac{1}{2} \cdot 60°\right)$ **b.** $\frac{\csc 60°}{2}$

43. a. $\cos\left(\frac{1}{2} \cdot 60°\right)$ **b.** $\sqrt{\frac{1+\cos 60°}{2}}$

44. a. $\tan(2 \cdot 60°)$ **b.** $\frac{2\tan 60°}{1-\tan^2 60°}$

45. a. $\cos\left(\frac{\pi}{2} - \frac{\pi}{6}\right)$ **b.** $\cos \frac{\pi}{2}\cos \frac{\pi}{6} + \sin \frac{\pi}{2}\sin \frac{\pi}{6}$

46. It can be shown that the exact value for $\sin 3°$ is

$$\frac{1}{16}\left[\sqrt{30} + \sqrt{10} - \sqrt{6} - \sqrt{2}\right]$$
$$-\frac{1}{8}\sqrt{20 + 4\sqrt{5} - 2\sqrt{15} - 10\sqrt{3}}$$

Use your calculator to verify this exact value.

47. It can be shown that $\sin 15° = \frac{1}{\sqrt{6} + \sqrt{2}}$ and $\cos 15° = \frac{2+\sqrt{3}}{\sqrt{6}+\sqrt{2}}$.
Simplify each of these exact values.

48. Verify, using the exact values in Problem 47, that $\cos^2 15° + \sin^2 15° = 1$
49. Find the equations of the lines passing through the given points with the given angles of inclination. Write your answers in the form $y = m(x - h) + k$.
  a. $30°$; $(2, 3)$ **b.** $120°$; $(1, 4)$
  c. $45°$; $(9, -5)$ **d.** $135°$; $(-3, -8)$
50. Find the approximate equation of the line passing through each point with the given angle of inclination. Write your equation in the form $y = mx + b$, and round all numbers and coefficients to two decimal places.
  a. $23°$; $(0.81, 0.53)$ **b.** $83°$; $(-1.45, 0.61)$
  c. $138°$; $(1, 8)$ **d.** $175°$; $(-3, -2)$
51. In terms of the angle of inclination ϕ, describe the following lines:
  a. A line that is *rising* from left to right
  b. A line that is *falling* from left to right
  c. A horizontal line
  d. A vertical line
52. Use the definition of tangent to show that the definition of slope based on angle of inclination is the same as the following definition of slope from algebra: *The slope of the line passing through the points $P_1(x_1, y_1)$ and $P_2(x_2, y_2)$ is*

$$m = \frac{y_2 - y_1}{x_2 - x_1}$$

53. Let $P(x, y)$ be any point in the plane. Show that $P(r\cos\theta, r\sin\theta)$ is a representation for P, where θ is the standard-position angle formed by drawing ray $\overrightarrow{OP}$.
54. Let A and B be any two points on a unit circle where A is a point on the terminal side of an angle α and B is a point on the terminal side of an angle β. Find the slope of the line passing through A and B.
55. Fill in the values in the given table.

θ	1	0.5	0.1	0.01	0.001	0.0001
$\sin\theta$						

As θ approaches 0, written $\theta \to 0$, make a conjecture about the values of $\sin\theta$.

56. Fill in the values in the given table.

θ	1	0.5	0.1	0.01	0.001	0.0001
$\dfrac{\sin\theta}{\theta}$						

As θ approaches 0, written $\theta \to 0$, make a conjecture about the values of $\dfrac{\sin\theta}{\theta}$.

LEVEL 3

57. The position of a piston, connecting rod, and crankshaft in an internal combustion engine are related as shown in Figure 5.31. It can be shown that

$$d = k + r - r\cos(2\pi vt) - \sqrt{k^2 - r^2\sin^2(2\pi vt)}$$

where

d = depth of piston stroke
k = length of connecting rod
r = radius of crankshaft
v = velocity of crankshaft in rev/min
t = time in minutes

Find the position of a piston after 0.06 second if the connecting rod is 15 cm and the radius of the crankshaft is 5.0 cm. Assume that the crankshaft is turning at 600 rev/min.

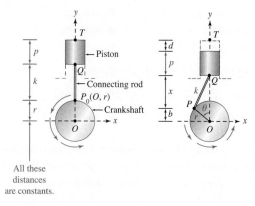

All these distances are constants.

Figure 5.31 Piston in an internal combustion engine

58. Find the position of a piston after $\frac{1}{2}$ second if the crankshaft (see Problem 57) is turning at 60 rev/min and the radius is 4.0 in. Assume that the connecting rod is 8.0 inches long.

59. a. If θ is in Quadrant I, then $\theta + \pi$ is in Quadrant III with a reference angle θ. Use this fact and the reduction principle to show that

$$\sin(\theta + \pi) = -\sin\theta$$

   b. Show that $\sin(\theta + \pi) = -\sin\theta$ if θ is in Quadrant II.
   c. Show that $\sin(\theta + \pi) = -\sin\theta$ if θ is in Quadrant III.
   d. Show that $\sin(\theta + \pi) = -\sin\theta$ if θ is in Quadrant IV.

60. Show that $\cos(\theta + \pi) = -\cos\theta$ for any angle θ.

5.4 Graphs of the Trigonometric Functions

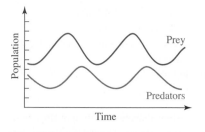

Figure 5.32 Prey-predator biological model

In a classic study,* Lotka considered the interdependence of two species, the first of which serves as food for the other. If we consider the populations of foxes and rabbits in a certain region, we might find that their populations go up and down in cycles (but out of phase with one another), as shown in Figure 5.32.

Suppose it was found that the population of the rabbits changed according to the formula

$$y = 400 + 100\cos\frac{\pi x}{4}$$

where y is the number of rabbits after time x (measured in months). What is the graph representing the rabbit population? If $x = 1$, then

$$y = 400 + 100\cos\frac{\pi x}{4}$$

$$= 400 + 100\cos\frac{\pi(1)}{4}$$

$$\approx 470.7106781$$

Thus, the point $(1, 470.7107)$ is a point on the graph, since x and $y = 470.7107$.† If we repeat this calculation for a number of x values, we develop a table of values, as shown in Table 5.4.

Graph these points on a Cartesian coordinate system, giving careful attention to the units on the x- and y-axes (it is not practical to choose the same units on both axes). The points are plotted in Figure 5.33**a**, and these points are connected in Figure 5.33**b**.

*Elements of Mathematical Biology by Alfred Lotka. New York: Dover Publications, 1956.
†If you obtained 499.990605, you need to change from degree mode to radian mode. Remember, if units are not specified, radians are understood.

The graphs of the cosine and sine functions look like waves. We will consider various properties associated with such graphs. The **amplitude** is half the distance between the maximum and minimum functional values. The **cycle** is one complete repetition of the wave (i.e., one set of y-values). The **period** is the horizontal length of the cycle. The **frequency** tells how many cycles occur in a given unit of time and is defined to be the reciprocal of the period. For example, if the horizontal axis is a function of time, then the frequency might be expressed in *cycles per year*, as in the case of rabbit populations, or in *cycles per second*, as in the broadcasting of radio waves.

TABLE 5.4	Values Satisfying the Equation $y = 400 + 100 \cos \dfrac{\pi x}{4}$
x	y
0	500
1	470.7107
2	400
3	329.2893
4	300
5	329.2893
6	400
7	470.7107
8	500
9	470.7101
10	400
11	329.2893
12	300

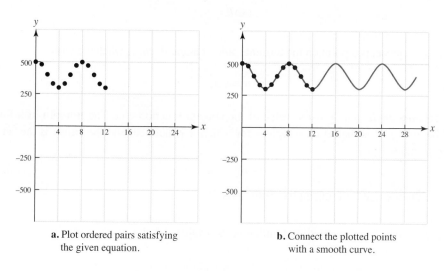

a. Plot ordered pairs satisfying the given equation.

b. Connect the plotted points with a smooth curve.

Figure 5.33 Graph of $= 400 + 100 \cos \dfrac{\pi x}{4}$

EXAMPLE 1   Finding amplitude, period, and frequency

Suppose we are given the calculator graph shown in Figure 5.34. Find the period, amplitude, and frequency.

Solution   Calculator graphs do not show the scale on the screen, nor do they label the axes. We mark the appropriate elements of Figure 5.34 in Figure 5.35.

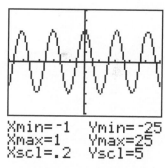

Figure 5.34 Calculator graph of a periodic function

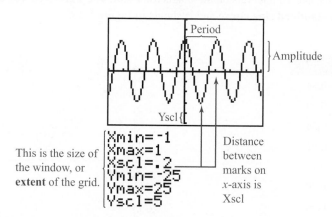

Figure 5.35 Reading a calculator graph output

We note here that the x-scale is 0.2 unit per tick mark, and the period is measured from any point to the next point horizontally where the cycle repeats. In this case, the period is 0.4 unit.

The y-scale is 5 units per tick mark, so each tick mark represents 5 units, and the amplitude is the maximum height of the wave. In this example, it is 15 units.

The frequency is the reciprocal of the period, so in this example, the frequency is $1/0.4 = 2.5$ or $2\frac{1}{2}$ cycles per unit.   ■

We now turn to graphing the standard trigonometric functions by plotting points. The process will then be generalized so we can graph the functions without too many calculations concerning points.

To plot points, we need to relate the trigonometric functions to the Cartesian coordinate system. For example, we may wish to write $y = \tan x$, but in so doing, we change the meaning of x and y as they were used in the last chapter. Recall from the *definition* of the trigonometric functions that

$$\tan \theta = \frac{y}{x}$$

where (x, y) represents a point on the terminal side of angle θ. *Now*, when we write

$$y = \tan x$$

x represents the *angle*, and y is the value of the tangent function at x. The **domain** of a function defined by $y = f(x)$ is the set of all values for which x is defined, and the **range** is the set of all resulting y-values.

Graphs of the Standard Trigonometric Functions

Sine Function

To graph $y = \sin x$, begin by plotting familiar (exact) values for the sine function:

x = real number	0	$\frac{\pi}{2}$	π	$\frac{3\pi}{2}$	2π
$y = \sin x$	0	1	0	-1	0

Exact values are used here, but a calculator could be used to generate other values. If you are graphing with a calculator or computer, you will need to specify a "window" or "frame" of values. If you are graphing with a pencil and paper, you may find it convenient to choose 12 intervals on the x-axis for π units, and 10 intervals on the y-axis for 1 unit. The smooth curve that connects these points is called the **standard sine curve** and is shown in Figure 5.36.

x	y
-1	-0.84
0	0.00
1	0.84
2	0.91
3	0.14
4	-0.76
5	-0.96
6	-0.28
7	0.66

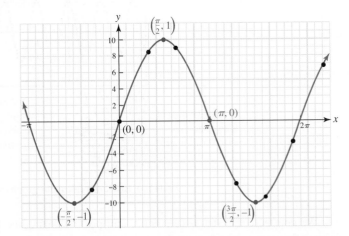

Figure 5.36 Graph of $y = \sin x$ by plotting points (see table at the left for plotted values)

HINT: Sine is positive in Quadrants I $\left(0 < x < \frac{\pi}{2}\right)$ and II $\left(\frac{\pi}{2} < x < \frac{3\pi}{2}\right)$ and negative in Quadrants III $\left(\pi < x < \frac{3\pi}{2}\right)$ and IV $\left(\frac{3\pi}{2} < x < 2\pi\right)$.

The domain for x is all real numbers, so what about values other than $0 \leq x \leq 2\pi$? Using the reduction principle (which was discussed on page 310), we know that

$$\sin x = \sin(x + 2\pi) = \sin(x - 2\pi) = \sin(x + 4\pi) = \cdots$$

More generally,

$$\sin(\theta + 2n\pi) = \sin \theta$$

for any integer n. In other words, the values of the sine function repeat themselves after 2π. We describe this by saying that the standard sine function is **periodic with period 2π**. The sine curve is shown in Figure 5.37. Notice that even though the domain of the sine function is all real numbers, the range is restricted to values between -1 and 1 (inclusive).

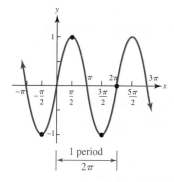

Amplitude: 1
Period: 2π
Domain: all real x, written $\mathbb{R}$
Range: $-1 \leq y \leq 1$

☠ Remember the shape of this graph, its amplitude, period, domain, and range. ☠

Figure 5.37 Graph of $y = \sin x$

The sine curve shown in Figure 5.37 is a particularly important part of trigonometry, so it is worthwhile to spend some time learning how to quickly and easily sketch not only the standard sine curve but also the sine curve with variations in its amplitude and period. To accomplish this, notice the section in Figure 5.37 labeled "1 period." This section of the graph (reading from left to right) begins at $(0, 0)$, goes *up* to $\left(\frac{\pi}{2}, 1\right)$, and then *down* to $\left(\frac{3\pi}{2}, -1\right)$, passing through $(\pi, 0)$, and then goes back up to $(2\pi, 0)$, which completes one period. This graph shows that the range of the sine function is $-1 \leq y \leq 1$. We summarize a technique for sketching the sine curve (see Figure 5.38) called **framing the sine curve**.

FRAMING A SINE CURVE

The standard sine function

$$y = \sin x$$

has domain $\mathbb{R}$ (all real numbers) and range $-1 \leq y \leq 1$ and is periodic with period 2π. One period of this curve can be sketched by framing, as follows:

Step 1 *Start* at the origin $(0, 0)$.

Step 2 *Height* of this frame is 2 units: 1 unit up and 1 unit down from the starting point.

Step 3 *Length* of this frame is 2π units (about 6.28) from the starting point.

Step 4 The curve is now framed. Plot *five* critical points within the frame:

    a. Two endpoints (along the axis)

    b. Midpoint (along axis)

    c. Two quarterpoints (up first, then down)

Step 5 Draw the curve through the critical points, remembering the shape of the sine curve.

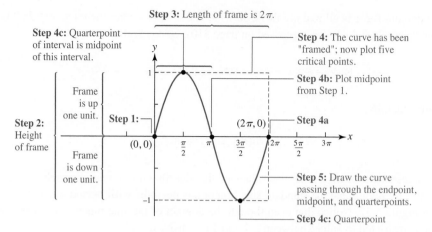

Step 3: Length of frame is 2π.

Step 4c: Quarterpoint of interval is midpoint of this interval.

Step 4: The curve has been "framed"; now plot five critical points.

Step 4b: Plot midpoint from Step 1.

Frame is up one unit.

Step 1:

Step 2: Height of frame

$(0, 0)$

Frame is down one unit.

$(2\pi, 0)$

Step 4a

Step 5: Draw the curve passing through the endpoint, midpoint, and quarterpoints.

Step 4c: Quarterpoint

Figure 5.38 Procedure for framing a sine curve (five steps)

The graph of $y = \sin x$ has a starting point of $(0, 0)$, but other sine curves may have other starting points, as illustrated by the following example.

EXAMPLE 2   Drawing one cycle of a standard sine curve

Draw one period of a standard sine curve using $(3, 1)$ as the starting point for building a frame.

Solution   Plot the point $(3, 1)$ as the starting point. Draw the standard frame as shown in Figure 5.39.

Remember, when framing a curve, first plot the starting point, and then count squares on your graphing paper to build the frame.

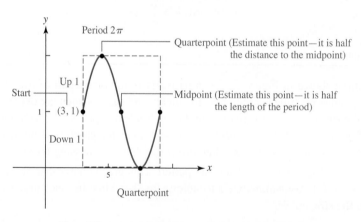

Period 2π

Quarterpoint (Estimate this point—it is half the distance to the midpoint)

Start

Up 1

Midpoint (Estimate this point—it is half the length of the period)

$(3, 1)$

Down 1

Quarterpoint

Figure 5.39 Graph of one period of a sine curve by framing

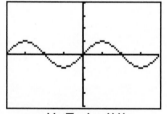

\Y₁ᗺsin(X)
Xmin=⁻6.152285…
Xmax=6.1522856…
Xscl=1.5707963…
Ymin=⁻4
Ymax=4
Yscl=1

Figure 5.40 Calculator graph for Example 2

☐ COMPUTATIONAL WINDOW

Curve sketching by plotting points can efficiently be done on a calculator. Many people believe that it is not necessary to *think* if you are using a calculator, that the "machine" will do all the work. However, this is not true; with a calculator, your attention must be focused on different matters. For example, in this section, we are concerned with graphing $y = \sin x$. The first task is to decide on the domain and the range. If you know what the curve looks like, then it is easy to make an intelligent choice about the domain and range. However, if you do not know anything about the curve, you may have to make several attempts at setting the domain and range before you obtain a satisfactory graph. Without the work of this section before graphing a sine curve on a calculator, the task would be difficult. On almost all graphing calculators, the scale is fixed automatically by pressing ZOOM and choosing the *trigonometric* scale. Notice that the "strange" scale for the x-axis is an approximation for $\pi/2$ and the Xmin and Xmax values were automatically set by choosing the trigonometric scale. Practice graphing a sine curve using your calculator. An example of what this graph should look like is shown in Figure 5.40.

The following example illustrates how you might be misled in obtaining a correct graph if you rely solely on your calculator.

EXAMPLE 3   Sketching a curve by plotting points

Graph $f(x) = \dfrac{\sin x}{x}$.

Solution   We note that f is not defined when $x = 0$. Set up a table of values, and plot the corresponding points on a coordinate system, as shown in Figure 5.41. Draw an open circle at $x = 0$, since the graph does not exist at that point.

x	y
-6	-0.047
-5	-0.192
-4	-0.189
-3	0.047
-2	0.455
-1	0.841
1	0.841
2	0.455
3	0.047
4	-0.189
5	-0.192
6	-0.047

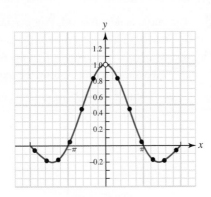

Figure 5.41 Graph of $y = \dfrac{\sin x}{x}$

A calculator graph for this function is shown in Figure 5.42.

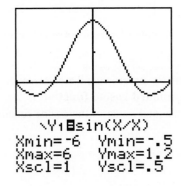

Figure 5.42 Calculator graph of $y = \dfrac{\sin x}{x}$

Even though the calculator graph is straightforward, you will notice there is no indication that there is a "hole" at $x = 0$. ∎

Cosine Function

We can graph the **cosine curve** by plotting points, as we did with the sine curve. The details of plotting points are left as an exercise. The cosine curve $y = \cos x$ is "framed" as described in Figure 5.43.

Notice that the only difference in the steps for graphing a cosine and a sine curve is in step 4; the procedures for building the frame are identical.

FRAMING A COSINE CURVE

The standard cosine function

$$y = \cos x$$

has domain $\mathbb{R}$ (all real numbers) and range $-1 \leq y \leq 1$ and is periodic with period 2π. One period of this can be sketched by framing, as follows:

Step 1 *Start* at the origin $(0, 0)$.

Step 2 *Height* of this frame is 2 units: 1 unit up and 1 unit down from the starting point.

Step 3 *Length* of this frame is 2π units (about 6.28) from the starting point.

Step 4 The curve is now framed. Plot *five* critical points within the frame:

   a. Two endpoints (at the top corners of the frame)

   b. Midpoint (at the bottom of the frame)

   c. Two quarterpoints (along the axis)

Step 5 Draw the curve through the critical points, remembering the shape of the cosine curve.

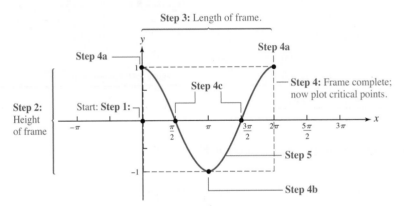

Figure 5.43 Procedure for framing a cosine curve (five steps)

Since values for x greater than 2π or less than 0 are coterminal with those already considered, we see that **the period of the standard cosine function is 2π.** The cosine curve is shown in Figure 5.44. Notice that the domain and range of the cosine function are the same as they are for the sine function.

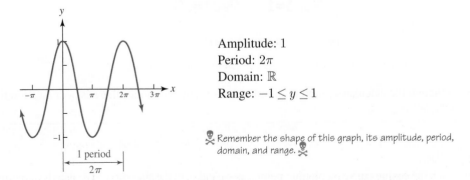

Amplitude: 1
Period: 2π
Domain: $\mathbb{R}$
Range: $-1 \leq y \leq 1$

☠ Remember the shape of this graph, its amplitude, period, domain, and range. ☠

Figure 5.44 Graph of $y = \cos x$

Tangent Function

By setting up a table of values and plotting points (the details are left as an exercise), we notice that $y = \tan x$ does not exist at $\frac{\pi}{2}$, $\frac{3\pi}{2}$, or $\frac{\pi}{2} + n\pi$ for any integer n. The lines $x = \frac{\pi}{2}$, $x = \frac{3\pi}{2}$, ..., $x = \frac{\pi}{2} + n\pi$ for which the tangent is not defined are **vertical asymptotes**. The procedure for graphing the **tangent curve** is summarized in Figure 5.45. Notice that this procedure differs from that of the sine and cosine in several ways: the starting point is in the center of the frame; the period is π (instead of 2π); and, as before, the details for the particular curve are listed in step 4.

FRAMING A TANGENT CURVE

The standard tangent function

$$y = \tan x$$

has domain $\mathbb{R}$ (all real numbers), where $x \ne \frac{\pi}{2} + n\pi$ (for any integer n) and range $\mathbb{R}$. The standard tangent function is periodic with period π. One period of this can be sketched by framing, as follows:

Step 1 *Start* at the origin $(0, 0)$; for the tangent curve this is the **center** of the frame.

Step 2 *Height* of this frame is 2 units: 1 unit up and 1 unit down from the starting point.

Step 3 *Length* of this frame is π units (about 3.14) and is drawn so that it is $\frac{\pi}{2}$ (about 1.57) units on each side of the starting point.

Step 4 The curve is now framed. Plot the asymptotes and plot *three* critical points within the frame:

    a. Extend the vertical sides of the frame; these are the asymptotes.

    b. Midpoint (this is the starting point from step 1).

    c. Two quarterpoints (down first, then up. To help you remember this, we note as you read from left to right the curve is increasing—up, up, and away.)

Step 5 Draw the curve through the critical points, remembering the shape of the tangent curve.

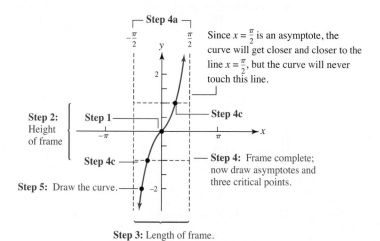

Figure 5.45 Procedure for framing a tangent curve (five steps)

The tangent curve is indicated in Figure 5.46. Even though the curve repeats for values of x greater than 2π or less than 0, it also repeats after it has passed through an interval with length π. For this reason, $\tan(\theta + n\pi) = \tan\theta$ for any integer n, and we see that **the standard tangent function has a period of π**. The domain of the tangent function is restricted so that multiples of π added to $\frac{\pi}{2}$ are excluded because the tangent is not defined for these values. The range, however, is unrestricted; it is the set of all real numbers.

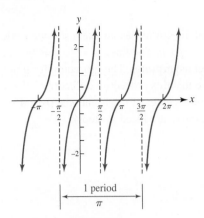

Amplitude: none
Period: π
Domain: $\mathbb{R}, x \neq \frac{\pi}{2} + n\pi$ for any integer n
Range: $\mathbb{R}$

☠ Remember the shape of this graph, its amplitude, period, domain, and range. ☠

Figure 5.46 Graph of $y = \tan x$

Reciprocal Functions

The graphs of the other three trigonometric functions could be done in the same fashion as sine, cosine, and tangent. Instead, however, we make use of the reciprocal relationships and graph them as shown in the following example.

EXAMPLE 4   Graph of the standard secant function using reciprocals

Sketch $y = \sec x$ by sketching the reciprocal of $y = \cos x$. This is called the **secant curve.**

Solution   Begin by sketching $y = \cos x$ (dashed curve in Figure 5.47). Wherever $\cos x = 0$, $\sec x$ is undefined; draw asymptotes at these points. Now, plot points by finding the reciprocals of the ordinates of points previously plotted. When $y = \cos x = \frac{1}{2}$, for example, the reciprocal is

$$y = \sec x = \frac{1}{\cos x} = \frac{1}{1/2} = 2$$

The completed graph is shown in Figure 5.47.

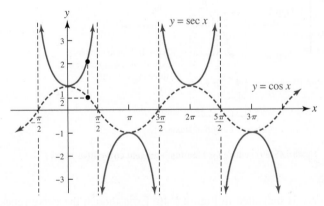

Figure 5.47 Graph of $y = \sec x$

The graphs of the other reciprocal trigonometric functions—namely, the **cosecant curve** and the **cotangent curve**, are shown in Figure 5.48.

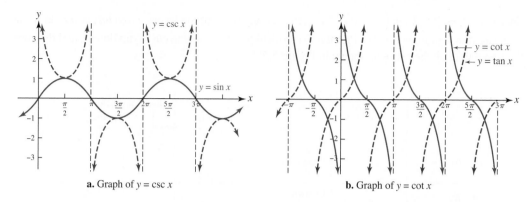

a. Graph of $y = \csc x$ **b.** Graph of $y = \cot x$

Figure 5.48 Graphs of the reciprocal functions

Graphs of the General Trigonometric Functions

You may have noticed that the graphs of $y = \cos x$ and $y = \sin x$ have the same shape, except one has been shifted to the right, as shown in Figure 5.49.

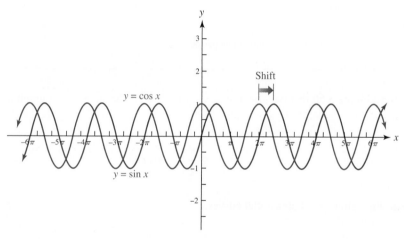

Figure 5.49 Comparison between standard cosine and sine graphs

In this section, we will discuss variations of the standard cosine, sine, and tangent graphs. These variations will include not only shifting, as shown in Figure 5.49, but also changes in amplitude and period. Since the cosine and sine graphs are identical (except for a possible shift), we call either of these curves a **sinusoidal** curve. That is, the graphs of both the cosine and sine functions are referred to as *sinusoidal* curves.

Translations and Phase Shift

We begin our discussion by considering a **phase shift** (illustrated in Figure 5.49), which means that a given curve is shifted to the left or the right. At the same time, we will also consider shifting the curve up or down. We will call these horizontal and vertical shifts **translations**.

In the case of the graph of the trigonometric functions, we will simply determine (h, k) and then build a frame at (h, k) rather than at the origin. If you can draw one cycle of the graph of a trigonometric function, then you can draw the entire function. For this reason, in this text, we will often graph only one cycle. The graph we will draw is the one enclosed in the standard frame of length equal to one period.

EXAMPLE 5   **Graphing one cycle of a translated sine curve**

Graph one period of $y = \sin\left(x + \frac{\pi}{2}\right)$.

Solution   First, $(h,\ k) = \left(-\frac{\pi}{2},\ 0\right)$. This is the starting point of the frame. *From this point*, draw the frame as shown in Figure 5.50a. Plot the five critical points for a sine curve (up/down), and then use the frame to draw the curve as shown in Figure 5.50b.

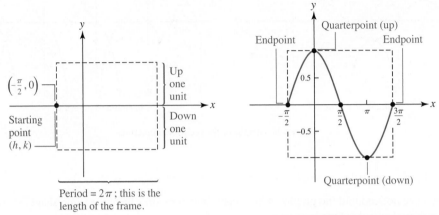

a. Drawing the frame. This
step is the same for both
sine and cosine curves.

b. Plot critical points and draw
one period of the curve.

Figure 5.50 Graph of one period of $y = \sin\left(x + \dfrac{\pi}{2}\right)$

Notice from Figure 5.50 that the graph of $y = \sin\left(x + \frac{\pi}{2}\right)$ is the same as the graph of $y = \cos x$. Thus,

$$\sin\left(x + \frac{\pi}{2}\right) = \cos x$$

This confirms the definition of sinusoidal curves.

EXAMPLE 6   Graph one cycle of a translated cosine function

Graph one period of $y - 2 = \cos\left(x - \frac{\pi}{6}\right)$.

Solution   Notice that $(h,\ k) = \left(\frac{\pi}{6},\ 2\right)$. Frame the curve, plot the five critical points for a cosine curve (down/up), and connect the points as shown in Figure 5.51.

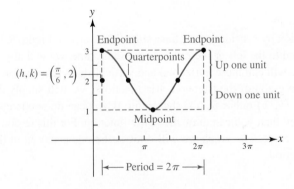

Figure 5.51 Graph of one period of $y - 2 = \cos\left(x - \dfrac{\pi}{6}\right)$

EXAMPLE 7   Graphing one cycle of a translated tangent function

Sketch one period of $y + 3 = \tan\left(x + \frac{\pi}{3}\right)$.

Solution   Notice that $(h,\ k) = \left(-\frac{\pi}{3},\ -3\right)$, and remember that this is the center of the frame and that the period of the tangent is π, as shown in Figure 5.52. Plot the midpoint and quarterpoints for the frame of a tangent curve (up, up, and away).

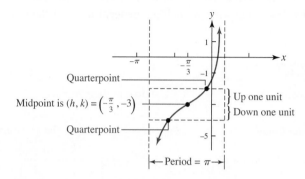

Figure 5.52 Graph of one period of $y + 3 = \tan\left(x + \dfrac{\pi}{3}\right)$

Variations in Amplitude and Period

We will now discuss two additional changes for the function defined by $y = f(x)$. The first, $y = af(x)$, changes the scale on the y-axis; the second, $y = f(bx)$, changes the scale on the x-axis.

For a function $y = af(x)$, it is clear that the y-value is a times the corresponding value of $f(x)$, which means that $f(x)$ is stretched or shrunk in the y-direction by the factor of a. For example, if $y = f(x) = \cos x$, then $y = 3f(x) = 3\cos x$ is the graph of $\cos x$ that has been stretched so that the high point is at 3 units and the low point is at -3 units (see Figure 5.53). In general, given $y = af(x)$, where f represents a trigonometric function, $2\,|a|$ gives the height of the frame for f. For sine and cosine curves, $|a|$ is the *amplitude* of the function. When $a = 1$, the amplitude is 1, so $y = \sin x$ and $y = \cos x$ are said to have amplitude 1.

EXAMPLE 8   Graphing a trigonometric curve with an arbitrary amplitude

Graph one period of $y = 3\cos x$.

Solution   From the starting point $(0,\ 0)$, draw a frame with amplitude $a = 3$ and period 2π, as shown in Figure 5.53. *After* drawing the frame, plot the five critical points for a cosine curve (down/up) using the frame as a guide.

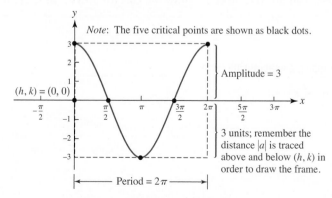

Figure 5.53 Graph of one period of $y = 3\cos x$

For a function $y = f(bx)$, $b > 0$, b affects the scale on the x-axis. Recall that $y = \sin x$ has a period of 2π $[f(x) = \sin x$, so $b = 1]$. A function $y = \sin 2x$ $[f(x) = \sin x$ and $f(2x) = \sin 2x]$ must complete one period as $2x$ varies from 0 to 2π. This means that one period is completed as x varies from 0 to π. For each value of x, the result is doubled *before* we find the sine of that number. In general, the period of $y = \sin bx$ is $2\pi/b$, and the period of $y = \cos bx$ is $2\pi/b$. Since the period of $\tan x$ is π, however, $y = \tan bx$ has a period of π/b. Therefore, when framing the curve, use $2\pi/b$ for the period of the sine and cosine and π/b for the period of the tangent.

We summarize by describing what we call the **general form** of the sine, cosine, and tangent curves.

GENERAL TRIGONOMETRIC CURVES

One period for the cosine, sine, and tangent are shown in their frames.

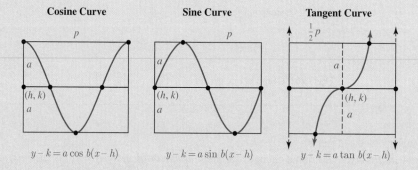

| Cosine Curve | Sine Curve | Tangent Curve |

$$y - k = a \cos b(x - h) \qquad y - k = a \sin b(x - h) \qquad y - k = a \tan b(x - h)$$

Step 1 Algebraically, put the equation into one of the general forms shown under the graphs.

Step 2 Identify (by inspection) the following values: (h, k), a, and b.

Calculate p: $p = \dfrac{2\pi}{b}$ for cosine and sine

$p = \dfrac{\pi}{b}$ for tangent.

Step 3 Draw the frame.

  a. Plot (h, k); this is the starting point.

  b. Draw **amplitude**, $|a|$; the *height of the frame* is $2|a|$; up a units from (h, k) and down a units from (h, k).*

  c. Draw **period**, p; the *length of the frame* is p. Remember, $p = \dfrac{2\pi}{b}$ for cosine and sine, and $p = \dfrac{\pi}{b}$ for tangent.

Step 4 Locate the critical values using the frame as a guide, and then sketch the appropriate graphs. **You do NOT need to know the coordinates of these critical values.**

EXAMPLE 9   Graphing a general sine curve

Graph $y + 1 = 2 \sin \frac{2}{3}\left(x - \frac{\pi}{2}\right)$.

Solution

$$(h, k) = \left(\tfrac{\pi}{2}, -1\right), \ a = 2, \ b = \tfrac{2}{3}; \ \text{so } p = \frac{2\pi}{\tfrac{2}{3}} = 3\pi$$

---

*For now, assume $a > 0$; $a < 0$ is considered in Chapter 4.

Now plot $(h,\ k)$ and frame the curve. Then plot the five critical points (two endpoints, the midpoint, and two quarterpoints). Finally, after sketching one period, draw the other periods as shown in Figure 5.54.

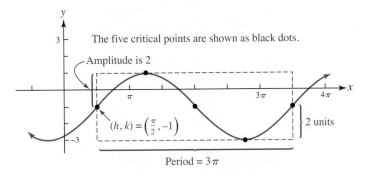

Figure 5.54 **Graph of** $y + 1 = 2\sin\dfrac{2}{3}\left(x - \dfrac{\pi}{2}\right)$

EXAMPLE 10   Graphing a general cosine curve

Graph $= 3\,\cos\left(2x + \frac{\pi}{2}\right) - 2.$

Solution   Rewrite in general form to obtain $y + 2 = 3\,\cos 2\left(x + \frac{\pi}{4}\right)$. ☠ Notice that you must factor out the coefficient of x in the argument of cosine. ☠ Note $(h,\ k) = \left(-\frac{\pi}{4},\ -2\right)$; $a = 3$, $b = 2$, and $p = 2\pi/2 = \pi$. Plot $(h,\ k)$, and frame the cosine curve (down/up) as shown in Figure 5.55.

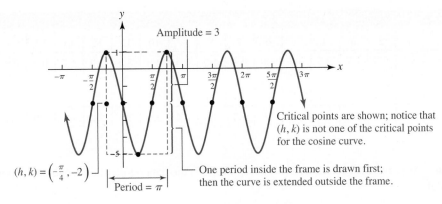

Figure 5.55 **Graph of** $y = 3\cos\left(2x + \dfrac{\pi}{2}\right) - 2$

EXAMPLE 11   Graphing a general tangent curve

Graph $y - 2 = 3\tan\frac{1}{2}\left(x - \frac{\pi}{3}\right).$

Solution $(h,\ k) = \left(\frac{\pi}{3},\ 2\right)$, $a = 3$, $b = \frac{1}{2}$, so $p = \dfrac{\pi}{\frac{1}{2}} = 2\pi$. Plot $(h,\ k)$ and frame the curve, as shown in Figure 5.56. Plot the midpoint and the quarterpoints for a tangent curve (up, up, and away).

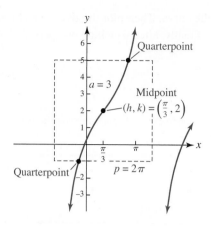

Figure 5.56 Graph of $y - 2 = 3\tan\dfrac{1}{2}\left(x - \dfrac{\pi}{3}\right)$

■

Graphing from Data Points

Sometimes we need to identify a trigonometric curve by looking at data points, as shown in the following example.

EXAMPLE 12   Graphing a curve from given data points

The flywheel on a lawn mower has 16 evenly spaced cooling fins. If we rotate the engine through two complete revolutions, we can generate 32 data points, with the first coordinate representing time and the second representing the depth of the piston. Determine an equation of the curve generated by the lawn-mower-engine data as shown in Table 5.5.

TABLE 5.5	Data Points for a Lawn Mower Engine															
x	0	1	2	3	4	5	6	7	8	9	10	11	12	13	14	15
y	0	0.2	0.8	1.7	2.5	3.4	4.0	4.4	4.5	4.4	4.0	3.5	2.2	1.7	0.9	0.2

x	16	17	18	19	20	21	22	23	24	25	26	27	28	29	30	31	32
y	0	0.2	0.8	1.7	2.5	3.4	4.0	4.4	4.5	4.4	4.0	3.5	2.2	1.7	0.9	0.2	0

Solution   As previously noted, we plot (in Figure 5.57) the data points and recognize the points as sinusoidal.

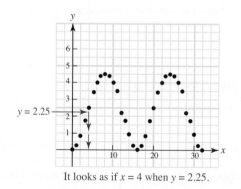

It looks as if $x = 4$ when $y = 2.25$.

Figure 5.57 Data points

The graph seems to have a period of 16 and an amplitude of 2.25. This means that for the general sine curve

$$y - k = a \sin b(x - h)$$

we have $a = 2.25 = \frac{9}{4}$; period $p = \frac{2\pi}{b} = 16$ so that $b = \frac{\pi}{8}$. Finally, we need to find an appropriate starting point, (h, k). There are many points we might choose, but we select that y-value halfway between the maximum and minimum, that is, $y = 2.25$. Next, we approximate the intersection of the line $y = 2.25 = \frac{9}{4}$ and the graph: it appears to be $(4, 2.25)$.

$$y - \frac{9}{4} = \frac{9}{4} \sin \frac{\pi}{8}(x - 4)$$

The graph is shown in Figure 5.58.

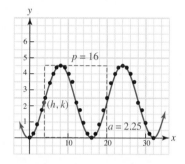

Figure 5.58 Graph of $y - \dfrac{9}{4} = \dfrac{9}{4} \sin \dfrac{\pi}{8}(x - 4)$

A tuning fork is a small two-pronged steel instrument that, when struck, vibrates at a particular frequency so that it produces a certain fixed tone in a perfect pitch. In the demonstration shown in Figure 5.59, a person strikes a tuning fork, which, in turn, is connected to an oscilloscope that allows us to "see" the sound. You will, no doubt, notice that the picture is sinusoidal.

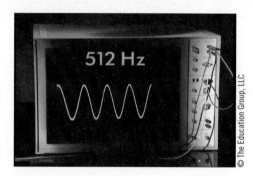

© The Education Group, LLC

© The Education Group, LLC

Figure 5.59 Looking at sound waves using an oscilloscope*

EXAMPLE 13 Sound frequency

MODELING APPLICATION

A middle C tuning fork vibrates at 264 Hz, with an amplitude of 0.02 cm. Write an equation and graph this sound wave for a middle C.

---

*Hz is an abbreviation for the unit of measurement called a *hertz*, which means "cycles per unit of time." It is named after Heinrich Hertz (1857–1894), who discovered radio waves. An amplitude of 0.02 cm means that the prong vibrates to the right and left a maximum of 0.02 cm (where we assume, as usual, that right is positive and left is negative).

Solution

Step 1: *Understand the problem.* We model this as a sinusoidal curve. This means we are looking for either a sine or a cosine equation of the form $y = a \cos bx$ or $y = a \sin bx$.

Step 2: *Devise a plan.* If we assume that x is the time, then a is the amplitude, and the period is $p = 2\pi/b$. Also, if $x = 0$, then the time is 0, and there is no vibration, so we select, as a model, the equation $y = a \sin bx$, since $y = 0$ for $x = 0$.

Step 3: *Carry out the plan.* Since a is the amplitude, $a = 0.02$. The period is the reciprocal of the frequency, so

$$p = \frac{1}{264} \qquad \text{and} \qquad b = \frac{2\pi}{p} = \frac{2\pi}{\frac{1}{264}} = 528\pi$$

The desired equation is

$$y = 0.02 \cos 528\,\pi x$$

The graph is shown in Figure 5.60. You will need to pay attention to the scale when graphing this function.

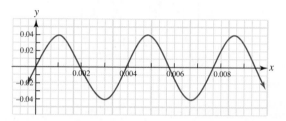

Figure 5.60 Graph of sound waves for a middle C tuning fork

Step 4: *Look back.* This graph looks correct, since we know that waves can be modeled as a sinusoidal curve.

PROBLEM SET 5.4

LEVEL 1

Draw a quick sketch of each curve in Problems 1–3 from memory.

1. **a.** $y = \cos x$ **b.** $y = \sec x$
2. **a.** $y = \sin x$ **b.** $y = \csc x$
3. **a.** $y = \tan x$ **b.** $y = \cot x$

State the amplitude and period in Problems 4–5.

4. a.

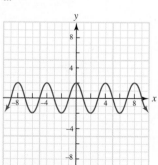

b.

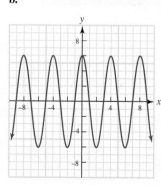

5. a.

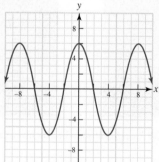

b.

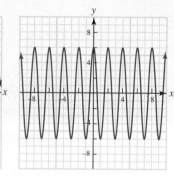

Give the amplitude, period, and frequency for the graphs in Problems 6–7.

6. a.   **b.**

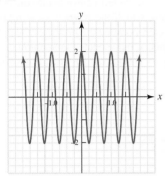   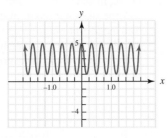

7. a.   **b.**

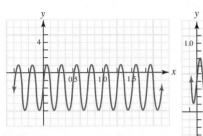   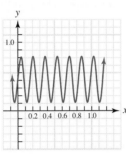

Plot the given point in Problems 8–11; then draw a frame using the given point as a starting point to draw one period of the requested curve.

8. a. $(\pi, 1)$; sine curve   **b.** $(\pi, 1)$; cosine curve
  c. $(\pi, 1)$; tangent curve

9. a. $\left(-\frac{\pi}{2}, 2\right)$; cosine curve   **b.** $\left(-\frac{\pi}{2}, 2\right)$; tangent curve
  c. $\left(-\frac{\pi}{2}, 2\right)$; sine curve

10. a. $\left(-\frac{\pi}{4}, -2\right)$; tangent curve   **b.** $\left(-\frac{\pi}{4}, -2\right)$; sine curve
  c. $\left(-\frac{\pi}{4}, -2\right)$; cosine curve

11. a. $(-1, -2)$; cosine curve   **b.** $(-1, -2)$; sine curve
  c. $(-1, -2)$; tangent curve

Graph one cycle of each function in Problems 12–17.

12. a. $y = \sin(x - 1)$ **b.** $y = \sin x - 1$
  c. How do the graphs differ?

13. a. $y = \cos\left(x - \frac{\pi}{3}\right)$ **b.** $y = \cos x - \frac{\pi}{3}$
  c. How do the graphs differ?

14. a. $y = 3\sin x$ **b.** $y = \sin 3x$
  c. How do the graphs differ?

15. a. $y = 2\cos x$ **b.** $y = \cos 2x$
  c. How do the graphs differ?

16. a. $y = \frac{1}{2}\sin x$ **b.** $y = \sin\frac{1}{2}x$
  c. How do the graphs differ?

17. a. $y = 4\tan x$ **b.** $y = \tan 4x$
  c. How do the graphs differ?

State the period and amplitude for the graphs in Problems 18–23. Also give a possible equation for the given (h, k) for each of these calculator graphs.

18. $(h, k) = (0, 0)$ **19.** $(h, k) = (0, 4)$

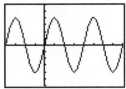   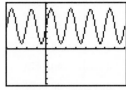

```
Xmin=-6.283185...        Xmin=-6.283185...
Xmax=12.566370...        Xmax=12.566370...
Xscl=1.5707963...        Xscl=1.5707963...
Ymin=-6                  Ymin=-6
Ymax=6                   Ymax=8
Yscl=1                   Yscl=1
```

20. $(h, k) = (\pi, 0)$ **21.** $(h, k) = \left(\frac{\pi}{2}, 0\right)$

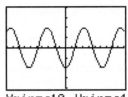   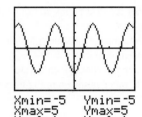

```
Xmin=-10  Ymin=-4        Xmin=-5   Ymin=-5
Xmax=10   Ymax=4         Xmax=5    Ymax=5
Xscl=1    Yscl=1         Xscl=1    Yscl=1
```

22. $(h, k) = (0, 20)$ **23.** $(h, k) = (-3, -10)$

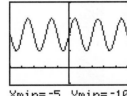   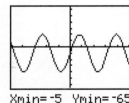

```
Xmin=-5   Ymin=-10       Xmin=-5   Ymin=-65
Xmax=5    Ymax=40        Xmax=5    Ymax=65
Xscl=1    Yscl=10        Xscl=1    Yscl=10
```

LEVEL 2

Graph one cycle of each function in Problems 24–29.

24. $y - 2 = \sin\left(x - \frac{\pi}{2}\right)$ **25.** $y + 1 = \cos\left(x + \frac{\pi}{3}\right)$

26. $y - 1 = \cos 2\left(x - \frac{\pi}{4}\right)$ **27.** $y - 3 = \tan\left(x + \frac{\pi}{6}\right)$

28. $y + 2 = \sin 3\left(x + \frac{\pi}{6}\right)$ **29.** $y - 1 = \tan 2\left(x - \frac{\pi}{4}\right)$

Specify the period and amplitude for each graph in Problems 30–41. Also graph each curve.

30. $y = \frac{1}{2}\cos\left(x + \frac{\pi}{6}\right)$ **31.** $y = 2\sin 2\pi x$

32. $y = 2\sin\left(x - \frac{\pi}{4}\right)$ **33.** $y = 3\cos 3\pi x$

34. $y = \sin(4x + \pi)$ **35.** $y = \sin(3x + \pi)$

36. $y = \tan\left(2x - \frac{\pi}{2}\right)$ **37.** $y = \tan\left(\frac{x}{2} + \frac{\pi}{3}\right)$

38. $y = 2\cos(3x + 2\pi) - 2$ **39.** $y = 4\sin\left(\frac{1}{2}x + 2\right) - 1$

40. $y = \sqrt{2}\cos x - 1$ **41.** $y = \sqrt{3}\sin\left(\frac{1}{3}x - \sqrt{\frac{1}{3}}\right)$

42. The current I (in amperes) in a certain circuit (for some convenient unit of time) generates the following set of data points:

Time	Height	Time	Height
0	−60.000	11	40.14784
1	−58.6889	12	48.54102
2	−54.8127	13	54.81273
3	−48.5410	14	58.68886
4	−40.1478	15	60.00000
5	−30.000	16	58.68886
6	−18.5410	17	54.81273
7	−6.27171	18	48.54102
8	6.27171	19	40.14784
9	18.54102	20	30.00000
10	30.00000		

Plot the data points and draw a smooth curve passing through these points.

43. Determine a possible equation of a curve that is generated by the data in Problem 42.

44. Suppose a point P on a waterwheel with a 30-ft radius is d units from the water as shown in Figure 5.61.

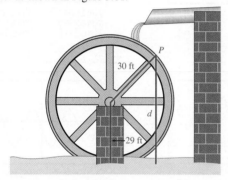

Figure 5.61 Waterwheel

If it turns at 6 revolutions per minute, the height of the point P above the water is given by the following set of data points:

Time	Height	Time	Height
0	−1.000	11	4.729
1	4.729	12	19.729
2	19.729	13	38.270
3	38.271	14	38.270
4	53.271	15	59.000
5	59.000	16	53.271
6	53.271	17	38.271
7	38.271	18	19.730
8	19.729	19	4.730
9	4.729	20	−1.000
10	−1.000		

Plot the data points and draw a smooth curve passing through these points.

45. Determine a possible equation of a curve that is generated by the data points for the waterwheel in Problem 44.

46. The pumping action of the heart consists of two phases: the systolic phase, in which blood rushes from the left ventricle into the aorta, and the diastolic phase, during which the heart muscle relaxes. The length of time between peaks is called the period of the pulse, and the pulse rate is the number of pulse beats in one minute. The graph of the blood pressure as a function of time is shown in Figure 5.62.

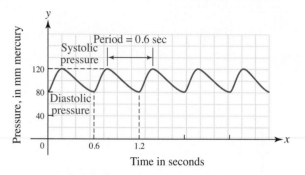

Figure 5.62 Blood pressure

  a. What is the amplitude?
  b. What is the pulse rate?

47. PROBLEMS FROM CALCULUS In *Calculus for the Life Sciences* by Rodolfo De Sapio, the motion of a human arm is analyzed. Consider a rhythmically moving arm as shown in Figure 5.63. The upper arm RO rotates back and forth about the point R, and the position of the arm is measured by the angle y. The graph shows the relationship between the angle and the time t, in seconds. Write an equation for this graph.

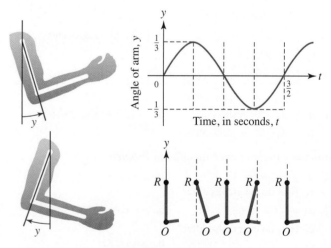

Figure 5.63 Arm motion

48. An E above middle C tuning fork vibrates at 330 Hz, with an amplitude of 0.02 cm. Write an equation, and graph this sound wave for an E above middle C.

49. A G above middle C tuning fork vibrates at 396 Hz, with an amplitude of 0.04 cm. Write an equation, and graph this sound wave for a G above middle C.

50. A spring at rest is shown in Figure 5.64. If the spring is pulled to position −3 and released, the spring will contract and stretch again for the first two seconds according to the table of values. Graph these data points, and then write an appropriate equation to model these data.

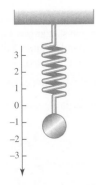

Figure 5.64 Spring at rest

Time	Position
0	−3.0
0.1	−2.6
0.2	−1.5
0.3	0.0
0.4	1.5
0.5	2.6
0.6	3
0.7	2.6
0.8	1.5
0.9	0
1.0	−1.5

Time	Position
1.1	−2.6
1.2	−3.0
1.3	−2.6
1.4	−1.5
1.5	0.0
1.6	1.5
1.7	2.6
1.8	3.0
1.9	2.6
2.0	1.5

51. A pendulum consisting of a 9-inch rod with a weight attached at one end is shown in Figure 5.65. If the rod is displaced to $60°$ and released, the angle through which the pendulum swings can be measured for the first 20 seconds after release. We measure swing to the right as a positive angle and swing to the left as a negative angle. After finding values for $t = 0$ to 10 in one-second intervals, we obtain the additional measurements as shown in the rightmost columns (specifically to help you answer this question). Graph these data points, and then write an appropriate equation to model the data.

Time	Angle
0	1.05
1	−0.32
2	−0.85
3	0.85
4	0.32
5	−1.05
6	0.33
7	0.84
8	−0.85
9	−0.32
10	1.05

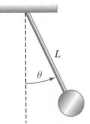

Figure 5.65 Motion of a pendulum

Time	Position
0.83	0.01
1.67	−1.05
2.50	0.00
3.33	1.05

52. Problem 51 is an important application from physics. If $t = 0$ for initial angle $\theta_0 = \frac{\pi}{3}$ (that is, $60°$), the table value is 1.047. This is an

approximate value. Explain where you think the number 1.047 comes from.

In physics, the equation for pendulum motion (which is derived in calculus) is

$$\theta = \theta_0 \cos\left(t\sqrt{\frac{g}{L}}\right)$$

where θ is the angle, θ_0 is the initial angle, t is the time in seconds, g is a constant due to gravity, and L is the length of the pendulum in inches. Do you think that the period depends on the length?

53. In physics, it is shown that the velocity for the pendulum in Problem 51 is given by the equation

$$V = -\frac{\pi}{3}\sqrt{\frac{32}{0.75}} \sin\left(\sqrt{\frac{32}{0.75}}\,t\right)$$

where V is in ft/s. Graph this function for two complete cycles. If you have a calculator, draw a graph representing the first 20 seconds.

LEVEL 3

54. The graph of the sound wave equation

$$y = 0.04\ \sin(880\pi x)$$

is shown in Figure 5.66. Use a graphing calculator with $0 \le x \le 2$ to graph this equation.

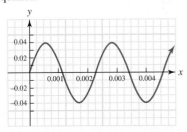

Figure 5.66 Graph of $y = 0.04 \sin(880\,\pi x)$

55. In this problem, we use the same equation for all of the graphs but have changed only the x-scale. These are all graphs of the curve

$$y = 0.04\ \sin(880\pi x)$$

(see Problem 54). Use a graphing calculator to match the equation and the graph.

a. $0 \le x \le 3.1$ **b.** $0 \le x \le 1$
c. $0 \le x \le 4$ **d.** $0 \le x \le 1.5$
e. $0 \le x \le 3$

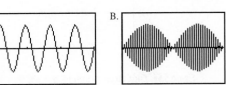

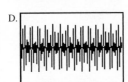

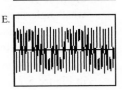

56. Take a piece of paper and wrap it around a candle, as shown in Figure 5.67. Make a perpendicular cut at A and a slanted cut at B as shown in Figure 5.67. Unroll the paper. Describe what you think the edges A and B will look like. Then perform the experiment to see whether your guess was correct. Describe what you found.

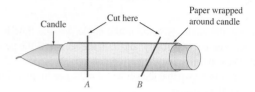

Figure 5.67 Candle experiment

57. Draw a figure to match this verbal description. Draw a Cartesian coordinate system (be sure to label the x- and y-axes); label the origin as point O. Draw a unit circle with center at $C(-1, 0)$. Draw another set of coordinate axes with origin at the point C and label these axes x' and y'. Let θ be an angle drawn with the vertex at C, initial side the positive x-axis, and the point $P(x', y')$ the intersection of the terminal side of this angle and the unit circle. Let $\overline{PQ}$ be the perpendicular drawn to the x'-axis and $\overline{PR}$ the perpendicular drawn to the y'-axis. Finally, let S be the intersection of the line determined by the terminal side and the y-axis.

a. Show $\sin\theta = \left|\overline{PQ}\right|$.

b. Show $\cos\theta = \left|\overline{PR}\right|$.

c. Show $\tan\theta = \left|\overline{SO}\right|$.

58. In Problem 57, it was discovered that $\sin\theta = \left|\overline{PQ}\right|$. Use this fact to help you sketch $y = \sin\theta$. For example, when $\theta = \frac{\pi}{4}$, draw this angle, and then measure $\overline{PQ}$ and *plot this height* at the location marked by $\theta = \frac{\pi}{4}$ on the θ-axis as shown in Figure 5.68. As P makes one revolution, it is easy to quickly plot the points on the curve $y = \sin\theta$. It is also easy to note the relationship between the unit circle definition of sine and what we call the *graph of the sine curve*.

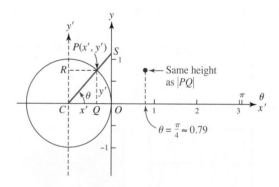

Figure 5.68 Unit circle graph of $y = \sin\theta$

59. In Problem 57, it was discovered that $\tan\theta = \left|\overline{SO}\right|$. Use this fact to help you sketch $y = \tan\theta$. For example, when $\theta = \frac{\pi}{4}$, draw this angle as shown in Figure 5.69 and then measure $\overline{SO}$ and *plot this height* at $\theta = \frac{\pi}{4}$ on the θ-axis as shown here. As P makes one revolution, notice that $\overline{SO}$ does not exist at $\theta = \frac{\pi}{2}$ and $\frac{3\pi}{2}$.

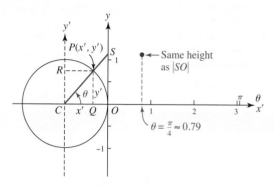

Figure 5.69 Unit circle graph of $y = \tan\theta$

60. In Problem 57, it was discovered that $\cos\theta = \left|\overline{PR}\right|$. Use this fact to help you sketch $y = \cos\theta$. To do this, rotate the unit circled described in Problem 57 by 90°, as shown in Figure 5.70. For example, when $\theta = \frac{\pi}{4}$, draw this angle, and then measure $\overline{PR}$ and plot this height at $\theta = \frac{\pi}{4}$ on the θ-axis. As P makes one revolution, it is easy to quickly plot the points on the curve $y = \cos\theta$.

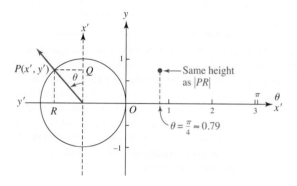

Figure 5.70 Unit circle graph of $y = \cos\theta$

5.5 Inverse Trigonometric Functions

Suppose you need to design parking spaces along a street. If the city council wants angle parking, at what angle should the parking spaces be laid out?

When you check out the physical situation, suppose you determine that city codes require that each parking space be a rectangle 8 ft wide and 18 ft long. Also suppose that the street width allows only 5 ft for the smaller side of the triangle at the top of the rectangle, as shown in Figure 5.71.

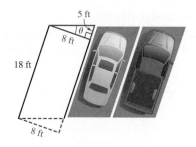

Figure 5.71 **Parking lot problem**

We know (from the definition of tangent) that $\tan\theta = \frac{x}{y}$ and for this problem (see Figure 5.71) $x = 5$ and $y = 8$, so that

$$\tan\theta = \frac{5}{8}$$

We would like to find the angle whose tangent is $\frac{5}{8} = 0.625$, which you are asked to find in Problem 44. But first, we begin by finding the angle θ whose sine is 0.625.

Inverse Sine Function

In Section 2.7, the notion of inverse functions was introduced. In this section, that idea is applied to the trigonometric functions. Recall that a function f has an inverse only when it is one-to-one on its domain—that is, when each number y in the range of f corresponds to exactly one number x in the domain. For example, this condition is satisfied by the natural logarithm because $\ln x$ does not decrease on its domain $x > 0$, and its inverse is the exponential function $y = e^x$. The trigonometric functions are not one-to-one, so their inverses do not exist. However, if we restrict the domains of the trigonometric functions, then the inverses exist.

Let us consider the sine function first. Figure 5.72 shows the graph of $y = \sin x$. Note the different x-values that lead to the same y-value; thus, the sine function is not a one-to-one function.

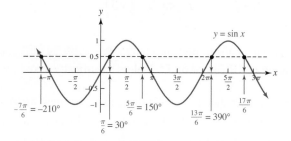

Figure 5.72 **Graph of** $y = \sin x$

We know, however, that the sine function is strictly increasing on the closed interval $\left[-\frac{\pi}{2}, \frac{\pi}{2}\right]$ and hence one-to-one. If we restrict $\sin x$ to this interval, it does have an inverse, as shown in Figure 5.73.

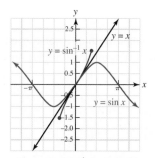

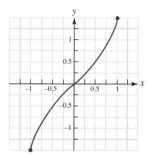

a. The graph of $\sin^{-1}x$ is obtained by reflecting the part of the sine on $\left[-\frac{\pi}{2}, \frac{\pi}{2}\right]$ about $y = x$.

b. The graph of the inverse sine function $y = \sin^{-1} x$

Figure 5.73 **Inverse sine function**

Note that the domain for this inverse function is $-1 \leq x \leq 1$. The stated restriction on the range—namely, $-\pi/2 \leq y \leq \pi/2$—tells us the angle y must be in Quadrant I (measured from 0 to $\pi/2$) or in Quadrant IV (measured as a negative angle from $-\pi/2$ to 0).

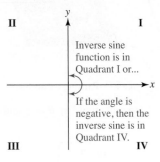

We now state the definition of $y = \sin^{-1} x$.

INVERSE SINE FUNCTION

The **inverse sine function** is defined by

$$y = \sin^{-1} x \text{ if and only if } x = \sin y \text{ and } -\tfrac{\pi}{2} \leq y \leq \tfrac{\pi}{2}$$

STOP The function $\sin^{-1} x$ is NOT the reciprocal of $\sin x$. To denote the reciprocal, write $(\sin x)^{-1}$. To **evaluate** $\sin^{-1} x$ means to find the angle θ so that $\sin \theta = x$.

EXAMPLE 1   Evaluating inverse sine functions

Evaluate:

a. $\sin^{-1} 0.625$ **b.** $\sin^{-1}\left(\tfrac{1}{2}\sqrt{3}\right)$ **c.** $\sin^{-1}\left(-\tfrac{1}{2}\sqrt{3}\right)$

Solution

a. We begin by using a calculator to find $\sin^{-1} 0.625$. Press: [sin⁻¹] [0.625] to find [0.675131532937]; in degrees, we find 38.68218745 or 38.7°. Unless specifically asked for degrees, we give the angles in radians.

b. We need to understand more than pressing a few calculator steps, so to understand the inverse sine function, we let $\theta = \sin^{-1}\left(\tfrac{1}{2}\sqrt{3}\right)$. Remember, the inverse sine function represents the measurement of an angle. Therefore, we denote it by θ, and we find the angle or real number θ with sine equal to $\tfrac{1}{2}\sqrt{3}$ so that $-\pi/2 \leq \theta \leq \pi/2$. From the memorized table of exact values, we know $\sin(\pi/3) = \tfrac{1}{2}\sqrt{3}$, and since $\pi/3$ is between $-\pi/2$ and $\pi/2$, we have

$$\sin^{-1}\left(\frac{1}{2}\sqrt{3}\right) = \frac{\pi}{3}$$

c. You may want to work with reference angles when finding inverse trigonometric functions. That is, because the table of exact values shows first-quadrant angles only, we work with the reference angle. Let

$$\overline{\theta} = \sin^{-1}\frac{1}{2}\sqrt{3}$$

reference angle   absolute value of the given number

so $\bar{\theta} = \frac{\pi}{3}$ the reference angle (from part **b**). Now place θ in the appropriate quadrant. The sine is negative in both the third and fourth quadrants, but because $-\pi/2 \le \theta \le \pi/2$, we see that $\theta = -\pi/3$.　　　　　　　　　　　　　　　　　　　　　　　　　　　　　　　　　■

Inverse Trigonometric Functions

The following table summarizes the relationship between a given trigonometric function, its inverse function, and the notation we use for the inverse.

Given Function	Inverse	Other Notations
$y = \cos x$	$x = \cos y$	$y = \cos^{-1} x;\ y = \arccos x$
$y = \sin x$	$x = \sin y$	$y = \sin^{-1} x;\ y = \arcsin x$
$y = \tan x$	$x = \tan y$	$y = \tan^{-1} x;\ y = \arctan x$
$y = \sec x$	$x = \sec y$	$y = \sec^{-1} x;\ y = \text{arcsec}\, x$
$y = \csc x$	$x = \csc y$	$y = \csc^{-1} x;\ y = \text{arccsc}\, x$
$y = \cot x$	$x = \cot y$	$y = \cot^{-1} x;\ y = \text{arccot}\, x$

Given this notation for the inverse trigonometric functions, we now need to determine the domain and range for each inverse trigonometric function. For example, by restricting $\tan x$ to the open interval $\left(-\frac{\pi}{2}, \frac{\pi}{2}\right)$ where it is one-to-one, we can define the inverse tangent function as follows.

INVERSE TANGENT FUNCTION

The **inverse tangent function** is defined by
$$y = \tan^{-1} x \text{ if and only if } x = \tan y \text{ and } -\tfrac{\pi}{2} < y < \tfrac{\pi}{2}$$

The graph of $y = \tan^{-1} x$ is shown in Figure 5.74.

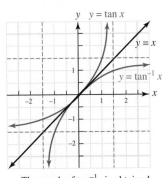
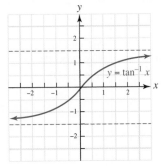

a. The graph of $\tan^{-1} x$ is obtained by reflecting the part of the tangent graph on $\left(-\frac{\pi}{2}, \frac{\pi}{2}\right)$ above the line $y = x$.

b. The graph of $\tan^{-1} x$

Figure 5.74 Graph of the inverse tangent

Since the definition of the inverse tangent function requires that $-\frac{\pi}{2} < y < \frac{\pi}{2}$, we can say, as we did with the inverse sine function, that it is defined in Quadrants I and IV. However, as we turn to the definition for the inverse cosine, we see that since the cosine function is positive in *both* Quadrants I and IV, and since we need one positive quadrant for cosine, we note the change in the quadrants as we state the definition for the inverse cosine function.

INVERSE COSINE FUNCTION

The **inverse cosine function** is defined by

$$y = \cos^{-1} x \text{ if and only if } x = \cos y \text{ and } 0 \le y \le \pi$$

EXAMPLE 2   Evaluating inverse trigonometric functions

Evaluate the given functions.

a. $\sin^{-1}\left(\dfrac{-\sqrt{2}}{2}\right)$ **b.** $\sin^{-1} 0.21$ **c.** $\cos^{-1} 0$ **d.** $\tan^{-1}\left(\dfrac{1}{\sqrt{3}}\right)$

Solution

a. $\sin^{-1}\left(\dfrac{-\sqrt{2}}{2}\right) = -\dfrac{\pi}{4}$;   *Think:* $x = \dfrac{-\sqrt{2}}{2}$ negative, so y is in Quadrant IV; the reference

angle is the angle whose sine is $\dfrac{\sqrt{2}}{2}$; it is $\frac{\pi}{4}$, so in Quadrant IV the

angle is $-\dfrac{\pi}{4}$.

b. $\sin^{-1} 0.21 \approx 0.2115750$ By calculator; be sure to use radian mode and inverse sine (not reciprocal).

c. $\cos^{-1} 0 = \dfrac{\pi}{2}$ Memorized exact value.

d. $\tan^{-1}\left(\dfrac{1}{\sqrt{3}}\right) = \dfrac{\pi}{6}$ *Think:* $x = \dfrac{1}{\sqrt{3}}$ positive, so y is in Quadrant I; the reference angle

is   the same as the value of the inverse tangent in Quadrant I. ∎

Definitions and graphs of the other fundamental inverse trigonometric functions are done similarly. We summarize in Table 5.6.

TABLE 5.6	Directory of Curves (part IV)			
Function	**Graph**	**Inverse**	**Domain*/Range**	**Inverse Graph**
$y = \sin x$	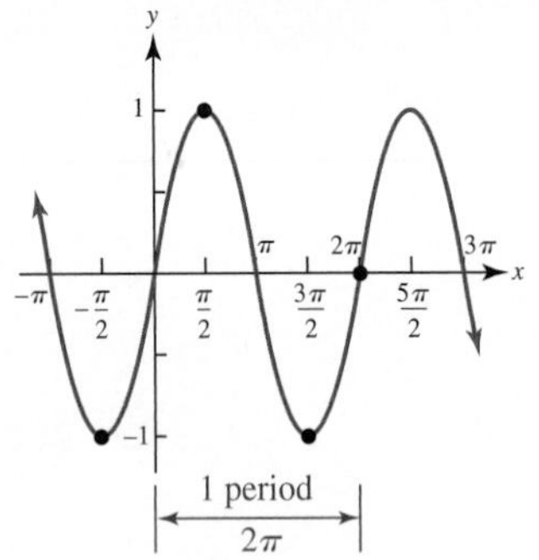	$y = \sin^{-1} x$	$D\colon -1 \le x \le 1$ $R\colon -\dfrac{\pi}{2} \le y \le \dfrac{\pi}{2}$ Pos: Quad I Neg: Quad IV Zero: $x = 0$	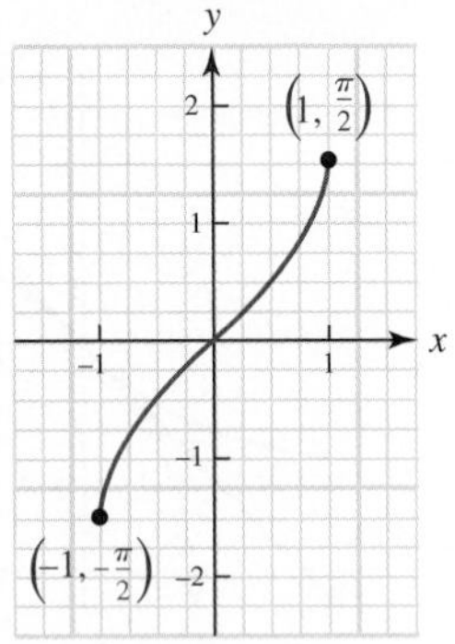
$y = \cos x$	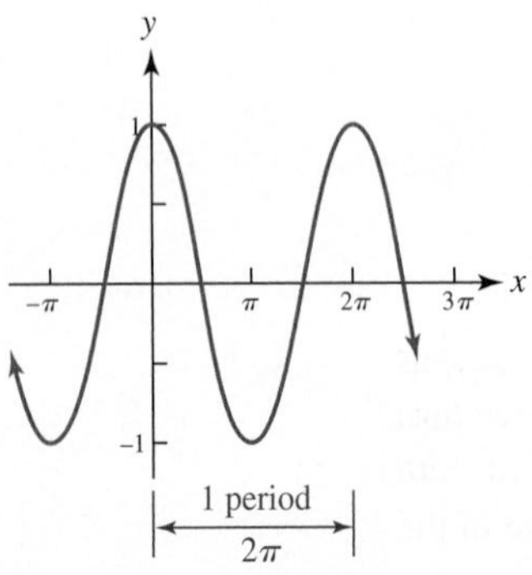	$y = \cos^{-1} x$	$D\colon -1 \le x \le 1$ $R\colon 0 \le y \le \pi$ Pos: Quad I Neg: Quad II Zero: $x = \dfrac{\pi}{2}$	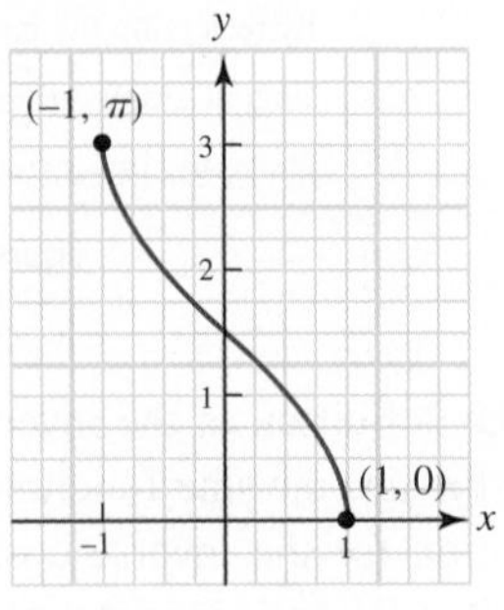

TABLE 5.6	Directory of Curves (part IV) (*Continued*)

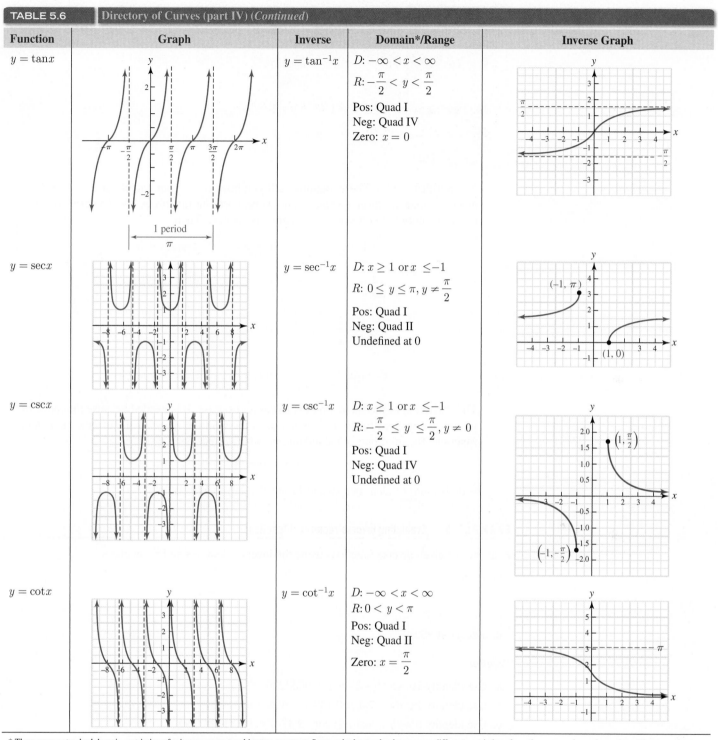

Function	Graph	Inverse	Domain*/Range	Inverse Graph
$y = \tan x$		$y = \tan^{-1}x$	$D: -\infty < x < \infty$ $R: -\dfrac{\pi}{2} < y < \dfrac{\pi}{2}$ Pos: Quad I Neg: Quad IV Zero: $x = 0$	
$y = \sec x$		$y = \sec^{-1}x$	$D: x \geq 1$ or $x \leq -1$ $R: 0 \leq y \leq \pi, y \neq \dfrac{\pi}{2}$ Pos: Quad I Neg: Quad II Undefined at 0	
$y = \csc x$		$y = \csc^{-1}x$	$D: x \geq 1$ or $x \leq -1$ $R: -\dfrac{\pi}{2} \leq y \leq \dfrac{\pi}{2}, y \neq 0$ Pos: Quad I Neg: Quad IV Undefined at 0	
$y = \cot x$		$y = \cot^{-1}x$	$D: -\infty < x < \infty$ $R: 0 < y < \pi$ Pos: Quad I Neg: Quad II Zero: $x = \dfrac{\pi}{2}$	

* There are no standard domain restrictions for inverse secant and inverse cosecant. Some calculus textbooks may use different restrictions from the ones we have chosen here. However, the ones we have used are the ones that are most convenient to use with standard calculators and the definitions used in most trigonometry texts.

Evaluating Inverses of the Reciprocal Functions

To evaluate arcsecant, arccosecant, or arccotangent functions when their exact values are not known, we need to use the reciprocal identities. First find the reciprocal, then find the inverse of the reciprocal. That is, find the inverse of the reciprocal function. This procedure is summarized in the form of some identities known as the *inverse identities*.*

---

*In Section 5.2, we presented identities numbered 1–8; we continue with that numbering in this section.

INVERSE IDENTITIES

9. $\cot^{-1} x = \begin{cases} \tan^{-1}\frac{1}{x} & \text{if } x \text{ is positive} \\ \tan^{-1}\frac{1}{x} + \pi & \text{if } x \text{ is negative} \\ \frac{\pi}{2} & \text{if } x = 0 \end{cases}$

10. $\sec^{-1} x = \cos^{-1}\frac{1}{x}$ if $x \geq 1$ or $x \leq -1$

11. $\csc^{-1} x = \sin^{-1}\frac{1}{x}$ if $x \geq 1$ or $x \leq -1$

》 IN OTHER WORDS These identities tell you how to use your calculator to evaluate the inverse cotangent, inverse secant, and inverse cosecant functions. We can derive the first part of identity 9. Let $\theta = \cot^{-1} x$ where x is positive. Then,

$$\cot\theta = x \qquad\qquad 0 < x < \tfrac{\pi}{2} \qquad \textit{Definition of } \cot^{-1}x$$

$$\frac{1}{\tan\theta} = x \qquad\qquad 0 < x < \tfrac{\pi}{2} \qquad \textit{Reciprocal identity}$$

$$\tan\theta = \frac{1}{x} \qquad\qquad 0 < x < \tfrac{\pi}{2} \qquad \textit{Solve for } \tan\theta.$$

$$\theta = \tan^{-1}\frac{1}{x} \qquad\qquad 0 < x < \tfrac{\pi}{2} \qquad \textit{Definition of } \tan^{-1}\theta$$

The derivation of the second part is left as an exercise. It is needed because the range of the inverse tangent function does not coincide with the range of the inverse cotangent function. The third part of the identity is obvious since $\cot\frac{\pi}{2} = 0$.

The derivations of identities 10 and 11 are also left as exercises.

EXAMPLE 3   **Evaluating inverse reciprocal functions**

Evaluate the given inverse functions using the inverse identities and a calculator.

   a. $\sec^{-1}(-3)$

   b. $\csc^{-1} 7.5$

   c. $\cot^{-1} 2.4747$

   d. $\operatorname{arccot}(-4.852)$

Solution

   a. Use identity 10: $\sec^{-1}(-3) \approx 1.910633236$

   b. Use identity 11: $\csc^{-1} 7.5 \approx 0.1337315894$

   c. Use identity 9 with x positive: $\cot^{-1} 2.4747 \approx 0.3840267299$

   d. Use identity 9 with x negative: $\operatorname{arccot}(-4.852) \approx 2.938338095$ ∎

Functions of Inverse Functions

A word of caution is in order regarding the inverse trigonometric functions, especially when you are using a calculator. Recall that for a function f with domain D and range R, then

$$f^{-1}[f(a)] = a \quad \text{for all } a \text{ in } D$$

$$f[f^{-1}(b)] = b \quad \text{for all } b \text{ in } R$$

This means that, for example, for the function $\cos x$ with restricted domain $0 \leq x \leq \pi$ and range $-1 \leq \cos x \leq 1$, we have

$$\cos^{-1}(\cos x) = x \quad \text{for every } x \text{ in the interval } [0, \pi]$$

$$\cos(\cos^{-1} x) = x \quad \text{for every } x \text{ in the interval } [-1, 1]$$

Similarly,

$$\sin^{-1}(\sin x) = x \quad \text{for every } x \text{ in the interval } \left[-\frac{\pi}{2}, \frac{\pi}{2}\right]$$

$$\sin(\sin^{-1} x) = x \quad \text{for every } x \text{ in the interval } [-1, 1]$$

and

$$\tan^{-1}(\tan x) = x \quad \text{for every } x \text{ in the interval } \left(-\frac{\pi}{2}, \frac{\pi}{2}\right)$$

$$\tan(\tan^{-1} x) = x \quad \text{for all real } x$$

EXAMPLE 4 Evaluate functions involving inverse trigonometric functions

Evaluate (correct to two decimal places):

a. $\cos^{-1}(\cos 2.2)$ **b.** $\cos^{-1}(\cos 4)$

c. $\cos(\cos^{-1} 2.2)$ **d.** $\cos(\cos^{-1} 0.22)$

e. $\sin^{-1}(\sin 2.2)$ **f.** $\sin^{-1}[\sin(-1)]$

g. $\sin(\sin^{-1} 0.46)$ **h.** $\sin(\sin^{-1} 2.46)$

i. $\tan\left(\tan^{-1} \sqrt{3}\right)$ **j.** $\tan^{-1}\left(\tan \sqrt{3}\right)$

Solution

a. $\cos^{-1}(\cos 2.2) = 2.2$ since 2.2 is on the interval $[0, \pi]$.
b. $\cos^{-1}(\cos 4) \approx 2.28$ *Note*: It is not 4 because 4 is not on the interval $[0, \pi]$.
c. $\cos(\cos^{-1} 2.2)$ is not defined. *Note*: 2.2 is not on the interval $[-1, 1]$.
d. $\cos(\cos^{-1} 0.22) = 0.22$
e. $\sin^{-1}(\sin 2.2) \approx 0.94$ *Note*: 2.2 is not on the interval $\left[-\frac{\pi}{2}, \frac{\pi}{2}\right]$.
f. $\sin^{-1}[\sin(-1)] = -1$ since -1 is on the interval $\left[-\frac{\pi}{2}, \frac{\pi}{2}\right]$.
g. $\sin(\sin^{-1} 0.46) = 0.46$ since 0.46 is on the interval $[-1, 1]$.
h. $\sin(\sin^{-1} 2.46)$ is not defined. *Note:* 2.46 is not on the interval $[-1, 1]$.
i. $\tan\left(\tan^{-1} \sqrt{3}\right) = \sqrt{3} \approx 1.73$, since $\tan(\tan^{-1} x) = x$ for all real numbers.
j. $\tan^{-1}\left(\tan \sqrt{3}\right) \approx -1.41$ *Note*: $\sqrt{3}$ not on the interval $\left(-\frac{\pi}{2}, \frac{\pi}{2}\right)$. ■

EXAMPLE 5 Evaluation by using a fundamental identity

Evaluate $\cot[\csc^{-1}(-3)]$ without tables or a calculator.

Solution Let $\theta = \csc^{-1}(-3)$.   Negative value, so θ is in Quadrant IV.

Then $\csc \theta = -3$. We want to find $\cot[\csc^{-1}(-3)] = \cot \theta$

$$\cot \theta = \pm\sqrt{\csc^2 \theta - 1} \qquad \text{Pythagorean identity}$$

$$= -\sqrt{(-3)^2 - 1} \qquad \text{\small csc }\theta = -3\text{; chose the negative value for the radical because }\theta \\ \text{\small is in Quadrant IV and cotangent is negative in Quadrant IV.}$$

$$= -\sqrt{8} \qquad \text{Simplify.}$$

$$= -2\sqrt{2}$$

EXAMPLE 6   Deriving a formula involving an inverse function

Find $\sin(\cos^{-1} x)$ as a function of x.

Solution

$$\sin(\cos^{-1} x) = \sin \theta \qquad \text{\small Let }\theta = \cos^{-1}x\text{ so that }\cos\theta = x.$$

$$= \pm\sqrt{1 - \cos^2 \theta} \qquad \text{Pythagorean identity}$$

$$= \sqrt{1 - x^2} \qquad \text{\small Substitute }\cos\theta = x,\text{ and choose the positive value} \\ \text{\small for the radical because }\theta\text{ is in Quadrant I or II.}$$

EXAMPLE 7   Solving an equation involving inverse functions

Solve $\cos^{-1} x = \tan^{-1} x$ using the indicated method:

a. Numerically   **b.** Graphically   **c.** Algebraically

Solution

a. Numerically; build a table of values for $y_1 = \cos^{-1} x$ and for $y_2 = \tan^{-1} x$ and look to see where they are equal.

X	Y1	Y2
.7	.7954	.61073
.71	.7813	.61741
.72	.76699	.62402
.73	.75247	.63058
.74	.73773	.63707
.75	.72273	.6435
.76	.70748	.64987
.77	.69196	.65618
.78	.67613	.66243
.79	.65999	.66861
.8	.6435	.67474

\Y1■cos⁻¹(X)
\Y2■tan⁻¹(X)

It looks like $y_1 = y_2$ near $x \approx 0.8$.

b. Graph $y_1 = \cos^{-1} x$ and $y_2 = \tan^{-1} x$ and look for the point of intersection.

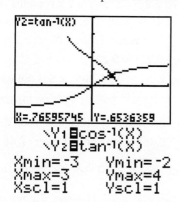

X=.76595745   Y=.6536359

\Y1■cos⁻¹(X)
\Y2■tan⁻¹(X)
Xmin=-3   Ymin=-2
Xmax=3   Ymax=4
Xscl=1   Yscl=1

It looks like the graphs cross near $x \approx 0.77$.

c. To find an exact answer, we proceed algebraically. Because inverse cosine is defined in Quadrants I and II and the inverse tangent is defined in Quadrants I and IV, the intersection point must be in Quadrant I.

$$\cos^{-1} x = \tan^{-1} x \qquad \text{Given equation}$$

$$\cos(\cos^{-1} x) = \cos(\tan^{-1} x) \qquad \text{Take the cosine of both angles.}$$

$$x = \cos \theta \qquad \text{\small cos}(\cos^{-1}x) = x\text{ in Quadrant I.}$$

Now, we need to simplify $\cos\theta$ where $\theta = \tan^{-1} x$. Then, $\tan\theta = x$

$$1 + \tan^2\theta = 1 + x^2$$

$$\sec^2\theta = 1 + x^2$$

$$\sec\theta = \sqrt{1 + x^2} \qquad \textit{Quad I}$$

$$\cos\theta = \frac{1}{\sqrt{1 + x^2}}$$

We now use this value for cosine to return to the solution of the main equation:

$$x = \cos\theta \qquad \textit{Continuation of the above equation}$$

$$x = \frac{1}{\sqrt{1 + x^2}} \qquad \textit{By substitution}$$

$$x^2 = \frac{1}{1 + x^2}$$

$$x^2 + x^4 = 1$$

$$x^4 + x^2 - 1 = 0$$

$$x^2 = \frac{-1 \pm \sqrt{1^2 - 4(1)(-1)}}{2(1)} \qquad \textit{Quadratic formula}$$

$$x^2 = \frac{-1 + \sqrt{5}}{2} \qquad \textit{Positive since } x^2 \textit{ is nonnegative}$$

$$x = \sqrt{\frac{\sqrt{5} - 1}{2}} \qquad \textit{Positive values since x is in Quad I}$$

The exact value is approximately 0.786153778 (by calculator), so the three methods are all approximately the same.

PROBLEM SET 5.5

1. **a.** What is the range for the inverse cosine?
   b. In which quadrants is the inverse cosine defined?
2. **a.** In which quadrants is the inverse sine defined?
   b. What is the range for the inverse sine?
3. **a.** What is the range for the inverse tangent?
   b. In which quadrants is the inverse tangent defined?

State the keys you press on your calculator to evaluate each function in Problems 4–6.

4. **a.** $y = \cos^{-1} x$ **b.** $y = \sec^{-1} x$
5. **a.** $y = \sin^{-1} x$ **b.** $y = \csc^{-1} x$
6. **a.** $y = \tan^{-1} x$ **b.** $y = \cot^{-1} x$
7. Let $c(x) = \cos x$ for $0 \le x \le \pi$. Using the graph in Figure 5.75, find the functional values for c and c^{-1} as requested.

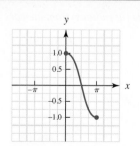

Figure 5.75 Graph of $y = c(x)$

a. $c\left(\frac{\pi}{4}\right)$ **b.** $c\left(\frac{\pi}{2}\right)$

c. $c(\pi)$ **d.** $c\left(\frac{3\pi}{4}\right)$

e. $c^{-1}(1)$ **f.** $c^{-1}(0)$

g. $c^{-1}(0.5)$ **h.** $c^{-1}(-1)$

i. Using this graph, draw the graph of $c^{-1}(x)$ by plotting points.

8. Let $s(x) = \sin x$ for $-\frac{\pi}{2} \le x \le \frac{\pi}{2}$. Using the graph shown in Figure 5.76, find the functional values for s and s^{-1} as requested.

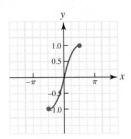

Figure 5.76 Graph of $y = s(x)$

a. $s\left(\frac{\pi}{4}\right)$ **b.** $s\left(\frac{\pi}{2}\right)$

c. $s\left(-\frac{\pi}{4}\right)$ **d.** $s\left(-\frac{\pi}{2}\right)$

e. $s^{-1}(1)$ **f.** $s^{-1}(0)$

g. $s^{-1}(0.5)$ **h.** $s^{-1}(-1)$

i. Using this graph, draw the graph of $s^{-1}(x)$ by plotting points.

9. Let $t(x) = \tan x$ for $-\frac{\pi}{2} < x \le \frac{\pi}{2}$. Using the graph shown in Figure 5.77, find the functional values for t and t^{-1} as requested.

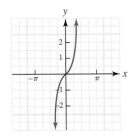

Figure 5.77 Graph of $y = t(x)$

a. $t\left(\frac{\pi}{4}\right)$ **b.** $t\left(\frac{\pi}{8}\right)$

c. $t(0)$ **d.** $t\left(-\frac{\pi}{4}\right)$

e. $t^{-1}(1)$ **f.** $t^{-1}(21)$

g. $t^{-1}(0.5)$ **h.** $t^{-1}(-1)$

i. Using this graph, draw the graph of $t^{-1}(x)$ by plotting points.

Obtain the given angle (in radians) from memory in Problems 10–15.

10. a. $\tan^{-1}\left(\frac{\sqrt{3}}{3}\right)$ **b.** $\cos^{-1} 1$

c. $\sin^{-1}\frac{1}{2}$ **d.** $\sin^{-1} 1$

11. a. $\cos^{-1}\left(\frac{\sqrt{2}}{2}\right)$ **b.** $\cot^{-1} 1$

c. $\sin^{-1}(-1)$ **d.** $\sin^{-1} 0$

12. a. $\cot^{-1}\sqrt{3}$ **b.** $\cos^{-1}\left(\frac{\sqrt{3}}{2}\right)$

c. $\tan^{-1} 1$ **d.** $\sin^{-1}\left(\frac{\sqrt{2}}{2}\right)$

13. a. $\tan^{-1}\sqrt{3}$ **b.** $\sin^{-1}(-1)$

c. $\sin^{-1}\left(-\frac{\sqrt{3}}{2}\right)$ **d.** $\cot^{-1}(-1)$

14. a. $\arccos(-1)$ **b.** $\operatorname{arccot}\left(-\sqrt{3}\right)$

c. $\arctan(-1)$ **d.** $\arcsin\left(-\frac{1}{2}\sqrt{3}\right)$

15. a. $\arctan\left(-\frac{1}{3}\sqrt{3}\right)$ **b.** $\arctan 0$

c. $\arccos\frac{1}{2}$ **d.** $\arcsin\left(\frac{1}{2}\sqrt{3}\right)$

Use a calculator in Problems 16–17 to find the values (in radians correct to the nearest hundredth).

16. a. $\arcsin 0.20846$ **b.** $\cos^{-1} 0.83646$

c. $\arctan 1.1156$ **d.** $\cot^{-1}(-0.08097)$

17. a. $\operatorname{arccot} 3.451$ **b.** $\sec^{-1} 4.315$

c. $\operatorname{arccsc} 2.461$ **d.** $\sec^{-1} 2.894$

Use a calculator in Problems 18–19 to find the values (to the nearest degree).

18. a. $\sin^{-1} 0.3584$ **b.** $\arccos 0.9455$

c. $\cos^{-1} 0.3584$ **d.** $\arcsin(-0.4696)$

19. a. $\cot^{-1}(-2)$ **b.** $\operatorname{arccsc} 3.945$

c. $\sec^{-1}(-6)$ **d.** $\arctan(-3)$

WHAT IS WRONG, *if anything, with each statement in Problems 20–25? Explain your reasoning.*

20. In $y = \cos x$, the angle is y.

21. In $y = \cos^{-1} x$, the angle is y.

22. $\tan^{-1}(-2.5)$ is in Quadrant II.

23. $\cot^{-1}(-2.5)$ is in Quadrant IV.

24. The domain in $y = \sin x$ is $[-1, 1]$.

25. The domain in $y = \sin^{-1} x$ is $[-1, 1]$.

Evaluate the expressions in Problems 26–31 without using a calculator or tables.

26. a. $\cot(\operatorname{arccot} 1)$ **b.** $\arccos\left(\cos\frac{\pi}{6}\right)$

27. a. $\sin\left(\sin^{-1}\frac{1}{3}\right)$ **b.** $\tan^{-1}\left(\tan\frac{\pi}{15}\right)$

28. a. $\cos\left(\cos^{-1}\frac{2}{3}\right)$ **b.** $\sin^{-1}\left(\sin\frac{2\pi}{15}\right)$

29. a. $\cot^{-1}(\cot 35°)$ **b.** $\tan(\tan^{-1} 0.4163)$

30. a. $\cos\left(\tan^{-1}\sqrt{3}\right)$ **b.** $\sin\left(\cos^{-1}\frac{2}{3}\right)$

31. a. $\cos\left[\sin^{-1}\left(-\frac{1}{3}\right)\right]$ **b.** $\sin\left[\cos^{-1}\left(-\frac{1}{3}\right)\right]$

LEVEL 2

Find a formula for each expression in Problems 32–37.

32. $\cos(\sin^{-1} x)$ **33.** $\sin(\csc^{-1} x)$

34. $\cos(\sec^{-1} x)$ **35.** $\sec(\tan^{-1} x)$

36. $\tan(\sec^{-1} x)$ **37.** $\csc(\cot^{-1} x)$

Solve the equations for $x \ge 0$ in Problems 38–43.

38. $\cos^{-1} x = \sin^{-1} x$

39. $\sec^{-1} = \tan^{-1} x$

40. $\sin^{-1} x = \tan^{-1} x$

41. $\cos^{-1} x = \cot^{-1} x$

42. $\sin^{-1}(\sin^{-1} x) = \sin^{-1}(\cos^{-1} x)$

43. $\cos^{-1}(\cos^{-1} x) = \cos^{-1}(\sin^{-1} x)$

44. At the beginning of this section, we answered a question about angle parking. Find the angle correct to three decimal places. Find the exact value of the angle.

45. Find the parking angle for the problem at the beginning of this section if the city ordinances require 10 ft for the smaller side of the triangle at the top of the rectangle shown in Figure 5.71.

Suppose a belt is connecting two pulleys of different sizes, as shown in Figure 5.78.

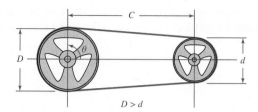

Figure 5.78 Two pulleys

The angle θ (measured in radians) is necessary to determine the length of the belt. It can be shown that

$$\theta = \cos^{-1}\left(\frac{D-d}{2C}\right)$$

where D is the diameter of the larger pulley, d is the diameter of the smaller pulley, and C is the distance between the pulleys. Find the angle θ (rounded to the nearest tenth) in Problems 46–47.

46. $D = 10$ in., $d = 8$ in., and $C = 14$ in.

47. $D = 12$ in., $d = 4$ in., and $C = 2$ ft

48. In Problem 46, the length of the belt is found by the formula

$$L = 2C\sin\theta + \pi D + (d-D)\theta$$

Find the length of the belt in Problem 46.

49. In Problem 47, the length of the belt is found by the formula

$$L = 2C\sin\theta + \pi D + (d-D)\theta$$

Find the length of the belt in Problem 47.

Graph each curve in Problems 50–53.

50. $y + 2 = \tan^{-1}x$ **51.** $y - 1 = \arcsin x$

52. $y = \arcsin(x-2)$ **53.** $y = \sin^{-1}(x+1)$

Solve the equations in Problems 54–57 on the interval $[0, 1]$ using the graphical method. Give your answer correct to two decimal places.

54. $1.2(x - 0.5)^2 = \cos^{-1}x$

55. $1.4(x - 0.5)^2 = \sin^{-1}x$

56. $\tan^{-1}x = 0.5\cos 3x$

57. $\dfrac{1}{\tan^{-1}x + \sin^{-1}x} = \sin 2x$

58. Prove identity 11:

$$\csc^{-1}x = \sin^{-1}\frac{1}{x} \text{ if } x \geq 1 \text{ or } x \leq -1$$

59. Prove identity 10:

$$\sec^{-1}x = \cos^{-1}\frac{1}{x} \text{ if } x \geq 1 \text{ or } x \leq -1$$

60. Prove identity 9 for $x < 0$:

$$\cot^{-1}x = \tan^{-1}\frac{1}{x} + \pi$$

5.6 Right Triangles

One of the most important uses of trigonometry is for solving triangles. Recall from geometry that every triangle has three sides and three angles, which are called the six *parts* of the triangle. We say that a **triangle is solved** when all six parts are known and listed. Typically, three parts will be given, or known, and it will be our task to find the other three parts to solve the triangle. Sometimes, however, we will be looking for only one part of the triangle.

A triangle is usually labeled as shown in Figure 5.79.

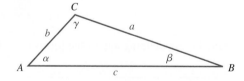

Figure 5.79 Correctly labeled triangle

The vertices are labeled A, B, and C, with the sides opposite those vertices labeled a, b, and c, respectively. The angles are labeled α, β, and γ, respectively. In this section, we assume that γ denotes the right angle and c the **hypotenuse** of a right triangle.

Right Triangle Definition of the Trigonometric Functions

According to the definition of the trigonometric functions, the angles under consideration must be in standard position. This requirement is sometimes inconvenient, so we use the ratio definition to create a special case that applies to any acute angle θ of a right triangle. Notice from Figure 5.79, θ might be α or β, but it would not be γ since γ is a right angle. Also notice that the hypotenuse is one of the sides of both acute angles. The other side making up the angle is called the **adjacent side**. Thus, side a is adjacent to β and side b is adjacent to α. The third side of the triangle (the one that is not part of the angle) is called the **opposite side**. Thus, side a is opposite α and side b is opposite β. Finally, note that in any right triangle,

$$c^2 = a^2 + b^2 \quad \text{Pythagorean theorem}$$

RIGHT TRIANGLE DEFINITION OF THE TRIGONOMETRIC FUNCTIONS

Let θ be an acute angle in a right triangle.

Then, the **right triangle definition of the trigonometric functions** is:

$$\cos\theta = \frac{\text{ADJACENT SIDE}}{\text{HYPOTENUSE}} \quad \sin\theta = \frac{\text{OPPOSITE SIDE}}{\text{HYPOTENUSE}} \quad \tan\theta = \frac{\text{OPPOSITE SIDE}}{\text{ADJACENT SIDE}}$$

Adjacent side

The secant, cosecant, and cotangent are defined as the reciprocals of the cosine, sine, and tangent, respectively.

EXAMPLE 1   Trigonometric functions for a given triangle

Given a triangle with sides 3, 4, and 5, as shown in Figure 5.80.

a. Find $\cos\alpha$, $\sin\alpha$, and $\tan\alpha$.

b. Find $\cos\beta$, $\sin\beta$, and $\tan\beta$.

c. Find α, β, and γ (correct to the nearest degree).

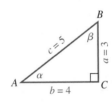

Figure 5.80 $\triangle ABC$

Solution
Note: Since $3^2 + 4^2 = 5^2$, we see that $\triangle ABC$ is a right triangle.

a. $\cos\alpha = \frac{4}{5} = 0.8$; $\sin\alpha = \frac{3}{5} = 0.6$; $\tan\alpha = \frac{3}{4} = 0.75$

b. $\cos\beta = \frac{3}{5} = 0.6$; $\sin\beta = \frac{4}{5} = 0.8$; $\tan\beta = \frac{4}{3} \approx 1.33$

c. Since $\cos\alpha = 0.8$, then $\alpha = \cos^{-1} 0.8 \approx 37°$.

Note: You could also have found $\sin^{-1} 0.6$ or $\tan^{-1} 0.75$; you might wish to verify that these expressions represent the same angle.

Since $\cos\beta = 0.6$, we see $\beta = \cos^{-1} 0.6 \approx 53°$.

Finally, since $\triangle ABC$ is a right triangle, $\gamma = 90°$. ∎

Trigonometric Substitutions

In calculus, when integrating functions containing any of the forms $(a > 0)$:

$$\sqrt{a^2 - u^2}, \quad \sqrt{u^2 - a^2}, \quad \text{and} \quad \sqrt{u^2 + a^2}$$

trigonometry is used in what is called a *trigonometric substitution*. These substitutions are based on the right triangle definition of the trigonometric functions. For example, to simplify the expression $\sqrt{a^2 - u^2}$, make the substitution $u = a\sin\theta$, where θ is an angle in a right triangle.

$$\sqrt{a^2 - u^2} = \sqrt{a^2 - (a\sin\theta)^2} \qquad \text{Substitute the value of } u.$$

$$= \sqrt{a^2 - a^2\sin^2\theta}$$

$$= \sqrt{a^2(1 - \sin^2\theta)}$$

$$= \sqrt{a^2\cos^2\theta} \qquad \text{Fundamental identity}$$

$$= a\,|\cos\theta| \qquad a > 0, \text{ but cosine may not be, hence absolute value.}$$

$$= a\cos\theta \qquad \theta \text{ is an angle in a right triangle, so it is in Quadrant I and positive.}$$

Another way of writing this substitution is to note that $\sin\theta = \frac{u}{a}$ so that $\theta = \sin^{-1}\frac{u}{a}$.

EXAMPLE 2   Trigonometric substitution

Make a trigonometric substitution in the expression $\dfrac{du}{u^2\sqrt{4 - u^2}}$ by making the substitution $u = 2\sin\theta$, $-\frac{\pi}{2} < \theta < \frac{\pi}{2}$, where du is a single variable representing $2\cos\theta$.

Solution

$$\frac{du}{u^2\sqrt{4 - u^2}} = \frac{2\cos\theta}{4\sin^2\theta\sqrt{4 - 4\sin^2\theta}} \qquad \text{Substitute given values for } u \text{ and } du.$$

$$= \frac{2\cos\theta}{4\sin^2\theta\sqrt{4(1 - \sin^2\theta)}} \qquad \text{Factor.}$$

$$= \frac{2\cos\theta}{4\sin^2\theta\sqrt{4\cos^2\theta}} \qquad \text{Fundamental identity}$$

$$= \frac{2\cos\theta}{4\sin^2\theta(2\cos\theta)} \qquad \sqrt{4\cos^2\theta} = 2|\cos\theta| = 2\cos\theta \text{ since cosine is positive when } -\frac{\pi}{2} < \theta < \frac{\pi}{2}.$$

$$= \frac{1}{4\sin^2\theta}$$

$$= \frac{1}{4}\csc^2\theta$$

By the way, one of the fundamental operations in calculus is integration, and a standard technique with this operation is called *trigonometric substitution*, which is illustrated in Example 2. These transformations used in calculus are summarized in Table 5.7.

Historical Note

From the time of the Egyptian rope stretchers in the 6th century BC to celestial navigation of spacecraft in the 20th century, we have used triangles and trigonometry to measure inaccessible distances.

TABLE 5.7   Trigonometric Substitutions

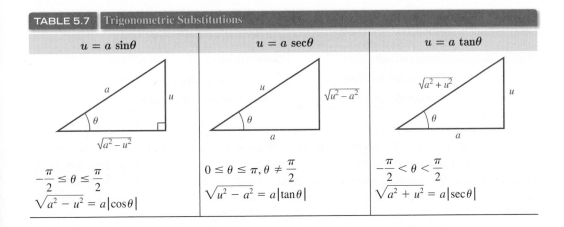

$u = a\sin\theta$	$u = a\sec\theta$	$u = a\tan\theta$
$-\dfrac{\pi}{2} \le \theta \le \dfrac{\pi}{2}$	$0 \le \theta \le \pi, \theta \ne \dfrac{\pi}{2}$	$-\dfrac{\pi}{2} < \theta < \dfrac{\pi}{2}$
$\sqrt{a^2 - u^2} = a|\cos\theta|$	$\sqrt{u^2 - a^2} = a|\tan\theta|$	$\sqrt{a^2 + u^2} = a|\sec\theta|$

EXAMPLE 3   Trigonometric functions using a reference triangle

Using the reference triangle for the substitution $u = a \tan \theta$, write all six trigonometric functions.

Solution   Use the reference triangle (third column of Table 5.7) to find

$$\cos \theta = \frac{a}{\sqrt{u^2 + a^2}}; \qquad \sin \theta = \frac{u}{\sqrt{u^2 + a^2}}; \qquad \tan \theta = \frac{u}{a};$$

The reciprocal functions are $\sec \theta = \dfrac{\sqrt{u^2 + a^2}}{a}$; $\csc \theta = \dfrac{\sqrt{u^2 + a^2}}{u}$; $\cot \theta = \dfrac{a}{u}$.

■

EXAMPLE 4   Proving an inverse identity using right triangles

Prove $\sin^{-1} x + \cos^{-1} x = \frac{\pi}{2}$ for $0 < x < 1$.

Solution   Consider the right triangle shown in Figure 5.81. Label the length of the side opposite α, x, and the length of the hypotenuse 1. Then

$$\sin \alpha = \frac{x}{1} = x \quad \text{so } \alpha = \sin^{-1} x \qquad \cos \beta = \frac{x}{1} = x \quad \text{so } \beta = \cos^{-1} x$$

Since the sum of the angles of any triangle is π and since C is a right angle, we see that the sum of the measures of angles α and β is $\pi/2$:

$$\alpha + \beta = \frac{\pi}{2}$$

$$\sin^{-1} x + \cos^{-1} x = \frac{\pi}{2}$$

■

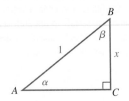

Figure 5.81 Right triangle

Solving Triangles

When solving triangles, we work with angles using degree measure, and we note that measurements cannot be exact. In "real life," measurements are made with a certain number of digits of precision, and results are not claimed to have more digits of accuracy than the least accurate number in the input data for multiplication or division, or the least precision for addition or subtraction. This limitation is sometimes confusing when comparing linear measurements for lengths of sides of a triangle with angle measurement in that same triangle. The digits known to be correct in a number obtained by a measurement are called **significant digits**. In this book, we assume the following relationship between **accuracy** of the measurements of sides and angles.

Accuracy in Sides	Equivalent Accuracy in Angles
Two significant digits	Nearest degree
Three significant digits	Nearest tenth of a degree
Four significant digits	Nearest hundredth of a degree

We do not want the focus of solving triangles to be on significant digits, but instead the focus is on the trigonometry. Just remember, your answers should not be more accurate than any of the given measurements. Another important rounding principle is often overlooked: ☠ Do not round at the beginning or in the middle of the problem. You should round only when you are stating your answers. ☠

EXAMPLE 5   Solving a right triangle given a side and an angle

Solve the right triangle with $a = 50.0$ and $\alpha = 35°$. (*Note:* We write $\alpha = 35°$ to mean the measure of angle α is $35°$.)

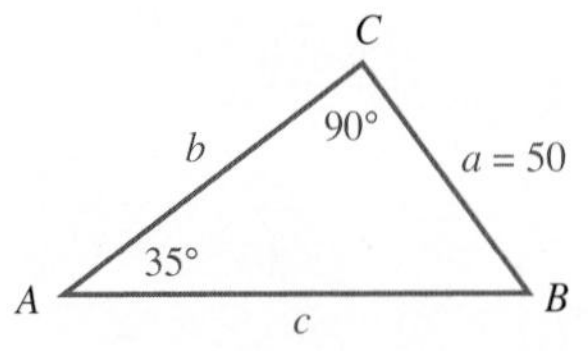

Solution   Begin by drawing the triangle (not necessarily to scale), as shown in Figure 5.82.

$\alpha = 35°$ *Given.*

$\beta = 55°$ *Since $\alpha + \beta = 90°$ for any right triangle with right angle C.*

$\gamma = 90°$ *Given.*

$a = 50$ *Given.*

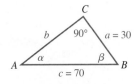

Figure 5.82 Given one side and one angle

Note that we write the answer to the nearest unit because the given angle is measured to the nearest degree, and we use the *least accurate* of the given measurements to determine the accuracy of the answer.

b: $\tan 35° = \dfrac{50}{b}$ *Definition of tangent*

$\qquad b = \dfrac{50}{\tan 35°}$ *Solve for b.*

$\qquad \approx 71.40740034$

$\qquad b = 71$ *Round answer to two significant digits.*

c: $\sin 35° = \dfrac{50}{c}$ *Definition of sine*

$\qquad c = \dfrac{50}{\sin 35°}$ *Solve for c.*

$\qquad \approx 87.17233978$

$\qquad c = 87$ *Again, round to two significant digits.* ∎

EXAMPLE 6   Solving a right triangle given two sides

Solve the right triangle with $a = 30.0$ and $c = 70.0$.

Solution   Begin by drawing the triangle as shown in Figure 5.83.

α : $\sin \alpha = \dfrac{30.0}{70.0}$ *Definition of sine*

$\qquad \alpha = \sin^{-1} \dfrac{30}{70}$ *Definition of arcsine*

$\qquad \approx 25.37693353$

$\qquad \alpha = 25.4°$ *Three significant digits*

β : $\beta = 180° - \alpha - \gamma$

$\qquad \approx 64.62306647°$

Figure 5.83 Given two sides

$$\beta = 64.6° \qquad \text{Since } \alpha + \beta = 90° \text{ for any right triangle with right angle } C$$

$$\gamma = 90° \qquad \text{Given (right triangle).}$$

$$a = 30.0 \qquad \text{Given.}$$

$$b: \qquad a^2 + b^2 = c^2 \qquad \text{Right triangle}$$

$$(30.0)^2 + b^2 = (70.0)^2$$

$$b^2 = 4,000$$

$$b = 20\sqrt{10}$$

$$\approx 63.2455532$$

$$b = 63.2 \qquad \text{Round answer to three significant digits.}$$

$$c = 70.0 \qquad \text{Given.}$$

The solution of right triangles is necessary in a variety of situations. The first one we will consider concerns an observer looking at an object. The **angle of elevation** is the acute angle measured up from a horizontal line to the line of sight, whereas the **angle of depression** is the acute angle measured down from the horizontal line of sight.

EXAMPLE 7   Angle of elevation to find an inaccessible height

The angle of elevation of a tree from a point on the ground 42 ft from its base is $33°$ (see Figure 5.84). Find the height of the tree.

Solution
The angle of elevation is $33°$; let $h =$ height of the tree. Then,

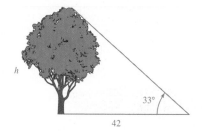

Figure 5.84 Height of a tree

$$\tan 33° = \frac{h}{42} \qquad \text{Definition of tangent}$$

$$h = 42 \tan 33°$$

$$\approx 27.27511891$$

The tree is 27 ft tall.

EXAMPLE 8   Angle of depression to find an inaccessible distance   `MODELING APPLICATION`

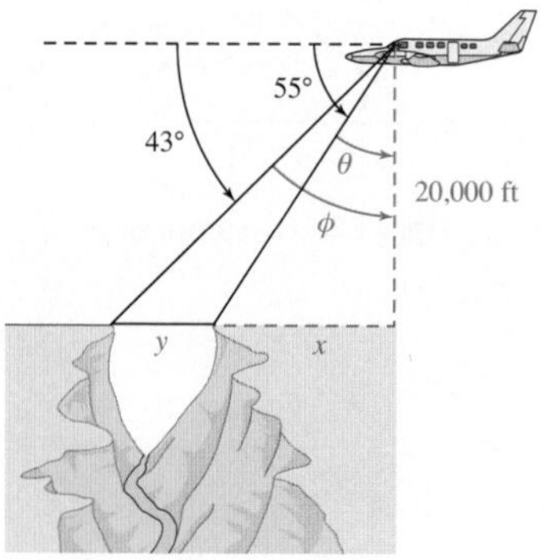

How might you use an aircraft to determine the distance across a canyon?

Solution

Step 1: *Understand the problem.* We will need to assume that the airplane is flying at a constant altitude and that it is possible to measure angles of depression to the two sides of the canyon.

Step 2: *Devise a plan.* Suppose the angles of depression to the two sides of a canyon are $43°$ and $55°$, as shown in Figure 5.85. If the altitude of the plane is 20,000 ft, how far is it across the canyon?

Step 3: *Carry out the plan.* Label parts x, y, θ, and ϕ, as shown in Figure 5.85.

$$\theta = 90° - 55° = 35° \qquad \text{and} \qquad \phi = 90° - 43° = 47°$$

Figure 5.85 Distance across a canyon from the air

First find x:

$$\tan 35° = \frac{x}{20{,}000}$$

$$x = 20{,}000\tan 35°$$

$$\approx 14{,}004$$

Next, find $x + y$:

$$\tan 47° = \frac{x+y}{20{,}000}$$

$$x + y = 20{,}000\tan 47°$$

$$\approx 21{,}447$$

Thus, $y \approx 21{,}447 - 14{,}004 = 7{,}443$. Rounding to two significant digits, we find the distance across the canyon to be 7,400 ft.

Step 4: *Look back.* Does 7,400 ft seem reasonable? That would be a bit further than a mile, and many canyons are that wide, so it seems to be a reasonable distance. ∎

Bearing

Another application of the solution of right triangles involves the **bearing** of a line, which is defined as an acute angle made with a north-south line. When giving the bearing of a line, first write N or S to determine whether to measure the angle from the north or from the south side of a point on the line. Then give the measure of the angle followed by E or W, denoting which side of the north–south line you are measuring. Some examples are shown in Figure 5.86.

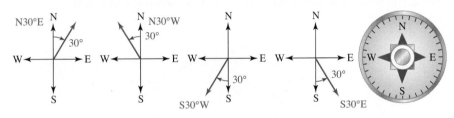

Figure 5.86 Measuring angles using bearing

EXAMPLE 9   Bearing to find an inaccessible distance

To find the width $\overline{AB}$ of a canyon from your present location to a landmark on the other side (see Figure 5.87a), measure 100 m from A in the direction of N42.6°W to locate point C (Figure 5.87b). Now, use a transit to determine that the bearing of $\overline{CB}$ is N73.5°E. Find the width of the canyon if point B is situated so that $\angle BAC = 90.0°$.

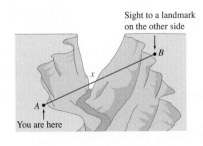

a. First sight to a landmark

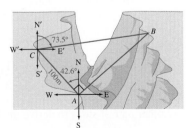

b. Steps in building a right triangle

Figure 5.87 Finding the distance across an inaccessible distance

© Photos.com

Historical Note

You may have seen surveyors using an instrument called a transit, or perhaps you have wondered how one could decide on the angle of elevation. A transit is a telescope that allows for accurate line-of-sight measurements. After fixing the line of sight, you can read directions from both horizontal and vertical circles.

Solution   Let $\theta = \angle BCA$ in Figure 5.87b.

$$\angle BCE' = 16.5° \qquad \text{Complementary angles}$$

$$\angle ACS' = 42.6° \qquad \text{Alternate interior angles}$$

$$\angle E'CA = 47.4° \qquad \text{Complementary angles}$$

$$\theta = \angle BCA = \angle BCE' + \angle E'CA = 16.5° + 47.4° = 63.9°$$

Since $\tan\theta = \dfrac{|AB|}{|AC|}$,

$$|AB| = |AC|\tan\theta$$

$$= 100\tan\theta$$

$$= 100\tan 63.9°$$

$$\approx 204.1253967$$

The canyon is 204 m across.   ∎

Areas

In elementary school you learned that the area, K, of a triangle is $K = \frac{1}{2}bh$.[*] You sometimes need to find the area of a triangle given the measurements for the sides and angles but not for the height, and, therefore, you need trigonometry. Consider $\triangle ABC$ as shown in Figure 5.88.

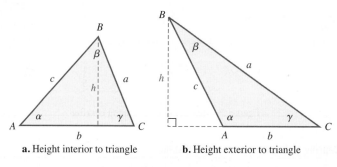

a. Height interior to triangle   **b.** Height exterior to triangle

Figure 5.88 Arbitrary triangles

Notice that the given triangle does not need to be a right triangle, nor does the height of the triangle need to be on the interior of the triangle.

$$K = \frac{1}{2}bh \qquad \text{and} \qquad \sin\alpha = \frac{h}{c} \quad \text{or} \quad h = c\sin\alpha$$

Thus,

$$K = \frac{1}{2}bc\sin\alpha$$

We could use any other pair of sides to derive the following area formulas.

---

[*]In elementary school, you no doubt used A for area. In trigonometry, we use K for area because we have already used A to represent a vertex of a triangle.

AREA OF A TRIANGLE (SAS)

The area of a triangle for which two sides and an included angle are known is found by using one of the following formulas:

$$K = \tfrac{1}{2}bc\sin\alpha \qquad K = \tfrac{1}{2}ac\sin\beta \qquad K = \tfrac{1}{2}ab\sin\gamma$$

>> IN OTHER WORDS The area of a triangle is equal to half the product of the lengths of two of the sides times the sine of the included acute angle.

EXAMPLE 10   Finding the area of a triangle

Find the area of $\triangle ABC$ where $\alpha = 18.4°$, $b = 154$ ft, and $c = 211$ ft.

Solution

$$K = \tfrac{1}{2}(154)(211)\sin 18.4° \approx 5{,}128.349903$$

To three significant digits, the area is $5{,}130$ ft^2. ∎

EXAMPLE 11   Area enclosed by the Pentagon

The largest ground area covered by any administrative building is that of the Pentagon in Arlington, Virginia. If the radius of the circumscribed circle is 783.5 ft, find the ground area enclosed by the exterior walls of the building.

Master Sgt. Ken Hammond, U.S. Air Force/U.S. Department of Defense

Solution   The Pentagon forms what is called a *regular pentagon*, which is a five-sided polygon with sides of equal length. Label the vertices as shown in Figure 5.89.

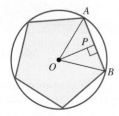

Figure 5.89 Schematic for the Pentagon

The area we seek is 5 times the area of $\triangle AOB$. We know that $|\overline{AO}| = |\overline{BO}| = 783.5$, and we also know that $\angle AOB = 72°$ (1/5 of 360°). The area of $\triangle AOB$ is

$$K \approx \frac{1}{2}(783.5)(783.5)\sin 72° \qquad K = \frac{1}{2}|\overline{AO}||\overline{BO}|\sin\angle AOB$$

$$\approx 291{,}913.6018$$

Next, find the total ground area:

$$5K \approx 5(291{,}913.6018) = 1{,}459{,}568.009$$

The ground area enclosed is about 1.460 million square feet. By the way, since the Pentagon is five stories high, it actually has a floor area of about 6.5 million square feet. ∎

PROBLEM SET 5.6

LEVEL 1

Solve each of the right triangles (γ is a right angle) in Problems 1–14.

1. $a = 80$; $\beta = 60°$
2. $b = 37$; $\alpha = 69°$
3. $b = 90$; $\beta = 13°$
4. $a = 49$; $\alpha = 45°$
5. $c = 75.4$; $\alpha = 62.5°$
6. $c = 418.7$; $\beta = 61.05°$
7. $\beta = 57.4°$; $a = 70.0$
8. $\alpha = 56.00°$; $b = 2{,}350$
9. $\beta = 32.17°$; $c = 343.6$
10. $b = 3{,}200$; $c = 7{,}700$
11. $b = 3{,}100$; $c = 3{,}500$
12. $a = 145$; $b = 240$
13. $a = 26.6$; $b = 10.84$
14. $a = 85.3$; $b = 125.5$

15. The most powerful lighthouse is on the coast of Brittany, France, and is 50 m tall. Suppose you are in a boat just off the coast, as shown in Figure 5.90. Determine your distance from the base of the lighthouse if $\theta = 12°$.

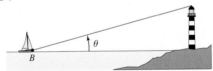

Figure 5.90 Distance from a boat to the *Créac'h d'Ouessant* on the coast of France

16. Find the distance $|\overline{PA}|$ across the river shown in Figure 5.91 if you know that $\theta = 32.5°$ and $|\overline{DA}| = 50$ ft.

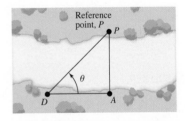

Figure 5.91 Distance across a river

17. The angle of elevation of a building from a point on the ground 30 m from its base is 38°. Find the height of the building.

18. From a cliff 150 m above the shoreline, the angle of depression of a ship is 37°. Find the distance of the ship from a point directly below the observer.

LEVEL 2

19. Using the reference triangle for the substitution $u = a \sin \theta$, write all six trigonometric functions.
20. Using the reference triangle for the substitution $u = a \sec \theta$, write all six trigonometric functions.
21. In a correctly labeled right triangle $\triangle ABC$, if $|AC| = |BC| = 5$ then evaluate:
    a. $\cos \alpha$, $\sin \alpha$, and $\tan \alpha$ b. $\cos \beta$, $\sin \beta$, and $\tan \beta$
22. In a correctly labeled right triangle $\triangle ABC$, if $|AB| = 1$ and $|BC| = \frac{1}{3}\sqrt{3}$, then evaluate:
    a. $\cos \alpha$, $\sin \alpha$, and $\tan \alpha$ b. $\cos \beta$, $\sin \beta$, and $\tan \beta$
23. From a police helicopter flying at 1,000 ft, a stolen car is sighted at an angle of depression of 71°. Find the distance of the car from a point directly below the helicopter.
24. Find the height of the Barrington Space Needle if the angle of elevation at 1,000 ft from a point on the ground directly below the top is 58.15°.
25. To get the most energy from the sun, suppose you need to adjust the height, h, of a 12-ft solar heater panel (see Figure 5.92) so that the angle θ is a right angle. What is h when the angle of elevation of the sun is 42°?

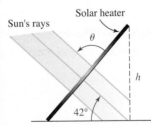

Figure 5.92 Solar heater panel

26. To get the most energy from the sun, suppose you need to adjust the height, of a 12-ft solar heater panel (see Figure 5.92) so that the angle θ is a right angle. What is h when the angle of elevation of the sun is 38°?

27. Suppose you want to plant a flowering plant that needs the direct rays of the sun, and you would like to know how close to a 6.0-ft fence this will be possible (see Figure 5.93). After consulting an almanac, you find that the lowest angle of elevation of the sun during the plant's flowering season is $35°20'$. How close to the fence (rounded to the nearest inch) should the plant be placed?

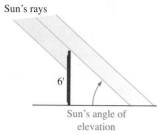

Sun's rays

6'

Sun's angle of elevation

Figure 5.93 Tracking the sun's rays

28. Rework Problem 27, assuming that the angle of elevation of the sun is $31°40'$.

29. Suppose you are constructing rafters with pitch angle $\theta = 35.0°$. If $x = 20.0$ ft, what is the height, h, as shown in Figure 5.94?

x x

h

θ θ

Figure 5.94 Rafter cross section

30. Suppose you are constructing rafters (see Figure 5.94) with a pitch angle of $40.0°$. If $h = 4$ ft 6 in., what is the length, x, of the rafters (rounded to the nearest inch)?

31. Suppose you are constructing a shed, and you need to cut a triangular piece of siding by scribing a pitch angle, θ. If the dimensions are shown in Figure 5.95, what is the proper angle for θ (to the nearest degree)?

Roof

Front   2 ft

12 ft

Figure 5.95 Shed roof angle

32. The angle of elevation of the top of the Great Pyramid of Khufu (or Cheops) from a point on the ground 351 ft from a point directly below the top is $52°$. Find the height of the pyramid.

© André Klaassen/ShutterStock, Inc.

33. To find the east–west boundary of a piece of land, a surveyor must divert her path from point C on the boundary by proceeding due south for 300 ft to point A. Point B, which is due east of point C, is now found to be in the direction of N49°E from point A. What is the distance $\left|\overline{CB}\right|$?

34. To find the distance across a river that runs east–west, a surveyor located points P and Q on a north–south line on opposites sides of the river.

She then paced out 150 ft from Q due east to a point R. Next, she determined that the bearing of RP is N58°W. How far is it across the river?

35. On top of the Empire State Building is a TV tower. From a point 1,000 ft from a point on the ground directly below the top of the tower, the angle of elevation to the bottom of the tower is $51.34°$ and to the top of the tower is $55.81°$. What is the length of the tower?

36. If the Empire State Building and the Sears Tower were situated 1,000 ft apart, the angle of depression from the top of the Sears Tower to the top of the Empire State Building would be $11.53°$, and the angle of depression to the foot of the Empire State Building would be $55.48°$. Find the heights of the buildings.

37. In the 1977 movie *Close Encounters of the Third Kind*, a scene showed star Richard Dreyfuss approaching Devil's Tower in Wyoming. He could have determined his distance from Devil's Tower by first stopping at a point P and estimating the angle P, as shown in Figure 5.96.

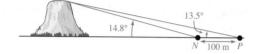

$13.5°$

$14.8°$

N 100 m P

Figure 5.96 Devil's Tower

After moving 100 m toward Devil's Tower, he could have estimated the angle N, as shown in the figure. How far away from Devil's Tower is point N?

38. What is the height of Devil's Tower in Problem 37?

39. To find the distance across a river, a surveyor locates points P and Q on opposite sides of the river. Next, the surveyor measures 100 ft from point Q in the direction S35°E to point R. Then the surveyor determines that point P is now in the direction N25.0°E from point R and that $\angle PQR$ is a right angle. Find the distance across the river.

40. To find the boundary of a piece of land, a surveyor must divert his path from a point A on the boundary for 500 ft in the direction S50°E. He then determines that the bearing of a point B located directly south of A is S40°W. Find the distance $\left|\overline{AB}\right|$.

41. The distance from the earth to the sun is 92.9 million miles and the angle formed between Venus, the earth, and the sun (as shown in Figure 5.97) is $47.0°$. Find the distance from the sun to Venus.

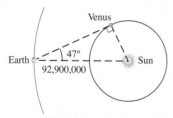

Venus

$47°$

Earth   Sun

92,900,000

Figure 5.97 Planetary distances

42. Use the information in Problem 41 to find the distance from earth to Venus.

43. To determine the height of the building shown in Figure 5.98, we select a point P and find that the angle of elevation is $59.64°$. We then move out a distance of 325.4 ft (on a level plane) to point Q and find that the angle of elevation is now $41.32°$. Find the height h of the building.

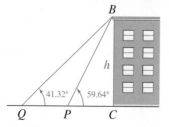

B

h

$41.32°$ $59.64°$

Q P C

Figure 5.98 Height of a building

44. Using Figure 5.98, let the angle of elevation at P be α and at Q be β, and let the distance from P to Q be d. If h is the height of the building, show that

$$h = \frac{d \tan\alpha \tan\beta}{\tan\alpha - \tan\beta}$$

45. Find the radius of the circle inscribed in the Pentagon in Arlington, Virginia. (See Example 11.)

46. A 6.0-ft person is casting a shadow of 4.2 ft. What time of the morning is it if the sun rose at 6:15 A.M. and is directly overhead at 12:15 P.M.?

47. A level lot has the dimensions shown in Figure 5.99. What is the total cost of treating the area for poison oak if the fee is \$45/acre (1 acre = 43,560 ft²)?

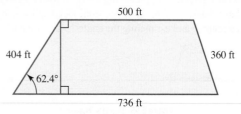

Figure 5.99 Area of a level lot

PROBLEMS FROM CALCULUS *Make the appropriate trigonometric substitutions to simplify each of the algebraic expressions in Problems 48–51.*

48. $\dfrac{du}{\sqrt{u^2 + a^2}}$

where du is the single variable $a \sec^2\theta$.

49. $\dfrac{\sqrt{u^2 - 25}}{u}\,du$

where du is the single variable $5\,\sec\theta\tan\theta$.

50. $\dfrac{x^2 dx}{\sqrt{x^2 + 5}}$

where dx is the single variable $\sqrt{5}\sec^2\theta$.

51. $\dfrac{dx}{x^2\sqrt{x^2 - 16}}$

where dx is the single variable $4\,\sec\theta\tan\theta$, and θ is in the first quadrant.

52. Prove $\tan^{-1}x + \cot^{-1}x = \frac{\pi}{2}, x \ne 0$.

53. Prove $\sec^{-1}x + \csc^{-1}x = \frac{\pi}{2}, x > 1$.

Find the area of the interior of the regular polygons inscribed in a circle of radius 1 in. in Problems 54–57.

54.

55.

56.

57.

LEVEL 3

In Problems 58–59, draw a unit circle, and label the points as shown in the given figure.

   a. What are the coordinates of A and P?
   Find:

   b. $|\overline{OA}|$ **c.** $|\overline{PA}|$

   d. $|\overline{OB}|$ **e.** Area of $\triangle AOB$.

58.

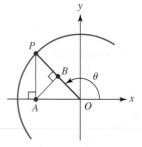

59.

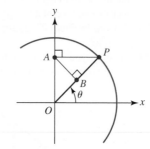

60. The volume of a cone with a circular base is given in geometry as

$$V = \frac{\pi r^2 h}{3}$$

where r is the radius of the base and h is the height of the cone. If you do not know the height, but do know the angle of elevation (see Figure 5.100), then the formula is

$$V = \frac{1}{3}\pi r^3 \tan\alpha$$

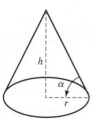

Figure 5.100 Volume of a cone

   a. Derive this formula.

   b. If sand is dropped from the end of a conveyor belt, the sand will fall in a conical heap such that the angle of elevation α is about 33°. Find the volume of sand when the radius is exactly 10 ft.

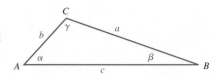

Figure 5.101 Correctly labeled triangle

5.7 Oblique Triangles

In the previous section, we solved right triangles. We now extend that study to triangles with no right angles. Such triangles are called **oblique triangles**. Consider a triangle labeled as in Figure 5.101, and notice that γ is now unrestricted.

In general, we will be given three parts of a triangle and be asked to find the remaining three parts. But can you do so given *any* three parts? Consider the possibilities:

SSS   Three sides are given.

SAS   Two sides and an included angle are given.

SSA   Two sides and the angle opposite one of them are given.

ASA   Two angles and an included side are given. AAS or SAA are considered to be the same as ASA, because if we know two angles, then we know all three angles (since the sum of the angles of a triangle is $180°$).

AAA   Three angles are given.

We will consider these possibilities, one at a time, in this and the next section. At the end of the next section, we will again summarize all these possibilities.

SSS

To solve a triangle given SSS, it is necessary for the sum of the lengths of the two smaller sides to be greater than the length of the largest side. We use a generalization of the Pythagorean theorem called the **law of cosines**.

LAW OF COSINES

In any $\triangle ABC$,

$$c^2 = a^2 + b^2 - 2ab\cos\gamma$$

The proof of the law of cosines involves the distance formula. Let γ be an angle in standard position with A on the positive x-axis, as shown in Figure 5.102.

The coordinates of the vertices are as follows:

$C(0, 0)$ Since C is in standard position

$A(b, 0)$ Since A is on the positive x-axis a distance of b units from the origin.

$B(a\cos\gamma, a\sin\gamma)$ Let $B = (x, y)$; then by definition of the trigonometric functions, $\cos\gamma = x/a$ and $\sin\gamma = y/a$. Thus, $x = a\cos\gamma$ and $y = a\sin\gamma$.

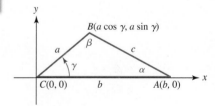

Figure 5.102 $\triangle ABC$ with angle γ in standard position

Use the formula for the distance between the point $A(b, 0)$ and the point $B(a\cos\gamma, a\sin\gamma)$:

$$c^2 = (a\cos\gamma - b)^2 + (a\sin\gamma - 0)^2 \qquad \text{Distance formula}$$

$$= a^2\cos^2\gamma - 2ab\cos\gamma + b^2 + a^2\sin^2\gamma \qquad \text{Algebraically simplify.}$$

$$= a^2(\cos^2\gamma + \sin^2\gamma) + b^2 - 2ab\cos\gamma \qquad \text{Common factor } a^2$$

$$= a^2 + b^2 - 2ab\cos\gamma \qquad \cos^2\gamma + \sin^2\gamma = 1$$

We can show that this law of cosines is a generalization of the Pythagorean theorem. This means that if $\gamma = 90°$:

$$c^2 = a^2 + b^2 - 2ab\cos\gamma \qquad \text{Law of cosines}$$

$$c^2 = a^2 + b^2 - 2ab\cos 90° \qquad \text{Given a right triangle with } \gamma = 90°$$

$$c^2 = a^2 + b^2 \qquad \text{This is the Pythagorean theorem.}$$

By letting A and B, respectively, be in standard position, we can also show that

$$a^2 = b^2 + c^2 - 2bc \cos\alpha \qquad \text{and} \qquad b^2 = a^2 + c^2 - 2ac \cos\beta$$

We call these the *alternate forms* of the law of cosines. To find the angle when you are given three sides, solve for α, β, or γ. We now summarize all the alternate forms for the law of cosines.

LAW OF COSINES ALTERNATE FORMS

In any $\triangle ABC$,

$$a^2 = b^2 + c^2 - 2bc \cos\alpha \qquad \text{or} \qquad \cos\alpha = \frac{b^2 + c^2 - a^2}{2bc}$$

$$b^2 = a^2 + c^2 - 2ac \cos\beta \qquad \text{or} \qquad \cos\beta = \frac{a^2 + c^2 - b^2}{2ac}$$

$$c^2 = a^2 + b^2 - 2ab \cos\gamma \qquad \text{or} \qquad \cos\gamma = \frac{a^2 + b^2 - c^2}{2ab}$$

EXAMPLE 1   Finding a part of a triangle given three sides (SSS)

What is the smallest angle of a triangular patio whose sides measure 25, 18, and 21 ft?

Solution   If γ represents the smallest angle, then c (the side opposite γ) must be the smallest side, so $c = 18$. Then,

$$\cos\gamma = \frac{a^2 + b^2 - c^2}{2ab} \qquad \text{\textit{Law of cosines}}$$

$$\cos\gamma = \frac{25^2 + 21^2 - 18^2}{2(25)(21)} \qquad \text{\textit{Substitute known values.}}$$

$$\gamma = \cos^{-1}\left[\frac{25^2 + 21^2 - 18^2}{2(25)(21)}\right] \qquad \text{\textit{Solve for }} \gamma.$$

$$\approx 45.03565072° \qquad \text{\textit{Evaluate (by calculator).}}$$

To two significant digits, the answer is $45°$.   ∎

Historical Note

Nicholas Copernicus (1473–1543) is probably best known as the astronomer who revolutionized the world with his heliocentric theory of the universe, but in his book *De revolutionibus orbium coelestium*, he also developed a substantial amount of trigonometry. This book was published in the year of his death; in fact, the first copy off the press was rushed to him as he lay on his deathbed. It was on Copernicus' work that his student Rheticus based his ideas, which soon brought trigonometry into full use. In a two-volume work, *Opus palatinum de triangulis*, Rheticus used and calculated elaborate tables for all six trigonometric functions.

SAS

The next possibility for solving oblique triangles is that of being given two sides and an included angle. It is necessary that the given angle be less than $180°$. Again, use the law of cosines for this possibility.

EXAMPLE 2   Solving an SAS triangle

Find c, where $a = 52.0$, $b = 28.3$, and $\gamma = 28.5°$.

Solution

$$c^2 = a^2 + b^2 - 2ab \cos\gamma \qquad \text{\textit{Law of cosines}}$$

$$= (52.0)^2 + (28.3)^2 - 2(52.0)(28.3)\cos 28.5° \qquad \text{\textit{Substitute known values.}}$$

$$\approx 918.355474 \qquad \text{\textit{Evaluate (by calculator).}}$$

Thus, $c \approx 30.30438044$; to three significant digits, $c = 30.3$.   ∎

EXAMPLE 3   Distance between planes

The Riddler just committed the crime of the century and left 30 minutes ago, traveling 460 mi/h in his Learjet with a heading of 120° (measured clockwise from North). At this instant, Batman is located 150 miles from this location on a bearing of 140°. How far apart are the two planes?

Solution   Draw the relative positions of the planes, as shown in Figure 5.103.

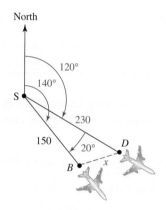

Figure 5.103 Batman and the Riddler

We seek the distance x, and from the law of cosines,

$$x^2 = \left|\overline{SB}\right|^2 + \left|\overline{SD}\right|^2 - 2\left|\overline{SB}\right|\left|\overline{SD}\right|\cos(\angle DSB) \qquad \text{Law of cosines}$$

$$= 150^2 + 230^2 - 2(150)(230)\cos 20° \qquad \text{Substitute known values.}$$

$$\approx 10{,}561.20917 \qquad \text{Evaluate (by calculator)}$$

Thus, $x \approx 102.7677438$ or 100 miles (to two significant digits). ∎

SSA

Suppose we know two sides and an angle that is not included. If we use the law of cosines, the resulting equation is second degree, which will require the quadratic formula.

EXAMPLE 4   Solving an SSA triangle

Solve the triangle where $a = 3.0$, $b = 2.0$, and $\alpha = 110°$.

Solution   Begin by drawing the triangle, as shown in Figure 5.104.

$$a^2 = b^2 + c^2 - 2bc \cos\alpha \qquad \text{Law of cosines}$$

$$(3.0)^2 = (2.0)^2 + c^2 - 2(2.0)c \cos 110° \qquad \text{Substitute known values.}$$

$$c^2 - 4\cos 110° \, c - 5 = 0 \qquad \text{Rewrite in quadratic equation form.}$$

$$c = \frac{-(-4\cos 110°) \pm \sqrt{(-4\cos 110°)^2 - 4(1)(-5)}}{2(1)} \qquad \text{The variable is } c, \text{ and we use the quadratic formula where } A = 1, \; B = -4\cos 110°, \text{ and } C = -5.$$

$$\approx 1.654316212, \; -3.022396785 \qquad \text{Use a calculator.}$$

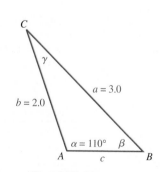

Figure 5.104 SSA triangle

Even though the quadratic formula gives two solutions, we need to reject the negative value (since c represents a length). We can now use the positive value to find the angles (using a calculator):

$$\cos \beta = \frac{a^2 + c^2 - b^2}{2ac} \qquad \text{Law of cosines}$$

$$\approx \frac{9 + 1.654316212^2 - 4}{2(3)(1.654316212)} \qquad \begin{array}{l}\text{Substitute known values.}\\ a = 3.0,\, b = 2.0,\, c = 1.654316212\end{array}$$

$$\approx 0.7794521661 \qquad \text{Use a calculator.}$$

Therefore, $\beta \approx 38.78955642°$. We state the solution (to the required degree of accuracy): $a = 3.0$, $b = 2.0$, $c = 1.7$; $\alpha = 110°$, $\beta = 39°$, $\gamma = 31°$. ∎

It may have occurred to you when working Example 4 that, since we are using the quadratic formula, it is certainly possible to obtain two solutions. We consider this possibility with the next example.

EXAMPLE 5   Solve an SSA triangle for which there are two solutions

Solve the triangle where $a = 1.50$, $b = 2.00$, and $\alpha = 40.0°$.

Solution

$$a^2 = b^2 + c^2 - 2bc \cos \alpha \qquad \text{Law of cosines}$$

$$(1.50)^2 = (2.00)^2 + c^2 - 2(2.00)c \cos 40° \qquad \text{Substitute known values.}$$

$$c^2 - 4\cos 40° c + 1.75 = 0 \qquad \begin{array}{l}\text{Rewrite in quadratic equation}\\ \text{form.}\end{array}$$

$$c = \frac{4\cos 40° \pm \sqrt{(-4\cos 40°)^2 - 4(1)(1.75)}}{2(1)} \qquad \begin{array}{l}\text{The variable is } c \text{, and we use the}\\ \text{quadratic formula where } A = 1,\\ B = -4 \cos 40°, \text{ and } C = 1.75.\end{array}$$

$$\approx 2.30493839,\, 0.7592393826 \qquad \text{Use a calculator.}$$

How can this be? How can we have two solutions? Look at Figure 5.105. There are two possible values of c and consequently two triangles that satisfy the conditions of this problem.

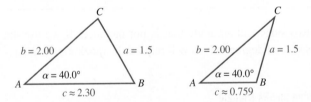

Figure 5.105 Two possible triangles

Since there are two triangles, we must now double our work (answers are shown to the correct number of significant digits):

Solution I		Solution II	
$a = 1.50$	Given	$a = 1.50$	Given
$b = 2.00$	Given	$b = 2.00$	Given
$c_1 = 2.30$	1st solution	$c_2 = 0.759$	2nd solution

☠ Do not work with rounded values when finding the other parts. This means that you should use the calculator output without rounding, not the rounded values shown here. ☠

Solution I		Solution II	
$\alpha = 40.0°$	Given	$\alpha = 40.0°$	Given
$\beta = \cos^{-1}\left(\dfrac{a^2 + c^2 - b^2}{2ac}\right)$	Law of cosines	$\beta = \cos^{-1}\left(\dfrac{a^2 + c^2 - b^2}{2ac}\right)$	Law of cosines
$\approx 58.98696953°$	1st solution	$\approx 121.0130305°$	2nd solution
$\beta_1 = 59.0°$	3 significant digits	$\beta_2 = 121.0°$	3 significant digits
$\gamma_1 = 81.0°$	Use $\gamma = 180° - \alpha - \beta$	$\gamma_2 = 19.0°$	Use $\gamma = 180° - \alpha - \beta$

Since there is more than one solution in Example 5, this possibility (namely, two sides with an angle that is not included) is sometimes referred to as the **ambiguous case**, because under certain circumstances two triangles result. However, as long as you solve this problem by the law of cosines, you simply need to interpret the results of the quadratic formula to decide whether there are one or two solutions.

◼ COMPUTATIONAL WINDOW   _ ◻ ✕

Many calculators will solve quadratic equations with built-in features (check your owner's manual). However, it is easy to write a program for using the quadratic formula to solve quadratic equations. For Example 5, input the A, B, and C values as follows:

SAMPLE CALCULATOR		
1	STO→	A
−4	STO→	B
1.75	STO→	C

The output for this program is: 2.30493839, 0.7592393826

EXAMPLE 6   Solving an SSA triangle with no solution

Solve the triangle where $a = 3.0$, $b = 1.5$, and $\beta = 40°$.

Solution   We see that we are given two sides and an angle that is not included. There *may* be two solutions, so we use the law of cosines.

$$b^2 = a^2 + c^2 - 2ac\cos\beta \qquad \text{Law of cosines}$$

$$(1.5)^2 = (3.0)^2 + c^2 - 2(3.0)c\cos 40° \qquad \text{Substitute known values.}$$

$$c^2 - 6\cos 40°\, c + 6.75 = 0 \qquad \text{Rewrite in quadratic equation form.}$$

$$c = \frac{6\cos 40° \pm \sqrt{(-6\cos 40°)^2 - 4(1)(6.75)}}{2(1)}$$

The variable is c, and we use the quadratic formula where $A = 1$, $B = -6\cos 40°$, and $C = 6.75$.

Since the discriminant (the number under the radical) is negative, we see there is no solution. Look at Figure 5.106 to see why there is no solution. The side opposite the $40°$ angle is not as long as the height of the triangle with a $40°$ angle and a side of length 3.

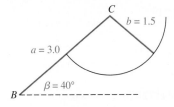

Figure 5.106 Triangle with no solution

ASA

Suppose that two angles and a side are given. For such a triangle to be formed, the angles must be positive, and the sum of the two given angles must be less than $180°$. Also, the length of the given side must be greater than 0. In this case, we state and prove a result called the **law of sines**.

LAW OF SINES

In any $\triangle ABC$,

$$\frac{\sin \alpha}{a} = \frac{\sin \beta}{b} = \frac{\sin \gamma}{c}$$

» IN OTHER WORDS This equation says that you can use any of the following equations to solve a triangle: $\dfrac{\sin \alpha}{a} = \dfrac{\sin \beta}{b}$, $\dfrac{\sin \alpha}{a} = \dfrac{\sin \gamma}{c}$, $\dfrac{\sin \beta}{b} = \dfrac{\sin \gamma}{c}$.

To prove this law, we use the formulas for the area of a triangle (see p. 367):

$$\frac{1}{2}bc \sin \alpha = \frac{1}{2}ac \sin \beta = \frac{1}{2}ab \sin \gamma$$

Multiply by 2 and then divide by abc to obtain:

$$\frac{\sin \alpha}{a} = \frac{\sin \beta}{b} = \frac{\sin \gamma}{c}$$

EXAMPLE 7   Solving a triangle given two angles (ASA)

Solve the triangle in which $a = 20$, $\alpha = 38°$, and $\beta = 121°$.

Solution   We are given AAS, but since knowing two angles ensures that we know all three angles, we call this ASA and draw the triangle as shown in Figure 5.107.

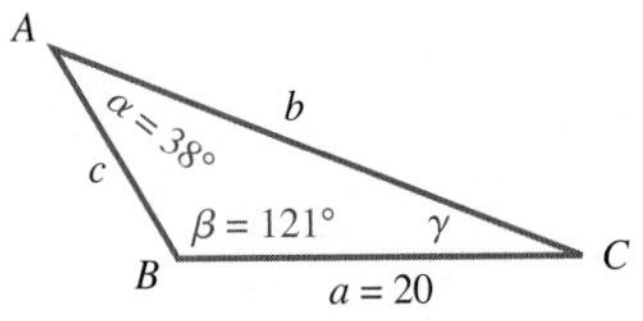

Figure 5.107 ASA triangle

$\alpha = 38°$ Given.

$\beta = 121°$ Given.

$\gamma = 21°$ Since $\alpha + \beta + \gamma = 180°$, $\gamma = 180° - 38° - 121° = 21°$.

$a = 20$ Given.

b: Use the law of sines:

$$\frac{\sin 38°}{20} = \frac{\sin 121°}{b}$$ Side b associated with opposite angle

$$b = \frac{20 \sin 121°}{\sin 38°}$$ Solve for b.

$$\approx 27.8454097$$ By calculator

$$b = 28$$ State answer to two significant digits.

c: Use the law of sines

$$\frac{\sin 38°}{20} = \frac{\sin 21°}{c}$$ Side b associated with opposite angle

$$c = \frac{20 \sin 21°}{\sin 38°}$$ Solve for c.

$$\approx 11.64172078$$ By calculator

$$c = 12$$ State answer to two significant digits.

EXAMPLE 8   Finding the length of a footbridge

Two points A and B on opposite sides of a river are the endpoints for a footbridge.

To find the length of this proposed bridge, a point C is located 100 ft from point A. It is then determined that $\alpha = 58°$ and $\gamma = 49°$. Find the length of the footbridge.

Solution   We draw a picture, as shown in Figure 5.108. We know $\alpha = 58°$ and $\gamma = 49°$, so $\beta = 73°$. We also know $b = 100$, so use the law of sines:

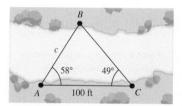

Figure 5.108 Length of a footbridge

$$\frac{\sin \beta}{b} = \frac{\sin \gamma}{c} \qquad \text{Law of sines}$$

$$\frac{\sin 73°}{100} = \frac{\sin 49°}{c} \qquad \text{Substitute known values.}$$

$$c = \frac{100 \sin 49°}{\sin 73°} \qquad \text{Multiply both sides by 100c, and divide both sides by } \sin 73°.$$

$$\approx 78.91935866 \qquad \text{By calculator}$$

To two significant digits, the length of the footbridge is 79 ft. ∎

EXAMPLE 9   Finding a rate by using trigonometry

A boat traveling at a constant rate due west passes a buoy that is 1.0 kilometer from a lighthouse. The lighthouse is N30°W of the buoy. After the boat has traveled for a half-hour, its bearing to the lighthouse is N74°E. How fast is the boat traveling?

Solution   The angle at the lighthouse (see Figure 5.109) is $180° - 60° - 16° = 104°$.

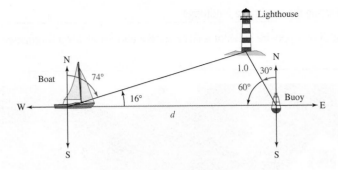

Figure 5.109 Find the boat's rate of travel

Use the law of sines:
$$\frac{\sin 104°}{d} = \frac{\sin 16°}{1.0}$$

$$d = \frac{\sin 104°}{\sin 16°}$$

$$\approx 3.520189502$$

After one hour, the distance is $2d$, so the rate of the boat is about 7.0 km/h. ∎

AAA

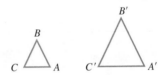

Figure 5.110 AAA implies similar triangles (no solution)

Another possible case supposes that three angles are given. However, from what we know of similar triangles (see Figure 5.110), they have the same shape but are not necessarily the same size.

This means their corresponding angles have equal measure, so we conclude that the triangle cannot be solved without knowing the length of at least one side.

Summary

An important skill to be learned from this section is the ability to select the proper trigonometric law when given a particular problem. Refer to Table 5.8.

TABLE 5.8	Procedure for Solving Triangles	
Given	**Conditions on the given information**	**Method of Evaluation**
	More than one side given:	**Law of cosines**
SSS	**a.** The sum of the lengths of the two smaller sides is less than or equal to the length of the largest side.	No solution
	b. The sum of the lengths of the two smaller sides is greater than the length of the largest side.	Law of cosines
SAS	**a.** The given angle is greater than or equal to 180°.	No solution
	b. The given angle is less than 180°.	Law of cosines
SSA	**a.** The given angle is greater than or equal to 180°.	No solution
	b. There are no solutions, one solution, or two solutions, as determined by the quadratic formula.	Law of cosines
	More than one angle given:	**Law of sines**
ASA or	**a.** The sum of the given angles is greater than or equal to 180°.	No solution
AAS	**b.** The sum of the given angles is less than 180°.	Law of sines
AAA	No unique triangle	No solution

EXAMPLE 10   **Determining the time to reach a destination**   MODELING APPLICATION

An airplane is 100 mi N40°E of a Loran station, traveling due west at 240 mi/h. How long will it take (to the nearest minute) before the plane is 90 miles from the Loran station?

Solution

Step 1:   *Understand the problem.* To understand the problem, we begin by drawing a picture and labeling the parts as shown in Figure 5.111.

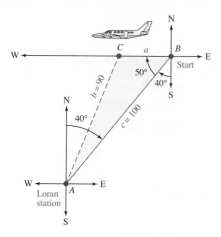

Figure 5.111 Time to reach a specified location

Step 2:   *Devise a plan.* Note that we are given more than one side, so we use the law of cosines. (We do not want you to think that every problem in this section is an application of the law of sines.) Furthermore, we note that we have SSA, so we expect to use the quadratic formula.

Step 3:   *Carry out the plan.* We find $\beta = 50°$, so we have

$$b^2 = a^2 + c^2 - 2ac\cos\beta \qquad \text{Law of cosines}$$

$$90^2 = a^2 + 100^2 - 2a(100)\cos 50° \qquad \text{Substitute known values.}$$

$$a^2 - 200\cos 50° \, a + 1,900 = 0 \qquad \text{Write as a quadratic.}$$

$$a = \frac{200\cos 50° \pm \sqrt{(-200\cos 50°)^2 - 4(1)(1,900)}}{2} \qquad \text{Quadratic formula}$$

$$\approx 111.5202587, \ 17.0372632 \qquad \text{By calculator}$$

We see that there are two solutions. The one shown in Figure 5.111 is $a \approx 17.04$. To find the times, we use the distance formula $d = rt$, for the two distances found in the quadratic formula and a given rate of 240 mi/h.

Solution I		Solution II	
$t = \dfrac{d}{r}$	Distance formula	$t = \dfrac{d}{r}$	
$t_1 = \dfrac{17.0372632}{240}$	Substitute known values	$t_2 = \dfrac{111.5202587}{240}$	
≈ 0.0709885967	By calculator	≈ 0.4646677447	

To convert these times (which are in hours) to minutes, multiply each by 60, and round to the nearest minute:

$$t_1 = 4 \text{ min} \qquad t_2 = 28 \text{ min}$$

The plane will be 90 mi from the Loran station in 4 min (location shown in Figure 5.111), and again in 28 min.

Step 4:   *Look back.* Does this answer make sense? Can you draw the location of the plane at this time?

Some problems can be worked by either the law of cosines or the law of sines. Consider Example 11, where we use both solutions with the goal of showing that even though both methods may work, one method is preferable.

EXAMPLE 11   Comparing solutions by laws of cosine and sine

Suppose we know the sides of a triangle are $a = 6.5$, $b = 8.3$, and $c = 12.2$. Also suppose we know $\alpha = 29.94°$. Find γ to the nearest tenth of a degree.

Solution   The triangle is shown in Figure 5.112.

Method I: Use the law of cosines.

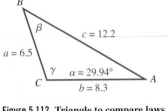

$$\cos \gamma = \frac{a^2 + b^2 - c^2}{2ab} \qquad \text{Law of cosines}$$

$$= \frac{6.5^2 + 8.3^2 - 12.2^2}{2(6.5)(8.3)} \qquad \text{Substitute known values.}$$

$$\approx -0.3493975904 \qquad \text{By calculator}$$

$$\gamma = \cos^{-1}(-0.3493975904) \qquad \text{Definition of inverse cosine}$$

$$\approx 110.4504735° \qquad \text{By calculator}$$

Figure 5.112 Triangle to compare laws of cosine and sine

To the nearest tenth of a degree, the solution is $\gamma = 110.5°$.

Method II: Use the law of sines.

$$\frac{\sin \alpha}{a} = \frac{\sin \gamma}{c} \qquad \text{Law of sines}$$

$$\frac{\sin 29.9°}{6.5} = \frac{\sin \gamma}{12.2} \qquad \text{Substitute known values.}$$

$$\sin \gamma = \frac{12.2 \sin 29.94°}{6.5} \qquad \text{Multiply both sides by 12.2.}$$

$$\gamma = \sin^{-1}(0.9367588433) \qquad \text{Definition of inverse sine}$$

$$\approx 69.51418175° \qquad \text{By calculator}$$

To the nearest tenth of a degree, the solution is 69.5°.

The solutions do not agree. Why? Remember, when we solved equations, we made a point of distinguishing the solution of an equation from the principal value. In method II, $\gamma = 69.5°$. Recall that the sine function is positive in the second quadrant as well as the first quadrant. Therefore, γ may be an obtuse angle as well as an acute angle.[*] In the second quadrant, the angle γ may be found by

$$\gamma = 180° - 69.51418175° = 110.4858183°$$

---

[*]Situations sometimes arise when a rough sketch is not sufficient to decide which value from the law of sines is appropriate. For this reason, we suggest that you use the law of cosines whenever possible. If you follow the recommendations given in Table 58, you will be led to the most appropriate method of solution.

From the problem as stated, we clearly want the obtuse solution, so from the law of cosines, $\gamma = 110.5°$ (to the nearest tenth of a degree). ∎

Area of a Triangle

In the previous section, right triangles were used to find the area of a triangle when two sides and an included angle are known. Suppose, however, that we know the angles, but only one side. If we know only side a, then we can use the formula $K = \frac{1}{2}ab\sin\gamma$ and the law of sines:

$$\frac{\sin\alpha}{a} = \frac{\sin\beta}{b} \quad \text{so} \quad b = \frac{a\sin\beta}{\sin\alpha}$$

Therefore,

$$K = \tfrac{1}{2}ab\sin\gamma = \tfrac{1}{2}a\frac{a\sin\beta}{\sin\alpha}\sin\gamma = \frac{a^2\sin\beta\sin\gamma}{2\sin\alpha}$$

The other area formulas shown in the following box can be similarly derived.

AREA OF A TRIANGLE (ASA)

The area of a triangle for which two angles and an included side are known is found by using one of the following formulas:

$$K = \frac{a^2\sin\beta\sin\gamma}{2\sin\alpha} \qquad K = \frac{b^2\sin\alpha\sin\gamma}{2\sin\beta} \qquad K = \frac{c^2\sin\alpha\sin\beta}{2\sin\gamma}$$

If three sides (but none of the angles) are known, you will need yet another formula to find the area of a triangle. The formula is derived from the law of cosines; this derivation is left as an exercise. The result is known as **Heron's** (or **Hero's**) formula.

AREA OF A TRIANGLE (SSS)

The area of a triangle for which three sides are known is found by using the following formula:

$$K = \sqrt{s(s-a)(s-b)(s-c)} \quad \text{where} \quad s = \tfrac{1}{2}(a+b+c)$$

EXAMPLE 12   Finding the area of an SSS triangle

Find the area of a triangle having sides 43 ft, 89 ft, and 120 ft.

Solution   Let $a = 43$, $b = 89$, and $c = 120$. Then $s = \tfrac{1}{2}(43 + 89 + 120) = 126$. Thus,

$$K = \sqrt{126(126 - 43)(126 - 89)(126 - 120)} \approx 1{,}523.704696$$

To two significant digits, the area is $1{,}500 \text{ ft}^2$. ∎

PROBLEM SET 5.7

Solve △ABC in Problems 1–6. Draw a representative triangle for each problem.

1. $a = 7.0$; $b = 8.0$; $c = 2.0$. Find α.
2. $a = 10.0$; $b = 4.0$; $c = 8.0$. Find γ.
3. $a = 12$; $b = 6.0$; $c = 15$. Find the smallest angle.
4. $a = 18$; $b = 25$; $\gamma = 30°$. Find c.
5. $a = 15$; $b = 8.0$; $\gamma = 38°$. Find c.
6. $b = 14$; $c = 12$; $\alpha = 82°$. Find a.

Label each given triangle in Problems 7–12 as SSS, SAS, SSA, ASA, AAS, or AAA. State whether you should use the law of sines or the law of cosines to solve the triangle, and give the number of significant digits in the solution. If there is no solution, so state.

7. **a.** $a = 15.8$; $b = 14.8$, $c = 5.4$
   b. $a = 5.8$; $b = 14.8$, $\gamma = 48°$
8. **a.** $b = 15.50$; $c = 9.30$; $\beta = 38.4°$
   b. $c = 6.45$; $\beta = 51.8°$; $\gamma = 16.5°$
9. **a.** $a = 3.5$; $\alpha = 122.8°$; $\beta = 12.6°$
   b. $\alpha = 85.2°$; $\beta = 21.4°$; $\gamma = 73.4°$
10. **a.** $b = 8.5$; $c = 9.3$; $\beta = 55.2°$
    b. $b = 45.3$; $c = 110.3$; $\gamma = 75.4°$
11. **a.** $b = 45.3$; $c = 56.4$; $\gamma = 75.4°$
    b. $a = 7.43$; $b = 8.52$; $\alpha = 35.00°$
12. **a.** $b = 45.3$; $c = 36.1$; $\alpha = 121°$
    b. $c = 28.36$; $\beta = 28.6°$; $\gamma = 102.3°$

Solve △ABC in Problems 13–29; if two triangles are possible, give both solutions. If the triangle cannot be solved, tell why. Draw a representative triangle for each problem.

13. $a = 14.2$; $b = 16.3$; $\beta = 115.0°$
14. $a = 5.0$; $b = 4.0$; $\alpha = 125°$
15. $a = 10$; $\alpha = 48°$; $\beta = 62°$
16. $a = 30$; $\beta = 50°$; $\gamma = 100°$
17. $b = 40$; $\alpha = 50°$; $\gamma = 60°$
18. $c = 53$; $\alpha = 82°$; $\beta = 19°$
19. $c = 115$; $\beta = 81.0°$; $\gamma = 64.0°$
20. $\beta = 85°$; $\gamma = 24°$; $b = 223$
21. $a = 41.0$; $\alpha = 45.2°$; $\beta = 21.5°$
22. $b = 58.3$; $\alpha = 120°$; $\gamma = 68.0°$
23. $b = 82.5$; $c = 52.2$; $\gamma = 32.1°$
24. $a = 10.2$; $b = 11.8$; $\alpha = 47.0°$
25. $a = 123$; $b = 225$; $c = 351$
26. $b = 5.2$; $c = 3.4$; $\alpha = 54.6°$
27. $a = 214$; $b = 3.20 \times 10^2$; $\gamma = 14.8°$
28. $a = 140$; $b = 85.0$; $c = 105$
29. $a = 36.9$; $b = 20.45$; $\gamma = 90.0°$

Find the area of each triangle whose base and height are given in Problems 30–41. Draw a triangle, and if no triangle is formed, so state.

30. $b = 15.6$; $h = 2.51$
31. $c = 6.81$; $h = 4.00$
32. $c = 12\sqrt{2}$; $h = 3\sqrt{3}$ (two sig figs)
33. $a = 15$; $b = 8.0$; $\gamma = 38°$
34. $b = 14$; $c = 12$; $\alpha = 82°$

35. $b = 40$; $\alpha = 50°$; $\gamma = 60°$
36. $a = b = c = 5$ (exact answer)
37. $a = 7.0$; $b = 8.0$; $c = 2.0$
38. $a = 11$; $b = 9.0$; $c = 8.0$
39. $a = 12.0$; $b = 9.00$; $\alpha = 52.0°$
40. $a = 10.2$; $b = 11.8$; $\alpha = 47.0°$
41. $b = 82.5$; $c = 52.2$; $\gamma = 32.1°$

42. A vertical tower is located on a hill whose inclination is 6°. From a point P located 100 ft down the hill from the base of the tower, the angle of elevation to the top of the tower is 28°. What is the height of the tower?

43. New York City is approximately 210 miles N9°E of Washington, D.C., and Buffalo, New York, is N49°W of Washington, D.C. How far is Buffalo from New York City if the distance from Buffalo to Washington, D.C., is approximately 285 miles?

44. An artillery-gun observer must determine the distance to a target at a point T. The observer knows that the target is 5.20 miles from point I on a nearby island. The observer knows that the gun location at H is 4.30 miles from point I. If $\angle HIT$ is 68.4°, how far is the gun from the target? (See Figure 5.113.)

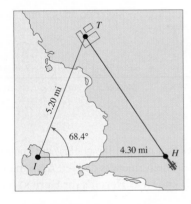

Figure 5.113 Distance to a target

45. A television antenna is attached to the gable of a roof with pitch and dimensions as shown in Figure 5.114. How long is the cable from point P to point Q?

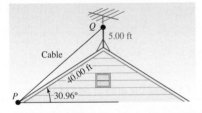

Figure 5.114 Guy wires

46. The Galactic Empire's computers on the Death Star are monitoring the positions of the invading forces. At a particular instant, two observation points 2,500 m apart make a fix on Luke's incoming spacecraft, which is between the observation points and in the same vertical plane. If the angle of elevation from the first observation point is 3.00° and the angle of elevation from the second is 1.90°, find the distance from Luke's spacecraft to each of the observation points.

47. At 500.0 ft in the direction the Tower of Pisa is leaning, the angle of elevation is 20.24° (see Figure 5.115). If the tower leans at an angle of 5.45° from the vertical, what is the length of the tower?

Figure 5.115 Leaning Tower

48. What is the angle of elevation of the Leaning Tower of Pisa (described in Problem 47) if you measure from a point 500 ft in the direction exactly opposite from the way it is leaning?

49. A weight is supported by ropes that are attached to both ends of a 12-ft balance beam (as shown in Figure 5.116). What are the angles that are formed between the balance beam and the ropes?

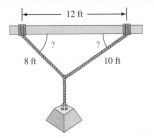

Figure 5.116 Supported weight

50. A dime, a penny, and a quarter are placed on a table so that they just touch one another, as shown in Figure 5.117. Let D, and P, and Q be the respective centers. Solve $\triangle DPQ$. HINT: If you measure the coins, the diameters are 1.75 cm, 2.00 cm, and 2.50 cm.

Figure 5.117 Three coins

51. What is the perimeter of a square inscribed in a circle of radius 5.0 in.?

52. The Pentagon in Arlington, Virginia, is a regular pentagon constructed inside a circle with radius 783.5 ft (see Figure 5.118). What is the perimeter of the Pentagon?

Figure 5.118 The Pentagon

53. In going from Newport Beach to Catalina, a boat traveling at a constant 15 mi/h travels for 60 min before the crew discovers that they need to correct course by 15° (see Figure 5.119). Assume Avalon is located 26 mi S45°W of Newport Beach.

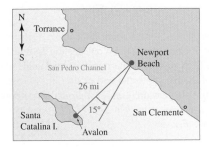

Figure 5.119 Catalina Island

How long will it take for the boat to reach Avalon?

54. What was the original heading of the boat in Problem 53, and what was the correct original heading? After traveling for an hour along the incorrect heading, what was the necessary new heading to reach Avalon?

55. A plane flying from Anchorage to Fairbanks (a distance of about 250 mi) has been traveling at 180 mi/h for 30 min before the pilot discovers that she is off course by 10.0° (see Figure 5.120). To reach Fairbanks at the scheduled time, the pilot must increase the airspeed. Find the necessary airspeed that enables the plane to reach Fairbanks on time.

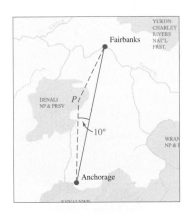

Figure 5.120 North to Alaska

56. From a blimp, the angle of depression to the top of the Eiffel Tower is 23.2° and to the bottom is 64.6°. After flying over the tower at the same height and at a distance of 1,010 ft from the first location, you determine that the angle of depression to the top of the tower is now 31.4°. What is the height of the Eiffel Tower, given that these measurements are in the same vertical plane, as shown in Figure 5.121?

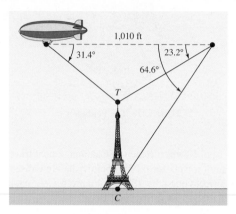

Figure 5.121 Eiffel Tower

57. A ski lift is planned for the south slope of Mt. Frissell in Connecticut. Point B is located directly below the top (point T), as shown in Figure 5.122. A surveyor determines that the angle of elevation from the start of the lift, S, to the end of the lift, T, to be 34.06°. Next, 1.000×10^3 ft is measured on level ground to determine point P. If the angle of elevation to T at P is 27.77°, what is the approximate height of Mt. Frissell?

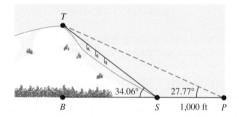

Figure 5.122 Mt. Frissell

58. How does a sundial work? Consider a sundial labeled as shown in Figure 5.123.

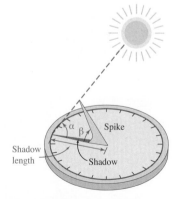

Figure 5.123 A sundial

To work properly, the angle (β in Figure 5.123) of the spike of a sundial must be the same as the latitude where the sundial is used.

Suppose on a certain day in Columbia, South Carolina (latitude 34.0°), the angle of elevation of the sun (α in Figure 5.123) is 68.0° at 1:00 P.M. If the length of the spike is 10.0 in., what is the length of the shadow?

LEVEL 3

59. A harpsichord is a keyboard instrument whose strings are plucked to make music. They are usually very beautifully finished (see Figure 5.124).

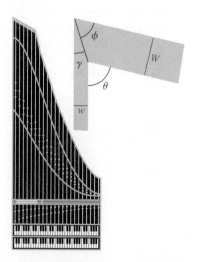

Courtesy of Harpsichord Clearing House [www.harpsicord.com]

Figure 5.124 A harpsichord

A harpsichord maker, reproducing a particular historical instrument, needs to join two pieces of wood of different thicknesses to give a neat miter at a certain angle. The pieces of wood have thickness w and W, and the necessary angle at the joint is θ. Find the angles γ and ϕ, as shown in Figure 5.125, at which the two pieces must be cut, in terms of w, W, and θ. HINT: $\theta = \gamma + \phi$

Figure 5.125 Harpsichord construction

60. The volume of a cylinder is found by the formula

$$V = (\text{HEIGHT})(\text{AREA OF THE BASE})$$

$$= h(\pi r^2)$$

$$= \pi r^2 h$$

The volume of a slice cut from a cylinder (see Figure 5.126) is found by the formula

$$V = hr^2 \left(\frac{\theta}{2} - \sin\frac{\theta}{2}\cos\frac{\theta}{2} \right)$$

for θ measured in radians such that $0 \le \theta \le \pi$.

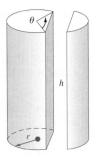

Figure 5.126 Cylinder

a. Derive this formula.
b. Find the volume (to the nearest cubic inch) of a slice cut from a 3.0-ft log with a 6.0-in. radius and a central angle of 90.0°.

CHAPTER 5 SUMMARY AND REVIEW

M *athematics is a type of thought which seems ingrained in the human mind, . . .*

J. W. A. Young (1907)

Take some time getting ready to work the review problems in this section. First, look back at the definition and property boxes. You will maximize your understanding of this chapter by working the problems in this section only after you have studied the material.

SELF TEST   *All of the answers for this self test are given in the back of the book.*

1. Complete the table of exact values.

degrees:	$-45°$	a.	b.	$210°$	$90°$
radians:	c.	π	$\dfrac{-4\pi}{3}$	d.	e.
sine:	f.	g.	h.	i.	j.
cosine:	k.	l.	m.	n.	o.
tangent:	p.	q.	r.	s.	t.

2. Evaluate the given functions (exact values).

a. $\sin^{-1}\left(-\dfrac{\sqrt{3}}{2}\right)$

b. $\cos^{-1}\dfrac{1}{2}$

c. $\tan^{-1}(-1)$

d. $\cot^{-1}\left(-\sqrt{3}\right)$

e. $\tan(\sec^{-1}3)$

f. $\sin\left[\cot^{-1}\left(-\dfrac{3}{4}\right)\right]$

g. $\cos^{-1}(\cos 1.5)$

h. $\cos(\cos^{-1}1.5)$

3. a. State and prove one of the reciprocal identities.
   b. State and prove one of the ratio identities.
   c. State and prove one of the Pythagorean identities.
4. Using the definition of the trigonometric functions, find their values for an angle α whose terminal side passes through $\left(\sqrt{7}, 3\right)$.
5. From memory, draw a quick sketch of each of the following. Label axes and scale.
   a. $y = \sin x$
   b. $y = \cot x$
   c. $y = \csc x$
6. Graph each curve.
   a. $y = \sqrt{3}\cos\frac{1}{2}x$
   b. $y + 2 = 2\sin(3x + \pi)$
   c. $y + 2 = \frac{1}{2}\tan\left(x - \frac{\pi}{4}\right)$

7. For the triangle shown in Figure 5.127, state (one form for each is sufficient):
   a. law of sines
   b. law of cosines
   c. If $s = 46$ cm, $t = 121$ cm, and $u = 92$ cm, find the area of $\triangle STU$.

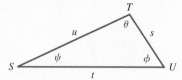

Figure 5.127 $\triangle STU$

8. The most westerly city in the continental United States is Ferndale, California, and it is located at 124°W longitude, 41°N latitude. How far (to the nearest 100 mi) is Ferndale from the equator if you assume that the earth's radius is approximately 3,950 mi?

9. Graton is 7.0 mi S40°W of Sebastopol. If I leave Sebastopol at noon and travel due west at 2.0 mi/h, when will I be exactly 6.0 mi from Graton?

10. A mine shaft is dug into the side of a sloping hill as shown in Figure 5.128.

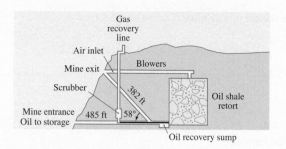

Figure 5.128 Mine shaft

A shaft is dug horizontally for 485 ft. Next, a turn is made so that the angle of elevation of the second shaft is 58.0°, thus forming a 58.0° angle between the shafts. The shaft is then continued for 382 ft before exiting.

a. How far is it along a straight line from the entrance to the exit, assuming that all tunnels are in a single plane?

b. If the slope of the hill follows the line from the entrance to the exit, what is the angle of elevation from entrance to exit?

🛑 STUDY HINTS *Compare your solutions and answers to the self test.*

Practice for Calculus—*Supplementary Problems Cumulative Review Chapters 1–5*

Evaluate each expression in Problems 1–4.

1. $x^3 - 2x + 3$ for $x = -2$
2. $-|a| - |b|$ for $a = -1$, $b = -3$
3. $f(x) = 4x^2 - 6x + 1$ for $x = 1.5$
4. $g(x) = |5 - x|$ for $x = 8$

Solve each equation in Problems 5–20.

5. $12x^2 - 20x - 25 = 0$
6. $x^2 - 125x - 31{,}250 = 0$
7. $\dfrac{1}{x+1} + \dfrac{1}{x} = 1$
8. $\dfrac{x-1}{x+2} + \dfrac{x-4}{x+3} = \dfrac{61}{12}$
9. $x - \sqrt{3-x} = 3$
10. $x - \sqrt{2x-3} = 1$
11. $4(x^2 - x - 2)^3 (2x - 1) = 0$
12. $5(x^2 - x - 6)^4 (2x - 1) = 0$
13. $\ln x = \ln 5 - 3\ln 2 + \frac{1}{4}\ln 3$
14. $1 = 2\log x - \log 1{,}000$
15. $e^{2x+1} = 8.5$
16. $83.4 = 261 e^{-t}$
17. $a^2 = 23^2 + 16^2 - 2(23)(16)\cos 15°$ (a positive, 2 sig figs)
18. $b^2 = 11^2 + 4.5^2 - 2(11)(4.5)\cos 16°$ (b positive, 2 sig figs)
19. $12^2 = 9^2 - 2(9)c\cos 128°$ (approximate answer)
20. $(16.3)^2 = (14.2)^2 + a^2 - 2(14.2)a\cos 115°$ (approximate answer)

Factor the expressions in Problems 21–26.

21. $\sin^2\theta + 5\sin\theta + 6$
22. $\tan^2\theta + 7\tan\theta + 12$
23. $\cos^2\theta - 1$
24. $\sec^2\theta - 4$
25. $3\sin^2\omega - 5\sin\omega - 2$
26. $6\cos^2\delta + 2\cos\delta - 4$

In Problems 27–30, find $\dfrac{f(x+h) - f(x)}{h}$ *and* **simplify**.

27. $f(x) = 3x^2 + 2x$
28. $f(x) = 5$
29. $f(x) = \dfrac{2x-3}{x-2}$
30. $f(x) = \dfrac{3x-1}{x+2}$

Find all vertical and horizontal asymptotes of the graph of the functions given in Problems 31–40. Finally, sketch the graph.

31. $f(x) = 4 + \dfrac{2x}{x-3}$
32. $g(x) = x - \dfrac{x}{4-x}$
33. $h(x) = x^2 + \dfrac{2}{x}$
34. $f(x) = \dfrac{x^2 + 2x - 3}{x+1}$
35. $y = \dfrac{3x}{\sqrt{9-x^2}}$
36. $y = \dfrac{5x}{\sqrt{16-x^2}}$
37. $y = \tan 2x$
38. $y = 3\sec x$
39. $y = 0.25\cot x$
40. $y = 0.5\csc x$

Graph each of the functions in Problems 41–48.

41. $y = \cos(5x - 3)$
42. $y + 1 = \sin(2x + 1)$
43. $y - 2 = \tan(2x + \frac{\pi}{2})$
44. $x = 4\cos\theta$, $y = 4\sin\theta$, $0° \le \theta \le 360°$
45. $x = 4\cos T$, $y = 2\cos T$, $90° \le T \le 270°$
46. $x = 2 - \sin T$, $y = 2 - \cos T$, $0° \le T \le 360°$
47. $f(x) = \begin{cases} x^2 - 3x & \text{if } x \le 1 \\ \sqrt{x} & \text{if } x > 1 \end{cases}$
48. $f(x) = \begin{cases} \sin x & \text{if } x \le 0 \\ \tan x & \text{if } 0 < x \le 1 \end{cases}$

In Problems 49–54, express each function as the composite of two functions. That is, express the given function f as $f(x) = g[u(x)]$ and then state the inner function u and the outer function g.

49. $f(x) = \sin(3x^2 + 4)$
50. $f(x) = \tan(x^3 - 2x^2)$
51. $f(x) = e^{\sin x}$
52. $f(x) = \sin e^x$
53. $f(x) = \ln|\sin x|$
54. $f(x) = \sin(\ln x)$

55. A rectangle is inscribed in a semicircle of diameter 16 cm. Express the area of the rectangle as a function of the width w of the rectangle.

56. Melissa commutes 30 mi to work each day, partly on a highway and partly in city traffic. On the highway, she doubles her city speed for just 15 min. If the entire trip takes an hour, how fast is she able to average in city traffic?

57. Shannon drives 56 mi to his job in the city every day, partly on the highway and partly on city streets. Off the highway, traffic crawls along at a third of his highway speed for 10 min to complete the journey. If the trip takes an hour, how fast is Shannon able to average in the city?

58. PROBLEMS FROM CALCULUS It is shown in calculus that for small values of x (called a *neighborhood* of the origin), the graph of $y = \sin x$ and the line $y = x$ (where x is measured in radians) are almost the same. This means, for example,

$$\sin 0.05 \approx 0.05$$

Illustrate this fact graphically.

59. To measure the span of the Rainbow Bridge in Utah, a surveyor selected two points P and Q, on opposite ends of the bridge. From point Q, the surveyor measured 500 ft in the direction N38.4°E to point R. Point P was then determined to be in the direction S67.5°W. What is the span of the Rainbow Bridge if all the preceding measurements are in the same plane and $\angle PQR$ is a right angle?

60. A fire is simultaneously spotted from two observation towers that are located 10,520 ft apart, as shown in Figure 5.129. The first observation post reports the fire at N54.8°E, and the second post reports the fire at N48.2°W. How far is the fire from each observation location? Assume that all the measurements are in the same plane and that the towers are on an east–west line.

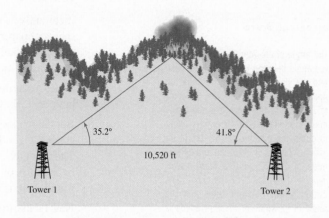

Figure 5.129 Triangulation for the location of a fire

CHAPTER 5 Group Research Projects

Working in small groups is typical of most work environments, and this book seeks to develop skills with group activities. At the end of each chapter, we present a list of suggested projects, and even though they could be done as individual projects, we suggest that these projects be done in groups of three or four students.

G5.1 a. Find $\tan^{-1} 1 + \tan^{-1} 2 + \tan^{-1} 3$

using a calculator. Make a conjecture about the exact value.

b. Prove the conjecture from part a using right triangles as follows. You may use the figure shown in Figure 5.130, and may assume that $\triangle ABC$, $\triangle ABD$, and $\triangle DEF$ are all right triangles with lengths of sides as shown in the figure.

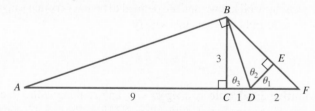

Figure 5.130 $\theta_1 + \theta_2 + \theta_3 = \pi$

G5.2 SUM OF CUBES Thirty-Seven-Year-Old Puzzle

For more than 37 years I have tried, on and off, to solve a problem that appeared in *Popular Science*. All right, I give up. What's the answer? In the September 1939 issue (page 14) the following letter appeared:

Have enjoyed monkeying with the problems you print, for the last couple of years.

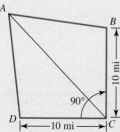

A man always drives at the same speed (his car probably has a governor on it). He makes it from A direct to C in 30 minutes; from A through B to C in 35 minutes; and from A through D to C in 40 minutes. How fast does he drive?

D.R.C., Sacramento, Calif.

It certainly looked easy, and I started to work on it. By 3 A.M. I had filled a lot of sheets of paper, both sides, with notations. In subsequent issues of *Popular Science*, all I could find on the subject was the following, in the November 1939 issue:

"In regard to the problem submitted by D.R.C. of Sacramento, Calif. in the September issue. According to my calculations, the speed of the car must be 38.843 miles an hour. W.L.B., Chicago, Ill."

But the reader gave no hint of how he had arrived at that figure. It may or may not be correct. Over the years, I probably have shown the problem to more than 500 people. The usual reaction was to nod the head knowingly and start trying to find out what the hypotenuse of the right triangle is. But those few who remembered the formula then found there is little you can do with it when you know the hypotenuse distance.

Now I'd like to challenge *Popular Science*. If your readers cannot solve this little problem (and prove the answer), how about your finding the answer, if there is one, editors?

R.F. Davis, Sun City, Ariz.*

The preceding letter to the editor appeared in the February 1977 issue of *Popular Science*. (Reprinted with permission from *Popular Science*. © 1977.) See if you can solve the puzzle.

G5.3 𝕳𝕚𝕤𝕥𝕠𝕣𝕚𝕔𝕒𝕝 𝕼𝕦𝕖𝕤𝕥 If we define the angle of elevation of a pyramid as the angle between the edge of the base and the slant height, we find that the major Egyptian pyramids are about 44° or 52°. Write a paper explaining why the Egyptians built pyramids using these angles of elevation.

References: Billard, Jules B., ed. *Ancient Egypt.* Washington, D.C.: National Geographic Society, 1978.

Edward, I. E. S. *The Pyramids of Egypt.* New York: Penguin Books, 1961.

Smith, Arthur F. "Angles of Elevation of the Pyramids of Egypt." *The Mathematics Teacher,* February 1982, pp. 124–127.

G5.4 Pythagorean triplets are well known and are defined to be integers that satisfy the Pythagorean theorem—namely a, b, and c integers that satisfy

$$c^2 = a^2 + b^2$$

We define a **primitive quadruple** as integers a, b, and c, along with θ (measured in degrees) that satisfy the law of cosines. For example, if $a = 7$, $b = 55$, $c = 97$, and $\theta = 120°$, then

$$97^2 = 57^2 + 55^2 - 2(57)(55)\cos 120°$$

Find another primitive quadruple. See if you can generate some procedures for finding primitive quadruples.

G5.5 Write a paper on the relationship between music and sine curves.

G5.6 Show that the following inequality holds for every right triangle:

$$\frac{a+b}{\sqrt{2}} \le c < a+b$$

𝕳𝕚𝕤𝕥𝕠𝕣𝕚𝕔𝕒𝕝 𝕹𝕠𝕥𝕖

The angle of elevation for a pyramid is the angle between the edge of the base and the slant height, the line from the apex of the pyramid to the midpoint of any side of the base. It is the maximum possible ascent for anyone trying to climb the pyramid to the top. In an article, "Angles of Elevation to the Pyramids of Egypt," in *The Mathematics Teacher* (February 1982, pp. 124–127), author Arthur F. Smith notes that the angle of elevation of these pyramids is either about 44° or 52°. Why did the Egyptians build pyramids using these angles of elevation?

Arctan $\frac{3}{\pi} \approx 43.7°$

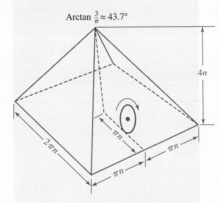

4n

Smith states that (according to Kurt Mendelssohn, *The Riddle of the Pyramids.* New York: Praeger Publications, 1974) Egyptians might have measured long horizontal distances by means of a circular drum with some convenient diameter such as 1 cubit. The circumference would then have been π cubits. To design a pyramid of convenient and attractive proportions, the Egyptians used a 1:4 ratio for the rise relative to one revolution of the drum. Smith then shows that the angle of elevation of the slant height is

$$\tan^{-1} \frac{4}{\pi} \approx 51.9°$$

If a small angle of elevation was desired (as in the case of the Red Pyramid), a 1:3 ratio might have been used. In that case, the angle of elevation of the slant height is

$$\tan^{-1} \frac{3}{\pi} \approx 43.7°$$

---

*Darrell Huff, master of the pocket calculator and frequent *Popular Science* contributor [February 1976, June 1976, December 1974] has come up with a couple of solutions. We'd like to see what you readers can do before we publish his answers.

G5.7 **Inscribed circle problem**. Consider $\triangle ABC$, where $s = \frac{1}{2}(a + b + c)$. The lines that bisect each angle of $\triangle ABC$ meet at a point that is the center of a circle called the inscribed circle. (See Figure 5.131.)

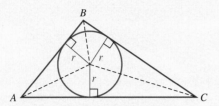

Figure 5.131 $\triangle ABC$ with inscribed circle

Show that the radius, r, of the inscribed circle is

$$r = \sqrt{\frac{(s - a)(s - b)(s - c)}{s}}$$

G5.8 Circumscribed circle problem. Consider $\triangle ABC$ and the circle passing through points A, B, and C. Such a circle is called the circumscribed circle. (See Figure 5.132.)

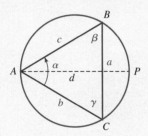

Figure 5.132 $\triangle ABC$ with circumscribed circle

Show that the diameter, d, of the circumscribed circle satisfies

$$d = \frac{a}{\sin \alpha}$$

HINT: Let O be the center of the circle and consider $\angle BOD$, where D is on $\overline{BC}$ so that $\overline{OD} \perp \overline{BC}$. Then, use the fact from geometry that $\angle BOC = 2\alpha$.

G5.9 Suppose that $\triangle ABC$ is an equilateral triangle so that each side has length 3 cm. Divide each side into two segments of lengths 1 cm and 2 cm as shown in Figure 5.133.
a. Calculate the area of $\triangle ABC$ using exact values.
b. Calculate the area of $\triangle DEF$ using exact values.

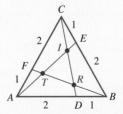

Figure 5.133 Area of $\triangle DEF$ and $\triangle ABC$

G5.10 Journal Problem (*The Mathematics Teacher*, November 1958, pp. 554−546)

Howard Eves published an article, "Pappus's Extension of the Pythagorean Theorem." In this article, he points out that if you combine Propositions 12 and 13 of Book II and Proposition 47 of Book I of Euclid's *Elements*, you can derive the law of cosines. Do some research and write a paper showing how this can be done.

Chapter Objectives

The material in this chapter is previewed in the following list of objectives. After completing this chapter, review this list again, and then complete the self test.

6.1
- 6.1 Solve first-degree trigonometric equations.
- 6.2 Solve second-degree trigonometric equations by factoring or by using the quadratic formula.

6.2
- 6.3 State and prove the eight fundamental identities.
- 6.4 Prove identities using algebraic simplification and the eight fundamental identities.
- 6.5 Prove a given identity by using various "tricks of the trade."

6.3
- 6.6 Use the cofunction identities.
- 6.7 Use the opposite-angle identities.

6.4
- 6.8 Use the double-angle identities.
- 6.9 Use the half-angle identities.
- 6.10 Use the product and sum identities.

6.5
- 6.11 Change rectangular-form complex numbers to trigonometric form.
- 6.12 Change trigonometric-form complex numbers to rectangular form.
- 6.13 Multiply and divide complex numbers.
- 6.14 State De Moivre's theorem and the **n**th root theorem.
- 6.15 Find all roots of a complex number.
- 6.16 Use the addition laws.

Trigonometric Equations and Identities

6

*P*ure mathematics is the world's best game.
It is more absorbing than chess, more of a gamble than poker, and lasts longer than
Monopoly. It's free. It can be played anywhere—Archimedes did it in a bathtub.
It is dramatic, challenging, endless, and full of surprises.

—Richard J. Trudeau (1976)

Chapter Sections

6.1 Trigonometric Equations
Principal Values
Complete Solution
Restricted Solutions

6.2 Proving Identities
Procedure for Proving Identities
Disproving an Identity

6.3 Addition Laws
Difference of Angles Identity
Cofunction Identities
Opposite-Angle Identities
Graphing with Opposites
Sum and Difference Angle Identities

6.4 Miscellaneous Identities
Double-Angle Identities
Half-Angle Identities
Product and Sum Identities

6.5 De Moivre's Theorem
Trigonometric Form of Complex Numbers
Multiplication and Division of Complex Numbers
Powers of Complex Numbers—de Moivre's Theorem

Chapter 6 Summary and Review
Self Test
Practice for Calculus
Chapter 6 Group Research Projects

▶ CALCULUS PERSPECTIVE

Equation solving is an important skill to learn in mathematics, and solving trigonometric equations is particularly important in calculus. The power of trigonometry is in using the eight fundamental identities introduced in the previous chapter to derive a variety of other useful identities.

To understand the content of this chapter, it is necessary to know the difference between an *equation* and an *identity*. The **solution** of a trigonometric equation has the same meaning as the solution of any open equation, namely, the value(s) that make a given equation true. Remember, **to solve an equation** means to find all replacements for the variable that makes the equation true. An open equation that is true for *all* values in the domain of the variable is called an **identity**.

6.1 Trigonometric Equations

In Section 5.5, we introduced the notion of inverse trigonometric functions. In this section, we will reconsider this problem in the context of solving equations.

Suppose we wish to solve $\cos x = \frac{1}{2}$. This is an *open* equation because there is a variable, whereas $\cos \frac{\pi}{3} = \frac{1}{2}$ is a *true* equation (no variable), and $\cos \frac{\pi}{4} = \frac{1}{2}$ is a *false* equation (no variable). To **solve** the open equation $\cos x = \frac{1}{2}$, we may mean the following:

1. Find all values for which $x = \cos^{-1}\frac{1}{2}$ is true; we call this the **principal-value solution**. These are the values you obtain when using a calculator. Using a calculator, we find

$$x = \cos^{-1}\frac{1}{2} \approx 1.05$$

2. Find all values that make the equation true within some given domain, for example, $0 \le x < 2\pi$; we call this a **restricted solution**. From the graph we see that the restricted solution is

$$x = \cos^{-1}\frac{1}{2} = \frac{\pi}{3}, \ \frac{5\pi}{3}$$

3. Find all values that make the equation true; we call this the **complete solution**.* This solution (from Section 3.3) is

$$x = \frac{\pi}{3} \pm 2\pi k, \qquad x = \frac{5\pi}{3} \pm 2\pi k \qquad \text{for any integer } k.$$

We now consider procedures for solving trigonometric equations.

EXAMPLE 1   Distinguish the types of solutions for a trig equation

Solve $\sin x = \frac{1}{2}$ using

a. principal value-solution (calculator solution to two decimal places)

b. exact answer restricted to $0 \le x < 2\pi$

c. complete solution (exact values)

d. Show these solutions graphically.

Solution

a. Principal value solution (calculator);

$$x = \sin^{-1}\frac{1}{2} \approx 0.5235987756$$

Rounded to two decimal places, $x = 0.52$.

b. Answer restricted to $0 \le x < 2\pi$. From $\sin x = \frac{1}{2}$ we note that sine is positive, and we know that sine is positive in Quadrants I and II, as shown in Figure 6.1.

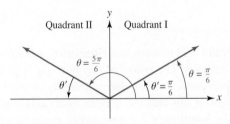

Figure 6.1 If $\theta' = \frac{\pi}{6}$, then $\sin\theta = \frac{1}{2}$ in Quadrants I and II

---

*In this book we will assume radian measure of angles (not degree) unless specifically requested.

Find the angles less than one revolution $(0 \leq x < 2\pi)$ whose reference angle is $\pi/6$:

$$x = \frac{\pi}{6}, \frac{5\pi}{6}$$

c. Since sine is periodic with period 2π, we see that there are infinitely many values x so that $\sin x = 1/2$, each found by adding multiples of 2π to the answers obtained in part **b.**

$$x = \frac{\pi}{6}, \frac{\pi}{6} + 2\pi, \frac{\pi}{6} + 4\pi, \cdots$$

and

$$x = \frac{5\pi}{6}, \frac{5\pi}{6} + 2\pi, \frac{5\pi}{6} + 4\pi, \cdots$$

We summarize by writing

$$x = \frac{\pi}{6} + 2k\pi; \frac{5\pi}{6} + 2k\pi$$

where k is any integer.

d. The graphs for the solutions are shown in Figure 6.2.

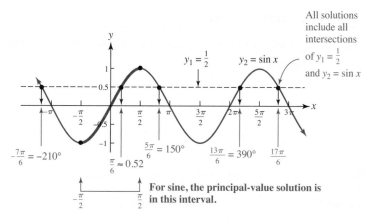

Figure 6.2 Graphical solution for $\sin x = \dfrac{1}{2}$

When solving a trigonometric equation, you must distinguish the unknown from the angle and the angle from the function. Consider $\cos(2x + 1)$:

$$\cos(2x + 1)$$

The unknown is x.

The function is cosine.

The angle, or argument, is $2x + 1$.

We can now state a procedure for solving trigonometric equations.

TRIGONOMETRIC EQUATIONS

The procedure for solving trigonometric equations is summarized:

Step 1 Solve for a single trigonometric function. You may use trigonometric identities, as well as the procedure for solving linear or quadratic equations.

    a. *Linear equation:* $Ax + B = 0, A \neq 0$

    Procedure: Isolate the variable: $x = -\dfrac{B}{A}$

(continues)

TRIGONOMETRIC EQUATIONS *(continued)*

b. *Quadratic equation:* $Ax^2 + Bx + C = 0, A \neq 0$

Procedure: Factor, if possible. Use the **zero-factor theorem**; that is, set each factor equal to zero and solve the resulting linear equations. If it is not factorable, use the **quadratic formula**:

$$x = \frac{-B \pm \sqrt{B^2 - 4AC}}{2A}$$

Step 2 Solve for the angle. You will use the definition of the inverse trigonometric functions for this step.

Step 3 Solve for the unknown.

Principal Values

The basic types of equation are linear and quadratic.

EXAMPLE 2   Linear trigonometric equations

Solve the equations for the principal values of θ. Compare the solutions of these similar problems.

a. $5 \sin(\theta + 1) = \frac{1}{2}$ **b.** $\sin 5(\theta + 1) = \frac{1}{2}$

Solution

a. $5 \sin(\theta + 1) = \dfrac{1}{2}$ Given equation

$\sin(\theta + 1) = \dfrac{1}{10}$ Step 1, solve for the trigonometric function.

$\theta + 1 = \sin^{-1} 0.1$ Step 2, solve for the angle; first use the definition of an inverse trigonometric function.

$\theta = -1 + \sin^{-1} 0.1$ Step 3, solve for the unknown; subtract 1 from both sides. This is the exact answer.

≈ -0.8998326 Use a calculator to find an approximate answer. Use radians (real numbers) unless otherwise specified.

b. $\sin 5(\theta + 1) = \dfrac{1}{2}$ Given equation. Note that step 1 is done—this is already solved for the trigonometric function.

$5(\theta + 1) = \sin^{-1} \dfrac{1}{2}$ Step 2, solve for the angle.

$5(\theta + 1) = \dfrac{\pi}{6}$ Principal value of $\sin^{-1} \dfrac{1}{2} = \dfrac{\pi}{6}$

$5\theta + 5 = \dfrac{\pi}{6}$ Step 3, solve for the unknown; simplify.

$5\theta = -5 + \dfrac{\pi}{6}$ Subtract 5 from both sides.

$\theta = -1 + \dfrac{\pi}{30}$ Divide both sides by 5. This is the exact answer.

≈ -0.8952802 Calculator approximation

Notice that these two answers are *not* the same.

EXAMPLE 3   Quadratic trigonometric equation (factoring)

Find the principal values of θ for $15 \cos^2 \theta - 2 \cos \theta - 8 = 0$.

Solution

$$15 \cos^2 \theta - 2 \cos \theta - 8 = 0 \qquad \text{\small\textit{Given equation}}$$
$$(3 \cos \theta + 2)(5 \cos \theta - 4) = 0 \qquad \text{\small\textit{Factor, if possible.}}$$

This is solved by setting each factor equal to zero (zero-factor theorem):

$$3 \cos \theta + 2 = 0 \qquad\qquad 5 \cos \theta - 4 = 0$$

$$\cos \theta = -\frac{2}{3} \qquad\qquad \cos \theta = \frac{4}{5}$$

$$\theta = \cos^{-1}\left(-\frac{2}{3}\right) \qquad\qquad \theta = \cos^{-1}\frac{4}{5}$$

$$\approx 2.300524 \qquad\qquad\qquad \approx 0.6435011$$

∎

EXAMPLE 4   Quadratic trigonometric equation (quadratic formula)

Find the principal values of θ for $\tan^2 \theta - 5 \tan \theta - 4 = 0$.

Solution

Because $\tan^2 \theta - 5 \tan \theta - 4$ cannot be factored, use the quadratic formula.

$$\tan^2 \theta - 5 \tan \theta - 4 = 0 \qquad \text{\small\textit{Given equation}}$$

$$\tan \theta = \frac{5 \pm \sqrt{25 - 4(1)(-4)}}{2} \qquad \text{\small\textit{Step 1, solve for the function.}}$$
$$\text{\small\textit{Use the quadratic formula.}}$$

$$\tan \theta = \frac{5 \pm \sqrt{41}}{2} \qquad \text{\small\textit{Simplify.}}$$

$$\theta = \tan^{-1}\left(\frac{5 \pm \sqrt{41}}{2}\right) \qquad \text{\small\textit{Step 2, solve for the angle. This}}$$
$$\text{\small\textit{is the exact solution.}}$$

$$\approx \tan^{-1}(5.7015621), \tan^{-1}(-0.7015621) \qquad \text{\small\textit{Two values, one with + and}}$$
$$\approx 1.3971718, -0.6117736 \qquad\qquad\qquad \text{\small\textit{one with -}}$$

∎

Complete Solution

To find the complete general solution of a trigonometric equation, first find the reference angle. You can find it by using the table of exact values or a calculator. Then find two values less than one revolution using the reference angle:

> For $y = \arccos x$ or $y = \text{arcsec } x$:
>> If x is positive, then the angle y (inverse function) is in Quadrants I and IV.
>> If x is negative, then the angle y (inverse function) is in Quadrants II and III.
> For $y = \arcsin x$ or $y = \text{arccsc } x$:
>> If x is positive, then the angle y (inverse function) is in Quadrants I and II.
>> If x is negative, then the angle y (inverse function) is in Quadrants III and IV.
> For $y = \arctan x$ or $y = \text{arccot } x$:
>> If x is positive, then the angle y (inverse function) is in Quadrants I and III.
>> If x is negative, then the angle y (inverse function) is in Quadrants II and IV.

The previous information is summarized in Figure 6.3.

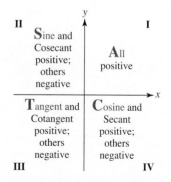

Figure 6.3 Signs of the trigonometric functions

After you have found the two values less than one revolution, the entire solution is found by using the period of the function:

> For arccosine, arcsine, arcsecant, and arccosecant: **Add multiples of 2π.**
> For arctangent and arccotangent: **Add multiples of π.**

EXAMPLE 5   General linear trigonometric equations

a. Solve $\cos\theta = \frac{1}{2}$.   **b.** Solve $\cos\theta = -\frac{1}{2}$.

Solution

a. Solve for θ: $\theta = \cos^{-1}\dfrac{1}{2}$. First, find the reference angle, $\overline{\theta}$:

$$\overline{\theta} = \arccos\left|\frac{1}{2}\right| = \frac{\pi}{3}$$

Compare $\theta = \cos^{-1}\frac{1}{2}$ with $y = \arccos\ x$, and note that $x = \frac{1}{2}$, so x is positive. The cosine is positive in Quadrants I and IV. Angles less than one revolution with a reference angle of $\frac{\pi}{3}$ are $\frac{\pi}{3}$ (Quadrant I) and $\frac{5\pi}{3}$ (Quadrant IV). The solution is infinite, so to find all solutions, add multiples of 2π to these values:

$$\theta = \frac{\pi}{3} + 2k\pi, \qquad \theta = \frac{5\pi}{3} + 2k\pi \qquad \text{for any integer } k.$$

b. Compare with part **a**. The only difference here is that x is negative, so the angle is in Quadrants II and III (that is, quadrants in which cosine is negative). In Quadrant II, the angle whose reference angle is $\frac{\pi}{3}$ is $\frac{2\pi}{3}$, and in Quadrant III, the angle whose reference angle is $\frac{\pi}{3}$ is $\frac{4\pi}{3}$, so to find all solutions, add multiples of 2π to these values:

$$\theta = \frac{2\pi}{3} + 2k\pi$$

$$\theta = \frac{4\pi}{3} + 2k\pi \qquad \text{for any integer } k.$$

Restricted Solutions

With restricted solutions, give the values from the complete solution that fall within the restricted domain.

EXAMPLE 6   Approximate values of a linear trigonometric function

Solve $\sin\theta = \frac{2}{\pi}$ for $0 \le \theta < 2\pi$.

Solution

First find the reference angle: $\overline{\theta} = \sin^{-1}\left|\frac{2}{\pi}\right| \approx 0.6901071$. Compare $\overline{\theta} = \sin^{-1}\frac{2}{\pi}$ with $y = \arcsin\ x$, and note that x is positive. The sine function is positive in Quadrants I and II, so

$$\theta \approx 0.6901071 \qquad \text{Quadrant I, } \theta = \overline{\theta}$$
$$\theta \approx 2.4514856 \qquad \text{Quadrant II, } \theta = \pi - \overline{\theta}$$

EXAMPLE 7   Trigonometric equation with a double angle

Solve $\sin 2\theta = \dfrac{\sqrt{3}}{2}$ for $0 \le \theta < 2\pi$.

Solution

Note: The angle in this problem is 2θ, which is called a double angle. A triple angle would be 3θ. Since we are looking for $0 \le \theta < 2\pi$, we see that the double-angle restriction is found by multiplying all components of this inequality by 2: $0 \le 2\theta < 4\pi$.

$$\sin 2\theta = \frac{\sqrt{3}}{2} \qquad \text{\textit{Given equation}}$$

$$2\theta = \sin^{-1}\left(\frac{\sqrt{3}}{2}\right) \qquad \text{\textit{Definition of inverse sine}}$$

$$2\theta = \frac{\pi}{3}, \frac{2\pi}{3}, \frac{7\pi}{3}, \frac{8\pi}{3} \qquad \text{\textit{The sine is positive in Quadrants I and II.}}$$

Quadrant I $\frac{\pi}{3}+2\pi$ $\frac{2\pi}{3}+2\pi$

Quadrant II (sine is positive in Quadrants I and II)

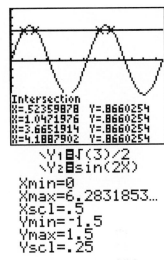

Mentally solve each equation below for θ

$$2\theta = \frac{\pi}{3} \quad 2\theta = \frac{2\pi}{3} \quad 2\theta = \frac{7\pi}{3} \quad 2\theta = \frac{8\pi}{3} \qquad \text{\textit{These are the values for } } 0 \le 2\theta < 4\pi.$$

$$\theta = \frac{\pi}{6} \quad \theta = \frac{\pi}{3} \quad \theta = \frac{7\pi}{6} \quad \theta = \frac{4\pi}{3} \qquad \text{\textit{These are the values for } } 0 \le \theta < 2\pi.$$

Find the approximate values 0.52, 1.05, 3.67, 4.19 (which all satisfy $0 \le \theta < 2\pi$), and check them graphically as shown in Figure 6.4. Note that the calculator solution uses the variable x instead of θ and gives an approximate answer.

Figure 6.4 Calculator solution of $\sin 2x = \dfrac{\sqrt{3}}{2}$

EXAMPLE 8   Trigonometric equation that has no solution

Solve $\cos 3\theta = 1.2862$.

Solution

The solution is empty because $-1 \le \cos 3\theta \le 1$. The graphical solution (see Figure 6.5) clearly shows that there is no solution:

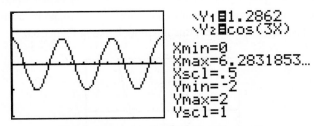

Figure 6.5 Graphs of $y = \cos 3x$ and $y = 1.2862$

EXAMPLE 9   Trigonometric equation that is also a radical equation

Solve the equation $\sin x - 1 = \sqrt{1 - \sin^2 x}$ for $0 \le x < 2\pi$.

Solution

To avoid radicals, we square both sides of the given equation:

$$\sin x - 1 = \sqrt{1 - \sin^2 x} \qquad \text{Given}$$

$$(\sin x - 1)^2 = \left(\sqrt{1 - \sin^2 x}\right)^2 \qquad \text{Square both sides; check required.}$$

$$\sin^2 x - 2\sin x + 1 = 1 - \sin^2 x \qquad \text{Simplify.}$$

$$2\sin^2 x - 2\sin x = 0 \qquad \text{Obtain a 0 on one side.}$$

$$\sin^2 x - \sin x = 0 \qquad \text{Divide both sides by 2.}$$

$$\sin(\sin x - 1) = 0 \qquad \text{Common factor}$$

$$\sin x = 0 \qquad \text{or} \quad \sin x - 1 = 0 \qquad \text{Zero-factor theorem}$$

$$x = \sin^{-1} 0 \qquad\qquad x = \sin^{-1} 1$$

$$= 0 \text{ or } \pi \qquad\qquad = \frac{\pi}{2} \qquad \text{These solutions are those for which } 0 \le x < 2\pi.$$

Check (because there might be extraneous roots).

Check $x = 0$:   Check $x = \pi$:   Check $x = \dfrac{\pi}{2}$:

$$\sin 0 - 1 \overset{?}{=} \sqrt{1 - \sin^2 0} \qquad \sin \pi - 1 \overset{?}{=} \sqrt{1 - \sin^2 \pi} \qquad \sin \frac{\pi}{2} - 1 \overset{?}{=} \sqrt{1 - \sin^2 \frac{\pi}{2}}$$

$$0 - 1 \overset{?}{=} \sqrt{1 - 0^2} \qquad\qquad 0 - 1 \overset{?}{=} \sqrt{1 - 0^2} \qquad\qquad 1 - 1 \overset{?}{=} \sqrt{1 - 1^2}$$

$$-1 \neq 1 \qquad\qquad\qquad -1 \neq 1 \qquad\qquad\qquad 0 = 0$$

The root is $\dfrac{\pi}{2}$.

■

▣ COMPUTATIONAL WINDOW

EXAMPLE 10   Trigonometric equation without an algebraic solution

Solve $\cos 2x + x^2 - 2 = 0$ for $0 \le x < 2\pi$.

Solution

There is no easy algebraic solution for this equation, so we can solve this equation numerically and graphically.

Graphical solution: We graph $y = \cos 2x + x^2 - 2$ and look to see where $y = 0$ (that is, where the graph crosses the x-axis). On many calculators, this is found by using the ▮ ROOT ▮ feature in the graph window. We show what this might look like in Figure 6.6.

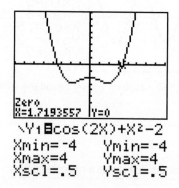

Figure 6.6 Positive root of $\cos 2x + x^2 - 2 = 0$

From the graph, we see that the root that satisfies $0 \le x < 2\pi$ is approximately $x = 1.7193557$. If your calculator does not have a ▮ ROOT ▮ feature, then you might be able to find it using the ▮ TRACE ▮, as shown in Figure 6.7.

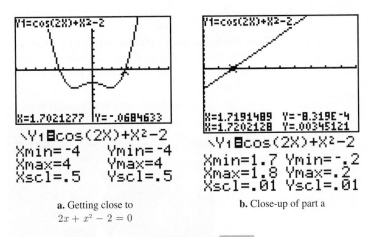

a. Getting close to
$2x + x^2 - 2 = 0$

b. Close-up of part a

Figure 6.7 Finding a root using TRACE

Using the TRACE, we see that the graph crosses the x-axis at $x \approx 1.7$ and at $x \approx -1.7$. If we want a more accurate approximation, we can use the ZOOM function to take a closer look in the neighborhood of $x = 1.7$, as shown in Figure 6.7**b**, where we see that the root is between $x = 1.7191489$ (y is negative) and $x = 1.7202128$ (y is positive). We estimate the solution to be $x \approx 1.72$

Numerical solution: We calculate $y = \cos 2x + x^2 - 2$ for values of x between 1.7 and 1.8, as shown in Figure 6.8.

X	Y1
1.7	-.0768
1.71	-.0374
1.72	.00259
1.73	.04316
1.74	.08432
1.75	.12604
1.76	.16835

X=1.72

X	Y1
1.716	-.0135
1.717	-.0095
1.718	-.0055
1.719	-.0014
1.72	.00259
1.721	.00662
1.722	.01066

X=1.719

Figure 6.8 Tabular approach for finding a root

From this table of values, we see that the root is between $x = 1.71$ (y is negative) and $x = 1.72$ (y is positive). We now calculate values between $x = 1.716$ and $x = 1.722$ to refine the table (as shown in Figure 6.8 at the right). From this table of values, we see that the root is between $x = 1.719$ (y is negative) and $x = 1.720$ (y is positive). From the table, we estimate the solution to be $x \approx 1.72$. This process could continue to any desired degree of accuracy. ∎

PROBLEM SET 6.1

LEVEL 1

Solve each equation in Problems 1–10.

a. Give the principal calculator value(s), rounded to two decimal places.
b. Give the exact restricted solution.
c. Give the complete exact solution.

1. $\cos x = \frac{1}{2}$
(Part **b** restriction is $0 \le x < \pi$).

2. $\cos x = \frac{1}{2}$
(Part **b** restriction is $0° \le x < 180°$).

3. $\cos x = -\frac{1}{2}$
(Part **b** restriction is $0 \le x < \pi$).

4. $\cos x = -\frac{1}{2}$
(Part **b** restriction is $0 \le x < 2\pi$).

5. $\tan x = -1$
(Part **b** restriction is $0 \le x < \pi$).

6. $\tan x = 1$
(Part **b** restriction is $0° \le x < 360°$).

7. $\sec x = 2$
(Part **b** restriction is $0 \le x < 2\pi$).

8. $\sec x = -2$
(Part **b** restriction is $0 < x < \pi$).

9. $\sin x = -\frac{\sqrt{2}}{2}$
(Part **b** restriction is $-\frac{\pi}{2} < x < \frac{\pi}{2}$).

10. $\sin x = \frac{\sqrt{2}}{2}$
(Part **b** restriction is $0 \le x < 2\pi$).

Solve each equation in Problems 11–20.

  a. Principal (calculator) values (two decimal places)
  b. Exact values such that $0 \le x < 2\pi$
  c. Complete exact solution

11. $\cos 2x = \frac{1}{2}$ **12.** $\sin 2x = -\frac{\sqrt{3}}{2}$

13. $\cos 2x = -\frac{1}{2}$ **14.** $\sin 2x = -\frac{\sqrt{2}}{2}$

15. $(\sec x)(\tan x) = 0$ **16.** $(\sin x)(\tan x) = 0$

17. $(\csc x)(\cos x) = 0$ **18.** $(\sin x)(\cos x) = 0$

19. $(\csc x - 2)(2\cos x - 1) = 0$ **20.** $(\sec x - 2)(2\sin x - 1) = 0$

LEVEL 2

Solve the equations in Problems 21–26 (exact answers) for
$0 \le x < 2\pi$.

21. $\tan^2 x = \tan x$ **22.** $\tan^2 x = \sqrt{3}\tan x$

23. $\sin^2 x = \frac{1}{2}$ **24.** $\cos^2 x = \frac{1}{2}$

25. $2\sin x \cos x = \sin x$ **26.** $\sqrt{2}\cos x \sin x = \sin x$

Find all solutions for each equation in Problems 27–32.

27. $\tan x = -\sqrt{3}$ **28.** $\cos x = -\frac{\sqrt{3}}{2}$

29. $\sin x = -\frac{\sqrt{2}}{2}$ **30.** $\sin x = 0.3907$

31. $\cos x = 0.2924$ **32.** $\tan x = 1.376$

Solve each equation in Problems 33–50 for principal values to two decimal places.

33. $\cos^2 x - 1 - \cos x = 0$ **34.** $\sin^2 x - \sin x - 2 = 0$

35. $\tan^2 x - 3\tan x + 1 = 0$ **36.** $\csc^2 x - \csc x - 1 = 0$

37. $\sec^2 x - \sec x - 1 = 0$ **38.** $\sin 2x + 2\cos x \sin 2x = 0$

39. $1 - \sin x = 1 - \sin^2 x$ **40.** $\cos x = 2\sin x \cos x$

41. $\cos(3x + 1) = \frac{1}{2}$ **42.** $\cos 3x + 1 = \sqrt{2}$

43. $\tan(2x + 1) = \sqrt{3}$ **44.** $\tan 2x + 1 = \sqrt{3}$

45. $\sin 2x + 1 = \sqrt{3}$ **46.** $1 - 2\sin^2 x = \sin x$

47. $2\cos^2 x - 1 = \cos x$ **48.** $2\cos x \sin x + \cos x = 0$

49. $\sin^2 3x + \sin 3x + 1 = 1 - \sin^2 3x$

50. $\sin^2 3x + \sin 3x = 1 - \cos^2 3x$

51. A tuning fork vibrating at 264 Hz (frequency $f = 264$) with an amplitude of 0.0050 cm produces C on the musical scale and can be described by an equation of the form

$$y = 0.0050\sin(528\pi x)$$

Find the smallest positive value of x (correct to four decimal places) for which $y = 0.0020$.

52. In a certain electrical circuit, the electromotive force V (in volts) and the time t (in seconds) are related by an equation of the form

$$V = \cos 2\pi t$$

Find the smallest positive value for t (correct to three decimal places) for which $V = 0.400$.

53. A lighthouse has a light that rotates once every 15 sec, as measured while you are sitting on the deck outside your kitchen. You know that the perpendicular distance between your location and the lighthouse is 350 ft.

It can be shown that if a person is standing at a point d ft from your location that the time between when you see the light and that person sees the light is given by

$$d = 350\tan\frac{2\pi t}{15}$$

How long after you see the light will your neighbor see it if she is 100 ft from your location?

54. The orbit of a particular satellite is shown in Figure 6.9.

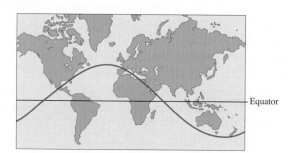

Figure 6.9 Orbital path of a satellite

This orbit can be described by the equation

$$y = 4,000 \sin\left(\frac{\pi}{45}t + \frac{5\pi}{18}\right)$$

where t is the time (in minutes) and y is the distance (in kilometers) from the equator. Find the times at which the satellite crosses the equator during the first hour and a half (that is, for $0 \leq t \leq 90$).

LEVEL 3

Solve the trigonometric equations in Problems 55–60 for $0 \leq x < 2\pi$. Give the exact values.

55. $\cos 2x - 1 = \sin 2x$ **56.** $\sin 2x - 1 = \cos 2x$

57. $\sin x + \cos x = 1$ **58.** $\sin x + \cos x = -1$

59. $x - \cos x = 0$ (answer rounded to the nearest tenth)

60. $\sin 2x + x^2 = 0$ (answer rounded to the nearest tenth)

6.2 Proving Identities

In Section 5.2, we considered eight fundamental identities that are used to simplify and change the form of a variety of trigonometric equations. These identities are repeated in Table 6.1 for easy reference. You might also wish to review some alternate forms of these identities; for example, $\sec\theta = 1/\cos\theta$ can also be written as $\cos\theta \sec\theta = 1$.

STOP

These are, indeed, fundamental. You should take whatever time you need to memorize them, as well as the alternative forms.

TABLE 6.1	Fundamental Identities
Fundamental Identities	**Alternative forms**
Reciprocal Identities	
1. $\sec\theta = \dfrac{1}{\cos\theta}$	$\sec\theta\cos\theta = 1$ $\cos\theta = \dfrac{1}{\sec\theta}$
2. $\csc\theta = \dfrac{1}{\sin\theta}$	$\csc\theta\sin\theta = 1$ $\sin\theta = \dfrac{1}{\csc\theta}$
3. $\cot\theta = \dfrac{1}{\tan\theta}$	$\cot\theta\tan\theta = 1$ $\tan\theta = \dfrac{1}{\tan\theta}$
Ratio Identities	
4. $\tan\theta = \dfrac{\sin\theta}{\cos\theta}$	$\tan\theta\cos\theta = \sin\theta$ $\cos\theta = \dfrac{\sin\theta}{\tan\theta}$
5. $\cot\theta = \dfrac{\cos\theta}{\sin\theta}$	$\cot\theta\sin\theta = \cos\theta$ $\sin\theta = \dfrac{\cos\theta}{\cot\theta}$
Pythagorean Identities	
6. $\cos^2\theta + \sin^2\theta = 1$	$\cos^2\theta = 1 - \sin^2\theta$ $\sin^2\theta = 1 - \cos^2\theta$
7. $1 + \tan^2\theta = \sec^2\theta$	$\tan^2\theta = \sec^2\theta - 1$ $\sec^2\theta - \tan^2\theta = 1$
8. $\cot^2\theta + 1 = \csc^2\theta$	$\cot^2\theta = \csc^2\theta - 1$ $\csc^2\theta - \cot^2\theta = 1$

EXAMPLE 1   **Graphical verification for a fundamental identity**

Show $\tan\theta = \dfrac{\sin\theta}{\cos\theta}$ graphically.

Solution

Graph $y = \tan\theta$; if you are using a graphing calculator, you will graph Y1 = tan X, which is shown in Figure 6.10**a**.

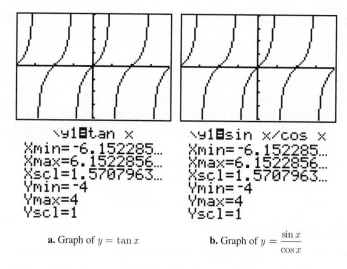

a. Graph of $y = \tan x$ **b.** Graph of $y = \dfrac{\sin x}{\cos x}$

Figure 6.10 Graphical verification that $\tan \theta = \dfrac{\sin \theta}{\cos \theta}$ **is an identity**

Now, graph Y2 = sin X/cos X, which is shown in Figure 6.10**b**. Notice that even though the functions being graphed are different, the graphs are identical. This means that the equation is an identity.

In addition to these fundamental identities, you will need to remember from arithmetic and algebra the proper procedure for simplifying expressions, especially the concepts of common denominators and complex fractions. You will combine this algebraic simplification with a knowledge of these fundamental identities, as shown by the next example.

EXAMPLE 2   Algebraically simplifying a trigonometric expression

Simplify $\dfrac{\sin \theta + \dfrac{\cos^2 \theta}{\sin \theta}}{\dfrac{\cos \theta}{\sin \theta}}$, and leave your answer in terms of sine and cosine functions only.

Solution

$$\frac{\sin \theta + \dfrac{\cos^2 \theta}{\sin \theta}}{\dfrac{\cos \theta}{\sin \theta}} = \frac{\dfrac{\sin^2 \theta}{\sin \theta} + \dfrac{\cos^2 \theta}{\sin \theta}}{\dfrac{\cos \theta}{\sin \theta}}$$

Common denominator; algebraic form:
$$\frac{a + \dfrac{b^2}{a}}{\dfrac{b}{a}} = \frac{\dfrac{a^2}{a} + \dfrac{b^2}{a}}{\dfrac{b}{a}}$$

$$= \frac{\sin^2 \theta + \cos^2 \theta}{\sin \theta} \cdot \frac{\sin \theta}{\cos \theta}$$

Dividing rational repressions; algebraic form:
$$\frac{\dfrac{a^2}{a} + \dfrac{b^2}{a}}{\dfrac{b}{a}} = \frac{a^2 + b^2}{a} \cdot \frac{a}{b}$$

$$= \frac{1}{\sin \theta} \cdot \frac{\sin \theta}{\cos \theta}$$

Fundamental identity: $\sin^2 \theta + \cos^2 \theta = 1$

$$= \frac{1}{\cos \theta}$$

Reduce rational expressions.

Procedure for Proving Identities

Suppose you are given a trigonometric equation such as

$$\tan \theta + \cot \theta = \sec \theta \csc \theta$$

and are asked to show that it is an identity, a process we term **proving an identity**. You must be careful not to treat this problem as though you are solving an algebraic equation. When asked to **prove** that a given equation is an identity, do *not* start with the given expression, since you cannot assume that it is true.

PROVING TRIGONOMETRIC IDENTITIES

The procedure for proving trigonometric identities is to *begin* with what you know is true and *end* with the given identity.

Step 1   Reduce the left-hand side of the equation to the right-hand side by using algebra and the fundamental identities.

Step 2   If step 1 does not give identical forms, start over and reduce the right-hand side equation to the left-hand side using algebra and the fundamental identities.

Step 3   If neither step 1 nor step 2 can be accomplished, then start over and reduce both sides *independently* to the same expression.

Step 4   If none of the above steps prove that the equation is an identity, try verifying graphically that it is an identity.

EXAMPLE 3   Proving an identity

Prove that $\tan \theta + \cot \theta = \sec \theta \csc \theta$.

Solution
Begin with either the left side or the right side (but begin with only *one* side). We begin with the left side:

$$\tan \theta + \cot \theta = \frac{\sin \theta}{\cos \theta} + \frac{\cos \theta}{\sin \theta}$$

Substitute one or more of the fundamental identities; in this case, we are substituting $\tan \theta = \dfrac{\sin \theta}{\cos \theta}$ and $\cot \theta = \dfrac{\cos \theta}{\sin \theta}$.

$$= \frac{\sin^2 \theta + \cos^2 \theta}{\cos \theta \sin \theta}$$

Common denominator to add rational expressions. The algebraic form is: $\dfrac{a}{b} + \dfrac{b}{a} = \dfrac{a^2 + b^2}{ab}$

$$= \frac{1}{\cos \theta \sin \theta}$$

Fundamental identity; in this case, we substituted $\sin^2 \theta + \cos^2 \theta = 1$.

$$= \frac{1}{\cos \theta} \cdot \frac{1}{\sin \theta}$$

Break up the product.

$$= \sec \theta \csc \theta$$

Fundamental identities; here we used two reciprocal identities.

Since the left-hand side (eventually) equals the right-hand side, we say that we have proved that the given equation is an identity.

■

Usually it is easier to begin with what seems to be the more complicated side (rather than always doing step 1 followed by step 2, . . .), and try to reduce it to the simpler side. If both sides seem equally complex, you might change all the functions to cosines and sines and then simplify, as we did in Example 3.

EXAMPLE 4   Proving an identity by combining fractions

Prove that $\dfrac{1}{1 + \cos \lambda} + \dfrac{1}{1 - \cos \lambda} = 2 \csc^2 \lambda$ is an identity.

Solution
We begin with the more complicated side (which, in this example, is the left side):

$$\frac{1}{1+\cos\lambda} + \frac{1}{1-\cos\lambda} = \frac{(1-\cos\lambda)+(1+\cos\lambda)}{(1+\cos\lambda)(1-\cos\lambda)}$$

Common denominators; algebraic form:

$$\frac{1}{1+b} + \frac{1}{1-b} = \frac{(1-b)+(1+b)}{(1+b)(1-b)}$$

$$= \frac{2}{1-\cos^2\lambda}$$

Simplify numerator and denominator.

$$= \frac{2}{\sin^2\lambda}$$

Fundamental identity

$$= 2 \cdot \frac{1}{\sin^2\lambda}$$

Multiplication of fractions

$$= 2\csc^2\lambda$$

Fundamental identity

EXAMPLE 5   Proving an identity by separating a fraction into a sum

Prove that $\dfrac{\sec 2\beta + \cot 2\beta}{\sec 2\beta} = 1 + \csc 2\beta - \sin 2\beta$ is an identity.

Solution

Begin with the left-hand side. When working with a fraction consisting of a single term as a denominator, it is often helpful to separate the fraction into the sum of several fractions.

$$\frac{\sec 2\beta + \cot 2\beta}{\sec 2\beta} = \frac{\sec 2\beta}{\sec 2\beta} + \frac{\cot 2\beta}{\sec 2\beta}$$

Break up fractions.

$$= 1 + \cot 2\beta \cdot \frac{1}{\sec 2\beta}$$

Note $\dfrac{\sec 2\beta}{\sec 2\beta} = 1$.

$$= 1 + \frac{\cos 2\beta}{\sin 2\beta} \cdot \cos 2\beta$$

Fundamental identities

$$= 1 + \frac{\cos^2 2\beta}{\sin 2\beta}$$

Multiply fractions.

$$= 1 + \frac{1 - \sin^2 2\beta}{\sin 2\beta}$$

Fundamental identity

$$= 1 + \frac{1}{\sin 2\beta} - \frac{\sin^2 2\beta}{\sin 2\beta}$$

Break up fractions.

$$= 1 + \csc 2\beta - \sin 2\beta$$

Fundamental identity and reduce the fraction.

EXAMPLE 6   Proving an identity by using a conjugate

Prove that $\dfrac{\cos\alpha}{1-\sin\alpha} = \dfrac{1+\sin\alpha}{\cos\alpha}$ is an identity.

Solution

Sometimes, when there is a binomial in the numerator or denominator, the identity may be proved by multiplying the numerator and denominator by the conjugate of the binomial. That is, multiply both the numerator and the denominator of the left side by $1 + \sin\alpha$. (This does not change the value of the expression because you are multiplying by 1.)

$$\frac{\cos\alpha}{1-\sin\alpha} = \frac{\cos\alpha}{1-\sin\alpha} \cdot \frac{1+\sin\alpha}{1+\sin\alpha}$$

Multiply by 1.

$$= \frac{\cos\alpha(1+\sin\alpha)}{1-\sin^2\alpha}$$

Multiply fractions.

$$= \frac{\cos\alpha(1+\sin\alpha)}{\cos^2\alpha} \qquad \text{Fundamental identities}$$

$$= \frac{1+\sin\alpha}{\cos\alpha} \qquad \text{Reduce fraction.}$$

EXAMPLE 7   Proving an identity by factoring and reducing

Prove that $\dfrac{\sec^2 2\gamma - \tan^2 2\gamma}{\sec 2\gamma + \tan 2\gamma} = \dfrac{\cos 2\gamma}{1+\sin 2\gamma}$ is an identity.

Solution

Sometimes the identity can be proved by factoring and reducing.

$$\frac{\sec^2 2\gamma - \tan^2 2\gamma}{\sec 2\gamma + \tan 2\gamma} = \frac{(\sec 2\gamma + \tan 2\gamma)(\sec 2\gamma - \tan 2\gamma)}{\sec 2\gamma + \tan 2\gamma}$$

$$= \sec 2\gamma - \tan 2\gamma \qquad \text{If an expression seems simplified, but it is not in the form you want, consider changing all the functions to cosines and sines.}$$

$$= \frac{1}{\cos 2\gamma} - \frac{\sin 2\gamma}{\cos 2\gamma}$$

$$= \frac{1-\sin 2\gamma}{\cos 2\gamma} \qquad \text{Subtract fractions.}$$

$$= \frac{1-\sin 2\gamma}{\cos 2\gamma} \cdot \frac{1+\sin 2\gamma}{1+\sin 2\gamma} \qquad \text{Multiply by 1.}$$

$$= \frac{1-\sin^2 2\gamma}{\cos 2\gamma(1+\sin 2\gamma)} \qquad \text{Multiply fractions.}$$

$$= \frac{\cos^2 2\gamma}{\cos 2\gamma(1+\sin 2\gamma)} \qquad \text{Fundamental identity}$$

$$= \frac{\cos 2\gamma}{1+\sin 2\gamma} \qquad \text{Reduce fractions.}$$

EXAMPLE 8   Proving an identity by multiplying by 1

Prove that $\dfrac{-2\sin\theta\cos\theta}{1-\sin\theta-\cos\theta} = 1+\sin\theta+\cos\theta$ is an identity.

Solution

Sometimes, when there is a fraction on one side, the identity can be proved by multiplying the other side by 1, where 1 is written so that the desired denominator is obtained. Thus, for this example,

$$1+\sin\theta+\cos\theta = (1+\sin\theta+\cos\theta) \cdot \frac{1-\sin\theta-\cos\theta}{1-\sin\theta-\cos\theta} \qquad \text{Multiply by 1.}$$

$$= \frac{(1+\sin\theta+\cos\theta)(1-\sin\theta-\cos\theta)}{1-\sin\theta-\cos\theta} \qquad \text{Multiply fractions.}$$

$$= \frac{1-\sin\theta-\cos\theta+\sin\theta-\sin^2\theta-\sin\theta\cos\theta+\cos\theta-\cos\theta\sin\theta-\cos^2\theta}{1-\sin\theta-\cos\theta}$$

$$= \frac{1-(\sin^2\theta+\cos^2\theta)-2\sin\theta\cos\theta}{1-\sin\theta-\cos\theta} \qquad \text{Combine terms.}$$

$$= \frac{-2\sin\theta\cos\theta}{1-\sin\theta-\cos\theta} \qquad \begin{array}{l}\text{Fundamental identity}\\ \sin^2\theta+\cos^2\theta=1 \\ \text{and } 1-1=0\end{array}$$

In summary, there is not a single method that is best for proving identities. However, the following hints should help:

> **TEN COMMANDMENTS FOR PROVING IDENTITIES**
>
> 1. Avoid the introduction of radicals.
> 2. If one side contains one function only, write all the trigonometric functions on the other side in terms of that function.
> 3. Change all trigonometric functions to sines and cosines and simplify. (See Example 3.)
> 4. Simplify by combining fractions. (See Example 4.)
> 5. If the denominator of a fraction consists of only one term, break up the fraction. (See Example 5.)
> 6. Multiply by the conjugate of either the numerator or the denominator. (See Example 6.)
> 7. Factoring is sometimes helpful. (See Example 7.)
> 8. Multiplying one side or the other by the number 1 is sometimes helpful. (See Example 8.)
> 9. If there are squares of functions, look for alternative forms of the Pythagorean identities.
> 10. Keep your destination in sight. Watch where you are going, and know when you are finished.

Disproving an Identity

If you suspect that a given equation is not an identity and want to prove that hypothesis, you may find a **counterexample**. That is, if you can find *one replacement* of the variable for which the functions are defined that will make the equation false, then you have proved that the equation is not an identity. In such a case, we say you have **disproved an identity**.

EXAMPLE 9   Finding a counterexample

A common mistake for trigonometry students is to write $\sin 2\theta$ as $2 \sin \theta$. Show that

$$\sin 2\theta = 2 \sin \theta$$

is not an identity by finding a counterexample.

Solution
Let $\theta = \frac{\pi}{6}$. Then, by substitution,

$$\sin\left(2 \cdot \frac{\pi}{6}\right) = 2 \sin \frac{\pi}{6}$$

$$\frac{\sqrt{3}}{2} = 2\left(\frac{1}{2}\right) \qquad \textit{False equation}$$

Since this is a false equation, we have found a counterexample, namely, $x = \frac{\pi}{6}$.   ■

◼ COMPUTATIONAL WINDOW   ⬓⬓⬓

It is easy to show that a given equation, such as the one given in Example 9, is not an identity by using a calculator. Graph $Y1 = \sin 2X$ and $Y2 = 2 \sin X$. These graphs are shown in Figure 6.11**a**.

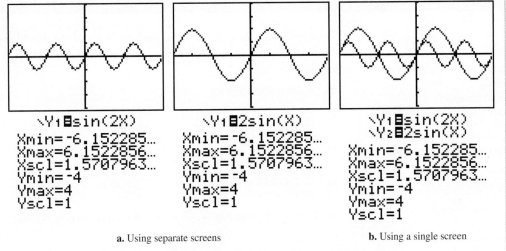

a. Using separate screens   **b.** Using a single screen

Figure 6.11 Calculator graphs can be used to disprove an identity.

You can see that these graphs are not the same, so the equation is not an identity. You can also graph both curves at the same time, as shown in Figure 6.11**b**.

EXAMPLE 10 Deciding whether an equation is an identity

Is the equation $\cos^2 x - \sin^2 x = \sin x$ an identity?

Solution

If you suspect that an equation might not be an identity, you can compare the graphs of both the left and right sides, or you can try to prove that it is not an identity by counterexample. Suppose we let $\theta = \frac{\pi}{6}$. Then, by substitution,

$$\cos^2 \frac{\pi}{6} - \sin^2 \frac{\pi}{6} = \sin \frac{\pi}{6} \qquad \textit{Substitute selected value.}$$

$$\left(\frac{\sqrt{3}}{2}\right)^2 - \left(\frac{1}{2}\right)^2 = \frac{1}{2} \qquad \textit{Evaluate the trigonometric functions.}$$

$$\frac{3}{4} - \frac{1}{4} = \frac{1}{2} \qquad \textit{True equation}$$

The fact that the value $\theta = \frac{\pi}{6}$ satisfies the equation does not answer the question. We have found a root for the equation, but we have not shown that the equation is an identity. Remember, for an equation to be an identity, it must be true for *all* values of the variable in the domain, not just one or two. *One* counterexample disproves the identity, but *one* true value does **not** prove an identity. In fact, if we solve this equation (the details are left for the reader), we find $\theta = \frac{\pi}{6} + 2k\pi$, $\frac{5\pi}{6} + 2k\pi$, $\frac{3\pi}{2} + 2k\pi$, k any integer. If you choose any of these values, the equation will be satisfied. If you choose any other value—for example, $\theta = \frac{\pi}{3}$—the equation is false.

$$\cos^2 \frac{\pi}{3} - \sin^2 \frac{\pi}{3} = \sin \frac{\pi}{3} \qquad \textit{Substitute selected value.}$$

$$\left(\frac{1}{2}\right)^2 - \left(\frac{\sqrt{3}}{2}\right)^2 = \frac{\sqrt{3}}{2} \qquad \textit{Evaluate the trigonometric functions.}$$

$$\frac{1}{4} - \frac{3}{4} = \frac{\sqrt{3}}{2} \qquad \textit{False equation}$$

We now have found a counterexample, and the identity is disproved.   ■

PROBLEM SET 6.2

LEVEL 1

1. IN YOUR OWN WORDS What is an equation? What is an identity?
2. IN YOUR OWN WORDS Distinguish the procedures for solving an equation and for proving that an equation is an identity.
3. IN YOUR OWN WORDS State the eight fundamental identities.
4. IN YOUR OWN WORDS List some of the procedures you would use in proving identities.

Use the fundamental identities to write the remaining five trigonometric values as functions of u in Problems 5–8. Assume u > 0, but pay attention to the given quadrant for the angle θ.

5. $\cos\theta = \dfrac{u}{3},\ 0 < \theta < \dfrac{\pi}{2}$

6. $\sin\theta = -3u,\ \dfrac{3\pi}{2} < \theta < 2\pi$

7. $\sin\theta = \dfrac{u}{\sqrt{3}},\ 0 < \theta < \dfrac{\pi}{2}$

8. $\cos\theta = -\dfrac{2u}{5},\ \dfrac{\pi}{2} < \theta < \pi$

Prove that the equations in Problems 9–28 are identities.

9. $\sec\theta = \sec\theta\sin^2\theta + \cos\theta$

10. $\tan\theta = \cot\theta\tan^2\theta$

11. $\dfrac{\sin\theta\cos\theta + \sin^2\theta}{\sin\theta} = \cos\theta + \sin\theta$

12. $\tan^2\theta - \sin^2\theta = \tan^2\theta\sin^2\theta$

13. $\cot^2\theta\cos^2\theta = \cot^2\theta - \cos^2\theta$

14. $\tan\theta + \cot\theta = \sec\theta\csc\theta$

15. $\dfrac{\cos\alpha + \tan\alpha\sin\alpha}{\sec\alpha} = 1$

16. $\dfrac{1 - \sec^2\beta}{\sec^2\beta} = -\sin^2\beta$

17. $(\sec\theta - \cos\theta)^2 = \tan^2\theta - \sin^2\theta$

18. $1 - \sin 2\theta = \dfrac{1 - \sin^2 2\theta}{1 + \sin 2\theta}$

19. $\sin\lambda = \dfrac{\sin^2\lambda + \sin\lambda\cos\lambda + \sin\lambda}{\sin\lambda + \cos\lambda + 1}$

20. $\dfrac{1 + \cos 2\lambda\sec 2\lambda}{\tan 2\lambda + \sec 2\lambda} = \dfrac{2\cos 2\lambda}{\sin 2\lambda + 1}$

21. $\sin 3\beta\cos 3\beta(\tan 3\beta + \cot 3\beta) = 1$

22. $(\sin\alpha + \cos\alpha)^2 + (\sin\alpha - \cos a)^2 = 2$

23. $\csc\beta - \cos\beta\cot\beta = \sin\beta$

24. $\dfrac{1 + \cot^2\gamma}{1 + \tan^2\gamma} = \cot^2\gamma$

25. $\dfrac{\sin^2\omega - \cos^2\omega}{\sin\omega + \cos\omega} = \sin\omega - \cos\omega$

26. $\tan^2\phi + \sin^2\phi + \cos^2\phi = \sec^2\phi$

27. $\dfrac{\tan\phi + \cot\phi}{\sec\phi\csc\phi} = 1$

28. $1 + \sin^2\theta = 2 - \cos^2\theta$

Find a counterexample in Problems 29–34 to show that each equation is not an identity. Be sure you do not choose a value that makes any of the functions undefined.

29. $\sec^2 x - 1 = \sqrt{3}\tan x$

30. $2\cos 2\theta\sin 2\theta = \sin 2\theta$

31. $\sin\theta = \cos\theta$

32. $\cos^2\theta - 3\sin\theta + 3 = 0$

33. $\sin^2\theta + \cos\theta = 0$

34. $2\sin^2\theta - 2\cos^2\theta = 1$

Graphically decide whether the equations in Problems 35–38 are identities. For each problem, show both the graphs and the input functions.

35. $\cos^4 x = \sin^3 x + \cos^2 x - \sin x$

36. $\dfrac{\tan^2 x - 2\tan x}{2\tan x - 4} = \dfrac{1}{2}\tan x$

37. $\dfrac{\cos^2 x - \cos x\sec x}{\cos^2 x\sec x - \sin^2 x\csc^2 x} = \cos x - 1$

38. $\dfrac{\sin x}{\csc x} + \dfrac{\cos x}{\sec x} = 1$

39. Prove or disprove that $\dfrac{\sin 2\theta}{2} = \sin\theta$. It is an identity, by dividing:

$$\dfrac{\sin \cancel{2}\theta}{\cancel{2}} = \sin\theta$$

40. Prove or disprove that the equation

$$\cos(\alpha - \beta) = \cos\alpha - \cos\beta$$

is an identity. It is an identity because of the distributive property.

LEVEL 2

Prove that the equations in Problems 41–56 are identities.

41. $\dfrac{1 + \tan\alpha}{1 - \tan\alpha} = \dfrac{\sec^2\alpha + 2\tan\alpha}{2 - \sec^2\alpha}$

42. $(\cot x + \csc x)^2 = \dfrac{\sec x + 1}{\sec x - 1}$

43. $\dfrac{\sin^3 x - \cos^3 x}{\sin x - \cos x} = 1 + \sin x\cos x$

44. $\dfrac{1 - \cos\theta}{1 + \cos\theta} = \left(\dfrac{1 - \cos\theta}{\sin\theta}\right)^2$

45. $\dfrac{(\sec^2\gamma + \tan^2\gamma)^2}{\sec^4\gamma - \tan^4\gamma} = 1 + 2\tan^2\gamma$

46. $(\sec\gamma + \csc\gamma)^2 = \dfrac{1 + 2\sin\gamma\cos\gamma}{\cos^2\gamma\sin^2\gamma}$

47. $\dfrac{1}{\sec\theta + \tan\theta} = \sec\theta - \tan\theta$

48. $\dfrac{1 + \tan^3\theta}{1 + \tan\theta} = \sec^2\theta - \tan\theta$

49. $\dfrac{1 - \sec^3\theta}{1 - \sec\theta} = \tan^2\theta + \sec\theta + 2$

50. $\dfrac{\tan\theta}{\cot\theta} - \dfrac{\cot\theta}{\tan\theta} = \sec^2\theta - \csc^2\theta$

51. $\sqrt{(3\cos\theta - 4\sin\theta)^2 + (3\sin\theta + 4\cos\theta)^2} = 5$

52. $\dfrac{\csc\theta + 1}{\cot^2\theta + \csc\theta + 1} = \dfrac{\sin^2\theta + \sin\theta\cos\theta}{\sin\theta + \cos\theta}$

53. $\dfrac{\cos^4\theta - \sin^4\theta}{(\cos^2\theta - \sin^2\theta)^2} = \dfrac{\cos\theta}{\cos\theta + \sin\theta} + \dfrac{\sin\theta}{\cos\theta - \sin\theta}$

54. $\dfrac{2\tan^2\theta + 2\tan\theta\sec\theta}{\tan\theta + \sec\theta - 1} = \tan\theta + \sec\theta + 1$

55. $\dfrac{\cos^2\theta - \cos\theta\csc\theta}{\cos^2\theta\csc\theta - \cos\theta\csc^2\theta} = \sin\theta$

56. $(\cos\alpha\cos\beta\tan\alpha + \sin\alpha\sin\beta\cot\beta)\csc\alpha\sec\beta = 2$

57. If a light ray passes from one medium to another of greater or lesser density, the ray will bend.

© Leslie Garland Picture Library/Alamy Images

A measure of this tendency is called the *index of refraction*, R, and is defined by

$$R = \frac{\sin \alpha}{\sin \beta}$$

For example, if the angle of incidence, α, of a light ray entering a diamond is $30.0°$ and the angle of refraction, β, is $11.9°$, then the diamond's index of refraction is:

$$R = \frac{\sin 30.0°}{\sin 11.9°} \approx 2.424781044$$

Since the index of refraction for any particular substance is a constant, we can compare our calculated number to the known index number to determine whether a particular stone is an imitation. The index of refraction of an imitation diamond is rarely greater than 2. If a stone is a diamond and the angle of incidence is $45.0°$, what should be its refracted angle?

$$R = \frac{\sin \alpha}{\sin \beta}, \text{ where } R \approx 2.42, a = 45°$$

Solve for β, and substitute known values to find the refracted angle β:

$$\sin \beta = \frac{\sin \alpha}{R}$$

$$\beta = \sin^{-1}\left(\frac{\sin \alpha}{R}\right) \approx 16.95°$$

To the nearest tenth, the refracted angle is $17.0°$. If a stone is a topaz, with an index of refraction of 1.63 and an angle of incidence is $20.0°$, what should be its refracted angle (correct to the nearest tenth of a degree)?

58. If an anthracite garnet has an index of refraction of 1.82 and an angle of incidence of $35.0°$, what should be its refracted angle (correct to the nearest tenth of a degree)? (See Problem 57.)

59. Journal Problem (*The Mathematics Student Journal* of the National Council of Teachers of Mathematics, April 1973, 20(4).) For $0 \le \theta < \frac{\pi}{2}$, solve

$$\left(\frac{16}{81}\right)^{\sin^2 \theta} + \left(\frac{16}{81}\right)^{1-\sin^2 \theta} = \frac{26}{27}$$

60. How many solutions are there for the equation

$$\frac{x}{100} = \sin x$$

6.3 Addition Laws

When proving identities, it is sometimes necessary to simplify the functional values of the sum or difference of two angles. If α and β represent any two angles,

$$\cos(a - \beta) \ne \cos \alpha - \cos \beta$$

For example, if $\alpha = 60°$ and $\beta = 30°$; then

$$\cos(60° - 30°) = \cos 60° - \cos 30° \qquad \text{Substitute values.}$$
$$\cos 30° = \cos 60° - \cos 30° \qquad \text{Parentheses first}$$
$$\frac{\sqrt{3}}{2} = \frac{1}{2} - \frac{\sqrt{3}}{2} \qquad \text{This is a false equation.}$$

Thus, by counterexample, $\cos(60° - 30°) \ne \cos 60° - \cos 30°$.

In this section, we discuss twelve more identities. Remember the first eight identities (listed in Table 6.1). The next three identities were listed in Section 5.5.

Difference of Angles Identity

We have shown what $\cos(\alpha - \beta)$ is not; therefore, it seems reasonable to ask what it is, which leads us to the next identity.

DIFFERENCE OF ANGLES

$$\cos(\alpha - \beta) = \cos\alpha\cos\beta + \sin\alpha\sin\beta$$

To prove this identity, find the length of any chord with corresponding arc intercepted by the central angle θ, where θ is in standard position (see Figure 6.12).

Let A be the point $(1, 0)$ and P be the point on the intersection of the terminal side of angle θ and the unit circle. This means that the coordinates of P are $(\cos\theta, \sin\theta)$. Now find the length of the chord $\overline{AP}$ by using the distance formula:

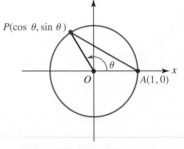

$$
\begin{aligned}
|\overline{AP}| &= \sqrt{(1 - \cos\theta)^2 + (0 - \sin\theta)^2} && \text{Distance formula} \\
&= \sqrt{1 - 2\cos\theta + \cos^2\theta + \sin^2\theta} \\
&= \sqrt{1 - 2\cos\theta + 1} && \sin^2\theta + \cos^2\theta = 1 \\
&= \sqrt{2 - 2\cos\theta}
\end{aligned}
$$

Figure 6.12 Length of a chord

Next, apply this result to a chord determined by any two angles α and β, as shown in Figure 6.13.

Let P_α and P_β be the points on the unit circle determined by the angles α and β, respectively. By the previous result,

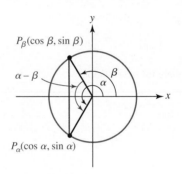

$$\left|P_\alpha P_\beta\right| = \sqrt{2 - 2\cos(\alpha - \beta)} \qquad \text{Substitute } \theta = \alpha - \beta.$$

Figure 6.13 Distance between points

But, we could also have found this distance using the distance formula:

$$
\begin{aligned}
\left|P_\alpha P_\beta\right| &= \sqrt{(\cos\beta - \cos\alpha)^2 + (\sin\beta - \sin\alpha)^2} \\
&= \sqrt{(\cos^2\beta - 2\cos\alpha\cos\beta + \cos^2\alpha + \sin^2\beta - 2\sin\alpha\sin\beta + \sin^2\beta} \\
&= \sqrt{(\cos^2\beta + \sin^2\beta) + (\cos^2\alpha + \sin^2\alpha) - 2(\cos\alpha\cos\beta + \sin\alpha\sin\beta)} \\
&= \sqrt{2 - 2(\cos\alpha\cos\beta + \sin\alpha\sin\beta)}
\end{aligned}
$$

Finally, equate these quantities since they both represent the distance between P_α and P_β:

$$
\begin{aligned}
\sqrt{2 - 2\cos(\alpha - \beta)} &= \sqrt{2 - 2(\cos\alpha\cos\beta + \sin\alpha\sin\beta)} \\
2 - 2\cos(\alpha - \beta) &= 2 - 2(\cos\alpha\cos\beta + \sin\alpha\sin\beta) \\
-2\cos(\alpha - \beta) &= -2(\cos\alpha\cos\beta + \sin\alpha\sin\beta) \\
\cos(\alpha - \beta) &= \cos\alpha\cos\beta + \sin\alpha\sin\beta
\end{aligned}
$$

We can use this identity to find the exact values of functions of angles that are multiples of $15°$.

EXAMPLE 1   Exact value of a cosine of a multiple of 15°

Find the exact value of $\cos 345°$.

Solution

$$\cos 345° = \cos 15°$$ Reduction principle; cosine is positive in Quadrant IV.

$$= \cos(45° - 30°)$$ Since $45° - 30° = 15°$

$$= \cos 45° \cos 30° + \sin 45° \sin 30°$$ Difference of angles identity

$$= \left(\frac{1}{2}\sqrt{2}\right)\left(\frac{1}{2}\sqrt{3}\right) + \left(\frac{1}{2}\sqrt{2}\right)\left(\frac{1}{2}\right)$$ Exact values

$$= \frac{\sqrt{6} + \sqrt{2}}{4}$$ Simplify.

It is also possible to derive the exact value of $\cos 15°$ by using a geometrical argument involving right triangles, which is given as a problem in the problem set.

Cofunction Identities

Even though this difference of angles identity is helpful in making evaluations (as in Example 1), its value lies in the fact that it is true for *any* choice of α and β. By making some particular choices for α and β, we find several useful special cases for this identity.

EXAMPLE 2 Derive a cofunction identity

Prove $\cos\left(\dfrac{\pi}{2} - \theta\right) = \sin\theta$.

Solution

This proof is based on the difference of angles identity.

$$\cos(\alpha - \beta) = \cos\alpha\cos\beta + \sin\alpha\sin\beta$$ Difference of angles identity

$$\cos\left(\frac{\pi}{2} - \theta\right) = \cos\frac{\pi}{2}\cos\theta + \sin\frac{\pi}{2}\sin\theta$$ Let $\alpha = \frac{\pi}{2}$ and $\beta = \theta$.

$$\cos\left(\frac{\pi}{2} - \theta\right) = 0 \cdot \cos\theta + 1 \cdot \sin\theta$$ Exact values: $\cos\frac{\pi}{2} = 0, \sin\frac{\pi}{2} = 1$

$$\cos\left(\frac{\pi}{2} - \theta\right) = \sin\theta$$ Simplify.

Example 2 is one of three identities known as the **cofunction identities.*

COFUNCTION IDENTITIES

For any real number (or angle) θ:

12. $\cos\left(\dfrac{\pi}{2} - \theta\right) = \sin\theta$

13. $\sin\left(\dfrac{\pi}{2} - \theta\right) = \cos\theta$

14. $\tan\left(\dfrac{\pi}{2} - \theta\right) = \cot\theta$

Identity 12 was proved in Example 2; the proof of identity 13 depends on identity 12, and is shown below:

$$\cos\theta = \cos\left[\frac{\pi}{2} - \left(\frac{\pi}{2} - \theta\right)\right] = \sin\left(\frac{\pi}{2} - \theta\right)$$

---

*Note that the numbering here begins with identity 12 because we already have eight fundamental identities (1–8) and three inverse identities (9–11).

Identities involving the tangent are usually proved after proving similar identities for cosine and sine. The fundamental identity $\tan \theta = \sin\theta/\cos\theta$ is applied first, allowing us then to use the appropriate identities for cosine and sine. This process is illustrated with the following derivation of identity 14.

$$\tan\left(\frac{\pi}{2} - \theta\right) = \frac{\sin\left(\frac{\pi}{2} - \theta\right)}{\cos\left(\frac{\pi}{2} - \theta\right)} = \frac{\cos\theta}{\sin\theta} = \cot\theta$$

The cofunction identities allow us to change a trigonometric function to the cofunction of its complement.

EXAMPLE 3   Writing a trigonometric function in terms of its cofunction

Write each function in terms of its cofunction.

   a. $\sin 38°$ **b.** $\cos 53°$ **c.** $\cot 4°$ **d.** $\tan \frac{\pi}{6}$ **e.** $\sec \alpha$ **f.** $\cos\left(\frac{\pi}{2} - \beta\right)$

Solution

   a. $\sin 38° = \cos(90° - 38°) = \cos 52°$
   b. $\cos 53° = \sin(90° - 53°) = \sin 37°$
   c. $\cot 4° = \tan(90° - 4°) = \tan 86°$

   d. $\tan\dfrac{\pi}{6} = \cot\left(\dfrac{\pi}{2} - \dfrac{\pi}{6}\right)$

            $= \cot\left(\dfrac{3\pi}{6} - \dfrac{\pi}{6}\right)$

            $= \cot\dfrac{2\pi}{6}$

            $= \cot\dfrac{\pi}{3}$

   e. $\sec\alpha = \csc\left(\frac{\pi}{2} - \alpha\right)$

   f. $\cos\left(\frac{\pi}{2} - \beta\right) = \sin\beta$

■

Opposite-Angle Identities

Suppose the given angle in Example 3 is larger than 90°; for example,

$$\cos 125° = \sin(90° - 125°) = \sin(-35°)$$

In order to further simplify this expression, we need the following **opposite-angle identities**.

OPPOSITE-ANGLE IDENTITIES

For any real number (or angle) θ:

   15. $\cos(-\theta) = \cos\theta$
   16. $\sin(-\theta) = -\sin\theta$
   17. $\tan(-\theta) = -\tan\theta$

Using identity 16, we see $\sin(-35°) = -\sin 35°$.

     The proof for identity 15 depends on the difference of angles identity with the following choice; let $\alpha = 0$ and $\beta = \theta$:

$$\cos(0 - \theta) = \cos 0 \cos\theta + \sin 0 \sin\theta = 1 \cdot \cos\theta + 0 \cdot \sin\theta = \cos\theta$$

Thus, $\cos(-\theta) = \cos\theta$. The proofs of identities 16 and 17 are similar and are left as exercises.

The opposite-angle identities remind us of the definition of even and odd functions. If $c(x) = \cos x$, then $c(-x) = \cos(-x) = \cos x = c(x)$, so cosine is an even function. Recall, this means that the graph of $y = \cos x$ is symmetric with respect to the y-axis. Also, if $s(x) = \sin x$, then $s(-x) = \sin(-x) = -\sin x = -s(x)$, so sine is an odd function. This means that the graph of $y = \sin x$ is symmetric with respect to the x-axis. Is tangent an even or an odd function? Consider the following example.

EXAMPLE 4   Finding the zeros of a trigonometric function

Find the zeros of the function $f(x) = \cos x + x^2$.

Solution
Since $\cos x$ is positive in the first quadrant $\left(0 < x < \frac{\pi}{2}\right)$ and x^2 is positive in the first quadrant, we see $f > 0$ in the first quadrant. For $x \geq \frac{\pi}{2}$, we see $x^2 > 1$ and since $-1 \leq \cos \leq 1$, we see $f > 0$ for all $x > 0$. Finally, since f is an even function, the graph is symmetric with respect to the y-axis, so we see that $f > 0$ for all x. Thus, there are no zeros. ∎

EXAMPLE 5   Writing a trigonometric function as a function of a positive angle

Write each as a function of a positive angle or number:

a. $\cos(-19°)$ **b.** $\sin(-19°)$ **c.** $\tan(-2)$

Solution
a. $\cos(-19°) = \cos 19°$ **b.** $\sin(-19°) = -\sin 19°$ **c.** $\tan(-2) = -\tan 2$ ∎

EXAMPLE 6   Writing a trigonometric function in terms of its cofunction

Write each given function in terms of its cofunction:

a. $\cos 125°$ **b.** $\sin 102°$ **c.** $\cot 2.5$

Solution
a. $\cos 125° = \sin(90° - 125°) = \sin(-35°) = -\sin 35°$
b. $\sin 102° = \cos(90° - 102°) = \cos(-12°) = \cos 12°$
c. $\cot 2.5 = \tan\left(\frac{\pi}{2} - 2.5\right) \approx \tan(-0.9292) = -\tan 0.9292$ ∎

Sometimes we use opposite-angle identities together with other identities. For example, you know that

$$a - b \quad \text{and} \quad b - a$$

are opposites. This means that $(a - b) = -(b - a)$. In trigonometry, we often see angles such as $\frac{\pi}{2} - \theta$ and want to write $\theta - \frac{\pi}{2}$. This means that $\frac{\pi}{2} - \theta = -\left(\theta - \frac{\pi}{2}\right)$. In particular,

$$\cos\left(\frac{\pi}{2} - \theta\right) = \cos\left[-\left(\theta - \frac{\pi}{2}\right)\right] = \cos\left(\theta - \frac{\pi}{2}\right)$$

EXAMPLE 7   Using an opposite-angle identity

Write the given functions using the opposite-angle identities:

a. $\sin\left(\frac{\pi}{2} - \theta\right)$ **b.** $\tan\left(\frac{\pi}{2} - \theta\right)$ **c.** $\cos(\pi - \theta)$

Solution

a. $\sin\left(\frac{\pi}{2} - \theta\right) = \sin\left[-\left(\theta - \frac{\pi}{2}\right)\right] = -\sin\left(\theta - \frac{\pi}{2}\right)$

b. $\tan\left(\frac{\pi}{2} - \theta\right) = \tan\left[-\left(\theta - \frac{\pi}{2}\right)\right] = -\tan\left(\theta - \frac{\pi}{2}\right)$

c. $\cos\left(\pi - \theta\right) = \cos\left[-\left(\theta - \pi\right)\right] = \cos\left(\theta - \pi\right)$

Graphing with Opposites

The procedure for graphing $y = -\cos x$ is identical to the procedure for graphing $y = \cos x$, except that, after the frame is drawn, the endpoints are at the bottom of the frame instead of at the top of the frame, and the midpoint is at the top, as shown in Figure 6.14.

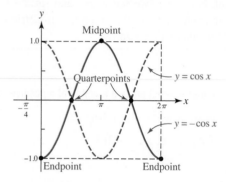

Figure 6.14 One period of $y = -\cos x$

EXAMPLE 8   Graphing a sine using an opposite-angle identity

Graph $y = \sin(-x)$.

Solution

First use an opposite-angle identity, if necessary; $\sin(-x) = -\sin x$. To graph $y = -\sin x$, build the frame as before; the endpoints and midpoints are the same, but the quarterpoints are reversed, as shown in Figure 6.15.

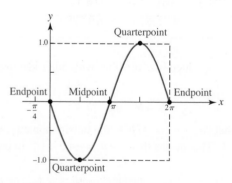

Figure 6.15 One period of $y = -\sin x$

Sum and Difference Angle Identities

The difference-of-angles identity proved at the beginning of this section is one of six identities known as the **addition laws**. Since subtraction can easily be written as a sum, the designation *addition laws* refers to both addition and subtraction.

ADDITION LAWS

18. $\cos(\alpha + \beta) = \cos\alpha\cos\beta - \sin\alpha\sin\beta$

19. $\cos(\alpha - \beta) = \cos\alpha\cos\beta + \sin\alpha\sin\beta$

20. $\sin(\alpha + \beta) = \sin\alpha\cos\beta + \cos\alpha\sin\beta$

21. $\sin(\alpha - \beta) = \sin\alpha\cos\beta - \cos\alpha\sin\beta$

22. $\tan(\alpha + \beta) = \dfrac{\tan\alpha + \tan\beta}{1 - \tan\alpha\tan\beta}$

23. $\tan(\alpha - \beta) = \dfrac{\tan\alpha - \tan\beta}{1 + \tan\alpha\tan\beta}$

EXAMPLE 9   Using an addition law

Write $\cos\left(\dfrac{2\pi}{3} + \theta\right)$ as a function of θ only.

Solution

$$\cos\left(\frac{2\pi}{3} + \theta\right) = \cos\frac{2\pi}{3}\cos\theta - \sin\frac{2\pi}{3}\sin\theta \qquad \text{Addition law}$$

$$= \left(-\frac{1}{2}\right)\cos\theta - \left(\frac{\sqrt{3}}{2}\right)\sin\theta \qquad \text{Exact values}$$

$$= -\frac{1}{2}\left(\cos\theta + \sqrt{3}\sin\theta\right) \qquad \text{Common factor}$$

EXAMPLE 10   Evaluation using an addition law

Evaluate $\dfrac{\tan 18° - \tan 40°}{1 + \tan 18°\tan 40°}$ using a calculator (two-place accuracy).

Solution

To evaluate this expression, you can do a lot of calculator arithmetic, or as the example title suggests, you can use an addition law:

$$\frac{\tan 18° - \tan 40°}{1 + \tan 18°\tan 40°} = \tan(18° - 40°)$$

$$= \tan(-22°)$$

$$\approx -0.4040262258$$

The result (to two-place accuracy) is -0.40.

EXAMPLE 11   Proving an addition law

Prove $\cos(\alpha + \beta) = \cos\alpha\cos\beta - \sin\alpha\sin\beta$

Solution

$$\cos(\alpha + \beta) = \cos[\alpha - (-\beta)] \qquad \text{Definition of subtraction}$$

$$= \cos\alpha\cos(-\beta) + \sin\alpha\sin(-\beta) \qquad \text{Difference-of-angles identity}$$

$$= \cos\alpha\cos\beta - \sin\alpha\sin\beta \qquad \text{Opposite-angle identities}$$

EXAMPLE 12   Use addition laws to evaluate a trigonometric expression

Evaluate $\cos\left(\sin^{-1}\frac{4}{5} + \cos^{-1}\frac{3}{5}\right)$ without tables or calculator.

Solution

Let α and β be first-quadrant angles (both inverse sine and inverse cosine are positive in the first quadrant) so that

$$\alpha = \sin^{-1}\frac{4}{5} \quad \text{or} \quad \sin\alpha = \frac{4}{5} \quad \text{and} \quad \beta = \cos^{-1}\frac{3}{5} \quad \text{or} \quad \cos\beta = \frac{3}{5}$$

We now need to evaluate

$$\cos\left(\sin^{-1}\frac{4}{5} + \cos^{-1}\frac{3}{5}\right) = \cos(\alpha + \beta)$$

$$= \cos\alpha\cos\beta - \sin\alpha\sin\beta$$

$$= \left(\sqrt{1 - \sin^2\alpha}\right)\cos\beta - \sin\alpha\left(\sqrt{1 - \cos^2\beta}\right)$$

$$= \left(\sqrt{1 - \left(\frac{4}{5}\right)^2}\right)\left(\frac{3}{5}\right) - \left(\frac{4}{5}\right)\left(\sqrt{1 - \left(\frac{3}{5}\right)^2}\right)$$

$$= \frac{3}{5}\sqrt{1 - \frac{16}{25}} - \frac{4}{5}\sqrt{1 - \frac{9}{25}}$$

$$= \frac{3}{5}\left(\frac{3}{5}\right) - \frac{4}{5}\left(\frac{4}{5}\right)$$

$$= -\frac{7}{25}$$

PROBLEM SET 6.3

LEVEL 1

WHAT IS WRONG, *if anything, with each statement in Problems 1–8? Explain your reasoning.*

1. $\sin(30° + 45°) = \sin 30° + \sin 45°$
2. $\cos(2 \cdot 45°) = \cos 45° + \cos 45°$
3. $\tan(30° + 60°) = \tan 90°$
4. $\tan(30° + 60°) = \tan 30° + \tan 60°$
5. $\sin\left(\frac{\pi}{2} - \theta\right) = \sin\frac{\pi}{2} - \sin\theta$
6. $\sin\left(\frac{\pi}{2} - \theta\right) = \sin\left(\theta - \frac{\pi}{2}\right)$
7. $\cos(-45°) = -\cos 45°$
8. $\sin(-60°) = -\sin 60°$

Change each of the expressions in Problems 9–10 to functions of θ only.

9. **a.** $\sin(\theta + 45°)$ **b.** $\cos(\theta - 45°)$ **c.** $\tan(\theta - 45°)$

10. **a.** $\cos\left(\frac{\pi}{6} + \theta\right)$ **b.** $\cos\left(\frac{\pi}{3} - \theta\right)$ **c.** $\tan\left(\frac{\pi}{4} + \theta\right)$

Write each function in Problems 11–12 in terms of its cofunction by using the cofunction identities.

11. **a.** $\sin 15°$ **b.** $\tan 62°$ **c.** $\cos\frac{5\pi}{6}$

12. **a.** $\sin 38°$ **b.** $\cos\frac{\pi}{6}$ **c.** $\cot\frac{2\pi}{3}$

Write each function in Problems 13–14 as a function of a positive angle.

13. **a.** $\sin(-23°)$ **b.** $\cos(-57°)$ **c.** $\tan(-29°)$
14. **a.** $\tan(-54°)$ **b.** $\cos(-19°)$ **c.** $\sin(-83°)$

Evaluate each of the expressions in Problems 15–20 to four decimal places.

15. $\sin 158° \cos 92° - \cos 158° \sin 92°$
16. $\sin 18° \cos 23° + \cos 18° \sin 23°$
17. $\cos 30° \cos 48° + \sin 30° \sin 48°$

18. $\cos 114° \cos 85° + \sin 114° \sin 85°$

19. $\dfrac{\tan 32° + \tan 18°}{1 - \tan 32° \tan 18°}$

20. $\dfrac{\tan 59° - \tan 25°}{1 + \tan 59° \tan 25°}$

Find the exact values for the cosine, sine, and tangent of each of the angles in Problems 21–26.

21. $15°$

22. $165°$

23. $-15°$

24. $195°$

25. $75°$

26. $105°$

Graph each function in Problems 27–32.

27. $y = -2 \cos x$

28. $y = -3 \sin x$

29. $y = \tan(-x)$

30. $y = \cos(-2x)$

31. $y - 1 = \cos(2\pi - x)$

32. $y - 2 = \cos(\pi - x)$

33. Decide whether each of the given functions is even, odd, or neither.

  a. $f(x) = \sec x$

  b. $g(x) = \dfrac{\sin x}{x}$

  c. $h(x) = \dfrac{x}{\cos x}$

  d. $F(x) = \sin x \tan x$

34. Write each as a function of $\theta - \frac{\pi}{3}$.

  a. $\cos\left(\frac{\pi}{3} - \theta\right)$ **b.** $\sin\left(\frac{\pi}{3} - \theta\right)$ **c.** $\tan\left(\frac{\pi}{3} - \theta\right)$

35. Write each as a function of $\alpha - \beta$.

  a. $\cos(\beta - \alpha)$ **b.** $\sin(\beta - \alpha)$ **c.** $\tan(\beta - \alpha)$

LEVEL 2

Prove the identities in Problems 36–41.

36. $\sin(\alpha - \beta) = \sin \alpha \cos \beta - \cos \alpha \sin \beta$

37. $\tan(\alpha - \beta) = \dfrac{\tan \alpha - \tan \beta}{1 + \tan \alpha \tan \beta}$

38. $\dfrac{\cos 5\theta}{\sin \theta} - \dfrac{\sin 5\theta}{\cos \theta} = \dfrac{\cos 6\theta}{\sin \theta \cos \theta}$

39. $\dfrac{\sin 6\theta}{\sin 3\theta} - \dfrac{\cos 6\theta}{\cos 3\theta} = \sec 3\theta$

40. $\sin(\alpha + \beta)\cos \beta - \cos(\alpha + \beta)\sin \beta = \sin \alpha$

41. $\cos(\alpha - \beta)\cos \beta - \sin(\alpha - \beta)\sin \beta = \cos \alpha$

Evaluate each expression in Problems 42–45 without using tables or a calculator.

42. $\sin\left[\sin^{-1}\left(\frac{4}{5}\right) - \cos^{-1}\left(\frac{3}{5}\right)\right]$

43. $\cos\left[\sin^{-1}\left(\frac{3}{5}\right) + \cos^{-1}\left(\frac{4}{5}\right)\right]$

44. $\cos\left[\cos^{-1}\left(\frac{1}{4}\right) - \sin^{-1}\left(\frac{1}{3}\right)\right]$

45. $\sin\left[\cos^{-1}\left(\frac{2}{3}\right) + \sin^{-1}\left(\frac{1}{2}\right)\right]$

46. Show (without a calculator) that

$$\tan^{-1}\left(\frac{1}{2}\right) + \tan^{-1}\left(\frac{1}{3}\right) = \frac{\pi}{4}$$

47. Find a formula for $\cos 2\theta$ by writing this as $\cos(\theta + \theta)$ and then using the appropriate addition law.

48. Find a formula for $\sin 2\theta$ by writing this as $\sin(\theta + \theta)$ and then using the appropriate addition law.

LEVEL 3

Let $\triangle ABC$ be a triangle labeled with angles α, β, and γ. Prove or disprove the identities in Problems 49–52 concerning interior angles in a triangle.

49. $\tan(\alpha + \beta) = -\tan \gamma$

50. $\sin \alpha = \cos\left(\frac{1}{2}\alpha + \frac{3}{2}\beta + \frac{1}{2}\gamma\right)$

51. $\sin\left(\dfrac{\alpha + \beta}{2}\right) = \cos\left(\dfrac{\gamma}{2}\right)$

52. $\tan \alpha + \tan \beta + \tan \gamma = \tan \alpha \tan \beta \tan \gamma$

53. In Example 1, we found the exact value of $\cos 15°$ by using the difference-of-angles identity. In this problem, we find the exact value of $\cos 15°$ geometrically. Draw isosceles $\triangle ABC$ with $|\overline{AC}| = |\overline{BC}| = 2$ and $\angle ACB = 150°$, as shown in Figure 6.16. Next, construct right $\triangle ABC$. Use this figure and the definition of the trigonometric functions to find the exact value of $\cos 15°$.

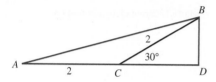

Figure 6.16 Finding $\cos 15°$

PROBLEMS FROM CALCULUS *An important expression, called a difference quotient, is defined to be*

$$\frac{f(x + h) - f(x)}{h}$$

Prove each identity in Problems 54–55.

54. Let $f(x) = \sin x$; prove that the difference quotient

$$\frac{\sin(x + h) - \sin x}{h}$$

  is equivalent to

$$\cos x\left(\frac{\sin h}{h}\right) - \sin x\left(\frac{1 - \cos h}{h}\right)$$

55. Let $f(x) = \cos x$; prove that the difference quotient

$$\frac{\cos(x + h) - \cos x}{h}$$

  is equivalent to

$$-\sin x\left(\frac{\sin x}{h}\right) - \cos x\left(\frac{1 - \cos h}{h}\right)$$

56. Journal Problem (*Mathematics Teacher*, May 1989, p. 384) Prove the addition law

$$\tan(\alpha + \beta) = \frac{\tan \alpha + \tan \beta}{1 - \tan \alpha \tan \beta}$$

using a geometrical argument for the illustration given in Figure 6.17.

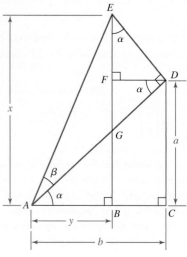

Figure 6.17 Proof of an addition law

57. Solve $\tan^{-1} x = \cot^{-1} x$

58. Solve

$$\tan^{-1} 1 = \tan^{-1}(1 - x) + \tan^{-1} x$$

59. In Section 5.6, we proved

$$\sin^{-1} x + \cos^{-1} x = \frac{\pi}{2}$$

for $0 < x < 1$. Using the addition formula, show that this identity actually holds on the closed interval $[-1, 1]$.

60. Prove

$$\tan^{-1} x + \tan^{-1} y = \tan^{-1}\left(\frac{x + y}{1 - xy}\right)$$

6.4 Miscellaneous Identities

In this section, we consider several additional identities that will enable you to write trigonometric expressions and equations in a variety of ways.

Double-Angle Identities

A special case of the addition laws is now considered (see Problems 47–48 of Section 6.3).

$$\begin{aligned}
\cos 2\theta &= \cos(\theta + \theta) &&\text{Addition law for } \cos(\alpha + \beta)\\
&= \cos\theta\cos\theta - \sin\theta\sin\theta \\
&= \cos^2\theta - \sin^2\theta \\
\sin 2\theta &= \sin(\theta + \theta) &&\text{Addition law for } \sin(\alpha + \beta)\\
&= \sin\theta\cos\theta + \cos\theta\sin\theta \\
&= 2\sin\theta\cos\theta \\
\tan 2\theta &= \tan(\theta + \theta) &&\text{Addition law for } \tan(\alpha + \beta)\\
&= \frac{\tan\theta + \tan\theta}{1 - \tan\theta\tan\theta} \\
&= \frac{2\tan\theta}{1 - \tan^2\theta}
\end{aligned}$$

These identities are known as the **double-angle identities**.

DOUBLE-ANGLE IDENTITIES

24. $\cos 2\theta = \cos^2\theta - \sin^2\theta$

25. $\sin 2\theta = 2\sin\theta\cos\theta$

26. $\tan 2\theta = \dfrac{2\tan\theta}{1 - \tan^2\theta}$

» IN OTHER WORDS Identity 24 is often combined with the identity $\cos^2\theta + \sin^2\theta = 1$ to write it in terms of cosines or in terms of sines:

$$\cos 2\theta = 2\cos^2\theta - 1 \quad \text{or} \quad \cos 2\theta = 1 - 2\sin^2\theta$$

EXAMPLE 1   Using a double-angle identity for a triple angle

Write $\cos 3\theta$ in terms of $\cos \theta$.

Solution

$$
\begin{aligned}
\cos 3\theta &= \cos(2\theta + \theta) &&\text{\textit{Write the triple angle as a sum.}}\\
&= \cos 2\theta \cos \theta - \sin 2\theta \sin \theta &&\text{\textit{Addition law (identity 18)}}\\
&= (\cos^2 \theta - \sin^2 \theta)\cos \theta - (2\sin \theta \cos \theta)\sin \theta &&\text{\textit{Double-angle identities}}\\
&= \cos^3 \theta - \sin^2 \theta \cos \theta - 2\sin^2 \theta \cos \theta &&\text{\textit{Distributive property}}\\
&= \cos^3 \theta - 3\sin^2 \theta \cos \theta &&\text{\textit{Combine similar terms.}}\\
&= \cos^3 \theta - 3(1 - \cos^2 \theta)\cos \theta &&\text{\textit{Fundamental identity}}\\
&= \cos^3 \theta - 3\cos \theta + 3\cos^3 \theta &&\text{\textit{Distributive property}}\\
&= 4\cos^3 \theta - 3\cos \theta &&\text{\textit{Combine similar terms.}}
\end{aligned}
$$

We continue with an example that you might find in a calculus text.

EXAMPLE 2   Finding a maximum value

Find the dimensions of the largest rectangle that can be inscribed in a semicircle of radius 1.

Solution
We begin by drawing a figure as shown in Figure 6.18.

The area of the rectangle is found, and then we apply a double-angle identity:

$$
\begin{aligned}
A &= 2xy\\
&= 2\cos \theta \sin \theta\\
&= \sin 2\theta
\end{aligned}
$$

The maximum value of $\sin 2\theta$ is when $2\theta = 90°$ or when $\theta = 45°$. Thus,

$$
\begin{aligned}
x &= \cos \theta & y &= \sin \theta\\
&= \cos 45° & &= \sin 45°\\
&= \frac{\sqrt{2}}{2} & &= \frac{\sqrt{2}}{2}
\end{aligned}
$$

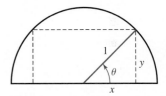

Figure 6.18 What is the largest possible rectangle?

Thus, the length of the rectangle is $2x = \sqrt{2}$ and the width is $y = \frac{1}{2}\sqrt{2}$.

EXAMPLE 3   Evaluation by using a double-angle identity

Evaluate $\dfrac{2\tan \frac{\pi}{16}}{1 - \tan^2 \frac{\pi}{16}}$ correct to four decimal places.

Solution
You could do this with a great deal of calculator work, or you can notice that this is the right-hand side of identity 26, so

$$
\frac{2\tan \frac{\pi}{16}}{1 - \tan^2 \frac{\pi}{16}} = \tan\left(2 \cdot \frac{\pi}{16}\right) = \tan \frac{\pi}{8} \approx 0.4142135624
$$

The requested value is 0.4142.

EXAMPLE 4 **Maximum range for a projectile** `MODELING APPLICATION`

Suppose that an object is propelled upward at an angle θ to the horizontal with an initial velocity of 224 ft/s. Find the angle θ for which the range is a maximum, and then find that maximum downrange point of impact. (Ignore air resistance.)

Solution

Step 1: *Understand the problem.* We have previously considered the path of a projectile.

Recall parametric equations for the path of a projectile (Example 6, Section 5.2):

$$x = (v_0 \cos\theta)t \qquad y = (v_0 \sin\theta)t - 16t^2$$

Step 2: *Devise a plan.* To answer this question, we know that the maximum range will be measured when the projectile hits the ground—that is, when $y = 0$. Our plan is to use the equation for the y-value to solve for t. Then, we will use this value for t in the x-value equation. Finally, we will use a double-angle to find the maximum range.

Step 3: *Carry out the plan.* We begin with the y-value equation.

$(v_0 \sin\theta)t - 16t^2 = y$	Given equation
$t[v_0 \sin\theta - 16t] = 0$	Let $y = 0$ and factor the left side.
$v_0 \sin\theta - 16t = 0$	Set each factor equal to 0; $t = 0$ and $v_0 \sin\theta - 16t = 0$.
$v_0 \sin\theta = 16t$	Add $16t$ to both sides.
$t = \dfrac{v_0 \sin\theta}{16}$	Divide both sides by 16.

Thus, the range (which is measured by the variable x) is

$x = (v_0 \cos\theta)t$	Given equation
$\quad = (v_0 \cos\theta)\left(\dfrac{v_0 \sin\theta}{16}\right)$	Write t as a function of θ.
$\quad = \dfrac{1}{16} v_0^2 \cos\theta \sin\theta$	Simplify.
$\quad = \dfrac{1}{32} v_0^2 \sin 2\theta$	Double-angle identity $\sin 2\theta = 2\sin\theta\cos\theta$

We can now answer the question. The maximum range (measured by x) will occur when $\sin 2\theta$ is the greatest, and we know that is when $\sin 2\theta = 1$, or when $2\theta = 90°$ and $\theta = 45°$. Finally, we find

$x = \dfrac{1}{32} v_0^2 \sin 2\theta$	Given equation
$\quad = \dfrac{1}{32}(224)^2 \sin 90°$	Substitute known values for this problem.
$\quad = 1{,}568$	Simplify.

The answer is that the projectile will impact 1,568 ft downrange if the angle of elevation is set at $45°$, and, furthermore, this is the maximum downrange distance.

Step 4: *Look back.* We check this path using a calculator, as shown in Figure 6.19.

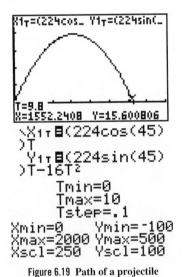

```
X1T=(224cos_ Y1T=(224sin(_
```

```
T=9.8
X=1552.2408  Y=15.600806
\X1T■(224cos(45)
)T
 Y1T■(224sin(45)
)T-16T²
    Tmin=0
    Tmax=10
    Tstep=.1
Xmin=0     Ymin=-100
Xmax=2000 Ymax=500
Xscl=250  Yscl=100
```

Figure 6.19 Path of a projectile

The next example exposes a common misunderstanding.

EXAMPLE 5   Graphing by using a double-angle identity

Graph $y = \sin^2 x$.

Solution

Since $\sqrt{x^2} = |x|$, some students *incorrectly* draw the graph of $y = |\sin x|$ to graph $y = \sin^2 x$, as shown in Figure 6.20**c**.

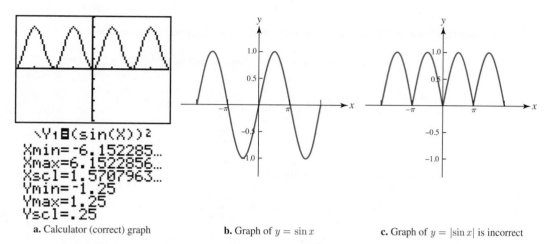

a. Calculator (correct) graph   **b.** Graph of $y = \sin x$ **c.** Graph of $y = |\sin x|$ is incorrect

Figure 6.20 Comparison of the graphs of $y = \sin^2 x$ (by calculator), $y = \sin x$, and $y = |\sin x|$

To correctly graph $y = \sin^2 x$ (other than by plotting points or by using a calculator as shown in Figure 6.20**a**), consider a double-angle identity:

$$\cos 2x = 1 - 2\sin^2 x \qquad \text{Double-angle identity}$$

$$2\sin^2 x = 1 - \cos 2x \qquad \text{Solve for } \sin^2 x$$

$$\sin^2 x = \frac{1}{2} - \frac{1}{2}\cos 2x$$

We see that the curve $y = \sin^2 x$ is a cosine curve with $a = -\frac{1}{2}$, $b = 2$, and $(h, k) = \left(0, \frac{1}{2}\right)$, as shown in Figure 6.21. Note that the period, p, is found by $p = (2\pi)/b = (2\pi)/2 = \pi$. The graph of $y = |\sin x|$ is shown as a dashed curve. Note that $y = \sin^2 x$ and $y = |\sin x|$ are not the same.

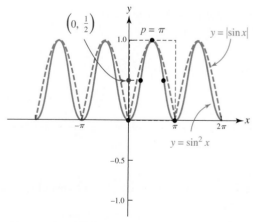

Figure 6.21 Graph of $y = \sin^2 x$

EXAMPLE 6   Trigonometric functions of double angles, given a single angle

If $\cos\theta = \frac{3}{5}$, and θ is in Quadrant IV, find $\cos 2\theta,\ \sin 2\theta,\ $ and $\tan 2\theta$.

Solution
Since $\cos 2\theta = 2\cos^2\theta - 1$,

$$\cos 2\theta = 2\left(\frac{3}{5}\right)^2 - 1 = -\frac{7}{25}$$

For the other functions of 2θ, we need to also know $\sin\theta$. To find this value, we use

$$\sin\theta = -\sqrt{1 - \cos^2\theta} \qquad \text{\small Negative since sine is negative in Quadrant IV}$$

$$= -\sqrt{1 - \left(\frac{3}{5}\right)^2} \qquad \text{\small Substitute given value of } \cos\theta.$$

$$= -\frac{4}{5} \qquad \text{\small Simplify.}$$

Now, we finish with the other double-angle formulas:

$$\sin 2\theta = 2\sin\theta\cos\theta = 2\left(-\frac{4}{5}\right)\left(\frac{3}{5}\right) = -\frac{24}{25}$$

$$\tan 2\theta = \frac{\sin 2\theta}{\cos 2\theta} = \frac{-\frac{24}{25}}{-\frac{7}{25}} = \frac{24}{7}$$

■

EXAMPLE 7   Solving an equation using a double-angle identity

Solve $\cos 2x - \cos x = 0$ for $0 \le x < 2\pi$.

Solution

$$\cos 2x - \cos x = 0$$
$$(\cos^2 x - \sin^2 x) - \cos x = 0$$
$$\cos^2 x - (1 - \cos^2 x) - \cos x = 0$$
$$2\cos^2 x - 1 - \cos x = 0$$
$$(2\cos x + 1)(\cos x - 1) = 0$$

$$\cos x = -\frac{1}{2} \qquad\qquad \cos x = 1$$

$$x = \frac{2\pi}{3},\ \frac{4\pi}{3} \qquad x = 0$$

■

Half-Angle Identities

Sometimes, as in Example 7, we want to write $\cos 2\phi$ in terms of cosines, and at other times, we want to write it in terms of sines:

$$\cos 2\phi = 2\cos^2\phi - 1 \quad \text{and} \quad \cos 2\phi = 1 - 2\sin^2\phi$$

These forms of the double-angle identity lead us to what are called the **half-angle identities**.

We begin by letting $2\phi = \theta$ so that $\phi = \frac{1}{2}\theta$.* We will show the derivation of the first half-angle identity (identity 27) and leave the second (identity 28) as an exercise.

$$\cos 2\phi = 2\cos^2\phi - 1 \qquad \text{\textit{Start with double-angle identity.}}$$

$$\cos\theta = 2\cos^2\frac{1}{2}\theta - 1 \quad \text{\textit{Substitute }} 2\phi = \theta.$$

$$2\cos^2\frac{1}{2}\theta = 1 + \cos\theta \qquad \text{\textit{Solve for }} \cos\tfrac{1}{2}\theta.$$

$$\cos^2\frac{1}{2}\theta = \frac{1 + \cos\theta}{2} \qquad \text{\textit{Divide both sides by 2.}}$$

$$\cos\frac{1}{2}\theta = \pm\sqrt{\frac{1 + \cos\theta}{2}} \quad \text{\textit{Square root property}}$$

The sign $+$ or $-$ is chosen according to the quadrant in which $\frac{1}{2}\theta$ is located.

☠ The formula requires either $+$ or $-$, but not both. This use of $\pm$ is different from the use of $\pm$ in algebra. For example, when using $\pm$ in the quadratic formula, we are indicating *two* possible correct roots. In this trigonometric identity, we obtain *one* correct sign, $+$ or $-$ (but not both), depending on the quadrant of $\frac{1}{2}\theta$. ☠

Read this sentence again! It is essential that you understand this important difference between the $\pm$ sign as it is used in algebra and the $\pm$ as it is used here. Study the following paragraph until you are sure you understand it. Ask someone if you are not sure.

HALF-ANGLE IDENTITIES

27. $\cos\dfrac{1}{2}\theta = \pm\sqrt{\dfrac{1 + \cos\theta}{2}}$

28. $\sin\dfrac{1}{2}\theta = \pm\sqrt{\dfrac{1 - \cos\theta}{2}}$

29. $\tan\dfrac{1}{2}\theta = \dfrac{\sin\theta}{1 + \cos\theta}$

To help you remember the correct sign between identities 27 and 28, remember "sinus is minus." To prove the last half-angle identity (identity 29), you can use

$$\tan\frac{1}{2}\theta = \frac{\sin\frac{1}{2}\theta}{\cos\frac{1}{2}\theta}$$

and then use identities 27 and 28 before simplifying. You can also use a geometric derivation, which is left for the problem set.

EXAMPLE 8   Using a half-angle identity

Find the exact value of $\cos\dfrac{9\pi}{8}$.

Solution

$$\cos\frac{9\pi}{8} = \cos\left(\frac{1}{2}\cdot\frac{9\pi}{4}\right) \qquad \text{\textit{Think of }} \frac{9\pi}{8} \text{ \textit{as a half-angle.}}$$

$$= -\sqrt{\frac{1 + \cos\frac{9\pi}{4}}{2}} \qquad \text{\textit{Choose a negative sign because }} \frac{9\pi}{8} \text{ \textit{is in Quadrant III, and the}}$$

$$\text{\textit{cosine is negative in Quadrant III.}}$$

---

*Remember from algebra that $\dfrac{1}{2}\theta$ and $\dfrac{\theta}{2}$ mean the same thing, so $\cos\dfrac{1}{2}\theta$ and $\cos\dfrac{\theta}{2}$ mean the same. However, do not confuse $\cos\dfrac{1}{2}\theta$ with $\dfrac{\cos\theta}{2}$; they are different and cannot be interchanged.

$$= -\sqrt{\frac{1 + \cos\frac{\pi}{4}}{2}} \qquad \text{Reduction principle}$$

$$= -\sqrt{\frac{1 + \frac{\sqrt{2}}{2}}{2}} \qquad \text{Exact value}$$

$$= -\sqrt{\frac{2 + \sqrt{2}}{4}} \qquad \text{Simplify.}$$

$$= -\frac{1}{2}\sqrt{2 + \sqrt{2}}$$

∎

EXAMPLE 9   Trigonometric functions given a double angle

If $\cot 2\theta = \frac{3}{4}$, find $\cos\theta$, $\sin\theta$, and $\tan\theta$, where 2θ is in Quadrant III.

Solution
We need to find $\cos 2\theta$ so that we can use it in the half-angle identities. To do this, we find $\tan 2\theta = 4/3$. Next, find $\sec 2\theta$:

$$\sec 2\theta = \pm\sqrt{1 + \tan^2 2\theta} \qquad \text{Pythagorean identity}$$

$$= -\sqrt{1 + \left(\frac{4}{3}\right)^2} \qquad \text{Negative because } 2\theta \text{ is in Quadrant III; substitute } \tan 2\theta = 4/3.$$

$$= -\frac{5}{3} \qquad \text{Simplify.}$$

Finally, using a reciprocal relationship, write $\cos 2\theta = -\frac{3}{5}$. We are now ready to use the half-angle identities (we show them side by side):

$$\cos\theta = \pm\sqrt{\frac{1 + \cos 2\theta}{2}} \qquad\qquad \sin\theta = \pm\sqrt{\frac{1 - \cos 2\theta}{2}} \qquad \text{Half-angle identities}$$

$$\underset{\underset{\text{Quadrant II}}{\uparrow}}{=} -\sqrt{\frac{1 + \left(-\frac{3}{5}\right)}{2}} \qquad\qquad \underset{\underset{\text{Quadrant II}}{\uparrow}}{=} +\sqrt{\frac{1 - \left(-\frac{3}{5}\right)}{2}} \qquad \begin{array}{l}\text{Since } \pi < 2\theta < \frac{3\pi}{2}, \frac{\pi}{2} < \theta < \frac{3\pi}{4} \\ \text{so } \theta \text{ is in Quadrant II}\end{array}$$

$$= -\frac{1}{\sqrt{5}} \qquad\qquad\qquad = \frac{2}{\sqrt{5}} \qquad \text{Simplify.}$$

Now, we use these results for the tangent function:

$$\tan\theta = \frac{\sin\theta}{\cos\theta} = \frac{2/\sqrt{5}}{-1/\sqrt{5}} = -2$$

∎

EXAMPLE 10   Proving an identity using a double-angle identity

Prove that $\sin\theta = \dfrac{2\tan\frac{1}{2}\theta}{1 + \tan^2\frac{1}{2}\theta}$ is an identity.

Solution

When proving identities involving functions of different angles, you should write all the trigonometric functions in the problem as functions of a single angle.

$$\frac{2\tan\frac{1}{2}\theta}{1+\tan^2\frac{1}{2}\theta} = \frac{2\frac{\sin\frac{1}{2}\theta}{\cos\frac{1}{2}\theta}}{\sec^2\frac{1}{2}\theta}$$ *Start with the more complicated side.*
    Use the fundamental identity $\tan\theta = \dfrac{\sin\theta}{\cos\theta}$.

$$= 2\frac{\sin\frac{1}{2}\theta}{\cos\frac{1}{2}\theta} \cdot \cos^2\frac{1}{2}\theta$$ *Fundamental identity $\cos\theta = \dfrac{1}{\sec\theta}$*

$$= 2\sin\frac{1}{2}\theta\cos\frac{1}{2}\theta$$ *Algebraically simplify.*

$$= \sin 2\left(\frac{1}{2}\theta\right)$$ *Double-angle identity*

$$= \sin\theta$$ *Algebraically simplify.*

∎

Product and Sum Identities

It is sometimes convenient, or even necessary, to write a trigonometric sum as a product or a product as a sum. The next box states the **product-to-sum identities**.

PRODUCT-TO-SUM IDENTITIES

30. $2\cos\alpha\cos\beta = \cos(\alpha - \beta) + \cos(\alpha + \beta)$
31. $2\sin\alpha\sin\beta = \cos(\alpha - \beta) - \cos(\alpha + \beta)$
32. $2\sin\alpha\cos\beta = \sin(\alpha + \beta) + \sin(\alpha - \beta)$
33. $2\cos\alpha\sin\beta = \sin(\alpha + \beta) - \sin(\alpha - \beta)$

The proofs of the product-to-sum identities involve systems of equations and will therefore be delayed until Section 8.4. By making appropriate substitutions and again using systems (as shown in Section 8.4), identities 30–33 can be rewritten in a form known as the **sum-to-product identities**.

SUM-TO-PRODUCT IDENTITIES

34. $\cos x + \cos y = 2\cos\left(\dfrac{x+y}{2}\right)\cos\left(\dfrac{x-y}{2}\right)$

35. $\cos x - \cos y = -2\sin\left(\dfrac{x+y}{2}\right)\sin\left(\dfrac{x-y}{2}\right)$

36. $\sin x + \sin y = 2\sin\left(\dfrac{x+y}{2}\right)\cos\left(\dfrac{x-y}{2}\right)$

37. $\sin x - \sin y = 2\sin\left(\dfrac{x-y}{2}\right)\cos\left(\dfrac{x+y}{2}\right)$

EXAMPLE 11   Using product and sum identities

a. Write $2\sin 3\sin 1$ as the sum of two functions.
b. Write $\sin 35° + \sin 27°$ as a product.
c. Write $\sin 40° \cos 12°$ as a sum.

Solution

a. Let $\alpha = 3$ and $\beta = 1$ in identity 31:

$$2 \sin 3 \sin 1 = \cos(3 - 1) - \cos(3 + 1) = \cos 2 - \cos 4$$

b. Let $x = 35°$, $y = 27°$, $\dfrac{x + y}{2} = \dfrac{35° + 27°}{2} = 31°$, and $\dfrac{x - y}{2} = \dfrac{35° - 27°}{2} = 4°$ in identity 36,

$$\sin 35° + \sin 27° = 2 \sin 31° \cos 4°$$

c. Let $\alpha = 40°$ and $\beta = 12°$ in identity 32:

$$2 \sin 40° \cos 12° = \sin(40° + 12°) + \sin(40° - 12°) = \sin 52° + \sin 28°$$

But what about the coefficient of 2 in this problem? Since you know that the preceding is an *equation* that is true, you can divide both sides by 2 to obtain

$$\sin 40° \cos 12° = \frac{1}{2}(\sin 52° + \sin 28°)$$

EXAMPLE 12   Proving an identity using a sum-to-product identity

Prove that $\dfrac{\sin 7\theta + \sin 5\theta}{\cos 7\theta - \cos 5\theta} = -\cot\theta$ is an identity.

Solution

$$\frac{\sin 7\theta + \sin 5\theta}{\cos 7\theta - \cos 5\theta} = \frac{2 \sin\left(\frac{7\theta + 5\theta}{2}\right)\cos\left(\frac{7\theta - 5\theta}{2}\right)}{-2 \sin\left(\frac{7\theta + 5\theta}{2}\right)\sin\left(\frac{7\theta - 5\theta}{2}\right)} \qquad \text{Begin with the left side.}$$

$$= \frac{2 \sin 6\theta \cos\theta}{-2 \sin 6\theta \sin\theta} \qquad \text{Simplify parentheses.}$$

$$= -\frac{\cos\theta}{\sin\theta} \qquad \text{Simplify fraction.}$$

$$= -\cot\theta \qquad \text{Fundamental identity}$$

EXAMPLE 13   Finding the period and amplitude for a graph

What are the exact period and amplitude for the graph with equation $y = \sin x + \cos x$?

Solution

We begin by using a cofunction identity and then use a sum-to-product identity.

$$y = \sin x + \cos x \qquad \text{Given.}$$

$$= \sin x + \sin\left(\frac{\pi}{2} - x\right) \qquad \text{Cofunction identity (identity 13)}$$

$$= 2 \sin\left(\frac{x + \frac{\pi}{2} - x}{2}\right)\cos\left(\frac{x - \frac{\pi}{2} + x}{2}\right) \qquad \text{Sum-to-product identity (identity 36)}$$

$$= 2 \sin\frac{\pi}{4}\cos\left(\frac{2x - \frac{\pi}{2}}{2}\right) \qquad \text{Simplify.}$$

$$= 2 \sin\frac{\pi}{4}\cos\left(x - \frac{\pi}{4}\right)$$

$$= 2\left(\frac{\sqrt{2}}{2}\right)\cos\left(x - \frac{\pi}{4}\right)$$ Exact value of $\sin\frac{\pi}{4}$ is $\frac{\sqrt{2}}{2}$.

$$= \sqrt{2}\cos\left(x - \frac{\pi}{4}\right)$$ Simplify.

From this form, we see the amplitude is $\sqrt{2}$ and the period is 2π. ∎

EXAMPLE 14   Solving an equation using a trigonometric identity

Solve $\cos 5x - \cos 3x = 0$ for $0 \le 2 < 2\pi$.

Solution

$$\cos 5x - \cos 3x = 0$$ Given.

$$-2\sin\left(\frac{5x + 3x}{2}\right)\sin\left(\frac{5x - 3x}{2}\right) = 0$$ Use sum-to-product identity.

$$\sin 4x \sin x = 0$$ Divide both sides by -2, and simplify.

$\sin 4x = 0 \qquad\qquad \sin x = 0$ Use the zero-factor theorem.
$\quad 4x = \sin^{-1} 0 \qquad\quad x = \sin^{-1} 0$
$\qquad = 0, \pi, 2\pi, 3\pi, \qquad\quad = 0, \pi$
$\qquad\quad 4\pi, 5\pi, 6\pi, 7\pi$

$$x = 0, \frac{\pi}{4}, \frac{\pi}{2}, \frac{3\pi}{4}, \pi, \frac{5\pi}{4}, \frac{3\pi}{2}, \frac{7\pi}{4}$$

∎

PROBLEM SET 6.4

LEVEL 1

WHAT IS WRONG, *if anything, with each statement in Problems 1–8? Explain your reasoning.*

1. $\cos 2\theta = \cos^2\theta + \sin^2\theta = 1$

2. $\dfrac{\cos 2\theta}{2} = \cos\theta$

3. $\cos\dfrac{1}{2}\theta = \dfrac{\cos\theta}{2}$

4. $\cos(\theta + \phi) = \cos\theta + \cos\phi$

5. $\tan(45° + \theta) = 1 + \tan\theta$

6. $\tan\dfrac{1}{2}\theta = \dfrac{1 - \cos\theta}{\sin\theta}$

7. If $\cos\dfrac{\theta}{2} = \pm\sqrt{\dfrac{1 + \cos\theta}{2}}$, then choose $+$ if θ is in Quadrants I or IV and $-$ if θ is in Quadrants II or III.

8. $\dfrac{2\tan\frac{1}{2}\theta}{1 - \tan^2\frac{1}{2}\theta} = \dfrac{2}{1 + \tan\theta}$

Assume that θ is positive and less than one revolution. Give the possible quadrant(s) for the angles indicated in Problems 9–14.

9. a. $\frac{1}{2}\theta$ if θ is in Quadrant I   **b.** 2θ if θ is in Quadrant I

10. a. 2θ if θ is in Quadrant II   **b.** $\frac{1}{2}\theta$ if θ is in Quadrant II

11. a. $\frac{1}{2}\theta$ if θ is in Quadrant III   **b.** $\frac{1}{2}\theta$ if θ is in Quadrant IV

12. a. 2θ if θ is in Quadrant III and $\cos 2\theta = -\frac{5}{9}$

b. θ if 2θ is in Quadrant IV and $\tan\theta = -\frac{3}{5}$

13. a. $\frac{1}{2}\theta$ in $\cos\frac{1}{2}\theta = \sqrt{\dfrac{1 + \cos\theta}{2}}$

b. $\frac{1}{2}\theta$ in $\sin\frac{1}{2}\theta = -\sqrt{\dfrac{1 - \cos\theta}{2}}$

14. a. $\frac{1}{2}\theta$ in $\cos\frac{1}{2}\theta = -\sqrt{\dfrac{1 + \cos\theta}{2}}$

b. $\frac{1}{2}\theta$ in $\sin\frac{1}{2}\theta = \sqrt{\dfrac{1 - \cos\theta}{2}}$

Use double-angle or half-angle identities to evaluate using exact values in Problems 15–18.

15. a. $2\cos^2 22.5° - 1$ **b.** $\tan 22.5°$

16. a. $\sin 22.5°$ **b.** $\cos\dfrac{\pi}{8}$

17. a. $\dfrac{2\tan\frac{\pi}{8}}{1 - \tan^2\frac{\pi}{8}}$ **b.** $\sqrt{\dfrac{1 - \cos 60°}{2}}$

18. a. $1 - 2\sin^2 90°$ **b.** $-\sqrt{\dfrac{1 - \cos 420°}{2}}$

Write each given product as a sum (or difference) and each given sum as a product in Problems 19–24.

19. a. $2\sin 35°\sin 24°$ **b.** $2\cos 46°\cos 18°$
20. a. $2\cos 70°\sin 24°$ **b.** $2\sin 41°\cos 19°$
21. a. $\cos\theta\cos 3\theta$ **b.** $\sin 2\theta\cos 5\theta$
22. a. $\sin 43° + \sin 64°$ **b.** $\cos 79° - \cos 77°$
23. a. $\sin 15° + \sin 30°$ **b.** $\cos 25° - \cos 100°$
24. a. $\cos 6x + \cos 2x$ **b.** $\sin 3x - \sin 2x$

LEVEL 2

25. Write $\sin 3\theta$ as a function of $\sin\theta$.
26. Write $\tan 3\theta$ as a function of $\tan\theta$.

Find $\cos\theta$, $\sin\theta$, and $\tan\theta$ when θ is in Quadrant I and $\cot 2\theta$ is given in Problems 27–32.

27. $\cot 2\theta = -\dfrac{3}{4}$ **28.** $\cot 2\theta = \dfrac{1}{\sqrt{3}}$

29. $\cot 2\theta = -\dfrac{4}{3}$ **30.** $\cot 2\theta = 0$

31. $\cot 2\theta = -\dfrac{1}{\sqrt{3}}$ **32.** $\cot 2\theta = \dfrac{4}{3}$

Find the exact value of cosine, sine, and tangent of both 2θ and $\frac{1}{2}\theta$ in Problems 33–38.

33. $\sin\theta = \dfrac{3}{5}$; θ in Quadrant I

34. $\sin\theta = \dfrac{5}{13}$; θ in Quadrant II

35. $\tan\theta = -\dfrac{5}{12}$; θ in Quadrant IV

36. $\tan\theta = -\dfrac{3}{4}$; θ in Quadrant II

37. $\cos\theta = \dfrac{5}{9}$; θ in Quadrant I

38. $\cos\theta = -\dfrac{5}{13}$; θ in Quadrant III

Graph each function in Problems 39–42.

39. $y = \cos^2 x$ **40.** $y = |\cos x|$

41. $y = \dfrac{\sin x}{1 + \cos x}$ **42.** $y = \pm\sqrt{\dfrac{1 + \cos x}{2}}$

Solve each equation for $0 \le x < 2\pi$ in Problems 43–46.

43. $\cos 2x - \sin x = 0$ **44.** $\sin 2x - \cos x = 0$
45. $\sin 2x - \sin x = 0$ **46.** $\cos 2x - \cos x = 0$

Prove that each equation in Problems 47–52 is an identity.

47. $\sin\alpha = 2\sin\dfrac{\alpha}{2}\cos\dfrac{\alpha}{2}$ **48.** $\sin 2\theta\tan\theta = 2\sin^2\theta$

49. $\cos 4\theta = \cos^2 2\theta - \sin^2 2\theta$ **50.** $\tan\dfrac{3}{2}\beta = \dfrac{2\tan\frac{3}{4}\beta}{1 - \tan^2\frac{3}{4}\beta}$

51. $\sin 2\theta = \dfrac{2\tan\theta}{1 + \tan^2\theta}$ **52.** $\tan\dfrac{1}{2}\theta = \dfrac{1 - \cos\theta}{\sin\theta}$

53. Graph $y = 3\sin x + 3\cos x$. Is this graph sinusoidal? If so, what are the period and amplitude of this graph?

54. Graph $y = \sin x - \cos x$. Is this graph sinusoidal? If so, what are the period and amplitude of this graph?

55. A touch-tone phone is arranged into four rows and three columns, with each row and each column set at a different frequency, as shown in Figure 6.22.

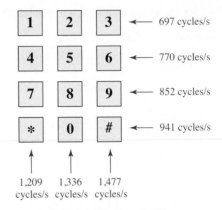

Figure 6.22 Touch-tone phone

When we press a particular button, the note that is sounded is the sum of two frequencies determined by the row and column. For example, if you press 1, then the sound emitted is

$$y = \sin 2\pi(697)x + \sin 2\pi(1,209)x$$

Write this sound as a product of sines and/or cosines.

56. Write the sound when you press the numeral 5 on a touch-tone phone as a product of sines and/or cosines. (See Problem 55.)

57. An airplane flying faster than the speed of sound is said to have a speed greater than Mach 1. In such a case, a sonic boom, created by sound waves that form a cone with vertex angle θ (see Figure 6.23), is heard. The Mach number is the ratio of the speed of the plane to the speed of sound and is denoted by M. It can be shown that, if $M > 1$, then $\sin\frac{1}{2}\theta = M^{-1}$.
   a. If $\theta = \frac{\pi}{6}$, find the Mach number to the nearest tenth.
   b. Find the exact Mach number for part **a**.
   c. Solve for θ; that is, suppose you know the Mach number, how can you determine the angle?

Figure 6.23 Pattern of sound waves

58. If a boat is moving at a constant rate that is faster than the water waves it produces, then the boat sends out waves in the shape of a cone with a vertex angle θ, as shown in Figure 6.24. If r represents the ratio of the speed of the boat to the speed of the wave and if $r > 1$, then $\sin \frac{1}{2}\theta = r^{-1}$.

a. If $\theta = \frac{\pi}{4}$, find r to the nearest tenth.

b. Find the exact value of r for part **a**.

c. Solve for θ; that is, suppose you know r, how can you determine the angle?

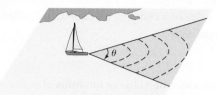

Figure 6.24 Pattern of boat waves

LEVEL 3

59. Suppose that an object is propelled upward at an angle θ to the horizontal with an initial velocity of 256 ft/s. Find the angle θ for which the range is a maximum, and then find that maximum downrange point of impact. (Ignore air resistance.)

60. Suppose that an object is propelled upward at an angle θ to the horizontal with an initial velocity of 384 ft/s. Find the angle θ for which the range is a maximum, and then find that maximum downrange point of impact. (Ignore air resistance.)

6.5 De Moivre's Theorem

Before studying this section, you might wish to review complex numbers (see Appendix B). In this section, we assume that the domain is the set of all complex numbers. Suppose we wish to solve

$$x^3 + 1 = 0$$

You might proceed by subtracting 1 from both sides, and then attempting to "take the cube root of both sides":

$$x^3 + 1 = 0 \qquad \textit{Given equation}$$
$$x^3 = -1 \qquad \textit{Subtract 1 from both sides.}$$
$$x = \sqrt[3]{-1} \qquad \textit{Cube root of both sides}$$
$$x = -1 \qquad \textit{Simplify.}$$

This "solution" is not correct because we note that this is a third-degree equation, and consequently by the fundamental theorem of algebra, we should expect three solutions. Let us try again, this time by factoring:

$$x^3 + 1 = 0 \qquad\qquad \textit{Given equation}$$
$$(x + 1)(x^2 - x + 1) = 0 \qquad\qquad \textit{Factor a sum of cubes.}$$
$$x + 1 = 0 \quad x^2 - x + 1 = 0 \qquad\qquad \textit{Set each factor equal to 0}$$
$$\textit{(zero-factor property).}$$

$$x = -1 \qquad\qquad x = \frac{-(-1) \pm \sqrt{(-1)^2 - 4(1)(1)}}{2(1)}$$
$$= \frac{1 \pm \sqrt{-3}}{2}$$
$$= \frac{1 \pm \sqrt{3}i}{2}$$
$$= \frac{1}{2} \pm \frac{\sqrt{3}}{2}i$$

The goal of this section is to develop techniques for solving such equations to find all the roots (not only the real roots but also those roots that are not real).

We begin this section by considering a graphical representation of a complex number. To give a graphic representation of complex numbers, such as

$$2 + 3i, \quad -i, \quad -3 - 4i, \quad 3i, \quad -2 + \sqrt{2}i, \quad \frac{3}{2} - \frac{5}{2}i$$

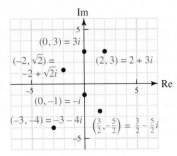

Figure 6.25 Complex plane

a two-dimensional coordinate system is used. The horizontal axis represents the **real axis** and the vertical axis the **imaginary axis**, so that the complex number $a + bi$ is represented by the ordered pair (a, b) in the usual manner, as shown in Figure 6.25.

The coordinate system in Figure 6.25 is called the **complex plane** or the **Gaussian plane**, in honor of Karl Friedrich Gauss.

The **absolute value** of a complex number z is graphically the distance between z and the origin (just as it is for real numbers). The absolute value of a complex number is also called the **modulus**. The distance formula leads to the following definition.

ABSOLUTE VALUE OR MODULUS

If $z = a + bi$ is a complex number, then the **absolute value** or **modulus** of z, denoted by $|z|$, is defined by

$$|z| = \sqrt{a^2 + b^2}$$

EXAMPLE 1   Finding the absolute value of a complex number

Find

a. $|3 + 4i|$ **b.** $\left| -2 + \sqrt{2}i \right|$ **c.** $|-3|$

Solution

a. $|3 + 4i| = \sqrt{3^2 + 4^2} = \sqrt{25} = 5$

b. $\left| -2 + \sqrt{2}i \right| = \sqrt{4 + 2} = \sqrt{6}$

c. You might suggest that -3 is a real number (which it is) and not a complex number (which is not correct). Recall, *all* real numbers are also complex numbers. We can write $-3 = -3 + 0i$.

$$|-3 + 0i| = \sqrt{9 + 0} = 3$$

Note: This is consistent with the absolute value of the real number -3, namely, $|-3| = 3$. ■

Trigonometric Form of Complex Numbers

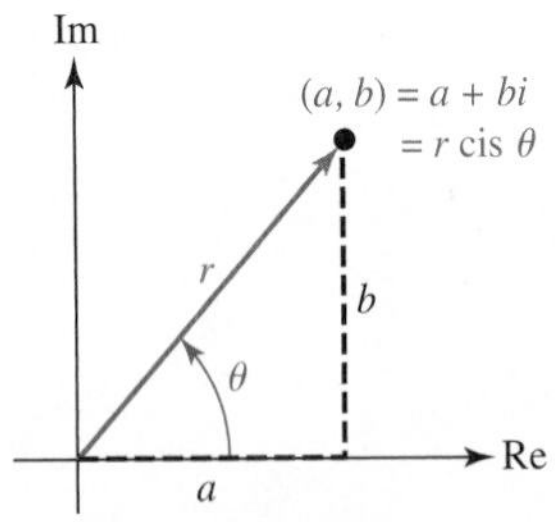

The form $a + bi$ is called the **rectangular form**, but another useful representation uses trigonometry. Consider the graphical representation of a complex number $a + bi$, as shown in Figure 6.26.

Let r be the distance from the origin to (a, b), and let θ be the angle the segment makes with the real axis. Then

$$r = \sqrt{a^2 + b^2}$$

Figure 6.26 Graphical representation of a complex number

and θ, called the **argument**, is chosen so that it is the smallest nonnegative angle the terminal side makes with the positive real axis.

From the definition of the trigonometric functions,

$$\cos \theta = \frac{a}{r} \qquad \sin \theta = \frac{b}{r} \qquad \tan \theta = \frac{b}{a}$$

$$a = r \cos \theta \qquad b = r \sin \theta$$

Thus,

$$a + bi = r\cos\theta + i(r\sin\theta) \qquad \text{Substitute } a = r\cos\theta \text{ and } b = r\sin\theta.$$
$$= r(\cos\theta + i\sin\theta) \qquad \text{Common factor}$$

Sometimes we write $r(\cos\theta + i\sin\theta)$ as $r\operatorname{cis}\theta$

TRIGONOMETRIC FORM OF A COMPLEX NUMBER

The **trigonometric form** of a complex number $z = a + bi$ is

$$r(\cos\theta + i\sin\theta) = r\operatorname{cis}\theta$$

where $r = \sqrt{a^2 + b^2}$; $\tan\theta = \frac{b}{a}(a \neq 0)$; $a = r\cos\theta$; $b = r\sin\theta$.
This representation is unique for $0° \leq 360°$ for all z except $0 + 0i$.

» IN OTHER WORDS This definition tells us how to change forms:
To change from rectangular form to trigonometric form:

Given $a + bi$. Find r and θ as follows:

$$r = \sqrt{a^2 + b^2} \quad \text{and} \quad \tan\theta = \frac{b}{a}$$

To change from trigonometric form to rectangular form:

Given $r\operatorname{cis}\theta$. Find a and b as follows:

$$a = r\cos\theta \quad \text{and} \quad b = r\sin\theta$$

The placement of θ in the proper quadrant is an important consideration because there are two values of $0° \leq \theta < 360°$ that will satisfy the relationship $\tan\theta = b/a$. For example, compare:

STOP Go back and read this definition and help window again. Write it down to help you remember it. Now, read the following paragraph SLOWLY!

$-1 + i$, a complex number in Quadrant II, where $a = -1$, $b = 1$, $\tan\theta = \frac{1}{-1} = -1$

$1 - i$, a complex number in Quadrant IV, where $a = 1$, $b = -1$, $\tan\theta = \frac{-1}{1} = -1$

Notice that the trigonometric equation is the same for both complex numbers, even though $-1 + i$ and $1 - i$ are in different quadrants. This consideration is even more important when you are doing the problem on a calculator because the usual sequence of steps for this example gives the result $-45°$, which is not true for either example, since $0° \leq \theta < 360°$ and $-45°$ is not within that domain. The entire process can be dealt with quite simply if we let $\overline{\theta}$ be the reference angle for θ. Find the reference angle

$$\overline{\theta} = \tan^{-1}|-1| = 45°$$

then decide the quadrant for the given complex number, and fix the argument accordingly. For $-1 + i$ in Quadrant II, we see $\theta = 135°$ (reference angle of $45°$ in Quadrant II); for $1 - i$ in Quadrant IV, we find $\theta = 315°$ (reference angle of $45°$ in Quadrant IV).

EXAMPLE 2　Changing rectangular form to trigonometric form

STOP Did you understand the material between the stop signs? If not, read it again, or ask your instructor to go over this with you.

Write each number in trigonometric form:

a. $-1 - \sqrt{3}i$ **b.** $6i$ **c.** -5 **d.** $4.310 + 5.516i$

Solution
a. $-1 - \sqrt{3}i$; $a = -1, b = -\sqrt{3}$ is in Quadrant III.

$$r = \sqrt{(-1)^2 + (-\sqrt{3})^2} = \sqrt{4} = 2$$

$$\overline{\theta} = \tan^{-1}\left|\frac{-\sqrt{3}}{-1}\right| = 60°; \text{ in Quadrant III}, \theta = 240°.$$

$$-1 - \sqrt{3} = 2\operatorname{cis}240°$$

b. $6i$; $a = 0$ and $b = 6$

Notice that $\tan \theta$ is not defined for $\theta = 90°$. By inspection,

$$6i = 6 \operatorname{cis} 90°$$

c. -5; $a = -5$, $b = 0$; by inspection, $-5 = 5 \operatorname{cis} 180°$.

d. $4.310 + 5.516i$; $a = 4.310$ and $b = 5.516$ is in Quadrant I.

$$r = \sqrt{(4.310)^2 + (5.516)^2} \approx \sqrt{49} = 7$$

$$\bar{\theta} = \tan^{-1} \left| \frac{5.516}{4.310} \right| \approx 52°. \text{ Thus, } 4.310 + 5.516i \approx 7 \operatorname{cis} 52°.$$

■

◼ COMPUTATIONAL WINDOW   ⊟◻☒

Many calculators have a built-in conversion for different representations of complex numbers. Most calculators call this *rectangular-polar* conversion, where $(x, y) \leftrightarrow (r, \theta)$ for $x + yi = r \operatorname{cis} \theta$. Check the owner's manual for your calculator.

EXAMPLE 3   **Changing trigonometric form to rectangular form**

Write each number in rectangular form.

   a. $4 \operatorname{cis} 330°$

   b. $5(\cos 38° + i \sin 38°)$

Solution

   a. $4 \operatorname{cis} 330° = 4(\cos 330° + i \sin 330°)$ **b.** $5(\cos 38° + i \sin 38°) = 5 \cos 38° + (5 \sin 38°)i$

$$= 4 \left[\frac{\sqrt{3}}{2} + i \left(-\frac{1}{2} \right) \right] \qquad\qquad \approx 3.94 + 3.08i$$

$$= 2\sqrt{3} - 2i$$

■

Multiplication and Division of Complex Numbers

The great advantage of the trigonometric form over the rectangular form is the ease with which you can multiply and divide complex numbers.

PRODUCTS AND QUOTIENTS

Let $z_1 = r_1 \operatorname{cis} \theta_1$ and $z_2 = r_2 \operatorname{cis} \theta_2$ be nonzero complex numbers. Then

$$z_1 z_2 = r_1 r_2 \operatorname{cis}(\theta_1 + \theta_2) \quad \text{and} \quad \frac{z_1}{z_2} = \frac{r_1}{r_2} \operatorname{cis}(\theta_1 - \theta_2)$$

To prove the product rule, multiply and use the addition laws.

$$z_1 z_2 = (r_1 \operatorname{cis} \theta_1)(r_2 \operatorname{cis} \theta_2)$$

$$= [r_1(\cos \theta_1 + i \sin \theta_1)][r_2(\cos \theta_2 + i \sin \theta_2)]$$

$$= r_1 r_2(\cos \theta_1 + i \sin \theta_1)(\cos \theta_2 + i \sin \theta_2)$$

$$= r_1 r_2(\cos \theta_1 \cos \theta_2 + i \cos \theta_1 \sin \theta_2 + i \sin \theta_1 \cos \theta_2 + i^2 \sin \theta_1 \sin \theta_2)$$

$$= r_1 r_2 [(\cos\theta_1 \cos\theta_2 - \sin\theta_1 \sin\theta_2) + i(\cos\theta_1 \sin\theta_2 + \sin\theta_1 \cos\theta_2)]$$

$$= r_1 r_2 [\cos(\theta_1 + \theta_2) + i\sin(\theta_1 + \theta_2)]$$

$$= r_1 r_2 \operatorname{cis}(\theta_1 + \theta_2)$$

The proof of the quotient form is similar and is left as a problem.

EXAMPLE 4   Simplifying expressions in complex form

Simplify:

a. $(5\operatorname{cis}38°)(4\operatorname{cis}75°)$ **b.** $\left(\sqrt{2}\operatorname{cis}188°\right)\left(2\sqrt{2}\operatorname{cis}310°\right)$

c. $(2\operatorname{cis}48°)^3$ **d.** $\dfrac{15(\cos48° + i\sin48°)}{5(\cos125° + i\sin125°)}$

Solution

a. $(5\operatorname{cis}38°)(4\operatorname{cis}75°) = 5 \cdot 4 \operatorname{cis}(38° + 75°)$

$$= 20\operatorname{cis}113°$$

b. $\left(\sqrt{2}\operatorname{cis}188°\right)\left(2\sqrt{2}\operatorname{cis}310°\right) = 2 \cdot 2 \operatorname{cis}(188° + 310°)$

$$= 4\operatorname{cis}498° \qquad \text{Arguments should be between } 0° \text{ and } 360°$$

$$= 4\operatorname{cis}138°$$

c. $(2\operatorname{cis}48°)^3 = (2\operatorname{cis}48°)(2\operatorname{cis}48°)^2$

$$= (2\operatorname{cis}48°)(4\operatorname{cis}96°)$$

$$= 8\operatorname{cis}144°$$

d. $\dfrac{15(\cos48° + i\sin48°)}{5(\cos125° + i\sin125°)} = \dfrac{15\operatorname{cis}48°}{5\operatorname{cis}125°}$

$$= 3\operatorname{cis}(48° - 125°)$$

$$= 3\operatorname{cis}(-77°)$$

$$= 3\operatorname{cis}283°$$

EXAMPLE 5   Simplifying a power

Simplify (expand) the expression $\left(1 - \sqrt{3}i\right)^5$.

Solution

First change to trigonometric form: $a = 1$, $b = -\sqrt{3}$; Quadrant IV

$r = \sqrt{1 + 3} = 2$ and $\overline{\theta} = \tan^{-1}\left|-\frac{\sqrt{3}}{1}\right| = 60°$; in Quadrant IV, $\theta = 300°$. Thus,

$$\left(1 - \sqrt{3}i\right)^5 = (2\operatorname{cis}300°)^5$$

$$= (2\operatorname{cis}300°)(2\operatorname{cis}300°)(2\operatorname{cis}300°)(2\operatorname{cis}300°)(2\operatorname{cis}300°)$$

$$= 2 \cdot 2 \cdot 2 \cdot 2 \cdot 2 \operatorname{cis}(300° + 300° + 300° + 300° + 300°)$$

$$= 2^5 \operatorname{cis}(5 \cdot 300°)$$

$$= 32\operatorname{cis}1{,}500°$$

$$= 32\operatorname{cis}60°$$

If we want the answer in rectangular form, we can now change back:

$$32\operatorname{cis}60° = 32\cos60° + i(32\sin60°)$$

$$= 16 + 16\sqrt{3}i$$

Powers of Complex Numbers—de Moivre's Theorem

As you can see from Example 5, multiplication in trigonometric form extends quite nicely to any positive integral power in a result called **de Moivre's theorem**. This theorem is proved by mathematical induction (see Appendix C).

> **DE MOIVRE'S THEOREM**
>
> If n is a natural number, then
>
> $$(r\operatorname{cis}\theta)^n = r^n \operatorname{cis} n\theta$$
>
> for a complex number $r\operatorname{cis}\theta = r(\cos\theta + i\sin\theta)$

Historical Note

Abraham de Moivre (1667–1754)

The great French mathematician Abraham de Moivre was born in France to a Protestant family at a time of serious religious persecution by Catholics at a time when Protestants were imprisoned for their beliefs. After his release from prison in 1688, he traveled to England and became a private tutor of mathematics, teaching out of the coffeehouses of London. In 1718, he was appointed to the commission set up by the Royal Society to review the controversy about the invention of calculus involving Leibniz and Newton. The society favored Newton, and de Moivre was appointed because of his friendship with Newton. The result, which today is called de Moivre's formula, appears in a paper de Moivre published in 1722 but is closely related to some work he published in 1707. Historian Howard Eves tells of a fable regarding de Moivre's death. According to the story, de Moivre noticed that each day he required a bit more sleep than on the previous day. When this progression reached twenty-four hours, de Moivre thought he would die. When it did, he did!

Although de Moivre's theorem is useful for powers, as illustrated by Example 5, its real usefulness is in finding the complex roots of numbers. Recall from algebra that $\sqrt[n]{r} = r^{1/n}$ is used to denote the principal nth root of r. However, $r^{1/n}$ is only *one* of the nth roots of r. How do you find *all* the nth roots of r? To find the principal root, you can use a calculator or logarithms along with the following theorem. The following **nth root theorem** follows directly from de Moivre's theorem.

> **nTH ROOT THEOREM**
>
> If n is any positive integer, then the nth roots of $r\operatorname{cis}\theta$ are given by
>
> $$\sqrt[n]{r}\operatorname{cis}\left(\frac{\theta + 360°k}{n}\right) \text{ or } \sqrt[n]{r}\operatorname{cis}\left(\frac{\theta + 2\pi k}{n}\right)$$
>
> for $k = 0, 1, 2, 3, \ldots, n-1$.

If $\theta = 0°$, the real root is positive and it is called the **principal nth root**. If $\theta = 180°$, the **principal nth root** is the negative real root.* Notice that the principal nth root of a positive or negative real number is the root that you obtain from your calculator. The proof of this theorem is left as a problem.

EXAMPLE 6   Finding the square roots of a number

Find the square roots of $-\frac{9}{2} + \frac{9}{2}\sqrt{3}i$.

Solution

First, change to trigonometric form.

$$r = \sqrt{\left(-\frac{9}{2}\right)^2 + \left(\frac{9}{2}\sqrt{3}\right)^2} = 9$$

$\overline{\theta} = \tan^{-1}\left|\dfrac{\frac{9}{2}\sqrt{3}}{-\frac{9}{2}}\right| = \tan^{-1}\sqrt{3} = 60°$; in Quadrant II, $\theta = 120°$. The square roots are

$$9^{1/2}\operatorname{cis}\left(\frac{120° + 360°k}{2}\right) = 3\operatorname{cis}(60° + 180°k)$$

$k = 0$: $3\operatorname{cis}60° = \dfrac{3}{2} + \dfrac{3}{2}\sqrt{3}i$

$k = 1$: $3\operatorname{cis}240° = -\dfrac{3}{2} - \dfrac{3}{2}\sqrt{3}i$

Check: $\left(\dfrac{3}{2} + \dfrac{3}{2}\sqrt{3}i\right)^2 = \dfrac{9}{4} + \dfrac{9}{2}\sqrt{3}i + \dfrac{9}{4}\cdot 3i^2 = -\dfrac{9}{2} + \dfrac{9}{2}\sqrt{3}i$

$\left(-\dfrac{3}{2} - \dfrac{3}{2}\sqrt{3}i\right)^2 = \dfrac{9}{4} + \dfrac{9}{2}\sqrt{3}i + \dfrac{9}{4}\cdot 3i^2 = -\dfrac{9}{2} + \dfrac{9}{2}\sqrt{3}i$

■

---

*Not all numbers have principal roots. Look at the following three examples; the first one has no principal roots, the second has a positive principal root ($\theta = 0°$), and the third has a negative principal root ($\theta = 180°$).

EXAMPLE 7 Finding the fifth roots of a number

Find the fifth roots of 32.

Solution

Begin by writing 32 in trigonometric form: $32 = 32 \operatorname{cis} 0°$. The fifth roots are found by

$$32^{1/5} \operatorname{cis}\left(\frac{0° + 360°k}{5}\right) = 2 \operatorname{cis} 72°k$$

$k = 0$: $2 \operatorname{cis} 0° = 2$ This is the principal *n*th root.
$k = 1$: $2 \operatorname{cis} 72° \approx 0.6180 + 1.90211i$
$k = 2$: $2 \operatorname{cis} 144° \approx -1.6180 + 1.1756i$
$k = 3$: $2 \operatorname{cis} 216° \approx -1.6180 - 1.1756i$
$k = 4$: $2 \operatorname{cis} 288° \approx 0.6180 - 1.9021i$

All other integral values for k repeat those listed here.   ■

If all the fifth roots of 32 are represented graphically, as shown in Figure 6.27, they all lie on a circle with radius 2 and are equally spaced.

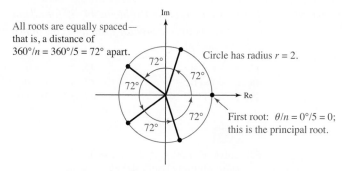

All roots are equally spaced— that is, a distance of $360°/n = 360°/5 = 72°$ apart.

Circle has radius $r = 2$.

First root: $\theta/n = 0°/5 = 0$; this is the principal root.

Figure 6.27 Graphical representation of the fifth roots of 32

If n is a positive integer, the nth roots of a complex number $a + bi = r \operatorname{cis} \theta$ are equally spaced on the circle of radius r centered at the origin. We return to the equation from the beginning of this section, this time using de Moivre's formula to solve.

EXAMPLE 8 Using de Moivre's formula to solve an equation

Solve $x^3 + 1 = 0$.

Solution

$x^3 = -1$, so we are looking for the cube roots of -1. Since $-1 = \operatorname{cis} 180°$, we have

$$1^{1/3} \operatorname{cis} \frac{180° + 360°k}{3} = \operatorname{cis}(60° + 120°k)$$

$k = 0$: $\operatorname{cis} 60° = \dfrac{1}{2} + \dfrac{\sqrt{3}}{2}i$

$k = 1$: $\operatorname{cis} 180° = -1$ This is the principal *n*th root.

$k = 2$: $\operatorname{cis} 300° = \dfrac{1}{2} - \dfrac{\sqrt{3}}{2}i$

Thus, the roots of $x^3 + 1 = 0$ are $\dfrac{1}{2} \pm \dfrac{\sqrt{3}}{2}i, -1$.   ■

PROBLEM SET 6.5

WHAT IS WRONG, *if anything, with each statement in Problems 1–6? Explain your reasoning.*

1. $\operatorname{cis} 90°$ does not exist because $\tan 90°$ is undefined.
2. All negative real numbers have $\theta = 180°$.
3. If a point is on the coordinate axis, then θ is $0°$, $90°$, $180°$, or $270°$.
4. If $\theta = 90°$, then the complex number is plotted on the negative imaginary axis.
5. $2 - 3i$ is a point in Quadrant II.
6. $4 \operatorname{cis} 250°$ is a point in Quadrant IV.

Plot the complex numbers in Problems 7–10, and find the modulus of each.

7. **a.** $3 + i$ **b.** $7 - i$ **c.** $3 + 2i$
8. **a.** $-3 - 2i$ **b.** $2 + 4i$ **c.** $5 + 6i$
9. **a.** $-2 + 5i$ **b.** $-5 + 4i$ **c.** $-3 - 2i$
10. **a.** $2 - 5i$ **b.** $4 - i$ **c.** $-1 + i$

Plot each number given in Problems 11–18, and then change to trigonometric form.

11. **a.** $1 + i$ **b.** $1 - i$ **c.** $\sqrt{3} - i$
12. **a.** $\sqrt{3} + i$ **b.** $1 - \sqrt{3}i$ **c.** $-1 - \sqrt{3}i$
13. **a.** -1 **b.** 2 **c.** -3
14. **a.** $2i$ **b.** $-4i$ **c.** $3i$
15. **a.** $-5i$ **b.** $-6i$ **c.** $-7i$
16. **a.** -5 **b.** -6 **c.** -7
17. **a.** $5.7956 - 1.5529i$ **b.** $1.5321 - 1.2856i$
18. **a.** $-0.6946 + 3.9392i$ **b.** $-2.0404 - 4.5647i$

Plot each number given in Problems 19–22, and then change to rectangular form. Use exact values whenever possible.

19. **a.** $2(\cos 45° + i \sin 45°)$ **b.** $\operatorname{cis} \dfrac{5\pi}{6}$

20. **a.** $3(\cos 60° + i \sin 60°)$ **b.** $5 \operatorname{cis} \dfrac{3\pi}{2}$

21. **a.** $6 \operatorname{cis} 247°$ **b.** $9 \operatorname{cis} 190°$
22. **a.** $2.5 \operatorname{cis} 300°$ **b.** $4.2 \operatorname{cis} 135°$

Perform the operations indicated in Problems 23–29.

23. **a.** $(2 \operatorname{cis} 60°)(3 \operatorname{cis} 150°)$ **b.** $(3 \operatorname{cis} 48°)(5 \operatorname{cis} 92°)$
24. **a.** $4(\cos 65° + i \sin 65°) \cdot 12(\cos 87° + i \sin 87°)$

  b. $\dfrac{5(\cos 315° + i \sin 315°)}{2(\cos 48° + i \sin 48°)}$

25. **a.** $(4 \operatorname{cis} 240°)(2 \operatorname{cis} 330°)$ **b.** $\dfrac{6 \operatorname{cis} 100°}{2 \operatorname{cis} 300°}$

26. **a.** $(\cos 210° + i \sin 210°)^5$ **b.** $(\cos 180° + i \sin 180°)^4$

27. **a.** $\dfrac{8 \operatorname{cis} 30°}{4 \operatorname{cis} 15°}$ **b.** $\dfrac{12 \operatorname{cis} 250°}{4 \operatorname{cis} 120°}$

28. **a.** $(2 \operatorname{cis} 50°)^3$ **b.** $(3 \operatorname{cis} 60°)^4$
29. **a.** $(1 + i)^6$ **b.** $\left(\sqrt{3} - i\right)^8$

Find the indicated roots of the given numbers in Problems 30–37. Leave your answers in trigonometric form.

30. square roots of $16 \operatorname{cis} 100°$ 31. fourth roots of $81 \operatorname{cis} 88°$
32. cube roots of $64 \operatorname{cis} 216°$ 33. cube roots of 27
34. cube roots of 8 35. fourth roots of $1 + i$
36. cube roots of $8 \operatorname{cis} 240°$ 37. fifth roots of $32 \operatorname{cis} 200°$

Find the indicated roots in Problems 38–43. Leave your answers in rectangular form. Show the roots graphically.

38. cube roots of 1 39. cube roots of -8
40. fourth roots of 1 41. cube roots of $4\sqrt{3} - 4i$
42. square roots of $\dfrac{25}{2} - \dfrac{25\sqrt{3}}{2}i$
43. fourth roots of $12.2567 + 10.2846$

Solve the equations in Problems 44–49 for all complex roots. Leave your answers in rectangular form and give approximate answers to four decimal places.

44. $x^3 - 1 = 0$ 45. $x^5 - 1 = 0$
46. $x^5 + 1 = 0$ 47. $x^6 - 1 = 0$
48. $x^4 + 16 = 0$ 49. $x^4 - 16 = 0$

50. Solve $x^4 + x^3 + x^2 + x + 1 = 0$.
    Hint: Multiply both sides by $(x - 1)$.
51. Solve $x^5 + x^4 + x^3 + x^2 + x + 1 = 0$.
    Hint: Multiply both sides by $(x - 1)$.
52. Find the cube roots of $\left(4\sqrt{2} + 4\sqrt{2}i\right)^2$.

53. Find the fifth roots of $\left(-16 + 16\sqrt{3}i\right)^3$.

PROBLEMS FROM CALCULUS *Problems 54–56 are found in most calculus books.*

54. Prove the *triangle inequality:*

$$\left| z_1 + z_2 \right| \le \left| z_1 \right| + \left| z_2 \right|$$

  for complex numbers $z_1 = a + bi$ and $z_2 = c + di$.
55. In calculus, it is shown that

$$e^{i\theta} = \cos \theta + i \sin \theta$$

  which is one of the most remarkable formulas in all of mathematics. This means, among other things, that

$$re^{i\theta} = r \operatorname{cis} \theta$$

  a. Explain the meaning for the following graffiti:

  We Are Number $-e^{-i\pi}$

  b. Plot $e^{i\pi/4}$.   **c.** Plot $2e^{-i\pi/6}$.   **d.** Plot $-e^{-i\pi}$.

56. If

$$\cos\theta = 1 - \frac{\theta^2}{2!} + \frac{\theta^4}{4!} - \frac{\theta^6}{6!} + \cdots + \frac{(-1)^n\theta^{2n}}{(2n)!} + \cdots$$

and

$$\sin\theta = \theta - \frac{\theta^3}{3!} + \frac{\theta^5}{5!} - \frac{\theta^7}{7!} + \cdots + \frac{(-1)^n\theta^{2n+1}}{(2n+1)!} + \cdots$$

and

$$e^{i\theta} = 1 + (i\theta) + \frac{(i\theta)^2}{2!} + \frac{(i\theta)^3}{3!} + \frac{(i\theta)^4}{4!} + \cdots + \frac{(i\theta)^n}{n!} + \cdots$$

show that $e^{i\theta} = \cos\theta + i\sin\theta$ This equation is called *Euler's formula.*

57. Complex numbers are used in the study of electrical currents. For example, the alternating current, in amperes, in an electric inductor is

$$I = \frac{E}{Z}$$

amperes, where is E the voltage and Z is the impedance. Suppose $E = 16\operatorname{cis}75°$ and $Z = 10 + 5i$. Find I, the current in amperes, in rectangular form (rounded to two decimal places).

58. Complex numbers are used in the study of electrical currents. For example, the alternating current, in amperes, in an electric inductor is

$$I = \frac{E}{Z}$$

amperes, where E is the voltage and Z is the impedance. Suppose $E = 12\operatorname{cis}27°$ and $Z = 3 + 2i$. Find I, the current in amperes, in rectangular form (rounded to two decimal places).

59. Prove that

$$\frac{r_1\operatorname{cis}\theta_1}{r_2\operatorname{cis}\theta_2} = \frac{r_1}{r_2}\operatorname{cis}(\theta_1 - \theta_2)$$

60. Prove that

$$\sqrt[n]{r}\operatorname{cis}\left(\frac{\theta + 360°k}{n}\right) = (r\operatorname{cis}\theta)^{1/n}$$

CHAPTER 6 SUMMARY AND REVIEW

C*alculators can only calculate—they cannot do mathematics.*

John A. Van de Walle

Take some time getting ready to work the review problems in this section. First, look back at the definition and property boxes. You will maximize your understanding of this chapter by working the problems in this section only after you have studied the material.

SELF TEST   *All of the answers for this self test are given in the back of the book.*

1. Evaluate using exact values.

a. $\dfrac{2\tan\frac{\pi}{6}}{1 - \tan^2\frac{\pi}{6}}$

b. $-\sqrt{\dfrac{1 + \cos 240°}{2}}$

2. a. If $\cos\theta = \frac{4}{5}$ and 2θ is in Quadrant IV, find the exact value of $\sin 2\theta$.

b. If $\cot 2\theta = -\frac{4}{3}$, find the exact value of $\tan\theta$ (θ is in Quadrant I).

3. a. Find the exact value of $\sin 105°$.

b. Write $\sin(x + h) - \sin x$ as a product.

4. a. Graph $y - 1 = 2\sin\left(\frac{\pi}{6} - \theta\right)$.

b. Graph $y + 1 = 2\cos(2 - \theta)$.

5. Solve for $0 \le \theta < 2\pi$.

a. $3\tan 2\theta - \sqrt{3} = 0$

b. $\sin^2\theta + 2\cos\theta = 1 + 3\cos\theta + \sin^2\theta$

6. Solve for $0 \le \theta < 2\pi$.

a. $3\cos^2\theta = 1 + \cos\theta$ **b.** $\tan^2 2\theta = 3\tan 2\theta$

Prove the identities in Problems 7–9.

7. a. $\dfrac{\csc^2\alpha}{1 + \cot^2\alpha} = 1$

b. $\dfrac{1 + \tan^2\theta}{\csc\theta} = \sec\theta\tan\theta$

8. a. $\dfrac{\sin^2\beta - \cos^2\beta}{\sin\beta + \cos\beta} = \sin\beta - \cos\beta$

b. $\cos^2\gamma\tan\gamma\csc\gamma\sec\gamma = 1$

9. a. $\dfrac{\sin 5\theta + \sin 3\theta}{\cos 5\theta - \cos 3\theta} = -\cot\theta$

b. $\cos(\alpha - \beta) = \cos\alpha\cos\beta + \sin\alpha\sin\beta$

10. a. Find the square roots of $\frac{7}{2}\sqrt{3} - \frac{7}{2}i$. Leave your answer in rectangular form.

b. Find the cube roots of i. Leave your answer in trigonometric form.

 STUDY HINTS *Compare your solutions and answers to the self test.*

Evaluate *each expression in Problems 1–4.*

1. a. $\sin\dfrac{\pi}{6}$ **b.** $\cos\dfrac{11\pi}{6}$

   c. $\sin\dfrac{11\pi}{6}$ **d.** $\tan\dfrac{5\pi}{6}$

   e. $\cos\left(-\dfrac{5\pi}{6}\right)$

2. $\cos 345°$ **3.** $-\sqrt{\dfrac{1+\cos 270°}{2}}$

4. $\sin 13° \cos 32° + \cos 13° \sin 32°$

Simplify *each expression in Problems 5–8.*

5. $\dfrac{1}{2}\left(\dfrac{x+1}{x-2}\right)^{-1/2}\left[\dfrac{(x-2)(1)-(x+1)(1)}{(x-2)^2}\right]$

6. $\dfrac{1}{2}\left(\dfrac{x-3}{x+2}\right)^{-1/2}\left[\dfrac{(x+2)(1)-(x-3)(1)}{(x+2)^2}\right]$

7. $\dfrac{1}{2}\left(\dfrac{2x-1}{x-1}\right)^{-1/2}\left[\dfrac{(x-1)(2)-(x+1)(2)}{(x-1)^2}\right]$

8. $\dfrac{1}{2}\left(\dfrac{3x-2}{x+1}\right)^{-1/2}\left[\dfrac{(x+1)(3)-(x+1)(3)}{(x+1)^2}\right]$

Factor *each expression in Problems 9–12.*

9. a. $3x^2 + 2x - 1$ **b.** $3\tan^2 x + 2\tan x - 1$
10. a. $x^3 - 1$ **b.** $\cot^3 x - 1$
11. a. $x^2 - x - 2$ **b.** $\sin^2 x - \sin x - 2$
12. a. $6x^2 + 29x - 5$ **b.** $6\sec^2 x + 29\sec x - 5$

Solve *each equation in Problems 13–22 for the domain* $[0, 2\pi)$.

13. $\cos^2 x - 3\sin x + 3 = 0$ **14.** $\sec^2 x - 1 = \sqrt{3}\tan x$
15. $\sin\left(x+\frac{\pi}{3}\right) + \sin\left(x-\frac{\pi}{3}\right) = 1$ **16.** $\sin\left(x+\frac{\pi}{6}\right) - \sin\left(x-\frac{\pi}{6}\right) = \frac{1}{2}$
17. $\ln(1+\cos\theta) + \ln(1-\cos\theta) = \ln(\sin^2\theta)$
18. $2^{\sin^2\theta} \cdot 2^{\cos^2\theta} = 2$ **19.** $\csc x(\ln e^{\sin x}) = 1$
20. $\dfrac{10^{\log(\cos x)}}{10^{\log(1-\sin x)}} = \sec x + \tan x$
21. $\sin x + \sin 3x = 0$ **22.** $\cos x = \sin 2x$

Use the trigonometric substitution $u = a\sin\theta$ *on* $\left[-\frac{\pi}{2}, \frac{\pi}{2}\right]$ *in Problems 23–26 to rewrite the algebraic expression in terms of θ and simplify.*

23. $\dfrac{1}{\sqrt{36-u^2}}$ **24.** $\dfrac{u}{\sqrt{9-u^2}}$

25. $\dfrac{(4-u^2)^2}{u^4}$ **26.** $\dfrac{\sqrt{49-u^2}}{u}$

Use the trigonometric substitution $u = a\tan\theta$ *on* $\left(-\frac{\pi}{2}, \frac{\pi}{2}\right)$ *in Problems 27–30 to rewrite the algebraic expression in terms of θ and simplify.*

27. $\sqrt{u^2+100}$ **28.** $\dfrac{1}{\sqrt{u^2+1}}$

29. $\dfrac{u^2}{u^2+9}$ **30.** $u^3(4+u^2)^{-3/2}$

Use the trigonometric substitution $u = a\sec\theta$ *on* $\left(0, \frac{\pi}{2}\right)$, $a>0$, *in Problems 31–34 to rewrite the algebraic expression in terms of θ and simplify.*

31. $\sqrt{u^2-64}$ **32.** $\dfrac{1}{(u^2-81)^2}$

33. $\dfrac{1}{u\sqrt{u^2-1}}$ **34.** $u^5(u^2-4)^{-3/2}$

The following list includes many of the equations that you will use in calculus. Following this list are **graphs** *numbered 35–60. Match each graph with an equation from this list.*

$y = Ax$	$y = Ax^2 + B$	$y = A\sin Bx$	$y = A\cos Bx$
$y = Ax + B$	$y = Ax^2 + x + B$	$y = A\sec Bx$	$y = A\csc Bx$
$y = Ax^2 + Bx$	$y = Ax^4 + B$	$y = A\tan Bx$	$y = A\cot Bx$
$y = Ax^3 + B$	$y = Ax^4 + Bx$	$y = A\sin B(x-2)$	$y = A\sin B(x+2)$
$y = Ax^4 + Bx^2$	$y = Ae^{Bx}$	$y = A\cos Bx - 4$	$y = A\cos Bx + 4$
$y = A\log B$	$y = A\ln B$	$y = \log_2 B$	$y = \dfrac{x+B}{x-A}$
$y = Ax^{1/3} + B$	$y = A(x-2)(x-3)(x+4) + B$		

35.

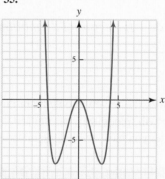

36.

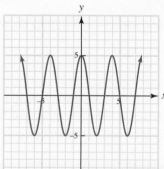

37.

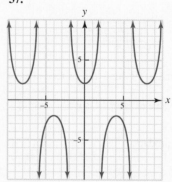

38.

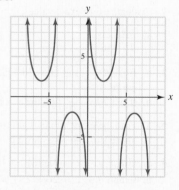

39.

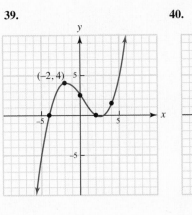

40.

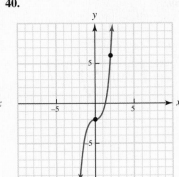

47.

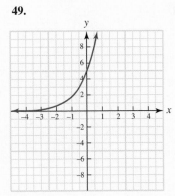

48.

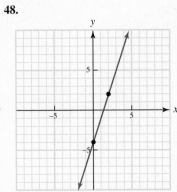

41.

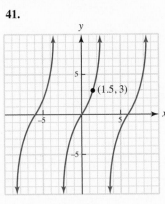

42.

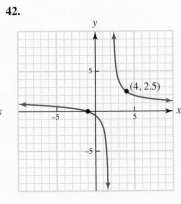

49.

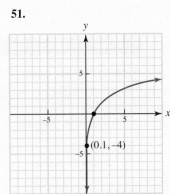

50.

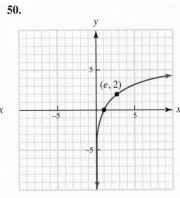

43.

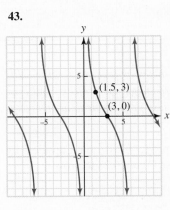

44.

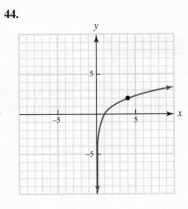

51.

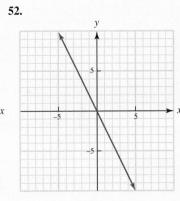

52.

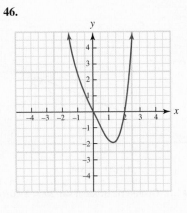

45.

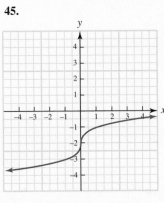

46.

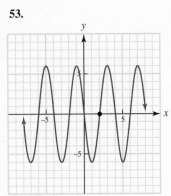

53.

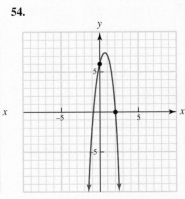

54.

55.

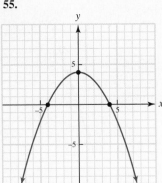

56.

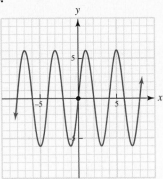

59.

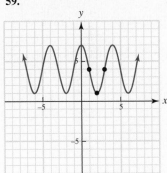

60.

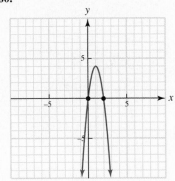

57.

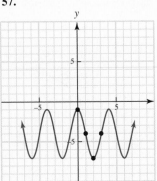

58.

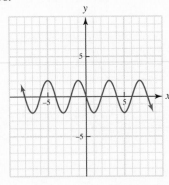

CHAPTER 6   Group Research Projects

Working in small groups is typical of most work environments, and this book seeks to develop skills with group activities. At the end of each chapter, we present a list of suggested projects, and even though they could be done as individual projects, we suggest that these projects be done in groups of three or four students.

G6.1 Journal Problem (*College Mathematics Journal*, January 1989, p. 51, "Classroom Capsules" by Roger B. Nelsen, Lewis and Clark College, Portland, OR). Use Figure 6.28 to prove

$$\sin 2\theta = 2\sin\theta\cos\theta$$

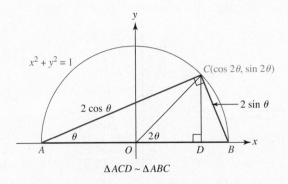

Figure 6.28 Geometric proof of $\sin 2\theta = 2\sin\theta\cos\theta$

G6.2 Use Figure 6.28 to prove

$$\cos 2\theta = 2\cos^2\theta - 1$$

G6.3 What are the precise times when the minute hand of an ordinary clock crosses the hour hand?

G6.4 Two dimes touch the side of a rectangle at the same point, but one is on the inside and the other is on the outside. (See Figure 6.29.) The coins are rolled in the plane along the perimeter of the rectangle until they come back to their initial positions. The height of the rectangle is twice the circumference of the dimes and its width is twice its height. How many revolutions will each coin make?

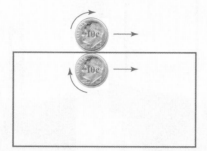

Figure 6.29 The sliding dimes

G6.5 Parking Lot Design A city engineer must design the parking lot along a certain street. She can use parallel or angle parking. Angle parking requires greater street width, as shown in Figure 6.30, but it allows more spaces along the curb. Let us assume that city code calls for a 10-ft width per parking space. This means that 100 ft of curb could accommodate ten parking spaces if parking is at 90°.

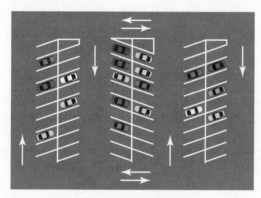

Figure 6.30 Parking lot

   a. Show that changing the angle from 90° to 45° increases the amount of curb space needed by 40%.

   b. Find the street width as a function of parking-space angle.

   c. If street size were of primary importance, what type of parking would you recommend? Write a report of your findings.

   d. If curb space were of primary importance, what type of parking would you recommend? Write a report of your findings.

G6.6 Historical Quest (*Mathematics Teacher,* March 1982, pp. 246–249.)

Clark Kimberling wrote in an article in the *Mathematics Teacher* that he talked to many people who knew Amalie Noether, and they were quick to mention two things about her. "One is that Noether's students seemed always to be flocked around her, following her as if following the Pied Piper, and the other is that she very seldom wrote or said anything that was not mathematical." Write a short biographical paper on Amalie Noether.

G6.7 The displacement of a particle that can be expressed in terms of equations of the form

$$y = a\cos(\omega t + h) \quad \text{or} \quad y = a\sin(\omega t + h)$$

where a, ω, and h are constants, is said to be in **simple harmonic motion**. However, many moving bodies do not move back and forth between precisely fixed limits because friction slows down their motion. Such motions are called **damped harmonic motions**.

a. Man on bridge

b. Bridge oscillates

c. Bridge collapses

Figure 6.31 Collapse of the Tacoma Narrows Bridge at Puget Sound, Washington

The period T of a harmonic motion is the time required to complete one round trip of the motion, and the frequency n is the number of oscillations or cycles per unit of time so that

$$n = \frac{1}{T} = \frac{\omega}{2\pi}$$

The effect of damping can be seen by looking at the graphs shown in Figure 6.32.

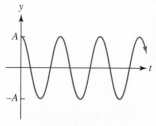
a. Simple harmonic motion

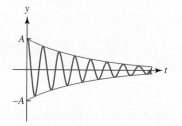
b. Damped harmonic motion

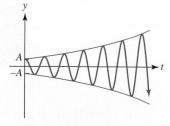
c. Resonance

Figure 6.32 Harmonic motion

Notice in Figure 6.32**b** that, for damped harmonic motion, the amplitude decreases toward zero. The reverse situation, when the amplitude increases to a maximum, is called **resonance** and has a graph as shown in Figure 6.32**c**. A striking example of the destructive force of resonance is the collapse of the Tacoma Narrows Bridge, which was opened on July 1, 1940. Early on the morning of November 7, 1940, the bridge began to vibrate torsionally in two segments with frequency 14 vib/min, with the main span eventually breaking. Write a paper on simple and damped harmonic motion and resonance.

G6.8 PROBLEMS FROM CALCULUS In calculus, it is shown that if a particle is moving in simple harmonic motion according to the equation

$$y = a\cos(\omega x + t)$$

then the velocity of the particle at any time x is given by

$$v = a\omega \sin(\omega x + t)$$

a. Graph this velocity curve, where the frequency is $\frac{1}{4}$ cycle per unit of time, and $t = \frac{\pi}{6}$, $a = \frac{120}{\pi}$.

b. Find a maximum value and a minimum value for the velocity.

Graph each of the given curves for $t \geq 0$, and classify each motion as harmonic motion, damped harmonic motion, resonance, neither, or both.

c. $y = 3\cos 4\pi x$

d. $y = 3x \sin 4\pi x$

e. $y = \frac{3}{x}\cos 3\pi x$

f. $y = \cos x + x \sin x$

g. $y = \sin x - x\cos x$

h. $y = 2\sin\left(5x - \frac{\pi}{2}\right) + \sin x$

i. $y = \frac{9}{4}\cos\frac{9}{4}x + \cos\frac{1}{4}x$

j. $y = \frac{13}{5}\cos x + \sin x - \frac{8}{5}\cos\frac{3}{2}x$

G6.9 Draw the unit circle $x^2 + y^2 = 1$. Draw the two lines passing through $(2, 3)$ that are tangent to the unit circle, as shown in Figure 6.33. What is the equation of the line passing through the points of tangency?

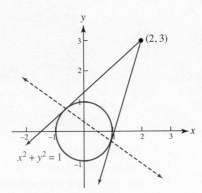

Figure 6.33 Equation of a line passing through the points of tangency

G6.10 Journal Problem (*Quantum*, July/August, 1995, p. 30, "Circuits and Symmetry," by Gary Haardeng-Pedersen). Consider the arrangement of the 12 identical resistors that make up the edges of a cube. What is the equivalent resistance between two corners that are diagonally opposite each other as shown in Figure 6.34?

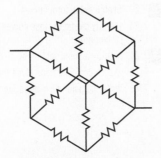

Figure 6.34 Electrical circuit

This question comes from the article. Read this article, and then answer this question. The article concludes with the following advice: "So—look for symmetry in physical situations. When you find it, use it to reduce the amount of algebra you need to solve the problem."

Chapter Objectives

The material in this chapter is reviewed in the following list of objectives. After completing this chapter, review this list again, and then complete the self test.

7.1 7.1 Graph parabolas.

  7.2 Find the equations of parabolas, given certain information about the graph.

  7.3 Parameterize parabolas.

  7.4 Find the tangent to a parabola at a given point.

  7.5 Solve applied problems involving parabolas.

7.2 7.6 Graph ellipses.

  7.7 Find the equations of ellipses, given certain information about the graph.

  7.8 Parameterize ellipses.

  7.9 Solve applied problems involving ellipses.

7.3 7.10 Graph hyperbolas.

  7.11 Find the equations of hyperbolas, given certain information about the graph.

  7.12 Parameterize hyperbolas.

  7.13 Solve applied problems involving hyperbolas.

  7.14 Describe the reflective property of parabolas, ellipses, and hyperbolas.

  7.15 Describe eccentricity for parabolas, ellipses, and hyperbolas.

  7.16 Know the definition and standard-form equations for the conic sections.

  7.17 Graph conic sections. Name the type of curve (by inspection) by looking at the graph of the curve.

7.4 7.18 Use the rotation of axes formulas and the amount of rotation formula to sketch rotated conic sections.

  7.19 Identify the rotated conic section by inspection of the equation.

7.5 7.20 Plot points in polar form and give both primary representations.

  7.21 Change from rectangular to polar form and from polar form to rectangular form.

7.6 7.22 Sketch polar-form curves.

  7.23 Identify cardioids, rose curves, and lemniscates by looking at the equation.

7.7 7.24 Identify a conic section by inspection of the polar-form equation.

  7.25 Sketch conic sections given the polar-form equation.

  7.26 Write the polar-form equation given some information about the curve.

Analytic Geometry

7

*T*he discovery of the conic sections, attributed to Plato, first threw open the higher species of form to the contemplation of geometers. But for this discovery, which was probably regarded in Plato's time and long after him, as the unprofitable amusement of a speculative brain, the whole course of practical philosophy of the present day, of the science of astronomy, of the theory of projectiles, of the art of navigation, might have run in a different channel; and the greatest discovery that has ever been made in the history of the world, the law of universal gravitation, with its innumerable direct and indirect consequences and applications to every department of human research and industry, might never to this hour have been elicited.

—J. J. Sylvester (1908)

Chapter Sections

7.1 Parabolas
Conic Sections
Standard-Form Parabolas with Vertex (0, 0)
Standard-Form Equations of Parabolas
Parabolic Reflectors

7.2 Ellipses
Definition of an Ellipse
Standard-Form Ellipse with Center (0, 0)
Standard-Form Equations of Ellipses
Eccentricity
Geometric Properties

7.3 Hyperbolas
Definition of a Hyperbola
Standard-Form Hyperbola with Center (0, 0)
Standard-Form Equations of Hyperbolas
Properties of Hyperbolas
Conic Section Summary

7.4 Rotations
The Rotation Concept
Rotation of Axes Formulas
Graphing Rotated Conics

7.5 Polar Coordinates
Plotting Points in Polar Coordinates
Primary Representations of a Point
Relationship between Polar and Rectangular Coordinates
Points on a Polar Curve

7.6 Graphing in Polar Coordinates
Graphing by Plotting Points
Cardioids
Symmetry and Rotations
Rose Curves
Lemniscates
Summary of Polar-Form Curves

7.7 Conic Sections in Polar Form
Polar-Form Parabola
Polar-Form Ellipse
Polar-Form Hyperbola
Rotated Conics in Polar Form

Chapter 7 Summary and Review
Self Test
Practice for Calculus
Chapter 7 Group Research Projects

▶CALCULUS PERSPECTIVE

The quotation of J. J. Sylvester (page 429) does not overstate the case for the importance of the conic sections in calculus. The conic sections consist of the circles (already considered), parabolas (those that open upward or downward have been considered), ellipses, and hyperbolas. In this chapter, we consider the standard and general forms of each of these curves.

The Greek mathematician Apollonius (ca. 262–190 B.C.) made a thorough investigation of the *conic sections* discussed in this chapter. His methods were expounded in an eight-volume work called *Conics*. His work was so modern it is sometimes judged to be an analytic geometry preceding the work of René Descartes (see Historical Note on page 20). A rudimentary knowledge of the fundamentals of the conic sections is required for the study of calculus, but beyond their use in more advanced mathematics, the conic sections are among the more common shapes used in architecture and design.

In this chapter, we introduce the conic sections and polar coordinates.

7.1 Parabolas

In Chapter 1, we considered parabolas that are functions. In this chapter, we consider a more general class of curves which include not only parabolas that are functions but also those that are not necessarily functions. The curves studied in this chapter satisfy the following general second-degree equation:

$$Ax^2 + Bxy + Cy^2 + Dx + Ey + F = 0$$

for any constants A, B, C, D, E, and F. This is called the **general form of a conic section**. If $A = B = C = 0$, then the equation is not quadratic but linear (first degree); but if at least one of A, B, or C is not zero, then the equation is quadratic and is known as a **conic section**.

Conic Sections

Today, we analyze the conic sections by looking at the general second-degree equation, but historically, they were studied by looking at the intersections of a double-napped right circular cone and a plane, and hence the name *conic section*. There are three general ways a plane can intersect a cone, as shown in Figure 7.1. There are several other special cases that we shall discuss later.

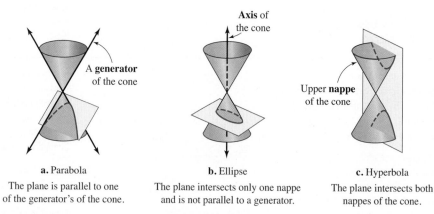

| **a.** Parabola | **b.** Ellipse | **c.** Hyperbola |
| The plane is parallel to one of the generator's of the cone. | The plane intersects only one nappe and is not parallel to a generator. | The plane intersects both nappes of the cone. |

Figure 7.1 Conic sections

Standard-Form Parabolas with Vertex (0, 0)

In Chapter 3, we said that the graph of the quadratic $y = ax^2$ is called a *standard-position parabola*. We now define a parabola, and then we show that the equation $y = ax^2$ is only one of the types of equations that represent parabolas.

> **PARABOLA**
>
> A **parabola** is the set of all points in the plane equidistant from a given point (called the **focus**) and a given line (called the **directrix**).

To see that this definition gives a curve with the same shape as the parabolas we considered in Chapter 3, we use a special type of graph paper, shown in Figure 7.2.

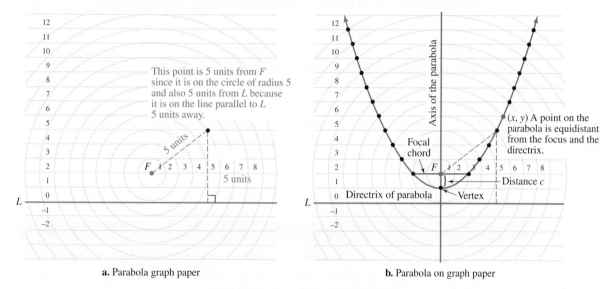

This point is 5 units from F since it is on the circle of radius 5 and also 5 units from L because it is on the line parallel to L 5 units away.

5 units

5 units

a. Parabola graph paper

Axis of the parabola

Focal chord

(x, y) A point on the parabola is equidistant from the focus and the directrix.

Directrix of parabola

Vertex

Distance c

b. Parabola on graph paper

Figure 7.2 Parabola graphed from definition

You can create such graph paper for yourself by starting with a clean sheet of paper. Draw a line and label it L, and then draw a point (not on L) and label it F. Next draw a circle with a given radius (say, 5, as shown in Figure 7.2**a**). Notice that we have drawn several concentric circles of radii 1, 2, 3, ... Finally, draw a line parallel to L and 5 units from L. You will notice in Figure 7.2**a** that we have drawn several such parallel lines, and we have them labeled 1, 2, 3, . . . to designate their distance from L. Mark the two points that are the intersection points of the circle with radius 5 and the line 5 units from L.

To sketch a parabola using the definition, plot points in the plane equidistant from the focus F and the directrix L. Draw a line through the focus and perpendicular to the directrix. This line is called the **axis of the parabola**. Let V be the point on this line halfway between the focus and the directrix. It is called the **vertex of the parabola**. Plot other points equidistant from F and L, as shown in Figure 7.2**b**.

In Figure 7.2**b**, let c be the distance from the vertex to the focus. Notice that the distance from the vertex to the directrix is also c. Consider the segment that passes through the focus perpendicular to the axis and with endpoints on the parabola. This segment has length $4c$ and is called the **focal chord**.

To obtain the equation of a parabola, first consider a special case—a parabola with focus $F(0, c)$ and directrix $y = -c$, where c is any positive number. This parabola must have its vertex at the origin (remember that the vertex is halfway between the focus and the directrix) and must open upward, as shown in Figure 7.3.

☠ This segment called the focal chord is crucial to graphing parabolas in this chapter. ☠

Let (x, y) be any point on the parabola. Then, from the definition of a parabola:

DISTANCE FROM (x, y) TO $(0, c)$ = DISTANCE FROM (x, y) TO DIRECTRIX

$$\sqrt{(x-0)^2 + (y-c)^2} = |y+c| \qquad \text{Distance formulas}$$
$$x^2 + (y-c)^2 = (y+c)^2 \qquad \text{Square both sides.}$$
$$x^2 + y^2 - 2cy + c^2 = y^2 + 2cy + c^2$$
$$x^2 = 4cy$$

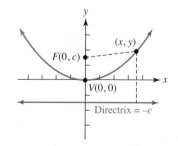

Figure 7.3 Graph of the parabola $x^2 = 4cy$

This is the equation of the parabola with vertex $(0, 0)$ and directrix $y = -c$. (Remember, we chose c to be a positive number.) You can repeat this argument for parabolas that have vertex at the origin and open downward, to the left, and to the right to obtain the results summarized in Table 7.1. A positive number c, the distance from the focus to the vertex, is assumed given. We call these the **standard-form parabola equations** with vertex $(0, 0)$.

» IN OTHER WORDS To sketch the graph of a standard-form parabola, solve the equation for the square term, and look at the side of the equation with the first-degree term:

Coefficient of 1st-degree term is positive (+) or negative (−).

↓

$+y$ means that the parabola opens upward
$-y$ means that the parabola opens downward
$+x$ means that the parabola opens rightward
$-x$ means that the parabola opens leftward

Notice that this is consistent with the directions on a number line:

positive y is up
negative y is down
positive x is right
negative x is left

TABLE 7.1   Standard-Form Parabola Equations with Vertex (0, 0)

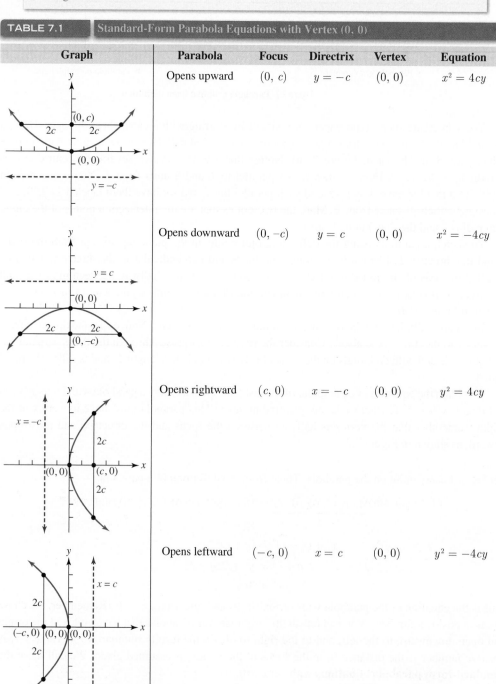

Graph	Parabola	Focus	Directrix	Vertex	Equation
	Opens upward	$(0, c)$	$y = -c$	$(0, 0)$	$x^2 = 4cy$
	Opens downward	$(0, -c)$	$y = c$	$(0, 0)$	$x^2 = -4cy$
	Opens rightward	$(c, 0)$	$x = -c$	$(0, 0)$	$y^2 = 4cy$
	Opens leftward	$(-c, 0)$	$x = c$	$(0, 0)$	$y^2 = -4cy$

We are now ready to sketch a parabola. In Chapter 3, we plotted a couple of points or found the y-intercept. In this chapter, we will use the focal chord to help us with the graph of a parabola. Remember, the *length of the focal chord is* $4c$, so we will plot the vertex, then the focus, and then we will mark the endpoints of the focal chord to determine the width of a parabola as shown in the following example.

EXAMPLE 1   Graphing a parabola that is a function

Graph $x^2 = 8y$.

Solution
This equation represents a parabola that opens upward. The vertex is $(0, 0)$, and we see (by inspection) that

$$4c = 8 \quad \text{Length of focal chord}$$
$$c = 2 \quad \text{Find } c.$$

Since $c = 2$, the focus is at $(0, 2)$. After the vertex $V(0, 0)$ and the focus $F(0, 2)$ are plotted, the only question is the width of the parabola. Since $4c = 8$, the focal chord has length 8, so we draw a segment of length 8 with the midpoint at F as shown in Figure 7.4. Use the vertex and the endpoints of the focal chord to sketch the parabola. ∎

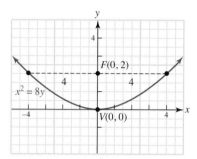

Figure 7.4 Graph of $x^2 = 8y$

EXAMPLE 2   Graphing a parabola that is not a function

Graph $y^2 = -12x$.

Solution
This equation represents a parabola that opens leftward. The vertex is $(0, 0)$, and the length of the focal chord is 12 (found by inspection).

$$4c = 12 \quad \text{Length of focal chord}$$
$$c = 3 \quad \text{Find } c.$$

The focus is 3 units to the *left* of the vertex; thus, the coordinates are $(-3, 0)$. The parabola is shown in Figure 7.5. ∎

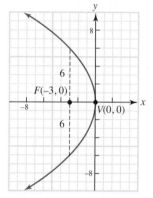

Figure 7.5 Graph of $y^2 = -12$

◼ COMPUTATIONAL WINDOW   ▫ ▢ ✕

It is easy to use a calculator to graph parabolas that are functions (open up/down), but graphing parabolas that are not functions (open right/left) requires some extra steps. Recall from Section 3.4 how we parameterize a curve. If you wish to use a calculator to graph a curve that is not a function, consider parameterizing the equation and using parametric form to graph the curve.

EXAMPLE 3   Parameterize the equation of a parabola

Graph $y^2 = -12x$ by considering a parameterization.

Solution
Let t be the variable that is squared. In this example, we let $y = t$; then

$$x = -\frac{1}{12}t^2 \quad \text{so that} \quad (x, y) = \left(-\frac{1}{12}t^2,\ t\right)$$

You can set up a table of values (as shown in the margin) or use a calculator to obtain the graph. The coefficient of the t^2 term dictates convenient choices for t.

Notice from the table that, because of our choices for t, the scaling numbers you might use for this graph are different from those shown in Figure 7.5 but nevertheless represent the same graph.

t	x	y
0	0	0
12	−12	12
−12	−12	−12
24	−48	24
−24	−48	−24
36	−108	36
−36	−108	−36

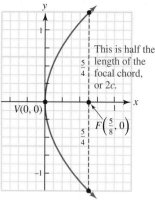

Figure 7.6 Graph of $2y^2 - 5x = 0$

EXAMPLE 4   Graphing with an equation not in standard form

Graph $2y^2 - 5x = 0$.

Solution

This equation is not in standard form; solve for the second-degree term:

$$y^2 = \frac{5}{2}x$$

The vertex is $(0, 0)$, and the focal chord has length $\frac{5}{2}$. We now need to find the focus; solve $4c = \frac{5}{2}$ to find $c = \frac{5}{8}$. Choose a scale to make it easy to plot the focus at $\left(\frac{5}{8}, 0\right)$ and the focal chord of length $\frac{5}{2}$ (so that is $\frac{5}{4}$ above and below F), as shown in Figure 7.6. ■

We will consider two basic types of problems in this chapter, as follows:

1. Given the equation, draw the graph; this is what we have done in Examples 1–4.
2. Given the graph (or information about the graph), write the equation. The following example is of this type.

EXAMPLE 5   Finding the equation of a parabola

Find the equation of the parabola with directrix $y = 4$ and focus at $F(0, -4)$.

Solution

This curve is a parabola that opens downward with vertex at the origin, as shown in Figure 7.7. The value for is found by inspection: $c = 4$. Thus, $4c = 16$. Since the equation is of the form $x^2 = -4cy$, the desired equation is (by substitution)

$$x^2 = -16y$$
■

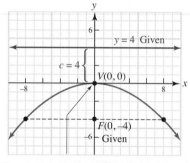

Figure 7.7 Find the equation of this parabola

Standard-Form Equations of Parabolas

The types of parabolas we have been considering are quite limited because we have assumed that the vertex is at the origin and the directrix is parallel to one of the coordinate axes. Suppose, however, that we consider a parabola with vertex at (h, k) and directrix parallel to one of the coordinate axes. In Section 3.2, we found the equation of a parabola that opens upward or downward and with vertex (h, k). We now extend this to parabolas that open right or left, as shown in Table 7.2.

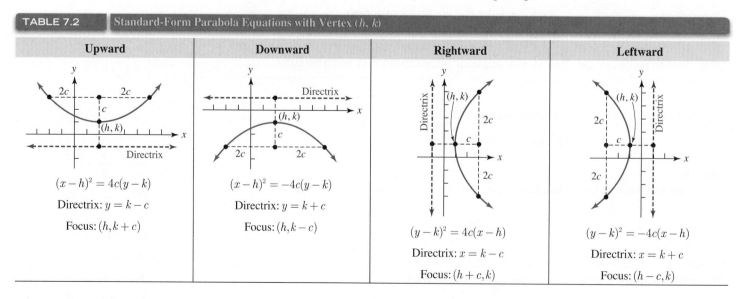

TABLE 7.2	Standard-Form Parabola Equations with Vertex (h, k)

Upward	Downward	Rightward	Leftward
$(x - h)^2 = 4c(y - k)$	$(x - h)^2 = -4c(y - k)$	$(y - k)^2 = 4c(x - h)$	$(y - k)^2 = -4c(x - h)$
Directrix: $y = k - c$	Directrix: $y = k + c$	Directrix: $x = k - c$	Directrix: $x = k + c$
Focus: $(h, k + c)$	Focus: $(h, k - c)$	Focus: $(h + c, k)$	Focus: $(h - c, k)$

To graph parabolas with vertex $(h, \ k)$, we use the idea of a translation introduced in Section 2.4 as summarized in the following box.

GRAPHING PARABOLAS

To graph a parabola, first make sure that one variable is second degree and the other is first degree.

Step 1 Associate together the terms involving the variable that is squared on one side of the equal sign and the other variable and constant on the other side. Note the direction that the parabola opens.

Step 2 Complete the square for the variable that is squared. That is, take one-half the first-degree coefficient, square it, and add it to both sides.

Step 3 Factor the completed square; factor the coefficient of the first-degree term.

Step 4 Determine the vertex by inspection and plot $(h, \ k)$.

Step 5 Determine the focus. That is, find c and plot this distance in the appropriate direction.

Step 6 Plot the endpoints of the focal chord and draw the parabola.

For the rest of the material in this chapter, you will need to remember the procedure for *completing the square*. The next two examples should provide enough detail to refresh your memory; however, if you need additional review, see Section 1.8.

EXAMPLE 6   Completing the square to graph a parabola (*x*-squared)

Sketch $x^2 + 4y + 8x + 4 = 0$.

Solution

We follow the steps in the graphing parabolas procedure.

$$x^2 + 4y + 8x + 4 = 0 \qquad \text{Given equation.}$$

Step 1:
$$x^2 + 8x = -4y - 4 \qquad \text{Negative } y, \text{ so parabola opens down.}$$

Step 2:
$$x^2 + 8x + \left(\frac{1}{2} \cdot 8\right)^2 = -4y - 4 + \left(\frac{1}{2} \cdot 8\right)^2 \qquad \text{Complete the square.}$$

$$x^2 + 8x + 16 = -4y + 12$$

Step 3:
$$(x + 4)^2 = -4(y - 3)$$

Step 4: The vertex is $(-4, \ 3)$. Plot this point.   By inspection

Step 5: $4c = 4$, so $c = 1$. Count out one unit down to find the focus, F.

Step 6: The length of the focal chord is 4 units, so plot the endpoints.
We draw the parabola as shown in Figure 7.8.

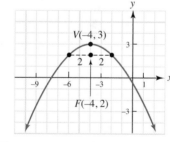

Figure 7.8 Graph of $x^2 + 4y + 8x + 4 = 0$

EXAMPLE 7   Completing the square to graph a parabola (*y*-squared)

Sketch $2y^2 + 6y + 5x + 10 = 0$.

Solution

This time we do not number the steps.

$$2y^2 + 6y + 5x + 10 = 0$$
$$2y^2 + 6y = -5x - 10$$
$$y^2 + 3y = -\frac{5}{2}x - 5$$

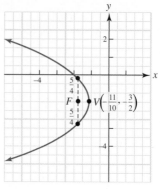

Figure 7.9 Graph of $2y^2 + 6y + 5x + 10 = 0$

$$y^2 + 3y + \left(\frac{3}{2}\right)^2 = -\frac{5}{2}x - 5 + \frac{9}{4}$$

$$\left(y + \frac{3}{2}\right)^2 = -\frac{5}{2}x - \frac{11}{4}$$

$$\left(y + \frac{3}{2}\right)^2 = -\frac{5}{2}\left(x + \frac{11}{10}\right)$$

The vertex is $\left(-\frac{11}{10}, -\frac{3}{2}\right)$, and $4c = \frac{5}{2}$, so $c = \frac{5}{8}$. Sketch the curve as shown in Figure 7.9. ∎

EXAMPLE 8   Equation of a parabola vertex not at origin

Find the equation of the parabola with focus at $(4, -3)$ and directrix the line $x + 2 = 0$.

Solution
Sketch the given information as shown in Figure 7.10.

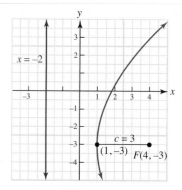

Figure 7.10 Given parabola

The vertex is $(1, -3)$ since this point must be equidistant from F and the directrix. Since the parabola opens to the right, we use the equation with form

$$(y - k)^2 = 4c(x - h)$$

We note $c = 3$ (the distance from the vertex to the directrix or the focus) and substitute this value, as well as $h = 1$ and $k = -3$ (from the vertex), to find

$$(y + 3)^2 = 12(x - 1)$$ ∎

In the next example, we return to a problem first introduced in Section 1.5. It might be interesting to mention that cannons existed nearly three centuries before enough was known about the behavior of a projectile to use them with accuracy. Indeed, even as late as World War II, the paths of projectiles were calculated by trial and error. A projectile does not travel in a straight line, it was discovered, because of the force of gravity. Newton's laws of motion can be used to prove that any projectile—a ball, an arrow, a bullet, a rock from a slingshot, and even water from the nozzle of your hose—travels a parabolic path. In fact, we modeled projectile motion using Newton's laws of motion and parametric equations in Section 5.2.

EXAMPLE 9   Modeling the path of a projectile   MODELING APPLICATION

In Section 1.5, we gathered data points for the path of a cannonball when we let x represent the horizontal distance and y the vertical height: $(64, 80)$, $(128, 128)$, $(192, 144)$, $(256, 128)$, $(320, 80)$, and $(384, 0)$. Find an equation to model this projectile.

Solution

Step 1:   *Understand the problem.* We begin by plotting the points as we did in Section 1.5.

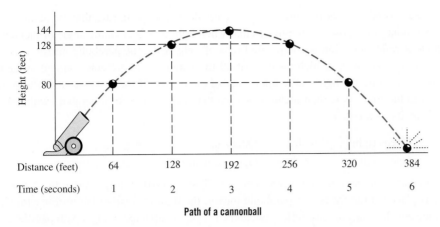

Path of a cannonball

We need to find an equation involving x and y that graphs this curve.

Step 2: *Devise a plan.* We note that the shape of this curve seems to be a parabola that opens down. If so, then the equation has the form $(x - h)^2 = -4c(y - k)$ with vertex (h, k). Our plan is to estimate the coordinates of the vertex and substitute them into this equation.

Step 3: *Carry out the plan.* By inspection, the vertex is $(h, k) = (192, 144)$, so we have

$$(x - h)^2 = -4c(y - k)$$
$$(x - 192)^2 = -4c(y - 144)$$

Since $(0, 0)$ satisfies the equation, we have

$$(0 - 192)^2 = -4c(0 - 144)$$
$$36,864 = 576c$$
$$64 = c$$

Thus, the path of the cannonball can be described by the equation

$$(x - 192)^2 = -256(y - 144)$$

Step 4: *Look back.* We can use a calculator (or graph it by hand) to graph this equation to verify that it passes through the data points. ∎

Parabolic Reflectors*

Parabolic curves are used in the design of lighting systems, telescopes, and radar antennas, mainly because of the reflective property illustrated in Figure 7.11.

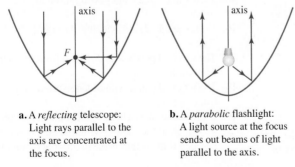

a. A *reflecting* telescope: Light rays parallel to the axis are concentrated at the focus.

b. A *parabolic* flashlight: A light source at the focus sends out beams of light parallel to the axis.

Figure 7.11 The reflective property of parabolas

As an illustration of the reflective property of parabolas, let us examine its application to reflecting telescopes (see Figure 7.11**a**). The eyepiece of such a telescope is placed at a point, F,

*Since these reflectors are three dimensional, the precise word is *paraboloidal*, but the most common usage refers to the cross-sectional shape, which is parabolic.

called the *focus* of a parabolic mirror. Light enters the telescope in rays that are parallel to the axis of the parabola. It is known from physics that when light is reflected, the angle of incidence equals the angle of reflection. Hence, the parallel rays of light strike the parabolic mirror so that they all reflect through the focus, which means that all the parallel rays are concentrated at the eyepiece, thereby maximizing the light-gathering ability of the mirror. If the distance from the vertex of the parabola to the focus is c, then it can be shown that the equation of a parabola with vertex $(0, 0)$ depends on its orientation.

Opens right:	$y^2 = 4cx$	Opens up:	$x^2 = 4cy$
Opens left:	$y^2 = -4cx$	Opens down:	$x^2 = -4cy$

Flashlights and automobile headlights (see Figure 7.11**b**) simply reverse the process: A light source is placed at the focus of a parabolic mirror, the light rays strike the mirror with an angle of incidence equal to the angle of reflection, and each ray is reflected along a path parallel to the axis, thus emitting a light beam of parallel rays.

Radar uses both of these properties. First, a pulse is transmitted from the focus to a parabolic surface. As with a reflecting telescope, parallel pulses are transmitted in this way. The reflected pulses then strike the parabolic surface and are sent back to be received at the focus.

PROBLEM SET 7.1

LEVEL 1

In Problems 1–6, use the definition of a parabola to sketch a parabola such that the distance between the focus and the directrix is the given number (see Figure 7.2).

1. 4
2. 6
3. 8
4. 10
5. 3
6. 7

WHAT IS WRONG, *if anything, with each statement in Problems 7–14? Explain your reasoning.*

7. If $x^2 = 10y$, then the parabola opens upward.
8. If $x^2 = -10y$, then the parabola opens downward.
9. If $y^2 = 10x$, then the parabola opens leftward.
10. If $y^2 = -10x$, then the parabola opens rightward.
11. To complete the square on

$$2x^2 + 12x = -3y + 6$$

add 9 to both sides.

12. If $(y - 2)^2 = -4(x - 3)$, then the vertex is at $(2, 3)$.
13. If $(x - 3)^2 = -8(y + 2)$, then the length of the focal chord is 8.
14. If $(y - 5)^2 = -4(x - 1)$, then the length of the focal chord is -1.

Graph the curves given by the equations in Problems 15–24.

15. $y^2 = 8x$
16. $4x^2 = 10y$
17. $3x^2 = -12y$
18. $2x^2 = -4y$
19. $3y^2 - 15x = 0$
20. $4x^2 + 3y = 12$
21. $(y + 3)^2 = 3(x - 1)$
22. $(x - 1)^2 = 3(y + 3)$
23. $(x - 1) = -2(y + 2)$
24. $(x + 3) = -3(y - 1)$

Parameterize and then graph the curves given by the equations in Problems 25–32.

25. $y^2 = -20x$
26. $y^2 = -12x$
27. $5y^2 + 15x = 0$
28. $2x^2 + 5y = 0$
29. $4y^2 + 3x = 12$
30. $5x^2 + 4y = 20$
31. $(y - 1)^2 = 2(x + 2)$
32. $(x + 2)^2 = 2(y - 1)$

LEVEL 2

Complete the squares to graph the curves given by the equations in Problems 33–38.

33. $y^2 - 4x + 10y + 13 = 0$
34. $2y^2 + 8y - 20x - 148 = 0$
35. $x^2 + 9y - 6x + 18 = 0$
36. $y^2 + 4x - 3y + 1 = 0$
37. $9x^2 + 6x + 18y - 23 = 0$
38. $9y^2 + 6y + 18x - 23 = 0$

Find the equations and sketch each curve satisfying the conditions given in Problems 39–48.

39. Directrix $x = 0$; focus at $(5, 0)$
40. Directrix $y = 0$; focus at $(0, -3)$
41. Directrix $x - 3 = 0$; vertex at $(-1, 2)$
42. Directrix $y + 4 = 0$; vertex at $(4, -1)$
43. Vertex at $(-2, -3)$; focus at $(-2, 3)$
44. Vertex at $(-3, 4)$; focus at $(1, 4)$
45. Vertex at $(-3, 2)$ and passing through $(-2, -1)$; axis parallel to the y-axis
46. Vertex at $(4, 2)$ and passing through $(-3, -4)$; axis parallel to the x-axis
47. The set of all points with distances from $(4, 3)$ that equal their distances from $(0, 3)$
48. The set of all points with distances from $(4, 3)$ that equal their distances from $(-2, 1)$
49. If the path of a baseball is parabolic and is 200 ft wide at the base and 50 ft high at the vertex, write the equation that gives the path of the baseball if the origin is the point of departure for the ball.

50. A radar antenna is constructed so that a cross section along its axis is a parabola with the receiver at the focus. Find the focus if the antenna is 12 m across and its depth is 4 m, as shown in Figure 7.12.

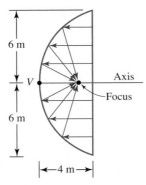

Figure 7.12 Radar antenna

51. A parabolic reflector (see Figure 7.12) is constructed so that a cross section along its axis is a parabola with the light source at the focus. Find the location of the focus if the reflector is 16 cm across and its depth is 8 cm.

52. Beams of light parallel to the axis of the parabolic mirror shown in Figure 7.13 strike the mirror and are reflected. Find the distance from the vertex to the point where the beams concentrate.

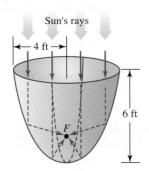

Figure 7.13 A parabolic mirror

53. One side of a storage yard is against a building. The other three sides of the rectangular yard are to be fenced with 36 ft of fencing. How long should the sides be to produce the greatest enclosed area with the given length of fence? What is the area obtained?

54. The sum of the length and the width of a rectangular area is 50 m. Find the greatest area possible, and find the dimensions of the figure.

55. A parabolic archway has the dimensions shown in Figure 7.14. Find the equation of the parabolic portion.

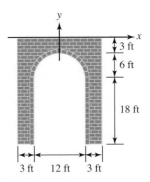

Figure 7.14 A parabolic archway

56. An arch has the shape of a parabola. If it is 72 ft wide at the base and 18 ft high at the center, how tall is it 12 ft from the center of the base?

LEVEL 3

57. Derive the equation of a parabola with focus $F(0, -c)$, where c is a positive number, and the directrix is the line $y = c$.

58. Derive the equation of a parabola with focus $F(c, 0)$, where c is a positive number, and the directrix is the line $x = -c$.

59. Derive the equation of a parabola with focus $F(-c, 0)$, where c is a positive number, and the directrix is the line $x = c$.

60. Show that the length of the focal chord for the parabola $y^2 = 4cx$ is $4c$ $(c > 0)$.

7.2 Ellipses

The second conic section we shall study in this chapter is called an *ellipse*.

Definition of an Ellipse

Some think of an ellipse as an "elongated circle," but in reality, a circle is a special type of ellipse. We begin with the definition.

ELLIPSE

An **ellipse** is the set of all points in the plane such that, for each point on the ellipse, the sum of its distances from two fixed points is a constant.

The fixed points are called the **foci** (plural of *focus*). To see what an ellipse looks like, we will imagine a special type of graph paper shown in Figure 7.15a. This paper is constructed by plotting two points, labeled F_1 and F_2, and then drawing concentric circles with radii $r = 1$, $r = 2$, $r = 3, \ldots$ emanating from each focus.

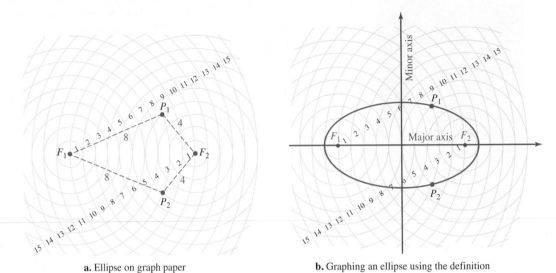

a. Ellipse on graph paper   **b.** Graphing an ellipse using the definition

Figure 7.15 Graphing an ellipse

We can use this special graph paper to help us graph an ellipse using the definition. We want to plot all points such that the sum of the distances from two fixed points is a constant, so we begin by picking a constant. This constant must be greater than the straight-line distance from F_1 to F_2. For this demonstration, we note that F_1 and F_2 in Figure 7.15 are 10 units apart, so let the given constant be 12. Plot all the points so that the sum of their distances from the foci is 12. This is easy using the special graph paper; if a point is 8 units from F_1, for example, then it is 4 units from F_2, so we look at the intersection of the circle with $r = 8$ from F_1 with the circle with $r = 4$ from F_2, as shown in Figure 7.15a. Continue this process by plotting the following points:

Distance from F_1	Distance from F_2
8	4
9	3
10	2
11	1

Do you see why 11 units is the maximum integer distance from F_1?

7	5
6	6
5	7
4	8
3	9
2	10
1	11

Do you see also why 11 units is the maximum integer distance from F_2?

There are, of course, other rational and irrational choices that could be made, but connecting these points as shown in Figure 7.15 will give us a good picture of an ellipse with constant 12 and a distance of 10 units between F_1 and F_2.

The line passing through F_1 and F_2 is called the **major axis**. The **center** is the midpoint of the segment $\overline{F_1F_2}$. The line passing through the center perpendicular to the major axis is called the **minor axis**. The ellipse is symmetric with respect to both the major and minor axes. Even though we know a line such as the major axis has no length, it is common to talk about the **length of the major axis**, which is the distance between the vertices along the major axis. This distance is $2a$. Similarly, we define the **length of the minor axis** to be $2b$.

Standard-Form Ellipse with Center (0, 0)

To find the equation of an ellipse, first consider a special case where the center is at the origin. Let the distance from the center to a focus be the positive number c; that is, let $F_1(-c, 0)$ and $F_2(c, 0)$ be the foci, as shown in Figure 7.16. Let the constant distance be $2a$.

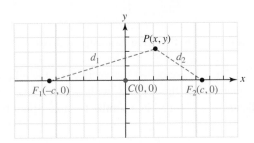

Figure 7.16 Graph using definition

Notice that the center of the ellipse shown in Figure 7.16 is $(0, 0)$. Let $P(x, y)$ be any point on the ellipse, and use the distance formula and the definition of an ellipse to derive the equation of this ellipse:

$$\text{SUM OF DISTANCES} = \text{FIXED GIVEN DISTANCE} \qquad \text{\textit{Definition of an ellipse}}$$

$$d_1 + d_2 = 2a$$

$$\sqrt{(x+c)^2 + (y-0)^2} + \sqrt{(x-c)^2 + (y-0)^2} = 2a \qquad \text{\textit{Distance formula}}$$

$$\sqrt{(x+c)^2 + y^2} = 2a - \sqrt{(x-c)^2 + y^2}$$

$$(x+c)^2 + y^2 = 4a^2 - 4a\sqrt{(x-c)^2 + y^2} + (x-c)^2 + y^2 \qquad \text{\textit{Square both sides.}}$$

$$x^2 + 2cx + c^2 + y^2 = 4a^2 - 4a\sqrt{(x-c)^2 + y^2} + x^2 - 2cx + c^2 + y^2$$

$$4a\sqrt{(x-c)^2 + y^2} = 4a^2 - 4cx$$

$$\sqrt{(x-c)^2 + y^2} = a - \frac{c}{a}x \qquad \text{\textit{Since } a \neq 0}$$

$$(x-c)^2 + y^2 = \left(a - \frac{c}{a}x\right)^2$$

$$x^2 - 2cx + c^2 + y^2 = a^2 - 2cx + \frac{c^2}{a^2}x^2$$

$$x^2 + y^2 = a^2 - c^2 + \frac{c^2}{a^2}x^2$$

$$x^2 - \frac{c^2}{a^2}x^2 + y^2 = a^2 - c^2$$

$$\left(1 - \frac{c^2}{a^2}\right)x^2 + y^2 = a^2 - c^2$$

$$\frac{a^2 - c^2}{a^2}x^2 + y^2 = a^2 - c^2$$

$$\frac{x^2}{a^2} + \frac{y^2}{a^2 - c^2} = 1 \qquad \text{\textit{Divide both sides by } } a^2 - c^2.$$

$$\frac{x^2}{a^2} + \frac{y^2}{b^2} = 1 \qquad \text{\textit{Let } } b^2 = a^2 - c^2.$$

The graph of the ellipse we are considering is shown in Figure 7.17.

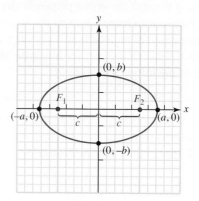

Figure 7.17   **Graph of standard-form ellipse with center (0, 0)**

The intercepts on the major axis are called the **vertices** of the ellipse. Notice also that since $a \neq c$, $a^2 - c^2 \neq 0$. The equation of the ellipse with major axis vertical, foci $F_1(0, c)$ and $F_2(0, -c)$, and constant distance $2a$ is found in a similar fashion. Simplifying as before, we find

$$\frac{y^2}{a^2} + \frac{x^2}{b^2} = 1$$

where $b^2 = a^2 - c^2$.

Notice from the graph shown in the margin that in both cases a^2 must be larger than both c^2 and b^2. Algebraically, we note that if it were not, a square number would be equal to a negative number, which is a contradiction in the set of real numbers. These equations are summarized in Table 7.3.

TABLE 7.3   Standard-Form Ellipse Equations with Center (0, 0)

Graph	Ellipse	Foci	Intercepts Major	Intercepts Minor	Center	Equation
	Horizontal	$(-c, 0)$ $(c, 0)$	$(a, 0)$ $(-a, 0)$	$(0, b)$ $(0, -b)$	$(0, 0)$	$\dfrac{x^2}{a^2} + \dfrac{y^2}{b^2} = 1$
	Vertical	$(0, c)$ $(0, -c)$	$(0, a)$ $(0, -a)$	$(b, 0)$ $(-b, 0)$	$(0, 0)$	$\dfrac{y^2}{a^2} + \dfrac{x^2}{b^2} = 1$

EXAMPLE 1   Graphing a horizontal ellipse

Graph $\dfrac{x^2}{9} + \dfrac{y^2}{4} = 1$.

Solution

The center of the ellipse is $(0, 0)$. The x-intercepts are places where the graph crosses the x-axis; that is, when $y = 0$:

$$\frac{x^2}{9} + \frac{0^2}{4} = 1$$
$$x^2 = 9$$
$$x = \pm 3$$

The y-intercepts y are found when $x = 0$:

$$\frac{0^2}{9} + \frac{y^2}{4} = 1$$
$$y^2 = 4$$
$$y = \pm 2$$

Sketch the ellipse using the four intercepts as shown in Figure 7.18. Even though you do not need to find the foci in order to graph the ellipse, you can find them if you are given the equation.

$$c^2 = a^2 - b^2$$
$$c^2 = 9 - 4$$
$$c^2 = 5$$
$$c = \pm\sqrt{5}$$

These distances are plotted on the major axis. ∎

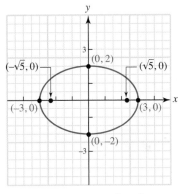

Figure 7.18 Graph of $\dfrac{x^2}{9} + \dfrac{y^2}{4} = 1$

EXAMPLE 2   Graphing a vertical ellipse

Graph $9x^2 + 4y^2 = 36$.

Solution

Divide both sides by 36 to put this equation into standard form:

$$9x^2 + 4y^2 = 36$$
$$\frac{9x^2}{36} + \frac{4y^2}{36} = \frac{36}{36}$$
$$\frac{x^2}{4} + \frac{y^2}{9} = 1$$

We see $a^2 = 9$ and $b^2 = 4$, which indicates an ellipse with the major axis vertical because the larger number is under the y^2. The x-intercepts are ± 2 and the y-intercepts are ± 3 (these are the vertices). The sketch is shown in Figure 7.19. ∎

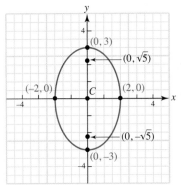

Figure 7.19 Graph of $\dfrac{x^2}{4} + \dfrac{y^2}{9} = 1$

▇ COMPUTATIONAL WINDOW   ▭▢☒

If you solve the equation of an ellipse for y, you will notice that an ellipse cannot be represented as a single function. The best way to graph an ellipse using a calculator is to write it in parametric form, as shown in the following example.

EXAMPLE 3   Parameterize the equation of an ellipse

Graph $2x^2 + 5y^2 = 10$ by considering a parameterization.

Solution

To parameterize an ellipse, you need to recall a fundamental identity from Section 5.2: $\cos^2\theta + \sin^2\theta = 1$. We begin by dividing both sides of the given equation by 10:

$$2x^2 + 5y^2 = 10$$

$$\frac{2x^2}{10} + \frac{5y^2}{10} = \frac{10}{10}$$

$$\frac{x^2}{5} + \frac{y^2}{2} = 1$$

$$\left(\frac{x}{\sqrt{5}}\right)^2 + \left(\frac{y}{\sqrt{2}}\right)^2 = 1$$

Let $\cos\theta = \frac{x}{\sqrt{5}}$ and $\sin\theta = \frac{y}{\sqrt{2}}$ so that $\cos^2\theta + \sin^2\theta = 1$. You can set up a table of values or use a calculator to obtain the graph as shown in Figure 7.20. Let $x = \sqrt{5}\,\cos\theta$ and $y = \sqrt{2}\,\sin\theta$. If you use a table of values, you need only consider values of θ between $0°$ and $90°$ because we know the ellipse is symmetric with respect to both the major and minor axes.

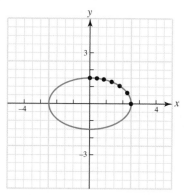

θ	x	y
0°	2.24	0
15°	2.16	0.37
30°	1.94	0.71
45°	1.58	1
60°	1.12	1.22
75°	0.58	1.37
90°	0	1.41

Figure 7.20 Graph of $x = \sqrt{5}\cos\theta, y = \sqrt{2}\sin\theta$

Standard-Form Equations of Ellipses

If the center of an ellipse is at (h, k), the equation can be written by completing the square, as shown in Table 7.4.

To graph ellipses with vertex (h, k), we use the idea of a translation introduced in Section 2.4 as summarized in the following box.

GRAPHING ELLIPSES

To graph an ellipse, first make sure that both variables are second degree and in standard form have coefficients with the same sign.

Step 1 Write the equation in standard form so there is a 1 on the right side and the coefficients of the square terms are also 1.

Step 2 Plot the center at (h, k).

Step 3 On the axis, plot $\pm$ the square root of the number under the x-axis in the x-direction and also $\pm$ the number under the y-axis in the y-direction. The longer axis is called the major axis and the smaller one the minor axis.

Step 4 Draw the ellipse using the four intercepts.

TABLE 7.4	Standard-Form Ellipse Equations with Vertex (h, k)

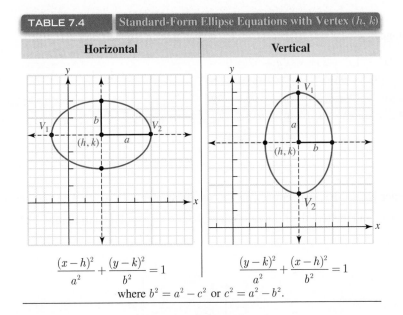

Horizontal	Vertical
$$\dfrac{(x-h)^2}{a^2} + \dfrac{(y-k)^2}{b^2} = 1$$	$$\dfrac{(y-k)^2}{a^2} + \dfrac{(x-h)^2}{b^2} = 1$$

where $b^2 = a^2 - c^2$ or $c^2 = a^2 - b^2$.

As before, the focus is found c units from the center on the major axis. The segment from the center to the vertex on the major axis is called a **semimajor axis** and has length a; the segment from the center to an intercept on the minor axis is called a **semiminor axis** and has length b. Use a and b to plot four points that are used to draw the ellipse.

EXAMPLE 4 Graphing an ellipse with center at (h, k)

Graph $16(x-3)^2 + 25(y-1)^2 = 400$.

Solution

We follow the steps in the graphing ellipses procedure.

$$16(x-3)^2 + 25(y-1)^2 = 400 \qquad \text{Given equation}$$

Step 1: $\quad \dfrac{(x-3)^2}{25} + \dfrac{(y-1)^2}{16} = 1 \qquad$ Divide both sides by 400.

Step 2: The center is $(3, 1)$; plot this point in Figure 7.21.

Step 3: Plot $\pm$ units from the center on the x-axis.

Plot $\pm$ units from the center on the y-axis.

Step 4: Draw the ellipse as shown in Figure 7.21.

Even though you do not need to find the foci in order to graph the ellipse, we can find them:

$$c^2 = a^2 - b^2$$
$$c^2 = 25 - 16$$
$$c = \pm 3$$

These distances are plotted on the major axis. ∎

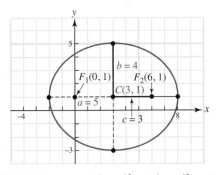

Figure 7.21 Graph of $16(x-3)^2 + 25(y-1)^2 = 400$

Many times the equation is given in general form, and we need to complete the squares to put the equation into standard form. The procedure is similar to that which we used with parabolas, except that for ellipses we need to complete the square for *both x and y*, as shown by the following example.

EXAMPLE 5 Graphing an ellipse by completing the square

Graph $3x^2 + 4y^2 + 24x - 16y + 52 = 0$.

Solution

Associate together the x and y terms (with the constant term isolated on the right):

$$3x^2 + 4y^2 + 24x - 16y + 52 = 0 \qquad \text{Given equation}$$
$$(3x^2 + 24x) + (4y^2 - 16y) = -52$$
$$3(x^2 + 8x \qquad) + 4(y^2 - 4y \qquad) = -52 \qquad \text{Factor coefficients.}$$
$$3(x^2 + 8x + 4^2) + 4(y^2 - 4y + 2^2) = -52 + 3 \cdot 16 + 4 \cdot 4 \qquad \text{Complete both squares.}$$
$$3(x + 4)^2 + 4(y - 2)^2 = 12 \qquad \text{Factor.}$$
$$\frac{(x + 4)^2}{4} + \frac{(y - 2)^2}{3} = 1 \qquad \text{Divide both sides by 12.}$$

Plot the center (h, k). By inspection, you can see the center is $(-4, 2)$. The vertices are ± 2 units from the center, and the length of the semiminor axis is $\sqrt{3}$, as shown in Figure 7.22.

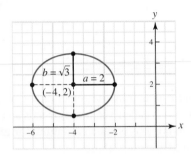

Figure 7.22 Graph of $3x^2 + 4y^2 + 24x - 16y + 52 = 0$

Eccentricity

We have seen that some ellipses are more circular and some more flat than others. A measure of the amount of flatness of an ellipse is called its **eccentricity**, which is defined as $\varepsilon = \frac{c}{a}$. Notice that

$$\varepsilon = \frac{c}{a} = \frac{\sqrt{a^2 - b^2}}{a} = \sqrt{\frac{a^2 - b^2}{a^2}} = \sqrt{1 - \left(\frac{b}{a}\right)^2}$$

Since $c < a$, ε is between 0 and 1. If $a = b$, then $\varepsilon = 0$ and the conic is a circle. If the ratio b/a is small, then the ellipse is very flat. Thus, for an ellipse, $0 \le \varepsilon < 1$ and ε measures the amount of roundness (or flatness) of the ellipse.

Consider a circle; that is, suppose $a = b$ so that $\varepsilon = 0$. In this case, let the *radius* $r = a = b$. We see that the equation of a circle with center (h, k) and radius r as derived in Section 1.4 is a special case of an ellipse:

$$(x - h)^2 + (y - k)^2 = r^2$$

EXAMPLE 6   Graphing a circle by completing the square

Graph $x^2 + y^2 + 6x - 14y + 22 = 0$.

Solution

Complete the squares in x and y:

$$x^2 + y^2 + 6x - 14y + 22 = 0$$
$$(x^2 + 6x) + (y^2 - 14y) = -22$$
$$(x^2 + 6x + 3^2) + (y^2 - 14y + 7^2) = -22 + 9 + 49$$
$$(x + 3)^2 + (y - 7)^2 = 36$$

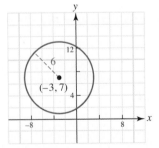

Figure 7.23 Graph of
$x^2 + y^2 + 6x - 14y + 22 = 0$

This is a circle with center at $(-3, 7)$ and radius 6, as shown in Figure 7.23.

We now turn to the problem of finding an equation if we are given a graph or some information about a graph.

EXAMPLE 7   Finding the equation of an ellipse

Find the equation of the ellipse with vertices at $(3, 2)$ and $(3, -4)$ and foci at $\left(3, \sqrt{5} - 1\right)$ and $\left(3, -\sqrt{5} - 1\right)$.

Solution
Most of the work for this type of problem is done by inspection. Plot these points as shown in Figure 7.24, and notice that the ellipse is vertical.

The center must be the midpoint of the segment connecting the vertices; this is the point $(3, -1)$. Use the usual variables to label the relevant distances: a is the distance from the center to a vertex, so $a = 3$; c is the distance from the center to a focus, so $c = \sqrt{5}$. We now calculate the b^2 value:

$$c^2 = a^2 - b^2$$
$$\left(\sqrt{5}\right)^2 = (3)^2 - b^2$$
$$b^2 = 9 - 5$$
$$= 4$$

The equation is

$$\frac{(y+1)^2}{9} + \frac{(x-3)^2}{4} = 1$$

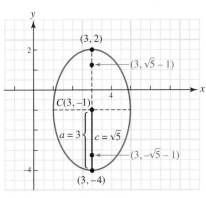

Figure 7.24 Given graph

EXAMPLE 8   Finding the equation of a curve

Find the equation of the set of all points the sum of whose distances from $(-3, 2)$ and $(5, 2)$ is equal to 16.

Solution
We note that this problem satisfies the definition of an ellipse. The given points are the foci and the center is the midpoint of the segment connecting the foci; this is the point $(1, 2)$. Since c is the distance from the center to a focus, we see that $c = 4$. The fixed distance (according to the definition of an ellipse) is $2a$, so we have $2a = 16$, which implies that $a = 8$. We can now calculate the value of b^2:

$$c^2 = a^2 - b^2$$
$$(4)^2 = (8)^2 - b^2$$
$$b^2 = 64 - 16$$
$$= 48$$

The desired equation is

$$\frac{(x-1)^2}{64} + \frac{(y-2)^2}{48} = 1$$

EXAMPLE 9   Finding the equation of a curve, given the eccentricity

Find the equation of the ellipse with foci $(-3, 6)$ and $(-3, 2)$ with $\varepsilon = \frac{1}{5}$.

Solution
By inspection, the ellipse is vertical and is centered at $(-3, 4)$ with $c = 2$. Since

☠ Just because $\frac{c}{a} = \frac{1}{5}$ you cannot assume
$c = 1$ and $a = 5$; all you know is that the ratio
of c to a is $\frac{1}{5}$. ☠

$$\varepsilon = \frac{c}{a} = \frac{1}{5}$$

and $c = 2$, we have

$$\frac{2}{a} = \frac{1}{5}$$

which implies $a = 10$. Since we know a and c, we can now find b:

$$c^2 = a^2 - b^2$$
$$(2)^2 = (10)^2 - b^2$$
$$b^2 = 100 - 4$$
$$= 96$$

Thus the equation is

$$\frac{(y-4)^2}{100} + \frac{(x+3)^2}{96} = 1$$

■

One of the most interesting applications of ellipses is their use in modeling planetary orbits (see Figure 7.25). The orbit of a planet can be described by an ellipse with the sun at one focus. The orbit is commonly identified by the length of its major axis $2a$ and its eccentricity ε. The **aphelion** is the point where a planet is farthest from the sun, and the **perihelion** is the point where a planet is closest to the sun. You might also wish to see the related Group Research Projects at the end of this chapter. The following modeling application uses this information.

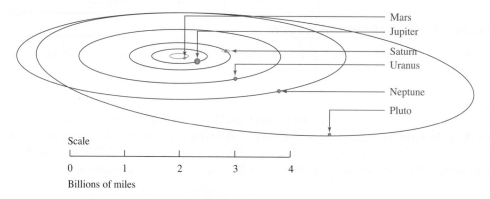

Figure 7.25 Planetary orbits

EXAMPLE 10   Modeling the orbit of the earth   MODELING APPLICATION

Find an equation that models the orbit of the earth around the sun.

Solution

Step 1:   *Understand the problem.* Before we can answer this question, we need additional information. We can consult an elementary astronomy book, an encyclopedia, or an almanac to find that the length of the major axis of the earth's orbit is 1.86×10^8 and the eccentricity of the earth is about $\frac{1}{62}$.

Step 2:   *Devise a plan.* Using this information, we have $\varepsilon = \frac{1}{62}$ and $2a = 1.86 \times 10^8$ miles or $a = 9.3 \times 10^7$. We will write the equation of the ellipse by first finding c, then using that information to find b^2, which will allow us to write the equation of the ellipse.

Step 3: *Carry out the plan.*

$$\varepsilon = \frac{c}{a}$$

$$\frac{1}{62} = \frac{c}{9.3 \times 10^7}$$

$$c = 1.5 \times 10^6$$

We can now find b^2:

$$c^2 = a^2 - b^2$$
$$b^2 = (9.3 \times 10^7)^2 - (1.5 \times 10^6)^2$$
$$= 8.649 \times 10^{15} - 2.25 \times 10^{12}$$
$$\approx 8.64675 \times 10^{15}$$

If we take the sun to be at one of the foci of the ellipse, we find the equation of the earth's orbit to be

$$\frac{x^2}{8.649 \times 10^{15}} + \frac{y^2}{8.64675 \times 10^{15}} = 1$$

Step 4: *Look back.* We see that the orbit is almost circular with the sun at the center, as shown in Figure 7.26.

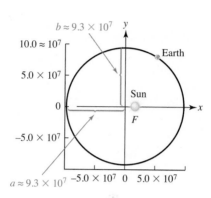

Figure 7.26 Earth's orbit around the sun

Geometric Properties

Like the parabola, the ellipse has some useful reflection properties.

REFLECTIVE PROPERTY

An elliptic mirror has the property that waves emanating from one focus are reflected toward the other focus.

This elliptic reflection principle is also used in a procedure called *lithotripsy* for disintegrating kidney stones. A patient is placed in a tub of water (see Figure 7.27) shaped like an ellipsoid (a three-dimensional elliptic figure) in such a way that the kidney stone is at one focus of the ellipsoid. A pulse generated at the other focus is then concentrated at the kidney stone.

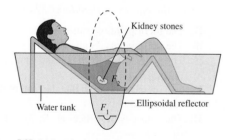

Figure 7.27 Lithotripsy uses the elliptic reflective property

The "whispering room" phenomenon is another application of this principle; it is found in many science museums and in famous buildings such as the old U.S. Capitol in Washington, D.C., and St. Paul's Cathedral in London (see Figure 7.28). If one person stands on a focus in St. Paul's Cathedral in London and whispers a secret message, another person standing on the other focus will clearly hear what is said, but anyone *not* near a focus will hear nothing. This is especially impressive if the foci are far apart. Christopher Wren completed this cathedral in 1711. It was the first Protestant cathedral and is still a major London landmark.

Figure 7.28 St. Paul's Cathedral

PROBLEM SET 7.2

Suppose the distance between two foci is 10 units. Use the definition to sketch the ellipse with the properties described in Problems 1–2.

1. The ellipse is the curve with the sum of the distances from the two foci 12 units.
2. The ellipse is the curve with the sum of the distances from the two foci 14 units.

WHAT IS WRONG, *if anything, with each statement in Problems 3–6? Explain your reasoning.*

3. If $\dfrac{x^2}{a^2} + \dfrac{y^2}{b^2} = 1$, then the intercepts are (a, b), $(a, -b)$, $(-a, b)$, and $(-a, -b)$.

4. If $\dfrac{x^2}{3} + \dfrac{y^2}{5} = 1$, then the vertices are $\left(\sqrt{3}, 0\right)$ and $\left(-\sqrt{3}, 0\right)$.

5. The eccentricity of an ellipse is less than 1.
6. The eccentricity of a circle is 1.

Sketch the curves in Problems 7–20. To check using a calculator, you may wish to convert to parametric form.

7. $x^2 + \dfrac{y^2}{9} = 1$

8. $4x^2 + 9y^2 = 36$

9. $5x^2 + 10y^2 = 7$

10. $3x^2 + y^2 = 5$

11. $(x-2)^2 + (y+3)^2 = 25$

12. $(x-1)^2 + (y-1)^2 = \frac{1}{4}$

13. $25x^2 + 16y^2 = 400$

14. $36x^2 + 25y^2 = 900$

15. $\dfrac{(x+3)^2}{81} + \dfrac{(y-1)^2}{49} = 1$

16. $\dfrac{(x-3)^2}{16} + \dfrac{(y-2)^2}{9} = 1$

17. $\dfrac{(x+2)^2}{25} + \dfrac{(y+2)^2}{9} = 1$

18. $\dfrac{(x+1)^2}{4} + \dfrac{(y-1)^2}{3} = 1$

19. $8(x-5)^2 + 8(y+2)^2 = 72$

20. $5(x+2)^2 + 3(y+4)^2 = 60$

Parameterize the equations and then graph the curves in Problems 21–26.

21. $\dfrac{x^2}{4} + \dfrac{y^2}{9} = 1$

22. $\dfrac{x^2}{25} + \dfrac{y^2}{36} = 1$

23. $3x^2 + 2y^2 = 6$

24. $16x^2 + 3y^2 = 48$

25. $(x+4)^2 + (y-2)^2 = 49$

26. $10(x-5)^2 + 6(y+2)^2 = 60$

Graph the curves in Problems 27–32.

27. The set of points 6 units from the point $(4, 5)$
28. The set of points 3 units from the point $(-2, 3)$
29. The set of points such that the sum of the distances from $(4, 0)$ and $(-4, 0)$ is 10
30. The set of points such that the sum of the distances from $(-4, 1)$ and $(2, 1)$ is 10
31. The ellipse with vertices at $(0, 7)$ and $(0, -7)$ and foci at $(0, 5)$ and $(0, -5)$
32. The ellipse with vertices at $(-6, 3)$ and $(4, 3)$ and foci at $(-4, 3)$ and $(2, 3)$

Find the equations of the curves in Problems 33–38, and then sketch each of the curves.

33. The set of points 6 units from $(-1, -4)$
34. The set of points such that the sum of the distances from $(-6, 0)$ and $(6, 0)$ is 20

35. The ellipse with vertices at $(4, 3)$ and $(4, -5)$ and foci at $(4, 2)$ and $(4, -4)$
36. The ellipse with foci at $(-4, -3)$ and $(2, -3)$ and eccentricity $\frac{4}{5}$
37. The circle with center at $(-1, 1)$ passing through $(2, 2)$
38. The ellipse with axis along the coordinate axis and minor intercept $(5, 0)$ passing through $\left(4, 3\sqrt{2}\right)$

Sketch the curves in Problems 39–50. You may wish to graph some of these by considering a parameterization.

39. $x^2 + 4x + y^2 + 6y - 12 = 0$
40. $9x^2 + 4y^2 - 18x + 16y - 11 = 0$
41. $16x^2 + 9y^2 + 96x - 36y + 36 = 0$
42. $x^2 + 4x + 16y + 4 = 0$
43. $y^2 + 6y + 25x + 159 = 0$
44. $3x^2 + 4y^2 + 6x - 8y + 4 = 0$
45. $144x^2 + 72y^2 - 72x + 48y - 100 = 0$
46. $4y^2 + x^2 - 16y + 4x - 8 = 0$
47. $x^2 + y^2 - 10x - 14y - 70 = 0$
48. $x^2 + y^2 - 4x + 10y + 15 = 0$
49. $4x^2 + y^2 + 24x + 4y + 16 = 0$
50. $x^2 + 9y^2 - 4x - 18y - 14 = 0$
51. If the length of the major axis of the earth's orbit is $186{,}000{,}000$ miles and its eccentricity is $1/62$, how far is the earth from the sun when it is at aphelion and at perihelion?
52. If the length of the semimajor axis of the orbit of Mars is 1.4×10^8 miles and the eccentricity is about 0.093, determine the greatest and least distance of Mars from the sun, correct to two decimal places.
53. If the planet Mercury is 28 million miles from the sun at perihelion, and the eccentricity of its orbit is $1/5$, how long is the major axis of Mercury's orbit?
54. The moon's orbit is elliptical with the earth at one focus. The point at which the moon is farthest from the earth is called the *apogee*, and the point at which it is closest is called the *perigee*. If the moon is $199{,}000$ miles from the earth at apogee and the length of the major axis of its orbit is $378{,}000$ miles, what is the eccentricity of the moon's orbit?
55. A stone tunnel is to be constructed such that the opening is a semielliptic arch as shown in Figure 7.29.

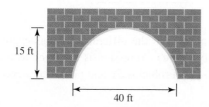

Figure 7.29 Semielliptic arch

Write the equation for the height of the tunnel as a function of the distance from the center of the tunnel.

56. It is necessary to know the height at 4-ft intervals from the center for the tunnel in Figure 7.29. That is, how high is the tunnel at 4, 8, 12, 16, and 20 ft from the center? (Answer to the nearest tenth of a foot.)

LEVEL 3

57. IN YOUR OWN WORDS In the text, we defined the eccentricity. Figure 7.30 shows some ellipses with the same vertices but different eccentricities.

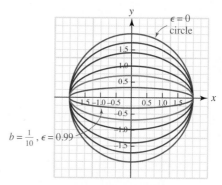

Figure 7.30 Circles with different eccentricities

Write a short paper exploring the notion of the eccentricity of an ellipse.
58. Derive the equation of the ellipse with foci at $(0, c)$ and $(0, -c)$ and constant distance $2a$. Let $b^2 = a^2 - c^2$. Show all of your work.
59. A segment through a focus parallel to a directrix and cut off by the ellipse is called the *focal chord*. Show that the length of the focal chord of the standard ellipse with center $(0, 0)$ is $2b^2/a$.
60. If we are given an ellipse with foci at $(-c, 0)$ and $(c, 0)$, where $c > 0$, and vertices at $(-a, 0)$ and $(a, 0)$, where $a > 0$, we define the *directrices* of the ellipse as the lines $x = a/\varepsilon$ and $x = -a/\varepsilon$. Show that an ellipse (see Figure 7.31) is the set of all points with distances from $F(c, 0)$ equal to ε times their distances from the line $x = a/\varepsilon$.

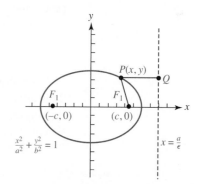

Figure 7.31 Directrix for an ellipse

7.3 Hyperbolas

The last of the conic sections to be considered has a definition similar to that of the ellipse.

Definition of a Hyperbola

The definition of a hyperbola is very similar to the definition of an ellipse, the only change being that the word "sum" is changed to "difference."

> **HYPERBOLA**
>
> A **hyperbola** is the set of all points in a plane such that, for each point on the hyperbola, the difference of its distances from two fixed points is a constant.

The fixed points are called the **foci**. A hyperbola with foci at F_1 and F_2, where the given constant is 8, is shown in Figure 7.32 with the same graph paper that we used for the ellipse.

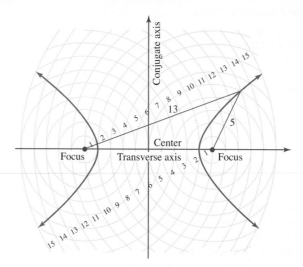

Figure 7.32 Graph of a hyperbola from the definition

The line passing through the foci is called the **transverse axis**. The **center** is the midpoint of the segment connecting the foci. The line passing through the center perpendicular to the transverse axis is called the **conjugate axis**. The hyperbola is symmetric with respect to both the transverse and the conjugate axes.

Standard-Form Hyperbola with Center (0, 0)

If you use the definition, you can derive the equation for a hyperbola with foci at $(-c,\ 0)$ and $(c,\ 0)$, and with constant distance $2a$.

If $(x,\ y)$ is any point on the curve, then

$$\left| \sqrt{(x+c)^2 + (y-0)^2} - \sqrt{(x-c)^2 + (y-0)^2} \right| = 2a$$

The absolute value is necessary here to make sure the difference is nonnegative (we know $2a$ is nonnegative). The procedure for simplifying this expression is the same as that shown for the ellipse, so the details are left as a problem (see Problems 59 and 60). After several steps, you should obtain

$$\frac{x^2}{a^2} - \frac{y^2}{c^2 - a^2} = 1$$

If $b^2 = c^2 - a^2$, then

$$\frac{x^2}{a^2} - \frac{y^2}{b^2} = 1$$

☠ Notice that $c^2 = a^2 - b^2$ for the ellipse and that $c^2 = a^2 + b^2$ for the hyperbola. For the ellipse, it is necessary that $a^2 > b^2$, but for the hyperbola, there is no restriction on the relative sizes of a and b. ☠

which is the standard-form equation.

Repeat the argument for a hyperbola with foci $(0,\ c)$ and $(0,\ -c)$, and you will obtain the other standard-form equation for a hyperbola with a vertical transverse axis. Standard-form hyperbolas are summarized in Table 7.5.

TABLE 7.5 Standard-Form Hyperbola Equations with Center (0, 0)

Graph	Ellipse	Foci	Vertices	Pseudo	Center	Equation
	Horizontal	$(-c, 0)$ $(c, 0)$	$(a, 0)$ $(-a, 0)$	$(0, b)$ $(0, -b)$	$(0, 0)$	$\dfrac{x^2}{a^2} - \dfrac{y^2}{b^2} = 1$
	Vertical	$(0, c)$ $(0, -c)$	$(0, a)$ $(0, -a)$	$(b, 0)$ $(-b, 0)$	$(0, 0)$	$\dfrac{y^2}{a^2} - \dfrac{y^2}{b^2} = 1$

As with the other conic sections, we will sketch a hyperbola by determining some information about the curve directly by inspection of the equation. The points of intersection of the hyperbola with the transverse axis are called the **vertices**. For

$$\frac{x^2}{a^2} - \frac{y^2}{b^2} = 1 \qquad \text{and} \qquad \frac{y^2}{a^2} - \frac{x^2}{b^2} = 1$$

notice that the vertices for the first equation occur at $(a, 0)$ and $(a, 0)$. These are found by letting $y = 0$ and solving for x. The vertices for the second equation occur at $(0, a)$ and $(0, -a)$. The number $2a$ is the **length of the transverse axis**. The hyperbola does not intersect the conjugate axis, but if we plot the points $(0, b)$, $(0, -b)$ and $(-b, 0)$, $(b, 0)$, respectively, we determine a segment on the conjugate axis whose length is called the **length of the conjugate axis**.

The procedure we will use to sketch a hyperbola is illustrated in the next example.

EXAMPLE 1 Procedure for sketching a hyperbola

Graph $9x^2 - 4y^2 - 36 = 0$.

Solution

	$9x^2 - 4y^2 - 36 = 0$	*Given equation*
Step 1:	$9x^2 - 4y^2 = 36$	*Add 36 to both sides.*
	$\dfrac{x^2}{4} - \dfrac{y^2}{9} = 1$	*Divide both sides by 36.*

Step 2: The center is $(0, 0)$; plot this point in Figure 7.33**a**.

Step 3: Plot the vertices: ± 2 units from the center on the x-axis. Plot the pseudovertices ± 3 units from the center on the y-axis.

Step 4: Draw a **central rectangle**. That is, draw lines through the vertices and pseudovertices parallel to the axes of the hyperbola. Draw the diagonal lines passing through the corners of the central rectangle, as shown in Figure 7.33b. These lines are called the **slant asymptotes**.

Step 5: Draw the hyperbola using the slant asymptotes and the vertices.

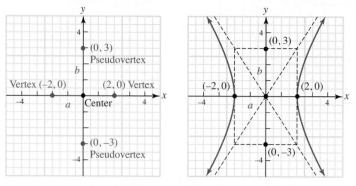

Figure 7.33 Graph of $9x^2 - 4y - 36 = 0$

To write a parameterization of a hyperbola, we begin with the left side of the standard-form equation of a hyperbola:

$$\frac{x^2}{a^2} - \frac{y^2}{b^2} = \left(\frac{x}{a}\right)^2 - \left(\frac{y}{b}\right)^2 \qquad \text{Let } \frac{x}{a} = \frac{1}{\cos\theta} \text{ and } \frac{y}{b} = \frac{\sin\theta}{\cos\theta}.$$

$$= \left(\frac{1}{\cos\theta}\right)^2 - \left(\frac{\sin\theta}{\cos\theta}\right)^2 \qquad \text{Substitute parametric values.}$$

$$= \frac{1}{\cos^2\theta} - \frac{\sin^2\theta}{\cos^2\theta}$$

$$= \frac{1 - \sin^2\theta}{\cos^2\theta}$$

$$= \frac{\cos^2\theta}{\cos^2\theta}$$

$$= 1$$

We see that the parameterization $x = \dfrac{a}{\cos\theta}$ and $y = b\tan\theta$ gives the standard-position horizontal hyperbola. Similarly, it can be shown that the parameterization for the standard-position vertical hyperbola is $x = b\tan\theta$ and $y = \dfrac{a}{\cos\theta}$.

EXAMPLE 2   Parameterize the equation of a hyperbola

Parameterize the hyperbola $\dfrac{x^2}{4} - \dfrac{y^2}{9} = 1$.

Solution

Let $x = \dfrac{2}{\cos\theta}$ and $y = \dfrac{3\sin\theta}{\cos\theta}$. We can verify that this represents the correct equation by eliminating the parameter:

$$\frac{x^2}{4} - \frac{y^2}{9} = \frac{\left(\dfrac{2}{\cos\theta}\right)^2}{4} - \frac{\left(\dfrac{3\sin\theta}{\cos\theta}\right)^2}{9}$$

$$= \frac{4}{\cos^2\theta} \cdot \frac{1}{4} - \frac{9\sin^2\theta}{\cos^2\theta} \cdot \frac{1}{9}$$

$$= \frac{1}{\cos^2 \theta} - \frac{\sin^2 \theta}{\cos^2 \theta}$$
$$= \frac{1 - \sin^2 \theta}{\cos^2 \theta}$$
$$= 1$$

■

Standard-Form Equations of Hyperbolas

If the center of a hyperbola is at (h, k), the equations shown in Table 7.6 are obtained.

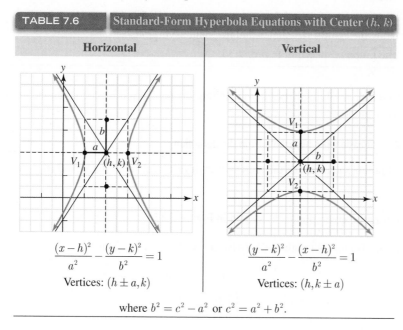

TABLE 7.6	Standard-Form Hyperbola Equations with Center (h, k)
Horizontal	**Vertical**
$\dfrac{(x - h)^2}{a^2} - \dfrac{(y - k)^2}{b^2} = 1$	$\dfrac{(y - k)^2}{a^2} - \dfrac{(x - h)^2}{b^2} = 1$
Vertices: $(h \pm a, k)$	Vertices: $(h, k \pm a)$

where $b^2 = c^2 - a^2$ or $c^2 = a^2 + b^2$.

To graph hyperbolas with vertex (h, k), we use the idea of a translation introduced in Section 2.4 as summarized in the following box.

GRAPHING HYPERBOLAS

To graph a hyperbola, first make sure that both variables are second degree and, in standard form, have coefficients with different signs.

Step 1 Write the equation in standard form so there is a 1 on the right side and the coefficients of the square terms are also 1.

Step 2 Plot the center at (h, k).

Step 3 Plot the vertices and the pseudovertices.

Step 4 Draw the central rectangle and the slant asymptotes.

Step 5 Draw the hyperbola using the slant asymptotes and the vertices.

To sketch a hyperbola when we are given a general-form equation, we complete the square in both the variables x and y to write the equation in standard form for easy graphing.

EXAMPLE 3   Completing the squares to sketch a hyperbola

Graph $16x^2 - 9y^2 - 128x - 18y + 103 = 0$.

Solution

First, we must write this equation so that it is in standard form. To do this, we complete the squares in both x and y:

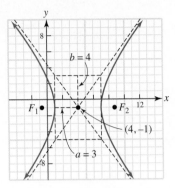

Figure 7.34 Graph of $16x^2 - 9y^2 - 128x - 18y + 103 = 0$

Step 1:

$$16x^2 - 9y^2 - 128x - 18y + 103 = 0$$
$$16(x^2 - 8x \quad) - 9(y^2 + 2y \quad) = -103$$
$$16(x^2 - 8x + 4^2) - 9(y^2 + 2y + 1^2) = -103 + 16 \cdot 16 - 9 \cdot 1$$
$$16(x - 4)^2 - 9(y + 1)^2 = 144$$
$$\frac{(x - 4)^2}{9} - \frac{(y + 1)^2}{16} = 1$$

☠ Watch the signs on the second part with the negative sign. ☠

Now, we follow the steps for plotting a hyperbola:

Step 2: Plot $(4, -1)$.

Step 3: Plot the vertices and the pseudovertices: $a = \pm 3$, $b = \pm 4$.

Step 4: Draw the central rectangles and the slant asymptotes.

Step 5: The graph is shown in Figure 7.34. ∎

The second task involving hyperbolas is to find the equation if we are given the graph (or information about the graph).

EXAMPLE 4   Equation of a hyperbola from the definition

Find the set of points such that the difference of their distances from $(6, 2)$ and $(6, -5)$ is always 3.

Solution
Begin by plotting the given points, as shown in Figure 7.35.

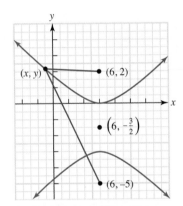

Figure 7.35 Plot the given points

We see the center is $\left(6, -\frac{3}{2}\right)$ with $c = \frac{7}{2}$. Also $2a = 3$, so $a = \frac{3}{2}$. Thus,

$$c^2 = a^2 + b^2$$
$$\left(\frac{7}{2}\right)^2 = \left(\frac{3}{2}\right)^2 + b^2$$
$$b^2 = \frac{49}{4} - \frac{9}{4}$$
$$= 10$$

Thus, the desired equation for the vertical hyperbola is

$$\frac{\left(y + \frac{3}{2}\right)^2}{\frac{9}{4}} - \frac{(x - 6)^2}{10} = 1$$

This can also be written as $\dfrac{4(y + 1.5)^2}{9} - \dfrac{(x - 6)^2}{10} = 1$. ∎

Properties of Hyperbolas

The eccentricity of the hyperbola and parabola is defined by the same equation that was used for the ellipse, namely,

$$\varepsilon = \frac{c}{a}$$

Remember, for the ellipse, $0 \le \varepsilon < 1$; however, for the hyperbola, $c > a$, so $\varepsilon > 1$, and for the parabola, $c = a$, so $\varepsilon = 1$.

EXAMPLE 5   Equation of a hyperbola given the foci and eccentricity

Find the equation of the conic with foci at $(-3, 2)$ and $(5, 2)$ with eccentricity 1.5.

Solution

We note that the eccentricity is greater than 1, so the conic is a hyperbola. We plot the foci, note that $c = 4$, and then we find the midpoint of the segment connecting the foci:

$$\left(\frac{-3+5}{2}, \frac{2+2}{2} \right) = (1,\ 2)$$

This point is the center of the hyperbola. Also, since

$$\varepsilon = \frac{c}{a}$$

$$\frac{3}{2} = \frac{4}{a}$$

$$a = \frac{8}{3}$$

Since $c^2 = a^2 + b^2$, we have $4^2 = \left(\frac{8}{3}\right)^2 + b^2$ or $b^2 = \frac{80}{9}$.
Thus, the equation is

$$\frac{(x-1)^2}{\frac{64}{9}} - \frac{(y-2)^2}{\frac{80}{9}} = 1$$

$$\frac{9(x-1)^2}{64} - \frac{9(y-2)^2}{80} = 1$$

There is also a useful reflection property of hyperbolas. Suppose an aircraft has crashed somewhere in the desert. A device in the wreckage emits a "beep" at regular intervals. Two observers, located at listening posts a known distance apart, time a beep. It turns out that the time difference between the two listening posts multiplied by the velocity of sound gives the value $2a$ for a hyperbola on which the airplane is located. A third listening post will determine two more hyperbolas in a similar fashion, and the airplane must be at the intersection of these hyperbolas. This is illustrated in Figure 7.36.

To transmit motion to a skew shaft, gears need to be fashioned with blades curved as hyperbolas. Such gears are called *hyperboloidal gears.*

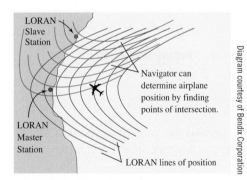

Figure 7.36 **LORAN navigation**

To transmit motion to a skew shaft, gears need to be fashioned with blades curved as hyperbolas. Such gears are called *hyperboloidal gears.*

Conic Section Summary

We have now considered the graphs of equations of the form

$$Ax^2 + Bxy + Cy^2 + Dx + Ey + F = 0$$

Geometrically, they represent the intersection of a plane and a cone, usually resulting in a parabola, ellipse, or hyperbola. If $B \neq 0$, then the conic is rotated, which we consider in Section 7.4.

There are certain positions of the plane as it intersects the cone that are called **degenerate conics** (see Figure 7.37). If the plane is situated so that one of its generators lies in the plane, then a line results and is called a **degenerate parabola**. If the plane intersects at the vertex of the upper and lower nappes, a point results and is called a **degenerate ellipse**. And, finally, for a **degenerate hyperbola**, visualize the plane situated so that the axis of the cone lies in the plane, resulting in a pair of intersecting lines.

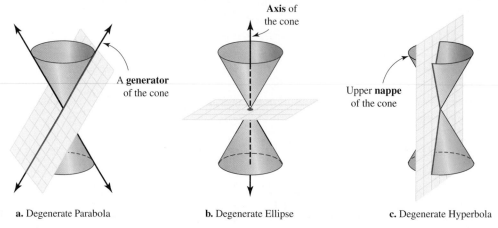

a. Degenerate Parabola   **b.** Degenerate Ellipse   **c.** Degenerate Hyperbola

Figure 7.37 Degenerate conic sections

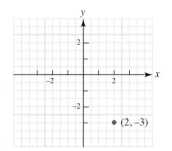

Figure 7.38 Degenerate ellipse is a point

EXAMPLE 6   Degenerate ellipse

Graph $\dfrac{(x-2)^2}{4} + \dfrac{(y+3)^2}{9} = 0$.

Solution

There is only one point that satisfies this equation—namely, $(2, -3)$. This is an example of a degenerate ellipse. Notice that, except for the zero, the equation has the "form of an ellipse." The graph is shown in Figure 7.38.   ∎

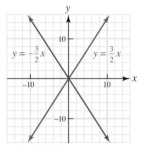

Figure 7.39 Degenerate hyperbola is a pair of lines

EXAMPLE 7   Degenerate hyperbola

Graph $\dfrac{x^2}{4} - \dfrac{y^2}{9} = 0$.

Solution

This equation has the "form of a hyperbola," but because of the zero, it cannot be put into standard form. You can, however, treat this in factored form:

$$\frac{x^2}{4} - \frac{y^2}{9} = 0$$

$$\left(\frac{x}{2} - \frac{y}{3}\right)\left(\frac{x}{2} + \frac{y}{3}\right) = 0$$

$$\frac{x}{2} - \frac{y}{3} = 0 \qquad \text{and} \qquad \frac{x}{2} + \frac{y}{3} = 0$$

The graph of each of these lines is shown in Figure 7.39. These lines *together* are considered a single curve called a *degenerate hyperbola*.   ∎

It is important to be able to recognize the curve by inspection of the equation before you begin. The first thing to notice is whether there is an xy-term. Remember, in this section, we assume $B = 0$ (that is, there is no xy-term); the case in which $B \neq 0$ is considered in the following section. Table 7.7 summarizes the method we will use to recognize a conic section.

TABLE 7.7	Recognizing a Conic Section by Looking at the General-Form Equation			
Type of Curve	**Degree**			**General Form Equation**
	Equation	**In x**	**In y**	$Ax^2 + Cy^2 + Dz + Ey + F = 0$
Line	first	first	first	$A = C = 0$
Parabola	second	first	second	$A = 0$ and $C \neq 0$
Parabola	second	second	first	$A \neq 0$ and $C = 0$
Ellipse	second	second	second	A and C have the same sign
Circle	second	second	second	$A = C \neq 0$
Hyperbola	second	second	second	A and C have opposite signs

This recognition table does not distinguish degenerate cases. This means that the test may indicate that the curve is an ellipse but, in fact, it may turn out to be a single point. Remember, too, that a circle is a special case of the ellipse.

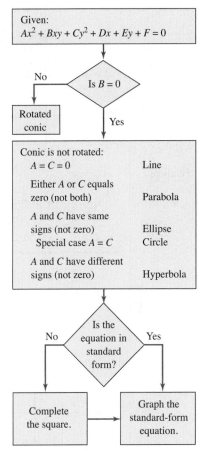

EXAMPLE 8   Identifying conic sections

Identify each of the given conics.

a. $4x^2 + 9y^2 = 36$ **b.** $4x^2 + 9y = 36$ **c.** $9x^2 - 16y^2 = 144$

d. $9x - 16y = 144$ **e.** $25x^2 + 25y^2 = 1$ **f.** $25x^2 + 4y^2 = 0$

g. $25x^2 - 4y^2 + 3x - 5y + 6 = 0$ **h.** $25x^2 + 4y^2 + 50x - 8y = 0$

Solution

To identify each conic, compare the given form with the form

$$Ax^2 + Bxy + Cy^2 + Dx + Ey + F = 0$$

a. $4x^2 + 9y^2 = 36$; second degree in both x and y, with $A = 4$ and $C = 9$ both positive; the graph is an ellipse.

b. $4x^2 + 9y = 36$; no y^2 term ($C = 0$), so the graph is a parabola ($A = 4$).

c. $9x^2 - 16y^2 = 144$; second degree in both x and y with $A = 9$ and $C = -16$ opposite in sign; thus, the graph is a hyperbola.

d. $9x - 16y = 144$; no x^2 and y^2 terms ($A = 0$, $C = 0$); thus, the graph is a line.

e. $25x^2 + 25y^2 = 1$; $A = C = 25$, so the graph is an ellipse (in fact, it is also a circle).

f. $25x^2 + 4y^2 = 0$; A and C have the same sign, so the graph is an ellipse. In fact, we notice the 0, so this is a degenerate ellipse, and the graph is the point $(0, 0)$.

g. $25x^2 - 4y^2 + 3x - 5y + 6 = 0$; A and C have opposite signs, so the graph is a hyperbola.

h. $25x^2 + 4y^2 + 50x - 8y = 0$; A and C have the same sign, so the graph is an ellipse. ∎

The procedure illustrated in Example 8 tells us what curve to expect. Complete the square, when necessary, to put the equation into standard form. Table 7.8 summarizes the standard-form and parametric-form equations for lines and the conic sections. We are given that (x, y) is any point on the curve, t is a parameter on $(-\infty, \infty)$, θ is a parameter on $[0°, 360°)$, and the other constants are as they have been defined in this chapter; in particular, remember $c > 0$.

You should remember the shapes of these conic sections, as summarized in Table 7.8, so we add these curves to our Library of Curves in Appendix F.

TABLE 7.8 Directory of Curves (Part V)

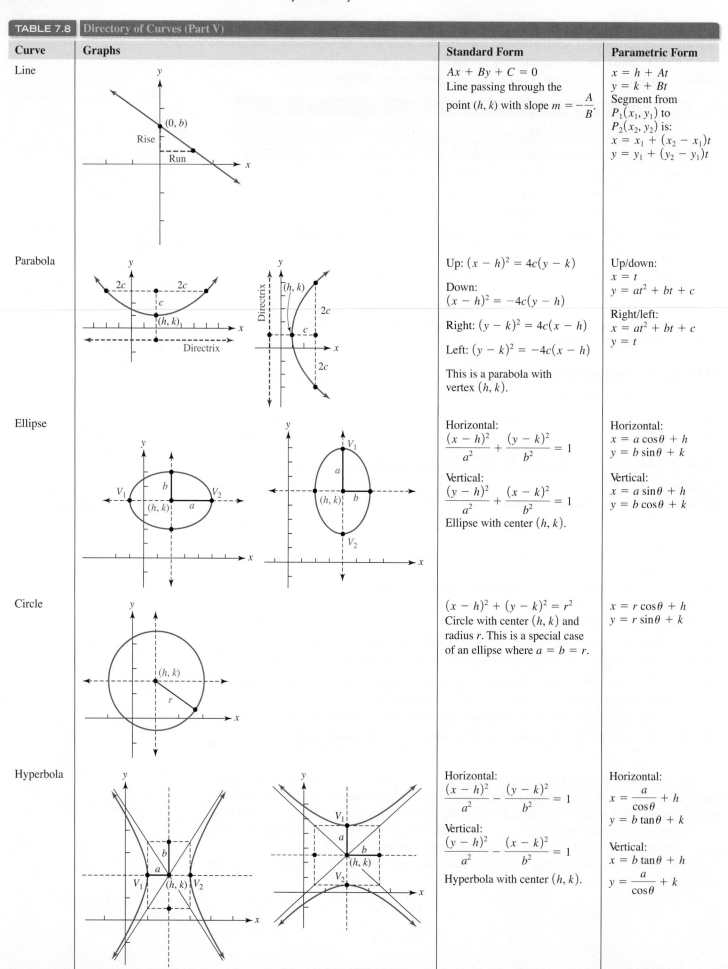

Curve	Graphs	Standard Form	Parametric Form
Line		$Ax + By + C = 0$ Line passing through the point (h, k) with slope $m = -\dfrac{A}{B}$.	$x = h + At$ $y = k + Bt$ Segment from $P_1(x_1, y_1)$ to $P_2(x_2, y_2)$ is: $x = x_1 + (x_2 - x_1)t$ $y = y_1 + (y_2 - y_1)t$
Parabola		Up: $(x - h)^2 = 4c(y - k)$ Down: $(x - h)^2 = -4c(y - h)$ Right: $(y - k)^2 = 4c(x - h)$ Left: $(y - k)^2 = -4c(x - h)$ This is a parabola with vertex (h, k).	Up/down: $x = t$ $y = at^2 + bt + c$ Right/left: $x = at^2 + bt + c$ $y = t$
Ellipse		Horizontal: $\dfrac{(x - h)^2}{a^2} + \dfrac{(y - k)^2}{b^2} = 1$ Vertical: $\dfrac{(y - h)^2}{a^2} + \dfrac{(x - k)^2}{b^2} = 1$ Ellipse with center (h, k).	Horizontal: $x = a\cos\theta + h$ $y = b\sin\theta + k$ Vertical: $x = a\sin\theta + h$ $y = b\cos\theta + k$
Circle		$(x - h)^2 + (y - k)^2 = r^2$ Circle with center (h, k) and radius r. This is a special case of an ellipse where $a = b = r$.	$x = r\cos\theta + h$ $y = r\sin\theta + k$
Hyperbola		Horizontal: $\dfrac{(x - h)^2}{a^2} - \dfrac{(y - k)^2}{b^2} = 1$ Vertical: $\dfrac{(y - h)^2}{a^2} - \dfrac{(x - k)^2}{b^2} = 1$ Hyperbola with center (h, k).	Horizontal: $x = \dfrac{a}{\cos\theta} + h$ $y = b\tan\theta + k$ Vertical: $x = b\tan\theta + h$ $y = \dfrac{a}{\cos\theta} + k$

PROBLEM SET 7.3

LEVEL 1

Suppose the distance between two foci is 10 units. Use the definition to sketch the hyperbola with the properties described in Problems 1–2.

1. The hyperbola is the curve with the difference of the distances from the two foci 8 units.
2. The hyperbola is the curve with the difference of the distances from the two foci 6 units.

WHAT IS WRONG, *if anything, with each statement in Problems 3–6? Explain your reasoning.*

3. The center of the graph of
$$\frac{(y-2)^2}{9} - \frac{(x+3)^2}{16} = 1$$
is $(2, -3)$.
4. $3x^2 + 4y^2 = 6$ is the equation of a circle.
5. $4x^2 - 9y^2 = -6$ is the equation of an ellipse.
6. To complete the square on
$$x^2 - y^2 + 10y + 14 = 0$$
add 25 to both sides.
7. Write out a definition for each of the following curves.
   a. circle   b. parabola
   c. ellipse   d. hyperbola
8. Explain how you recognize each of the following by looking at the equation only.
   a. parabola   b. circle
   c. ellipse   d. hyperbola
9. If $Ax^2 + Cy^2 + Dx + Ey + F = 0$, then list conditions on the constants to assure that the indicated graph results.
   a. a hyperbola   b. a circle
   c. a parabola   d. an ellipse
10. What is the eccentricity for each of the given curves?
    a. ellipse   b. hyperbola
    c. circle   d. parabola

Graph the curves in Problems 11–18.

11. $x^2 - y^2 = 1$ 12. $x^2 - y^2 = 4$
13. $\dfrac{x^2}{9} - \dfrac{y^2}{4} = 1$ 14. $\dfrac{x^2}{4} - \dfrac{y^2}{9} = 1$
15. $36y^2 - 25x^2 = 900$ 16. $3y^2 = 4x^2 + 12$
17. $81(x-1)^2 - 4(y+2)^2 = 36$ 18. $4(y+1)^2 - 16(x-2)^2 = 1$

Write a parameterization for the equations in Problems 19–24 and then sketch the curve.

19. $x^2 - y^2 = 9$ 20. $y^2 - x^2 = 9$
21. $3x^2 - 4y^2 = 12$ 22. $3x^2 - 4y^2 = 5$
23. $\dfrac{(x+4)^2}{8} - \dfrac{(y+2)^2}{5} = 1$ 24. $\dfrac{(y+2)^2}{25} - \dfrac{(x+1)^2}{16} = 1$

First identify and then graph each curve in Problems 25–36.

25. $2x - y - 8 = 0$ 26. $2x + y - 10 = 0$
27. $4x^2 - 16y = 0$ 28. $\dfrac{(x-3)^2}{4} - \dfrac{(y+2)}{6} = 1$
29. $\dfrac{(x-3)^2}{9} - \dfrac{(y+2)^2}{25} = 1$ 30. $\dfrac{x-3}{9} + \dfrac{y-2}{25} = 1$
31. $(x+3)^2 + (y-2)^2 = 0$ 32. $9(x+3)^2 + 4(y-2) = 0$
33. $x^2 + y^2 - 3y = 0$ 34. $x^2 + 8(y-12)^2 = 16$
35. $x^2 + 64(y+4)^2 = 16$ 36. $9(x+3)^2 - 4(y-2)^2 = 0$

LEVEL 2

Find the equations of the curves in Problems 37–40.

37. The hyperbola with vertices $(0, 5)$ and $(0, -5)$, and foci at $(0, 7)$ and $(0, -7)$
38. The set of points such that the difference of their distances from $(-6, 0)$ and is 10
39. The set of points such that the difference of their distances from $(4, -3)$ and is $(-4, -3)$ is 6
40. The hyperbola with vertices at $(4, 4)$ and $(4, 8)$ and foci at $(4, 3)$ and $(4, 9)$

Identify and sketch the curves in Problems 41–58. You may use a parameterization if you wish.

41. $5(x-2)^2 - 2(y+3)^2 = 10$
42. $4(x+4)^2 - 3(y+3)^2 = -12$
43. $3x^2 - 4y^2 + 12x + 80y = 88$
44. $9x^2 - 18x - 11 = 4y^2 + 16y$
45. $4y^2 - 8y + 4 = 3x^2 - 6x$
46. $x^2 - 4x + y^2 + 6y - 12 = 0$
47. $x^2 - y^2 = 2x + 4y - 3$
48. $3x^2 - 5y^2 + 24x + 20y - 2 = 0$
49. $4x^2 - 3y^2 - 24y - 112 = 0$
50. $y^2 - 4x + 2y + 21 = 0$
51. $9x^2 + 2y^2 - 48y + 270 = 0$
52. $x^2 + 4x + 12y + 64 = 0$
53. $y^2 - 6y - 4x + 5 = 0$
54. $100x^2 - 7y^2 + 98y - 368 = 0$
55. $x^2 + y^2 + 2x - 4y - 20 = 0$
56. $4x^2 + 12x + 4y^2 + 8y + 1 = 0$
57. $x^2 - 4y^2 - 6x - 8y - 11 = 0$
58. $9x^2 + 25y^2 - 54x - 200y + 256 = 0$

LEVEL 3

59. Derive the equation of the hyperbola with foci at $(-c, 0)$ and $(c, 0)$ and constant distance $2a$. Let $b^2 = c^2 - a^2$. Show all your work.
60. Derive the equation of the hyperbola with foci at $(0, c)$ and $(0, -c)$ and constant distance $2a$. Let $b^2 = c^2 - a^2$. Show all your work.

7.4 Rotations

All of the curves called conic sections can be characterized by the general second-degree equation

$$Ax^2 + Bxy + Cy^2 + Dx + Ey + F = 0$$

Notice that up to now the xy-term has not appeared (that is, $B = 0$). The presence of this term indicates that the conic has been rotated.

It is important to be able to recognize the curves by inspection of the equation before we begin. The first thing to notice is whether there is an xy-term. Next, note the type of conic (line, parabola, ellipse, circle, or hyperbola). You can use Table 7.9 in the last section if $B = 0$. If $B \neq 0$, then we use the *discriminant* of the general second-degree equation as shown in Table 7.9.

TABLE 7.9	Recognizing a Rotated Conic Section ($B \neq 0$)
Discriminant	**Type of Curve**
$B^2 - 4AC < 0$	Ellipse
$B^2 - 4AC = 0$	Parabola
$B^2 - 4AC > 0$	Hyperbola

EXAMPLE 1   Identifying a conic equation from its equation

Identify each of the given curves.

a. $x^2 + 4xy + 4y^2 = 9$ **b.** $2x^2 + 3xy + y^2 = 25$
c. $x^2 + xy + y^2 - 8x - 8y = 0$ **d.** $xy = 5$

Solution
All of these parts have $B \neq 0$; therefore, we proceed as follows:
a. $B^2 - 4AC = 16 - 4(1)(4) = 0$; parabola
b. $B^2 - 4AC = 9 - 4(2)(1) > 0$; hyperbola
c. $B^2 - 4AC = 1 - 4(1)(1) < 0$; ellipse
d. $B^2 - 4AC = 1 - 4(0)(0) > 0$; hyperbola ∎

The Rotation Concept

To graph a conic that has been rotated, we need to introduce a new coordinate axis for which the curve is in standard position. How can we do this?

This question can be answered by considering how you were able to read the preceding paragraph. What did you do? Probably you turned the book until the paragraph was right-side up. We do the same for a rotated conic.

If we rotate the axis through an angle θ ($0 < \theta < 90°$), the relationship between the old coordinates (x, y) and the new coordinates (x', y') can be found by considering Figure 7.40.

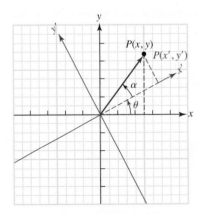

Figure 7.40 Rotation of axes

Rotation of Axes Formulas

Let O be the origin and P be a point with coordinates (x, y) relative to the old coordinate system and (x', y') relative to the newly rotated coordinate system. Let θ be the amount of rotation, and let α be the angle between the x'-axis and $|\overline{OP}|$. Then, using the definition of sine and cosine, we have

$$x = |\overline{OP}| \cos(\theta + \alpha) \qquad x' = |\overline{OP}| \cos\alpha$$
$$y = |\overline{OP}| \sin(\theta + \alpha) \qquad y' = |\overline{OP}| \sin\alpha$$

Using the addition laws (Section 6.4) we find:

$$
\begin{aligned}
x &= |\overline{OP}| \cos(\theta + \alpha) \\
&= |\overline{OP}| (\cos\theta\cos\alpha - \sin\theta\sin\alpha) \\
&= |\overline{OP}| \cos\theta\cos\alpha - |\overline{OP}| \sin\theta\sin\alpha \\
&= (|\overline{OP}| \cos\alpha)\cos\theta - (|\overline{OP}| \sin\alpha)\sin\theta \\
&= x'\cos\theta - y'\sin\theta
\end{aligned}
$$

and
$$
\begin{aligned}
y &= |\overline{OP}| \sin(\theta + \alpha) \\
&= |\overline{OP}| (\sin\theta\cos\alpha + \cos\theta\sin\alpha) \\
&= |\overline{OP}| \sin\theta\cos\alpha + |\overline{OP}| \cos\theta\sin\alpha \\
&= (|\overline{OP}| \cos\alpha)\sin\theta + (|\overline{OP}| \sin\alpha)\cos\theta \\
&= x'\sin\theta + y'\cos\theta
\end{aligned}
$$

We summarize this in the following property box.

ROTATION OF AXES FORMULAS

$$x = x'\cos\theta - y'\sin\theta$$
$$y = x'\sin\theta + y'\cos\theta$$

It is important that we rotate the axes "the right amount." That is, the new axes should be rotated the same amount as the given conic so that the conic will be in standard position after the rotation. To find out how much to rotate the axes, we substitute the values of x and y from the rotation of axes formulas into the general equation for the conic sections:

$$Ax^2 + Bxy + Cy^2 + Dx + Ey + F = 0$$

We obtain (after a lot of simplifying; see Problem 59)

$$(A\cos^2\theta + B\cos\theta\sin\theta + C\sin^2\theta)x'^2 + [B(\cos^2\theta - \sin^2\theta) + 2(C - A)\sin\theta\cos\theta]x'y'$$
$$+ (A\sin^2\theta - B\sin\theta\cos\theta + \cos^2\theta)y'^2 + (D\cos\theta + E\sin\theta)x' + (-D\sin\theta + E\cos\theta)y' + F = 0$$

This looks terrible, but we want to choose θ so that there is no $x'y'$-term. That is,

$$B(\cos^2\theta - \sin^2\theta) + 2(C - A)\sin\theta\cos\theta = 0$$
$$B\cos 2\theta + (C - A)\sin 2\theta = 0$$
$$B\cos 2\theta = (A - C)\sin 2\theta$$
$$\frac{\cos 2\theta}{\sin 2\theta} = \frac{A - C}{B}$$
$$\cot 2\theta = \frac{A - C}{B}$$

Simplifying, we obtain the following formula used for the angle of rotation. (Remember that $B \neq 0$ or there would have been no rotation in the first place.)

AMOUNT OF ROTATION FORMULA

If θ is chosen so that

$$\cot 2\theta = \frac{A - C}{B}$$

then substitution into the rotation of axes formula will give the equation of a conic that does not have a rotation.

Notice that we required $0 < \theta < 90°$, so 2θ is in Quadrant I or Quadrant II. This means that if $\cot 2\theta$ is positive, then 2θ must be in Quadrant I; if $\cot 2\theta$ is negative, then 2θ is in Quadrant II. The procedure is to solve

$$\cot 2\theta = \frac{A - C}{B}$$

and then use that result in the formulas:

$$x = x' \cos\theta - y' \sin\theta$$
$$y = x' \sin\theta + y' \cos\theta$$

Finally, substitute these formulas into the given equation to find the equation for the curve relative to the rotated axes.

EXAMPLE 2   Finding the amount of rotation for a given conic

Find the appropriate rotation so that the given curve will be in standard position relative to the rotated axes. Also find the x and y values in the new coordinate system.

a. $xy = 6$ **b.** $7x^2 - 6\sqrt{3}xy + 13y^2 - 16 = 0$ **c.** $x^2 - 4xy + 4y^2 + 5\sqrt{5}y - 10 = 0$

Solution

a. For $xy = 6$, $A = 0$, $B = 1$, and $C = 0$; $\cot 2\theta = \dfrac{A - C}{B} = \dfrac{0 - 0}{1} = 0$

Thus, $2\theta = 90°$, $\theta = 45°$.

$$x = x' \cos\theta - y' \sin\theta \qquad y = x' \sin\theta + y' \cos\theta$$
$$= x'\left(\frac{1}{\sqrt{2}}\right) - y'\left(\frac{1}{\sqrt{2}}\right) \qquad = x'\left(\frac{1}{\sqrt{2}}\right) + y'\left(\frac{1}{\sqrt{2}}\right)$$
$$= \frac{1}{\sqrt{2}}(x' - y') \qquad = \frac{1}{\sqrt{2}}(x' + y')$$

b. For $7x^2 - 6\sqrt{3}xy + 13y^2 - 16 = 0$, $A = 7$, $B = -6\sqrt{3}$, $C = 13$;

$$\cot 2\theta = \frac{A - C}{B} = \frac{7 - 13}{-6\sqrt{3}} = \frac{1}{\sqrt{3}}$$

Thus, $2\theta = 60°$, $\theta = 30°$.

$$x = x' \cos\theta - y' \sin\theta \qquad y = x' \sin\theta + y' \cos\theta$$
$$= x'\left(\frac{\sqrt{3}}{2}\right) - y'\left(\frac{1}{2}\right) \qquad = x'\left(\frac{1}{2}\right) + y'\left(\frac{\sqrt{3}}{2}\right)$$
$$= \frac{1}{2}\left(\sqrt{3}x' - y'\right) \qquad = \frac{1}{2}\left(x' + \sqrt{3}y'\right)$$

c. For $x^2 - 4xy + 4y^2 + 5\sqrt{5}y - 10 = 0$; $A = 1$, $B = -4$, and $C = 4$;

$$\cot 2\theta = \frac{A - C}{B} = \frac{1 - 4}{-4} = \frac{3}{4}$$

This does not give us an exact value for θ, so we need to use the following identities from trigonometry:

$$\tan 2\theta = \frac{1}{\cot 2\theta} \qquad \sec 2\theta = \pm\sqrt{1 + \tan^2 2\theta}$$
$$= \frac{4}{3} \qquad = \sqrt{1 + \frac{16}{9}} \qquad \text{\textit{First quadrant, so we choose positive value.}}$$
$$= \frac{5}{3}$$
$$\cos 2\theta = \frac{3}{5}$$

Finally,

$$\cos\theta = \pm\sqrt{\frac{1+\cos 2\theta}{2}} \qquad\qquad \sin\theta = \pm\sqrt{\frac{1-\cos 2\theta}{2}}$$

$$= \sqrt{\frac{1+\frac{3}{5}}{2}} \quad \text{\scriptsize First quadrant, choose +.} \qquad\qquad = \sqrt{\frac{1-\frac{3}{5}}{2}} \quad \text{\scriptsize First quadrant, choose +.}$$

$$= \frac{2}{\sqrt{5}} \qquad\qquad\qquad\qquad\qquad\qquad = \frac{1}{\sqrt{5}}$$

Hence, $\quad x = x'\cos\theta - y'\sin\theta \qquad\qquad y = x'\sin\theta + y'\cos\theta$

$$= x'\left(\frac{2}{\sqrt{5}}\right) - y'\left(\frac{1}{\sqrt{5}}\right) \qquad\qquad = x'\left(\frac{1}{\sqrt{5}}\right) + y'\left(\frac{2}{\sqrt{5}}\right)$$

$$= \frac{1}{\sqrt{5}}(2x' - y') \qquad\qquad\qquad = \frac{1}{\sqrt{5}}(x' + 2y')$$

To find the rotation (or the slope of the rotated axes), we first find

$$m = \tan\theta = \frac{\sin\theta}{\cos\theta} = \frac{\dfrac{1}{\sqrt{5}}}{\dfrac{2}{\sqrt{5}}} = \frac{1}{2}$$

Recall (Section 5.3) that the slope of the x'-axis is the tangent of the angle of inclination. Since θ is the angle of inclination, we draw the x'-axis so that it has a rise of one unit and a run of two units (from $m = \frac{1}{2}$). Notice that we found the rotated axes without ever consulting a calculator or a table of values. If you do want to know the angle of rotation in Example 2, use the inverse tangent function on your calculator to find $\theta = \tan^{-1}\frac{1}{2} \approx 26.6°$.

Graphing Rotated Conics

We now state a procedure for graphing rotated conics.

GRAPH A ROTATED CONIC

The procedure for graphing a rotated conic section follows these steps:

Step 1 Determine the nature of the conic by calculating $B^2 - 4AC$.

Step 2 Find the angle of rotation.

Step 3 Find x and y in the new coordinate system.

Step 4 Substitute the values found in Step 3 into the given equation and simplify.

Step 5 Sketch the resulting equation relative to the new x'- and y'-axes. You will have to complete the square if it is not centered at the origin.

EXAMPLE 3   Graphing a rotated hyperbola

Graph $xy = 6$.

Solution

The graph of this equation is a hyperbola, since $B^2 - 4AC = 1 - 0 > 0$. From Example 2a, $\theta = 45°$, and $x = \frac{1}{\sqrt{2}}(x' - y')$ and $y = \frac{1}{\sqrt{2}}(x' + y')$. Substitute these values into the original equation:

$$xy = 6$$

$$\left[\frac{1}{\sqrt{2}}(x' - y')\right]\left[\frac{1}{\sqrt{2}}(x' + y')\right] = 6$$

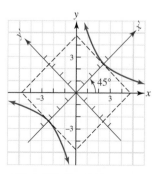

Figure 7.41 Graph of $xy = 6$

$$x'^2 - y'^2 = 12 \qquad \text{Simplify (see Problem 1) for details.}$$
$$\frac{x'^2}{12} - \frac{y'^2}{12} = 1$$

The sketch is shown in Figure 7.41.   ∎

EXAMPLE 4   Graphing a rotated ellipse

Graph $7x^2 - 6\sqrt{3}xy + 13y^2 - 16 = 0$.

Solution

The graph of this equation is an ellipse, since $B^2 - 4AC = 36(3) - 4(7)(13) < 0$. From Example 2b, $\theta = 30°$, $x = \frac{1}{2}\left(\sqrt{3}x' - y'\right)$, and $y = \frac{1}{2}\left(x' + \sqrt{3}y'\right)$. Substitute this into the original equation:

$$7x^2 - 6\sqrt{3}xy + 13y^2 - 16 = 0$$

$$7\left[\frac{1}{2}\left(\sqrt{3}x' - y'\right)\right]^2 - 6\sqrt{3}\left[\frac{1}{2}\left(\sqrt{3}x' - y'\right)\right]\left[\frac{1}{2}\left(x' + \sqrt{3}y\right)\right] + 13\left[\frac{1}{2}\left(x' + \sqrt{3}y'\right)\right]^2 - 16 = 0$$

Simplify (see Problem 2) for details: $\dfrac{x'^2}{4} + \dfrac{y'^2}{1} = 1$

The sketch is shown in Figure 7.42.   ∎

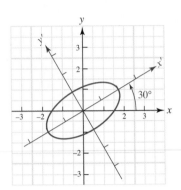

Figure 7.42 Graph of
$7x^2 - 6\sqrt{3}xy + 13y^2 - 16 = 0$

EXAMPLE 5   Graph of a rotated parabola

Graph $x^2 - 4xy + 4y^2 + 5\sqrt{5}y - 10 = 0$.

Solution

The graph of this equation is a parabola, since $B^2 - 4AC = 16 - 4(1)(4) = 0$. From Example 2c, the rotation is given by $\tan\theta = \frac{1}{2}$, and $x = \frac{1}{\sqrt{5}}(2x' - y')$ and $y = \frac{1}{\sqrt{5}}(x' + 2y')$. Substitute this into the original equation:

$$x^2 - 4xy + 4y^2 + 5\sqrt{5}y - 10 = 0$$

$$\left[\frac{1}{\sqrt{5}}(2x' - y')\right]^2 - 4\left[\frac{1}{\sqrt{5}}(2x' - y')\right]\left[\frac{1}{\sqrt{5}}(x' + 2y')\right] + 4\left[\frac{1}{\sqrt{5}}(x' + 2y')\right]^2 + 5\sqrt{5}\left[\frac{1}{\sqrt{5}}(x' + 2y')\right] - 10 = 0$$

Simplify (see Problem 3) for details: $y'^2 + 2y' = -(x' - 3)$
Complete the square to find: $(y' + 1)^2 = -(x' - 3)$
The sketch is shown in Figure 7.43.   ∎

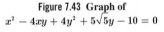

Vertex (3, –1)
measured on
rotated axes

Figure 7.43 Graph of
$x^2 - 4xy + 4y^2 + 5\sqrt{5}y - 10 = 0$

To graph the general second-degree equation

$$Ax^2 + Bxy + Cy^2 + Dx + Ey + F = 0$$

follow the steps shown in the box.

GRAPHING A CONIC

To graph a general second-degree equation

$$Ax^2 + Bxy + Cy^2 + Dx + Ey + F = 0$$

Step 1   Identify the curve (see the following flowchart).

Step 2   Rotate the axes; if $B = 0$, go to the next step.

Step 3   Translate the axes.

Step 4   Graph the standard-form equation.

(continues)

GRAPHING A CONIC *(continued)*

These steps are summarized in the following flowchart.

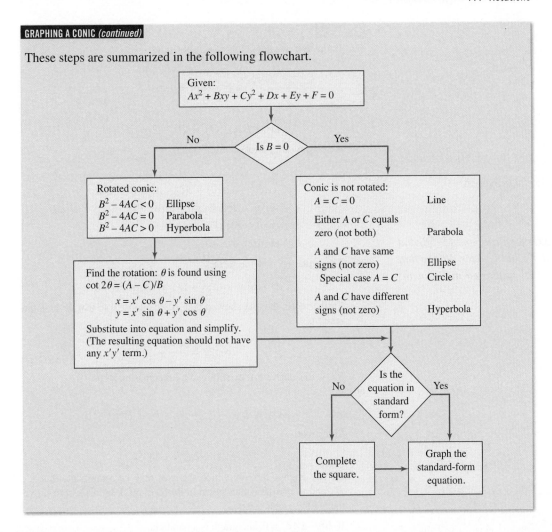

PROBLEM SET 7.4

LEVEL 1

Because of the amount of arithmetic and algebra involved, many steps were left out of the examples in this section. In Problems 1–3, fill in the details left out of the indicated example.

1. Example 3   2. Example 4   3. Example 5

Find the x and y substitutions (using exact values) for a given (x', y') and the angle θ given in Problems 4–6.

4. $60°$ 5. $-30°$ 6. $-45°$

Find the x and y substitutions (correct to two decimal places) for a given (x', y') and the angle θ given in Problems 7–9.

7. $15°$ 8. $28°$ 9. $83°$

10. IN YOUR OWN WORDS Describe a process for sketching any conic section.

Identify the curves given in Problems 11–30.

11. $xy = 12$
12. $xy = -14$
13. $8x^2 - 4xy + 5y^2 = 36$
14. $8x^2 + 4xy + 5y^2 - 12 = 0$
15. $5x^2 - 26xy + 5y^2 + 72 = 0$
16. $6x^2 + 4\sqrt{3}xy + y^2 + x - y = 3$
17. $3x^2 + 2\sqrt{3}xy + y^2 + 16x - 16\sqrt{3}y = 0$
18. $3x^2 - 2\sqrt{3}xy + y^2 + 24x + 24\sqrt{3}y = 0$
19. $x^2 + 2xy + y^2 + 12\sqrt{2}x - 6 = 0$
20. $3x^2 - 10xy + 3y^2 - 32 = 0$
21. $2x^2 - 4\sqrt{3}xy + 3y^2 - 2x + 5 = 0$
22. $x^2 + 4xy + 4y^2 + 10\sqrt{5}x = 9$
23. $5x^2 - 4xy + 8y^2 = 36$
24. $5x^2 - 15xy + 5y^2 + 18 = 0$
25. $5x^2 - 3xy + y^2 + 65x - 25y + 203 = 0$
26. $6x^2 - 5xy - 2y^2 + 2x - 3 = 0$
27. $17x^2 - 12xy + 8y^2 - 80 = 0$
28. $24x^2 + 16\sqrt{3}xy + 8y^2 - x + \sqrt{3}y - 8 = 0$
29. $16x^2 - 24xy + 9y^2 - 60x - 80y + 100 = 0$
30. $x^2 + 2\sqrt{3}xy + 3y^2 + 2\sqrt{3}x + 2y - 16 = 0$

Find the appropriate rotation so that the curves given by the equations in Problems 31–44 will be in standard position relative to the rotated axes. Also find the x and y values in the new coordinate system.

31. $xy = 8$
32. $xy = 10$
33. $xy = -4$
34. $xy = -1$
35. $13x^2 - 10xy + 13y^2 - 72 = 0$

36. $23x^2 + 26\sqrt{3}xy - 3y^2 - 144 = 0$

37. $3x^2 - 10xy + 3y^2 - 32 = 0$

38. $5x^2 - 8xy + 5y^2 - 9 = 0$

39. $x^2 + 4xy + 4y^2 + 10\sqrt{5}x = 9$

40. $10x^2 + 24xy + 17y^2 - 9 = 0$

41. $3xy - 4y^2 - 18 = 0$

42. $24x^2 + 16\sqrt{3}xy + 8y^2 - x + \sqrt{3}y - 8 = 0$

43. $13x^2 - 6\sqrt{3}xy + 7y^2 + \left(16\sqrt{3} - 8\right)x + \left(-16 - 8\sqrt{3}\right)y + 16 = 0$

44. $21x^2 + 10\sqrt{3}xy + 31y^2 - 72x - 16\sqrt{3}x - 72\sqrt{3}y + 16 = 0$

LEVEL 2

First find the equation of the rotated axes for Problems 45–58, and then graph the curve using the rotated x'- and y'-axes. Notice that preliminary work for these problems was requested in Problems 31–44.

45. $xy = 8$

46. $xy = 10$

47. $xy = -4$

48. $xy = -1$

49. $13x^2 - 10xy + 13y^2 - 72 = 0$

50. $23x^2 + 26\sqrt{3}xy - 3y^2 - 144 = 0$

51. $3x^2 - 10xy + 3y^2 - 32 = 0$

52. $5x^2 - 8xy + 5y^2 - 9 = 0$

53. $x^2 + 4xy + 4y^2 + 10\sqrt{5}x = 9$

54. $10x^2 + 24xy + 17y^2 - 9 = 0$

55. $3xy - 4y^2 - 18 = 0$

56. $24x^2 + 16\sqrt{3}xy + 8y^2 - x + \sqrt{3}y - 8 = 0$

57. $13x^2 - 6\sqrt{3}xy + 7y^2 + \left(16\sqrt{3} - 8\right)x + \left(-16 - 8\sqrt{3}\right)y + 16 = 0$

58. $21x^2 + 10\sqrt{3}xy + 31y^2 - 72x - 16\sqrt{3}x - 72\sqrt{3}y + 16y + 16 = 0$

LEVEL 3

59. In the text, we noted that if we substitute

$$x = x'\cos\theta - y'\sin\theta$$
$$y = x'\sin\theta + y'\cos\theta$$

into

$$Ax^2 + Bxy + Cy^2 + Dx + Ey + F = 0$$

we obtain (after a lot of simplifying)

$$(A\cos^2\theta + B\cos\theta\sin\theta + C\sin^2\theta)x'^2$$
$$+ [B(\cos^2\theta - \sin^2\theta) + 2(C - A)\sin\theta\cos\theta]x'y'$$
$$+ (A\sin^2\theta - B\sin\theta\cos\theta + C\cos^2\theta)y'^2$$
$$+ (D\cos\theta + E\sin\theta)x'$$
$$+ (-D\sin\theta + E\cos\theta)y' + F = 0$$

This equation is of the form

$$A'x^2 + B'xy + C'y^2 + D'x + E'y + F' = 0$$

Furthermore, the text showed that, if we choose θ so that $B' = 0$, the result is

$$\cos 2\theta = \frac{A - C}{B}$$

Show this derivation by filling in all the missing steps.

60. Let

$$Ax^2 + Bxy + Cy^2 + Dx + Ey + F = 0$$

Show that

$$B^2 - 4AC = B'^2 - 4A'C'$$

for any angle θ through which the axes may be rotated and that A', B', and C' are the values given in Problem 59. Use this fact to prove that (if the graph exists):

If $B^2 - 4AC = 0$, the graph is a parabola.
If $B^2 - 4AC < 0$, the graph is an ellipse.
If $B^2 - 4AC > 0$, the graph is a hyperbola.

7.5 Polar Coordinates

Up to now, the only coordinate system we have used is a Cartesian coordinate system. In this section, we introduce a *polar coordinate system*, in which points are located by specifying their distance from a fixed point and their direction in relation to a fixed line. We use the term **rectangular coordinates** to refer to Cartesian coordinates to help you remember that Cartesian coordinates are plotted in a rectangular fashion.

Plotting Points in Polar Coordinates

In a **polar coordinate system**, we fix a point O, called the **pole**, and draw a horizontal ray (half-line), called the **radial axis**, from the point O. Then each point in the plane is represented by an ordered pair $P(r, \theta)$ where θ (the **polar angle**) measures the angle from the radial axis, and r (the **radial distance**) measures the directed distance from the pole to the point P. Both r and θ can be any real number.

When plotting points, rotate the radial axis through an angle θ, as shown in Figure 7.44. You might find it helpful to rotate your pencil as the axis—the tip points in the positive direction and the eraser in the negative direction. If θ is measured in a counterclockwise direction, it is positive; if θ is measured in a clockwise direction, it is negative. Next, plot r on the radial axis (the pencil).

Notice that any real number can be plotted on this real number line in the direction of the tip if the number is positive and in the direction of the eraser if it is negative. Plotting points seems easy, but it is necessary that you completely understand this process. Study each part of Example 1 and make sure you understand how each point is plotted. We assume an orientation provided by an x-axis and a y-axis with the pole at the origin, and the rotation of the radial axis is measured from the positive x-axis. We consider that points can be plotted using either rectangular or polar coordinates, and that the system we are using in a particular example must be either specified or obvious from the context.

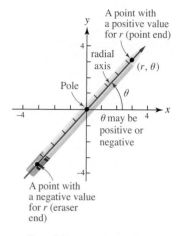

Figure 7.44 Polar-form points

EXAMPLE 1   Plotting polar-form points

Plot each of the following polar-form points:

$$A\left(4,\ \frac{\pi}{3}\right),\ B\left(-4,\ \frac{\pi}{3}\right),\ C\left(3,\ -\frac{\pi}{6}\right),\ D\left(-3,\ -\frac{\pi}{6}\right),\ E(-3,\ 3),\ F(-3,\ -3),$$

$$G(-4,\ -2),\ H\left(5,\ \frac{3\pi}{2}\right),\ I\left(-5,\ \frac{\pi}{2}\right),\ J\left(5,\ -\frac{\pi}{2}\right).$$

Solution
Points A, B, C, and D illustrate the basic ideas of plotting polar-form points.

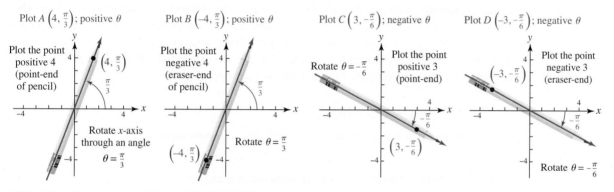

Points E, F, G, H, I, and J illustrate common situations that can sometimes be confusing. Make sure you take time with each example.

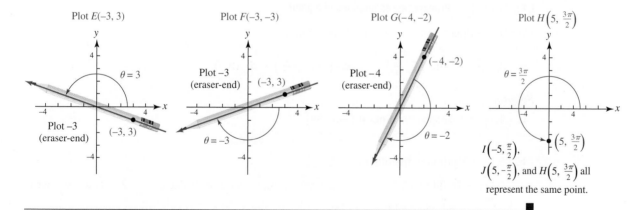

Note that the point $(-r,\ \theta)$ can be obtained by *reflecting* $(r,\ \theta)$ through the origin, whereas the point $(r,\ -\theta)$ can be obtained by reflecting $(r,\ \theta)$ in the x-axis. These features of the polar coordinate system are illustrated in Figure 7.45.

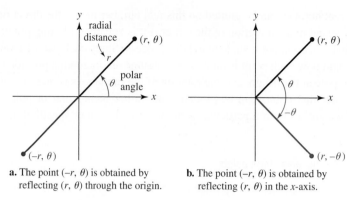

a. The point $(-r, \theta)$ is obtained by reflecting (r, θ) through the origin.

b. The point $(-r, \theta)$ is obtained by reflecting (r, θ) in the x-axis.

Figure 7.45 Plotting points using reflections

Primary Representations of a Point

Notice from Example 1 that ordered pairs in polar form are not associated in a one-to-one fashion with points in the plane. Indeed, given any point in the plane, there are infinitely many ordered pairs of polar coordinates associated with that point. If you are given a point $P(r, \theta)$ other than the pole in polar coordinates, then $(-r, \theta + \pi)$ also represents P. In addition, there are infinitely many other representations of P—namely, multiples of 2π added to these representations. We call (r, θ) and $(-r, \theta + \pi)$ the *primary representations of the point* if the angles θ and $\theta + \pi$ are between 0 and 2π. If the angle is not between 0 and 2π, then you can add or subtract a multiple of 2π so that it is between 0 and 2π. We summarize in the following definition box.

☠ Remember to simplify, so the second component represents a nonnegative angle less than one revolution. ☠

> **PRIMARY REPRESENTATIONS OF A POINT IN POLAR FORM**
>
> Every nonzero point in polar form has two **primary representations**:
> $$(r, \theta) \quad \text{and} \quad (-r, \pi + \theta)$$
> where the second component in each case is between 0 and 2π.

You still need to recognize when a representation is primary, and you will be leaving your answers in this form.

EXAMPLE 2   Primary representations of a point

Give both primary representations of each point:

a. $\left(3, \frac{\pi}{4}\right)$ **b.** $\left(5, \frac{5\pi}{4}\right)$ **c.** $\left(-6, -\frac{2\pi}{3}\right)$ **d.** $(9, 5)$ **e.** $(9, 7)$

Solution

a. $\left(3, \frac{\pi}{4}\right)$ is primary; the other is $\left(-3, \frac{5\pi}{4}\right)$

   Change the sign of r and add π: $\theta = \frac{\pi}{4} + \pi = \frac{5\pi}{4}$.

b. $\left(5, \frac{5\pi}{4}\right)$ is primary; the other is $\left(-5, \frac{\pi}{4}\right)$.

   $\frac{5\pi}{4} + \pi = \frac{9\pi}{4}$, but $\left(-5, \frac{9\pi}{4}\right)$ is not a primary representation because $\frac{9\pi}{4} > 2\pi$. Thus, we write $\left(-5, \frac{9\pi}{4}\right)$ and then add a multiple of 2π to obtain the correct primary representation of $\left(-5, \frac{\pi}{4}\right)$.

c. $\left(-6, -\frac{2\pi}{3}\right)$ has primary representations $\left(-6, \frac{4\pi}{3}\right)$ and $\left(6, \frac{\pi}{3}\right)$.

d. $(9, 5)$ is primary; the other is $(-9, 5 - \pi)$; a point such as $(-9, 5 - \pi)$ may be approximated by $(-9, 1.86)$.

e. $(9, 7)$ has primary representations $(9, 7 - 2\pi)$, or $(9, 0.72)$, and $(-9, 7 - \pi)$, or $(-9, 3.86)$.

EXAMPLE 3 Find polar-form coordinates

A portion of the floor of the Paris Observatory is reproduced in Figure 7.46. Can you locate Washington, D.C.?

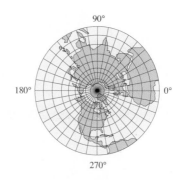

Figure 7.46 Location of Washington, D.C.?

Solution

There are infinitely many polar forms, but only one actual location. We will find the primary representations for this city in polar form. First, we locate the point showing the approximate location of Washington, D.C. (See arrow in Figure 7.47.)

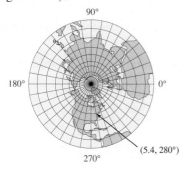

(5.4, 280°)

Figure 7.47 Location of Washington, D.C.

Begin by finding r; we count the concentric circles and see that Washington, D.C., is located between the 5th and 6th circles—estimate $r \approx 5.4$. Next, we notice from the rays marked 0°, 90°, 180°, and 270° that each such ray represents 10°. From this, we can estimate $\theta \approx 280°$.

Now for the two primary representations, we have $(r, \theta) = (5.4, 280°)$ and $(-r, \theta + 180°) = (-5.4, 460°)$ or $(-5.4, 100°)$. The two primary representations are $(5.4, 280°)$ and $(-5.4, 100°)$. ■

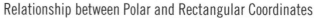

Relationship between Polar and Rectangular Coordinates

The relationship between polar and rectangular coordinates can be found by using the definition of trigonometric functions (see Figure 7.48).

CHANGING COORDINATES

The procedure for changing from one coordinate system to another:

Step 1 To change from *polar to rectangular* form, use the formulas:
$$x = r \cos\theta \qquad y = r \sin\theta$$

Step 2 To change from *rectangular to polar* form, use the formulas:
$$r = \sqrt{x^2 + y^2} \qquad \overline{\theta} = \tan^{-1}\left|\frac{y}{x}\right|, x \neq 0$$

where $\overline{\theta}$ is the reference angle for θ. Place θ in the proper quadrant by noting the signs of x and y. If $x = 0$, then $\overline{\theta} = \frac{\pi}{2}$.

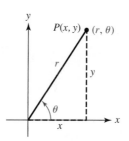

Figure 7.48 Conversions

EXAMPLE 4 Converting from polar to rectangular coordinates

Change the polar coordinates $\left(-3, \frac{7\pi}{4}\right)$ to rectangular coordinates.

Solution

$$x = -3\cos\frac{7\pi}{4} = -3\left(\frac{\sqrt{2}}{2}\right) = -\frac{3\sqrt{2}}{2}$$

$$y = -3\sin\frac{7\pi}{4} = -3\left(-\frac{\sqrt{2}}{2}\right) = \frac{3\sqrt{2}}{2}$$

The rectangular coordinates are $\left(-\frac{3\sqrt{2}}{2}, \frac{3\sqrt{2}}{2}\right)$.

EXAMPLE 5   Converting from rectangular to polar coordinates

Write both primary representations of the polar-form coordinates for the point with rectangular coordinates $\left(\frac{5\sqrt{3}}{2}, -\frac{5}{2}\right)$.

Solution

$$r = \sqrt{\left(\frac{5\sqrt{3}}{2}\right)^2 + \left(-\frac{5}{2}\right)^2} = \sqrt{\frac{75}{4} + \frac{25}{4}} = 5$$

Note that θ is in Quadrant IV because x is positive and y is negative.

$$\bar{\theta} = \tan^{-1}\left|\frac{-\frac{5}{2}}{\frac{5\sqrt{3}}{2}}\right| = \tan^{-1}\left(\frac{1}{\sqrt{3}}\right) = \frac{\pi}{6}; \text{ thus, } \theta = \frac{11\pi}{6} \text{ (Quadrant IV)}.$$

The polar form coordinates are $\left(5, \dfrac{11\pi}{6}\right)$ and $\left(-5, \dfrac{5\pi}{6}\right)$.

■ COMPUTATIONAL WINDOW   _□✕

Check the owner's manual for your calculator. Many have keys for converting between polar and rectangular coordinates.

Points on a Polar Curve

The **graph** of an equation in polar coordinates is the set of all points P having polar coordinates (r, θ) that satisfy the given equation. Circles, lines through the origin, and rays emanating from the origin have particularly simple equations in polar coordinates.

EXAMPLE 6   Graphing circles, rays, and lines

Graph: **a.** $r = 6$ **b.** $\theta = \frac{\pi}{6}, r \geq 0$ **c.** $\theta = \frac{\pi}{6}$

Solution

  a. The graph is the set of all points (r, θ) such that the first component is 6 for any angle θ. This is a circle with radius 6 centered at the origin, as shown in Figure 7.49**a**.
  b. The graph is the closed half-line (ray) that emanates from the origin and makes an angle of $\frac{\pi}{6}$ with the positive x-axis. This graph is shown in Figure 7.49**b**.
  c. This is the line through the origin that makes an angle of $\frac{\pi}{6}$ with the positive x-axis, as shown in Figure 7.49**c**.

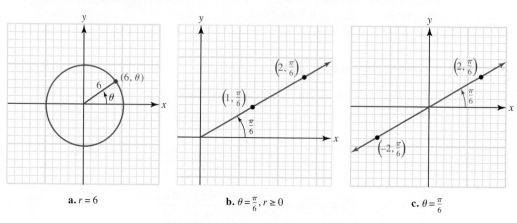

a. $r = 6$ **b.** $\theta = \frac{\pi}{6}, r \geq 0$ **c.** $\theta = \frac{\pi}{6}$

Figure 7.49 Graphing circles, rays, and lines

As with other equations, we begin graphing polar-form curves by plotting some points. However, you must first be able to recognize whether a point in polar form satisfies a given equation.

EXAMPLE 7   Verifying that polar coordinates satisfy an equation

Show that each of the given points lies on the polar graph whose equation is

$$r = \frac{2}{1 - \cos\theta}$$

a. $\left(2, \frac{\pi}{2}\right)$ **b.** $\left(-2, \frac{3\pi}{2}\right)$ **c.** $(-1, 2\pi)$

Solution
Begin by substituting the given coordinates into the equation.

a. $2 \overset{?}{=} \dfrac{2}{1 - \cos\frac{\pi}{2}} = \dfrac{2}{1 - 0} = 2$; $\left(2, \frac{\pi}{2}\right)$ is on the curve because it satisfies the equation.

b. $-2 \overset{?}{=} \dfrac{2}{1 - \cos\frac{3\pi}{2}} = \dfrac{2}{1 - 0} = 2$; this equation is **not** true. Although the equation is not satisfied, we *cannot* say that the point is not on the curve. Indeed, we see from part **a** that it is on the curve because $\left(-2, \frac{3\pi}{2}\right)$ and $\left(2, \frac{\pi}{2}\right)$ name the same point! So even if one primary representation of a point does not satisfy the equation, we must still check the other primary representation of the point to see whether it satisfies the equation.

c. The primary representation of $(-1, 2\pi)$ is $(-1, 0)$.

$$-1 = \frac{2}{1 - \cos 0}$$

is undefined. Check the other representation of the same point—namely, $(1, \pi)$:

$$1 \overset{?}{=} \frac{2}{1 - \cos\pi} = \frac{2}{1 - (-1)} = 1$$

Thus, the point is on the curve. ∎

We will discuss the graphing of polar-form curves in the next section, but sometimes a polar-form equation can be graphed by changing it to rectangular form.

EXAMPLE 8   Polar-form graphing by changing to rectangular form

Graph the given polar-form curves by changing to rectangular form.

a. $r = 3\cos\theta$ **b.** $r\cos\theta = 4$ **c.** $r = \dfrac{6}{2\sin\theta + \cos\theta}$

Solution

a.

$r = 3\cos\theta$	*Given equation*
$r^2 = 3r\cos\theta$	*Multiply by r.*
$x^2 + y^2 = 3x$	*Because $x = r\cos\theta$ and $r^2 = x^2 + y^2$*
$x^2 - 3x + y^2 = 0$	*Subtract $3x$ from both sides.*
$\left[x^2 - 3x + \left(\dfrac{3}{2}\right)^2\right] + y^2 = \dfrac{9}{4}$	*Complete the square.*
$\left(x - \dfrac{3}{2}\right)^2 + y^2 = \dfrac{9}{4}$	*Factor.*

We see this is a circle with center at $\left(\frac{3}{2}, 0\right)$ and radius $\frac{3}{2}$. The graph is shown in Figure 7.50a.

b. Because $x = r\cos\theta$, we see that the given equation can be written $x = 4$, which is a vertical line (Figure 7.50**b**).

c.
$$r = \frac{6}{2\sin\theta + \cos\theta} \qquad \textit{Given equation}$$

$$2r\sin\theta + r\cos\theta = 6 \qquad \textit{Multiply both sides by } (2\sin\theta + \cos\theta).$$

$$2y + x = 6 \qquad \textit{Substitute } y = r\sin\theta \text{ and } x = r\cos\theta.$$

$$y = -\frac{1}{2}x + 3 \qquad \textit{Solve for } y.$$

We see this is a line with y-intercept 3 and slope $-\frac{1}{2}$ (Figure 7.50**c**).

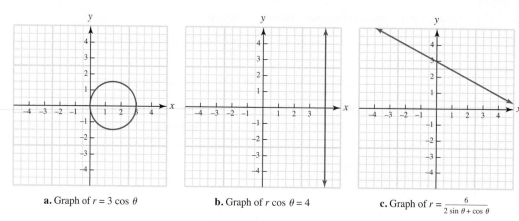

a. Graph of $r = 3\cos\theta$

b. Graph of $r\cos\theta = 4$

c. Graph of $r = \dfrac{6}{2\sin\theta + \cos\theta}$

Figure 7.50 Polar-form graphs

PROBLEM SET 7.5

1. IN YOUR OWN WORDS Describe a procedure for plotting points in polar form.

2. IN YOUR OWN WORDS Show the derivation of the polar-to-rectangular-form conversion equations.

3. What are the primary representations of a point, and why is it necessary to consider two primary representations?

4. What is the procedure for verifying whether a given ordered pair satisfies an equation?

Plot each of the polar-form points given in Problems 5–8. Give both primary representations and the rectangular coordinates of the point.

5. a. $\left(4, \frac{\pi}{4}\right)$ **b.** $\left(6, \frac{\pi}{3}\right)$ **c.** $\left(5, \frac{2\pi}{3}\right)$

6. a. $\left(3, -\frac{\pi}{6}\right)$ **b.** $\left(\frac{3}{2}, -\frac{5\pi}{6}\right)$ **c.** $(-4, 4)$

7. a. $(1, 3\pi)$ **b.** $(1, 2)$ **c.** $\left(-2, -\frac{3\pi}{2}\right)$

8. a. $(2, -\pi)$ **b.** $(-2, \pi)$ **c.** $(0, -3)$

Plot the given rectangular-form points in Problems 9–12 and give both primary representations in polar form.

9. a. $(5, 5)$ **b.** $\left(-1, \sqrt{3}\right)$ **c.** $\left(2, -2\sqrt{3}\right)$

10. a. $(1, 1)$ **b.** $(3, -3)$ **c.** $(-2, -2)$

11. a. $(3, 7)$ **b.** $(-3, 7)$ **c.** $\left(3, -3\sqrt{3}\right)$

12. a. $(3, 0)$ **b.** $(-3, 0)$ **c.** $\left(\sqrt{3}, -1\right)$

13. Name the country or island in Figure 7.51 in which each of the named points is located.

a. $(5.5, 260°)$ **b.** $(7.5, 78°)$
c. $(-2, 140°)$ **d.** $(-4, 80°)$

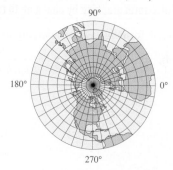

Figure 7.51 Map on the floor of the Paris Observatory

14. Name possible coordinates for each of the following cities (see Figure 7.51).

a. Miami, Florida **b.** Los Angeles, California
c. Mexico City, Mexico **d.** London, England

Write each equation given in Problems 15–22 in rectangular form.

15. $r = 4\sin\theta$ **16.** $r = 16$
17. $r = \sec\theta$ **18.** $r = 2\cos\theta$

19. $r^2 = \dfrac{2}{1 + \sin^2 \theta}$ **20.** $r^2 = \dfrac{2}{3\cos^2 \theta - 1}$

21. $r = 1 - \sin \theta$ **22.** $r = 4\tan \theta$

Sketch the graph of each equation given in Problems 23–30.

23. $r = \frac{3}{2}$ **24.** $r = \frac{3}{2}, \; 0 \le \theta < 2$

25. $r = \sqrt{2}, \; 0 \le \theta < 2$ **26.** $r = 4$

27. $\theta = 1$ **28.** $\theta = 1, \; r \ge 0$

29. $\theta = \frac{\pi}{6}, \; r < 0$ **30.** $\theta = -\frac{\pi}{3}$

Tell whether each of the points given in Problems 31–38 lies on the curve

$$r = \dfrac{5}{1 - \sin \theta}$$

31. $\left(10, \frac{\pi}{6}\right)$ **32.** $\left(5, \frac{\pi}{2}\right)$

33. $\left(-10, \frac{5\pi}{6}\right)$ **34.** $\left(-5, \frac{\pi}{2}\right)$

35. $\left(-\frac{10}{3}, \frac{5\pi}{6}\right)$ **36.** $\left(\frac{10}{3}, \frac{\pi}{6}\right)$

37. $\left(20 + 10\sqrt{3}, \frac{\pi}{3}\right)$ **38.** $\left(-10, \frac{\pi}{3}\right)$

Tell whether each of the points given in Problems 39–46 lies on the curve

$$r = 2(1 - \cos \theta)$$

39. $\left(1, \frac{\pi}{3}\right)$ **40.** $\left(1, -\frac{\pi}{3}\right)$

41. $\left(-1, \frac{\pi}{3}\right)$ **42.** $\left(-2, \frac{\pi}{2}\right)$

43. $\left(2 + \sqrt{2}, \frac{\pi}{4}\right)$ **44.** $\left(-2 - \sqrt{2}, \frac{\pi}{4}\right)$

45. $\left(0, \frac{\pi}{4}\right)$ **46.** $\left(0, -\frac{2\pi}{3}\right)$

Find three distinct ordered pairs satisfying each of the equations given in Problems 47–56. Give both primary representations for each point.

47. $r^2 = 9\cos \theta$ **48.** $r^2 = 9\cos 2\theta$

49. $r = 3\theta$ **50.** $r = 5\theta$

51. $r = 3$ **52.** $\theta = 3$

53. $r = 2 - 3\sin \theta$ **54.** $r = 2(1 + \cos \theta)$

55. $\dfrac{r}{1 - \sin \theta} = 2$ **56.** $r = \dfrac{8}{1 - 2\cos \theta}$

57. The equation $r = 1 - \sin \theta$ is used by entomologists to describe the "dance" of a certain species of bee. The dance tells other bees the direction and distance of a food source. Plot points to graph the path of this bee's dance. (For more information about this topic, see the Group Research Projects at the end of this chapter.)

58. The equation $r = 1 - \cos \theta$ is used by entomologists to describe the "dance" of a certain species of bee. The dance tells other bees the direction and distance of a food source. Plot points to graph the path of this bee's dance. (For more information about this topic, see the Group Research Projects at the end of this chapter.)

LEVEL 3

59. a. What is the distance between the polar-form points $\left(3, \frac{\pi}{3}\right)$ and $\left(7, \frac{\pi}{4}\right)$? Explain why you cannot use the distance formula for these ordered pairs.

    b. What is the distance between the polar-form points (r_1, θ_1) and (r_2, θ_2)?

    c. Find an equation for a circle of radius a and polar-form center (R, α).

60. Show that the graph of the polar equation $r = a \sin \theta + b \cos \theta$ is a circle. Find its center and radius.

7.6 Graphing in Polar Coordinates

In the last section, we found that the graph of the polar equation $r = a$ is a circle centered at the origin with radius a, and $\theta = a$ is a line passing through the origin. In this section, we discuss techniques for graphing polar-form equations and categorize several important types of polar-form curves.

Graphing by Plotting Points

We have examined polar forms for lines and circles, and in this section, we shall examine curves that are more easily represented in polar coordinates than in rectangular coordinates. We begin with a simple **spiral**, which is a plane curve that winds around a fixed center point at a continuously increasing or decreasing distance from that fixed point.

EXAMPLE 1   **A spiral by plotting points**

Graph $r = \theta$ for $\theta \ge 0$.

Solution

Set up a table of values. Notice that as θ increases, r must also increase.

θ	r
1	1
2	2
3	3
4	4
5	5
6	6

Choose a value of θ and then find a corresponding r so that (r, θ) satisfies the equation.

Plot each of these points and connect them, as shown in Figure 7.52.[*]

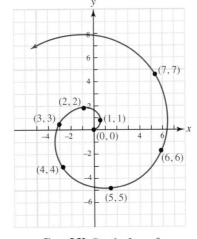

Figure 7.52 Graph of $r = \theta$

---

[*]Many books use what is called *polar graph paper*, but such paper is not really necessary. It also obscures the fact that polar curves and rectangular curves are both plotted on a Cartesian coordinate system, only with a different meaning attached to the ordered pairs. You can estimate the angles as necessary without polar graph paper. Remember, just as in graphing rectangular curves, the key is not in plotting many points but in recognizing the type of curve and then plotting a few key points.

In Example 1, the polar equation $r = \theta$ has exactly one value of r for each value of θ. Thus, the relationship given by $r = \theta$ is a function of θ, and we can write the polar form $r = f(\theta)$, where $f(\theta) = \theta$. A function of the form $r = f(\theta)$, where θ is a polar angle and r is the corresponding radial distance, is called a **polar function**.

Cardioids

Next, we examine a class of polar curves called **cardioids** because of their heartlike shape.

EXAMPLE 2   A cardioid by plotting points

Graph $r = 2(1 - \cos \theta)$.

Solution
Construct a table of values by choosing values for θ and approximating the corresponding values for r. Do not forget that even though we pick θ and find a value for r, we still represent the table values as ordered pairs of the form (r, θ). The points are connected as shown in Figure 7.53.

θ	r
0	0
1	0.9193954
2	2.832294
3	3.979985
4	3.307287
5	1.432676
6	0.079659

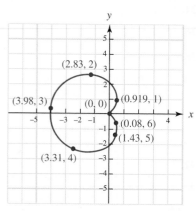

Figure 7.53 Graph of $r = 2(1 - \cos \theta)$

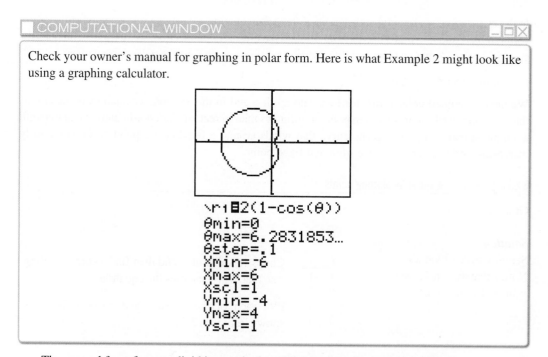

Check your owner's manual for graphing in polar form. Here is what Example 2 might look like using a graphing calculator.

The general form for a cardioid in standard position is given in the following box.

CARDIOID

The equation of a **standard-position cardioid** is

$$r = a(1 - \cos\theta)$$

In general, a cardioid in standard position can be completely determined by plotting four particular points:

θ	$r = a(1 - \cos\theta)$	Ordered Pair Notation
0	$r = a(1 - \cos 0) = a(1 - 1) = 0$	$(0, 0)$
$\frac{\pi}{2}$	$r = a\left(1 - \cos\frac{\pi}{2}\right) = a(1 - 0) = a$	$\left(a, \frac{\pi}{2}\right)$
π	$r = a(1 - \cos\pi) = a(1 + 1) = 2a$	$(2a, \pi)$
$\frac{3\pi}{2}$	$r = a\left(1 - \cos\frac{3\pi}{2}\right) = a(1 - 0) = a$	$\left(a, \frac{3\pi}{2}\right)$

These reference points are all you need when graphing other standard-position cardioids, because they will all have the same shape as the one shown in Figure 7.53.

Symmetry and Rotations

When sketching a polar graph, it is often useful to determine whether the graph has been rotated or whether it has any **polar-form symmetry**. If an angle α is subtracted from θ in a polar-form equation, it has the effect of rotating the curve (see Problem 59).

POLAR-FORM ROTATIONS

The polar graph of $r = f(\theta - \alpha)$ is the same as the polar graph of $r = f(\theta)$, only rotated through an angle α. If α is positive, the **rotation** is counterclockwise, and if α is negative, the **rotation** is clockwise.

EXAMPLE 3   Rotated cardioid

Graph $r = 3 - 3\cos\left(\theta - \frac{\pi}{6}\right)$.

Solution
Recognize this as a cardioid with $a = 3$ and a rotation of $\frac{\pi}{6}$. Plot the four points shown in Figure 7.54 and draw the cardioid.

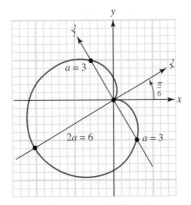

Figure 7.54 Graph of $r = 3 - 3\cos\left(\theta - \dfrac{\pi}{6}\right)$

If the rotation is $180° = \pi$, the equation simplifies considerably. Consider

$r = 3 - 3\cos(\theta - \pi)$ Cardioid with 180° rotation
$= 3 - 3[\cos\theta\cos\pi + \sin\theta\sin\pi]$ Addition law for cosine

$$= 3 - 3[\cos\theta(-1) + \sin\theta(0)] \qquad \text{Substitute exact values.}$$
$$= 3 - 3(-1)\cos\theta \qquad\qquad\qquad \text{Simplify.}$$
$$= 3 + 3\cos\theta$$

Compare this with Example 3 and you will see that the only difference is a 180° rotation instead of a 30° rotation. This means that, whenever you graph an equation of the form

$$r = a(1 + \cos\theta) \qquad \text{Note the plus sign.}$$

instead of $r = a(1 - \cos\theta)$, it is a standard-form cardioid with a 180° rotation. Similarly,

$$r = a(1 - \sin\theta)$$

is a standard-form cardioid with a 90° rotation.

$$r = a(1 + \sin\theta)$$

is a standard-form cardioid with a 270° rotation. ∎

These curves, along with other polar-form curves, are graphed and summarized in Table 7.10 on page 483.

The cardioid is only one of the polar-form curves we will consider. Before sketching other curves, we will consider symmetry. There are three important kinds of **polar symmetry**, which are described in the following box and demonstrated in Figure 7.55.

POLAR-FORM SYMMETRY

A polar-form graph $r = f(\theta)$ is symmetric with respect to . . .	. . . if the equation $r = f(\theta)$ is unchanged when (r, θ) is replaced by . . .
x-axis	$(r, -\theta)$
y-axis	$(r, \pi - \theta)$
origin	$(-r, \theta)$

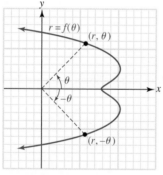

a. Symmetry with respect to the x-axis

b. Symmetry with respect to the y-axis

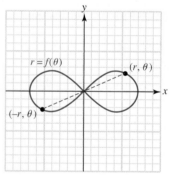

c. Symmetry with respect to the origin

Figure 7.55 Symmetry in polar form

Limaçons

We will illustrate symmetry in polar-form curves by graphing a curve called a **limaçon**.

EXAMPLE 4   Graphing a limaçon using symmetry

Graph $r = 3 + 2\cos\theta$.

Solution
Let $f(\theta) = 3 + 2\cos\theta$.

Symmetry with respect to the x-axis

$$f(-\theta) = 3 + 2\cos(-\theta) = 3 + 2\cos\theta = f(\theta)$$

Yes; it is symmetric, so it is enough to graph f for θ between 0 and π.

Symmetry with respect to the y-axis

$$f(\pi - \theta) = 3 + 2\cos(\pi - \theta)$$
$$= 3 + 2[\cos \pi \cos \theta + \sin \pi \sin \theta]$$
$$= 3 - 2\cos \theta$$

After checking the other primary representation, we find that it is not symmetric with respect to the y-axis.

Symmetry with respect to the origin

$$-r \neq f(\theta) \qquad \text{and} \qquad r \neq f(\theta + \pi),$$

so the graph is **not symmetric** with respect to the origin.

The graph is shown in Figure 7.56; note that we sketch the top half of the graph (for $0 \leq \theta \leq \pi$) by plotting points and then complete the sketch by reflecting the graph in the x-axis. Because $\cos \theta$ steadily decreases from its largest value of 1 at $\theta = 0$ to its smallest value -1 at $\theta = \pi$, the radial distance $r = 3 + 2\cos \theta$ will also steadily decrease as θ increases from 0 to π. The largest value of r is $r = 3 + 2(1) = 5$ at $\theta = 0$, and its smallest value is $r = 3 + 2(-1) = 1$ at $\theta = \pi$.

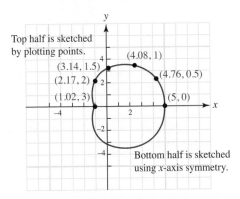

Figure 7.56 Graph of $r = 3 + 2\cos \theta$

The graph of any polar equation of the general form

$$r = b \pm a \cos \theta \qquad \text{or} \qquad r = b \pm a \sin \theta$$

is called a **limaçon** (derived from the Latin word "limax," which means "slug"). The special case where $a = b$ is the *cardioid*. Figure 7.57 shows four different kinds of limaçons that can occur. Note how the appearance of the graph depends on the ratio b/a. We have discussed cases II and III in Examples 2 and 4. Case I (the "inner loop" case) and case IV (the "convex") case are examined in the problem set.

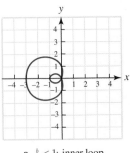

a. $\frac{b}{a} < 1$; inner loop

Case I:

$(r = 1 - 2\cos \theta)$

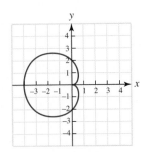

b. $\frac{b}{a} = 1$; cardioid

Case II:

$(r = 2 - 2\cos \theta)$

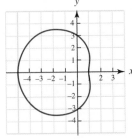

c. $1 < \frac{b}{a} < 2$; dimple

Case III:

$(r = 3 - 2\cos \theta)$

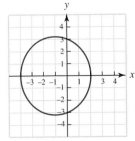

d. $\frac{b}{a} \geq 2$; convex

Case IV:

$(r = 3 - \cos \theta)$

Figure 7.57 Limaçon $r = b \pm a \cos \theta$ or $r = b \pm a \sin \theta$

Just as with the cardioid, we designate a standard-form limaçon and consider the others as rotations, as shown in the box.

LIMAÇON

$r = b - a\cos$	**Standard form**
$r = b - a\sin\theta$	90° rotation
$r = b + a\cos\theta$	180° rotation
$r = b + a\sin\theta$	270° rotation

Rose Curves

There are several polar-form curves known as **rose curves**, which consist of several loops called *leaves* or *petals*.

EXAMPLE 5   Graphing a four-leaved rose

Graph $r = 4\cos 2\theta$.

Solution
Let $f(\theta) = 4\cos 2\theta$.

Symmetry with respect to the *x*-axis

$$f(-\theta) = 4\cos 2(-\theta) = 4\cos 2\theta = f(\theta);   \textbf{Yes}, \text{ symmetric}$$

Symmetry with respect to the *y*-axis

$$f(\pi - \theta) = 4\cos 2(\pi - \theta) = 4\cos(2\pi - 2\theta) = 4\cos(-2\theta) = 4\cos 2\theta$$
$$= f(\theta);   \textbf{Yes}, \text{ symmetric}$$

Symmetry with respect to the origin
If $-r = 4\cos 2\theta$, then the other primary representation is

$$r = 4\cos 2(\theta + \pi) = 4\cos(2\theta + 2\pi) = 4\cos 2\theta;   \textbf{Yes}, \text{ symmetric}$$

Because of this symmetry, we shall sketch the graph of $r = f(\theta)$ for θ between 0 and $\frac{\pi}{2}$ and then use symmetry to complete the graph.

When $\theta = 0$, $r = 4$, and as θ increases from 0 to $\frac{\pi}{4}$, the radial distance decreases from 4 to 0. Then, as θ increases from $\frac{\pi}{4}$ to $\frac{\pi}{2}$, r becomes negative and decreases from 0 to -4. A table of values is given.

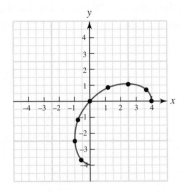

θ	$f(\theta)$
0	4
0.2	3.684244
0.4	2.786827
0.8	−0.116798
1.0	−1.664587
1.2	−2.949575
1.3	−3.427555
1.4	−3.768889

The next part of the graph is obtained by first reflecting in the y-axis.

The last part is found by reflecting in the x-axis, as shown in Figure 7.58.

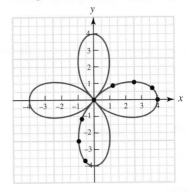

Figure 7.58 Graph of $r = 4\cos 2\theta$

In general, $r = a\cos n\theta$ is the equation of a rose curve in which each petal has length a. If n is an even number, the rose has $2n$ petals; if n is odd, the number of petals is n. The tips of the petals are equally spaced on a circle of radius a. Equations of the form $r = a\sin n\theta$ are handled as rotations.

ROSE CURVE

The **standard-position rose curve** has equation

$$r = a\cos n\theta$$

EXAMPLE 6   Graphing a rose curve with 8 petals and a rotation

Graph $r = 5\sin 4\theta$.

Solution
We begin by finding the amount of rotation.

$$r = 5\sin 4\theta \qquad \text{Given equation}$$

$$= 5\cos\left(\frac{\pi}{2} - 4\theta\right) \qquad \text{Cofunctions of complementary angles}$$

$$= 5\cos\left(4\theta - \frac{\pi}{2}\right) \qquad \text{Remember, } \cos(-\theta) = \cos\theta.$$

$$= 5\cos 4\left(\theta - \frac{\pi}{8}\right) \qquad \text{Common factor of 4}$$

Recognize this as a rose curve rotated $\pi/8$. There are $2(4) = 8$ petals of length 5. The petals are a distance of $\pi/4$ (one revolution $= 2\pi$ divided by the number of petals) apart. The graph is shown in Figure 7.59.

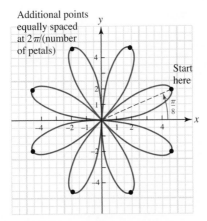

Additional points equally spaced at $2\pi/$(number of petals)

Start here

Figure 7.59 Graph of $r = 5\sin 4\theta$

EXAMPLE 7   Find the equation of a polar-form curve

What is the equation of the curve shown in Figure 7.60?

Solution
By comparing this graph with the one we did for Example 6, we see that the curve is a rose curve with 8 leaves, and since 8 is an even number, we know the equation has the form

$$r = a\cos n\theta$$

where $n = 4$. Furthermore, we see that the length of each leaf is 5 units, so $a = 5$. Thus, the desired equation is

$$r = 5\cos 4\theta$$

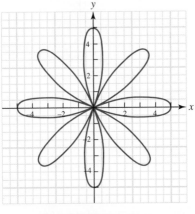

Figure 7.60 Eight-leaf rose

Lemniscates

The last general type of polar-form curve we will consider is called a **lemniscate**.

> **LEMNISCATE**
>
> The **standard-position lemniscate** has equation
>
> $$r^2 = a^2 \cos 2\theta$$

EXAMPLE 8   Graphing a lemniscate

Graph $r^2 = 9\cos 2\theta$.

Solution

As before, when graphing a curve for the first time, begin by checking symmetry and plotting points. For this example, note that you obtain two values for r when solving this quadratic equation. For example, if $\theta = 0$, then $\cos 2\theta = 1$ and $r^2 = 9$, so $r = 3$ or -3.

Symmetry with respect to the *x*-axis:
$9\cos[2(-\theta)] = 9\cos 2\theta$, so $r^2 = 9\cos 2\theta$ is not affected when θ is replaced by $-\theta$; **yes**, symmetric with respect to the x-axis.

Symmetry with respect to the *y*-axis:
$9\cos[2(\pi - \theta)] = 9\cos(2\pi - 2\theta) = 9\cos(-2\theta) = 9\cos 2\theta$; **yes**, symmetric with respect to the y-axis.

Symmetry with respect to the origin:
$(-r)^2 = r^2$ so $r^2 = 9\cos 2\theta$ is not affected when r is replaced by $-r$; **yes**, symmetric with respect to the origin.

Note that because $r^2 \geq 0$, the function $\cos 2\theta$ is defined only for those values of θ for which $\cos 2\theta \geq 0$; that is,

$$-\frac{\pi}{4} \leq \theta \leq \frac{\pi}{4}; \quad \frac{3\pi}{4} \leq \theta \leq \frac{5\pi}{4}; \quad \ldots$$

We begin by restricting our attention to the interval $0 \leq \theta \leq \frac{\pi}{4}$, as shown in Figure 7.61**a**. Note that $\cos 2\theta$ decreases steadily from 3 to 0 as θ varies from 0 to $\frac{\pi}{4}$.

A second step is to use symmetry to reflect the curve in the x-axis (Figure 7.61**b**):

Finally, obtain the rest of the graph by reflecting the curve in the y-axis (Figure 7.61**c**):

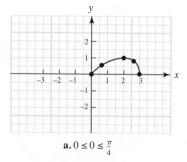

a. $0 \leq \theta \leq \frac{\pi}{4}$

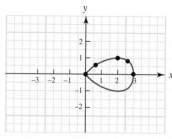

b. *x*-reflection

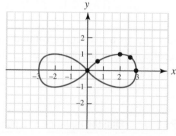

c. Completed curve

Figure 7.61 Graph of $r^2 = 9\cos 2\theta$

Summary of Polar-Form Curves

We conclude this section by summarizing in Table 7.10 the special types of polar-form curves we have examined. There are many others, some of which are represented in the problems and others which are found in the Library of Curves in Appendix F.

TABLE 7.10 | Directory of Curves (Part VI)

LIMAÇONS $r = b \pm a \cos\theta$ or $r = b \pm a \sin\theta$

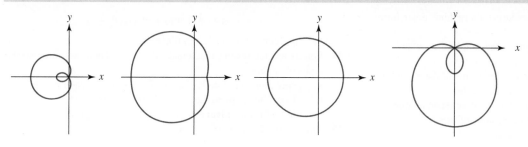

$r = b - a\cos\theta, \frac{b}{a} < 1$
standard form
with inner loop

$r = b - a\cos\theta, 1 < \frac{b}{a} < 2$
standard form
with a dimple

$r = b - a\cos\theta, \frac{b}{a} \geq 2$
standard form,
convex

$r = b - a\sin\theta, \frac{b}{a} < 1$
$\frac{\pi}{2}$ rotation
with inner loop

CARDIOIDS $r = a(1 \pm \cos\theta)$ and $r = a(1 \pm \sin\theta)$ Limaçons in which $a = b$

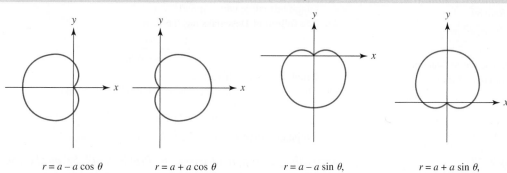

$r = a - a\cos\theta$
standard form

$r = a + a\cos\theta$
π rotation

$r = a - a\sin\theta,$
$\frac{\pi}{2}$ rotation

$r = a + a\sin\theta,$
$\frac{3\pi}{2}$ rotation

ROSE CURVES
$r = a\cos n\theta$ or $r = a\sin n\theta$

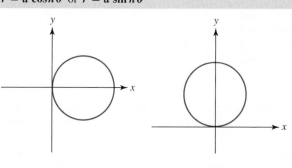

$r = a\cos\theta$
standard form,
one (circular) leaf

$r = a\sin\theta$
$\frac{\pi}{2}$ rotation,
one (circular) leaf

LEMNISCATES
$r^2 = a^2\cos 2\theta$ or $r^2 = a^2\sin 2\theta$

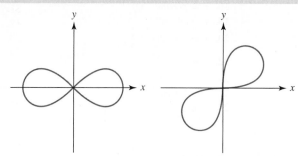

$r^2 = a^2\cos 2\theta$
standard form

$r^2 = a^2\sin 2\theta$
$\frac{\pi}{4}$ rotation

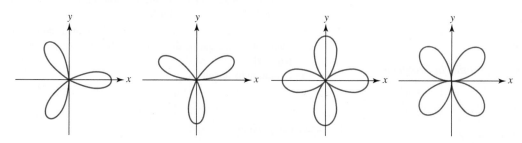

$r = a\cos 3\theta$
standard form,
three petals

$r = a\sin 3\theta$
$\frac{\pi}{6}$ rotation,
three petals

$r = a\cos 2\theta$
standard form,
four petals

$r = a\sin 2\theta$
$\frac{\pi}{4}$ rotation,
four petals

PROBLEM SET 7.6

LEVEL 1

1. IN YOUR OWN WORDS Describe a procedure for graphing polar-form curves.
2. IN YOUR OWN WORDS Discuss symmetry in the graph of a polar-form function.
3. IN YOUR OWN WORDS Compare and contrast the forms of the equations for limaçons, cardioids, rose curves, and lemniscates.
4. Identify each of the curves as a cardioid, rose curve (state number of petals), lemniscate, limaçon, circle, line, or none of these.

   a. $r^2 = 9\cos 2\theta$ **b.** $r = 2\sin\frac{\pi}{6}$
   c. $r = 3\sin 3\theta$ **d.** $r = 3\theta$
   e. $r = 2 - 2\cos\theta$ **f.** $\theta = \frac{\pi}{6}$
   g. $r^2 = \sin 2\theta$ **h.** $r - 2 = 4\cos\theta$

5. Identify each of the curves as a cardioid, rose curve (state number of petals), lemniscate, limaçon, circle, line, or none of these.

   a. $r = 2\sin 2\theta$ **b.** $r^2 = 2\cos 2\theta$
   c. $r = 5\cos 60°$ **d.** $r = 5\sin 8\theta$
   e. $r\theta = 3$ **f.** $r^2 = 9\cos\left(2\theta - \frac{\pi}{4}\right)$
   g. $r = \sin 3\left(\theta + \frac{\pi}{6}\right)$ **h.** $\cos\theta = 1 - r$

6. Identify each of the curves as a cardioid, rose curve (state number of petals), lemniscate, limaçon circle, line, or none of these.

   a. $r = 2\cos 2\theta$ **b.** $r = 4\sin 30°$
   c. $r + 2 = 3\sin\theta$ **d.** $r + 3 = 3\sin\theta$
   e. $\theta = 4$ **f.** $\theta = \tan\frac{\pi}{4}$
   g. $r = 3\cos 5\theta$ **h.** $r\cos\theta = 2$

Graph the polar-form curves in Problems 7–28.

7. $r = 3,\ 0 \le \theta \le \frac{\pi}{2}$ 8. $r = -1,\ 0 \le \theta \le \pi$
9. $\theta = -\frac{\pi}{4},\ 0 \le r \le 4$ 10. $\theta = \frac{\pi}{4},\ 1 \le r \le 2$
11. $r = \theta + 1,\ 0 \le \theta \le \pi$ 12. $r = \theta - 1,\ 0 \le \theta \le \pi$
13. $r = 2\theta,\ \theta \ge 0$ 14. $r = \frac{\theta}{2},\ \theta \ge 0$
15. $r = 2\cos 2\theta$ 16. $r = 3\cos 3\theta$
17. $r = 5\sin 3\theta$ 18. $r^2 = 9\cos 2\theta$
19. $r = 3\cos 3\left(\theta - \frac{\pi}{3}\right)$ 20. $r = 2\cos 2\left(\theta + \frac{\pi}{3}\right)$
21. $r = \sin\left(2\theta + \frac{2\pi}{3}\right)$ 22. $r = \cos\left(2\theta + \frac{\pi}{3}\right)$
23. $r = 2 + \cos\theta$ 24. $r = 3 + \sin\theta$
25. $r = 1 + \sin\theta$ 26. $r = 1 + \cos\theta$
27. $r\cos\theta = 1$ 28. $r\sin\theta = 2$

LEVEL 2

Sketch the graph of the polar functions in Problems 29–34.

29. $f(\theta) = \sin 2\theta,\ 0 \le \theta \le \frac{\pi}{2}$; rose petal
30. $f(\theta) = |\sin 2\theta|,\ 0 \le \theta \le 2\pi$; four-leaf rose
31. $f(\theta) = 2|\cos\theta|,\ 0 \le \theta \le 2\pi$
32. $f(\theta) = 4|\sin\theta|,\ 0 \le \theta \le 2\pi$
33. $f(\theta) = \sqrt{|\cos\theta|},\ 0 \le \theta \le 2\pi$; lazy eight
34. $f(\theta) = \sqrt{\cos 2\theta},\ 0 \le \theta \le 2\pi$; lemniscate

Graph the set of points (r, θ) so that the inequalities in Problems 35–42 are satisfied.

35. $2 \le r \le 3,\ 0 \le \theta < 2\pi$ 36. $0 \le r \le 4,\ 0 \le \theta \le \frac{\pi}{2}$
37. $0 \le r \le 4,\ 0 \le \theta \le \pi$ 38. $r \ge 2,\ \frac{\pi}{2} \le \theta \le \pi$
39. $0 \le \theta \le \frac{\pi}{4},\ r \ge 0$ 40. $r > 1,\ 0 \le \theta < 2\pi$
41. $0 \le \theta \le \frac{\pi}{4},\ 1 \le r \le 2$ 42. $0 \le \theta \le \frac{\pi}{4},\ r \ge 1$

43. Show that the polar equations

$$r = \cos\theta + 1 \text{ and } r = \cos\theta - 1$$

have the same graph in the xy-plane.

44. Spirals are interesting mathematical curves. There are three special types of spirals:

   a. A **spiral of Archimedes** has the form $r = a\theta$; graph $r = 2\theta$.
   b. A **hyperbolic spiral** has the form $r\theta = a$; graph $r\theta = 2$.
   c. A **logarithmic spiral** has the form $r = a^k$; graph $r = 2^\theta$.

45. The **strophoid** is a curve of the form

$$r = a\cos 2\theta\sec\theta$$

Graph this curve where $a = 2$.

46. The **bifolium** has the form

$$r = a\sin\theta\cos^2\theta$$

Graph this curve where $a = 1$.

47. The **folium of Descartes** has the form

$$r = \frac{3a\sin\theta\cos\theta}{\sin^3\theta + \cos^3\theta}$$

Graph the curve where $a = 2$.

48. The curve known as the **ovals of Cassini** has the form

$$r^4 + b^4 - 2b^2r^2\cos 2\theta = k^4$$

Graph the curve where $b = 2.6,\ k = 3$.

Graph the given pair of curves in Problems 49–54 on the same coordinate axes. The first equation uses (x, y) as rectangular coordinates and the second uses (r, θ) as polar coordinates.

49. $y = \cos x$ and $r = \cos\theta$
50. $y = \sin x$ and $r = \sin\theta$
51. $y = \tan x$ and $r = \tan\theta$
52. $y = \sec x$ and $r = \sec\theta$
53. $y = \csc x$ and $r = \csc\theta$
54. $y = \cot x$ and $r = \cot\theta$

LEVEL 3

55. Show that the curve $r = f(\theta)$ is symmetric with respect to the x-axis if the equation is unaffected when r is replaced by $-r$ and θ is replaced by $\pi - \theta$.
56. Show that the curve $r = f(\theta)$ is symmetric with respect to the y-axis if the equation is unaffected when r is replaced by $-r$ and θ is replaced by $-\theta$.
57. Show that the curve $r = f(\theta)$ is symmetric in the line $y = x$ if the equation is unaffected when θ is replaced by $\frac{\pi}{2} - \theta$.
58. State and prove a result regarding the relationship of the graph of $r = f(\theta)$ to that of $r = f(\theta - \theta_0)$ for $0 < \theta_0 < \frac{\pi}{2}$.
59. **a.** Show that if the polar curve is $r = f(\theta)$ rotated about the pole through an angle α, the equation for the new curve is $r = f(\theta - \alpha)$.

   b. Use a rotation to sketch

$$r = 2\sec\left(\theta - \frac{\pi}{3}\right)$$

60. Sketch the graph of

$$r = \frac{\theta}{\cos\theta} \quad \text{for} \quad 0 \le \theta \le \frac{\pi}{2}$$

In particular, show that the graph has a vertical asymptote at $\theta = \frac{\pi}{2}$.

7.7 Conic Sections in Polar Form

We have previously noted that the eccentricity ε can be used to distinguish one conic section from another. Up to now, we let the origin lie at the center of the conic or else we translated to to a point $(h,\ k)$. However, there are many important applications of conics that put a focus at the origin (or pole). In this section, we give an alternate definition of the conic section that puts a focus at the pole and also uses ε to distinguish one conic from another.

CONIC SECTION DEFINITION

The set of points in a plane such that the distance from a fixed point (called the **focus**) is in constant ratio to its distance from a fixed line (called the **directrix**) is called a **conic**. Moreover, the conic is

    a **parabola** if $\varepsilon = 1$;

    an **ellipse** if $0 < \varepsilon < 1$;

    a **hyperbola** if $\varepsilon > 1$.

The single equation for all of these conics is

$$r = \frac{\varepsilon p}{1 - \varepsilon \cos \theta}$$

The graphs that follow from this definition are shown in Figure 7.62.

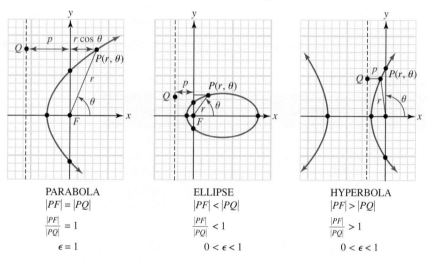

PARABOLA	ELLIPSE	HYPERBOLA
$\|PF\| = \|PQ\|$	$\|PF\| < \|PQ\|$	$\|PF\| > \|PQ\|$
$\dfrac{\|PF\|}{\|PQ\|} = 1$	$\dfrac{\|PF\|}{\|PQ\|} < 1$	$\dfrac{\|PF\|}{\|PQ\|} > 1$
$\epsilon = 1$	$0 < \epsilon < 1$	$0 < \epsilon < 1$

Figure 7.62 Polar-form graphs of the conic sections

Polar-Form Parabola

We can show that this definition is equivalent to our previous definitions for the conics. For example, if $\varepsilon = 1$, then

$$r = \frac{p}{1 - \cos \theta}$$
$$r(1 - \cos \theta) = p$$
$$r - r \cos \theta = p$$

To transform this equation into rectangular form, we need to remember the conversion formulas: $r = \sqrt{x^2 + y^2}$ and $x = r \cos \theta$.

$$\sqrt{x^2 + y^2} - x = p$$
$$\sqrt{x^2 + y^2} = x + p$$
$$x^2 + y^2 = (x + p)^2$$
$$x^2 + y^2 = x^2 + 2px + p^2$$
$$y^2 = 2px + p^2$$
$$y^2 = 2p\left(x + \frac{p}{2}\right)$$

This is the equation of a parabola with vertex (in rectangular coordinates) $(h, k) = \left(-\frac{p}{2}, 0\right)$ and focal chord length $2p$. The standard polar equations are summarized next. We can similarly derive equations for parabolas that open downward, left, or to the right. These graphs are shown in Figure 7.63.

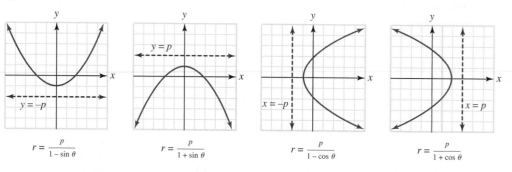

$$r = \frac{p}{1 - \sin\theta} \qquad r = \frac{p}{1 + \sin\theta} \qquad r = \frac{p}{1 - \cos\theta} \qquad r = \frac{p}{1 + \cos\theta}$$

Figure 7.63 Standard-position polar-form parabolas $(p > 0)$. If $p < 0$, rotate parabola through an angle of π.

STANDARD POLAR-FORM PARABOLAS

Direction	Equation	Focus	Directrix Rectangular Form*	Polar-form Vertex
Upward	$r = \dfrac{p}{1 - \sin\theta}$	$(0, 0)$	$y = -p$	$\left(\dfrac{p}{2}, \dfrac{3\pi}{2}\right)$
Downward	$r = \dfrac{p}{1 + \sin\theta}$	$(0, 0)$	$y = p$	$\left(\dfrac{p}{2}, \dfrac{\pi}{2}\right)$
Rightward	$r = \dfrac{p}{1 - \cos\theta}$	$(0, 0)$	$x = -p$	$\left(\dfrac{p}{2}, \pi\right)$
Leftward	$r = \dfrac{p}{1 + \cos\theta}$	$(0, 0)$	$x = p$	$\left(\dfrac{p}{2}, 0\right)$

EXAMPLE 1   Graphing a polar-form parabola

Describe and sketch the graph of the equation

$$r = \frac{4}{3 - 3\cos\theta}$$

Solution

We multiply the given equation by 1 to put the equation into standard form:

$$r = \frac{4}{3 - 3\cos\theta} \cdot \frac{\frac{1}{3}}{\frac{1}{3}} = \frac{\frac{4}{3}}{1 - \cos\theta}$$

---

*We could, of course, state these equations of lines in polar form. For example, $y = -p$ is $r\sin\theta = -p$, but we prefer writing the equation of the directrix in rectangular form.

By inspection, you can now see (Figure 7.64) that the parabola opens to the right and that $p = \frac{4}{3}$. Thus, the vertex is

$$\left(\frac{p}{2}, \pi\right) = \left(\frac{2}{3}, \pi\right)$$

Plot the vertex and the line $x = -\frac{4}{3}$, as shown in the figure.

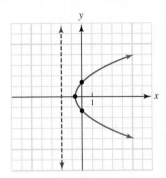

Figure 7.64 Parabola in polar form

You can plot other points that are easy to calculate, such as the points where $\theta = \frac{\pi}{2}$ and $\theta = \pi$ or $\theta = \frac{3\pi}{2}$. ∎

Most graphing calculators will graph curves with polar-form equations. Most require that you first change to polar-form mode.

EXAMPLE 2   Finding a polar-form equation of a parabola

Find a polar-form equation for the parabola with focus at the origin and vertex at the polar-form point $(3, \pi)$.

Solution
The vertex $(3, \pi)$ is on the x-axis and to the left of the focus (the pole). Thus, the parabola opens to the right and has a polar-form equation

$$r = \frac{p}{1 - \cos\theta}$$

where p is the distance from the focus to the directrix. Because the vertex $(3, \pi)$ is halfway between the focus and the directrix, we must have $p = 6$, so that the required equation is

$$r = \frac{6}{1 - \cos\theta}$$

■

Polar-Form Ellipse

Next, we shall examine polar characterizations for the ellipse with eccentricity ε. Consider an ellipse with one focus F at the origin of a polar coordinate plane. Assume that the corresponding directrix L is the vertical line $x = p$, or in polar-form, $r\cos\theta = p$ $(p > 0)$ and that the ellipse has eccentricity ε. Then, if $P(r, \theta)$ is a polar-form point on the ellipse, we have

$$\varepsilon = \frac{\text{DISTANCE FROM } P \text{ TO } F}{\text{DISTANCE FROM } P \text{ TO } L} = \frac{r}{p - r\cos\theta}$$

This relationship is shown in Figure 7.65.

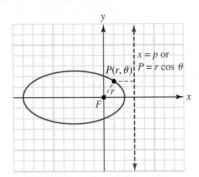

Figure 7.65 Polar-form representation of an ellipse

Solving for r, we find the ellipse has the polar equation

$$r = \frac{\varepsilon p}{1 + \varepsilon \cos\theta} \qquad \text{Because } 0 \le \varepsilon < 1, \varepsilon\cos\theta \ne -1.$$

Similarly, if the directrix is $x = -p$, the equation is $r = \dfrac{\varepsilon p}{1 - \varepsilon \cos\theta}$, and if the directrix is $y = p$ or $y = -p$, the corresponding equations are, respectively,

$$r = \frac{\varepsilon p}{1 + \varepsilon \sin\theta} \qquad \text{and} \qquad r = \frac{\varepsilon p}{1 - \varepsilon \sin\theta}$$

These four possibilities are summarized in Figure 7.66.

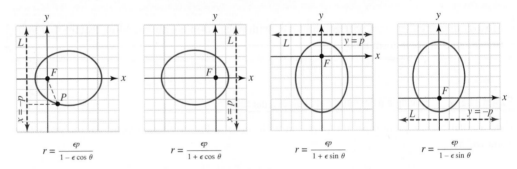

Figure 7.66 Forms for the equation of an ellipse ($\varepsilon < 1$) in standard polar form

STANDARD POLAR-FORMS ELLIPSES ($\varepsilon < 1$)

Direction	Equation	Focus	Directrix Rectangular Form*
Vertical	$r = \dfrac{\varepsilon p}{1 + \varepsilon \sin\theta}$	$(0, 0)$	$y = p$
Vertical	$r = \dfrac{\varepsilon p}{1 - \varepsilon \sin\theta}$	$(0, 0)$	$y = -p$
Horizontal	$r = \dfrac{\varepsilon p}{1 + \varepsilon \cos\theta}$	$(0, 0)$	$x = p$
Horizontal	$r = \dfrac{\varepsilon p}{1 - \varepsilon \cos\theta}$	$(0, 0)$	$x = -p$

*We could, of course, state these equations of lines in polar form. For example, $y = -p$ is $r\sin\theta = -p$, so we prefer writing the equation of the directrix in rectangular form.

EXAMPLE 3   Describing the graph of an equation in polar form

Discuss the graph of the polar-form equation $r = \dfrac{2}{2 - \cos\theta}$.

Solution

Begin by writing the equation in standard form (multiply by 1): $r = \dfrac{1}{1 - \frac{1}{2}\cos\theta}$. This form involves

a cosine and has $\varepsilon = \frac{1}{2} < 1$, so by comparing with the forms in Figure 7.66, we see that the graph must be a horizontal ellipse. The form also tells us that $\varepsilon p = 1$, so $p = 2$ and the directrix is $x = -2$. The focus F_1 closer to the directrix is at the pole, and the vertices occur where $\theta = 0$ and $\theta = \pi$. For $\theta = 0$, we obtain $r = 2$, and for $\theta = \pi$, $r = \frac{2}{3}$, so the vertices are the polar points $V_1\left(\frac{2}{3}, \pi\right)$ and $V_2(2, 0)$. Since the focus $F_1(0, 0)$ is $\frac{2}{3}$ unit to the left of $V_1\left(\frac{2}{3}, \pi\right)$, F_2 is the polar point $\left(\frac{4}{3}, 0\right)$. The center of the ellipse is midway between the foci, at $\left(\frac{2}{3}, 0\right)$, so the minor axis is the vertical line passing through this point. This is the line $x = \frac{2}{3}$, or, in polar form,

$$r\cos\theta = \frac{2}{3} \qquad \text{or} \qquad r = \frac{\frac{2}{3}}{\cos\theta}$$

To find the endpoints of the minor axis, we need to solve simultaneously the equations for the axis with the equation of the ellipse, namely,

$$\frac{2}{2 - \cos\theta} = \frac{\frac{2}{3}}{\cos\theta}$$

$$2\cos\theta = \frac{2}{3}(2 - \cos\theta)$$

$$\frac{8}{3}\cos\theta = \frac{4}{3}$$

$$\cos\theta = \frac{1}{2}$$

so that $\theta = \frac{\pi}{3}$ and $\frac{5\pi}{3}$. Solving for r, we find $r = \frac{4}{3}$, which gives the vertices $\left(\frac{4}{2}, \frac{\pi}{3}\right)$ and $\left(\frac{4}{3}, \frac{5\pi}{3}\right)$. The graph is shown in Figure 7.67.

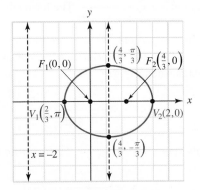

Figure 7.67 Ellipse in polar form

Polar-Form Hyperbola

Formulas for hyperbolas in polar coordinates are obtained in essentially the same way as polar formulas for ellipses. The four different cases that can occur for hyperbolas in standard form are summarized in Figure 7.68.

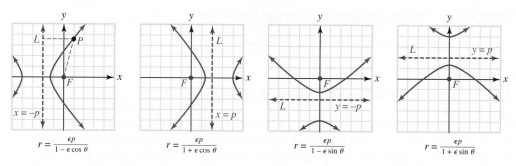

Figure 7.68 Forms for the equation of a hyperbola ($\varepsilon > 1$) in standard polar form

STANDARD POLAR-FORM HYPERBOLAS ($\varepsilon > 1$)			
Direction	**Equation**	**Focus**	**Directrix** Rectangular Form
Vertical	$r = \dfrac{\varepsilon p}{1 + \varepsilon \sin\theta}$	$(0, 0)$	$y = p$
Vertical	$r = \dfrac{\varepsilon p}{1 - \varepsilon \sin\theta}$	$(0, 0)$	$y = -p$
Horizontal	$r = \dfrac{\varepsilon p}{1 + \varepsilon \cos\theta}$	$(0, 0)$	$x = p$
Horizontal	$r = \dfrac{\varepsilon p}{1 - \varepsilon \cos\theta}$	$(0, 0)$	$x = -p$

EXAMPLE 4   Describing the graph of a hyperbola in polar form

Discuss the graph of the polar equation $r = \dfrac{5}{3 + 4\sin\theta}$.

Solution

The standard form of the equation is $r = \dfrac{\frac{5}{3}}{1 + \frac{4}{3}\sin\theta}$. The form tells us that the eccentricity is $\varepsilon = \frac{4}{3}$, and because $\varepsilon > 1$, the graph is a hyperbola. We also see that the transverse axis is the y-axis, and because $\varepsilon p = \frac{5}{3}$, we have $p = \frac{5}{4}$. Thus, the graph has one focus F_1 at the pole and directrix $y = p = \frac{5}{4}$. The corresponding vertex occurs when $\theta = \frac{\pi}{2}$:

$$r = \frac{5}{3 + 4\sin\frac{\pi}{2}} = \frac{5}{7}$$

so the polar coordinates of this vertex V_1 are $\left(\frac{5}{7}, \frac{\pi}{2}\right)$. The opposite vertex occurs where $\theta = \frac{3\pi}{2}$ and

$$r = \frac{5}{3 + 4\sin\frac{3\pi}{2}} = -5$$

so the point is $V_2\left(-5, \frac{3\pi}{2}\right)$. Because the vertex V_1 is located $\frac{5}{7}$ unit above $F_1(0, 0)$, we find the other focus, F_2, will be located 5 units above the vertex V_1. Thus, F_2 is the polar point $\left(\frac{40}{7}, \frac{\pi}{2}\right)$. The graph is shown in Figure 7.69.

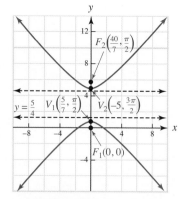

Figure 7.69 Polar-form hyperbola

Rotated Conics in Polar Form

It is easy (compared with what you needed to do in Section 7.4) to rotate the conic sections when the equation is given in polar form. Remember that, if the argument of a trigonometric function is $(\theta - \alpha)$, then the curve is rotated through an angle α.

EXAMPLE 5   Sketching a rotated ellipse in polar form

Graph $r = \dfrac{16}{5 - 3\cos\left(\theta - \frac{\pi}{6}\right)}$.

Solution

First, determine the type of conic section by writing the equation in standard form:

$$r = \frac{16}{5 - 3\cos\left(\theta - \frac{\pi}{6}\right)} = \frac{\frac{16}{5}}{1 - \frac{3}{5}\cos\left(\theta - \frac{\pi}{6}\right)}$$

This ellipse $\left(\varepsilon = \frac{3}{5}\right)$ has been rotated $\frac{\pi}{6}$. First, rotate the conic section, and then plot points on the rotated axis using the equation

$$r' = \frac{\frac{16}{5}}{1 - \frac{3}{5}\cos\theta'}$$

If $\theta' = 0$, then $r' = 8$; plot $(8, 0)$ on the rotated axes.
If $\theta' = \frac{\pi}{2}$, then $r' = \frac{16}{5}$; plot $\left(\frac{16}{5}, \frac{\pi}{2}\right)$ on the rotated axes.
If $\theta' = \pi$, then $r' = 2$; plot $(2, \pi)$ on the rotated axes.
If $\theta' = \frac{3\pi}{2}$, then $r' = \frac{16}{5}$; plot $\left(\frac{16}{5}, \frac{3\pi}{2}\right)$ on the rotated axes.

The graph is shown in Figure 7.70.

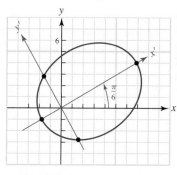

Figure 7.70 Rotated conic section

PROBLEM SET 7.7

LEVEL 1

1. IN YOUR OWN WORDS Outline a procedure for recognizing a conic section by looking at its polar-form equation.
2. IN YOUR OWN WORDS Discuss how the eccentricity can be used to identify a conic section.

Identify the conic in each of Problems 3–20. In each case, write the standard form of the equation and give ε.

3. $r = \dfrac{3}{1 - \cos\theta}$

4. $r = \dfrac{5}{1 + \sin\theta}$

5. $r = \dfrac{5}{3 + 5\sin\theta}$

6. $r = \dfrac{6}{2 - 5\sin\theta}$

7. $r = \dfrac{6}{5 - 4\cos\theta}$

8. $r = \dfrac{16}{8 + \cos\theta}$

9. $r = \dfrac{4}{1 - \sin\theta}$

10. $r = \dfrac{3}{4 - 3\cos\theta}$

11. $r(1 - 2\cos\theta) = 4$

12. $r(1 + 3\sin\theta) = 5$

13. $r(2 - 2\cos\theta) = 8$

14. $r(4 - 4\sin\theta) = 32$

15. $r = \dfrac{3}{1 - \cos\left(\theta - \frac{\pi}{4}\right)}$

16. $r = \dfrac{4}{2 - 2\cos\left(\theta - \frac{\pi}{3}\right)}$

17. $r = \dfrac{5}{4 - 2\cos\left(\theta - \frac{\pi}{6}\right)}$

18. $r = \dfrac{2}{2 - \cos(\theta + 1)}$

19. $r = \dfrac{\sqrt{2}}{1 + \cos(\theta + 0.5)}$

20. $r = \dfrac{\sqrt{3}}{1 + \sin(\theta - 1)}$

21. Solve the equation $r = \dfrac{\varepsilon p}{1 - \varepsilon\cos\theta}$ for ε.

22. Solve the equation $r = \dfrac{\varepsilon p}{1 + \varepsilon\cos\left(\theta + \frac{\pi}{2}\right)}$ for ε.

23. Solve the equation $r = \dfrac{\varepsilon p}{1 - \varepsilon\cos\theta}$ for θ.

24. Solve the equation $r = \dfrac{\varepsilon p}{1 + \varepsilon\sin\left(\theta + \frac{\pi}{2}\right)}$ for θ.

LEVEL 2

Sketch the graph of the polar-form parabola in Problems 25–46.

25. $r = \dfrac{3}{1 - \cos\theta}$

26. $r = \dfrac{5}{1 + \sin\theta}$

27. $r = \dfrac{6}{1 + \cos\theta}$

28. $r = \dfrac{4}{1 - \sin\theta}$

29. $r = \dfrac{-9}{1 + \sin\theta}$

30. $r = \dfrac{-2}{1 - \cos\theta}$

31. $r = \dfrac{5}{3 + 5\cos\theta}$

32. $r = \dfrac{6}{2 - 5\sin\theta}$

33. $r = \dfrac{5}{5 - 4\cos\theta}$

34. $r = \dfrac{16}{8 + \cos\theta}$

35. $r = \dfrac{8}{2 - 2\cos\theta}$

36. $r = \dfrac{9}{3 + 3\cos\theta}$

37. $r = \dfrac{4}{6 + \cos\theta}$

38. $r = \dfrac{4}{2 + 3\cos\theta}$

39. $r = \dfrac{5}{1 - 2\sin\theta}$

40. $r = \dfrac{-3}{2 - \sin\theta}$

41. $r = \dfrac{3}{1 - \cos\left(\theta - \frac{\pi}{4}\right)}$

42. $r = \dfrac{4}{2 - 2\cos\left(\theta - \frac{\pi}{3}\right)}$

43. $r = \dfrac{5}{4 - 2\cos\left(\theta - \frac{\pi}{6}\right)}$

44. $r = \dfrac{2}{2 - \cos(\theta + 1)}$

45. $r = \dfrac{\sqrt{2}}{1 + \cos(\theta + 0.5)}$

46. $r = \dfrac{\sqrt{3}}{1 - \sin(\theta - 1)}$

Find a polar equation with its focus at the pole and with the given property or equation in Problems 47–56.

47. $x^2 + y^2 = 16$

48. $9x^2 + 9y^2 = 1$

49. vertex at the polar-form point $(4, 0)$

50. vertex at the polar-form point $(2, \pi)$

51. parabola with directrix at $y = -4$

52. parabola with directrix at $x = 3$

53. $y^2 = 4x + 4$

54. $x^2 = 1 - 2y$

55. $4x^2 = y + \frac{1}{16}$

56. $4x^2 = y - 1$

57. Assume that $\varepsilon = \frac{1}{2}$ in the standard polar-form ellipse formula (page 488). Convert the polar-form equation to rectangular form and show that it is an ellipse.

58. Assume that $\varepsilon = \frac{3}{2}$ in the standard polar-form hyperbola formula (page 490). Convert the polar-form equation to rectangular form and show it is a hyperbola.

59. An **epitrochoid** is a curve with the parametric equations

$$x = a\cos\theta + b\cos k\theta, \qquad y = a\sin\theta + b\sin k\theta$$

Sketch the graph of an epitrochoid for $a = 4$, $b = 3$, and $k = 5$.

60. A **witch of Agnesi** is a curve with the parametric equations

$$x = a\cot\theta, \qquad y = a\sin^2\theta$$

Sketch the graph of a witch of Agnesi for $a = 2$.

CHAPTER 7 SUMMARY AND REVIEW

*A*ll mathematicians share . . . a sense of amazement over the infinite depth and the mysterious beauty and usefulness of mathematics.

Martin Gardner

Take some time getting ready to work the review problems in this section. First, look back at the definition and property boxes. You will maximize your understanding of this chapter by working the problems in this section only after you have studied the material.

SELF TEST   *All of the answers for this self test are given in the back of the book.*

1. Write the following standard-form equations:
   a. A parabola opening downward with vertex (h, k)
   b. A vertical ellipse centered at (h, k)
   c. A horizontal hyperbola centered at (h, k)
   d. A cardioid with no rotation
   e. A lemniscate that has been rotated $\pi/4$
   f. A four-leaved rose with no rotation

2. If $Ax^2 + Cy^2 + Dx + Ey + F = 0$, how do you recognize each of the following curves?
   a. A line; state the standard-form equation.
   b. A parabola that opens upward or downward; state the standard-form equation.
   c. An ellipse; state both standard-form equations.
   d. A circle; state the standard form equation.
   e. A parabola that opens right or left; state the standard-form equation.
   f. A hyperbola; state both standard-form equations.

Graph the curves in Problems 3–4.

3. a. $x + 5 = 2(y - 3)^2$ **b.** $(x - 2)^2 + (y + 1)^2 = 9$
   c. $\dfrac{x^2}{4} + \dfrac{y^2}{16} = 1$ **d.** $\dfrac{y^2}{9} - \dfrac{x^2}{1} = 1$

4. a. $r = 2\cos 2\theta$ **b.** $r = 2 + 2\cos\theta$
   c. $r^2 = 25\cos 2\theta$ **d.** $r = \tan 45°$

5. a. Graph the parametric equations $x = 3\cos\theta$, $y = 5\sin\theta$.
   b. Write $x = 4y^2 + 3y + 1$ in parametric form, and then sketch the curve.

6. Write each equation in parametric form, and then sketch each curve.
   a. $\dfrac{x^2}{16} - \dfrac{y^2}{4} = 1$ **b.** $\dfrac{(x + 2)^2}{5} + \dfrac{(y - 3)^2}{10} = 1$

Find the equations of the curves described in Problems 7–10.

7. The set of points with the sum of the distances from $(-3, 4)$ and $(-7, 4)$ equal to 12

8. The set of points with the difference of distances from $(-3, 4)$ and $(-7, 4)$ equal to 2

9. The set of points whose distance from $(-3, 4)$ is equal to the distance from the line $x = 2$

10. The set of points whose distance from $(-3, 4)$ is equal to 7

 STUDY HINTS *Compare your solutions and answers to the self test.*

For the functions given in Problems 1–10, match the function with one of the graphs labeled A–X.

1. **a.** Linear function; $f(x) = 3x - 5$
    b. Quadratic function; $f(x) = 2x^2 + 3x - 5$
2. **a.** Cubic function; $f(x) = 2x^2 + 3x - x^3 - 5$
    b. Quartic function; $f(x) = x^4 - x^2$
3. **a.** Rational function; $f(x) = \dfrac{x^4 - 16x^2}{x^2 - 16}$

    b. Radical function: $f(x) = \sqrt{x^2 + x - 12}$
4. **a.** Exponential function (common); $f(x) = 10^{2x+1}$
    b. Exponential function (natural); $f(x) = e^{-x}$
5. **a.** Logarithmic (common) function; $f(x) = \log\frac{1}{x}$
    b. Logarithmic (natural) function; $f(x) = \ln x$

6. **a.** Trigonometric function; $f(x) = \sin(x+1)$
    b. Trigonometric function; $f(x) = \tan x$
7. **a.** Cardioid; $r = 2 + 2\cos\theta$
    b. Circle; $r = \sin(\theta + 1)$
8. **a.** Four-leaved rose; $r = \cos 2\theta$
    b. Lemniscate; $r^2 = \sin\theta$
9. **a.** Vertical hyperbola; $\dfrac{(y-3)^2}{9} - \dfrac{(x-2)^2}{25} = 1$

    b. Spiral; $r = e^{\theta/10}$
10. **a.** Horizontal ellipse; $\dfrac{(x-2)^2}{5} + \dfrac{(y-3)^3}{2} = 1$

    b. Parabola (not a function); $x = \frac{1}{4}y^2 + 3y + 1$

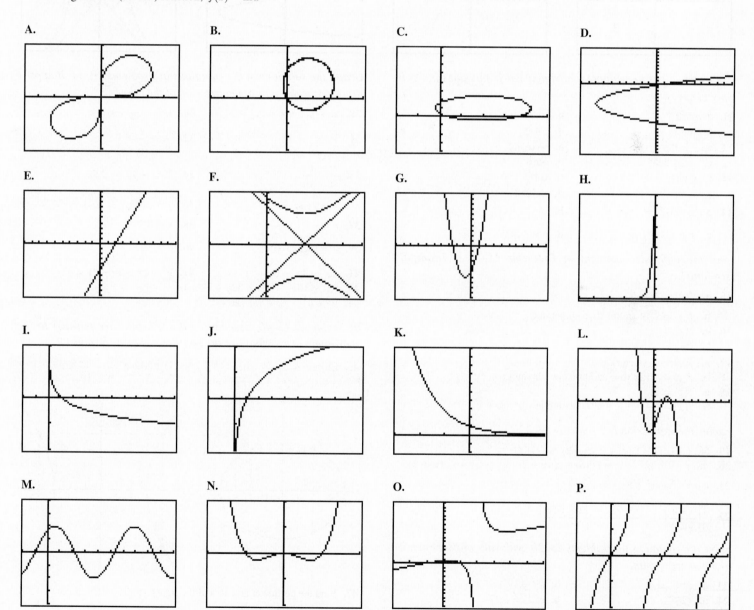

A.   B.   C.   D.

E.   F.   G.   H.

I.   J.   K.   L.

M.   N.   O.   P.

Q.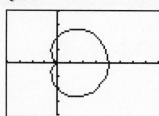

R.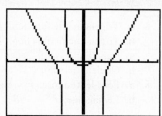

S.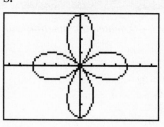

T.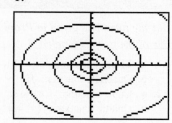

U.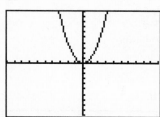

V.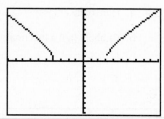

W.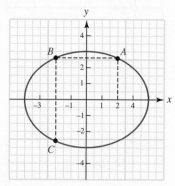

X.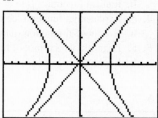

*Use exact values to **evaluate** the trigonometric functions in Problems 11–14.*

11. a. $\sin 60°$ **b.** $\cos(-60°)$
   c. $\tan\frac{\pi}{3}$ **d.** $\cot\frac{5\pi}{3}$
12. a. $\cos 135°$ **b.** $-\cot 45°$
   c. $-\tan 225°$ **d.** $\sec 0°$
13. a. $\csc 2\pi$ **b.** $\cot\frac{3\pi}{2}$
   c. $\tan \pi$ **d.** $\csc 240°$
14. a. $\sec 210°$ **b.** $\tan\left(-\frac{11\pi}{3}\right)$
   c. $\sin(-330°)$ **d.** $\cos(-300°)$

*Find the difference quotient in Problems 15–18 and **simplify** (see hints).*

15. a. $f(x) = 2x + 5$
   b. $f(x) = \sqrt{5x}$ (rationalize numerator)
16. a. $f(x) = 3x^2 - 5x$
   b. $f(x) = -\frac{5}{x}$
17. a. $f(x) = e^x$ (factor)
   b. $f(x) = \sin 2x$ (use addition law for sine)
18. a. $f(x) = 51$
   b. $f(x) = \log x$ (use multiplication law of logs)

***Factor** the expression in Problems 19–22.*

19. $3(2x-1)^2(x-1)(2) + 2(2x-2)^3(x-1)(1)$
20. $4(5x+2)^3(3x^2+2x+1)^2(5) + 2(5x+2)^4(3x^2+2x+1)(6x+2)$
21. $6xe^{3x^2+4}\sin x^2 + 2x\cos x^2\left(e^{3x^2+5}\right)$
22. $\frac{1}{2}\left(\frac{x-1}{x-2}\right)^{-1/2}\left[\frac{(x-2)(1)-(x-1)(1)}{(x-2)^2}\right]$

***Solve** the equations in Problems 23–28, with your answer rounded to the nearest tenth.*

23. a. $8^{2x+3} = 4$ **b.** $e^{4x} = \frac{1}{10}$
24. a. $125^{2x+1} = 25$ **b.** $e^{2x+1} = 5.474$
25. a. $\frac{1}{2} = e^{0.0032x}$ **b.** $\log_2 85.2 = x$
26. $\frac{x+1}{x+3} = \frac{2x-1}{2x+1}$ **27.** $\sqrt{2x+3}+3 = 3\sqrt{x+1}$
28. $(2x+3)^2(4x-1)^3 + 2(2x+3)(4x-1)^3(2) + 3(2x+3)^2(4x-1)^2 = 0$

***Graph** the solutions of the equations in Problems 29–44. If appropriate, consider a parameterization of the curve.*

29. $2x - y + 8 = 0$ **30.** $4x^2 - 16y = 0$
31. $\frac{x}{9} - \frac{y}{16} = 1$ **32.** $\frac{x^2}{9} - \frac{y}{16} = 1$
33. $\frac{x^2}{9} - \frac{y^2}{16} = 1$ **34.** $25x^2 + 9y^2 = 225$
35. $y = 2\cos\left(x - \frac{\pi}{6}\right)$ **36.** $y = -3\tan(2x - 4)$
37. $y = -e^x$ **38.** $y = 5^{\sin x}$
39. $y = \log(x - 6)$ **40.** $y = \frac{x}{\sqrt{x^2 - 16}}$
41. $4x^2 - 3y^2 - 6y - 111 = 0$ **42.** $y^2 - 4x + 2y + 21 = 0$
43. $9x^2 + 16y^2 - 90x - 32y + 97 = 0$
44. $x^2 + y^2 + 2x - 4y - 20 = 0$

In Problems 45–46, assume the dashed lines are parallel to the coordinate axes. What are the coordinates of A, B, and C?

45. Ellipse: $\frac{x^2}{16} + \frac{y^2}{9} = 1$ **46.** Hyperbola: $\frac{x^2}{9} - \frac{y^2}{16} = 1$

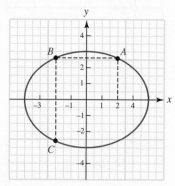

47. Find the perimeter and area of the right triangle with vertices $(-1, 3)$, $(-1, 8)$, and $(11, 8)$.
48. Find the standard-form equation of the tangent to $y = 2x^2 + 4x$ at $x = 2$.
49. Find the equation of the parabola with vertex at $(6, 3)$ and directrix $x = 1$.

50. Find the equation of the parabola with vertex at $(4, 2)$ and passing through $(-3, -4)$ with axis parallel to the y-axis.
51. Find the equation of the set of points 8 units from the point $(-1, -2)$.
52. Find the equation of the ellipse with foci at $(2, 3)$ and $(-1, 3)$, with eccentricity $\frac{3}{5}$.
53. Find the equation of the hyperbola with vertices at $(-3, 0)$ and $(3, 0)$, and foci at $(5, 0)$ and $(-5, 0)$.
54. Find constants A and B so that

$$\tan\left(x + \frac{\pi}{3}\right) = \frac{A + \tan x}{1 + B \tan x}$$

55. Find constants A and B so that

$$\sin^3 x = A \sin 3x + B \sin x$$

56. A bus charter company offers a travel club the following arrangements: If no more than 100 people go on a certain tour, the cost will be $500 per person, but the cost per person will be reduced by $4 for each person in excess of 100 who takes the tour.
    a. Express the total revenue R obtained by the charter company as a function of the number of people who go on the tour.
    b. Sketch the graph of R. Estimate the number of people that results in the greatest total revenue for the charter company.
57. Suppose the number of worker-hours (in hundreds) required to distribute new telephone books to x percent of the households in a certain rural community is given by the function

$$f(x) = \frac{6x}{3 - x}$$

    a. What is the domain of the function f?
    b. For what values of x does $f(x)$ have a practical interpretation in this context?
    c. How many worker-hours were required to distribute new telephone books to the first 50% of the households?

d. How many worker-hours were required to distribute new telephone books to the entire community?
e. What percentage of the households in the community had received new telephone books by the time 150 worker-hours had been expended?

58. A mural 7 feet high is hung on a wall in such a way that its lower edge is 5 feet higher than the eye of an observer standing 12 feet from the wall. (See Figure 7.71.) Find $\tan \theta$, where θ is the angle subtended by the mural at the observer's eye.

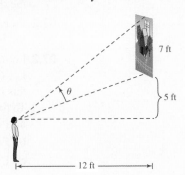

7 ft

5 ft

12 ft

Figure 7.71 Estimating an angle

59. Two jets bound for Los Angeles leave New York 30 min apart. The first travels 550 mi/h, while the second goes 650 mi/h. How long will it take the second plane to pass the first?
60. An open box with a square base is to be built for $48. The sides of the box will cost $3 per square ft and the base will cost $4 per square ft. Express the volume of the box as a function of the length of its base.

CHAPTER 7 Group Research Projects

Working in small groups is typical of most work environments, and this book seeks to develop skills with group activities. At the end of each chapter, we present a list of suggested projects, and even though they could be done as individual projects, we suggest that these projects be done in groups of three or four students.

G7.1 Investigate the topic of conic sections. Build models and/or find three-dimensional models for the conic sections. What did the Greeks know of the conic sections?

G7.2 Prepare a portfolio of photographs or clippings from magazines that show examples of conic sections.

G7.3 Let $L: Ax + By + C = 0$ be any nonvertical line and let $P(x_0, y_0)$ be any point not on the line. Let Q be the point of intersection of the line through P which is perpendicular to L. The goal of this problem is to find a formula for $|\overline{PQ}|$. That is, what is the distance from P to the line L? To find this formula, follow the steps outlined next.

  a. Find the slope L. Let L' be the line through P parallel to L. Find the slope of L'.

  b. Find the equation of L'.

  c. Draw $\overline{RS}$ so that it passes through $(0, 0)$ and is parallel to $\overline{PQ}$. Let R be the point of intersection of the lines $\overline{RS}$ and L' and S be the point of intersection of the lines $\overline{RS}$ and L. Find the equation of the line passing through R and S.

  d. Find the coordinates of point R.

  e. Find the coordinates of point S.

  f. Use the distance formula to find $|\overline{RS}|$

  g. State a formula for the distance from P to L.

G7.4 𝔥𝔦𝔰𝔱𝔬𝔯𝔦𝔠𝔞𝔩 𝔔𝔲𝔢𝔰𝔱 Write an essay on the difficulties women in the history of mathematics faced in competing in a discipline dominated by men. Include a list of female mathematicians in the history of mathematics.

G7.5 PROBLEM FROM CALCULUS In Example 9, Section 7.1, we modeled the path of a cannonball by the equation

$$(x - 192)^2 = -256(y - 144)$$

Relate this equation to the angle of the cannon and the initial velocity of the cannonball as described by Example 6, Section 5.2. You can provide an empirical answer (that is, an answer found by trial-and-error experimentation).

G7.6 Consider a person A who fires a rifle at a distant gong labeled B. Assuming the ground is flat, where must you stand to hear the sound of the gun and the sound of the gong simultaneously? *Hint:* To answer this question, let d be the distance sound travels in the length of time it takes the bullet to travel from the gun to the gong. Show that the person who hears the sounds simultaneously must stand on a branch of a hyperbola (the one nearest the target) so that the difference of the distances from A to B is d.

G7.7 Ships at sea locate their positions using the LOng RAnge Navigation system known as LORAN. In this system, a master station sends signals that can be received by ships at sea. To fix the position of a particular ship, a secondary (slave) sending station also emits signals that can be received by the ship. Since the ship monitoring the two signals will usually be nearer one of the two stations, there will be a difference in the distances that the two signals travel. Because $d = rt$, there will be a slight time difference between the signals. Suppose the distance between the master and slave stations is 100 mi and suppose that the difference in the arrival of the time signals is 300 μsec. (*Note:* a μsec is a microsecond—that is, one-millionth of a second. Also, suppose that signals travel at 980 ft/μsec. Write an equation for the path of the ship.

G7.8 Journal Problem (*Nature*, May 31, 2001) Ungless and others investigated the nature of the "waggle dance" of the domestic honeybee (*Apis melliferra*). Their work is extended from data collected by the German ethologist Karl von Frisch who discovered much of what we know today about the honeybee.

Polar coordinates can be applied to the manner in which bees communicate the distance and source of a food supply. The dance is done on the vertical hanging honeycomb inside the hive (see Figure 7.72).

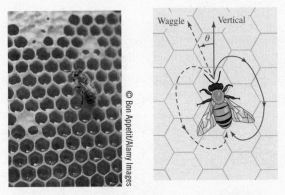

Figure 7.72 Beehive and waggle dance of a bee

The vertical direction represents the sun's direction, and the angle θ represents the direction of the food source. Write a paper describing how the bees communicate with this polar-form "waggle dance."

G7.9 𝕳𝖎𝖘𝖙𝖔𝖗𝖎𝖈𝖆𝖑 𝕼𝖚𝖊𝖘𝖙 In 1986, Halley's Comet returned for our once-every-76-year view. By using mathematics and a knowledge of the conic sections, we can predict the comet's path as well as its exact arrival time. Do some research on planetary orbits (see Figure 7.73) and write a paper considering, but not limited to, questions such as:

If the mass of the earth is 6.58×10^2 tons, the sun 2.2×10^{27} tons, and the moon 8.1×10^{19} tons, calculate the orbit of the moon. Assume that the length of the semimajor axis of the earth is 9.3×10^7 miles, the period of the earth 365.25 days, and the period of the moon 27.3 days.

Courtesy of the Observatories of the Carnegie Institution of Washington

Figure 7.73 Halley's Comet with orbit detail

For each problem you model, be sure to state the assumptions you are making, as shown in the preceding question.

G7.10 Do some research to find the eccentricity of the planetary orbits (see Figure 7.25). Which planet has an orbit that is most closely circular, and which planet has the least circular orbit? Using the results of your research, plot the planetary orbits. (Use Example 10, Section 7.2 as a model for your work.)

Chapter Objectives

The material in this chapter is previewed in the following list of objectives. After completing this chapter, review this list again, and then complete the self test.

8.1 8.1 Classify a sequence as arithmetic, geometric, Fibonacci, or none of these.

 8.2 Be able to generate a sequence from a general term.

8.2 8.3 Determine whether a given sequence converges.

 8.4 State and use the bounded, monotonic convergence property.

8.3 8.5 Use summation notation with series.

 8.6 Know and use the formula for arithmetic and geometric series.

 8.7 Find the sum of an infinite geometric series that converges.

 8.8 Work applied problems with series.

8.4 8.9 Solve a system of equations by graphing.

 8.10 Solve a system of equations by substitution.

 8.11 Solve a system of equations by linear combinations (adding).

8.5 8.12 Be able to apply the elementary row operations to a system of equations written in matrix form.

 8.13 Solve a system using Gauss–Jordan elimination.

 8.14 Solve systems of equations with more equations than unknowns, including inconsistent systems.

 8.15 Solve systems of equations with more unknowns than equations, including those with parametric solutions.

 8.16 Solve applied system problems.

8.6 8.17 Perform matrix operations.

 8.18 Find the inverse of a matrix (if it exists).

 8.19 Solve a system of equations by using an inverse matrix.

8.7 8.20 Graph the solution of a system of linear inequalities.

 8.21 Graph the solution of a system of nonlinear inequalities.

Sequences, Systems, and Matrices

8

We can see that the development of mathematics is a process of conflict among the many contrasting elements: the concrete and the abstract, the particular and the general, the formal and the material, the finite and the infinite, the discrete and the continuous, . . .

—A. D. Aleksandrov (1963)

Chapter Sections

8.1 Sequences
Arithmetic Sequences
Geometric Sequences
Fibonacci-Type Sequences

8.2 Limit of a Sequence
Convergence
Limit Properties
Bounded, Monotonic Sequences

8.3 Series
Summation Notation
Arithmetic Series
Geometric Series
Infinite Geometric Series
Summary of Sequence and Series Formulas

8.4 Systems of Equations
Graphing Method
Substitution Method
Linear Combination (Addition) Method

8.5 Matrix Solution of a System of Equations
Definition of a Matrix
Matrix Form of a System of Equations
Elementary Row Operations and Pivoting
Gauss–Jordan Elimination
Applications

8.6 Inverse Matrices
Matrix Operations
Zero-One Matrices
Algebraic Properties of Matrices
Inverse Property
Systems of Equations

8.7 Systems of Inequalities
Half-planes
Graphing Systems of Inequalities
Nonlinear Systems

Chapter 8 Summary and Review
Self Test
Practice for Calculus
Chapter 8 Group Research Projects

▶ CALCULUS PERSPECTIVE

Sequences of letters, symbols, or numbers are the basis of pattern recognition. Sequences are defined mathematically in this chapter, and several special sequences are studied. The sum of the numbers of a sequence—a series—is similarly explored, and special notation is introduced to simplify the work. The limit of a sequence as the number of terms increases is of particular interest in calculus, so we introduce some preliminary concepts in this chapter. A second very important tool in mathematics, the notion of a matrix, is also introduced in this chapter. We are motivated by our desire to solve systems of equations, but we soon learn that matrices have many other applications and uses. If we are to solve real-life examples, we need to be able to solve very complicated systems, and for the most part, these systems are too difficult to solve without some sort of technological assistance, so using technology will also be part of the development of this chapter.

8.1 Sequences

Patterns and proof are two of the cornerstones upon which mathematics is founded. Consider a function whose domain is the set of counting numbers, namely, $N = \{1, 2, 3, 4, \ldots, n, \ldots\}$. Such a function is called an **infinite sequence.** If we denote this function by s (for sequence), then the number $s(1)$ is the *first term*, $s(2)$ is the *second term*, $\ldots$, and $s(23)$ is the *twenty-third term*. The number $s(n)$ is the *nth term* or *general term* of the sequence. In the context of sequences, we use subscripts instead of the usual functional notation. Thus, we write s_1 to mean $s(1)$, s_2 to mean $s(2)$, $\ldots$, and s_{23} to mean $s(23)$.

Arithmetic Sequences

Perhaps the simplest pattern is the counting numbers themselves: 1, 2, 3, 4, 5, 6, ... A list of numbers having a first number, a second number, a third number, and so on, is called a **sequence.**

| 1 | 2 | 3 | 4 | 5 |

Arithmetic Growth Is Linear

The numbers in a sequence are called the **terms** of the sequence. The sequence of counting numbers is formed by adding 1 to each term to obtain the next term. Your math assignments may well have been identified by some sequence: "Do the multiples of 3 from 3 to 30," that is, 3, 6, 9, . . . , 27, 30. This sequence is formed by adding 3 to each term. Sequences obtained by adding the same number to each term to obtain the next term are called *arithmetic sequences* or *arithmetic progressions.*

> **ARITHMETIC SEQUENCE**
>
> An **arithmetic sequence** is a sequence whose consecutive terms differ by the same real number, called the **common difference.**

EXAMPLE 1   Arithmetic sequences

Show that each sequence is arithmetic, and find the missing term.

a. 1, 4, 7, 10, 13, ___, . . .

b. 20, 14, 8, 2, −4, −10, ___, . . .

c. $a_1,\ a_1 + d,\ a_1 + 2d,\ a_1 + 3d,\ a_1 + 4d,$ ___, . . .

Solution

a. Look for a common difference by subtracting each term from the succeeding term:

$$4 - 1 = 3, \qquad 7 - 4 = 3, \qquad 10 - 7 = 3, \qquad 13 - 10 = 3$$

If the difference between each pair of consecutive terms of the sequence is the same number, then that number is the common difference; in this case, it is 3. To find the missing term, simply add the common difference. The next term is

$$13 + 3 = 16$$

☠ Do not check only the first difference. All the differences must be the same. ☠

b. The common difference is -6. The next term is found by adding the common difference:

$$-10 + (-6) = -16$$

c. The common difference is d, so the next term is

$$(a_1 + 4d) + d = a_1 + 5d$$

■

Example 1c leads us to a formula for arithmetic sequences:

a_1 is the first term of an arithmetic sequence.

a_2 is the second term of an arithmetic sequence, and

$$a_2 = a_1 + d$$

a_3 is the third term of an arithmetic sequence, and

$$a_3 = a_2 + d = (a_1 + d) + d = a_1 + 2d$$

a_4 is the fourth term of an arithmetic sequence, and

$$a_4 = a_3 + d = (a_1 + 2d) + d = a_1 + 3d$$

a_{43} is the 43rd term of an arithmetic sequence, and

$$a_{43} = a_1 + \underset{\text{One less than the term number}}{\underline{42}}d$$

⋮

This pattern leads to the following formula.

> **GENERAL TERM OF AN ARITHMETIC SEQUENCE**
>
> The **general term** of an arithmetic sequence $a_1, a_2, a_3, \ldots, a_n, \ldots$ with common difference d is
> $$a_n = a_1 + (n-1)d$$

EXAMPLE 2   Finding an arithmetic sequence, given a general term

If $a_n = 26 - 6n$, list the sequence.

Solution

$$a_1 = 26 - 6(1) = 20 \qquad \text{a_1 means evaluate $26 - 6n$ for $n = 1$.}$$
$$a_2 = 26 - 6(2) = 14$$
$$a_3 = 26 - 6(3) = 8$$
$$a_4 = 26 - 6(4) = 2$$

The sequence is 20, 14, 8, 2, ... The graph of this sequence is shown in Figure 8.1. Arithmetic sequences appear linear, but because of the limited domain (counting numbers), the graph appears as a set of discrete points.

■

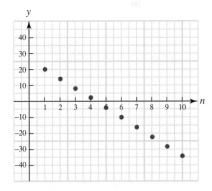

Figure 8.1 Graph of $a_n = 26 - 6n$

Geometric Sequences

A second type of sequence is the *geometric sequence* or *geometric progression*. If each term is *multiplied* by the same number (instead of *added* to the same number) to obtain successive terms, a geometric sequence is formed.

Geometric Growth Is Exponential

> **GEOMETRIC SEQUENCE**
>
> A **geometric sequence** is a sequence whose consecutive terms have the same quotient, called the **common ratio**.

If the sequence is geometric, the number obtained by dividing any term into the following term of that sequence will be the same nonzero number. No term of a geometric sequence may be zero.

EXAMPLE 3   Geometric sequences

Show that each sequence is geometric, and find the common ratio.

  a. 2, 4, 8, 16, 32, ____,

  b. $10, 5, \frac{5}{2}, \frac{5}{4}, \frac{5}{8},$ ____, $\ldots$

  c. $g_1, g_1 r, g_1 r^2, g_1 r^3, g_1 r^4,$ ____, $\ldots$

Solution

  a. First, verify that there is a common ratio:

$$\frac{4}{2} = 2, \quad \frac{8}{4} = 2, \quad \frac{16}{8} = 2, \quad \frac{32}{16} = 2$$

  The common ratio is 2, so to find the next term, multiply the common ratio by the preceding term. The next term is

$$32(2) = 64$$

  b. The common ratio is $\frac{1}{2}$ (be sure to check *each* ratio). The next term is found by multiplication:

$$\frac{5}{8}\left(\frac{1}{2}\right) = \frac{5}{16}$$

  c. The common ratio is r, and the next term is

$$g_1 r^4 (r) = g_1 r^5$$

___

As with arithmetic sequences, we denote the terms of a geometric sequence by using a special notation. Let $g_1, g_2, g_3, \ldots, g_n, \ldots$ be the terms of a *geometric* sequence. Example 3c leads us to a formula for geometric sequences.

$$g_2 = r g_1$$
$$g_3 = r g_2 = r(r g_1) = r^2 g_1$$
$$g_4 = r g_3 = r(r g_2) = r(r^2 g_1) = r^3 g_1$$

$$g_5 = rg_4 = r(r^3 g_1) = r^4 g_1$$
$$\vdots$$

one less than the term number

$$g_{92} = \overbrace{r^{91}}^{} g_1$$

Look for patterns to find the following formula.

GENERAL TERM OF A GEOMETRIC SEQUENCE

For a geometric sequence $g_1,\ g_2,\ g_3,\ \ldots,\ g_n,\ \ldots$, with common ratio r, the **general term** is

$$g_n = g_1 r^{n-1}$$

EXAMPLE 4   **Finding a geometric sequence, given a general term**

List the sequence generated by $g_n = 50(2)^{n-1}$.

Solution

$$g_1 = 50(2)^{1-1} = 50$$
$$g_2 = 50(2)^{2-1} = 100$$
$$g_3 = 50(2)^{3-1} = 200$$
$$g_4 = 50(2)^{4-1} = 400$$

The sequence is 50, 100, 200, 400, ...
The graph of this sequence is shown in Figure 8.2.
Geometric sequences appear exponential, but because
of the domain, the graph appears as a set of discrete points.

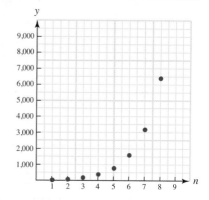

Figure 8.2 Graph of $g_n = 50(2)^{n-1}$

Fibonacci-Type Sequences

Even though our attention is focused on arithmetic and geometric sequences, it is important to realize there can be other types of sequences.

© Eric Isselée/ShutterStock, Inc.

   The next type of sequence came about, oddly enough, by looking at the birth patterns of rabbits. In the 13th century, Leonardo Fibonacci wrote a book, *Liber Abaci*, in which he discussed the advantages of the Hindu-Arabic numerals over Roman numerals. In this book, one problem was to find the number of rabbits alive after a given number of generations. Let us consider what he did with this problem.

EXAMPLE 5   **Fibonacci's problem**   **MODELING APPLICATION**

Suppose a pair of rabbits will produce a new pair of rabbits in their second month and thereafter will produce a new pair every month. The new rabbits will do exactly the same. Start with one pair. How many pairs will there be in 10 months?

**Leonardo Fibonacci
(ca. 1175–1250)**

© Karl Smith Library

Fibonacci, also known as Leonardo da Pisa, visited a number of Eastern and Arabic cities, where he became interested in the Hindu-Arabic numeration system we use today. He wrote *Liber Abaci*, in which he strongly advocated the use of the Hindu-Arabic numeration system. In his book *In Mathematical Circles*, Howard Eves states: "Fibonacci sometimes signed his work with the name Leonardo Bigollo. Now *bigollo* has more than one meaning; it means both 'traveler' and 'blockhead.' In signing his work as he did, Fibonacci may have meant that he was a great traveler, for so he was. But a story has circulated that he took pleasure in using this signature because many of his contemporaries considered him a blockhead (for his interest in the new numbers), and it pleased him to show these critics what a blockhead could accomplish."

Solution

Step 1: *Understand the problem.* We can begin to understand the problem by looking at the following chart:

Number of Months	Number of Pairs	Pairs of Rabbits (the pairs shown in color are ready to reproduce in the next month)
Start	1	
1	1	
2	2	
3	3	
4	5	
5	8	
⋮	⋮	Same pair (rabbits never die)

Step 2: *Devise a plan.* We look for a pattern with the sequence 1, 1, 2, 3, 5, 8, . . . ; it is not arithmetic and it is not geometric. It looks as if (after the first two months) each new number can be found by adding the two previous terms.

Step 3: *Carry out the plan.*
$$1 + 1 = 2$$
$$1 + 2 = 3$$
$$2 + 3 = 5$$
$$3 + 5 = 8$$
$$5 + 8$$

Do you see the pattern? The sequence is 1, 1, 2, 3, 5, 8, 13, 21, 34, 55, 89, ...

Step 4: *Look back.* Using this pattern, Fibonacci was able to compute the number of pairs of rabbits alive after 10 months (it is the 10th term after the first 1): 89. He could also compute the number of pairs of rabbits after the first year or any other interval. Without a pattern, the problem would indeed be a difficult one.   ∎

FIBONACCI SEQUENCE

A **Fibonacci-type sequence** is a sequence in which any two numbers form the first and second terms, and subsequent terms are found by adding the previous two terms. *The* **Fibonacci sequence** is that sequence for which the first two terms are both 1.

If we state this definition symbolically with first term s_1, second term s_2, and general term s_n, we have a formula for the general term.

GENERAL TERM OF A FIBONACCI SEQUENCE

For a Fibonacci sequence $s_1, s_2, s_3, \ldots, s_n, \ldots$, the **general term** is given by

$$s_n = s_{n-1} + s_{n-2}$$

for $n \geq 3$.

EXAMPLE 6   Fibonacci-type sequences

a. If $s_1 = 5$ and $s_2 = 2$, list the first five terms of this Fibonacci-type sequence.

b. If s_1 and s_2 represent any first numbers, list the first eight terms of this Fibonacci-type sequence.

Solution

a. 5, 2, $\underset{5+2}{7}$, 9, $\underset{7+9}{16}$ with $\overset{2+7}{9}$

b. s_1 and s_2 are given.

$$
\begin{aligned}
s_3 &= s_2 + s_1 \\
s_4 &= s_3 + s_2 = (s_1 + s_2) + s_2 = s_1 + 2s_2 \\
s_5 &= s_4 + s_3 = (s_1 + 2s_2) + (s_1 + s_2) = 2s_1 + 3s_2 \\
s_6 &= s_5 + s_4 = (2s_1 + 3s_2) + (s_1 + 2s_2) = 3s_1 + 5s_2 \\
s_7 &= s_6 + s_5 = (3s_1 + 5s_2) + (2s_1 + 3s_2) = 5s_1 + 8s_2 \\
s_8 &= s_7 + s_6 = (5s_1 + 8s_2) + (3s_1 + 5s_2) = 8s_1 + 13s_2
\end{aligned}
$$

Look at the coefficients in the algebraic simplification at the right, and notice that the Fibonacci sequence 1, 1, 2, 3, 5, 8, . . . is part of the construction of any Fibonacci-type sequence regardless of the terms s_1 and s_2. ∎

Historically, there has been much interest in the Fibonacci sequence. It is used in botany, zoology, business, economics, statistics, operations research, archaeology, architecture, education, and sociology. There is even an official Fibonacci Association.

An example of Fibonacci numbers occurring in nature is illustrated by a sunflower. The seeds are arranged in spiral curves as shown in Figure 8.3.

© Priscilla R. Steele/ShutterStock, Inc.

Figure 8.3 **The arrangement of the pods (phyllotaxy) of a sunflower illustrates a Fibonacci sequence.**

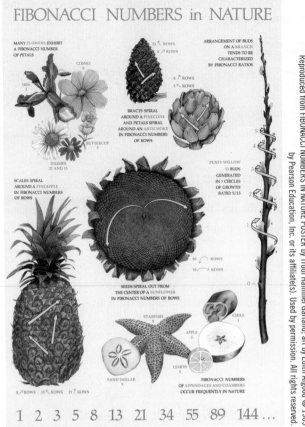

If we count the number of counterclockwise spirals (13 and 21 in Figure 8.3), they are successive terms in the Fibonacci sequence. This is true of all sunflowers and, indeed, of the seed head of any composite flower such as the daisy or aster. These patterns are mentioned in the Group Research Projects at the end of this chapter.

EXAMPLE 7   Classifying sequences

Classify the given sequences as arithmetic, geometric, Fibonacci-type, or none of the above. Find the next term for each sequence, and give its general term if it is arithmetic, geometric, or Fibonacci-type.

   a. 15, 30, 60, 120, . . .

   b. 15, 30, 45, 60, . . .

   c. 15, 30, 45, 75, . . .

   d. 15, 20, 26, 33, . . .

   e. 3, 3, 3, 3, . . .

   f. 15, 30, 90, 360, . . .

Solution

   a. 15, 30, 60, 120, . . . does not have a common difference, but a common ratio of 2, so this is a geometric sequence. The next term is $120(2) = 240$. The general term is $g_n = 15(2)^{n-1}$.

   b. 15, 30, 45, 60, . . . has a common difference of 15, so this is an arithmetic sequence. The next term of the sequence is $60 + 15 = 75$. The general term is $a_n = 15 + (n-1)15 = 15 + 15n - 15 = 15n$.

c. 15, 30, 45, 75, . . . does not have a common difference or a common ratio. Next, we check for a Fibonacci-type sequence by adding successive terms: $15 + 30 = 45$; $30 + 45 = 75$, so this is a Fibonacci-type sequence. The next term is $45 + 75 = 120$. The general term is $s_n = s_{n-1} + s_{n-2}$, where $s_1 = 15$ and $s_2 = 30$.

d. 15, 20, 26, 33, . . . does not have a common difference or a common ratio. Since $15 + 20 \neq 26$, we see it is not a Fibonacci-type sequence. We do see a pattern, however, when looking at the differences:

$$20 - 15 = 5; \qquad 26 - 20 = 6; \qquad 33 - 26 = 7$$

The next difference is 8. Thus, the next term is $33 + 8 = 41$.

e. 3, 3, 3, 3, . . . has a common difference of 0 and the common ratio is 1, so it is both arithmetic and geometric. The next term is 3. The general term is $a_n = 3 + (n - 1)0 = 3$ or $g_n = 3(1)^{n-1} = 3$.

f. 15, 30, 90, 360, . . . does not have a common difference or a common ratio. We see a pattern when looking at the ratios:

$$\frac{30}{15} = 2; \qquad \frac{90}{30} = 3; \qquad \frac{360}{90} = 4$$

The next ratio is 5. Thus, the next term is $360(5) = 1,800$. ∎

General terms can be given for sequences that are not arithmetic, geometric, or Fibonacci-type. Consider the following example.

EXAMPLE 8 Generating a sequence from a general term

Find the first four terms for the sequences with the given general terms. If the general term defines an arithmetic, geometric, or Fibonacci-type sequence, so state.

a. $s_n = n^2$; this is also written as $\{n^2\}$.

b. $\{(-1)^n - 5n\}$

c. $s_n = s_{n-1} + s_{n-2}$, where $s_1 = -4$ and $s_2 = 6$

d. $\{2n\}$

e. $s_n = 2n + (n - 1)(n - 2)(n - 3)(n - 4)$

Solution

a. Since $s_n = n^2$, we find $s_1 = (1)^2 = 1$; $s_2 = (2)^2 = 4$, $s_3 = (3)^2 = 9, \ldots$.
The sequence is 1, 4, 9, 16, ...

b. Since $s_n = (-1)^n - 5n$, we find

$$s_1 = (-1)^1 - 5(1) = -6$$
$$s_2 = (-1)^2 - 5(2) = -9$$
$$s_3 = (-1)^3 - 5(3) = -16$$
$$s_4 = (-1)^4 - 5(4) = -19$$

The sequence is $-6, -9, -16, -19, \ldots$.

c. $s_n = s_{n-1} + s_{n-2}$ is the form of a Fibonacci-type sequence. Using the two given terms, we find:

$$s_1 = -4 \qquad \textit{Given.}$$
$$s_2 = 6 \qquad \textit{Given.}$$
$$s_3 = s_2 + s_1 = 6 + (-4) = 2$$
$$s_4 = s_3 + s_2 = 2 + 6 = 8$$

The sequence is $-4, 6, 2, 8, \ldots$.

d. $s_n = 2n$; $s_1 = 2(1) = 2$; $s_2 = 2(2) = 4$; $s_3 = 2(3) = 6$; . . . .
The sequence is 2, 4, 6, 8, . . . ; this is an arithmetic sequence.

e. $s_n = 2n + (n - 1)(n - 2)(n - 3)(n - 4)$

$$s_1 = 2(1) + (1 - 1)(1 - 2)(1 - 3)(1 - 4) = 2$$
$$s_2 = 2(2) + (2 - 1)(2 - 2)(2 - 3)(2 - 4) = 4$$
$$s_3 = 2(3) + (3 - 1)(3 - 2)(3 - 3)(3 - 4) = 6$$
$$s_4 = 2(4) + (4 - 1)(4 - 2)(4 - 3)(4 - 4) = 8$$

The sequence is 2, 4, 6, 8, ... ∎

Examples 8d and 8e show that if only a finite number of successive terms are known and no general term is given, then a *unique* general term cannot be given. That is, if we are given the sequence

$$2, 4, 6, 8, . . .$$

the next term is probably 10 (if we are thinking of the general term of Example 8d), but it *may* be something different. In Example 8e,

$$s_5 = 2(5) + (5 - 1)(5 - 2)(5 - 3)(5 - 4) = 34$$

This gives the unlikely sequence 2, 4, 6, 8, 34, ... In general, you are looking for the simplest general term; nevertheless, you must remember that answers are not unique *unless the general term is given.*

PROBLEM SET 8.1

LEVEL 1

WHAT IS WRONG, *if anything, with each statement in Problems 1–4?*

1. The notation s_n is sequence notation for a function s, where $s(n) = s_n$ whose domain is the set of real numbers.
2. If the general term, s_n, of a sequence is replaced by d_n, then the common difference is d, and if it is replaced by r_n then the common ratio is r.
3. In the notation $s_n = s_{n-1} + s_{n-2}$, if $n = 23$, then $s_{n-1} = 22$ and $s_{n-2} = 21$.
4. The sequence $\{5\}$ has only one term.

In Problems 5–26,

  a. Classify the sequences as arithmetic, geometric, Fibonacci, or none of these.
  b. If the sequence is arithmetic, give d; if geometric, give r; if Fibonacci, give the first two terms; and if none of these, state a pattern using your own words.
  c. Supply the next term.

5. 2, 4, 6, 8, ___, . . .
6. 2, 4, 8, 16, ___, . . .
7. 2, 4, 6, 10, ___, . . .
8. 5, 15, 25, ___, . . .
9. 5, 15, 45, ___, . . .
10. 5, 15, 20, ___, . . .
11. 1, 5, 25, ___, . . .
12. 25, 5, 1, ___, . . .
13. 21, 20, 18, 15, 11, ___, . . .
14. 8, 6, 7, 5, 6, 4, ___, . . .

15. 2, 5, 8, 11, 14, ___, . . .
16. 3, 6, 12, 24, 48, ___, . . .
17. 5, −15, 45, −135, 405, ___, . . .
18. 10, 10, 10, ___, . . .
19. 2, 5, 7, 12, ___, . . .
20. 3, 6, 9, 15, ___, . . .
21. 1, 8, 27, 64, 125, ___, . . .
22. 8, 12, 18, 27, ___, . . .
23. $\frac{1}{2}, \frac{1}{3}, \frac{2}{3}, \frac{1}{4}, \frac{3}{4}, \frac{1}{5}, \frac{2}{5}, \frac{3}{5}, \frac{4}{5}, \frac{1}{6}$, ___, . . .
24. $\frac{1}{10}, \frac{1}{5}, \frac{3}{10}, \frac{2}{5}, \frac{1}{2}$, ___, . . .
25. $\frac{4}{3}, 2, 3, 4\frac{1}{2}$, ___, . . .
26. $\frac{7}{12}, \frac{2}{3}, \frac{3}{4}, \frac{5}{6}$, ___, . . .

LEVEL 2

In Problems 27–40,

  a. Find the first three terms of the sequences whose nth terms are given.
  b. Classify the sequence as arithmetic (give d), geometric (give r), both, or neither.

27. $s_n = 4n - 3$
28. $s_n = -3 + 3n$
29. $s_n = 2 - n$
30. $s_n = 7 - 3n$
31. $s_n = \frac{2}{n}$
32. $s_n = 1 - \frac{1}{n}$
33. $s_n = \dfrac{n-1}{n+1}$
34. $s_n = \dfrac{10}{2^{n-1}}$
35. $\left\{\frac{1}{2}n(n+1)\right\}$
36. $\left\{\frac{1}{4}n^2(n+1)^2\right\}$
37. $\{-5\}$
38. $\left\{\frac{2}{3}\right\}$
39. $\left\{(-1)^{n+1}\right\}$
40. $\{(-1)^n(n+1)\}$

Classify the situations in Problems 41–46 as arithmetic, geometric, or neither. Do NOT answer these questions, simply classify. (Do not worry, you will be given the opportunity to answer these questions later.)

41. Suppose that a teacher obtains a job with a starting salary of $25,000 and receives a $500 raise every year thereafter. What will the teacher's salary be in 10 years?

42. Suppose that an auto worker obtains a job with a starting salary of $30,000 and receives an 8% raise every year thereafter. What will the auto worker's salary be in 10 years?

43. Suppose that a teacher obtains a job with a starting salary of $25,000 and receives a $3\frac{1}{3}\%$ raise every year thereafter. What will the teacher's salary be in 10 years?

44. A grocery clerk must stack 30 cases of canned fruit, each containing 24 cans. He decides to display the cans by stacking them in a pyramid where each row after the bottom row contains one less can. Is it possible to use all the cans and end up with a top row of only one can?

45. Suppose that a chain letter asks you to send copies to 10 of your friends. If everyone carries out the directions and sends the chain letter, and if nobody receives more than one chain letter, how many people will be involved in 10 mailings?

46. March Madness begins with 32 teams playing an elimination tournament. After each game, the winner is paired with another winner, until one team is declared the best. What is the total number of games played in the tournament?

Find the requested terms in Problems 47–54.

47. Find the 15th term of the sequence
$$s_n = 4n - 3$$

48. Find the 69th term of the sequence
$$s_n = 7 - 3n$$

49. Find the 20th term of the sequence
$$s_n = (-1)^n (n + 1)$$

50. Find the 3rd term of the sequence
$$s_n = (-1)^{n+1} 5^{n+1}$$

51. Find the first five terms of the sequence where $s_1 = 2$ and $s_n = 3s_{n-1}$, $n \geq 2$.

52. Find the first five terms of the sequence where $s_1 = 3$ and $s_n = \frac{1}{3} s_{n-1}$, $n \geq 2$.

53. Find the first five terms of the sequence where $s_1 = 1$, $s_2 = 1$, and $s_n = s_{n-1} + s_{n-2}$, $n \geq 3$.

54. Find the first five terms of the sequence where $s_1 = 1$, $s_2 = 2$, and $s_n = s_{n-1} + s_{n-2}$, $n \geq 3$.

LEVEL 3

55. Suppose that $g_1, g_2, g_3, \ldots$ is a geometric sequence with common ratio r. Find the common ratio for the sequence
$$g_1^2, g_2^2, g_3^2, \ldots$$

56. Suppose that $g_1, g_2, g_3, \ldots$ is a geometric sequence of positive terms with positive common ratio. Show that the sequence
$$\log g_1, \log g_2, \log g_3, \ldots$$
is an arithmetic sequence. What is the common difference?

57. Suppose that $a_1, a_2, a_3, \ldots$ is an arithmetic sequence with common difference d. Show that the sequence
$$e^{a_1}, e^{a_2}, e^{a_3}, \ldots$$
is a geometric sequence; find the common ratio.

58. The sum of three terms of an arithmetic sequence is 570. Find the middle term.

59. If the first term and the common difference of an arithmetic sequence are the same, and we know that the product of the first three terms of this sequence is 1,053,696, what is the middle term?

60. Fill in the blanks so that
$$\_\_\_, 8, \_\_\_, \_\_\_, 27, \_\_\_, \ldots$$
is
a. an arithmetic sequence.
b. a geometric sequence.
c. a sequence that is neither arithmetic nor geometric, for which you are able to write a general term.

8.2 Limit of a Sequence

It is often desirable to examine the behavior of a given sequence $\{s_n\}$ as n gets arbitrarily large. For example, consider the sequence

$$s_n = \frac{n}{n+1}$$

> ☠ Even though we write s_n, remember this is a function, $f(n) = \frac{n}{n+1}$, where the domain is the set of nonnegative integers. ☠

Because $s_1 = \frac{1}{2}$, $s_2 = \frac{2}{3}$, $s_3 = \frac{3}{4}$, $\ldots$, we can plot the terms of this sequence where the domain is on a number line, as shown in Figure 8.4a, or the sequence can be plotted in two dimensions, as shown in Figure 8.4b.

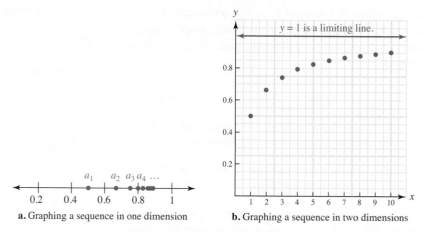

a. Graphing a sequence in one dimension

b. Graphing a sequence in two dimensions

Figure 8.4 Graphing the sequence $s_n = \dfrac{n}{n+1}$

Convergence

By looking at either graph in Figure 8.4, we see that it appears the terms of the sequence are approaching 1. In general, if the terms of the sequence approach the number L as n increases without bound, we say that the sequence *converges to the limit* L and write

$$L = \lim_{n \to \infty} s_n$$

For instance, in our example, we would expect

$$\lim_{n \to \infty} \frac{n}{n+1} = 1$$

This limiting behavior is analogous to the continuous case (discussed in Section 2.8) and may be defined informally as follows, even though a precise definition needs to wait for calculus.

CONVERGENT SEQUENCE

The sequence $\{s_n\}$ **converges** to the number L, written

$$L = \lim_{n \to \infty} s_n$$

which means that the terms of the sequence $\{s_n\}$ can be made as close to L as may be desired by making n sufficiently large. If $\{s_n\}$ does not converge, then we say that the sequence **diverges**.

A geometric interpretation of this definition is shown in Figure 8.5.

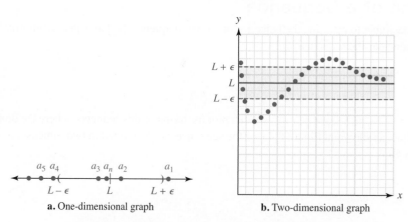

a. One-dimensional graph

b. Two-dimensional graph

Figure 8.5 Geometric interpretation of a converging sequence

In one dimension, if the terms of the sequence eventually end up in some interval $(L - \varepsilon, \, L + \varepsilon)$, then we say the sequence converges. In two dimensions, if the terms of the sequence eventually end up in some vertical band $(L - \varepsilon, \, L + \varepsilon)$, then we say the sequence converges. Note that if

$$L = \lim_{n \to \infty} s_n$$

the numbers s_n may be practically anywhere at first (that is, for "small" n), but eventually, the a_n must "cluster" near the limiting value L.

Limit Properties

We have the following useful result, which you will spend some time proving in calculus, but for now, we simply state them as properties.

LIMIT PROPERTIES

If $\lim_{n \to \infty} s_n = L$ and $\lim_{n \to \infty} t_n = M$, then

Linearity rule $\lim_{n \to \infty} (a s_n + b t_n) = aL + bM$

Product rule $\lim_{n \to \infty} (s_n t_n) = LM$

Quotient rule $\lim_{n \to \infty} \dfrac{s_n}{t_n} = \dfrac{L}{M}$ provided $M \neq 0$

Root rule $\lim_{n \to \infty} \sqrt[m]{s_n} = \sqrt[m]{L}$ provided $\sqrt[m]{s_n}$ is defined for all n and $\sqrt[m]{L}$ exists.

EXAMPLE 1   Convergent sequences

Find the limit of each of these convergent sequences:

a. $\left\{ \dfrac{100}{n} \right\}$ **b.** $\left\{ \dfrac{2n^2 + 5n - 7}{n^3} \right\}$ **c.** $\left\{ \dfrac{3n^4 + n - 1}{5n^4 + 2n^2 + 1} \right\}$

Solution

a. As n grows arbitrarily large, $100/n$ gets smaller and smaller. Thus,

$$\lim_{n \to \infty} \frac{100}{n} = 0$$

A graphical representation is shown in Figure 8.6.

b. We cannot use the quotient rule for sequences because neither the limit in the numerator nor the one in the denominator exists. However,

$$\frac{2n^2 + 5n - 7}{n^3} = \frac{2}{n} + \frac{5}{n^2} - \frac{7}{n^3}$$

and by using the linearity rule, we find that

$$
\begin{aligned}
\lim_{n \to \infty} \frac{2n^2 + 5n - 7}{n^3} &= \lim_{n \to \infty} \frac{2}{n} + 5 \lim_{n \to \infty} \frac{1}{n^2} - 7 \lim_{n \to \infty} \frac{1}{n^3} \\
&= 0 + 0 + 0 \\
&= 0
\end{aligned}
$$

A graph is shown in Figure 8.7a.

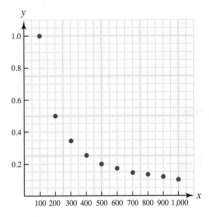

Figure 8.6 Graphical representation of $s_n = 100/n$

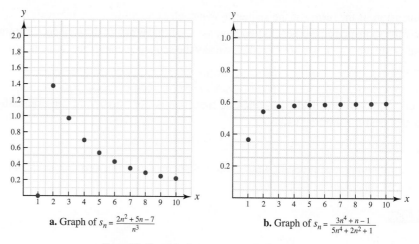

a. Graph of $s_n = \frac{2n^2 + 5n - 7}{n^3}$ **b.** Graph of $s_n = \frac{3n^4 + n - 1}{5n^4 + 2n^2 + 1}$

Figure 8.7 Graphical representations of sequences

c. Divide the numerator and denominator by n^4 to obtain

$$\lim_{n \to \infty} \frac{3n^4 + n - 1}{5n^4 + 2n^2 + 1} = \lim_{n \to \infty} \frac{3 + \frac{1}{n^3} - \frac{1}{n^4}}{5 + \frac{2}{n^2} + \frac{1}{n^4}} = \frac{3}{5}$$

A graph of this sequence is shown in Figure 8.7b.   ∎

EXAMPLE 2   Divergent sequences

Show that the following sequences diverge:

a. $\left\{ (-1)^n \right\}$ **b.** $\left\{ \dfrac{n^5 + n^3 + 2}{7n^4 + n^2 + 3} \right\}$

Solution

a. The sequence defined by $\{(-1)^n\}$ is $-1, 1, -1, 1, \ldots$, and this sequence diverges by oscillation because the nth term is always either 1 or -1. Thus, a_n cannot approach one specific number L as n grows large.

b. $\displaystyle\lim_{n \to \infty} \frac{n^5 + n^3 + 2}{7n^4 + n^2 + 3} = \frac{1 + \frac{1}{n^2} + \frac{2}{n^5}}{\frac{7}{n} + \frac{1}{n^3} + \frac{3}{n^5}}$

The numerator tends toward 1 as $n \to \infty$, and the denominator approaches 0. Hence, the quotient increases without bound, and the sequence must diverge.   ∎

If $\displaystyle\lim_{n \to \infty} s_n$ does not exist because the numbers s_n become arbitrarily large as $n \to \infty$, we write $\displaystyle\lim_{n \to \infty} s_n = \infty$. We summarize this more precisely in the following box.

LIMIT TO INFINITY

The **limit notation** $\displaystyle\lim_{n \to \infty} s_n$ is a **limit to infinity** and is informally defined as follows:

$\displaystyle\lim_{n \to \infty} s_n = \infty$ means that for any real number m, we have

$s_n > m$ for all sufficiently large n.

$\displaystyle\lim_{n \to \infty} t_n = -\infty$ means that for any real number M, we have $b_n < M$ for all sufficiently large n.

Rewriting the answer to Example 2b in this notation,

$$\lim_{n \to \infty} \frac{n^5 + n^3 + 2}{7n^4 + n^2 + 3} = \infty$$

Also notice that $\lim_{n \to \infty} (-5n) = -\infty$ (that is, decreases without bound), whereas $\lim_{n \to \infty} (-1)^n$ does not exist. Thus, the answer to Example 2a is neither ∞ nor $-\infty$.

EXAMPLE 3   Determining the convergence or divergence of a sequence

Determine the convergence or divergence of the sequence $\left\{ \sqrt{n^2 + 3n} - n \right\}$.

Solution
Consider

$$\lim_{n \to \infty} \left(\sqrt{n^2 + 3n} - n \right)$$

It would not be correct to apply the linearity property for sequences (because neither $\lim_{n \to \infty} \sqrt{n^2 + 3n}$ nor $\lim_{n \to \infty} n$ exists). It is also not correct to use this as a reason to say that the limit does not exist. You might even try some values of n to guess that there is some limit. In order to find the limit, however, we shall rewrite the general term algebraically as follows:

$$\sqrt{n^2 + 3n} - n = \left(\sqrt{n^2 + 3n} - n \right) \frac{\sqrt{n^2 + 3n} + n}{\sqrt{n^2 + 3n} + n}$$

$$= \frac{n^2 + 3n - n^2}{\sqrt{n^2 + 3n} + n}$$

$$= \frac{3n}{\sqrt{n^2 + 3n} + n} \cdot \frac{\frac{1}{n}}{\frac{1}{n}}$$

$$= \frac{3}{\frac{\sqrt{n^2 + 3n}}{n} + 1}$$

$$= \frac{3}{\sqrt{\frac{n^2 + 3n}{n^2}} + 1}$$

$$= \frac{3}{\sqrt{1 + \frac{3}{n}} + 1}$$

Hence, $\lim_{n \to \infty} \left(\sqrt{n^2 + 3n} - n \right) = \lim_{n \to \infty} \frac{3}{\sqrt{1 + \frac{3}{n}} + 1} = \frac{3}{2}$ ∎

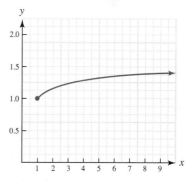

Figure 8.8 Graph of $y = \sqrt{x^2 + 3x} - x$, $x \geq 1$

The graph of a sequence consists of a succession of isolated points, but we can compare the graph of the sequence with the graph of the related function. For Example 3, this graph is shown in Figure 8.8.

The only difference between $\lim_{n \to \infty} s_n = L$ and $\lim_{x \to \infty} f(x) = L$ is that n is required to be an integer. This is stated in the hypothesis of the following property (which is proved in a calculus course).

LIMIT OF A SEQUENCE

Suppose f is a function such that $s_n = f(n)$ for $n = 1, 2, \ldots$. If $\lim_{x \to \infty} f(x)$ exists and $\lim_{x \to \infty} f(x) = L$ the sequence $\{ s_n \}$ converges and $\lim_{x \to \infty} s_n = L$.

Be sure you read this property correctly. In particular, note that it *does not* say that if $\lim\limits_{n \to \infty} s_n = L$, then $\lim\limits_{x \to \infty} f(x) = L$ (see Figure 8.9).

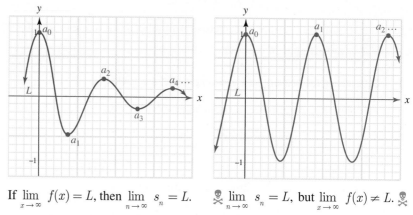

If $\lim\limits_{x \to \infty} f(x) = L$, then $\lim\limits_{n \to \infty} s_n = L$.　☠ $\lim\limits_{n \to \infty} s_n = L$, but $\lim\limits_{x \to \infty} f(x) \neq L$. ☠

Figure 8.9 Graphical comparison

Bounded, Monotonic Sequences

A sequence $\{s_n\}$ is said to be **increasing** if $s_1 < s_2 < s_3 < \ldots < s_{k-1} < s_k < \ldots$ and **nondecreasing** if $s_1 \leq s_2 \leq \ldots \leq s_{k-1} \leq s_k \leq \ldots$. However, the sequence is **decreasing** if $s_1 > s_2 > s_3 > \ldots > s_{k-1} > s_k > \ldots$ and **nonincreasing** if $s_1 \geq s_2 \geq s_3 \geq \ldots \geq s_{k-1} \geq s_k \geq \ldots$. Finally, it is **bounded above** by M if $s_n \leq M$ for $n = 1, 2, 3, \ldots$ and **bounded below** by m if $m \leq s_n$ for $n = 1, 2, 3, \ldots$.

Moreover, a sequence is referred to as **monotonic** if it is nondecreasing or nonincreasing and **strictly monotonic** if it is increasing or decreasing. It is **bounded** if it is bounded both above and below.

In general, it is difficult to tell whether a given sequence converges or diverges, but thanks to the following property, it is easy to make this determination if we know the sequence is monotonic.

> **BMCP**
>
> The **bounded, monotonic convergence property**, sometimes called the BMCP, states that a monotonic sequence $\{s_n\}$ converges if it is bounded and diverges otherwise.

For the following informal argument, we will assume that $\{s_n\}$ is a nondecreasing sequence. You might wish to see whether you can give a similar informal argument for the increasing case. Because the terms of the sequence satisfy $s_1 \leq s_2 \leq s_3 \leq \ldots$, we know that the sequence is bounded from below by s_1 and that the graph of the corresponding points (n, s_n) will be rising in the plane. Two cases can occur, as shown in Figure 8.10.

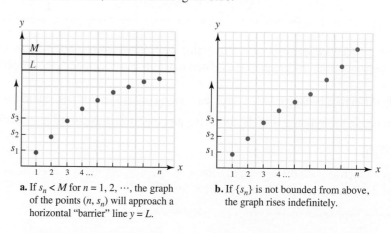

a. If $s_n < M$ for $n = 1, 2, \cdots$, the graph of the points (n, s_n) will approach a horizontal "barrier" line $y = L$.

b. If $\{s_n\}$ is not bounded from above, the graph rises indefinitely.

Figure 8.10 Graphical possibilities for the bounded, monotonic convergence property

Suppose the sequence $\{s_n\}$ is also bounded from above by a number M, so that $s_1 \leq s_n \leq M$ for $n = 1, 2, \ldots$. Then the graph of the points (n, s_n) must continually rise (because the sequence is monotonic), and yet, it must stay below the line $y = M$. The only way this can happen is for the graph to approach a "barrier" line $y = L$ (where $L \leq M$), and we have $\lim\limits_{n \to \infty} s_n = L$, as shown in Figure 8.10a. However, if the sequence is not bounded from above, the graph will rise indefinitely (Figure 8.10b) and the terms in the sequence $\{s_n\}$ cannot approach any finite number L. A formal proof of this property is given in most calculus textbooks.

EXAMPLE 4   Convergence using BMCP

Show that the sequence $\left\{ \dfrac{1 \cdot 3 \cdot 5 \cdot \cdots \cdot (2n - 1)}{2 \cdot 4 \cdot 6 \cdot \cdots \cdot (2n)} \right\}$ converges.

Solution
The first few terms of this sequence are

$$s_1 = \frac{1}{2} \quad s_2 = \frac{1 \cdot 3}{2 \cdot 4} = \frac{3}{8} \quad s_3 = \frac{1 \cdot 3 \cdot 5}{2 \cdot 4 \cdot 6} = \frac{5}{16}$$

Because $\frac{1}{2} > \frac{3}{8} > \frac{5}{16}$, it appears that the sequence is decreasing (that is, it is monotonic). We can prove this by showing that $s_{n+1} < s_n$, or equivalently, $\dfrac{s_{n+1}}{s_n} < 1$. (Note that $s_n \neq 0$ for all n.)

$$\frac{s_{n+1}}{s_n} = \frac{\dfrac{1 \cdot 3 \cdot 5 \cdots [2(n+1) - 1]}{2 \cdot 4 \cdot 6 \cdots [2(n+1)]}}{\dfrac{1 \cdot 3 \cdot 5 \cdots (2n - 1)}{2 \cdot 4 \cdot 6 \cdots (2n)}}$$

$$= \frac{1 \cdot 3 \cdot 5 \cdots (2n + 1)}{2 \cdot 4 \cdot 6 \cdots (2n + 2)} \cdot \frac{2 \cdot 4 \cdot 6 \cdots (2n)}{1 \cdot 3 \cdot 5 \cdots (2n - 1)}$$

$$= \frac{2n + 1}{2n + 2}$$

$$< 1$$

for any $n > 0$. Hence, for all $s_{n+1} < s_n$ for all n, and $\{s_n\}$ is a decreasing sequence. Because $s_n > 0$ for all n, it follows that $\{s_n\}$ is bounded below by 0. Applying the bounded, monotonic convergence property, we see that $\{s_n\}$ converges, but the bounded, monotonic convergence property tells us nothing about the limit. ∎

Technically, the sequence $\{s_n\}$ is monotonic only when its terms are either always nonincreasing or always nondecreasing, but the bounded, monotonic convergence property also applies to sequences whose terms are *eventually* monotonic. That is, it can be shown that the sequence $\{s_n\}$ converges if it is bounded and there exists an integer N such that $\{s_n\}$ is monotonic for all $n > N$. This modified form of the bounded, monotonic convergence property is illustrated in the following example.

EXAMPLE 5   Convergence of a sequence that is eventually monotonic

Show that the sequence $\left\{ \dfrac{\ln n}{\sqrt{n}} \right\}$ converges.

Solution
We will apply the bounded, monotonic convergence property. Some initial values are not decreasing:

n	$\dfrac{\ln n}{\sqrt{n}}$	n	$\dfrac{\ln n}{\sqrt{n}}$
1	0	20	0.67
2	0.49	30	0.62
3	0.63	40	0.58
4	0.69	50	0.55
5	0.72	100	0.46
6	0.73	1,000	0.22
7	0.73	10,000	0.09
8	0.74	100,000	0.04
9	0.73	1,000,000	0.01
10	0.73	10,000,000	0.00

The succession of numbers suggests that the sequence increases at first and then gradually decreases. To verify this behavior algebraically requires calculus, but if we use the numerical approach, it appears that the terms are bounded by 0.8 and that, for $n > 8$, the sequence is decreasing.

The BMCP is an extremely valuable theoretical tool. For example, in Chapter 4, 251, we defined the number e by the limit

$$\lim_{n \to \infty} \left(1 + \frac{1}{n}\right)^n = e$$

but to do so, we assumed that this limit exists. We now claim this assumption is warranted, because it turns out that the sequence $\left\{\left(1 + \frac{1}{n}\right)^n\right\}$ is increasing and bounded from above by 3. Thus, the bounded, monotonic convergence property assures us that the sequence converges, and this in turn guarantees the existence of the limit.

PROBLEM SET 8.2

LEVEL 1

1. IN YOUR OWN WORDS What do we mean by the limit of a sequence?
2. IN YOUR OWN WORDS Contrast an infinite limit and a limit to infinity.
3. IN YOUR OWN WORDS What is meant by the bounded, monotonic convergence property?
4. IN YOUR OWN WORDS Explain the difference and/or similarities between the limit of a function and the limit of a sequence.

Write out the first five terms (beginning with $n = 1$) of the sequences given in Problems 5–18. Approximate Problems 11–14 to the nearest hundredth.

5. $\{1 + (-1)^n\}$

6. $\left\{\left(\dfrac{-1}{2}\right)^{n+2}\right\}$

7. $\left\{\dfrac{\cos 2n\pi}{n}\right\}$

8. $\left\{n \sin \dfrac{n\pi}{2}\right\}$

9. $\left\{\dfrac{3n+1}{n+2}\right\}$

10. $\left\{\dfrac{n^2 - n}{n^2 + n}\right\}$

11. $\left\{\tan^{-1} n\right\}$

12. $\left\{\sin\left(\dfrac{\pi}{2} + \dfrac{1}{n}\right)\right\}$

13. $\left\{\dfrac{\ln n}{n^{1/n}}\right\}$

14. $\left\{\dfrac{\ln n}{\ln 2n}\right\}$

15. $\{s_n\}$ where $s_1 = 256$ and $s_n = \sqrt{a_{n-1}}$ for $n \geq 2$.
16. $\{s_n\}$ where $s_1 = 10$ and $s_n = \frac{1}{n} s_{n-1}$ for $n \geq 2$.
17. $\{s_n\}$ where $s_1 = -1$ and $s_n = n + s_{n-1}$ for $n \geq 2$.
18. $\{s_n\}$ where $s_1 = 1$ and $s_n = (s_{n-1})^2 + s_{n-1} + 1$ for $n \geq 2$.

Compute the limit of the convergent sequences in Problems 19–44.

19. $\left\{\dfrac{5n+8}{n}\right\}$

20. $\left\{\dfrac{5n}{n+7}\right\}$

21. $\left\{5 + 0.1^n\right\}$

22. $\left\{\left(-\dfrac{1}{2}\right)^n\right\}$

23. $\left\{\dfrac{n+(-1)^n}{n}\right\}$ **24.** $\left\{\dfrac{2n+1}{3n-4}\right\}$

25. $\left\{\dfrac{4-7n}{8+n}\right\}$ **26.** $\left\{\dfrac{8n^2+800n+5,000}{2n^2-1,000n+2}\right\}$

27. $\left\{\dfrac{100n+7,000}{n^2-n-1}\right\}$ **28.** $\left\{\dfrac{8n^2+6n+4,000}{n^3+1}\right\}$

29. $\left\{\dfrac{n^3-6n^2+85}{2n^3-5n+170}\right\}$ **30.** $\left\{\dfrac{2n}{n+7\sqrt{n}}\right\}$

31. $\left\{\dfrac{8n-500\sqrt{n}}{2n+800\sqrt{n}}\right\}$ **32.** $\left\{\dfrac{3\sqrt{n}}{5\sqrt{n}+\sqrt[4]{n}}\right\}$

33. $\left\{\dfrac{\sin n}{n}\right\}$ **34.** $\left\{\dfrac{\ln n}{n^2}\right\}$

35. $\left\{2^{5/n}\right\}$ **36.** $\left\{n^{3/n}\right\}$

37. $\left\{\left(1+\dfrac{3}{n}\right)^n\right\}$ **38.** $\left\{\left(1+\dfrac{4}{n}\right)^n\right\}$

39. $\left\{\left(1-\dfrac{2}{n}\right)^n\right\}$ **40.** $\left\{\left(1-\dfrac{5}{n}\right)^n\right\}$

41. $\left\{\sqrt{n^2+n}-n\right\}$ **42.** $\left\{\sqrt{n+5\sqrt{n}}-\sqrt{n}\right\}$

43. $\left\{\ln n-\ln(n+1)\right\}$ **44.** $\left\{\ln 3n-\ln(2n+1)\right]$

LEVEL 2

Show that each sequence given in Problems 45–50 converges either by showing it is increasing with an upper bound or decreasing with a lower bound.

45. $\left\{\dfrac{3n-2}{n}\right\}$ **46.** $\left\{\dfrac{4n+5}{n}\right\}$

47. $\left\{\dfrac{n}{2^n}\right\}$ **48.** $\left\{\dfrac{3n-7}{2^n}\right\}$

49. $\left\{\ln\left(\dfrac{n+1}{n}\right)\right\}$ **50.** $\left\{\sqrt[n]{n}\right\}$

Explain why each sequence in Problems 51–54 diverges.

51. $\left\{1+(-1)^n\right\}$ **52.** $\left\{\cos n\pi\right\}$

53. $\left\{\dfrac{n^3-7n+5}{100n^2+219}\right\}$ **54.** $\left\{\sqrt{n}\right\}$

55. PROBLEM FROM CALCULUS Suppose a particle of mass m moves back and forth along a line segment of length $|a|$. In classical mechanics, the particle can move at any speed, and thus, its energy can be any positive number. However, quantum mechanics replaces this continuous model of the particle's behavior with one in which the particle's energy level can have only certain discrete values, say, $E_1, E_2, \ldots$

Specifically, it can be shown that the nth term in this quantum sequence has the value

$$E_n=\frac{n^2h^2}{8ma^2}\qquad n=0, 1, 2, \ldots$$

where h is a physical constant known as *Planck's constant* ($h\approx 6.63\times 10^{-27}$ erg/s). List the first four values of E_n for a particle with mass 8 mg moving along a segment of length 100 cm.

56. The Fibonacci rabbit problem
Consider Fibonacci's rabbit problem of Example 5, Section 8.1. Let s_n denote the number of adult pairs of rabbits in her "colony" at the end of n months.

a. Explain why $s_1=1$, $s_2=1$, $s_3=2$, $s_4=3$ and in general

$$s_{n+1}=s_{n-1}+s_n \text{ for } n=2, 3, 4, \ldots$$

b. The *growth rate* of the colony during the $(n+1)$th month is

$$r_n=\frac{s_{n+1}}{s_n}$$

Compute r_n for $n=1, 2, 3, \ldots, 10$.

c. Assume that the growth rate sequence $\{r_n\}$ defined in part **b** converges, and let

$$L=\lim_{n\to\infty} r_n$$

Use the recursion formula in part **a** to show that

$$\frac{s_{n+1}}{s_n}=1+\frac{s_{n-1}}{s_n}$$

and conclude that L must satisfy the equation

$$L=1+\frac{1}{L}$$

Use this information to compute L.

LEVEL 3

In Problems 57–60, use the fact that $\lim\limits_{n\to\infty} s_n=L$ means $|s_n-L|$ is arbitrarily small when n is sufficiently large.

57. If $\lim\limits_{n\to\infty}\dfrac{n}{n+1}=1$, find N so that $\left|\dfrac{n}{n+1}-1\right|<0.01$ if $n>N$.

58. If $\lim\limits_{n\to\infty}\dfrac{2n}{n+3}=2$, find N so that $\left|\dfrac{2n+1}{n+3}-2\right|<0.01$ if $n>N$.

59. If $\lim\limits_{n\to\infty}\dfrac{n^2+1}{n^3}=0$, find N so that $\left|\dfrac{n^2+1}{n^3}\right|<0.001$ if $n>N$.

60. If $\lim\limits_{n\to\infty} e^{-n}=0$, find N so that $\dfrac{1}{e^n}<0.001$ if $n>N$.

8.3 Series

If the terms of a sequence are added, the expression is called a **series**. We first consider a finite sequence along with its associated sum. Note that a capital letter is used to indicate the sum.

> **FINITE SERIES**
>
> The indicated sum of the terms of a finite sequence
>
> $$s_1, \ s_2, \ s_3, \dots, s_n$$
>
> is called a **finite series** and is denoted by
>
> $$S_n = s_1 + s_2 + s_3 + \cdots + s_n$$

EXAMPLE 1 Finding the sum of the terms of a sequence

a. Find S_4, where $s_n = 26 - 6n$. **b.** Find S_3, where $s_n = (-1)^n n^2$.

Solution

a.
$$S_4 = s_1 + s_2 + s_3 + s_4$$
$$= \overbrace{[26 - 6(1)]}^{s_1} + \overbrace{[26 - 6(2)]}^{s_2} + \overbrace{[26 - 6(3)]}^{s_3} + \overbrace{[26 - 6(4)]}^{s_4}$$
$$= 20 + 14 + 8 + 2$$
$$= 44$$

b.
$$S_3 = s_1 + s_2 + 3_3$$
$$= \overbrace{[(-1)^1(1)^2]}^{s_1} + \overbrace{[(-1)^2(2)^2]}^{s_2} + \overbrace{[(-1)^3(3)^2]}^{s_3}$$
$$= -1 + 4 + (-9)$$
$$= -6$$

The terms of the sequence in Example 1b alternate in sign: $-1, 4, -9, 16, \dots$. A factor of $(-1)^n$ or $(-1)^{n+1}$ in the general term will cause the signs of the terms to alternate, creating a series called an **alternating series**.

Summation Notation

Before we continue to discuss finding the sum of the terms of a sequence, we need a handy notation, called **summation notation**. In Example 1a, we wrote

$$S_4 = s_1 + s_2 + s_3 + s_4$$

Using summation notation or, as it is sometimes called, **sigma notation**, we could write this sum using the Greek letter Σ:

$$S_4 = \sum_{k=1}^{4} s_k = s_1 + s_2 + s_3 + s_4$$

The sigma notation evaluates the expression (s_k) immediately following the sigma (Σ) sign, first for $k = 1$, then for $k = 2$, then for $k = 3$, and finally for $k = 4$, and then adds these numbers. That is, the expression is evaluated for *consecutive counting numbers*, starting with the value of k listed at the bottom of the sigma ($k = 1$) and ending with the value of k listed at the top of the sigma ($k = 4$). For example, consider $s_k = 2k$ with $k = 1, 2, 3, 4, 5, 6, 7, 8, 9, 10$. Then

This is the last natural number in the domain. It is called the upper limit.

$$\downarrow$$

$$\sum_{k=1}^{10} 2k\} \leftarrow \text{This is the function being evaluated. It is called the general term.}$$

$$\uparrow$$

This is the first natural number in the domain. It is called the lower limit.

Thus, $\displaystyle\sum_{k=1}^{10} 2k = 2(1)+2(2)+2(3)+2(4)+2(5)+2(6)+2(7)+2(8)+2(9)+2(10) = 110$. The words **evaluate** and **expand** are both used to mean write out an expression in summation notation, and then sum the resulting terms, if possible.

EXAMPLE 2   Evaluate an expression with summation notation

a. Evaluate: $\displaystyle\sum_{k=3}^{6}(2k+1)$ **b.** Expand: $\displaystyle\sum_{k=3}^{n}\frac{1}{2^k}$

Solution

a.

$$\sum_{k=3}^{6}(2k+1) = \underbrace{(2\cdot 3+1)}_{\text{Evaluate the expression } 2k+1 \text{ for } k=3} + \overbrace{(2\cdot 4+1)}^{k=4} + \underbrace{(2\cdot 5+1)}_{k=5} + \underbrace{(2\cdot 6+1)}_{k=6}$$

$$= 7 + 9 + 11 + 13$$
$$= 40$$

b. $\displaystyle\sum_{k=3}^{n}\frac{1}{2^k} = \frac{1}{2^3}+\frac{1}{2^4}+\frac{1}{2^5}+\frac{1}{2^6}+\cdots+\frac{1}{2^{n-1}}+\frac{1}{2^n}$ ∎

If the summation notation is given, then we can evaluate or expand to write out the terms as a sum. Summation notation is also important to simplify the way we write expressions. It is quite common to have a sum, such as the one shown in Example 2b, and to rewrite this sum using summation notation to simplify the expression.

EXAMPLE 3   Using summation notation

Write each expression using summation notation.

a. $2 + 4 + 6 + 8 + \cdots + 2n$ **b.** $2 + 4 + 8 + 16 + \cdots + 2^n$

c. $1 + 2 + 4 + 8 + 16 + \cdots + 2^n$ **d.** $8 + 16 + 32 + \cdots + 2^{n-2}$

e. $a_1 + a_2 + a_3 + \cdots + a_n$ **f.** $a_1 + (a_1 + d) + (a_1 + 2d) + \cdots + [a_1 + (n-1)d]$

g. $g_1 + g_2 + g_3 + \cdots + g_n$ **h.** $g_1 + g_1 r + g_1 r^2 + \cdots + g_1 r^{n-1}$

Solution

a. $2 + 4 + 6 + 8 + \cdots + 2n = \displaystyle\sum_{k=1}^{n} 2k$ **b.** $2 + 4 + 8 + 16 + \cdots + 2^n = \displaystyle\sum_{k=1}^{n} 2^k$

c. $1 + 2 + 4 + 8 + 16 + \cdots + 2^n = \displaystyle\sum_{k=0}^{n} 2^k$ **d.** $8 + 16 + 32 + \cdots + 2^{n+2} = \displaystyle\sum_{k=3}^{n+2} 2^k$

Note: Answers for these problems are not unique; for example, we could also write

$$8 + 16 + 32 + \cdots + 2^{n-2} = \sum_{k=5}^{n} 2^{k-2}$$

This opens the door for infinitely many other representations for the given sum using summation notation. For this reason, the variable k is sometimes called a **dummy variable** because it serves as a placeholder only. The test to see whether a summation is correct is to write it out without summation notation; if it generates the correct series, then it is a correct representation. Generally, we write the simplest or most obvious summation representation.

e. $a_1 + a_2 + a_3 + \cdots + a_n = \displaystyle\sum_{k=1}^{n} a_k$

f. Recognize that this is an arithmetic sequence for which we know the formula for the general term.

$$a_1 + (a_1 + d) + (a_1 + 2d) + \cdots + [a_1 + (n-1)d] = \sum_{k=1}^{n} [a_1 + (k-1)d]$$

g. $g_1 + g_2 + g_3 + \cdots + g_n = \displaystyle\sum_{k=1}^{n} g_k$

h. Recognize that this is a geometric sequence for which we know the formula for the general term.

$$g_1 + g_1 r + g_1 r^2 + \cdots + g_1 r^{n-1} = \sum_{k=1}^{n} g_1 r^{k-1}$$

Arithmetic Series

An **arithmetic series** is the sum of the terms of an arithmetic sequence. Let us consider a rather simple-minded example. How many blocks are shown in the stack in Figure 8.11?

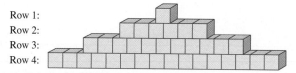

Row 1:
Row 2:
Row 3:
Row 4:

Figure 8.11 How many blocks?

We can answer this question by simply counting the blocks: There are 34 blocks. Somehow it does not seem like this is what we have in mind with this question. Suppose we ask a better question: How many blocks are in a similar building with n rows?

We notice that the number of blocks (counting from the top) in each row forms an arithmetic series:

$$1 + 6 + 11 + 16 + \cdots$$

Look for a pattern:

Denote one row by $A_1 = 1$ block
Two rows: $A_2 = 1 + 6 = 7$
Three rows: $A_3 = 1 + 6 + 11 = 18$
Four rows: $A_4 = 1 + 6 + 11 + 16 = 34$ (shown in Figure 8.11)

What about 10 rows?

$$A_{10} = 1 + 6 + 11 + 16 + 21 + 26 + 31 + 36 + 41 + 46$$

Instead of adding all these numbers directly, let us try an easier way. Write down A_{10} twice, once counting from the top and once counting from the bottom:

$$A_{10} = 1 + 6 + 11 + 16 + 21 + 26 + 31 + 36 + 41 + 46$$
$$\updownarrow \quad \updownarrow \quad \updownarrow \quad \updownarrow \quad \updownarrow \quad \updownarrow \quad \updownarrow \quad \updownarrow \quad \updownarrow \quad \updownarrow$$
$$A_{10} = 46 + 41 + 36 + 31 + 26 + 21 + 16 + 11 + 6 + 1$$

Add these equations:

$$2A_{10} = 47 + 47 + 47 + 47 + 47 + 47 + 47 + 47 + 47 + 47$$

Number of terms

$$2A_{10} = \overbrace{10}\ \underbrace{(47)} = 470$$

Repeated term

$$A_{10} = 10\ \underbrace{\left(\tfrac{47}{2}\right)} = 235 \qquad \text{Divide both sides by 2.}$$

Average of 1st and last terms

This pattern leads us to a formula for n terms. We note that the number of blocks is an arithmetic sequence with $a_1 = 1$, $d = 5$; thus, since $a_n = a_1 + (n-1)d$, we have for the blocks in Figure 8.11 a stack starting with 1 and ending (in the nth row) with

$$a_n = 1 + (n-1)5 = 1 + 5n - 5 = 5n - 4$$

Thus, Number of blocks in n rows   Number of rows

$$\downarrow \qquad\qquad\qquad\qquad \downarrow$$

$$A_n \qquad\qquad\qquad = n\underbrace{\left[\frac{1 + (5n - 4)}{2}\right]}$$

Average of 1st and last terms

$$= n\left[\frac{5n - 3}{2}\right]$$

$$= \frac{1}{2}(5n^2 - 3n)$$

This formula can be used for the number of blocks for any number of rows. Looking back, we see

$$n = 1:\quad A_1 = \frac{1}{2}[5(1)^2 - 3(1)] = 1$$

$$n = 4:\quad A_4 = \frac{1}{2}[5(4)^2 - 3(4)] = 34 \qquad \text{(Figure 8.11)}$$

$$n = 10:\quad A_{10} = \frac{1}{2}[5(10)^2 - 3(10)] = 235$$

If we carry out these same steps for A_n, where $a_n = a_1 + (n-1)d$, we derive the following formula for the sum of the terms of an arithmetic sequence.

ARITHMETIC SERIES FORMULA

The sum of the terms of an arithmetic sequence a_1, a_2, a_3, . . . , a_n with common difference d is

$$A_n = \sum_{k=1}^{n} a_k = n\left(\frac{a_1 + a_n}{2}\right) \quad \text{or} \quad A_n = \frac{n}{2}[2a_1 + (n-1)d]$$

» IN OTHER WORDS The first form tells us that the sum of n terms of an arithmetic sequence is n times the average of the first and last terms.

The last part of the formula for A_n is used when the last term is not explicitly stated or known. To derive this formula, we know $a_n = a_1 + (n-1)d$ so

$$A_n = n\left(\frac{a_1 + a_n}{2}\right)$$

$$= n\left(\frac{a_1 + [a_1 + (n-1)d]}{2}\right)$$

$$= \frac{n}{2}[a_1 + a_1 + (n-1)d]$$

$$= \frac{n}{2}[2a_1 + (n-1)d]$$

EXAMPLE 4 Arithmetic series

In a classroom of 35 students, each student "counts off" by threes (i.e., 3, 6, 9, 12, . . .). What is the sum of the students' numbers?

Solution

We recognize the sequence 3, 6, 9, 12, . . . as an arithmetic sequence, with the first term $a_1 = 3$ and the common difference $d = 3$. The sum of these numbers is denoted by A_{35} since there are 35 students "counting off":

$$A_{35} = \frac{35}{2}\left[2(3) + (35-1)3\right] = 1{,}890$$

Geometric Series

A **geometric series** is the sum of the terms of a geometric sequence. To motivate a formula for a geometric series, we once again consider an example. Suppose Charlie Brown receives a chain letter, and he is to copy this letter and send it to six of his friends.

You may have heard that chain letters "do not work." Why not? Consider the number of people who could become involved with this chain letter if we assume that all recipients carry out his or her task and do not break the chain. The first mailing would consist of six letters. The two mailings involve 42 letters since the second mailing of 36 letters is added to the total: $6 + 36 = 42$.

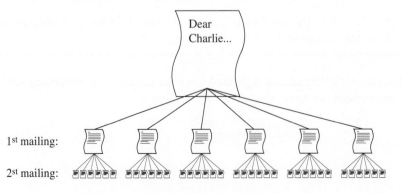

The number of letters in each successive mailing is a number in the geometric sequence

$$6, 36, 216, 1296, \ldots \text{ or } 6, 6^2, 6^3, 6^4, \ldots$$

How many people receive letters with 11 mailings, assuming that no person receives a letter more than once? To answer this question, consider the series associated with a geometric sequence. We begin with a pattern:

Denote one mailing by: $G_1 = 6$
Two mailings: $G_2 = 6 + 6^2$
Three mailings: $G_3 = 6 + 6^2 + 6^3$
$$\vdots$$
Eleven mailings: $G_{11} = 6 + 6^2 + \cdots + 6^{11}$

We could probably use a calculator to find this sum, but we are looking for a formula, so we try something different. In fact, this time we will work out the general formula. Let

$$G_n = g_1 + g_2 + g_3 + \cdots + g_n$$
$$G_n = g_1 + g_1 r + g_1 r^2 + \cdots + g_1 r^{n-1}$$

Multiply both sides of the latter equation by r:

$$r G_n = g_1 r + g_1 r^2 + g_1 r^3 + \cdots + g_1 r^n$$

Notice that, except for the first and last terms, all the terms in the expansions for G_n and $r G_n$ are the same, so that if we subtract one equation from the other, we have

$$G_n - r G_n = g_1 - g_1 r^n$$
$$(1 - r)G_n = g_1(1 - r^n) \qquad \textit{We solve for } G_n.$$
$$G_n = \frac{g_1(1 - r^n)}{1 - r} \quad \text{if } r \neq 1$$

For Charlie Brown's chain letter problem, $g_1 = 6$, $n = 11$, and $r = 6$, so we find

$$G_{11} = \frac{6(1 - 6^{11})}{1 - 6} = \frac{6}{5}(6^{11} - 1) = 435,356,466$$

This is more than the number of people in the United States! The number of letters in only two more mailings would exceed the number of men, women, and children in the whole world.

GEOMETRIC SERIES FORMULA

The sum of the terms of a geometric sequence $g_1, g_2, g_3, \ldots, g_n$ with common ratio r (where $r \neq 1$) is

$$G_n = \frac{g_1(1 - r^n)}{1 - r}$$

EXAMPLE 5   Salary problem

Suppose some eccentric millionaire offered to hire you for a month (say, 31 days) and offered you the following salary choice. She will pay you $500,000 per day or else will pay you $1 for the first day, $2 for the second day, $4 for the third day, and so on for the 31 days. Which salary should you accept?

Solution
If you are paid $500,000 per day, your salary for the 31 days is

$$\$500,000(31) = \$15,500,000$$

Now, if you are paid using the doubling scheme, your salary is (in dollars)

$$1 + 2 + 4 + 8 + \cdots + \overset{\text{31st day}}{\overline{\text{last day}}} \quad \text{or} \quad 2^0 + 2^1 + 2^2 + 2^3 + \cdots + 2^{30}$$

We see that this is the sum of the geometric sequence, where $g_1 = 1$ and $r = 2$. We are looking for G_{31}:

$$G_{31} = \frac{1(1 - 2^{31})}{1 - 2} = -(1 - 2^{31}) = 2^{31} - 1$$

Using a calculator, we find this to be $2,147,483,647. You should certainly accept the doubling scheme.

EXAMPLE 6   Finding sums

Find the indicated sums:

a. $128 - 192 + 288 - \cdots + 1,458$ **b.** $10 + 17 + 24 + \cdots + 136 + 143$
c. The first 50 odd numbers

Solution

a. The terms form a geometric sequence, where $g_1 = 128$, $g_n = 1,458$, and $r = -\frac{192}{128} = -1.5$. We need to know the number of terms, so we use the formula for a geometric sequence:

$$g_n = g_1 r^{n-1} \qquad \text{Geometric sequence formula}$$
$$1,458 = 128(-1.5)^{n-1} \qquad \text{Substitute given values.}$$
$$11.390625 \approx (-1)^{n-1}(1.5)^{n-1} \qquad \text{Divide both sides by 128 and use property of exponents.}$$
$$n - 1 = \log_{1.5} 11.390625 \qquad \text{Disregard the } -1 \text{ factor for the moment, and solve the exponential equation.}$$
$$n - 1 = 6 \qquad \text{Evaluate logarithm.}$$
$$n = 7$$

For $n = 7$, we see that $(-1)^{7-1} = (-1)^6 = 1$. Thus, $n = 7$. We can now compute the sum:

$$G_7 = \frac{g_1(1 - r^n)}{1 - r} \qquad \text{Geometric series formula}$$

$$= \frac{128[1 - (-1.5)^7]}{1 - (-1.5)} \qquad \text{Substitute given values.}$$

$$= 926$$

b. The terms form an arithmetic sequence, where $a_1 = 10$, $a_n = 143$, and $d = 17 - 10 = 7$. We need to find n, so we use the formula for the general term of an arithmetic sequence:

$$a_n = a_1 + (n - 1)d \qquad \text{Arithmetic sequence formula}$$
$$143 = 10 + (n - 1)7 \qquad \text{Substitute given values.}$$
$$143 = 3 + 7n$$
$$140 = 7n$$
$$20 = n$$

We can now compute the sum:

$$A_n = \frac{n}{2}[2a_1 + (n - 1)d] \qquad \text{Arithmetic series formula}$$

$$= \frac{20}{2}[2(10) + (19)(7)] \qquad \text{Substitute given values.}$$

$$= 10(20 + 133)$$
$$= 1,530$$

c. The sum of the first 50 odd numbers is the sum of the arithmetic sequence $1, 3, 5, \ldots$. For this problem, $a_1 = 1$, $d = 2$, and $n = 50$. Then

$$A_n = \frac{n}{2}[2a_1 + (n-1)d]$$
$$= \frac{50}{2}[2(1) + (50-1)2]$$
$$= 2,500$$

■

EXAMPLE 7   March Madness   MODELING APPLICATION

The NCAA men's basketball tournament has 64 teams. How many games are necessary to determine a champion?

Solution

Step 1: *Understand the problem.* Most tournaments are formed by drawing an elimination schedule similar to the one shown in Figure 8.12. This is sometimes called a *two-team elimination tournament.*

Step 2: *Devise a plan.* We could obtain the answer to the question by direct counting, but instead, we will find a general solution, working backward. We know there will be 1 championship game and 2 semifinal games; continuing to work backward, there are 4 quarterfinal games, . . . :

$$1 + 2 + 2^2 + 2^3 + \cdots$$

We recognize this as a geometric series. We note that since there are 64 teams, the first round has $64/2 = 32$ games. Also, we know that $2^5 = 32$, so we must find

$$1 + 2 + 2^2 + 2^3 + 2^4 + 2^5$$

Step 3: *Carry out the plan.* We note that $g_1 = 1$, $r = 2$, and $n = 6$:

$$G_6 = \frac{1(1 - 2^6)}{1 - 2} = 2^6 - 1 = 63$$

Thus, the NCAA playoffs will require 63 games.

Step 4: *Look back.* We see not only that the NCAA tournament has 63 games for a playoff tournament, but, in general, that if there are 2^n teams in a tournament, there will be $2^n - 1$ games. However, do not forget to check by estimation or by using common sense. Each game eliminates one team, and so 63 teams must be eliminated to crown a champion.

■

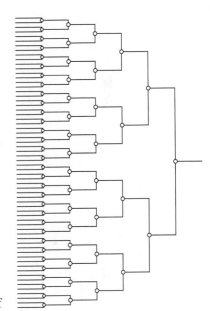

Figure 8.12 NCAA playoffs

Infinite Geometric Series

We have just found a formula for the sum of the first n terms of a geometric sequence. Sometimes it is also possible to find the sum of an entire infinite geometric series. Suppose an infinite geometric series

$$g_1 + g_2 + g_3 + g_4 + \cdots$$

is denoted by G. This is an **infinite series**.

The **partial sums** are defined by

$$G_1 = g_1; \qquad G_2 = g_1 + g_2; \qquad G_3 = g_1 + g_2 + g_3 + \cdots$$

Consider the partial sums for an infinite geometric series, with $g_1 = \frac{1}{2}, r = \frac{1}{2}$. The geometric sequence is $\frac{1}{2}, \frac{1}{4}, \frac{1}{8}, \frac{1}{16}, \ldots$ The first few partial sums can be found as follows:

$$G_1 = \frac{1}{2}; \; G_2 = \frac{1}{2} + \frac{1}{4} = \frac{3}{4}; \; G_3 = \frac{1}{2} + \frac{1}{4} + \frac{1}{8} = \frac{7}{8}$$

Does this series have a sum if you add *all* its terms? It does seem that as you take more terms of the series, the sum is closer and closer to 1. Analytically, we can check the partial sums on a calculator or spreadsheet. Geometrically, you can see (Figure 8.13) that if the terms are laid end-to-end as lengths on a number line, each term is half the remaining distance to 1.

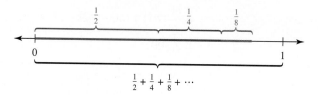

Figure 8.13 Series $\frac{1}{2} + \frac{1}{4} + \frac{1}{8} + \cdots$

It appears that the partial sums are getting closer to 1 as n becomes larger. We *can* find the sum of an infinite geometric sequence. Consider

$$G_n = \frac{g_1(1 - r^n)}{1 - r} = \frac{g_1 - g_1 r^n}{1 - r} = \frac{g_1}{1 - r} - \frac{g_1}{1 - r} r^n$$

Now, g_1, r, and $1 - r$ are fixed numbers. If $|r| < 1$, then r^n approaches 0 as n grows and thus G_n approaches $\frac{g_1}{1 - r}$.

INFINITE GEOMETRIC SERIES FORMULA

If $g_1, g_2, g_3, \ldots, g_n \ldots$ is an infinite geometric sequence with a common ratio r such that $|r| < 1$, then its sum is denoted by G and is found by

$$G = \frac{g_1}{1 - r}$$

If $|r| \geq 1$, the infinite geometric series has no sum.

EXAMPLE 8   Pendulum swing

The path of each swing of a pendulum is 0.85 as long as the path of the previous swing (after the first). If the path of the tip of the first swing is 36 in. long, how far does the tip of the pendulum travel before it eventually comes to rest?

Solution

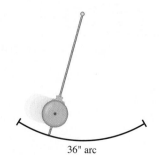

36" arc

$$\text{TOTAL DISTANCE} = 36 + 36(0.85) + 36(0.85)^2 + \cdots \qquad \textit{Infinite geometric series: } g_1 = 36; r = 0.85$$

$$= \frac{36}{1 - 0.85}$$

$$= 240$$

The tip of the pendulum travels 240 in.

EXAMPLE 9   Accumulation of medication in a body

A patient is given an injection of 10 units of a certain drug every 24 hr. The drug is eliminated exponentially so that the fraction that remains in the patient's body after t days is $f(t) = e^{-t/5}$. If the treatment is continued indefinitely, approximately how many units of the drug will eventually be in the patient's body just before an injection?

Solution

Of the original dose of 10 units, only $10e^{-1/5}$ are left in the patient's body after the first day (just before the second injection). That is,

$$S_1 = 10e^{-1/5}$$

The medication in the patient's body after 2 days consists of what remains from the first two doses. Of the original dose, only $10e^{-2/5}$ units are left (because 2 days have elapsed) and of the second dose, $10e^{-1/5}$ units remain:

$$S_2 = 10e^{-1/5} + 10e^{-2/5}$$

Similarly, for n days,

$$S_n = 10e^{-1/5} + 10e^{-2/5} + \cdots + 10e^{-n/5}$$

The amount S of medication in the patient's body in the long run is the limit of S_n as $n \to \infty$. That is,

$$S = \lim_{n \to \infty} S_n$$

$$= \sum_{k=1}^{\infty} 10e^{-k/5} \qquad \text{Geometric series with } a = 10e^{-1/5} \text{ and } r = e^{-1/5}.$$

$$= \frac{10e^{-1/5}}{1 - e^{-1/5}}$$

$$\approx 45.166556$$

We see that about 45 units remain in the patient's body. ∎

Summary of Sequence and Series Formulas

We conclude this section by repeating the important formulas related to sequences and series. Given a sequence of numbers, a *series* arises by considering the *sum* of the terms of the sequence. The formulas for the *general term of a sequence* and for the *sum* (of terms of the sequence—that is, a *series*) are given in Table 8.1.

TABLE 8.1	Distinguishing a Series from a Sequence		
Type	**Definition**	**Notation**	**Formula**
Sequences	A list of numbers having a first term, a second term, ...	s_n	
Arithmetic	A sequence with a common difference, d	a_n	$a_n = a_1 + (n - 1)d$
Geometric	A sequence with a common ratio, r	g_n	$g_n = g_1 r^{n-1}$
Fibonacci	A sequence with first two terms given and subsequent terms the sum of the two previous terms		$s_n = s_{n-1} + s_{n-2}, n \geq 3$
Series	The indicated sum of terms of a sequence	S_n	
Arithmetic	Sum of the terms of an arithmetic sequence: $A_n = \sum_{k=1}^{n} a_k = a_1 + a_2 + a_3 + \cdots + a_n$	A_n	$A_n = n\left(\dfrac{a_1 + a_n}{2}\right)$ or $A_n = \dfrac{n}{2}[2a_1 + (n - 1)d]$
Geometric	Sum of the terms of a geometric sequence: $G_n = \sum_{k=1}^{n} g_k = g_1 + g_2 + g_3 + \cdots + g_n$	G_n	$G_n = \dfrac{g_1(1 - r^n)}{1 - r}, r \neq 1$
	Sum of the terms of an infinite geometric sequence: $G = g_1 + g_2 + g_3 + \cdots$	G	$G = \dfrac{g_1}{1 - r}, |r| < 1$

☠ As a final note, remember that a sequence is a mere succession of terms, while a series is a sum of such terms. Do not confuse the two concepts. For example, a sequence of terms may converge, but the series of the same terms may diverge:

$$\left\{1+\frac{1}{2^n}\right\} \text{ is the } sequence \ \frac{3}{2}, \frac{5}{4}, \frac{9}{8}, \frac{17}{16}, \ldots, \text{ which converges to 1.}$$

$$\sum_{k=1}^{\infty}\left(1+\frac{1}{2^k}\right) \text{ is the } series \ \frac{3}{2}+\frac{5}{4}+\frac{9}{8}+\cdots, \text{ which diverges. ☠}$$

PROBLEM SET 8.3

LEVEL 1

WHAT IS WRONG, *if anything, with each statement in Problems 1–4? Explain your reasoning.*

1. $1 + 2 + 3 + 4 + \cdots$ is an infinite arithmetic sequence.

2. $\sum_{k=1}^{6} 5 = 6 \cdot 5$

3. We use s_n to represent a sequence and S_n to represent a series.

4. We use s_n to represent the general term of any sequence, and we use a_n if we know the sequence is arithmetic.

Find the requested values in Problems 5–10.

5. S_5 when $s_n = 15 - 3n$
6. S_8 when $s_n = 5n$
7. S_4 when $s_n = 5 \cdot 2^n$
8. S_6 when $s_n = (-1)^n$
9. S_7 when $s_n = (-1)^n$
10. S_3 when $s_n = 8 \cdot 5^n$

Evaluate the expressions in Problems 11–16.

11. $\sum_{k=2}^{6} k$
12. $\sum_{k=1}^{4} k^2$
13. $\sum_{k=2}^{5} (10 - 2k)$
14. $\sum_{k=0}^{4} 3(-2)^k$
15. $\sum_{k=0}^{3} 2(3^k)$
16. $\sum_{k=1}^{3} (-1)^k (k^2 + 1)$

Find an expression for the general term of each of the sequences in Problems 17–22, and then write the expression using summation notation.

17. $3, 7, 11, 15, \ldots$
18. $151, 142, 133, \ldots$
19. $3, 6, 12, \ldots$
20. $7, 21, 63, \ldots$
21. $x - 5b, x - 3b, x - b, \ldots$
22. $xyz, xy, \dfrac{xy}{z}, \ldots$

Find the sum of each series in Problems 23–28.

23. $35 + 46 + 57 + \cdots + 123 + 134$
24. $17 + 9 + 1 + \cdots + (-55)$
25. $5 + 15 + 45 + \cdots + 3,645$
26. $-12 - 36 - 108 - \cdots - 2,916$
27. $3,125 + 2,500 + 2,000 + \cdots + 1,024$
28. $-29 - 22 - 15 - \cdots + 20$

If possible, find the sum of the infinite geometric series in Problems 29–34.

29. $1 + \dfrac{1}{2} + \dfrac{1}{4} + \cdots$
30. $1 + \dfrac{1}{3} + \dfrac{1}{9} + \cdots$
31. $1 + \dfrac{3}{4} + \dfrac{9}{16} + \cdots$
32. $1 + \dfrac{3}{2} + \dfrac{9}{4} + \cdots$
33. $1,000 + 500 + 250 + \cdots$
34. $-20 + 10 - 5 + \cdots$

LEVEL 2

The game of pool uses 15 balls numbered from 1 to 15 (see Figure 8.14). In the game of rotation, a player attempts to "sink" a ball in a pocket of the table and receives the number of points on the ball. Answer the questions in Problems 35–38.

Figure 8.14 Pool balls

35. How many points would a player who "runs the table" receive? (To "run the table" means to sink all the balls.)

36. Suppose Missy sinks balls 1 through 8 and Shannon sinks balls 9 through 15. What are their respective scores?

37. Suppose Missy sinks the even-numbered balls and Shannon sinks the odd-numbered balls. What are their respective scores?

38. Suppose we consider a game of "super pool," which has 30 consecutively numbered balls on the table. How many points would a player receive to "run the table"?

39. The *Peanuts* cartoon (p. 522) expresses a common feeling regarding chain letters. Consider the total number of letters sent after a particular mailing:

   1st mailing: 6
   2nd mailing: 6 + 36 = 42
   3rd mailing: 6 + 36 + 216 = 258

   Determine the total number of letters sent in five mailings of the chain letter.

40. How many blocks would be needed to build a stack like the one shown in Figure 8.15 if the bottom row has 28 blocks?

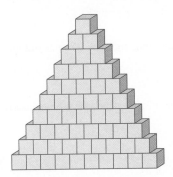

Figure 8.15 **How many blocks?**

41. Repeat Problem 40 if the bottom row has 87 blocks.
42. Repeat Problem 40 if the bottom row has 100 blocks.
43. A pendulum is swung 20 cm and is allowed to swing freely until it eventually comes to rest. Each subsequent swing of the bob of the pendulum is 90% as far as the preceding swing. How far will the bob travel before coming to rest?
44. The initial swing of the tip of a pendulum is 25 cm. If each swing of the tip is 75% of the preceding swing, how far does the tip travel before eventually coming to rest?
45. A flywheel is brought to a speed of 375 revolutions per minute (rpm) and allowed to slow and eventually come to rest. If, in slowing, it rotates three-fourths as fast each subsequent minute, how many revolutions will the wheel make before returning to rest?
46. A rotating flywheel is allowed to slow to a stop from a speed of 500 rpm. While slowing, each minute it rotates two-thirds as many times as in the preceding minute. How many revolutions will the wheel make before coming to rest?
47. Advertisements say that a new type of superball will rebound to 9/10 of its original height. If it is dropped from a height of 10 ft, how far, based on the advertisements, will the ball travel before coming to rest?
48. A tennis ball is dropped from a height of 10 ft. If the ball rebounds 2/3 of its height on each bounce, how far will the ball travel before coming to rest?
49. Rework Example 5, but assume that the starting salary is 1¢ instead of $1. That is, you are paid 1¢ for the first day, 2¢ for the second day, 4¢ for the third day, and so on.
50. A culture of bacteria increases by 100% every 24 hours. If the original culture contains 1 million bacteria ($a_0 = 1$ million), find the number of bacteria present after 10 days.
51. Use Problem 50 to find a formula for the number of bacteria present after d days.
52. How many games are necessary for a two-team elimination tournament with 32 teams?
53. Games like *Wheel of Fortune* and *Jeopardy* have one winner and two losers. A three-team game tournament is illustrated by Figure 8.16. If *Jeopardy* has a "Tournament of Champions" consisting of 27 players, what is the necessary number of games?

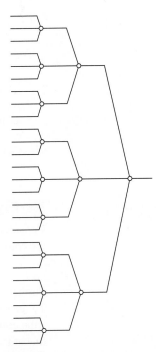

Figure 8.16 **A three-team tournament**

54. How many games are necessary for a three-team elimination tournament with 729 teams?
55. A patient is given an injection of 20 units of a certain drug every 24 hours. The drug is eliminated exponentially, so the fraction that remains in the patient's body after t days is $f(t) = e^{-t/2}$. If the treatment is continued indefinitely, approximately how many units of the drug will eventually be in the patient's body?
56. Each January 1, the administration of a certain private college adds six new members to its board of trustees. If the fraction of trustees who remain active for at least t years is $f(t) = e^{-0.2t}$, approximately how many active trustees will the college have on December 31, in the long run? Assume that there were six members on the board on January 1 of the present year.

LEVEL 3

57. Suppose you were hired for a job paying $21,000 per year and were given the following options:
OPTION A: Annual salary increase of $1,440
OPTION B: Semiannual salary increase of $360
OPTION C: Quarterly salary increase of $90
OPTION D: Monthly salary increase of $10
Write the arithmetic series for the total amount of money earned in 10 years under each of these options. Which is the best option?
58. Find three distinct numbers whose sum is 9 so that these numbers form an arithmetic sequence and their squares form a geometric sequence. Can you find a second possibility?
59. The sum of an infinite geometric series is 30. Each term is exactly 4 times the sum of the remaining terms. Determine the series with these properties.

60. The infinite series

$$\lim_{n \to \infty} \sum_{k=1}^{n} \frac{1}{k^p} = \frac{1}{1^p} + \frac{1}{2^p} + \frac{1}{3^p} + \cdots$$

is called a *p*-**series**. Look at the sequence of partial sums and state whether you think the given infinite *p*-series has a limit.

a. $p = 1$ **b.** $p = 2$
c. $p = 3$ **d.** $p = 0.5$

Can you make a conjecture (guess) about the values of p for which the infinite *p*-series has a limit?

8.4 Systems of Equations

Consider the following puzzle problem dealing with sequences: The nonzero numbers a, b, and c are consecutive terms of an arithmetic sequence so that

$$a + b + c = 60$$

Furthermore, $4a$, $8b$, and $25c$ are consecutive terms of a geometric sequence. Find both sequences. To answer this question, we let d denote the common difference for the arithmetic sequence and r the common ratio for the geometric series. We are led to several equations, all of which must be true at the same time, or *simultaneously*.

$$a + b + c = 60$$
$$a = b - d$$
$$c = b + d$$
$$\frac{8b}{r} = 4a \text{ or } 8b = (4a)r$$
$$25c = (8b)r$$

We see that we have listed five equations, and there are five unknowns. We call these five equations a **system of equations**, and a **solution** is a set of values for the variables $(a,\ b,\ c,\ d,$ and $r)$ so that all equations are true *at the same time*—that is, simultaneously. The **simultaneous solution** of a system of equations is the intersection of the solution sets of the individual equations. We use a brace to show that we are looking for a simultaneous solution. If all the equations in a system are linear, it is called a **linear system**. It is not necessary for the number of equations and the number of variables to be the same. We will solve this equation in Example 7 of this section.

We begin by reviewing those methods of solving linear systems that you encountered in your previous algebra courses; then we generalize first to *nonlinear systems* and, later in this chapter, to more complicated linear systems.

Graphing Method

Since the graph of each equation in a system of linear equations in two variables is a line, the solution set is the intersection of two lines. In two dimensions, two lines must be related to each other in one of three possible ways:

1. They intersect at a single point.

2. The graphs are parallel lines. In this case, the solution set is empty, and the system is called *inconsistent*. In general, any system that has an empty solution set is referred to as an **inconsistent system**.

3. The graphs are the same line. In this case, infinitely many points are in the solution set, and any solution of one equation is also a solution of the other. Such a system is called a **dependent system**.

» IN OTHER WORDS Graph the equations and check for the intersection point. If the graphs do not intersect, then the equations represent an inconsistent system. If they coincide, then they represent a dependent system. When we solve a system by graphing the equations and then looking for points of intersection, we call it the **graphing method**.

EXAMPLE 1   Solving systems of equations by graphing

Solve the given systems by graphing:

a. $\begin{cases} 2x - 3y = -8 \\ x + y = 6 \end{cases}$ b. $\begin{cases} 2x - 3y = -8 \\ 4x - 6y = 0 \end{cases}$ c. $\begin{cases} 2x - 3y = -8 \\ y = \dfrac{2}{3}x + \dfrac{8}{3} \end{cases}$

Solution

a. Graph the given lines.

$$2x - 3y = -8$$

$$3y = 2x + 8$$

$$y = \frac{2}{3}x + \frac{8}{3}$$

Graph is shown in blue.

$$x + y = 6$$

$$y = -x + 6$$

Graph is shown in red.

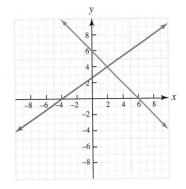

Look at the point(s) of intersection. The solution appears to be $(2, 4)$, which can be verified by direct substitution into both the given equations.

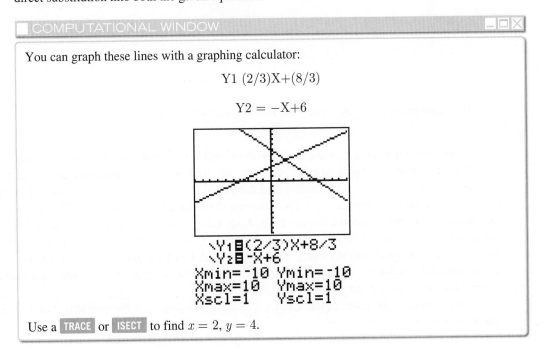

COMPUTATIONAL WINDOW

You can graph these lines with a graphing calculator:

$$Y1\ (2/3)X + (8/3)$$

$$Y2 = -X + 6$$

Use a TRACE or ISECT to find $x = 2$, $y = 4$.

b. The graph of $\begin{cases} 2x - 3y = -8 \\ 4x - 6y = 0 \end{cases}$ is shown in Figure 8.17.

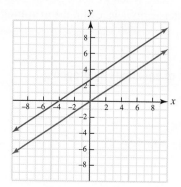

Figure 8.17 Inconsistent system

Notice that these lines are parallel; you can show this analytically by noting that the slopes of the lines are the same. Since they are distinct parallel lines, there is not a point of intersection. This is an *inconsistent system*.

c. The graph of $\begin{cases} 2x - 3y = -8 \\ y = \dfrac{2}{3}x + \dfrac{8}{3} \end{cases}$ is shown in Figure 8.18.

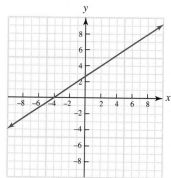

Figure 8.18 Dependent system

The equations represent the same line. This is a *dependent system*.

In your previous algebra courses, it was probably sufficient simply to state that the system was dependent. However, to simply say that there are infinitely many solutions does not adequately describe the situation. Look at Example 1 and the graph of the system in Figure 8.18. A solution is any point on the line—that is, any point satisfying the equation $2x - 3y = -8$. This means we can choose *any* value for y and calculate the corresponding value for x. The value chosen for y is a *parameter*, and we usually represent this parameter by t. Thus, if $y = t$, then $2x - 3t = -8$ so that $x = \frac{3}{2}t - 4$. We now can say that the solution for part **c** of Example 1 is $\left(\frac{3}{2}t - 4, t\right)$. We will discuss parametric solutions to systems of equations later in this chapter. In this course, if a system is dependent, we will give the solutions in parametric form.

If one or more of the given equations of a system is not linear, then the system is called a **nonlinear system**.

EXAMPLE 2   Solving a nonlinear system by graphing

Solve by graphing: $\begin{cases} x - y = 3 \\ 6x - y = x^2 + 7 \end{cases}$

Solution

This is a nonlinear system, but the solution is, nevertheless, the intersection point(s) of the graphs. The graph of $x - y = 3$ is found by writing $y = x - 3$ and graphing a y-intercept of -3 and a slope of 1. For $6x - y = x^2 + 7$, we solve for y and complete the square:

$$6x - y = x^2 + 7$$
$$y + 7 = -x^2 + 6x$$
$$y + 7 - 9 = -(x^2 - 6x + 3^2) \qquad \text{Complete the square.}$$
$$y - 2 = -(x - 3)^2$$

This is a parabola that opens downward with vertex at $(3, 2)$. The graph is shown in Figure 8.19.

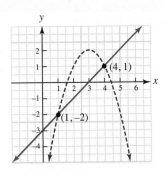

Figure 8.19 Graph of nonlinear system

By inspection, the solution appears to be the points $(4, 1)$ and $(1, -2)$. To check an answer, be sure that every member of the solution satisfies all the equations of the system. ∎

EXAMPLE 3   Salary problem revisited

In the previous section, we solved the following salary problem: Suppose some eccentric millionaire offered to hire you for a month (say, 31 days) and offered you the following salary choice. She will pay you $0.5 million per day or else will pay you $1 for the first day, $2 for the second day, $4 for the third day, and so on for the 31 days. The answer showed that, without a doubt, you should accept the doubling scheme. We now ask a new question: For how many days would the salary offers be the same?

Solution

If we work x days, then solve the equation

$$500{,}000x = 2^x - 1$$

We solve this graphically as a system of equations:

$$\begin{cases} y = 500{,}000x \\ y = 2^x - 1 \end{cases}$$

The graph is shown in Figure 8.20.

Figure 8.20 Notice that straight salary is a straight line.

The point of intersection looks like (to the nearest day) $x = 24$. Let's compare salaries:

x	Straight Salary	Geometric Salary
23	$500,000(23) = 11,500,000$	$G_{23} = 2^{23} - 1 = 8,388,607$
24	$500,000(24) = 12,000,000$	$G_{24} = 2^{24} - 1 = 16,777,215$

That is, if you work 23 or fewer days, take the straight salary; if you work 24 or more days, take the geometric salary.

Substitution Method

The graphing method can give solutions as accurate as the graphs you can draw, and consequently, it is adequate for many applications. However, more exact methods are often needed.

In general, given a system, the procedure for solving is to write a simpler equivalent system. Two systems are said to be **equivalent** if they have the same solution set. In this book, we limit ourselves to finding only real roots. There are several ways to go about writing equivalent systems. The first nongraphical method we consider comes from the substitution property of real numbers and leads to a **substitution method** for solving systems.

SUBSTITUTION METHOD

To solve a system of equations with two equations and two unknowns by substitution:

Step 1 *Solve* one of the equations for one of the variables.

Step 2 *Substitute* the expression that you obtain into the other equation.

Step 3 *Solve* the resulting equation.

Step 4 *Substitute* that solution into either of the original equations to find the value of the other variable.

Step 5 *State* the solution.

EXAMPLE 4   Solving a linear system using substitution

Solve: $\begin{cases} 2p + 3q = 5 \\ q = -2p + 7 \end{cases}$

Solution
Since $q = -2p + 7$, substitute $-2p + 7$ for q in the other equation:

$$2p + 3q = 5$$
$$2p + 3(-2p + 7) = 5$$
$$2p - 6p + 21 = 5$$
$$-4p = -16$$
$$p = 4$$

Now, substitute 4 for p in either of the given equations:

$$q = -2p + 7$$
$$= -2(4) + 7$$
$$= -1$$

The solution is $(p, q) = (4, -1)$. If the variables are not x and y, then you must also show the variables along with the ordered pair. This establishes which variable is associated with which number—namely, $p = 4$ and $q = -1$.

■ COMPUTATIONAL WINDOW ▭☐✕

If you are graphing with variables other than x and y, you will need to reassign them. For this example, we let $x = p$ and $y = q$:

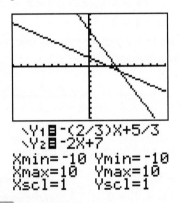

```
\Y₁☐-(2/3)X+5/3
\Y₂☐-2X+7
Xmin=-10  Ymin=-10
Xmax=10   Ymax=10
Xscl=1    Yscl=1
```

Many calculators have an [ISECT] function to find: x = 4   y = -1

EXAMPLE 5   Solving a nonlinear system using substitution

Solve: $\begin{cases} x - y + 7 = 0 \\ y = x^2 + 4x + 3 \end{cases}$

Solution
Solve one of the equations for one of the variables. The second equation is solved for y, so substitute $x^2 + 4x + 3$ for y in the first equation:

$$x - y + 7 = 0$$
$$x - (x^2 + 4x + 3) + 7 = 0$$
$$-x^2 - 3x + 4 = 0$$
$$x^2 + 3x - 4 = 0$$
$$(x + 4)(x - 1) = 0$$
$$x = -4, 1$$

If $x = -4$, then we use the second equation to find: $y = (-4)^2 + 4(-4) + 3 = 3$; if $x = 1$, then $y = (1)^2 + 4(1) + 3 = 8$.

The solution is $(-4, 3)$ and $(1, 8)$. If you have a graphing calculator, you can check this solution by graphing.

■

EXAMPLE 6   Solving a system with two nonlinear equations

Solve: $\begin{cases} y = x^2 - x - 1 \\ 4x = 7 + 2y - y^2 \end{cases}$

Solution
Substitute $x^2 - x - 1$ for y in the second equation:

$$4x = 7 + 2y - y^2$$
$$4x = 7 + 2(x^2 - x - 1) - (x^2 - x - 1)^2$$
$$4x = 7 + 2x^2 - 2x - 2 - x^4 + 2x^3 + x^2 - 2x - 1$$
$$4x = 4 + 3x^2 - 4x - x^4 + 2x^3$$
$$x^4 - 2x^3 - 3x^2 + 8x - 4 = 0$$

To solve this equation, you can use technology or you can use synthetic division with the rational root theorem. The possible rational roots are ± 1, ± 2, ± 4.

	1	−2	−3	8	−4
1	1	−1	−4	4	0
1	1	0	−4	0	

We continue with the factorization of the fourth-degree equation:

$$x^4 - 2x^3 - 3x^2 + 8x - 4 = 0 \qquad \text{Given equation}$$
$$(x - 1)(x^3 - x^2 - 4x + 4) = 0 \qquad \text{From first line of synthetic division}$$
$$(x - 1)(x - 1)(x^2 - 4) = 0 \qquad \text{From second line of synthetic division}$$
$$(x - 1)^2(x - 2)(x + 2) = 0 \qquad \text{Factor difference of squares.}$$
$$x = 1,\, 2,\, -2$$

The roots of this fourth-degree equation are 1 (multiplicity 2), 2, −2. We now use the first equation to find the corresponding y-values.

$$y = x^2 - x - 1$$

If $x = 1$:	$y = (1)^2 - 1 - 1 = -1$
If $x = 2$:	$y = (2)^2 - 2 - 1 = 1$
If $x = -2$:	$y = (-2)^2 - (-2) - 1 = 5$

The solution is $(1, -1)$, $(2, 1)$, $(-2, 5)$.

■

EXAMPLE 7   Puzzle sequence problem with five variables

The nonzero numbers, a, b, and c are consecutive terms of an arithmetic sequence so that

$$a + b + c = 60$$

Furthermore, $4a$, $8b$, and $25c$ are consecutive terms of a geometric sequence. Find both sequences.

Solution

Let d denote the common difference for the arithmetic sequence, and let r denote the common ratio for the geometric series. We are led to the following system of equations:

$$\begin{cases} a + b + c = 60 \\ a = b - d \\ c = b + d \\ 8b = (4a)r \\ 25c = (8b)r \end{cases}$$

Substitute the second and third equations into the first equation to obtain

$$a + b + c = 60$$
$$\uparrow \qquad \uparrow$$
$$b - d \qquad b + d$$
$$(b - d) + b + (b + d) = 60$$
$$3b = 60$$
$$b = 20$$

Next, substitute $b = 20$ into the fourth equation to obtain

$$8b = (4a)r$$
$$8(20) = 4ar$$
$$a = \frac{40}{r}$$

Also, substitute $b = 20$ into the fifth equation to obtain

$$25c = (8b)r$$
$$25c = 8(20)r$$
$$c = \frac{32r}{5}$$

We now return to the first equation with these newfound values:

$$20$$
$$\downarrow$$
$$a + b + c = 60 \qquad \text{First equation}$$
$$\uparrow \qquad \uparrow$$
$$\frac{40}{r} \qquad \frac{32r}{5}$$

$$\frac{40}{r} + 20 + \frac{32r}{5} = 60 \qquad \text{Substitute values.}$$
$$200 + 100r + 32r^2 = 300r \qquad \text{Multiply both sides by } 5r.$$
$$32r^2 - 200r + 200 = 0 \qquad \text{Solve quadratic equation.}$$
$$8(r - 5)(4r - 5) = 0 \qquad \text{Factor.}$$
$$r = 5, \frac{5}{4} \qquad \text{Zero factor theorem}$$

We see if $r = 5$, $a = 8$, and $c = 32$; and if $r = \frac{5}{4}$, $a = 31$, and $c = 8$; so in either case we find the arithmetic sequence to be 8, 20, 32, . . . or 32, 20, 8, . . . and the geometric sequence to be 32, 160, 800, . . . or 128, 160, 200, ... ∎

Linear Combination (Addition) Method

A third method for solving systems is called the **linear combination method** (or, as it is often called, the **addition method**). It involves substitution and the idea that if equal quantities are added to equal quantities, the resulting equation is equivalent to the original system. In general, addition will not simplify the system unless the numerical coefficients of one or more terms are opposites. However, you can often force them to be opposites by multiplying one or both of the given equations by nonzero constants.

LINEAR COMBINATIONS

To solve a system of equations by linear combinations:

Step 1   *Multiply* one or both of the equations by a constant or constants so that the coefficients of one of the variables become opposites.

Step 2   *Add* corresponding members of the equations to obtain a new equation in a single variable.

Step 3   *Solve* the derived equation for that variable.

Step4   *Substitute* the value of the found variable into either of the original equations, and solve for the second variable.

Step 5   *State* the solution.

»   IN OTHER WORDS Multiply one or both of the equations by a constant or constants so that the coefficients of one of the variables are opposites, and then add the equations to eliminate the variable.

EXAMPLE 8   Solving systems using linear combinations

Solve the given systems by the linear combination method.

a. $\begin{cases} 3x + 5y = -2 \\ 2x + 3y = 0 \end{cases}$ **b.** $\begin{cases} y^2 - 5xy = 3 \\ 5xy + 4 = y^2 \end{cases}$

Solution

a. Multiply **both sides** of the first equation by 2 and **both sides** of the second equation by -3. This procedure, denoted as shown below, forces the coefficients of x to be opposites:

$$\begin{array}{r} 2 \\ -3 \end{array} \begin{cases} 3x + 5y = -2 \\ 2x + 3y = 0 \end{cases}$$

This means you should add the equations.

$$\begin{array}{c} \downarrow \\ + \end{array} \begin{cases} 6x + 10y = -4 \\ -6x - 9y = 0 \end{cases}$$

$$y = -4$$

Mentally, add the equations.

If $y = -4$, then $2x + 3y = 0$ means $2x + 3(-4) = 0$, which implies that $x = 6$. The solution is $(6, -4)$.

b. Arrange the equations so similar terms are aligned.

$$\begin{cases} y^2 - 5xy = 3 \\ y^2 - 5xy = 4 \end{cases}$$

Multiply both sides of the first equation by -1, and then add the equations:

$$+ \begin{cases} -y^2 + 5xy = -3 \\ \underline{y^2 - 5xy = 4} \\ \quad\quad 0 = 1 \end{cases}$$

This equation is false, so we see this is an inconsistent system. ∎

PROBLEM SET 8.4

LEVEL 1

Solve the systems in Problems 1–8 by graphing.

1. $\begin{cases} x - y = 2 \\ 2x + 3y = 9 \end{cases}$
2. $\begin{cases} 3x - 4y = 16 \\ -x + 2y = -6 \end{cases}$

3. $\begin{cases} y = 3x + 1 \\ x - 2y = 8 \end{cases}$
4. $\begin{cases} 2x - 3y = 12 \\ -4x + 6y = 18 \end{cases}$

5. $\begin{cases} y = \frac{3}{5}x + 2 \\ 3x - 5y = -15 \end{cases}$
6. $\begin{cases} y = \frac{2}{3}x - 7 \\ 2x + 3y = 3 \end{cases}$

7. $\begin{cases} y = x^2 - 6x + 11 \\ x + y = -7 \end{cases}$
8. $\begin{cases} y = x^2 - 4x \\ y = x^2 - 4x + 8 \end{cases}$

Solve the systems in Problems 9–16 by the substitution method.

9. $\begin{cases} y = 3 - 2x \\ 3x + 2y = -17 \end{cases}$
10. $\begin{cases} 5x - 2y = -19 \\ x = 3y + 4 \end{cases}$

11. $\begin{cases} 4y + 5x = 2 \\ y = \frac{5}{4}x + 2 \end{cases}$
12. $\begin{cases} \frac{x}{3} - y = 7 \\ x + \frac{y}{2} = 7 \end{cases}$

13. $\begin{cases} 3t_1 + 5t_2 = 1{,}541 \\ t_2 = 2t_1 + 160 \end{cases}$
14. $\begin{cases} x = -7y - 3 \\ 2x + 5y = 3 \end{cases}$

15. $\begin{cases} \alpha = 3\beta - 4 \\ 5\alpha - 4\beta = -9 \end{cases}$
16. $\begin{cases} \gamma + 3\delta = 0 \\ \gamma = 5\delta + 16 \end{cases}$

Solve the systems in Problems 17–24 using linear combinations (that is, by the addition method).

17. $\begin{cases} x + y = 16 \\ x - y = 10 \end{cases}$
18. $\begin{cases} x + y = 560 \\ x - y = 490 \end{cases}$

19. $\begin{cases} 6r - 4s = 10 \\ 2s = 3r - 5 \end{cases}$
20. $\begin{cases} 3u + 2v = 5 \\ 4v = 10 - 6u \end{cases}$

21. $\begin{cases} 3a_1 + 4a_2 = -9 \\ 5a_1 + 7a_2 = -14 \end{cases}$
22. $\begin{cases} 5s_1 + 2s_2 = 23 \\ 2s_1 + 7s_2 = 34 \end{cases}$

23. $\begin{cases} 2\alpha + 5\beta = 7 \\ 2\alpha + 6\beta = 14 \end{cases}$
24. $\begin{cases} 5\gamma + 4\delta = 5 \\ 15\gamma - 2\delta = 8 \end{cases}$

LEVEL 2

Solve the systems in Problems 25–36 for all real solutions, using any suitable method.

25. $\begin{cases} x + y = 7 \\ x - y = -1 \end{cases}$
26. $\begin{cases} x - y = 8 \\ x + y = 2 \end{cases}$

27. $\begin{cases} x = \frac{3}{4}y - 2 \\ 3y - 4x = 5 \end{cases}$
28. $\begin{cases} 100x - y = 0 \\ 50x + y = 300 \end{cases}$

29. $\begin{cases} 12x - 5y = -39 \\ y = 2x + 9 \end{cases}$
30. $\begin{cases} y = 2x - 1 \\ y = -3x - 9 \end{cases}$

31. $\begin{cases} y = \frac{2}{3}x - 5 \\ y = -\frac{4}{3}x + 7 \end{cases}$
32. $\begin{cases} x + y = a \\ x - y = b \end{cases}$

33. $\begin{cases} x - y = c \\ x - 2y = d \end{cases}$
34. $\begin{cases} x + y = 2a \\ x - y = 2b \end{cases}$

35. $\begin{cases} 3x^2 + 4y^2 = 12 \\ x^2 + y^2 = -8 \end{cases}$
36. $\begin{cases} 3x + 4y^2 = 19 \\ x^2 + y^2 = 5 \end{cases}$

37. Suppose some eccentric millionaire offered to hire you for a month (say, 31 days) and offered you the following salary choice. She will pay you $500 per day or else will pay you 1¢ for the first day, 2¢ for the second day, 4¢ for the third day, and so on. Obviously, if you worked only one day you would take the $500 salary, and if you worked 31 days you would take the doubling salary. For which day of the month does the choice change?

38. Find two numbers whose sum is 9 and whose product is 18.

39. Find two numbers whose difference is 4 and whose product is -3.

40. Find the lengths of the legs of a right triangle whose area is 60 ft^2 and whose hypotenuse is 17 ft.

41. Find the length and width of a rectangle whose area is 60 ft^2 with a diagonal of length 13 ft.

In Section 6.4, we stated some trigonometric identities whose proofs were delayed because their proofs require systems of equations. Problems 42–45 now request those proofs.

42. Use the system
$$\begin{cases} \cos(x + y) = \cos x \cos y - \sin x \sin y \\ \cos(x - y) = \cos x \cos y + \sin x \sin y \end{cases}$$
to prove the identity
$$2\cos x \cos y = \cos(x + y) + \cos(x + y)$$

43. Use the system
$$\begin{cases} \cos(x + y) = \cos x \cos y - \sin x \sin y \\ \cos(x - y) = \cos x \cos y + \sin x \sin y \end{cases}$$
to prove the identity
$$2\sin x \sin y = \cos(x - y) - \cos(x + y)$$

44. Use the system
$$\begin{cases} \sin(x + y) = \sin x \cos y + \cos x \sin y \\ \sin(x - y) = \sin x \cos y - \cos x \sin y \end{cases}$$
to prove the identity
$$2\sin x \cos y = \sin(x + y) + \sin(x - y)$$

45. Use the system
$$\begin{cases} \cos(x+y) = \cos x \cos y - \sin x \sin y \\ \cos(x-y) = \cos x \cos y + \sin x \sin y \end{cases}$$
to prove the identity
$$2\cos x \sin y = \sin(x+y) - \sin(x-y)$$

LEVEL 3

PROBLEM FROM CALCULUS *In calculus, a method for optimizing a function is called the method of Lagrange multipliers, which requires solving a system of equations. Solve each system in Problems 46–49 for variables x, y, and z. These systems of equations were found in* Calculus *by Bradley/Smith (Englewood Cliffs: NJ, Prentice Hall, 1995), pp. 806–811.*

46. $\begin{cases} -2x = \lambda \\ -2y = \lambda \\ x+y = 1 \end{cases}$

47. $\begin{cases} -4x = \lambda \\ -2y = \lambda \\ -2z = \lambda \\ x+y+z = 1 \end{cases}$

48. $\begin{cases} 20x^{-3/5}y^{3/5} = \lambda \\ 30x^{2/5}y^{-2/5} = \lambda \\ x+y-150 = 0 \end{cases}$

49. $\begin{cases} 2x = \lambda + 2x\mu \\ 2y = 2\lambda + 2y\mu \\ 2z = \lambda - \mu \\ x+2y+z = 10 \\ z = x^2 + y^2 \end{cases}$

Solve Problems 50 and 51 for solutions between 0 and 6.

50. $\begin{cases} \sin x + \cos y = 1 \\ \sin x - \cos y = 0 \end{cases}$

51. $\begin{cases} \cos x - \sin y = 1 \\ \cos x + \sin y = 0 \end{cases}$

Solve Problems 52 and 53 for exact solutions.

52. $\begin{cases} 2^x + 2^y = 12 \\ 4^x + 4^y = 80 \end{cases}$

53. $\log x + \log y = 71$
$2\log x - \log y = 2$

Find the solution to Problems 54 and 55 with answers correct to the nearest tenth.

54. $\begin{cases} y^2 - 5xy = 1 \\ y^2 - 4xy = 2y \end{cases}$

55. $\begin{cases} \dfrac{x^2}{64} + \dfrac{y^2}{16} = 1 \\ \dfrac{x^2}{9} - \dfrac{y^2}{4} = 1 \end{cases}$

56. If the parabola $(x-h)^2 = y - k$ passes through the points $(-2, 6)$ and $(-4, 2)$, find (h, k).

57. If the parabola $-3(x-h)^2 = y - k$ passes through the points $(2, 5)$ and $(-1, -4)$, find (h, k).

58. **Historical Quest** The following problem was written by Leonhard Euler: "Two persons owe conjointly 29 pistoles*; they both have money, but neither of them enough to enable him, singly, to discharge this common debt." The first debtor says therefore to the second, "If you give me $\frac{2}{3}$ of your money, I can immediately pay the debt." The second answers that he also could discharge the debt if the other would give him $\frac{3}{4}$ of his money. Determine how many pistoles each had.

59. **Historical Quest** The Babylonians knew how to solve systems of equations as early as 1600 B.C. One tablet, called the Yale Tablet, shows a system equivalent to
$$\begin{cases} xy = 600 \\ (x+y)^2 - 150(x-y) = 100 \end{cases}$$
Find a positive solution for this system correct to the nearest tenth.

60. **Historical Quest** The Louvre Tablet from the Babylonian civilization is dated about 1500 B.C. It shows a system equivalent to
$$\begin{cases} xy = 1 \\ x+y = a \end{cases}$$
Solve this system for x and y in terms of a.

*A pistole is a unit of money.

8.5 Matrix Solution of a System of Equations

One of the most common types of problems to which we can apply mathematics in a variety of different disciplines is to the solution of systems of equations. In fact, common real-world problems require the simultaneous solution of systems involving 3, 4, 5, or even 20 or 100 unknowns. The methods of the previous section will not suffice, and, in practice, techniques that will allow computer or calculator help in solving systems are common. In this section, we introduce a way of solving large systems of equations in a general way so that we can handle the solution of a system of m equations with n unknowns.

Definition of a Matrix

Consider a system with unknowns x_1 and x_2. We use **subscripts** 1 and 2 to denote the unknowns, instead of using variables x and y, because we want to be able to handle n unknowns, which we can easily denote as $x_1, x_2, x_3, \ldots, x_n$; if we continued by using $x, y, z, \ldots$ for systems in general, we would soon run out of letters. Here is the way we will write a general system of two equations with two unknowns.

$$\begin{cases} a_{11}x_1 + a_{12}x_2 = b_1 \\ a_{21}x_1 + a_{22}x_2 = b_2 \end{cases}$$

The coefficients of the unknowns use **double subscripts** to denote their position in the system; a_{11} is used to denote the numerical coefficient of the first variable in the first row; a_{12} denotes the numerical coefficient of the second variable in the first row; and so on. The constants are denoted by b_1 and b_2.

We now separate the parts of this system of equations into *rectangular arrays* of numbers. An **array** of numbers is called a **matrix**. A matrix is denoted by enclosing the array in large brackets.

Let [A] be the matrix (array) of coefficients: $\begin{bmatrix} a_{11} & a_{12} \\ a_{21} & a_{22} \end{bmatrix}$

Let [X] be the matrix of unknowns: $\begin{bmatrix} x_1 \\ x_2 \end{bmatrix}$

Let [B] be the matrix of constants: $\begin{bmatrix} b_1 \\ b_2 \end{bmatrix}$

We will write the system of equations as a *matrix equation*, $[A][X] = [B]$, but before we do this, we will do some preliminary work with matrices.

Matrices are classified by the number of (horizontal) **rows** and (vertical) **columns**. The numbers of rows and columns of a matrix need not be the same; but if they are, the matrix is called a **square matrix**.

The **order** or **dimension** of a matrix is given by an expression $m \times n$ (pronounced "m by n"), where m is the number of rows and n is the number of columns. For example, the matrix [A] shown above is of order 2×2, and matrices (plural for matrix) [X] and [B] have order 2×1.

Matrix Form of a System of Equations

We write a system of equations in the form of an **augmented matrix**. The matrix refers to the matrix of coefficients, and we *augment* (add to, or affix) this matrix by writing the constant terms at the right of the matrix (separated by a vertical line):

$$\begin{cases} a_{11}x_1 + a_{12}x_2 = b_1 \\ a_{21}x_1 + a_{22}x_2 = b_2 \end{cases} \quad \text{in matrix form is} \quad \begin{bmatrix} a_{11} & a_{12} & | & b_1 \\ a_{21} & a_{22} & | & b_2 \end{bmatrix}$$

EXAMPLE 1   **Writing systems in augmented form**

Write the augmented matrix, and then give the order of the resulting system.

a. $\begin{cases} 2x + y = 3 \\ 3x - y = 2 \\ 4x + 3y = 7 \end{cases}$ **b.** $\begin{cases} 5x - 3y + z = -3 \\ 2x + 5z = 14 \end{cases}$ **c.** $\begin{cases} x_1 - 3x_3 + x_5 = -3 \\ x_2 + x_4 = -1 \\ x_3 + x_5 = 7 \\ x_1 + x_2 - x_3 + 4x_4 = -8 \\ x_1 + x_2 + x_3 + x_4 + x_5 = 8 \end{cases}$

Solution

Note that some coefficients are negative and some are zero.

a. $\begin{bmatrix} 2 & 1 & | & 3 \\ 3 & -1 & | & 2 \\ 4 & 3 & | & 7 \end{bmatrix}$; Order 3×3 **b.** $\begin{bmatrix} 5 & -3 & 1 & | & -3 \\ 2 & 0 & 5 & | & 14 \end{bmatrix}$; Order 2×4

c. $\begin{bmatrix} 1 & 0 & -3 & 0 & 1 & | & -3 \\ 0 & 1 & 0 & 1 & 0 & | & -1 \\ 0 & 0 & 1 & 0 & 1 & | & 7 \\ 1 & 1 & -1 & 4 & 0 & | & -8 \\ 1 & 1 & 1 & 1 & 1 & | & 8 \end{bmatrix}$; Order 5×6

EXAMPLE 2   Writing systems in algebraic form

Write a system of equations (use x_1, x_2, x_3, . . .) that has the given augmented matrix.

a. $\begin{bmatrix} 2 & 1 & -1 & -3 \\ 3 & -2 & 1 & 9 \\ 1 & -4 & 3 & 17 \end{bmatrix}$ b. $\begin{bmatrix} 1 & 0 & 0 & 3 \\ 0 & 1 & 0 & -2 \\ 0 & 0 & 1 & 21 \end{bmatrix}$ c. $\begin{bmatrix} 1 & 0 & 0 & 5 \\ 0 & 1 & 0 & 12 \\ 0 & 0 & 1 & -3 \\ 0 & 0 & 0 & 4 \end{bmatrix}$ d. $\begin{bmatrix} 1 & 0 & 0 & -7 \\ 0 & 1 & 0 & 3 \\ 0 & 0 & 1 & -1 \\ 0 & 0 & 0 & 0 \end{bmatrix}$

Solution

a. $\begin{cases} 2x_1 + x_2 - x_3 = -3 \\ 3x_1 - 2x_2 + x_3 = 9 \\ x_1 - 4x_2 + 3x_3 = 17 \end{cases}$ b. $\begin{cases} x_1 = 3 \\ x_2 = -2 \\ x_3 = 21 \end{cases}$ c. $\begin{cases} x_1 = 5 \\ x_2 = 12 \\ x_3 = -3 \\ 0 = 4 \end{cases}$ d. $\begin{cases} x_1 = -7 \\ x_2 = 3 \\ x_3 = -1 \\ 0 = 0 \end{cases}$

■

The goal of this section is to solve a system of m equations with n unknowns. We have already looked at systems of two equations with two unknowns. In previous courses, you may have solved three equations with three unknowns. Now, however, we want to be able to solve problems with two equations and five unknowns, or three equations and two unknowns, or systems with any number of linear equations and unknowns. The procedure for this section—*Gauss–Jordan elimination*—is a general method for solving all these types of systems. We write the system in augmented matrix form (as in Example 1) and then carry out a process that transforms the matrix until the solution is obvious. Look back at Example 2—the solution to part **b** is obvious. Part **c** shows $0 = 4$ in the last equation, so this system has no solution (0 cannot equal 4), and part **d** shows $0 = 0$ (which is true for all replacements of the variable), which means that the solution is found by looking at the other equations (namely, $x_1 = -7$, $x_2 = 3$, and $x_3 = -1$). The terms with nonzero coefficients in these examples (that is, parts **b**, **c**, and **d**) are arranged on a diagonal, and such a system is said to be in **diagonal form**.

Elementary Row Operations and Pivoting

What process will allow us to transform a matrix into diagonal form? We begin with some steps called **elementary row operations**. Elementary row operations change the *form* of a matrix, but the new form represents an equivalent system. Matrices that represent equivalent systems are called **equivalent matrices**; we now introduce the elementary row operations, which allow us to write equivalent matrices. Let us work with a system consisting of three equations and three unknowns (any size will work the same way).

System Format	Matrix Format	Calculator Format
$\begin{cases} 2x - 2y + 4z = 14 \\ x - y - 2z = -9 \\ 3x + 2y + z = 16 \end{cases}$	$\begin{bmatrix} 2 & -2 & 4 & 14 \\ 1 & -1 & -2 & -9 \\ 3 & 2 & 1 & 16 \end{bmatrix}$	$[2\ -2\ \ 4\ \ 14]$ $[1\ -1\ -2\ -9]$ $[3\ \ 2\ \ 1\ \ 16]$

For the discussion, we call this matrix A, and denote it by $[A]$.

◼ COMPUTATIONAL WINDOW   ▭◻✕

Check your owner's manual for the matrix operations. Most will have a ▮MATRIX▮ key and then you will need to name the matrix. We assume your calculator will allow for at least three matrices (which we will call [A], [B], and [C]). After naming the matrix, you will need to input the order (most will handle up to order 6×6). After inputting a matrix, it is a good idea to recall it to make sure it is input correctly.

Elementary Row Operation 1: RowSwap

Interchanging two equations is equivalent to interchanging two rows in the matrix format, and certainly, if we do this, the solution to the system will be the same:

System Format	Matrix Format	Calculator Format
$\begin{cases} x - y - 2z = -9 \\ 2x - 2y + 4z = 14 \\ 3x + 2y + z = 16 \end{cases}$	$\begin{bmatrix} 1 & -1 & -2 & -9 \\ 2 & -2 & 4 & 14 \\ 3 & 2 & 1 & 16 \end{bmatrix}$	$\begin{bmatrix} 1 & -1 & -2 & -9 \end{bmatrix}$ $\begin{bmatrix} 2 & -2 & 4 & 14 \end{bmatrix}$ $\begin{bmatrix} 3 & 2 & 1 & 16 \end{bmatrix}$

In this example, we interchanged the first and the second rows of the matrix. If we denote the original matrix as $[A]$, then we indicate the operation of interchanging the first and second rows of matrix A by RowSwap($[A], 1, 2$).

Elementary Row Operation 2: Row+

Because adding the entries of one equation to the corresponding entries (similar terms) of another equation will not change the solution to a system of equations, the second elementary row operation is called row addition. (This is the step called linear combinations in Section 8.1.) In terms of matrices, we see that this operation corresponds to adding one row to another:

System Format	Matrix Format	Calculator Format
$\begin{cases} 2x - 2y + 4z = 14 \\ x - y - 2z = -9 \\ 3x + 2y + z = 16 \end{cases}$	$\begin{bmatrix} 2 & -2 & 4 & 14 \\ 1 & -1 & -2 & -9 \\ 3 & 2 & 1 & 16 \end{bmatrix}$	$\begin{bmatrix} 2 & -2 & 4 & 14 \end{bmatrix}$ $\begin{bmatrix} 1 & -1 & -2 & -9 \end{bmatrix}$ $\begin{bmatrix} 3 & 2 & 1 & 16 \end{bmatrix}$

Add row 1 to row 3:

System Format	Matrix Format	Calculator Format
$\begin{cases} 2x - 2y + 4z = 14 \\ x - y - 2z = -9 \\ 5x + 5z = 30 \end{cases}$	$\begin{bmatrix} 2 & -2 & 4 & 14 \\ 1 & -1 & -2 & -9 \\ 5 & 0 & 5 & 30 \end{bmatrix}$	$\begin{bmatrix} 2 & -2 & 4 & 14 \end{bmatrix}$ $\begin{bmatrix} 1 & -1 & -2 & -9 \end{bmatrix}$ $\begin{bmatrix} 5 & 0 & 5 & 30 \end{bmatrix}$

Notice that only row 3 changes; we call the row being added to (that is, the row that is being changed) the **target row**. We indicate this operation by Row+($[A]1, 3$).

$\uparrow$
target row

Elementary Row Operation 3: *Row

Multiplying or dividing both sides of an equation by any nonzero number does not change the simultaneous solution, so, in matrix format, the solution will not be changed if any row is multiplied or divided by a nonzero constant. In this context, we call the constant a **scalar**. For example, we can multiply both sides of the first equation of the original system by $\frac{1}{2}$. (Note that dividing both sides of an equation by 2 can be considered as multiplying both sides by $\frac{1}{2}$.)

System Format	Matrix Format	Calculator Format
$\begin{cases} 2x - 2y + 4z = 14 \\ x - y - 2z = -9 \\ 3x + 2y + z = 16 \end{cases}$	$\begin{bmatrix} 2 & -2 & 4 & 14 \\ 1 & -1 & -2 & -9 \\ 3 & 2 & 1 & 16 \end{bmatrix}$	$\begin{bmatrix} 2 & -2 & 4 & 14 \end{bmatrix}$ $\begin{bmatrix} 1 & -1 & -2 & -9 \end{bmatrix}$ $\begin{bmatrix} 3 & 2 & 1 & 16 \end{bmatrix}$

Multiply the first row by $\frac{1}{2}$ to obtain

System format	Matrix format	Calculator format
$\begin{cases} x - y + 2z = 7 \\ x - y - 2z = -9 \\ 3x + 2y + z = 16 \end{cases}$	$\begin{bmatrix} 1 & -1 & 2 & 7 \\ 1 & -1 & -2 & -9 \\ 3 & 2 & 1 & 16 \end{bmatrix}$	$\begin{array}{l}[1\ -1\ \ 2\ \ 7] \\ [1\ -1\ -2\ -9] \\ [3\ \ \ 2\ \ 1\ 16]\end{array}$

We indicate this operation by *Row$(\frac{1}{2}, [A], 1)$.

$$\underset{\text{scalar}}{\uparrow} \quad \underset{\text{target row}}{\uparrow}$$

Elementary Row Operation 4: *Row+

When solving systems, more often than not, we need to multiply both sides of an equation by a scalar before adding to make the coefficients opposites. Elementary row operation 4 combines row operations 2 and 3 so that this can be accomplished in one step. Let us return to the original system:

System Format	Matrix Format	Calculator Format
$\begin{cases} 2x - 2y + 4z = 14 \\ x - y - 2z = -9 \\ 3x + 2y + z = 16 \end{cases}$	$\begin{bmatrix} 2 & -2 & 4 & 14 \\ 1 & -1 & -2 & -9 \\ 3 & 2 & 1 & 16 \end{bmatrix}$	$\begin{array}{l}[2\ -2\ \ 4\ 14] \\ [1\ -1\ -2\ -9] \\ [3\ \ \ 2\ \ 1\ 16]\end{array}$

We can change this system by multiplying the second equation by -2 and adding the result to the first equation. In matrix terminology, we say that we multiply the second row by -2 and add it to the first row. Denote this by *Row+$(-2, [A], 2, 1)$.

$$\underset{\text{scalar}}{\uparrow} \quad \underset{\text{target row}}{\uparrow}$$

System Format	Matrix Format	Calculator Format
$\begin{cases} 8z = 32 \\ x - y - 2z = -9 \\ 3x + 2y + z = 16 \end{cases}$	$\begin{bmatrix} 0 & 0 & 8 & 32 \\ 1 & -1 & -2 & -9 \\ 3 & 2 & 1 & 16 \end{bmatrix}$	$\begin{array}{l}[0\ \ 0\ \ 8\ 32] \\ [1\ -1\ -2\ -9] \\ [3\ \ 2\ \ 1\ 16]\end{array}$

Once again, multiply the second row, this time by -3, and add it to the third row. Wait! Why -3? Where did that come from? The idea is the same one we used in the linear combination method—we use a number that will give a zero coefficient to the x in the third equation.

System Format	Matrix Format	Calculator Format
$\begin{cases} 8z = 32 \\ x - y - 2z = -9 \\ 5y + 7z = 43 \end{cases}$	$\begin{bmatrix} 0 & 0 & 8 & 32 \\ 1 & -1 & -2 & -9 \\ 0 & 5 & 7 & 43 \end{bmatrix}$	$\begin{array}{l}[0\ \ 0\ \ 8\ \ 32] \\ [1\ -1\ -2\ -9] \\ [0\ \ 5\ \ 7\ \ 43]\end{array}$

Note that the multiplied row is not changed; instead, the changed row is the one to which the multiplied row is added. We call the original row the **pivot row**. Note also that we did not work with the original matrix $[A]$ but rather with the previous answer, so we indicate this by

$$\overset{\text{privot row}}{\underset{\underset{\text{scalar}}{\uparrow} \quad \underset{\text{target row}}{\uparrow}}{\text{*Row+}(-2, [\text{Ans}], 2, 1)}}$$

There you have it! You can carry out these four elementary operations until you have a system for which the solution is obvious, as illustrated by the following example.

EXAMPLE 3 Using matrix methods for solving a system

Solve using system, matrix, and calculator format: $\begin{cases} 2x - 5y = 5 \\ x - 2y = 1 \end{cases}$

Solution

System Format	Matrix Format	Calculator Format	Operation Performed
$\begin{cases} 2x - 5y = 5 \\ x - 2y = 1 \end{cases}$	$[A] = \begin{bmatrix} 2 & -5 & 5 \\ 1 & -2 & 1 \end{bmatrix}$	$[A] = \begin{bmatrix} 2 & -5 & 5 \\ 1 & -2 & 1 \end{bmatrix}$	Given system.

Swap the rows to obtain a 1 in the first row first column of the matrix format:

System Format	Matrix Format	Calculator Format	Operation Performed
$\begin{cases} x - 2y = 1 \\ 2x - 5y = 5 \end{cases}$	$\begin{bmatrix} 1 & -2 & 1 \\ 2 & -5 & 5 \end{bmatrix}$	$\begin{bmatrix} 1 & -2 & 1 \\ 2 & -5 & 5 \end{bmatrix}$	RowSwap($[A], 1, 2$) *This matrix is called [Ans] because it is the result of the previous matrix operation.*

Obtain a zero below the 1 in the first column:

System Format	Matrix Format	Calculator Format	Operation Performed
$\begin{cases} x - 2y = 1 \\ -y = 3 \end{cases}$	$\begin{bmatrix} 1 & -2 & 1 \\ 0 & -1 & 3 \end{bmatrix}$	$\begin{bmatrix} 1 & -2 & 1 \\ 0 & -1 & 3 \end{bmatrix}$	*Row+($-2, [Ans], 1, 2$) *This matrix is now referred to as [Ans].*

Multiply the last row by -1:

System Format	Matrix Format	Calculator Format	Operation Performed
$\begin{cases} x - 2y = 1 \\ y = -3 \end{cases}$	$\begin{bmatrix} 1 & -2 & 1 \\ 0 & 1 & -3 \end{bmatrix}$	$\begin{bmatrix} 1 & -2 & 1 \\ 0 & 1 & -3 \end{bmatrix}$	*Row($-1, [Ans], 2$)

Finally, multiply the last row by 2 and add it to the first row:

System Format	Matrix Format	Calculator Format	Operation Performed
$\begin{cases} x = -5 \\ y = -3 \end{cases}$	$\begin{bmatrix} 1 & 0 & -5 \\ 0 & 1 & -3 \end{bmatrix}$	$\begin{bmatrix} 1 & 0 & -5 \\ 0 & 1 & -3 \end{bmatrix}$	*Row+($[Ans], 2, 1$)

The solution $(-5, -3)$ is now obvious. ∎

As you study Example 3, first look at how the elementary row operations led to a system equivalent to the first—but one for which the solution is obvious. Next, try to decide *why* a particular row operation was chosen when it was. Many students quickly learn the elementary row operations, but then use a series of (almost random) steps until the obvious solution results. This often works but is not very efficient. The steps chosen in Example 3 illustrate an efficient method of using the elementary row operations to determine a system whose solution is obvious. Let us restate the elementary row operations as part of the following definition.

There are four *elementary row operations* for producing equivalent matrices:

1. **RowSwap** Interchange any two rows.
2. **Row+** Row addition—add a row to any other row.
3. ***Row** Scalar multiplication—multiply (or divide) all the elements of a row by the same nonzero real number.
4. ***Row+** Multiply all the entries of a row (*pivot row*) by a nonzero real number and add each resulting product to the corresponding entry of another specified row (*target row*).

 This operation changes only the target row.

These elementary row operations are used together in a process called **pivoting**.

PIVOTING

The process known as **pivoting** means to carry out the following two steps.

Step 1 Divide all entries in the row in which the pivot appears (called the **pivoting row**) by the nonzero pivot element so that the pivot entry becomes a 1. This uses elementary row operation 3, *Row.

Step 2 Obtain zeros above and below the pivot element by using elementary row operation 4, *Row+.

Gauss–Jordan Elimination

You are now ready to see the method worked out by Gauss and Jordan; it is known as **Gauss–Jordan elimination**. It efficiently uses the elementary row operations to diagonalize the matrix. That is, the first pivot is the first entry in the first row, first column; the second is the entry in the second row, second column; and so on until the solution is obvious. A **pivot** element is an element that is used to eliminate elements above and below it in a given column by using elementary row operations.

GAUSS–JORDAN ELIMINATION

Step 1 Select as the first pivot the element in the first row, first column, and pivot.

Step 2 The next pivot is the element in the second row, second column; pivot.

Step 3 Repeat the process until you arrive at the last row or until the pivot element is a zero. If it is a zero and you can interchange that row with a row below it, so that the pivot element is no longer a zero, do so and continue. If it is zero and you cannot interchange rows so that it is not a zero, continue with the next row. The final matrix is called the **row-reduced form**.

EXAMPLE 4 Gauss–Jordan elimination

Solve: $\begin{cases} x + 2y - z = 0 \\ 2x + 3y - 2z = 3 \\ -x - 4y + 3z = -2 \end{cases}$

Karl Friedrich Gauss (1777–1855)

Historical Note

Karl Gauss, described in a previous historical note as one of history's greatest mathematicians, was credited with the process for solving equations described in this section. This method was often referred to as *Gaussian elimination*. Recently, however, it became known that another mathematician, Camille Jordan, who is known for his work in algebra and a branch of mathematics called Galois theory, also developed, independently from Gauss, the solution of systems of equations outlined in this section. For this reason, we call the procedure the Gauss–Jordan elimination. It has been reported that when Gauss was 3 years old, he corrected an error in his father's payroll calculations. By the time Gauss was 21, he had contributed more to mathematics than most mathematicians do in a lifetime.

Solution

We solve this system by choosing the steps according to the Gauss–Jordan method.

$$[A] = \begin{bmatrix} 1 & 2 & -1 & | & 0 \\ 2 & 3 & -2 & | & 3 \\ -1 & -4 & 3 & | & -2 \end{bmatrix}$$ This matrix represents the given system.

First pivot is Row 1, Column 1.

$$\rightarrow \begin{bmatrix} 1 & 2 & -1 & | & 0 \\ 0 & -1 & 0 & | & 3 \\ 0 & -2 & 2 & | & -2 \end{bmatrix}$$ *Row+(-2, [A], 1, 2)

*Row+(1, [Ans], 1, 3)

$$\rightarrow \begin{bmatrix} 1 & 2 & -1 & | & 0 \\ 0 & 1 & 0 & | & -3 \\ 0 & -2 & 2 & | & -2 \end{bmatrix}$$ *Row(-1, [Ans], 2)

Second pivot is Row 2, Column 2.

$$\rightarrow \begin{bmatrix} 1 & 0 & -1 & | & 6 \\ 0 & 1 & 0 & | & -3 \\ 0 & 0 & 2 & | & -8 \end{bmatrix}$$ *Row+(-2, [Ans], 2, 1)

*Row+(2, [Ans], 2, 3)

$$\rightarrow \begin{bmatrix} 1 & 0 & -1 & | & 6 \\ 0 & 1 & 0 & | & -3 \\ 0 & 0 & 1 & | & -4 \end{bmatrix}$$ *Row(0.5, [Ans], 3)

Third pivot is Row 3, Column 3.

$$\rightarrow \begin{bmatrix} 1 & 0 & 0 & | & 2 \\ 0 & 1 & 0 & | & -3 \\ 0 & 0 & 1 & | & -4 \end{bmatrix}$$ *Row+(1, [Ans], 3, 1)

The solution $(2, -3, -4)$ is found by inspection since this matrix represents the system

$$\begin{cases} 1x + 0y + 0z = 2 \\ 0x + 1y + 0z = -3 \\ 0x + 0y + 1z = -4 \end{cases}$$

To check this answer, substitute the x-, y-, and z-values into the original equations:

$$\begin{cases} 2 + 2(-3) - (-4) = 0 \quad \checkmark \\ 2(2) + 3(-3) - 2(-4) = 3 \quad \checkmark \\ -(2) - 4(-3) + 3(-4) = -2 \quad \checkmark \end{cases}$$

■

X Y Z

WHO NEEDS EM

$$\begin{bmatrix} 1 & 2 & -1 & : & 0 \\ 2 & 3 & -2 & : & 3 \\ -1 & -4 & 3 & : & -2 \end{bmatrix}$$

β

Courtesy of Patrick J. Boyle

EXAMPLE 5   **Parametric solution for a system of equations**

Solve: $\begin{cases} 3x - 2y + 4z = 8 \\ x - y - 2z = 5 \\ 4x - 3y + 2z = 13 \end{cases}$

Solution

$$\text{Let } [A] = \begin{bmatrix} 3 & -2 & 4 & | & 8 \\ 1 & -1 & -2 & | & 5 \\ 4 & -3 & 2 & | & 13 \end{bmatrix} \quad \text{This matrix represents the given system.}$$

$$\rightarrow \begin{bmatrix} 1 & -1 & -2 & | & 5 \\ 3 & -2 & 4 & | & 8 \\ 4 & -3 & 2 & | & 13 \end{bmatrix} \quad \text{RowSwap}([A], 1, 2)$$

First pivot is Row 1, Column 1.

$$\rightarrow \begin{bmatrix} 1 & -1 & -2 & | & 5 \\ 0 & 1 & 10 & | & -7 \\ 0 & 1 & 10 & | & -7 \end{bmatrix} \quad \begin{matrix} \text{*Row+}(-3, [\text{Ans}], 1, 2) \\ \text{*Row+}(-4, [\text{Ans}], 1, 3) \end{matrix}$$

Second pivot is Row 2, Column 2.

$$\rightarrow \begin{bmatrix} 1 & 0 & 8 & | & -2 \\ 0 & 1 & 10 & | & -7 \\ 0 & 0 & 0 & | & 0 \end{bmatrix} \quad \begin{matrix} \text{*Row+}(1, [\text{Ans}], 2, 1) \\ \\ \text{*Row+}(-1, [\text{Ans}], 2, 3) \end{matrix}$$

This is equivalent to the system $\begin{cases} 1x + 0y + 8z = -2 \\ 0x + 1y + 10z = -7 \\ 0x + 0y + 0z = 0 \end{cases}$

Because the third equation $(0 = 0)$ is true for all values of x, y, and z, we focus on the first and second equations; there are two equations with three unknowns. This is a *dependent system.* In your previous courses, you probably finished by saying the system is dependent. In this course, as well as in more advanced mathematics, we will find the set of all values that makes the system true. By looking at the solved system, we see that z appears in the first two equations:

$$\begin{cases} x + 8z = -2 \\ y + 10z = -7 \end{cases} \quad \text{or} \quad \begin{cases} x = -2 - 8z \\ y = -7 - 10z \end{cases}$$

We can pick *any* value, say, $z = 2$; then

$$x = -2 - 8z = -2 - 8(2) = -18 \qquad y = -7 - 10z = -7 - 10(2) = -27$$

Thus, a solution is $(-18, -27, 2)$. But why did we pick $z = 2$? What if we picked $z = 0$? We would find another solution, namely,

$$x = -2 - 8(0) = -2 \qquad y = -7 - 10(0) = -7$$

for a solution $(-2, -7, 0)$. But we do not want only two solutions, we want *all* solutions, so we use the concept of a *parameter,* which is an arbitrary constant used to distinguish various specific cases. For this example, let $z = t$ (the variable t is the parameter) to find

$$x = -2 - 8t$$
$$y = -7 - 10t$$

This gives a complete **parametric solution** $(-2 - 8t, -7 - 10t, t)$. ∎

The real beauty of Gauss–Jordan elimination is that it works with all sizes of linear systems. Consider the following example consisting of five equations and five unknowns.

EXAMPLE 6   **Gauss–Jordan elimination for five equations with five unknowns**

Solve: $\begin{cases} x_1 - 3x_3 + x_5 = -3 \\ x_2 + x_4 = -1 \\ x_3 + x_5 = 7 \\ x_1 + x_2 - x_3 + 4x_4 = -8 \\ x_1 + x_2 + x_3 + x_4 + x_5 = 8 \end{cases}$

Solution

$$[A] = \left[\begin{array}{ccccc|c} 1 & 0 & -3 & 0 & 1 & -3 \\ 0 & 1 & 0 & 1 & 0 & -1 \\ 0 & 0 & 1 & 0 & 1 & 7 \\ 1 & 1 & -1 & 4 & 0 & -8 \\ 1 & 1 & 1 & 1 & 1 & 8 \end{array}\right]$$

This matrix represents the given system.

First pivot is Row 1, Column 1.

$$\rightarrow \left[\begin{array}{ccccc|c} 1 & 0 & -3 & 0 & 1 & -3 \\ 0 & 1 & 0 & 1 & 0 & -1 \\ 0 & 0 & 1 & 0 & 1 & 7 \\ 0 & 1 & 2 & 4 & -1 & -5 \\ 0 & 1 & 4 & 1 & 0 & 11 \end{array}\right]$$

*Row+$(-1, [A], 1, 4)$

*Row+$(-1, [Ans], 1, 5)$

Second pivot is Row 2, Column 2.

$$\rightarrow \left[\begin{array}{ccccc|c} 1 & 0 & -3 & 0 & 1 & -3 \\ 0 & 1 & 0 & 1 & 0 & -1 \\ 0 & 0 & 1 & 0 & 1 & 7 \\ 0 & 0 & 2 & 3 & -1 & -4 \\ 0 & 0 & 4 & 0 & 0 & 12 \end{array}\right]$$

*Row+$(-1, [Ans], 2, 4)$

*Row+$(-1, [Ans], 2, 5)$

Third pivot is Row 3, Column 3.

$$\rightarrow \left[\begin{array}{ccccc|c} 1 & 0 & 0 & 0 & 4 & 18 \\ 0 & 1 & 0 & 1 & 0 & -1 \\ 0 & 0 & 1 & 0 & 1 & 7 \\ 0 & 0 & 0 & 3 & -3 & -18 \\ 0 & 0 & 0 & 0 & -4 & -16 \end{array}\right]$$

*Row+$(3, [Ans], 3, 1)$

*Row+$(-2, [Ans], 3, 4)$

*Row+$(-4, [Ans], 3, 5)$

Fourth pivot is Row 4, Column 4.

$$\rightarrow \left[\begin{array}{ccccc|c} 1 & 0 & 0 & 0 & 4 & 18 \\ 0 & 1 & 0 & 1 & 0 & -1 \\ 0 & 0 & 1 & 0 & 1 & 7 \\ 0 & 0 & 0 & 1 & -1 & -6 \\ 0 & 0 & 0 & 0 & -4 & -16 \end{array}\right]$$

*Row$\left(\dfrac{1}{3}, [Ans], 4\right)$

$$\rightarrow \left[\begin{array}{ccccc|c} 1 & 0 & 0 & 0 & 4 & 18 \\ 0 & 1 & 0 & 0 & 1 & 5 \\ 0 & 0 & 1 & 0 & 1 & 7 \\ 0 & 0 & 0 & 1 & -1 & -6 \\ 0 & 0 & 0 & 0 & -4 & -16 \end{array}\right]$$

*Row+$(-1, [Ans], 4, 2)$

Fifth pivot is Row 5, Column 5

$$\rightarrow \begin{bmatrix} 1 & 0 & 0 & 0 & 4 & | & 18 \\ 0 & 1 & 0 & 0 & 1 & | & 5 \\ 0 & 0 & 1 & 0 & 1 & | & 7 \\ 0 & 0 & 0 & 1 & -1 & | & -6 \\ 0 & 0 & 0 & 0 & 1 & | & 4 \end{bmatrix}$$

$*\text{Row}+\left(-\dfrac{1}{4},[\text{Ans}],4,2\right)$

$$\rightarrow \begin{bmatrix} 1 & 0 & 0 & 0 & 0 & | & 2 \\ 0 & 1 & 0 & 0 & 0 & | & 1 \\ 0 & 0 & 1 & 0 & 0 & | & 3 \\ 0 & 0 & 0 & 1 & 0 & | & -2 \\ 0 & 0 & 0 & 0 & 1 & | & 4 \end{bmatrix}$$

$*\text{Row}+(-4,[\text{Ans}],5,1)$
$*\text{Row}+(-1,[\text{Ans}],5,2)$
$*\text{Row}+(-1,[\text{Ans}],5,3)$
$*\text{Row}+(1,[\text{Ans}],5,4)$

The solution is $(x_1,\ x_2,\ x_3,\ x_4,\ x_5) = (2,\ 1,\ 3,\ -2,\ 4)$. ∎

It should not take more than the preceding two examples to convince you of two facts: (1) the Gauss–Jordan procedure is powerful, and (2) it is computationally difficult, even when using a calculator. For real-life examples, you will need to rely on computer methods. If the number of equations is the same as the number of unknowns, and if you have a calculator, the matrix inverse method introduced in the next section will help with the computational tedium of Gauss–Jordan. However, we will need to rely on Gauss–Jordan elimination when the number of equations is not the same as the number of unknowns. We will illustrate with several examples.

Systems with More Equations than Unknowns

Much of your previous work with systems was centered on solving systems in which the number of unknowns and number of equations were equal. In the real world, however, most of the time the number of unknowns will not equal the number of equations. We begin by considering an example with three equations and two unknowns.

EXAMPLE 7   Three equations and two unknowns

Solve $\begin{cases} 2x + y = 3 \\ 3x - y = 2 \\ 4x + 3y = 7 \end{cases}$

Solution

Let $[A] = \begin{bmatrix} 2 & 1 & | & 3 \\ 3 & -1 & | & 2 \\ 4 & 3 & | & 7 \end{bmatrix}$.

The first Gauss–Jordan step is to multiply the first row by $\frac{1}{2}$ to obtain a 1 in row 1, column 1. However, if we assume that we do not have a calculator, we might wish to minimize the amount of arithmetic dealing with fractions. We can simplify our work if we perform one or two valid elementary row operations *before* we begin Gauss–Jordan. This step will give us a 1 in the first column, which we can then use as a pivot.

$$\begin{bmatrix} 2 & 1 & | & 3 \\ 3 & -1 & | & 2 \\ 4 & 3 & | & 7 \end{bmatrix} \xrightarrow[*\text{Row}+(-1,[A],1,2)]{} \begin{bmatrix} 2 & 1 & | & 3 \\ 1 & -2 & | & -1 \\ 4 & 3 & | & 7 \end{bmatrix} \xrightarrow[\text{RowSwap}([\text{Ans}],1,2)]{} \begin{bmatrix} 1 & -2 & | & -1 \\ 2 & 1 & | & 3 \\ 4 & 3 & | & 7 \end{bmatrix}$$

We now begin the Gauss–Jordan process.

$$\begin{bmatrix} 1 & -2 & -1 \\ 2 & 1 & 3 \\ 4 & 3 & 7 \end{bmatrix} \rightarrow \begin{bmatrix} 1 & -2 & -1 \\ 0 & 5 & 5 \\ 0 & 11 & 11 \end{bmatrix}$$

First pivot is Row 1, Column 1.

*Row+$(-2,[\text{Ans}],1,2)$
*Row+$(-4,[\text{Ans}],1,3)$

Second pivot is Row 2, Column 2.

$$\rightarrow \begin{bmatrix} 1 & -2 & -1 \\ 0 & 1 & 1 \\ 0 & 11 & 11 \end{bmatrix} \rightarrow \begin{bmatrix} 1 & 0 & 1 \\ 0 & 1 & 1 \\ 0 & 0 & 0 \end{bmatrix}$$

*Row$\left(\frac{1}{5},[\text{Ans}],2\right)$

*Row+$(2,[\text{Ans}],2,1)$

*Row+$(-11,[\text{Ans}],2,3]$

The final matrix is equivalent to the system $\begin{cases} x = 1 \\ y = 1 \\ 0 = 0 \end{cases}$. The solution is $(1,\ 1)$.

When there are more equations than unknowns, the row-reduced form will yield some open equations in the system followed by some true or false equations. There will be a solution when the open equations are followed by true equations ($0 = 0$ is true). However, if the open equations are followed by at least one false equation (such as $1 = 0$), then there will be no solution for the system; that is, it is an inconsistent system.

EXAMPLE 8   More equations than unknowns; an inconsistent system

Solve: $\begin{cases} 2x + y = 3 \\ 3x - y = 2 \\ x - 2y = 4 \end{cases}$

Solution

This looks very much like the previous example. Let $[A]$ be the initial matrix:

$$[A] = \begin{bmatrix} 2 & 1 & 3 \\ 3 & -1 & 2 \\ 1 & -2 & 4 \end{bmatrix} \xrightarrow[\text{RowSwap}([A],1,3)]{} \begin{bmatrix} 1 & -2 & 4 \\ 3 & -1 & 2 \\ 2 & 1 & 3 \end{bmatrix} \begin{array}{c} \text{*Row+}(-3,[\text{Ans}],1,2) \\ \rightarrow \\ \text{*Row+}(-2,[\text{Ans}]1,3) \end{array} \begin{bmatrix} 1 & -2 & 4 \\ 0 & 5 & -10 \\ 0 & 5 & -5 \end{bmatrix}$$

$$\xrightarrow[\text{*Row}\left(\frac{1}{5},[\text{Ans}],2\right)]{} \begin{bmatrix} 1 & -2 & 4 \\ 0 & 1 & -2 \\ 0 & 5 & -5 \end{bmatrix} \begin{array}{c} \text{*Row+}(2,[\text{Ans}],2,1) \\ \rightarrow \\ \text{*Row+}(-5,[\text{Ans}],2,3) \end{array} \begin{bmatrix} 1 & 0 & 0 \\ 0 & 1 & -2 \\ 0 & 0 & 5 \end{bmatrix}$$

The final matrix is equivalent to the system $\begin{cases} x = 0 \\ y = -2 \\ 0 = 5 \end{cases}$. Since $0 \neq 5$ (that is, $0 = 5$ is false), the system is inconsistent (it has no solution).

Systems with More Unknowns than Equations

You may need a parameter when there are more unknowns than equations. For example, the system

$$\begin{cases} 5x - 3y + 12z = 13 \\ 2x + 5z = 4 \end{cases}$$

has more unknowns than equations so it is a *dependent system*. We can use the Gauss–Jordan method to find a parametric solution to this system.

EXAMPLE 9   Two equations with three unknowns

Solve $\begin{cases} 5x - 3y + 12z = 13 \\ 2x + 5z = 4 \end{cases}$

Solution

Let $[A] = \begin{bmatrix} 5 & -3 & 12 & | & 13 \\ 2 & 0 & 5 & | & 4 \end{bmatrix}$. Before we begin the Gauss–Jordan procedure, we obtain a 1 in the first column by multiplying the second row by -2 and adding it to the first row.

$$\begin{bmatrix} 5 & -3 & 12 & | & 13 \\ 2 & 0 & 5 & | & 4 \end{bmatrix} \xrightarrow{\text{*Row+}(-2,[A],2,1)} \begin{bmatrix} 1 & -3 & 2 & | & 5 \\ 2 & 0 & 5 & | & 4 \end{bmatrix} \xrightarrow{\text{*Row+}(-2,[\text{Ans}],1,2)} \begin{bmatrix} 1 & -3 & 2 & | & 5 \\ 0 & 6 & 1 & | & -6 \end{bmatrix}$$

$$\xrightarrow{\text{*Row}\left(\frac{1}{6},[\text{Ans}],2\right)} \begin{bmatrix} 1 & -3 & 2 & | & 5 \\ 0 & 1 & \frac{1}{6} & | & -1 \end{bmatrix} \xrightarrow{\text{*Row+}(3,[\text{Ans}],2,1)} \begin{bmatrix} 1 & 0 & \frac{5}{2} & | & 2 \\ 0 & 1 & \frac{1}{6} & | & -1 \end{bmatrix}$$

This is equivalent to the system $\begin{cases} x + \frac{5}{2}z = 2 \\ y + \frac{1}{6}z = -1 \end{cases}$. Notice that one of these equations is in terms of x and z and the other is in terms of y and z. This means that if we choose z to parameterize, it will be easy to find the corresponding values of x and y. Since we can choose the parameter, we make a choice that will simplify the arithmetic. Let t be a parameter, and let $z = 6t$. Then, from the equivalent system, we see

$$x + \frac{5}{2}z = 2 \qquad\qquad y + \frac{1}{6}z = -1$$
$$x = 2 - \frac{5}{2}z \qquad\qquad y = -1 - \frac{1}{6}z$$
$$= 2 - \frac{5}{2}(6t) \qquad\qquad y = -1 - \frac{1}{6}(6t)$$
$$= 2 - 15t \qquad\qquad = -1 - t$$

This gives the parametric solution (x, y, z) $(2 - 15t, -1 - t, 6t)$. What does this mean? It means that for *every* choice of t, we obtain a solution of the original system. For example,

if $t = 0$ $(x, y, z) = (2, -1, 0)$ satisfies both equations
if $t = 1$ $(x, y, z) = (-13, -2, 6)$ satisfies both equations
if $t = -1$ $(x, y, z) = (17, 0, -6)$ satisfies both equations
$\vdots$ $\vdots$

The parametric representation for a dependent system is not unique. For Example 9, what if we let $z = -6t$? Then the parametric solution

$$(x, y, z) = (2 + 15t, -1 + t, -6t)$$

is not identical to the one obtained in Example 9. However, every choice for t that makes this parametric solution true will also simultaneously satisfy the original system. In more advanced work, techniques are developed for showing that different parametric representations are equivalent.

Applications

Our discussion concludes with two applied problems.

EXAMPLE 10 Mixture problem

A rancher has to mix three types of feed for her cattle. The following analysis shows the amounts per bag (100 lb) of grain:

Grain	Protein	Carbohydrates	Sodium
A	7 lb	88 lb	1 lb
B	6 lb	90 lb	1 lb
C	10 lb	70 lb	2 lb

How many bags of each type of grain should she mix to provide 71 lb of protein, 854 lb of carbohydrates, and 12 lb of sodium?

Solution

Let a, b, and c be the number of bags of grains A, B, and C, respectively, that are needed for the mixture. Then

Grain	Protein	Carbohydrates	Sodium
A	$7a$ lb	$88a$ lb	a lb
B	$6b$ lb	$90b$ lb	b lb
C	$10c$ lb	$70c$ lb	$2c$ lb
Total needed:	71 lb	854 lb	12 lb

Thus, $\begin{cases} 7a + 6b + 10c = 71 \\ 88a + 90b + 70c = 854 \\ a + b + 2c = 12 \end{cases}$ Let $[A] = \begin{bmatrix} 7 & 6 & 10 & | & 71 \\ 88 & 90 & 70 & | & 854 \\ 1 & 1 & 2 & | & 12 \end{bmatrix}$.

We show the steps in Gauss–Jordan elimination:

$$\begin{bmatrix} 7 & 6 & 10 & | & 71 \\ 88 & 90 & 70 & | & 854 \\ 1 & 1 & 2 & | & 12 \end{bmatrix} \xrightarrow{\text{RowSwap}([A],1,3)} \begin{bmatrix} 1 & 1 & 2 & | & 12 \\ 88 & 90 & 70 & | & 854 \\ 7 & 6 & 10 & | & 71 \end{bmatrix} \begin{array}{c} *\text{Row}+(-7,[\text{Ans}],1,3) \\ \to \\ *\text{Row}+(-88,[\text{Ans}],1,2) \end{array} \begin{bmatrix} 1 & 1 & 2 & | & 12 \\ 0 & 2 & -106 & | & -202 \\ 0 & -1 & -4 & | & -13 \end{bmatrix}$$

$$\xrightarrow[*\text{Row}(1/2,[\text{Ans}],2)]{} \begin{bmatrix} 1 & 1 & 2 & | & 12 \\ 0 & 1 & -53 & | & -101 \\ 0 & -1 & -4 & | & -13 \end{bmatrix} \begin{array}{c} *\text{Row}+(-1,[\text{Ans}],2,1) \\ \to \\ *\text{Row}+(1,[\text{Ans}],2,3) \end{array} \begin{bmatrix} 1 & 0 & 55 & | & 113 \\ 0 & 1 & -53 & | & -101 \\ 0 & 0 & -57 & | & -114 \end{bmatrix}$$

$$\xrightarrow[*\text{Row}(-1/57,[\text{Ans}],3)]{} \begin{bmatrix} 1 & 0 & 55 & | & 113 \\ 0 & 1 & -53 & | & -101 \\ 0 & 0 & 1 & | & 2 \end{bmatrix} \begin{array}{c} *\text{Row}+(-55,[\text{Ans}],3,1) \\ \to \\ *\text{Row}+(53,[\text{Ans}],3,2) \end{array} \begin{bmatrix} 1 & 0 & 0 & | & 3 \\ 0 & 1 & 0 & | & 5 \\ 0 & 0 & 1 & | & 2 \end{bmatrix}$$

Mix three bags of grain A, five bags of grain B, and two bags of grain C. ∎

EXAMPLE 11 Equation of a parabola passing through known points

Suppose the equation of a certain parabola has the form $y = ax^2 + bx + c$. Find the equation of the parabola passing through $(-1, -6)$, $(2, 9)$, and $(-2, -3)$.

Solution

If a curve passes through a point, then the coordinates of that point must satisfy the equation:

$$(-1,-6): \quad -6 = a(-1)^2 + b(-1) + c$$
$$(2,9): \quad 9 = a(2)^2 + b(2) + c$$
$$(-2,-3): \quad -3 = a(-2)^2 + b(-2) + c$$

In standard form, this system is written $\begin{cases} a - b + c = -6 \\ 4a + 2b + c = 9 \\ 4a - 2b + c = -3 \end{cases}$. We show the Gauss–Jordan elimination steps:

$$[A] = \begin{bmatrix} 1 & -1 & 1 & | & -6 \\ 4 & 2 & 1 & | & 9 \\ 4 & -2 & 1 & | & -3 \end{bmatrix} \begin{array}{c} *\text{Row}+(-4,[A],1,2) \\ \rightarrow \\ *\text{Row}+(-4,[\text{Ans}],1,3) \end{array} \begin{bmatrix} 1 & -1 & 1 & | & -6 \\ 0 & 6 & -3 & | & 33 \\ 0 & 2 & -3 & | & 21 \end{bmatrix}$$

$$\begin{array}{c} \rightarrow \\ *\text{Row}\left(\frac{1}{6},[\text{Ans}],2\right) \end{array} \begin{bmatrix} 1 & -1 & 1 & | & -6 \\ 0 & 1 & -\frac{1}{2} & | & \frac{11}{2} \\ 0 & 2 & -3 & | & 21 \end{bmatrix} \begin{array}{c} *\text{Row}+(1,[\text{Ans}],2,1) \\ \rightarrow \\ *\text{Row}+(-2,[\text{Ans}],2,3) \end{array} \begin{bmatrix} 1 & 0 & \frac{1}{2} & | & -\frac{1}{2} \\ 0 & 1 & -\frac{1}{2} & | & \frac{11}{2} \\ 0 & 0 & -2 & | & 10 \end{bmatrix}$$

$$\begin{array}{c} \rightarrow \\ *\text{Row}\left(-\frac{1}{2},[\text{Ans}],3\right) \end{array} \begin{bmatrix} 1 & 0 & \frac{1}{2} & | & -\frac{1}{2} \\ 0 & 1 & -\frac{1}{2} & | & \frac{11}{2} \\ 0 & 0 & 1 & | & -5 \end{bmatrix} \begin{array}{c} *\text{Row}\left(-\frac{1}{2},[\text{Ans}],3,1\right) \\ \rightarrow \\ *\text{Row}\left(\frac{1}{2},[\text{Ans}],3,2\right) \end{array} \begin{bmatrix} 1 & 0 & 0 & | & 2 \\ 0 & 1 & 0 & | & 3 \\ 0 & 0 & 1 & | & -5 \end{bmatrix}$$

Thus, $a = 2$, $b = 3$, and $c = -5$, so the desired equation is $y = 2x^2 + 3x - 5$. ∎

PROBLEM SET 8.5

LEVEL 1

1. Write each system in augmented matrix form.

   a. $\begin{cases} 4x + 5y = -16 \\ 3x + 2y = 5 \end{cases}$

   b. $\begin{cases} x + y + z = 4 \\ 3x + 2y + z = 7 \\ x - 3y + 2z = 0 \end{cases}$

   c. $\begin{cases} x_1 + 2x_2 - 5x_3 + x_4 = 5 \\ x_1 - 3x_3 + 6x_4 = 0 \\ x_3 - 3x_4 = -15 \\ x_2 - 5x_3 + 5x_4 = 2 \end{cases}$

2. Write a system of equations that has the given augmented matrix.

   a. $\begin{bmatrix} 6 & 7 & 8 & | & 3 \\ 1 & 2 & 3 & | & 4 \\ 0 & 1 & 3 & | & 4 \end{bmatrix}$

   b. $\begin{bmatrix} 1 & 0 & 0 & | & 3 \\ 0 & 1 & 2 & | & 4 \end{bmatrix}$

   c. $\begin{bmatrix} 1 & 0 & 0 & | & 32 \\ 0 & 1 & 0 & | & 27 \\ 0 & 0 & 1 & | & -5 \\ 0 & 0 & 0 & | & 3 \end{bmatrix}$

Given the matrices in Problems 3–4, perform elementary row operations to obtain a 1 in the row 1, column 1 position. Answers may vary.

3. **a.** $[A] = \begin{bmatrix} 3 & 1 & 2 & | & 1 \\ 0 & 2 & 4 & | & 5 \\ 1 & 3 & -4 & | & 9 \end{bmatrix}$ **b.** $[B] = \begin{bmatrix} -2 & 3 & 5 & | & -8 \\ 1 & 0 & 2 & | & 2 \\ 0 & 1 & 0 & | & 5 \end{bmatrix}$

4. **a.** $[C] = \begin{bmatrix} 2 & 4 & 10 & | & -12 \\ 6 & 3 & 4 & | & 6 \\ 10 & -1 & 0 & | & 1 \end{bmatrix}$ **b.** $[D] = \begin{bmatrix} 5 & 20 & 15 & | & 6 \\ 7 & -5 & 3 & | & 2 \\ 12 & 0 & 1 & | & 4 \end{bmatrix}$

Given the matrices in Problems 5–6, perform elementary row operations to obtain zeros under the 1 in the first column. Answers may vary.

5. **a.** $[A] = \begin{bmatrix} 1 & 2 & -3 & | & 0 \\ 0 & 3 & 1 & | & 4 \\ 2 & 5 & 1 & | & 6 \end{bmatrix}$ **b.** $[B] = \begin{bmatrix} 1 & 3 & -5 & | & 6 \\ -3 & 4 & 1 & | & 2 \\ 0 & 5 & 1 & | & 3 \end{bmatrix}$

6. **a.** $[C] = \begin{bmatrix} 1 & 2 & 4 & | & 1 \\ -2 & 5 & 0 & | & 2 \\ -4 & 5 & 1 & | & 3 \end{bmatrix}$ **b.** $[D] = \begin{bmatrix} 1 & 5 & 3 & | & 2 \\ 2 & 3 & -1 & | & 4 \\ 3 & 2 & 1 & | & 0 \end{bmatrix}$

Given the matrices in Problems 7 and 8, perform elementary row operations to obtain a 1 in the second row, second column without changing the entries in the first column. Answers may vary.

7. a. $[A] = \begin{bmatrix} 1 & 3 & 5 & | & 2 \\ 0 & 2 & 6 & | & -8 \\ 0 & 3 & 4 & | & 1 \end{bmatrix}$　**b.** $[B] = \begin{bmatrix} 1 & 5 & -3 & | & 5 \\ 0 & 3 & 9 & | & -15 \\ 0 & 2 & 1 & | & 5 \end{bmatrix}$

8. a. $[C] = \begin{bmatrix} 1 & 4 & -1 & | & 6 \\ 0 & 5 & 1 & | & 3 \\ 0 & 4 & 6 & | & 5 \end{bmatrix}$　**b.** $[D] = \begin{bmatrix} 1 & 3 & -2 & | & 0 \\ 0 & 4 & 2 & | & 9 \\ 0 & 3 & 6 & | & 1 \end{bmatrix}$

Given the matrices in Problems 9 and 10, perform elementary row operations to obtain a zero (or zeros) above and below the 1 in the second column without changing the entries in the first column. Answers may vary.

9. a. $[A] = \begin{bmatrix} 1 & 5 & -3 & | & 2 \\ 0 & 1 & 4 & | & 5 \\ 0 & 3 & 4 & | & 2 \end{bmatrix}$　**b.** $[B] = \begin{bmatrix} 1 & 3 & 6 & | & 12 \\ 0 & 1 & -2 & | & -5 \\ 0 & -2 & 2 & | & 6 \end{bmatrix}$

10. a. $[C] = \begin{bmatrix} 1 & 6 & -3 & 4 & | & 1 \\ 0 & 1 & 7 & 3 & | & 0 \\ 0 & 3 & 4 & 0 & | & -2 \\ 0 & -2 & 3 & 1 & | & 0 \end{bmatrix}$　**b.** $[D] = \begin{bmatrix} 1 & 5 & -1 & 2 & | & 8 \\ 0 & 1 & 5 & 2 & | & 0 \\ 0 & 1 & 4 & 0 & | & 5 \\ 0 & 2 & -3 & 1 & | & 7 \end{bmatrix}$

Given the matrices in Problems 11 and 12, perform elementary row operations to obtain a 1 in the third row, third column without changing the entries in the first two columns. Answers may vary.

11. a. $[A] = \begin{bmatrix} 1 & 0 & 4 & | & 5 \\ 0 & 1 & -3 & | & 6 \\ 0 & 0 & 5 & | & 10 \end{bmatrix}$　**b.** $[B] = \begin{bmatrix} 1 & 0 & 4 & | & -5 \\ 0 & 1 & 3 & | & 6 \\ 0 & 0 & 8 & | & 12 \end{bmatrix}$

12. a. $[C] = \begin{bmatrix} 1 & 0 & -2 & 1 & | & 1 \\ 0 & 1 & 6 & 2 & | & 0 \\ 0 & 0 & 4 & 0 & | & -2 \\ 0 & 0 & 2 & 1 & | & 0 \end{bmatrix}$　**b.** $[D] = \begin{bmatrix} 1 & 0 & -8 & 2 & | & 8 \\ 0 & 1 & -1 & 3 & | & 2 \\ 0 & 0 & 2 & 0 & | & 10 \\ 0 & 0 & -2 & 1 & | & 6 \end{bmatrix}$

Given the matrices in Problems 13 and 14, perform elementary row operations to obtain zeros above and below the 1 in the third column without changing the entries in the first or second columns. Answers may vary.

13. a. $[A] = \begin{bmatrix} 1 & 0 & -1 & | & 5 \\ 0 & 1 & 2 & | & 6 \\ 0 & 0 & 1 & | & 4 \end{bmatrix}$　**b.** $[B] = \begin{bmatrix} 1 & 0 & -3 & | & -2 \\ 0 & 1 & 4 & | & 5 \\ 0 & 0 & 1 & | & 3 \end{bmatrix}$

14. a. $[C] = \begin{bmatrix} 1 & 0 & -1 & 4 & | & 1 \\ 0 & 1 & 4 & 3 & | & 0 \\ 0 & 0 & 1 & 0 & | & -2 \\ 0 & 0 & -3 & 1 & | & 0 \end{bmatrix}$　**b.** $[D] = \begin{bmatrix} 1 & 0 & -8 & 2 & | & 8 \\ 0 & 1 & 4 & 2 & | & 0 \\ 0 & 0 & 1 & 0 & | & 2 \\ 0 & 0 & -1 & 1 & | & 7 \end{bmatrix}$

LEVEL 2

Solve the systems in Problems 15–46 by the Gauss–Jordan method. Give parametric solutions for dependent systems, but some systems do not have solutions.

15. $\begin{cases} x - y = 2 \\ 2x + 3y = 9 \end{cases}$

16. $\begin{cases} 3x - 4y = 16 \\ -x + 2y = -6 \end{cases}$

17. $\begin{cases} 3x - 2y = 12 \\ 6x - 4y = 24 \end{cases}$

18. $\begin{cases} 4x - 4y = 8 \\ -8x + 8y = -16 \end{cases}$

19. $\begin{cases} x + y = 2 \\ x - z = 1 \\ -y + z = 1 \end{cases}$

20. $\begin{cases} x + 5z = 9 \\ y + 2z = 2 \\ 2x + 3z = 4 \end{cases}$

21. $\begin{cases} x + 2z = 13 \\ 2x + y = 8 \\ -2y + 9z = 41 \end{cases}$

22. $\begin{cases} 4x + y = -2 \\ 3x + 2z = -9 \\ 2y + 3z = -5 \end{cases}$

23. $\begin{cases} 5x + z = 9 \\ x - 5z = 7 \\ x + y - z = 0 \end{cases}$

24. $\begin{cases} x + y = -2 \\ y + z = 2 \\ x - y - z = -1 \end{cases}$

25. $\begin{cases} 2x + y = 0 \\ 3x - 2y = -7 \\ x - 3y = -1 \end{cases}$

26. $\begin{cases} 2x - 3y = 5 \\ 9x - 7y = 4 \\ x + y = 4 \end{cases}$

27. $\begin{cases} x + 2z = 9 \\ x + y = 13 \\ 2y + z = 8 \end{cases}$

28. $\begin{cases} 3x + 3z = y + 2 \\ x - z = 0 \\ x + y = z \end{cases}$

29. $\begin{cases} x + 2z = 9 \\ 2x - 4y = 8 \\ x + 3y = -11 \end{cases}$

30. $\begin{cases} x + 2z = 0 \\ 3x - y + 2z = 0 \\ 4x + y = 6 \end{cases}$

31. $\begin{cases} x - y = 1 \\ x + z = 1 \\ y - z = 1 \end{cases}$

32. $\begin{cases} 3x - y + 2z = 3 \\ y - 4x + z = 5 \\ 2x - y + 5z = 2 \end{cases}$

33. $\begin{cases} x + 2y - z = 3 \\ z - 3x - 6y = 1 \\ 4y + 2x - z = 2 \end{cases}$

34. $\begin{cases} x + y = 3 \\ x + 2y + z = -2 \end{cases}$

35. $\begin{cases} y - 2x = -2 \\ 2x - y + 4z = -14 \end{cases}$

36. $\begin{cases} y - x = 8 \\ y + 2x = 1 \\ x - 3y = 3 \end{cases}$

37. $\begin{cases} x + 2y = 9 \\ x + y = 5 \\ 3x - y = 26 \end{cases}$

38. $\begin{cases} x + 3y - z = -2 \\ x - y + 2z = -4 \\ 2x + y - 3z = 3 \end{cases}$

39. $\begin{cases} 2x + y - z = 1 \\ x + 3y - 2z = 4 \\ 3x - 2y + z = -2 \end{cases}$

40. $\begin{cases} 2x - y + z = -1 \\ x + 2y + z = 4 \\ x - y - 4z = -8 \end{cases}$

41. $\begin{cases} w + z = 5 \\ x + y = 3 \\ y + z = 3 \\ 2w + y = 10 \end{cases}$

42. $\begin{cases} x + z = 3 \\ w + y = x \\ x + y = 0 \\ y + z = 1 \end{cases}$

43. $\begin{cases} 2x + 2y - z + w = 20 \\ 3y - x + 3z = -10 \\ 3x - 2z - w = 21 \\ y + z + 2w = 0 \end{cases}$

44. $\begin{cases} x + y + z + w = 3 \\ x - y - w = 0 \\ y + 2z + w = 0 \\ 2x - z + w = 3 \end{cases}$

45. $\begin{cases} 3x - 2y = 13 \\ 4x + 5y = 2 \\ x + y = 1 \\ -2x - 3y = 0 \end{cases}$

46. $\begin{cases} x - 6y = 8 \\ 2x - y = 5 \\ x + y = 1 \\ 4x + 5y = 3 \end{cases}$

47. Find the equation of the parabola of the form $y = ax^2 + bx + c$ that passes through $(0, 5)$, $(-1, 2)$, and $(3, 26)$.

48. Find the equation of the parabola of the form $y = ax^2 + bx + c$ that passes through $(1, -2)$, $(-2, -14)$, and $(3, -4)$.

49. In the last chapter, we saw that the equation of a circle has the form

$$x^2 + y^2 + Ax + By + C = 0$$

Find the equation of the circle passing through $(-3, 4)$, $(4, 5)$, and $(1, -4)$.

50. In the last chapter, we saw that the equation of a circle has the form

$$x^2 + y^2 + Ax + By + C = 0$$

Find the equation of the circle passing through $(2, 2)$, $(-2, -6)$, and $(5, 1)$.

51. To control a certain type of crop disease, it is necessary to use 23 gal of chemical A and 34 gal of chemical B. The dealer can order commercial spray I, each container of which holds 5 gal of chemical A and 2 gal of chemical B, and commercial spray II, each container of which holds 2 gal of chemical A and 7 gal of chemical B. How many containers of each type of commercial spray should be used to obtain exactly the right proportion of chemicals needed?

52. To manufacture a certain alloy, it is necessary to use 33 kg (kilograms) of metal A and 56 kg of metal B. It is cheaper for the manufacturer if she buys and mixes an alloy, each bar of which contains 3 kg of metal A and 5 kg of metal B, along with another alloy, each bar of which contains 4 kg of metal A and 7 kg of metal B. How much of the two alloys should she use to produce the alloy desired?

53. A candy maker mixes chocolate, milk, and mint extract to produce three kinds of candy (I, II, and III) with the following proportions:
I: 7 lb chocolate, 5 gal milk, 1 oz mint extract
II: 3 lb chocolate, 2 gal milk, 2 oz mint extract
III: 4 lb chocolate, 3 gal milk, 3 oz mint extract
If 67 lb of chocolate, 48 gal of milk, and 32 oz of mint extract are available, how much of each kind of candy can be produced?

54. Using the data from Problem 53, how much of each type of candy can be produced with 62 lb of chocolate, 44 gal of milk, and 32 oz of mint extract?

LEVEL 3

Solve the systems in Problems 55–60. Each system has infinitely many solutions. Give the parametric solution.

55. $\begin{cases} x + 2y = 1 \\ 4y + 3z = 3 \\ 3z - 2x = 1 \end{cases}$

56. $\begin{cases} x - 2y = 1 \\ 3x + z = 2 \\ 6y + z = -1 \end{cases}$

57. $\begin{cases} 2x - y + z = 2 \\ 2x - 4y = 32 \end{cases}$

58. $\begin{cases} x - 2y + z = -3 \\ 3x - 7z = -6 \end{cases}$

59. $\begin{cases} x - 2y + 3z = 3 \\ 2x + 3y - z = 5 \\ x + 5y - 4y = 2 \end{cases}$

60. $\begin{cases} x + y + 3z = -1 \\ x + 3y - 3z = 3 \\ 2x + 5y - 3z = 4 \end{cases}$

8.6 Inverse Matrices

The availability of computer software and calculators that can carry out matrix operations has considerably increased the importance of a matrix solution to systems of equations. This process involves the inverse of a matrix. If we let $[A]$ be the matrix of coefficients of a system of equations, $[X]$ the matrix of unknowns, and $[B]$ the matrix of constants, we can then represent the system of equations by the **matrix equation**

$$[A][X] = [B]$$

If we can define an inverse matrix, denoted by $[A]^{-1}$, we should be able to solve the *system* by finding

$$[X] = [A]^{-1}[B]$$

To understand this simple process, we need to understand what it means for matrices to be inverses and develop a basic algebra for matrices. Even though these matrix operations are difficult with a pencil and paper, they are easy with the aid of computer software or a calculator that does matrix operations.

☠ Note that $[A]^{-1}$ does not mean $\frac{1}{[A]}$ but rather means the inverse of matrix $[A]$. ☠

Matrix Operations

The next box gives a definition of matrix equality along with the fundamental matrix operations.

MATRIX OPERATIONS

Equality
$[M] = [N]$ if and only if matrices $[M]$ and $[N]$ are the same order and the corresponding entries are the same.

Addition

$[M] + [N] = [S]$ if and only if $[M]$ and $[N]$ are the same order and the entries of $[S]$ are found by adding the corresponding entries of $[M]$ and $[N]$.

Multiplication by a scalar

$c[M] = [cM]$ is the matrix in which each entry of $[M]$ is multiplied by the scalar (real number) c.

Subtraction

$[M] - [N] = [D]$ if and only if $[M]$ and $[N]$ are the same order and the entries of $[D]$ are found by subtracting the entries of $[N]$ from the corresponding entries of $[M]$.

Multiplication

Let $[M]$ be an $m \times r$ matrix and $[N]$ an $r \times n$ matrix. The product matrix $[M][N] = [P]$ is an $m \times n$ matrix. The entry in the ith row and jth column of $[M][N]$ is *the sum of the products formed by multiplying each entry of the ith row of $[M]$ by the corresponding entry in the jth column of $[N]$.*

All of these definitions, except multiplication, are straightforward, so we will consider multiplication separately after Example 1. If an addition or multiplication cannot be performed because of the order of the given matrices, the matrices are said to be **nonconformable**.

EXAMPLE 1   Matrix addition, subtraction, and scalar multiplication

Let $[A] = [5 \quad 2 \quad 1]$, $[B] = [4 \quad 8 \quad -5]$, $[C] = \begin{bmatrix} 7 & 3 & 2 \\ 5 & -4 & -3 \end{bmatrix}$, $[D] = \begin{bmatrix} 4 & -2 & 1 \\ -3 & 3 & -1 \\ 2 & 4 & -1 \end{bmatrix}$,

$[E] = \begin{bmatrix} 3 & 4 & -1 \\ 2 & 0 & 5 \\ -4 & 2 & 3 \end{bmatrix}$. Find: **a.** $[A] + [B]$ **b.** $[A] + [C]$ **c.** $[D] + [E]$ **d.** $(-5)[C]$ **e.** $2[A] - 3[B]$

Solution

a. $[A] + [B] = [5 \quad 2 \quad 1] + [4 \quad 8 \quad -5]$
$= [5+4 \quad 2+8 \quad 1+(-5)]$
$= [9 \quad 10 \quad -4]$

b. $[A] + [C]$ is not defined because $[A]$ and $[C]$ are nonconformable.

c. $[D] + [E] = \begin{bmatrix} 4 & -2 & 1 \\ -3 & 3 & -1 \\ 2 & 4 & -1 \end{bmatrix} + \begin{bmatrix} 3 & 4 & -1 \\ 2 & 0 & 5 \\ -4 & 2 & 3 \end{bmatrix}$

$= \begin{bmatrix} 7 & 2 & 0 \\ -1 & 3 & 4 \\ -2 & 6 & 2 \end{bmatrix}$ Add entry by entry.

d. $(-5)[C] = (-5) \begin{bmatrix} 7 & 3 & 2 \\ 5 & -4 & -3 \end{bmatrix}$

$= \begin{bmatrix} -35 & -15 & -10 \\ -25 & 20 & 15 \end{bmatrix}$ Multiply each entry by -5.

e. $2[A] - 3[B] = 2[5 \quad 2 \quad 1] + (-3)[4 \quad 8 \quad -5]$
$= [10 \quad 4 \quad 2] + [-12 \quad -24 \quad 15]$
$= [-2 \quad -20 \quad 17]$

Matrix operations are particularly easy using a calculator that handles matrices. Check with the owner's manual or the website for the calculator you are using.

We illustrate matrix multiplication for matrices of different order (size).

$$\underbrace{\begin{bmatrix} 2 & 3 & 4 & 5 \end{bmatrix}}_{1 \times \underset{\uparrow}{4} \; matrix} \begin{bmatrix} a \\ b \\ c \\ d \end{bmatrix} \! \! \underbrace{\phantom{\begin{bmatrix} a \\ b \\ c \\ d \end{bmatrix}}}_{\underset{\uparrow}{4} \times 1 \; matrix} = \underbrace{\begin{bmatrix} 2a + 3b + 4c + 5d \end{bmatrix}}_{1 \times 1 \; matrix}$$

To be conformable, these numbers must be the same.

$$\underbrace{\begin{bmatrix} 2 & 3 & 4 & 5 \end{bmatrix}}_{1 \times \underset{\uparrow}{4} \; matrix} \begin{bmatrix} 1 \\ -3 \\ 0 \\ 2 \end{bmatrix} \! \! \underbrace{\phantom{\begin{bmatrix} 1 \\ -3 \\ 0 \\ 2 \end{bmatrix}}}_{\underset{\uparrow}{4} \times 1 \; matrix} = \underbrace{\begin{bmatrix} 2(1) + 3(-3) + 4(0) + 5(2) \end{bmatrix}}_{1 \times 1 \; matrix \; answer} = \begin{bmatrix} 3 \end{bmatrix}$$

Same

$$\underbrace{\begin{bmatrix} 5 & 3 & 2 \end{bmatrix}}_{1 \times \underset{\uparrow}{3} \; matrix} \begin{bmatrix} 1 \\ 2 \\ 3 \\ 4 \end{bmatrix} \! \! \underbrace{\phantom{\begin{bmatrix} 1 \\ 2 \\ 3 \\ 4 \end{bmatrix}}}_{\underset{\uparrow}{4} \times 1 \; matrix \; answer} \qquad \text{not comformable}$$

Not the same

$$\underbrace{\begin{bmatrix} 2 & 3 & 4 & 5 \end{bmatrix}}_{1 \times \underset{\uparrow}{4} \; matrix} \begin{bmatrix} 1 & 2 \\ -3 & 1 \\ 0 & -2 \\ 2 & 3 \end{bmatrix} \! \! \underbrace{\phantom{\begin{bmatrix} 1 & 2 \\ -3 & 1 \\ 0 & -2 \\ 2 & 3 \end{bmatrix}}}_{\underset{\uparrow}{4} \times 2 \; matrix} = \begin{bmatrix} \underbrace{2\,(1) + 3\,(-3) + 4\,(0) + 5\,(2)}_{(\,row\,1,\,column\,1\,)\;entry} & \underbrace{2\,(2) + 3\,(1) + 4\,(-2) + 5\,(3)}_{(\,row\,1,\,column\,2\,)\;entry} \end{bmatrix}$$

$$\underbrace{\phantom{\begin{bmatrix} 2\,(1) + 3\,(-3) & 2\,(2) + 3\,(1) \end{bmatrix}}}_{1 \times 2 \; matrix \; answer}$$

$$= \begin{bmatrix} 2 - 9 + 0 + 10 & 4 + 3 - 8 + 15 \end{bmatrix}$$
$$= \begin{bmatrix} 3 & 14 \end{bmatrix}$$

Same: Answer is a 1 × 2 matrix.

$$\underbrace{\begin{bmatrix} 2 & -1 & 2 \end{bmatrix}}_{1 \times 3 \; matrix} \begin{bmatrix} 4 & 2 & -1 \\ 1 & 0 & 2 \\ 3 & -1 & 3 \end{bmatrix}$$

$$\underbrace{\phantom{\begin{bmatrix} 4 & 2 & -1 \\ 1 & 0 & 2 \\ 3 & -1 & 3 \end{bmatrix}}}_{3 \times 3 \; matrix}$$

$$= \begin{bmatrix} \underbrace{2\,(4) + (-1)\,(1) + 2\,(3)}_{(\,row\,1,\,column\,1\,)\;entry} & \underbrace{2\,(2) + (-1)\,(0) + 2\,(-1)}_{(\,row\,1,\,column\,2\,)\;entry} & \underbrace{2\,(-1) + (-1)\,(2) + 2\,(3)}_{(\,row\,1,\,column\,3\,)\;entry} \end{bmatrix}$$

$$\underbrace{\phantom{\begin{bmatrix} 2\,(4) + (-1)\,(1) & 2\,(2) + (-1)\,(0) & 2\,(-1) + (-1)\,(2) \end{bmatrix}}}_{1 \times 3 \; matrix \; answer}$$

The first step (above) is written out to demonstrate what is happening. Your work, however, should look like this:

$$[2 \quad -1 \quad 2] \begin{bmatrix} 4 & 2 & -1 \\ 1 & 0 & 2 \\ 3 & -1 & 3 \end{bmatrix} = [8 - 1 + 6 \quad 4 + 0 - 2 \quad -2 - 2 + 6]$$

$$= [13 \quad 2 \quad 2]$$

■ COMPUTATIONAL WINDOW ▁□✕

A principal advantage of using matrix multiplication to solve systems of equations is the ease with which matrix multiplication can be done using a calculator that handles matrices. You should check the following calculations using a calculator.

EXAMPLE 2   Matrix multiplication

Find $[A][B]$ and $[B][A]$.

a. Let $[A] = \begin{bmatrix} 1 & 2 & 3 & 4 \\ 5 & 6 & 7 & 8 \end{bmatrix}$ and $[B] = \begin{bmatrix} -3 & 1 & -2 \\ 0 & -1 & 5 \\ -4 & 3 & -1 \\ 2 & 3 & -2 \end{bmatrix}$.

b. Let $[A] = \begin{bmatrix} 3 & -1 & 4 \\ 2 & 1 & 0 \\ -1 & 3 & 2 \end{bmatrix}$ and $[B] = \begin{bmatrix} 5 & 1 & -1 \\ 2 & 3 & -2 \\ 0 & 3 & 4 \end{bmatrix}$.

Solution

a. $[A][B] = \begin{bmatrix} 1 & 2 & 3 & 4 \\ 5 & 6 & 7 & 8 \end{bmatrix} \begin{bmatrix} -3 & 1 & -2 \\ 0 & -1 & 5 \\ -4 & 3 & -1 \\ 2 & 3 & -2 \end{bmatrix}$ *Mental step:*

$$= \begin{bmatrix} 1(-3) + 2(0) + 3(-4) + 4(2) & 1(1) + 2(-1) + 3(3) + 4(3) & 1(-2) + 2(5) + 3(-1) + 4(-2) \\ 5(-3) + 6(0) + 7(-4) + 8(2) & 5(1) + 6(-1) + 7(3) + 8(3) & 5(-2) + 6(5) + 7(-1) + 8(-2) \end{bmatrix}$$

$$= \begin{bmatrix} -7 & 20 & -3 \\ -27 & 44 & -3 \end{bmatrix}$$

$$[B][A] = \begin{bmatrix} -3 & 1 & -2 \\ 0 & -1 & 5 \\ -4 & 3 & -1 \\ 2 & 3 & -2 \end{bmatrix} \begin{bmatrix} 1 & 2 & 3 & 4 \\ 5 & 6 & 7 & 8 \end{bmatrix}$$

$[B]$ and $[A]$ are not conformable. Note that $[A][B] \neq [B][A]$.

b. $[A][B] = \begin{bmatrix} 3 & -1 & 4 \\ 2 & 1 & 0 \\ -1 & 3 & 2 \end{bmatrix} \begin{bmatrix} 5 & 1 & -1 \\ 2 & 3 & -2 \\ 0 & 3 & 4 \end{bmatrix}$ *Mental step:*

$$= \begin{bmatrix} 3(5) + (-1)(2) + 4(0) & 3(1) + (-1)(3) + 4(3) & 3(-1) + (-1)(-2) + 4(4) \\ 2(5) + (1)(2) + 0(0) & 2(1) + (1)(3) + 0(3) & 2(-1) + (1)(-2) + 0(4) \\ (-1)(5) + 3(2) + 2(0) & (-1)(1) + 3(3) + 2(3) & (-1)(-1) + 3(-2) + 2(4) \end{bmatrix}$$

$$= \begin{bmatrix} 13 & 12 & 15 \\ 12 & 5 & -4 \\ 1 & 14 & 3 \end{bmatrix}$$

$$[B][A] = \begin{bmatrix} 5 & 1 & -1 \\ 2 & 3 & -2 \\ 0 & 3 & 4 \end{bmatrix} \begin{bmatrix} 3 & -1 & 4 \\ 2 & 1 & 0 \\ -1 & 3 & 2 \end{bmatrix} = \begin{bmatrix} 18 & -7 & 18 \\ 14 & -5 & 4 \\ 2 & 15 & 8 \end{bmatrix}$$

Once again, note that $[A][B] \neq [B][A]$. ∎

EXAMPLE 3   Matrix form of a system of equations

Let $[A] = \begin{bmatrix} 1 & 2 & 3 \\ 4 & -1 & 5 \\ 3 & 2 & -1 \end{bmatrix}$, $[X] = \begin{bmatrix} x \\ y \\ z \end{bmatrix}$ and $[B] = \begin{bmatrix} 3 \\ 16 \\ 5 \end{bmatrix}$. What is $[A][X] = [B]$?

Solution

$$[A][X] = \begin{bmatrix} x + 2y + 3z \\ 4x - y + 5z \\ 3x + 2y - z \end{bmatrix}$$ so $[A][X] = [B]$ is a matrix equation representing the

system $\begin{cases} x + 2y + 3z = 3 \\ 4x - y + 5z = 16 \\ 3x + 2y - z = 5 \end{cases}$ ∎

Zero-One Matrices

A **zero-one matrix** (sometimes called a *communication matrix*) is a square matrix in which the entries symbolize the occurrence of some facet or event with a 1 and the nonoccurrence with a 0. Consider the following example.

EXAMPLE 4   Diplomatic relations   **MODELING APPLICATION**

Use matrix notation to summarize the following information: The United States has diplomatic relations with Russia and with Mexico but not with Cuba. Mexico has diplomatic relations with the United States and Russia but not with Cuba. Russia has diplomatic relations with the United States, Mexico, and Cuba. Finally, Cuba has diplomatic relations with Russia, but not with the United States and not with Mexico. Note that a country is not considered to have diplomatic relations with itself.

a. Write a matrix showing direct communication possibilities.

b. Write a matrix showing the channels of communication that are open to the various countries if they are willing to speak through an intermediary.

Solution

    Step 1:   *Understand the problem.* Part **a** is fairly easy to understand, so we will do this first:

	U.S.	Russia	Cuba	Mexico
U.S.	0	1	0	1
$[A] =$ Russia	1	0	1	1
Cuba	0	1	0	0
Mexico	1	1	0	0

For part **b**, we want to find a matrix that will tell us the number of ways the countries can communicate through an intermediary. For example, the United States can talk to Cuba by talking through Russia. A country can also communicate with itself (to test security, perhaps) if it does so through an intermediary.

Step 2: *Devise a plan.* We begin with a simplified example. Let

$$[B] = \begin{array}{c} \\ \text{U.S.} \\ \text{Russia} \\ \text{Cuba} \end{array} \begin{array}{ccc} \text{U.S.} & \text{Russia} & \text{Cuba} \\ \begin{bmatrix} 0 & 1 & 0 \\ 1 & 0 & 1 \\ 0 & 1 & 0 \end{bmatrix} \end{array}$$

Consider the product: $[B]^2 = \begin{bmatrix} 0 & 1 & 0 \\ 1 & 0 & 1 \\ 0 & 1 & 0 \end{bmatrix} \begin{bmatrix} 0 & 1 & 0 \\ 1 & 0 & 1 \\ 0 & 1 & 0 \end{bmatrix}$

$$= \begin{bmatrix} 0+1+0 & 0+0+0 & 0+1+0 \\ 0+0+0 & 1+0+1 & 0+0+0 \\ 0+1+0 & 0+0+0 & 0+1+0 \end{bmatrix}$$

$$= \begin{bmatrix} 1 & 0 & 1 \\ 0 & 2 & 0 \\ 1 & 0 & 1 \end{bmatrix}$$

Can we attach any meaning to this product? Note that the 1 in the first row, first column means that the United States can communicate with the United States one way (via Russia). The 0 in the first row, second column means that the United States cannot communicate with Russia through an intermediary. The 1 in the first row, third column means that the United States can communicate with Cuba 1 way (via Russia). This explanation corresponds to the real-life possibilities. What do you think the 2 in row 2, column 2 means? It means that Russia can communicate with Russia via one intermediary in two ways (via United States and via Cuba).

Since this seems to make sense, we now look at the same product for the matrix $[A]$.

$$[A]^2 = \begin{bmatrix} 0 & 1 & 0 & 1 \\ 1 & 0 & 1 & 1 \\ 0 & 1 & 0 & 0 \\ 1 & 1 & 0 & 0 \end{bmatrix} \begin{bmatrix} 0 & 1 & 0 & 1 \\ 1 & 0 & 1 & 1 \\ 0 & 1 & 0 & 0 \\ 1 & 1 & 0 & 0 \end{bmatrix}$$

$$= \begin{bmatrix} 0+1+0+1 & 0+0+0+1 & 0+1+0+0 & 0+1+0+0 \\ 0+0+0+1 & 1+0+1+1 & 0+0+0+0 & 1+0+0+0 \\ 0+1+0+0 & 0+0+0+0 & 0+1+0+0 & 0+1+0+0 \\ 0+1+0+0 & 1+0+0+0 & 0+1+0+0 & 1+1+0+0 \end{bmatrix}$$

$$= \begin{bmatrix} 2 & 1 & 1 & 1 \\ 1 & 3 & 0 & 1 \\ 1 & 0 & 1 & 1 \\ 1 & 1 & 1 & 2 \end{bmatrix}$$

Step 3: *Carry out the plan.* The plan could list all possibilities by considering one possibility at a time but instead we will use matrix multiplication to find $[A]^2$; we will then verify that this is the desired matrix. This means, for example, that Mexico can talk to Cuba one way (row 4, column 3). Mexico talks to Cuba using Russia as an intermediary.

Step 4: *Look back.* We see that the United States can communicate with Cuba (entry in the first row, third column) through an intermediary. The zeros indicate that Russia and Cuba cannot communicate through intermediaries (they can communicate only directly). Matrix $[A]^3$ tells us how many ways the countries can communicate if they use two intermediaries.

Algebraic Properties of Matrices

Properties for an algebra of matrices can also be developed. The $m \times n$ **zero matrix**, denoted by $[0]$, is the matrix with m rows and n columns in which each entry is 0. The **identity matrix** *for multiplication*, denoted by $[I_n]$, is the square matrix with n rows and n columns consisting of a 1 in each position on the **main diagonal** (entries $m_{11}, m_{22}, m_{33}, \ldots, m_{nn}$ and zeros elsewhere:

$$[I_2] = \begin{bmatrix} 1 & 0 \\ 0 & 1 \end{bmatrix}, \qquad [I_3] = \begin{bmatrix} 1 & 0 & 0 \\ 0 & 1 & 0 \\ 0 & 0 & 1 \end{bmatrix}, \qquad [I_4] = \begin{bmatrix} 1 & 0 & 0 & 0 \\ 0 & 1 & 0 & 0 \\ 0 & 0 & 1 & 0 \\ 0 & 0 & 0 & 1 \end{bmatrix}$$

☠ $[M]^{-1} \neq \dfrac{1}{[M]}$ ☠

Notice that identity matrices are zero-one matrices. The **additive inverse** of a matrix $[M]$ is denoted by $[-M]$ and is defined by $(-1)[M]$; the **multiplicative inverse** of a matrix $[M]$ is denoted by $[M]^{-1}$ if it exists. Table 8.2 summarizes the properties of matrices. Assume that $[M]$, $[N]$, and $[P]$ all have order $n \times n$, which forces them to be conformable for the given operations. If the context makes the order of the identity obvious, we sometimes write $[I]$ to denote the identity matrix; that is, we assume that $[I]$ means $[I_n]$.

TABLE 8.2	Properties of Matrices	
Property	**Addition**	**Multiplication**
Commutative	$[M] + [N] = [N] + [M]$	$[M][N] \neq [N][M]$
Associative	$([M] + [N] + [P]) = [M] + ([N] + [P])$	$([M][N])[P] = [M]([N][P])$
Identity	$[M] + [0] = [0] + [M]$	$[I][M] = [M][I] = [M]$
Inverse	$[M] + [-M] = [-M] + [M] = [0]$	$[M][M]^{-1} = [M]^{-1}[M] = [I]$
Distributive	$[M]([N] + [P]) = [M][N] + [M][P]$	
Distributive	$([N] + [P])[M] = [N][M] + [P][M]$	

Inverse Property

The property from this list that is particularly important for us in solving systems of equations is the inverse property. There are two unanswered questions about the inverse property. Given a *square* matrix $[M]$, when does $[M]^{-1}$ exist? And if it exists, how do you find it?

> **INVERSE OF A MATRIX**
>
> If $[A]$ is a square matrix and if there exists a matrix $[A]^{-1}$ such that
>
> $$[A]^{-1}[A] = [A][A]^{-1} = [I]$$
>
> where $[I]$ is the identity matrix for multiplication, then $[A]^{-1}$ is called the **inverse** of $[A]$ for multiplication.

Usually, in the context of matrices, when we talk simply of the inverse of $[A]$ we mean the inverse of $[A]$ for multiplication, and when we talk of the identity matrix, we mean the identity matrix for multiplication.

EXAMPLE 5   Showing that given matrices are inverses

Show that [A] and [B] are inverses, given:

a. $[A] = \begin{bmatrix} 2 & 1 \\ 3 & 2 \end{bmatrix}$; $[B] = \begin{bmatrix} 2 & -1 \\ -3 & 2 \end{bmatrix}$

b. $[A] = \begin{bmatrix} 0 & 1 & 2 \\ -1 & 1 & 2 \\ 1 & -2 & -5 \end{bmatrix}$; $[B] = \begin{bmatrix} 1 & -1 & 0 \\ 3 & 2 & 2 \\ -1 & -1 & -1 \end{bmatrix}$

Solution

a. We must show that $[A][B] = [I]$ and $[B][A] = [I]$.

$$[A][B] = \begin{bmatrix} 2 & 1 \\ 3 & 2 \end{bmatrix} \begin{bmatrix} 2 & -1 \\ -3 & 2 \end{bmatrix} = \begin{bmatrix} 4-3 & -2+2 \\ 6-6 & -3+4 \end{bmatrix} = \begin{bmatrix} 1 & 0 \\ 0 & 1 \end{bmatrix} = [I]$$

$$[B][A] = \begin{bmatrix} 2 & -1 \\ -3 & 2 \end{bmatrix} \begin{bmatrix} 2 & 1 \\ 3 & 2 \end{bmatrix} = \begin{bmatrix} 4-3 & 2-2 \\ -6+6 & -3+4 \end{bmatrix} = \begin{bmatrix} 1 & 0 \\ 0 & 1 \end{bmatrix} = [I]$$

Since $[A][B] = [B][A] = [I]$, we see that $[B] = [A]^{-1}$.

b. $[A][B] = \begin{bmatrix} 0 & 1 & 2 \\ -1 & 1 & 2 \\ 1 & -2 & -5 \end{bmatrix} \begin{bmatrix} 1 & -1 & 0 \\ 3 & 2 & 2 \\ -1 & -1 & -1 \end{bmatrix}$

$$= \begin{bmatrix} 0+3-2 & 0+2-2 & 0+2-2 \\ -1+3-2 & 1+2-2 & 0+2-2 \\ 1-6+5 & -1-4+5 & 0-4+5 \end{bmatrix}$$

$$= \begin{bmatrix} 1 & 0 & 0 \\ 0 & 1 & 0 \\ 0 & 0 & 1 \end{bmatrix} = [I]$$

$[B][A] = \begin{bmatrix} 1 & -1 & 0 \\ 3 & 2 & 2 \\ -1 & -1 & -1 \end{bmatrix} \begin{bmatrix} 0 & 1 & 2 \\ -1 & 1 & 2 \\ 1 & -2 & -5 \end{bmatrix}$

$$= \begin{bmatrix} 0+1+0 & 1-1+0 & 2-2+0 \\ 0-2+2 & 3+2-4 & 6+4-10 \\ 0+1-1 & -1-1+2 & -2-2+5 \end{bmatrix}$$

$$= \begin{bmatrix} 1 & 0 & 0 \\ 0 & 1 & 0 \\ 0 & 0 & 1 \end{bmatrix} = [I]$$

Since $[A][B] = [I] = [B][A]$, then $[B] = [A]^{-1}$. ∎

If a given matrix has an inverse, we say it is **nonsingular**. The unanswered question, however, is how to *find* an inverse matrix.

If you have a calculator that does matrix operations, use a MATRIX function and then press [A]⁻¹ to find the inverse of a matrix [A]. We will show this step in the margin as we find the inverse matrix in the next two examples.

EXAMPLE 6   Inverse of a 2 × 2 matrix

Find the inverse of $[A] = \begin{bmatrix} 1 & 2 \\ 1 & 4 \end{bmatrix}$.

Solution

Find a matrix $[B]$, if it exists, so that $[A][B] = [I]$; since we do not know $[B]$, let its entries be variables: $[B] = \begin{bmatrix} x_1 & x_2 \\ y_1 & y_2 \end{bmatrix}$. Then,

$$[A][B] = [I]$$

$$\begin{bmatrix} 1 & 2 \\ 1 & 4 \end{bmatrix} \begin{bmatrix} x_1 & x_2 \\ y_1 & y_2 \end{bmatrix} = \begin{bmatrix} 1 & 0 \\ 0 & 1 \end{bmatrix}$$

$$\begin{bmatrix} x_1 + 2y_1 & x_2 + 2y_2 \\ x_1 + 4y_1 & x_2 + 4y_2 \end{bmatrix} = \begin{bmatrix} 1 & 0 \\ 0 & 1 \end{bmatrix}$$

By definition of equality of matrices, we see that

$$\begin{cases} x_1 + 2y_1 = 1 \\ x_1 + 4y_1 = 0 \end{cases} \quad \text{and} \quad \begin{cases} x_2 + 2y_2 = 0 \\ x_2 + 4y_2 = 1 \end{cases}$$

Solve each of these systems simultaneously to find $x_1 = 2$, $x_2 = -1$, $y_1 = -\frac{1}{2}$, $y_2 = \frac{1}{2}$. Thus, the inverse is $[B] = \begin{bmatrix} x_1 & x_2 \\ y_1 & y_2 \end{bmatrix} = \begin{bmatrix} 2 & -1 \\ -\dfrac{1}{2} & \dfrac{1}{2} \end{bmatrix}$

Input MATRIX and then press [A]⁻¹ to obtain the output:

$$\begin{bmatrix} 2 & -1 \\ -0.5 & 0.5 \end{bmatrix}$$

EXAMPLE 7   Inverse of a 3 × 3 matrix

Find the inverse for $[A] = \begin{bmatrix} 1 & -1 & 0 \\ 3 & 2 & 2 \\ -1 & -1 & -1 \end{bmatrix}$.

Solution

We need to find a matrix $\begin{bmatrix} x_1 & x_2 & x_3 \\ y_1 & y_2 & y_3 \\ z_1 & z_2 & z_3 \end{bmatrix}$ so that

$$\begin{bmatrix} 1 & -1 & 0 \\ 3 & 2 & 2 \\ -1 & -1 & -1 \end{bmatrix} \begin{bmatrix} x_1 & x_2 & x_3 \\ y_1 & y_2 & y_3 \\ z_1 & z_2 & z_3 \end{bmatrix} = \begin{bmatrix} 1 & 0 & 0 \\ 0 & 1 & 0 \\ 0 & 0 & 1 \end{bmatrix}$$

The definition of equality of matrices gives rise to three systems of equations.

$$\begin{cases} x_1 - y_1 + 0z_1 = 1 \\ 3x_1 + 2y_1 + 2z_1 = 0 \\ -x_1 - y_1 - z_1 = 0 \end{cases} \quad \begin{cases} x_2 - y_2 + 0z_2 = 0 \\ 3x_2 + 2y_2 + 2z_2 = 1 \\ -x_2 - y_2 - z_2 = 0 \end{cases} \quad \begin{cases} x_3 - y_3 + 0z_3 = 0 \\ 3x_3 + 2y_3 + 2z_3 = 0 \\ -x_3 - y_3 - z_3 = 1 \end{cases}$$

We could solve these as three separate systems using Gauss–Jordan elimination; however, all the steps would be identical because the coefficients are the same in each system. Therefore, suppose we augment the matrix of the coefficients by the *three* columns of constants and do all three at once. Write the augmented matrix as $[B]$ $[A \mid I]$.

$$[B] = \left[\begin{array}{ccc|ccc} 1 & -1 & 0 & 1 & 0 & 0 \\ 3 & 2 & 2 & 0 & 1 & 0 \\ -1 & -1 & -1 & 0 & 0 & 1 \end{array}\right] \quad \underset{\substack{*\text{Row}+(-3,[B],1,2) \\ *\text{Row}+(1,[\text{Ans}],1,3)}}{\longrightarrow} \quad \left[\begin{array}{ccc|ccc} 1 & -1 & 0 & 1 & 0 & 0 \\ 0 & 5 & 2 & -3 & 1 & 0 \\ 0 & -2 & -1 & 1 & 0 & 1 \end{array}\right] \quad \underset{*\text{Row}+(2,[\text{Ans}],3,2)}{\longrightarrow}$$

$$\left[\begin{array}{ccc|ccc} 1 & -1 & 0 & 1 & 0 & 0 \\ 0 & 1 & 0 & -1 & 1 & 2 \\ 0 & -2 & -1 & 1 & 0 & 1 \end{array}\right] \quad \underset{\substack{*\text{Row}+(1,[\text{Ans}],2,1) \\ *\text{Row}+(2,[\text{Ans}],2,3)}}{\longrightarrow} \quad \left[\begin{array}{ccc|ccc} 1 & 0 & 0 & 0 & 1 & 2 \\ 0 & 1 & 0 & -1 & 1 & 2 \\ 0 & 0 & -1 & -1 & 2 & 5 \end{array}\right] \quad \underset{*\text{Row}(-1,[\text{Ans}],3)}{\longrightarrow}$$

$$\left[\begin{array}{ccc|ccc} 1 & 0 & 0 & 0 & 1 & 2 \\ 0 & 1 & 0 & -1 & 1 & 2 \\ 0 & 0 & 1 & 1 & -2 & -5 \end{array}\right]$$

Now, if we relate this to the original three systems, we see that the inverse matrix is

$$\begin{bmatrix} 0 & 1 & 2 \\ -1 & 1 & 2 \\ 1 & -2 & -5 \end{bmatrix}$$

■ COMPUTATIONAL WINDOW ▭□✕

Use the **MATRIX** function and input $[A]$. Then press $[A]^{-1}$ to find:

$$\begin{bmatrix} 0 & 1 & 2 \\ -1 & 1 & 2 \\ 1 & -2 & -5 \end{bmatrix}$$

By studying Example 7, we are led to a procedure for finding the inverse of a nonsingular matrix.

INVERSE OF A MATRIX

Follow this procedure to finding the inverse of a *square* matrix A:

Step 1 Augment [A] with [I]; that is, write [A | I], where [I] is the identity matrix of the same order as [A].

Step 2 Perform elementary row operations using Gauss–Jordan elimination to change the matrix [A] into the identity matrix [I], if possible.

Step 3 If at any time you obtain all zeros in a row or column to the left of the dividing line, then there will be no inverse.

Step 4 If steps 1 and 2 can be performed, the result in the augmented part is the inverse of [A].

EXAMPLE 8   Inverse may not exist

Find the inverse, if possible, of $[A] = \begin{bmatrix} 1 & 2 \\ 0 & 0 \end{bmatrix}$.

Solution

Write the augmented matrix [A | I]:

$$\left[\begin{array}{cc|cc} 1 & 2 & 1 & 0 \\ 0 & 0 & 0 & 1 \end{array}\right]$$

We want to make the left-hand side look like the corresponding identity matrix. This is impossible because there are no elementary row operations that will put it into the required form. Thus, there is no inverse. If you use a calculator to try to find the inverse of a **singular** matrix (one that does not have an inverse), you will obtain an error message.

EXAMPLE 9   Inverse of a 2 × 2 matrix using Gauss–Jordan

Use matrix methods to find the inverse of $[A] = \begin{bmatrix} 1 & 2 \\ 1 & 4 \end{bmatrix}$. Compare with Example 6.

Solution

Begin with the augmented matrix [B] = [A | I].

$$\left[\begin{array}{cc|cc} 1 & 2 & 1 & 0 \\ 1 & 4 & 0 & 1 \end{array}\right] \rightarrow \left[\begin{array}{cc|cc} 1 & 2 & 1 & 0 \\ 0 & 2 & -1 & 1 \end{array}\right] \rightarrow \left[\begin{array}{cc|cc} 1 & 2 & 1 & 0 \\ 0 & 1 & -\frac{1}{2} & \frac{1}{2} \end{array}\right] \rightarrow \left[\begin{array}{cc|cc} 1 & 0 & 2 & -1 \\ 0 & 1 & -\frac{1}{2} & \frac{1}{2} \end{array}\right]$$

The inverse is on the right of the dividing line: $\begin{bmatrix} 2 & -1 \\ -\frac{1}{2} & \frac{1}{2} \end{bmatrix}$; this result agrees with Example 6.

When a matrix has fractional entries, it is often rewritten with a fractional coefficient and integer entries to simplify the arithmetic. For example, we could rewrite this inverse matrix as

$$[A]^{-1} = \frac{1}{2}\begin{bmatrix} 4 & -2 \\ -1 & 1 \end{bmatrix}$$

EXAMPLE 10 Inverse of a 3 × 3 matrix using Gauss–Jordan

Find the inverse, if possible, of $[A] = \begin{bmatrix} 0 & 1 & 2 \\ 2 & -1 & 1 \\ -1 & 1 & 0 \end{bmatrix}$.

Solution

Write the augmented matrix $[\ A\ |\ I\ ]$ and make the left-hand side look like the corresponding identity matrix (if possible):

$$\begin{bmatrix} 0 & 1 & 2 & | & 1 & 0 & 0 \\ 2 & -1 & 1 & | & 0 & 1 & 0 \\ -1 & 1 & 0 & | & 0 & 0 & 1 \end{bmatrix} \rightarrow \begin{bmatrix} -1 & 1 & 0 & | & 0 & 0 & 1 \\ 2 & -1 & 1 & | & 0 & 1 & 0 \\ 0 & 1 & 2 & | & 1 & 0 & 0 \end{bmatrix} \rightarrow \begin{bmatrix} 1 & -1 & 0 & | & 0 & 0 & -1 \\ 2 & -1 & 1 & | & 0 & 1 & 0 \\ 0 & 1 & 2 & | & 1 & 0 & 0 \end{bmatrix}$$

$$\rightarrow \begin{bmatrix} 1 & -1 & 0 & | & 0 & 0 & -1 \\ 0 & 1 & 1 & | & 0 & 1 & 2 \\ 0 & 1 & 2 & | & 1 & 0 & 0 \end{bmatrix} \rightarrow \begin{bmatrix} 1 & 0 & 1 & | & 0 & 1 & 1 \\ 0 & 1 & 1 & | & 0 & 1 & 2 \\ 0 & 0 & 1 & | & 1 & -1 & -2 \end{bmatrix} \rightarrow \begin{bmatrix} 1 & 0 & 0 & | & -1 & 2 & 3 \\ 0 & 1 & 0 & | & -1 & 2 & 4 \\ 0 & 0 & 1 & | & 1 & -1 & -2 \end{bmatrix}$$

Thus, $[A]^{-1} = \begin{bmatrix} -1 & 2 & 3 \\ -1 & 2 & 4 \\ 1 & -1 & -2 \end{bmatrix}$ ∎

Historical Note

© Karl Smith Library

James Sylvester (1814–1897)

A second mathematician responsible for much of what we study in this section is James Sylvester. He was a contemporary and good friend of Cayley (see Historical Note on page 560). Cayley would try out many of his ideas on Sylvester, and Sylvester would react and comment on Cayley's work. Sylvester originated many of the ideas that Cayley used and even suggested the term *matrix* to Cayley. He founded the *American Journal of Mathematics* while he taught in the United States at Johns Hopkins University (from 1877 to 1883).

Systems of Equations

In Example 3, we saw how a system of equations can be written in matrix form. We can now see how to solve a system of linear equations by using the inverse. Consider a system of n linear equations with n unknowns whose matrix of coefficients [A] has an inverse $[A]^{-1}$:

$$\begin{aligned} [A][X] &= [B] && \text{Given system} \\ [A]^{-1}[A][X] &= [A]^{-1}[B] && \text{Multiply both sides by } [A]^{-1}. \\ ([A]^{-1}[A])[X] &= [A]^{-1}[B] && \text{Associative property} \\ [I][X] &= [A]^{-1}[B] && \text{Inverse property} \\ [X] &= [A]^{-1}[B] && \text{Identity property} \end{aligned}$$

> **» IN OTHER WORDS** To solve a system of equations, look to see whether the number of equations is the same as the number of unknowns. If so, find the inverse of the matrix of the coefficients (if it exists) and multiply it (on the right) by the matrix of the constants.

EXAMPLE 11 Solving a system using the inverse matrix method

Solve: $\begin{cases} y + 2z = 0 \\ 2x - y + z = -1 \\ y - x = 1 \end{cases}$

Solution

Write in matrix form: $[A] = \begin{bmatrix} 0 & 1 & 2 \\ 2 & -1 & 1 \\ -1 & 1 & 0 \end{bmatrix}$, $[X] = \begin{bmatrix} x \\ y \\ z \end{bmatrix}$ $[B] = \begin{bmatrix} 0 \\ -1 \\ 1 \end{bmatrix}$.

From Example 10, $[A]^{-1} = \begin{bmatrix} -1 & 2 & 3 \\ -1 & 2 & 4 \\ 1 & -1 & -2 \end{bmatrix}$. Thus,

$$[X] = [A]^{-1}[B] = \begin{bmatrix} -1 & 2 & 3 \\ -1 & 2 & 4 \\ 1 & -1 & -2 \end{bmatrix} \begin{bmatrix} 0 \\ -1 \\ 1 \end{bmatrix} = \begin{bmatrix} 1 \\ 2 \\ -1 \end{bmatrix}$$

Therefore, the solution to the system is $(x, y, z) = (1, 2, -1)$.

COMPUTATIONAL WINDOW   ▬□✕

The calculator, of course, is what really makes this inverse method worthwhile. It makes the solution of any system *that has the same number of equations and variables* almost trivial. For this example, simply enter $[A]$ and $[B]$ and then press [A]⁻¹[B] . The output is:

$$\begin{bmatrix} 1 \\ 2 \\ -1 \end{bmatrix}$$

That is all there is! Does this make the method worthwhile?

The method of solving a system by using the inverse matrix is efficient if you know the inverse. Unfortunately, *finding* the inverse for one system is usually more work than using another method to solve the system. However, certain applications yield the same system over and over, and the only thing to change is the constants. In this case, you should think about the inverse method. And, finally, computers and calculators can find approximations for inverse matrices quite easily, so this method becomes the method of choice if you have access to this technology.

PROBLEM SET 8.6

LEVEL 1

WHAT IS WRONG, *if anything, with each statement in Problems 1–6?*
Explain your reasoning.

1. $[M]$ is singular if it has an inverse.
2. For conformable matrices $[A]$ and $[B]$, $[A] + [B] = [B] + [A]$.
3. For conformable matrices $[A]$ and $[B]$, $[A][B] = [B][A]$.
4. For conformable matrices,

$$([A] + [B])^2 = [A]^2 + [B]^2$$

5. $[A]^{-1} = \dfrac{1}{[A]}x$, provided $[A]^{-1}$ exists.

6. If $[A][X] = [B]$, then $[X] = [B][A]^{-1}$.

In Problems 7–14, find the indicated matrices if possible.

$$[A] = \begin{bmatrix} 1 & 2 \\ 4 & 0 \\ -1 & 3 \\ 2 & 1 \end{bmatrix} \qquad [B] = \begin{bmatrix} 4 & 2 \\ -1 & 3 \end{bmatrix} \qquad [C] = \begin{bmatrix} 1 & 0 & 0 & 0 \\ 0 & 1 & 0 & 0 \\ 0 & 0 & 1 & 0 \\ 0 & 0 & 0 & 1 \end{bmatrix}$$

$$[D] = \begin{bmatrix} 4 & 1 & 3 & 6 \\ -1 & 0 & -2 & 3 \end{bmatrix} \qquad [E] = \begin{bmatrix} 1 & 0 & 2 \\ 3 & -1 & 2 \\ 4 & 1 & 0 \end{bmatrix}$$

$$[F] = \begin{bmatrix} 1 & 4 & 0 \\ 3 & -1 & 2 \\ -2 & 1 & 5 \end{bmatrix} \qquad [G] = \begin{bmatrix} 8 & 1 & 6 \\ 3 & 5 & 7 \\ 4 & 9 & 2 \end{bmatrix}$$

7. a. $[E] + [F]$		**b.** $2[E] - [G]$
8. a. $[E][F]$		**b.** $[E][G]$
9. a. $[F][G]$		**b.** $[G][F]$
10. a. $[A][B]$		**b.** $[B][D]$
11. a. $[B]^2$		**b.** $[C]^3$
12. a. $[E]([F] + [G])$		**b.** $[E][F] + [E][G]$
13. a. $([B] + [C])[A]$		**b.** $[B][A] + [C][A]$
14. a. $([E][F])[G]$		**b.** $[E]([F][G])$

Write $[A][X] = [B]$, *if possible, for the matrices given in Problems 15–16.*

15. $[A] = \begin{bmatrix} 1 & 2 & 4 \\ -3 & 2 & 1 \\ 2 & 0 & 1 \end{bmatrix}$,

$[X] = \begin{bmatrix} x \\ y \\ z \end{bmatrix}$, and $[B] = \begin{bmatrix} 13 \\ 11 \\ 0 \end{bmatrix}$.

16. $[A] = \begin{bmatrix} 4 & 1 & 0 \\ 3 & -1 & 2 \\ 2 & 3 & 1 \end{bmatrix}$,

$[X] = \begin{bmatrix} x \\ y \\ z \end{bmatrix}$, and $[B] = \begin{bmatrix} 2 \\ 11 \\ -1 \end{bmatrix}$.

Show that the given matrices are inverses in Problems 17–18.

17. $[A] = \begin{bmatrix} 2 & 7 \\ 1 & 4 \end{bmatrix}$; $[B] = \begin{bmatrix} 4 & -7 \\ -1 & 2 \end{bmatrix}$

18. $[A] = \begin{bmatrix} -16 & -2 & 7 \\ 7 & 1 & -3 \\ -3 & 0 & 1 \end{bmatrix}$;

$[B] = \begin{bmatrix} 1 & 2 & -1 \\ 2 & 5 & 1 \\ 3 & 6 & -2 \end{bmatrix}$

LEVEL 2

Find the inverse of each matrix in Problems 19–24, if it exists.

19. $\begin{bmatrix} 4 & -7 \\ -1 & 2 \end{bmatrix}$

20. $\begin{bmatrix} 8 & 6 \\ -2 & 4 \end{bmatrix}$

21. $\begin{bmatrix} 1 & 0 & 2 \\ 2 & 1 & 0 \\ 0 & -2 & 9 \end{bmatrix}$

22. $\begin{bmatrix} 6 & 1 & 20 \\ 1 & -1 & 0 \\ 0 & 1 & 3 \end{bmatrix}$

23. $\begin{bmatrix} 1 & 0 & 0 & 1 \\ 0 & 2 & 0 & 0 \\ 0 & 0 & 0 & 1 \\ 2 & 0 & 1 & 0 \end{bmatrix}$

24. $\begin{bmatrix} 0 & 1 & 2 & 0 \\ 0 & 0 & 0 & 1 \\ 1 & 1 & 3 & 0 \\ 2 & 4 & 0 & 0 \end{bmatrix}$

Solve the systems in Problems 25–52 by solving the corresponding matrix equation with an inverse, if possible.

Problems 25–30 use the inverse found in Problem 19.

25. $\begin{cases} 4x - 7y = -2 \\ -x + 2y = 1 \end{cases}$

26. $\begin{cases} 4x - 7y = -65 \\ -x + 2y = 18 \end{cases}$

27. $\begin{cases} 4x - 7y = 48 \\ -x + 2y = -13 \end{cases}$

28. $\begin{cases} 4x - 7y = 2 \\ -x + 2y = 3 \end{cases}$

29. $\begin{cases} 4x - 7y = 5 \\ -x + 2y = 4 \end{cases}$

30. $\begin{cases} 4x - 7y = -3 \\ -x + 2y = 8 \end{cases}$

Problems 31–36 use the inverse found in Problem 20.

31. $\begin{cases} 8x + 6y = 12 \\ -2x + 4y = -14 \end{cases}$

32. $\begin{cases} 8x + 6y = 16 \\ -2x + 4y = 18 \end{cases}$

33. $\begin{cases} 8x + 6y = -6 \\ -2x + 4y = -26 \end{cases}$

34. $\begin{cases} 8x + 6y = -28 \\ -2x + 4y = 18 \end{cases}$

35. $\begin{cases} 8x + 6y = -26 \\ -2x + 4y = 12 \end{cases}$

36. $\begin{cases} 8x + 6y = -36 \\ -2x + 4y = -2 \end{cases}$

Problems 37–42 all use the same inverse.

37. $\begin{cases} 2x + 3y = 9 \\ x - 6y = -3 \end{cases}$

38. $\begin{cases} 2x + 3y = 2 \\ x - 6y = 16 \end{cases}$

39. $\begin{cases} 2x + 3y = 2 \\ x - 6y = -14 \end{cases}$

40. $\begin{cases} 2x + 3y = 9 \\ x - 6y = 42 \end{cases}$

41. $\begin{cases} 2x + 3y = -22 \\ x - 6y = 49 \end{cases}$

42. $\begin{cases} 2x + 3y = 12 \\ x - 6y = -24 \end{cases}$

Problems 43–48 use the inverse found in Problem 21.

43. $\begin{cases} x + 2z = 7 \\ 2x + y = 16 \\ -2y + 9z = -3 \end{cases}$

44. $\begin{cases} x + 2z = 4 \\ 2x + y = 0 \\ -2y + 9z = 19 \end{cases}$

45. $\begin{cases} x + 2z = 4 \\ 2x + y = 0 \\ -2y + 9z = 31 \end{cases}$

46. $\begin{cases} x + 2z = 7 \\ 2x + y = 1 \\ -2y + 9z = 28 \end{cases}$

47. $\begin{cases} x + 2z = 12 \\ 2x + y = 0 \\ -2y + 9z = 10 \end{cases}$

48. $\begin{cases} x + 2z = 5 \\ 2x + y = 8 \\ -2y + 9z = 9 \end{cases}$

Problems 49–52 use the inverse found in Problem 22.

49. $\begin{cases} 6x + y + 20z = 27 \\ x - y = 0 \\ y + 3z = 4 \end{cases}$

50. $\begin{cases} 6x + y + 20z = 14 \\ x - y = 1 \\ y + 3z = 1 \end{cases}$

51. $\begin{cases} 6x + y + 20z = -12 \\ x - y = 6 \\ y + 3z = -7 \end{cases}$

52. $\begin{cases} 6x + y + 20z = 57 \\ x - y = 1 \\ y + 3z = 5 \end{cases}$

LEVEL 3

53. If we take the first few rows of Pascal's triangle and arrange them into a lower triangular matrix, we form what is called a *Pascal matrix*:

$$[P] = \begin{bmatrix} 1 & 0 & 0 & 0 & 0 \\ 1 & 1 & 0 & 0 & 0 \\ 1 & 2 & 1 & 0 & 0 \\ 1 & 3 & 3 & 1 & 0 \\ 1 & 4 & 6 & 4 & 1 \end{bmatrix}$$

   a. Find the inverse of this matrix.
   b. Make a general statement about the inverse of any order Pascal's matrix.

54. If we add Pascal's matrix (Problem 53) and the identity matrix, we find $[P] + [I]$. For the order shown in Problem 53,

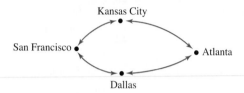

$$[P] + [I] = \begin{bmatrix} 2 & 0 & 0 & 0 & 0 \\ 1 & 2 & 0 & 0 & 0 \\ 1 & 2 & 2 & 0 & 0 \\ 1 & 3 & 3 & 2 & 0 \\ 1 & 4 & 6 & 4 & 2 \end{bmatrix}$$

Find the inverse of this matrix.

55. How many channels of communication are open to each country in Example 4 if the countries are willing to speak to each other through two intermediaries?

56. Consider the map of airline routes shown in Figure 8.21.

Kansas City

San Francisco

Atlanta

Dallas

Figure 8.21 Travel routes for four cities

a. Fill in the blanks in the following zero-one matrix representing the airline routes:

$$[T] = \begin{array}{c} \\ SF \\ D \\ A \\ KC \end{array} \begin{array}{cccc} SF & D & A & KC \\ \left[\begin{array}{cccc} & & & \\ & & & \\ & & & \\ & & & \end{array} \right] \end{array}$$

b. Write a matrix showing the number of routes among these cities if you make exactly one intermediate stop. Use this matrix to state in how many ways you can travel from Kansas City to San Francisco making exactly one intermediate stop.

c. Write a matrix showing the number of routes among these cities if you make exactly two intermediate stops. Use this matrix to state in how many ways you can travel from San Francisco to Kansas City making two intermediate stops.

57. The Seedy Vin Company produces Riesling, Charbono, and Rosé wines. There are three procedures for producing each wine (the procedures for production affect the cost of the final product). One procedure allows an outside company to bottle the wine; a second allows the wine to be produced and bottled at the winery; a third allows the wine to be estate bottled. The amount of wine produced by the Seedy Vin Company by each method is shown by the matrix

$$[W] = \begin{array}{c} \\ \\ \\ \end{array} \begin{array}{ccc} \text{Riesling} & \text{Charbono} & \text{Rosé} \\ \left[\begin{array}{ccc} 2 & 1 & 3 \\ 4 & 3 & 6 \\ 1 & 2 & 4 \end{array} \right] & \begin{array}{l} \text{outside bottling} \\ \text{produced and bottled at winery} \\ \text{estate bottled} \end{array} \end{array}$$

Suppose the cost for each method of production is given by the matrix

$$[C] = [\text{outside} \quad \text{winery} \quad \text{estate}]$$
$$= [\ 1 \quad\quad 4 \quad\quad 6\ ]$$

Suppose the production cost of a unit of each type of wine is given by the matrix

$$[D] = \begin{bmatrix} 40 \\ 60 \\ 30 \end{bmatrix} \begin{array}{l} \text{Riesling} \\ \text{Charbono} \\ \text{Rosé} \end{array}$$

a. Find the cost of producing each of the three types of wine.
   Hint: This is $[C][W]$.
b. Find the dollar amount for producing each unit of the different types of wine for these methods of production.
   Hint: This is $[W][D]$.
c. Find $([C][W])[D]$ and $[C]([W][D])$. Give a verbal interpretation for these products.

58. If $[M]$ and $[N]$ are nonsingular, then $[M][N]$ is nonsingular. Show that the inverse of $[M][N]$ is

$$([M][N])^{-1} = [N]^{-1}[M]^{-1}$$

59. If $[M]$ is nonsingular, show that if $[M][N] = [M][P]$, then $[N] = [P]$.
60. Prove that if $[M]$ is nonsingular and $[M][N] = [M]$, then $[N] = [I_3]$ for square matrices of order 3.

8.7 Systems of Inequalities

In previous sections, we have discussed the simultaneous solution of a system of equations. In this section, we discuss the graphical solution of a simultaneous **system of inequalities**. The solution of a *system of inequalities* refers to the intersection of the solutions of the individual inequalities in the system. The procedure for graphing the solution of a system of inequalities is to graph the solution of the individual inequalities and then shade the intersection of all the graphs. We begin with a quick review of graphing half-planes (which you probably did in a previous course).

Half-planes

Once you know how to graph a line, you can also graph linear inequalities. Linear functions divide a plane into three parts, as shown in Figure 8.22.

The two parts labeled I and II are called **half-planes**. The third part is called the **boundary** and is the line separating the half-planes. The solution of a first-degree inequality in two unknowns is the

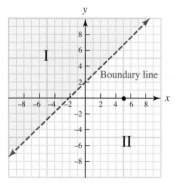

Figure 8.22 Half-planes

set of all ordered pairs that satisfy the given inequality. This solution set is a half-plane. The following table offers some examples of first-degree inequalities with two unknowns, along with some associated terminology.

Example	Inequality Symbol	Boundary Included	Terminology
$3x - y > 5$	$>$	no	**open half-plane**
$3x - y < 5$	$<$	no	open half-plane
$3x - y \geq 5$	$\geq$	yes	**closed half-plane**
$3x - y \leq 5$	$\leq$	yes	closed half-plane

We can now summarize the procedure for graphing a first-degree inequality in two unknowns.

PROCEDURE FOR GRAPHING A LINEAR INEQUALITY

To graph a linear inequality:

Step 1 **Graph the boundary.**

Replace the inequality symbol with an equality symbol and draw the resulting line. This is the boundary line. Use a solid line when the boundary is included ($\leq$ or $\geq$) and a dashed line when the boundary is not included ($<$ or $>$).

Step 2 **Test a point.**

Choose any point in the plane, called a **test point**, that is not on the boundary line; the point $(0, 0)$ is usually the simplest choice.

If this test point makes the *inequality* true, shade in the half-plane that contains the test point. That is, the shaded plane is the solution set.[*]

If the test point makes the *inequality* false, shade in the other half-plane for the solution.

This process may sound complicated, but if you know how to draw lines from equations, you will not find this difficult.

EXAMPLE 1   Half-plane with boundary included

Graph $3x - y \geq 5$.

Solution

Note that the inequality symbol is $\geq$, so the boundary is included.

Step 1 Graph the boundary; draw the (solid) line corresponding to

$$3x - y = 5 \qquad \text{Replace the inequality symbol with an equality symbol.}$$
$$y = 3x - 5$$

The y-intercept is -5 and the slope is 3; the boundary line is shown in Figure 8.23.

Step 2 Choose a test point; we choose $(0, 0)$. Plot $(0, 0)$ in Figure 8.23 and note that it lies in one of the half-planes determined by the boundary line. We now check this test point with the given *inequality*:

$$3x - y \geq 5$$
$$3(0) - (0) \geq 5 \qquad \text{You can usually test this in your head.}$$
$$0 \geq 5 \qquad \text{This is false.}$$

Therefore, shade the half-plane that does *not* contain $(0, 0)$, as shown in Figure 8.23. ∎

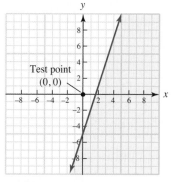

Figure 8.23 Graph of $3x - y \geq 5$

---

[*]A highlighter pen does a nice job of shading in your work.

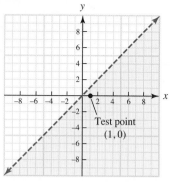

Figure 8.24 Graph of $y < x$

EXAMPLE 2   Half-plane with boundary not included

Graph $y < x$.

Solution

Note that the inequality symbol is $<$, so the boundary line is not included.

Step 1 Draw the (dashed) boundary line, $y = x$, as shown in Figure 8.24.

Step 2 Choose a test point. We can't pick $(0, 0)$ because $(0, 0)$ is on the boundary line. Since we must choose some point not on the boundary, we choose $(1, 0)$:

$$y < x$$
$$0 < 1 \qquad \text{This is true.}$$

Therefore, shade the half-plane that contains the test point, as shown in Figure 8.24. ∎

Graphing Systems of Inequalities

EXAMPLE 3   Graphical solution of a system of inequalities

Solve: $\begin{cases} y > -20x + 110 \\ x \geq 0 \\ y \geq 0 \\ x \leq 8 \end{cases}$

Solution

Graph each half-plane, but instead of shading, mark the appropriate regions with arrows. For this example, the two inequalities

$$x \geq 0$$
$$y \geq 0$$

give the first quadrant, which is marked with arrows as shown in Figure 8.25a. Next, graph $y = -20x + 110$ and use a test point, say, $(0, 0)$, to determine the half-plane. Finally, draw $x = 8$ and mark it with arrows to show that $x \leq 8$. The last step is to shade the intersection, as shown in Figure 8.25b.

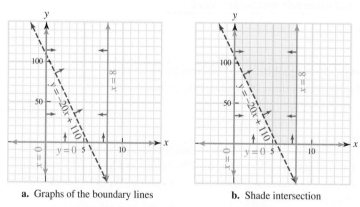

a. Graphs of the boundary lines   **b.** Shade intersection

Figure 8.25 Graph of a system of inequalities

∎

EXAMPLE 4   Graphical solution of a system of linear inequalities

Graph the solution of the system: $\begin{cases} 2x + y \leq 3 \\ x - y > 5 \\ x \geq 0 \\ y \geq -10 \end{cases}$

Solution
The graph of the individual inequalities and their intersection is shown in Figure 8.26.

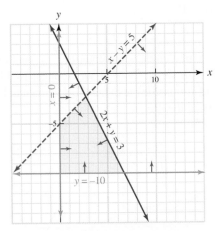

Figure 8.26 System of inequalities

Note the use of arrows to show the solutions of the individual inequalities. This device replaces the use of a lot of shading, which can be confusing if there are many inequalities in the system. In this book, we show the intersection (solution) in color. You can show the intersection in your work in a variety of different ways, but a color highlighter works well. ■

Nonlinear Systems

Nonlinear systems can be solved graphically using the same technique—namely, looking at the intersection of the individual solutions of each inequality. You might wish to review graphing quadratic inequalities in Section 1.9.

EXAMPLE 5   Graphical solution of a nonlinear system

Sketch the graph: $\begin{cases} y - 1 \ge (x - 3)^2 \\ x - 1 > \dfrac{1}{2}(y - 5)^2 \end{cases}$

Solution
As with linear inequalities, first sketch the associated equations. Second, determine the individual graphs and their intersections. The solution is the intersection of the shaded portions (the color portion in Figure 8.27).

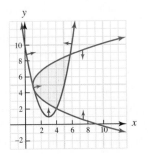

Figure 8.27 System of inequalities

■

EXAMPLE 6   Graphical solution of a nonlinear system

Sketch the graph: $\begin{cases} (x-2)^2 + (y+1)^2 \leq 9 \\ x - 3 < -(y+1)^2 \end{cases}$

Solution

The first inequality is the interior of a circle with center $(2, -1)$ and radius 3. The boundary of the second inequality is a parabola with vertex $(3, -1)$, and it opens to the left. The intersection is the set of points inside the circle and also inside the parabola, as shown in Figure 8.28. Check by choosing some points within and outside the region.

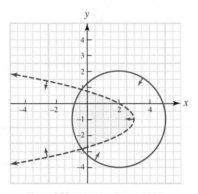

Figure 8.28 System of inequalities

Systems of inequalities are sometimes used in applied problems to define a set of **feasible values** from which other values satisfying certain conditions, called **constraints**, can be chosen. We illustrate with the following example.

EXAMPLE 7   Feasible values in crop production   **MODELING APPLICATION**

A farmer has 100 acres on which to plant two crops, corn and wheat, and the problem is to maximize the profit. There are several considerations; the first is the expense:

	Cost per acre	
Expenses	**Corn**	**Wheat**
Seed	$12	$40
Fertilizer	$58	$80
Planting/care/harvesting	$50	$90
Total	$120	$210

After the harvest, the farmer must store the crops while awaiting proper market conditions. Each acre yields an average of 110 bushels of corn or 30 bushels of wheat. The limitations of resources are as follows:

Available capital: $15,000
Available storage facilities: 4,000 bushels
Net profit for wheat is $2.00/bu.
Net profit for corn is $1.30/bu.

Graph the set of feasible values satisfying these conditions.

Solution

Step 1: *Understand the problem.* First, you might try to solve this problem by using your intuition.

Plant 100 acres in wheat

Production:	100 acres @ 30 bu/acre = 3,000 bushels
Net profit:	3,000 bu × $2.00/bu = $6,000
Costs:	100 acres @ $210/acre = $21,000
	These costs are more than the $15,000 available, so the farmer cannot plant 100 acres in wheat.

Plant 100 acres in corn

Production:	100 acres @ 110 bu/acre = 11,000 bushels
Net profit:	11,000 bu × $1.30/bu = $14,300
Costs:	100 acres @ $120/acre = $12,000
	These costs can be met with the available $15,000, but the production of 11,000 bushels exceeds the available storage capacity of 4,000 bu.

Clearly, some mix of wheat and corn is necessary. Let us build a mathematical model.

Step 2: *Devise a plan.* Let $x =$ number of acres to be planted in corn; $y =$ number of acres to be planted in wheat. With a mathematical model, we must make certain assumptions. For this example, we assume:

$x \geq 0$	The number of acres of corn cannot be negative.	☠ These first two assumptions (constraints) will apply in almost every linear programming model. ☠
$y \geq 0$	The number of acres of wheat cannot be negative.	
$x + y \leq 100$	The amount of available land is 100 acres. We do not assume that $x + y = 100$ because it might be more profitable to leave some land unplanted.	
EXPENSES $\leq$ 15,000	The total expenses cannot exceed $15,000. We also know:	

$$\text{EXPENSES} = \text{EXPENSE FOR CORN} + \text{EXPENSE FOR WHEAT}$$
$$= 120x + 210y$$

Thus, this constraint (in terms of x and y) is

$$120x + 210y \leq 15,000$$

TOTAL YIELD $\leq$ 4,000   The total yield cannot exceed the storage capacity of 4,000 bushels. We also know:

$$\text{TOTAL YIELD} = \text{YIELD FOR CORN} + \text{YIELD FOR WHEAT}$$
$$= 110x + 30y$$

Thus, this constraint (in terms of x and y) is

$$110x + 30y \leq 4,000$$

We have defined the following set of feasible values.

$$\begin{cases} x \geq 0 \\ y \geq 0 \\ x + y \leq 100 \\ 120x + 210y \leq 15,000 \\ 110x + 30y \leq 4,000 \end{cases}$$

Step 3: *Carry out the plan.* We graph the constraints as shown in Figure 8.29.

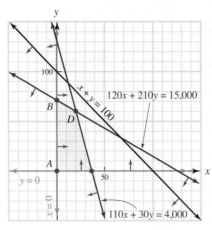

Figure 8.29 Farmer problem

Step 4: *Look back.* Notice from Figure 8.29 that some of the constraints could be eliminated from the problem and everything else would remain unchanged. For example, the boundary $x + y = 100$ was not necessary in finding the maximum value of P. Such a condition is said to be a **superfluous constraint**.

PROBLEM SET 8.7

LEVEL 1

WHAT IS WRONG, *if anything, with each statement in Problems 1–4?* *Explain your reasoning.*

1. The linear inequality $2x + 5y < 2$ does not have a boundary line because the inequality is $<$.
2. A good test point for the linear inequality $y \geq x$ is $(0, 0)$.
3. The test point $(-2, 4)$ satisfies the inequality $5x + 2y \leq 9$.
4. The test point $(-2, 4)$ satisfies the inequality $4x < 3y$.

Graph the first-degree inequalities in two unknowns in Problems 5–22.

5. $y \leq 2x + 1$ 6. $y \geq 5x - 3$
7. $y > 5x - 3$ 8. $y \geq -2x + 3$
9. $y > 3x - 3$ 10. $y < -2x + 5$
11. $x \geq y$ 12. $y > x$
13. $3x \leq 2y$ 14. $2x < 3y$
15. $y \geq 0$ 16. $x \leq 0$
17. $x - 3y \geq 9$ 18. $6x - 2y < 0$
19. $3x + 2y > 1$ 20. $2x - 3y < 1$
21. $x + 2y + 4 \leq 0$ 22. $2x - 3y + 6 \leq 0$

Graph the solution of each system given in Problems 23–38.

23. **a.** $\begin{cases} x \geq 0 \\ y \leq 0 \end{cases}$ **b.** $\begin{cases} x \geq 0 \\ y \geq 0 \end{cases}$

24. **a.** $\begin{cases} x \leq 0 \\ y \leq 0 \end{cases}$ **b.** $\begin{cases} x \leq 0 \\ y \geq 0 \end{cases}$

25. $\begin{cases} x \geq 0 \\ y \geq 0 \\ x < 5 \\ y \leq 6 \end{cases}$ 26. $\begin{cases} x \geq 0 \\ y \geq 0 \\ x < 500 \\ y \leq 1,000 \end{cases}$

27. $\begin{cases} -10 < x \\ x < 6 \\ 3 \leq 2y \\ y \leq 4 \end{cases}$ 28. $\begin{cases} -4 \leq x \\ -2 \geq x \\ -5 \leq y \\ 9 \geq y \end{cases}$

29. $\begin{cases} y \geq \frac{3}{4}x - 4 \\ y \leq -\frac{3}{4}x + 11 \\ x \geq 6 \end{cases}$ 30. $\begin{cases} y \geq \frac{3}{2}x + 3 \\ y \leq \frac{3}{2}x + 6 \\ 3 \leq y \leq 6 \end{cases}$

31. $\begin{cases} x \geq 0 \\ y \geq 0 \\ x + y \leq 9 \\ 2x - 3y \geq -6 \\ x - y \leq 3 \end{cases}$ 32. $\begin{cases} x \geq 0 \\ y \geq 0 \\ x + y \leq 8 \\ y \leq 4 \\ x \leq 6 \end{cases}$

33. $\begin{cases} 5x + 2y \leq 30 \\ 5x + 2y \geq 20 \\ x \geq 0 \\ y \geq 0 \end{cases}$ 34. $\begin{cases} 5x - 2y + 30 \geq 0 \\ 5x - 2y + 20 \leq 0 \\ x \leq 0 \\ y \geq 0 \end{cases}$

35. $\begin{cases} 2x + y \leq 8 \\ y \leq 5 \\ x - y \leq 2 \\ 3x - y \geq 5 \end{cases}$ 36. $\begin{cases} 2x + 3y \leq 30 \\ 3x + 2y \geq 20 \\ x \geq 0 \\ y \geq 0 \end{cases}$

37. $\begin{cases} 2x - 3y + 30 \geq 0 \\ 3x - 2y + 20 \leq 0 \\ x \leq 0 \\ y \geq 0 \end{cases}$ 38. $\begin{cases} x + y - 10 \leq 0 \\ x + y + 4 \geq 0 \\ x - y \leq 6 \\ y - x \leq 4 \end{cases}$

LEVEL 2

Graph the solution of each system given in Problems 39–50.

39. $\begin{cases} y \le -(x-2)^2 - 3 \\ y \ge -x - 4 \end{cases}$ **40.** $\begin{cases} y \ge \frac{1}{2}(x-3)^2 \\ y \le x + 1 \end{cases}$

41. $\begin{cases} y \ge x^2 \\ 5x - 4y + 26 \ge 0 \end{cases}$ **42.** $\begin{cases} y \le 4 - x^2 \\ y \ge \frac{1}{2}x \end{cases}$

43. $\begin{cases} x^2 + y^2 \le 25 \\ y \ge x \end{cases}$ **44.** $\begin{cases} x^2 + y^2 \ge 16 \\ 4y \le x \end{cases}$

45. $\begin{cases} x^2 + y^2 \le 36 \\ |x| > 2 \end{cases}$ **46.** $\begin{cases} x^2 + y^2 \le 25 \\ |y| < 2 \end{cases}$

47. $\begin{cases} x^2 + y^2 \le 16 \\ x^2 + y^2 + 2y \ge 8 \end{cases}$ **48.** $\begin{cases} x^2 + y^2 \le 16 \\ x^2 + y^2 - 6x \ge 16 \end{cases}$

49. $\begin{cases} -1 \le x - y \le 1 \\ y < -x^2 + 3 \end{cases}$ **50.** $\begin{cases} y < \frac{1}{2}(x-2)^2 \\ -1 \le x < 4 \end{cases}$

51. Suppose the farmer in Example 7 contracted to have the grain stored at a neighboring farm, and the contract calls for at least 4,000 bushels to be stored. Sketch the set of feasible solutions if the other conditions in Example 2 remain the same.

52. Your broker tells you of two investments she thinks are worthwhile. She advises that you buy a new issue of Pertec stock, which should yield 20% over the next year, and then to balance your account, she advises Campbell Municipal Bonds with a 10% yearly yield. The bond-to-stock ratio should not be greater than 3 to 1. If you have no more than $100,000 to invest and do not want to invest more than $70,000 in Pertec or less than $20,000 in bonds, sketch the set of feasible solutions.

53. A convalescent hospital wishes to provide, at a minimum cost, a diet that has a minimum of 200 g of carbohydrates, more than 100 g of protein, and more than 20 g of fat per day. These requirements can be met with two foods, A and B:

Food	Carbohydrates	Protein	Fats
A	10 g	2 g	3 g
B	5 g	5 g	4 g

If food A costs $0.29 per gram and food B costs $0.15 per gram, how many grams of each food should be purchased for each patient per day to meet the minimum requirements at the lowest cost?

54. The following carbohydrate information is given on the side of the respective cereal boxes (for 1 oz of cereal with $\frac{1}{2}$ cup of whole milk):

Cereal	Starch/carbohydrates	Sucrose/sugar
Kellogg's Corn Flakes	23 g	7 g
Post Honeycombs	14 g	15 g

Sketch the set of feasible solutions if at least 366 g starch and >119 g sucrose are consumed by these two cereals.

LEVEL 3

Graph the solution of each system given in Problems 55–60.

55. $\begin{cases} (x-4)^2 + (y-2)^2 \ge 25 \\ (x-2)^2 + (y-6)^2 \le 16 \end{cases}$ **56.** $\begin{cases} \dfrac{x^2}{4} + \dfrac{y^2}{9} \le 1 \\ \dfrac{x^2}{9} + \dfrac{y^2}{4} \ge 1 \end{cases}$

57. $\begin{cases} x^2 + y^2 - 12 \le 6x + 4y \\ x^2 + y^2 + 2x \le 11 - 4y \end{cases}$ **58.** $\begin{cases} x^2 + y^2 + 4x \le 6y + 3 \\ x + y^2 - 6y + 7 \le 0 \end{cases}$

59. $\begin{cases} x^2 - 4y - 8 < 4x - y^2 \\ x + y \ge 1 \end{cases}$ **60.** $\begin{cases} y^2 - x + 4y < -4 \\ 4x^2 - 12x - 8y + 1 \le 0 \end{cases}$

CHAPTER 8 SUMMARY AND REVIEW

*M*athematics is loved by many, disliked by a few, admired and respected by all. Because of their immense power and reliability, mathematical methods inspire confidence in persons who comprehend them and awe in those who do not.

Hollis R. Cooley (1963)

Take some time getting ready to work the review problems in this section. First, look back at the definition and property boxes. You will maximize your understanding of this chapter by working the problems in this section only after you have studied the material.

SELF TEST   *All of the answers for this self test are given in the back of the book.*

1. For each given sequence,
   i. Classify as arithmetic, geometric, both, or neither.
   ii. If arithmetic, give d; if geometric, give r; if neither, state a pattern using your own words.
   iii. Supply the next term.
   iv. State the general term (if possible).
   a. $6, 11, 16, \ldots$
   b. $35, 46, 57, \ldots$
   c. $1, \frac{1}{2}, \frac{1}{4}, \ldots$
   d. $4, 6, 10, 16, \ldots$
   e. $3, -1, \frac{1}{3}, -\frac{1}{9}, \ldots$

2. i. State the first four terms of the given sequence.
   ii. Either find the limit or show that the sequence diverges.

   a. $\left\{\dfrac{e^n}{n!}\right\}$

   b. $\left\{\dfrac{3n^2 - n + 1}{(1 - 2n)n}\right\}$

   c. $\left\{\left(1 + \dfrac{1}{n}\right)^n\right\}$

3. a. Find the sum of the first ten terms of the sequence
   $$a + b, \ a^2 + ab, \ a^3 + a^2b, \ldots$$

   b. Find the sum of the first ten terms of the sequence
   $$P, \ P(1+r), \ P(1+r)^2, \ P(1+r)^3, \ldots$$

   c. Find the sum of the infinite series $2{,}000$
   $$1{,}000 + 500 + 250 + \cdots$$

   d. Find the sum of the infinite series
   $$\frac{1}{8} + \frac{1}{16} + \frac{1}{32} + \cdots$$

4. Evaluate:

   a. $\displaystyle\sum_{k=0}^{5} 3^k$

   b. $\displaystyle\sum_{k=1}^{8} 3$

   c. $\displaystyle\sum_{k=1}^{10} 2(3)^{k-1}$

   d. $\displaystyle\sum_{k=1}^{100} [5 + (k-1)4]$

Solve the systems in Problems 5–9 by the indicated method.

5. By graphing: $\begin{cases} 2x - y = 2 \\ 3x - 2y = 1 \end{cases}$

6. By addition: $\begin{cases} x + 3y = 3 \\ 4x - 6y = -6 \end{cases}$

7. By substitution: $\begin{cases} x + y = 1 \\ x^2 + y^2 = 13 \end{cases}$

8. By Gauss–Jordan: $\begin{cases} 2x + y = 1 \\ x + z = 1 \\ 2x - y = 1 \end{cases}$

9. By using an inverse matrix: $\begin{cases} 2x + 7y = 13 \\ x + 4y = 8 \end{cases}$

10. Graph the solution of the system: $\begin{cases} 2x - y + 2 \le 0 \\ 2x - y + 12 \ge 0 \\ 3x + 2y + 10 > 0 \\ 3x + 2y - 18 < 0 \end{cases}$

STOP STUDY HINTS *Compare your solutions and answers to the self test.*

Simplify *the expressions in Problems 1–4.*

1. $\dfrac{(3 - x^2)(4) - (4x - 7)(-2x)}{(3 - x^2)^2}$

2. $\dfrac{(2x^2 + x - 3)(6x) - (3x^2 + 5)(4x + 1)}{(2x^2 + x - 3)^2}$

3. $\dfrac{\frac{1}{2}t^{-1/2}\cos t - t^{1/2}(-\sin t)}{\cos^2 t}$

4. $\dfrac{1}{4}\left(\dfrac{x}{1 - 3x}\right)^{-3/4}\left[\dfrac{(1 - 3x)(1) - x(-3)}{(1 - 3x)^2}\right]$

Factor *the expressions in Problems 5–8.*

5. $-24x^{-4}y^4 + 32x^{-3}y^3$
6. $2(2x + 1)(5x^2 + 4) + (2x + 1)^2(10x)$
7. $16\tan^3 x - 2$
8. $4\cos^2 x \sin x + 4\cos x \sin x - 3\sin x$

Evaluate *the expressions in Problems 9–12.*

9. $f(x) = \dfrac{x - 1}{x + 1}$ for $x = 0.999$

10. $|x^2 - \pi^2|$ for $x = \pi$

11. $\displaystyle\lim_{n \to \infty}\left(\dfrac{2n^2 + 5n - 7}{n^3}\right)$

12. $\dfrac{8}{\sqrt{3}}\left[\dfrac{u(u^2 + 1)^{3/2}}{4} - \dfrac{u\sqrt{u^2 + 1}}{8}\right]$ for $u = \sqrt{3x}$

Solve *the equations in Problems 13–20.*

13. $\dfrac{x(9x^2 - 2)}{\sqrt{3x^2 - 1}} = 0$

14. $\sqrt{2x - 1} + \sqrt{x + 4} = 6$

15. $2\sin^2 x - 2\cos^2 x = 1$

16. $I = \frac{E}{K}(1 - e^{-Rt/L})$ for L

17. a. $\left(\frac{1}{2}\right)^x = \frac{1}{8}$ b. $\left(\frac{1}{2}\right)^x = 8$

18. a. $8^x = 300$ b. $2^x = 1{,}000$

19. $6^{3x+2} = 200$

20. $\ln 2 + \frac{1}{2}\ln 3 + 4\ln 5 = \ln x$

In Problems 39–44 of the first cumulative review (page 153) we asked for the y-components of the points P, Q, and R for an arbitrary function f as shown in Figure 8.30.

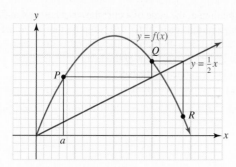

Figure 8.30 Finding coordinates

In this course, we have considered many types of functions, in particular those listed in Problems 21–27. Find the coordinates P or Q for the given functions. **Graph** *the given function showing the points P, Q, and R as well as the line $y = \frac{1}{2}x$.*

21. Point P for the trigonometric function: $f(x) = 2\cos\frac{x}{\pi}, 1 < a < 1.5$.

22. Point P for the cubic function: $f(x) = 2 - 0.2(x-3)^3, 1 < a < 2$.

23. Point Q for the rational function: $f(x) = \frac{x-3}{x+2}, -8 < a < -6$.

24. Point Q for the exponential function: $f(x) = 1 + e^{-x}, -2 < a < -1$.

25. Point Q for the quadratic function: $f(x) = 3x - x^2, 0 < a < 0.5$.

26. Point Q for the radical function: $f(x) = \sqrt{x+2}, -2 < a < -1$.

27. Point Q for the logarithmic function: $f(x) = \log x^2, -1.5 < a < -1.0$

In Problems 28–33, find $\dfrac{f(x+h) - f(x)}{h}$.

28. $f(x) = (2x-1)^2$

29. $f(x) = \dfrac{1}{x+1}$

30. $f(x) = \sqrt{3x}$

31. $f(x) = \sin x$

32. $f(x) = \ln x$

33. $f(x) = e^{2x}$

Prove that each of the equations given in Problems 34–38 is an identity.

34. $\cos^4\theta - \sin^4\theta = \cos 2\theta$

35. $\sec\left(\frac{\pi}{2} - \theta\right) = \csc\theta$

36. $\sec(-\theta) = \sec\theta$

37. $\dfrac{\tan(\alpha+\beta) - \tan\beta}{1 + \tan(\alpha+\beta)\tan\beta} = \tan\alpha$

38. $\cos(\alpha-\beta)\cos\beta - \sin(\alpha-\beta)\sin\beta = \cos\alpha$

Solve the systems in Problems 39–46.

39. a. $\begin{cases} 2x + y = 1 \\ 3x + 2y = 4 \end{cases}$
b. $\begin{cases} \dfrac{2}{w} + \dfrac{1}{z} = 1 \\ \dfrac{3}{w} + \dfrac{2}{z} = 4 \end{cases}$

40. a. $\begin{cases} 2x + y = 1 \\ 5x + 2y = 1 \end{cases}$
b. $\begin{cases} \dfrac{2}{w} + \dfrac{1}{z} = 1 \\ \dfrac{1}{w} + \dfrac{1}{z} = 1 \end{cases}$

41. a. $\begin{cases} 2x + 3y = 1 \\ 4x - 6y = -5 \end{cases}$
b. $\begin{cases} 2\sqrt{w} + 3\sqrt{z} = 1 \\ 4\sqrt{w} - 6\sqrt{z} = -5 \end{cases}$

42. a. $\begin{cases} 5x + 3y = 5 \\ 4x - 6y = 2 \end{cases}$
b. $\begin{cases} 5\sqrt{w} + 3\sqrt{z} = 5 \\ 4\sqrt{w} - 6\sqrt{z} = 2 \end{cases}$

43. $\begin{cases} 3x - 2y + z = 13 \\ x - 5y + 2z = 24 \\ 2x + 3y - z = -11 \end{cases}$

44. $\begin{cases} 2x - 2y + 3z = 7 \\ x - 3y + 2z = 13 \\ 3x - 5y + 5z = 20 \end{cases}$

45. $\begin{cases} x^2 + y^2 = 3 \\ x^2 - y^2 = 1 \end{cases}$

46. $\begin{cases} 2x + y = 1 \\ w + y - 3z = 0 \\ x + y - z = 0 \\ w + x - z = -1 \end{cases}$

Graph the solution of each system of inequalities in Problems 47–50.

47. $\begin{cases} 2x + 7y \geq 420 \\ 2x + 2y \leq 500 \\ x > 50 \\ y < 100 \end{cases}$

48. $\begin{cases} 2x + 3y \geq 600 \\ x + 2y \geq 360 \\ x - 2y \leq 480 \\ x \geq 0 \\ y \geq 0 \end{cases}$

49. $\begin{cases} -2^x \leq y \leq 2^x \\ -2 \leq x \leq 2 \end{cases}$

50. $\begin{cases} -0.5 \leq \sin x \leq 0.5 \\ -1 \leq x \leq 5 \end{cases}$

51. If water at temperature B is surrounded by air at temperature A, it will gradually cool so that the temperature T, t minutes later, is given by

$$T = A + (B - A)e^{-kt}$$

This is Newton's law of cooling. Assume you draw water for a bath, and the water temperature is $120°F$ and the room temperature is $75°F$. Draw a graph showing the cooling of the water for the first half-hour if we assume $k = 0.08$.

52. Find constants a and b that guarantee that the graph of the function defined by

$$f(x) = \frac{ax + 5}{3 - bx}$$

will have a vertical asymptote at $x = 5$ and a horizontal asymptote at $y = -3$.

53. Radium has a half-life of approximately 1,600 years so that the decay rate is -0.00043322. If we start with 100 milligrams (mg) of radium, use the growth/decay formula $A = A_0 e^{rt}$ to draw a graph representing the amount A left after t years.

54. Santa Rosa and The Nut Tree are approximately 60 miles apart. A pilot flying from Santa Rosa finds that the plane is $15.0°$ off course after 20 miles. How far is the plane from The Nut Tree at this time?

55. Mr. T, whose eye level is 6 ft, is standing on a hill with an inclination of $5°$ and needs to throw a rope to the top of a nearby building. If the angle of elevation to the top of the building is $20°$ and the angle of depression to the base of the building is $11°$, how tall is the building?

56. The lengths of two sides of an isosceles triangle are 12 ft, and the angle between those two sides is θ with α the measure of each of the base angles.

a. Express the area of the triangle as a function of $\alpha/2$.

b. Express the area of the triangle as a function of α.

57. If $g_1, g_2, g_3, \ldots$ is a geometric sequence with common ratio $r > 0$ and $g_1 > 0$, show that the sequence

$$\log g_1, \log g_2, \log g_3, \ldots$$

is an arithmetic sequence and find the common difference.

58. If $a_1, a_2, a_3 \ldots$ is an arithmetic sequence with common difference $d > 0$ and $a_1 > 0$, show that the sequence

$$10^{a_1}, 10^{a_2}, 10^{a_3}, \ldots$$

is a geometric sequence, and find the common ratio.

59. In Example 6, Section 5.2, we modeled projectile motion with the parametric equations

$$x = (v_0 \cos\theta)t, \qquad y = h_0 + (v_0 \sin\theta)t - 16t^2$$

A boy standing at the edge of a cliff throws a ball upward at a 30° angle with an initial speed of 64 ft/s. Suppose that when the ball leaves the boy's hand, it is 48 ft above the ground at the base of the cliff.

a. What is the time of flight of the ball?

b. What is the range for the ball?

60. Find the sum of the infinite series

$$[\tan^{-1} 1 - \tan^{-2} 2] + [\tan^{-1} 2 - \tan^{-1} 3] + \cdots [\tan^{-1} n - \tan^{-1}(n+1)] + \cdots$$

CHAPTER 8 Group Research Projects

Working in small groups is typical of most work environments, and this book seeks to develop skills with group activities. At the end of each chapter, we present a list of suggested projects, and even though they could be done as individual projects, we suggest that these projects be done in groups of three or four students.

G8.1 Joan Ross and Marc Ross tell a story that, at a recent meeting of the Michigan Council of Teachers of Mathematics, someone pointed out that the U.S. budget is so large that if all the trees in the United States were cut down and the wood made into dollar bills, the total value would not add up to the 2007 U.S. budget. Write a paper to refute or substantiate this claim.

G8.2 Use a calculator to find the inverse of a matrix to solve a system of equations has only recently been possible for the majority of students. Do you see this as a help or a hindrance? Write a 500-word essay on the *specific* use of calculators to help you with the material in this course. From your viewpoint, discuss the advantages and disadvantages of incorporating calculators (in particular, graphing calculators and calculators that handle matrix operations) into the *learning* of mathematics.

G8.3 Square *ABCD* has sides of length a. Square *EFGH* is formed by connecting the midpoints of the sides of the first square, as shown in Figure 8.31. Assume that the pattern of shaded regions of the square is continued indefinitely. Write the total area of the shaded regions as a function of a.

Figure 8.31 Area of shaded regions

G8.4 What forms in nature represent mathematical formulas and patterns? The spiral is a key to understanding organic nature. The book *The Curves of Life* by Theodore Andrea Cook (New York: Dover Publications, 1979), states, "It may be said that with very few exceptions the spiral formation is intimately connected with the phenomena of life and growth. When it is found in inorganic phenomena the logarithmic spiral is again connected with those forms of energy which are most closely compared with the energy we describe as life and growth, such, for instance, as the mathematical definition of electrical phenomena, or the spiral nebulae of the astronomer." There are five (and only five) regular solids in mathematics: the cube, tetrahedron, octahedron, icosahedron, and dodecahedron. All of these solids can be found in nature. Figure 8.32 shows skeletons of marine animals called *radiolaria*. These are only some of the examples of mathematics in nature.

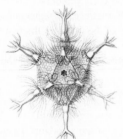

Figure 8.32 Radiolaria as examples of the regular solids in nature

Prepare a paper or an exhibit showing the five regular solids (see Figure 8.32) or describing spiral shells, crystals, ellipsoidal stones or eggs, spiraling sunflower seed pods, branch distributions illustrating Fibonacci series, snowflake patterns, body bones acting as levers, or the ratio of food consumed to body size.

G8.5 Suppose that we fit a band tightly around the earth at the equator. Now suppose that you remove it and cut it at one place, then splice in an additional piece 10 ft long. If you now replace it on the equator, it would fit more loosely, would it not? Assuming that the slack is uniform all the way around the equator, make a guess that the band would be loose enough to:

A. Walk under?

B. Crawl under?

C. Slip your hand under?

D. Slip a piece of paper under?

E. Not even enough to get the sheet of paper under?

Write a paper based on this problem.

G8.6 Journal Problem (*Function*, Vol. 7, No. 3, 1983.) Solve for x:

$$\sqrt{x-1} - \sqrt{x - \sqrt{3x}} = \frac{1}{2}x$$

G8.7 Journal Problem (*Parabola*, Vol. 17, No. 1, 1981.) Given the polynomial

$$P(x) = p_0 x^n + p_1 x^{n-1} + \cdots + p_n$$

determine the coefficients p_k for each of the following cases:

a. $P(x)$ is linear, $P(0) = 15$, and the numerical value of $P(x)$ is 3 when $x = 2$.

b. $P(x)$ is quadratic, $P(0) = 32$ and $P(2^t) = 0$ has roots $t = 1$ and $t = 3$.

G8.8 Journal Problem (*Function*, April 1980, Vol. 4, No., problem by Ravi Sidhu.) Give the meaning and value of the continued fraction

$$1 + \cfrac{1}{1 + \cfrac{1}{1 + \cfrac{1}{1 + \cfrac{1}{\ddots}}}}$$

Write the continued fraction using functional notation.

G8.9 Journal Problem (*Greece*, 1984.) Find the range of the function

$$f(x) = \frac{\sqrt{x^2 + 1} + x - 1}{\sqrt{x^2 + 1} + x + 1}$$

and show that f is an odd function.

G8.10 Journal Problem (*Quantum*, January/February 1996; problem by Andrey Yegorov.)

Solve the equation

$$5x^2 - 2xy + 2y^2 - 2x - 2y + 1 = 0$$

Chapter Objectives

The material in this chapter is previewed in the following list of objectives. After completing this chapter, review this list again, and then complete the self test.

9.1
9.1 Define and sketch vectors.

9.2 Carry out vector operations.

9.3 Find a unit vector with given conditions.

9.4 Find a force that is a resultant of other forces.

9.2
9.5 Be able to locate objects or points in three dimensions; be able to sketch coordinate planes.

9.6 Graph a plane in space. Write the equation of a plane.

9.7 Know the equation of a sphere; be able to find the center and radius of a sphere by completing the square.

9.8 Define and sketch cylinders.

9.9 Define and sketch quadric surfaces.

9.10 Recognize a quadric surface by looking at the equation.

9.3 9.11 Find the standard representation of a vector in $\mathbb{R}^3$.

9.12 Find the magnitude of a vector.

9.13 Find a unit vector in a given direction.

9.14 Find a vector that is parallel to a given vector.

9.15 Define dot product; be able to find the dot product of vectors.

9.16 Find the angle between two vectors.

9.17 Know the orthogonal vector property.

9.18 Find both the scalar projection and the vector projection of a vector **v** onto a vector **w**.

9.19 Define work; find the amount of work as a dot product.

9.4 9.20 Define cross product; be able to find the cross product of vectors.

9.21 State and use the geometric property of cross product.

9.22 Use the right-hand rule.

9.23 Find the magnitude of a cross product.

9.24 Use triple scalar product to find the volume of a parallelepiped.

9.25 Use vectors to find the torque.

9.5 9.26 Find the parametric form of a line in $\mathbb{R}^3$.

9.27 Find the symmetric form of a line in $\mathbb{R}^3$.

9.28 Determine whether given lines are intersecting, parallel, or skew.

9.29 Find the direction cosines of a vector.

9.30 Find the equation of a plane; know the relationship between normal vectors and planes.

9.31 Find the equation of a line orthogonal to a given plane.

9.32 Find the equation of a plane containing three given points.

9.33 Find the equation of a line parallel to the intersection of two given planes.

9.34 Find the equation of a plane containing two intersecting lines.

Vectors and Solid Analytic Geometry

9

I *have absolutely no idea of what paleontology is; and if someone would spend an hour with me, or an hour a day for a week, or an hour a day for a year, teaching it to me, my soul would be richer. In that sense, I was doing the same thing for my colleague, the paleontologist. I was telling him what mathematics is. That, I think is important. All educable human beings should know what mathematics is because their souls would grow by that. They would enjoy life more, they would understand life more, they would have greater insight. They should, in that sense, understand all human activity such as paleontology and mathematics.*

—Paul Halmos (1982)

Chapter Sections

9.1 Vectors in the Plane
Introduction
Vector Operations
Standard Representation of Vectors in the Plane

9.2 Quadric Surfaces and Graphing in Three Dimensions
Three-Dimensional Coordinate System
Graphs in $\mathbb{R}^3$
Quadric Surfaces

9.3 The Dot Product
Vectors in $\mathbb{R}^3$
Definition of Dot Product
Angle Between Vectors
Projections
Work as a Dot Product

9.4 The Cross Product
Definition of Cross Product
Geometric Interpretation of Cross Product
Properties of Cross Product
Torque

9.5 Lines and Planes in Space
Lines in $\mathbb{R}^3$
Planes in $\mathbb{R}^3$

Chapter 9 Summary and Review
Self Test
Practice for Calculus
Final Review—Set I
Final Review—Set II
Chapter 9 Group Research Projects

▶ CALCULUS PERSPECTIVE

Up to now, in this book, we have lived in the two-dimensional world called the Cartesian plane. However, it is necessary to move from that flat world into the three-dimensional world that we inhabit. Our primary tool for this transition is the notion of a vector. We will see in this chapter that vectors can be defined not only in two but also in three dimensions. Further work in mathematics will extend vectors into n dimensions. The last third of most calculus books concerns itself with vectors and vector operations. Many applications of mathematics involve quantities that have both magnitude and direction, such as force, velocity, acceleration, and momentum. Vectors are an important tool in mathematics, and in this section, we introduce you to the terminology and notation for vector representation.

9.1 Vectors in the Plane

Introduction

A **vector** in a plane can be thought of as a directed line segment, an "arrow" with **initial point** P and **terminal point** Q. The direction of the vector is that of the arrow, and its magnitude is represented by the arrow's length, as shown in Figure 9.1a. We shall indicate such a vector by writing **PQ** in boldface print, but in your work, you may write an arrow over the designated points: $\overrightarrow{PQ}$. The order of letters you write down is important: **PQ** means that the vector is from P to Q, but **QP** means that the vector is from Q to P. The first letter is always the initial point, and the second letter, the terminal point, for a vector always points to the right. We shall denote the magnitude (length) of a vector by $\| \mathbf{PQ} \|$. Two vectors are regarded as **equal** (or **equivalent**) if they have the same magnitude and the same direction, even if they do not coincide.

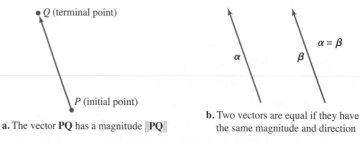

a. The vector **PQ** has a magnitude $\|\mathbf{PQ}\|$

b. Two vectors are equal if they have the same magnitude and direction

Figure 9.1 Vectors in a plane

A vector with magnitude 0 is called a **zero** (or **null**) **vector** and is denoted by **0**. The **0** vector has no specific direction, and we shall adopt the convention of assigning it any direction that may be convenient in a particular problem.

Figure 9.2 A concept car, the Vector Biturbo

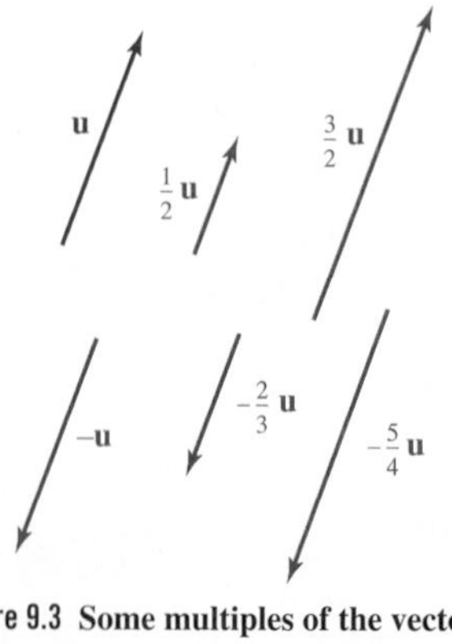

Vector Operations

If **v** is a vector other than **0**, then any vector **w** that is parallel to **v** is called a **scalar multiple** of **v** and satisfies $\mathbf{w} = s\mathbf{v}$ for some nonzero number s. A **scalar** quantity is one that has only magnitude and, in the context of vectors, is used to describe a real number. The scalar multiple $s\mathbf{v}$ has length $|s|$ times that of **v**; it points in the same direction as **v** if $s > 0$ and the opposite direction if $s < 0$ (see Figure 9.3). Notice that, for any distinct points P and Q, $\mathbf{PQ} = -\mathbf{QP}$. For the zero vector **0**, we define $s\mathbf{0} = \mathbf{0}$ for any scalar s.

Figure 9.3 Some multiples of the vector u

Physical experiments indicate that force and velocity vectors can be added (or **resolved**) according to a **triangular rule** displayed in Figure 9.4**a**, and we use this rule as our definition of vector addition. In particular, to add the vector **v** to the vector **u**, we place the end (initial point) of **v** at the tip (terminal point) of **u** and define the sum, also called the **resultant u + v**, to be the vector that extends from the initial point of **u** to the terminal point of **v**.

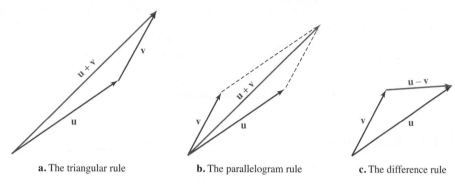

a. The triangular rule   **b.** The parallelogram rule   **c.** The difference rule

Figure 9.4 Adding and subtracting vectors

Equivalently, the **parallelogram rule** for **u + v** is the diagonal of the parallelogram formed with sides **u** and **v**, as shown in Figure 9.4**b**. The **difference u − v** is just the vector **w** that satisfies **v + w = u**, and it may be found by placing the initial points of **u** and **v** together and extending a vector from the terminal point of **v** to the terminal point of **u** (see Figure 9.4**c**).

A vector **OQ** with initial point at the origin O of a coordinate plane can be uniquely represented by specifying the coordinates of its terminal point Q. If Q has coordinates (a, b), we denote the vector **OQ** by $\langle a, b \rangle$ where the pointed brackets $\langle \ \ \rangle$ are used to distinguish the *vector* **OQ** $= \langle a, b \rangle$ from the *point* (a, b). We call this vector representation **component form**.

The vector **PQ** with initial point $P(c, d)$ and terminal point $Q(a, b)$ can be denoted in a similar fashion. Using analytic geometry (see Problem 56), it can be shown that **PQ** equals vector **OR** with initial point at the origin $(0, 0)$ and terminal point $R(a − c, b − d)$, as shown in Figure 9.5. Notice that **PQ** $= \langle a − c, b − d \rangle$.

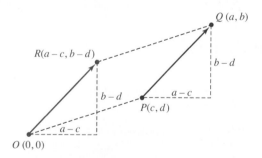

Figure 9.5 Representation of PQ

Vector operations are easily represented when vectors are given in component form. In particular, we have

$$\langle a_1, b_1 \rangle = \langle a_2, b_2 \rangle \qquad \text{If and only if } a_1 = a_2 \text{ and } b_1 = b_2$$
$$k\langle a, b \rangle = \langle ka, kb \rangle \qquad \text{For constant } k$$
$$\langle a, b \rangle + \langle c, d \rangle = \langle a + c, b + d \rangle$$
$$\langle a, b \rangle - \langle c, d \rangle = \langle a - c, b - d \rangle$$

These formulas may be verified by analytic geometry. For instance, the rule for multiplication by a scalar may be obtained by using the relationships in Figure 9.6**a**, and Figure 9.6**b** can be used to obtain the rule for vector addition.

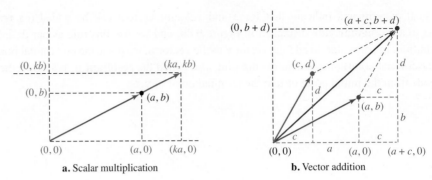

Figure 9.6 Vector operations

EXAMPLE 1   Vector operations

For the vectors $\mathbf{u} = \langle 2, -3 \rangle$ and $\mathbf{v} = \langle -1, 7 \rangle$, find

  a. $\mathbf{u} + \mathbf{v}$

  b. $\frac{3}{4}\mathbf{u}$

  c. $3\mathbf{u} - \frac{1}{2}\mathbf{v}$

Solution

  a. $\mathbf{u} + \mathbf{v} = \langle 2, -3 \rangle + \langle -1, 7 \rangle$
$$= \langle 2 + (-1), -3 + 7 \rangle$$
$$= \langle 1, 4 \rangle$$

  b. $\dfrac{3}{4}\mathbf{u} = \dfrac{3}{4}\langle 2, -3 \rangle$

$$= \left\langle \dfrac{3}{4}(2), \dfrac{3}{4}(-3) \right\rangle$$

$$= \left\langle \dfrac{3}{2}, \dfrac{-9}{4} \right\rangle$$

  c. $3\mathbf{u} - \dfrac{1}{2}\mathbf{v} = 3\langle 2, -3 \rangle - \dfrac{1}{2}\langle -1, 7 \rangle$

$$= \langle 6, -9 \rangle + \left\langle \dfrac{1}{2}, -\dfrac{7}{2} \right\rangle \quad \text{Scalar multiplication}$$

$$= \left\langle 6 + \dfrac{1}{2}, -9 - \dfrac{7}{2} \right\rangle \quad \text{Add vectors.}$$

$$= \left\langle \dfrac{13}{2}, -\dfrac{25}{2} \right\rangle \quad \text{Simplify.}$$

    In general, an expression of the form $a\mathbf{u} + b\mathbf{v}$ is called a **linear combination** of the vectors $\mathbf{u}$ and $\mathbf{v}$. Note that if $\mathbf{u} = \langle u_1, u_2 \rangle$ and $\mathbf{v} = \langle v_1, v_2 \rangle$, then

$$a\mathbf{u} + b\mathbf{v} = a\langle u_1, u_2 \rangle + b\langle v_1, v_2 \rangle = \langle au_1 + bv_1, au_2 + bv_2 \rangle$$

Vector addition and multiplication of a vector by a scalar behave very much like ordinary addition and multiplication (see Appendix A for a review of these properties of addition and multiplication in the set of real numbers).

VECTOR OPERATIONS

For any vectors **u**, **v**, and **w** in the plane and scalars s and t,

Commutativity of vector addition	$\mathbf{u} + \mathbf{v} = \mathbf{v} + \mathbf{u}$
Associativity of vector addition	$(\mathbf{u} + \mathbf{v}) + \mathbf{w} = \mathbf{u} + (\mathbf{v} + \mathbf{w})$
Associativity of scalar multiplication	$(st)\mathbf{u} = s(t\mathbf{u})$
Identity for addition	$\mathbf{u} + \mathbf{0} = \mathbf{u}$
Inverse property for addition	$\mathbf{u} + (-\mathbf{u}) = \mathbf{0}$
Vector distributivity	$(s + t)\mathbf{u} = s\mathbf{u} + t\mathbf{u}$
Scalar distributivity	$s(\mathbf{u} + \mathbf{v}) = s\mathbf{u} + s\mathbf{v}$

Each vector property can be established by using a corresponding property of real numbers (see Appendix A). For example, to prove associativity of vector addition, let $\mathbf{u} = \langle u_1, u_2 \rangle$, $\mathbf{v} = \langle v_1, v_2 \rangle$, and $\mathbf{w} = \langle w_1, w_2 \rangle$. Then

$$
\begin{aligned}
(\mathbf{u} + \mathbf{v}) + \mathbf{w} &= (\langle u_1, u_2 \rangle + \langle v_1, v_2 \rangle) + \langle w_1, w_2 \rangle \\
&= \langle u_1 + v_1, u_2 + v_2 \rangle + \langle w_1, w_2 \rangle \\
&= \langle (u_1 + v_1) + w_1, (u_2 + v_2) + w_2 \rangle \\
&= \langle u_1 + (v_1 + w_1), u_2 + (v_2 + w_2) \rangle \qquad \textit{Associativity of addition for the real numbers} \\
&= \langle u_1, u_2 \rangle + (\langle v_1, v_2 \rangle + \langle w_1, w_2 \rangle) \\
&= \mathbf{u} + (\mathbf{v} + \mathbf{w})
\end{aligned}
$$

You are asked to prove the other six properties in the problem set.

EXAMPLE 2   Vector proof of a geometric property

Show that the line segment joining the midpoints of two sides of a triangle is parallel to the third side and has half its length.

Solution

Consider $\triangle ABC$ and let P and Q be the midpoints of sides $\overline{AC}$ and $\overline{BC}$, respectively, as shown in Figure 9.7.

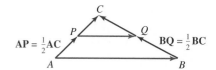

Figure 9.7 **Vector proof of a geometric property**

Given: $\mathbf{AP} = \frac{1}{2}\mathbf{AC}$ and $\mathbf{BQ} = \frac{1}{2}\mathbf{BC}$.
Prove: $\mathbf{PQ}$ is parallel to $\mathbf{AB}$ and $\| \mathbf{PQ} \| = \frac{1}{2}\| \mathbf{AB} \|$, which means that we must establish the vector equation $\mathbf{PQ} = \frac{1}{2}\mathbf{AB}$. Toward this end, we begin by noting that $\mathbf{AB}$ can be expressed as the following vector sum:

$$
\begin{aligned}
\mathbf{AB} &= \mathbf{AP} + \mathbf{PQ} + \mathbf{QB} \\
&= \frac{1}{2}\mathbf{AC} + \mathbf{PQ} - \mathbf{BQ} \qquad && \mathbf{AP} = \frac{1}{2}\mathbf{AC} \text{ and } \mathbf{QB} = -\mathbf{BQ} \\
&= \frac{1}{2}(\mathbf{AB} + \mathbf{BC}) + \mathbf{PQ} - \frac{1}{2}\mathbf{BC} \qquad && \mathbf{AC} = (\mathbf{AB} + \mathbf{BC}) \text{ and } \mathbf{BQ} = \frac{1}{2}\mathbf{BC} \\
&= \frac{1}{2}\mathbf{AB} + \frac{1}{2}\mathbf{BC} + \mathbf{PQ} - \frac{1}{2}\mathbf{BC} \\
&= \frac{1}{2}\mathbf{AB} + \mathbf{PQ} \\
\frac{1}{2}\mathbf{AB} &= \mathbf{PQ} \qquad && \textit{Subtract } \frac{1}{2}\mathbf{AB} \textit{ from both sides.}
\end{aligned}
$$

Therefore, the theorem is proved.

When a vector **u** is represented in component form $\mathbf{u} = \langle u_1, u_2 \rangle$, its length is given by the formula

$$\| \mathbf{u} \| = \sqrt{u_1^2 + u_2^2}$$

This is a simple application of the Pythagorean theorem as shown in Figure 9.8**a**. Another important relationship involving the length of vectors is the *triangle inequality*

$$\| \mathbf{u} + \mathbf{v} \| \leq \| \mathbf{u} \| + \| \mathbf{v} \|$$

for any vectors **u** and **v**. Equality will occur precisely when **u** and **v** are multiples of one another (that is, when **u** and **v** have the same direction). To establish the inequality, we observe that **u** and **v** are two sides of a triangle in the plane, the third side has length $\| \mathbf{u} + \mathbf{v} \|$ and is "shorter than the sum $\| \mathbf{u} \| + \| \mathbf{v} \|$ of the lengths of the other two sides," as shown in Figure 9.8**b**.

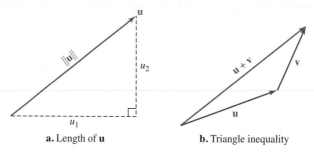

a. Length of **u**   **b.** Triangle inequality

Figure 9.8 Geometric representation of two vector properties

In component form, two vectors are equal if their components are equal, as illustrated by the following example.

EXAMPLE 3   Equal vectors in component form

If $\mathbf{u} = \langle 8, -2 \rangle$ and $\mathbf{v} = \langle -3, 5 \rangle$, find s and t so that $s\mathbf{u} + t\mathbf{v} = \mathbf{w}$, where $\mathbf{w} = \langle 2, 8 \rangle$.

Solution

$$\begin{aligned}
s\mathbf{u} + t\mathbf{v} &= s\langle 8, -2 \rangle + t\langle -3, 5 \rangle \\
&= \langle 8s, -2s \rangle + \langle -3t, 5t \rangle \\
&= \langle 8s - 3t, -2s + 5t \rangle
\end{aligned}$$

Thus, if $s\mathbf{u} + t\mathbf{v} = \mathbf{w}$, we see

$$\begin{cases} 8s - 3t = 2 \\ -2s + 5t = 8 \end{cases}$$

Solving this system, we find $(s, t) = (1, 2)$. ∎

EXAMPLE 4   Using vectors in a velocity problem   MODELING APPLICATION

A special boat, *The Earthrace*, draws attention wherever it goes (see Figure 9.9).

A river 4 mi wide flows south with a current of 5 mi/h. What speed and heading should a motorboat assume to travel directly across the river from east to west in 20 min?

Solution

Step 1:   *Understand the problem.* Begin by drawing a diagram, as shown in Figure 9.10.

Step 2:   *Devise a plan.* Let **B** be the velocity vector of the boat in the direction of the angle θ. If the river's current has velocity **C**, the given information tells us that $\| \mathbf{C} \| = 5$ mi/h and that **C** points directly south. Moreover, because the boat is to cross the river from east

Figure 9.9 *Earthrace* high-speed boat

to west in 20 min (that is, $\frac{1}{3}$ h), its *effective velocity* after compensating for the current is a vector **V** that points west and has magnitude. We will calculate $\| \mathbf{V} \|$ to find the effective velocity and then use this information to find the magnitude and direction of **B**.

Step 3:   *Carry out the plan.*

$$\| \mathbf{V} \| = \frac{\text{WIDTH OF THE RIVER}}{\text{TIME OF CROSSING}} = \frac{4 \text{ mi}}{\frac{1}{3} \text{ h}} = 12 \text{ mi/h}$$

The effective velocity **V** is the resultant of **B** and **C**; that is, $\mathbf{V} = \mathbf{B} + \mathbf{C}$. Because **V** and **C** act in perpendicular directions, we can determine **B** by referring to the right triangle with sides $\| \mathbf{V} \| = 12$ and $\| \mathbf{C} \| = 5$ with hypotenuse $\| \mathbf{B} \|$. We find that

$$\| \mathbf{B} \| = \sqrt{\| \mathbf{V} \|^2 + \| \mathbf{C} \|^2} = \sqrt{12^2 + 5^2} = 13$$

The direction of the velocity vector **V** is given by the angle θ in Figure 9.10, and we find that

$$\tan \theta = \frac{5}{12} \quad \text{so that} \quad \theta = \tan^{-1}\left(\frac{5}{12}\right) \approx 0.3948$$

Thus, the boat should travel at 13 mi/h in a direction of approximately 0.3948.

Step 4:   *Look back.* In navigation, it is common to specify direction in degrees rather than radians; this is 22.6° north of west. This answer makes sense by looking at Figure 9.10.

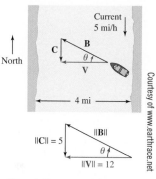

Figure 9.10 A velocity problem

A **unit vector** is just a vector with length 1, and a **direction vector** for a given vector **v** is a unit vector **u** that points in the same direction as **v**. Such a vector can be found by simply dividing **v** by its length $\| \mathbf{v} \|$; that is, $\mathbf{u} = \dfrac{\mathbf{v}}{\| \mathbf{v} \|}$.

EXAMPLE 5   **Finding a direction vector**

Find a direction vector for the vector $\mathbf{v} = \langle 2, -3 \rangle$.

Solution
The vector **v** has length (magnitude) $\| \mathbf{v} \| = \sqrt{2^2 + (-3)^2} = \sqrt{13}$. Thus, the required direction vector is the unit vector

$$\mathbf{u} = \frac{\mathbf{v}}{\| \mathbf{v} \|} = \frac{\langle 2, -3 \rangle}{\sqrt{13}} = \frac{\sqrt{13}}{13} \langle 2, -3 \rangle = \left\langle \frac{2\sqrt{13}}{13}, \frac{-3\sqrt{13}}{13} \right\rangle$$

Standard Representation of Vectors in the Plane

The unit vectors $\mathbf{i} = \langle 1, 0 \rangle$ and $\mathbf{j} = \langle 0, 1 \rangle$ point in the directions of the positive x- and y-axes, respectively, and are called **standard basis vectors**. Any vector $\mathbf{v} = \langle v_1, v_2 \rangle$ in the plane can be expressed as a linear combination of the vectors **i** and **j**, because

$$\mathbf{v} = \langle v_1, v_2 \rangle = v_1 \langle 1, 0 \rangle + v_2 \langle 0, 1 \rangle = v_1 \mathbf{i} + v_2 \mathbf{j}$$

This is called the **standard representation** of the vector **v**, and it can be shown that the representation is unique in the sense that if $\mathbf{v} = a\mathbf{i} + b\mathbf{j}$, then $a = v_1$ and $b = v_2$, as shown in Figure 9.11. In this context, the scalars v_1 and v_2 are called the **horizontal and vertical components** of **v**, respectively.

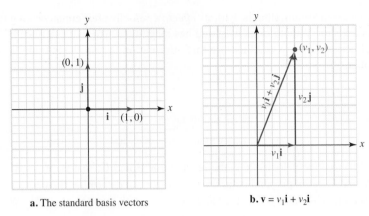

a. The standard basis vectors   **b.** $\mathbf{v} = v_1\mathbf{i} + v_2\mathbf{i}$

Figure 9.11 Standard representation of vectors in the plane

EXAMPLE 6   Finding the standard representation of a vector

If $\mathbf{u} = 3\mathbf{i} + 2\mathbf{j}$, $\mathbf{v} = -2\mathbf{i} + 5\mathbf{j}$, and $\mathbf{w} = \mathbf{i} - 4\mathbf{j}$, what is the standard representation of the vector $2\mathbf{u} + 5\mathbf{v} - \mathbf{w}$?

Solution
Using properties of vectors, we find that

$$
\begin{aligned}
2\mathbf{u} + 5\mathbf{v} - \mathbf{w} &= 2(3\mathbf{i} + 2\mathbf{j}) + 5(-2\mathbf{i} + 5\mathbf{j}) - (\mathbf{i} - 4\mathbf{j}) \\
&= [2(3) + 5(-2) - 1]\mathbf{i} + [2(2) + 5(5) - (-4)]\mathbf{j} \\
&= -5\mathbf{i} + 33\mathbf{j}
\end{aligned}
$$

EXAMPLE 7   A vector connecting two points

Find the standard representation of the vector $\mathbf{PQ}$, for the points $P(3, -4)$ and $Q(-2, 6)$.

Solution
The component form of $\mathbf{PQ}$ is

$$\mathbf{PQ} = \langle (-2) - 3,\ 6 - (-4) \rangle = \langle -5,\ 10 \rangle$$

This means that $\mathbf{PQ}$ has the standard representation $\mathbf{PQ} = -5\mathbf{i} + 10\mathbf{j}$.

EXAMPLE 8   Computing a resultant force   MODELING APPLICATION

Two forces $\mathbf{F}_1$ and $\mathbf{F}_2$ act on the same body. This example investigates the additional force $\mathbf{F}_3$ that must be applied to keep the body at rest.

Solution

Step 1: *Understand the problem.* Two forces $\mathbf{F}_1$ and $\mathbf{F}_2$ act on the same body. It is known, for example, that $\mathbf{F}_1$ has magnitude 3 newtons and acts in the direction of $-\mathbf{i}$, whereas $\mathbf{F}_2$ has magnitude 2 newtons and acts in the direction of the unit vector

$$\mathbf{u} = \frac{3}{5}\mathbf{i} - \frac{4}{5}\mathbf{j}$$

What additional force $\mathbf{F}_3$ must be applied in order to keep the body at rest? (See Figure 9.12.)

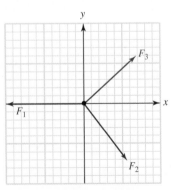

Figure 9.12 Keeping a body at rest

Step 2: *Devise a plan.* We will calculate $\mathbf{F}_1$ and $\mathbf{F}_2$, and we want to find $\mathbf{F}_3 = a\mathbf{i} + b\mathbf{j}$ so that $\mathbf{F}_1 + \mathbf{F}_2 + \mathbf{F}_3 = \mathbf{0}$.

Step 3: *Carry out the plan.* According to the given information, we have

$$\mathbf{F}_1 = 3(-\mathbf{i}) = -3\mathbf{i} \quad \text{and} \quad \mathbf{F}_2 = 2\left(\frac{3}{5}\mathbf{i} - \frac{4}{5}\mathbf{j}\right) = \frac{6}{5}\mathbf{i} - \frac{8}{5}\mathbf{j}$$

and we want to find $\mathbf{F}_3 = a\mathbf{i} + b\mathbf{j}$ so that $\mathbf{F}_1 + \mathbf{F}_2 + \mathbf{F}_3 = \mathbf{0}$. Substituting into this vector equation, we obtain

$$(-3\mathbf{i}) + \left(\frac{6}{5}\mathbf{i} - \frac{8}{5}\mathbf{j}\right) + (a\mathbf{i} + b\mathbf{j}) = 0\mathbf{i} + 0\mathbf{j}$$

By combining terms on the left, we find that

$$\left(-3 + \frac{6}{5} + a\right)\mathbf{i} + \left(-\frac{8}{5} + b\right)\mathbf{j} = 0\mathbf{i} + 0\mathbf{j}$$

Because the standard representation is unique, we must have

$$-3 + \frac{6}{5} + a = 0 \quad \text{and} \quad -\frac{8}{5} + b = 0$$

$$a = \frac{9}{5} \qquad\qquad b = \frac{8}{5}$$

The required force is $\mathbf{F}_3 = \frac{9}{5}\mathbf{i} + \frac{8}{5}\mathbf{j}$. This is a force of magnitude:

$$\| \mathbf{F}_3 \| = \sqrt{\left(\frac{9}{5}\right)^2 + \left(\frac{8}{5}\right)^2} = \frac{1}{5}\sqrt{145} \text{ newtons}$$

which acts in the direction of the unit vector

$$\mathbf{v} = \frac{\mathbf{F}_3}{\| \mathbf{F}_3 \|} = \frac{5}{\sqrt{145}}\left(\frac{9}{5}\mathbf{i} + \frac{8}{5}\mathbf{j}\right) = \frac{9}{\sqrt{145}}\mathbf{i} + \frac{8}{\sqrt{145}}\mathbf{j}$$

Step 4: *Look back.* If we consider these three vectors with the ones shown in Figure 9.12, we see that the result is reasonable. ∎

PROBLEM SET 9.1

Sketch each vector given in Problems 1–4 assuming that its initial point is at the origin.

1. $3\mathbf{i} + 4\mathbf{j}$
2. $-2\mathbf{i} - 3\mathbf{j}$
3. $-\frac{1}{2}\mathbf{i} + \frac{5}{2}\mathbf{j}$
4. $-2(-\mathbf{i} + 2\mathbf{j})$

The initial point P and terminal point Q of a vector are given in Problems 5–8. Sketch each vector and then write it in component form.

5. $P(3, -1), Q(7, 2)$
6. $P(5, -2), Q(5, 8)$
7. $P(3, 4), Q(-2, 4)$
8. $P\left(\frac{1}{2}, 6\right), Q(-3, -2)$

*Express each vector in **PQ** Problems 9–12 in standard form and also find its length.*

9. $P(-1, -2), Q(1, -2)$
10. $P(5, 7), Q(6, 8)$
11. $P(-4, -3), Q(0, -1)$
12. $P(3, -5), Q(2, 8)$

Find a direction vector for each of the vectors given in Problems 13–16.

13. $\mathbf{i} + \mathbf{j}$
14. $\frac{1}{2}\mathbf{i} + \frac{1}{4}\mathbf{j}$
15. $3\mathbf{i} - 4\mathbf{j}$
16. $-4\mathbf{i} + 7\mathbf{j}$

Let $\mathbf{u} = \langle -3, 4 \rangle$ and $\mathbf{v} = \langle 1, -1 \rangle$. Find s and t so that $s\mathbf{u} + t\mathbf{v} = \mathbf{w}$ for the given vector in Problems 17–20.

17. $\mathbf{w} = \langle 6, 0 \rangle$ **18.** $\mathbf{w} = \langle 0, -3 \rangle$
19. $\mathbf{w} = \langle -2, 1 \rangle$ **20.** $\mathbf{w} = \langle 8, 11 \rangle$

Suppose $\mathbf{u} = 3\mathbf{i} - 4\mathbf{j}$, $\mathbf{v} = 4\mathbf{i} - 3\mathbf{j}$, and $\mathbf{w} = \mathbf{i} + \mathbf{j}$. Express each of the expressions in Problems 21–24 in standard form.

21. $2\mathbf{u} + 3\mathbf{v} - \mathbf{w}$ **22.** $\frac{1}{2}(\mathbf{u} + \mathbf{v}) - \frac{1}{4}\mathbf{w}$
23. $\|\mathbf{v}\|\mathbf{u} + \|\mathbf{u}\|\mathbf{v}$ **24.** $\|\mathbf{u}\|\|\mathbf{v}\|\mathbf{w}$

Find all real numbers x and y that satisfy the vector equations given in Problems 25–28.

25. $(x - y - 1)\mathbf{i} + (2x + 3y - 12)\mathbf{j} = \mathbf{0}$
26. $x\mathbf{i} - 4y^2\mathbf{j} = (5 - 3y)\mathbf{i} + (10 - 7x)\mathbf{j}$
27. $(x^2 + y^2)\mathbf{i} + y\mathbf{j} = 20\mathbf{i} + (x + 2)\mathbf{j}$
28. $(y - 1)\mathbf{i} + y\mathbf{j} = (\log x)\mathbf{i} + [\log 2 + \log(x + 4)]\mathbf{j}$

In Problems 29–32, find a unit vector $\mathbf{u}$ with the given characteristics.

29. $\mathbf{u}$ makes an angle of $30°$ with the positive x-axis.
30. $\mathbf{u}$ has the same direction as the vector $2\mathbf{i} - 3\mathbf{j}$.
31. $\mathbf{u}$ has the direction opposite that of $-4\mathbf{i} + \mathbf{j}$.
32. $\mathbf{u}$ has the direction of the vector from $P(-1, 5)$ to $Q(7, -3)$.

In Problems 33–36, let $\mathbf{u} = 4\mathbf{i} - \mathbf{j}$, $\mathbf{v} = \mathbf{i} + 2\mathbf{j}$, and $\mathbf{w} = -3\mathbf{i} + 4\mathbf{j}$.

33. Find a unit vector in the same direction as $\mathbf{u} + \mathbf{v}$.
34. Find a vector of length 3 with the same direction as $\mathbf{u} - 2\mathbf{v} + 2\mathbf{w}$.
35. Find the terminal point of the vector $5\mathbf{i} + 7\mathbf{j}$ if the initial point is $(-2, 3)$.
36. Find the initial point of the vector $-\mathbf{i} + 2\mathbf{j}$ if the terminal point is $(-1, -2)$.

LEVEL 2

37. Use vectors to find the coordinates of the midpoint of the line segment joining the points $P(-3, -8)$ and $Q(9, -2)$. What point is located $\frac{5}{6}$ of the distance from P to Q?

38. If $\|\mathbf{v}\| = 3$ and $-3 \leq r \leq 1$, what are the possible values of $\|r\mathbf{v}\|$?
39. Show that $\mathbf{v} = (\cos\theta)\mathbf{i} + (\sin\theta)\mathbf{j}$ is a unit vector for any angle θ.
40. If $\mathbf{u}$ and $\mathbf{v}$ are nonzero vectors and

$$r = \frac{\|\mathbf{u}\|}{\|\mathbf{v}\|}$$

what is $\|r\mathbf{v}\|$?

41. If $\mathbf{u}$ and $\mathbf{v}$ are nonzero vectors with $\|\mathbf{u}\| = \|\mathbf{v}\|$, does it follow that $\mathbf{u} = \mathbf{v}$? Explain.
42. If $\mathbf{u} = 2\mathbf{i} - 3\mathbf{j}$ and $\mathbf{v} = x\mathbf{i} + y\mathbf{j}$, describe the set of points in the plane whose coordinates (x, y) satisfy $\|\mathbf{v} - \mathbf{u}\| \leq 2$.
43. Let $\mathbf{u}_0 = x_0\mathbf{i} + y_0\mathbf{j}$ for constants x_0 and y_0, and let $\mathbf{u} = x\mathbf{i} + y\mathbf{j}$. Describe the set of all points in the plane whose coordinates satisfy
    a. $\|\mathbf{u} - \mathbf{u}_0\| = 1$
    b. $\|\mathbf{u} - \mathbf{u}_0\| \leq r$
44. Let $\mathbf{u} = 3\mathbf{i} - \mathbf{j}$ and $\mathbf{v} = -6\mathbf{i} + 2\mathbf{j}$. Show that there are no numbers a, b for which $a\mathbf{u} + b\mathbf{v} = 2\mathbf{i} + 5\mathbf{j}$.
45. Suppose $\mathbf{u}$ and $\mathbf{v}$ are a pair of nonzero, nonparallel vectors. Find numbers a, b, c such that

$$a\mathbf{u} + b(\mathbf{u} - \mathbf{v}) + c(\mathbf{u} + \mathbf{v}) = \mathbf{0}$$

46. Suppose $\mathbf{u}$ and $\mathbf{v}$ are perpendicular vectors, with $\|\mathbf{u}\| = 2$ and $\|\mathbf{v}\| = 4$. Sketch the vector $c\mathbf{u} + (1 - c)\mathbf{v}$ for the cases where $c = 0$, $c = \frac{1}{4}$, $c = \frac{1}{2}$, $c = \frac{3}{4}$, and $c = 1$. In general, if the initial point of a vector $c\mathbf{u} + (1 - c)\mathbf{v}$ with $0 \leq c \leq 1$ is at the origin, where is its terminal point?
47. Two forces $\mathbf{F}_1 = 3\mathbf{i} + 4\mathbf{j}$ and $\mathbf{F}_2 = 3\mathbf{i} - 7\mathbf{j}$ act on an object. What additional force should be applied to keep the body at rest?
48. Three forces $\mathbf{F}_1 = \mathbf{i} - 2\mathbf{j}$, $\mathbf{F}_2 = 3\mathbf{i} - 7\mathbf{j}$, and $\mathbf{F}_3 = \mathbf{i} + \mathbf{j}$ act on an object. What additional force $\mathbf{F}_4$ should be applied to keep the body at rest?
49. A rock is thrown into the air at an angle of $40°$ with the horizontal and with an initial velocity of 60 ft/s. Find the vertical and horizontal components of the velocity vector.
50. A river 2.1 mi wide flows south with a current of 3.1 mi/h. What speed and heading should a motorboat assume to travel across the river from east to west in 30 min?

LEVEL 3

51. Four forces act on an object: $\mathbf{F}_1$ has magnitude 10 lb and acts at an angle of $\frac{\pi}{6}$ measured counterclockwise from the positive x-axis; $\mathbf{F}_2$ has magnitude 8 lb and acts in the direction of the vector $\mathbf{j}$; $\mathbf{F}_3$ has magnitude 5 lb and acts at an angle of $4\pi/3$ measured counterclockwise from the positive x-axis. What must the fourth force $\mathbf{F}_4$ be to keep the object at rest?
52. Use vector methods to show that the diagonals of a parallelogram bisect each other.
53. In a triangle, let $\mathbf{u}$, $\mathbf{v}$, and $\mathbf{w}$ be the vectors from each vertex to the midpoint of the opposite side. Use vector methods to show that $\mathbf{u} + \mathbf{v} + \mathbf{w} = \mathbf{0}$.
54. Two nonzero vectors $\mathbf{u}$ and $\mathbf{v}$ are said to be **linearly independent** in the plane if they are not parallel.
    a. If $\mathbf{u}$ and $\mathbf{v}$ have this property and $a\mathbf{u} = b\mathbf{v}$ for constants a, b, show that $a = b = 0$.
    b. Show that the standard representation of a vector is unique. That is, if the vector $\mathbf{u}$ has the representation $\mathbf{u} = a_1\mathbf{i} + b_1\mathbf{j}$ and $\mathbf{u} = a_2\mathbf{i} + b_2\mathbf{j}$ is another such representation, then $a_1 = a_2$ and $b_1 = b_2$.
55. Prove that the medians of a triangle intersect at a single point by completing the following argument (see Figure 9.13).

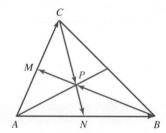

Figure 9.13 Median intersection

    a. Let M and N be the midpoints of sides $\overline{AC}$ and $\overline{AB}$, respectively. Show that

$$\mathbf{CN} = \frac{1}{2}\mathbf{AB} - \mathbf{AC} \quad \text{and} \quad \mathbf{BM} = \frac{1}{2}\mathbf{AC} - \mathbf{AB}$$

    b. Let P be the point where medians $\overline{BM}$ and $\overline{CN}$ intersect. Show that there are constants r and s such that

$$\mathbf{CP} = r\left(\tfrac{1}{2}\mathbf{AB} - \mathbf{AC}\right) \quad \text{and} \quad \mathbf{BP} = s\left(\tfrac{1}{2}\mathbf{AC} - \mathbf{AB}\right)$$

Note that $\mathbf{CP} + \mathbf{PB} = \mathbf{CB}$. Use this relationship to prove that $r = s = \frac{2}{3}$. Explain why this shows that any pair of medians meet at a point located $\frac{2}{3}$ the distance from each vertex to the midpoint of the opposite side. Why does this show that *all three* medians meet at a single point?

c. The *centroid* of a triangle is the point where the medians meet. Show that if a triangle has vertices (x_1, y_1), (x_2, y_2), (x_3, y_3), then the centroid has coordinates

$$\left(\frac{x_1 + x_2 + x_3}{3}, \frac{y_1 + y_2 + y_3}{3} \right)$$

56. Show that the vector with initial point $P(c, d)$ and terminal point $Q(a, b)$ has the component form

$$\mathbf{PQ} = \langle a - c, b - d \rangle.$$

See Figure 9.5.

57. a. Prove the commutativity property for vectors.
   b. Prove the associative property for scalar multiplication.
58. a. Prove the identity property for vectors.
   b. Prove the inverse property of addition for vectors.
59. a. Prove vector distributivity.
   b. Prove scalar distributivity.
60. 𝕳𝖎𝖘𝖙𝖔𝖗𝖎𝖈𝖆𝖑 𝕼𝖚𝖊𝖘𝖙 The history of mathematics has many more white males recorded than it does females or minorities. In previous problems, you have been asked to do research in mathematical contributions of some of these groups. In this chapter, we highlight Dr. Ellen Ochoa (see below), the first Hispanic female astronaut. For this Quest, investigate contributions of Hispanic mathematicians.

9.2 Quadric Surfaces and Graphing in Three Dimensions

Because we live in a three-dimensional world, it is essential that we be able to visualize surfaces in three dimensions. To that end, we consider ordered triplets and their representation in three-dimensional space, as well as develop some knowledge of simple three-dimensional surfaces—namely, the quadric surfaces. In the next section, we will consider vectors in three dimensions to describe motion in space efficiently.

Three-Dimensional Coordinate System

Our next goal is to see how analytic geometry and vector methods can be applied in space. We have already considered ordered pairs and a two-dimensional coordinate system. We shall denote this two-dimensional system by $\mathbb{R}^2$. Because we exist in at least a three-dimensional world, it is also important to consider a three-dimensional system. We call this *three-space* and denote it by $\mathbb{R}^3$. We introduce a coordinate system to three-space by choosing three mutually perpendicular axes to serve as a frame of reference. The orientation of our reference system will be *right-handed* in the sense that if you stand at the origin with your arms stretched out in the direction of the positive x-axis and y-axis, your right arm will be the one along the x-axis and your left arm will be stretched straight ahead along the y-axis, as shown in Figure 9.14. Your head will then point in the direction of the positive z-axis.

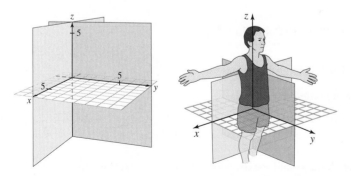

Figure 9.14 A "right-handed" rectangular coordinate system for $\mathbb{R}^3$

In order to orient yourself to a three-dimensional coordinate system, think of the x-axis and y-axis as lying in the plane of the floor and the z-axis as a line perpendicular to the floor. All the graphs we have drawn in the first nine chapters of this book would now be drawn on the floor.

𝕳𝖎𝖘𝖙𝖔𝖗𝖎𝖈𝖆𝖑 𝕹𝖔𝖙𝖊

Ellen Ochoa (1958–)

Born in 1958, Ellen Ochoa was the first Hispanic female astronaut. Since 1991, she has flown four spaceflights and is a specialist in the operation of robotic arms. When we start studying lines and planes in space, we can't help but think of applications of mathematics to spaceflight and space exploration. Ochoa graduated from Grossmont High School, received a BS from San Diego State University, and MS and PhD degrees from Stanford University. She is a co-inventor of three patents for an optical inspection system, an optical object recognition method, and a method for noise removal in images. She believes a space station is "critical . . . to human exploration in space, a transportation mode to new frontiers."

If you orient yourself in a room (your classroom, for example), as shown in Figure 9.15, you may notice some important planes. Assume the room is 25 ft × 30 ft × 8 ft and fix the origin at a front corner (where the board hangs).

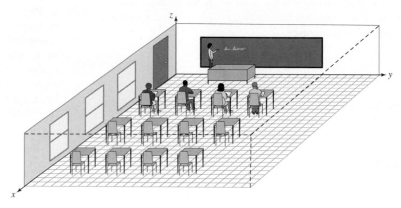

Figure 9.15 A typical classroom

Floor: **xy-plane**; equation is $z = 0$.

Ceiling: Plane parallel to the xy-plane; equation is $z = 8$.

Front wall: **yz-plane**; equation is $x = 0$.

Back wall: Plane parallel to the yz-plane; equation is $x = 30$.

Left wall: **xy-plane**; equation is $y = 0$.

Right wall: Plane parallel to the xz-plane; equation is $y = 25$.

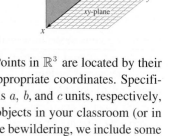

The xy-, xz-, and yz-planes are called the **coordinate planes**. Points in $\mathbb{R}^3$ are located by their position in relation to the three coordinate planes and are given appropriate coordinates. Specifically, the point P is assigned coordinates (a, b, c) to indicate that it is a, b, and c units, respectively, from the yz-, xz-, and xy-planes. Name the coordinates of several objects in your classroom (or in Figure 9.15). Because working in three dimensions can sometimes be bewildering, we include some **drawing lessons** to help you along; these include plotting points in $\mathbb{R}^3$, drawing vertical planes, graphing surfaces, sketching a plane in one octant, observer viewpoint, and sketching a plane in $\mathbb{R}^3$. The following example gives us our first drawing lesson, and it would be a good idea to duplicate it on your own paper.

Drawing Lesson: Plotting Points in $\mathbb{R}^3$

EXAMPLE 1   Points in three dimensions

Graph the following ordered triplets:

  a. $(3, 4, -5)$ **b.** $(10, 20, 10)$ **c.** $(-12, 6, 12)$ **d.** $(-12, -18, 6)$ **e.** $(20, -10, 18)$

Solution

  a. Step 1: Sketch the x- and y-axis, adding tick marks. Outline the xy-plane.

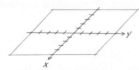

Step 2: Sketch the z-axis, adding tick marks. Use dashed segments for hidden parts.

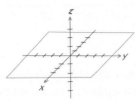

Step 3: Plot x-distance and y-distance on the xy-plane. Darken segments from each along grid lines. Colored pencil or highlighter may help you visualize the figure.

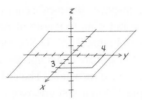

Step 4: Plot z-distance, using the unit size from the z-axis. Lightly sketch a grid on the xy-plane, using tick marks as guides.

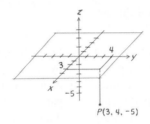

The other points are plotted similarly as shown.

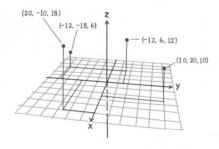

Plotting points and sketching objects in three dimensions is a skill that takes some practice. Do not be discouraged with your first attempts. Work with a pencil and make preliminary sketches. If the perspective does not look right, erase one or more of your preliminary lines and draw some new lines. For the final step, you might wish to use colored pencils or highlighter pens to enhance the three-dimensional effect of your drawing.

In Example 1, we measured distances in the x-, y-, and z-directions. We will, however, also need to measure distances between points in $\mathbb{R}^3$. The formula for distance in $\mathbb{R}^2$ easily extends to $\mathbb{R}^3$.

DISTANCE FORMULA

The distance $\left| P_1 P_2 \right|$ between $P_1(x_1, y_1, z_1)$ and $P_2(x_2, y_2, z_2)$ is

$$\left| P_1 P_2 \right| = \sqrt{(x_2 - x_1)^2 + (y_2 - y_1)^2 + (z_2 - z_1)^2}$$

This formula will be derived in Section 9.3. The distance between $(10, 20, 10)$ and $(-12, 6, 12)$ is

$$d = \sqrt{(-12-10)^2 + (6-20)^2 + (12-10)^2} = \sqrt{684} = 6\sqrt{19}$$

Graphs in $\mathbb{R}^3$

The **graph of an equation** in $\mathbb{R}^3$ is the collection of all points (x, y, z) whose coordinates satisfy a given equation. This graph is called a **surface**. You are not expected to spend a great deal of time graphing three-dimensional surfaces, but the drawing lessons in this section should help. You may also have access to a computer program to help you look at graphs in three dimensions.

Planes

We shall obtain equations for planes in space after we discuss vectors in space. However, in beginning to visualize objects in $\mathbb{R}^3$, we do not want to ignore planes because they are so common (for example, the walls, ceiling, and floor in Figure 9.15). In Section 9.5, we will show that the graph of $ax + by + cz = d$ is a **plane** if a, b, c, and d are real numbers (not all zero). The following drawing lesson will help you sketch planes in three dimensions. The task described begins with the xy-plane and point $P(3, 4, -5)$ from Example 1. Then you are asked to draw (on your paper) the planes $x = 2$ and $y = 0$.

Drawing Lesson: Drawing Vertical Planes

Begin with the xy-plane from Example 1, and then draw the line on $x = 2$ on the xy-plane. Now through each endpoint, draw a segment parallel to the z-axis. Then connect the endpoints. When you are finished, your drawing should look like this:

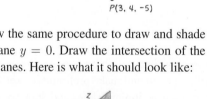

Shade the plane $x = 2$ where it is not hidden by the xy-plane. Erase hidden parts of both planes, and use your eraser to dash hidden parts of the axis. When you are finished, your drawing should look like this:

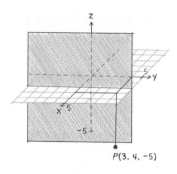

$P(3, 4, -5)$

Follow the same procedure to draw and shade the plane $y = 0$. Draw the intersection of the two planes. Here is what it should look like:

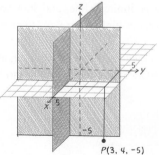

$P(3, 4, -5)$

Use colored pencils or highlighters to distinguish individual planes. Here is an example of a finished drawing:

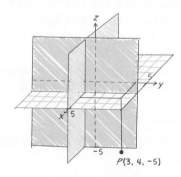

$P(3, 4, -5)$

The following example will also help you sketch planes.

EXAMPLE 2   Graphing planes

Graph the planes defined by the given equations.

a. $x = 4$

b. $y + z = 5$

c. $x + 3y + 2z = 6$

Solution

To graph a plane, find some ordered triplets satisfying the equation. The best ones to use are often those that fall on a coordinate axis (the intercepts).

a. When two variables are missing, then the plane is parallel to one of the coordinate planes, as shown in Figure 9.16**a**.

b. When one of the variables is missing from an equation of a plane, then that plane is parallel to the axis corresponding to the missing variable; in this case it is parallel to the x-axis. Draw the line $y + z = 5$ on the yz-plane, and then complete the plane, as shown in Figure 9.16**b**.

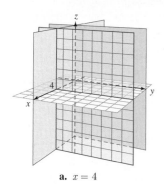 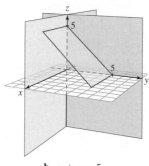 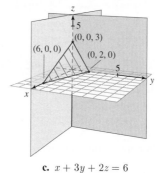

a. $x = 4$	**b.** $y + z = 5$	**c.** $x + 3y + 2z = 6$

Figure 9.16 Graphs of planes

c. Let $x = 0$ and $y = 0$; then $z = 3$; plot the point $(0, 0, 3)$.
Let $x = 0$ and $z = 0$; then $y = 2$; plot the point $(0, 2, 0)$.
Let $y = 0$ and $z = 0$; then $x = 6$; plot the point $(6, 0, 0)$.
Use these points to draw the intersection lines (called **trace lines**) of the plane you are graphing with each of the coordinate planes. The result is shown in Figure 9.16**c**. ∎

Spheres

A **sphere** (see Figure 9.17) is defined as the collection of all points located a fixed distance (the **radius**) from a fixed point (the **center**).

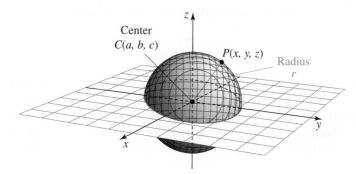

Figure 9.17 Sphere

In particular, if $P(x,\ y,\ z)$ is a point on the sphere with radius r and center $C(a,\ b,\ c)$, then the distance from C to P is r. Thus,

$$r = \sqrt{(x-a)^2 + (y-b)^2 + (z-c)^2}$$

If you square both sides of this equation, you can see that it is equivalent to the equation of a sphere displayed in the following box. Conversely, if the point $(x,\ y,\ z)$ satisfies an equation of this form, it must lie on a sphere with center $(a,\ b,\ c)$ and radius r.

EQUATION OF A SPHERE

The graph of the equation

$$(x-a)^2 + (y-b)^2 + (z-c)^2 = r^2$$

is a sphere with center $(a,\ b,\ c)$ and radius r, and any sphere has an equation of this form. This is called the **standard form of the equation of a sphere** (or simply *standard-form sphere*).

EXAMPLE 3   Center and radius of a sphere from a given equation

Show that the graph of the equation $x^2 + y^2 + z^2 + 4x - 6y - 3 = 0$ is a sphere, and find its center and radius.

Solution
By completing the square in both variables x and y, we have

$$(x^2 + 4x) + (y^2 - 6y) + z^2 = 3$$
$$(x^2 + 4x + 2^2) + (y^2 - 6y + (-3)^2) = 3 + 4 + 9$$
$$(x+2)^2 + (y-3)^2 + z^2 = 16$$

Comparing this equation with the standard form, we see that it is the equation of a sphere with center $(-2, 3, 0)$ and radius 4. The graph is shown in Figure 9.18. Note that you cannot see the bottom half of the sphere because your view is obstructed by the xy-plane.

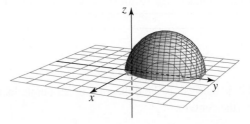

Figure 9.18 Sphere with center $(-2, 3, 0)$ and radius 4

Cylinders

A **cross-section** of a surface in space is a curve obtained by intersecting the surface with a plane. If parallel planes intersect a given surface in congruent cross-sectional curves, the surface is called a **cylinder**. We define a cylinder with **principal cross-sections C and generating line L** to be the surface obtained by moving lines parallel to L along the boundary of the curve C, as shown in Figure 9.19. In this context, the curve C is called a **directrix** of the cylinder, and L is the **generatrix**.

We shall deal primarily with cylinders in which the directrix is a conic section and the generatrix L is one of the coordinate axes. Such a cylinder is often named for the type of conic section in its principal cross-sections and is described by an equation involving only two of the variables x, y, z. In this case, the generating line L is parallel to the coordinate axis of the missing variable. Thus,

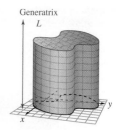
Figure 9.19 Cylinder

$x^2 + y^2 = 5$ is a **circular cylinder** with L parallel to the z-axis.

$y^2 - z^2 = 9$ is a **hyperbolic cylinder** with L parallel to the x-axis.

$x^2 + 2z^2 = 25$ is an **elliptic cylinder** with L parallel to the y-axis.

The graphs of these cylinders are shown in Figure 9.20.

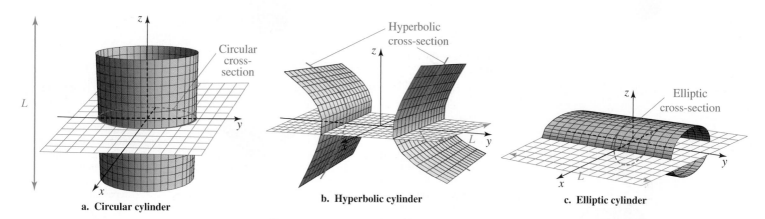

a. Circular cylinder   **b. Hyperbolic cylinder**   **c. Elliptic cylinder**

Figure 9.20 Cylinders

Quadric Surfaces

Spheres and elliptic, parabolic, and hyperbolic cylinders are examples of **quadric surfaces**. In general, such a surface is the graph of an equation of the form

$$Ax^2 + By^2 + Cz^2 + Dxy + Exz + Fyz + Gx + Hy + Iz + J = 0$$

Quadric surfaces may be thought of as the generalizations of the conic sections in $\mathbb{R}^3$. The **trace** of a curve is found by setting one of the variables equal to a constant and then graphing the resulting curve. If $x = k$ (k a constant), the resulting curve is drawn in the plane $x = k$, which is parallel to the yz-plane; and if $z = k$, the curve is drawn in the plane $z = k$, parallel to the xy-plane. Table 9.1 shows the quadric surfaces and is the final installment in our Directory of Curves.

EXAMPLE 4   Identifying and sketching a quadric surface

Identify and sketch the surface with equation $9x^2 - 16y^2 + 144z = 0$.

Solution

Look at Table 9.1 on page 599 and note that the equation is second degree in x and y but first degree in z. This means it is an elliptic paraboloid or a hyperbolic paraboloid. Solve the equation for z:

$$9x^2 - 16y^2 + 144z = 0$$
$$144z = 16y^2 - 9x^2$$
$$z = \frac{y^2}{9} - \frac{x^2}{16}$$

TABLE 9.1 Directory of Curves (part VII)—Quadric Surfaces

Surface	Description	Surface	Description

Elliptic cone

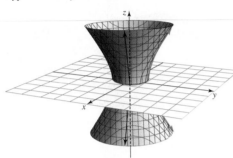

The trace in the xy-plane is a point; in planes parallel to the xy-plane, it is an ellipse. Traces in the xz- and yz-planes are intersecting lines; in planes parallel to these, they are hyperbolas:

$$z^2 = \frac{x^2}{a^2} + \frac{y^2}{b^2}$$

The axis of this cone is the z-axis.

Circular cone

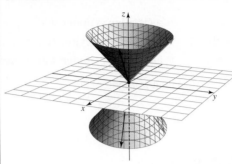

The circular cone is a special kind of elliptic cone for which $a = b = r$.

$$z^2 = \frac{x^2}{r^2} + \frac{y^2}{r^2}$$

The axis of this cone is the z-axis.

Hyperboloid of one sheet

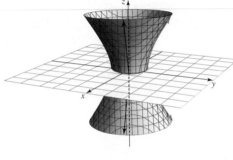

The trace in the xy-plane is an ellipse; to the xz- and yz-planes the traces are hyperbolas:

$$\frac{x^2}{a^2} + \frac{y^2}{b^2} - \frac{z^2}{c^2} = 1$$

One negative on equation; the axis of this one is the z-axis.

Hyperboloid of two sheets

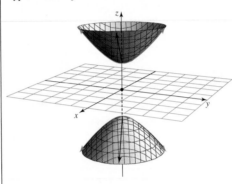

There is no trace in the xy-plane. In planes parallel to the xy-plane, which intersect the surface, the traces are ellipses. Traces in the xz- and yz-planes are hyperbolas.

$$\frac{x^2}{a^2} + \frac{y^2}{b^2} - \frac{z^2}{c^2} = -1$$

Two negatives; the axis of this one is the z-axis.

Ellipsoid

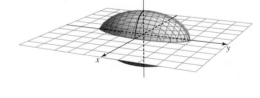

The trace in the coordinate planes are ellipses.

$$\frac{x^2}{a^2} + \frac{y^2}{b^2} + \frac{z^2}{c^2} = 1$$

Sphere

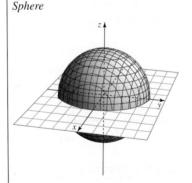

The sphere is a special kind of ellipsoid for which $a = b = c = r$

$$x^2 + y^2 + z^2 = r^2$$

Paraboloid

The trace in the xy-plane is a point; in planes parallel to the xy-plane, it is an ellipse. Traces in the xz- and yz-planes are parabolas

$$z = \frac{x^2}{a^2} + \frac{y^2}{b^2}$$

Axis is z-axis (z first degree). If $a = b = r$, then the cross-section is a circle, and it is called a *circular paraboloid*.

Hyperbolic paraboloid

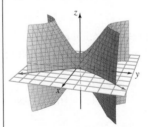

The trace in the xy-plane is two intersecting lines; in the plane parallel to the xy-plane, the traces are hyperbolas. Traces in the xz-plane and yz-planes are parabolas.

$$z = \frac{y^2}{b^2} - \frac{x^2}{a^2}$$

This is also called a *saddle*.

We recognize this as a hyperbolic paraboloid. Next, we take cross-sections of $9x^2 - 16y^2 + 144z = 0$:

Cross-Section	Chosen Value	Equation	Description
xy-plane	$z = 0$	$\dfrac{y^2}{9} - \dfrac{x^2}{16} = 0$	Two intersecting lines
Parallel to the xy-plane	$z = 4$	$\dfrac{y^2}{36} - \dfrac{x^2}{64} = 1$	Hyperbola
xz-plane	$y = 0$	$z = -\dfrac{x^2}{16}$	Parabola opens downward
Parallel to the xz-plane	$y = 10$	$z - \dfrac{100}{9} = -\dfrac{x^2}{16}$	Parabola opens downward
yz-plane	$x = 0$	$z = \dfrac{y^2}{9}$	Parabola opens upward
Parallel to the yz-plane	$x = 5$	$z + \dfrac{25}{16} = \dfrac{y^2}{9}$	Parabola opens upward

These traces are shown in Figure 9.21.

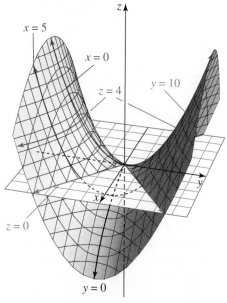

Figure 9.21 Graph of quadric surface

PROBLEM SET 9.2

LEVEL 1

Plot the points in Problems 1–4.

1. $A(2, -3, 8)$
2. $B(8, 5, -3)$
3. $C(5, 7, 5)$
4. $D(-8, 0, 3)$

In Problems 5–8, plot the points P and Q in $\mathbb{R}^3$, and find the distance $|\overrightarrow{PQ}|$.

5. $P(3, -4, 5)$, $Q(1, 5, -3)$
6. $P(0, 3, 0)$, $Q(-2, 5, -7)$
7. $P(-3, -5, 8)$, $Q(3, 6, -7)$
8. $P(5, 7, -3)$, $Q(2, -4, 3)$

In Problems 9–12, find the standard-form equation of the sphere with the given center C and radius r.

9. $C(0, 0, 0)$, $r = 1$
10. $C(-3, 5, 7)$, $r = 2$
11. $C(0, 4, -5)$, $r = 3$
12. $C(-2, 3, -1)$, $r = \sqrt{5}$

Find the center and radius of each sphere whose equations are given in Problems 13–16.

13. $x^2 + y^2 + z^2 - 2y + 2z - 2 = 0$
14. $x^2 + y^2 + z^2 + 4x - 2z - 8 = 0$
15. $x^2 + y^2 + z^2 - 6x + 2y - 2z + 10 = 0$
16. $x^2 + y^2 + z^2 - 2x - 4y + 8z + 17 = 0$

Follow the steps shown in the drawing lesson to plot the planes in Problems 17–20.

17. Sketch the planes $x = 0$, $y = 0$, and $z = 0$.
18. Sketch the planes $x = 4$, $y = 0$, and $z = 0$.
19. Sketch the planes $x = 0$, $y = 2$, and $z = 0$.
20. Sketch the planes $x = 0$, $y = -3$, and $z = 0$.

In Problems 21–30, match the equation with its graph (A–L).

21. $x^2 = z^2 + y^2$

22. $z^2 = \dfrac{x^2}{4} + \dfrac{y^2}{9}$

23. $x^2 + y^2 + z^2 = 9$

24. $y = x^2 + z^2$

25. $x = \dfrac{y^2}{25} + \dfrac{z^2}{16}$

26. $y = \dfrac{z^2}{4} - \dfrac{x^2}{9}$

27. $\dfrac{x^2}{2} + \dfrac{y^2}{4} + \dfrac{z^2}{9} = 1$

28. $\dfrac{x^2}{9} + \dfrac{y^2}{16} - \dfrac{z^2}{4} = 1$

29. $y^2 + z^2 - x^2 = -1$

30. $\dfrac{x^2}{4} - \dfrac{y^2}{9} + \dfrac{z^2}{9} = -1$

A.

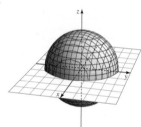

B.

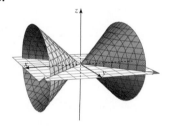

C. **D.**

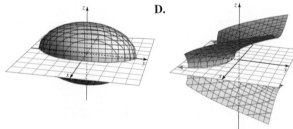

E. **F.**

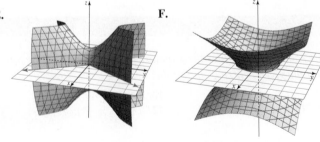

G. **H.**

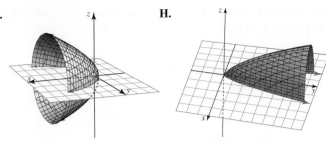

I. **J.**

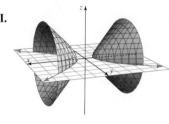

K.

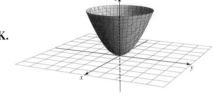

L.

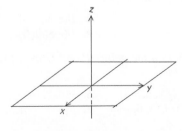

LEVEL 2

The vertices A, B, and C of a triangle in space are given in Problems 31–34. Find the lengths of the sides of the triangle and determine whether it is a right triangle, an isosceles triangle, or neither.

31. $A(3, -1, 0)$, $B(7, 1, 4)$, $C(1, 3, 4)$
32. $A(1, 1, 1)$, $B(3, 3, 2)$, $C(3, -3, 5)$
33. $A(1, 2, 3)$, $B(-3, 2, 4)$, $C(1, -4, 3)$
34. $A(2, 4, 3)$, $B(-3, 2, -4)$, $C(-6, 8, -10)$

35. **Drawing Lesson: Surfaces**

Sketch the graph of $z = \dfrac{1}{1 + x^2 + y^2}$ by reproducing the following steps on your own paper.

Step 1: Draw the xy-plane in three dimensions, adding the z-axis.

Step 2: Draw a trace in one of the coordinate planes (in this case, the plane is $x = 0$). If necessary, adjust the z-scale to show the trace more clearly.

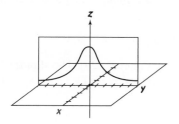

Step 3: Draw a trace in another coordinate plane (in this case, the plane $y = 0$).

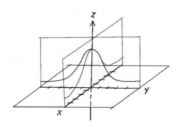

Step 4: Draw several additional trace curves to reveal the contours of the surface.

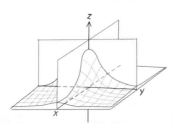

Step 5: Erase all hidden lines. Use highlighters to color the surface and the xy-plane.

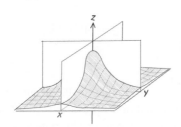

Follow the steps shown in Problem 35 to sketch the surfaces in Problems 36–39.

36. Sketch the graph of $y = \dfrac{1}{1 + x^2 + z^2}$.

37. Sketch the graph of $x = \dfrac{1}{1 + y^2 + z^2}$.

38. Sketch the graph of $x^2 + y^2 + z^2 = 1$.

39. Sketch the graph of $y = x^2 + y^2$.

In Problems 40–51, sketch the graph of each equation in $\mathbb{R}^3$.

40. $2x + y + 3z = 6$ **41.** $x = 4$
42. $x + 2y + 5z = 10$ **43.** $x + y + z = 1$
44. $3x - 2y - z = 6$ **45.** $y = x^2$
46. $z = x - 1$ **47.** $x^2 + y^2 = z$
48. $z = e^y$ **49.** $y = \ln x$
50. $2z + 5y^2 = 1$ **51.** $z = \sin y$

In Problems 52–57, identify the quadric surface and describe the traces. Sketch the graph.

52. $9x^2 + 4y^2 + z^2 = 1$ **53.** $\dfrac{x^2}{4} + \dfrac{y^2}{9} - z^2 = 1$

54. $\dfrac{x^2}{9} - y^2 - z^2 = 1$ **55.** $z^2 = x^2 + \dfrac{y^2}{4}$

56. $z = x^2 + \dfrac{y^2}{4}$ **57.** $x^2 - 2y^2 = 9z$

58. Find an equation for a sphere, given that the endpoints of a diameter of the sphere are $(1, 2, -3)$ and $(-2, 3, 3)$.

LEVEL 3

59. Let $P(3, 2, -1)$, $Q(-2, 1, c)$ and $R(c, 1, 0)$ be points in $\mathbb{R}^3$. For what values of c (if any) is $\triangle PQR$ a right triangle?

60. Find the point P that lies $\frac{2}{3}$ of the distance from the point $A(-1, 3, 9)$ to the midpoint of the line segment joining points $B(-2, 3, 7)$ and $C(4, 1, -3)$.

9.3 The Dot Product

In this chapter, we have introduced vectors in $\mathbb{R}^2$ and then considered graphing in three dimensions. We now join these topics by considering three-dimensional vectors.

Vectors in $\mathbb{R}^3$

A vector in $\mathbb{R}^3$ may be thought of as a directed line segment (an "arrow") in space. The vector $\mathbf{P_1 P_2}$ with initial point $P_1(x_1, y_1, z_1)$ and terminal point $P_2(x_2, y_2, z_2)$ has the component form

$$\mathbf{P_1 P_2} = \langle x_2 - x_1, \ y_2 - y_1, \ z_2 - z_1 \rangle$$

Vector addition and multiplication of a vector by a scalar are defined for vectors in $\mathbb{R}^3$ in essentially the same way as these operations were defined for vectors in $\mathbb{R}^2$. In addition, the properties of vector algebra listed in Section 9.1 apply to vectors in $\mathbb{R}^3$ as well as to those in $\mathbb{R}^2$.

For example, we observed that each vector in $\mathbb{R}^2$ can be expressed as a unique linear combination of the standard basis vectors $\mathbf{i}$ and $\mathbf{j}$. This representation can be extended to vectors in $\mathbb{R}^3$ by adding a vector $\mathbf{k}$ defined to be the unit vector in the direction of the positive z-axis. In component form, we have in $\mathbb{R}^3$:

$$\mathbf{i} = \langle 1,0,0 \rangle \qquad \mathbf{j} = \langle 0,1,0 \rangle \qquad \mathbf{k} = \langle 0,0,1 \rangle$$

We call these the **standard basis vectors in** $\mathbb{R}^3$ (see Figure 9.22a). The **standard representation** of the vector with initial point at the origin O and terminal point $Q(a_1, a_2, a_3)$ is $\mathbf{OQ} = a_1\mathbf{i} + a_2\mathbf{j} + a_3\mathbf{k}$ as shown in Figure 9.22b.

a. The standard basis vectors in $\mathbb{R}^3$

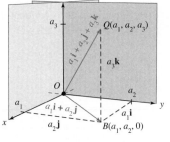

b. Vector from O to Q

Figure 9.22 Standard representation of vectors in $\mathbb{R}^3$

The vector $\mathbf{PQ}$ with initial point $P(x_0, y_0, z_0)$ and terminal point $Q(x_1, y_1, z_1)$ has the standard representation

$$\mathbf{PQ} = (x_1 - x_0)\mathbf{i} + (y_1 - y_0)\mathbf{j} + (z_1 - z_0)\mathbf{k}$$

EXAMPLE 1   Standard representation of a vector in $\mathbb{R}^3$

Find the standard representation of the vector $\mathbf{PQ}$ with initial point $P(-1, 2, 2)$ and terminal point $Q(3, -2, 4)$.

Solution
We have

$$\begin{aligned}\mathbf{PQ} &= [3 - (-1)]\mathbf{i} + [-2 - 2]\mathbf{j} + [4 - 2]\mathbf{k} \\ &= 4\mathbf{i} - 4\mathbf{j} + 2\mathbf{k}\end{aligned}$$

By referring to Figure 9.22b, we can also derive a formula for the length of a vector, which in turn can be used to prove the distance formula (stated in the previous section) between any two points in $\mathbb{R}^3$. Specifically, note that $\triangle OBQ$ in Figure 9.22b is a right triangle with hypotenuse of length $\|\mathbf{OQ}\|$ and legs $\|\mathbf{BQ}\| = |a_3|$ and $\|\mathbf{OB}\| = \sqrt{a_1^2 + a_2^2}$; by applying the Pythagorean theorem, we conclude that the vector $\mathbf{OQ} = a_1\mathbf{i} + a_2\mathbf{j} + a_3\mathbf{k}$ has length

$$\begin{aligned}\|\mathbf{OQ}\| &= \sqrt{\|\mathbf{OB}\|^2 + \|\mathbf{BQ}\|^2} \\ &= \sqrt{\left(a_1^2 + a_2^2\right) + a_3^2} \\ &= \sqrt{a_1^2 + a_2^2 + a_3^2}\end{aligned}$$

Moreover, if $A(a_1, a_2, a_3)$ and $B(b_1, b_2, b_3)$ are any two points in $\mathbb{R}^3$, the distance between them is the length of the vector $\mathbf{AB}$. We find

$$\begin{aligned}\mathbf{AB} &= \mathbf{OB} - \mathbf{OA} \\ &= (b_1 - a_1)\mathbf{i} + (b_2 - a_2)\mathbf{j} + (b_3 - a_3)\mathbf{k}\end{aligned}$$

so that the distance between A and B is given by

$$\| \mathbf{AB} \| = \sqrt{(b_1 - a_1)^2 + (b_2 - a_2)^2 + (b_3 - a_3)^2}$$

> **MAGNITUDE OF A VECTOR**
>
> The **magnitude**, or length, of the vector $\mathbf{v} = a_1\mathbf{i} + a_2\mathbf{j} + a_3\mathbf{k}$ is
>
> $$\| \mathbf{v} \| = \sqrt{a_1^2 + a_2^2 + a_3^2}$$

EXAMPLE 2   Magnitude of a vector

Find the magnitude of the vector $\mathbf{v} = 2\mathbf{i} - 3\mathbf{j} + 5\mathbf{k}$ and the distance between the points $A(1, -1, -4)$ and $B(-2, 3, 8)$.

Solution

$\| \mathbf{v} \| = \sqrt{2^2 + (-3)^2 + 5^2} = \sqrt{38}$ and

$$\| \overrightarrow{AB} \| = \sqrt{(-2 - 1)^2 + [3 - (-1)]^2 + [8 - (-4)]^2} = 13$$

As in $\mathbb{R}^2$, if $\mathbf{v}$ is a given nonzero vector in $\mathbb{R}^3$, then a unit vector $\mathbf{u}$ that points in the same direction as $\mathbf{v}$ is

$$\mathbf{u} = \frac{\mathbf{v}}{\| \mathbf{v} \|}$$

EXAMPLE 3   Unit vector in a given direction

Find a unit vector that points in the direction of the vector $\mathbf{PQ}$ from $P(-1, 2, 5)$ to $Q(0, -3, 7)$.

Solution

$$\mathbf{PQ} = (0 + 1)\mathbf{i} + (-3 - 2)\mathbf{j} + (7 - 5)\mathbf{k} \qquad \| \mathbf{PQ} \| = \sqrt{1^2 + (-5)^2 + 2^2}$$
$$= \mathbf{i} - 5\mathbf{j} + 2\mathbf{k} \qquad\qquad\qquad = \sqrt{30}$$

Thus,

$$\mathbf{u} = \frac{\mathbf{PQ}}{\| \mathbf{PQ} \|}$$
$$= \frac{\mathbf{i} - 5\mathbf{j} + 2\mathbf{k}}{\sqrt{30}}$$
$$= \frac{1}{\sqrt{30}}\mathbf{i} - \frac{5}{\sqrt{30}}\mathbf{j} + \frac{2}{\sqrt{30}}\mathbf{k}$$

As in $\mathbb{R}^2$, two vectors in $\mathbb{R}^3$ are **parallel** if they are multiples of one another, because parallel vectors point either in the same direction or in opposite directions. Nonzero vectors $\mathbf{u}$ and $\mathbf{v}$ are parallel if and only if $\mathbf{u} = s\mathbf{v}$ for some nonzero scalar s.

EXAMPLE 4   Parallel vectors

A vector $\mathbf{PQ}$ has initial point $P(1, 0, -3)$ and length 3. Find Q so that $\mathbf{PQ}$ is parallel to $\mathbf{v} = 2\mathbf{i} - 3\mathbf{j} + 6\mathbf{k}$.

Solution

Let Q have coordinates (a_1, a_2, a_3). Then,

$$\mathbf{PQ} = (a_1 - 1)\mathbf{i} + (a_2 - 0)\mathbf{j} + (a_3 + 3)\mathbf{k}$$
$$= (a_1 - 1)\mathbf{i} + a_2\mathbf{j} + (a_3 + 3)\mathbf{k}$$

Because $\mathbf{PQ}$ is parallel to $\mathbf{v}$, we have $\mathbf{PQ} = \mathbf{v}$ for some scalar s; that is,

$$(a_1 - 1)\mathbf{i} + a_2\mathbf{j} + (a_3 + 3)\mathbf{k} = s(2\mathbf{i} - 3\mathbf{j} + 6\mathbf{k})$$

This implies

$$a_1 - 1 = 2s \qquad a_2 = -3s \qquad a_3 + 3 = 6s$$
$$a_1 = 2s + 1 \qquad\qquad\qquad a_3 = 6s - 3$$

Because $\mathbf{PQ}$ has length 3, we have

$$3 = \sqrt{(a_1 - 1)^2 + a_2^2 + (a_3 + 3)^2}$$
$$= \sqrt{[(2s + 1) - 1]^2 + (-3s)^2 + [(6s - 3) + 3]^2}$$
$$= \sqrt{4s^2 + 9s^2 + 36s^2}$$
$$= \sqrt{49s^2}$$
$$= 7\,|s|$$

Thus, $s = \pm\dfrac{3}{7}$ and

$$a_1 = 2s + 1 \qquad a_2 = -3s \qquad a_3 = 6s - 3$$
$$= 2\left(\pm\frac{3}{7}\right) + 1 \qquad = -3\left(\pm\frac{3}{7}\right) \qquad = 6\left(\pm\frac{3}{7}\right) - 3$$
$$= \frac{13}{7}, \frac{1}{7} \qquad = -\frac{9}{7}, \frac{9}{7} \qquad = -\frac{3}{7}, -\frac{39}{7}$$

Two points satisfy the conditions for the required terminal point Q: $\left(\frac{13}{7}, -\frac{9}{7}, -\frac{3}{7}\right)$ and $\left(\frac{1}{7}, \frac{9}{7}, -\frac{39}{7}\right)$. ∎

Definition of Dot Product

The **dot (scalar) product** and the **cross (vector) product** are two important vector operations. We shall examine the cross product in the next section. The dot product is also known as a scalar product because it is a product of vectors that gives a scalar (that is, real number) as a result. Sometimes the dot product is called the **inner product**.

DOT PRODUCT

The **dot product** of vectors $\mathbf{v} = a_1\mathbf{i} + a_2\mathbf{j} + a_3\mathbf{k}$ and $\mathbf{w} = b_1\mathbf{i} + b_2\mathbf{j} + b_3\mathbf{k}$ is the scalar denoted by $\mathbf{v} \cdot \mathbf{w}$ and given by

$$\mathbf{v} \cdot \mathbf{w} = a_1 b_1 + a_2 b_2 + a_3 b_3$$

The dot product of two vectors $\mathbf{v} = a_1\mathbf{i} + a_2\mathbf{j}$ and $\mathbf{w} = b_1\mathbf{i} + b_2\mathbf{j}$ in a plane is given by a similar formula with $a_3 = b_3 = 0$: $\mathbf{v} \cdot \mathbf{w} = a_1 b_1 + a_2 b_2$.

EXAMPLE 5   Dot product

Find the dot product of $\mathbf{v} = -3\mathbf{i} + 2\mathbf{j} + \mathbf{k}$ and $\mathbf{w} = 4\mathbf{i} - \mathbf{j} + 2\mathbf{k}$.

Solution

$\mathbf{v} \cdot \mathbf{w} = -3(4) + 2(-1) + 1(2) = -12$. ∎

EXAMPLE 6   Dot product in component form

If $\mathbf{v} = \langle 4, -1, 3 \rangle$ and $\mathbf{w} = \langle -1, -2, 5 \rangle$ find the dot product, $\mathbf{v} \cdot \mathbf{w}$.

Solution

$\mathbf{v} \cdot \mathbf{w} = 4(-1) + (-1)(-2) + 3(5) = 13$. ∎

Before we can apply the dot product to geometric and physical problems, we need to know how it behaves algebraically. A number of important general properties of the dot product are listed in the following box.

DOT PRODUCT PROPERTIES

If $\mathbf{u}$, $\mathbf{v}$, and $\mathbf{w}$ are vectors in $\mathbb{R}^2$ or $\mathbb{R}^3$ and c is a scalar, then:

Magnitude of a vector	$\mathbf{v} \cdot \mathbf{v} = \| \mathbf{v} \|^2$
Zero product	$\mathbf{0} \cdot \mathbf{v} = 0$
Commutativity	$\mathbf{v} \cdot \mathbf{w} = \mathbf{w} \cdot \mathbf{v}$
Scalar multiple	$c(\mathbf{v} \cdot \mathbf{w}) = (c\mathbf{v}) \cdot \mathbf{w} = \mathbf{v} \cdot (c\mathbf{w})$
Distributivity	$\mathbf{u} \cdot (\mathbf{v} + \mathbf{w}) = \mathbf{u} \cdot \mathbf{v} + \mathbf{u} \cdot \mathbf{w}$

To show where these properties come from, let $\mathbf{u} = a_1\mathbf{i} + a_2\mathbf{j} + a_3\mathbf{k}$, $\mathbf{v} = b_1\mathbf{i} + b_2\mathbf{j} + b_3\mathbf{k}$, and $\mathbf{w} = c_1\mathbf{i} + c_2\mathbf{j} + c_3\mathbf{k}$. We begin with the magnitude of a vector property.

$$\| \mathbf{v} \|^2 = \left(\sqrt{a_1^2 + a_2^2 + a_3^2} \right)^2$$
$$= a_1^2 + a_2^2 + a_3^2$$
$$= \mathbf{v} \cdot \mathbf{v}$$

The zero product and scalar multiple properties are similarly shown. For the distributive property, we have

$$\mathbf{u} \cdot (\mathbf{v} + \mathbf{w}) = (a_1\mathbf{i} + a_2\mathbf{j} + a_3\mathbf{k}) \cdot [(b_1 + c_1)\mathbf{i} + (b_2 + c_2)\mathbf{j} + (b_3 + c_3)\mathbf{k}]$$
$$= a_1(b_1 + c_1) + a_2(b_2 + c_2) + a_3(b_3 + c_3)$$
$$= a_1b_1 + a_1c_1 + a_2b_2 + a_2c_2 + a_3b_3 + a_3c_3$$

Also,

$$\mathbf{u} \cdot \mathbf{v} + \mathbf{u} \cdot \mathbf{w} = (a_1b_1 + a_2b_2 + a_3b_3) + (a_1c_1 + a_2c_2 + a_3c_3)$$
$$= a_1b_1 + a_1c_1 + a_2b_2 + a_2c_2 + a_3b_3 + a_3c_3$$

Since these are the same, we see

$$\mathbf{u} \cdot (\mathbf{v} + \mathbf{w}) = \mathbf{u} \cdot \mathbf{v} + \mathbf{u} \cdot \mathbf{w}$$

The commutative property for the dot product can be established in a similar fashion.

Angle Between Vectors

The angle between two nonzero vectors $\mathbf{v}$ and $\mathbf{w}$ is defined to be the angle θ with $0 \le \theta \le \pi$ that is formed when the vectors are in standard position (initial points at the origin), as shown in Figure 9.23.

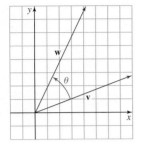

Figure 9.23 The angle between two vectors

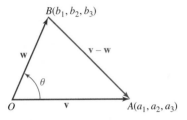

Figure 9.24 Angle between two vectors

The angle between vectors plays an important role in certain applications and may be computed by using the following formula involving the dot product.

ANGLE BETWEEN VECTORS

If θ is the angle between the nonzero vectors $\mathbf{v}$ and $\mathbf{w}$, then

$$\cos\theta = \frac{\mathbf{v}\cdot\mathbf{w}}{\|\mathbf{v}\|\,\|\mathbf{w}\|}$$

We can derive this formula as follows. Suppose $\mathbf{v}=a_1\mathbf{i}+a_2\mathbf{j}+a_3\mathbf{k}$ and $\mathbf{w}=b_1\mathbf{i}+b_2\mathbf{j}+b_3\mathbf{k}$, and consider $\triangle AOB$ with vertices at the origin O and the points $A(a_1,\ a_2,\ a_3)$ and $B(b_1,\ b_2,\ b_3)$, as shown in Figure 9.24.

Note that sides $\overline{OA}$ and $\overline{OB}$ have lengths $\|\mathbf{v}\|=\sqrt{a_1^2+a_2^2+a_3^3}$ and $\|\mathbf{w}\|=\sqrt{b_1^2+b_2^2+b_3^2}$, respectively, and that side $\overline{AB}$ has length $\|\mathbf{v}-\mathbf{w}\|=\sqrt{(a_1-b_1)^2+(a_2-b_2)^2+(a_3-b_3)^2}$. Next, we use the law of cosines.

$$a^2 = b^2 + c^2 - 2bc\cos\theta \qquad \text{Law of cosines}$$

$$\|\mathbf{v}-\mathbf{w}\|^2 = \|\mathbf{v}\|^2 + \|\mathbf{w}\|^2 - 2\|\mathbf{v}\|\|\mathbf{w}\|\cos\theta \qquad \text{See Figure Figure 9.24.}$$

$$\cos\theta = \frac{\|\mathbf{v}\|^2 + \|\mathbf{w}\|^2 - \|\mathbf{v}-\mathbf{w}\|^2}{2\|\mathbf{v}\|\|\mathbf{w}\|} \qquad \text{Solve for } \cos\theta.$$

$$= \frac{a_1^2+a_2^2+a_3^2+b_1^2+b_2^2+b_3^2-[(a_1-b_1)^2+(a_2-b_2)^2+(a_3-b_3)^2]}{2\|\mathbf{v}\|\|\mathbf{w}\|}$$

$$= \frac{2a_1b_1+2a_2b_2+2a_3b_3}{2\|\mathbf{v}\|\|\mathbf{w}\|}$$

$$= \frac{\mathbf{v}\cdot\mathbf{w}}{\|\mathbf{v}\|\|\mathbf{w}\|}$$

EXAMPLE 7   Angle between two given vectors

Let $\triangle ABC$ be the triangle with vertices $A(1,\ 1,\ 8)$, $B(4,\ -3,\ -4)$, and $C(-3,\ 1,\ 5)$. Find the angle formed at A.

Solution
Draw $\triangle ABC$ and label the angle formed at A as α as shown in Figure 9.25.

The angle α is the angle between vectors $\mathbf{AB}$ and $\mathbf{AC}$, where

$$\mathbf{AB}=(4-1)\mathbf{i}+(-3-1)\mathbf{j}+(-4-8)\mathbf{k} \qquad \mathbf{AC}=(-3-1)\mathbf{i}+(1-1)\mathbf{j}+(5-8)\mathbf{k}$$
$$= 3\mathbf{i}-4\mathbf{j}-12\mathbf{k} \qquad\qquad\qquad = -4\mathbf{i}-3\mathbf{k}$$

Thus,

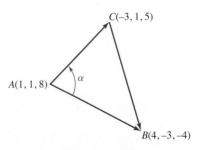

Figure 9.25 Angle in a triangle

$$\cos\alpha = \frac{\mathbf{AB}\cdot\mathbf{AC}}{\|\mathbf{AB}\|\|\mathbf{AC}\|}$$

$$= \frac{3(-4)+(-4)(0)+(-12)(-3)}{\sqrt{3^2+(-4)^2+(-12)^2}\,\sqrt{(-4)^2+(-3)^2}}$$

$$= \frac{24}{\sqrt{169}\sqrt{25}}$$

$$= \frac{24}{65}$$

and the required angle is $\alpha=\cos^{-1}\left(\frac{24}{65}\right)\approx 1.19$ (or about $68°$). ∎

The formula for the angle between vectors is often used in conjunction with the dot product. If we multiply both sides of the formula by $\| \mathbf{v} \| \| \mathbf{w} \|$ we obtain the following alternate form for the dot product formula.

DOT PRODUCT (ALTERNATE)

$$\mathbf{v} \cdot \mathbf{w} = \| \mathbf{v} \| \| \mathbf{w} \| \cos\theta$$

where θ is the angle $(0 \leq \theta \leq \pi)$ between the vectors $\mathbf{v}$ and $\mathbf{w}$.

Two vectors are said to be **perpendicular**, or **orthogonal**, if the angle between them is $\theta = \frac{\pi}{2}$. The following theorem provides a useful condition for orthogonality.

ORTHOGONAL VECTORS

Nonzero vectors $\mathbf{v}$ and $\mathbf{w}$ are **orthogonal** if and only if

$$\mathbf{v} \cdot \mathbf{w} = 0$$

If the vectors are orthogonal, then the angle between them is $\frac{\pi}{2}$; therefore,

$$\mathbf{v} \cdot \mathbf{w} = \| \mathbf{v} \| \| \mathbf{w} \| \cos\frac{\pi}{2} = 0$$

Conversely, if $\mathbf{v} \cdot \mathbf{w} = 0$ and $\mathbf{v}$ and $\mathbf{w}$ are nonzero vectors, then $\cos\theta = 0$, so that $\theta = \frac{\pi}{2}$ and the vectors are orthogonal.

EXAMPLE 8   Orthogonal vectors

Determine which (if any) of the following vectors are orthogonal:

$$\mathbf{u} = 3\mathbf{i} + 7\mathbf{j} - 2\mathbf{k} \qquad \mathbf{v} = 5\mathbf{i} - 3\mathbf{j} - 3\mathbf{k} \qquad \mathbf{w} = \mathbf{j} - \mathbf{k}$$

Solution

$$
\begin{aligned}
\mathbf{u} \cdot \mathbf{v} &= 3(5) + 7(-3) + (-2)(-3) = 0; && \text{orthogonal vectors.} \\
\mathbf{u} \cdot \mathbf{w} &= 3(0) + 7(1) + (-2)(-1) = 9; && \text{not orthogonal vectors.} \\
\mathbf{v} \cdot \mathbf{w} &= 5(0) + (-3)(1) + (-3)(-1) = 0; && \text{orthogonal vectors.}
\end{aligned}
$$

■

Projections

Let $\mathbf{v}$ and $\mathbf{w}$ be two vectors in $\mathbb{R}^2$ drawn so that they have a common initial point, as shown in Figure 9.26.* If we drop a perpendicular from the head of $\mathbf{v}$ to the line determined by $\mathbf{w}$, we determine a vector called the **vector projection of v onto w**, which we have labeled $\mathbf{u}$ in Figure 9.26. The **scalar projection of v onto w** (also called the **component of v along w**) is the length of the vector projection, so it is denoted by $\| \mathbf{u} \|$. Let θ be the acute angle between $\mathbf{v}$ and $\mathbf{w}$. Then, by definition of cosine we have

$$\| \mathbf{u} \| = \| \mathbf{v} \| \cos\theta$$

$$= \| \mathbf{v} \| \left(\frac{\mathbf{v} \cdot \mathbf{w}}{\| \mathbf{v} \| \| \mathbf{w} \|} \right)$$

$$= \frac{\mathbf{v} \cdot \mathbf{w}}{\| \mathbf{w} \|} \qquad \textit{Cosine is positive because } \theta \textit{ is acute.}$$

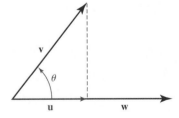

Figure 9.26 Projection of v onto w

If θ is obtuse, its cosine is negative and $\| \mathbf{u} \| = -\| \mathbf{v} \| \cos\theta$. To find a formula for the vector projection, we note that it is the scalar component of $\mathbf{v}$ in the direction of $\mathbf{w}$. If θ is acute, then the vector projection has length $\| \mathbf{v} \| \cos\theta$ and has direction $\mathbf{w}/\| \mathbf{w} \|$ (the unit vector of $\mathbf{w}$). If the

---

*Even though Figure 9.26 is drawn in $\mathbb{R}^2$, the projection formula applies to $\mathbb{R}^3$ as well.

angle is obtuse, the vector projection has length $-\|\mathbf{v}\|\cos\theta$ and has direction $-\mathbf{w}/\|\mathbf{w}\|$. In either case,

$$\mathbf{u} = (\|\mathbf{v}\|\cos\theta)\frac{\mathbf{w}}{\|\mathbf{w}\|}$$

$$= \frac{\mathbf{v}\cdot\mathbf{w}}{\|\mathbf{w}\|}\left(\frac{\mathbf{w}}{\|\mathbf{w}\|}\right)$$

$$= \frac{\mathbf{v}\cdot\mathbf{w}}{\|\mathbf{w}\|^2}\mathbf{w}$$

$$= \left(\frac{\mathbf{v}\cdot\mathbf{w}}{\mathbf{w}\cdot\mathbf{w}}\right)\mathbf{w}$$

☠ Note that $\left(\frac{\mathbf{v}\cdot\mathbf{w}}{\mathbf{w}\cdot\mathbf{w}}\right)\mathbf{w}$ is not the same as $\left(\frac{\mathbf{v}}{\mathbf{w}}\right)\mathbf{w}$; you cannot "cancel" the vector $\mathbf{w}$. Remember, $\mathbf{v}\cdot\mathbf{w}$ and $\mathbf{w}\cdot\mathbf{w}$ are real numbers, whereas $\mathbf{v}/\mathbf{w}$ is not defined. ☠

PROJECTIONS

Scalar projection of $\mathbf{v}$ onto $\mathbf{w}$ (a number): $\left|\dfrac{\mathbf{v}\cdot\mathbf{w}}{\|\mathbf{w}\|}\right|$

Vector projection of $\mathbf{v}$ in the direction of $\mathbf{w}$ (a vector): $\left(\dfrac{\mathbf{v}\cdot\mathbf{w}}{\mathbf{w}\cdot\mathbf{w}}\right)\mathbf{w}$

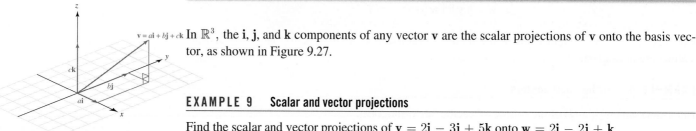

Figure 9.27 Basis vectors are vector projections

In $\mathbb{R}^3$, the $\mathbf{i}$, $\mathbf{j}$, and $\mathbf{k}$ components of any vector $\mathbf{v}$ are the scalar projections of $\mathbf{v}$ onto the basis vector, as shown in Figure 9.27.

EXAMPLE 9   Scalar and vector projections

Find the scalar and vector projections of $\mathbf{v} = 2\mathbf{i} - 3\mathbf{j} + 5\mathbf{k}$ onto $\mathbf{w} = 2\mathbf{i} - 2\mathbf{j} + \mathbf{k}$.

Solution
We first find the vector projection:

$$\text{vector projection of } \mathbf{v} \text{ onto } \mathbf{w} = \left(\frac{\mathbf{v}\cdot\mathbf{w}}{\mathbf{w}\cdot\mathbf{w}}\right)\mathbf{w}$$

$$= \left(\frac{2(2)+(-3)(-2)+5(1)}{2^2+(-2)^2+1^2}\right)(2\mathbf{i}-2\mathbf{j}+\mathbf{k})$$

$$= \frac{15}{9}(2\mathbf{i}-2\mathbf{j}+\mathbf{k})$$

$$= \frac{10}{3}\mathbf{i} - \frac{10}{3}\mathbf{j} + \frac{5}{3}\mathbf{k}$$

To find the scalar projection, we can find the length of the vector projection, or we can use the scalar projection formula (which is usually easier than finding the length directly):

$$\text{scalar projection of } \mathbf{v} \text{ onto } \mathbf{w} = \left|\frac{\mathbf{v}\cdot\mathbf{w}}{\|\mathbf{w}\|}\right|$$

$$= \left|\frac{15}{3}\right|$$

$$= 5$$

Work as a Dot Product

One important application of the dot product and projections occurs in physics when calculating the amount of work done by a constant force. When a constant force of magnitude F is applied to an object through a distance d, the **work** performed is defined to be the product of force and distance. This means that the work, W, done by a constant force directed along the line of the motion of the object is $W = Fd$ (Figure 9.28**a**). We now consider a force acting in some other direction (Figure 9.28**b**).

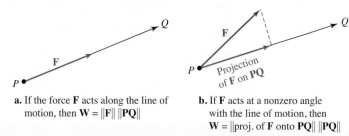

a. If the force **F** acts along the line of motion, then $\mathbf{W} = \|\mathbf{F}\| \|\mathbf{PQ}\|$

b. If **F** acts at a nonzero angle with the line of motion, then $\mathbf{W} = \|\text{proj. of } \mathbf{F} \text{ onto } \mathbf{PQ}\| \|\mathbf{PQ}\|$

Figure 9.28 Work W as a dot product

Physics experiments indicate that when a force **F** moves an object along the line from point P to point Q, the work performed is given by the product

$$W = [\text{SCALAR COMPONENT OF } \mathbf{F} \text{ ALONG } \mathbf{PQ}][\text{DISTANCE MOVED BY THE OBJECT}]$$

The distance moved by the object is the length of the *displacement vector* **PQ**, and we find that

Scalar component of **F** along **PQ**

$$W = \frac{\mathbf{F} \cdot \mathbf{PQ}}{\|\mathbf{PQ}\|} \|\mathbf{PQ}\| = \mathbf{F} \cdot \mathbf{PQ}$$

Distance moved by the object

WORK AS A DOT PRODUCT

An object that moves along a line with displacement **PQ** against a constant force **F** performs

$$W = \mathbf{F} \cdot \mathbf{PQ}$$

units of work.

EXAMPLE 10 Work performed by a constant force

Suppose that the wind is blowing with a force **F** of magnitude 500 lb in the direction of N30°E over a boat's sail. How much work does the wind perform in moving the boat in a northerly direction a distance of 100 ft? Give your answer in foot-pounds.

Solution

We see that $\|\mathbf{F}\| = 500$ lb and is in the direction of N30°E, as shown in Figure 9.29.

The displacement direction is $\mathbf{PQ} = 100\mathbf{j}$, so $\|\mathbf{PQ}\| = 100$ ft. Thus, $\mathbf{F} = 500\cos 60°\mathbf{i} + 500\sin 60°\mathbf{j} = 250\mathbf{i} + 250\sqrt{3}\mathbf{j}$. Thus, the work performed is

$$W = \mathbf{F} \cdot \mathbf{PQ} = 100\left(250\sqrt{3}\right) = 25{,}000\sqrt{3}$$

Thus, the work is approximately 43,300 ft-lb.

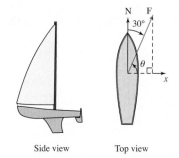

Side view   Top view

Figure 9.29 Work performed

In the U.S. system of measurements, work is typically expressed in *foot-pounds*, *inch-pounds*, or *foot-tons*. In the International System (SI), work is expressed in newton-meters (called *joules*), and in the centimeter-gram-second (CGS) system, the basic unit of work is the dyne-centimeter (called an *erg*). These units are shown in Table 9.2.*

TABLE 9.2	Common Units of Work and Force		
Mass	**Distance**	**Unit of force**	**Work**
kg	m	newton (N)	joule
g	cm	dyne (dyn)	erg
slug	ft	pound	ft-lb

PROBLEM SET 9.3

LEVEL 1

1. IN YOUR OWN WORDS Discuss how to find a dot product, and describe an application of dot product.
2. IN YOUR OWN WORDS Describe and contrast scalar and vector projections.

Find the standard representation of the vector **PQ**, *and then find* $\| \mathbf{PQ} \|$ *in Problems 3–6.*

3. $P(1, -1, 3)$, $Q(-1, 1, 4)$ 4. $P(0, 2, 3)$, $Q(2, 3, 0)$
5. $P(1, 1, 1)$, $Q(-3, -3, -3)$ 6. $P(3, 0, -4)$, $Q(0, -4, 3)$

Find the dot product $\mathbf{v} \cdot \mathbf{w}$ *in Problems 7–10.*

7. $\mathbf{v} = \langle 3, -2, 4 \rangle$; $\mathbf{w} = \langle 2, -1, -6 \rangle$
8. $\mathbf{v} = \langle 2, -6, 0 \rangle$; $\mathbf{w} = \langle 0, -3, 7 \rangle$
9. $\mathbf{v} = 2\mathbf{i} + 3\mathbf{j} - \mathbf{k}$; $\mathbf{w} = -3\mathbf{i} + 5\mathbf{j} + 4\mathbf{k}$
10. $\mathbf{v} = 3\mathbf{i} - \mathbf{j}$; $\mathbf{w} = 2\mathbf{i} + 5\mathbf{j}$

State whether the given pairs of vectors in Problems 11–14 are orthogonal.

11. $\mathbf{v} = \mathbf{i}$; $\mathbf{w} = \mathbf{k}$
12. $\mathbf{v} = \mathbf{j}$; $\mathbf{w} = -\mathbf{k}$
13. $\mathbf{v} = 3\mathbf{i} - 2\mathbf{j}$; $\mathbf{w} = 6\mathbf{i} + 9\mathbf{j}$
14. $\mathbf{v} = 4\mathbf{i} - 5\mathbf{j} + \mathbf{k}$; $\mathbf{w} = 8\mathbf{i} + 10\mathbf{j} - 2\mathbf{k}$

Evaluate the expressions given in Problems 15–18.

15. $\| \mathbf{i} + \mathbf{j} + \mathbf{k} \|$
16. $\| \mathbf{i} - \mathbf{j} + \mathbf{k} \|$
17. $\| 2\mathbf{i} + \mathbf{j} - 3\mathbf{k} \|^2$
18. $\| 2(\mathbf{i} - \mathbf{j} + \mathbf{k}) - 3(2\mathbf{i} + \mathbf{j} - \mathbf{k}) \|^2$

Let $\mathbf{v} = \mathbf{i} - 2\mathbf{j} + 2\mathbf{k}$ *and* $\mathbf{w} = 2\mathbf{i} + 4\mathbf{j} - \mathbf{k}$; *and find the vector or scalar requested in Problems 19–22.*

19. $2\mathbf{v} - 3\mathbf{w}$
20. $\| \mathbf{v} \| \mathbf{w}$
21. $\| 2\mathbf{v} - 3\mathbf{w} \|$
22. $\| \mathbf{v} - \mathbf{w} \| (\mathbf{v} + \mathbf{w})$

Determine whether each vector in Problems 23–26 is parallel to $\mathbf{T} = 2\mathbf{i} - 3\mathbf{j} + 5\mathbf{k}$.

23. $\mathbf{u} = \langle -4, 6, -10 \rangle$
24. $\mathbf{v} = \langle 4, 0, 10 \rangle$
25. $\mathbf{w} = \left\langle 1, -\frac{3}{2}, 2 \right\rangle$
26. $\mathbf{v} = \left\langle -\frac{1}{2}, \frac{3}{2}, 2 \right\rangle$

Let $\mathbf{v} = 3\mathbf{i} - 2\mathbf{j} + \mathbf{k}$ *and* $\mathbf{w} = \mathbf{i} + \mathbf{j} - \mathbf{k}$. *Carry out the indicated operations in Problems 27–30 to find a vector or a number.*

27. $(\mathbf{v} + \mathbf{w}) \cdot (\mathbf{v} - \mathbf{w})$
28. $(\mathbf{v} \cdot \mathbf{w})\mathbf{w}$
29. $(\| \mathbf{v} \| \mathbf{w}) \cdot (\| \mathbf{w} \| \mathbf{v})$
30. $\dfrac{2\mathbf{v} + 3\mathbf{w}}{\| 3\mathbf{v} + 2\mathbf{w} \|}$

---

*Some of the measurements and units in this table may not be familiar to you. For example, a *slug* is a unit of measurement in physics defined to be "the unit of mass that is accelerated at the rate of one foot per second per second when acted on by a force of one pound weight." If you have not studied physics, this may be meaningless to you right now, but remember that our focus here is not on the units of measurement, but on how to calculate *work* using scalar multiplication of vectors. However, if you are interested, a *dyne* is defined to be "the force required to accelerate a mass of one centimeter per second per second to a mass of one gram" and a *newton* is defined to be "the unit of force required to accelerate a mass of one kilogram one meter per second per second." Thus, one newton is equal to 100,000 dynes.

Find the angle between the vectors given in Problems 31–34. Round to the nearest degree.

31. $v = i + j + k$; $w = i - j + k$
32. $v = 2i + k$; $w = j - 3k$
33. $v = 2j + k$; $w = i - 2k$
34. $v = 4i - j + k$; $w = 2i + 3j + 5k$

*Find the vector and scalar projections of **v** onto **w** in Problems 35–38.*

35. $v = i + j + k$; $w = 2k$
36. $v = i + 2k$; $w = -3j$
37. $v = 2i - 3j$; $w = 2j - 3k$
38. $v = i + j - 2k$; $w = i + j + k$

LEVEL 2

Find two distinct unit vectors that are orthogonal to each pair of vectors given in Problems 39–42.

39. $v = i + k$; $w = i - 2k$
40. $v = i + j - k$; $w = -i + j + k$
41. $v = 2i + j + 2k$; $w = -i + 2j - k$
42. $v = 2j + 3k$; $w = i + j + k$
43. Find a unit vector that points in the direction opposite to $v = 2i + 3j - 2k$.
44. Find a vector that points in the same direction as $v = i + 2j - k$ and has length one-third unit.
45. Find a number a that guarantees that the vectors $3i - 2j + k$ and $2i + aj - 2ak$ will be orthogonal.
46. Find x if the vectors $v = 3i - xj + 2k$ and $w = xi + j - 2k$ are to be orthogonal.
47. Find the angles between the vector $2i + j - k$ and each of the coordinate axes. The cosines of these angles (to the nearest degree) are known as the **direction cosines**.
48. Find the cosine of the angle between the vectors $v = i - j + 2k$ and $w = 2i + j - k$. Then find the vector projection of v onto w.
49. Let $v = 4i - j + k$ and $w = 2i + 3j - k$. Find:
   a. $v \cdot w$
   b. $\cos\theta$, where θ is the angle between v and w
   c. a scalar s such that v is orthogonal to $v - sw$
   d. a scalar s such that $sv + w$ is orthogonal to w
50. Let $v = 2i - 3j + 6k$ and $w = 4i + 3k$. Find
   a. $v \cdot w$
   b. $\cos\theta$, where θ is the angle between v and w
   c. a scalar s such that v is orthogonal to $v - sw$
   d. scalar s such that $sv + w$ is orthogonal to w
51. Find the work done by the constant force $F = 2i + 3j + k$ when it moves a particle along the line from $P(1, 0, -1)$ to $Q(3, 1, 2)$.
52. Find the work performed when a force $F = \frac{6}{7}i - \frac{2}{7}j + \frac{6}{7}k$ is applied to an object moving along the line from $P(-3, -5, 4)$ to $Q(4, 9, 11)$.

53. A boy mowing the lawn wants to impress his dad by calculating the amount of work he is doing (see Figure 9.30).

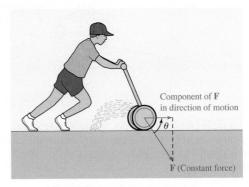

Figure 9.30 Work mowing the lawn

The *displacement*, d, the distance the lawn mower is moved, is 2,450 ft and the angle in which the boy is pressing down on the handle is 40° with a force of 48.0 lb. What is the amount of work done by the boy?

54. A dad pulls his child in a wagon (see Figure 9.31) for 150 ft by exerting a constant force of 25 lb along the handle. How much work is done?

Figure 9.31 Work pulling a wagon

55. PROBLEMS FROM CALCULUS Fred and his son Sam are pulling a heavy log along flat horizontal ground by ropes attached to the front of the log. The ropes are 8 ft long. Fred holds his rope 2 ft above the log and 1 ft to the side, and Sam holds his end 1 ft above the log and 1 ft to the opposite side, as shown in Figure 9.32.

Figure 9.32 Work pulling a log

If Fred exerts a force of 30 lb and Sam exerts a force of 20 lb, what is the resultant force on the log?

56. PROBLEMS FROM CALCULUS Find the force required to keep a 5,000-lb van from rolling downhill if it is parked on a 10° slope.

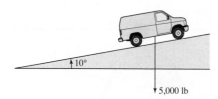

57. PROBLEMS FROM CALCULUS A block of ice is dragged 20 ft across a floor, using a force of 50 lb. Find the work done if the direction of the force is inclined θ to the horizontal, where

a. $\theta = \frac{\pi}{6}$ **b.** $\theta = \frac{\pi}{4}$

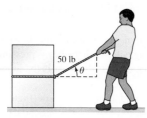

58. PROBLEMS FROM CALCULUS Suppose that the wind is blowing with a 1,000-lb magnitude force **F** in the direction of N30°E over a boat's sail. How much work does the wind perform in moving the boat an easterly direction a distance of 50 ft? Give your answer in foot-pounds.

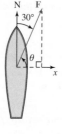

Side view   Top view

LEVEL 3

59. a. Show that

$$(\mathbf{v}+\mathbf{w})\cdot(\mathbf{v}+\mathbf{w})=\|\mathbf{v}\|^2 + \|\mathbf{w}\|^2 +2(\mathbf{v}\cdot\mathbf{w})$$

   b. Use part **a** to prove the **triangle inequality**:

$$\|\mathbf{v}+\mathbf{w}\|\leq\|\mathbf{v}\|+\|\mathbf{w}\|$$

   Hint: Note that

$$\|\mathbf{v}+\mathbf{w}\|^2= (\mathbf{v}+\mathbf{w})\cdot(\mathbf{v}+\mathbf{w})$$

60. The Cauchy-Schwarz inequality in $\mathbb{R}^3$ states that for any vectors **v** and **w**

$$|\mathbf{v}\cdot\mathbf{w}| \leq \|\mathbf{v}\|\|\mathbf{w}\|$$

   a. Prove the Cauchy-Schwarz inequality.
      Hint: Use the formula for the angle between vectors.
   b. Use the Cauchy-Schwarz inequality to prove the triangle inequality.

9.4 The Cross Product

We have considered two types of products involving vectors. The first was *scalar multiplication*, which is the multiplication of a vector by a number (which is called a *scalar*). In the last section, we defined a product of vectors that produces a number (i.e., scalar); this product is known as *scalar product* (because the answer is a scalar) or *dot product* since the notation we use for this product is a dot. We are now ready to consider a third product of vectors that produces a vector. It is called the *vector product* or *cross product* because we use the multiplication cross ($\times$) to denote this type of product.

Definition of Cross Product*

The cross product is sometimes called **vector product** because the result is a vector. In other applications, it is called the **outer product**.

> **CROSS PRODUCT**
>
> If $\mathbf{v} = a_1\mathbf{i} + a_2\mathbf{j} + a_3\mathbf{k}$ and $\mathbf{w} = b_1\mathbf{i} + b_2\mathbf{j} + b_3\mathbf{k}$, the **cross product**, written $\mathbf{v}\times\mathbf{w}$, is the vector
>
> $$\mathbf{v}\times\mathbf{w} = (a_2b_3 - a_3b_2)\mathbf{i} + (a_3b_1 - a_1b_3)\mathbf{j} + (a_1b_2 - a_2b_1)\mathbf{k}$$

---

*Determinants (see Appendix E) are necessary for this section.

This definition seems very strange, indeed, until we show that these terms can be obtained by using a determinant

$$\mathbf{v} \times \mathbf{w} = \begin{vmatrix} \mathbf{i} & \mathbf{j} & \mathbf{k} \\ a_1 & a_2 & a_3 \\ b_1 & b_2 & b_3 \end{vmatrix}$$

We can verify this is the formula for cross product by expanding this determinant about the first row. (A brief discussion of the basic properties of determinants is included in Appendix E.)

$$\mathbf{v} \times \mathbf{w} = \begin{vmatrix} \mathbf{i} & \mathbf{j} & \mathbf{k} \\ a_1 & a_2 & a_3 \\ b_1 & b_2 & b_3 \end{vmatrix} = \begin{vmatrix} a_2 & a_3 \\ b_2 & b_3 \end{vmatrix} \mathbf{i} \underset{\uparrow}{-} \begin{vmatrix} a_1 & a_3 \\ b_1 & b_3 \end{vmatrix} \mathbf{j} + \begin{vmatrix} a_1 & a_2 \\ b_1 & b_2 \end{vmatrix} \mathbf{k}$$

j is in row 1, column 2, so do not forget the negative sign here.

$$= (a_2 b_3 - a_3 b_2)\mathbf{i} - (a_1 b_3 - a_3 b_1)\mathbf{j} + (a_1 b_2 - a_2 b_1)\mathbf{k}$$
$$= (a_2 b_3 - a_3 b_2)\mathbf{i} + (a_3 b_1 - a_1 b_3)\mathbf{j} + (a_1 b_2 - a_2 b_1)\mathbf{k}$$

The definition requires a basis of $\mathbf{i}$, $\mathbf{j}$, and $\mathbf{k}$ and is, therefore, a definition that makes sense only in $\mathbb{R}^3$.

EXAMPLE 1   Cross product

Find $\mathbf{v} \times \mathbf{w}$ where $\mathbf{v} = 2\mathbf{i} - \mathbf{j} + \mathbf{k}$ and $\mathbf{w} = 7\mathbf{j} - 4\mathbf{k}$.

Solution

$$\mathbf{v} \times \mathbf{w} = \begin{vmatrix} \mathbf{i} & \mathbf{j} & \mathbf{k} \\ 2 & -1 & 3 \\ 0 & 7 & -4 \end{vmatrix}$$

Do not forget a minus here ($\mathbf{j}$ is in the negative position).
↓
$$= [(-1)(-4) - 3(7)]\mathbf{i} - [2(-4) - 0(3)]\mathbf{j} + [2(7) - 0(-1)]\mathbf{k}$$
$$= -17\mathbf{i} + 8\mathbf{j} + 14\mathbf{k}$$
∎

Properties of determinants can also be used to establish properties of the cross product. For instance, the following computation shows that the cross product is *not* commutative. This property is sometimes called **anticommutativity**.

$$\mathbf{v} \times \mathbf{w} = \begin{vmatrix} \mathbf{i} & \mathbf{j} & \mathbf{k} \\ a_1 & a_2 & a_3 \\ b_1 & b_2 & b_3 \end{vmatrix} = - \begin{vmatrix} \mathbf{i} & \mathbf{j} & \mathbf{k} \\ b_1 & b_2 & b_3 \\ a_1 & a_2 & a_3 \end{vmatrix} = -(\mathbf{w} \times \mathbf{v})$$

The cross product has interesting properties. We have already seen that $\mathbf{v} \times \mathbf{w} = -(\mathbf{w} \times \mathbf{v})$. This and other properties are listed in the following box.

CROSS PRODUCT

If $\mathbf{u}$, $\mathbf{v}$, and $\mathbf{w}$ are vectors in $\mathbb{R}^3$ and s and t are scalars, then several properties can be derived:

Scalar distributivity	$(s\mathbf{v}) \times (t\mathbf{w}) = st(\mathbf{v} \times \mathbf{w})$
Vector distributivity*	$\mathbf{u} \times (\mathbf{v} + \mathbf{w}) = (\mathbf{u} \times \mathbf{v}) + (\mathbf{u} \times \mathbf{w})$
	$(\mathbf{u} + \mathbf{v}) \times \mathbf{w} = (\mathbf{u} \times \mathbf{w}) + (\mathbf{v} \times \mathbf{w})$
Anticommutativity	$\mathbf{v} \times \mathbf{w} = -(\mathbf{w} \times \mathbf{v})$
Parallel vectors	$\mathbf{v} \times \mathbf{v} = \mathbf{0}$; if $\mathbf{w} = s\mathbf{v}$, then $\mathbf{v} \times \mathbf{w} = \mathbf{0}$
Zero product	$\mathbf{v} \times \mathbf{0} = \mathbf{0} \times \mathbf{v} = \mathbf{0}$

---

*Properly, it is the distributive property of vectors for cross product over addition, which, for convenience, we shorten to vector distributivity.

To derive these properties, let

$$\mathbf{u} = a_1\mathbf{i} + a_2\mathbf{j} + a_3\mathbf{k}, \mathbf{v} = b_1\mathbf{i} + b_2\mathbf{j} + b_3\mathbf{k}, \text{ and } \mathbf{w} = c_1\mathbf{i} + c_2\mathbf{j} + c_3\mathbf{k}$$

Scalar distributivity and *vector distributivity* are proved by using the definition of cross product and the corresponding properties of real numbers. For **zero factors** we find:

$$\mathbf{v} \times \mathbf{v} = \begin{vmatrix} \mathbf{i} & \mathbf{j} & \mathbf{k} \\ b_1 & b_2 & b_3 \\ b_1 & b_2 & b_3 \end{vmatrix} = 0 \quad \text{Property of determinants (two rows the same)}$$

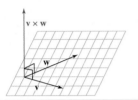

Figure 9.33 Vector product

Geometric Interpretation of Cross Product

We will show with the following property that the vector $(\mathbf{v} \times \mathbf{w})$ is orthogonal to both the vectors $\mathbf{v}$ and $\mathbf{w}$. The only way this can occur is in a three-dimensional setting. Any two distinct nonzero vectors in $\mathbb{R}^3$ that are not parallel (that is, not scalar multiples of one another) can be arranged to determine a plane. Then, according to the geometric property of cross product, the vector product *must* be orthogonal to this plane, as shown in Figure 9.33.

> **GEOMETRIC PROPERTY OF CROSS PRODUCT**
>
> If $\mathbf{v}$ and $\mathbf{w}$ are nonzero vectors in $\mathbb{R}^3$ that are not multiples of one another, then $\mathbf{v} \times \mathbf{w}$ is orthogonal to both $\mathbf{v}$ and $\mathbf{w}$.

☠ This property is one of the most important properties when working in three dimensions. We will use this property extensively in the next section. ☠

We will show $(\mathbf{v} \times \mathbf{w})$ is orthogonal to $\mathbf{v}$ and leave the proof that $\mathbf{v} \times \mathbf{w}$ is orthogonal to $\mathbf{w}$ as an exercise. Let $\mathbf{v} = a_1\mathbf{i} + a_2\mathbf{j} + a_3\mathbf{k}$ and $\mathbf{w} = b_1\mathbf{i} + b_2\mathbf{j} + b_3\mathbf{k}$. Then,

$$\mathbf{v} \times \mathbf{w} = \begin{vmatrix} \mathbf{i} & \mathbf{j} & \mathbf{k} \\ a_1 & a_2 & a_3 \\ b_1 & b_2 & b_3 \end{vmatrix} = (a_2 b_3 - a_3 b_2)\mathbf{i} - (a_1 b_3 - a_3 b_1)\mathbf{j} + (a_1 b_2 - a_2 b_1)\mathbf{k}$$

To show this vector is orthogonal to $\mathbf{v}$, we find $\mathbf{v} \cdot (\mathbf{v} \times \mathbf{w})$:

$$\mathbf{v} \cdot (\mathbf{v} \times \mathbf{w}) = a_1(a_2 b_3 - a_3 b_2) - (a_2(a_1 b_3 - a_3 b_1) + a_3(a_1 b_2 - a_2 b_1)$$
$$= a_1 a_2 b_3 - a_1 a_3 b_2 - a_1 a_2 b_3 + a_2 a_3 b_1 + a_1 a_3 b_2 - a_2 a_3 b_1$$
$$= 0$$

EXAMPLE 2   A vector orthogonal to two given vectors

Find a nonzero vector that is orthogonal to both $\mathbf{v} = -2\mathbf{i} + 3\mathbf{j} - 7\mathbf{k}$ and $\mathbf{w} = 5\mathbf{i} + 9\mathbf{k}$.

Solution
The cross product $\mathbf{v} \times \mathbf{w}$ is orthogonal to both $\mathbf{v}$ and $\mathbf{w}$.

$$\mathbf{v} \times \mathbf{w} = \begin{vmatrix} \mathbf{i} & \mathbf{j} & \mathbf{k} \\ -2 & 3 & -7 \\ 5 & 0 & 9 \end{vmatrix}$$
$$= (27 + 0)\mathbf{i} - (-18 + 35)\mathbf{j} + (0 - 15)\mathbf{k}$$
$$= 27\mathbf{i} - 17\mathbf{j} - 15\mathbf{k}$$

∎

Because both $\mathbf{v} \times \mathbf{w}$ and $\mathbf{w} \times \mathbf{v}$ are orthogonal to the plane determined by $\mathbf{v}$ and $\mathbf{w}$, and because $(\mathbf{v} \times \mathbf{w}) = -(\mathbf{w} \times \mathbf{v})$, we see that one points up from the given plane and the other points down. To see which is which, we state the **right-hand rule**, which is described in Figure 9.34.

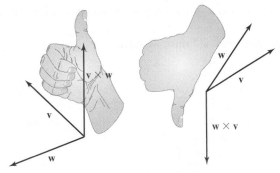

Figure 9.34 Right-hand rule

If you place the palm of your right hand along $\mathbf{v}$ and curl your fingers toward $\mathbf{w}$, then your fingers are pointing in the direction of $\mathbf{v}$, and your thumb points in the direction of $\mathbf{v} \times \mathbf{w}$.

EXAMPLE 3   Right-hand rule

Use the right-hand rule to verify each of the following cross products.

$$
\begin{array}{lll}
\mathbf{i} \times \mathbf{j} = \mathbf{k} & \mathbf{j} \times \mathbf{i} = -\mathbf{k} & \mathbf{k} \times \mathbf{i} = \mathbf{j} \\
\mathbf{i} \times \mathbf{k} = -\mathbf{j} & \mathbf{j} \times \mathbf{k} = \mathbf{i} & \mathbf{k} \times \mathbf{j} = -\mathbf{i} \\
\mathbf{i} \times \mathbf{i} = 0 & \mathbf{j} \times \mathbf{j} = 0 & \mathbf{k} \times \mathbf{k} = 0
\end{array}
$$

Solution

$\mathbf{i} \times \mathbf{j}$ Place the palm of your right hand along $\mathbf{i}$ and curl your fingers toward $\mathbf{j}$. The answer is $\mathbf{k}$.

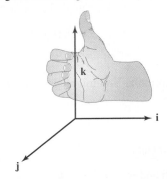

$\mathbf{j} \times \mathbf{i}$ Place the palm of your right hand along $\mathbf{j}$ and curl your fingers toward $\mathbf{i}$. The answer is $-\mathbf{k}$.

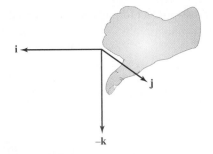

$\mathbf{i} \times \mathbf{i}$ Can't use right-hand rule. However, we know

$$
\mathbf{i} \times \mathbf{i} = \begin{vmatrix} \mathbf{i} & \mathbf{j} & \mathbf{k} \\ 1 & 0 & 0 \\ 1 & 0 & 0 \end{vmatrix} = 0
$$

The other parts are left for you to verify. ∎

Properties of Cross Product

Next, we will find the magnitude of the cross product of two vectors. The magnitude of the cross product of two vectors $\| \mathbf{v} \times \mathbf{w} \|$ is equal to the area of the parallelogram having $\mathbf{v}$ and $\mathbf{w}$ as adjacent sides, as shown in Figure 9.35. Let $\mathbf{v} = a_1\mathbf{i} + a_2\mathbf{j} + a_3\mathbf{k}$ and $\mathbf{w} = b_1\mathbf{i} + b_2\mathbf{j} + b_3\mathbf{k}$.

To see this, note that because $h = \| \mathbf{w} \| \sin\theta$, we have

$$\begin{aligned}
\text{AREA} &= (\text{BASE})(\text{HEIGHT}) && \text{Area formula} \\
&= \| \mathbf{v} \| (\| \mathbf{w} \| \sin\theta) && \text{Substitute given values.} \\
&= \| \mathbf{v} \| \| \mathbf{w} \| \sqrt{1 - \cos^2\theta} && \text{From } \sin^2\theta + \cos^2\theta = 1 \\
&= \| \mathbf{v} \| \| \mathbf{w} \| \sqrt{1 - \left[\frac{\mathbf{v} \cdot \mathbf{w}}{\| \mathbf{v} \| \| \mathbf{w} \|} \right]^2} && \text{Angle between vectors } \cos\theta = \frac{\mathbf{v} \cdot \mathbf{w}}{\| \mathbf{v} \| \| \mathbf{w} \|} \\
&= \| \mathbf{v} \| \| \mathbf{w} \| \frac{\sqrt{\| \mathbf{v} \|^2 \| \mathbf{w} \|^2 - (\mathbf{v} \cdot \mathbf{w})^2}}{\| \mathbf{v} \| \| \mathbf{w} \|} && \text{Common denominator} \\
&= \sqrt{\| \mathbf{v} \|^2 \| \mathbf{w} \|^2 - (\mathbf{v} \cdot \mathbf{w})^2} \\
&= \sqrt{\left(a_1^2 + a_2^2 + a_3^2 \right)\left(b_1^2 + b_2^2 + b_3^2 \right) - \left(a_1 b_1 + a_2 b_2 + a_3 b_3 \right)^2} \\
&= \sqrt{(a_2 b_3 - a_3 b_2)^2 + (a_3 b_1 - a_1 b_3)^2 + (a_1 b_2 - a_2 b_1)^2} \\
&= \| \mathbf{v} \times \mathbf{w} \|
\end{aligned}$$

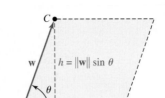

Figure 9.35 Area of a parallelogram

MAGNITUDE OF CROSS PRODUCT

If $\mathbf{v}$ and $\mathbf{w}$ are nonzero vectors in $\mathbb{R}^3$ with θ the angle between $\mathbf{v}$ and $\mathbf{w}$ $(0 \leq \theta \leq \pi)$, then

$$\| \mathbf{v} \times \mathbf{w} \| = \| \mathbf{v} \| \| \mathbf{w} \| \sin\theta$$

Here is an example that uses this formula. You might also note that even though $\mathbf{v} \times \mathbf{w} \neq \mathbf{w} \times \mathbf{v}$, it is true that $\| \mathbf{v} \times \mathbf{w} \| = \| \mathbf{w} \times \mathbf{v} \|$.

EXAMPLE 4   Angle between two vectors

Find the angle between $\mathbf{v} = \mathbf{i} + 2\mathbf{j}$ and $\mathbf{w} = \mathbf{i} - \mathbf{j} + \mathbf{k}$.

Solution

We will use the magnitude of cross-product formula, so we first find some preliminary values:

$$\| \mathbf{v} \| = \sqrt{1^2 + 2^2} \qquad\qquad \| \mathbf{w} \| = \sqrt{1^2 + (-1)^2 + 1^2}$$
$$= \sqrt{1 + 4} \qquad\qquad\qquad = \sqrt{1 + 1 + 1}$$
$$= \sqrt{5} \qquad\qquad\qquad\quad = \sqrt{3}$$

$$\mathbf{v} \times \mathbf{w} = \begin{vmatrix} \mathbf{i} & \mathbf{j} & \mathbf{k} \\ 1 & 2 & 0 \\ 1 & -1 & 1 \end{vmatrix} \qquad \| \mathbf{v} \times \mathbf{w} \| = \sqrt{2^2 + (-1)^2 + (-3)^2}$$
$$= (2 + 0)\mathbf{i} - (1 - 0)\mathbf{j} + (-1 - 2)\mathbf{k} \qquad\quad = \sqrt{4 + 1 + 9}$$
$$= 2\mathbf{i} - \mathbf{j} - 3\mathbf{k} \qquad\qquad\qquad\qquad\quad = \sqrt{14}$$

Thus,

$$\| \mathbf{v} \times \mathbf{w} \| = \| \mathbf{v} \| \| \mathbf{w} \| \sin \theta$$

$$\sin \theta = \frac{\| \mathbf{v} \times \mathbf{w} \|}{\| \mathbf{v} \| \| \mathbf{w} \|}$$

$$\sin \theta = \frac{\sqrt{14}}{\sqrt{5}\sqrt{3}}$$

$$\theta = \sin^{-1} \sqrt{\frac{14}{15}}$$

$$\approx 75°$$

EXAMPLE 5   Area of a triangle

Find the area of the triangle with vertices $P(-2, 4, 5)$, $Q(0, 7, -4)$, and $R(-1, 5, 0)$.

Solution
Draw this triangle as shown in Figure 9.36.

Then ΔPQR has half the area of the parallelogram determined by the vectors $\mathbf{PQ}$ and $\mathbf{PR}$; that is, the triangle has area

$$A = \frac{1}{2} \| \mathbf{PQ} \times \mathbf{PR} \|$$

First find

$$\mathbf{PQ} = (0 + 2)\mathbf{i} + (7 - 4)\mathbf{j} + (-4 - 5)\mathbf{k} = 2\mathbf{i} + 3\mathbf{j} - 9\mathbf{k}$$
$$\mathbf{PR} = (-1 + 2)\mathbf{i} + (5 - 4)\mathbf{j} + (0 - 5)\mathbf{k} = \mathbf{i} + \mathbf{j} - 5\mathbf{k}$$

and compute the cross product:

$$\mathbf{PQ} \times \mathbf{PR} = \begin{vmatrix} \mathbf{i} & \mathbf{j} & \mathbf{k} \\ 2 & 3 & -9 \\ 1 & 1 & -5 \end{vmatrix}$$

$$= (-15 + 9)\mathbf{i} - (-10 + 9)\mathbf{j} + (2 - 3)\mathbf{k}$$

$$= -6\mathbf{i} + \mathbf{j} - \mathbf{k}$$

Thus, the triangle has area

$$A = \frac{1}{2} \| \mathbf{PQ} \times \mathbf{PR} \|$$

$$= \frac{1}{2} \sqrt{(-6)^2 + 1^2 + (-1)^2}$$

$$= \frac{1}{2} \sqrt{38}$$

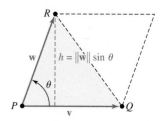

Figure 9.36 Area of a triangle

The cross product can also be used to compute the volume of a parallelepiped in $\mathbb{R}^3$. Consider the parallelepiped determined by three nonzero vectors $\mathbf{u}$, $\mathbf{v}$, and $\mathbf{w}$ that do not all lie in the same plane, as shown in Figure 9.37.

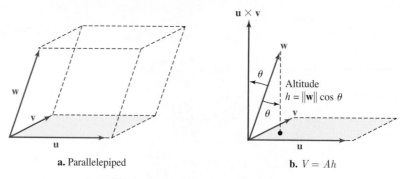

Figure 9.37 Computing the volume of a parallelepiped

It is known from solid geometry that this parallelogram has volume $V = Ah$, where A is the area of the face determined by **u** and **v**, and h is the altitude from the tip of **w** to this face.

The face determined by **u** and **v** is a parallelogram with area $A = \| \mathbf{u} \times \mathbf{v} \|$, and we know that the cross-product vector $\mathbf{u} \times \mathbf{v}$ is perpendicular to both **u** and **v** and hence to the face determined by **u** and **v**. From the alternative form of the dot product, we have

$$(\mathbf{u} \times \mathbf{v}) \cdot \mathbf{w} = \| \mathbf{u} \times \mathbf{v} \| \, \| \mathbf{w} \| \cos \theta$$

where θ is the angle between $\mathbf{u} \times \mathbf{v}$ and **w**. Thus, the parallelepiped has altitude

$$h = | \, \| \mathbf{w} \| \cos \theta \, | = \left| \frac{(\mathbf{u} \times \mathbf{v}) \cdot \mathbf{w}}{\| \mathbf{u} \times \mathbf{v} \|} \right|$$

and the volume is given as

$$V = Ah = \| \mathbf{u} \times \mathbf{v} \| \left| \frac{(\mathbf{u} \times \mathbf{v}) \cdot \mathbf{w}}{\| \mathbf{u} \times \mathbf{v} \|} \right| = | (\mathbf{u} \times \mathbf{v}) \cdot \mathbf{w} |$$

The combined operation $(\mathbf{u} \times \mathbf{v}) \cdot \mathbf{w}$ is called the **triple scalar product** of **u**, **v**, and **w**.

EXAMPLE 6   Volume of a parallelepiped

Find the volume of the parallelepiped determined by the vectors

$$\mathbf{u} = \mathbf{i} - 2\mathbf{j} + 3\mathbf{k} \qquad \mathbf{v} = -4\mathbf{i} + 7\mathbf{j} - 11\mathbf{k} \qquad \mathbf{w} = 5\mathbf{i} + 9\mathbf{j} - \mathbf{k}$$

Solution
We first find the cross product.

$$\mathbf{u} \times \mathbf{v} = \begin{vmatrix} \mathbf{i} & \mathbf{j} & \mathbf{k} \\ 1 & -2 & 3 \\ -4 & 7 & -11 \end{vmatrix}$$
$$= (22 - 21)\mathbf{i} - (-11 + 12)\mathbf{j} + (7 - 8)\mathbf{k}$$
$$= \mathbf{i} - \mathbf{j} - \mathbf{k}$$

Thus,

$$V = | (\mathbf{u} \times \mathbf{v}) \cdot \mathbf{w} |$$
$$= | (\mathbf{i} - \mathbf{j} - \mathbf{k}) \cdot (5\mathbf{i} + 9\mathbf{j} - \mathbf{k}) |$$
$$= | 5 - 9 + 1 |$$
$$= 3$$

The volume of the parallelepiped is 3 cubic units.

Computationally, there is an easier way to work Example 6.

TRIPLE SCALAR PRODUCT

If $\mathbf{u} = a_1\mathbf{i} + a_2\mathbf{j} + a_3\mathbf{k}$, $\mathbf{v} = b_1\mathbf{i} + b_2\mathbf{j} + b_3\mathbf{k}$, and $\mathbf{w} = c_1\mathbf{i} + c_2\mathbf{j} + c_3\mathbf{k}$, then the triple scalar product can be found by evaluating the determinant

$$(\mathbf{u} \times \mathbf{v}) \cdot \mathbf{w} = \begin{vmatrix} a_1 & a_2 & a_3 \\ b_1 & b_2 & b_3 \\ c_1 & c_2 & c_3 \end{vmatrix}$$

We can use this formula to rework Example 6.

$$(\mathbf{u} \times \mathbf{v}) \cdot \mathbf{w} = \begin{vmatrix} 1 & -2 & 3 \\ -4 & 7 & -11 \\ 5 & 9 & -1 \end{vmatrix} = -3$$

Thus, the volume is $|-3| = 3$.

We summarize the area and volume vector formulas in the following box.

AREAS AND VOLUMES

Let $\mathbf{u}$, $\mathbf{v}$, and $\mathbf{w}$ be nonzero vectors that do not all lie in the same plane. Then,

Area of a parallelogram: $\qquad A = \|\mathbf{u} \times \mathbf{v}\|$

Area of a triangle: $\qquad A = \dfrac{1}{2}\|\mathbf{u} \times \mathbf{v}\|$

Volume of a parallelepiped: $\qquad V = |(\mathbf{u} \times \mathbf{v}) \cdot \mathbf{w}|$

In the problem set, we have included several exercises involving the triple scalar product and the triple vector product $\mathbf{u} \times \mathbf{v} \times \mathbf{w}$. In case you wonder why we neglect the product $(\mathbf{u} \cdot \mathbf{v}) \times \mathbf{w}$, notice that such a product makes no sense, because $\mathbf{u} \cdot \mathbf{v}$ is a *scalar* and the cross product is an operation involving only vectors. Thus, the product

$$\mathbf{u} \cdot \mathbf{v} \times \mathbf{w} \quad \text{must mean} \quad \mathbf{u} \cdot (\mathbf{v} \times \mathbf{w})$$

Torque

A useful physical application of the cross product involves **torque**. Suppose the force $\mathbf{F}$ is applied to the point Q. Then the torque of $\mathbf{F}$ around P is defined as the cross product of the "arm" vector $\mathbf{PQ}$ with the force $\mathbf{F}$, as shown in Figure 9.38.

The torque, $\mathbf{T}$, of $\mathbf{F}$ at Q about P is

$$\mathbf{T} = \mathbf{PQ} \times \mathbf{F}$$

The magnitude of the torque, $\|\mathbf{T}\|$, provides a measure of the tendency of the vector arm $\mathbf{PQ}$ to rotate counterclockwise about an axis perpendicular to the plane determined by $\mathbf{PQ}$ and $\mathbf{F}$.

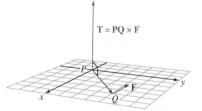

Figure 9.38 Torque as a cross product

EXAMPLE 7 Torque on the hinge of a door

Figure 9.39 shows a half-open door that is 3 ft wide. A horizontal force of 30 lb is applied at the edge of the door. Find the torque of the force about the hinge on the door.

Solution

We represent the force by $\mathbf{F} = -30\mathbf{i}$. Because the door is half open, it makes an angle of $\frac{\pi}{4}$ with the horizontal, and we can represent the door (i.e., the "arm" $\mathbf{PQ}$) by the vector

$$\mathbf{PQ} = 3\left(\cos\frac{\pi}{4}\mathbf{i} + \sin\frac{\pi}{4}\mathbf{j}\right) = 3\left(\frac{\sqrt{2}}{2}\mathbf{i} + \frac{\sqrt{2}}{2}\mathbf{j}\right) = \frac{3\sqrt{2}}{2}\mathbf{i} + \frac{3\sqrt{2}}{2}\mathbf{j}$$

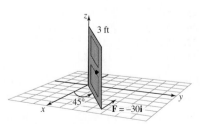

Figure 9.39 Swinging door

The torque can now be found:

$$T = PQ \times F = \begin{vmatrix} i & j & k \\ \dfrac{3\sqrt{2}}{2} & \dfrac{3\sqrt{2}}{2} & 0 \\ -30 & 0 & 0 \end{vmatrix} = 45\sqrt{2}k$$

The magnitude of the torque ($45\sqrt{2}$ ft-lb) is a measure of the tendency of the door to rotate about its hinges. ∎

PROBLEM SET 9.4

LEVEL 1

Find $v \times w$ *for the vectors given in Problems 1–10.*

1. $v = i$; $w = j$
2. $v = k$; $w = k$
3. $v = 3i + 2k$; $w = 2i + j$
4. $v = i - 3j$; $w = i + 5k$
5. $v = 3i - 2j + 4k$; $w = i + 4j - 7k$
6. $v = 5i - j + 2k$; $w = 2i + j - 3k$
7. $v = 3i - j + 2k$; $w = 2i + 3j - 4k$
8. $v = -j + 4k$; $w = 5i + 6k$
9. $v = i - 6j + 10k$; $w = -i + 5j - 6k$
10. $v = \cos\theta i + \sin\theta j$; $w = -\sin\theta i + \cos\theta j$

Find $\sin\theta$ *where* θ *is the angle between* v *and* w *in Problems 11–16.*

11. $v = i + k$; $w = i + j$
12. $v = i + j$; $w = i + j + k$
13. $v = j + k$; $w = i + k$
14. $v = i + j$; $w = j + k$
15. $v = i + 2j + 3k$; $w = 4i + 5j + 6k$
16. $v = \cos\omega i - \sin\omega j$; $w = \sin\omega i + \cos\omega j$

Find a unit vector that is orthogonal to both v *and* w *in Problems 17–20.*

17. $v = 2i + k$; $w = i - j - k$
18. $v = j - 3k$; $w = -i + j + k$
19. $v = i + j + k$; $w = 3i + 12j - 4k$
20. $v = 2i - 2j + k$; $w = 4i + 2j - 3k$

Find the area of the parallelogram determined by the vectors in Problems 21–24.

21. $3i + 4j$ and $i + j - k$ 22. $2i - j + 2k$ and $4i - 3j$
23. $4i - j + k$ and $2i + 3j - k$ 24. $2i + 3k$ and $2j - 3k$

Find the area of $\triangle PQR$ *in Problems 25–28.*

25. $P(0, 1, 1)$, $Q(1, 1, 0)$, $R(1, 0, 1)$
26. $P(1, 0, 0)$, $Q(2, 1, -1)$, $R(0, 1, -2)$
27. $P(1, 2, 3)$, $Q(2, 3, 1)$, $R(3, 1, 2)$
28. $P(-1, -1, -1)$, $Q(1, -1, -1)$, $R(-1, 1, -1)$

Determine whether each product in Problems 29–32 is a scalar or a vector or does not exist. Explain your reasoning.

29. a. $u \times (v \cdot w)$ b. $u \cdot (v \times w)$
30. a. $u \times (v \times w)$ b. $u \cdot (v \cdot w)$
31. a. $(u \times v) \cdot (u \times w)$ b. $(u \times v) \times (u \times w)$
32. a. $(u \times v) \cdot w$ b. $|(u \times v) \cdot w|$

In Problems 33–36, find the volume of the parallelepiped determined by vectors u, v, *and* w.

33. $u = i + j$; $v = j + 2k$; $w = 3k$
34. $u = j + k$; $v = 2i + j + 2k$; $w = 5i$
35. $u = i + j + k$; $v = i - j - k$; $w = 2i + 3k$
36. $u = 2i + j - k$; $v = 3i + k$; $w = j + k$

LEVEL 2

37. IN YOUR OWN WORDS Contrast dot and cross products of vectors, including a discussion of some of their properties.
38. IN YOUR OWN WORDS What is the right-hand rule?
39. IN YOUR OWN WORDS Describe the meaning of $|(u \times v) \cdot w|$.
40. IN YOUR OWN WORDS
    a. If $u \times w = v \times w$, does it follow that $u = v$?
    b. If $u \cdot w = v \cdot w$, does it follow that $u = v$?
    c. If both $u \times w = v \times w$ and $u \cdot w = v \cdot w$, does it follow that $u = v$?
41. Find a number s that guarantees that the vectors i, $i + j + k$, and $i + 2j + sk$ will be parallel to the same plane.
42. Find a number t that guarantees the vectors $i + j$, $2i - j + k$, and $i + j + tk$ will be parallel to the same plane.
43. Find the angle between the vector $2i - j + k$ and the plane determined by the points $P(1, -2, 3)$, $Q(-1, 2, 3)$, and $R(1, 2, -3)$.
44. Let $u = i + j$, $v = 2i - j + k$, and $w = 3i$. Compute $(u \times v) \times w$ and $u \times (v \times w)$. What does this say about the associativity of cross product?
45. Show that $(au) \times (bv) = ab(u \times v)$ and scalars a and b.
46. A 40-lb child sits on a seesaw, 3 ft from the fulcrum, as shown in Figure 9.40. What torque is exerted when the child is 2 ft above the horizontal?

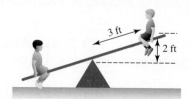

Figure 9.40 Seesaw torque

47. IN YOUR OWN WORDS One end of a 2-ft lever pivots about the origin in the yz-plane, as shown in Figure 9.41.

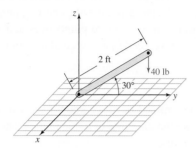

Figure 9.41 Finding the torque

If a vertical force of 40 lb is applied at the end of the lever, what is the torque of the lever about the pivot point (the origin) when the lever makes an angle of $30°$ with the xy-plane?

48. PROBLEMS FROM CALCULUS A 3-lb weight hangs from a rope at the end of Q of a 5-ft stick PQ that is held at an angle of $60°$ to the horizontal, as shown in Figure 9.42. What is the torque about the point P due to the weight?

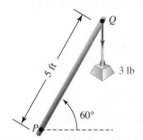

Figure 9.42 Finding the torque

LEVEL 3

49. Prove the determinant formula for evaluating a triple scalar product.
50. a. Show that the vectors **u**, **v**, and **w** are coplanar (all in the same plane) if $\mathbf{u} \cdot (\mathbf{v} \times \mathbf{w}) = 0$ or $(\mathbf{u} \times \mathbf{v}) \cdot \mathbf{w} = 0$
b. Verify that the vectors $\mathbf{u} = \mathbf{i} + 3\mathbf{j} + \mathbf{k}$, $\mathbf{v} = 2\mathbf{i} - \mathbf{j} - \mathbf{k}$, and $\mathbf{w} = 7\mathbf{j} + 3\mathbf{k}$ are coplanar.
51. Show that the triangle with vertices $P(x_1, y_1)$, $Q(x_2, y_2)$, $R(x_3, y_3)$ has area $A = \frac{1}{2}D$, where

$$D = \begin{vmatrix} x_1 & y_1 & 1 \\ x_2 & y_2 & 1 \\ x_3 & y_3 & 1 \end{vmatrix}$$

52. Using the properties of determinants, show that

$$\mathbf{u} \cdot (\mathbf{v} \times \mathbf{w}) = (\mathbf{u} \times \mathbf{v}) \cdot \mathbf{w}$$

for any vectors **u**, **v**, and **w**.

53. Let A, B, C, and D be four points that do not lie in the same plane (see Figure 9.43).

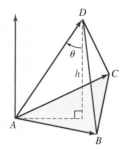

Figure 9.43 Four noncollinear points

It can be shown that the volume of the tetrahedron with vertices A, B, C, and D satisfies

$$\begin{bmatrix} \text{VOLUME OF} \\ \text{TETRAHEDRAON } ABCD \end{bmatrix} = \frac{1}{3} \begin{pmatrix} \text{AREA OF} \\ \triangle ABC \end{pmatrix} \begin{pmatrix} \text{ALTITUDE FROM} \\ D \text{ TO } \triangle ABC \end{pmatrix}$$

Show that the volume is given by

$$V = \frac{1}{6} |(\mathbf{AB} \times \mathbf{AC}) \cdot \mathbf{AD}|$$

54. Show that if **u**, **v**, and **w** are vectors in $\mathbb{R}^3$ with $\mathbf{u} + \mathbf{v} + \mathbf{w} = 0$, then

$$\mathbf{u} \times \mathbf{v} = \mathbf{v} \times \mathbf{w} = \mathbf{w} \times \mathbf{u}$$

55. Suppose **u**, **v**, and **w** are nonzero vectors in $\mathbb{R}^3$ with $\mathbf{u} \times \mathbf{v} = \mathbf{w}$ and $\mathbf{u} \cdot \mathbf{v} = 0$. Show that

$$\mathbf{v} = s(\mathbf{w} \times \mathbf{u}) \text{ and } \mathbf{u} = t(\mathbf{v} \times \mathbf{w})$$

for scalars s and t.
56. Show that

$$(c\mathbf{u}) \times (d\mathbf{v}) = cd(\mathbf{u} \times \mathbf{v})$$

57. Show that

$$\tan\theta = \frac{\| \mathbf{v} \times \mathbf{w} \|}{\mathbf{v} \cdot \mathbf{w}}$$

where $\theta \left(0 \leq \theta \leq \frac{\pi}{2}\right)$ is the angle between **v** and **w**.
58. Show that, if all three vertices of a triangle have integer coordinates, the area of the triangle is at least 0.5.
59. Let **u**, **v**, and **w** be nonzero nonplanar vectors in $\mathbb{R}^3$. Show that

$$|\mathbf{u} \cdot (\mathbf{v} \times \mathbf{w})| = |\mathbf{v} \cdot (\mathbf{u} \times \mathbf{w})|$$

What other triple scalar products involving **u**, **v**, and **w** have the same absolute values?
60. Let **a**, **b**, and **c** be vectors in space. Show

$$\mathbf{a} \times (\mathbf{b} \times \mathbf{c}) = (\mathbf{c} \cdot \mathbf{a})\mathbf{b} - (\mathbf{b} \cdot \mathbf{a})\mathbf{c}$$

This is called the "cab – bac" formula.

9.5 Lines and Planes in Space

As we move from $\mathbb{R}^2$ to $\mathbb{R}^3$, we look for patterns. As we have seen, the number of components of points corresponds to the dimension. In $\mathbb{R}$, we locate a point with one component, in $\mathbb{R}^2$ we use ordered pairs, and in $\mathbb{R}^3$ we use triplets. Circles in $\mathbb{R}^2$ are two points in $\mathbb{R}$ and spheres in $\mathbb{R}^3$. However, lines in $\mathbb{R}^2$ generalize to planes in $\mathbb{R}^3$, so we need to pay special attention to developing lines in $\mathbb{R}^3$.

Lines in $\mathbb{R}^3$

As in the plane, a line in space is completely determined once we know one of its points and its direction. We used the concept of slope to measure the direction of a line in the plane, but in space, it is more convenient to specify direction with vectors.

Suppose L is a line in space that contains $Q(x_0, y_0, z_0)$ whose location is determined by the vector $\mathbf{v} = A\mathbf{i} + B\mathbf{j} + C\mathbf{k}$. We say that L is **aligned with v**, as shown in Figure 9.44.

We also say that the line has **direction numbers** A, B, and C and denote these direction numbers by $[A, B, C]$. If $P(x, y, z)$ is any point on L, then the vector $\mathbf{QP}$ is parallel to $\mathbf{v}$ and must satisfy the vector equation $\mathbf{QP} = t\mathbf{v}$ for some number t. If we introduce coordinates and use the standard representation, we can rewrite this vector equation as

$$(x - x_0)\mathbf{i} + (y - y_0)\mathbf{j} + (z - z_0)\mathbf{k} = t[A\mathbf{i} + B\mathbf{j} + C\mathbf{k}]$$

By equating components on both sides of this equation, we find that the coordinates of P must satisfy the linear system

$$x - x_0 = tA \qquad y - y_0 = tB \qquad z + z_0 = tC$$

where t is a real number.

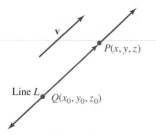

Figure 9.44 *L is aligned with v*

☠ The parameter t will be a different value for each point (x, y, z). ☠

PARAMETRIC FORM OF A LINE IN $\mathbb{R}^3$

If L is a line that contains the point (x_0, y_0, z_0) and is aligned with the vector $\mathbf{v} = A\mathbf{i} + B\mathbf{j} + C\mathbf{k}$, then the point (x, y, z) is on L if and only if its coordinates satisfy

$$x - x_0 = tA \qquad y - y_0 = tB \qquad z + z_0 = tC$$

for some number t.

Turning things around, if we are given the equation of a line with direction numbers $[A, B, C]$, then $\mathbf{v} = A\mathbf{i} + B\mathbf{j} + C\mathbf{k}$ is the **vector aligned with L**.

To graph a line in space, follow the steps in the drawing lesson shown in Problem 17.

EXAMPLE 1   Parametric equations of a line in space

Find the parametric equations for the line that contains the point $(3, 1, 4)$ and is aligned with the vector $\mathbf{v} = -\mathbf{i} + \mathbf{j} - 2\mathbf{k}$. Find where this line passes through the coordinate planes and sketch the line.

Solution

The direction numbers are $[-1, 1, -2]$ and $x_0 = 3$, $y_0 = 1$, $z_0 = 4$, so the line has the parametric form

$$x = 3 - t \qquad y = 1 + t \qquad z = 4 - 2t$$

This line will intersect the xy-plane when $z = 0$; solve

$$0 = 4 - 2t \quad \text{implies} \quad t = 2$$

If $t = 2$, then $x = 3 - 2 = 1$ and $y = 1 + 2 = 3$. This is the point $(1, 3, 0)$. Similarly, the line intersects the xz-plane at $(4, 0, 6)$ and the yz-plane at $(0, 4, -2)$. Plot these points and draw the line, as shown in Figure 9.45.

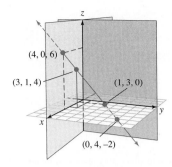

Figure 9.45 Graph of a line in space

In the special case where none of the direction numbers A, B, or C is 0, we can solve each of the parametric-form equations for t to obtain the following **symmetric equations** for a line.

> **SYMMETRIC FORM OF A LINE IN $\mathbb{R}^3$**
>
> If L is a line that contains the point (x_0, y_0, z_0) and is aligned with the vector
>
> $$\mathbf{v} = A\mathbf{i} + B\mathbf{j} + C\mathbf{k}$$
>
> (A, B, and C nonzero numbers), then the point (x, y, z) is on L if and only if its coordinates satisfy
>
> $$\frac{x - x_0}{A} = \frac{y - y_0}{B} = \frac{z - z_0}{C}$$

EXAMPLE 2   Symmetric form of the equation of a line in space

Find symmetric equations for the line L through the points $P(-1, 3, 7)$ and $Q(4, 2, -1)$. Find the points of intersection with the coordinate planes and sketch the line.

Solution
The required line passes through P and is aligned with the vector

$$\mathbf{PQ} = [4 - (-1)]\mathbf{i} + [2 - 3]\mathbf{j} + [-1 - 7]\mathbf{k} = 5\mathbf{i} - \mathbf{j} - 8\mathbf{k}$$

Thus, the direction numbers of the line are $[5, -1, -8]$, and we can choose either P or Q as (x_0, y_0, z_0). Choosing P, we obtain

$$\frac{x + 1}{5} = \frac{y - 3}{-1} = \frac{z - 7}{-8}$$

Next, we find points of intersection with the coordinate planes:

$$xy\text{-plane: } z = 0, \text{ so } \frac{x + 1}{5} = \frac{7}{8} \text{ implies } x = \frac{27}{8} \text{ and } \frac{y - 3}{-1} = \frac{7}{8} \text{ implies } y = \frac{17}{8}.$$

The point of intersection of line with the xy-plane is $\left(\frac{27}{8}, \frac{17}{8}, 0\right)$. Similarly, the other intersections are xz-plane: $(14, 0, -17)$; yz-plane: $\left(0, \frac{14}{5}, \frac{27}{5}\right)$. The graph is shown in Figure 9.46.

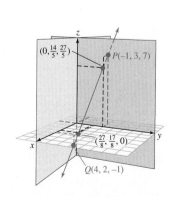

Figure 9.46 Graph of $\dfrac{x + 1}{5} = \dfrac{y - 3}{-1} = \dfrac{z - 7}{-8}$

You might remember that two lines in $\mathbb{R}^2$ must intersect if their slopes are different (because they cannot be parallel), but two lines in space may have different direction numbers and still not intersect. In this case, the lines are said to be **skew**. The situation in $\mathbb{R}^3$ is different from that in $\mathbb{R}^2$ because even though lines with different direction numbers cannot be parallel, there is still enough "room" in space for the lines to lie in parallel planes and be aligned with vectors that are not parallel. This situation is shown in Figure 9.47.

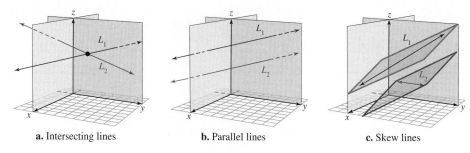

a. Intersecting lines   **b.** Parallel lines   **c.** Skew lines

Figure 9.47 Lines in space may intersect, be parallel, or be skew.

EXAMPLE 3   Skew lines in space

Determine whether the following pair of lines intersect, are parallel, or are skew.

$$L_1: \frac{x-1}{2} = \frac{y+1}{1} = \frac{z-2}{4} \quad \text{and} \quad L_2: \frac{x+2}{4} = \frac{y}{-3} = \frac{z+1}{1}$$

Solution

Note that L_1 has direction numbers $[2, 1, 4]$ (that is, L_1 is aligned with $2\mathbf{i} + \mathbf{j} + 4\mathbf{k}$) and L_2 has direction numbers $[4, -3, 1]$. If we solve $\langle 2, 1, 4 \rangle = t \langle 4, 3, 1 \rangle$ for t, we find no possible solution for any value of t. This implies that the lines are not parallel. (See Figure 9.48.)

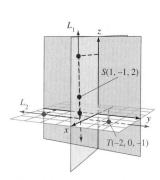

Figure 9.48 Graphs of L_1 and L_2

Next, we determine whether the lines intersect or are skew. Note that $S(1, -1, 2)$ lies on L_1 and $T(-2, 0, -1)$ lies on L_2. The lines intersect if and only if there is a point P that lies on both lines. To determine this, we write the equations of the lines in parametric form. We use a different parameter for each line so that points on one line do not depend on the value of the parameter of the other line:

L_1:	$x = 1 + 2s$	$y = -1 + s$	$z = 2 + 4s$
L_2:	$x = -2 + 4t$	$y = -3t$	$z = -1 + t$
Thus:	$x = 1 + 2s = -2 + 4t$	$y = -1 + s = -3t$	$z = 2 + 4s = -1 + t$
or:	$2s - 4t = -3$	$s + 3t = 1$	$4s - t = -3$

This is equivalent to the system of linear equations

$$\begin{cases} 2s - 4t = -3 \\ s + 3t = 1 \\ 4s - t = -3 \end{cases}$$

Any solution of this system must correspond to a point of intersection of L_1 and L_2, and if no solution exists, then L_1 and L_2 are skew. Because this is a system of three equations with two unknowns, we first solve the first two equations simultaneously to find $s = -\frac{1}{2}$, $t = \frac{1}{2}$. Because $s = -\frac{1}{2}$ and $t = \frac{1}{2}$ do not satisfy the third equation, it follows that L_1 and L_2 do not intersect, so they must be skew. ∎

EXAMPLE 4   Intersecting lines

Show that the lines

$$L_1: \frac{x-1}{2} = \frac{y+1}{1} = \frac{z-2}{4} \quad \text{and} \quad L_2: \frac{x+2}{4} = \frac{y}{-3} = \frac{z-\frac{1}{2}}{-1}$$

intersect and find the point of intersection. (See Figure 9.49.)

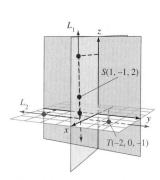

Figure 9.49 Graph of L_1 and L_2

$(0, -\frac{3}{2}, 0)$

Solution

L_1 has direction numbers $[2, 1, 4]$ and L_2 has direction numbers $[4, -3, -1]$. Because there is no t for which $[2, 1, 4] = t[4, -3, -1]$, the lines are not parallel. Express the lines in parametric form:

$$L_1: \qquad x = 1 + 2s \qquad\qquad y = -1 + s \qquad\qquad z = 2 + 4s$$

$$L_2: \qquad x = -2 + 4t \qquad\qquad y = -3t \qquad\qquad z = \frac{1}{2} - t$$

$$\text{Thus:} \qquad x = 1 + 2s = -2 + 4t \qquad y = -1 + s = -3t \qquad z = 2 + 4s = \frac{1}{2} - t$$

$$\text{or: } 2s - 4t = -3 \qquad\qquad s + 3t = 1 \qquad\qquad 4s + t = -\frac{3}{2}$$

Solving the first two equations simultaneously, we find $s = -\frac{1}{2}$, $t = \frac{1}{2}$. This solution satisfies the third equation, namely,

$$4\left(-\frac{1}{2}\right) + \frac{1}{2} = -\frac{3}{2}$$

To find the coordinates of the point of intersection, substitute $s = -\frac{1}{2}$ into the parametric-form equations for L_1 (or substitute $t = \frac{1}{2}$ into L_2) to obtain

$$x_0 = 1 + 2\left(-\frac{1}{2}\right) = 0$$

$$y_0 = -1 + \left(-\frac{1}{2}\right) = -\frac{3}{2}$$

$$z_0 = 2 + 4\left(-\frac{1}{2}\right) = 0$$

Thus, the lines intersect at $P\left(0, -\frac{3}{2}, 0\right)$.

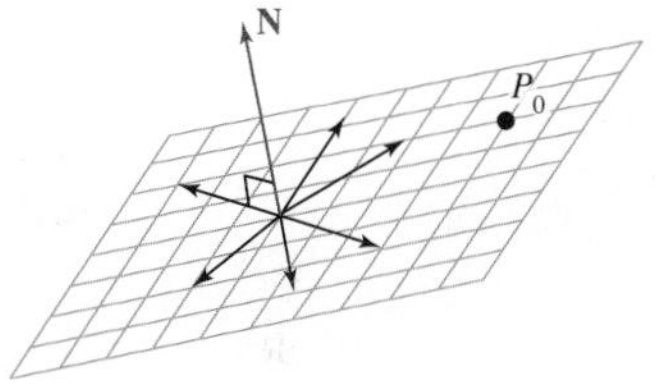

■

Planes in $\mathbb{R}^3$

Planes in space can also be characterized by vector methods. In particular, any plane is completely determined once we know one of its points and its orientation—that is, the "direction it faces." A common way to specify the direction of a plane is by means of a vector $\mathbf{N}$ that is orthogonal to every vector in the plane, as shown in Figure 9.50. Such a vector is called a **normal** to the plane.

Figure 9.50 A plane through P with a normal vector N

EXAMPLE 5   Obtain the equation for a plane

Find an equation for the plane that contains the point $Q(3, -7, 2)$ and is normal to the vector $\mathbf{N} = 2\mathbf{i} + \mathbf{j} - 3\mathbf{k}$.

Solution

The normal vector $\mathbf{N}$ is orthogonal to every vector in the plane. In particular, if $P(x, y, z)$ is any point in the plane, then $\mathbf{N}$ must be orthogonal to the vector

$$\mathbf{QP} = (x - 3)\mathbf{i} + (y + 7)\mathbf{j} + (z - 2)\mathbf{k}$$

Because the dot (or scalar) product of two orthogonal vectors is 0, we have

$$\mathbf{N} \cdot \mathbf{QP} = 2(x - 3) + (1)(y + 7) + (-3)(z - 2) = 0$$
$$2x - 6 + y + 7 - 3z + 6 = 0$$
$$2x + y - 3z + 7 = 0$$

Therefore, $2x + y - 3z + 7 = 0$ is the equation of the plane.

■

By generalizing the approach illustrated in Example 5, we can show that the plane that contains the point (x_0, y_0, z_0) and has normal vector $\mathbf{N} = A\mathbf{i} + B\mathbf{j} + C\mathbf{k}$ must have the Cartesian equation

$$A(x - x_0) + B(y - y_0) + C(z - z_0) = 0$$

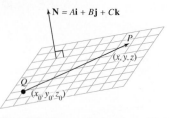

Figure 9.51 The graph of a plane with attitude numbers $[A, B, C]$

This is called the **point-normal form** of the equation of a plane. By rearranging terms, we can rewrite this equation in the form $Ax + By + Cz + D = 0$. This is called the **standard form** of the equation of a plane. The numbers $[A, B, C]$ are called **attitude numbers** of the plane (see Figure 9.51).

Notice from Figure 9.51 that *attitude numbers of a plane are the same as direction numbers of a normal line.* This means that you can find normal vectors to a plane *by inspecting the equation of the plane.*

EXAMPLE 6   Relationship between normal vectors and planes

Find normal vectors to the planes.

a. $5 + 7y - 3z = 0$ **b.** $x - 5y + \sqrt{2}z = 6$ **c.** $3x - 7z = 10$

Solution

a. A normal to the plane $5 + 7y - 3z = 0$ is $\mathbf{N} = 5\mathbf{i} + 7\mathbf{j} - 3\mathbf{k}$

b. For $x - 5y + \sqrt{2}z = 6$ the normal is $\mathbf{N} = \mathbf{i} - 5\mathbf{j} + \sqrt{2}\mathbf{k}$

c. For $3x - 7z = 10$ it is $\mathbf{N} = 3\mathbf{i} - 7\mathbf{k}$. ∎

EQUATIONS OF A PLANE

A plane with normal $\mathbf{N} = A\mathbf{i} + B\mathbf{j} + C\mathbf{k}$ that contains the point (x_0, y_0, z_0) has the following equations:

Point-normal form: $A(x - x_0) + B(y - y_0) + C(z - z_0) = 0$
Standard form: $Ax + By + Cz + D = 0$

for some constants A, B, C, and D.

EXAMPLE 7   Equation of a line orthogonal to a given plane

Find an equation of the line that passes through the point $Q(2, -1, 3)$ and is orthogonal to the plane $3x - 7y + 5z + 55 = 0$. Where does the line intersect the plane? (See Figure 9.52.)

Plane
$3x - 7y + 5z = -55$

$N = 3i - 7j + 5k$

Line
$\dfrac{x-2}{3} = \dfrac{y+1}{-7} = \dfrac{z-3}{5}$

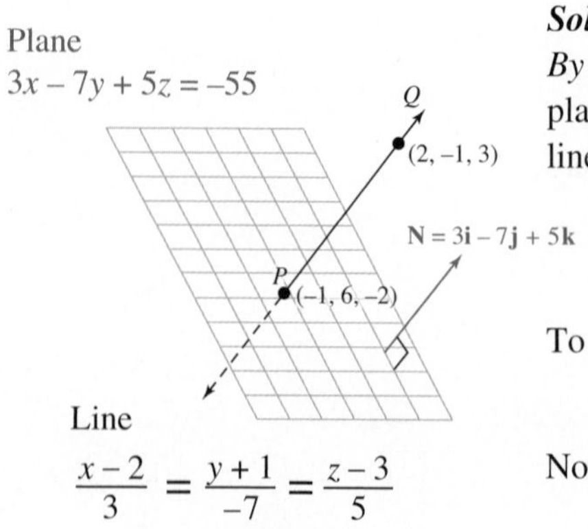

Figure 9.52 Graph of a plane

Solution

By inspection of the equation of the plane, we see that $\mathbf{N} = 3\mathbf{i} - 7\mathbf{j} + 5\mathbf{k}$ is a normal vector to the plane. Because the required line is also orthogonal to the plane, it must be parallel to $\mathbf{N}$. Thus, the line contains the point $Q(2, -1, 3)$ and has direction numbers $[3, -7, 5]$, so that its equation is

$$\frac{x-2}{3} = \frac{y+1}{-7} = \frac{z-3}{5}$$

To find the point where this line intersects the plane, we rewrite it in parametric form:

$$x = 2 + 3t, \quad y = -1 - 7t, \quad \text{and} \quad z = 3 + 5t$$

Now, substitute into the equation of the plane:

$$3(2 + 3t) - 7(-1 - 7t) + 5(3 + 5t) = -55$$
$$6 + 9t + 7 + 49t + 15 + 25t = -55$$
$$83t = -83$$
$$t = -1$$

Thus, the point of intersection is found by substituting $t = -1$ for x, y, and z:

$$x = 2 + 3(-1) = -1$$
$$y = -1 - 7(-1) = 6$$
$$z = 3 + 5(-1) = -2$$

The point of intersection is $(-1, 6, -2)$.

∎

EXAMPLE 8   Equation of a plane containing three given points

Find the standard-form equation of a plane containing $P(-1, 2, 1)$, $Q(0, -3, 2)$, and $R(1, 1, -4)$. (See Figure 9.53.)

Solution

Because a normal $\mathbf{N}$ to the required plane is orthogonal to the vectors $\mathbf{PQ}$ and $\mathbf{PR}$, we find $\mathbf{N}$ by computing the cross product $\mathbf{N} = \mathbf{PQ} \times \mathbf{PR}$.

$$\mathbf{PQ} = (0 + 1)\mathbf{i} + (-3 - 2)\mathbf{j} + (2 - 1)\mathbf{k} = \mathbf{i} - 5\mathbf{k} + \mathbf{k}$$
$$\mathbf{PR} = (1 + 1)\mathbf{i} + (1 - 2)\mathbf{j} + (-4 - 1)\mathbf{k} = 2\mathbf{i} - \mathbf{j} - 5\mathbf{k}$$

$$\mathbf{N} = \mathbf{PQ} \times \mathbf{PR} = \begin{vmatrix} \mathbf{i} & \mathbf{j} & \mathbf{k} \\ 1 & -5 & 1 \\ 2 & -1 & -5 \end{vmatrix}$$
$$= (25 + 1)\mathbf{i} - (-5 - 2)\mathbf{j} + (-1 + 10)\mathbf{k}$$
$$= 26\mathbf{i} + 7\mathbf{j} + 9\mathbf{k}$$

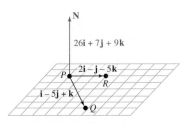

Figure 9.53 Plane passing through three points

We can now find the equation of the plane using this normal vector and any point in the plane. We will use the point P:

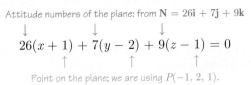

Attitude numbers of the plane: from $\mathbf{N} = 26\mathbf{i} + 7\mathbf{j} + 9\mathbf{k}$

$$26(x + 1) + 7(y - 2) + 9(z - 1) = 0$$

Point on the plane; we are using $P(-1, 2, 1)$.

Thus, the equation of the plane is

$$26x + 26 + 7y - 14 + 9z - 9 = 0$$
$$26x + 7y + 9z + 3 = 0$$

∎

EXAMPLE 9   Line parallel to the intersection of two planes

Find the equation of a line passing through $(-1, 2, 3)$ that is parallel to the line of intersection of the planes $3x - 2y + z = 4$ and $x + 2y + 3z = 5$. (See Figure 9.54.)

Solution

By inspection, we see that the normals to the given planes are $\mathbf{N}_1 = 3\mathbf{i} - 2\mathbf{j} + \mathbf{k}$ and $\mathbf{N}_2 = \mathbf{i} + 2\mathbf{j} + 3\mathbf{k}$. The desired line is perpendicular to both of these normals, so the aligned vector is found by computing with the cross product:

$$\mathbf{N}_1 \times \mathbf{N}_2 = \begin{vmatrix} \mathbf{i} & \mathbf{j} & \mathbf{k} \\ 3 & -2 & 1 \\ 1 & 2 & 3 \end{vmatrix}$$
$$= -(-6 - 2)\mathbf{i} - (9 - 1)\mathbf{j} + (6 + 2)\mathbf{k}$$
$$= -8\mathbf{i} - 8\mathbf{j} + 8\mathbf{k}$$

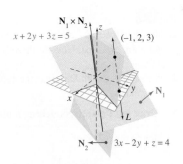

Figure 9.54 Given planes

The direction of this vector is $\langle -8, -8, 8 \rangle = -8\langle 1, 1, -1 \rangle$. The equation of the desired line is

$$\frac{x+1}{1} = \frac{y-2}{1} = \frac{z-3}{-1}$$

■

Example 9 can also be used to find the equation of the line of intersection of the two planes. Instead of using the given point $(-1, 2, 3)$, you will first need to find a point in the intersection and then proceed, using the steps of Example 9. We conclude this section by finding the equation of a plane containing two given (nonparallel) lines.

EXAMPLE 10   Equation of a plane containing two intersecting lines

Find the standard-form equation of the plane determined by the intersecting lines

$$\frac{x-2}{3} = \frac{y+5}{-2} = \frac{z+1}{4} \quad \text{and} \quad \frac{x+1}{2} = \frac{y}{-1} = \frac{z-16}{5}$$

Solution

Proceeding as in Example 4, we find that the lines intersect at $(-19, 9, -29)$. The aligned vectors for these two lines are $\mathbf{v}_1 = 3\mathbf{i} - 2\mathbf{j} + 4\mathbf{k}$ and $\mathbf{v}_2 = 2\mathbf{i} - \mathbf{j} + 5\mathbf{k}$, and the normal to the desired plane is orthogonal to both $\mathbf{v}_1$ and $\mathbf{v}_2$. (See Figure 9.55.)

Thus, we take the normal to be the cross product:

$$\mathbf{N} = \mathbf{v}_1 \times \mathbf{v}_2$$
$$= \begin{vmatrix} \mathbf{i} & \mathbf{j} & \mathbf{k} \\ 3 & -2 & 4 \\ 2 & -1 & 5 \end{vmatrix}$$
$$= (-10+4)\mathbf{i} - (15-8)\mathbf{j} + (-3+4)\mathbf{k}$$
$$= -6\mathbf{i} - 7\mathbf{j} + \mathbf{k}$$

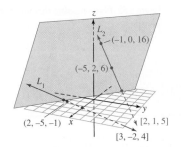

Figure 9.55 Plane determined by lines

The point of intersection $P(-19, 9, -29)$ is certainly in the plane, so we can use $(2, -5, -1)$, $(-1, 0, 16)$, or $(-19, 9, -29)$ to obtain

$$-6(x-2) - 7(y+5) + 1(z+1) = 0$$
$$-6x + 12 - 7y - 35 + z + 1 = 0$$
$$-6x - 7y + z = 22$$
$$6x + 7y - z + 22 = 0$$

■

PROBLEM SET 9.5

LEVEL 1

1. IN YOUR OWN WORDS Contrast the parametric and symmetric forms of the equation of a line. How do these forms compare with the equation of a plane?

2. IN YOUR OWN WORDS Describe the relationship between normal vectors and planes.

3. **Drawing Lesson: Sketching a plane in one octant**

Sketch the graph of

$$4(x - 1) + 6y + 2(z - 4) = 0$$

by reproducing the following steps on your own paper.

Step 1: Draw the three coordinate axes.

Step 2: Draw the three coordinate planes.

Step 3: Plot the points where the plane passes through each of the coordinate axes.

If $y = z = 0$, then

$$4(x - 1) + 6(0) + 2(0 - 4) = 0$$
$$4x - 4 + 0 + 8 - 8 = 0$$
$$x = 3$$

Plot $(3, 0, 0)$.

If $x = y = 0$, then

$$4(0 - 1) + 6(0) + 2(z - 4) = 0$$
$$-4 + 0 + 2z - 8 = 0$$
$$z = 6$$

Plot $(0, 0, 6)$.

If $x = z = 0$, then

$$4(0 - 1) + 6y + 2(0 - 4) = 0$$
$$-4 + 6y - 8 = 0$$
$$y = 2$$

Plot $(0, 2, 0)$.

Connect those three points:

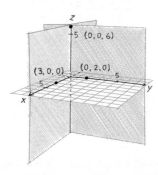

Step 4: Use highlighters or pencils to color the planes in order to add depth to the figure.

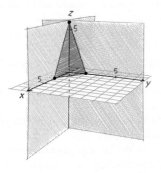

4. **Drawing Lesson: Observer Viewpoint**

Sketch the graph of

$$2(x - 1) - 2(y - 1) + (z - 2) = 0$$

by reproducing the following steps on your own paper. Complete steps 1 and 2 as shown in the previous problem. If one or more of the coordinate points are negative, change the orientation (that is, the vantage point of your eye as you are looking at the sketch).

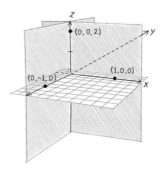

Notice the orientation has changed.

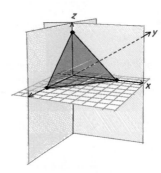

Follow the steps shown in Problems 3 and 4 to plot the points in Problems 5–8. Write each equation for a plane in standard form. Sketch each plane.

5. $4(x - 5) + 3(y - 4) + 2(z - 7) = 0$
6. $5(x - 2) - 3(y + 4) + 4(z - 1) = 0$
7. $3(x - 4) - 2(y - 7) + 2(z + 4) = 0$
8. $2(x + 3) - 4(y - 5) + 8(z - 4) = 0$

Find the parametric and symmetric equations for the line(s) passing through the given points with the properties described in Problems 9–16.

9. $(1, -1, -2)$; parallel to $3\mathbf{i} - 2\mathbf{j} + 5\mathbf{k}$
10. $(1, 0, -1)$; parallel to $3\mathbf{i} + 4\mathbf{j}$
11. $(1, -1, 2)$; through $(2, 1, 3)$
12. $(2, 2, 3)$; through $(1, 3, -1)$
13. $(1, -3, 6)$; parallel to

$$\frac{x - 5}{1} = \frac{y + 2}{-3} = \frac{z}{-5}$$

14. $(1, 0, -4)$; parallel to $x = -2 + 3t, y = 4 + t, z = 2 + 2t$
15. $(3, -1, 2)$; parallel to the xy-plane and the yz-plane
16. $(-1, 1, 6)$; perpendicular to $3x + y - 2z = 5$

17. **Drawing Lesson: Sketching a line in $\mathbb{R}^3$**

Sketch the graph of

$$\frac{x-2}{2} = \frac{y}{10} = \frac{z-3}{-3}$$

by reproducing the following steps on your own paper.

Step 1: Draw the three coordinate axes.

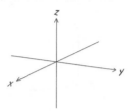

Step 2: Draw the three coordinate planes.

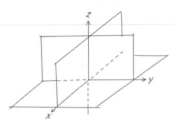

Step 3: Plot points where the line intersects each coordinate plane. For this example, we see, by inspection, that the line passes through $(1, 0, 3)$; plot this point. Next, if $x = 0$

$$\frac{0-1}{2} = \frac{y}{10} \text{ so that } y = -5$$

$$\frac{0-1}{2} = \frac{z-3}{-3} \text{ so that } z = \frac{9}{2}$$

Plot the point $\left(0, -5, \frac{9}{2}\right)$. Finally, if $z = 0$,

$$\frac{x-1}{2} = \frac{0-3}{-3} \text{ so that } y = 10$$

Plot the point $\left(3, 10, \frac{9}{2}\right)$. Draw the line passing through the given points; used dashed lines for hidden parts.

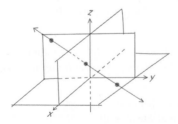

Step 4: Use highlighter or pencils to color the planes in order to add depth to the figure:

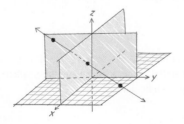

Find the points of intersection of each line in Problems 18–21 with each of the coordinate planes. Follow the steps shown in Problem 17 to graph each line in $\mathbb{R}^3$.

18. $x = 6 - 2t$, $y = 1 + t$, $z = 3t$

19. $x = 6 + 3t$, $y = 2 - t$, $z = 2t$

20. $\dfrac{x-4}{4} = \dfrac{y+3}{3} = \dfrac{z+2}{1}$

21. $\dfrac{x+1}{1} = \dfrac{y+2}{2} = \dfrac{z-6}{3}$

In Problems 22–25, tell whether the two lines intersect, are parallel, are skew, or coincide. If they intersect, give the point of intersection.

22. $\dfrac{x-4}{2} = \dfrac{y-6}{-3} = \dfrac{z+2}{5}; \dfrac{x}{4} = \dfrac{y+2}{-6} = \dfrac{z-3}{10}$

23. $\dfrac{x-3}{2} = \dfrac{y-1}{-1} = \dfrac{z-4}{1}; \dfrac{x+2}{3} = \dfrac{y-3}{-1} = \dfrac{z-2}{1}$

24. $x = 4 - 2s$, $y = 6s$, $z = 7 - 4s$;
$x = 5 + t$, $y = 1 - 3t$, $z = -3 + 2t$

25. $x = 1 + 4s$, $y = 2 + 3s$, $z = 2 + 2s$;
$x = 1 + 4t$, $y = -6t$, $z = -3 - t$

Find an equation for the plane that contains the point P and has the normal $\mathbf{N}$ vector given in Problems 26–31.

26. $P(-1, 3, 5)$; $\mathbf{N} = 2\mathbf{i} + 4\mathbf{j} - 3\mathbf{k}$

27. $P(0, -7, 1)$; $\mathbf{N} = -\mathbf{i} + \mathbf{k}$

28. $P(0, -3, 0)$; $\mathbf{N} = -2\mathbf{j} + 3\mathbf{k}$

29. $P(1, 1, -1)$; $\mathbf{N} = -\mathbf{i} - 2\mathbf{j} + 3\mathbf{k}$

30. $P(0, 0, 0)$; $\mathbf{N} = \mathbf{k}$

31. $P(0, 0, 0)$; $\mathbf{N} = \mathbf{i}$

32. Find two unit vectors parallel to the line

$$\frac{x-3}{4} = \frac{y-1}{2} = \frac{z+1}{1}$$

33. Find two unit vectors parallel to the line

$$\frac{x-1}{2} = \frac{y+2}{4} = \frac{z+5}{1}$$

34. Find two unit vectors perpendicular to the plane $2x + 4y - 3z = 4$.

35. Find two unit vectors perpendicular to the plane $5x - 3y + 2z = 15$.

LEVEL 2

36. Show that the vector $\mathbf{v} = 3\mathbf{i} - 4\mathbf{j} + \mathbf{k}$ is orthogonal to the line that passes through the points $P(0, 0, 1)$ and $Q(2, 1, -1)$.

37. Show that the vector $\mathbf{v} = 7\mathbf{i} + 4\mathbf{j} + 3\mathbf{k}$ is orthogonal to the line passing through the points $R(-2, 2, 7)$ and $S(3, -3, 2)$.

38. Find two unit vectors that are parallel to the line of intersection of the planes $x + y = 1$ and $x - 2z = 3$.

39. Find two unit vectors that are parallel to the line of intersection of the planes $x + y + z = 3$ and $x - y + z = 1$.

40. Find an equation for the line of intersection of the planes $x + y - z = 4$ and $2x - y + 3z = 1$.

41. Find an equation for the line of intersection of the planes $x + 2y + z = 3$ and $x + y - 2z = 4$.

42. Find an equation for the plane that passes through $P(1, -1, 2)$ and is normal to $\mathbf{PQ}$, where $Q(2, 1, 3)$.

43. Find an equation for the line that passes through the point $(1, -5, 3)$ and is orthogonal to the plane $2x - 3y + z = 1$.

44. Find an equation for the plane that contains the point $(2, 1, -1)$ and is orthogonal to the line

$$\frac{x-3}{3} = \frac{y+1}{5} = \frac{z}{2}$$

45. Find a plane that passes through the point $(1, 2, -1)$ and is parallel to the plane $2x - y + 3z = 1$.

46. Show that the line

$$\frac{x-1}{2} = \frac{y+1}{3} = \frac{z-2}{4}$$

is parallel to the plane $x - 2y + z = 6$.

47. Find the point where the line

$$\frac{x-1}{2} = \frac{y+1}{-1} = \frac{z}{3}$$

intersects the plane $3x + 2y - z = 5$.

48. The *angle* between two planes is defined to be the acute angle between their normal vectors. Find the angle between the planes $x + 2y - 4z = 3$ and $x - 2y + z = 2$, rounded to the nearest degree.

49. Find the equation of the line that passes through the point $P(2, 3, 1)$ and is parallel to the line of intersection of the planes $x + 2y - 3z = 4$ and $x - 2y + z = 0$.

50. Find the equation of the line that passes through the point $P(0, 1, -1)$ and is parallel to the line of intersection of the planes $2x + y - 2z = 5$ and $3x - 6y - 2z = 7$.

51. Find a vector that is parallel to the line of intersection of the planes $2x + 3y = 0$ and $3x - y + z = 1$.

52. Find the equation of the line of intersection of the planes $3x + y - z = 5$ and $x - 6y - 2z = 10$.

53. Find the equation of the line of intersection of the planes $2x - y + z = 8$ and $x + y - z = 5$.

54. Let $\mathbf{v} = 2\mathbf{i} + \mathbf{j}$ and $\mathbf{w} = 2\mathbf{i} - \mathbf{j} + 3\mathbf{k}$. Find the direction cosines and the direction angles of $\mathbf{v} \times \mathbf{w}$.

55. Find the direction cosines of a vector determined by the line of intersection of the planes $x + y + z = 3$ and $2x + 3y - z = 4$.

56. What can be said about the lines

$$\frac{x-x_0}{a_1} = \frac{y-y_0}{b_1} = \frac{z-z_0}{c_1}$$

and

$$\frac{x-x_0}{a_2} = \frac{y-y_0}{b_2} = \frac{z-z_0}{c_2}$$

in the case where $a_1 a_2 + b_1 b_2 + c_1 c_2 = 0$?

57. Show that the shortest distance from a point P to a line L is given by the formula

$$d = \frac{\| \mathbf{v} \times \mathbf{QP} \|}{\| \mathbf{v} \|}$$

where Q is any point on L.

58. In Figure 9.56, $\mathbf{N}$ is normal to the plane p and L is a line that intersects p.

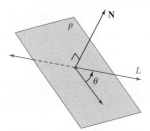

Figure 9.56 Find θ

Assume $\mathbf{N} = a\mathbf{i} + b\mathbf{j} + c\mathbf{k}$ and that L is given by $x = x_0 + At$, $y = y_0 + Bt$, $z = z_0 + Ct$.

a. Find the angle θ between L and the plane p.

b. Find the angle (to the nearest degree) between the plane $x + y + z = 10$ and the line

$$\frac{x-1}{2} = \frac{y+3}{3} = \frac{z-2}{-1}$$

59. Show that a plane with x-intercept a, y-intercept b, and z-intercept c has the equation

$$\frac{x}{a} + \frac{y}{b} + \frac{z}{c} = 1$$

assuming a, b, and c are all nonzero.

60. Suppose planes and p_1 and p_2 intersect. If $\mathbf{v}_1$ and $\mathbf{w}_1$ are vectors on p_1 and $\mathbf{v}_2$ and $\mathbf{w}_2$ are on plane p_2, then show that

$$(\mathbf{v}_1 \times \mathbf{w}_1) \times (\mathbf{v}_2 \times \mathbf{w}_2)$$

is aligned with the line of intersection of the planes.

CHAPTER 9 SUMMARY AND REVIEW

W*ith me everything turns into mathematics.*

René Descartes

Take some time to get ready to work the review problems in this section. First, look back at the definition and property boxes. You will maximize your understanding of this chapter by working the problems in this section only after you have studied the material.

SELF TEST   *All of the answers for this self test are given in the back of the book.*

1. Given $\mathbf{v} = 2\mathbf{i} - 3\mathbf{j} + \mathbf{k}$ and $\mathbf{w} = 3\mathbf{i} - 2\mathbf{j}$. Find each of the following vectors.
   a. $2\mathbf{v} + 3\mathbf{w}$ b. $\|\mathbf{v}\|^2 - \|\mathbf{w}\|^2$
   c. $\mathbf{v} \cdot \mathbf{w}$ d. $\mathbf{v} \times \mathbf{w}$
2. For $\mathbf{v} = 2\mathbf{i} - 5\mathbf{j} + \mathbf{k}$ and $\mathbf{w} = \mathbf{j} - 3\mathbf{k}$, find
   a. vector projection of $\mathbf{v}$ onto $\mathbf{w}$
   b. scalar projection of $\mathbf{w}$ onto $\mathbf{v}$
3. Given $\mathbf{u} = 2\mathbf{i} - 3\mathbf{j} + \mathbf{k}$, $\mathbf{v} = \mathbf{i} + \mathbf{j} - 2\mathbf{k}$, and $\mathbf{w} = 3\mathbf{i} + 5\mathbf{k}$. In each of the following cases, either perform the indicated computation or explain why it cannot be computed.
   a. $(\mathbf{u} \times \mathbf{v}) \cdot \mathbf{w}$ b. $(\mathbf{u} \cdot \mathbf{v}) \times \mathbf{w}$
   c. $(\mathbf{u} \times \mathbf{v}) \times \mathbf{w}$ d. $(\mathbf{u} \cdot \mathbf{v}) \cdot \mathbf{w}$
4. Find the equation of the lines with the given properties.
   a. The parametric form of a line through the points $P(-1, 4, -3)$ and $Q(0, -2, 1)$
   b. The parametric form of a line of intersection of the planes $2x + 3y + z = 2$ and $y - 3z = 5$
   c. The symmetric form of a line that contains the points $R(1, 2, -1)$ and $S(-3, -1, 2)$
5. Find the equation of the planes with the given properties.
   a. The plane that contains the point $P(1, 1, 3)$ and is normal to the vector $\mathbf{v} = 2\mathbf{i} + 3\mathbf{k}$
   b. The plane that contains the points $P(0, 2, -1)$, $Q(1, -3, 5)$, and $R(3, 0, -2)$

6. Find the direction cosines and the direction angles of the vector $\mathbf{u} = -2\mathbf{i} + 3\mathbf{j} + \mathbf{k}$. Round to the nearest degree.
7. In each case, determine whether the lines intersect, are parallel, or are skew. If they intersect, find the point of intersection.
   a. $x = 2t - 3$, $y = t$, $z = 2t$ and $\dfrac{x+2}{3} = \dfrac{y-3}{5}$; $z = 3$
   b. $\dfrac{x-7}{5} = \dfrac{y-6}{4} = \dfrac{z-8}{5}$ and $\dfrac{x-8}{6} = \dfrac{y-6}{4} = \dfrac{z-9}{6}$
8. Let $\mathbf{u} = 2\mathbf{i} + \mathbf{j}$, $\mathbf{v} = \mathbf{i} - \mathbf{j} - \mathbf{k}$, and $\mathbf{w} = 3\mathbf{i} + \mathbf{k}$.
   a. Find the volume of the parallelepiped determined by these vectors.
   b. Find a positive number A that guarantees that the tetrahedron determined by $A\mathbf{u}$, $A\mathbf{v}$, and $\mathbf{w}$ has volume that is twice the volume of the original tetrahedron.
9. An airplane flies at 200 mi/h parallel to the ground at an altitude of 10,000 ft. If the plane flies due south and the wind is blowing toward the northeast at 50 mi/h, what is the ground speed of the plane (that is, effective speed)?
10. A girl pulls a sled 50 ft on level ground with a rope inclined at an angle of 30° with the horizontal (the ground). If she applies 3 lb of tension to the rope, how much work is performed on the sled?

STOP   STUDY HINTS *Compare your solutions and answers to the self test.*

Simplify the expressions in Problems 1–6.

1. $\dfrac{(6 - x^2)(5) - (5x - 4)(-2x)}{(6 - x^2)^2}$ 2. $(3x^2 - 2)^4$

3. $\dfrac{\cos x(\cos x) - \sin x(-\sin x)}{\cos^2 x}$ 4. $\sqrt{1 + \left(\dfrac{4y^4 - 1}{4y^2}\right)^2}$

5. $\left(\dfrac{e^x + e^{-x}}{2}\right)^2 - \left(\dfrac{e^x - e^{-x}}{2}\right)^2$

6. $\dfrac{(-r\sin t)(-r\sin t) - (r\cos t)(-r\cos t)}{[(-r\sin t)^2 + (r\cos t)^2]^{3/2}}$

Factor the expressions in Problems 7–10.

7. $-3x^{-2}y^3 + 3x^{-3}y^2$
8. $-3(x + y)^{-4}(x - y)^2 + 2(x + y)^{-3}(x - y)$
9. $3(x^2 + 3x + 4)^2(3x^4 + 6)^2(4x + 3) + 2(x^2 + 3x + 4)^3(3x^4 + 6)(12x^3)$
10. $\dfrac{(1 - 4x)^5(\sec^2 7x)(7) - (\tan^2 7x)(5)(1 - 4x)^4(-4)}{(1 - 4x)^{10}}$

Evaluate $\lim\limits_{\Delta x \to 0} \dfrac{f(x + \Delta x) - f(x)}{\Delta x}$ *for the function f in*

Problems 11–16.

11. $f(x) = 3x + 5$ 12. $f(x) = 2x^2 - 3x + 10$

13. $f(x) = \sqrt{2x}$

14. $f(x) = x(x+1)$

15. $f(x) = \dfrac{x+1}{x-2}$

16. $f(x) = \dfrac{4}{x}$

Solve the equations in Problems 17–22 correct to two decimal places.

17. $\dfrac{4x-5}{2\sqrt{x-3}\sqrt{2x+1}} = 0$

18. $\cos 3\theta + \cos \theta = 0$

19. $10^{x+3} = 214$

20. $\dfrac{1}{2} = e^{-0.000425t}$

21. $\dfrac{10}{3}x^{-1/3} - \dfrac{10}{32}x^{2/3} = 0$

22. $\ln x - \dfrac{1}{2}\ln 3 = \dfrac{1}{2}\ln(x+6)$

Graph the functions in Problems 23–28.

23. $y = \dfrac{6x^2 - 11x}{2x - 1}$

24. $y = \dfrac{2x^2 - 5x - 3}{3x^2 - 7x - 6}$

25. $y - 2 = \sin\left(x - \dfrac{\pi}{2}\right)$

26. $y = -\dfrac{3}{5}\sqrt{x^2 - 5}$

27. $y = \sqrt{\dfrac{x+3}{x^2-4}}$

28. $y = 2 + 3t,\ y = 4 - 5t$

In Problems 29–36, graph the solutions. If appropriate, consider a parameterization of the curve.

29. $2x + y - 10 = 0$

30. $x^2 + y^2 - 6y = 0$

31. $y = x^2 - 10x + 23$

32. $x^2 < y$

33. $\dfrac{x^2}{49} + \dfrac{y^2}{25} = 1$

34. $\dfrac{y^2}{16} - \dfrac{x^2}{9} = 1$

35. $y^2 + 4x + 4y = 0$

36. $4x = 4y^2 + 4y + 3$

Solve the systems in Problems 37–42.

37. a. $\begin{cases} x + 2y = 3 \\ 3x + 4y = 3 \end{cases}$

b. $\begin{cases} \dfrac{1}{x} + \dfrac{2}{y} = 3 \\ \dfrac{3}{x} + \dfrac{4}{y} = 3 \end{cases}$

38. a. $\begin{cases} x - 6 = y \\ 4x + y = 9 \end{cases}$

b. $\begin{cases} \sqrt{x} - 6 = \sqrt{y} \\ 4\sqrt{x} + \sqrt{y} = 9 \end{cases}$

39. $\begin{cases} 3x - 2y + 17 = 0 \\ 5x + 4y + z = 8 \\ 3x + z + 2 = 0 \end{cases}$

40. $\begin{cases} x_1 + 2x_2 - x_4 = 0 \\ x_1 - 3x_3 - x_4 = 10 \\ x_2 - x_1 + x_3 = -2 \\ 2x_4 - x_1 = 3 \end{cases}$

41. $\begin{cases} 2x - y + z = 4 \\ 3x - 2y + 2z = 31 \\ x - y + 3z = 2 \end{cases}$

42. $\begin{cases} 6x - y - 2z = 7 \\ 5x - 4y - 5z = 5 \\ x + 3y + 3z = 4 \end{cases}$

43. Estimate the coordinates of each point on the graph in Figure 9.57 where a tangent line would be horizontal.

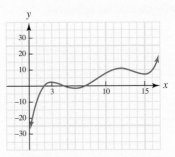

Figure 9.57 Graph of a function

44. Graph $\begin{cases} y \geq 0 \\ 3x + 2y > -3 \\ x - y < 0 \end{cases}$

45. Find the equation and graph the parabola with directrix $y - 3 = 0$ and focus $(-3, -2)$.

46. Find the area of each of the following plane figures:
 a. the circle $x^2 + y^2 = 9$
 b. the rectangle with vertices $(0, 0)$, $(3, 0)$, $(3, 2)$, and $(0, 2)$

47. a. Find the sum of the first ten terms of the sequence

$$5,\ 20,\ 80,\ \ldots$$

 b. Evaluate $\displaystyle\sum_{k=1}^{100}(3k - 1)$.

48. Find equations, in both parametric and symmetric forms, of the lines passing through $A(1, -2, 3)$, $B(4, -1, 2)$. Find two additional points on each line.

49. Find the area of the parallelogram determined by $3\mathbf{i} - 4\mathbf{j}$ and $-\mathbf{i} - \mathbf{j} + \mathbf{k}$.

50. Find the equation of the plane satisfying the given conditions.
 a. the xy-plane
 b. the plane parallel to the xz-plane passing through $(4, 3, 7)$
 c. the plane through $(-1, 4, 5)$ and perpendicular to a line with direction numbers $\langle 4, 4, -3 \rangle$
 d. the plane through $(4, -3, 2)$ and parallel to the plane $5x - 2y + 3z - 10 = 0$
 e. the plane passing through $P(4, 1, 3)$, $Q(-4, 2, 1)$, and $R(1, 0, 2)$

51. Find $\|\mathbf{v}\|$, $\mathbf{v} - \mathbf{w}$, $2\mathbf{v} + 3\mathbf{w}$, $\mathbf{v} \cdot \mathbf{w}$, and $\mathbf{v} \times \mathbf{w}$.
 a. $\mathbf{v} = 3\mathbf{i} - 2\mathbf{j} + \mathbf{k},\ \mathbf{w} = 4\mathbf{i} + \mathbf{j} - 3\mathbf{k}$
 b. $\mathbf{v} = \mathbf{i},\ \mathbf{w} = \mathbf{j}$

52. Find the points of intersection of the line $x = 4 - 3t$, $y = 2 + t$, $z = 5t$ with each of the coordinate planes.

53. Find the points of intersection of the line

$$\frac{x-3}{2} = \frac{y-6}{-2} = \frac{z+2}{4}$$

with each of the coordinate planes.

54. A charter flight has signed up 100 travelers. They are told that if they can sign up an additional 25 persons, they can save $78 each. What is the cost per person if 100 persons make the trip?

55. A closed-top box is to be made from the 10 in. by 16 in. piece of cardboard by cutting out squares of equal size as shown in Figure 9.58 and folding along the dashed lines, tucking the two extra flaps inside. Let x be the length of the side of each square.

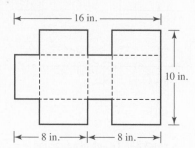

Figure 9.58 Building a box

    a. Write the volume as a function of x.
    b. What is the domain of x?
    c. Graph the volume $V(x)$.
    d. Estimate the dimensions of the box of largest volume.

56. A projectile is fired from ground level at an angle of $40°$ with muzzle velocity of 110 ft/s. Find the time of flight and the range.

57. Find a constant c that guarantees that the graph of the equation

$$x^2 + xy + cy = 4$$

will have a y-intercept of $(0, -5)$. What are the x-intercepts of the graph?

58. In Figure 9.59, ship A is at point P at noon and sails due east at 9 km/h. Ship B arrives at point P at 1:00 P.M. and sails at 7 km/h along a course that forms an angle of $60°$ with the course of ship A. Find a formula for the distance $s(t)$ separating the ships t hours after noon. Approximately how far apart (to the nearest kilometer) are the ships at 4:00 P.M.?

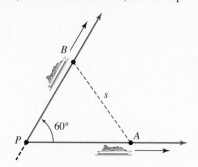

Figure 9.59 Distance between ships

59. A manufacturer estimates that when the price for each unit is p dollars, the profit will be $N = -p^2 + 14p - 50$ million dollars. Sketch the graph of the profit formula and answer these questions.

    a. For what values of p is this a profitable operation? (That is, when is $N > 0$?)

    b. What price results in maximum profit? What is the maximum profit?

60. The vectors $\mathbf{u}$, $\mathbf{v}$, and $\mathbf{w}$ are said to be *linearly independent* in $\mathbb{R}^3$ if the only solution to the equation $a\mathbf{u} + b\mathbf{v} + c\mathbf{w} = \mathbf{0}$ is $a = b = c = 0$. Otherwise, the vectors are *linearly dependent*. Determine whether the vectors $\mathbf{u} = -\mathbf{i} + 2\mathbf{k}$, $\mathbf{v} = 2\mathbf{i} - \mathbf{j} + 3\mathbf{k}$, $\mathbf{w} = \mathbf{i} + 3\mathbf{j} - 2\mathbf{k}$ are linearly independent or dependent.

FINAL REVIEW SET I

Match the equations in Problems 1–26 with the correct calculator graph labeled A–Z.

1. $y = mx + b$
2. $y = ax^3$
3. $y = e^x$
4. $y = x^4 - 3$
5. $y = \dfrac{x - a}{x - b}$
6. $y = \dfrac{(x - a)^2}{x - b}$
7. $y = a \sin bx$
8. $y = a \cos bx$
9. $y = a \tan bx$
10. $y = a \cot bx$
11. $y = a \sec bx$
12. $y = a \csc bx$
13. $x = t, y = 2t^2 + 3$
14. $x = 25t^2 + 3, y = t, t \geq 0$
15. $r^2 = a^2 \sin 2\theta$
16. $r^2 = a^2 \cos 2\theta$
17. $r = a \cos 2\theta$
18. $r = \sin 4\theta$
19. $r = a \sin 3\theta$
20. $r = a \sin \theta$
21. $y = \log x$
22. $y = -\log x$
23. $r = a - a \cos \theta$
24. $y = 2 \cos t, y = 3 \sin t$
25. $y = ax^4 + bx^3 + cx^2 + x$
26. $y = ax^5 + bx^4 + cx^3 + dx^2 + x$

A.

B.

C.

D.

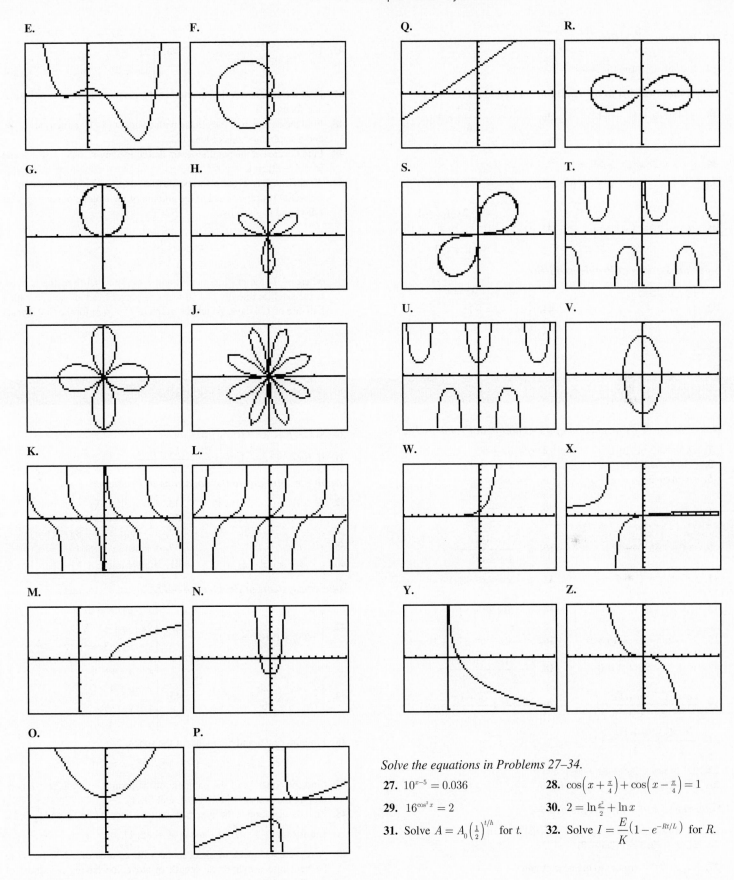

E.   F.   Q.   R.

G.   H.   S.   T.

I.   J.   U.   V.

K.   L.   W.   X.

M.   N.   Y.   Z.

O.   P.

Solve the equations in Problems 27–34.

27. $10^{x-5} = 0.036$ **28.** $\cos\left(x + \frac{\pi}{4}\right) + \cos\left(x - \frac{\pi}{4}\right) = 1$

29. $16^{\cos^2 x} = 2$ **30.** $2 = \ln\frac{e^2}{2} + \ln x$

31. Solve $A = A_0\left(\frac{1}{2}\right)^{t/h}$ for t.   **32.** Solve $I = \frac{E}{K}\left(1 - e^{-Rt/L}\right)$ for R.

33. If $f(x) = \dfrac{5}{x-2}$, find $\dfrac{f(x+h) - f(x)}{h}$.

34. If $f(x) = e^x$, find $\dfrac{f(x+h) - f(x)}{h}$.

Graph the solutions to the equations in Problems 35–42.

35. $x^2 + 16(y+4)^2 = 64$

36. $\dfrac{(x-3)^2}{4} - \dfrac{y+2}{6} = 1$

37. $y^2 - 6y - 4x + 5 = 0$

38. $12x^2 - 4y^2 + 24x - 8y + 4 = 0$

39. $y - 2 = 2\cos\left(x - \frac{\pi}{4}\right)$

40. $x = t^2 \sin t, y = t\cos t$ for $0 \le t \le 2\pi$

41. $f(x) = 3\sin 2\left(x - \frac{\pi}{6}\right) - 1$

42. $f(x) = \begin{cases} x^2 - 3x & \text{if } x \le 1 \\ \frac{1}{2}(3x - 7) & \text{if } 1 < x < 3 \\ \frac{3x-1}{x+5} & \text{if } x \ge 3 \end{cases}$

Solve the systems in Problems 43–46.

43. $\begin{cases} 2x + y = 3 \\ -\dfrac{3}{2}x - \dfrac{1}{2}y = 1 \end{cases}$

44. $\begin{cases} 2x + 2y + 3z = 1 \\ 2x - z = -11 \\ 3y + 2z = 6 \end{cases}$

45. $\begin{cases} y = 2x^2 - 12x + 18 \\ y + 6x = x^2 + 10 \end{cases}$

46. $\begin{cases} 6x_1 - 3x_2 + x_4 = -12 \\ 2x_2 + 4x_3 - x_4 = 1 \\ 3x_1 + 2x_2 + 2x_3 = -3 \end{cases}$

47. Find the standard form equation of the parabola with vertex $(-3, 5)$ and focus $(-3, 1)$.

48. Find the equation of the ellipse with center at $(4, 1)$, a focus at $(5, 1)$, and a semimajor axis of length 2.

49. Find the area of the parallelogram determined by $\mathbf{v} = 2\mathbf{i} - 3\mathbf{j} - \mathbf{k}$ and $\mathbf{w} = \mathbf{i} - 2\mathbf{j} + \mathbf{k}$.

50. A satellite has a radioisotope power supply. The power output P (in watts) is proportional to the amount of the isotope remaining after t days and is given by

$$P = P_0\left(\frac{1}{2}\right)^{t/h}$$

where P_0 is the original output and h is the half-life of the isotope. If the satellite initially has 50 watts of power from an isotope with a half-life of 174 days, sketch the graph of $P = f(t)$ for the first 2 years $(0 \le t \le 730)$.

FINAL REVIEW SET II

Name the curve by inspection of the equation in Problems 1–23.

1. $xy = 5$

2. $y = 3x + 4$

3. $r^2 = 9\cos 2\theta$

4. $r = 2 + 2\sin\theta$

5. $r = 4\cos 5\theta$

6. $r = 5\sin\frac{\pi}{4}$

7. $r = \dfrac{3}{1 - \cos\theta}$

8. $r = \dfrac{4}{1 - 2\cos\theta}$

9. $r(4 - 4\sin\theta) = 32$

10. $r = \dfrac{\sqrt{3}}{1 + \sin(\theta - 1)}$

11. $(x-1)(y+1) = 7$

12. $y = 2\log(x - 3)$

13. $y = e^{2x} + 3$

14. $y = \sin\left(2x - \frac{\pi}{2}\right)$

15. $x^2 - 2y^2 + 3x + 4y = 0$

16. $x^2 - 2x + y^2 + 5y + 10 = 0$

17. $x^2 + 5x - 4y + 3 = 0$

18. $\dfrac{(x-4)^2}{2} + \dfrac{(y+1)^2}{4} = 1$

19. $\dfrac{(x-4)^2}{2} + \dfrac{(y+1)}{4} = 1$

20. $\dfrac{(x-4)}{2} + \dfrac{(y+1)^2}{4} = 1$

21. $\dfrac{(x-4)}{2} + \dfrac{(y+1)}{4} = 1$

22. $2x^2 - 4xy + 2y^2 + 3x + 4y - 5 = 0$

23. $x^2 + 2xy - y^2 + 2x + 6y + 11 = 0$

Solve the equations in Problems 24–27.

24. $1 - \sin x = \cos 2x$

25. $8^{5x+2} = 16$

26. $2\ln x - \frac{1}{2}\ln 9 = \ln 3(x - 2)$

27. $\frac{1}{2} = e^{-0.000425t}$ (round to the nearest unit)

28. Solve $I = \dfrac{E}{R}(1 - e^{-Rt/L})$ for t.

29. Solve $P = P_0 e^{rt}$ for r.

30. Solve $T = A + (B - A)10^{-kt}$ for k.

31. If $f(x) = (3x + 1)$, find $\dfrac{f(x+h) - f(x)}{h}$.

Graph the solutions to the equations in Problems 32–41.

32. $(x+1)^2 + (y-2)^2 = 25$

33. $5(x+3)^2 + 9(y-2)^2 = 45$

34. $x(x - y) = y(y - x) - 1$

35. $8y^2 - x - 32y + 31 = 0$

36. $x^2 + y^2 = 4x + 2y - 3$

37. $x^2 - 4y^2 - 6x - 8y - 11 = 0$

38. $x^{2/3} + y^{2/3} = 1$

39. $x = \sin t, y = \cos t$

40. $y - 3 = \sin(\pi - x)$

41. $f(x) = \tan\left(x + \frac{\pi}{4}\right) - 2$

Solve the systems in Problems 42–45.

42. $\begin{cases} q + d - 147 \\ 0.25q + 0.10d = 24.15 \end{cases}$

43. $\begin{cases} x + y + z = 6 \\ 2x - y + z = 3 \\ x - 2y + 3z = -12 \end{cases}$

44. $\begin{cases} x^2 + y^2 = 25 \\ 7x - y = 25 \end{cases}$

45. $\begin{cases} x - y - 8 \le 0 \\ x - y + 4 \ge 0 \\ x + y \ge -5 \\ y + x \le 4 \end{cases}$

46. Find the center and the radius of the sphere

$$4x^2 + 4y^2 + 4z^2 - 12y + 4z + 1 = 0$$

47. Find the equation of the parabola of the form $y = ax^2 + bx + c$ containing the points $(1, 1)$, $(2, 0)$, and $(4, 4)$.

48. Find the equation of the hyperbola with vertices at $(-3, 1)$ and $(-5, 1)$, and foci at $\left(-4 - \sqrt{6}, 1\right)$ and $\left(-4 + \sqrt{6}, 1\right)$.

49. A rectangular piece of sheet metal with an area of 1,800 in.² is to be bent into a cylindrical length of stovepipe having a volume of 1,200 in.³. What are the dimensions of the sheet metal?

50. A closed box with a square base is to have a volume of 250 cubic meters. The material for the top and bottom of the box costs $2 per square meter, and the material for the sides costs $1 per square meter. Express the construction cost of the box as a function of the length of its base.

Working in small groups is typical of most work environments, and this book seeks to develop skills with group activities. At the end of each chapter, we present a list of suggested projects, and even though they could be done as individual projects, we suggest that these projects be done in groups of three or four students. Since this is the last set of group research projects in the book, these are slightly different. They can be used to help you prepare for a calculus course. In each case, write a paper on the given topic.

G9.1 The box problem An open box is to be made from an $x \times y$ piece of cardboard by cutting out squares of equal size from the four corners and bending up the flaps to form the sides. The calculus problem is, "What are the dimensions of the box of largest volume that can made in this way?" For this problem, find the volume of the box, the variable, and the domain for the variable.

G9.2 The oil can problem Suppose you must design an oil can in the shape of a right circular cylinder with a cone lid, and you are told that the can must hold V quarts of oil. The calculus problem is, "What dimensions should you specify so that your can will require the least amount of metal for its construction?" For this problem, find the surface area of the oil can, the variable, and the domain for the variable.

G9.3 The fence problem You wish to fence a rectangular pasture and you have a roll of fencing material containing x ft. The calculus problem is, "What are the dimensions of a rectangular pasture of greatest area that can be enclosed with x ft of available fencing material?" For this problem, find the area of the enclosure, the variable, and the domain for this variable.

G9.4 The conical tank problem Water is flowing into a cone-shaped tank at a constant rate. The tank is h ft tall, has an R-ft radius at the top, and takes 1 hr to fill. How much water is contained in the tank after m minutes?

G9.5 The tangent problem Consider the problem of drawing a tangent line to a given curve. Be careful about your definitions and explanations.

G9.6 The limit problem Write a paper on the concept of limit, as it was presented in this text. Research some calculus books for a precise definition of a limit. Reconcile the intuitive notion of limit with the definition as presented in a calculus book.

G9.7 The continuity problem Write a paper on the concept of continuity.

G9.8 The area problem Write a paper on finding the area of a closed region.

G9.9 Your history problem Write a paper about your mathematical history, and in particular include your experiences in taking this course. Include a prospective about your expectations when you enroll in a calculus course.

G9.10 Your future problem Write a paper about what you expect to do after your graduate from college. Does this choice require calculus as a prerequisite? How do you think what you have learned in this course will be of value in your future career?

Appendices

A: Field Properties
Properties of Equality
Properties of Real Numbers

B: Complex Numbers

C: Mathematical Induction

D: Binomial Theorem
Binomial Theorem

E: Determinants and Cramer's Rule
Determinants
Properties of Determinants

F: Library of Curves and Surfaces

G: Answers

Appendix A: Field Properties

To study mathematics, it is worthwhile to state the assumptions used in this course. It is not our intent to rigorously develop the material in the book, but it is important to present some of the terminology you will find in the book.

Properties of Equality

In Table 1.1, we presented the sets of numbers used in this text, and in Section 1.1 we stated the property of comparison (trichotomy), which requires that for any two real numbers then exactly one of the following is true: (1) $a = b$, (2) $a > b$, (3) $a < b$. In addition to the property of comparison, four properties of equality are used in mathematics.

PROPERTIES OF EQUALITY

Let a, b, and c be real numbers.

REFLEXIVE PROPERTY: $a = a$

SYMMETRIC PROPERTY: If $a = b$, then $b = a$.

TRANSITIVE PROPERTY: If $a = b$ and $b = c$, then $a = c$.

SUBSTITUTION PROPERTY: If $a = b$, then a may be replaced by b (or b by a) throughout any statement without changing the truth or falsity of the statement.

EXAMPLE 1   Practice recognizing the properties of equality

Identify the property of equality illustrated by each statement:

a. $(a + b)(c + d) = (a + b)(c + d)$

b. If $(a + b)(c + d) = (a + b)c + (a + b)d$

and $(a + b)c + (a + b)d = ac + bc + ad + bd,$

then

$(a + b)(c + d) = ac + bc + ad + bd.$

c. If $a(b + c) = ab + ac,$ then $ab + ac = a(b + c).$

d. If $a = 2$ and $(a + 3)(a + 4) = 30,$ then $(2 + 3)(2 + 4) = 30.$

Solution

a. reflexive

b. transitive

c. symmetric

d. substitution

Properties of Real Numbers

In developing the material in this book, we assumed some properties of the real numbers. When we are adding real numbers, the result is called the **sum** and the numbers added are called **terms**. When we are multiplying real numbers, the result is called the **product** and the numbers multiplied are called the **factors**. The result from subtraction is called the **difference**; the result from division is the **quotient**. The real numbers, together with the relation of equality and the operations of addition and multiplication, satisfy what are called **field properties**.

FIELD

A **field** is a set along with two defined operations satisfying the following eleven properties. Let a, b, and c be real numbers.

	Addition Properties	*Multiplication Properties*
CLOSURE:	$a + b$ is a unique real number.	ab is a unique real number.
COMMUTATIVE:	$a + b = b + a$	$ab = ba$
ASSOCIATIVE:	$(a + b) + c = a + (b + c)$	$(ab)c = a(bc)$
IDENTITY:	There exists a unique real number zero, denoted by 0, such that $a + 0 = 0 + a = a$	There exists a unique real number one, denoted by 1, such that $a \cdot 1 = 1 \cdot a = a$
INVERSE:	For each real number a, there is a unique real number $-a$ such that $a + (-a) = (-a) + a = 0$	For each *nonzero* real number a there is a unique real number $\frac{1}{a}$ such that $a\left(\frac{1}{a}\right) = \left(\frac{1}{a}\right)a = 1$
DISTRIBUTIVE:	$a(b + c) = ab + ac$	

A set is said to be **closed** with respect to a particular operation if it satisfies the closure property, as shown in Example 2.

EXAMPLE 2   Checking the closure property

Is the set $\{-1, 0, 1\}$ closed for

a. multiplication

b. addition

Solution

a. Check all possible products:

$$(-1)(-1) = 1;\ (0)(1) = 0;\ (-1)(0) = 0;\ (0)(0) = 0;\ (-1)(1) = -1;\ (1)(1) = 1$$

Since we have exhausted all possibilities, we see that the set is closed for multiplication.

b. $1 + 1 = 2$ is not in the set, so the set is not closed for addition. ∎

EXAMPLE 3   Finding inverses

Find the additive and multiplicative inverse for $\sqrt{2} + 1$.

Solution

Additive inverse Find a number I so that

$$\sqrt{2} + 1 + I = 0$$
$$I = -\sqrt{2} - 1$$

Multiplicative inverse Find a number I so that

$$\left(\sqrt{2} + 1\right)I = 1$$

$$I = \frac{1}{\sqrt{2} + 1} \cdot \frac{\sqrt{2} - 1}{\sqrt{2} - 1}$$

$$= \frac{\sqrt{2} - 1}{2 - 1}$$

$$= \sqrt{2} - 1$$

∎

EXAMPLE 4   Practice identifying field properties

Identify the field property illustrated.

a. $4 \cdot 3$ is a real number.

b. $\frac{1}{4}(4 \cdot 3) = \left(\frac{1}{4} \cdot 4\right)3$

c. $\left(\frac{1}{4} \cdot 4\right)3 = 1 \cdot 3$

d. $1 \cdot 3 = 3$

e. $4 + (3 + 5) = (3 + 5) + 4$

f. $4 + (3 + 5) = (4 + 3) + 5$

Solution

a. Closure property for multiplication

b. Associative property for multiplication

c. Inverse for multiplication

d. Identity for multiplication

e. Commutative property for addition

f. Associative property for addition

∎

PROBLEM SET **APPENDIX A**

LEVEL 1

1. IN YOUR OWN WORDS Explain the identity property.
2. IN YOUR OWN WORDS Explain the inverse property.

Identify the property of equality or field property illustrated by Problems 3–18.

3. $14x + 8x = 14x + 8x$
4. $14x + 8x = (14 + 8)x$
5. $10y + 4y = 10y + 4y$
6. $10y + 4y = (10 + 4)y$
7. $(14 + 8)x = 22x$
8. If $14x + 8x = (14 + 8)x$, and $(14 + 8)x = 22x$, then $14x + 8x = 22x$
9. $(10 + 4)y = 14y$
10. If $10y + 4y = (10 + 4)y$, and $(10 + 4)y = 14y$, then $10y + 4y = 14y$
11. If $14x + 8y = 22$ and $y = b$, then $14x + 8b = 22$
12. 4π is a real number.
13. If $22 = 14x + 8y$, then $14x + 8y = 22$.
14. $\frac{1}{6}$ is a real number.
15. $\pi \cdot \frac{1}{\pi} = 1$
16. $5x(a + b) = 5xa + 5xb$
17. $\dfrac{\sqrt{3}+2}{\sqrt{5}+1} = \dfrac{\sqrt{3}+2}{\sqrt{5}+1} \cdot \dfrac{\sqrt{5}-1}{\sqrt{5}-1}$
18. $(a + b)5x = 5x(a + b)$

LEVEL 2

19. Is the set $\{0, 1\}$ closed for addition?
20. Is the set $\{0, 1\}$ closed for multiplication?
21. Is the set $\{-1, 0, 1\}$ closed for subtraction?
22. Is the set $\{-1, 0, 1\}$ closed for nonzero division?
23. Is the set $\{0, 3, 6, 9, 12, \ldots\}$ closed for addition?
24. Is the set $\{0, 3, 6, 9, 12, \ldots\}$ closed for multiplication?
25. Is the set $\{0, 3, 6, 9, 12, \ldots\}$ closed for subtraction?
26. Is the set $\{0, 3, 6, 9, 12, \ldots\}$ closed for nonzero division?
27. What is the additive inverse of the real number $1 + \sqrt{3}$?
28. What is the additive inverse of the real number $2 + \sqrt{3}$?
29. What is the multiplicative inverse of the real number $1 + \sqrt{3}$?
30. What is the multiplicative inverse of the real number $2 + \sqrt{3}$?
31. What is the additive inverse of the real number $\pi/3 + 1$?
32. What is the additive inverse of the real number $\frac{\sqrt{2}}{2} + 1$?
33. What is the multiplicative inverse of the real number $\pi/3 + 1$?
34. What is the multiplicative inverse of the real number $\frac{\sqrt{2}}{2} + 1$?
35. Let a be the process of putting on a shirt; let b be the process of putting on a pair of socks; and let c be the process of putting on a pair of shoes. Let $\star$ be the operation of "followed by."
    a. Is $\star$ commutative for $\{a, b, c\}$?
    b. Is $\star$ associative for $\{a, b, c\}$?
36. Consider the set $\mathbb{N}$ of natural numbers and an operation $\lhd$, which means "select the first of the two." That is,
    $$4 \lhd 3 = 4; \ 3 \lhd 4 = 3;$$
    $$5 \lhd 7 = 5; \ 6 \lhd 6 = 6$$
    a. Is $\lhd$ associative for $\mathbb{N}$? Give reasons.
    b. Is $\lhd$ commutative for $\mathbb{N}$? Give reasons.

37. Let $\downarrow$ mean "select the smaller number" and $\rightarrow$ mean "select the second of the two." Is $\rightarrow$ distributive over $\downarrow$ in the set of natural numbers, $\mathbb{N}$?
38. Let $\downarrow$ mean "select the smaller number" and $\rightarrow$ mean "select the second of the two." Is $\downarrow$ distributive over $\rightarrow$ in the set of natural numbers, $\mathbb{N}$?
39. IN YOUR OWN WORDS Is the operation of subtraction on $\mathbb{R}$ associative?
40. IN YOUR OWN WORDS Is the operation of division on the set of nonzero real numbers associative?
41. IN YOUR OWN WORDS Find an example showing that the operation of subtraction on $\mathbb{R}$ is not commutative.
42. IN YOUR OWN WORDS Find an example showing that the operation of division on the set of nonzero real numbers is not commutative.

Consider a soldier facing in a given direction (say north). Let us denote "left face" by ℓ, "right face" by r, "about face" by a, and "stand fast" by f. (The element f means "don't move from your present position." It does not mean "return to your original position.") Then we define

$$H = \{\ell, r, a, f\}$$

and an operation $\star$ meaning "followed by." Thus, $\ell \star \ell = a$ means "left face" followed by "left face" and is the same as the single command "about face." Use this information in Problems 43–46.

43. Complete the following table.

44. Is the set closed for H the operation of $\star$?
45. Does the set H and operation $\star$ satisfy the associative property? Give reasons.
46. Does the set H and operation $\star$ satisfy the commutative property? Give reasons.

LEVEL 3

47. Which of the field properties are satisfied by the set $\mathbb{N}$ (set of natural numbers)?
48. Which of the field properties are satisfied by the set $\mathbb{W}$ (set of whole numbers)?
49. Which of the field properties are satisfied by the set $\mathbb{Z}$ (set of integers)?
50. Which of the field properties are satisfied by the set $\mathbb{Q}$ (set of rationals)?

51. Which of the field properties are satisfied by the set $\mathbb{Q}'$ (set of irrationals)?

52. Which of the field properties are satisfied by the set $\mathbb{R}$ (set of reals)?

53. Which of the field properties are satisfied by the $H = \{1, -1, i, -i\}$, where i is the number with the property that $i^2 = -1$?

54. Which of the field properties are satisfied by the set of nonnegative multiples of 5, namely,

$$F = \{0, 5, 10, 15, 20, 25, \ldots\}?$$

Symmetries of a Square *Cut out a small square and label it as shown in Figure A.1.*

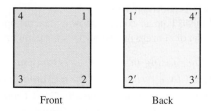

Front   Back

Figure A.1 Square construction

Be sure that 1 is in front of 1′, 2 is in front of 2′, 3 is in front of 3′, and 4 is in front of 4′. We will study certain symmetries of this square—that is, the results that are obtained when a square is moved around according to certain rules that we will establish. Hold the square with the front facing you and the 1 in the top right-hand corner as shown. This is called the basic position.

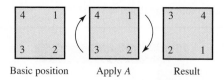

Basic position   Apply A Result

Now rotate the square 90° clockwise so that 1 moves into the position formerly held by 2 and so that 4 and 3 end up on top. We use the letter A to denote this rotation of the square. That is, A indicates a clockwise rotation of the square through 90°. Other symmetries can be obtained similarly according to Table A.1. You should be able to tell how each of the results in the table was found. Do this before continuing with the problem. We now have a set of elements: {A, B, C, D, E, F, G, H}. We must define an operation that combines a pair of these symmetries. Define an operation ★, which means "followed by." Consider, for example, A ★ B: Start with the basic position, apply A, followed by B (without returning to basic position) to obtain the result shown:

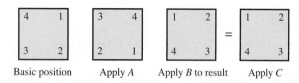

Basic position   Apply A Apply B to result   Apply C

TABLE A.1	Symmetries of a square	
Element	*Description*	*Result*
A	90° clockwise rotation	$\begin{smallmatrix}3&4\\2&1\end{smallmatrix}$
B	180° clockwise rotation	$\begin{smallmatrix}2&3\\1&4\end{smallmatrix}$
C	270° clockwise rotation	$\begin{smallmatrix}1&2\\4&3\end{smallmatrix}$
D	360° clockwise rotation	$\begin{smallmatrix}4&1\\3&2\end{smallmatrix}$
E	Flip about a **horizontal** line through the middle of the square	$\begin{smallmatrix}3'&2'\\4'&1'\end{smallmatrix}$
F	Flip about a **vertical** line through the middle of the square	$\begin{smallmatrix}1'&4'\\2'&3'\end{smallmatrix}$
G	Flip along a line drawn from **upper** left to lower right	$\begin{smallmatrix}4'&3'\\1'&2'\end{smallmatrix}$
H	Flip along a line drawn from **lower** left to upper right	$\begin{smallmatrix}2'&1'\\3'&4'\end{smallmatrix}$

Thus, we say $A \star B = C$, meaning "A followed by B is the same as applying the single element C." Use this information in Problems 55–60.

55. Complete a table for the operation ★ and the set $\{A, B, C, D, E, F, G, H\}$.

56. Is the set closed for ★?

57. Is the set associative for ★?

58. Is the set commutative for ★?

59. Does the set have an identity for ★?

60. Does the inverse property hold for the set and the operation ★?

Appendix B: Complex Numbers

To find the roots of certain equations, you must sometimes consider the square root of a negative number. Since the set of real numbers does not allow for such a possibility, we need to define a number that is *not a real number*. This number is denoted by the symbol i.

IMAGINARY UNIT

The number i, called the **imaginary unit**, is defined as a number with the following properties:

$$i^2 = -1 \quad \text{and} \quad \sqrt{-a} = i\sqrt{a} \quad \text{if } a > 0$$

With this number available, the square root of any negative number can be written as the product of a real number and the number i:

$$\sqrt{-1} = i\sqrt{1} = i \qquad \text{STOP} \quad \text{You should remember this.}$$

In particular,

$$\sqrt{-4} = i\sqrt{4} = 2i \qquad \textit{Check: } (2i)^2 = 4i^2 = 4(-1) = -4$$

So, by definition $\sqrt{-4} = 2i$. Also, note that

$$\sqrt{-13} = i\sqrt{13}$$ If you write this as $\sqrt{13}\ i$, make sure that the i is not included under the radical.

⚠ $\sqrt{-b}$ is simplified if $b < 0$.

$$\sqrt{-b} = i\sqrt{b} \text{ if } b > 0$$

Now we consider numbers of the form $a + bi$.

If $b = 0$, then $a + bi = a + 0i = a$ is a *real* number.
If $a = 0$ and $b = 1$, then $a + bi = 0 + 1i = i$ is the *imaginary unit*.
If $a = 0$ and $b \neq 0$, then $a + bi = 0 + bi$ is called a *pure imaginary number*.

Numbers of the form $a + bi$ are called *complex numbers*.

COMPLEX NUMBERS

The set of numbers of the form

$$a + bi$$

where a and b are real numbers and i is the imaginary unit, is called the set of **complex numbers**.

A complex number is **simplified** when it is written in the form $a + bi$, where the real numbers a and b are separately written in simplified form:

$5 + 3i$, $3 - i$, and $\sqrt{2} - \sqrt{3}i$ are all in simplified form.

$\sqrt{-9}$ is not in simplified form; write $\sqrt{-9} = 3i$

7 is a complex number in simplified form, since it is not necessary to write this as $7 + 0i$.

EXAMPLE 1   Solve an equation with nonreal solutions

Solve $x^2 + 1 = 0$.

Solution

$$x^2 + 1 = 0 \qquad \text{Given}$$
$$x^2 = -1 \qquad \text{Subtract 1 from both sides.}$$
$$x = \pm\sqrt{-1} \qquad \text{Square root property}$$
$$x = \pm i \qquad \text{Definition of } i$$

Check: If $x = i$, then   Check: If $x = -i$,

$$x^2 + 1 = 0 \quad \text{Given} \qquad\qquad (-i)^2 + 1 = i^2 + 1$$
$$i^2 + 1 = 0 \quad \text{Check value } x = i. \qquad\qquad = -1 + 1$$
$$-1 + 1 = 0 \quad i^2 = -1 \qquad\qquad\qquad\quad = 0 \qquad x = -i \text{ also checks}$$
$$0 = 0 \quad \text{True equation, so } x = i \text{ is a solution.}$$

Our next consideration is to carry out operations with complex numbers.

EQUALITY AND OPERATIONS WITH COMPLEX NUMBERS

Let $a + bi$ and $c + di$ be any complex numbers.

Equality $a + bi = c + di$ if and only if $a = c$ and $b = d$
Addition $(a + bi) + (c + di) = (a + c) + (b + d)i$
Subtraction $(a + bi) - (c + di) = (a - c) + (b - d)i$
Distributive $c(a + bi) = ca + cbi$

EXAMPLE 2   Addition and subtraction of complex numbers

Simplify:

a. $(3 + 4i) + (2 + 3i)$ **b.** $(2 - i) - (3 - 2i)$ **c.** $(6 + 2i) + (2 - 2i)$

d. $(4 + 3i) - (4 - 2i)$ **e.** $\dfrac{2 - \sqrt{-8}}{2}$

Solution

a. $(3 + 4i) + (2 + 3i) = 5 + 7i$

b. $(2 - i) - (3 - 2i) = -1 + i$

c. $(6 + 2i) + (2 - 2i) = 8$

d. $(4 + 3i) - (4 - 2i) = 5i$

e.
$$\frac{2 - \sqrt{-8}}{2} = \frac{2 - \sqrt{8}i}{2}$$
$$= \frac{2 - 2\sqrt{2}i}{2}$$
$$= \frac{2\left(1 - \sqrt{2}i\right)}{2}$$
$$= 1 - \sqrt{2}i$$

Notice that the operations of addition and subtraction of complex numbers conform to the usual way you handle the addition and subtraction of binomials. Multiplication is defined similarly. Since $i^2 = -1$, multiply in the usual way and replace i^2 by -1 wherever it occurs.

Now consider the complex numbers $a + bi$ and $c + di$. If multiplication of complex numbers is handled in the same manner as binomials, we have

$$(a + bi)(c + di) = ac + adi + bci + bdi^2$$
$$= ac + adi + bci - bd$$
$$= (ac - bd) + (ad + bc)i$$

We use this as the definition of multiplication of complex numbers.

MULTIPLICATION OF COMPLEX NUMBERS

If $a + bi$ and $c + di$ are any two complex numbers, then

$$(a + bi)(c + di) = (ac - bd) + (ad + bc)i$$

It is not necessary (or even desirable) to memorize this definition, since two complex numbers are handled as you would any binomials.

EXAMPLE 3   Multiplying complex numbers

Simplify:

   a. $(2 + 3i)(4 + 2i)$ **b.** $(4 - 3i)(2 - i)$ **c.** $(3 - 2i)(3 + 2i)$

Solution

  a. $(2 + 3i)(4 + 2i) = 8 + 16i + 6i^2$ Usual binomial multiplication.
$$= 8 + 16i - 6 \qquad \text{Write } i^2 = (-1).$$
$$= 2 + 16i$$

  b. $(4 - 3i)(2 - i) = 8 - 10i + 3i^2$
$$= 5 - 10i$$

  c. $(3 - 2i)(3 + 2i) = 9 + 6i - 6i - 4i^2$
$$= 9 - 4i^2$$
$$= 13$$

You must be very careful when multiplying complex numbers in radical form so that you correctly use the laws of radicals.

             If a and b are both positive: $\sqrt{a}\sqrt{b} = \sqrt{ab}$

Pay attention!   If a and b are both negative: $\sqrt{a}\sqrt{b} = -\sqrt{ab}$

This means that when working with complex numbers you should *first* write them in the form $a + bi$ and *then* perform the arithmetic.

EXAMPLE 4   Multiplying complex numbers with radicals

Simplify:

  a. $\sqrt{-4}\sqrt{-4}$

  b. $\left(2 - \sqrt{-4}\right)\left(4 + \sqrt{-9}\right)$

  c. $\left(\sqrt{7} - \sqrt{-16}\right)^2$

Solution

  a. $\sqrt{-4}\sqrt{-4} = \left(\sqrt{4}i\right)\left(\sqrt{4}i\right)$ It is WRONG to write $\sqrt{-4}\sqrt{-4} = \sqrt{16} = 4$ because $\sqrt{a}\sqrt{b} = -\sqrt{ab}$ if a and b are both negative.
$$= 4i^2$$
$$= -4$$

  b. $\left(2 - \sqrt{-4}\right)\left(4 + \sqrt{-9}\right) = (2 - 2i)(4 + 3i)$ Write in terms of i.
$$= 8 - 2i - 6i^2 \qquad \text{Multiply.}$$
$$= 14 - 2i \qquad \text{Simplify.}$$

  c. $\left(\sqrt{7} - \sqrt{-16}\right)^2 = \left(\sqrt{7} - 4i\right)^2$
$$= 7 - 8\sqrt{7}i + 16i^2$$
$$= 7 - 8\sqrt{7}i - 16$$
$$= -9 - 8\sqrt{7}i$$

Example 3c gives a clue for dividing complex numbers. When a complex number is multiplied by its conjugate, the result is a real number. The numbers $a + bi$ and $a - bi$ are called **complex conjugates**. In general,

$$(a + bi)(a - bi) = a^2 - b^2 i^2 = a^2 + b^2$$

which is a real number since a and b are real numbers. Thus, for division

$$\frac{a + bi}{c + di} = \frac{a + bi}{c + di} \cdot \frac{c - di}{c - di} \qquad \text{Multiply by 1.}$$

$$= \frac{(ac + bd) + (bc - ad)i}{c^2 - d^2 i^2} \qquad \text{Expand.}$$

$$= \frac{ac + bd}{c^2 + d^2} + \frac{bc - ad}{c^2 + d^2} i \qquad \text{Simplify.}$$

DIVISION OF COMPLEX NUMBERS

If $a + bi$ and $c + di$ are any two complex numbers (c and d are not both zero), then

$$\frac{a + bi}{c + di} = \frac{ac + bd}{c^2 + d^2} + \frac{bc - ad}{c^2 + d^2} i$$

Notice that the result is in the form of a complex number, but instead of memorizing this definition, simply remember: *To divide complex numbers, you multiply both the numerator and the denominator by the conjugate of the denominator.*

EXAMPLE 5   Division of complex numbers

Simplify:

a. $\dfrac{7 + i}{2 + i}$ **b.** $\dfrac{15 - \sqrt{-25}}{2 - \sqrt{-1}}$

Solution

a. $\dfrac{7 + i}{2 + i} = \dfrac{7 + i}{2 + i} \cdot \dfrac{2 - i}{2 - i}$ Multiply by 1.

$= \dfrac{14 - 5i - i^2}{4 - i^2}$ Simplify.

$= \dfrac{15 - 5i}{5}$ $i^2 = -1$ and simplify.

$= \dfrac{5(3 - i)}{5}$ Common factor.

$= 3 - i$ Simplify.

b. $\dfrac{15 - \sqrt{-25}}{2 - \sqrt{-1}} = \dfrac{15 - 5i}{2 - i}$

$= \dfrac{15 - 5i}{2 - i} \cdot \dfrac{2 + i}{2 + i}$

$= \dfrac{30 + 5i - 5i^2}{4 - i^2}$

$= \dfrac{35 + 5i}{5}$

$= \dfrac{5(7 + i)}{5}$

$= 7 + i$

PROBLEM SET **APPENDIX B**

LEVEL 1

1. IN YOUR OWN WORDS What is the number i?
2. IN YOUR OWN WORDS What does it mean to *simplify* a complex number?
3. Is every real number a complex number? Is every complex number a real number?

WHAT IS WRONG, *if anything, with each statement in Problems 4–9? Explain.*

4. $-i^2 = i^2 = -1$
5. $\sqrt{i} = -1$
6. $\sqrt{-2}\sqrt{-3} = \sqrt{6}$
7. $\sqrt{a}\sqrt{b} = \sqrt{ab}$
8. $\sqrt{-2}\sqrt{-2} = \sqrt{(-2)(-2)} = \sqrt{4} = 2$
9. $\left(2+\sqrt{-4}\right)\left(2-\sqrt{-4}\right) = 4 - 4 = 0$

Simplify each of the expressions in Problems 10–26.

10. **a.** $\sqrt{-36}$ **b.** $\sqrt{-100}$
11. **a.** $\sqrt{-49}$ **b.** $\sqrt{-8}$
12. **a.** $\sqrt{-20}$ **b.** $\sqrt{-24}$
13. **a.** $\dfrac{-3\sqrt{-144}}{5}$ **b.** $\dfrac{-6\sqrt{-4}}{8}$
14. **a.** $\dfrac{3\sqrt{-49}}{7}$ **b.** $\dfrac{-2\sqrt{-24}}{8}$
15. **a.** $i(5-2i)$ **b.** $i(2+3i)$
16. **a.** $(3+3i)+(5+4i)$ **b.** $(2-i)(2+i)$
17. **a.** $(8-2i)(8+2i)$ **b.** $(3-4i)(3+4i)$
18. **a.** $-i^2$ **b.** $-i^3$
19. **a.** $-i^5$ **b.** i^7
20. **a.** i^{15} **b.** $-i^{18}$
21. **a.** $-i^9$ **b.** i^{10}
22. **a.** $-i^{2008}$ **b.** i^{2009}
23. **a.** $2+\sqrt{2}-4+\sqrt{-2}$ **b.** $6-\sqrt{3}-8+\sqrt{-3}$
24. **a.** $\left(3+\sqrt{3}\right)+\left(5-\sqrt{-9}\right)$ **b.** $\left(5-\sqrt{2}\right)+\left(2-\sqrt{-4}\right)$
25. **a.** $(5+3i)(2+6i)$ **b.** $(2-7i)-(3-2i)$
26. **a.** $(5-4i)-(5-9i)$ **b.** $(2-3i)-(4+5i)$

LEVEL 2

Simplify each expression in Problems 27–43.

27. **a.** $(1-3i)^2$ **b.** $(6-2i)^2$
28. **a.** $(3+4i)^2$ **b.** $(2-5i)^2$
29. **a.** $\left(\sqrt{2}+3i\right)^2$ **b.** $\left(\sqrt{5}-3i\right)^2$
30. **a.** $\left(6-\sqrt{-1}\right)\left(2+\sqrt{-1}\right)$ **b.** $\left(2-\sqrt{-4}\right)\left(3-\sqrt{-4}\right)$
31. **a.** $\left(3-\sqrt{-3}\right)\left(3+\sqrt{-3}\right)$ **b.** $\left(2-\sqrt{-3}\right)\left(2+\sqrt{-3}\right)$
32. **a.** $\left(2-\sqrt{-2}\right)\left(3-\sqrt{-3}\right)$ **b.** $\left(4-\sqrt{-9}\right)\left(5+\sqrt{-16}\right)$
33. **a.** $\dfrac{4-2i}{3+i}$ **b.** $\dfrac{5+3i}{4-i}$
34. **a.** $\dfrac{1+3i}{1-2i}$ **b.** $\dfrac{3-2i}{5+i}$
35. **a.** $\dfrac{-3}{1+i}$ **b.** $\dfrac{5}{4-i}$
36. **a.** $\dfrac{2}{i}$ **b.** $\dfrac{3}{-i}$
37. **a.** $\dfrac{-2i}{3+i}$ **b.** $\dfrac{3i}{5-2i}$
38. **a.** $\dfrac{-i}{2-i}$ **b.** $\dfrac{1-6i}{1+6i}$
39. **a.** $\dfrac{2+7i}{2-7i}$ **b.** $\dfrac{-1+4i}{1+2i}$
40. **a.** $\dfrac{\sqrt{-1}+1}{\sqrt{-1}-1}$ **b.** $\dfrac{10-\sqrt{-25}}{2-\sqrt{-1}}$
41. **a.** $\dfrac{4-\sqrt{-4}}{1+\sqrt{-1}}$ **b.** $\dfrac{2-\sqrt{-9}}{1+\sqrt{-1}}$
42. $(1.9319+0.5176i)(2.5981+1.5i)$
43. $\dfrac{4.5963+3.8567i}{1.9696+0.3473i}$
44. Find the value of $x^2 + 25$ when $x = 5i$.
45. Find the value of $x^2 + 3$ when $x = \sqrt{3}i$.
46. Solve $x^2 + 2 = 0$.
47. Solve $x^2 + 4 = 0$.
48. Find the value of $x^2 + 5$ when $x = 1 + 2i$.
49. Find the value of $x^2 - 2x + 5$ when $x = 1 - 2i$.
50. Evaluate $x^3 - 11x^2 + 40x - 50$ when $x = 3 + i$.
51. Evaluate $2x^3 - 11x^2 + 14x + 10$ when $x = 3 - i$.
52. Solve $x^3 - 1 = 0$. *Hint*: Find three solutions; begin by factoring.
53. Solve $x^3 + 1 = 0$. *Hint*: Find three solutions; begin by factoring.

LEVEL 3

54. Simplify:
$$\left[\left(2+\sqrt{3}\right)+\left(4-\sqrt{2}\right)i\right]\left[\left(2+\sqrt{3}\right)-\left(4-\sqrt{2}\right)i\right]$$

55. Simplify:
$$\left[\left(4-\sqrt{2}\right)+\left(3-\sqrt{2}\right)i\right]\left[\left(1+\sqrt{2}\right)+\left(2-\sqrt{2}\right)i\right]$$

56. WHAT IS WRONG, if anything, with the following "proof"?

$$\sqrt{-1} = \sqrt{-1} \qquad \text{Property of equality}$$

$$\sqrt{\frac{1}{-1}} = \sqrt{\frac{-1}{1}} \qquad -1 = \frac{1}{-1} \text{ and } -1 = \frac{-1}{1}$$

$$\frac{\sqrt{1}}{\sqrt{-1}} = \frac{\sqrt{-1}}{\sqrt{1}} \qquad \text{Property of radicals}$$

$$\frac{1}{i} = \frac{i}{1} \qquad \sqrt{1} = 1 \text{ and } \sqrt{-1} = i$$

$$\frac{1}{i} \cdot \frac{i}{i} = i \qquad \text{Multiplication by 1 and } \frac{i}{1} = i$$

$$\frac{i}{i^2} = i \qquad \text{Multiplying fractions}$$

$$\frac{i}{-1} = i \qquad i^2 = -1$$

$$-i = i \qquad \text{Impossible!}$$

Let $z_1 = a_1 + b_1i$, $z_2 = a_2 + b_2i$, and $z_3 = a_3 + b_3i$ in Problems 57–60.

57. Prove the distributive law for complex numbers. That is, prove

$$z_1(z_2 + z_3) = z_1z_2 + z_1z_3$$

58. Prove the commutative laws for complex numbers. That is, prove

$$z_1 + z_2 = z_2 + z_1, \quad z_1z_2 = z_2z_1$$

59. Prove the associative laws for complex numbers. That is, prove

$$z_1 + (z_2 + z_3) = (z_1 + z_2) + z_3, \quad z_1(z_2z_3) = (z_1z_2)z_3$$

60. IN YOUR OWN WORDS Show that the field properties are satisfied for the operations of addition and multiplication over the set of complex numbers.

Appendix C: Mathematical Induction

Mathematical induction is an important method of proof in mathematics that allows us to prove results involving the set of positive integers. Do not confuse mathematical induction with the scientific method or the inductive logic used in the experimental sciences. Mathematical induction is a form of deductive logic in which conclusions are inescapable.

The first step in establishing a result by mathematical induction is often the observation of a pattern. Let us begin with a simple example: Find the sum of 100 consecutive odd numbers. We can show this pattern geometrically:

$1 = 1$ □ This is 1^2.

$1 + 3 = 4$ This is 2^2.

$1 + 3 + 5 = 9$ This is 3^2.

$1 + 3 + 5 + 7 = 16$ This is 4^2.
$1 + 3 + 5 + 7 + 9 = 25$ This is 5^2.

Do you see a pattern? It appears that the sum of the first n odd numbers is n^2 since we see that the sum of the first 3 odd numbers is 3^2, of the first 4 odd numbers is 4^2, and so on. You now wish to *prove* deductively that

$$1 + 3 + 5 + \cdots + \underbrace{(2n - 1)}_{n\text{th odd number}} = n^2$$

is true for all positive integers n. How can you proceed? You can use a method called *mathematical induction*, which is used to prove certain propositions about the positive integers. The proposition is denoted by $P(n)$. For example, in the above statement, let

$$P(n): \quad 1 + 3 + 5 + \cdots + (2n - 1) = n^2$$

This means

$n = 1;\ P(1):$	$1 = 1^2$
$n = 2;\ P(2):$	$1 + 3 = 2^2$
$n = 3;\ P(3):$	$1 + 3 + 5 = 3^2$
$n = 4;\ P(4):$	$1 + 3 + 5 + 7 = 4^2$
	$\vdots$
$n = 100;\ P(100):$	$1 + 3 + 5 + \cdots + 199 = 100^2$
	$\vdots$
$n = x - 1;\ P(x - 1):$	$1 + 3 + 5 + \cdots + (2x - 3) = (x - 1)^2$
$n = x;\ P(x):$	$1 + 3 + 5 + \cdots + (2x - 1) = x^2$
$n = x + 1;\ P(x + 1):$	$1 + 3 + 5 + \cdots + (2x + 1) = (x + 1)^2$

Now, you want to show that $P(n)$ is true for *all* n (n a positive integer).

MATH INDUCTION

Principle of mathematical induction (PMI): If a given proposition $P(n)$ is true for $n = 1$ and if the truth of the proposition for $n = k$ implies the truth of the proposition for $n = k + 1$, then the proposition $P(n)$ is true for all positive integers.

>> IN OTHER WORDS This gives us the following procedure for proof by mathematical induction:

1. Prove the proposition $P(1)$ is true.
2. Assume the proposition $P(k)$ is true.
3. Prove the proposition $P(k + 1)$ is true.
4. Conclude the proposition $P(n)$ is true for all positive integers n.

Students often have a certain uneasiness when they first use the principle of mathematical induction as a method of proof. Suppose we use this principle with a stack of dominoes, as shown in the cartoon in the margin.

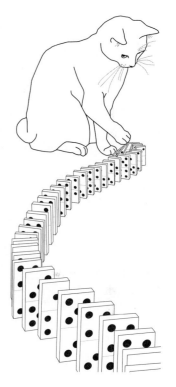

How can the cat in the cartoon be certain of knocking over all the dominoes? The cat would have to be able to knock over the first one. He would have to have the dominoes arranged so that *if* the kth domino falls, then the next one, the $(k + 1)$st, will also fall. That is, each domino is set up so that if it falls, it causes the next one to fall. We have set up a kind of "chain reaction" here. The first domino falls; this knocks over the next one (the second domino); the second one knocks over the next one (the third domino); the third one knocks over the next one; this continues until all the dominoes are knocked over.

We return to the example to prove (for all positive integers n)

$$1 + 3 + 5 + \cdots + (2n - 1) = n^2$$

We refer to this as the *proposition* or as the *original statement*.

Step 1 Prove $P(1)$ is true: $1 = 1^2$ is true.

Step 2 Assume the proposition $P(k)$ is true:

$$1 + 3 + 5 + \cdots + (2k - 1) = k^2$$

This is found by substituting k for n is the original statement. We refer to this statement as the *hypothesis*.

Step 3 Prove $P(k + 1)$ true. In other words,

TO PROVE: $1 + 3 + 5 + \cdots + [2(k + 1) - 1] = (k + 1)^2$

This is found by substituting $(k + 1)$ for n in the original statement you want to prove. Next, simplify the TO PROVE statement so you will know when you are finished with step 3 of the proof.

$1 + 3 + 5 + \cdots + [2(k + 1) - 1] = (k + 1)^2$	Substitute.
$1 + 3 + 5 + \cdots + [2k + 2 - 1] = (k + 1)^2$	Simplify.
$1 + 3 + 5 + \cdots + (2k + 1) = (k + 1)^2$	This is what needs to be proved.

The procedure for step 3 is to begin with the hypothesis (from step 2) and *prove* $1 + 3 + 5 + \cdots + (2k + 1) = (k + 1)^2$.

STATEMENTS	REASONS
$1 + 3 + 5 + \cdots + (2k - 1) = k^2$	By hypothesis (step 2)
$1 + 3 + 5 + \cdots + (2k - 1) + (2k + 1) = k^2 + (2k + 1)$	Add $(2k + 1)$ to both sides.
$1 + 3 + 5 + \cdots + (2k - 1) + (2k + 1) = k^2 + 2k + 1$	Associative
$1 + 3 + 5 + \cdots + (2k - 1) + (2k + 1) = (k + 1)^2$	Factoring (distributive)

Step 4 The proposition is true for all positive integers by PMI.

EXAMPLE 1   Mathematical induction; sum of terms

Prove or disprove: $2 + 4 + 6 + \cdots + 2n = n(n + 1)$.

Solution

We begin by using mathematical induction. Since we do not know that this is a true proposition, we proceed one step at a time.

Step 1   Prove $P(1)$ is true

$$2 = 1(1 + 1)$$
$$2 = 2$$

This is true.

Step 2   Assume $P(k)$; HYPOTHESIS: $2 + 4 + 6 + \cdots + 2k = k(k + 1)$

Step 3   Prove $P(k + 1)$: That is, prove the proposition is true when $n = k + 1$.

TO PROVE: $2 + 4 + 6 + \cdots + 2(k + 1) = (k + 1)(k + 2)$

STATEMENTS	REASONS
$2 + 4 + 6 + \cdots + 2k = k(k + 1)$	By hypothesis (step 2)
$2 + 4 + 6 + \cdots + 2k + (2k + 1) = k(k + 1) + 2(k + 1)$	Add $2(k + 1)$ to both sides.
$= (k + 1)(k + 2)$	Factoring

Step 4   The proposition is true for all positive integers by PMI.   ∎

EXAMPLE 2   Mathematical induction; divisibility

Prove or disprove: $n^3 + 2n$ is divisible by 3.

Solution

Step 1   Prove $P(1)$: $1^3 + 2 \cdot 1 = 3$, which is divisible by 3.

Step 2   Assume $P(k)$:   HYPOTHESIS: $k^3 + 2k$ is divisible by 3.

Step 3   Prove $P(k + 1)$: TO PROVE: $(k + 1)^3 + 2(k + 1)$ is divisible by 3.

STATEMENTS	REASONS
$(k + 1)^3 + 2(k + 1) = k^3 + 3k^2 + 3k + 1 + 2k + 2$	Distributive, associative, and commutativity properties
$= (3k^2 + 3k + 3) + (k^3 + 2k)$	Commutative and associative properties
$= 3(k^2 + k + 1) + (k^3 + 2k)$	Distributive property
$3(k^2 + k + 1)$ is divisible by 3.	Definition of divisibility by 3
$k^3 + 2k$ is divisible by 3.	Hypothesis
$(k + 1)^3 + 2(k + 1)$ is divisible by 3.	Both terms are divisible by 3 and therefore the sum is also.

Step 4   The proposition is true for all positive integers by PMI.   ∎

Example 3 shows that *even though* you make an assumption in step 2, it is not going to help if the proposition is not true.

EXAMPLE 3   Mathematical induction; failed step

Prove or disprove: $n + 1$ is prime.

Solution

Step 1   Prove $P(1)$: $1 + 1 = 2$ is a prime.

Step 2   Assume $P(k)$: HYPOTHESIS: $k + 1$ is a prime.

Step 3 $P(k+1)$: TO PROVE: $(k+1)+1$ is a prime.

This is not possible since $(k+1)+1 = k+2$, which is not prime whenever k is an even positive integer.

Step 4 Any conclusions? You cannot conclude that the statement is false, only that induction does not work. But this statement is in fact false, and a counterexample is $n = 3$, since $n + 1 = 4$ is not prime. ∎

EXAMPLE 4 Mathematical induction; skipped step

Prove or disprove: $1 \cdot 2 \cdot 3 \cdot 4 \cdots \cdot n < 0$.

Solution

Students often slip into the habit of skipping either the first or second step in a proof by mathematical induction. This is dangerous, and it is important to check every step. Suppose a careless person did not verify the first step.

Step 2 Assume $P(k)$: HYPOTHESIS: $1 \cdot 2 \cdot 3 \cdots \cdot k < 0$

Step 3 Prove $P(k+1)$: $1 \cdot 2 \cdot 3 \cdots \cdot k \cdot (k+1) < 0$

$1 \cdot 2 \cdot 3 \cdots \cdot k < 0$ by hypothesis

$k + 1$ is positive since k is a positive integer. Then, since we know that the product of a negative and a positive is negative,

$$\underbrace{1 \cdot 2 \cdot 3 \cdots \cdot k}_{\text{Negative}} \cdot \underbrace{(k+1)}_{\text{Positive}} < 0$$

Step 3 is proved.

Step 4 The proposition is not true for all positive integers since the first step, $1 < 0$, does not hold. ∎

PROBLEM SET APPENDIX C

LEVEL 1

State and prove or disprove each proposition in Problems 1–12 for $n = 1$.

1. $1 + 2 + 3 + \cdots + n = \dfrac{n(n+1)}{2}$

2. $1 + 4 + 9 + \cdots + n^2 = \dfrac{n(n+1)(2n+1)}{6}$

3. $1 + 8 + 27 + \cdots + n^3 = \dfrac{n^2(n+1)^2}{4}$

4. $5 + 9 + 13 + \cdots + (4n+1) = n(2n+3)$

5. $3 + 9 + 15 + \cdots + (6n-3) = 3n^2$

6. $4 + 16 + 36 + \cdots + 4n^2 = \dfrac{2n(n+1)(2n+1)}{3}$

7. $n^2 + n$ is even.

8. $n + 1$ is odd.

9. $n^3 - n + 3$ is divisible by 3.

10. $n^5 - n$ is divisible by 5.

11. $\left(\dfrac{2}{3}\right)^{n+1} < \left(\dfrac{2}{3}\right)^n$

12. $1 + 2n \le 3^n$

In Problems 13–24, state and simplify the expression for $n = k$ and $n = k + 1$.

13. $1 + 2 + 3 + \cdots + n = \dfrac{n(n+1)}{2}$

14. $1 + 4 + 9 + \cdots + n^2 = \dfrac{n(n+1)(2n+1)}{6}$

15. $1 + 8 + 27 + \cdots + n^3 = \dfrac{n^2(n+1)^2}{4}$

16. $5 + 9 + 13 + \cdots + (4n+1) = n(2n+3)$

17. $3 + 9 + 15 + \cdots + (6n-3) = 3n^2$

18. $4 + 16 + 36 + \cdots + 4n^2 = \dfrac{2n(n+1)(2n+1)}{3}$

19. $n^2 + n$ is even.

20. $n + 1$ is odd.

21. $n^3 - n + 3$ is divisible by 3.

22. $n^5 - n$ is divisible by 5.

23. $\left(\dfrac{2}{3}\right)^{n+1} < \left(\dfrac{2}{3}\right)^n$

24. $1 + 2n \le 3^n$

PROBLEMS FROM CALCULUS *Problems 25–27 are given as a preliminary formula for the process of integration in* Calculus *by Gerald Bradley and Karl Smith, p. 263. Use mathematical induction to prove each statement.*

25. $1 + 2 + 3 + \cdots + n = \dfrac{n(n+1)}{2}$

26. $1^2 + 2^2 + 3^2 + \cdots + n^2 = \dfrac{n(n+1)(2n+1)}{6}$

27. $1^3 + 2^3 + 3^3 + \cdots + n^3 = \dfrac{n^2(n+1)^2}{4}$

In each of Problems 28–38, prove that the given formula is true for all positive integers n.

28. $5 + 9 + 13 + \cdots + (4n + 1) = n(2n + 3)$

29. $3 + 9 + 15 + \cdots + (6n - 3) = 3n^2$

30. $4 + 16 + 36 + \cdots + 4n^2 = \dfrac{2n(n+1)(2n+1)}{3}$

31. $1 + 4 + 7 + \cdots + (3n - 2) = \dfrac{n(3n-1)}{2}$

32. $2 + 7 + 12 + \cdots + (5n - 3) = \dfrac{n(5n-1)}{2}$

33. $1 + 9 + 25 + \cdots + (2n - 1)^2 = \dfrac{n(2n-1)(2n+1)}{3}$

34. $2 + 6 + 12 + \cdots + n(n + 1) = \dfrac{n(n+1)(n+2)}{3}$

35. $3 + 8 + 15 + \cdots + n(n + 2) = \dfrac{n(n+1)(2n+7)}{6}$

36. $3 + 9 + 27 + \cdots + 3^n = \dfrac{3^{n+1} - 3}{2}$

37. $5 + 25 + 125 + \cdots + 5^n = \dfrac{5^{n+1} - 5}{4}$

38. $r + r^2 + r^3 + \cdots + r^n = \dfrac{r^{n+1} - r}{r - 1}$

LEVEL 2

Define $b^{n+1} = b^n \cdot b$ *and* $b^0 = 1$. *Use this definition to prove the properties of exponents in Problems 39–42 for all positive integers* n.

39. $b^m \cdot b^n = b^{m+n}$

40. $(b^m)^n = b^{mn}$

41. $(ab)^n = a^n b^n$

42. $\left(\dfrac{a}{b}\right)^n = \dfrac{a^n}{b^n}$

Prove that the statements in Problems 43–53 are true for every positive integer n.

43. $n^2 + n$ is even.

44. $n^5 - n$ is divisible by 5.

45. $n^3 - n + 3$ is divisible by 3.

46. $n(n + 1)(n + 2)$ is divisible by 6.

47. $10^{n+1} + 3 \cdot 10^n + 5$ is divisible by 9.

48. $(1 + n)^2 \geq 1 + n^2$

49. $\left(\dfrac{2}{3}\right)^{n+1} < \left(\dfrac{2}{3}\right)^n$

50. $1 + 2n \leq 3^n$

51. $2^n > n$

52. $\dfrac{1}{2} + \dfrac{1}{3} + \dfrac{1}{4} + \dfrac{1}{5} + \cdots + \dfrac{1}{n+1} < n$

53. $1 + 2 + 3 + \cdots + n < \dfrac{(2n+1)^2}{8}$

54. Prove the generalized distributive property
$$a(b_1 + b_2 + \cdots + b_n) = ab_1 + ab_2 + \cdots + ab_n$$

55. Prove: $g_1 + g_1 r + g_1 r^2 + \cdots + g_1 r^{n-1} = \dfrac{g_1(1 - r^n)}{1 - r}$

LEVEL 3

56. Notice the following:
$$1^3 = 1^2$$
$$1^3 + 2^3 = 3^2$$
$$1^3 + 2^3 + 3^3 = 6^2$$
$$1^3 + 2^3 + 3^3 + 4^3 = 10^2$$

Make a conjecture based on this pattern and then prove your conjecture.

57. Notice the following:
$$1 = 1$$
$$1 + 4 = 5$$
$$1 + 4 + 7 = 12$$
$$1 + 4 + 7 + 10 = 22$$

Make a conjecture based on this pattern and then prove your conjecture.

58. Notice the following:
$$2 = 2$$
$$2 + 2 \cdot 3 = 8$$
$$2 + (2 \cdot 3) + (2 \cdot 3^2) = 26$$
$$2 (2 \cdot 3) + (2 \cdot 3^2) + (2 \cdot 3^3) = 80$$

Make a conjecture based on this pattern and then prove your conjecture.

59. Notice the following:
$$(-1)^1 = -1$$
$$(-1) + (-1)^2 = 0$$
$$(-1) + (-1)^2 + (-1)^3 = -1$$
$$(-1) + (-1)^2 + (-1)^3 + (-1)^4 = 0$$

Make a conjecture based on this pattern and then prove your conjecture.

60. Prove the generalized triangle inequality:
$$|a_1 + a_2 + \cdots + a_n| \leq |a_1| + |a_2| + \cdots + |a_n|$$

Appendix D: Binomial Theorem

In calculus, you will need to raise a binomial to a variety of different powers, and we need an easy and efficient way of doing this. We begin with an example.

EXAMPLE 1   **Expand (multiply out) the expression $(a + b)^8$.**   **MODELING APPLICATION**

Solution

Step 1: *Understand the problem.* We could begin by using the definition of 8th power and the distributive property:

$$(a+b)^8 = \underbrace{(a+b)(a+b)\cdots(a+b)}_{8 \text{ factors of } a+b}$$

However, we soon see that this is too lengthy to complete directly.

Step 2: *Devise a plan.* We will consider a pattern of successive powers; that is, consider $(a+b)^n$ for $n = 0, 1, 2, \ldots$.

Step 3: *Carry out the plan.* We begin by actually doing the multiplications:

$$
\begin{array}{lll}
n = 0: & (a+b)^0 = & 1 \\
n = 1: & (a+b)^1 = & 1 \cdot a + 1 \cdot b \\
n = 2: & (a+b)^2 = & 1 \cdot a^2 + 2 \cdot ab + 1 \cdot b^2 \\
n = 3: & (a+b)^3 = & 1 \cdot a^3 + 3 \cdot a^2 b + 3 \cdot ab^2 + 1 \cdot b^3 \\
n = 4: & (a+b)^4 = & 1 \cdot a^4 + 4 \cdot a^3 b + 6 \cdot a^2 b^2 + 4 \cdot ab^3 + 1 \cdot b^4 \\
\vdots & \vdots &
\end{array}
$$

First, ignore the coefficients (shown in color) and focus on the variables:

$$
\begin{array}{lllll}
(a+b)^1: & a & b \\
(a+b)^2: & a^2 & ab & b^2 \\
(a+b)^3: & a^3 & a^2 b & ab^2 & b^3 \\
(a+b)^4: & a^4 & a^3 b & a^2 b^2 & ab^3 & b^4 \\
\vdots & & & \vdots
\end{array}
$$

Do you see a pattern? As you read from left to right, the powers of a decrease and the powers of b increase. Note that the sum of the exponents for each term is the same as the original exponent:

$$(a+b)^n: \quad a^n b^0 \quad a^{n-1} b^1 \quad a^{n-2} b^2 \quad \cdots \quad a^{n-r} b^r \quad \cdots \quad a^2 b^{n-2} \quad a^1 b^{n-1} \quad a^0 b^n$$

Next, consider the numerical coefficients (shown in color):

$$
\begin{array}{l}
(a+b)^0: \qquad\qquad\quad 1 \\
(a+b)^1: \qquad\qquad 1 \quad 1 \\
(a+b)^2: \qquad\quad 1 \quad 2 \quad 1 \\
(a+b)^3: \qquad 1 \quad 3 \quad 3 \quad 1 \\
(a+b)^4: \quad 1 \quad 4 \quad 6 \quad 4 \quad 1 \\
\qquad\qquad \vdots \qquad\qquad \vdots
\end{array}
$$

Do you see the pattern? We continue this pattern in Figure D.2. It is called *Pascal's Triangle* after Blaise Pascal.

Historical Note

© Karl Smith Library

Blaise Pascal (1623–1662)

Described as "the greatest 'might-have-been' in the history of mathematics," Pascal was a person of frail health, and because he needed to conserve his energy, he was forbidden to study mathematics. This aroused his curiosity and forced him to acquire most of his knowledge of the subject by himself. At 18, he had invented one of the first calculating machines. However, at 27, because of his health, he promised God that he would abandon mathematics and spend his time in religious study. Three years later he broke this promise and wrote *Traite du triangle arithmetique*, in which he investigated what we today call Pascal's triangle. The very next year he was almost killed when his runaway horse jumped an embankment. He took this to be a sign of God's displeasure with him and again gave up mathematics—this time permanently.

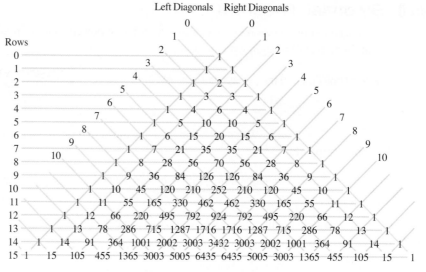

Figure D.2 Pascal's triangle

To find $(a + b)^8$, we look at the 8th row of Pascal's triangle for the coefficients to complete the product:

$$(a + b)^8 = a^8 + 8a^7b + 28a^6b^2 + 56a^5b^3 + 70a^4b^4 + 56a^3b^5 + 28a^2b^6 + 8ab^7 + b^8$$

Step 4: *Look back.* Does this pattern seem correct? We have verified the pattern directly for $n = 0, 1, 2, 3,$ and 4. With a great deal of algebraic work, you can verify by direct multiplication that the pattern checks for $(a + b)^8$. ∎

Binomial Theorem

The pattern we discovered for $(a + b)^8$ is a very important theorem in mathematics. It is called the **binomial theorem**. The difficulty in stating this theorem is in relating the coefficients of the expansion to Pascal's triangle. We write $\begin{pmatrix} 4 \\ 2 \end{pmatrix}$ to represent the number in row 4, diagonal 2 of Pascal's triangle (see Figure D.2). We see

$$\begin{pmatrix} 4 \\ 2 \end{pmatrix} = 6 \qquad \begin{pmatrix} 6 \\ 4 \end{pmatrix} = 15 \qquad \begin{pmatrix} 5 \\ 3 \end{pmatrix} = 10 \qquad \begin{pmatrix} 15 \\ 7 \end{pmatrix} = 6,435$$

We summarize this pattern in the following box.

> **BINOMIAL THEOREM**
>
> For any positive integer n,
>
> $$(a + b)^n = \begin{pmatrix} n \\ 0 \end{pmatrix} a^n + \begin{pmatrix} n \\ 1 \end{pmatrix} a^{n-1}b + \begin{pmatrix} n \\ 2 \end{pmatrix} a^{n-2}b^2 + \cdots + \begin{pmatrix} n \\ n-1 \end{pmatrix} ab^{n-1} + \begin{pmatrix} n \\ n \end{pmatrix} b^n$$
>
> where $\begin{pmatrix} n \\ r \end{pmatrix}$ is the number in the nth row, rth diagonal of Pascal's triangle.

The proof of the binomial theorem is by mathematical induction, and the procedure is lengthy, so we leave the proof for the problem set. You are led through the steps and are asked to fill in the details in Problems 57–59. Pascal's triangle is efficient for finding the numerical coefficients for exponents that are relatively small, as shown in Figure D.1.

 You should actually FIND these numbers using Figure D.2. Spend some time studying this theorem. There is a lot of notation here, and you should make sure you understand what it means.

Historical Note

The binomial theorem was first proved for real exponents in 1665 by one of the greatest mathematicians of all time, Sir Isaac Newton (1642–1727). Note that Newton was 23 years old at the time. At 18 he entered Trinity College, Cambridge, and remained there until 1696. It was when the university closed for a year because of bubonic plague that he invented not only the binomial theorem but also the calculus, because he found himself home for a year and needed to keep himself occupied.

EXAMPLE 2   Binomial expansion of a difference

Expand $(x - 2y)^4$.

Solution

In this example, let $a = x$ and $b = -2y$, and look at row 4 of Pascal's triangle for the coefficients.

$$(a + b)^4 = a^4 + 4a^3b + 6a^2b^2 + 4ab^3 + b^4 \qquad \text{Binomial theorem, } n = 4$$

$$(x - 2y)^4 = x^4 + 4x^3(-2y) + 6x^2(-2y)^2 + 4x(-2y)^3 + (-2y)^4 \quad a = x, b = -2y$$

$$= x^4 - 8x^3y + 24x^2y^2 - 32xy^3 + 16y^4$$

If the power of the binomial is very large, then Pascal's triangle is not efficient so the next step is to find a formula for $\binom{n}{r}$. This formula is found by using a notation called **factorial notation**.

FACTORIAL NOTATION

The symbol

$$n! = n(n - 1)(n - 2) \cdot \cdots \cdot 3 \cdot 2 \cdot 1$$

is called n **factorial** (n is a natural number). Also, we define $0! = 1$ and $1! = 1$.

EXAMPLE 3   Factorial values

Find $n!$ for $n = 0, 1, 2, 3, 4, 5, 6, 7, 8, 9, 10$.

Solution

$0! = 1;$ $1! = 1;$ $2! = 2 \cdot 1 = 2;$ $3! = 3 \cdot 2 \cdot 1 = 6;$

$4! = 4 \cdot 3 \cdot 2 \cdot 1 = 24$ $5! = 5 \cdot 4 \cdot 3 \cdot 2 \cdot 1 = 120$ $6! = 6 \cdot 5 \cdot 4 \cdot 3 \cdot 2 \cdot 1 = 720$

$7! = 7 \cdot 6 \cdot 5 \cdot 4 \cdot 3 \cdot 2 \cdot 1 = 5{,}040$ $8! = 8 \cdot 7 \cdot 6 \cdot 5 \cdot 4 \cdot 3 \cdot 2 \cdot 1 = 40{,}320$ $9! = 9 \cdot 8 \cdot 7 \cdot 6 \cdot 5 \cdot 4 \cdot 3 \cdot 2 \cdot 1 = 362{,}880$

$10! = 10 \cdot 9 \cdot 8 \cdot 7 \cdot 6 \cdot 5 \cdot 4 \cdot 3 \cdot 2 \cdot 1 = 3{,}628{,}800$

In finding the values in Example 3, notice that $3! = 3 \cdot 2!$, $4! = 4 \cdot 3!$, $5! = 5 \cdot 4!$, ...

FACTORIAL PROPERTY

For any counting number n, $n! = n(n - 1)!$.

EXAMPLE 4   Evaluating expressions involving factorial notation

Evaluate:

   a. $8! - 4!$ **b.** $(8 - 4)!$ **c.** $\dfrac{8!}{4!}$ **d.** $\left(\dfrac{8}{4}\right)!$ **e.** $\binom{8}{4}!$ **f.** $\dfrac{500!}{498!}$

Solution

   a. $8! - 4! = 40{,}320 - 24 = 40{,}296$ **b.** $(8 - 4)! = 4! = 24$

   c. $\dfrac{8!}{4!} = \dfrac{40{,}320}{24} = 1{,}680$ **d.** $\left(\dfrac{8}{4}\right)! = 2! = 2$

e. $\begin{pmatrix} 8 \\ 4 \end{pmatrix}! = 70!$

When working with factorials, the numbers can become very large quickly, so it is necessary to leave your answers in factorial form. Even if you have a calculator, you will probably find that most factorials are beyond the capacity of your calculator. However, just to give you an idea about the size of a fairly innocent number such as 70!, we did calculate it using a computer:

$70! = 11,978,571,669,969,891,796,072,783,721,689,098,736,458,938,142,546,425,$
$857,555,362,864,628,009,582,789,845,319,680,000,000,000,000,000$
$= 1.20 \times 10^{100}$

f. $\dfrac{500!}{498!} = \dfrac{500 \cdot 499 \cdot 498!}{498!}$ *Factorial property*

$= 249,500$

> These numbers are too large to do the calculation with a calculator.

Now, we can state a formula for $\begin{pmatrix} n \\ r \end{pmatrix}$ using factorial notation.

BINOMIAL COEFFICIENT

The symbol $\begin{pmatrix} n \\ r \end{pmatrix}$ is defined for integers r and n such that $0 \le r \le n$.

$$\begin{pmatrix} n \\ r \end{pmatrix} = \frac{n!}{r!(n-r)!}$$

is called the **binomial coefficient** n, r, and is sometimes pronounced as "n choose r."

EXAMPLE 5 Binomial coefficients by formula

Use the binomial coefficient formula to find: **a.** $\begin{pmatrix} 6 \\ 4 \end{pmatrix}$ **b.** $\begin{pmatrix} 52 \\ 3 \end{pmatrix}$ **c.** $\begin{pmatrix} n \\ n-1 \end{pmatrix}$

Solution

a. $\begin{pmatrix} 6 \\ 4 \end{pmatrix} = \dfrac{6!}{4!(6-4)!}$

$= \dfrac{6!}{4!2!}$

$= \dfrac{6 \cdot 5 \cdot 4!}{4! \cdot 2 \cdot 1}$

$= 15$

If the numbers are small (as with this example), you can check your answer by looking at Pascal's triangle. For this example, look at row 6, diagonal 4 of Pascal's triangle.

b. $\begin{pmatrix} 52 \\ 3 \end{pmatrix} = \dfrac{52!}{3!(52-3)!}$

$= \dfrac{52!}{3!49!}$

$= \dfrac{52 \cdot 51 \cdot 50 \cdot 49!}{3 \cdot 2 \cdot 49!}$

$= \dfrac{52 \cdot 51 \cdot 50}{3 \cdot 2}$

$= 22,100$

c. $\begin{pmatrix} n \\ n-1 \end{pmatrix} = \dfrac{n!}{(n-1)![n-(n-1)]!}$ *Binomial coefficient formula*

$= \dfrac{n!}{(n-1)!1!}$

$= \dfrac{n(n-1)!}{(n-1)!}$ *Factorial property*

$= n$

EXAMPLE 6   Binomial expansion for a large n

Find: $(x + y)^{15}$

Solution
The power is rather large, so use the binomial theorem and the formula for the coefficients.

$$(x+y)^{15} = \binom{15}{0}x^{15} + \binom{15}{1}x^{14}y + \binom{15}{2}x^{13}y^2 + \cdots + \binom{15}{14}xy^{14} + \binom{15}{15}y^{15}$$

$$= \frac{15!}{0!15!}x^{15} + \frac{15!}{1!14!}x^{14}y + \frac{15!}{2!13!}x^{13}y^2 + \cdots + \frac{15!}{14!1!}xy^{14} + \frac{15!}{15!0!}y^{15}$$

$$= x^{15} + 15x^{14}y + 105x^{13}y^2 + \cdots + 15xy^{14} + y^{15}$$

∎

EXAMPLE 7   Finding a particular binomial coefficient

Find the coefficient of the term x^2y^{10} in the expansion of $(x + 2y)^{12}$.

Solution
We note that $n = 12$, $r = 10$, $a = x$, and $b = 2y$; we then look at the rth term in the binomial expansion:

$$\binom{12}{10}x^2(2y)^{10} = \frac{12!}{10!2!}(2)^{10}x^2y^{10}$$

$$= 66(1,024)x^2y^{10}$$

$$= 67,584x^2y^{10}$$

∎

EXAMPLE 8   Birth orders   MODELING APPLICATION

If a family has 5 children, in how many different birth orders could the parents have a 3–boy, 2–girl family?

Solution

Step 1: *Understand the problem.* Part of understanding the problem might involve estimation. For example, if a family has 1 child, there are 2 possible orders (B or G). If a family has 2 children, there are 4 orders (BB, BG, GB, GG); for 3 children, 8 orders; for 4 children, 16 orders; and for 5 children, a total of 32 orders. This means, for example, that an answer of 140 possible orders is an unreasonable answer.

Step 2: *Devise a plan.* You might begin by enumeration:

BBBGG, BBGBG, BBGGB, . . .

This would seem to be too tedious. Instead, rewrite this as a simpler problem and look for a pattern.

1 child:	B ← one way	**2 children:**	BB ← one way	**3 children:**	BBB ← one way
	G ← one way		BG ⎫ ← two ways		BBG ⎫
			GB ⎭		BGB ⎬ ← three ways
					GBB ⎭
			GG ← one way		BGG ⎫
					GBG ⎬ ← three ways
					GGB ⎭
					GGG ← one way

Look at the possibilities:

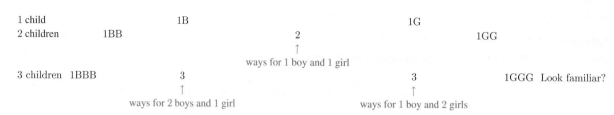

Look at Pascal's triangle in Figure D.2; for 5 children, look at row 5.

Step 3:   *Carry out the plan.*

They could have 3 boys and 2 girls in a total of 10 ways.

Step 4:   *Look back.* We predicted that there are 32 ways a family could have 5 children; let's sum the number of possibilities we found in carrying out the plan to see it totals 32:

$$1 + 5 + 10 + 10 + 5 + 1 = 32$$

PROBLEM SET **APPENDIX D**

LEVEL 1

WHAT IS WRONG, *if anything, with each statement in Problems 1–4? Explain your reasoning.*

1. $(a + b)^2 = a^2 + b^2$
2. $(a + b)^3 = a^3 + b^3$
3. $(2x - 3y)^5$ has six terms.
4. $(H + T)^5$ can be used to find the number of ways of obtaining three heads and two tails by looking at the coefficient of HT.

Evaluate the expressions in Problems 5–8 using both Pascal's triangle and the binomial coefficient formula.

5. **a.** $\begin{pmatrix} 8 \\ 1 \end{pmatrix}$ **b.** $\begin{pmatrix} 9 \\ 5 \end{pmatrix}$

    c. $\begin{pmatrix} 15 \\ 0 \end{pmatrix}$ **d.** $\begin{pmatrix} 7 \\ 5 \end{pmatrix}$

6. **a.** $\begin{pmatrix} 30 \\ 30 \end{pmatrix}$ **b.** $\begin{pmatrix} 32 \\ 31 \end{pmatrix}$

    c. $\begin{pmatrix} 18 \\ 3 \end{pmatrix}$ **d.** $\begin{pmatrix} 14 \\ 2 \end{pmatrix}$

7. **a.** $\dfrac{14!}{11!}$ **b.** $\dfrac{10!}{4!6!}$

    c. $6! - 3!$ **d.** $(6 - 3)!$

8. **a.** $\dfrac{20!}{3!(20 - 3)!}$ **b.** $\dfrac{52!}{3!(52 - 3)!}$

In Problems 9 and 10, expand without using the binomial theorem and then verify your result using the binomial theorem.

9. $(x + 3)^4$ 10. $(a + b)^6$

In Problems 11–24, expand using the binomial theorem.

11. $(a + b)^7$ 12. $(x - 1)^5$
13. $(x - 1)^9$ 14. $(x - y)^6$
15. $(x - 2)^6$ 16. $(x - 3)^5$
17. $(2x + 3y)^4$ 18. $(x - 2y)^8$
19. $\left(\frac{1}{2}x + y^3\right)^3$ 20. $(x^{-2} + y^{-2})^4$
21. $(x^{1/2} + y^{1/2})^4$ 22. $(1 + x)^{10}$
23. $(1 - x)^8$ 24. $(1 - 2y)^6$

LEVEL 2

Find the coefficient of the given term in the expansion of the given binomial in Problems 25–32.

25. a^5b^6 in $(a - b)^{11}$ 26. a^4b^7 in $(a + b)^{11}$
27. $x^{10}y^4$ in $(x + y)^{14}$ 28. $x^{10}y^5$ in $(x - y)^{15}$
29. x^{12} in $(x - 1)^{16}$ 30. y^8 in $(y + 1)^{12}$
31. a^7b in $(a - 2b)^8$ 32. a^4b^4 in $(a + 2b)^8$

Find the first four terms in the expansion of the given binomial in Problems 33–40.

33. $(x - y)^{15}$ 34. $(x + 2y)^{16}$
35. $\left(x + \sqrt{2}\right)^8$ 36. $(x - 2y)^{12}$
37. $(x - 3y)^{10}$ 38. $\left(x + \sqrt{3}\right)^9$
39. $(x^2 + 5k)^{11}$ 40. $(z^3 - k^2)^7$
41. What is the constant term in the expansion of $(9x^{-1} + x^2/3)^6$?
42. What is the constant term in the expansion of $(4y^{-2} + y^3/2)^5$?

Find the last three terms in the expansion of the given binomial in Problems 43–48.

43. $(x^{2/3} + y^{1/3})^{15}$

44. $(2x^{1/5} + 3y^{3/5})^{10}$

45. $\left(q^{-2} - \sqrt{2}r^{-1}\right)^{6}$

46. $(xy^{-1} - 2y^{-2})^{11}$

47. $\left(2m^{-1} + \frac{1}{3}m^{-2}\right)^{9}$

48. $(3r - 2r^{-1})^{12}$

49. If a family has five children, in how many ways could the parents have two boys and three girls?

50. If a family has six children, in how many ways could the parents have three boys and three girls?

51. If a family has seven children, in how many ways could the parents have four boys and three girls?

52. Suppose a coin is tossed eight times. How many outcomes of five heads and three tails are possible?

53. Suppose a coin is tossed 10 times. How many outcomes of four heads and six tails are possible?

54. Suppose a coin is tossed nine times. How many different outcomes of two heads and seven tails are possible?

LEVEL 3

55. Show that

$$\binom{n-1}{r-1} + \binom{n-1}{r} = \binom{n}{r}$$

56. Prove:

$$\binom{k}{r} + \binom{k}{r-1} = \binom{k+1}{r}$$

Problems 57–59 will lead you through the induction proof of the binomial theorem. Let n be any positive integer. To prove:

$$(a+b)^n = \binom{n}{0}a^n + \binom{n}{1}a^{n-1}b + \cdots + \binom{n}{n-1}ab^{n-1} + \binom{n}{n}b^n$$

57. Prove it true for $n = 1$.

58. Assume it true for $n = k$. Fill in this statement of the hypothesis.

59. Prove it true for $n = k + 1$. That is,

TO PROVE:

$$(a+b)^{k+1} = a^{k+1} + \cdots + \left[\binom{k}{r} + \binom{k}{r-1}\right]a^{k-r+1}b^r + \cdots + b^{k+1}$$

Hint: To prove this you will need to use the results of Problem 56.

60. WHAT IS WRONG, if anything, with the following "proof" that $1 = 2$?

$$(a+b)^n = a^n + na^{n-1}b + \frac{n(n-1)}{2!}a^{n-2}b^2 + \cdots + nab^{n-1} + b^n$$

Binomial theorem

Let $n = 0$. Then,

$$(a+b)^0 = a^0 + 0 + 0 + \cdots + 0 + b^0 \qquad \text{Substitution}$$
$$1 = 1 + 0 + 0 + \cdots + 0 + 1$$
$$1 = 2 \qquad \text{Simplify.}$$

Appendix E: Determinants and Cramer's Rule

Suppose we decide to look for a *formula* solution to a system of equations. This formula method for solving a system of equations is known as *Cramer's rule*. Consider

$$\begin{cases} a_{11}x_1 + a_{12}x_2 = b_1 \\ a_{21}x_1 + a_{22}x_2 = b_2 \end{cases}$$

Solve this system by using linear combinations.

$$\begin{aligned} a_{22} & \begin{cases} a_{11}x_1 + a_{12}x_2 = b_1 \\ a_{21}x_1 + a_{22}x_2 = b_2 \end{cases} \\ -a_{12} & \\ & + \begin{cases} a_{11}a_{22}x_1 + a_{12}a_{22}x_2 = b_1a_{22} \\ -a_{12}a_{21}x_1 - a_{12}a_{22}x_2 = -b_2a_{12} \end{cases} \\ & a_{11}a_{22}x_1 - a_{12}a_{21}x_1 = b_1a_{22} - b_2a_{12} \end{aligned}$$

If $a_{11}a_{22} - a_{12}a_{21} \neq 0$, then

$$x_1 = \frac{b_1a_{22} - b_2a_{12}}{a_{11}a_{22} - a_{12}a_{21}}$$

The difficulty with using this formula is that it is next to impossible to remember. To help with this matter, we introduce the concept of determinants.

Determinants

A matrix is an array of numbers, and associated with each square matrix is a real number called its **determinant**.

DETERMINANT OF ORDER 2

If A is the 2×2 matrix $\begin{bmatrix} a & b \\ c & d \end{bmatrix}$, then the **determinant** of A is defined to be the number $ad - bc$.

☠ Do not confuse $|A|$ with $[A]$; the symbol $|A|$ is a determinant or a real number, and $[A]$ is an array or a matrix of real numbers. ☠

Some notations for this determinant are det A, $\begin{vmatrix} a & b \\ c & d \end{vmatrix}$, and $|A|$. We will generally use $|A|$.

EXAMPLE 1   Evaluating determinants of order 2

Evaluate:

a. $\begin{vmatrix} 4 & -2 \\ -1 & 3 \end{vmatrix}$ **b.** $\begin{vmatrix} 2 & 2 \\ 2 & 2 \end{vmatrix}$ **c.** $\begin{vmatrix} 1 & 0 \\ 0 & 1 \end{vmatrix}$ **d.** $\begin{vmatrix} \sin\theta & \cos\theta \\ -\cos\theta & \sin\theta \end{vmatrix}$

e. $\begin{vmatrix} a_{11} & a_{12} \\ a_{21} & a_{22} \end{vmatrix}$ **f.** $\begin{vmatrix} b_1 & a_{12} \\ b_2 & a_{22} \end{vmatrix}$ **g.** $\begin{vmatrix} a_{11} & b_1 \\ a_{21} & b_2 \end{vmatrix}$

Solution

a. $\begin{vmatrix} 4 & -2 \\ -1 & 3 \end{vmatrix} = 4 \cdot 3 - (-2)(-1)$ **b.** $\begin{vmatrix} 2 & 2 \\ 2 & 2 \end{vmatrix} = 2 \cdot 2 - 2 \cdot 2$

$\qquad\qquad = 12 - 2 \qquad\qquad\qquad\qquad\qquad = 0$

$\qquad\qquad = 10$

c. $\begin{vmatrix} 1 & 0 \\ 0 & 1 \end{vmatrix} = 1 \cdot 1 - 0 \cdot 0$ **d.** $\begin{vmatrix} \sin\theta & \cos\theta \\ -\cos\theta & \sin\theta \end{vmatrix} = \sin^2\theta + \cos^2\theta$

$\qquad\qquad = 1 \qquad\qquad\qquad\qquad\qquad\qquad = 1$

e. $\begin{vmatrix} a_{11} & a_{12} \\ a_{21} & a_{22} \end{vmatrix} = a_{11}a_{22} - a_{12}a_{21}$ **f.** $\begin{vmatrix} b_1 & a_{12} \\ b_2 & a_{22} \end{vmatrix} = b_1 a_{22} - b_2 a_{12}$

g. $\begin{vmatrix} a_{11} & b_1 \\ a_{21} & b_2 \end{vmatrix} = b_2 a_{11} - b_1 a_{21}$

■

The solution to the general system

$$\begin{cases} a_{11}x_1 + a_{12}x_2 = b_1 \\ a_{21}x_1 + a_{22}x_2 = b_2 \end{cases}$$

can now be stated using determinant notation:

$$x_1 = \frac{\begin{vmatrix} b_1 & a_{12} \\ b_2 & a_{22} \end{vmatrix}}{\begin{vmatrix} a_{11} & a_{12} \\ a_{21} & a_{22} \end{vmatrix}} \qquad x_2 = \frac{\begin{vmatrix} a_{11} & b_1 \\ a_{21} & b_2 \end{vmatrix}}{\begin{vmatrix} a_{11} & a_{12} \\ a_{21} & a_{22} \end{vmatrix}}$$

Notice that the denominator of both variables is the same. This determinant is called the **determinant of the coefficients** for the system and is denoted by $|D|$. The numerators are also found by looking at $|D|$. For x_1, replace the first column of $|D|$ by the constant numbers b_1 and b_2, respectively. Denote this by $|D_1|$. Similarly, let $|D_2|$ be the determinant formed by replacing the coefficients of x_2 in $|D|$ by the constant numbers b_1 and b_2, respectively. The solution to the system can now be stated with a result called **Cramer's rule**.

CRAMER'S RULE (TWO UNKNOWNS)

Let $|D|$ be the determinant of the coefficients ($|D| \neq 0$) of a system of linear equations, and let $|D_1|$ and $|D_2|$ be the determinants where the coefficients of x_1 and x_2 are replaced by b_1 and b_2, respectively. Then,

$$x_1 = \frac{|D_1|}{|D|} \quad \text{and} \quad x_2 = \frac{|D_2|}{|D|}$$

EXAMPLE 2   Solving a system of equations using Cramer's rule

Solve $\begin{cases} 2x - 3y = -8 \\ x + y = 6 \end{cases}$

Solution

$$x_1 = \frac{\begin{vmatrix} -8 & -3 \\ 6 & 1 \end{vmatrix}}{\begin{vmatrix} 2 & -3 \\ 1 & 1 \end{vmatrix}} \quad x_2 = \frac{\begin{vmatrix} 2 & -8 \\ 1 & 6 \end{vmatrix}}{\begin{vmatrix} 2 & -3 \\ 1 & 1 \end{vmatrix}}$$

$$= \frac{-8 + 18}{2 + 3} \qquad = \frac{12 + 8}{2 + 3}$$

$$= 2 \qquad\qquad = 4 \qquad\qquad \text{The solution is } (2, 4).$$

One advantage of Cramer's rule is the ease with which its form can be used when working with more complicated solutions.

EXAMPLE 3   Solving a difficult system of equations using Cramer's rule

Solve $\begin{cases} 14x - 3y = 1 \\ 5x + 7y = -2 \end{cases}$

Solution

$$x_1 = \frac{\begin{vmatrix} 1 & -3 \\ -2 & 7 \end{vmatrix}}{\begin{vmatrix} 14 & -3 \\ 5 & 7 \end{vmatrix}} \quad x_2 = \frac{\begin{vmatrix} 14 & 1 \\ 5 & -2 \end{vmatrix}}{\begin{vmatrix} 14 & -3 \\ 5 & 7 \end{vmatrix}}$$

$$= \frac{7 - 6}{98 + 15} \qquad = \frac{-28 - 5}{98 + 15}$$

$$= \frac{1}{113} \qquad\qquad = \frac{-33}{113} \qquad\qquad \text{The solution is } \left(\frac{1}{113}, \frac{-33}{113} \right).$$

Cramer's rule also works for independent and inconsistent systems. For example, if $a_{11}a_{22} - a_{12}a_{21} = 0$ and the numerator is not zero, the system is *inconsistent*. If, however, either of the numerators is 0 when $a_{11}a_{21} - a_{12}a_{21} = 0$, then the system is *dependent*.

Properties of Determinants

The strength of Cramer's rule is that it can be applied to linear systems of n equations with n unknowns. To accomplish this application, we need to define the determinant of a 3×3 matrix. We need a preliminary observation. If we delete the first row and first column of the matrix,

$$A = \begin{bmatrix} a_{11} & a_{12} & a_{13} \\ a_{21} & a_{22} & a_{23} \\ a_{31} & a_{32} & a_{33} \end{bmatrix}$$

We obtain the 2×2 matrix $\begin{bmatrix} a_{22} & a_{23} \\ a_{32} & a_{33} \end{bmatrix}$. The determinant of this 2×2 matrix is referred to as the **minor** associated with entry a_{11}, which is the element in the first row and first column. Similarly,

The minor of a_{12} of $A = \begin{bmatrix} a_{11} & a_{12} & a_{13} \\ a_{21} & a_{22} & a_{23} \\ a_{31} & a_{32} & a_{33} \end{bmatrix}$ is $\begin{vmatrix} a_{21} & a_{23} \\ a_{31} & a_{33} \end{vmatrix}$.

The minor of a_{13} of $A = \begin{bmatrix} a_{11} & a_{12} & a_{13} \\ a_{21} & a_{22} & a_{23} \\ a_{31} & a_{32} & a_{33} \end{bmatrix}$ is $\begin{vmatrix} a_{21} & a_{22} \\ a_{31} & a_{32} \end{vmatrix}$.

EXAMPLE 4   Finding minors

Consider $\begin{vmatrix} 2 & -4 & 0 \\ 1 & -3 & -1 \\ 6 & 5 & 3 \end{vmatrix}$ and note that the a_{11} entry is 2; the a_{21} entry is 1; and the a_{31} entry is 6. Find the minor of 2, 1, and 6.

Solution
The minor of 2 is $\begin{vmatrix} -3 & -1 \\ 5 & 3 \end{vmatrix} = (-3)(3) - (5)(-1) = -4$

The minor of 1 is $\begin{vmatrix} -4 & 0 \\ 5 & 3 \end{vmatrix} = (-4)(3) - (5)(0) = -12$

The minor of 6 is $\begin{vmatrix} -4 & 0 \\ -3 & -1 \end{vmatrix} = (-4)(-1) - (-3)(0) = 4$

The **cofactor** of an entry a_{ij} is $(-1)^{i+j}$ times the minor of the a_{ij} entry. This says that if the sum of the row and column numbers is even, the cofactor is the same as the minor. If the sum of the row and the column numbers of an entry is odd, the cofactor of that entry is the opposite of its minor.

EXAMPLE 5   Finding cofactors

Find the cofactor of the a_{45} entry (which is a 2) for the determinant:

$$\begin{vmatrix} 1 & 32 & -3 & 4 & 5 \\ -6 & 7 & 8 & 9 & 10 \\ 11 & -10 & 1 & -8 & -7 \\ -6 & -5 & -9 & 3 & 2 \\ 1 & 0 & 3 & 5 & 12 \end{vmatrix}$$

Solution

Locate the a_{45} entry mentally; delete the entries in the 4th row, 5th column:

$$\begin{vmatrix} 1 & 32 & -3 & 4 & 5 \\ -6 & 7 & 8 & 9 & 10 \\ 11 & -10 & 1 & -8 & -7 \\ -6 & -5 & -9 & 3 & 2 \\ 1 & 0 & 3 & 5 & 12 \end{vmatrix}$$

Since $4 + 5$ is odd, the sign is negative; therefore, the cofactor of 2 is "$-$."

$$-\begin{vmatrix} 1 & 32 & -3 & 4 \\ -6 & 7 & 8 & 9 \\ 11 & -10 & 1 & -8 \\ 1 & 0 & 3 & 5 \end{vmatrix}$$

We know how to evaluate a 2×2 determinant, but we have not yet evaluated higher-order determinants, such as the one shown in Example 5.

DETERMINANT OF ORDER n

A **determinant of order** n is a real number whose value is the sum of the products obtained by multiplying each element of a row (or column) by its cofactor.

EXAMPLE 6   Evaluating determinants of order 3

Evaluate $\begin{vmatrix} 1 & 4 & -1 \\ -2 & 0 & 2 \\ 3 & 1 & 2 \end{vmatrix}$ by *expanding* it about the first row.

Solution

$$\begin{vmatrix} 1 & 4 & -1 \\ -2 & 0 & 2 \\ 3 & 1 & 2 \end{vmatrix} = 1\begin{vmatrix} 0 & 2 \\ 1 & 2 \end{vmatrix} - 4\begin{vmatrix} -2 & 2 \\ 3 & 2 \end{vmatrix} + (-1)\begin{vmatrix} -2 & 0 \\ 3 & 1 \end{vmatrix}$$

$$= 1(0 - 2) - 4(-4 - 6) + (-1)(-2 - 0)$$

$$= -2 + 40 + 2$$

$$= 40$$

The value of the determinant in Example 6 is the same regardless of the row or column that is chosen for evaluation. Try at least one other row or column of Example 6 to show that you obtain the same value of 40.

The method of evaluating determinants by rows or columns is not very efficient for higher-order determinants. The following theorem considerably simplifies the work in evaluating determinants.

DETERMINANT REDUCTION

If $|A'|$ is a determinant obtained from a determinant $|A|$ by multiplying any row by a constant k and adding the result to any other row (entry by entry), then $|A'| = |A|$. The same results holds for columns.

EXAMPLE 7 Evaluating a determinant using reduction

Expand $\begin{vmatrix} 1 & -2 & -5 \\ 2 & -1 & 0 \\ -4 & 5 & 6 \end{vmatrix}$ by using determinant reduction.

Solution

We wish to obtain some row or column with two zeros. Add twice the second column to the first column.

$$\begin{vmatrix} 1 & -2 & -5 \\ 2 & -1 & 0 \\ -4 & 5 & 6 \end{vmatrix} = \begin{vmatrix} -3 & -2 & -5 \\ 0 & -1 & 0 \\ 6 & 5 & 6 \end{vmatrix}$$

Next, expand about the second row (do not even bother to write down the products that are zero):

$$\begin{vmatrix} -3 & -2 & -5 \\ 0 & -1 & 0 \\ 6 & 5 & 6 \end{vmatrix} = \underset{\substack{\uparrow \\ \text{position sign}}}{+} \quad \underset{\substack{\uparrow \\ \text{entry}}}{(-1)} \begin{vmatrix} -3 & -5 \\ 6 & 6 \end{vmatrix}$$

Now that we have a 2×2 determinant, we finish off the multiplication:

$$+(-1)\begin{vmatrix} -3 & -5 \\ 6 & 6 \end{vmatrix} = -(-18 + 30) = -12$$

■

EXAMPLE 8 Evaluating a higher-order determinant

Evaluate $\begin{vmatrix} 2 & -3 & 2 & 5 & 0 \\ 4 & 2 & -1 & 4 & 0 \\ 5 & 1 & 0 & -2 & 0 \\ 6 & 2 & 3 & 6 & 0 \\ 3 & 4 & 6 & 1 & -2 \end{vmatrix}$ using determinant reduction.

Solution

First, notice that all the entries in the fifth column except one are zeros, so begin by expanding about the fifth column. The nonzero entry is located in position a_{55} and $5 + 5 = 10$, which is even, so the leading coefficient is $+1$:

$$\begin{vmatrix} 2 & -3 & 2 & 5 & 0 \\ 4 & 2 & -1 & 4 & 0 \\ 5 & 1 & 0 & -2 & 0 \\ 6 & 2 & 3 & 6 & 0 \\ 3 & 4 & 6 & 1 & -2 \end{vmatrix} = +(-2)\begin{vmatrix} 2 & -3 & 2 & 5 \\ 4 & 2 & -1 & 4 \\ 5 & 1 & 0 & -2 \\ 6 & 2 & 3 & 6 \end{vmatrix}$$

Next, obtain a row or column with all entries zero, except one. We work toward obtaining zeros in column 3.

$$= -2\begin{vmatrix} 10 & 1 & 0 & 13 \\ 4 & 2 & -1 & 4 \\ 5 & 1 & 0 & -2 \\ 6 & 2 & 3 & 6 \end{vmatrix}$$

$$= -2 \begin{vmatrix} 10 & 1 & 0 & 13 \\ 4 & 2 & -1 & 4 \\ 5 & 1 & 0 & -2 \\ 18 & 8 & 0 & 18 \end{vmatrix}$$ Expand along the third column.

$$= -2 \left[-(-1) \begin{vmatrix} 10 & 1 & 13 \\ 5 & 1 & -2 \\ 18 & 8 & 18 \end{vmatrix} \right]$$ Now, we work toward obtaining zeros in the middle row.

$$= (-2) \begin{vmatrix} 5 & 1 & 15 \\ 0 & 1 & 0 \\ -22 & 8 & 34 \end{vmatrix}$$

$$= (-2) \left[+(1) \begin{vmatrix} 5 & 15 \\ -22 & 34 \end{vmatrix} \right]$$

$$= -2(170 + 330)$$

$$= -1,000$$

∎

We can now state Cramer's rule for n linear equations with n unknowns.

$$\begin{cases} a_{11}x_1 + a_{12}x_2 + a_{13}x_3 + \cdots + a_{1n}x_n = b_1 \\ a_{21}x_1 + a_{22}x_2 + a_{23}x_3 + \cdots + a_{2n}x_n = b_2 \\ a_{31}x_1 + a_{32}x_2 + a_{33}x_3 + \cdots + a_{3n}x_n = b_3 \\ \qquad\qquad\qquad \vdots \\ a_{n1}x_1 + a_{n2}x_2 + a_{n3}x_3 + \cdots + a_{nn}x_n = b_n \end{cases}$$

The unknowns are x_1, x_2, x_3, . . . , x_n; a_{ij} are the coefficients and b_1, b_2, b_3, . . . , b_n are the constants. When written in this form, the system is said to be in **standard form**.

CRAMER'S RULE

Let $|D|$ be the determinant of the coefficients ($|D| \neq 0$) of a system of n linear equations with n unknowns. Let $|D_i|$ be the determinant where the coefficients of x_i have been replaced by the constants b_1, b_2, b_3, . . . , b_n, respectively. Then,

$$x_i = \frac{|D_i|}{|D|}$$

As before, if $|D_i|$ and $|D|$ are zero, the system is dependent; if at least one of the $|D_i|$ is not zero when $|D|$ is zero, the system is inconsistent.

EXAMPLE 9 Using Cramer's rule to solve a 3 × 3 system of equations

Solve $\begin{cases} x - 2y - 5z = -12 \\ 2x - y = 7 \\ 5y + 6z = 4x + 1 \end{cases}$

Solution

Rewrite the system in standard form:

$$\begin{cases} x - 2y - 5z = -12 \\ 2x - y = 7 \\ -4x + 5y + 6z = 1 \end{cases}$$

Now, we find

$$\left|D_1\right| = \begin{vmatrix} -12 & -2 & -5 \\ 7 & -1 & 0 \\ 1 & 5 & 6 \end{vmatrix} \quad \left|D_2\right| = \begin{vmatrix} 1 & -12 & -5 \\ 2 & 7 & 0 \\ -4 & 1 & 6 \end{vmatrix} \quad \left|D_3\right| = \begin{vmatrix} 1 & -2 & -12 \\ 2 & -1 & 7 \\ -4 & 5 & 1 \end{vmatrix}$$

$$= \begin{vmatrix} -26 & -2 & -5 \\ 0 & -1 & 0 \\ 36 & 5 & 6 \end{vmatrix} \quad = \begin{vmatrix} 1 & -12 & -5 \\ 2 & 7 & 0 \\ -3 & -11 & 1 \end{vmatrix} \quad = \begin{vmatrix} -3 & -2 & -26 \\ 0 & -1 & 0 \\ 6 & 5 & 36 \end{vmatrix}$$

$$= (-1)\begin{vmatrix} -26 & -5 \\ 36 & 6 \end{vmatrix} \quad = \begin{vmatrix} -14 & -67 & 0 \\ 2 & 7 & 0 \\ -3 & -11 & 1 \end{vmatrix} \quad = (-1)\begin{vmatrix} -3 & -26 \\ 6 & 36 \end{vmatrix}$$

$$= -(-156 + 180) \quad = +(1)\begin{vmatrix} -14 & -67 \\ 2 & 7 \end{vmatrix} \quad = -(-108 + 156)$$

$$= -24 \quad\quad\quad = -98 + 134 \quad\quad = -48$$

$$\quad\quad\quad\quad\quad = 36$$

Finally, $|D| = \begin{vmatrix} 1 & -2 & -5 \\ 2 & -1 & 0 \\ -4 & 5 & 6 \end{vmatrix} = -12$ (from Example 7), so

$$x = \frac{|D_1|}{|D|} \quad y = \frac{|D_2|}{|D|} \quad z = \frac{|D_3|}{|D|}$$

$$= \frac{-24}{-12} \quad = \frac{36}{-12} \quad = \frac{-48}{-12}$$

$$= 2 \quad\quad = -3 \quad\quad = 4$$

The solution is $(x, y, z) = (2, -3, 4)$. ∎

PROBLEM SET **APPENDIX E**

LEVEL 1

Evaluate the determinants in Problems 1–14.

1. $\begin{vmatrix} -5 & -3 \\ 4 & 2 \end{vmatrix}$

2. $\begin{vmatrix} \pi & 3 \\ 0 & 1 \end{vmatrix}$

3. $\begin{vmatrix} \sqrt{3} & \sqrt{5} \\ \sqrt{5} & \sqrt{3} \end{vmatrix}$

4. $\begin{vmatrix} \sin\theta & -\cos\theta \\ \cos\theta & \sin\theta \end{vmatrix}$

5. $\begin{vmatrix} \tan\theta & 1 \\ -1 & \tan\theta \end{vmatrix}$

6. $\begin{vmatrix} \cos x & \sin x \\ \sin y & \cos y \end{vmatrix}$

7. $\begin{vmatrix} 1 & -2 & 3 \\ 0 & 4 & 0 \\ 3 & -1 & -3 \end{vmatrix}$

8. $\begin{vmatrix} 3 & 1 & 0 \\ -2 & 4 & 0 \\ -3 & 5 & -4 \end{vmatrix}$

9. $\begin{vmatrix} 2 & -3 & 1 \\ 1 & 14 & -3 \\ 3 & -12 & -1 \end{vmatrix}$

10. $\begin{vmatrix} 2 & -1 & -3 \\ 1 & 3 & 14 \\ 3 & 3 & 12 \end{vmatrix}$

11. $\begin{vmatrix} 2 & 4 & 3 \\ -2 & 3 & -2 \\ 4 & 3 & 5 \end{vmatrix}$

12. $\begin{vmatrix} 6 & 3 & -3 \\ 2 & 0 & 5 \\ 3 & 5 & -2 \end{vmatrix}$

13. $\begin{vmatrix} 4 & 8 & 5 \\ 3 & 2 & 3 \\ 5 & 5 & 4 \end{vmatrix}$

14. $\begin{vmatrix} 3 & 7 & 1 \\ 2 & 4 & 3 \\ 5 & 6 & 2 \end{vmatrix}$

LEVEL 2

Evaluate the determinants in Problems 15–24.

15. $\begin{vmatrix} 5 & 2 & 6 & -11 \\ -3 & 0 & 3 & 1 \\ 4 & 0 & 0 & 6 \\ 5 & 0 & 0 & -1 \end{vmatrix}$

16. $\begin{vmatrix} 4 & 3 & 2 & 1 \\ -5 & 0 & 0 & 0 \\ 11 & -4 & 0 & 0 \\ 9 & 6 & 3 & -5 \end{vmatrix}$

17. $\begin{vmatrix} 3 & 1 & -1 & 2 \\ 4 & 0 & 3 & 0 \\ 2 & 4 & 3 & -3 \\ 6 & 1 & 4 & 0 \end{vmatrix}$

18. $\begin{vmatrix} 2 & 1 & 3 & 1 \\ 6 & 3 & -3 & 2 \\ 2 & 0 & 5 & 1 \\ 3 & 5 & -2 & -1 \end{vmatrix}$

19. $\begin{vmatrix} 1 & 2 & 3 & 4 \\ 8 & -1 & 5 & 7 \\ 2 & 4 & 6 & 8 \\ -1 & 5 & 3 & 7 \end{vmatrix}$

20. $\begin{vmatrix} 8 & 1 & -7 & 5 \\ -1 & 2 & 2 & 3 \\ -7 & 8 & 15 & 4 \\ -3 & 6 & 6 & 9 \end{vmatrix}$

21. $\begin{vmatrix} -3 & 4 & 5 & 8 & -9 \\ 0 & 0 & 0 & 0 & 6 \\ 3 & 0 & 0 & -1 & 4 \\ 3 & 0 & 0 & -1 & 4 \\ 0 & 0 & 4 & 0 & 9 \end{vmatrix}$

22. $\begin{vmatrix} 3 & 4 & -1 & -2 & 0 \\ 1 & -2 & 3 & 0 & 1 \\ 0 & 4 & -2 & 1 & 0 \\ 5 & 0 & -3 & 0 & 0 \\ 2 & 1 & -2 & 2 & 0 \end{vmatrix}$

23. $\begin{vmatrix} 2 & -1 & 0 & 1 & -1 \\ 3 & 0 & 0 & 2 & 0 \\ 2 & 1 & 3 & 1 & 3 \\ 0 & 0 & 0 & -2 & 0 \\ 1 & 3 & -1 & 2 & 1 \end{vmatrix}$

24. $\begin{vmatrix} 2 & 1 & 3 & 0 & 0 \\ -6 & 1 & 5 & 2 & -1 \\ 1 & 4 & -5 & 9 & 3 \\ 3 & 2 & 5 & 7 & 2 \\ 2 & -3 & -2 & 4 & 6 \end{vmatrix}$

Use Cramer's rule to solve the systems in Problems 25–47.

25. $\begin{cases} x + y = 1 \\ 3x + y = -5 \end{cases}$

26. $\begin{cases} 4x + 3y = 5 \\ 3x + 2y = 2 \end{cases}$

27. $\begin{cases} 2x - 3y - 5 = 0 \\ 3x - 5y + 2 = 0 \end{cases}$

28. $\begin{cases} 2x - 3y - 4 = 0 \\ y = \frac{3}{2}x - 8 \end{cases}$

29. $\begin{cases} 2x + y = \alpha \\ x - 3y = \beta \end{cases}$

30. $\begin{cases} ax + by = 1 \\ bx + ay = 0 \end{cases}$

31. $\begin{cases} ax - by = \gamma \\ cx + dy = \delta \end{cases}$

32. $\begin{cases} ax + by = \alpha \\ cx + dy = \beta \end{cases}$

33. $\begin{cases} 2x + 3y = \cos^2 45° \\ x - y = -\sin^2 45° \end{cases}$

34. $\begin{cases} 3x - y = \sec^2 30° \\ x - y = \tan^2 30° \end{cases}$

35. $\begin{cases} x + 3y = \csc^2 60° \\ x - 2y = \cot^2 60° \end{cases}$

36. $\begin{cases} x + y = \sec^2 45° \\ 2x - y = \tan^2 45° \end{cases}$

37. $\begin{cases} 3x^2 + 4y^2 = 16 \\ x^2 + y^2 = 5 \end{cases}$

38. $\begin{cases} 13x^2 + 12y^2 = 169 \\ x^2 + y^2 = 12 \end{cases}$

39. $\begin{cases} x + y + z = 4 \\ x + 3y + 2z = 4 \\ x - 2y + z = 7 \end{cases}$

40. $\begin{cases} x + y + z = 3 \\ 2x + z = -1 \\ y = 5 \end{cases}$

41. $\begin{cases} 2x + 2y + 3z = 1 \\ 2x - z = -11 \\ 3y + 2z = 6 \end{cases}$

42. $\begin{cases} x + 2y + z = 1 \\ x - 3y - 2z = 2 \\ 3x - 2y + z = 3 \end{cases}$

43. $\begin{cases} 2x - y + z = 4 \\ 3x - 2y + 2z = 3 \\ x - y + 3z = 2 \end{cases}$

44. $\begin{cases} 5x - 3y + 2z = 10 \\ 4x + 2y - 3z = 4 \\ 3x + y + 4z = -8 \end{cases}$

45. $\begin{cases} w + x - y + z = 7 \\ 2w + y - 3z = 1 \\ 2x - z + w = 4 \\ y - w + z = -4 \end{cases}$

46. $\begin{cases} s + 3t - 2u = 4 \\ u + 2x - t = -5 \\ x - u - t = 0 \\ s - 2x = -1 \end{cases}$

47. $\begin{cases} 2x - y + v - w = -4 \\ 3x + 2v = 0 \\ x + y + 3z + w + 3v = 5 \\ -2w = -6 \\ x + 3y - z + 2w + v = 10 \end{cases}$

Use the determinant equation given in Problem 56 to find the equations of the lines passing through the points given in Problems 48–51.

48. $(-3, -2)$, $(1, -3)$

49. $(4, -5)$, $(7, -8)$

50. $(1, 5)$, $(-2, -3)$

51. $(8, 3)$, $(-5, -2)$

Find the absolute value of the determinant expression given in Problem 57 to find the areas of the triangles with vertices given in Problems 52–55.

52. $(1, 1)$, $(-2, -3)$, $(11, -3)$

53. $(-2, 12)$, $(5, 6)$, $(-3, -9)$

54. $(-8, 0)$, $(12, 10)$, $(4, -5)$

55. $(6, 2)$, $(-2, -3)$, $(2, 5)$

LEVEL 3

56. Prove that $\begin{vmatrix} x & y & 1 \\ x_1 & y_1 & 1 \\ x_2 & y_2 & 1 \end{vmatrix} = 0$ is the equation of the line passing through (x_1, y_1) and (x_2, y_2).

57. Prove that the area of a triangle with vertices at (x_1, y_2), (x_2, y_2), and (x_3, y_3) is the absolute value of

$$\frac{1}{2} \begin{vmatrix} x_1 & y_1 & 1 \\ x_2 & y_2 & 1 \\ x_3 & y_3 & 1 \end{vmatrix}$$

58. If r_1, r_2, r_3, and r_4 are the fourth roots of 1, show that

$$\begin{vmatrix} r_1 & r_2 & r_3 & r_4 \\ r_2 & r_3 & r_4 & r_1 \\ r_3 & r_4 & r_1 & r_2 \\ r_4 & r_1 & r_2 & r_3 \end{vmatrix} = 0$$

For Problems 59 and 60 let

$$A = \begin{vmatrix} a_{11} & a_{12} & a_{13} \\ a_{21} & a_{22} & a_{23} \\ a_{31} & a_{32} & a_{33} \end{vmatrix}$$

59. Show that you obtain the same result if you expand along the first or third row.

60. Show that you obtain the same result if you expand along the second row or second column.

Appendix F: Library of Curves and Surfaces

We have placed summaries of important curves and surfaces throughout this book and have called them "Directory of Curves" (Parts I–VII). Those curves are repeated here, along with additional curves in alphabetical order. This library will be valuable to you in calculus and in your future studies.

For some of these curves, the area enclosed (A) and the arc length s are given. Rectangular curves can be translated to (h, k). Polar-form curves can be rotated through an angle α. The particular values for the constants used for each graph are shown in parentheses.

Absolute value function

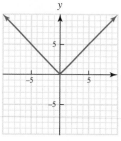

$$y = a|x| \quad (a = 1)$$

Archimedean spiral (*See* Spiral of Archimedes)

Astroid

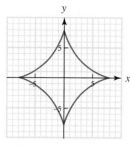

$$x^{2/3} + y^{2/3} = a^{2/3} \quad (a = 8)$$
$$A = \frac{3}{8}\pi a^2 \quad s = 6a$$

This curve is described by a point P on a circle of radius $a/4$ as it rolls on the inside of a circle of radius a.

Bifolium

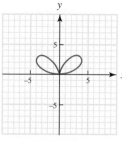

$$(x^2 + y^2)^2 = ax^2 y \quad (a = 12)$$
$$r = a\sin\theta\cos^2\theta$$

Cardioid

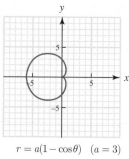

$$r = a(1 - \cos\theta) \quad (a = 3)$$
$$A = \frac{3}{2}\pi a^2 \quad s = 8a$$

This is the curve described by a point P of a circle of radius a as it rolls on the outside of a fixed circle of radius a. This curve is also a special case of a limaçon and epicycloid.

Cassinian curves (*See* Ovals of Cassini)

Catenary (*See* Hyperbolic cosine)

Circle

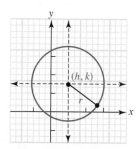

$$x^2 + y^2 = r^2 \quad (r = 5)$$
$$A = \pi r^2 \quad s = 2\pi r$$

Circular cone

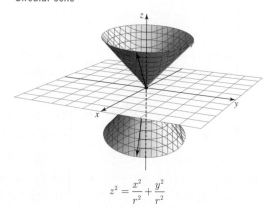

$$z^2 = \frac{x^2}{r^2} + \frac{y^2}{r^2}$$

Cissoid of Diocles

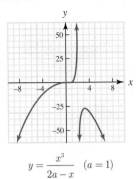

$$y = \frac{x^3}{2a - x} \quad (a = 1)$$

This curve is used in the problem of *duplication of a cube* (i.e., finding the side of a cube that is twice the volume of a given cube).

Cochleoid (or Oui-ja board curve)

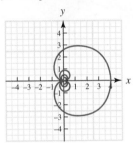

$$\tan\left(\frac{ay}{x^2 + y^2}\right) = \frac{y}{x} \quad (a = 4)$$

Conic section
(*See* Circle, Ellipse, Hyperbola, and Parabola)

Cosecant function

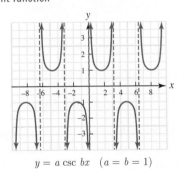

$$y = a \csc\, bx \quad (a = b = 1)$$

Cosine function

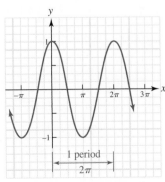

$$y = a \cos\, bx \quad (a = b = 1)$$

Cotangent function

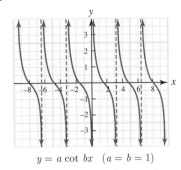

$$y = a \cot\, bx \quad (a = b = 1)$$

Cube root function

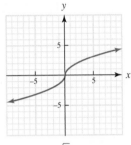

$$y = a\sqrt[3]{x} \quad (a = 1)$$

Cubical parabola

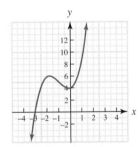

$$y = ax^3 + bx^2 + cx + d$$
$$(a = 1,\ b = 3,\ c = 1,\ d = 4)$$

Cubic function

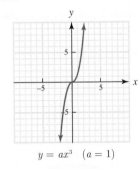

$$y = ax^3 \quad (a = 1)$$

Curate cycloid (*See* Trochoid)

Cycloid

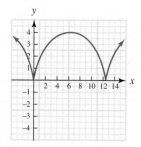

$$\begin{cases} x = a\theta - b\sin\theta \\ y = a - b\cos\theta \end{cases} \quad (a = b = 2)$$

or

$$x = a\cos^{-1}\left(\frac{a-y}{a}\right) \mp \sqrt{2ay - y^2}$$

For one arch, $A = 3\pi a^2 \quad s = 8a$

Deltoid

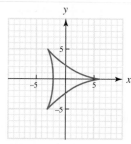

$$\begin{cases} x = 2a\cos\phi + a\cos 2\phi \\ y = 2a\sin\phi - a\sin 2\phi \end{cases} \quad (a = 2)$$

Ellipse

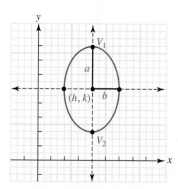

$$\frac{(x-h)^2}{a^2} + \frac{(y-k)^2}{b^2} = 1 \quad (a = 2, b = 3)$$

$$A = \pi ab \quad s \approx 2\pi\sqrt{\frac{a^2+b^2}{2}}$$

Ellipsoid

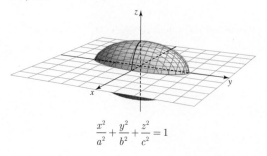

$$\frac{x^2}{a^2} + \frac{y^2}{b^2} + \frac{z^2}{c^2} = 1$$

Elliptic cone

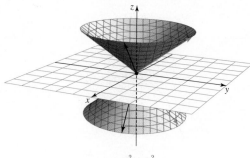

$$z^2 = \frac{x^2}{a^2} + \frac{y^2}{b^2}$$

Elliptic paraboloid

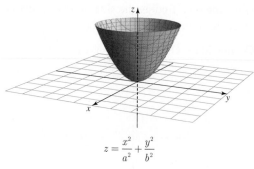

$$z = \frac{x^2}{a^2} + \frac{y^2}{b^2}$$

Epicycloid

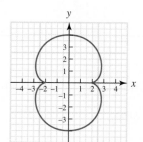

$$\begin{cases} x = (a+b)\cos\theta - b\cos\left(\frac{a+b}{b}\right)\theta \\ y = (a+b)\sin\theta - b\sin\left(\frac{a+b}{b}\right)\theta \end{cases}$$
$$(a = 2, b = 1)$$

This is the curve described by a point P on a circle of radius b as it rolls on the outside of a circle of radius a.

Evolute of an ellipse

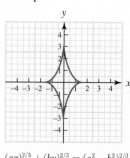

$$(ax)^{2/3} + (by)^{2/3} = (a^2 - b^2)^{2/3}$$
$$(a = 2, b = 1)$$

Exponential curve

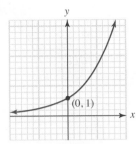

Type I: $b > 1$
$$y = b^{ax} \quad (b = e,\ a = 1)$$

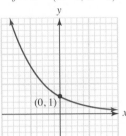

Type II: $0 < b < 1$
$$y = b^{ax} \quad (b = e,\ a = 1)$$

Folium of Descartes

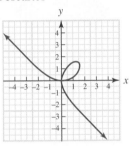

$$x^3 + y^3 = 3axy \quad (a = 1)$$

$$A = \frac{3}{2}a^2$$

Four-leaved rose

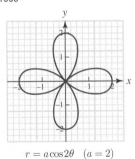

$$r = a\cos 2\theta \quad (a = 2)$$

Gamma function

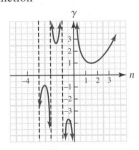

$$\Gamma(n) = \int_0^\infty x^{n-1}e^{-x}dx \quad (n > 0)$$

Hyperbola

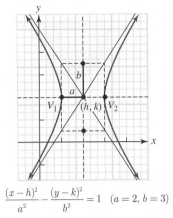

$$\frac{(x-h)^2}{a^2} - \frac{(y-k)^2}{b^2} = 1 \quad (a = 2,\ b = 3)$$

Hyperbolic cosecant

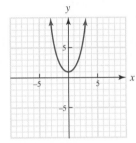

$$y = \operatorname{csch} x = \frac{1}{\sinh x} = \frac{2}{e^x - e^{-x}}$$

Hyperbolic cosine (Catenary)

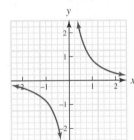

$$y = \cosh x = \frac{e^x + e^{-x}}{2}$$

Hyperbolic cotangent

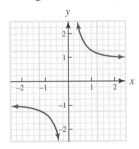

$$y = \coth x = \frac{1}{\tanh x} = \frac{e^x + e^{-x}}{e^x - e^{-x}}$$

Hyperbolic paraboloid

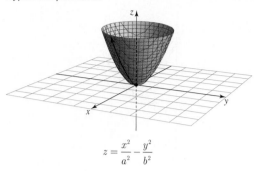

$$z = \frac{x^2}{a^2} - \frac{y^2}{b^2}$$

Hyperbolic secant

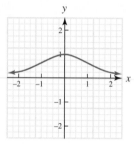

$$y = \operatorname{sech} x = \frac{1}{\cosh x} = \frac{2}{e^x + e^{-x}}$$

Hyperbolic sine

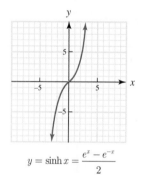

$$y = \sinh x = \frac{e^x - e^{-x}}{2}$$

Hyperbolic spiral (*See* Spiral)

Hyperbolic tangent

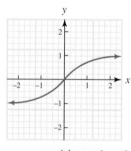

$$y = \tanh x = \frac{\sinh x}{\cosh x} = \frac{e^x - e^{-x}}{e^x + e^{-x}}$$

Hyperboloid of one sheet

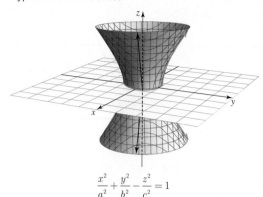

$$\frac{x^2}{a^2} + \frac{y^2}{b^2} - \frac{z^2}{c^2} = 1$$

Hyperboloid of two sheets

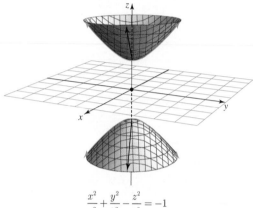

$$\frac{x^2}{a^2} + \frac{y^2}{b^2} - \frac{z^2}{c^2} = -1$$

Hypocycloid

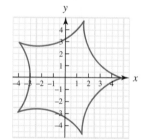

$$\begin{cases} x = (a-b)\cos\phi + b\cos\left(\dfrac{a-b}{b}\right)\phi \\ y = (a-b)\sin\phi - b\sin\left(\dfrac{a-b}{b}\right)\phi \end{cases}$$

This curve is described by a point P on a circle of radius b as it rolls on the inside of a circle of radius a.

Hypocycloid with four cusps (*See* Astroid)

Hypocycloid with three cusps (*See* Deltoid)

Identity function

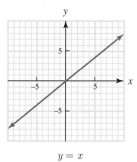

$$y = x$$

Inverse cosecant

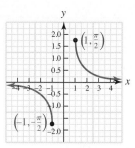

$$y = \sec^{-1} x$$

Inverse cosine

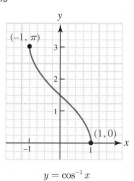

$$y = \cos^{-1} x$$

Inverse cotangent

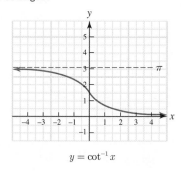

$$y = \cot^{-1} x$$

Inverse secant

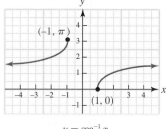

$$y = \sec^{-1} x$$

Inverse sine

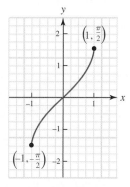

$$y = \sin^{-1} x$$

Inverse tangent

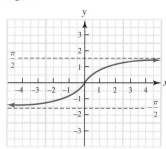

$$y = \tan^{-1} x$$

Involute of a circle

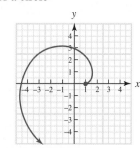

$$\begin{cases} x = a(\cos\phi + \phi\sin\phi) \\ y = a(\sin\phi - \phi\cos\phi) \end{cases} \quad (a = 1)$$

Lemniscate (or Lemniscate of Bernoulli)

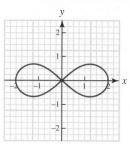

$$(x^2 + y^2)^2 = a^2(x^2 - y^2) \quad (a = 2)$$
$$r^2 = a^2 \cos 2\theta$$

Limaçon (or Limaçon of Pascal)

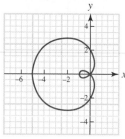

$$r = b - a\cos\theta \quad (a = 3, \, b = 2)$$

Line

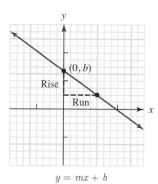

$$y = mx + b$$

Lituus (spiral)

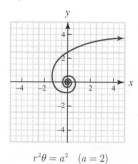

$$r^2\theta = a^2 \quad (a = 2)$$

Logarithmic curve

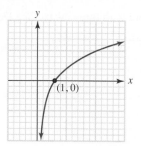

Type I: $b > 1$
$$y = \log_b x \quad (b = 10)$$

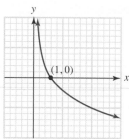

Type II: $0 < b < 1$
$$y = \log_b x \quad \left(b = \frac{1}{2}\right)$$

One-leaved rose (*See* Circle)

Oui-ja board curve (*See* Cochleoid)

Ovals of Cassini

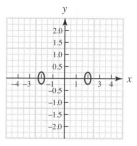

$$(x^2 + y^2 + a^2)^2 - 4a^2x^2 = k^2$$
$$(a = 2, \, k = 1)$$

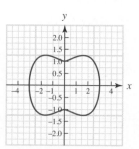

$$(x^2 + y^2 + a^2)^2 - 4a^2x^2 = k^2$$
$$(a = 2, \, k = 5)$$

These curves are sections of a torus on planes parallel to the axis of the torus.

Parabola

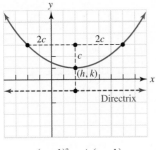

$$(y-k)^2 = 4c(x-h)$$

Probability curve

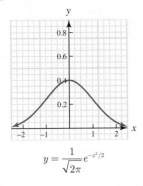

$$y = \frac{1}{\sqrt{2\pi}} e^{-x^2/2}$$

Prolate cycloid (*See* Trochoid)

Quadratic function (standard)

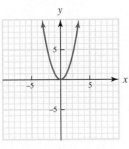

$$y = ax^2 \quad (a=1)$$

Quadratrix of Hippias

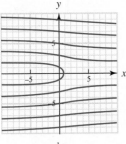

$$y = ax \tan \frac{\pi by}{2} \quad (a=b=1)$$

Reciprocal function (standard)

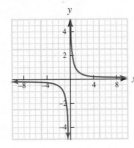

$$y = \frac{a}{x} \quad (a=1)$$

Reciprocal squared function (standard)

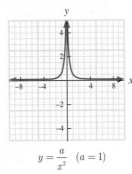

$$y = \frac{a}{x^2} \quad (a=1)$$

Rose curves
(Indexed under number of leaves)

Secant function

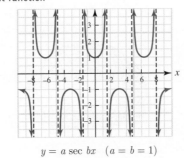

$$y = a \sec bx \quad (a=b=1)$$

Semicubical parabola

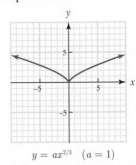

$$y = ax^{2/3} \quad (a=1)$$

Serpentine curve

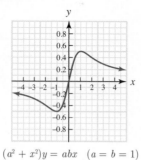

$$(a^2 + x^2)y = abx \quad (a = b = 1)$$

Sine function

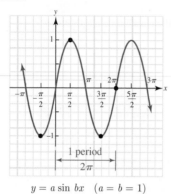

$$y = a \sin bx \quad (a = b = 1)$$

Sinusoid curve (general sine function)

$$y = a \sin(bx + c) \text{ or } y = a \cos(bx + c)$$

Sphere

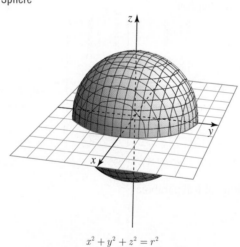

$$x^2 + y^2 + z^2 = r^2$$

Spiral (or Spiral of Archimedes)

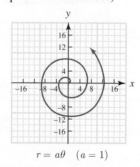

$$r = a\theta \quad (a = 1)$$

Spiral, hyperbolic

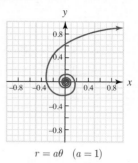

$$r = a\theta \quad (a = 1)$$

Spiral, logarithmic

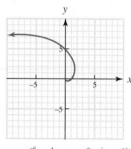

$$r = e^{a\theta} \text{ or } \ln r = a\theta \quad (a = 1)$$

Spiral, parabolic

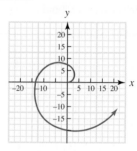

$$(r - a)^2 = 4ak\theta \quad (a = k = 1)$$

Square root function

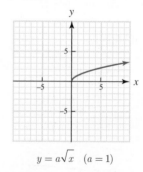

$$y = a\sqrt{x} \quad (a = 1)$$

Square root reciprocal (standard)

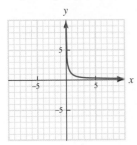

$$y = \frac{1}{\sqrt{x}}$$

Strophoid

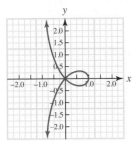

$$y^2 = x^2 \frac{a - x}{a + x} \quad (a = 1)$$

Tangent function

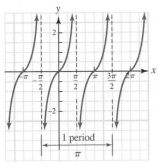

1 period

π

$$y = a \tan bx \quad (a = b = 1)$$

Three-leaved rose

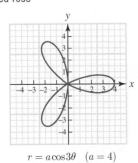

$$r = a \cos 3\theta \quad (a = 4)$$

Tractrix

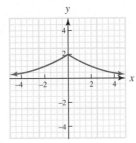

$$\begin{cases} x = a\left(\ln \cot \frac{1}{2}\phi - \cos\phi \right) \\ y = a \sin\phi \quad (a = 2) \end{cases}$$

Trochoid

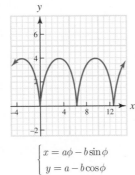

$$\begin{cases} x = a\phi - b\sin\phi \\ y = a - b\cos\phi \\ \quad (a = b = 1) \end{cases}$$

Two-leaved rose (*See* Lemniscate)

Witch of Agnesi

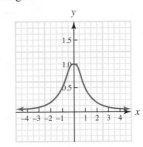

$$y = \frac{a^3}{x^2 + a^2} \quad (a = 1)$$
$$\text{or} \begin{cases} x = a \cot \theta \\ y = a \sin^2 \theta \end{cases}$$

Appendix G: Answers

CHAPTER 1

1. a. $\mathbb{Z}, \mathbb{Q}, \mathbb{R}$ **b.** $\mathbb{Q}, \mathbb{R}$
  c. $\mathbb{N}, \mathbb{W}, \mathbb{Z}, \mathbb{Q}, \mathbb{R}$ **d.** $\mathbb{Q}', \mathbb{R}$
3. a. $\mathbb{Q}', \mathbb{R}$ **b.** $\mathbb{Q}', \mathbb{R}$
  c. $\mathbb{Q}', \mathbb{R}$ **d.** $\mathbb{Q}, \mathbb{R}$
5. a. $\mathbb{Q}, \mathbb{R}$ **b.** $\mathbb{Q}, \mathbb{R}$
  c. $\mathbb{Q}', \mathbb{R}$ **d.** $\mathbb{Q}', \mathbb{R}$
7. a. irrational **b.** rational
  c. rational **d.** rational
9. a. irrational **b.** irrational
  c. rational **d.** irrational
11. a. rational **b.** rational
13. a. $<$ **b.** $<$
15. a. $>$ **b.** $<$
17. a. $>$ **b.** $>$
19. a. $<$ **b.** $>$
21. F **23.** F
25. F
27. a. $\pi - 2$ **b.** $5 - \pi$
  c. x^2 **d.** $x^2 + 3$
29. a. $2\pi - 6$ **b.** $7 - 2\pi$
  c. $x^4 + x^2$ **d.** $x^6 + 4$
31. a. x (or y) **b.** x
  c. y
33. a. 7 **b.** 109
35. a. 73 **b.** $3 - \sqrt{5}$
37. a. $\dfrac{8}{15}$ **b.** $\sqrt{2}$

39.

41.

43. a.

$(3, 7)$

b.

$(-4, -1)$

c.

$[-2, 6]$

d.

$(-3, 0]$

45. a.

$(-\infty, -3]$

b.

$[-\pi, \infty)$

c.

$(-\infty, 0)$

d.

$(2, \infty)$

47. a.

$-3 \le x \le 2$

b.

$-2 < x < 2$

c.

$x \le 3$

d.

$x \ne 6$

49. a.

$-2 < x \le 0$ or $3 < x < 5$

b.

$x \ne 2$

c.

$x < -3$ or $0 < x \le 3$

d.

$-5 \le x < -1$ or $0 < x \le 5$

51. a. F **b.** T
53. a. F **b.** F

1. a. 1 **b.** 13
  c. 125 **d.** 35
3. a. 1 **b.** 16
  c. 4 **d.** 81
5. a. $3x^2 + 3x + 1$ **b.** $3x^2 - x - 7$
  c. $4x - 7$ **d.** $x^2 + 4x - 9$
7. a. $-3x^2 + 13x - 9$ **b.** $x^2 + 7x + 9$
  c. 13 **d.** $x^2 - 4x - 1$
9. a. $x^2 + 3x + 2$ **b.** $y^2 + y - 6$
  c. $x^2 - x - 2$ **d.** $y^2 - y - 6$
11. a. $25x^2 - 16$ **b.** $9y^2 - 4$
  c. $a^2 + 4a + 4$ **d.** $b^2 - 4b + 4$
13. a. $2x^2 - 6x + 8$ **b.** $8x - 22$
  c. $3x^3 + 8x^2 - 9x + 2$ **d.** $2x^3 + 5x^2 - 8x - 5$

15. a. $2x^3 - 3x^2 - 8x - 3$ **b.** $6x^3 + 17x^2 - 4x - 3$
  c. $x^3 - 3x^2 - 10x + 24$ **d.** $4x^3 + 4x^2 - 9x - 9$
17. a. $m(e + i + y)$ **b.** $(a - b)(a + b)$
  c. not factorable **d.** $(a - b)(a^2 + ab + b^2)$
19. a. $(a + b)^3$ **b.** $(p - q)^3$
  c. $(d - c)^3$ **d.** $xy(x + y)$
21. a. $(3x + 1)(x - 2)$ **b.** $(3y - 2)(2y - 1)$
  c. $b(4a - 1)(2a + 3)$ **d.** $2(s - 8)(s + 3)$
23. a. $(x - y + 1)(x - y - 1)$ **b.** $4(x + 1)(x + 2)$
  c. $5(a - 1)(5a + 1)$ **d.** $3(p - 2)(3p + 2)$
25. a. $(a + b + x + y)(a + b - x - y)$
  b. $(-3)(2m - 1)$
  c. $(2x - 3)(x + 2)$ **d.** $(3x + 1)(x - 4)$
27. a. $(4x - 3)(x + 4)$ **b.** $(9x + 2)(x - 5)$
  c. $(9x - 2)(x - 6)$ **d.** not factorable
29. F **31.** F
33. F **35.** F
37. $-2x^{-3}(4x - 5)(2x - 5)^{-3}$

39. $\dfrac{(7x + 11)^2}{(x^2 + 3)^2}(7x^2 - 22x + 63)$

41. $(x + 1)^2(x - 2)^3(7x - 2)$
43. $3(2x - 1)^2(3x + 2)^2(12x + 1)$
45. $2(x + 5)^3(x^2 - 2)^2(5x^2 + 15x - 4)$

  1.3 Equations of Lines, page 27

1.
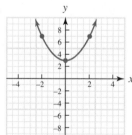

3.
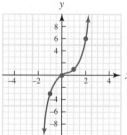

5.
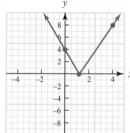

7.
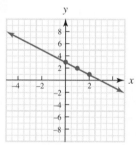

9. a.
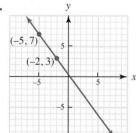
  b. $m = \dfrac{-4}{3}$

11. a.
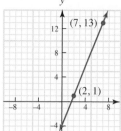
  b. $m = \dfrac{12}{5}$

13. a.
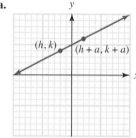
  b. $m = -1$

15. a.
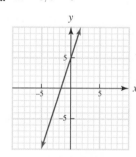
  b. $m = 1$

23. F **25.** F
27. F
29. a. $m = 3; b = 5$ **b.** $m = -4; b = 3$

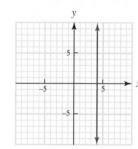 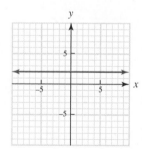

31. a. no slope, no y-intercept **b.** $m = 0; b = 2$

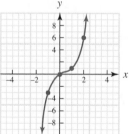 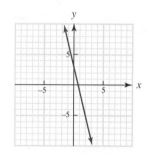

33. a. $m = \dfrac{5}{4}; b = -2$ **b.** $m = \dfrac{1}{3}; b = \dfrac{2}{3}$

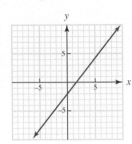
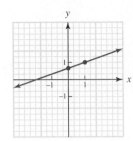

35. a. $5x - y + 6 - 0$ **b.** $y = 0$
37. a. $3x - y - 3 = 0$ **b.** $x + y - 1 = 0$
39. a. $x - 4 = 0$ **b.** $x - 2y + 2 = 0$
41. $2x + 3y - 16 = 0$ **43.** $2x + y + 4 = 0$
45. $x + y - 35 = 0$
47. $x - 2y + 80 = 0$; cost \$237

49. a.   **b.** $273°$

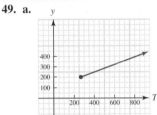

51. $2x + 5y - 19 = 0$

1.4 Distance and Symmetry, page 35

3. a. $5; \left(\dfrac{13}{2}, 3\right)$ **b.** $13; \left(7, \dfrac{13}{2}\right)$

   c. $-5x; \left(5x, \dfrac{7}{2}x\right)$

5. a. $\sqrt{37}; \left(\dfrac{7}{2}, 2\right)$ **b.** $\sqrt{29}; \left(\dfrac{5}{2}, -1\right)$

   c. $5x; \left(5x, \dfrac{7}{2}x\right)$

7.   **9.**

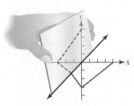
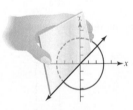

11.   **13.**

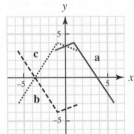
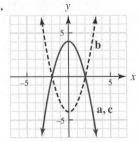

15. none   **17.** none

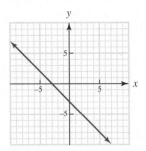
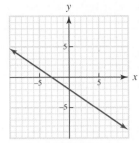

19. y-axis   **21.** x-axis, y-axis, origin

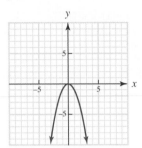
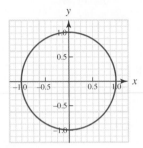

23. y-axis   **25.** y-axis

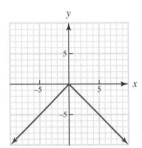
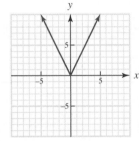

27. y-axis   **29.** none
31. x-axis, y-axis, origin   **33.** none
35. x-axis, y-axis, origin   **37.** origin
39. y-axis
41. $(x - 5)^2 + (y + 1)^2 = 16$ **43.** $(x - 2)^2 + (y - 1)^2 = 26$

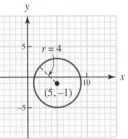
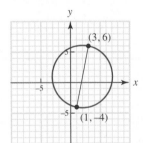

45. $(x - a)^2 + (y - b)^2 = b^2$ **47.**

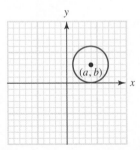
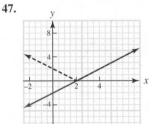

49. $(x - 2)^2 + (y - 3)^2 = 49$ **51.** $\left(2 + 4\sqrt{3}, 0\right), \left(2 - 4\sqrt{3}, 0\right)$

1.5 Linear Inequalities and Coordinate Systems, page 41

3. a. 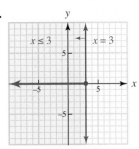 **b.**

5. a. 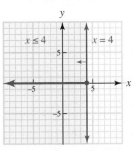 **b.**

7. a. 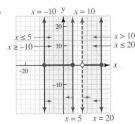 **b.**

9. a. 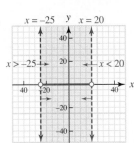 **b.**

11. a. **b.**

13. F **15.** F
17. F **19.** F
21. a. $x = -2$ **b.** $x > -2$
    c. $x < -2$
23. a. $x = 20$ **b.** $x > 20$
    c. $x < 20$
25. a. $x = \pm 3$ **b.** $x < -3$ or $x > 3$
    c. $-3 < x < 3$
27. a. $x = -3, 8$ **b.** $x < -3$ or $x > 8$
    c. $-3 < x < 8$

29. a. $x = -3, 1, 4$ **b.** $-3 < x < 1$ or $x > 4$
    c. $x < -3$ or $-1 < x < 4$
31. a. $x = -5, -1, 8$
    b. $-5 < x < -1$ or $x > 8$
    c. $x < -5$ or $-1 < x < 8$
33. a. $x = 0, 8$ **b.** $x > 8$
    c. $x < 0$ or $0 < x < 8$
35. $[7, \infty)$ **37.** $(-\infty, -41]$
39. $(-\infty, 2)$ **41.** $(-\infty, -1)$
43. $(7, \infty)$ **45.** $(2, 3]$
47. $\left(-\dfrac{8}{5}, 0 \right)$ **49.** $[-4, 2)$
51. $(-7, -4)$ **53.** $\left[-\dfrac{15}{2}, 4 \right)$
55. $8 \le P \le 40$

1.6 Absolute Value Equations and Inequalities, page 48

1. F **3.** T
5. F
7. a. ± 5 **b.** no values
9. a. $-6, 24$ **b.** no values
11. a. $-2, \dfrac{2}{5}$ **b.** $-3, \dfrac{19}{3}$
13. a. $x = 1$ **b.** $x > 1$
    c. $x < 1$
15. a. $x = -4, 0$ **b.** $-4 < x < 0$
    c. $x < -4$ or $x > 0$
17. a. $x = -2, 2$ **b.** $-2 < x < 2$
    c. $x < -2$ or $x > 2$
19. a. $x = 0, 1, 2$ **b.** $x < 0$ or $1 < x < 2$
    c. $0 < x < 1$ or $x > 2$
21. a. $-2 < x < 2$ **b.** no values
23. $x < -4$ or $x > 10$ **25.** $x \le -7$ or $x \ge 1$
27. $-3 \le x \le 10$
29. $3 < x < 8$ **31.** $x < -\dfrac{1}{5}$ or $x > \dfrac{3}{5}$
33. $x \le -25$ or $x \ge -6$ **35.** $|x - 10| = 3$
37. $|5 - 40| = d$ **39.** $|k + 4| \ge 3$
41. If $|x - 5| < 1$, then $|4x - 20| < 4$.
43. If $|x - 5| < \delta$, then $|4x - 20| < \varepsilon$.
45. If $|x - 2| < 0.0006$, then $\left| \dfrac{x^2 - 2x + 2}{x - 4} + 1 \right| < 0.001$.
47. $|m - 3.7| \le 0.04$ **49.** $|m - 9.15| < 0.15$
51. $|t - 120| \le 30$ **53.** $|p - 0.11| \le 0.01$
55. $|a - 6{,}500| \le 1{,}500$

1.7 What Is Calculus?, page 56

3. $\dfrac{1}{3}$ **5.** $\dfrac{3}{11}$

7.

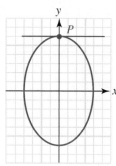

9.

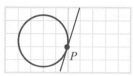

11.

13. 2   **15.** 0
17. 3   **19.** 3
21. 6 or 7   **23.** 6
25. 5   **27.** 4.1
29. 4   **31.** 19
33. 12.61   **35.** 12
37. 0.2   **39.** 0.23607

41. $\dfrac{1}{4}$ **43.** $-\dfrac{1}{8}$

45. -0.2381 **47.** $-\dfrac{1}{4}$

49. π **51.** 5
53. 3.1875   **55.** 0.3984375

1.8   Equations for Calculus, page 64

1. F   **3.** F
5. F   **7.** F
11. a. $-1, 2$ **b.** $-4, -3$

13. a. $-3, -\dfrac{1}{3}, \dfrac{1}{2}$ **b.** $-\dfrac{3}{2}, -\dfrac{1}{3}, 2$

15. a. $-a \pm 4$ **b.** $-a \pm \sqrt{b}$
17. $-5, 3$ **19.** $-9, 2$

21. $0, \dfrac{5}{6}$ **23.** $-\dfrac{1}{2}, \dfrac{4}{5}$

25. $-\dfrac{3}{4}$ (mult. 2)   **27.** $\dfrac{-2 \pm \sqrt{17}}{3}$

29. $-3 \pm i$; no solution in $\mathbb{R}$ **31.** $\dfrac{-1 \pm \sqrt{1 + 8w}}{4}$

33. $\dfrac{-1 \pm \sqrt{3y - 5}}{3}$ **35.** $\dfrac{1 \pm t}{2}$

37. $\dfrac{-1 \pm \sqrt{8y - 47}}{4}$ **39.** $3 \pm \sqrt{4y - y^2}$

41. $-1.98, 2.02$ **43.** $-0.26, 0.69$
45. $-2.87, 4.87$
47. $A = 5$; $B = 0$; $C = 3$; $D = 0$; $E = -3$; $F = 8$

49. $s = 16$; $t = -16$; $u = -\dfrac{5}{7}$

51. $m = \pm 3$; $n = -10$; $p = 0$
53. The horizontal distance is 192 ft.
55. The object is in the air for 4 seconds.

1.9   Inequalities for Calculus, page 70

1. F   **3.** T, provided $a \neq 0$
5. $(-1, 0)$ **7.** $(-\infty, 2] \cup [5, \infty)$
9. $(-7, 3)$ **11.** $(-\infty, -2] \cup [3, \infty)$

13. $\left(-\infty, \dfrac{1}{3}\right) \cup (4, \infty)$ **15.** $(-\infty, -4] \cup [0, 3]$

17. $(-2, 0)$ **19.** $(-4, 3]$

21. $[-3, 2] \cup [4, \infty)$ **23.** $\left(-\infty, -\dfrac{5}{2}\right) \cup \left(-1, \dfrac{7}{3}\right)$

25. $[-3, 1] \cup [5, \infty)$ **27.** $(-\infty, -3) \cup (-1, 2)$
29. $[0, 3) \cup (3, \infty)$ **31.** $(-\infty, 1) \cup (2, \infty)$

33. $[-1, 2]$ **35.** $\left(-\dfrac{1}{5}, 1\right]$

37. $(-\infty, 1] \cup [5, \infty)$ **39.** $\left(-4, \dfrac{3}{2}\right)$

41. $\left[\dfrac{2}{3}, \dfrac{3}{2}\right]$ **43.** $\left(-\infty, \dfrac{2}{5}\right] \cup \left(\dfrac{5}{2}, \infty\right)$

45. $(-\infty, -17] \cup [20, \infty)$ **47.** $(-\infty, 0) \cup \left(\dfrac{1}{2}, 5\right]$

49. $(-\infty, -2) \cup (0, 3)$
51. a. $(-1, 0) \cup (2, \infty)$
     b. $(-\infty, 1) \cup (0, 2)$

Chapter 1 Self Test, page 72

1. b. $8x^2 + 10x - 3$ **c.** $x^3 - 6x^2 + 12x - 8$
     d. $5x^3 + 2x^2 + 4$
2. b. $(2x + 3)(3x - 2)$
     c. $(x - 3)(x + 3)(x - 1)(x + 1)$
     d. $(2x - 1)(x + 1)(4x^2 + 2x + 1)(x^2 - x + 1)$
3. b. 5 **c.** $\sqrt{50} - 6$
     d. $x^2 + 2\pi$

4. b.

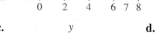

     c.   **d.**

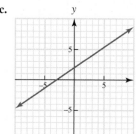
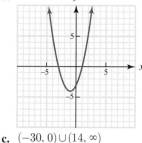

5. b. $-3, 5$ **c.** $(-30, 0) \cup (14, \infty)$
     d. $[-2, 1) \cup [5, \infty)$
6. a. $2x - y + 1 = 0$ **b.** $y - 16 = 0$

7. $\dfrac{5 \pm \sqrt{21}}{2}$ **8.** $0, \pm\sqrt{2}$

9. y-axis
10. a. $(1, 1)$; $m = 1$ **b.** $\left(\sqrt{2}, 0\right)$, $m = 0$

Practice for Calculus, page 72

1. $x^2 + 2$ **3.** -9
5. 19   **7.** $-x^3 - 3x^2 + 4x$

9. $10x^4 + 13x^3 - 66x^2 - 65x - 15$

11. $-5x^3 - 4x^2 + 37x + 16$ **13.** $5(2x - 5)(x + 2)$

15. $(x - 1)(x + 1)(x - 5)(x + 5)$

17. $3(x - 1)(x + 1)(2x + 1)$

19.
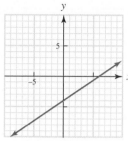

21.
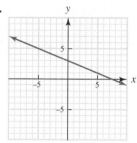

23.
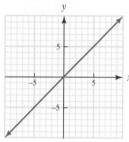

25. $x \le -1$ **27.** $1 \le x \le 10$

29. $-3.0001 \le x \le -2.999$

31.
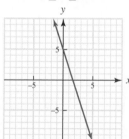

33.
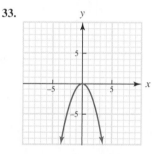

35. $5x + y - 17 = 0$ **37.** $3x + 2y - 22 = 0$

39. $5, \dfrac{7}{3}$ **41.** $(-\infty, -1) \cup (0, 2)$

43. no real values   **45.** no real values

47. $\left(-\dfrac{3}{2}, \dfrac{13}{2}\right)$ **49.** $\left[-\dfrac{1}{2}, 3\right]$

51. $\dfrac{3 \pm \sqrt{5}}{2}$ **53.** no

55. $(-\infty, -2) \cup [-1, 2)$

57. a. x-axis   **b.** $x = 0$

59. a. none   **b.** $-4, 1, 3$

CHAPTER 2

2.1 Problem Solving, page 90

7. Statement (form: $p \to q$): If you break the law, then you will go to jail. Converse (form $q \to p$): If you go to jail, then you broke the law. Inverse (form not $p \to$ not q): If you don't break the law, then you will not go to jail. Contrapositive (form not $q \to$ not p): If you don't go to jail, then you did not break the law.

9. Statement (form: $p \to q$): If a polygon has three sides, then it is a triangle. Converse (form $p \to q$): If it is a triangle, then it has three sides. Inverse (form not $p \to$ not q): If it does not have three sides, then it is not a triangle. Contrapositive (form not $q \to$ not p): If is not a triangle, then it does not have three sides.

11. Statement (form: $p \to q$): If $5 + 10 = 15$, then $15 - 10 = 2$. Converse (form $q \to p$): If $15 - 10 = 2$, then $5 + 10 = 15$. Inverse (form not $p \to$ not q): If $5 + 10 \neq 15$, then $15 - 10 \neq 2$. Contrapositive (form not $q \to$ not p): If $15 - 10 \neq 2$, then $5 + 10 \neq 15$.

13. $A = bh$ **15.** $A = \dfrac{1}{2}pq$

17. $V = s^3$ **19.** $V = \dfrac{1}{3}\pi r^2 h$

21. 400   **23.** 23

25. 17, 18, 19, and 20   **27.** $P = 2\ell + 2w$; 350 ft

29. $P = 4s$; 2,160 ft; $A = s^2$; 291,600 ft²

31. $A = \pi r^2$; 225π in.²   **33.** $V = \dfrac{4}{3}\pi r^3$; $\dfrac{62,500\pi}{3}$ cm³

35. $V = \dfrac{1}{3}\pi r^2 h$; $\dfrac{250\pi}{3}$ ft³   **37.** 5 ft by 11 ft

39. The width of the new figure is any number greater than zero.

41. 4 hr   **43.** 6 mi/h; 8 mi/h

45. width greater than 3 and length greater than 6

47. 2 or 10

49. a. 1,454 ft   **b.** 11.2 sec

51. 16 sec   **53.** 14.8 ft

55. a. $123,456,789 \times 36 = 4,444,444,404$

b. 7,777,777,707   **c.** 9,999,999,909

57. 120 lilies   **59.** 50 cubits

2.2 Introduction to Functions, page 101

1. a. function; D: $\{5, 6, 7, 9\}$ **b.** not a function

c. function; D: $\mathbb{R}$ **d.** function; D: $\mathbb{R}$

3. a. function; D: {years the stock market has been in operation}

b. not a function

5. $y = 3x - 5$; D: $\mathbb{R}$ **7.** $y = \sqrt{5 - x}$; D: $x \le 5$

9. a. 4   **b.** 0

c. 25

11. a. -1 **b.** 9

c. -16 **d.** $5\sqrt{5} - 1$

13. a. $5t - 1$ **b.** $5p - 1$

c. $5t + 4$ **d.** $3t^2 + 6t + 4$

15. a. $2t^2 - 27t - 14$ **b.** $2t^2 - 8t - 24$

17. a. 5   **b.** $6t + 3h$

19. not a function; D: $-1 \le x \le 1$; R: $-3 \le y \le 3$

21. not a function; D: $x \ge -3$; R: $\mathbb{R}$

23. function; D: $-2 \le x \le 3$; R: $-8 \le y \le 4$

25. F   **27.** F

29. T   **31.** T

33. F   **35.** $8x + 4h$

37. $2x + h$ **39.** $2x + h - 1$

41. $\dfrac{3 + 2x}{4 - x}$ **43.** a is arbitrary; $b = 0$

45. 0

47. a. no   **b.** $E(x) = x$

49. a. 512   **b.** 192

c. 80   **d.** $64h + 16h^2$

e. $32xh + 16h^2$

f. $32t + 16h$ is the average speed over the interval between time t and $t + h$.

51. a. 64   **b.** 44

53. $A = \left(\dfrac{P}{4}\right)^2$ **55.** $V = 45x - \dfrac{1}{4}x^3$

57. $f(x) = \dfrac{1}{24}(x^2 + 51x + 144)$

59. $A(x) = \dfrac{4(x-1)^2}{\pi} + x^2$

2.3   Graph of a Function, page 112

7. a. not equal   **b.** equal
9. a. equal   **b.** not equal
11. a. $\mathbb{R}$ **b.** $x \ne 1$
13. a. $\mathbb{R}$ **b.** $\mathbb{R}$
15. $-2 \le x \le 1$
17. a. $(x_0,\ G(x_0))$ **b.** $(x_0 + h,\ G(x_0 + h))$
19. D: $[-4, 7]$; constant on $(-4, 1)$; decreasing on $(1, 3)$ and $(5, 7)$; increasing on $(3, 5)$; intercepts $(3, 0)$, $(6, 0)$, $(0, 5)$; turning points $(3, 0)$, $(5, 5)$
21. D: $[-5, 3) \cup (3, 0)$; constant on $(-5, -2)$; decreasing on $(-2, 0)$; increasing on $(0, \infty)$; intercepts $(-1.75, 0)$, $(1.75, 0)$, $(0, -3)$; turning point $(0, -3)$
23. D: $[-6, 6]$; increasing on $(-6, 0)$; decreasing on $(0, 6)$; intercepts $(-3, 0)$, $(3, 0)$, $(0, 5)$; turning point $(0, 5)$
25. D: $(-\infty, \infty)$; decreasing on $(-\infty, 0)$; increasing on $(0, \infty)$; intercepts $(-3, 0)$, $(3, 0)$, $(0, -9)$; turning point $(0, -9)$
27. D: $(-\infty, \infty)$; decreasing on $(-\infty, -1)$; increasing on $(-1, 0)$; decreasing on $(0, 1)$; increasing on $(1, \infty)$; intercepts $(-2, 0)$, $(2, 0)$, $(0, -8)$; turning points $(-1, -9)$, $(1, -9)$, $(0, -8)$
29. D: $(-\infty, 4) \cup (-4, 4) \cup (4, \infty)$; decreasing on $(-\infty, -4)$, $(-4, 4)$, $(4, \infty)$; intercept $(0, 0)$; no turning points

31. 1   **33.** $\dfrac{-1}{x(x+h)}$

35. D: $\mathbb{R}$; R: $\mathbb{R}$; neither   **37.** D: $\mathbb{R}$; R: $\mathbb{R}$; odd
39. D: $\mathbb{R}$; R: $\mathbb{R}$; odd   **41.** D: $x \ne 0$; R: $y \ne 0$; odd
43. D: $\mathbb{R}$; R: $y \ge -8$; even
45. D: $x \le -2$ or $x \ge 2$; R: $y \ge 0$; even
47. D: $x \le -4$, $x \ge 3$; R: $y \ge 0$; neither
49. D: $x \ne 0$; R: $y \ne 0$; odd
51. D: $x \le -\dfrac{5}{2}$, $x \ge \dfrac{5}{2}$; R: $\mathbb{R}$; not a function

53. $(3, 3)$, $\left(-\dfrac{1}{5}, 3\right)$ **55.** $(-1, 4)$

57. no point of intersection

2.4   Transformations of Functions, page 121

1. $y + 4 = x^2$ **3.** $y + 4 = (x+4)^2$
5. $y + 6 = |x+2|$ **7.** $y - 4 = \sqrt[3]{x}$
9. $y = \dfrac{1}{(x-4)^2}$

11.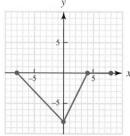

13.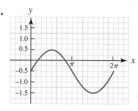

15.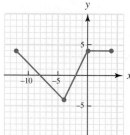

17.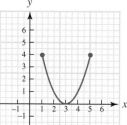

19.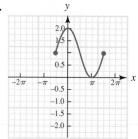

21.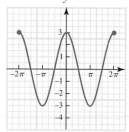

23.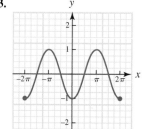

25.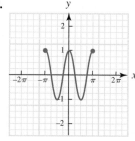

27.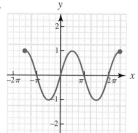

29.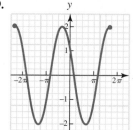

31.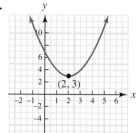

33.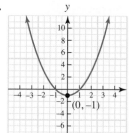

35.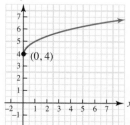

37.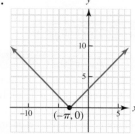

39.

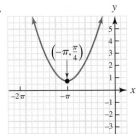

(−π, π/4)

41.

(3, −π)

43.

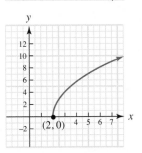

(2, 0)

45.

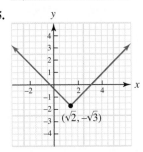

(√2, −√3)

47.

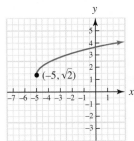

(−5, √2)

49.

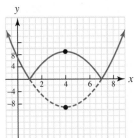

51.

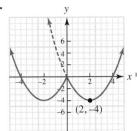

(2, −4)

53.

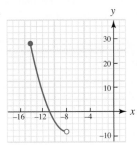

55.

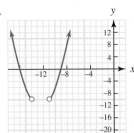

57.

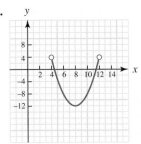

59.

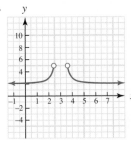

1. a. 1   **b.** 5.3
    c. π   **d.** $\frac{1}{2}$
    e. 5.3

3. a. 3   **b.** 7.3
    c. π + 2   **d.** $\frac{3}{2}$
    e. 3.3

5. a. 2   **b.** 10.6
    c. 2π   **d.** 0
    e. 0

7. a. 1   **b.** 1
    c. 1   **d.** −1
    e. −1

9. a. 2   **b.** 28.09
    c. π²   **d.** −1
    e. 5.3

11. a. 2   **b.** 4.3
    c. π − 1   **d.** $\frac{7}{2}$
    e. 8.3

13. a. 1   **b.** 5
    c. 3   **d.** −1
    e. −6

15. a. 2   **b.** 10.3
    c. π + 3   **d.** $-\frac{3}{2}$
    e. −11.3

17. F   **19.** F
21. F   **23.** T

25.

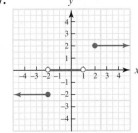

27.

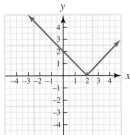

29.

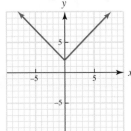

31.

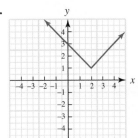

33.

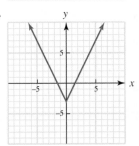

35.

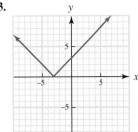

37.

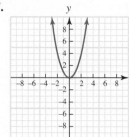

39.

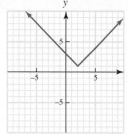

41. 1

43. −1

45. $-\left[-\dfrac{n}{500}\right]$

47. $t(x) = \$2.50 - \$0.25[-2x]$

49. $c(x) = \begin{cases} 1.00 & \text{if } 0 < x \le 3 \\ 1.00 - 0.35[3-x] & \text{if } x > 3 \end{cases}$

51.

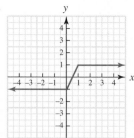

53.

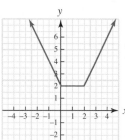

55.

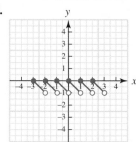

57.

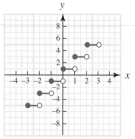

59.

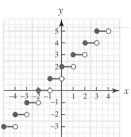

2.6 Composition and Operations of Functions, page 134

1. a. 43 **b.** 36
3. a. 25 **b.** 33
5. a. 0 **b.** 10,000
7. a. 4 **b.** $-\dfrac{5}{4}$
9. a. 0 **b.** does not exist
11. a. $-\dfrac{7}{4}$ **b.** $-\dfrac{5}{4}$
13. a. 3 **b.** 7
15. a. 4 **b.** not defined
17. F
19. F
21. $x^2 - 2x - 2$
23. x^2

---

25. $x^3 + 1$
27. $u(x) = x^2 + 1;\ g(x) = x^2$
29. $u(x) = 2x^2 - 1;\ g(x) = x^4$
31. $u(x) = 3x^2 + 4x - 5;\ g(x) = x^3$
33. $u(x) = 5x - 1;\ g(x) = \sqrt{x}$
35. $u(x) = x^2 - 4;\ g(x) = \sqrt[3]{x}$
37. $g(x) = x^2 - 1;\ u(x) = x^3 + \sqrt{x} + 5$

39. $(f + g)(x) = x^2 + 2x - 2$, D: $\mathbb{R}$; $(f - g)(x) = -x^2 + 2x - 4$, D: $\mathbb{R}$

$(fg)(x) = 2x^3 - 3x^2 + 2x - 3$, D: $\mathbb{R}$; $(f/g)(x) = \dfrac{2x-3}{x^2+1}$, D: $\mathbb{R}$

41. $(f + g)(x) = \dfrac{(x+1)(x-1)^2}{(x-2)}$, D: $\mathbb{R}$, $x \ne 2$;

$(f - g)(x) = \dfrac{-(x+1)(x^2 - 6x + 7)}{(x-2)}$, D: $\mathbb{R}$, $x \ne 2$;

$(fg)(x) = 2x^3 + x^2 - 4x - 3$, D: $\mathbb{R}$, $x \ne 2$;

$(f/g)(x) = \dfrac{2x-3}{(x-2)^2}$, D: $\mathbb{R}$, $x \ne -1, 2$

43. $(f \circ g)(x) = 2x^2 - 1$; $(g \circ f)(x) = 4x^2 - 12x + 10$
45. $(f \circ g)(x) = 2x^2 - 2x - 7$; $(g \circ f)(x) = 4x^2 - 14x + 10$
47. a. $36x^2 + 36x + 9$ **b.** $36x^2 + 36x + 9$
49. a. $x + 2$ **b.** $x + 2$

51. a. $\dfrac{16}{3}\pi$ **b.** $\dfrac{2}{3}\pi t^3$ **c.** $0 \le t \le \sqrt[3]{\dfrac{9}{\pi}}$

53. $f(x) = x + 1$
55. all iterate to 1
57. a. $\dfrac{2x+1}{x+1}$ **b.** $\dfrac{3x+2}{2x+1}$
c. $\dfrac{5x+3}{3x+2}$ **d.** yes

59. Calculator approximation: $\sqrt{5} \approx 2.2360679775$; it seems to be approaching this value.

2.7 Inverse Functions, page 140

1. F
3. F
5. a. inverses **b.** inverses
c. inverses **d.** not inverses
7. a. $\{(5, 4), (3, 6), (1, 7), (4, 2)\}$
b. $x - 3$ **c.** $\dfrac{1}{5}x$
d. no inverse
9. a. no inverse **b.** $\dfrac{1}{x} + 3$
11. $\dfrac{3x+6}{2-3x}$
13. a. 1 **b.** −1
c. 0 **d.** 2
15. a. 2 **b.** 4
c. 6 **d.** −1
17. D: $[-6, 6]$; R: $[-2, 2]$

19.

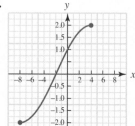

21.

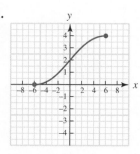

23.

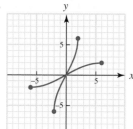

25. a.

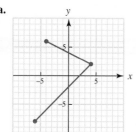

b.

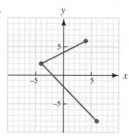

27.

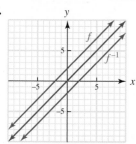

29.

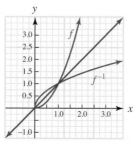

31.

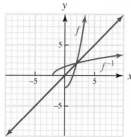

33.

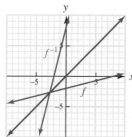

35.

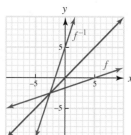

37.

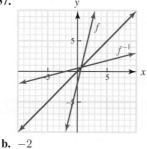

39. a. 5

b. -2

c. π

d. $\sqrt{3}$

41. 0

43. answers vary

45. $y = 0.3937x$

47.

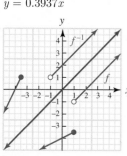

49. a.

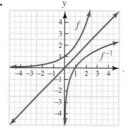

b. answers vary

51. inverses

53. not inverses

55. inverses

57. a. $y = \sqrt{x}$ on $[0, \infty)$

b. does not exist

59. a. $y = \sqrt{\dfrac{x}{2}}$ on $[8, 200]$

b. $y = -\sqrt{\dfrac{x}{2}}$ on $[2, 200]$

2.8   Limits and Continuity, page 150

1. $\lim\limits_{x \to 2} f(x) = 8$

3. $\lim\limits_{x \to 0^+} F(x) = 0$

5. a. continuous function; $[0, 24)$

   b. continuous function; $[0, 24)$

   c. not a continuous function; $(8, 16)$

   d. not a continuous function; $[1, 31]$

   e. not a continuous function; $[0, \infty)$

   f. not a continuous function; $(0, 70]$

7. a. 0

   b. 2

   c. 7

   d. 7

   e. 4

   f. 2

9. 0

11. -9

13. 8

15. $-\dfrac{1}{6}$

17. $-\dfrac{1}{9}$

19. continuous

21. not continuous

23. not continuous

25. 0

27. 1

29. no suspicious points and no points of discontinuity

31. suspicious point $x = 0, 1$; discontinuous at $x = 0$ and $x = 1$

33. suspicious point $x = 0$; discontinuous at $x = 0$

35. suspicious points $x = 0$ and $x = -1$; discontinuous at $x = 0$ and $x = -1$

37. suspicious point $x = 1$; no points of discontinuity

39. a. continuous

   b. discontinuous

   c. continuous

41. discontinuous

43. diverges

45. does not exist

47. 8

49. does not exist

51. $f(-1) < 0, f(3) > 0$; solution on $[-5, 5]$

53. $x = 0$ is a solution, which is on $[-5, 5]$.

55. $f(0) < 0, f(6) > 0$; solution on $[-10, 10]$.

57. a. $-32t + 40$

   b. 40 ft/s

   c. -56 ft/s

   d. 1.25 seconds

59. $a = 4, b = \dfrac{11}{2}$

Chapter 2 Self Test, page 152

1. a. $6x + 2$

   b. $6x + 4$

   c. 0

   d. 3

   e. $f^{-1}(x) = \dfrac{1}{3}(x - 2)$

2. a. $-\sqrt{17} \le x \le \sqrt{17}$

   b. $0 \le y \le \sqrt{34}$

   c. $u(x) = 34 - 2x^2$ and $g(x) = \sqrt{x}$

   d. 4

   e. 4

3. a. $(-\infty, -1) \cup \left[-\frac{1}{4}, \frac{1}{4}\right] \cup (1, \infty)$

b. $(-\infty, -4) \cup [-1, 1] \cup (4, \infty)$

c. $\left(\frac{1}{4}, 0\right), \left(-\frac{1}{4}, 0\right)$

d. $(0, 1), (0, -1)$ **e.** no

4. a. y-compression   **b.**

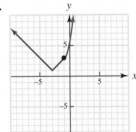

c. 2   **d.** $\frac{1}{2}(h + 2x - 4)$

e. $x - 2$

5. a. $\mathbb{R}$ **b.** $(-1, 0), (0, 1)$

c.   **d.** yes

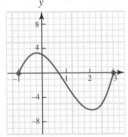

e.

6. a. yes   **b.** no, no

c. D: $[-1, 3]$; R: $[-12, 7]$

d.   **e.**

7. a. yes   **b.** no

c. $(1, 0), (-3, 0), (4, 0), (0, -1.2)$

d. $(1, 0), (3.1, -2.4), (-1.9, -5.5)$

e. -1.2

8. a. $g^{-1}(x) = \dfrac{5 - 2x}{x + 1}$ **b.** $-x^4 + 10x^2 - 20$

c. $\dfrac{x^2}{7 - x^2}$ **d.** f, even; g, neither

e. $[-5, 5]$

9. a. 13 hr; 2.3 hr; 1.8 hr; 14 workers; $C(H) = 25x\left(\dfrac{3x + 4}{2x - 5}\right)$

b. no

10. $V(x) = 8x - \dfrac{2}{3}x^3$

Practice for Calculus—Cumulative Review, page 153

1. $|x^2 - 4|$ **3.** $|2x - 5|$

5. $-2.001 \le x \le -1.999$ **7.** $-0.41\overline{6} \le x \le -0.25$

9. $-1.666\overline{3} \le x \le 1.667$ **11.** $0.49995 \le x \le 0.50005$

13. $\dfrac{5 \pm \sqrt{13}}{2}$ **15.** $-5, -4$

17. $-\dfrac{11}{2}, \dfrac{5}{2}$ **19.** $-3, \dfrac{13}{3}$

21. $\pm 2\sqrt{cy}$ **23.** $\pm\dfrac{1}{2}\sqrt{1 - 3y^2}$

25. $-(x - 1)(x + 1)(x^2 + x + 1)(x^2 - x + 1)$

27. $\dfrac{1}{225}(3x - 1)(3x + 1)(5x - 1)(5x + 1)$

29. $-8x^{-5}(3x - 2)^{-4}$

31. $x(2x^2 + 3)(x - 1)^2(x^2 + x + 1)^2(26x^3 + 27x - 8)$

33. 1   **35.** $5(2x + h)$

37. $\dfrac{-1}{x(x + h)}$ **39.** $P(0.1, 0.4), Q(0.4, 0.4)$

41. $P(a, f(a)); Q(a + h, f(a + h))$

43. $R(f(a + h), f(a + h))$

45. 5   **47.** -4

49. 1   **51.** $2x$

53. $u(x) = 3x^2 - 5x; g(x) = x^2$ **55.** $u(x) = 3x^4 - 1; g(x) = x^{3/2}$

57. $u(x) = x^2 + 3; g(x) = x^4 - x^{5/2}$

59. a. $\mathbb{R}, n \ne 0$ **b.** positive integers

c. 7 minutes   **d.** 12

e. time decreases as the number of trials increases

CHAPTER 3

3.1 Linear Functions, page 167

3. $7x - y - 9 = 0$ **5.** $11x + 30y - 13 = 0$

7. 55   **9.** 10

11. $\dfrac{8}{5}$ **13.** 4

15. $A(x_0, f(x_0))$ and $B(x_0 + h, f(x_0 + h))$; $m = \dfrac{f(x_0 + h) - f(x_0)}{h}$

17. $A(x_0, g(x_0))$ and $B(x_0 + h, g(x_0 + \Delta x))$; $m = \dfrac{g(x_0 + \Delta x) - g(x_0)}{\Delta x}$

19. 2   **21.** 4

23. -2 **25.** $y = 3x - 7$

27. $y = 9$

29. $12x - y - 16 = 0$

31. $x + 3y - 9 = 0$

33. $x = 0$

35. $x + 12y - 98 = 0$

37. a. 35 ft/s   **b.** 350 ft

39. a. P **b.** Q

   c. 7 sec

41. $y = \dfrac{3}{5}x$; the weight is 15.8 lb.

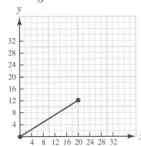

43. a. 0 mi/h   **b.** 180 mi

   c. 60 mi   **d.** 60 mi/h

45. a. increases

   b. Increases until the price reaches \$83, and then as the price moves below \$82, it decreases.

   c. increases

   d. The profit increases until the price reaches \$87, and then it decreases.

47. a. \$340,300 when the price is \$82 or \$83

   b. \$45,962.50 when the price is \$100

   c. There is more profit when revenue is maximized.

49. a.

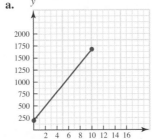

   b. 150   **c.** 150; same

51. The rate of depreciation is \$18,000/year.

53. \$6,250

55. $A(1,\ m + b)$; $B(m + b,\ m + b)$; $C(m + b,\ m^2 + mb + b)$

57. g satisfies neither   **59.** 3

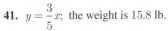

3.2   Quadratic Functions, page 175

1. a. c   **b.** a

3. a. a   **b.** c

5. F   **7.** F

9. T   **11.** F

13. F

15. a.   **b.**

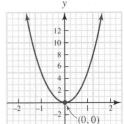

17. a.   **b.**

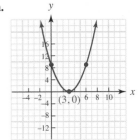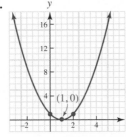

19.   **21.**

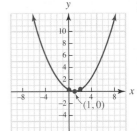

Answers in Problems 23–28 may vary.

23. $-2 \le x \le 2$; $0 \le y \le 200$ **25.** $0 \le x \le 20$; $0 \le y \le 10$

27. $-30 \le x \le 10$; $-30 \le y \le 100$

29. $A\left(1,\ \frac{1}{4}\right)$, $B\left(\frac{1}{4},\ \frac{1}{4}\right)$, $C\left(\frac{1}{4},\ \frac{1}{64}\right)$

31. $y = 0$;   **33.** $y = -4$;

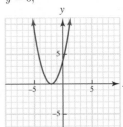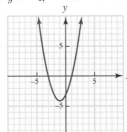

35. $y = 2$;   **37.** $y = -3$;

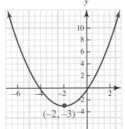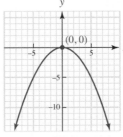

39. $y = 2$;

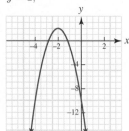

41. $y = -5$;

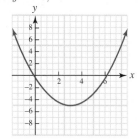

43. $y = 2$;

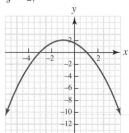

45. $\left(-\dfrac{b}{2a}, \ -\dfrac{b^2 - 4ac}{4a} \right)$

47. $A = -x^2 + 22x$

49. $A = 8x - \dfrac{1}{2}x^3$

51. $A = -2x^2 + 600x$

53. $A = 300y - 2y^2$

55. $A = \dfrac{1}{2}x\sqrt{x}$

57. $P = 4x + 4\sqrt{25 - x^2}$

59. $V = \dfrac{1}{3}\pi r \sqrt{576 - r^4}$

3.3   Optimization Problems, page 184

3. 2

5. 1

7. 40

9. 3

11. −2

13. −647

15. max, 12

17. max, 8

19. max, 8

21. 25

23. −9

25. $48

27. $2,437.50

29. max, 96; min, 51

31. max, 208; min, −135

33. max, 131; min, −125

35. F

37. T

39. 3,456 ft

41. 9 ft × 18 ft; 162 ft²

43. 8 units

45. a. 8,100 ft²; 90 ft × 90 ft

    b. 16,200 ft²; 90 ft × 180 ft

47. $4

49. $250

51. July 11

55. 5 in. × 14 in. × 35 in.

57. 2 ft × 6 ft

59. 168 in.³

3.4   Parametric Equations, page 192

1. F

3. T

5. F

7. $x = 4 + 2t, \ y = 5 + 3t$

9. $x = -5 + t, \ y = 2 + 3t$

11. $x = 2t, \ y = -t$

13. $x = -4 + at, \ y = 5$

15.

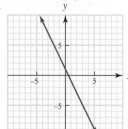

17.

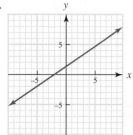

19.

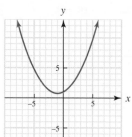

21.

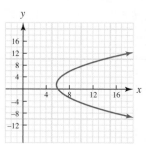

23.

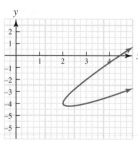

25. $x = 4t, \ y = 9t, \ 0 \le t \le 1$

27. $x = 4 - 4t, \ y = 9 - 9t, \ 0 \le t \le 1$

29. $x = 3 - t, \ y = (3 - t)^2, \ 0 \le t \le 3$

31. $x = t^2, \ y = t, \ 1 \le t \le 3$

33. $x = (3 - t)^3, \ y = 3 - t, \ 0 \le t \le 3$

35.

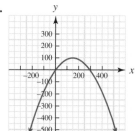

37.

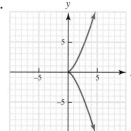

39.

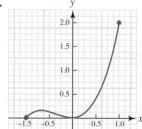

41.

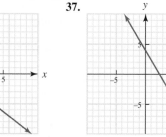

43.

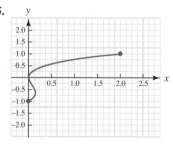

45.

47. $x = \dfrac{1}{2}t, \ y = 6\left(\dfrac{1}{2}t\right) = 3t$

49. $x = -2t, \ y = -12t + 1$

51. $x = 10t, \ y = 100t^2$

53. no

55. Graph approaches the x-axis as k increases.

57.

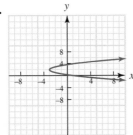

59.

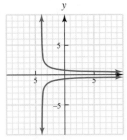

3.5 Graphing Polynomial Functions, page 201

1. a. 3   **b.** -9
   c. -6 **d.** $3x^2 - 3x + 5$

3. a. 1   **b.** 1
   c. 0   **d.** -3
   e. 2   **f.** 0
   g. -7 **h.** -7

   i. $x^3 + x^2 - 2x + \dfrac{-7}{x-1}$

5. $x^2 + 3x + 2$ **7.** $x^2 - 10x + 21$

9. $2x^2 + x + 6 + \dfrac{2}{x-2}$ **11.** $5x^3 - 5x^2 - 5x + 3 + \dfrac{-11}{x+3}$

13. a. yes; 4th degree; 3 zeros   **b.** yes; 1st degree; 1 zero
15. a. no; 2 zeros   **b.** no; 2 zeros
17. a. -70 **b.** -109
   c. -29 **d.** 875
19. a. 18   **b.** 0
   c. 0   **d.** 0
21. four turning points   **23.** four turning points
25. should be continuous

27. a. $x = -4, 2, 5$; positive on $(-4, 2)$ and on $(5, \infty)$; negative on $(-\infty, -4)$ and on $(2, 4)$

   b. $x = 1 \pm \sqrt{7}$; rising on $\left(-\infty, 1 - \sqrt{7}\right)$ and on $\left(1 + \sqrt{7}, \infty\right)$ falling on $\left(1 - \sqrt{7}, 1 + \sqrt{7}\right)$

   c. $x = 1$; concave down on $(-\infty, 1)$; concave up on $(1, \infty)$

29. a. $x = -6, -4, 3, 5$; positive on $(-\infty, -6)$, $(-4, 3)$ and on $(5, \infty)$; negative on $(-6, -4)$ and on $(3, 5)$

   b. $x = \dfrac{-1 \pm \sqrt{85}}{2}, -\dfrac{1}{2}$; on $\left(-\infty, \dfrac{-1 - \sqrt{85}}{2}\right)$ and on $\left(-\dfrac{1}{2}, \dfrac{-1 + \sqrt{85}}{2}\right)$; rising on $\left(\dfrac{-1 - \sqrt{85}}{2}, -\dfrac{1}{2}\right)$ and on $\left(\dfrac{-1 + \sqrt{85}}{2}, \infty\right)$.

   c. $x = \dfrac{-3 \pm \sqrt{255}}{6}$; concave up on $\left(-\infty, \dfrac{-3 - \sqrt{255}}{6}\right)$ and on $\left(\dfrac{-3 + \sqrt{255}}{6}, \infty\right)$; concave down on $\left(\dfrac{-3 - \sqrt{255}}{6}, \dfrac{-3 + \sqrt{255}}{6}\right)$

31. a. $x = -5, -1, 3, 6$; positive on $(-\infty, -5)$, $(-1, 3)$, and on $(6, \infty)$; negative on $(-5, -1)$ and on $(3, 6)$

   b. $x \approx -3.488, 0.941, 4.797$; falling on $(-\infty, -3.488)$ and on $(0.941, 4.797)$; rising on $(-3.488, 0.941)$ and on $(4.797, \infty)$

   c. $x \approx -1.644, 3.144$; concave up on $(-\infty, -1.644)$ and on $(3.1044, \infty)$ concave down on $(-1.644, 3.144)$

33.

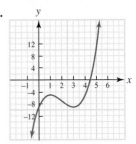

35.

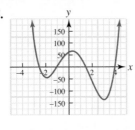

37.

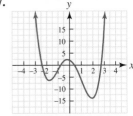

39.

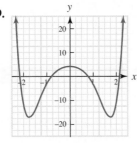

41.

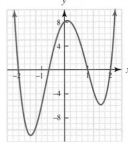

43.

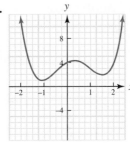

45.

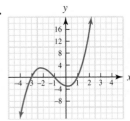

47.

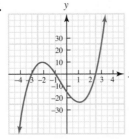

49.

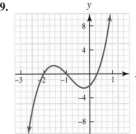

51.

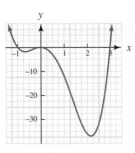

53.

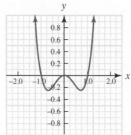

55.

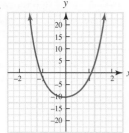

57.

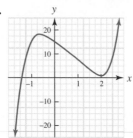

59.

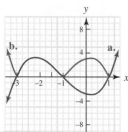

c.

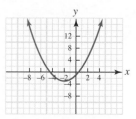

d.

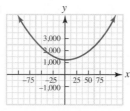

2. a.

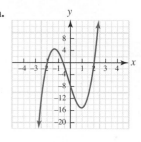

b.

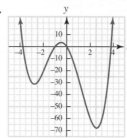

c.

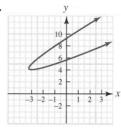

d.

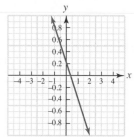

3. a.

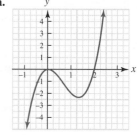

b. 0, 2

3.6 Polynomial Equations, page 212

1. a. $1, -4$ (mult. 2) **b.** $-2, 3$ (both mult. 2)
3. a. 1 (mult. 2), -1 (mult. 2), -6
b. $2, 3$ (mult. 3), -3 (mult. 3)
5. a. 0 (mult. 2), 4 (mult. 2) **b.** $0, -3$ (both mult. 2)
7. a. $0, 2$, or 4 pos; no neg **b.** 0 or 2 pos; 1 neg
9. a. 1 pos; 0 or 2 neg **b.** 0 or 2 pos; 1 neg
11. a. 0 or 2 pos; 0 or 2 neg **b.** 0 or 2 pos; 0 or 2 neg
17. no **19.** yes; $1 - i$
21. no **23.** 1
25. ± 5 **27.** $\pm 1, 2 \pm \sqrt{5}$
29. $\pm 3, \ \pm i\sqrt{3}$ **31.** $2, 3, \ -1, \ \pm i$
33. $-\frac{1}{2}, 1 \pm i\sqrt{2}$
35.

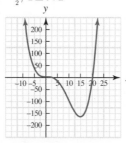

37.

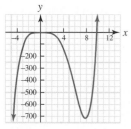

c. 3 **d.** one root has multiplicity 2

4.

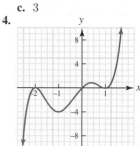

39. $2, \ -1 \pm i\sqrt{3}$ **41.** $\pm 3, \ \pm 3i$
43. $\pm\sqrt{5}, \ \pm i\sqrt{5}$ **45.** $\pm 2i, \ \pm i\sqrt{5}$
47. $\pm 2i, \pm 3i$ **49.** $1, 7, 3 \pm i$
51. $2 \pm \sqrt{5}; \frac{3}{2} \pm \frac{1}{2}\sqrt{11}i$ **59.** $0, -0.13$

5. a. 1 positive root; 2 or 0 negative roots; $\pm 1, \pm 2, \pm 3, \pm 4, \pm 6, \pm 12,$
$\pm\frac{1}{3}, \pm\frac{2}{3}, \pm\frac{4}{3}; 3, -\frac{1}{3}, -4$
b. 1 positive root; 2 or 0 negative roots; $\pm 1, \pm 2, \pm 4, \pm 8, \pm\frac{1}{3}, \pm\frac{2}{3},$
$\pm\frac{4}{3}; \pm\frac{8}{3}, \pm 2, -\frac{2}{3}$
6. a. 3 or 1 positive roots; 1 negative root; $\pm 1, \pm 5, \pm\frac{1}{2}, \pm\frac{1}{3}, \pm\frac{1}{6}, \pm\frac{5}{2},$
$\pm\frac{5}{3}, \pm\frac{5}{6}; x = 1$ (mult. 3), $-\frac{5}{6}$
b. 3 or 1 positive roots; 1 negative root; $\pm 1, \pm 2, \pm 13, \pm 26; 2, -13$
7. $(x - 13)(x + 11)(2x - 5)(3x - 17)$

Chapter 3 Self Test, page 214

1. a.

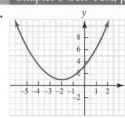

b.

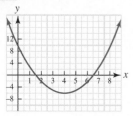

8. a.

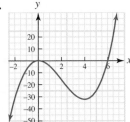

b.

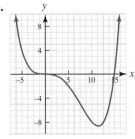

c.

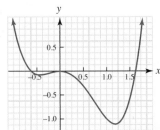

d.

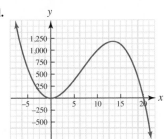

9. $\left(7, \sqrt{14}\right), \left(7, -\sqrt{14}\right)$ **10.** 49 ft

Practice for Calculus, Cumulative Review, page 214

1. $x = -\dfrac{3}{2}, \dfrac{5}{6}$ **3.** $x = \dfrac{25 \pm 5\sqrt{10}}{3}$

5. $x = \dfrac{-1 \pm \sqrt{3y-2}}{3}$ **7.** $-1.66667 \le x \le -1.66666$

9. $-1.55 \le x \le 1.45$ **11.** $x^{-1}y^{-2}(3x^2y^4 - 5y^6 + 1)$

13. $-2x^{-3}(4x-3)(2x-3)^{-3}$

15. $6(x^2+1)^2(3x+2)^3(5x^2+2x+2)$

17. 5   **19.** 0

21. $3x^2$ **23.** $2x - 5 + h$

25. $12x^2 - 6x + 4h^2 + 12xh - 3h$

27. $4x^3 + 6x^2h + 4xh^2 + h^3$ **29.** $y = x - 2$ on $[1, 3]$

31. $y = -\dfrac{1}{255}x^2 + \dfrac{4}{3}x$ on $[0, 180]$

33. $y = 2 + \dfrac{2}{3}(x-1)$ on $[2, 5]$ **35.** 3.75

37. 15   **39.** 6.25

41. 5   **43.** -5

45. -6.89 **47.** -1

49. -5 **51.** -7.5625

53. max 81 at $x = 3$; min -863 at $x = 1$

55. max 81 at $x = 3$; min $-1,375$ at $x = 5$

57. max 1,512 at $x = 6$; min $-206,712$ at $x = -6$

59. $5\sqrt{13}$; 73 ft

CHAPTER 4

4.1 Rational Functions, page 228

1.

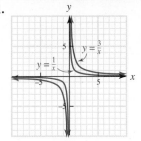

3.

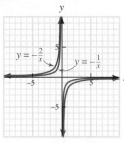

5.

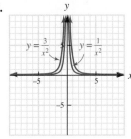

7.

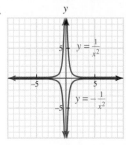

9.

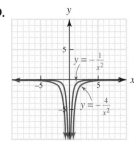

11.

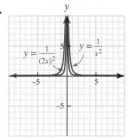

13.

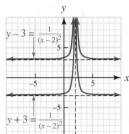

15.

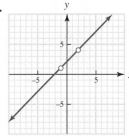

17.

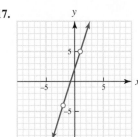

19.

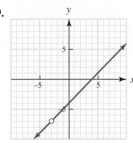

21.

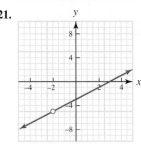

23.

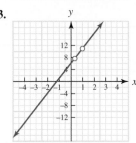

25. $x = 0, y = 0$

27. $x = 0, y = 2$

29. none

31. none

33.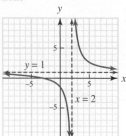

35.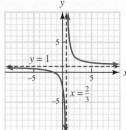

37.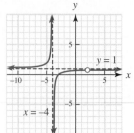

39.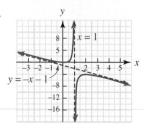

41.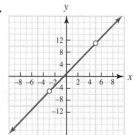

43.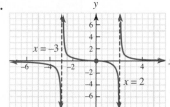

45.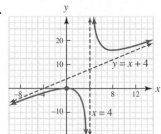

47.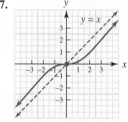

49.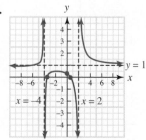

51.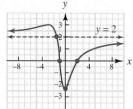

53.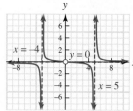

55.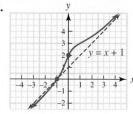

57. 6

59. 490 units

4.2 Radical Functions, page 235

5. F

7. F

9.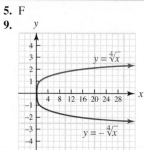

11.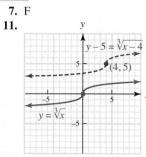

13.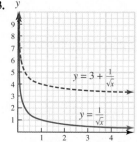

15.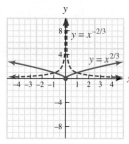

17.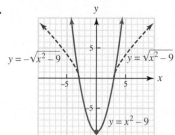

19.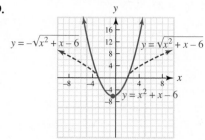

21.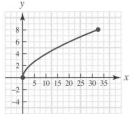

23.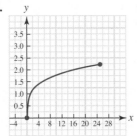

25.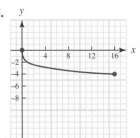

27.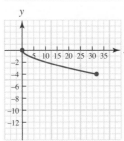

29.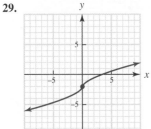

31.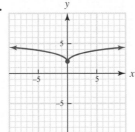

33.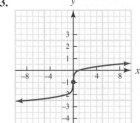

35.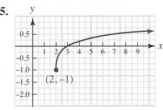

37.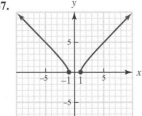

39.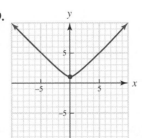

41.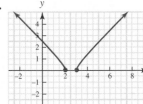

43.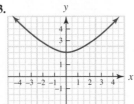

45.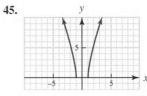

47.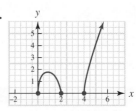

49.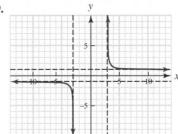

51.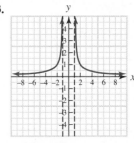

53.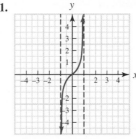

55.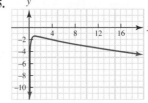

57. a. 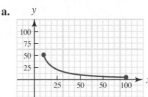   **b.**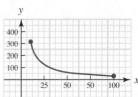

59. It will take 72 minutes.

4.3 Real Roots of Rational and Radical Equations, page 242

1. F

3. F

5. $4, -1$

7. $x \neq 0; \left\{\frac{4}{3}\right\}$

9. $x \neq 2; \varnothing; 2$ is extraneous

11. $y \neq 0, 1; \{4\}$

13. $x \neq \pm 2; \{-8\}$

15. $x \neq \pm 1; \left\{-\frac{1}{5}, 5\right\}$

17. $x \neq \pm 3; \{-1, 1\}$

19. $\{1\}$

21. $\{7\}$

23. $\{1\}$

25. $\{4\}; 1$ is extraneous

27. $\{3\}; 18$ is extraneous

29. $\varnothing$

31. $\left\{\frac{3}{2}\right\}$

33. $\{2\}; 0$ is an extraneous root

35. $\{5\}; 1$ is an extraneous root

37. $\{1, 5\}$

39. $\{0\}$

41. $\{12\}; 0$ is extraneous

43. $\{-1, 0\}$

45. $\varnothing; -3$ is extraneous

47. $\left\{\frac{2}{3}, -\frac{2}{3}, \frac{1}{2}, -\frac{1}{2}\right\}$

49. $\left\{ \pm \frac{\sqrt{3}}{3}, \pm \frac{\sqrt{2}}{2} \right\}$

51. $\{-3, -2, -1\}$

53. $w = \dfrac{1 + \sqrt{5}}{2}$

55. 3.1 feet

57. 15 cm

59. no inconsistency

4.4   Exponential Functions, page 252

1. a. 20.08553692   **b.** 0.1353352832
   c. 2.117000017

3. a. 1.172887932   **b.** 1.81759428
   c. 2.716923932

5. a. 3   **b.** −3
   c. undefined

7. a. −2   **b.** −2
   c. 2

9. a. 4   **b.** 4
   c. undefined   **d.** 4
   e. 4   **f.** undefined

11. a. 3   **b.** 3
   c. undefined   **d.** 3
   e. 3   **f.** undefined

13.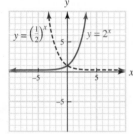

15.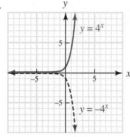

17.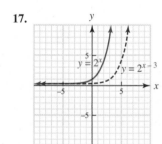

19.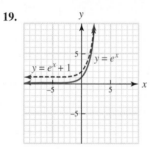

21.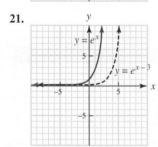

23. $5,427.43

25. $1,054.41

27. 6.14%

29.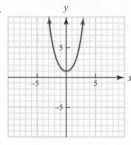

31. a. 1
   b. e^x
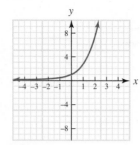

33. $459.54   **35.** $22,831.20

37. $136.17   **39.** $5,413.59

41.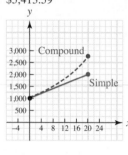

43.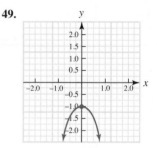

45. $2^{\sqrt{2}} \approx 2.7$

47.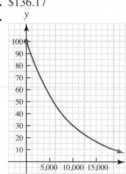

49.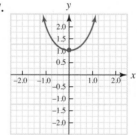

51.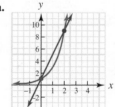

53. The theorem $(b^q)^p = b^{pq}$ applies only for nonnegative b.

55. Let $b = \frac{1}{2}$.

59. a. 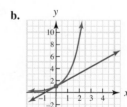   **b.**

   c. 1.1

4.5 Logarithmic Functions, page 260

1. a. $6 = \log_2 64$ **b.** $2 = \log 100$
3. a. $3 = \log_5 125$ **b.** $c = \log_b a$
5. a. $-2 = \log_{1/3} 9$ **b.** $-3 = \log_{1/2} 8$
7. a. $10^4 = 10,000$ **b.** $10^{-2} = 0.01$
9. a. $e^2 = e^2$ **b.** $e^3 = e^3$
11. a. $2^{-3} = \frac{1}{8}$ **b.** $2^5 = 32$
13. a. 2 **b.** 3
15. a. -1 **b.** 3
17. a. 0.6304 **b.** 0.0334
19. a. 4.8549 **b.** -1.3768
21. a. 0.8198 **b.** 2.8196
23. a. 2.5649 **b.** -1.8971
25. a. 3.4650 **b.** 2.8028
27. a. -1.2851 **b.** 0.8044
29. a. -7.4804 **b.** 2.1991
31. a. 111.7887 **b.** 83.2047

33. 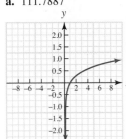 **35.**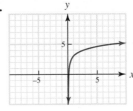

37. 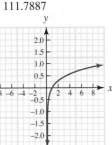 **39.**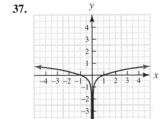

41. 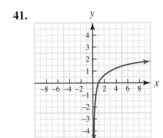 **43.**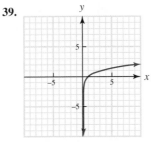

45. F **47.** F
49. F **51.** T
53. a. 3,572 **b.** 4,746
55. a. 3.89 **b.** 8.80
57. a. 30 days **b.** not possible
59.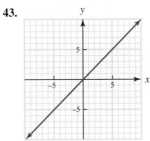

4.6 Logarithmic Equations, page 267

1. a. 2 **b.** 7
   c. 4 **d.** 3
3. a. 3 **b.** $2\sqrt{21}$
   c. $2\sqrt{7}$ **d.** $5\sqrt{2}$
5. a. $\pm 5\sqrt{5}$ **b.** $\pm 2\sqrt{3}$
   c. $\pm\sqrt{70}$ **d.** $\frac{7}{2}$
7. F **9.** F
11. F **13.** T
15. T **17.** 2
19. 3 **21.** 1
23. 93
25. a. 1.79175946923 **b.** $\ln 6$
   c. $\ln 6^2$
27. a. $\log 24$ **b.** $\log 2$
   c. $\ln \dfrac{x^2 y^3}{z^4}$ **d.** $\ln \dfrac{x^5 z^3}{y^2}$
29. a. $\log \dfrac{x^3(x-3)}{x+3}$ **b.** $\log \dfrac{x-3}{x^6(x+3)}$
   c. $\ln(x+2)$ **d.** $\ln(x+2)^{1/3}$
31. 0 **33.** 10^{100}
35. 25 **37.** $\sqrt{e}$
39. 5 **41.** $\log 18$
43. 1 **45.** 2, 4
47. $\sqrt{10}$ **49.** e^2
51. 5 **53.** 3
55. e^{10} **57.** $t = e^{(80-R)/27}$

4.7 Exponential Equations, page 276

1. a. 7 **b.** 5
3. a. -2 **b.** $-\frac{4}{3}$
5. a. $\dfrac{\log_5 4 - 3}{2}$ **b.** $\dfrac{\ln 25 - 1}{2}$
7. a. 1.6232492904 **b.** -1.63078414259
9. a. -1.29202967422 **b.** -2.48126242604
11. a. 0.861654166907 **b.** 2.30539742365
13. 0 **15.** -9.4520
17. 8.6653 **19.** ± 6.2261
21. ± 9.4476 **23.** ± 8.0241
25. 0.9031, 1 **27.** 0.6931
29. 0.5378 **31.** ± 2.2924
33. 2.3522 **35.** 3.1555
37. 0.0208 **39.** ± 1
41. $t = \dfrac{1}{r}\ln\left(\dfrac{P}{P_0}\right)$ **43.** $a = \dfrac{-100}{21}\ln\left(\dfrac{P}{14.7}\right)$
45. $t = h\log_{0.5}\left(\dfrac{A}{A_0}\right)$ **47.** $n = 1 - \dfrac{100}{11}\ln\left(\dfrac{P_n}{P_1}\right)$
49. $T = \dfrac{E}{R\ln(\eta A)}$ **51.** 10,000 years
53. 1 year 11 months **55.** 1,007,000
57. 7 days
59. a. $\dfrac{x^5}{2\cdot 3\cdot 4\cdot 5}; \dfrac{x^6}{2\cdot 3\cdot 4\cdot 5\cdot 6}$ **b.** 2.716667
   c. 1.6458333

Chapter 4 Self Test, page 277

1. a.

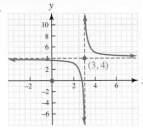

b.

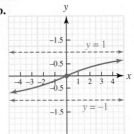

c.

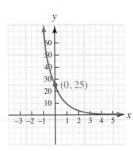

d.

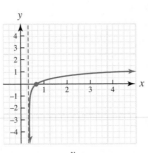

2. a.

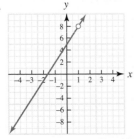

b.

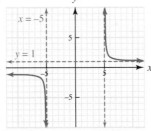

c.

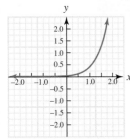

d.

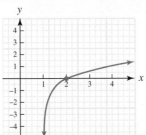

3. a. $\frac{5}{2} + e$ **b.** 3
    c. 546   **d.** 4.3

4. a. $\left\{ \frac{5}{3} \right\}$ **b.** {2}
    c. $\log 85$ **d.** $\left\{ \frac{101}{99} \right\}$

5. a. {9, 10}   **b.** {−2}
    c. $\dfrac{\ln 45 - 1}{3}$ **d.** $\left\{ 100\sqrt{5} \right\}$

6. a. 1.1178   **b.** 1,296
    c. 5   **d.** $\log_{(1+i)} \frac{A}{P}$

7. a. $a = 14.8$, $b = \ln 2.5$

8. a. 3 years 3 months   **b.** 10 years
    c. 14.1%

9. a. −0.02615649738   **b.** 101 hours

10. $A = A_0 e^{0.5941r}$

Practice for Calculus, Cumulative Review, page 278

1. $\dfrac{-4}{(x+h-2)(x-2)}$ **3.** $\dfrac{e^x(e^h - 1)}{h}$

5. $\dfrac{1}{h} \log\left(\dfrac{x+h}{x} \right)$ **7.** $\dfrac{16(2^{-h} - 1)}{h}$

9. $\dfrac{1}{h} \log\left(\dfrac{2+h}{2} \right)$ **11.** $\dfrac{e^6 + 1}{e^3}$

13. $-\dfrac{1}{4e^2}$

15. a. 3   **b.** 19

17. a. $e^3 - e^2$ **b.** $e^{11} - e^{10}$

19. a. $\ln 2$ **b.** $\ln \frac{10}{9}$

21. $(2x - 3)(4x + 5)$

23. $5x^{-2}y^{-1}z(x^4y^4 + 4y^2z + 20x^3z^2)$

25. a. polynomial   **b.** logarithmic
    c. polynomial   **d.** polynomial
    e. rational

27. $u(x) = 2x^2 + 1$; $g(x) = e^x$

29. $u(x) = x^2 + 3x$; $g(u) = u^2$

31. $u(x) = 3x^4 + 5x$; $g(u) = \log u$

33.

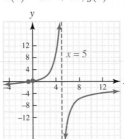

35.

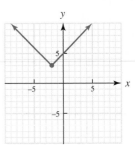

37.

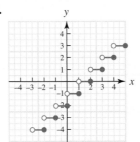

39.

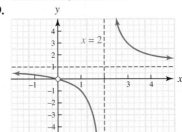

41.

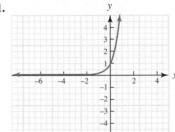

43.

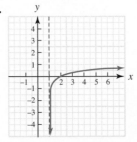

45.

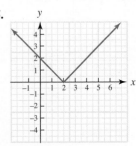

47.

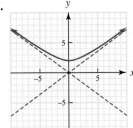

49. {4, 7}

51. {1}

53. {1}

55. $t = -20\ln\left(\frac{3}{8}\right)$

57. $n = \log_{(1+r)}\left(\dfrac{m}{m - Pi}\right)$

59. 23 hours

CHAPTER 5

5.1 Angles, page 292

1. F

3. F

5. F

7. a. $30° = \dfrac{\pi}{6}$

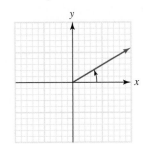

b. $90° = \dfrac{\pi}{2}$

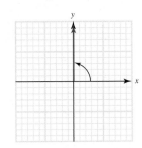

c. $45° = \dfrac{\pi}{4}$

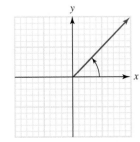

d. $60° = \dfrac{\pi}{3}$

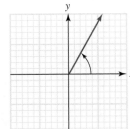

9. a. $\dfrac{\pi}{3} = 60°$

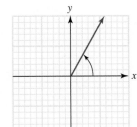

b. $\dfrac{\pi}{6} = 30°$

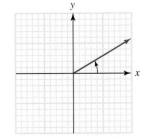

c. $\dfrac{\pi}{2} = 90°$

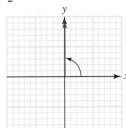

d. $\dfrac{\pi}{4} = 45°$

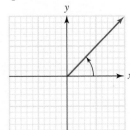

11. a. F   **b.** D
c. A   **d.** E
e. C   **f.** B

13.

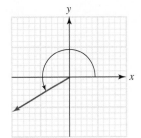

a. none   **b.** reference angle
c. coterminal   **d.** coterminal
e. equal

15.

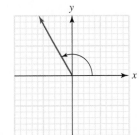

a. reference angle   **b.** none
c. coterminal   **d.** coterminal
e. equal

17.

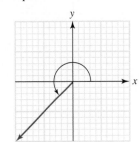

a. reference angle   **b.** none
c. coterminal   **d.** none
e. none

19. a. $270° = \dfrac{3\pi}{2}$

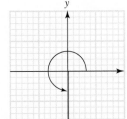

b. $480° = \dfrac{8\pi}{3}$

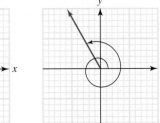

c. $40° = \dfrac{2\pi}{9}$

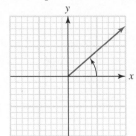

21. a. $300° = \dfrac{5\pi}{3}$

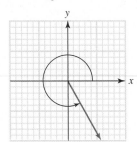

b. $-150° = -\dfrac{5\pi}{6}$

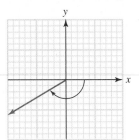

c. $85° = \dfrac{17\pi}{36}$

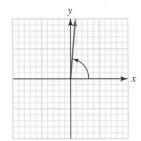

23. a. $120° \approx 2.09$

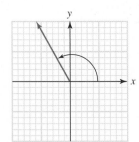

b. $-115° \approx -2.01$

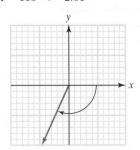

c. $100° \approx 1.75$

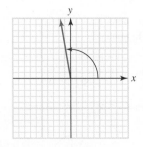

25. a. $350° \approx 6.11$

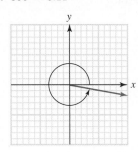

b. $525° \approx 9.16$

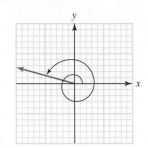

c. $-45° \approx -0.79$

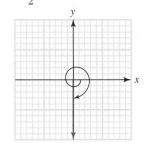

27. a. $5\pi = 900°$

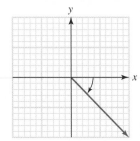

b. $-\dfrac{5\pi}{2} = -450°$

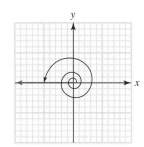

c. $\dfrac{11\pi}{6} = 330°$

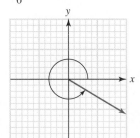

29. a. $\dfrac{2\pi}{3} = 120°$

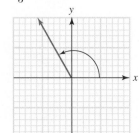

b. $\dfrac{4\pi}{3} = 240°$

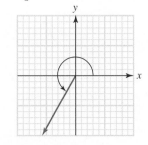

c. $\dfrac{11\pi}{3} = 660°$

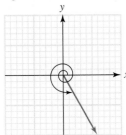

31. a. $\dfrac{2\pi}{9} = 40°$ **b.** $\dfrac{\pi}{2} = 90°$

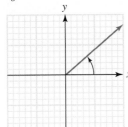   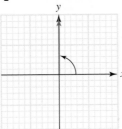

c. $\dfrac{5\pi}{3} = 300°$

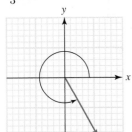

33. a. $-0.42 \approx -24°$ **b.** $0.4 \approx 23°$

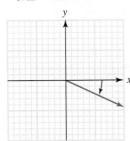   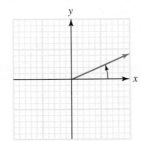

c. $7 \approx 401°$

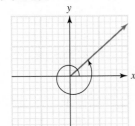

35. a. $\pi;\ 0$ **b.** $\dfrac{\pi}{6};\dfrac{\pi}{6}$

  c. $\pi;\ 0$ **d.** $7 - 2\pi;\ 7 - 2\pi$

  e. $4\pi - 7;\ 4\pi - 7$ **f.** $\dfrac{3\pi}{4};\dfrac{\pi}{4}$

37. a. 2.7168   **b.** 1.2832
  c. 0.7879
39. a. 4.8673   **b.** 3.1416
  c. 1.0472
41. a. $0° < \alpha < 90°$ **b.** $90° < \beta < 180°$
  c. $180° < \gamma < 270°$ **d.** $270° < \varphi < 360°$
43. a. Quad III or IV (including 270°)
  b. Quad IV   **c.** Quad I
  d. Quad I or II
45. a. $0 \le 2\theta < 2\pi$ **b.** $0 \le 3\theta < 3\pi$
  c. $0 \le \frac{1}{2}\theta < \frac{\pi}{2}$ **d.** $0 \le \frac{1}{3}\theta < \frac{\pi}{3}$
47. a. 1 m   **b.** 4.19 in.
49. 3.14 cm   **51.** 1.59 revolutions
53. 2,600 mi   **55.** 143°
57. 141.4 in.   **59.** 28,000 kilometers per hour

5.2 Fundamentals, page 303

1. $(\cos \alpha,\ \sin \alpha)$ **3.** $[\cos(\alpha + \pi),\ \sin(\alpha + \pi)]$
5. a. secant   **b.** tangent
  c. cosecant   **d.** cosine
  e. sine   **f.** cotangent
7. F   **9.** F
11. F
13. a. 0.6   **b.** 0.9
  c. -0.5
15. a. -0.4 **b.** -0.8
  c. 0.3
17. a. 0.6427876097   **b.** 0.342020143
  c. 2.9238044   **d.** 0.342020143326
  e. -0.9961946981 **f.** 0.3639702343
19. a. 0.6225146366   **b.** -0.1124755272
  c. -0.1154340956 **d.** -0.7071067812
  e. 0.7071067812   **f.** 1
21. a. 0.9092974268   **b.** -0.9364566873
  c. 4.637332055   **d.** -19.10732261
23. a. positive   **b.** negative
  c. positive   **d.** negative
25. a. positive   **b.** positive
  c. positive   **d.** negative
27. a. positive   **b.** negative
  c. negative   **d.** negative
29. a. I, IV   **b.** I, II
31. a. II, III   **b.** II, IV
33. a. III   **b.** II
35. a. I   **b.** III

41. $\sin \theta = \sin \theta$; $\cos \theta = \sqrt{1 - \sin^2 \theta}$; $\tan \theta = \dfrac{\pm \sin \theta}{\sqrt{1 - \sin^2 \theta}}$;

$\sec \theta = \dfrac{\pm 1}{\sqrt{1 - \sin^2 \theta}}$; $\csc \theta = \dfrac{1}{\sin \theta}$; $\cot \theta = \dfrac{\pm\sqrt{1 - \sin^2 \theta}}{\sin \theta}$

43. $\cos \theta = \dfrac{\pm 1}{\sqrt{1 + \tan^2 \theta}}$; $\sin \theta = \dfrac{\pm \tan \theta}{\sqrt{1 + \tan^2 \theta}}$; $\tan \theta = \tan \theta$;

$\sec \theta = \pm\sqrt{1 + \tan^2 \theta}$; $\csc \theta = \dfrac{\pm\sqrt{1 + \tan^2 \theta}}{\tan \theta}$; $\cot \theta = \dfrac{1}{\tan \theta}$

45. $\cos\theta = \dfrac{1}{\sec\theta}$; $\sin\theta = \dfrac{\pm\sqrt{\sec^2\theta - 1}}{\sec\theta}$; $\tan\theta = \pm\sqrt{\sec^2\theta - 1}$;

$\sec\theta = \sec\theta$; $\csc\theta = \dfrac{\pm\sec\theta}{\sqrt{\sec^2\theta - 1}}$; $\cot\theta = \dfrac{\pm 1}{\sqrt{\sec^2\theta - 1}}$

47. a. $\sin\theta = \frac{12}{13}$; $\tan\theta = \frac{12}{5}$; $\sec\theta = \frac{13}{5}$; $\csc\theta = \frac{13}{12}$; $\cot\theta = \frac{5}{12}$

    b. $\sin\theta = \frac{12}{13}$; $\tan\theta = -\frac{12}{5}$; $\sec\theta = -\frac{13}{5}$; $\csc\theta = \frac{13}{12}$; $\cot\theta = -\frac{5}{12}$

49. a. $\cos\theta = \frac{12}{13}$; $\sin\theta = \frac{5}{13}$; $\sec\theta = \frac{13}{12}$; $\csc\theta = \frac{13}{5}$; $\cot\theta = \frac{12}{5}$

    b. $\cos\theta = -\frac{12}{13}$; $\sin\theta = \frac{5}{13}$; $\sec\theta = -\frac{13}{12}$; $\csc\theta = \frac{13}{5}$; $\cot\theta = -\frac{12}{5}$

51.

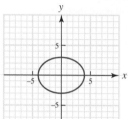

53.

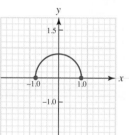

55.

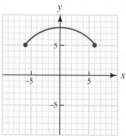

57. If the angle size is increased, then the height is increased but the range is decreased. The time of flight is increased.

59. Range of 2,800 ft in 12.5 sec.

5.3 Trigonometric Functions of Any Angle, page 310

1. a. 1   **b.** 1

    c. $\dfrac{\sqrt{3}}{2}$ **d.** $\dfrac{\sqrt{3}}{2}$

    e. 0   **f.** $\dfrac{\sqrt{3}}{3}$

    g. 0   **h.** $\dfrac{\sqrt{2}}{2}$

    i. 0   **j.** 1

3. a. $\sqrt{2}$ **b.** 1

    c. 0   **d.** undefined

    e. 2   **f.** $\sqrt{3}$

    g. 0   **h.** $\dfrac{2\sqrt{3}}{3}$

    i. $\dfrac{\sqrt{3}}{3}$ **j.** undefined

5. a. $\dfrac{1}{2}$ **b.** 1

    c. $\dfrac{1}{2}$ **d.** $\dfrac{1}{2}$

7. a. $-\dfrac{\sqrt{2}}{2}$ **b.** $\dfrac{1}{2}$

    c. $\dfrac{\sqrt{3}}{2}$ **d.** $-\sqrt{3}$

9. a. $-\dfrac{\sqrt{3}}{2}$ **b.** undefined

    c. $-\dfrac{1}{2}$ **d.** -1

11. $\cos\theta = \frac{3}{5}$; $\sin\theta = \frac{-4}{5}$; $\tan\theta = \frac{-4}{3}$; $\sec\theta = \frac{5}{3}$; $\csc\theta = \frac{-5}{4}$; $\cot\theta = \frac{-3}{4}$

13. $\cos\theta = \frac{-5}{13}$; $\sin\theta = \frac{12}{13}$; $\tan\theta = \frac{-12}{5}$; $\sec\theta = \frac{-13}{5}$; $\csc\theta = \frac{13}{12}$; $\cot\theta = \frac{-5}{12}$

15. $\cos\theta = -\frac{2}{13}\sqrt{13}$; $\sin\theta = -\frac{3}{13}\sqrt{13}$; $\tan\theta = \frac{3}{2}$; $\sec\theta = \frac{-\sqrt{13}}{2}$; $\csc\theta = \frac{-\sqrt{13}}{3}$; $\cot\theta = \frac{2}{3}$

17. F   **19.** F

21. T   **23.** F

25. F   **27.** F

29. $-\frac{1}{2}\sqrt{2}$ **31.** $-\dfrac{\sqrt{3}}{2}$

33. a. 0   **b.** -1

35. a. $\frac{3}{2}$ **b.** 1

37. a. 1   **b.** 1

39. a. 1   **b.** 1

41. a. $\sqrt{3}$ **b.** $\dfrac{2\sqrt{3}}{3}$

43. a. $\dfrac{\sqrt{3}}{2}$ **b.** $\dfrac{\sqrt{3}}{2}$

45. a. $\dfrac{1}{2}$ **b.** $\dfrac{1}{2}$

47. $\dfrac{\sqrt{6}-\sqrt{2}}{4}$; $\dfrac{\sqrt{6}+\sqrt{2}}{4}$

49. a. $y = \frac{1}{3}\sqrt{3}(x-2)+3$ **b.** $y = -\sqrt{3}(x-1)+4$

    c. $y = x-14$ **d.** $y = -x-11$

51. a. $0° < \phi < 90°$ **b.** $90° < \phi < 180°$

    c. $\phi = 0°$ **d.** $\phi = 90°$

55. As $\theta \to 0$, $\sin\theta \to 0$ and $\sin\theta \approx \theta$.

57. It is 9.3 cm from the top.

5.4 Graphs of the Trigonometric Functions, page 328

1. a.

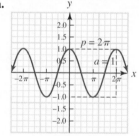

    b.

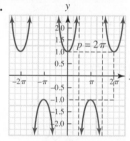

3. a.

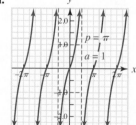

    b.

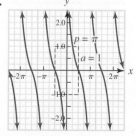

5. a. $a = 6, p = 8$ **b.** $a = 6, p = 2$

7. a. $a = 3, p = \frac{1}{4}, f = 4$

 b. 6 cycles for 1.0, so $a = 0.3, p = \frac{1}{6}, f = 6$

9. a.
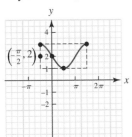

b.
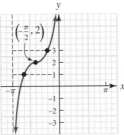

c.
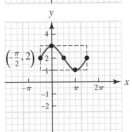

11. a.
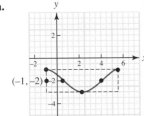

b.
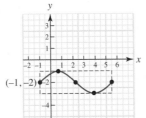

c.
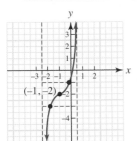

13. a.
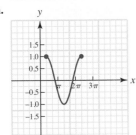

b.
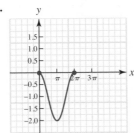

 c. The starting point changes.

15. a.
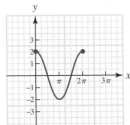

b.
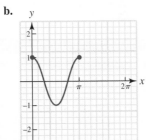

 c. The amplitude and period change.

17. a.
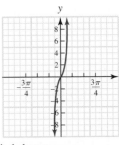

b.
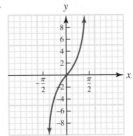

 c. The height of the frame and period change.

19. $a = 3, p = \frac{2\pi}{2} = \pi;\ y - 4 = 3 \sin 2x$

21. $a = 3, p = \pi;\ y = 3 \cos 2\left(x - \frac{\pi}{2}\right)$

23. $a = 30, p = 3;\ y + 10 = 30 \sin \frac{2\pi}{3}(x + 3)$

25.
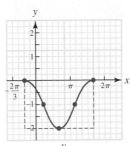

27.
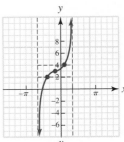

29.
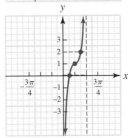

31.
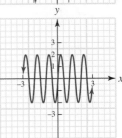

33.
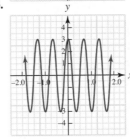

35.
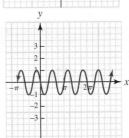

37.
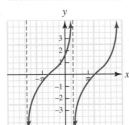

39.
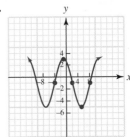

41.
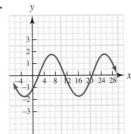

43. $y = 60 \sin \frac{\pi}{15}(x - 7.5)$

45. $y - 29 = 30 \cos \frac{\pi}{5}(x - 5)$ **47.** $y = \frac{1}{3} \sin \frac{4\pi}{3} t$

49. $y = 0.04 \sin 792 \pi x$ **51.** $y = \cos 1.9x$

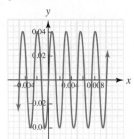

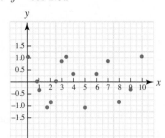

53.

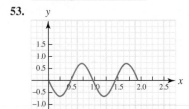

55. Using a calculator, we can duplicate the graphs. **a** is graph **E**; **b** is graph **E**; **c** is graph **D**; **d** is graph **C**; **e** is graph **A**.

59.

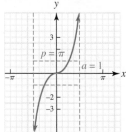

5.5 Inverse Trigonometric Functions, page 341

1. a. $0 \le y \le \pi$ **b.** I, II

3. a. $-\frac{\pi}{2} < y < \frac{\pi}{2}$ **b.** I, IV

5. a. $\sin^{-1} x$; [sin⁻¹] [x] [ENTER] or [x] [inv] [sin]

   b. $\csc^{-1} x$; [sin⁻¹] [x] [x⁻¹] [ENTER] or [x] [1/x] [inv] [sin]

7. a. 0.75   **b.** 0

   c. -1.0 **d.** -0.75

   e. 0   **f.** $\frac{\pi}{2}$

   g. $\frac{3\pi}{8}$ **h.** π

   i.

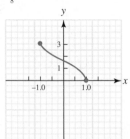

9. a. 1.0   **b.** 0.4

   c. 0.0   **d.** -1.0

   e. $\frac{\pi}{4}$ **f.** 1.1

   g. 0.5   **h.** $-\frac{\pi}{4}$

i.

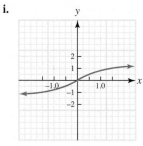

11. a. $\frac{\pi}{4}$ **b.** $\frac{\pi}{4}$

   c. $-\frac{\pi}{2}$ **d.** 0

13. a. $\frac{\pi}{3}$ **b.** $-\frac{\pi}{2}$

   c. $-\frac{\pi}{3}$ **d.** $\frac{3\pi}{4}$

15. a. $-\frac{\pi}{6}$ **b.** 0

   c. $\frac{\pi}{3}$ **d.** $\frac{\pi}{3}$

   a. 0.28   **b.** 1.34

   c. 0.42   **d.** 1.22

19. a. 153°   **b.** 15°

   c. 100°   **d.** $-72°$

21. T   **23.** F

25. T

27. a. $\frac{1}{3}$ **b.** $\frac{\pi}{15}$

29. a. 35°   **b.** 0.4163

31. a. $\dfrac{2\sqrt{2}}{3}$ **b.** $\dfrac{2\sqrt{2}}{3}$

33. $\sin(\csc^{-1} x) = \frac{1}{x}$ **35.** $\sec(\tan^{-1} x) = \sqrt{1 + x^2}$

37. $\csc(\cot^{-1} x) = \pm\sqrt{1 + x^2}$ **39.** $\frac{1}{2}\sqrt{2}$

41. 0   **43.** $\frac{1}{2}\sqrt{2}$

45. 51°   **47.** 1.4

49. 6 ft

51.

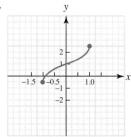

53.   **55.** 0.16

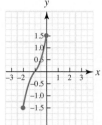

57. ± 0.56

5.6 Right Triangles, page 352

	α	β	γ	a	b	c
1.	30°	60°	90°	80	140	160
3.	77°	13°	90°	390	90	400
5.	62.5°	27.5°	90.0°	66.9	34.8	75.4
7.	32.6°	57.4°	90.0°	70.0	109	130
9.	57.83°	32.17°	90.00°	290.8	182.9	343.6
11.	28°	62°	90°	1,600	3,100	3,500
13.	67.8°	22.2°	90.0°	26.6	10.8	28.7

15. The boat is 240 m from the lighthouse.

17. The height of the building is 23 m.

19. $\cos\theta = \dfrac{\sqrt{a^2-u^2}}{a}$, $\sin\theta = \dfrac{u}{a}$, $\tan\theta = \dfrac{u}{\sqrt{a^2-u^2}}$; $\sec\theta = \dfrac{a}{\sqrt{a^2-u^2}}$,

$\csc\theta = \dfrac{a}{u}$, $\cot\theta = \dfrac{\sqrt{a^2-u^2}}{u}$

21. a. $\cos\alpha = \sin\alpha = \frac{1}{2}\sqrt{2}$; $\tan\alpha = 1$
   b. $\cos\beta = \sin\beta = \frac{1}{2}\sqrt{2}$; $\tan\beta = 1$

23. The car is 340 ft from a point directly below the helicopter.

25. The height of the solar panel should be 8.9 ft.

27. The plant can be no closer than 8 ft 6 in.

29. The height of the rafter should be 11.5 ft.

31. The pitch angle for the rafters is 9°.

33. The distance is about 350 ft.

35. The length of the tower is 222.0 ft.

37. The point N is 995 m from the base of Devil's Tower.

39. The distance across the river is 170 ft.

41. The distance from the sun to Venus is about 6.79×10^7 mi.

43. The height of the building is about 589.8 ft.

45. The radius of the inscribed circle is 633.9 ft.

47. The cost is $228.   **49.** $\tan^2\theta$

51. $\frac{1}{16}\cos^2\theta$

55. The area of the square is 2 in.²

57. The area of the polygon is about 2.94 in.²

59. a. $A(0 \sin\theta)$; $P(\cos\theta, \sin\theta)$ **b.** $\sin\theta$
   c. $\cos\theta$ **d.** $\sin^2\theta$
   e. $\frac{1}{2}\cos\theta \sin^3\theta$

5.7 Oblique Triangles, page 366

1. $\alpha = 54°$ **3.** $\alpha = 22°$

5. $c = 10$

7. a. SSS; law of cosines; 2 sig figs
   b. SAS; law of cosines; 2 sig figs

9. a. AAS; law of sines; 2 sig figs
   b. AAA; no solution because three angles do not yield a unique triangle.

11. a. SSA; law of cosines; 3 sig figs
   b. SSA; law of cosines; 3 sig figs

13. $a = 14.2$; $b = 16.3$; $c = 4.00$; $\alpha = 52.1°$; $\beta = 115.0°$; $\gamma = 12.9°$

15. $a = 10$; $b = 12$; $c = 13$; $\alpha = 48°$; $\beta = 62°$; $\gamma = 70°$

17. $a = 33$; $b = 40$; $c = 37$; $\alpha = 50°$; $\beta = 70°$; $\gamma = 60°$

19. $a = 73.4$; $b = 126$; $c = 115$; $\alpha = 35.0°$; $\beta = 81.0°$; $\gamma = 64.0°$

21. $a = 41.0$; $b = 21.2$; $c = 53.1$; $\alpha = 45.2°$; $\beta = 21.5°$; $\gamma = 113.3°$

23. $a = 98.2$; $a' = 41.6$; $b = 82.5$; $c = 52.2$; $\alpha = 90.8°$; $\alpha' = 25.0°$; $\beta = 57.1°$; $\beta' = 122.9°$; $\gamma = 32.1°$

25. No solution

27. $a = 214$; $b = 320$; $c = 126$; $\alpha = 25.8°$; $\beta = 139.4°$; $\gamma = 14.8°$

29. $a = 36.9$; $b = 20.5$; $c = 42.4$; $\alpha = 61.0°$; $\beta = 29.0°$; $\gamma = 90.0°$

31. 13.6 sq units   **33.** 37 sq units

35. 560 sq units   **37.** 6.4 sq units

39. 54 sq units   **41.** 2,150 sq units and 911 sq units

43. The distance is 249 mi.   **45.** The cable is 42.8 ft.

47. The length of the tower is 178.9 feet.

49. The angle with the 8-ft side is 56°, and with the 10-ft side it is 41°.

51. The perimeter of the square is $20\sqrt{2}$ in. $\approx$ 28 in.

53. It will take 49 min.

55. The pilot must increase the airspeed to 182 mi/h.

57. Mt. Frissell is about 2,382 ft high.

59. $\tan^{-1}\left(\dfrac{\sin\theta}{\frac{W}{w}+\cos\theta}\right)$.

Chapter 5 Self Test, page 369

1. a. $180°$ **b.** $-240°$
   c. $-\frac{\pi}{4}$ **d.** $\frac{7\pi}{6}$
   e. $\frac{\pi}{2}$ **f.** $-\frac{\sqrt{2}}{2}$
   g. 0 **h.** $\frac{\sqrt{3}}{2}$
   i. $-\frac{1}{2}$ **j.** 1
   k. $\frac{\sqrt{2}}{2}$ **l.** -1
   m. $-\frac{1}{2}$ **n.** $-\frac{\sqrt{3}}{2}$
   o. 0 **p.** -1
   q. 0 **r.** $-\sqrt{3}$
   s. $\frac{\sqrt{3}}{3}$ **t.** undefined

2. a. $-\frac{\pi}{3}$ **b.** $\frac{\pi}{3}$
   c. $-\frac{\pi}{4}$ **d.** $\frac{5\pi}{6}$
   e. $2\sqrt{2}$ **f.** $\frac{4}{5}$
   g. 1.5 **h.** does not exist

3. a. $\sec\theta = \dfrac{1}{\cos\theta}$; $\csc\theta = \dfrac{1}{\sin\theta}$; $\cot\theta = \dfrac{1}{\tan\theta}$
   b. $\tan\theta = \dfrac{\sin\theta}{\cos\theta}$; $\cot\theta = \dfrac{\cos\theta}{\sin\theta}$
   c. $\cos^2\theta + \sin^2\theta = 1$; $\tan^2\theta + 1 = \sec^2\theta$; $1 + \cot^2\theta = \csc^2\theta$

4. $\cos\alpha = \frac{\sqrt{7}}{4}$; $\sin\alpha = \frac{3}{4}$; $\tan\alpha = \frac{3}{\sqrt{7}}$; $\sec\alpha = \frac{4}{\sqrt{7}}$; $\csc\alpha = \frac{4}{3}$; $\cot\alpha = \frac{\sqrt{7}}{3}$

5. a.   **b.**

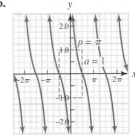

   c.

6. a.   **b.**

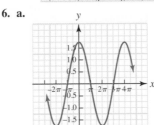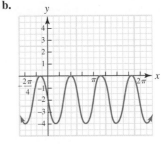

c.

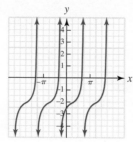

7. a. $\dfrac{\sin\theta}{t} = \dfrac{\sin\phi}{u} = \dfrac{\sin\psi}{s}$

b. $s^2 = t^2 + u^2 - 2tu\cos\psi$; $t^2 = s^2 + u^2 - 2su\cos\theta$;

$u^2 = s^2 + t^2 - 2st\cos\phi$

c. 1,900 cm²

8. Ferndale is about 2,900 mi above the equator.

9. 12:42 P.M. or 4:40 P.M.

10. a. 430 ft   **b.** 48.9°

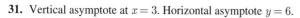
Practice for Calculus, Cumulative Review, page 370

1. −1   **3.** 1

5. $-\dfrac{5}{6}, \dfrac{5}{2}$ **7.** $\left\{\dfrac{1 \pm \sqrt{5}}{2}\right\}$

9. {3}   **11.** $\{2, -1, \tfrac{1}{2}\}$

13. $\dfrac{5}{8}\sqrt[4]{3}$ **15.** $\tfrac{1}{2}(\ln 8.5 - 1)$

17. 8.6   **19.** 6

21. $(\sin\theta + 3)(\sin\theta + 2)$ **23.** $(\cos\theta - 1)(\cos\theta + 1)$

25. $(3\sin\omega + 1)(\sin\omega - 2)$ **27.** $6x + 3h + 2$

29. $\dfrac{-1}{(x+h-2)(x-2)}$

31. Vertical asymptote at $x = 3$. Horizontal asymptote $y = 6$.

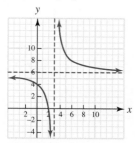

33. Vertical asymptote at $x = 0$. No horizontal asymptote.

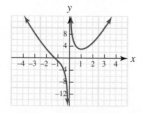

35. Vertical asymptotes at $x = \pm 3$. No horizontal asymptote.

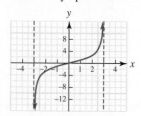

37. Vertical asymptotes at $x = \dfrac{\pi}{2} + n\pi$ for any interger n. No horizontal asymptote.

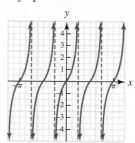

39. Vertical asymptotes at $x = n\pi$ for any integer n. No horizontal asymptote.

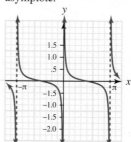

41.

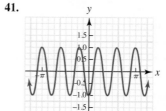

43.

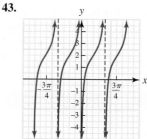

45.

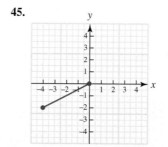

47.

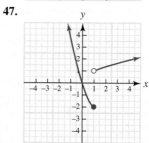

49. $u(x) = 3x^2 + 4$; $g(u) = \sin u$ **51.** $u(x) = \sin x$; $g(u) = e^u$

53. $u(x) = \sin x$; $g(u) = \ln|u|$ **55.** $A = 2w\sqrt{64 - w^2}$

57. 21 mi/h   **59.** 278 ft

CHAPTER 6

6.1 Trigonometric Equations, page 383

1. a. 1.05   **b.** $\dfrac{\pi}{3}$

c. $\dfrac{\pi}{3} + 2\pi k$; $\dfrac{5\pi}{3} + 2\pi k$

3. a. 2.09   **b.** $\dfrac{2\pi}{3}$

c. $\dfrac{2\pi}{3} + 2k\pi$, $\dfrac{4\pi}{3} + 2k\pi$

5. a. -0.79 **b.** $\dfrac{3\pi}{4}$

    c. $\dfrac{3\pi}{4} + k\pi,\ \dfrac{7\pi}{4} + k\pi$

7. a. 1.05 **b.** $\dfrac{\pi}{3},\ \dfrac{5\pi}{3}$

    c. $\dfrac{\pi}{3} + 2k\pi;\ \dfrac{5\pi}{3} + 2k\pi$

9. a. -0.79 **b.** $-\dfrac{\pi}{4}$

    c. $\dfrac{5\pi}{4} + 2k\pi,\ \dfrac{7\pi}{4} + 2k\pi$

11. a. 0.52 **b.** $\dfrac{\pi}{6},\ \dfrac{5\pi}{6},\ \dfrac{7\pi}{6},\ \dfrac{11\pi}{6}$

    c. $\dfrac{\pi}{6} + 2k\pi;\ \dfrac{5\pi}{6} + 2k\pi;\ \dfrac{7\pi}{6} + 2k\pi;\ \dfrac{11\pi}{6} + 2k\pi$

13. a. 1.05 **b.** $\dfrac{\pi}{3},\ \dfrac{2\pi}{3},\ \dfrac{4\pi}{3},\ \dfrac{5\pi}{3}$

    c. $\dfrac{\pi}{3} + 2k\pi;\ \dfrac{2\pi}{3} + 2k\pi;\ \dfrac{4\pi}{3} + 2k\pi;\ \dfrac{5\pi}{3} + 2k\pi$

15. a. 0 **b.** $0,\ \pi$

    c. $k\pi$

17. a. 1.57 **b.** $\dfrac{\pi}{2},\ \dfrac{3\pi}{2}$

    c. $\dfrac{\pi}{2} + k\pi$

19. a. $0.52,\ 1.05$ **b.** $\dfrac{\pi}{6},\ \dfrac{5\pi}{6},\ \dfrac{\pi}{3},\ \dfrac{5\pi}{3}$

    c. $\dfrac{\pi}{6} + 2k\pi;\ \dfrac{5\pi}{6} + 2k\pi;\ \dfrac{\pi}{3} + 2k\pi;\ \dfrac{5\pi}{3} + 2k\pi$

21. $0,\ \pi,\ \dfrac{\pi}{4},\ \dfrac{5\pi}{4}$ **23.** $\dfrac{\pi}{4},\ \dfrac{3\pi}{4},\ \dfrac{5\pi}{4},\ \dfrac{7\pi}{4}$

25. $0,\ \pi,\ \dfrac{\pi}{3},\ \dfrac{5\pi}{3}$ **27.** $\dfrac{2\pi}{3} + 2k\pi,\ \dfrac{5\pi}{3} + 2k\pi$

29. $\dfrac{5\pi}{4} + 2k\pi,\ \dfrac{7\pi}{4} + 2k\pi$

31. $1.2741 + 6.2832k,\ 5.0091 + 6.2832k$

33. 2.24 **35.** $0.36,\ 1.21$

37. 0.90 **39.** $0.00,\ 1.57$

41. 0.02 **43.** 0.02

45. 0.41 **47.** $0.00,\ 2.09$

49. $0.00,\ -0.17$ **51.** 0.0002

53. 0.66 sec   **55.** $0,\ \dfrac{3\pi}{4},\ \pi,\ \dfrac{7\pi}{4}$

57. $0,\ \dfrac{\pi}{2}$ **59.** 0.7

6.2   Proving Identities, page 392

3. $(1)\ \sec\theta = \dfrac{1}{\cos\theta};\ (2)\ \csc\theta = \dfrac{1}{\sin\theta};\ (3)\ \cot\theta = \dfrac{1}{\tan\theta};$

    $(4)\ \tan\theta = \dfrac{\sin\theta}{\cos\theta};\ (5)\ \cot\theta = \dfrac{\cos\theta}{\sin\theta};\ (6)\ \cos^2\theta + \sin^2\theta = 1;$

    $(7)\ 1 + \tan^2\theta = \sec^2\theta;\ (8)\ \cot^2\theta + 1 = \csc^2\theta$

5. $\cos\theta = \dfrac{u}{3};\ \sin\theta = \dfrac{\sqrt{9-u^2}}{3};\ \tan\theta = \dfrac{\sqrt{9-u^2}}{u};\ \sec\theta = \dfrac{3}{u};$

    $\csc\theta = \dfrac{3\sqrt{9-u^2}}{9-u^2};\ \cot\theta = \dfrac{u\sqrt{9-u^2}}{9-u^2}$

7. $\cos\theta = \dfrac{\sqrt{9-3u^2}}{3};\ \sin\theta = \dfrac{u}{\sqrt{3}};\ \tan\theta = \dfrac{u\sqrt{3-u^2}}{3-u^2};$

    $\sec\theta = \dfrac{\sqrt{9-3u^2}}{3-u^2};\ \csc\theta = \dfrac{\sqrt{3}}{u};\ \cot\theta = \dfrac{\sqrt{3-u^2}}{u}$

Proofs of identities may vary; hints (or first steps) are given here.

9. Change to sines and cosines.   **11.** Change to sines and cosines.

13. Common factor.   **15.** Change to sines and cosines.

17. Use fundamental identities.   **19.** Common factor.

21. Change to sines and cosines.   **23.** Change to sines and cosines.

25. Factor difference of squares.   **27.** Change to sines and cosines.

Problems 29–34, counterexamples vary.

35. The graphs are not the same.   **37.** The graphs are the same.

39. F   **41.** Multiply by $\dfrac{1 + \tan\alpha}{1 + \tan\alpha}$.

43. Factor difference of cubes.   **45.** Factor difference of squares.

47. Multiply by $\dfrac{\sec\theta - \tan\theta}{\sec\theta - \tan\theta}$.   **49.** Factor difference of cubes.

51. Simplify and use fundamental identities.

53. Factor difference of squares on left and then simplify the right.

55. Common factor.   **57.** $12.1°$

59. $\dfrac{\pi}{6},\ \dfrac{\pi}{3}$

6.3   Addition Laws, page 400

1. F   **3.** T

5. F   **7.** F

9. a. $(\sin\theta + \cos\theta)\dfrac{\sqrt{2}}{2}$ **b.** $(\cos\theta + \sin\theta)\dfrac{\sqrt{2}}{2}$

    c. $\dfrac{\tan\theta - 1}{1 + \tan\theta}$

11. a. $\cos 75°$ **b.** $\cot 28°$

    c. $-\sin\dfrac{\pi}{3}$

13. a. $-\sin 23°$ **b.** $\cos 57°$

    c. $-\tan 29°$

15. 0.9135 **17.** 0.9511

19. 1.1918

21. $\cos 15° = \dfrac{\sqrt{6} + \sqrt{2}}{4};\ \sin 15° = \dfrac{\sqrt{6} - \sqrt{2}}{4};\ \tan 15° = 2 - \sqrt{3}$

23. $\cos(-15°) = \dfrac{\sqrt{6} + \sqrt{2}}{4};\ \sin(-15°) = \dfrac{\sqrt{2} - \sqrt{6}}{4};\ \tan(-15°) = \sqrt{3} - 2$

25. $\cos 75° = \dfrac{\sqrt{6} - \sqrt{2}}{4};\ \sin 75° = \dfrac{\sqrt{6} + \sqrt{2}}{4};\ \tan 75° = 2 + \sqrt{3}$

27.

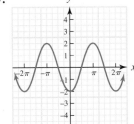

29.

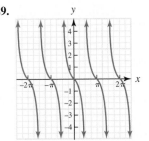

31.

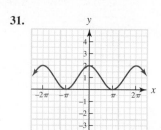

33. a. even   **b.** even
  c. odd   **d.** even
35. a. $\cos(\alpha - \beta)$ **b.** $-\sin(\alpha - \beta)$
  c. $-\tan(\alpha - \beta)$
37. Use the addition law for tangent.
39. Common denominator first.
41. Use the addition laws for cosine and sine.
43. $\dfrac{7}{25}$ **45.** $\dfrac{\sqrt{15}+2}{6}$
47. $\cos^2\theta - \sin^2\theta$
49. Let $\alpha + \beta = 180° - \gamma$ and then use the addition law of tangents; the statement is true.
51. First use cofunction of complementary angles and then simplify; the statement is true.
53. $\dfrac{\sqrt{2+\sqrt{3}}}{2}$ **55.** Use addition law for cosine.
57. 1
59. Let $\alpha = \sin^{-1}x$ and $\beta = \cos^{-1}x$ and use the addition law for cosine.

6.4 Miscellaneous Identities, page 411

1. F   **3.** F
5. F   **7.** F
9. a. Quadrant I   **b.** Quadrant I or II
11. a. Quadrant II   **b.** Quadrant II
13. a. Quadrant I or IV   **b.** Quadrant III or IV
15. a. $\dfrac{\sqrt{2}}{2}$ **b.** $\sqrt{2}-1$
17. a. 1   **b.** $\dfrac{1}{2}$
19. a. $\cos 11° - \cos 59°$ **b.** $\cos 28° + \cos 64°$
21. a. $\dfrac{1}{2}\cos 2\theta - \dfrac{1}{2}\cos 4\theta$ **b.** $\dfrac{1}{2}\sin 7\theta - \dfrac{1}{2}\sin 3\theta$
23. a. $2\sin 22.5° \cos 7.5°$ **b.** $2\sin 62.5° \cos 37.5°$
25. $3\sin\theta - 4\sin^3\theta$
27. $\cos\theta = \dfrac{\sqrt{5}}{5}$; $\sin\theta = \dfrac{2\sqrt{5}}{5}$; $\tan\theta = 2$
29. $\cos\theta = \dfrac{\sqrt{10}}{10}$; $\sin\theta = \dfrac{3\sqrt{10}}{10}$; $\tan\theta = 3$
31. $\cos\theta = \dfrac{1}{2}$; $\sin\theta = \dfrac{\sqrt{3}}{2}$; $\tan\theta = \sqrt{3}$
33. $\cos 2\theta = \dfrac{7}{25}$; $\sin 2\theta = \dfrac{24}{25}$; $\tan 2\theta = \dfrac{24}{7}$; $\cos\dfrac{1}{2}\theta = \dfrac{3}{10}\sqrt{10}$;
  $\sin\dfrac{1}{2}\theta = \dfrac{1}{10}\sqrt{10}$; $\tan\dfrac{1}{2}\theta = \dfrac{1}{3}$

35. $\cos 2\theta = \dfrac{119}{169}$; $\sin 2\theta = \dfrac{-120}{169}$; $\tan 2\theta = \dfrac{-120}{119}$; $\cos\dfrac{1}{2}\theta = \dfrac{5\sqrt{26}}{26}$;
  $\sin\dfrac{1}{2}\theta = \dfrac{\sqrt{26}}{26}$; $\tan\dfrac{1}{2}\theta = \dfrac{1}{5}$
37. $\cos 2\theta = \dfrac{-31}{81}$; $\sin 2\theta = \dfrac{20\sqrt{14}}{81}$; $\tan 2\theta = \dfrac{-20\sqrt{14}}{31}$; $\cos\dfrac{1}{2}\theta = \dfrac{\sqrt{7}}{3}$;
  $\sin\dfrac{1}{2}\theta = \dfrac{\sqrt{2}}{3}$; $\tan\dfrac{1}{2}\theta = \dfrac{\sqrt{14}}{7}$

39.   **41.**

43. $\dfrac{\pi}{6}, \dfrac{5\pi}{6}, \dfrac{3\pi}{2}$ **45.** $0, \pi, \dfrac{\pi}{3}, \dfrac{5\pi}{3}$
47. Use a double-angle identity.   **49.** Use a double-angle identity.
51. Write the tangent side as sines and cosines and simplify.
53. yes; $a = 3\sqrt{2}$; $p = 2\pi$ **55.** $2\sin(1{,}906\pi x)(\cos 512\pi x)$
57. a. $M \approx 3.9$ **b.** $2\sqrt{2+\sqrt{3}}$
  c. $\theta = 2\sin^{-1}M^{-1}$
59. 2,048 ft

6.5 De Moivre's Theorem, page 420

1. F   **3.** T
5. F
7. a. $\sqrt{10}$ **b.** $5\sqrt{2}$
  c. $\sqrt{13}$
9. a. $\sqrt{29}$ **b.** $\sqrt{41}$
  c. $\sqrt{13}$
11. a. $\sqrt{2}\,\text{cis}\,45°$ **b.** $\sqrt{2}\,\text{cis}\,315°$
  c. $2\,\text{cis}\,330°$
13. a. $\text{cis}\,180°$ **b.** $2\,\text{cis}\,0°$
  c. $3\,\text{cis}\,180°$
15. a. $5\,\text{cis}\,270°$ **b.** $6\,\text{cis}\,270°$
  c. $7\,\text{cis}\,270°$
17. a. $6\,\text{cis}\,345°$ **b.** $2\,\text{cis}\,320°$
19. a. $\sqrt{2}+\sqrt{2}i$ **b.** $-\dfrac{\sqrt{3}}{2}+\dfrac{1}{2}i$
21. a. $-2.3444 - 5.5230i$ **b.** $-8.8633 - 1.5628i$
23. a. $6\,\text{cis}\,210°$ **b.** $15\,\text{cis}\,140°$
25. a. $8\,\text{cis}\,210°$ **b.** $3\,\text{cis}\,160°$
27. a. $2\,\text{cis}\,15°$ **b.** $3\,\text{cis}\,130°$
29. a. $-8i$ **b.** $-128 + 128i\sqrt{3}$
31. $3\,\text{cis}\,22°, 3\,\text{cis}\,112°, 3\,\text{cis}\,202°, 3\,\text{cis}\,292°$
33. $3\,\text{cis}\,0°, 3\,\text{cis}\,120°, 3\,\text{cis}\,240°$
35. $\sqrt[8]{2}\,\text{cis}\,11.25°, \ \sqrt[8]{2}\,\text{cis}\,101.25°, \ \sqrt[8]{2}\,\text{cis}\,191.25°, \ \sqrt[8]{2}\,\text{cis}\,281.25°$
37. $2\,\text{cis}\,40°, 2\,\text{cis}\,112°, 2\,\text{cis}\,184°, 2\,\text{cis}\,256°, 2\,\text{cis}\,328°$
39. $-2, 1 \pm \sqrt{3}i$
41. $-0.6840 + 1.8794i, -1.2856 - 1.5321i, 1.9696 - 0.3473i$

43. $1.9696 + 0.3473i, -0.3473 + 1.9696i, -1.9696 - 0.3473i,$
$0.3473 - 1.9696i$

45. $1, 0.3090 + 0.9511i, -0.8090 + 0.5878i, -0.8090 - 0.5878i,$
$0.3090 - 0.9511i$

47. $\pm 1, \dfrac{1}{2} \pm \dfrac{\sqrt{3}}{2}i, -\dfrac{1}{2} \pm \dfrac{\sqrt{3}}{2}i$

49. $\pm 2, \pm 2i$

51. $-1, \dfrac{1}{2} \pm \dfrac{\sqrt{3}}{2}i, -\dfrac{1}{2} \pm \dfrac{\sqrt{3}}{2}i$

53. $8, 2.4721 \pm 7.6085, -6.4721 \pm 4.7023$

55. a. "We are number 1."

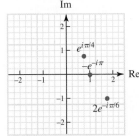

57. $0.95 + 1.07i$

Chapter 6 Self Test, page 421

1. a. $\sqrt{3}$ **b.** $-\dfrac{1}{2}$

2. a. $-\dfrac{24}{25}$ **b.** 3

3. a. $\dfrac{\sqrt{6} + \sqrt{2}}{4}$ **b.** $2\sin\dfrac{h}{2}\cos\left(x + \dfrac{h}{2}\right)$

4. a.

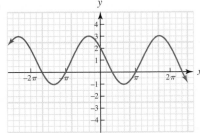

b.

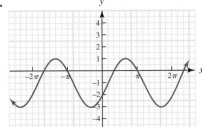

5. a. $\dfrac{\pi}{12}, \dfrac{7\pi}{12}, \dfrac{13\pi}{12}, \dfrac{19\pi}{12}$ **b.** π

6. a. $0.6957, 2.0200$

    b. $0, \dfrac{\pi}{2}, \pi, \dfrac{3\pi}{2}, 0.6245, 2.1953, 3.7661, 5.3369$

7. a. Use a Pythagorean identity. **b.** Use a Pythagorean identity.

8. a. Factor the numerator.
    b. Write as sines and cosines and simplify.

9. a. Use the sum-to-product identities.
    b. Let $P_\alpha(\cos\alpha, \sin\alpha)$ and $P_\beta(\cos\beta, \sin\beta)$ be two points on a unit circle. Calculate the distance between these points two ways: by finding the length of a chord and by using the distance formula.

10. a. $-2.5556 + 0.6848i, 2.5556 - 0.6848i$
    b. $\operatorname{cis}30°, \operatorname{cis}150°, \operatorname{cis}270°$

Practice for Calculus, Cumulative Review, page 422

1. a. $\dfrac{1}{2}$ **b.** $\dfrac{\sqrt{3}}{2}$

    c. $-\dfrac{1}{2}$ **d.** $-\dfrac{\sqrt{3}}{3}$

    e. $-\dfrac{\sqrt{3}}{2}$

3. $-\dfrac{\sqrt{2}}{2}$ **5.** $-\dfrac{3}{2}(x+1)^{-1/2}(x-2)^{3/2}$

7. $-2(2x-1)^{-1/2}(x-1)^{-3/2}$

9. a. $(3x-1)(x+1)$ **b.** $(3\tan x - 1)(\tan x + 1)$

11. a. $(x-2)(x+1)$ **b.** $(\sin x - 2)(\sin x + 1)$

13. $\dfrac{\pi}{2}$ **15.** $\dfrac{\pi}{2}$

17. True for all values of θ.   **19.** all $x; x \neq 0, \pi$

21. $0, \dfrac{\pi}{2}, \dfrac{3\pi}{2}, 2\pi$ **23.** $\dfrac{1}{6\cos\theta}$

25. $\cot^4 \theta$ **27.** $10\sec\theta$

29. $\sin^2 \theta$ **31.** $8\tan\theta$

33. $\dfrac{1}{\sec\theta\tan\theta}$ **35.** $y = Ax^4 + Bx^2$

37. $y = A\sec Bx$ **39.** $y = A(x-2)(x-3)(x+4) + B$

41. $y = A\tan Bx$ **43.** $y = A\cot Bx$

45. $y = Ax^{1/3} + B$ **47.** $y = Ax^4 + B$

49. $y = Ae^{Bx}$ **51.** $y = A\log B$

53. $y = A\sin B(x-2)$ **55.** $y = Ax^2 + B$

57. $y = A\cos Bx - 4$ **59.** $y = A\cos Bx + 4$

CHAPTER 7

7.1 Parabolas, page 438

1.

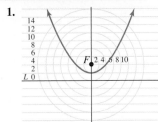

3.

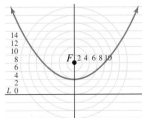

5.

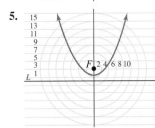

7. T

9. F

11. F

13. T

15.

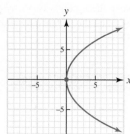

17.

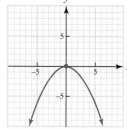

19.

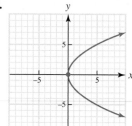

21.

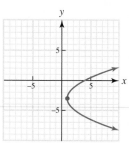

23.

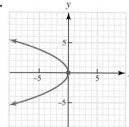

25.

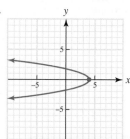

27.

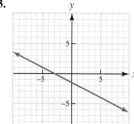

29.

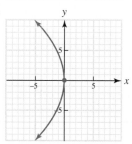

31.

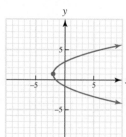

33.

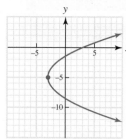

35.

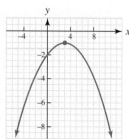

37.

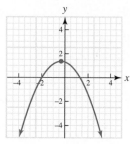

39. $y^2 = 10\left(x - \dfrac{5}{2}\right)$

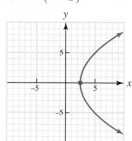

41. $(y-2)^2 = -16(x+1)$

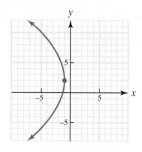
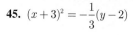

43. $(x+2)^2 = 24(y+3)$

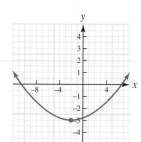

45. $(x+3)^2 = -\dfrac{1}{3}(y-2)$

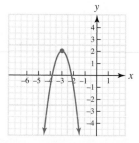

47. $x = 2$

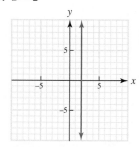

49. The domain is $[0,\ 200]$, and the equation is $(x-100)^2 = -200(y-50)$.

51. The focus is 2 cm from the vertex on the axis of the parabola.

53. The storage yard is 9 ft by 18 ft, with area 162 ft².

55. $x^2 = -6(y+3)$

7.2   Ellipses, page 450

1.

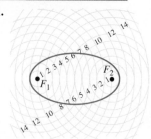

3. F

5. T

7.

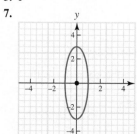

9.

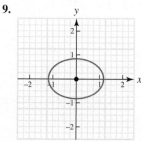

11.

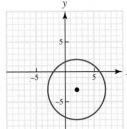

13.

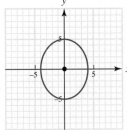

35. $\dfrac{(x-4)^2}{7} + \dfrac{(y+1)^2}{16} = 1$

37. $(x+1)^2 + (y-1)^2 = 10$

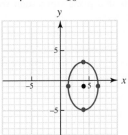

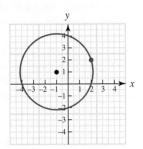

15.

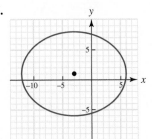

17.

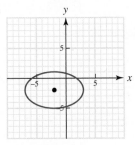

39.

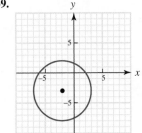

41.

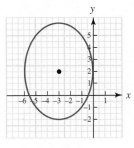

19.

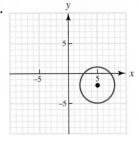

21. $(x, y) = (2\cos\theta, \ 3\sin\theta)$

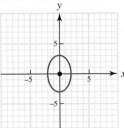

43.

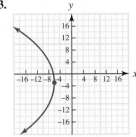

45.

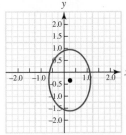

23. $(x, y) = \left(\sqrt{2}\cos\theta, \ \sqrt{3}\sin\theta\right)$

25. $(x, y) = (-4 + 7\cos\theta, \ 2 + 7\sin\theta)$

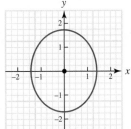

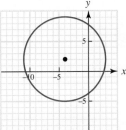

47.

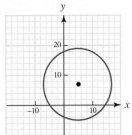

49.

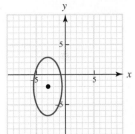

27.

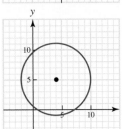

29.

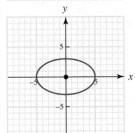

51. aphelion: 9.45×10^7 mi; perihelion: 9.15×10^7 mi

53. The length of the major axis ($2a$) is 7×10^7 mi.

55. $y = \dfrac{3}{4}\sqrt{400 - x^2}$

7.3 Hyperbolas, page 461

1.

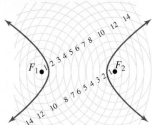

31.

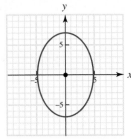

33.

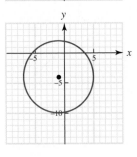

3. F   **5.** F

7. a. A circle is the set of all points in a plane a fixed distance from a fixed point.
  b. A parabola is the set of all points in a plane equidistant from a given point and a given line.
  c. An ellipse is the set of all points in a plane such that the sum of the distances from two given points is a constant.
  d. A hyperbola is the set of all points in a plane such that the difference of the distances from two given points is a constant.

9. a. A and C have opposite signs.
  b. $A = C \neq 0$
  c. $A = 0$ and $C \neq 0$ or $A \neq 0$ and $C = 0$
  d. A and C have the same sign

11.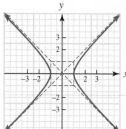

13.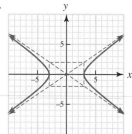

15.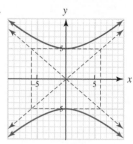

17.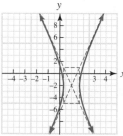

19. $x = \dfrac{3}{\cos\theta},\ y = 3\tan\theta$

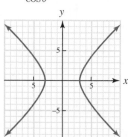

21. $x = \dfrac{2}{\cos\theta},\ y = \sqrt{3}\tan\theta$

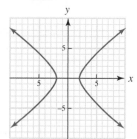

23. $x = \dfrac{\sqrt{5}}{\cos\theta} - 4,\ y = \sqrt{5}\tan\theta - 2$ **25.** line

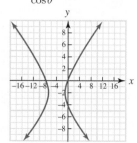   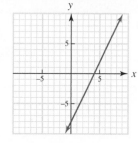

27. parabola

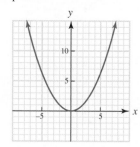

29. hyperbola

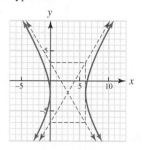

31. degenerate circle

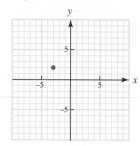

33. circle

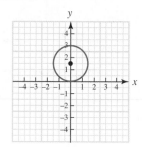

35. ellipse

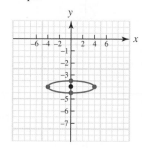

37. $\dfrac{y^2}{25} - \dfrac{x^2}{24} = 1$

39. $\dfrac{x^2}{9} - \dfrac{(y+3)^2}{7} = 1$

41. hyperbola

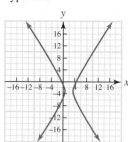

43. hyperbola

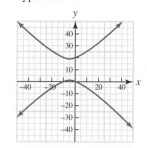

45. hyperbola

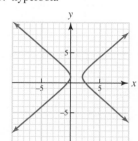

47. hyperbola

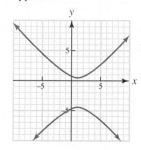

49. hyperbola

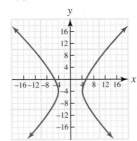

51. ellipse

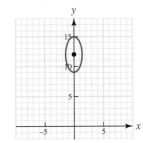

53. parabola

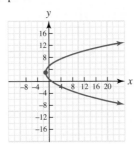

55. circle

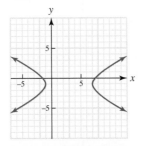

57. hyperbola

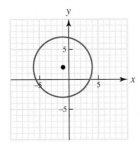

45. $\dfrac{x'^2}{16} - \dfrac{y'^2}{16} = 1$

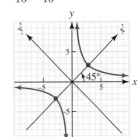

47. $\dfrac{y'^2}{8} - \dfrac{x'^2}{8} = 1$

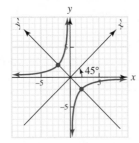

49. $\dfrac{x'^2}{9} + \dfrac{y'^2}{4} = 1$

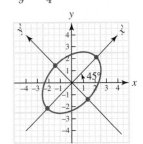

51. $\dfrac{y'^2}{4} - \dfrac{x'^2}{16} = 1$

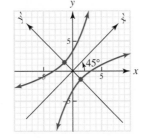

53. $(x'+1)^2 = 4\left(y' + \dfrac{7}{10}\right)$

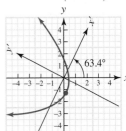

55. $\dfrac{x'^2}{36} - \dfrac{y'^2}{4} = 1$

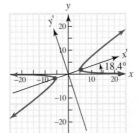

57. $\dfrac{(x'-2)^2}{4} + \dfrac{(y'-1)^2}{1} = 1$

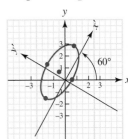

7.4 Rotations, page 467

5. $x = \frac{1}{2}\left(\sqrt{3}x' + y'\right),\; y = -\frac{1}{2}\left(x' - \sqrt{3}y'\right)$

7. $x = 0.97x' - 0.26y',\; y = 0.26x' + 0.97y'$

9. $x = 0.12x' - 0.99y',\; y = 0.99x' + 0.12y'$

11. hyperbola

13. ellipse

15. hyperbola

17. parabola

19. parabola

21. hyperbola

23. ellipse

25. ellipse

27. ellipse

29. parabola

31. $45°;\; x = \frac{1}{\sqrt{2}}(x' - y'),\; y = \frac{1}{\sqrt{2}}(x' + y')$

33. $45°;\; x = \frac{1}{\sqrt{2}}(x' - y'),\; y = \frac{1}{\sqrt{2}}(x' + y')$

35. $45°;\; x = \frac{1}{\sqrt{2}}(x' - y'),\; y = \frac{1}{\sqrt{2}}(x' + y')$

37. $45°;\; x = \frac{1}{\sqrt{2}}(x' - y'),\; y = \frac{1}{\sqrt{2}}(x' + y')$

39. $63.4°;\; x = \frac{1}{\sqrt{5}}(x' - 2y'),\; y = \frac{1}{\sqrt{5}}(2x' + y')$

41. $18.4°;\; x = \frac{1}{\sqrt{10}}(3x' - y'),\; y = \frac{1}{\sqrt{10}}(x' + 3y')$

43. $60°;\; x = \frac{1}{2}\left(x' - \sqrt{3}y'\right),\; y = \frac{1}{2}\left(\sqrt{3}x' + y'\right)$

7.5 Polar Coordinates, page 474

3. The primary representations are $(r,\,\theta)$ and $(-r,\,\theta + \pi)$, provided the second component in each case is between $0°$ and $360°$ or 0 and 2π. If the second component is not, then add or subtract a multiple of $360°$ or 2π to make it so.

5. a. $\left(4,\,\frac{\pi}{4}\right),\;\left(-4,\,\frac{5\pi}{4}\right);\;\left(2\sqrt{2},\,2\sqrt{2}\right)$

   b. $\left(6,\,\frac{\pi}{3}\right),\;\left(-6,\,\frac{4\pi}{3}\right);\;\left(3,\,3\sqrt{3}\right)$

   c. $\left(5,\,\frac{2\pi}{3}\right),\;\left(-5,\,\frac{5\pi}{3}\right);\;\left(-\frac{5}{2},\,\frac{5\sqrt{3}}{2}\right)$

7. a. $(1, \pi)$, $(-1, 0)$; $(-1, 0)$ **b.** $(1, 2)$, $(-1, 2+\pi)$; $(\cos 2, \sin 2)$
c. $\left(-2, \frac{\pi}{2}\right)$, $\left(2, \frac{3\pi}{2}\right)$; $(0, -2)$

9. a. $\left(5\sqrt{2}, \frac{\pi}{4}\right)$, $\left(-5\sqrt{2}, \frac{5\pi}{4}\right)$ **b.** $\left(2, \frac{2\pi}{3}\right)$, $\left(-2, \frac{5\pi}{3}\right)$
c. $\left(4, \frac{5\pi}{3}\right)$, $\left(-4, \frac{2\pi}{3}\right)$

11. a. $(7.62, 1.17)$, $(-7.62, 4.31)$ **b.** $(7.62, 1.98)$, $(-7.62, 5.12)$
c. $\left(6, \frac{5\pi}{3}\right)$, $\left(-6, \frac{2\pi}{3}\right)$

13. a. United States **b.** India
c. Greenland **d.** Canada
15. $x^2 + y^2 - 4y = 0$ **17.** $x = 1$
19. $x^2 + 2y^2 = 2$ **21.** $x^2 + y^2 = \sqrt{x^2 + y^2} - y$
23. **25.**

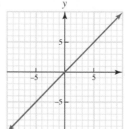

27. **29.**

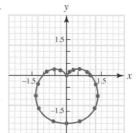 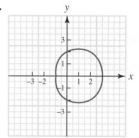

31. yes **33.** no
35. yes **37.** yes
39. yes **41.** no
43. no **45.** no

Answers to Problems 47–56 vary.

57.

59. a. 4.1751
b. $d = \sqrt{r_1^2 + r_2^2 - 2r_1 r_2 \cos(\theta_2 - \theta_1)}$
c. $a^2 = R^2 + r^2 - 2R r \cos(\theta - \alpha)$

7.6 Graphing in Polar Coordinates, page 484

5. a. four-leaved rose **b.** lemniscate
c. circle (r is a constant) **d.** sixteen-leaved rose
e. none (it is a spiral) **f.** lemniscate
g. three-leaved rose **h.** cardioid

7.

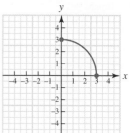

9.

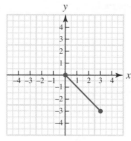

11.

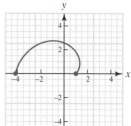

13.

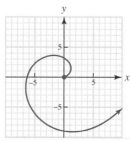

15.

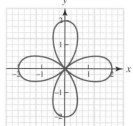

17.

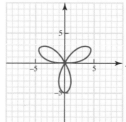

19.

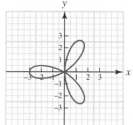

21.

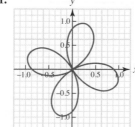

23.

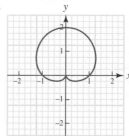

29.

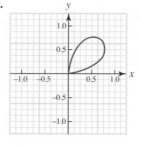

27.

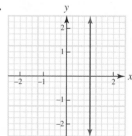

29.

31.

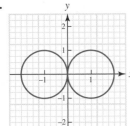

33.

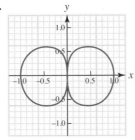

59. b.

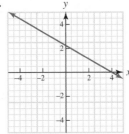

35.

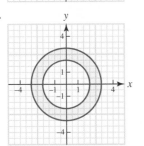

37.

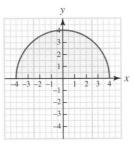

7.7 Conic Sections in Polar Form, page 491

3. parabola   **5.** hyperbola

7. ellipse   **9.** parabola

11. hyperbola   **13.** parabola

15. parabola   **17.** ellipse

19. parabola   **21.** $\varepsilon = \dfrac{r}{p + r\cos\theta}$

23. $\theta = \cos^{-1}\left(\dfrac{r - \varepsilon p}{\varepsilon r}\right)$

39.

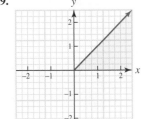

41.

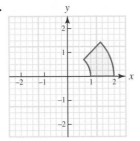

25.

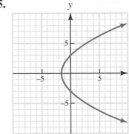

27.

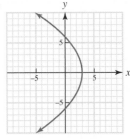

43.

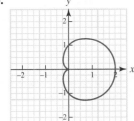

45.

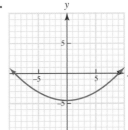

29.

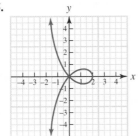

31.

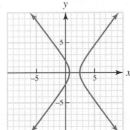

47.

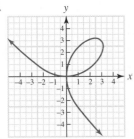

49.

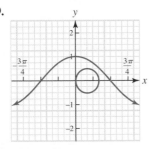

33.

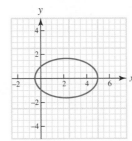

35.

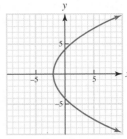

51.

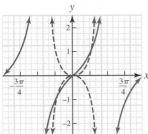

53.

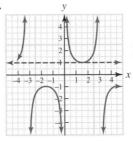

37.

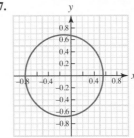

39.

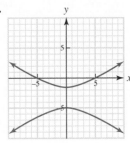

41.
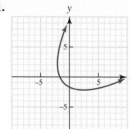

43.
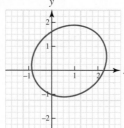

c.
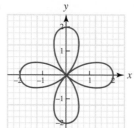

d.
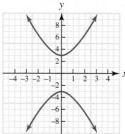

45.
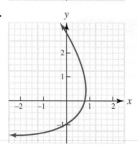

4. a.
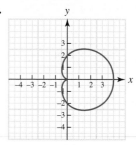

b.
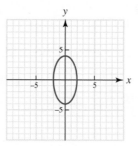

47. $r = 4$

49. $r = \dfrac{8}{1 + \cos\theta}$

51. $r = \dfrac{4}{1 - \sin\theta}$

53. $r = \dfrac{2}{1 - \cos\theta}$

c.
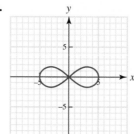

d.
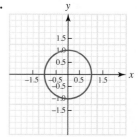

55. $r = \dfrac{1}{8 - 8\sin\theta}$

57. $3\left(x - \dfrac{p}{3}\right)^2 + 4y^2 = \dfrac{4p^2}{3}$

59.
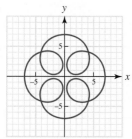

5. a.
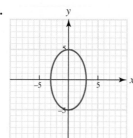

b. $x = 4t^2 + 4t + 1,\ y = t$
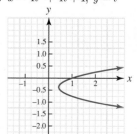

Chapter 7 Self Test, page 492

1. a. $(x - h)^2 = -4c(y - k)$ **b.** $\dfrac{(y - k)^2}{a^2} + \dfrac{(x - h)^2}{b^2} = 1$

c. $\dfrac{(x - h)^2}{a^2} - \dfrac{(y - k)^2}{b^2} = 1$ **d.** $r = a(1 - \cos\theta)$

e. $r^2 = a^2 \cos 2\left(\theta - \dfrac{\pi}{4}\right)$ **f.** $r = a\cos 2\theta$

2. a. $Ax + By + C = 0$ **b.** $(x - h)^2 = \pm 4c(y - k)$

c. $\dfrac{(x - h)^2}{a^2} + \dfrac{(y - k)^2}{b^2} = 1; \dfrac{(y - k)^2}{a^2} + \dfrac{(x - h)^2}{b^2} = 1$

d. $(x - h)^2 + (y - k)^2 = r^2$ **e.** $(y - k)^2 = \pm 4c(x - h)$

f. $\dfrac{(x - h)^2}{a^2} - \dfrac{(y - k)^2}{b^2} = 1; \dfrac{(y - k)^2}{a^2} - \dfrac{(x - h)^2}{b^2} = 1$

6. a. $x = \dfrac{4}{\cos\theta}$ and $y = \dfrac{2\sin\theta}{\cos\theta}$
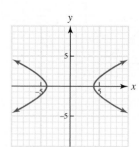

3. a.
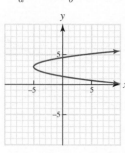

b.
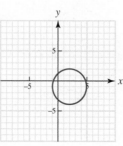

b. $x = \sqrt{5}\cos\theta - 2$
$y = \sqrt{10}\sin\theta + 3$
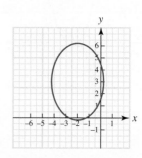

7. $\dfrac{(x+5)^2}{36} + \dfrac{(y-4)^2}{32} = 1$ **8.** $\dfrac{(x+5)^2}{1} = \dfrac{(y-4)^2}{3} = 1$

9. $(y-4)^2 = -10\left(x+\dfrac{1}{2}\right)$ **10.** $(x+3)^2 + (y-4)^2 = 49$

Practice for Calculus, Cumulative Review, page 493

1. a. E   **b.** G
3. a. U   **b.** V
5. a. I   **b.** J
7. a. Q   **b.** B
9. a. F   **b.** T

11. a. $\dfrac{\sqrt{3}}{2}$ **b.** $\dfrac{1}{2}$

   c. $\sqrt{3}$ **d.** $\dfrac{1}{\sqrt{3}}$

13. a. not defined   **b.** 0

   c. not defined   **d.** $-\dfrac{2}{\sqrt{3}}$

15. a. 2   **b.** $\dfrac{5}{\sqrt{5(x+h)}+\sqrt{5x}}$

17. a. $\dfrac{e^x(e^h-1)}{h}$ **b.** $\sin x\left(\dfrac{\cos h-1}{h}\right) + \cos x\left(\dfrac{\sin h}{h}\right)$

19. $2(x-1)(8x^3 - 12x^2 + 12x - 5)$
21. $2xe^{3x^2+5}(3\sin x^2 + \cos x^2)$
23. a. -1.2 **b.** -0.6
25. a. -216.6 **b.** 6.4
27. $\{3.0\}$

29.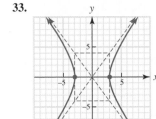

31.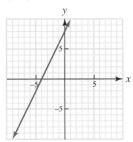

33.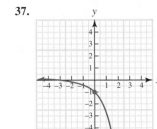

35.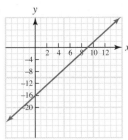

37.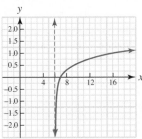

39.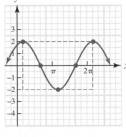

41.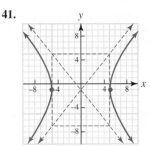

43.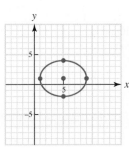

45. $A\left(2, \dfrac{3\sqrt{3}}{2}\right)$, $B\left(-2, \dfrac{3\sqrt{3}}{2}\right)$, $C\left(-2, -\dfrac{3\sqrt{3}}{2}\right)$

47. $P = 30$ units; $A = 30$ square units
49. $(y-3)^2 = 20(x-6)$ **51.** $(x+1)^2 + (y+2)^2 = 64$

53. $\dfrac{x^2}{9} - \dfrac{y^2}{16} = 1$ **55.** $A = -\dfrac{1}{4}$, $B = \dfrac{3}{4}$

57. a. $x \neq 3$ **b.** $[0, 100]$
   c. 120   **d.** 300
   e. 60%
59. 3 hours 15 minutes

CHAPTER 8

8.1   Sequences, page 508

1. F   **3.** F
5. a. arithmetic   **b.** $d = 2$
   c. 10
7. a. Fibonacci   **b.** $s_1 = 2$, $s_2 = 4$
   c. 16
9. a. geometric   **b.** $r = 3$
   c. 135
11. a. geometric   **b.** $r = 5$
   c. 125
13. a. none
   b. subtract 1, then subtract 2, then subtract 3, . . .
   c. 6
15. a. arithmetic   **b.** $d = 3$
   c. 17
17. a. geometric   **b.** $r = -3$
   c. $-1{,}215$
19. a. Fibonacci   **b.** $s_1 = 2$, $s_2 = 5$
   c. 19
21. a. none   **b.** perfect cubes
   c. 216
23. a. none
   b. $\dfrac{1}{2}$ followed by all not previously listed reduced fractions
   c. $\dfrac{5}{6}$
25. a. geometric   **b.** $r = \dfrac{3}{2}$
   c. $\dfrac{27}{4}$
27. a. 1, 5, 9   **b.** arithmetic; $d = 4$
29. a. 1, 0, -1 **b.** arithmetic; $d = -1$
31. a. 2, 1, $\dfrac{2}{3}$ **b.** neither
33. a. 0, $\dfrac{1}{3}, \dfrac{1}{2}$ **b.** neither
35. a. 1, 3, 6   **b.** neither
37. a. $-5, -5, -5$
   b. arithmetic, $d = 0$; and geometric, $r = 1$

39. a. $1, -1, 1$ **b.** geometric; $r = -1$
41. arithmetic   **43.** geometric
45. geometric   **47.** 57
49. 21   **51.** 2, 6, 18, 54, 162
53. 1, 1, 2, 3, 5   **55.** r^2
57. e^d **59.** 112

8.2   Limit of a Sequence, page 516

5. $0, 2, 0, 2, 0$ **7.** $1, \frac{1}{2}, \frac{1}{3}, \frac{1}{4}, \frac{1}{5}$
9. $\frac{4}{3}, \frac{7}{4}, 2, \frac{13}{6}, \frac{16}{7}$ **11.** $0.79, 1.11, 1.25, 1.33, 1.37$
13. $0, 0.49, 0.76, 0.98, 1.17$ **15.** $256, 16, 4, 2, \sqrt{2}$
17. $-1, 1, 4, 8, 13$ **19.** 5
21. 5   **23.** 1
25. -7 **27.** 0
29. $\frac{1}{2}$ **31.** 4
33. 0   **35.** 1
37. e^3 **39.** e^{-2}
41. $\frac{1}{2}$ **43.** 0
45. increasing; upper bound 3   **47.** decreasing; lower bound 0
49. decreasing; lower bound 0   **51.** diverges by oscillation
53. diverges because unbounded
55. $E_1 = 6.9 \times 10^{-59}, E_2 = 2.7 \times 10^{-58}, E_3 = 6.2 \times 10^{-58}, E_4 = 1.1 \times 10^{-57}$
57. $N = 100$ **59.** $N = 1{,}001$

8.3   Series, page 528

1. F   **3.** T
5. 30   **7.** 150
9. -1 **11.** 20
13. 12   **15.** 80
17. $\displaystyle\sum_{k=1}^{\infty}(4k-1)$ **19.** $\displaystyle\sum_{k=1}^{\infty}3(2)^{k-1}$
21. $\displaystyle\sum_{k=1}^{\infty}(x-7b+2bk)$ **23.** 845
25. 5,465   **27.** 11,529
29. 2   **31.** 4
33. 2,000   **35.** 120
37. 56; 64   **39.** 9,330
41. 3,828   **43.** 200 cm
45. 1,500 revolutions   **47.** 190 ft
49. $21,474,836.47   **51.** $P = 1{,}000{,}000 \cdot 2^d$
53. 13 games   **55.** 51 units
57. Option D is best (not only for 10 years but for any number of years).
59. $24, \frac{24}{5}, \frac{24}{25}, \ldots$

8.4   Systems of Equations, page 539

1. $(3, 1)$ **3.** $(-2, -5)$
5. inconsistent   **7.** inconsistent
9. $(23, -43)$ **11.** $\left(-\frac{3}{5}, \frac{5}{4}\right)$
13. $(t_1, t_2) = (57, 274)$ **15.** $(\alpha, \beta) = (-1, 1)$
17. $(13, 3)$ **19.** $(r, s) = \left(2t, 3t - \frac{5}{2}\right)$
21. $(a_1, a_2) = (-7, 3)$ **23.** $(\alpha, \beta) = (-14, 7)$
25. $(3, 4)$ **27.** inconsistent
29. $(3, 15)$ **31.** $(6, -1)$
33. $(2c - d, c - d)$ **35.** inconsistent
37. If you work 19 days, take the straight salary; if you work 20 days, take the geometric salary.
39. The numbers are 1 and -3 or 3 and -1.
41. 5 ft and 12 ft

47. $\left(\frac{1}{5}, \frac{2}{5}, \frac{2}{5}\right)$ **49.** $(1, 2, 5), (-2, -4, 20)$
51. $\left(\frac{\pi}{3}, \frac{7\pi}{6}\right), \left(\frac{\pi}{3}, \frac{11\pi}{6}\right), \left(\frac{5\pi}{3}, \frac{7\pi}{6}\right), \left(\frac{5\pi}{3}, \frac{11\pi}{6}\right)$
53. $(10^{73/3}, 10^{140/3})$
55. $(-5.4, 2.9), (5.4, -2.9), (-5.4, 2.9), (-5.4, -2.9)$
57. $(h, k) = (1, 8)$ **59.** $(34.7, 17.3)$

8.5   Matrix Solution of a System of Equations, page 554

1. a. $\begin{bmatrix} 4 & 5 & | & -16 \\ 3 & 2 & | & 5 \end{bmatrix}$ **b.** $\begin{bmatrix} 1 & 1 & 1 & | & 4 \\ 3 & 2 & 1 & | & 7 \\ 1 & -3 & 1 & | & 0 \end{bmatrix}$

c. $\begin{bmatrix} 1 & 2 & -5 & 1 & | & 5 \\ 1 & 0 & -3 & 6 & | & 0 \\ 0 & 0 & 1 & -3 & | & -15 \\ 0 & 1 & -5 & 5 & | & 2 \end{bmatrix}$

3. a. $\begin{bmatrix} 1 & 3 & -4 & | & 9 \\ 0 & 2 & 4 & | & 5 \\ 3 & 1 & 2 & | & 1 \end{bmatrix}$ **b.** $\begin{bmatrix} 1 & 0 & 2 & | & -8 \\ -2 & 3 & 5 & | & 9 \\ 0 & 1 & 0 & | & 5 \end{bmatrix}$

5. a. $\begin{bmatrix} 1 & 2 & -3 & | & 0 \\ 0 & 3 & 1 & | & 4 \\ 0 & 1 & 7 & | & 6 \end{bmatrix}$ **b.** $\begin{bmatrix} 1 & 3 & -5 & | & 6 \\ 0 & 13 & -14 & | & 20 \\ 0 & 5 & 1 & | & 3 \end{bmatrix}$

7. a. $\begin{bmatrix} 1 & 3 & 5 & | & 2 \\ 0 & 1 & 3 & | & -4 \\ 0 & 3 & 4 & | & 1 \end{bmatrix}$ **b.** $\begin{bmatrix} 1 & 5 & -3 & | & 5 \\ 0 & 1 & 3 & | & -5 \\ 0 & 2 & 1 & | & 5 \end{bmatrix}$

9. a. $\begin{bmatrix} 1 & 0 & -23 & | & -23 \\ 0 & 1 & 4 & | & 5 \\ 0 & 0 & -8 & | & -13 \end{bmatrix}$ **b.** $\begin{bmatrix} 1 & 0 & 12 & | & 27 \\ 0 & 1 & -2 & | & -5 \\ 0 & 0 & -2 & | & -4 \end{bmatrix}$

11. a. $\begin{bmatrix} 1 & 0 & 4 & | & 5 \\ 0 & 1 & -3 & | & 6 \\ 0 & 0 & 1 & | & 2 \end{bmatrix}$ **b.** $\begin{bmatrix} 1 & 0 & 4 & | & -5 \\ 0 & 1 & 3 & | & 6 \\ 0 & 0 & 1 & | & 1.5 \end{bmatrix}$

13. a. $\begin{bmatrix} 1 & 0 & 0 & | & 9 \\ 0 & 1 & 0 & | & -2 \\ 0 & 0 & 1 & | & 4 \end{bmatrix}$ **b.** $\begin{bmatrix} 1 & 0 & 0 & | & 7 \\ 0 & 1 & 0 & | & -7 \\ 0 & 0 & 1 & | & 3 \end{bmatrix}$

15. $(3, 1)$ **17.** $(2t + 4, 3t)$
19. $(2, 0, 1)$ **21.** $(3, 2, 5)$
23. $(2, -3, -1)$ **25.** inconsistent system
27. $(9, 4, 0)$ **29.** $\left(-2, -3, \frac{11}{2}\right)$
31. $\left(\frac{3}{2}, \frac{1}{2}, -\frac{1}{2}\right)$ **33.** inconsistent system
35. $(t + 1, 2t, -4)$ **37.** inconsistent
39. $\left(\frac{1}{2}, \frac{7}{2}, \frac{7}{2}\right)$ **41.** $(x, y, z, w) = (1, 2, 1, 4)$
43. $(x, y, z, w) = (4, 3, -5, 1)$ **45.** $(3, -2)$
47. $y = x^2 + 4x + 5$ **49.** $x^2 + y^2 - 2x - 2y - 23 = 0$
51. Mix 3 containers of spray I with 4 containers of spray II.
53. Produce 4 units of candy I, 5 units of candy II, and 6 units of candy III.
55. $\left(-\frac{1}{2} + 6t, \frac{3}{4} - 3t, 4t\right)$ **57.** $(2t - 4, t - 10, -3t)$
59. $\left(\frac{19}{7} - t, -\frac{1}{7} + t, t\right)$

8.6   Inverse Matrices, page 568

1. F   **3.** F
5. F

7. a. $\begin{bmatrix} 2 & 4 & 2 \\ 6 & -2 & 4 \\ 2 & 2 & 5 \end{bmatrix}$ **b.** $\begin{bmatrix} -6 & -1 & -2 \\ 3 & -7 & -3 \\ 4 & -7 & -2 \end{bmatrix}$

9. a. $\begin{bmatrix} 20 & 21 & 34 \\ 29 & 16 & 15 \\ 7 & 48 & 5 \end{bmatrix}$
b. $\begin{bmatrix} -1 & 37 & 32 \\ 4 & 14 & 45 \\ 27 & 9 & 28 \end{bmatrix}$

11. a. $\begin{bmatrix} 14 & 14 \\ -7 & 7 \end{bmatrix}$
b. $\begin{bmatrix} 1 & 0 & 0 & 0 \\ 0 & 1 & 0 & 0 \\ 0 & 0 & 1 & 0 \\ 0 & 0 & 0 & 1 \end{bmatrix}$

13. a. not conformable
b. not conformable

15. $\begin{cases} x + 2y + 4z = 13 \\ -3x + 2y + z = 11 \\ 2x + z = 0 \end{cases}$
19. $\begin{bmatrix} 2 & 7 \\ 1 & 4 \end{bmatrix}$

21. $\begin{bmatrix} 9 & -4 & -2 \\ -18 & 9 & 4 \\ -4 & 2 & 1 \end{bmatrix}$
23. $\begin{bmatrix} 1 & 0 & -1 & 0 \\ 0 & \frac{1}{2} & 0 & 0 \\ -2 & 0 & 2 & 1 \\ 0 & 0 & 1 & 0 \end{bmatrix}$

25. $(3, 2)$
27. $(5, -4)$
29. $(38, 21)$
31. $(3, -2)$
33. $(3, -5)$
35. $(-4, 1)$
37. $(3, 1)$
39. $(-2, 2)$
41. $(1, -8)$
43. $(5, 6, 1)$
45. $(-26, 52, 15)$
47. $(88, -176, -38)$
49. $(1, 1, 1)$
51. $(2, -4, -1)$

53. a. $\begin{bmatrix} 1 & 0 & 0 & 0 & 0 \\ -1 & 1 & 0 & 0 & 0 \\ 1 & -2 & 1 & 0 & 0 \\ -1 & 3 & -3 & 1 & 0 \\ 1 & -4 & 6 & -4 & 1 \end{bmatrix}$

b. It is the same as the Pascal's matrix except that the signs of the terms alternate.

55. $[A]^3 = \begin{bmatrix} 2 & 4 & 1 & 3 \\ 4 & 2 & 3 & 4 \\ 1 & 3 & 0 & 1 \\ 3 & 4 & 1 & 2 \end{bmatrix}$ For example, the United States can talk to Cuba through two intermediaries in one way, namely, United States to Mexico to Russia to Cuba.

57. a. Riesling costs 24; Charbono, 25; and Rosé, 51
b. Outside bottling, 230; produced and bottled at winery, 520; estate bottles, 280
c. [3,990]; It is the total cost of production of all three wines.

8.7 Systems of Inequalities, page 576

1. F
3. T
5.
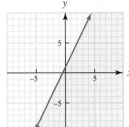
7.
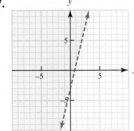

9.
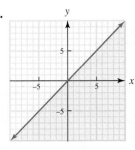
11.
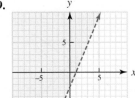

13.
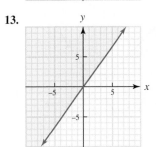
15.
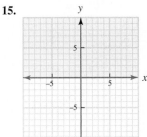

17.
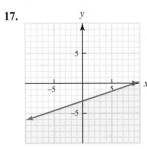
19.
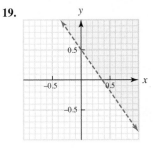

21.
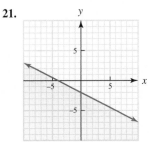

23. a.
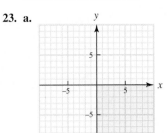
b.
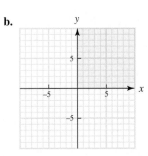

25.
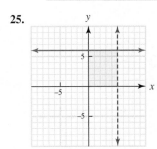
27.
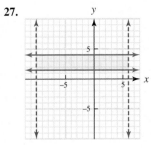

29.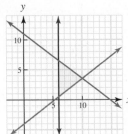

31.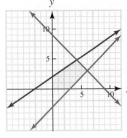

53.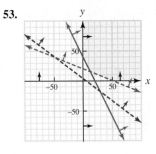

55.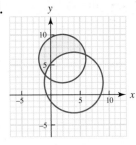

33.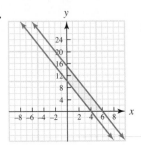

35.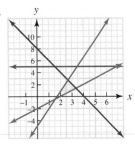

57.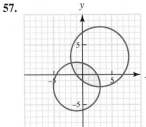

59.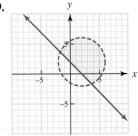

37.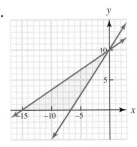

39.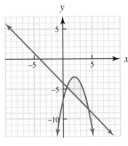

Chapter 8 Self-Test, page 578

1. **a.** arithmetic; $d = 5$; 21; $a_n = 1 + 5n$
   b. arithmetic; $d = 11$; 68; $a_n = 24 + 11n$
   c. geometric; $r = \frac{1}{2}$; $\frac{1}{8}$; $g_n = 2\left(\frac{1}{2}\right)^n$
   d. neither; add 2, then add 4, add 6, add 8, and so on; 24
   e. geometric; $r = -\frac{1}{3}$; $\frac{1}{27}$; $g_n = 3\left(-\frac{1}{3}\right)^{n-1}$

2. **a.** $e, \frac{1}{2}e^2, \frac{1}{6}e^3, \frac{1}{24}e^4$; $\displaystyle\lim_{n \to \infty} \frac{e^n}{n!} = 0$

   b. $-3, -\frac{11}{7}, -\frac{5}{3}, -\frac{15}{7}$; $\displaystyle\lim_{n \to \infty} \frac{3n^2 - n + 1}{(1 - 2n)n} = -\frac{3}{2}$

   c. $2, \frac{9}{4}, \frac{64}{27}, \frac{625}{256}$; $\displaystyle\lim_{n \to \infty} \left(1 + \frac{1}{n}\right)^n = e$

3. **a.** $G_{10} = \dfrac{(a+b)(1-a^{10})}{1-a}$ **b.** $G_{10} = \dfrac{P[1-(1-r)^{10}]}{1-P}$
   c. 2,000   **d.** $\frac{1}{4}$

4. **a.** 364   **b.** 24
   c. 59,048   **d.** 20,300

5. $(3, 4)$ 6. $(0, 1)$

41.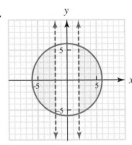

43.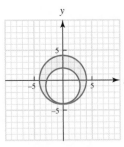

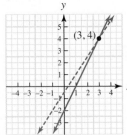

7. $(-2, 3), (3, -2)$ 8. $\left(\frac{1}{2}, 0, \frac{1}{2}\right)$
9. $(-4, 3)$

10.

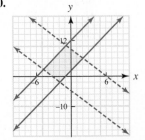

45.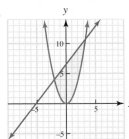

47.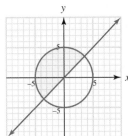

49.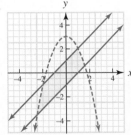

51.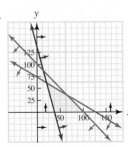

Practice for Calculus, Cumulative Review, page 578

1. $\dfrac{2(2x^2 - 7x + 6)}{(3 - x^2)^2}$

3. $\dfrac{\cos t + t\sin t}{2\sqrt{t}\cos^2 t}$

5. $8x^{-4}y^3(4x - 3y)$

7. $2(2\tan x - 1)(4\tan^2 x + 2\tan x + 1)$

9. $-\dfrac{1}{1{,}999}$

11. 0

13. $\pm\dfrac{\sqrt{2}}{3}$

15. $\dfrac{\pi}{3} + n\pi,\ \dfrac{4\pi}{3} + n\pi$

17. a. 3 **b.** -3

19. $\dfrac{1}{3}(\log_6 200 - 2)$

21. $P\left(a, 2\cos\frac{a}{\pi}\right)$

23. $Q\left(\dfrac{2a - 6}{a + 2}, \dfrac{-a - 12}{4a - 2}\right)$

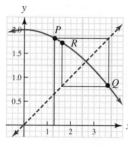

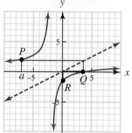

25. $Q(-a^2 + 6a,\ -a^4 + 12a^3 - 39a^2 + 18a)$

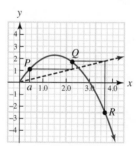

27. $Q(4\log a,\ 2\log(4\log a))$

29. $\dfrac{-1}{(x + h + 1)(x + 1)}$

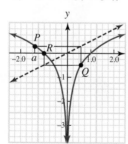
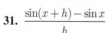

31. $\dfrac{\sin(x + h) - \sin x}{h}$

33. $\dfrac{e^{2x}(e^{2h} - 1)}{h}$

35. Write secant as cosine.

37. Use sum formula for tangents.

39. a. $(-2, 5)$ **b.** $\left(-\dfrac{1}{2}, \dfrac{1}{5}\right)$

41. a. $\left(-\dfrac{3}{8}, \dfrac{7}{12}\right)$ **b.** no real solution

43. $\left(\dfrac{17}{13} - t,\ -\dfrac{59}{13} + 5t, 13t\right)$

45. $\left(\sqrt{2}, 1\right), \left(\sqrt{2}, -1\right),$ $\left(-\sqrt{2}, 1\right), \left(-\sqrt{2}, -1\right)$

47.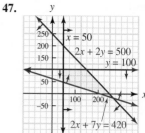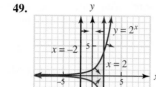

49.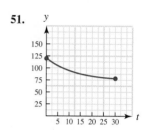

51.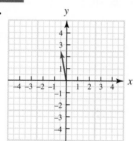

53.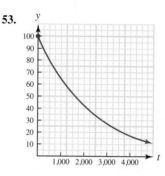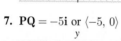

55. 31 feet

59. a. 3 seconds   **b.** 170 feet

CHAPTER 9

9.1   Vectors in the Plane, page 591

1.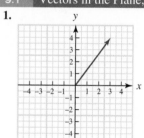

3.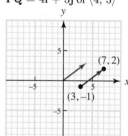

5. $\mathbf{PQ} = 4\mathbf{i} + 3\mathbf{j}$ or $\langle 4, 3 \rangle$

7. $\mathbf{PQ} = -5\mathbf{i}$ or $\langle -5, 0 \rangle$

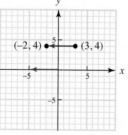

9. $\mathbf{PQ} = 2\mathbf{i}$
$\|\mathbf{PQ}\| = 2$

11. $\mathbf{PQ} = 4\mathbf{i} + 2\mathbf{j}$
$\|\mathbf{PQ}\| = 2\sqrt{5}$

13. $\dfrac{1}{2}\sqrt{2}(\mathbf{i} + \mathbf{j})$

15. $\dfrac{3}{5}\mathbf{i} - \dfrac{4}{5}\mathbf{j}$

17. $(s, t) = (6, 24)$

19. $(s, t) = (-1, -5)$

21. $17\mathbf{i} - 18\mathbf{j}$

23. $35\mathbf{i} - 35\mathbf{j}$

25. $(3, 2)$

27. $(2, 4), (-4, -2)$

29. $\dfrac{\sqrt{3}}{2}\mathbf{i} + \dfrac{1}{2}\mathbf{j}$

31. $\dfrac{4}{17}\sqrt{17}\mathbf{i} - \dfrac{1}{\sqrt{17}}\mathbf{j}$

33. $\dfrac{5}{26}\sqrt{26}\mathbf{i} + \dfrac{1}{26}\sqrt{26}\mathbf{j}$

35. $(3, 10)$

37. $M(3, -5); (7, -3)$

41. No

43. a. circle with center (x_0, y_0) and radius 1

   b. the set of all points on or interior to the circle with center (x_0, y_0) and radius r

45. $a = -2t$, $b = t$, $c = t$ **47.** $-6\mathbf{i} + 3\mathbf{j}$

49. $x \approx 45.96\,\text{ft/s}$, $y \approx 38.57\,\text{ft/s}$ **51.** $\mathbf{F}_4 = \left(\frac{5}{2} - 5\sqrt{3}\right)\mathbf{i} - \left(13 - \frac{5}{2}\sqrt{3}\right)\mathbf{j}$

9.2   Quadric Surfaces and Graphing in Three Dimensions, page 601

1–3.

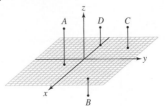

5.

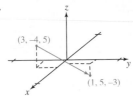

7.

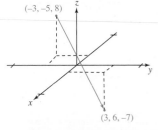

9. $x^2 + y^2 + z^2 = 1$ **11.** $x^2 + (y-4)^2 + (z+5)^2 = 9$

13. Center: $(0, 1, -1)$; $r = 2$ **15.** Center: $(3, -1, 1)$; $r = 1$

17.   **19.**

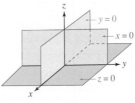

21. B   **23.** A

25. G   **27.** C

29. I

31. isosceles, but not a right triangle

33. neither right nor isosceles

35.   **37.**

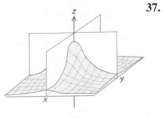

39.   **41.**

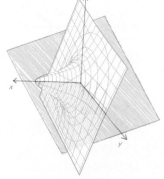

43.   **45.**

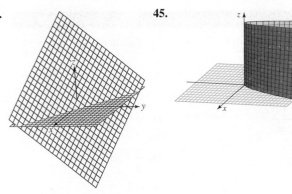

47.   **49.**

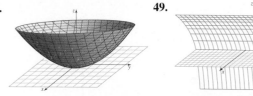

51.

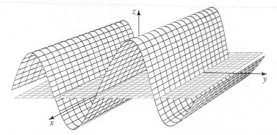

53. Hyperboloid on one sheet

xy-plane: ellipse

yz-plane: hyperbola

xz-plane: hyperbola

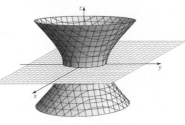

55. Elliptic cone parallel to the xy-plane; $z > 0$: ellipse parallel to the yz-plane: hyperbola parallel to the xz-plane: hyperbola

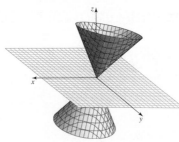

57. Hyperbolic paraboloid

xy-plane: hyperbola

yz-plane: parabola

xz-plane: parabola

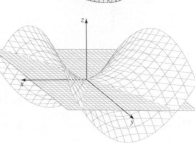

59. $\dfrac{3 \pm \sqrt{33}}{2}$, $\dfrac{17}{3}$

9.3 The Dot Product, page 612

3. $-2\mathbf{i} + 2\mathbf{j} + \mathbf{k}; 3$

5. $-4\mathbf{i} - 4\mathbf{j} - 4\mathbf{k}; 4\sqrt{3}$

7. -16

9. 5

11. orthogonal

13. orthogonal

15. $\sqrt{3}$

17. 14

19. $-4\mathbf{i} - 16\mathbf{j} + 7\mathbf{k}$

21. $\sqrt{321}$

23. yes

25. no

27. 11

29. 0

31. $71°$

33. $114°$

35. $1; \mathbf{k}$

37. $\frac{6}{13}\sqrt{13}; -\frac{12}{13}\mathbf{j} + \frac{18}{13}\mathbf{k}$

39. $\pm\mathbf{j}$

41. $\frac{\sqrt{2}}{2}\mathbf{i} + \frac{\sqrt{2}}{2}\mathbf{k}, -\frac{\sqrt{2}}{2}\mathbf{i} - \frac{\sqrt{2}}{2}\mathbf{k}$

43. $-\frac{2}{17}\sqrt{17}\mathbf{i} - \frac{3}{17}\sqrt{17}\mathbf{j} + \frac{2}{17}\sqrt{17}\mathbf{k}$ **45.** $a = \frac{3}{2}$

47. $35°, 66°, 114°$

49. a. 4

b. $\frac{2\sqrt{7}}{21}$

c. $\frac{9}{2}$

d. $-\frac{7}{2}$

51. 10

53. $90,100$ ft-lb

55. $48.5\mathbf{i} - 1.25\mathbf{j} + 10\mathbf{k}$

57. a. 500 ft-lb

b. $500\sqrt{2}$ ft-lb

9.4 The Cross Product, page 622

1. $\mathbf{k}$

3. $-2\mathbf{i} + 4\mathbf{j} + 3\mathbf{k}$

5. $-2\mathbf{i} + 25\mathbf{j} + 14\mathbf{k}$

7. $-2\mathbf{i} + 16\mathbf{j} + 11\mathbf{k}$

9. $-14\mathbf{i} - 4\mathbf{j} - \mathbf{k}$

11. $\frac{\sqrt{3}}{2}$

13. $\frac{\sqrt{3}}{2}$

15. $\frac{3\sqrt{33}}{77}$

17. $\frac{1}{14}\sqrt{14}\mathbf{i} + \frac{3}{14}\sqrt{14}\mathbf{j} - \frac{1}{7}\sqrt{14}\mathbf{k}$ **19.** $\frac{-16}{386}\sqrt{386}\mathbf{i} + \frac{7}{386}\sqrt{386}\mathbf{j} + \frac{9}{386}\sqrt{386}\mathbf{k}$

21. $\sqrt{26}$

23. $2\sqrt{59}$

25. $\frac{1}{2}\sqrt{3}$

27. $\frac{3}{2}\sqrt{3}$

29. a. does not exist

b. scalar

31. a. scalar

b. vector

33. 3

35. 9

41. 2

43. $130°$

47. $-40\sqrt{3}\mathbf{i}$

9.5 Lines and Planes in Space, page 630

5. $4 + 3y + 2z - 46 = 0$

7. $3x - 2y + 2z + 10 = 0$

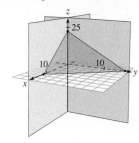

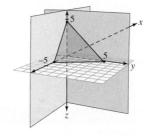

9. $x = 1 + 3t, y = -1 - 2t, z = -2 + 5t; \dfrac{x-1}{3} = \dfrac{y+1}{-2} = \dfrac{z+2}{5}$

11. $x = 1 + t, y = -1 + 2t, z = 2 + t; \dfrac{x-1}{1} = \dfrac{y+1}{2} = \dfrac{z-2}{1}$

13. $\dfrac{x-1}{1} = \dfrac{y+3}{-3} = \dfrac{z-6}{-5}; x = 1 + t, y = -3 - 3t, z = 6 - 5t$

15. $x = 3, y = -1 + t, z = 0$

19. $(0, 4, -4), (12, 0, 4), (6, 2, 0)$

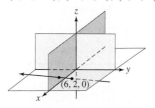

21. $(0, 0, 9), (0, 0, 9), (-3, -6, 0)$

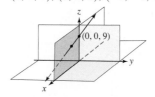

23. intersect at $(1, 2, 3)$

25. skew

27. $x - z + 1 = 0$

29. $x + 2y - 3z - 6 = 0$

31. $x = 0$

33. $\mathbf{u} = \pm\frac{1}{21}\sqrt{21}(2\mathbf{i} + 4\mathbf{j} + \mathbf{k})$

35. $\mathbf{u} = \pm\frac{1}{38}\sqrt{38}(5\mathbf{i} - 3\mathbf{j} + 2\mathbf{k})$

39. $\mathbf{N} = \pm\frac{1}{2}\sqrt{2}(\mathbf{i} - \mathbf{k})$

41. $\dfrac{x-5}{5} = \dfrac{y+1}{-3} = \dfrac{z}{1}$

43. $\dfrac{x-1}{2} = \dfrac{y+5}{-3} = \dfrac{z-3}{1}$

45. $2x - y + 3z + 3 = 0$

47. $(9, -5, 12)$

49. $\dfrac{x-2}{1} = \dfrac{y-3}{1} = \dfrac{z-1}{1}$

51. $3\mathbf{i} - 2\mathbf{j} - 11\mathbf{k}$

53. $\dfrac{y-\frac{2}{3}}{1} = \dfrac{z}{1}$ and $x = \dfrac{13}{3}$

55. $\alpha = 2.47$ or $142°$, $\beta = 0.94$ or $54°$, $\gamma = 1.37$ or $79°$

Chapter 9 Self Test, page 634

1. a. $13\mathbf{i} - 12\mathbf{j} + 2\mathbf{k}$

b. 1

c. 0

d. $2\mathbf{i} + 3\mathbf{j} + 5\mathbf{k}$

2. a. $-\frac{4}{5}\mathbf{j} + \frac{12}{5}\mathbf{k}$

b. $\frac{4}{5}\sqrt{10}$

3. a. 40

b. not defined

c. $25\mathbf{i} - 10\mathbf{j} - 15\mathbf{k}$

d. not defined

4. a. $x = t, y = -2 - 6t, z = 1 + 4t$

b. $x = -\frac{13}{2} - 10t, y = 5 + 6t, z = 2t$

c. $\dfrac{x-1}{4} = \dfrac{y-2}{3} = \dfrac{z+1}{-3}$

5. a. $2x + 3z - 11 = 0$

b. $17x + 19y + 13z - 25 = 0$

6. $122°, 37°, 74°$

7. a. skew

b. $(2, 2, 3)$

8. a. 6

b. $\sqrt{2}$

9. 168.4 mi/h

11. 130 ft-lb

Practice for Calculus, Cumulative Review, page 634

1. $\dfrac{5x^2 - 8x + 30}{(x^2 - 6)^2}$

3. $\sec^2\theta$

5. 1

7. $-3x^{-3}y^2(xy - 1)$

9. $9(x^2 + 3x + 4)^2(x^4 + 2)(20x^5 + 33x^4 + 32x^3 + 24x + 18)$

11. 3

13. $\dfrac{\sqrt{2x}}{2x}$

15. $\dfrac{-3}{(x-2)^2}$

17. no real values

19. -0.67

21. 10.67

23.
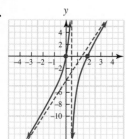

25.
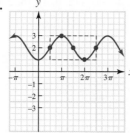

27.
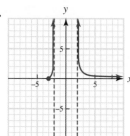

29.
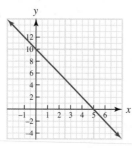

31.
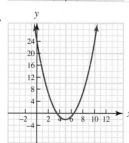

33.
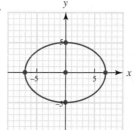

35.
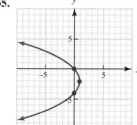

37. **a.** $(-3, 3)$　　　**b.** $\left(-\frac{1}{3}, \frac{1}{3}\right)$

39. $(-3, 4, 7)$　　**41.** $\left(-23, -\frac{125}{2}, -\frac{25}{2}\right)$

43. $(3, 3)$, $(6, -2)$, $(12, 10)$, $(15, 7)$

45. $x^2 + 6x + 10y + 4 = 0$

47. **a.** $1,747,625$　　　**b.** $15,050$

49. $\sqrt{74}$ square units

51. **a.** $\|\mathbf{v}\| = \sqrt{14}$; $\mathbf{v} - \mathbf{w} = -\mathbf{i} - 3\mathbf{j} + 4\mathbf{k}$; $2\mathbf{v} - 3\mathbf{w} = 18\mathbf{i} - \mathbf{j} - 7\mathbf{k}$;
$\mathbf{v} \cdot \mathbf{w} = 7$; $\mathbf{v} \times \mathbf{w} = 5\mathbf{i} + 13\mathbf{j} + 11\mathbf{k}$

　　b. $\|\mathbf{v}\| = 1$; $\mathbf{v} - \mathbf{w} = \mathbf{i} - \mathbf{j}$; $2\mathbf{v} - 3\mathbf{w} = 2\mathbf{i} + 3\mathbf{j}$; $\mathbf{v} \cdot \mathbf{w} = 0$;
$\mathbf{v} \times \mathbf{w} = \mathbf{k}$

53. $(0, 9, -8)$, $(9, 0, 10)$, $(4, 5, 0)$

55. **a.** $V(x) = x(8 - x)(10 - 2x)$

　　b. $[0, 5]$

　　c.
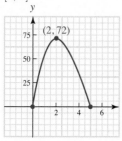

　　d. 72 occurs when $x = 2$; dimensions are $2 \times 6 \times 6$

57. $c = \frac{4}{5}$; $(2, 0)$ and $(-2, 0)$

59. **a.** $6 \le p \le 8$;
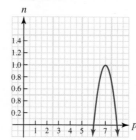

　　b. A price of \$7 yields a maximum profit of \$1,000,000.

Final Review, Set I, page 636

1. Q　　　　　　　　　**3.** W

5. X　　　　　　　　　**7.** B

9. L　　　　　　　　　**11.** U

13. O　　　　　　　　　**15.** S

17. I　　　　　　　　　**19.** H

21. A　　　　　　　　　**23.** F

25. E　　　　　　　　　**27.** $5 + \log 0.036$

29. $\frac{\pi}{3} + 2n\pi$, $\frac{2\pi}{3} + 2n\pi$, $\frac{4\pi}{3} + 2n\pi$, $\frac{5\pi}{3} + 2n\pi$

31. $t = -h \log_2\left(\frac{A}{A_0}\right)$　　**33.** $\dfrac{-5}{(x+h-2)(x-2)}$

35.
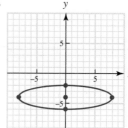

37.
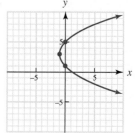

39.
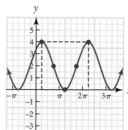

41.
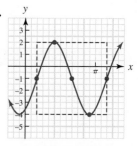

43. $(-5, 13)$　　　　　**45.** $(2, 2)$, $(4, 2)$

47. $(x + 3)^2 = -16(y - 5)$　　**49.** $\sqrt{35}$

Final Review, Set II, page 638

1. hyperbola　　　　　**3.** lemniscate

5. five-leaved rose　　**7.** parabola

9. parabola　　　　　**11.** hyperbola

13. exponential　　　　**15.** hyperbola

17. parabola　　　　　**19.** parabola

21. line

23. hyperbola

25. $-\frac{2}{15}$

27. 1,631

29. $r = \frac{1}{t}\ln\left(\dfrac{P}{P_0}\right)$

31. 3

33.

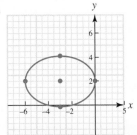

35.

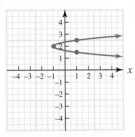

37.

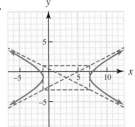

39.

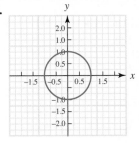

41.

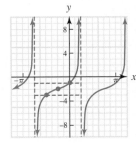

43. $(5, 4, -3)$

45.

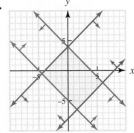

47. $y = x^2 - 4x + 4$

49. 18.8 in. × 63.7 in.

APPENDICES

Appendix A   Field Properties, page 643

3. reflexive property

5. reflexive property

7. closure for addition

9. closure for addition

11. substitution property

13. symmetric property

15. inverse property for multiplication

17. identity property for multiplication

19. no

21. no

23. yes

25. no

27. $-1 - \sqrt{3}$

29. $-\frac{1}{2} + \frac{1}{2}\sqrt{3}$

31. $-1 - \frac{\pi}{3}$

33. $\frac{3}{\pi + 3}$

35. a. no   **b.** yes

---

37. yes

43.

$\star$	ℓ	r	a	f
ℓ	a	f	r	ℓ
r	f	a	ℓ	r
a	r	ℓ	f	a
f	ℓ	r	a	f

45. yes

47. closure for $+$; closure for $\times$; associative for $+$; associative for $\times$; identity for $\times$ is 1; commutative for $+$; commutative for $\times$; distributive

49. $\mathbb{Z}$; closure for $+$; closure for $\times$; associative for $+$; associative for $\times$; identity for $+$ is 0; identity for $\times$ is 1; inverse for $+$; commutative for $+$; commutative for $\times$; distributive

51. associative for $+$; associative for $\times$; commutative for $+$; commutative for $\times$; distributive

53. H; closure for $\times$; associative for $+$; associative for $\times$; identity for $\times$ is 1; inverse for $\times$; commutative for $+$; commutative for $\times$; distributive

55.

$\star$	A	B	C	D	E	F	G	H
A	B	C	D	A	H	G	E	F
B	C	D	A	B	F	E	H	G
C	D	A	B	C	G	H	F	E
D	A	B	C	D	E	F	G	H
E	G	F	H	E	D	B	A	C
F	H	E	G	F	B	D	C	A
G	F	H	E	G	C	A	D	B
H	E	G	F	H	A	C	B	D

57. yes

59. yes

Appendix B   Complex Numbers, page 649

3. yes; no

5. F

7. F

9. F

11. a. $7i$

b. $2i\sqrt{2}$

13. a. $-\frac{36}{5}i$

b. $-\frac{3}{2}i$

15. a. $2 + 5i$

b. $-3 + 2i$

17. a. 68

b. 25

19. a. $-i$

b. $-i$

21. a. $-i$

b. -1

23. a. $-2 + \sqrt{2} + \sqrt{2}i$

b. $-2 - \sqrt{3} + \sqrt{3}i$

25. a. $-8 + 36i$

b. $-1 - 5i$

27. a. $-8 - 6i$

b. $32 - 24i$

29. a. $-7 + 6\sqrt{2}i$

b. $-4 - 6\sqrt{5}i$

31. a. 12

b. 7

33. a. $1 - i$

b. $1 + i$

35. a. $\frac{-3}{2} + \frac{3}{2}i$

b. $\frac{20}{17} + \frac{5}{17}i$

37. a. $-\frac{1}{5} - \frac{3}{5}i$

b. $-\frac{6}{29} + \frac{15}{29}i$

39. a. $\frac{-45}{53} + \frac{28}{53}i$

b. $\frac{7}{5} + \frac{6}{5}i$

41. a. $1 - 3i$

b. $-\frac{1}{2} - \frac{5}{2}i$

43. $2.6066 + 1.5497i$

45. 0

47. $\pm 2i$

49. 0

51. 0

53. $-1, \frac{1}{2} \pm \frac{\sqrt{3}}{2}$

55. $\left(-6 + 8\sqrt{2}\right) + \left(11 - 4\sqrt{2}\right)i$

Appendix C Mathematical Induction, page 653

1. T

3. T

5. T

7. F

9. T

11. T

13. $1 + 2 + 3 + \cdots + k = \frac{k(k+1)}{2}$; $1 + 2 + 3 + \cdots + (k+1) = \frac{(k+1)(k+2)}{2}$

15. $1 + 8 + 27 + \cdots + k^3 = \frac{k^2(k+1)^2}{4}$; $1 + 8 + 27 + \cdots + (k+1)^3 = \frac{(k+1)^2(k+2)^2}{4}$

17. $3 + 9 + 15 + \cdots + (6k-3) = 3k^2$; $3 + 9 + 15 + \cdots + (6k+3) = 3(k+1)^2$

19. $k^2 + k$ is even; $k^2 + 3k + 1$ is even

21. $k^3 - k + 3$ is divisible by 3; $k^3 + 3k^2 + 2k + 3$ is divisible by 3.

23. $\left(\frac{2}{3}\right)^{k+1} < \left(\frac{2}{3}\right)^k$; $\left(\frac{2}{3}\right)^{k+2} < \left(\frac{2}{3}\right)^{k+1}$

Appendix D Binomial Theorem, page 660

1. F

3. T

5. a. 8

b. 126

c. 1

d. 21

7. a. 2,184

b. 210

c. 714

d. 6

9. $x^4 + 12x^3 + 54x^2 + 108x + 81$

11. $a^7 + 7a^6b + 21a^5b^2 + 35a^4b^3 + 35a^3b^4 + 21a^2b^5 + 7ab^6 + b^7$

13. $x^9 - 9x^8 + 36x^7 - 84x^6 + 126x^5 - 126x^4 + 84x^3 - 36x^2 + 9x - 1$

15. $x^6 - 12x^5 + 60x^4 - 160x^3 + 240x^2 - 192x + 64$

17. $16x^4 + 96x^3y + 216x^2y^2 + 216xy^3 + 81y^4$

19. $\frac{1}{8}x^3 + \frac{3}{4}x^2y^3 + \frac{3}{2}xy^6 + y^9$

21. $x^2 + 4x^{3/2}y^{1/2} + 6xy + 4x^{1/2}y^{3/2} + y^2$

23. $1 - 8x + 28x^2 - 56x^3 + 70x^4 - 56x^5 + 28x^6 - 8x^7 + x^8$

25. 462

27. 1,001

29. 1,820

31. −16

33. $x^{15} - 15x^{14}y + 105x^{13}y^2 - 455x^{12}y^3 + \cdots$

35. $x^8 + 8\sqrt{2}x^7 + 56x^6 + 112\sqrt{2}x^5 + \cdots$

37. $x^{10} - 30x^9y + 405x^8y^2 - 3,240x^7y^3 + \cdots$

39. $x^{22} + 55x^{20}k + 1,375x^{18}k^2 + 20,625x^{16}k^3 + \cdots$

41. 10,935

43. $\cdots + 105x^{4/3}y^{13/3} + 15x^{2/3}y^{14/3} + y^5$

45. $\cdots + 60q^{-4}r^{-4} - 24\sqrt{2}q^{-2}r^{-5} + 8r^{-6}$

47. $\cdots + \frac{16}{243}m^{-16} + \frac{2}{729}m^{-17} + \frac{1}{19,683}m^{-18}$

49. 10

51. 35

53. 210

Appendix E Determinants and Cramer's Rule, page 668

1. 2

3. −2

5. $\sec^2\theta$

7. −48

9. −130

11. 4

13. 21

15. −204

17. −5

19. 0

21. 0

23. 24

25. $(-3, 4)$

27. $(31, 19)$

29. $\left(\frac{1}{7}(3\alpha + \beta), \frac{1}{7}(\alpha - 2\beta)\right)$

31. $\left(\frac{\gamma d + \delta b}{ad + bc}, \frac{\delta a - \gamma c}{ad + bc}\right)$

33. $\left(-\frac{1}{5}, \frac{3}{10}\right)$

35. $\left(\frac{11}{15}, \frac{1}{5}\right)$

37. $(2, 1), (2, -1), (-2, 1), (-2, -1)$

39. $(3, -1, 2)$

41. $(-4, 0, 3)$

43. $\left(5, \frac{15}{2}, \frac{3}{2}\right)$

45. $(w, x, y, z) = (3, 1, -2, 1)$

47. inconsistent

49. $x + y + 1 = 0$

51. $5x - 13y - 1 = 0$

53. 76.5 square units

55. 22 square units

Index

A

AAA, 355, 362
  procedure, 362
AAS, 355
  procedure, 362
Abel, Niels, 659
Abscissa, 20
Absolute value, 7, 414
  double property, 44
  equation, 9
    solving, 43
  equation property, 43
  function, 104
    graph, 670
  inequality, 9
    solving, 45
  inequality property, 45
  table of properties, 9
Abstraction, 81
Accuracy, 346
Achilles and the tortoise problem, 51
Addition
  complex numbers, 646
  laws, 399
  of matrices, 557
  method, 538
  property of inequality, 40
  vectors, 585
Additive
  identity, 641
  inverse of a matrix, 562
Adjacent side, 344
Agnesi, witch of, 679
AIDS, 272
Aleksandrov, A. D., 499
Alexandria, Egypt, 295
Algebra
  fundamental theorem, 206
  of matrices, 556

Algebraic
  expression, 12
  function, 101, 243
  proof, 386
Aligned, 624
Alpha (α), 284
Alternating series, 518
Ambiguous case, 359
Amplitude, 312
  general trigonometric curve, 324
Angle, 284
  between planes, 633
  of depression, 348
  of elevation, 348
  of inclination, 309
  initial side, 284
  negative, 284
  polar, 468
  positive, 284
  quadrantal, 284
  reference, 289
  standard-position, 284
  terminal side, 284
  vertex, 284
Answers, 681
Antenna, 437
Anticommutativity, 615
  cross product, 616
Aphelion, 448
Apogee, 451
Approximately equal to, 4
Aquarium of the Americas, 280
Arc, 290
  subtended, 291
Arc length, 290
  formula, 290
Arccosecant, 258
Arccosine, 258
Arccotangent, 258
Archimedes, 54, 206

Archimedian spiral, 670, 678
Arcsecant, 258
Arcsin, 258
Arctangent, 258
Area
　ASA, 365
　curved boundaries, 50
　infinite sum, 50
　parallelogram by vectors, 618
　problem, 85
　SAS, 365
　segment boundaries, 50
　SSS, 365
　of a triangle, 365
　triangle by vectors, 619
Argument, 255, 296, 414
　arm motion, 330
Arithmetic
　growth, 500
　progression, 500
　sequence, 500
　　formula for, 501
　series, 520
　　formula for, 521
Array, 541
　rectangular, 542
Arrhenius function, 205
Āryabhat, 296
ASA, 355
　area of a triangle, 365
　procedure, 362
Asimov, Isaac, 263
Associative property
　complex numbers, 650
　of matrixes, 562
　real numbers, 641
　scalar multiplication, 587
Astroid, 670
Asymptote, 319
　horizontal, 224
　hyperbola, 453
　oblique, 224
　parabola, 431
　slant, 224, 453
　tangent curve, 319
　vertical, 221, 319
Attitude numbers, 628
Augmented matrix, 541
Average, 47
　finite collection, 50
　rate of change of a function, 110
Axes
　conjugate, 452
　coordinate, 20

imaginary, 414
major, 441
minor, 441
radial, 468
real, 414
semimajor, 445
semiminor, 445
of symmetry, 32
transverse, 452, 453
x, 20
y, 20

B

Babylonians, 263
Bacon, Roger, 3
Bannaker, Benjamin, 350
Base, 12, 245
　natural, 251
Basis vectors, 589
　three dimensions, 604
Batman problem, 357
Bearing, 349
Becker, Melissa, 235, 371, 528
Bee waggle dance, 496
Beehive problem, 92
Bernoulli, lemniscate of, 676
Beta (β), 284
Bhaskara, 92
Bifolium, 484, 670
Binomial, 13
　theorem, 656
　　Pascal's triangle, 656
Blimp problem, 368
Blood pressure problem, 330
Boat wave, 413
Bound
　lower, 207
Boundary, 571
Bounded sequence, 514
Brache, Tycho, 263
Brahmagupta, 5, 92, 243
Broadway play problem, 281
Buckling problem, 91

C

Cain, Herman, 180
Calculator, *also see* Computational window
　cardioid, 476
　ellipse graphing, 443
　evaluating trigonometric functions, 298
　function graphing, 433
　graph input, 103

graphing critical values, 39
  parametric equations, 433
  periodic functions, 313
  polar graphing, 476, 487
  quadratic formula, 359
  reciprocal key, 299
  rectangular to trigonometric conversion, 416
  root feature, 383
  trigonometric scale, 316
  zoom, 316
Calculus, 50
  derivative, 50
  difference quotient, 401
  differential, 52
  distinguish from elementary mathematics, 50
  integral, 55
  invention of, 52
Candle experiment problem, 332
Cannonball path, 37, 437, 496
Cardan, Girolamo, 5
Cardioid, 476, 483
  calculator, 476
  equation, 477
  graph, 670
  rotated, 477
Carom paths, 155
Cartesian plane, 20
Cassini, ovals of, 676
Cassinian curve, 670
Catenary, 670, 673
Cauchy-Schwartz inequality, 614
Cayley, Arthur, 560
Ceiling function, 126
Center
  circle, 30, 598
  ellipse, 441
  hyperboloid, 452
Centimeter-gram-second, 612
Central rectangle, 453
Centroid, 593
CGS, 612
Chain letter problem, 522
Change of base theorem, 257
  proof, 267
Cheops pyramid, 353
Chinese mathematics, 5
Circle, 30
  center, 598
  graph, 670
  involute, 675
  polar-form, 472
  unit, 30

Circular
  cone, 600
  cylinder, 599
Circumference, 290
Cissoid of Diocleles, 671
Classroom, three-dimensional model, 594
Close Encounters of a Third Kind, 353
Closed half-plane, 571
Closure, 641
Cochleoid, 671
Coefficient, 13, 100
  determinate of, 662
  equal, 64
  leading, 13
Cofactor, 664
Cofunction
  cosine, 395
  identities, 395
  sine, 395
  tangent, 395
Column, 541
Common
  difference, 500
  logarithm, 256
  ratio, 502
Common logarithm, 256
Communication matrix, 560
Commutative property
  complex numbers, 650
  dot product, 607
  of matrices, 562
  real numbers, 641
  vector addition, 587
Comparison, property of, 5
Complement, 47
Completed solution, 376
Completing the square, 60
Complex number, 4
  addition, 646
  associative property, 650
  commutative property, 650
  conjugates, 648
  distributive property, 646
  product, 416
  rectangular form, 414
    change to trigonometric
      form, 415
  simplify, 417
  subtraction, 646
  trigonometric form, 415
    change to rectangular
      form, 415

Complex plane, 414
Component
  form, 585
  horizontal, 589
  of **v** along **w**, 610
  vertical, 589
Composite function, 129
Composition of functions, 129
Compound interest, 247
Compte, Auguste, 152
Computational window
  graphing lines, 531
  inverse matrix, 564
  ISECT key, 535
  matrices, 542
  matrix entry, 542
  matrix multiplication, 559
  matrix operations, 558
  systems, 559
    graphing, 535
    inverse matrices, 565
    matrices, 564
Concave
  down, 199
  up, 199
Conditional, 80
Cone, 600
  circular, 600
  elliptic, 600
  problem, 89
  volume of, 354
Conic section, 430
  alternate definition, 485
  degenerate, 458
  general form, 430
  recognition of types, 462
Conjugate, 648
  axis, 452
    length of, 453
  function, 95, 109
  term, 100
Constant, 13
  Planck's, 517
Constraints, 574
  superfluous, 576
Continuity, 145
  on a closed interval, 147
  definition, 146
  on an interval, 14
  on an open interval, 147
Continuous compounding, 251
Contract an expression, 265
Contrapositive, 80

Convergent sequence, 510
Converse, 80
Cooley, Hollis R., 577
Coordinate
  axes, 20
  plane, 20, 594
  right-handed, 593
  system, $\mathbb{R}^3$, 593
  three-dimensional, 593
  two dimensions, 20
    polar, 468
    rectangular, 468
Copernican theory, 263
Copernicus, Nicholas, 356
Cosecant
  curve, 320
    polar/rectangular form, 484
  function, 305
  graph, 671
Cosine(s)
  amplitude, 318
  compared to sine curve, 321
  curve, 317
    polar/rectangle form, 484
  direction, 613
  frame, 318
  function, 296
  general form, 318
  graph, 671
    from unit circle, 332
  law of, 355
  period, 318
  standard form, 318
  starting point, 318
Cost
  marginal, 161
Cotangent
  curve, 320
    polar/rectangular form, 484
  function, 305
  graph, 671
Coterminal angles, 284
Counterexample, 390
Counting numbers, 4
Craig, John, 263
Cramer's rule, 663
Critical value
  equation, 39
  inequality, 66
  rational inequality, 69
Cross product, 615
  distributive property, 616
  geometric property, 616

magnitude, 618
    parallel vector property, 616
    properties of, 616
    zero factor cross product, 616
Cross section, 599
Cube root
    function, 104
    graph, 671
Cubes, sum of, 216
Cubic
    function graph, 671
    parabola graph, 671
Cummings, Louise, 604
Curate cycloid, 671
Cycle, 312
Cycloid, 672
    curate, 671
    prolate, 677
Cylinder, 599
    circular, 599
    elliptic, 599
    volume of, 369

D

Damped harmonic motion, 426
Dantzia, Tobias, 219
De Moivre, Abraham, 418
De Moivre's Theorem, 418
    proof, 421
De Sapio, Rodolfo, 330
Death Star problem, 366
Decay formula, 271
Decreasing
    function, 109
    sequence, 514
Dedekind, Richard, 145
Deductive reasoning, 79
Degenerate conic, 458
Degree, 13, 100, 284
    change to radian, 288
    protractor, 285
    and radian table comparison, 288
Deleted point, 221
Delta (Greek letter lower case δ, capital Δ), 22, 48, 284
Deltoid, 672
Demand function, 178
Dependent
    system, 530, 663
        graph of, 532
    variable, 93
Depressed equation, 204
Depression, angle of, 348

Derivative, 50, 52
Derived equation, 240
Descartes, folium of, 673
Descartes, René, 634
Determinant, 662
    of the coefficients, 662
    definition, 662
    order, 665
Devil's Tower, 353
Diagonal
    form, 542
    main, 562
Diastolic phase, 330
Difference, 4, 641
    of angles identity, 393
    common, 500
    quotient, 97
    rule, 585
Differential calculus, 52
Dilation, 119
Dimension, 541
    one, 5
    two, 38
    three, 593
Diné (Navajo) Nation, 142
Diocles, cissoid of, 671
Diplomatic relations example, 560
Direction
    cosines, 613
    numbers, 624
    vector, 589
Directory of curves,
    Part I (basic curves), 104
    Part II (exponential curves), 246
    Part III (logarithmic curves), 259
    Part IV (trigonometric curves), 336
    Part V (conic sections), 460
    Part VI (polar curves), 483
    Part VII (quadratic surfaces), 600
Directrix, 430, 599
    ellipse, 451
    polar form, 485
Dirichlet, Lejeune, 93
Discontinuity, 146
Discriminant, 62
    general conic section, 462
Disproving an identity, 390
Distance
    formula
        one dimension, 5, 8
        two dimensions, 29
        in $\mathbb{R}^3$, 596
    point to a line, 496
    radial, 468

Distributive property, 646, 650
  complex numbers, 646
  cross product, 616
  dot product, 607
  matrices, 562
  real numbers, 641
  scalar product, 616
  vectors, 587
Diverge, 149
Divergent sequence, 510
Divine proportion, 238
Division, 4
  algorithm, 195
  long, 193
  by zero, 296
Domain, 20, 93, 314
  convention, 103
Dot product
  anticommutativity, 616
  definition, 607
    alternative, 609
  distributive property, 607
  properties, 607
Double
  absolute value, 44
  subscripts, 541
Double-angle identities, 402
Downward parabola, 170
Drawing lessons
  lines in $\mathbb{R}^3$, 632
  observer viewpoint, 631
  planes in one octant, 630
  plotting points in $\mathbb{R}^3$, 594
  surfaces, 603
  vertical planes, 596
Drive sprocket assembly, 294
Dune buggy problem, 234
Dyn, 612
Dyne, 612
Dyne-centimeter, 612

E

e, 251
Earth
  orbit, 448
  radius, 291
  radius problem, 295
Eccentricity
  ellipse, 446
  hyperbola, 456
  parabola, 456
Egyptian, 54
Eiffel Tower problem, 368

Eight-leafed rose, 481
Electrical current problem, 421
Electromotive force, 384
Elementary row operations, 542, 546
Elevation, angle of, 348
Elimination tournament, 525
Ellipse
  calculator graphing, 443
  definition, 439
  degenerate, 458
  focal chord, 451
  geometric properties, 449
  graphing procedure, 455
  polar-form, 485, 487
  reflective property, 449
  standard form, 442
  standard polar form, 488
  vertex, 442
Ellipsoid, 600, 672
Elliptic
  cone, 600, 672
  cylinder, 599
  paraboloid, 672
Empire State Building problem, 353
Epicycloid, 672
Epitrochoid, 492
Equal
  coefficients, 63
  functions, 105
  matrices, 556
  polynomials, 13
  vectors, 584
Equality, 646
  properties of, 640
Equation(s), 57
  absolute value, 9
    solving, 43
  depressed, 204
  derived, 240
  equal coefficients, 64
  exponential, 268
  false, 57, 376
  first degree, 5
  linear, 377
  logarithmic, 262
  matrices, 541
  open, 57, 376
  polynomial, 203
  principal value solution, 376
  procedure for solving, 377
  quadratic, 378
    formula, 378
  radical, 239
  rational, 236

restricted solution, 424
system of, 530
theory of, 204
true, 57, 376
Equivalent
  matrices, 542
  systems, 534
  vectors, 584
Eratosthenes, 295
Erg, 612
Euler, Leonhard, 177, 534
Euler's
  formula, 420
  number, 251
Evaluate, 80
  function, 296
  inverse trigonometric function, 334
  logarithm, 255
  summation, 519
  trigonometric functions, 298
Eves, Howard, 296, 395
Even function, 111
Evolute of an ellipse, 672
Exact
  interest, 248
  values, 306
    trigonometric functions, 306
Expand, 519
Exponent, 12, 245
Exponential, 268
  curve, 673
  directory of curves, 246
  function, 101, 245
  nonmatching bases, 274
  property of equality, 255
  quadratic formula, 275
  squeeze theorem, 244
Expression, 12
Extraneous root, 236

F

Factor, 15, 641
  theorem, 378
Factoring, 15
  complete, 16
  types, table of, 15
Falling object problem, 98
False equation, 57, 376
Farmer problem, 574
Feasible value, 574
Fenway Park, 304
Fermat, Pierre de, 164

Ferndale, CA, 370
Fibonacci
  Association, 505
  Leonardo, 503
  rabbit problem, 504, 517
  sequence, 504
    general term, 505
    in nature, 505
Field, 641
  properties, 640
Finite series, 518
First component, 20
First degree
  inequality, 571
  polynomial, 160
Focal chord
  parabola, 431
  ellipse, 451
Foci, (pl of *focus*), *see* Focus, 440
Focus, 485
  ellipse, 440
  hyperbola, 452
  parabola, 430
Folium of Descartes, 484, 673
Foot-pounds, 612
Four-leaved rose, 480, 673
Frame, 314
  cosine curve, 318
  sine curve, 315
  starting point, 318
  tangent curve, 319
Frequency, 312
Frog and tree problem, 92
Ft-lb, 612
Functional notation, 97
Function(s)
  absolute value, 104
  algebraic, 101, 243
  Arrhenius, 276
  ceiling, 126
  composite, 129
  constant, 95, 109
  cube root, 104
  cubic, 104
  decreasing, 109
  definition, 93
  demand, 178
  equal, 105
  even, 111
  exponential, 101, 245
  greatest integer, 125
  identity, 104
  increasing, 109

Function(s) (*Continued*)
  inverse, 137
  of an inverse function, 400
  limit of, 513
  linear, 100, 160
  notation, 97
  odd, 111
  one-to-one, 95
  piecewise, 123
  polar, 476
  polynomial, 100
  power, 100
  quadratic, 169
  quadric, 100, 104
  radical, 230
  rational, 100, 220
  reciprocal, 104
  reciprocal squared, 104
  square root, 104
  square root reciprocal, 104
  standard
    cubic, 104
    quadratic, 104
    reciprocal, 104
    reciprocal squared, 104
    square root reciprocal, 104
  transcendental, 243
  zero, 107
Fundamental theorem of algebra, 206
Fundamental identities, 298, 385
  table of, 385
Future value, 247

G

Galileo, Galilei, 132, 145
Gamma (γ), 284
  function graph, 673
Gardner, Martin, 492
Gauss, Karl, 206, 414
Gaussian
  elimination, 542
  plane, 414
Gauss-Jordan elimination, 542
Geese problem, 155
General form, *see* Particular form
  conic section, 430
  parabola, 171
General term
  arithmetic sequence, 501
  Fibonacci sequence, 505
  geometric sequence, 503
Generating line, 599
Generatrix, 599

Geometric
  growth, 501
  proof, 385
  property of cross product, 616
  sequence, 502
    formula, 503
  series, 522
    formula, 523
Germain, Sophie, 39
Golden ratio, 238, 243
Golden section, 238
Grant's tomb property, 260
Graph(s)
  critical values by calculator, 39
  directory of curves
    Part I (basic curves), 104
    Part II (exponential curves), 246
    Part III (logarithmic curves), 259
    Part IV (trigonometric curves), 336
    Part V (conic sections), 460
    Part VI (polar curves), 483
    Part VII (quadratic surfaces), 600
  drawing lessons
    lines in $\mathbb{R}^3$, 632
    observer viewpoint, 631
    planes, 630
    plotting points in $\mathbb{R}^3$, 594
    surfaces, 603
    vertical planes, 596
  of a function, 103
  library of curves, 670
  line, 23
  number, 5
  paper, 475
  in $\mathbb{R}^3$, 596
Graphing
  cardioid, 476
  double-angle identity, 405
  functions, calculator, 433
  method, 531
  polar form, 472
    change to rectangular-form, 473
  rectangular-form, 468
  sum-to-product identity, 410
Graton, CA, 370
Great pyramid, 353
Greatest integer function, 125
Greek letters, 284
Growth
  formula, 271
  geometric, 501
Gunter, Edmund, 296
Gutenberg, Beno, 205

H

Half-angle identities, 406, 407
Half-life, 272
  NA-22, 276
  U-234, 276
Half-plane, 571
  closed, 571
  open, 571
Halley's Comet, 497
Halmos, Paul, 277, 583
Harmonic motion, 426
  damped, 426
Harpsichord problem, 368
Heart pumping problem, 330
Height
  of cosine frame, 318
  of frame, 318
  of sine frame, 315
  of tangent frame, 319
Herbart, J. F., 282
Heron's formula, 365
Hero's formula, 365
Hindu-Arabic
  numerals, 503
  numeration system, 5
Historical notes
  Abel, Niels, 659
  Brache, Tycho, 263
  Cain, Herman, 180
  Caley, Arthur, 560
  Copernicus, 356
  de Moivre, Abraham, 418
  Descartes, Rene, 20
  early trigonometry books, 395
  Egyptian rope-stretchers, 345
  Euler, Leonard, 534
  Fibonacci, 504
  functions, 93
  Gauss, Karl, 206
  Germain, Sophie, 39
  Hindu-Arabic numeration system, 5
  Hypatia, 496
  Invention of calculus, 52
  Kovalevsky, Sofia, 190
  Lawrence-Neimark, Ruth, 160
  Leibniz, Gottfried, 52, 93
  logarithm invention, 255
  Napier, John, 258
  Newton, Isaac, 52
  Noether, Emmy, 604
  Ochoa, Ellen, 593
  Oresme, Nicole, 132
  Origin of trigonometry terminology, 296
  Pascal, Blaise, 655
  Pólya, George, 78
  Scott, Charlotte Angas, 604
  sequence vs series, 518
  solving triangles, 395
  Surveying, 349
  Sylvester, James, 567
  value of mathematics, 40
  women in mathematics, 496
Historical Quest problems
  angle of elevation for Egyptian pyramids, 372
  Bhaskara, 92
  black mathematicians, 74
  Brahmagupta, 92
    geese problem, 243
  frog and tree problem, 92
  function, 155
  history of calculus, 74
  Kadamba flower problem, 92
  monkey in a tree problem, 92
  Napier, John, 281
  nature of functions, 216
  Noether, Amalie, 425
  Pólya, George, 155
  quadratic formula, 74
  radius of earth, 295
  water lily problem, 92
  Weyl, Hermann, 155
  women in mathematics, 74
Horizontal
  asymptote, 224
  component, 589
  line, 22, 160
  test, 95
How to Solve It, 78
Hypatia, 496
Hyperbola, 673
  definition, 452
  degenerate, 458
  eccentricity, 456
  polar-form, 485, 490
  standard polar form, 490
  vertex, 453
Hyperbolic
  cosecant graph, 673
  cosine, 253
    graph, 673
  cylinder, 599
  graph, 673, 674
  paraboloid, 600, 674
  secant graph, 674
  sine, 253
    graph, 674
  spiral, 484, 674, 678
  tangent, 253

Hyperboloid
  one sheet, 600, 674
  two sheets, 600, 674
Hyperboloidal gears, 457
Hypocycloid, 674
  with four cusps, 675
  with three cusps, 675
Hypotenuse, 343

I

i vector, 604
Identities, 298
  fundamental, 298
  inverse, 338
  Pythagorean, 299
  ratio, 299
  reciprocal, 298
Identity
  addition, 587
  additive, 641
  difference of angles, 394
  disproving, 390
  double-angle, 402
  function, 104
    graph, 675
  fundamental, 385
  half-angle, 407
  matrix
    addition, 562
    multiplication, 562
  multiplicative, 641
  product-to-sum, 409
  property for matrices, 562
  proving, 386
    hints for, 390
    procedure, 387
  Pythagorean, 385
  ratio, 385
  real numbers, 641
  reciprocal, 385
  sum-to-product, 409
Imaginary axis, 414
Imaginary number(s)
  pure, 645
  unit, 645
Inclination, angle of, 309
Inconsistent
  system, 530, 663
    graph of, 532
Increasing
  function, 109
  sequence, 514

Independent variable, 93
Index of refraction, 393
Induction, mathematical, 78, 651
Inductive reasoning, 78
Inequality
  absolute value, 9
    solving, 45
  first degree, 571
  linear, 39, 570
  one variable, 39
  properties, 40
Infinite
  sequence, 500
  series, 525
  slope, 310
Infinity
  limit, 512
  symbol, 6
Inflation, 251
Initial
  point, 584
  side, 284
Inner product, 607
Integers, 4
Integral, 54
Integral calculus, 50, 55
Interest, 247
  compound, 247
  continuous, 251
  exact, 248
  ordinary, 248
  simple, 247
Intensity of sound, 261
Intercepts, 22, 38, 107
Intersecting lines, 626
Interval notation, 6
Inverse, 80
  addition, 587
  additive for matrices, 562
  cosecant, 258
    graph, 675
  cosine, 258
    graph, 675
  cotangent, 258
    graph, 675
  function, 136, 334
    definition, 137
  function table, 258
  identities, 338
  procedure for finding, 138
  property of matrices, 562
  real numbers, 641
  reciprocal, contrast, 338

secant, 258
  graph, 675
sine, 258
  graph, 675
tangent, 258, 338
  graph, 675
Inverted cone problem, 89
Involute of a circle, 675
Irrational number, 4
ISECT key, 535
Iteration, 133

J

j vector, 604
Jeopardy, 529
Joule, 612
Journal problems
  Function, radical inequality, 11
  Journal of Recreational Mathematics, strange
    calculator, 74
  Mathematical Gazette, factoring, 74
  Mathematics and Computer Education, factoring, 20
  Mathematics Monthly, $3x + 1$
    conjecture, 156
  Mathematics Olympiad, absolute value inequality, 155
  Mathematics Student Journal, functional dates, 216
  Mathematics Teacher, function composition with a
    calculator, 155
  Mathematics Teacher, max/min problems, 217
  Ontario Secondary School Mathematics Bulletin, functional
    notation puzzle, 216
  Parabola, computer exercise, 216
  Parabola, farmer Jones fence problems, 216
  Parabola, quadratic inequality, 74
  Quantum, quadratic inequality, 74
  School Sciences and Mathematics, functional
    composition, 135

K

k vector, 604
Kadamba flower problem, 92
Khufu Pyramid, 353
Kimberling, Clark, 425
Kovalevsky, Sofia, 190

L

Lambda (λ), 284
Latitude, 291
Law of cosines, 355
  compared to law of sines, 364
  proof of, 355

Law of sines, 360
  compared to law of cosines, 364
  proof of, 360
Lawrence-Neimark, Ruth, 160
Lazy-eight curve, 484
Leading coefficient, 13, 100
Leaf, 480
Learning curve, 274
Leibniz, Gottfried, 52, 93, 145
Lemniscate, 482, 483
  Bernoulli, 676
  graph of, 676, 679
  standard position, 482
Length
  of cosine frame, 318
  of frame, 318
  of sine frame, 316
  of tangent frame, 319
Letter graphing, 217
Liber Abaci, 503
Library of curve and surfaces, 670
Limaçon, 478, 483, 676
  standard form, 480
Limit, 52, 142
  of a function, 510
  by graphing, 142
  infinity, 512
  introduction, 50, 51
  intuitive notion, 141
  notation, 512
  of a polynomial, 144
  properties, 511
  of a sequence, 513
  by table, 144
Line(s)
  aligned, 624
  calculator graphing, 564
  generating, 599
  graph, 676
  graphing, 23
  horizontal, 22
  normal, 167
  parallel, 26, 626
  parametric form, 625
  perpendicular, 26
  polar-form, 472
  in $\mathbb{R}^3$, 625
  secant, 52, 164
  skew, 625
  slope-intercept form, 23
  symmetric form, 625
  symmetry, 31
  tangent, 52, 164
  vertical, 22, 160

Linear
  combination, 586
    method, 538
  equality, 570
  function, 100, 160
  independent, 592, 636
  inequality, 570
  system, 530
Linear equation, 377
Linear inequality
  one variable, 39
Linear polynomial, 13
Linearity rule, 511
Lithotripsy, 449
Lituus, 676
ln x (phonetically, "lon x"), 256
Logarithm, 255
  both sides theorem, 262
  change of base theorem, 257
  common, 256
  contract, 265
  directory of curves, 259
  evaluate, 255
  function, 258
  natural, 256
Logarithmic equations, 262
  graph, 676
  spiral, 280, 484, 678
Log of both sides theorem, 262
Log x, 256
Logical statements, 80
Long division, 193
Longitude, 291
LORAN, 457
Lotka, Alfred, 312
Lower bound, 207

M

Mach number, 412
MacLane, Sanders, 159
Magnitude, 605
  cross product, 618
  vector, 607
Main diagonal, 562
Major axis, 441
  length of, 441
Marginal cost, 161
Mathematical
  induction, 78, 651
  modeling, 81
Mathematical Circles, 296

Mathematics
  value of, 40
Matrices, 541
  addition, 557
  additive inverse, 562
  algebra of, 556
  associative property, 562
  augmented, 541
  calculator use, 542
  communication, 560
  commutative property, 562
  definition, 540
  diagonal form, 542
  dimension, 541
  equal, 556
  equation, 541, 556
  equivalent, 542
  identity property, 562
  inverse matrix
    additive, 562
    by calculator, 564
    multiplicative, 562
    operations by calculator, 558
  multiplication, 559
  multiplicative inverse, 562
  noncomformable, 557
  nonsingular, 563
  operations, 556
  order, 541
  Pascal, 569
  properties of, 562
  RowSwap, 543
  singular, 566
  square, 541
  subtraction, 557
  test point, 571
  times row, 543
  zero, 562
  zero-one, 560
Maximum, 170
Middle C, 328, 384
Mile, nautical, 291
Million dollar salary problem, 523
Minimum, 170
Minor, 664
  axis, 441
  length of, 441
Minute, 284
Modeling
  applications, 82
  example, 274
  formulating a math problem, 36
Modulus, 414

Monkey-in-a-tree problem, 92
Monomial, 13
Monotonic sequence, 514
Mt. Frissell problem, 368
Mu (μ), 284
Müller, Johann, 395
Multiplication, 4
  of inverse matrices, 562
  of matrices, 557
Multiplicative identity, 641
Multiplicity, 59

N

Napier, John, 263
NASA, 295
Natural
  base, 251
  logarithm, 256
  number, 4
Nautical mile, 291
Navajo (Diné) nation, 142
NCAA problem, 525
Negative
  angle, 284
  multiplication property of inequalities, 40
  number, 5
newton, 612
Newton, Isaac, 53, 206
  law of cooling, 579
  laws of motion, 302
Newton-Raphson method, 135
Noether, Amalie (Emmy), 425, 604
Nonconformable matrices, 557
Nondecreasing sequence, 514
Nonincreasing sequence, 514
Nonlinear system, 532
Nonsingular, 563
Normal, 627
  line, 167
North to Alaska problem, 367
nth power, 12
nth root theorem, 418
Null vector, 584
Number
  complex, 4
  counting, 4
  integer, 4
  irrational, 4
  line, 5
  natural, 4
  negative, 5

  positive, 5
  rational, 4
  real, 4
  of roots theorem, 206
  sets of, 4
  types of, 4
  whole, 4
Numerical expression, 12

O

Objectives
  Chapter 1, 2
  Chapter 2, 76
  Chapter 3, 158
  Chapter 4, 218
  Chapter 5, 282
  Chapter 6, 374
  Chapter 7, 428
  Chapter 8, 498
  Chapter 9, 582
Oblique asymptote, 224
Oblique triangles, 355
Ochoa, Ellen, 593
Odd function, 111
Oil reserves, 280
Omega (ω), 284
One, 641
One-dimensional coordinate system, 5
  compared with two dimensional, 38
One-leaved rose, 676
One-to-one
  correspondence, 5
  function, 95
Open
  equation, 57, 376
  half-plane, 571
Opposite, 5
  angle identities, 396
  side, 344
Optimization, 177
  procedure, 177
Orbit, planets, 448
Order, 4, 541
  determinant, 665
  of operations agreement, 5
Ordered pair, 20
Ordinary interest, 248
Ordinate, 20
Oresme, Nicole, 132
Orientation, 187
Origin, 5, 20

Orthogonal, 609
  vector property, 609
Osgood, W. F., 72
Oui-ja board curve, 671, 676
Outer product, 614
Ovals of Cassini, 484, 676

P

Pappus Extension of the Pythagorean
    Theorem, 373
Parabola, 169
  axes, 431
  cubic, 671
  definition, 430
  degenerate, 458
  downward, 170
  eccentricity, 456
  focal chord, 431
  general form, 171
  graph, 677
    from definition, 431
    procedure, 435
  at (h, k), 434
    position, 169, 430
  at origin, 432
  polar-form, 485, 490
  semicubical, 677
  standard form, 431
  types, 431
  upward, 170
  vertex, 431
Parabolic spiral, 678
Paraboloid, 600
Parallel
  lines, 26, 530, 626
  vector cross-product, 616
  vectors, 606
Parallelogram
  area of by vectors, 618
  rule, 585
Parameter, 187, 532
  eliminating, 301
  equation, 187
    calculator graphing, 433
  graphing, 301
Parametric
  graphing, 301
  line in $\mathbb{R}^3$, 625
  solution, 548
Parametrization, 190
Paris observatory floor, 471
Parthenon, 238, 243

Partial sum, 525
Pascal
  Blaise, 655
  limaçon of, 676
  matrix, 569
Pascal's triangle, 656
  binomial theorem, 656
Peanuts chain letter, 522
Pendulum problem, 331
Pentagon, 351
  problem, 367
  regular, 351
Perigee, 451
Perihelion, 448
Period, 312
    general trigonometric curve, 324
Periodic, 315
Perpendicular lines, 26
Perpendicular vectors, 609
Petal, 480
Petroleum reserves, 280
Phase shift, 321
Phi (φ), 284
Phyllotaxy, 505
Pi (π), 284
  divided by 3, 307
Piecewise function, 123
Pipeline problem, 91
Pisa, tower of, 92, 367
Pistol problem, 540
Piston position problem, 312
Pivot, 544
  row, 546
Pivoting, 546
Planck's constant, 517
Plane(s), 20, 596
  angle between, 633
  attitude numbers, 628
  coordinate, 20
  xy, 20
  xy-plane, 594
  xz-plane, 594
  yz-plane, 594
Planetary orbits, 448
Plot a point, 20
Plus/minus symbol as used in trigonometry, 407
PMI, 651
Point
  deleted, 221
  initial, 584
  polar-form plotting, 469
  primary representations, 470
  terminal, 584

Point-normal form, 628
Polar
  angle, 468
  calculator, 476, 487
  change to rectangular form, 473
  circle, 472
  coordinate system, 468
  ellipse, 485, 487
  function, 476
  graph paper, 475
  graphing, 472, 483
  hyperbola, 485, 490
  line, 472
  parabola, 485, 490
  Paris observatory floor, 471
  plotting points, 469
  pole, 468
  ray, 472
  rotation, 477
  summary, 483
  symmetry, 477, 478
Pole, 468
Pólya, George, 77, 78, 177
Polynomial, 13
  equation, 203
  first-degree, 160
  function, 118
  inequalities, 68
  limit of, 144
  linear, 13
  long division, 193
Pool ball problem, 528
Positive, 5
  angle, 284
Positive multiplication property of
    inequalities, 40
Power function, 100
Preditor-prey problem, 312
Present value, 247
Prey-preditor problem, 312
Price Is Right problem, 124
Primary representations of a point, 470
Primitive quadruple, 372
Principal, 247
  cross section, 599
  of mathematical induction, 651
  nth root, 418
  square root, 4
  value solution, 376
Probability curve, 677
Problem-solving
  guide, 84
  procedure, 78
  word problems, 84

Product, 4, 641
  complex numbers, 416
  cross, 615
  dot, 607
  inner, 607
  outer, 614
  rule, 511
  scalar, 584
  scalar of dot, 607
  to sum identities, 409
  vector, 614
Progression, 500
Projection, 609
  formulas, 610
  scalar, 610
  vector, 610
Projectile
  motion, 301
  problem, 404
Projects group, 74, 155, 215, 279, 371, 424, 495, 580, 638
Prolate cycloid, 677
Prostaphaeresis, 263
Protractor, 285
Proving an identity, 386
  hints for, 390
  procedure, 387
p-series, 530
Pseudovertex, 453
Pulley problem, 294, 343
Pure imaginary number, 645
Pyramid
  Cheops, 353
  Great, 353
  Khufu, 353
Pythagorean
  identities, 299, 385
  theorem, 29
  triplets, 372

Q

Quadrant, 20, 284
Quadrantal angle, 284
Quadratic
  equations, 378
    one variable, 58
  formula, 378
    calculator, 359
    exponential equations, 275
  functions, 100, 169
    graphs, 677
  inequality, 66
  polynomial, 13
  surfaces, 599
    table of, 600

Quadratix of Hippias, 677
Quotient, 4, 195, 641
  complex numbers, 416
  rule, 511

R

Rabbit problem, 504, 517
Radar, 438
  antenna, 437
Radial
  axis, 468
  distance, 468
Radian, 285
  to degree, 288
  protractor, 285
Radical
  equation, 239
  function, 230
Radioactive decay, 272
Radiolaria, 580
Radius, 30, 598
  of the earth, 295
Rafter problem, 353
Range, 93, 314
Rate problems, 86
Ratio
  common, 502
  definition, 305
  identities, 299, 385
Rational
  equation, 236
  expression, 239
  function, 100, 220
  number, 4
  table of types, 220
Ray, polar, 472
Real
  axis, 414
  line, 5
  numbers, 4
    associative property, 641
    closure, 641
    commutative property, 641
    properties of, 641
Reasoning
  deductive, 79
  inductive, 78
Reciprocal, 4
  calculator button, 299
  function, 298, 677
    squared, 677
  identity, 298, 385

Rectangular
  array, 541
  coordinates, 468
  form, 414
    change to trigonometric form, 415
Reduction principle, 308
Reference angle, 289
Reflecting telescope, 437
Reflection, 32, 469
  in the x-axis, 117
  in the y-axis, 117
Reflexive property, 640
  ellipse, 449
Refraction, index of, 393
Regiomontanus, 395
Regular pentagon, 351
Remainder, 195
  theorem, 197
Research projects, 74, 155, 215, 279, 371, 424, 495,
    580, 638
Resolved vectors, 585
Restricted solution, 424
Resultant vector, 585
Reviews
  Chapter 1, 72
  Chapter 2, 152
  Chapter 3, 214
  Chapter 4, 277
  Chapter 5, 369
  Chapter 6, 421
  Chapter 7, 492
  Chapter 8, 577
  Chapter 9, 634
Rheticus, 356
Richter, Charles, 261
Richter scale, 261
  research project, 280
Right-hand rule, 617
Right-handed coordinate system, 593
Right triangle definition of the trigonometric functions, 344
Roe, E. D., Jr., 214
Roman numerals, 503
Root, 57
  extraneous, 236
  location theorem, 148
  of multiplicity, 58, 207
  rule, 511
  theorem, 206
Rope stretchers, 345
Rose curve, 480, 483, 677
  eight-leaved, 481
  four-leaved, 480
  four leaves, 673

one leaf, 676
standard position, 481
three leaves, 679
two leaves, 679
Rotation
polar-form, 477, 490
Row, 541
operations, 542
reduced form, 546
target, 543
RowSwap, 543
Rules of divisibility, 74

S

SAA, 355
Saint Paul's Cathedral, 450
Salary millionaire problem, 523
SAS, 355
area, 365
procedure, 362
Satisfy an equation, 20, 473
Scalar multiplication, 584
associative, 587
distributive property, 587, 616
dot product, 607
matrices, 557
projection, 610
formula, 610
Scott, Charlotte Angas, 604
Sear's Tower, 353
Sebastopol, CA, 370
Secant
curve, 320
polar/rectangular form, 484
function, 296
graph, 677
line, 52, 164
Second, 284
component, 20
Selfridges theorem, 257
Semicubical parabola graph, 677
Semimajor axis, 445
Semiminor axis, 445
Sequence, 500
arithmetic, 500
formula, 503
bounded, 514
compared to series, 518
convergent, 510
decreasing, 514
divergent, 510
Fibonacci, 504
finite, 500

geometric, 502
formula, 503
graphing, 510
increasing, 514
infinite, 500
limit of, 513
monotonic, 514
nondecreasing, 514
nonincreasing, 514
summary, 527
Series, 518
alternating, 518
arithmetic, 520
compared to sequence, 518
finite, 518
geometric, 522
formula, 523
infinite, 525
Serpentine curve, 678
Shannon, 371, 528
Shift, 114
Sigma notation, 518
Significant digits, 346
Signs of trigonometric functions,
298, 379
Similar
terms, 13
triangles, 88, 304
property, 88
Simple
harmonic motion, 426
interest, 247
Simplify, 13
complex number, 417
Simultaneous solution, 530
Sine
amplitude, 315
compared to cosine curve, 321
curve, 314
polar/rectangular form, 484
frame, 315
function, 296, 305
general form, 324
graph, 678
from the unit circle, 323
period, 315
procedure for graphing, 316
properties of, 315
standard form, 314
starting point, 316
Singular matrix, 566
Sinusoidal curve, 321, 678
Skew lines, 625
Slant asymptote, 224, 453

Slope, 22
  of a curve, 50
  of a line, 50
  no, 321
  as a rate, 160
  of a tangent line, 165
Slope-intercept form, 23
Slug, 612
Smart Trig Class, A, 298
Solution, 533
  system of equations, 530
  system of inequalities, 570
Solar panel problem, 352
Soldier problem, 643
Solution
  complete solution, 376
  restricted, 424
  trigonometric, 375
Solve
  equation, 57, 376
  inequality, 40
  triangle, 343, 362
Sonic boom problem, 412
Sound wave frequency, 327
Sphere, 600, 678
  equation, 598
Spiral
  Archimedian, 670, 678
  hyperbolic, 674, 678
  logarithmic, 678
  parabolic, 678
Spring problem, 331
Square
  completing, 60
  matrix, 541
  root, 4
    function, 104, 678
    reciprocal function, 679
Squeeze theorem, 244
SSA, 355
  properties, 362
SSS, 355
  area, 365
  properties, 362
Stairway construction problem, 366
Standard
  basis vectors, 604
  form
    curve, *see* Particular curve
    plane, 628
  equation, 628
    sphere, 598

functions
  cubic, 104
  quadratic, 104
  reciprocal, 104
  reciprocal squared, 104
  square root reciprocal, 104
  representation, 589
    in $\mathbb{R}^3$, 604
  trigonometric curves
    cosine, 318
    sine, 314
    tangent, 319
Standard-position
  angle, 284
  parabola, 169
Street light problem, 88
Strictly monotonic, 514
Strophoid, 484, 679
Subscripts, 540
  double, 541
Substitution
  method, 534
  property, 79, 640
Subtended arc, 291
Subtraction, 4
  complex numbers, 646
Subtraction of matrices, 557
Sum, 4, 641
  of cubes problem, 216
  partial, 525
Sum and difference angle identities, 398
Summaries
  Chapter 1, 72
  Chapter 2, 152
  Chapter 3, 214
  Chapter 4, 277
  Chapter 5, 369
  Chapter 6, 421
  Chapter 7, 492
  Chapter 8, 577
  Chapter 9, 634
Summation
  evaluate, 519
  notation, 518
Sum-to-product identities, 409
Sundial problem, 368
Sunflower, 505
Superball problem, 529
Superfluous constraint, 576
Surface, 596
Suspicious point, 147
Sylvester, James, 429, 567

Symmetry
  axis of, 32
  polar-form, 477, 478
  with respect to a line, 31, 484
  x-axis, 484
  y-axis, 484
Symmetric
  form, 625
  property, 640
Symmetries
  of a square, 644
  table of, 664
Systems of equations, 530
  consistent, 530
  Cramer's rule, 663
  dependent, 530, 663
  equivalent, 534
  graphing
    by calculator, 535
    method, 531
  inconsistent, 530, 663
  linear, 530
  matrices by calculator, 559, 564
  matrix method, 567
  nonlinear, 530, 532
  simultaneous solution, 530
  solution, 530
  standard form, 667
  substitution method, 534
Systems of inequalities, 572
  solution, 570
Swap rows, 543
Systolic phase, 330

T

Table
  exact values, 308
Tangent
  amplitude, 320
  asymptotes, 319
  curve, 320
    polar/rectangular form, 484
  frame, 319
  function, 305
  general form, 324
  graph, 679
    from unit circle, 332
  standard form, 319
Tangent line, 52, 164
  slope of, 165
Target row, 543

Ten commandments for proving identities, 390
Terminal
  point, 584
  side, 284
Term(s), 12, 641
  similar, 13
Test point, 571
Tetrahedron, 623
Theory of equations, 204
Theta (θ), 284
Thirty-seven-year-old puzzle, 371
Three coin problem, 367
Three-dimensional coordinate system, 593
Three-leaved rose, 679
Three space, 593
Time, 270
Times
  row, 543
  row plus, 544
Torque, 621
Tournament, 525
  of champions, 529
  two-team elimination, 525
Tower of Pisa, 92
  problem, 367
Trace, 599
  lines, 597
Tractrix, 679
Transcendental, 101
  function, 243
Transitive property, 640
Transit, 349
Transitivity, 40
Translation, 114
  trigonometric curve, 321
Transverse axis, 452
  length of, 453
Triangle
  area by vectors, 619
  area of, 365
    SAS, 365
  correctly labeled, 370
  inequality, 9, 588
    generalized form, 654
  no solution, 359
  oblique, 355
  one solution, 357
  solve, 343
    procedure, 362
  two solutions, 358
Triangular rule, 585
Trichotomy, 5, 640

Trigonometric equation, 376
  complete solution, 376
  double-angle, 381
  graphing solution, 381
  procedure for solving, 377
  radical form, 381
  restricted solution, 424
  solution, 375
  tabular form solution, 383
Trigonometric form, 415
  change to rectangular form, 415
Trigonometric functions, 296
  evaluate, 298
  exact values, 308
  general form, 324
  ratio definition, 305
  right triangle definition, 344
  signs of, 298, 379
  substitutions, 344
    table, 345
  unit circle definition, 332
Trigonometry, 296
Trinomial, 13
Triple scalar product, 620
Trochoid, 679
Trudear, Richard, J., 375
True equation, 57, 376
Tuning fork, 327, 384
Turner, Bird, 604
Turning point, 109, 182
Two-dimensional coordinate system, 20, 38
  compared with one-dimensional, 38
Two-leaved rose, 679

Vector(s), 584
  addition, 585
  area of a triangle, 619
  basis, 589, 604
  component form, 585
  distributive property, 587, 615
  equal, 584
  equivalent, 584
  notation, 584
  null, 584
  operations, 586
  orthogonal, 609
  perpendicular, 609
  product, 614
  projection, 610
  resolved, 585
  resultant, 585
  standard representation, 589
  three-space, 593
  unit, 589
  zero, 584
Velocity, 161
Vertex, 170, 294
  ellipse, 442
  hyperbola, 453
  parabola, 431
Vertical
  asymptote, 221, 319
  change, 22
  components, 589
  line, 22, 160, 310
    test, 95
Volume of a cylinder, 369

U

Unit
  circle, 30
    definition of the trigonometric functions, 295
  imaginary, 645
  multiplicative identity, 641
  one, 641
  vector, 589
Upper and lower bound theorem, 207
Upper bound, 207
Upward parabola, 170

V

Van deWalle, John A., 421
Variable
  dependent, 93
  independent, 93

W

Waggle dance, 496
Water-lily problem, 92
Weierstraß, Karl, 145
Wheel of Fortune, 529
Whole number, 4
Witch of Agnesi, 492, 679
Women in mathematics, 496
Word problem(s)
  rate problems, 86
  solving procedure, 84
Work
  as a dot product, 613
  units of, 612
World population, 253
  growth example, 273

X

x-axis, 20
x-intercept, 22, 38, 107
xy-plane, 20, 594
xz-plane, 594
yz-plane, 594

Y

y-axis, 20
y-intercept, 22, 101
Young, J. W. A., 369

Z

Zeno, 51
  paradox, 51
Zero, 5, 641
  division by, 296
  factor cross product, 616
  factor theorem, 58
  of a function, 107
  matrix, 562
  of multiplicity k, 207
  product of a vector, 607
  vector, 584
Zero-one matrix, 560
Zoom, 369

FORMULAS

Distance between (x_1, y_1) and (x_2, y_2) is $d = \sqrt{(x_2 - x_1)^2 + (y_2 - y_1)^2}$.

Slope of the line passing through (x_1, y_1), and (x_2, y_2) is $m = \dfrac{y_2 - y_1}{x_2 - x_1}$.

Pythagorean theorem: A triangle with sides a and b, and hypotenuse c (the side opposite the right angle) is a right triangle if and only if $a^2 + b^2 = c^2$.

Quadratic formula: If $ax^2 + bx + c = 0$, $a \neq 0$, then $x = \dfrac{-b \pm \sqrt{b^2 - 4ac}}{2a}$.

The *discriminant* is $b^2 - 4ac$;
if $b^2 - 4ac < 0$, then there are *no real* solutions,
$b^2 - 4ac = 0$, then there is *one real* solution, and
$b^2 - 4ac > 0$, then there are *two real* solutions.

Factoring rules:
$$x^2 - y^2 = (x - y)(x + y)$$
$$x^3 - y^3 = (x - y)(x^2 + xy + y^2)$$
$$x^3 + y^3 = (x + y)(x^2 - xy + y^2)$$

Also see table of factoring techniques, Table 1.4, p. 15.

Growth/Decay formula: $A = A_0 e^{rt}$

Principle of mathematical induction: If a given proposition $P(n)$ is true for $P(1)$ and if the truth of $P(k)$ implies its truth for $P(k + 1)$, then $P(n)$ is true for all positive integers n.

Binomial expansion: $(a + b)^n = \displaystyle\sum_{k=0}^{n} \binom{n}{k} a^{n-k} b^k$

Factorial: $n! = n(n - 1)(n - 2) \cdots 3 \cdot 2 \cdot 1$; $\quad 0! = 1$; $\quad n! = n(n - 1)!$

Arithmetic sequence: $a_n = a_1 + (n - 1)d$ $\qquad$ **Geometric sequence:** $g_n = g_1 r^{n-1}$

Arithmetic series: $A_n = \dfrac{n}{2}[2a_1 + (n - 1)d]$ $\quad$ or $\quad$ $A_n = n\left(\dfrac{a_1 + a_n}{2}\right)$

Geometric series: $G_n = \dfrac{g_1(1 - r^n)}{1 - r}$; $\qquad G = \dfrac{g_1}{1 - r}$ if $|r| < 1$

Forms of a complex number z:
Rectangular form: $z = a + bi$
Polar form: $z = r(\cos\theta + i\sin\theta) = r\operatorname{cis}\theta$
Conversion formulas:
From rectangular to polar: $r = \sqrt{a^2 + b^2}$; $\bar{\theta} = \tan^{-1}\left|\dfrac{b}{a}\right|$, where $\bar{\theta}$ is the reference angle for θ.

From polar to rectangular: $a = r\cos\theta$, $b = r\sin\theta$

De Moivre's theorem: If n is a natural number, $(r\operatorname{cis}\theta)^n = r^n \operatorname{cis} n\theta$.

nth root theorem: If n is a positive integer, $(r\operatorname{cis}\theta)^{1/n} = \sqrt[n]{r}\operatorname{cis}\left[\frac{1}{n}(\theta + 360°k)\right]$, where $k = 0, 1, \cdots, n - 1$